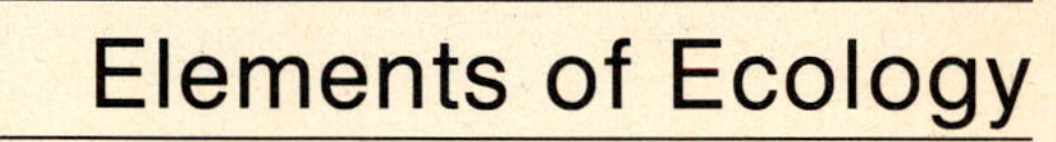

Elements of Ecology

ELEMENTS OF ECOLOGY

Second Edition

West Virginia University

ROBERT LEO SMITH

HARPER & ROW, PUBLISHERS, New York
Cambridge, Philadelphia, San Francisco,
London, Mexico City, São Paulo, Singapore, Sydney

Sponsoring Editor: Claudia M. Wilson
Project Editor: Donna DeBenedictis
Cover and Text Design: Hudson River Studio
Cover Photograph: Sunrise, Great Blue Heron, Texas City, Texas. © EKM-NEPENTHE.
Text Illustration: Robert Leo Smith, Jr. and Ned Smith
Text Rendition of Technical Art: Fineline Illustrations, Inc., and Robert Leo Smith, Jr.
Photo Research: Mira Schachne
Production: Debra Forrest Bochner
Compositor: Ruttle, Shaw & Wetherill, Inc.
Printer and Binder: R. R. Donnelley & Sons Company

Part-opening photograph credits: Hamlin, Stock, Boston (Part I); Thomas M. Smith (Part II); Menzel, Stock, Boston (Part III); Kirschenbaum, Stock, Boston (Part IV); Delevingne, Stock, Boston (Part V); © Schroeder/Eastwood, Southern Light (Part VI).

Figure credits start on facing page.

Previous edition was published under the title: *Elements of Ecology and Field Biology.*

Elements of Ecology, Second Edition

Library of Congress Cataloging-in-Publication Data
Smith, Robert Leo.
Elements of ecology.

Rev. ed. of: Elements of ecology and field biology/ Robert Leo Smith. © 1977.
Bibliography: p.
Includes index.
1. Ecology. 2. Biology—Field work. I. Smith, Robert Leo. Elements of ecology and field biology. II. Title.
QH541.S624 1986 574.5 85-17622
ISBN 0-06-046327-9

85 86 87 88 9 8 7 6 5 4 3 2 1

ACKNOWLEDGMENTS

I wish to thank the following for permission to adapt, reprint, or redraw the following figures from their publications. The list is long, and I trust I did not overlook anyone.

Figure 2.1. C. B. Cox, I. N. Healy, and P. D. Moore, *Biogeography: An Ecological and Evolutionary Approach,* 1st ed., p. 117. Copyright © 1973 Blackwell Scientific Publications. Used with permission.

Figure 2.2, Figure 4.3. A. Strahler, *The Earth Sciences,* Harper & Row, New York, pp. 52, 54. Copyright © 1971 Arthur Strahler. Reprinted with permission.

Figure 2.3. R. F. Flint, *Glacial and Quaternary Geology,* p. 545. Wiley, New York. Copyright © 1971 John Wiley and Sons. Used with permission.

Figure 2.4. H. S. Dybas and M. Lloyd, Isolation by habitat in two synchronized species of periodical cicadas (Homoptera, Cicadidae, *Magicada*), *Ecology,* vol. 43, p. 447. Copyright © 1962 Ecological Society of America. Used with permission.

Figure 2.6. T. Hunkapiller, H. Huang, L. Hood, and J. H. Campbell, The impact of modern genetics on evolutionary theory. In *Perspectives in Evolution,* ed. R. Milkman, pp. 166, 167. Sinauer Associates, Sunderland, MA. Copyright © 1982 Sinauer Associates. Used with permission.

Figure 2.7. L. Hood, J. H. Campbell, and S. R. C. Elgin, The organization, expression and evolution of antibody genes and other multigene families. Reproduced, with permission, from *Annual Review of Genetics,* vol. 9, p. 306. Copyright © 1975 Annual Reviews Inc.

Figure 2.10. P. T. Boag and P. R. Grant, Intense natural selection in a population of Darwin's finches, *Science,* vol. 214, p. 83. Copyright © 1981 American Association for the Advancement of Science. Used with permission.

Figure 3.3. R. W. Ficken, M. S. Ficken, and D. H. Morse, Competition and character displacement in two sympatric pine-dwelling warblers (*Dendroica,* Parulidae), *Evolution,* vol. 22, p. 310. Copyright © 1968 Society for the Study of Evolution.

Figure 3.4. L. E. Mettler and T. G. Gregg, *Population Genetics and Evolution,* p. 184. Prentice-Hall, Englewood Cliffs, NJ. Copyright © 1969 Prentice-Hall. Used with permission.

Figure 4.14. H. E. Landsberg, Man-made climatic changes, *Science,* vol. 170, p. 1721. Copyright © 1970 American Association for the Advancement of Science. Used with permission.

Figure 4.15. R. J. Kopec, Further observations of the urban heat island of a small city, *Bulletin American Meteorological Society,* vol. 51, p. 604. Copyright © 1970 American Meteorological Society. Used with permission.

Figure 5.1. D. Gates, *Energy Exchange and the Biosphere,* p. 13. Harper & Row, New York. Copyright © 1962 David M. Gates. Reprinted with permission.

Figure 5.2. B. McNab, The evolution of endothermy in the phylogeny of mammals, *American Naturalist,* vol. 112, p. 9. Copyright © 1978 University of Chicago Press. Used with permission.

Figure 5.5b. B. Heinrich, Heat exchange in relation to blood flow between thorax and abdomen in bumblebees, *Journal of Experimental Biology,* vol. 64, p. 564. Copyright © 1976 The Company of Biologists, Cambridge, England. Used with permission.

Figure 5.5c. C. R. Taylor, The desert gazelle: A parody resolved. In *Comparative Physiology of Desert Animals,* Symposium of Zoological Society of London, no. 31, G. M. O. Malory, ed. Copyright © 1972, Academic Press. Used with permission.

Figure 5.5d. K. Schmidt-Nielsen, *Animal Physiology: Adaptation and Environment,* p. 272. Cambridge University Press. Copyright © 1979 Cambridge University Press. Used with permission.

Figure 5.5e K. Schmidt-Nielsen, *Animal Physiology,* 3d. ed., p. 56. Prentice-Hall, Englewood Cliffs, NJ. Copyright © 1970 Prentice-Hall. Used with permission.

Figure 6.3. R. L. Dix and F. E. Smeins, The prairie, meadow, and marsh vegetation of Nelson County, North Dakota, *Canadian Journal of Botany,* vol. 45, p. 34. Copyright 1967 National Research Board of Canada. Used with permission.

Figure 7.2. W. Larcher, *Physiological Plant Ecology,* pp. 4, 15. Copyright © 1979 Springer-Verlag. Used with permission.

Figure 7.4. B. A. Hutchinson and D. R. Matt, The distribution of solar radiation within a deciduous forest, *Ecological Monographs,* vol. 47, p. 205. Copyright © 1977 Ecological Society of America. Used with permission.

Figure 7.12. R. J. Goss, C. E. Dinsmore, L. N. Grimes, and J. K. Rosen, Expression and suppression of circannual antler cycle in deer. In *Circannual Clocks: Annual Biological Rhythms,* ed. E. T. Pengelley, p. 407. Copyright © 1974 Academic Press. Used with permission.

Figure 8.5. H. Jenny, *The Soil Resource,* pp. 51, 54. Copyright © 1980 Springer-Verlag. Used with permission.

Figure 10.2. J. A. Wiens, Population responses to patchy environments. Reproduced, with permission, from *Annual Review of Ecology and Systematics,* vol. 7, p. 94. Copyright © 1976 by Annual Reviews Inc.

Figure 10.3. J. Wiens, Pattern and process in grassland bird communities, *Ecological Monographs,* vol. 43, p. 240. Copyright © 1973 Ecological Society of America. Used with permission.

Figure 10.10c. J. Hett and O. L. Loucks, Age structure models of balsam fir and eastern hemlock, *Journal of Ecology,* vol. 64, p. 1035. Copyright © 1976 Blackwell Scientific Publications. Used with permission.

Figure 10.10d. W. R. Hawthorn and P. B. Cavers, Population dynamics of the perennial herbs *Plantago major* and *P. Rugelli* Decne., *Journal of Ecology,* p. 516. Copyright © 1976 Blackwell Scientific Publications. Used with permission.

Figure 11.1b, Figure 11.3e. V. P. W. Lowe, Population dynamics of red deer (*Cervus elaphus* L) on Rhum, *Journal of Animal Ecology,* vol. 38, pp. 436, 437. Copyright © 1969, Blackwell Scientific Publications. Used with permission.

Figure 11.1d. D. W. Tinkle, J. W. Condon, and P. C. Rosen, Nesting season and success: Implications for the demography of painted turtles, *Ecology,* vol. 62, p. 1431. Copyright © 1981 Ecological Society of America. Used with permission.

Figure 11.1c. M. C. Baker, L. R. Mewaldt, and R. M. Stewart, Demography of white-crowned sparrows *(Zonotrichia leucophrys nuttali), Ecology,* vol. 62, p. 639. Copyright © 1981 Ecological Society of America. Used with permission.

Figure 11.1e. G. A. Polis and R. D. Farley, Population biology of a desert scorpion: Survivorship, microhabitat, and the evolution of life history strategy, *Ecology,* vol. 61, p. 623. Copyright © 1980 Ecological Society of America. Used with permission.

Figure 11.1f. O. Solbrig and B. B. Simpson, Components of regulation of a population of dandelion in Michigan, *Journal of Ecology,* vol. 62, p. 479. Copyright © 1974 Blackwell Scientific Publications. Used with permission.

Figure 11.4a. R. R. Scharitz and J. R. McCormick, Population dynamics of two competing plant species, *Ecology,* vol. 54, p. 729. Copyright © 1973 Ecological Society of America. Used with permission.

Figure 11.4c. N. C. West, K. H. Rea, and R. O. Harniss, Plant demographic studies in sagebrush-grass communities of southeastern Idaho, *Ecology,* vol. 60, p. 382. Copyright © 1979 Ecological Society of America. Used with permission.

Figure 11.4d. M. Yarranton and G. A. Yarranton, Demography of a jack pine stand, *Canadian Journal of Botany,* vol. 53, p. 311. Copyright © 1975 National Research Council of Canada. Used with permission.

Figure 11.5. C. Clark, *Population Growth and Land Use,* p. 43. Macmillan, London, and St. Martin's Press, New York. Copyright © 1967 Macmillan and St. Martin's Press. Used with permission.

Figure 11.6a and b. P. Williamson, Above-ground primary production of chalk grassland allowing for leaf death, *Journal of Ecology,* vol. 64, p. 1063. Copyright © 1976 Blackwell Scientific Publications. Used with permission.

Figure 11.6c. R. J. Reader, Contribution of overwintering leaves to the growth of three broad-leaved evergreen shrubs belonging to the Ericaceae family, *Canadian Journal of Botany,* vol. 56, p. 1251. Copyright © 1978 National Research Board of Canada. Used with permission.

Figure 12.2. V. Scheffer, The rise and fall of a reindeer herd, *Scientific Monthly,* vol. 73, p. 356. Copyright © 1951 American Association for the Advancement of Science. Used with permission.

Figure 12.5. F. A. Pitelka, Some characteristics of microtine cycles in the Arctic, *Proceedings 18th Biology Colloquium,* Oregon State College, pp. 79, 80. Copyright © 1957, Oregon University Press.

Figure 13.2 M. C. Dash and A. R. Hota, Density effects on survival, growth rate, and metamorphosis of *Rana tigrina* tadpoles, *Ecology,* vol. 61, p. 1027. Copyright © 1980 Ecological Society of America. Used with permission.

Figure 13.5. C. W. Fowler, Density dependence as related to life history strategy, *Ecology,* vol. 62, p. 607. Copyright © 1981 Ecological Society of America. Used with permission.

Figure 13.5, Figure 13.11. A. R. E. Sinclair, *The African buffalo,* pp. 140, 144. Copyright © 1977 University of Chicago Press. Reprinted with permission.

Figure 13.6. R. H. Tamarin, Dispersal, population regulation, and K-selection in field mice, *American Nat-*

uralist, vol. 112, p. 547. Copyright © 1978 University of Chicago Press. Reprinted with permission.

Figure 13.7. D. H. Janzen, Herbivores and the number of tree species in tropical forest, *American Naturalist,* vol. 104, p. 502. Copyright © 1970 University of Chicago Press. Reprinted with permission.

Figure 13.10. M. W. Fox, *Soul of the Wolf,* p. 61. Little, Brown Company, Boston. Copyright © 1980, M. J. Fox. Used with permission.

Figure 13.11b. J. R. Krebs, Territory and breeding density in the great tit *Parus major, Ecology,* vol. 52, p. 4. Copyright © 1971 Ecological Society of America. Used with permission.

Figure 13.13. S. D. Fretwell and H. L. Lucas, On territorial behavior and other factors influencing habitat distribution in birds, *Acta Biotheoretica,* vol. 19, p. 24. Copyright © 1969 Martin Nijhoff, Publishers, The Hague, Netherlands. Used with permission.

Figure 13.14. S. M. Smith, The "underworld" in a territorial adaptive strategy for floaters, *American Naturalist,* vol. 112, pp. 573, 576. Copyright © 1978 University of Chicago Press. Reprinted with permission.

Figure 14.2. M. L. Cody, A general theory of clutch size, *Evolution,* vol. 20, p. 179. Copyright © 1976 Society for the Study of Evolution. Used with permission.

Figure 15.3. H. C. Heller and D. Gates, Altitudinal zonation of chipmunks *(Eutamias)*: Energy budgets, *Ecology,* vol. 52, p. 424. Copyright © 1971 Ecological Society of America. Used with permission.

Figure 15.5. N. K. Wieland and F. A. Bazzaz, Physiological ecology of three codominant successional annuals, *Ecology,* vol. 56, p. 686. Copyright © 1975 Ecological Society of America. Used with permission.

Figure 15.6. M. A. Bowers and J. M. Brown, Body size and coexistence in desert rodents: Chance or community structure? *Ecology,* vol. 63, p. 396. Copyright © 1982 Ecological Society of America. Used with permission.

Figure 15.9. P. D. Putwain and J. L. Harper, Studies of dynamics of plant populations. 3. The influence of associated species on populations of *Rumex acetosa* and *R. acetosella* in grasslands, *Journal of Ecology,* vol. 58, p. 262. Copyright © 1970 Blackwell Scientific Publications. Used with permission.

Figure 15.10. E. R. Pianka, *Evolutionary Ecology,* 3d ed., p. 256. Harper & Row, New York. Copyright © 1983 Eric R. Pianka. Reprinted with permission.

Figure 15.12. P. Williamson, Feeding ecology of the red-eyed vireo *(Vireo olivaceous)* and associated foliage-gleaning birds, *Ecological Monographs,* vol. 41, p. 136. Copyright © 1971 Ecological Society of America. Used with permission.

Figure 15.14. B. J. Fox, Niche parameters and species richness, *Ecology,* vol. 62, p. 1418. Copyright © Ecological Society of America. Used with permission.

Figure 16.2b. M. P. Hassell, J. H. Lawton, and J. R. Beddington, The components of arthropod predation, *Journal of Animal Ecology,* vol. 45, p. 139. Copyright © 1976 Blackwell Scientific Publications. Used with permission.

Figure 16.3. M. P. Hassell and R. M. May, Aggregration in predators and insect parasites and its effects on stability, *Journal of Animal Ecology,* vol. 43, p. 576. Copyright © Blackwell Scientific Publications. Used with permission.

Figure 16.6. M. P. Hassell, Evaluation of predator or parasite response, *Journal of Animal Ecology,* vol. 35, p. 65. Copyright © 1966 Blackwell Scientific Publications. Used with permission.

Figure 16.7. D. H. Rusch, E. C. Meslow, P. D. Doerr, and L. B. Kieth, Response of great horned owl populations to changed prey densities, *Journal of Wildlife Management,* vol. 36, p. 291. Copyright © 1972, The Wildlife Society. Used with permission.

Figure 16.10. E. E. Werner and J. D. Hall, Optimal foraging and size selection of prey by the bluegill sunfish, *Ecology,* vol. 55, p. 1048. Copyright © 1974 Ecological Society of America. Used with permission.

Figure 16.11. N. B. Davies, Prey selection and social behavior in wagtails (Aves, Monticillidae), *Journal of Animal Ecology,* vol. 46, p. 48. Copyright © 1977 Blackwell Scientific Publications. Used with permission.

Figure 16.13. J. R. Krebs, Optimal foraging decision rules for predators. In *Behavioral Ecology: An Evolutionary Approach,* 1st ed., eds. J. R. Krebs and N. B. Davies, p. 42. Copyright © 1978 Blackwell Scientific Publications. Used with permission.

Figures 18.3, 18.4, 18.5. M. G. Barbour, J. H. Burk, and W. D. Pitts, *Terrestrial Plant Ecology,* p. 310. Benjamin/Cummins, Menlo Park, CA. Copyright © 1980 Benjamin/Cummins. Used with permission.

Figure 18.7. M. Kluge, Crassulacean acid metabolism (CAM): CO_2 and water economy. In *Water and Plant Life,* Ecological Studies no. 19, eds. O. L. Lange, L. Kappen, and E. D. Schulze, p. 317. Springer-

Verlag, New York. Copyright © 1972 Springer-Verlag. Used with permission.

Figure 18.8. O. Bjorkman, Comparative studies on photosynthesis in higher plants. In *Photophysiology,* vol. 8, ed. A. G. Giese, pp. 53, 56. Copyright © 1973 Academic Press. Used with permission.

Figure 18.9. J. D. Stout, K. R. Tate, and L. F. Molloy, Decomposition processes in New Zealand soils with particular respect to rates and pathways of plant degradation. In *The Role of Terrestrial and Aquatic Organisms in Decomposition Processes,* eds. J. M. Anderson and A. Macfayden, p. 98. Copyright © 1976 Blackwell Scientific Publications. Used with permission.

Figure 18.10. T. Fenchel and P. Harrison, The significance of bacterial grazing and mineral cycling for the decomposition of particulate detritus. In *The Role of Terrestrial and Aquatic Organisms in Decomposition Processes,* eds. J. M. Anderson and A. Macfayden, p. 296. Copyright © 1976 Blackwell Scientific Publications. Used with permission.

Figure 18.11. G. W. Saunders, Decomposition in fresh water. In *The Role of Terrestrial and Aquatic Organisms in Decomposition Processes,* eds. J. M. Anderson and A. Macfayden, p. 349. Copyright © 1976 Blackwell Scientific Publications. Used with permission.

Figure 19.1, Figure 19.2. D. F. Westlake, Primary production. In *The Functioning of Freshwater Ecosystems,* International Biological Programme no. 22, eds. E. D. LeCren and R. H. Lowe-McConnell, pp. 166, 179. Copyright © 1980 Cambridge University Press. Reprinted with permission.

Figure 19.4. J. R. Etherington, *Environment and Plant Ecology,* 2d ed., p. 355. Wiley, New York. Copyright © 1982 John Wiley and Sons. Used with permission.

Figure 19.5. F. B. Golley and H. Leith, Basis of organic production in the tropics. In *Tropical Ecology with Emphasis on Organic Production,* eds. P. M. Golley and F. B. Golley. University of Georgia Press, Athens, GA. Copyright © 1972 University of Georgia Press. Used with permission.

Figure 19.7. R. H. Whittaker, F. H. Bormann, G. E. Likens, and T. G. Siccama, The Hubbard Brook ecosystem study: Forest biomass and production, *Ecological Monographs,* vol. 44, p. 239. Copyright © 1974 Ecological Society of America. Used with permission.

Figure 19.8. G. D. Cooke, The pattern of autotrophic succession in laboratory microcosms, *Bioscience,* vol. 17, p. 719. Copyright © 1967 American Institute of Biological Sciences. Used with permission.

Figure 20.7. M. Lamotte, The structure and function of a tropical savanna ecosystem. In *Trends in Tropical Ecology,* eds. F. Golley and E. Medina, p. 216. Copyright © 1980 Springer-Verlag. Used with permission.

Figure 20.9. J. M. Teal, Energy flow in a salt marsh ecosystem of Georgia, *Ecology,* vol. 43, p. 622. Copyright © 1962 Ecological Society of America. Used with permission.

Figure 21.3. C. F. Baes, Jr., J. E. Goeller, J. S. Olson, and R. M. Rotty, Carbon dioxide and climate: The uncontrolled experiment, *American Scientist,* vol. 65, p. 311. Copyright © 1977, Sigma Xi. Used with permission.

Figure 21.7. W. W. Kellogg, R. D. Cadle, E. R. Allen, A. L. Lazrus, and E. A. Martell, The sulfur cycle, *Science,* vol. 175, p. 594. Copyright © 1972, American Association for the Advancement of Science. Used with permission.

Figure 21.11. D. W. Johnson et al., Nutrient cycling in the forests of the Pacific Northwest. In *Coniferous Forest Ecosystems in the Western United States,* ed. R. L. Edwards, p. 193. Dowden, Hutchinson & Ross, Stroudsburg, PA. Copyright © 1982 Dowden, Hutchinson & Ross. Used with permission.

Figures 21.14, 21.15. M. Williamson, *Island Populations,* pp. 98, 101. Oxford University Press, Oxford, England. Copyright © 1981, Oxford University Press. Used with permission.

Figure 22.6. J. Ranney et al., Importance of edge in the structure and dynamics of forest islands. In *Forest Island Dynamics in Man-dominated Landscapes,* eds. R. L. Burgess and D. M. Sharpe, p. 83. Copyright © 1981 Springer-Verlag. Used with permission.

Figure 22.7. R. F. Whitcomb et al., Effects of forest fragmentation on avifauna of the eastern deciduous forest. In *Forest Island Dynamics in Man-dominated Landscapes,* eds. R. L. Burgess and D. M. Sharpe, p. 143. Copyright © 1981 Springer-Verlag. Used with permission.

Figure 22.16. J. B. Levenson, Woodlots as biogeographic islands in southeastern Wisconsin. In *Forest Island Dynamics in Man-dominated Landscapes,* eds. R. L. Burgess and D. M. Sharpe, p. 32. Copyright © 1981 Springer-Verlag. Used with permission.

Figure 23.2. F. A. Bazzaz, Plant species diversity in old field successional ecosystems in southern Illinois, *Ecology,* vol. 56, pp. 486, 487. Copyright © 1975 Ecological Society of America. Used with permission.

Figure 23.3. F. H. Bormann and G. E. Likens, *Pattern and Process in a Forested Ecosystem,* p. 166.

Copyright © 1979 Springer-Verlag. Used with permission.

Figure 23.4. P. L. Marks, The role of pin cherry (*Prunus pensylvanica* L) in the maintenance of stability in the northern hardwoods system, *Ecological Monographs,* vol. 44, p. 75. Copyright © 1974 Ecological Society of America. Used with permission.

Figure 23.5. J. E. Ash and J. F. Barkham, Change and variability in the field layer of coppiced woodland in Norfolk, England, *Journal of Ecology,* vol. 64, p. 706. Copyright © 1976 Blackwell Scientific Publications. Used with permission.

Figure 23.7. D. G. Sprugel, Dynamic structure of wave generated *Abies balsamea* forests in northeastern United States, *Journal of Ecology,* vol. 64., p. 891. Copyright © 1976 Blackwell Scientific Publications. Used with permission.

Figure 24.3. W. K. Lavenroth, Grassland primary production: North American grasslands in perspective. In *Perspectives in Grassland Ecology,* ed. N. R. French, p. 10. Copyright © 1979 Springer-Verlag. Used with permission.

Figure 24.4. N. R. French, Principal subsystem interactions in grasslands. In *Perspectives in Grassland Ecology,* ed. N. R. French, p. 185. Copyright © 1979 Springer-Verlag. Used with permission.

Figure 24.5. A. I. Breymeyer, Trophic structure and relationships. In *Grasslands, Systems Analysis and Man,* International Biological Programme no. 19, eds. A. I. Breymeyer and G. Van Dyne, p. 803. Copyright © 1980 Cambridge University Press. Reprinted with permission.

Figure 24.6. F. E. Clark, C. Cole, and R. A. Bowman, Nutrient cycling. In *Grasslands, Systems Analysis and Man,* International Biological Programme no. 19, eds. A. I. Breymeyer and G. Van Dyne, p. 666. Copyright © 1980 Cambridge University Press. Reprinted with permission.

Figure 24.19. P. L. Johnson and W. O. Billings, The alpine vegetation of the Beartooth Plateau and its relation to cryopedogenic processes and patterns, *Ecological Monographs,* vol. 32, pp. 122, 123, 126, 129. Copyright © 1962 Ecological Society of America. Used with permission.

Figures 24.20, 24.21. F. S. Chapin III, P. C. Miller, W. D. Billings, and P. I. Coyne, Carbon and nutrient budgets and their control in coastal tundra. In *An Arctic Ecosystem,* eds. J. Brown, P. C. Miller, L. L. Tieszen, and F. L. Bunnell, pp. 471, 481. Dowden, Hutchinson & Ross, Stroudsburg, PA. Copyright © 1980 Dowden, Hutchinson & Ross. Used with permission.

Figure 24.23. D. E. Reichle, Advances in ecosystem analysis, *Bioscience,* vol. 25, p. 260. Copyright © 1975 American Institute of Biological Sciences. Used with permission.

Figure 25.3. K. W. Cummins, Structure and function of stream ecosystems, *Bioscience,* vol. 24, p. 633. Copyright © 1974 American Institute of Biological Sciences. Used with permission.

Figure 25.5. J. W. Newbold, R. V. O'Neill, J. W. Elwood, W. Van Winkle, Nutrient spiralling in streams: Implications for nutrients and invertebrate activity, *American Naturalist,* vol. 120, Figure 1, p. 630. Copyright © 1982 University of Chicago Press. Reprinted with permission.

Figure 25.9. G. E. Likens and F. H. Bormann, Linkages between terrestrial and aquatic ecosystems, *Bioscience,* vol. 24, p. 448. Copyright © 1974 American Institute of Biological Science. Used with permission.

Figure 25.10. P. H. Rich and R. G. Wetzel, Detritus in the lake ecosystem, *American Naturalist,* vol. 112, p. 68. Copyright © 1978 University of Chicago Press. Reprinted with permission.

Figure 25.11. M. Brylinsky, Estimating the productivity of lakes and reservoirs. In *The Functioning of Freshwater Ecosystems,* International Biological Programme no. 22, eds. E. D. LeCren and R. H. Lowe-McConnell, p. 433. Copyright © 1980 Cambridge University Press. Reprinted with permission.

Figure 25.12. H. L. Golterman and F. A. Kouwe, Chemical budgets and nutrient patterns. In *The Functioning of Freshwater Ecosystems,* International Biological Programme no. 22, eds. E. D. LeCren and R. H. Lowe-McConnell, p. 183. Copyright © 1982 Cambridge University Press. Reprinted with permission.

Figure 25.14. J. R. Gosselink and R. E. Turner, Role of hydrology in freshwater wetland ecosystems. In *Freshwater Wetlands,* eds. R. Good, D. F. Whigham, and R. L. Simpson, p. 73. Copyright © 1978 Academic Press. Used with permission.

Figure 25.16. D. F. Whigham, J. McCormick, R. E. Good, and R. L. Simpson, Biomass and primary production in freshwater tidal wetlands of the middle Atlantic Coast. In *Freshwater Wetlands,* eds. R. Good, D. F. Whigham, and R. L. Simpson, p. 20. Copyright © 1978 Academic Press. Used with permission.

Figure 25.17. W. Odum and M. Haywood, Decomposition of intertidal freshwater marsh plants. In *Freshwater Wetlands,* eds. R. Good, D. F. Whigham, and

R. L. Simpson, p. 92. Copyright © 1978 Academic Press. Used with permission.

Figure 25.18. T. Rosswall et al., Stordalen (Abisko) Sweden. In *Structure and Function of Tundra Ecosystems,* T. Rosswall and O. W. Heal, eds. Swedish Natural Science Research Council, Stockholm. Copyright © 1975 Swedish Natural Science Research Council. Used with permission.

Figure 26.1. J. W. Nybakken, *Marine Biology: An Ecological Approach,* p. 67. Harper & Row, New York. Copyright © 1982 J. W. Nybakken.

Figure 26.5. D. L. Correll, Estuarine productivity, *Bioscience,* vol. 28, p. 648. Copyright © 1978 American Institute Biological Sciences. Used with permission.

Figure 26.16. S. K. Eltringham, *Life in Mud and Sand,* p. 203, Crane, Russak, New York. Copyright © 1971 Crane, Russak. Used with permission.

Figure 26.17. © Tortel, 1977, DPI.

Figure 26.18. J. E. G. Raymont, *Plankton and Productivity in Oceans,* p. 547. Pergamon, New York. Copyright © 1963 Pergamon. Used with permission.

Figure VI.5. Reprinted by permission of Macmillan Publishing Company from R. H. Whittaker, *Communities and Ecosystems.* Copyright © 1970 R. H. Whittaker.

Table 14.1. S. T. Emlen and L. W. Oring, Ecology, sexual selection, and evolution of mating systems, *Science,* vol. 197, p. 217. Copyright © 1977 American Association for the Advancement of Science. Used with permission.

To Colleen Schoemer Smith and Mary Ward Smith

CONTENTS

PREFACE

This long overdue revision of *Elements of Ecology* is a result of a complete rethinking of the first edition of this text. As I contemplated the revision, looked closely at the first edition, and studied the comments of users and reviewers, I decided that *Elements* should be something more than just a shortened version of my larger text, *Ecology and Field Biology.* That book emphasizes the ecosystem approach; this book would emphasize an evolutionary approach.

My main objective when writing the first edition of *Elements* was to produce a more descriptive, yet concise, ecology text that would be useful in shorter and less rigorous ecology courses. That remains the principal objective of this revised edition. Although there are a number of short ecology texts on the market, they are, in my opinion, too short and lack rigor. They merely skim the outer periphery of ecology and fail to present the most current ideas in the field.

A second objective of this text is to integrate evolutionary ecology and ecosystem ecology. That is a tall order and how well I have succeeded can be judged by the reader.

A third objective is to present the most current concepts in ecology and not skirt about the more controversial issues.

In most fields other than ecology, a general or introductory text is expected to provide a broad survey of the subject. This is the approach of introductory texts in such subject areas as biology, chemistry, physics, psychology, and economics, among others. These texts have one feature in common: They provide an overview of the field. The approach of each text may be different, but the subject matter covered is similar. In other words, these subject areas all have well-defined, widely recognized sets of concepts and principles that make up introductory texts. This is not the case with ecology, however. A review of all the so-called general ecology texts reveals no standard set of concepts or coverage of subject matter areas. Many of the texts are highly idiosyncratic, each reflecting the author's own views of what constitutes general ecology. Many authors lean heavily toward population or evolutionary ecology or selected topics within these broad areas. Such approaches more aptly might be called selected topics in ecology.

Some authors remove all descriptive material and present only highly quantified material. Ecology, they say, must be shown as an exact science with certain underlying principles and laws as fundamental as the laws of *pure* physics. In my view, ecology should be quantitative, but it will never become an exact science with laws comparable to that of *pure* physics. Ecology cuts across the domain of too many "pure" sciences and involves such a variable natural world that it can never become an exact science. Two plus two will always equal four; 2H + O will combine to form H_2O; and the energy of a photon is directly proportional to the frequency of light, $E = h\nu$. But the Lotka-Volterra equations will never exactly predict or describe the growth and interactions of populations because neither they nor any other mathematical formulas can take into account the vagaries of nature and the individualistic behaviors of the organisms involved.

Writing an ecology text is becoming increasingly difficult. Back in the late 1960s, when I wrote the first edition of *Ecology and Field Biology,* modern ecology was just emerging and the number of journals was relatively small. Then came the information explosion and the phenomenal growth of ecology as a field of study in the 1970s. Ecology diverged into many different paths, many of which became increasingly divergent from each other.

The growth of ecosystem ecology, the emergence of theoretical ecology (especially population ecology and evolutionary ecology), the rapid growth of behavioral ecology, the development of chemical ecology, and the resurgence of physiological ecology have made writing a general ecology text a much more laborious, time-consuming, and yet challenging task. As most ecologists become more and more specialized, the author of a general ecology text must become more and more of a generalist. I find myself reading a wider range of journals than ever before as well as review volumes and monographs in many areas of ecology. Attempting to remain familiar with the many pathways of ecology becomes nearly a full-time job. But to ignore the wide range of ecological literature, to fail to maintain familiarity with the areas of ecology, would result in a very narrow text that would in no way reflect the current state of ecology. This expansion of ecology, revealing the need to provide a broad survey of ecology even in an introductory text, is the reason why I rewrote *Elements of Ecology.*

A general ecology text, then, should introduce students to the broad sweep of the field. It should include some discussion of the various types of ecosystems, or "habitats" as some wish to call them, as well as general principles. To fail to introduce the major ecosystems in a general ecology course is to deny beginning students an introduction both to major ecosystems and to a large body of important ecological literature. What differences exist between terrestrial ecosystems and aquatic ecosystems, streams and lakes, tropical rain forests and deciduous forests, and forests and grasslands? By that I do not mean differences in plant and animal life, the natural history of the ecosystems, but functional differences, the way they operate. Such knowledge should be a part of any ecologist's education.

Changes in the Second Edition

Simply by looking at the table of contents you will observe how much the second edition has changed from the first. Part I reviews the history of ecology, including a discussion of current trends and schisms. Too many young ecologists are unfamiliar with the roots of ecology. They don't know who the pioneers in ecology were or the contributions they made. They don't understand how ecology arrived where it is today. They fail to grasp that many of the ideas of modern ecology were actually formulated much earlier by ecologists who were ahead of their time and whose ideas either were not accepted or remained buried in literature. Because ecologists ignore older literature, they spend precious time rediscovering old concepts.

Part II introduces evolutionary ecology and establishes a theme that is carried through the text. Although the chapter titles are similar to those of the first edition, the contents have been rewritten to emphasize current evolutionary theory, including the possible role of gene behavior in influencing evolution outside of natural selection and gradualism versus punctuated equilibrium.

Part III logically follows. The emphasis here is on adaptation of organisms to the environment with a heavy emphasis on physiological ecology. Here, too, physical or abiotic components of the environment are discussed. Temperature, moisture, light, and soil are covered in separate chapters and a new one on fire has been added with an emphasis on its role in natural selection.

Part IV deals with population and evolutionary ecology. The subject matter has been divided into chapters of varying length; each deals with specific or related topics. I expanded population ecology considerably over the first edition to reflect the importance population and theoretical ecology have assumed in the past decade. Emphasized are new ideas in population regulation, life history strategies, foraging, competition, and coevolution.

Part V introduces the ecosystem. This part, too, has changed greatly. The two major functional processes in ecosystems, photosynthesis and decomposition, are treated in one chapter. The chapter on energy flow and food webs contains a discussion of current food web theory and the one on community structure treats the concept of island biogeography and edge to a greater extent than in other ecology texts. The chapter on succession has been completely rewritten to reflect the varying viewpoints on this phenomenon.

Part VI deals with marine, freshwater, and terrestrial ecosystems. This material, which comprised a major part of the first edition, has been condensed with an emphasis on comparative function rather than a description of their component populations. Description of grassland, rivers, ponds, and the like are available in many popular and semipopular books and other publications.

This edition of *Elements* introduces a number of innovations added to increase its pedagogical usefulness. Each chapter opens with an outline of material covered and a set of objectives that alerts readers to what they should gain from the chapter. Each chapter concludes with a summary and a set of review questions intended to reinforce the objectives for that chapter. My students have been using similar review questions for years with considerable success.

Three other innovations are boxed material, introductions to each Part, and Ecological Excursions. The boxed material, held to a minimum, presents information whose inclusion in the main text would have interrupted the flow, and yet expands the text discussion. The introductions to each Part discuss concepts essential to the material, yet do not fit well into any of the chapters themselves. Readings entitled "Ecological Excursions" are included at the end of certain chapters for added interest but with a purpose. Some demonstrate that ideas in ecology are relevant to real life situations (local age pyramids, graveyard ecology, boom and bust of local wildlife populations) and can provide material for good writing (Muir, Farb, Leopold). Others introduce readers to a sampling of classical ecological literature (Henderson, Errington), much of which escapes the notice of today's ecologists. Like field trips, these excursions down some side trails of ecology should be refreshing and stimulating.

An instructor's manual is also available. This manual is meant as an aid in an ecology course. But it can provide more. Because the manual can be a last-minute production from camera-ready copy, I am able to present more recent references to journal articles and books.

Acknowledgments

No textbook is the product of the author alone. Although the author writes the text, its content represents a distillation of the work of researchers who have spent long hours in the field and in the laboratory. They are the ones who through their published papers provide the raw material from which the text is formed. A text is a synthesis of that research, an attempt to bring ecological knowledge into a coherent whole.

Revisions of a textbook depend heavily on the input of users. I am grateful for the comments, criticisms, and suggestions offered by a number of instructors and students, as well as several anonymous reviewers who critically examined the first edition prior to revision. William E. Dunscombe of Union County College, Stanley I. Dodson of the University of Wisconsin, Glen C. Kroh of

Texas Christian University, I. Jack Stout of the University of Central Florida, Joseph J. Mahoney of Kean College of New Jersey, and James MacMahon of Utah State University provided detailed critiques of the new manuscript, all of which were extremely helpful, pointing out gaps, redundancies, and places of difficulty in the text.

I appreciate the constant encouragement, prodding, and, especially, patience of Claudia Wilson, biology editor at Harper & Row. When she sat down in my office to go over the revision, I am sure she was thinking of a project spanning about a year's time. What she got instead was a new book over three years later, which in turn has slowed revision of *Ecology and Field Biology.* Project Editor Donna DeBenedictis did an outstanding job of pulling manuscript, art, galleys, pages, and all that goes with it together. At times, we both felt that a hotline was needed between Morgantown and New York.

And I got a lot of family help. My wife Alice saw to it that I got back to the desk when I began to procrastinate. She compiled the index and helped with the permissions. My son Thomas Michael Smith, an ecologist variously with the University of Witswatersrand, the Oak Ridge National Laboratory, and soon (1986) with the Australian National University, Canberra, helps keep me modern and provided the opportunity to expand my ecological knowledge, especially of tropical savannas and succession. My other son Robert Leo Smith, Jr., a natural history and scientific illustrator, did most of the new graphs and all of the new drawings. A few of the illustrations taken from *Ecology and Field Biology* were done by Ned Smith (no relation) for the "green" first edition of that text. (See, for example, Figures 8.10, 20.2, 20.6, 21.4.) I would like to pay tribute to his memory. Ned, one of the country's outstanding wildlife artists, died this spring (1985) of a heart attack. The drawings he did for the first edition of *Ecology and Field Biology* contributed greatly to the success of that book.

Robert Leo Smith

PART I

INTRODUCTION

CHAPTER 1

What Is Ecology?

Outline

Objectives

Upon completion of this chapter, you should be able to:

1. Indicate the origin and development of various areas of ecology.
2. Briefly describe some of the major philosophical backgrounds and differences in the development of ecology.
3. Discuss the relationship of ecology to environmental science and resource management.
4. Identify some of the major figures in the development of ecology.

Ecology. For years the term was known and used only by those working in the field, a sort of a peripheral adjunct to biology, overshadowed, especially in the 1950s and 1960s, by molecular biology. Ecology was passé, scarcely recognized by the academic world. Then came the environmental movement of the late 1960s and early 1970s. Because it concerned environmental relationships, ecology became popular. The term appeared everywhere—in newspapers, magazines, and books. *Ecology* became a household word, and as such, it was used, misused, and abused.

Ecology was and still is equated with the environment. For example, over the past several years, I have been hearing seasonal radio advertisements urging the substitution of limestone chips for salt and cinders or sand for winter road treatment. The major selling points were that limestone is nonpolluting and good for the ecology. In a book review that appeared in a well-known biological journal was the statement: "If the heat dissipated by a large expenditure of energy goes into, say, warming a river, it can indeed dislocate the ecology." In each instance *ecology* was used when *environment, ecosystem,* or some equivalent word or phrase would have been correct.

Ecology, if one goes by the usual definition, is the study of the relationship between organisms and their environment. Even that definition can be faulted unless you consider the words *relationship* and *environment* in their fullest meaning. Environment includes not only the physical but also the biological conditions under which an organism lives; and relationships involve interactions with the physical world as well as interrelationships with members of other species and individuals of the same species.

The word *ecology* was coined by the German zoologist Ernest Hackel in 1866. He called it "Oekologie" and defined its scope as the study of the relationship of animals to their environment. Specifically, he was trying to relate animal morphology to Darwin's theory of evolution. The word did not come into general use, however, until it appeared in Warming's book *Plantesamfund: Grundtrak af den okologiske Plantegeografi* in 1895. Its modern meaning was largely fixed by that publication.

The term *ecology* comes from the Greek word *oikos,* which refers to the family household, and *logy,* meaning "the study of." Literally, ecology is the study of the household. It has the same root words as economics, or "management of the household." In effect, ecology could be considered as the study of the economics of nature. In fact, some economic concepts—such as resource allocation, cost-benefit ratios, and optimization theory—have crept into ecology.

The Roots of Ecology

Because of its hybrid nature, ecology is difficult to trace back to any definitive roots (Figure 1.1) as can be done with mathematics, chemistry, microbiology, and other sciences. One can argue that ecology goes back to the Greek scholar Theophrastus, a friend and associate of Aristotle who wrote about the interrelations between organisms and the environment. But ecology as we know it today has its strongest roots in plant geography and natural history, including the study of plants, birds, mammals, fish, and insects.

Although the study of natural history provided much of the knowledge upon which ecology was based, the real impetus came from the study and explorations of plant geographers. They discovered that although plants differed in various regions of the world, the vegetation assumed certain similarities and differences that demanded explanation. One of the early influential plant geographers was Carl Ludwig Willdenow (1765–1812). He pointed out that similar climates supported similar vegetation. Willdenow was a major influence on a wealthy young Prussian, Friedrich Heinrich Alexander von Humboldt (1769–1859). Sponsored by King Carlos IV of Spain—who provided the equipment needed to measure altitude, latitude, elevation, humidity, and temperature—young Humboldt spent 5 years exploring Latin America. He traveled in Mexico, Cuba, Venezuela, and Peru and explored the Orinoco and Amazon Rivers. Ultimately, he described his travels in a 30-volume work, *Voyage to the Equatorial Regions.* Of these, 14 volumes were de-

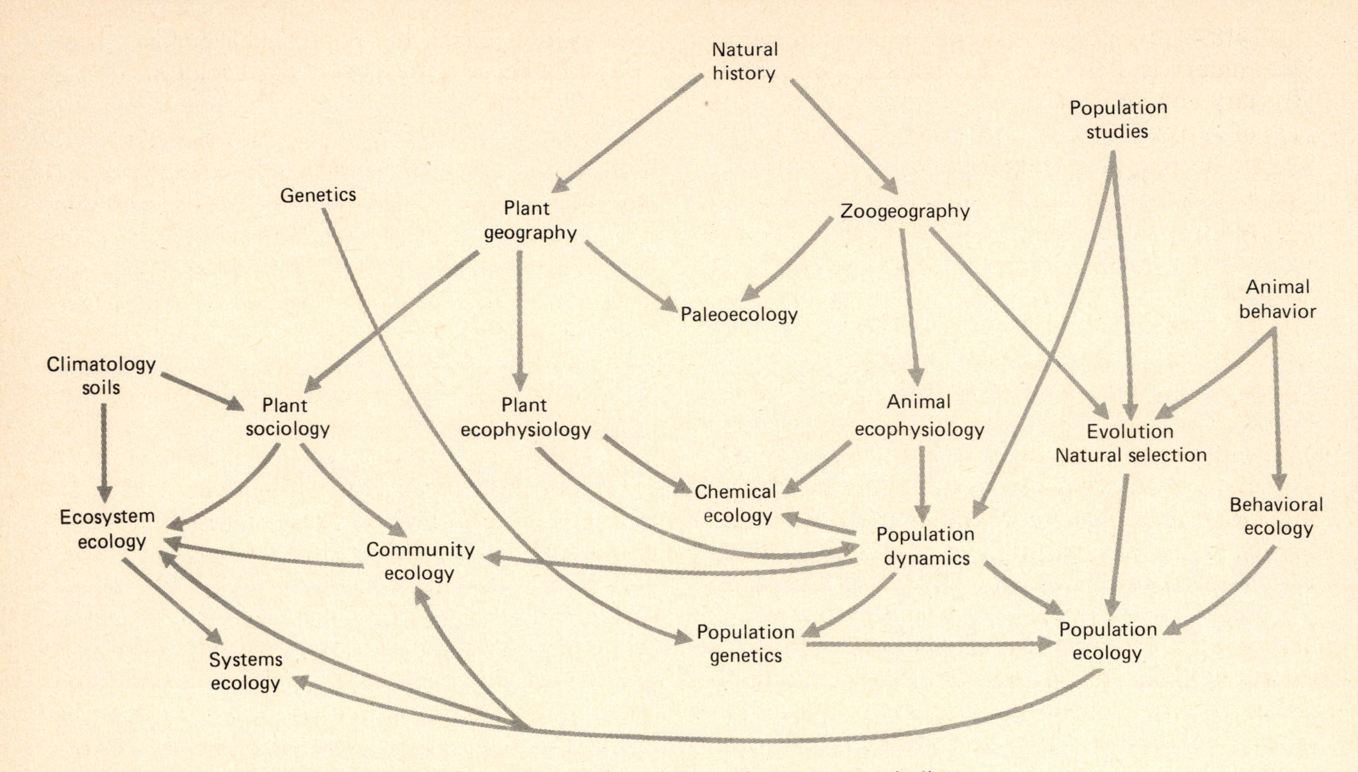

FIGURE 1.1 A suggested evolutionary pathway of modern ecology. Arrows indicate both the direction of development and the ancestry of various areas of ecology. The upper horizontal row makes up the parental stock. Natural history gave rise to plant geography, pioneered by Warming, Schimper, and Humboult, and to zoogeography, dominated by Alfred Wallace. Genetics began with Mendel. Population studies were stimulated by the ideas of Malthus. These areas gave rise to plant sociology, evolution, and natural selection. These further radiated into population ecology, community ecology, ecosystem ecology, and systems ecology. The diagram also emphasizes the wide range of ecology and shows why it is difficult, if not impossible, for ecology to evolve into a definitive science, such as physics or chemistry.

voted to plants. Humboldt described vegetation in terms of physiognomy, correlated vegetation types with environmental characteristics, and coined the term *association*.

The plants collected and the data obtained by such explorer naturalists as Humboldt provided the material for other, less traveled plant geographers. One was J. F. Schouw (1789–1852) at the University of Copenhagen, who studied the effects of major environmental factors on plant distribution. He emphasized the role of temperature on plant distribution and introduced the idea of naming plant associations after dominant species. The French taxonomist Alphonse De Candolle (1806–1893) stayed at home, but he made a major contribution to plant geography by using the sums of annual temperature as a basis for describing vegetational distribution. De Candolle's concept of climate classification was adopted and modified by the German climatologist Wladimir Köppen for his well-known Köppen's climate types.

Meanwhile, a contemporary, Anton Kerner

(1831–1898), was commissioned by the Hungarian government to describe the vegetation of eastern Hungary and Transylvania. He introduced the concept of vegetational change through time, or succession, in his delightful book *Plant Life of the Danube Basin,* still available today. In addition, Kerner pioneered the use of experimental transplant gardens to study the behavior of plants taken from different topographic elevations.

The work of these botanists became the basis for another generation of plant geographers, one of whom was Johannes Warming (1841–1924), who followed Schouw at the University of Copenhagen. Like Humboldt, Warming traveled to South America to study the tropical vegetation of Brazil. The outcome was a book on Brazilian vegetation notable for Warming's modern approach to plant ecology. Warming advanced the idea of life forms and the use of dominant plants to describe vegetational associations. He noted the influence of fire and time on vegetational change. But Warming contributed more. He wrote the first text on plant ecology, *Plantesamfund,* eventually translated into four languages, including English. In it Warming synthesized and unified plant morphology, physiology, taxonomy, and biogeography into a coherent whole. This book had a tremendous influence on the development of ecology.

At the same time, another botanist, Andreas Schimper (1856–1901), also traveled extensively in the tropics. The outcome of his studies was the book *Plant Geography on a Physiological Basis.* In it Schimper tried to explain regional differences in vegetation. In spite of the intended physiological approach, Schimper stressed morphology over physiology.

While Warming and Schimper were interested in the influence of environment on vegetational development, Jozef Paczoski (1864–1941) at the University of Poznan was studying vegetation-environmental interactions on a more local scale. He wrote a text in which he described how plants modify the environment by creating their own microenvironment. He introduced such concepts as shade tolerance, competition, succession, and the role of fire. In effect, he developed the field of plant sociology. Paczoski's impact on ecology was lessened, however, because his book, published in Slavic, was only belatedly discovered by ecologists outside the Slavic world.

These plant geographers and botanists built a solid foundation upon which modern ecology would rise. They influenced the development of European plant sociology, dominated by the Zurich-Montpellier school, led by Josias Braun-Blanquet, who introduced new methods of studying and quantifying plant associations. And these plant geographers stimulated the flowering of American plant ecology, dominated by F. E. Clements, with his emphasis on vegetational dynamics.

In Europe two of the pioneering plant ecologists were Christen Raunkiaer, who succeeded Warming as professor of botany at the University of Copenhagen, and A. E. Tansley of Great Britain. Raunkiaer's contributions were a life-form classification of plants that is still in use today (see page 437 and Mueller-Dombois and Ellenberg, 1974) and quantitative methods of sampling vegetation, the data from which could be treated statistically. Tansley, one of the giants of modern ecology, introduced both the term and the concept of *ecosystem.* He urged a more experimental approach to plant ecology in the field—especially in the areas of physiological ecology, life-history traits, and plant population dynamics—instead of descriptions of communities and supposed adaptations to habitats. He founded the British Ecological Society and headed a conservation movement in England. His views on research needed and future growth of ecology antedated by years the emergence of such work, which took place mostly in the 1970s.

Two early plant ecologists in the United States were J. M. Coulter of the University of Chicago and C. E. Bessey of the University of Nebraska. Their major contributions were the encouragement of their doctoral students, two of whom would shape the development of plant ecology. One was H. C. Cowles (1869–1939), a doctoral student under Coulter. A geologist turned botanist, Cowles in 1897 wrote his dissertation on "Ecological Relations of Sand Dune Flora of Northern Indiana." His geological background provided Cowles with an appreciation of change through time, and he applied that geological concept to ecological changes through time. This work marked the beginning of field studies on plant succession. Between 1898 and 1911 Cowles wrote a number of papers on sand dune succession in which he emphasized the dynamic nature of vegetation.

At the same time, another graduate student, F. E. Clements, working under Bessey at the University of Nebraska, studied the plant geography of that state. Dogmatic and convincing, Clements quickly became the major theorist of plant ecology in the United States. He described the vegetation of North America in terms of regional formations and associations as well as local variations. A classicist by training in his undergraduate days, Clements gave ecology a hierarchical framework, introduced innumerable terms (most happily extinct) and the idea of environmental indicators, described succession, and developed a theory of succession and development of plant associations in terms of an organism that still colors ecology today.

Early plant ecologists were concerned mostly with terrestrial plants. Another group of European biologists was interested in the relationship between aquatic organisms, plant and animal, and their environment—the natural history of fresh waters. Prominent among these biologists were A. Thienemann (1931) and F. A. Forel (1901). Thienemann viewed freshwater biology as ecological. He introduced the ideas of organic nutrient cycling and trophic feeding levels, using the terms *producers* and *consumers*. Forel, on the other hand, was more interested in the physical parameters of the freshwater habitats, particularly lakes. He described thermal stratification and internal seiches in lakes and introduced the term *limnology* for the study of freshwater life in his monograph on Lake Leman.

In the United States, S. A. Forbes (1887), an entomologist at the University of Illinois and the Illinois State Laboratory of Natural History (Illinois Natural History Survey), wrote a classic of ecology, "The Lake as a Microcosm," which first appeared in the *Bulletin of the Peoria Scientific Association*. This essay concerned the interrelations of life in a lake, particularly through food chains, and the role of natural selection in the regulation of numbers of predators and prey.

Wholly unrelated to limnology, but nevertheless destined to have an important influence on its future development and that of ecology, was the work of Edgar Transeau in an Illinois cornfield. Transeau was not an ecologist, much less a limnologist. He was interested in improving the productivity of agriculture by understanding the photosynthetic efficiency of the corn plant. His landmark paper "The Accumulation of Energy in Plants," which appeared in the *Ohio Journal of Science* in 1926, marked the beginning, far ahead of its time, of primary productivity and energy budgets.

Thienemann's concept of trophic levels and producers and consumers and Transeau's concept of energy budgets and primary production sparked the study of energy budgets of lakes by E. A. Birge and especially by C. Juday of the Wisconsin Natural History Survey. Juday, in a classic paper, "The Annual Energy Budget of an Inland Lake," published in *Ecology* in 1940, summarized the total energy relations of a system, not a single crop, over an entire year. The study involved not only the accumulation of energy by plants but also its movement through various feeding groups, including the decomposers.

The work of Juday and Birge influenced the research of a young limnologist at the University of Minnesota, R. A. Lindeman. He investigated the food relations of organisms found in Cedar Creek Bogs with the idea of integrating food cycle dynamics with the principles of community succession. Based on his studies Lindeman formulated his trophic-dynamic or "energy-available relationships" within the community. His 1942 paper "The Trophic-Dynamic Aspect of Ecology" represented a most significant advance in the development of modern ecology and marked the beginning of ecosystem ecology.

Lindeman's theory stimulated further pioneering work on energy flow and energy budgets by G. E. Hutchinson of Yale University and H. T. and E. P. Odum of the University of Georgia in the 1950s. Work on nutrient cycling was done by J. Ovington (1965) in England and Rodin and Bazilevic (1967) in Russia. The increasing ability to measure energy flows and nutrient cycling by means of radioactive tracers and to analyze large amounts of data with computers permitted the development of *systems ecology*, which is the application of general systems theory and methods to ecology.

As plant ecology was evolving out of plant geography, activities in other areas of natural history were assuming an important role in the development of ecology. One was the voyage of Charles Darwin on the *Beagle*, during which he collected numerous biological specimens, made detailed notes, and mentally framed his view of life on Earth. Influenced by the works of the geologist Charles Lyell, who proposed that Earth changed through time, Darwin noted how life, too, apparently changed through time. Working for years on his notes and collections, Darwin observed the relationships between organisms and environments, the similarities and dissimilarities of organisms within continental land masses and between continents. He attributed these differences to geological barriers separating inhabitants. He noted how successive groups of plants and animals, distinct yet obviously related, replaced one another.

Out of his years of study, Darwin came up with his theory of evolution and the origin of species. In developing his theory, Darwin was influenced by the writings of Thomas Malthus (1798). An economist, Malthus advanced the principle that populations grew in geometric fashion, doubling after some period of time. Experiencing such rapid growth, a population would outstrip its food supply. Ultimately, the population would be restrained by a "strong, constantly operating force—among plants and animals the waste of seed, sickness, and premature death. Among mankind misery and vice." From this concept Darwin developed the idea of "the survival of the fittest" as a mechanism of natural selection and evolution.

Meanwhile, unknown to Darwin, an Austrian monk, Gregor Mendel, was studying the transmission of inheritable characters from one generation of pea plants to another in his garden. The work of Mendel would have answered a number of Darwin's questions on the mechanism of inheritance and provided his theory of natural selection with the firm base it needed. Belatedly, Darwin's theory of evolution and Mendelian genetics were combined to form the study of evolution and adaptation, two central themes in ecology. The theoretical basis of the role of inheritance in evolution was advanced by Sewell Wright (1931); R. A. Fisher (1930); and J. Haldane (1932, 1954), who developed the field of *population genetics*, concerned with abstractions—gene frequency, selection coefficients, and population size.

The Malthusian concept of population growth and limitations stimulated the study of population dynamics. P. F. Verhulst (1838) first gave the mathematical basis of the nature of population growth under limiting conditions. Verhulst's work, also belatedly recognized, preceded the contributions of A. Lotka (1925) and V. Volterra (1926). Their theoretical contributions to the study of population growth, predation, and competition provided the foundation of *population ecology*, concerned with population growth (including natality and mortality), regulation (fluctuations, dispersal), and intraspecific and interspecific relations. The mathematical models of Lotka and Volterra were tested experimentally by the Russian biologist G. F. Gause (1934) with laboratory populations of protozoans and by the American Thomas Park (1954) with flour beetles. Many of the concepts of population genetics have been combined with ideas from population ecology to make up the field of *population biology*, which is concerned with the interactions of population dynamics and genetics in natural selection and evolution.

The ideas of natural selection, evolution, and population dynamics arose not out of plant ecology, in which the concepts of population dynamics have only recently been applied, but out of the areas of natural history related to animals. Early animal ecology developed later than plant ecology and along lines divorced from it. An early concept of the community, comparable to that advanced by plant ecologists, was presented for animals by the German zoologist Karl Mobius (1877). In the essay "An Oyster Bank Is a Bioconose, or a Social Community," he explained that the oyster bank, though dominated by one animal, the oyster, was really a complex community inhabited by many interdependent marine organisms, a number of them attached to oyster shells. Mobius (1877:123) noted that "Science possesses, as yet, no word by which such a community of living beings may be designated . . .

I propose the word bioconose for such a community." The word comes from the Greek, meaning "life having something in common."

However advanced the ideas of Mobius were for his time, the real beginnings of animal ecology can be traced to R. Hesse's *Tiergeographie auf Okologische Grundlage,* which appeared in 1924 and was translated into English in 1937 as *Ecological Animal Geography,* and Charles Elton's *Animal Ecology,* published in 1927. He defined animal ecology as the study of the sociology and economics of animals. Hesse and Elton had a very strong influence on the development of animal ecology.

Two pioneer animal ecologists in the United States were Charles Adams and Victor Shelford. Adams wrote the first text on animal ecology, *A Guide to the Study of Animals,* in 1913, and Shelford followed in the same year with *Animal Communities in North America.* The latter was a landmark work because Shelford stressed the relationship between plants and animals and emphasized the idea of ecology as a science of communities. The community concept was central to ecology until Tansley advanced the concept of the ecosystem.

The appearance in 1949 of the encyclopedic *Principles of Animal Ecology* by five ecologists from the University of Chicago—W. C. Allee, A. E. Emerson, Thomas Park, Orlando Park, and K. P. Schmidt—pointed the direction modern ecology would take, with its emphasis on trophic structure, energy budgets, population dynamics, natural selection, and evolution.

The discovery of the role of territory in bird life by H. E. Howard (1920) in England led to further studies by M. M. Nice in the 1930s in the United States. Out of such studies came the field of *behavioral ecology,* which was shaped in the 1940s and 1950s by two European ethologists, Konrad Lorenz of Germany and N. Tinbergen of Denmark. They developed the concepts of instinctive and aggressive behavior. Behavioral ecology has spawned a controversial offspring, *sociobiology,* which for one, holds that genetics rather than environment is the controlling force in behavioral development.

Paralleling the growth of community ecology was that of *physiological ecology.* It is concerned with responses of individual organisms to temperature, moisture, light, and other such stresses. Physiological ecology dates to Justic Liebig (1840), who studied the role of limited supplies of nutrients in the growth and development of plants. The idea of limiting factors was extended to a maximum one—too much of a good thing—by F. F. Blackman (1905). He also advanced the idea of factor interaction and studied the relationships among light, carbon dioxide, temperature and the rate of assimilation in plants. V. E. Shelford applied the concept of limiting factors to animals in a law of tolerance. It incorporated both the physiology of an organism and the environment. He suggested that organisms had both a negative and an optimal response to environmental conditions and that these responses had a role in the distribution of organisms. L. J. Henderson (1913), in a classic book, *The Fitness of the Environment,* explored the biological significance for life of the properties of matter.

Observations and later experimental work on such phenomena as the effects of plant exudates on the growth of associated species and the use of chemical defensive mechanisms by animals led to investigations on chemical substances in the natural world. These include such studies as the use of chemicals in animal recognition, trail making, courtship, and defense in both plants and animals as well as studies on the nature of the chemicals themselves. Such work has grown into the specialized field of *chemical ecology.*

Ecology has developed from so many roots that it still is and probably always will remain as Robert McIntosh (1980b) calls it, "a polymorphic discipline." Attempts to reduce ecology to a set of basic principles have not been successful. Ecology ranges over many diverse areas—marine, freshwater aquatic, terrestrial. It involves all taxonomic groups from bacteria and protozoa to mammals and forest trees, and it deals with all these at different levels: individual, populations, ecosystems. Any of these levels and groups may be studied from various points of view: behavioral, physiological, mathematical, chemical. As a result, ecology, by necessity, involves groups of specialists with little knowledge of or interest in other areas. Unfortunately, a few consider their own particular area alone to be the core of ecology.

The Development of Ecology

Paul Sears (1964) has called ecology a subversive science because it undercuts traditional thinking regarding economics and the relationship of humans to resources and environment. Ecology is also a divisive science because of strong philosophical differences that arise between groups. Some of the divisiveness has healed through time and been forgotten, but new issues arise. This divisiveness has had an important influence on the development of ecology.

PLANT ECOLOGY VERSUS ANIMAL ECOLOGY

The first split was the failure of plant and animal ecology to meet on a common ground. Ecology began with plants and later picked up animals. In England plant ecology was strongly influenced by A. E. Tansley and animal ecology by Charles Elton. Both areas were covered by one journal, sponsored by the newly formed British Ecological Society, *The Journal of Ecology*. In a few years Elton started the *Journal of Animal Ecology*. The approach to ecology, as reflected in the journals, differed. Plant ecology emphasized the plant association. Animal ecology, defined as the economics and sociology of animals, emphasized population dynamics. The two, plant and animal ecology, went their separate ways.

In America the split was less amicable. Early in the history of American ecology, a schism developed over the term *ecology*. Botanists decided at the Madison (Wisconsin) Botanical Congress in 1893 to drop the *o* from *oecology* and adopt an anglicized spelling. Zoologists refused to recognize the term *ecology*. The entomologist William Morton Wheeler complained that botanists had usurped the word, had eliminated the letter *o* improperly, and had distorted the science. He urged that zoologists drop the term *ecology* and adopt the word *ethology*. The schism was widened by a more fundamental difference in approach. Plant ecologists ignored any interaction between plants and animals. In effect, they viewed plants as growing in a world without parasitic insects and grazing herbivores. For years plant and animal ecologists went their separate ways.

Two Americans—Victor Shelford, an animal ecologist, and F. E. Clements, a plant ecologist—attempted to bring plant and animal ecology together by emphasizing how plants and animals interacted. In the book *Bio-Ecology,* they suggested that plants and animals be merged into broad biotic communities, or biomes. Clements, however, insisted that plants determined the presence of animals and never vice versa.

The force that integrated plant and animal ecology was the emergence of the "new ecology," as E. P. Odum (1964) called it, in the 1960s. Based on the ecosystem theory of Tansley, the energetic studies of Transeau and Juday, and the trophic-dynamic theory of Lindeman, the new ecology tied plants and animals together in energy-flow, nutrient cycling, and food chains. Animal ecologists, separated from plant ecologists by the gulf of Lotka-Volterra equations, found that barrier bridged when plant ecologists, led by J. L. Harper of England, began applying principles of population dynamics to plants.

Another important influence on the integration of plant ecology with animal ecology was the International Biological Program, known as the IBP. The organization of the program was stimulated by a growing concern over environmental problems facing the world. The program in general focused on the cooperative study and analysis of whole ecosystems, including tundra, coniferous forests, heathlands, deciduous forests, desert, and so on. The multiple goals of IBP included: (1) understanding the interactions of the many components of complex ecological systems; (2) exploiting this understanding to increase biological productivity; (3) increasing the capacity to predict the effects of environmental impacts; (4) enhancing the capacity to manage natural resources; and (5) advancing the knowledge of human genetic, physiological, and behavioral adaptations. IBP's greatest contribution was to increase our understanding of processes involved in the functioning of ecosystems—particularly photosynthesis and productivity, water and mineral cycling, decomposition, and the role of detritus. Although it may have lacked strong organization and coordinated direction in research, it did more than any other program to advance modern ecology and integrate plant and animal ecology.

THE ORGANISMIC GAP

Although the old ecology, with its broad divisions of plant ecology and animal ecology, disappeared, there arose a new divisiveness, which reaches back into the old ecology, dominated by F. E. Clements. He strongly influenced philosophical ideas in ecology in America and Great Britain.

Clements was influenced by the writings of Herbert Spenser, biologist, philosopher, and social Darwinist. Spenser proposed that human society was a self-evolving organism. Nutrition and growth in the human organism had its counterpart in human society in the form of profit in commerce. The circulatory and nervous systems had their counterparts in transportation and communication networks. Further, Spenser argued, as human society developed, it became more differentiated, more integrated, and advanced toward an equilibrium with environmental conditions. He saw human society as evolving from a homogeneous structure characteristic of primitive societies to a highly heterogeneous one characteristic of his own Victorian society. A part removed from a homogeneous society would not disturb it, but a part removed from a heterogeneous one could upset its natural balance. A progressive change in human society resulted from competition within societies—a Darwinian struggle for existence. Spenser cautioned against interference in the natural course of human succession.

These Spenserian ideas colored Clement's approach to plant communities. He viewed the plant community as an organism. Like an individual organism, vegetation moved through several stages of development, from colonizing bare ground to a mature, self-reproducing climax in balance with its climatically determined environment. The climax was the end, or goal, toward which all vegetation progressed. If disturbed, vegetation responded by retracing its developmental stages to the climax again. Clements (1916:3) wrote:

> The unit of vegetation, the climax formation, is an organic entity. As an organism, the formation arises, grows, matures, and dies. . . . The climax formation is the adult organism, the fully developed community, of which all initial and medial stages are but stages of development. Succession is the process of the reproduction of a formation, and this reproductive process can no more fail to terminate in the adult form of vegetation than it can in the case of the individual.

Clement's organismic approach to ecology was not lost on animal ecologists. The American zoologist and animal behaviorist William Morton Wheeler (already introduced), an international authority on ants and termites, advanced the idea that ant colonies behaved as organisms. They carried out such functions as food gathering, nutrition, self-defense, and reproduction. Like Clements, Wheeler was influenced by Herbert Spenser, but he went further. Wheeler developed a hierarchical organization of life: cells, organs, individuals, social groups, nations, all embraced by a greater ecological order.

Wheeler was also influenced by Lloyd Morgan, a biological philosopher. Morgan tried to bridge two competing views of life held by biologists of his time, vitalism versus mechanism, by substituting a theory of emergent evolution: nature evolves by sudden leaps. The new level was a synthesis of parts that left previous identity behind. It was then impossible to reduce a higher level to a lower one. $X + Y = Z$, not XY. Wheeler adapted Morgan's emergence theory to ecology. He proposed that the natural association had certain emergent properties as aggregations of organisms—predators and prey, parasites and hosts—that arose from lower levels of organization. All levels occurred together in an ecological community, or biocoenosis. The biocoenosis modified its component species through behavioral changes and new levels of integration. Everything in the biocoenosis was related to everything else. This view of a tight but orderly nature contrasted with the chaotic effect imposed on nature by humans.

This organismic, levels-of-hierarchy view of nature advanced by Clements and Wheeler captured the thinking of an influential group of ecologists at the University of Chicago—W. C. Allee, Thomas Park, Alfred Emerson, and Karl Schmidt—and Orlando Park of Northwestern University, the authors of *The Principles of Animal Ecology.* In the book they stated that the organismic concept of ecology was "one of the fruitful ideas contributed by biological science to modern civilization." The Chicago group attempted to extend the organismic concept, with its tight organization and levels of hierarchy, to human society. They promoted a philosophy based on ecology in which human society should succumb to forces of progressive evolution and worldwide integration and in which the interspecific (international) community was all-important. Their ideas, suggestive of a totalitarian society, died suddenly with World War II. The five quickly faded from the ecological scene.

Although the organismic concept dominated ecology until the 1960s, it was not accepted by all ecologists. One of the major early opponents was H. A. Gleason. In 1926 he published a paper in the *Bulletin of the Torrey Biotanical Club* titled "The Individualistic Concept of the Plant Association." In it he challenged the organismic concept and argued that the plant association was hardly an organism capable of self-reproduction. Instead, each community is unique. It arises randomly through environmental selection of seeds, spores, and other reproductive parts of plants that enter a particular area. He proposed that vegetation of an area results from "the fluctuating and fortuitous immigration of plants and an equally fluctuating variable environment. Each . . . species of plant is a law unto itself . . ." (1926:25) Gleason's ideas had little influence on ecology until the 1950s, when his individualistic concepts became the basis of a population and continuum approach to plant ecology.

The English ecologist A. E. Tansley, once enamored with the organismic concept, ultimately rejected it. Although he admitted to certain striking similarities between development of vegetation and development of individual organisms, vegetational development was quite different from the ontogeny of plants and animals. Vegetation, he allowed, might be called a quasi organism, but certainly not an organism or a complex organism.

Tansley went further. He rejected the concept of the biotic community as anthropomorphic. No social relationship exists among plants or between plants and animals, as the term connotes. In its place Tansley substituted the term *ecosystem,* in which plants and animals were components of a system that also included the effective physical factors shaping the system. Tansley also would not accept Clement's concept of the climax as the highest state of vegetation developed outside of human influence. Tansley lived in a part of the world in which vegetation was shaped by centuries of human interference. Europe and Britain had no pristine climatic climaxes. To Tansley vegetational succession was a reality, but disturbance—including that induced by humans—was the major force shaping vegetation. It destroyed preexisting ecosystems and at the same time formed new ones of a very different nature. Climaxes achieved by purely natural processes were not necessarily the ideal.

The views of Tansley were echoed in the United States by a geographer, R. Hartshore—geographers, too, had embraced the organismic idea—and by Hugh Raup, a forest ecologist and plant geographer. Raup argued that the community was not a product of long-term development in a relatively stable physical environment. Rather, vegetation was a product of repeated major disturbances caused by external forces. Vegetation responds through the various species' inherent abilities to adjust to environmental changes.

THE NEW ECOLOGY

Although the organismic concept was no longer viable by the 1960s, many of its philosophical and functional attributes live on in the new ecology. The new ecology, as defined by E. P. Odum (1964, 1971), is a "systems ecology," an "integrative discipline that deals with supraindividual levels of organization."

Ecosystems develop from youth to maturity. Each stage of development exhibits some of its own unique characteristics. Young ecosystems, for example, put energy into growth; mature ecosystems use energy for maintenance. Mature ecosystems are structurally and functionally more complex than young ecosystems. Interactions among populations and between plants and animals result in a hierarchical organization. This organization involves interacting components, which produce large functional wholes. The outcome is the emergence of new system properties that are not evident at the level below (Odum, 1971, 1982). These emergent properties account for most of the changes in species and growth forms that take place over time.

The basic philosophical argument of ecosystem ecology is that ecosystems are unique ecological entities, constrained by their own organization, which guides the evolution of their species (F. Smith, 1976). The system itself is independent of its species composition. One species can replace another without necessarily upsetting the function of the system. The approach of study is holistic because systems are considered too complex to study in bits. Since the whole is greater than the sum of its parts, ecosystems can be studied only as functional units.

An opposite view is held by theoretical (or population or evolutionary) ecology, which also emerged in the 1960s as part of the new ecology. Theoretical ecology holds that an ecosystem is the sum of its parts. By understanding how each part—the species, their numbers and characteristics—functions, one can discover how the whole system operates. Rather than guiding the evolution of species, the nature of ecosystems results from the evolution of species.

The new theoretical ecology, led largely by the late Robert MacArthur, was primarily animal ecology. But recently, theoretical ecology has been adopted by plant ecologists, especially in the field of plant population biology, led by J. L. Harper (1977).

In general, theoretical ecologists are antipathetic toward ecosystem ecology, which they regard as stagnant, barren of hypotheses, and unable to reach new and exciting dimensions. The bases of theoretical ecology are the mathematical models of Lotka and Volterra, especially their models of interspecific competition and predation, and the experimental work of G. Gause and T. Park. From that foundation theoretical ecologists have developed a body of theory relating to competition, population growth, life-history strategies, resource utilization, niche, coevolution, and the like.

If ecosystem ecology lacked theory and hypotheses, then theoretical ecology has been weighted down with too much theory that has become dogma and too many hypotheses that are untested or untestable in the field as well as experimental results involving one or two species populations that are not easily transferred to real-world situations. On the other hand, some of the theories and the field research stimulated by theoretical ecologists have provided new insights into the relationships among species, utilization of resources, and life-history patterns.

Although a division exists between ecosystem ecology and theoretical ecology, the latter is also divided within itself over certain issues. In the 1960s a bitter dispute arose over the importance of density dependence and density independence in the regulation of animal populations—whether animal populations were regulated by some external force, such as weather, or by some internal force, such as competition among members of the same species. For a number of years the dispute polarized a segment of ecology until the controversy died. A group of newer ecologists saw that population regulation involved both external and internal forces.

A current bitter dispute is over the role of interspecific competition in determining the structure and pattern of natural communities. A cornerstone of theoretical ecology is that competition is the major process shaping ecological communities. That dogma has been challenged by another group, which argues that although competition may be important, other forces are just as important or more so, particularly predation and natural disturbances. Yet the argument is considerably more complicated than that. It also involves a philosophical approach to research on the species structure of a community—for example, the value of manipulative field experiments versus natural experiments, in which one observes the outcome of ecological and evolutionary processes over time.

Ecosystem ecology and theoretical ecology may seem poles apart, yet much of the divisiveness is unwarranted. Neither side can lay sole claim to ecological truth. Certain features of ecosystems do mimic organisms. Ecosystems in a sense do mature. Young ecosystems use most of their energy for growth and biomass accumulation, and mature ecosystems expend most of their energy on maintenance. But ecosystems do so because individuals of the species involved also accumulate biomass as growth when young and use most of their energy intake for maintenance when mature. Energy and nutrients do flow through the ecosystem from one feeding level to another, but that takes place through individual organisms that make up the species population. Ecosystems probably do function independently of individual species. If one species is lost, its function may be taken over by another. Energy flow continues, even though species composition may shift. Because functionally ecosystems change little through time, evolution is constrained by functional limitations within ecosystems. In turn, ecosystems do emerge as products of the evolution of individual species.

Ecosystem ecology and theoretical ecology are closely related. Ecosystem ecology can gain from a better understanding of how each part functions—a reductionist's approach. Theoretical ecology can gain considerably from an understanding of how all the parts fit together to function as a whole.

MATHEMATICAL ECOLOGY

One feature that overrides both systems ecology and theoretical ecology is the increasing role of mathematics. Indeed, one mathematical ecologist, E. C. Pielou (1977), states that ecology is essentially a mathematical subject. She defines mathematical ecology as the construction and study of purely hypothetical models, in contrast to statistical ecology, which is the development and use of methods for interpreting observations.

Both systems ecologists and theoretical ecologists are concerned with models, but the models employed by each are very different. Systems ecologists attempt to model the functioning of ecosystems—energy flows and nutrient cycling and the response of systems to disturbance. They draw on general systems theory and techniques and concepts from engineering and depend heavily on computers to handle the complex interactions described. These models are intended to mimic natural systems and provide some insight, hopefully predictive, on how ecosystems function. Theoretical ecologists, in general, hold holistic systems models in disdain, describing them as low in predictive power and unable to yield general ecological principles. Theoretical models, supposedly more elegant, are based directly or indirectly on the Lotka-Volterra equations and on experimental models of single and two-species populations. These models, too, have severe limitations, since they rarely describe the phenomena. Many of them are little more than exercises in higher mathematics, with little regard for the basic natural history of the organisms involved.

Part of the problem arises from what R. P. McIntosh (1980b) calls "hit and run entrants" into ecology from other sciences, especially engineering, mathematics, and physics. They have left behind a trail of models that, as L. B. Slobodkin (1974) observed, "develop biological nonsense with mathematical certainty." E. C. Pielou (1981:17) has stated that "mathematical modeling is rapidly becoming an end unto itself. . . . As a result the whole body of ecological knowledge and theory has grown topheavy with models."

This does not mean that models have no place in ecology. They are in fact an important tool in explaining the functions of natural ecosystems. They are useful in forecasting both the future behavior of ecosystems and population fluctuations and in generating hypotheses. They can serve as standards of comparison, especially in population dynamics. But, as Pielou (1981:30) admonishes: "too much should not be expected of them. Modeling is only a part, a subordinate part, of ecological research," to be used, as Skellem (1972:27) warns, with "enlightenment and eternal vigilance on the part of both ecologists and mathematicians."

The Application of Ecology

In the 1970s ecology, popularly described as a new science, became involved in social, political, and economic issues. This involvement came about primarily because people became aware of the problems of pollution, overpopulation, and degraded environments. Although the public treated these environmental issues as if they were new, ecologists had been involved with such issues for years. The problem was that people were not listening. For example, in 1864 George Perkin Marsh called attention to the effects of poor land use on the human environment in his dramatic book *Man and Nature*. F. E. Clements in the 1930s urged that the Great Plains be managed as grazing land and not be broken by the plow, which they ultimately were, resulting in a dust bowl. Paul B. Sears' *Deserts on the March* (1935) was written in response to the dust bowl of the 1930s, and William Vogt's *Road to Survival* (1948) and Fairfield Osburn's *Our Plundered Planet* (1948) both called attention to the growing population-resource problem. Aldo Leopold's *A Sand County Almanac* (1949), which called for an ecological land ethic, was read largely by those interested in wildlife management until the 1970s, when it became the bible of the ecology movement.

In 1932 H. L. Stoddard introduced the idea of the role of fire in the control of plant succession in his book *The Bobwhite Quail*. Aldo Leopold expounded the application of ecological principles in the management of wildlife in his classic *Game Management* (1933). In *Forest Soils* (1954), H. L. Lutz and R. F. Chandler discussed nutrient cycles and their role in the forest ecosystem. And J. Kittredge pointed out the impact of forests on the environment in *Forest Influences* (1948).

Rachel Carson probably did more than anyone to bring environmental problems to the attention of the public. Since the publication of her book *Silent Spring* (1962), people have become more aware of the chemical poisons and other pollutants being recycled. Once castigated as more fiction than fact, Carson's predictions came only too true as carnivorous birds especially fell victim to pesticides. With a ban on the use of certain pesticides, hawks and fish-eating birds are gradually making a comeback.

Application of ecosystem and theoretical ecology to resource management has been most apparent in the past decade or so, even though economics often takes precedence over ecology. Forestry, once concerned mostly with silvics or autecology of commercial species of trees, now emphasizes biomass accumulation, nutrient cycling, the effects of timber harvesting on nutrient budgets, and the role of fire in forest ecosystems. Range management is interested in the functioning of grassland ecosystems, the effects of grazing intensities on above-ground and below-ground production by plants, and species structure. Wildlife management, once emphasizing only game species, now considers the entire wildlife spectrum, including species hunted and not hunted. The range of interest covers both the population ecology of wildlife and the maintenance and management of plant communities as wildlife habitat. In the past few years, wildlife management has developed an interest in population genetics, with an emphasis on the effects of hunting on the genetics of game species.

The National Environmental Protection Act (NEPA), enacted in 1969, required the prediction of the environmental effects of all projects financed wholly or in part by federal funds. Later, state and local environmental laws extended assessment of environmental impacts to cover more local situations. These laws required environmental impact statements (EISs) about many projects, ranging from industrial developments and power plants to dams and highway construction. EISs involve an interdisciplinary approach toward predicting environmental (as well as social and economic) consequences of planned actions, stimulating a need for applied ecologists. *Applied ecology* is concerned with predicting ecological consequences resulting from activities with goals other than biological ones and making recommendations to reduce, eliminate, or mitigate environmental damage (Ghiselin, 1981).

Generally, applied ecologists need to develop baseline studies describing existing conditions and later studies documenting the effects of changes on those conditions relative to predictions. Unfortunately, such studies become corporate secrets for various reasons, and important information is lost to ecologists and the public.

Review and Study Questions

1. Compare and contrast: (1) the organismic with the individualistic approach to ecology, (2) holism and reductionism, (3) ecosystem ecology and theoretical ecology. How are all of these related?
2. Make a list of the major figures in the history of ecology and note their contributions.
3. Why should there be disagreements among the various areas of ecology? How do these arguments affect the "holism" of ecology?
4. Has some familiarity with the history of ecology changed your original concept of ecology?

Note: The purpose of this chapter is to stimulate your interest in the history and background of ecology. Many of the ideas briefly introduced here will be expanded in the chapters ahead. The best way to study this chapter is to read more on the history of ecology and to review some of the original papers mentioned.

Suggested Readings

Brewer, R. 1960. A brief history of ecology. Part I. Prenineteenth century to 1910. Occas. Paper no. 1. Kalamazoo, Mich.: C. C. Adams Center for Ecological Studies.

Clements, F. E. 1916. *Plant succession: An analysis of the development of vegetation.* Publ. 242. Washington, D.C.: Carnegie Institute.

Egerton, F. N. 1976. Ecological studies and observations before 1900. In *Issues and ideas in America,* ed. B. J. Taylor and T. J. White, 311–351. Norman: University of Oklahoma Press.

———. 1977. A bibliographical guide to the history of general ecology and population ecology. *Hist. Sci.* 15:189–215.

Egerton, F. N., ed. 1977. *History of American ecology.* New York: Arno Press.

Eiseley, L. 1958. *Darwin's century: Evolution and the men who discovered it.* Garden City, N.Y.: Doubleday.

Elton, C. 1927. *Animal ecology.* London: Sedwick & Jackson.

Gleason, H. A. 1926. The individualistic concept of the plant association. *Bull. Torrey Bot. Club* 53:7–26.

———. 1927. Further views on the succession concept. *Ecology* 8:299–326.

———. 1975. Delving into the history of American ecology. *Bull. Ecol. Soc. America* 56:5–10.

Kendeigh, S. C. 1954. History and evaluation of various concepts of plant and animal communities in North America. *Ecology* 35:152–171.

Kormondy, E. J., ed. 1974. *Readings in ecology.* Englewood Cliffs, N.J.: Prentice-Hall. (Excerpts from a wide range of historical papers.)

Kormondy, E. J., and J. F. McCormick, eds. 1981. *Handbook of contemporary developments in world ecology.* Westport, Conn.: Greenwood Press.

Lindeman, R. 1942. The trophic-dynamic aspect of ecology. *Ecology* 23:399–418.

McIntosh, R. P. 1976. Ecology since 1900. In *Issues and ideas in America,* ed. B. J. Taylor and T. J. White, 353–372. Norman: University of Oklahoma Press.

———. 1980. The background and some current problems of theoretical ecology. *Synthese* 43:195–255.

Margalef, R. 1968. *Perspectives in ecological theory.* Chicago: University of Chicago Press.

Odum, E. P. 1964. The new ecology. *Bioscience* 14:14–16.

———. 1968. Energy flow in ecosystems: A historical review. *Amer. Zool.* 8:11–18.

Raup, H. 1964. Some problems in ecological theory. *J. Ecology* (Suppl.): 19–28.

Tansley, A. E. 1935. The use and abuse of vegetational concepts and terms. *Ecology* 16:284–307.

Van Dyne, G. 1969. *Ecosystem concept in natural resource management.* New York: Academic Press.

Worster, D. 1977. *Nature's economy.* San Francisco: Sierra Club Books.

PART II
NATURAL SELECTION AND SPECIATION

INTRODUCTION

Natural Selection

In 1831—on the 27th of December, to be exact—young Charles Darwin, 22 years old, shipped aboard the HMS *Beagle,* a surveying ship of the British Navy, as a naturalist. The *Beagle* sailed from Plymouth, England, to the eastern coast of South America, through the Strait of Magellan, past Tierra del Fuego, up the west coast to Peru, and then across the water to the Galapagos Islands. From there the ship sailed to the South Sea Islands, New Zealand, and Australia, across the Indian Ocean, around the Cape of Good Hope, and on to England. Five years later Darwin returned, his notebooks filled with observations, his boxes filled with specimens of rocks, plants, and animals, and his head filled with ideas on evolution. On this trip he explored the jungles and pampas of South America and climbed the Andes. On the pampas he unearthed the fossil remains of prehistoric creatures—the glyptodon, the megatherium, and the guanaco—and he noted the similarities and differences between these and such existing species as the armadillo, the sloth, and the llama. He wondered that two such similar forms, one living and the other extinct, should exist in exactly the same part of the world. About this, Darwin later wrote (1887): "The wonderful relationship in the same continent between the dead and the living will, I do not doubt, hereafter, throw more light on the appearance of organic beings on our earth, and their disappearance from it, than any other class of facts."

During frequent stops along the South American coast, Darwin noticed that although individuals of the same species were identical or nearly so in one locality, they were slightly different in another. The nearer the locations were, the more closely did the two populations resemble each other. The more distant they were, the greater was their divergence. So often did Darwin observe this subtle difference within so many different species that he was convinced it was a general rule.

Darwin pondered these similarities and differences between fossil and living representatives and between populations of animals separated by space. He eventually came to the conclusion that species gradually became modified with the passage of time, an idea already advanced by the French biologist Lamarck. In a similar manner, he reasoned, when a species consisting of a homogeneous group increased its range into new habitats, the organisms would evolve or change in different ways, resulting in slightly different races in each region.

But the questions "How did evolution occur?" and "What are its mechanisms?" still remained unexplained. Then, in October 1838 Darwin happened to read "An Essay on Population," by Malthus, which considered the relationship between population size and food supply. Malthus (1798: 9–10) wrote:

> Through the animal and vegetable kingdoms, nature has scattered the seeds of life abroad with the most profuse and liberal hand. She has been comparatively sparing in the room, and nourishment necessary to rear them. The germs of existence contained in this spot of earth, with ample food, and ample room to expand in, would fill millions of worlds in the course of a few thousand years. Necessity, that imperious all-pervading law of nature, restrains them within prescribed bounds. The race of plants, and the race of animals shrink under this great restrictive law. And the race of man cannot, by any efforts of reason, escape from it. Among plants and animals its effects are waste of seed, sickness, and premature death. Among mankind, misery and vice.

From this Darwin developed his concept of "the struggle for existence," which forms an important part of his hypothesis regarding the mechanism of evolution.

Darwin reasoned from his observations on the differences among living things that some variations were more advantageous than others, that some variations enabled the organisms to occupy an area or to survive. Since only a few animals would survive, Darwin argued, then those with more favorable variations would have better odds for survival. "Under these circumstances," wrote Darwin (1887: 120), "favorable variations would tend to be preserved, and unfavorable ones to be destroyed. The result of this would be the formation of a new species. Here then I had at last got a theory by which to work. . . ."

This process that resulted in a greater survival of individuals possessing advantageous characteristics over those with less advantageous ones Darwin called *natural selection*. As a result of it, favorable variations will be retained in the population and will increase. Through time, this will result in a population better adapted to its environment; if continued long enough, new and different species will evolve.

Although Darwin's theory of evolution is in essence accepted today, Darwin hesitated to publish it. He started to develop his hypotheses in 1837, and in 1842 he wrote a penciled 35-page abstract of his ideas, which he enlarged to 255 pages in 1844. He still withheld publication but discussed his ideas with two friends, the geologist Sir Charles Lyell and the botanist Dr. J. D. Hooker. In 1856 Lyell urged Darwin to publish his material, and by 1858 he had his book half completed.

Then in the summer of that year, Darwin received a letter from a fellow naturalist, Alfred Russel Wallace, who was exploring the Malay Archipelago. Its contents shocked Darwin, for with the letter, Wallace sent an essay entitled "On the Tendency of Varieties to Depart Indefinitely from the Original Type." This essay contained an excellent summary of Darwin's own theory of evolution by natural selection.

After much reflection by Darwin over the dilemma, and following the intervention of Lyell and Hooker, the two men, who had independently arrived at the same conclusion, presented their views jointly, through Lyell and Hooker, to the Linnaean Society of London on July 1, 1858, and subsequently published them in the society's journal.

Darwin pointed out that animals as well as plants produce more offspring than necessary to maintain the species. He concluded that the number of organisms must be held in check by their own struggle either with other individuals of the same species or different species or against "external nature."

Wrote Darwin (1884:119):

> Now, can it be doubted, from the struggle each individual has to obtain subsistence, that any minute variation in structure, habits, or instincts adapting that individual better to new conditions would tell upon its vigor and health? In the struggle it would have a better *chance* of surviving; and those of its offspring which inherited the variation, be it ever so slight, would also have a better *chance*. Yearly more are bred than can survive; the smallest grain in the balance, in the long run, must tell on which death shall fall, and which shall survive. Let this work of selection on one hand, and death on the other, go on for a thousand generations, who will pretend to affirm that it would produce no effect, when we remember what, in a few years, Blakewell ef-

fected in cattle, and Western in sheep, by this identical principle of selection?

Wallace stated much the same theory, except that he emphasized population control largely by food supply. Like Darwin, Wallace concluded that forms best adapted to their environment evolved through selection of the individual. Wallace (1858: 274) argued that if any environmental change made existence difficult, then of all the individuals composing the species, those forming the least numerous and most feebly organized variety would suffer first, and, were the pressure severe, must soon become extinct. The same causes continuing in action, the parent species would next suffer, would gradually diminish in numbers, and with a recurrence of similar unfavorable conditions might also become extinct. The superior variety would then alone remain, and on a return to favorable circumstances would rapidly increase in numbers and occupy the place of the extinct species and variety.

Darwin further expounded his theory of natural selection and evolution in the *Origin of Species*, published in November 1859. Darwin received, as he should, the major credit for arriving at a theory of evolution, while Wallace is best known for his fundamental studies of the distribution of animals.

Darwin's theory of evolution can be summarized briefly: Variation exists among individuals of sexually reproducing species that affects their chances of survival and their reproductive rates. In addition, many species have such a high potential rate of increase that if unchecked they would exhaust both food and living space. Since food and space are limited, those with the most advantageous variations will have the better chance to survive. Constant selection of the better adapted and the elimination of the less fit result in the evolutionary change of populations. This is "survival of the fittest."

Neither Wallace nor Darwin could explain adequately or fully understand the nature and origin of variation or how these variations were transmitted from parent to offspring, although they realized that they were inherited. Had they known, much of the storm of controversy over the theory might never have developed. Ironically, the answers to Darwin's most pressing questions on variation and a basis for understanding the mechanisms of evolution were being discovered in a monastery garden by a contemporary—an Austrian monk, Gregor Mendel. His findings on the laws of inheritance, presented as a lecture to the Natural History Society of Brunn in 1866 and published in the *Transactions* of the society, were unappreciated by the scientific world at that time and remained lost in the obscure journal until 1900, when he was discovered independently by three biologists: deVries, Correns, and Tschermak.

CHAPTER 2

Natural Selection and Evolution

Outline

Objectives

Upon completion of this chapter, you should be able to:

1. Briefly describe the changes in Earth through geological time.
2. Explain the differences among adaptation, natural selection, and evolution.
3. Explain the importance of mutation, recombination of genes, and multigene families as sources of genetic variation.
4. Distinguish among directional, disruptive, and stabilizing selection.
5. Tell what is meant by coevolution.
6. Discuss other influences on evolution that lie outside of natural selection.
7. Discuss group selection and kin selection.

The Evolution of Life: An Overview

To all of us Earth appears immutable. Waters of oceans wash the shores of continents that seemingly never change. Familiar mountain peaks have looked the same through centuries. Even local hills, rivers, and other topographic features have an air of permanence about them. But Earth was not always as it is today, nor will it be the same tomorrow, because changes are continually taking place, however slowly. Volcanoes erupt, new islands appear in oceans, beaches erode. Physical features of Earth—its oceans and land masses, its vegetation and animal life—have changed radically through time. These changes of the past have influenced life on Earth today.

Earth, geologists estimate, is some 4,600 million years old (Table 2.1). The first era, the Precambrian, which extended from 4,600 million years ago to 570 million years ago, saw the production of an atmosphere and a hydrosphere; the evolution of preliving compounds and, later, autotrophic forms of life; the internal reorganization of Earth; and the development of ocean basins and continents.

In the early Paleozoic, some 540 million years ago, three separate land masses existed: Asia, North America and Europe, and Gondwanaland, which included modern-day Africa, South America, Australia, New Zealand, and Antarctica as well as scattered fragments of land. During the Paleozoic, 420 million years ago, North America and Africa lay close together around the south pole, and the rest of Gondwanaland lay on the far side of the south pole, pointing toward the equator. Slowly the land mass moved northward, so that by the Carboniferous age, 340 million years ago, the whole of Africa had moved across the south pole and Antarctica lay in the region of the south pole. Glaciers covered southern South America, South Africa, India, and Australia; and Europe and North America lay along the equator. There the climate was warm, humid, and seasonless, and a large part of the area was covered with swamps and tropical rain forests. During the Permian these three blocks joined, raising the Caledonian, Ural, and Appalachian mountain ranges and forming a single land mass, Pangaea (Figure 2.1).

As Pangaea moved northward, it began to break apart slowly, 5 to 10 cm a year. The first break in the single land mass apparently took place in the mid-Mesozoic age, approximately 180 million years ago, when North America and Africa parted to form the first narrow strip of the Atlantic. Africa and South America were still connected by the end

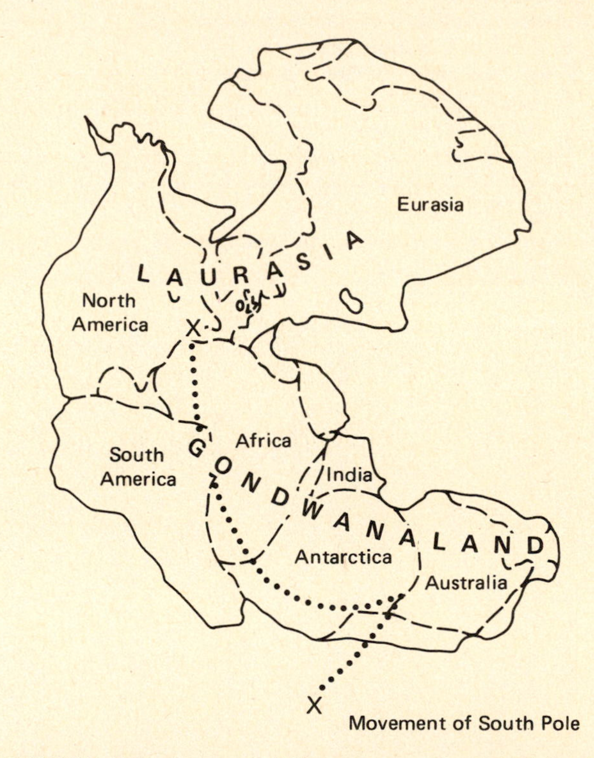

FIGURE 2.1 A map of Pangaea as it existed in the Permian and Triassic (without Southeast Asia because its position at this time is uncertain). The dashed lines indicate the outlines of present-day continental coastlines. Note that Laurasia, consisting of North America and Eurasia, is attached to Gondwanaland, the continents of Africa, South America, Antarctica, and Australia. The series of dots indicates the successive positions of the south pole from the Cambrian to the Jurassic. This provides some indication of the northward movement of the land masses over millions of years. (Redrawn from Cox, Healey, and Moore, 1973:177; Oxford, England: © Blackwell Scientific Publications.)

TABLE 2.1 GEOLOGICAL TIME SCALE.

Era	Period	Epoch	Age, Millions of Years	DOMINANT LIFE Plants	DOMINANT LIFE Animals
Cenozoic: The age of mammals	Quaternary	Recent	0.01	Agricultural plants	Domesticated animals
		Pleistocene	2		Ice Age—modern humans; mixture and then thinning out of mammalian faunas
	Tertiary	Pliocene	10	Herbaceous plants rise; forests spread	Culmination of mammals; radiation of apes
		Miocene	25	First extensive grasslands	
		Oligocene	35		Modernization of mammals; mammals become dominant
		Eocene	55		Mammals become conspicuous
		Paleocene	70		Expansion of mammals; extinction of dinosaurs
Mesozoic: The age of reptiles	Cretaceous		135	Angiosperms or flowering plants rise; gymnosperms decline	Dinosaurs reach peak; first snakes appear
	Jurassic		180	Cycads prevalent	First birds and mammals appear
	Triassic		230	Gymnosperms rise; seed ferns die out	First dinosaurs; reptiles prominent
Paleozoic	Permian		280	Conifers become forest trees; cycads important	Great expansion of primitive reptiles
	Carboniferous Pennsylvanian		310	Lepidodendron, sigillaria, and calamites dominant; the swamp forest	Age of cockroaches; first reptiles
	Mississippian		345	Lycopods and seed ferns abundant	Peak of crinoids and bryozoans
	Devonian		405	First spread of forests	First amphibians; insects and spiders
	Silurian		425	First known land plants	First land animals (scorpions)
	Ordovician		500	Algae, fungi, bacteria	Earliest known fishes; peak of trilobites
	Cambrian		600	Algae, fungi, bacteria; lichens on land	Trilobites and brachipods; marine invertebrates
Precambrian	Late			Algae, fungi, bacteria	First known fossils
	Early		4,500	Bacteria	No fossils found

of the Jurassic, but drift was taking place in the South Atlantic region. By the middle Cretaceous, about 100 million years ago, Africa and South America had split apart. By the late Cretaceous, southern Greenland had begun to separate from the British Isles and move northward. South America and Antarctica still clung together until sometime in the early Cenozoic, finally becoming separated in the Eocene. Africa had already become separated from Antarctica between the middle Jurassic and the middle Cretaceous. Thus, by the end of the Cretaceous, Gondwanaland (Figure 2.2) had broken up; the only intact land mass was North America/Eurasia, collectively known as Laurasia.

Until the lower Eocene, Laurasia remained intact. North America was connected to Europe by Greenland and Scandinavia. But in the mid-Eocene, the North Atlantic joined the Arctic Ocean, separating Laurasia into North America and Europe.

The formation, breakup, and northward drift of continents resulted in broad climatic changes and the formation of geological barriers that affected evolving plant and animal life. Between 2,700 million and 2,000 million years ago, prokaryote (lacking chromosomes and nuclei) bacteria and photosynthetic blue-green algae developed. Eukaryote (with characteristic nuclei and chromosomes) organisms evolved 1,800 million to 1,000 million years ago. The Precambrian era came to a close with the sudden rise of Metazoa, a diverse and complex assemblage of animals in ancient seas that ushered in the Paleozoic era.

The Paleozoic saw the evolution of the earliest known fish, the first amphibians, the first reptiles, the first insects, and the first land plants as well as the rise of the great coal forests. The latter, dominant on the North American and European continents that at that time lay on the equator, consisted of such tall trees as *Lepidodendron, Sigillaria,* and *Cordaites,* forerunners of coniferous trees.

The end of the Paleozoic witnessed the expansion of primitive reptiles that ushered in the Mesozoic era, the age of reptiles. The Mesozoic, between 230 million and 70 million years ago, saw the rise and fall of dinosaurs and an explosive evolutionary radiation of the angiosperms. Throughout the Mesozoic no distinct floral or faunal regions existed. The land, even though partially separated, was one, with no effective barriers to dispersal of plants and animals. Mountain ranges lay only about the edges of Pangaea, and although shallow seas invaded the land, the invasions were short-lived, geologically. The climate, too, was warm and equitable, mostly tropical to subtropical, even to the coast of Alaska. These conditions allowed the great reptiles and early mammals to roam freely over continental land masses.

But in the late Cretaceous, climatic conditions began to change as continental land masses drifted apart. The warm, subtropical middle Cretaceous climate was replaced in the late Cretaceous by a warm, temperate climate. The cooling marked the end of the great reptiles. As the Cretaceous period moved into the Paleocene period of the Cenozoic, greater changes took place. Between the lower Cretaceous and the Eocene, the single, connected land mass inhabited by gymnosperms and reptiles was replaced by a divided land mass inhabited by flowering plants and mammals. Continental drift effectively separated plant and animal populations and hastened the development of different floras and faunas. For example, when Gondwanaland broke apart, the resulting continental land masses carried with them similar flora. But Laurasia retained the conifers that had arisen from the *Cordaites*. Conifers could not spread across the hotter equatorial regions. Because of the colder climate in the northern part of Laurasia, a still different flora evolved in northern and central Eurasia.

Although plants achieved a worldwide distribution before the continents broke apart, mammals did not. They never achieved any significance in the fauna until the continents were well on their way to separation. One of the first groups of mammals, the marsupials, confronted competition from more advanced placental animals. Had the placentals achieved dominance before the breakup of continents, the marsupials might never have survived. But marsupials had spread to Antarctica and Australia before these land masses separated. Antarctica moved south, where the cold eliminated terrestrial mammalian life, while Australia, separate from

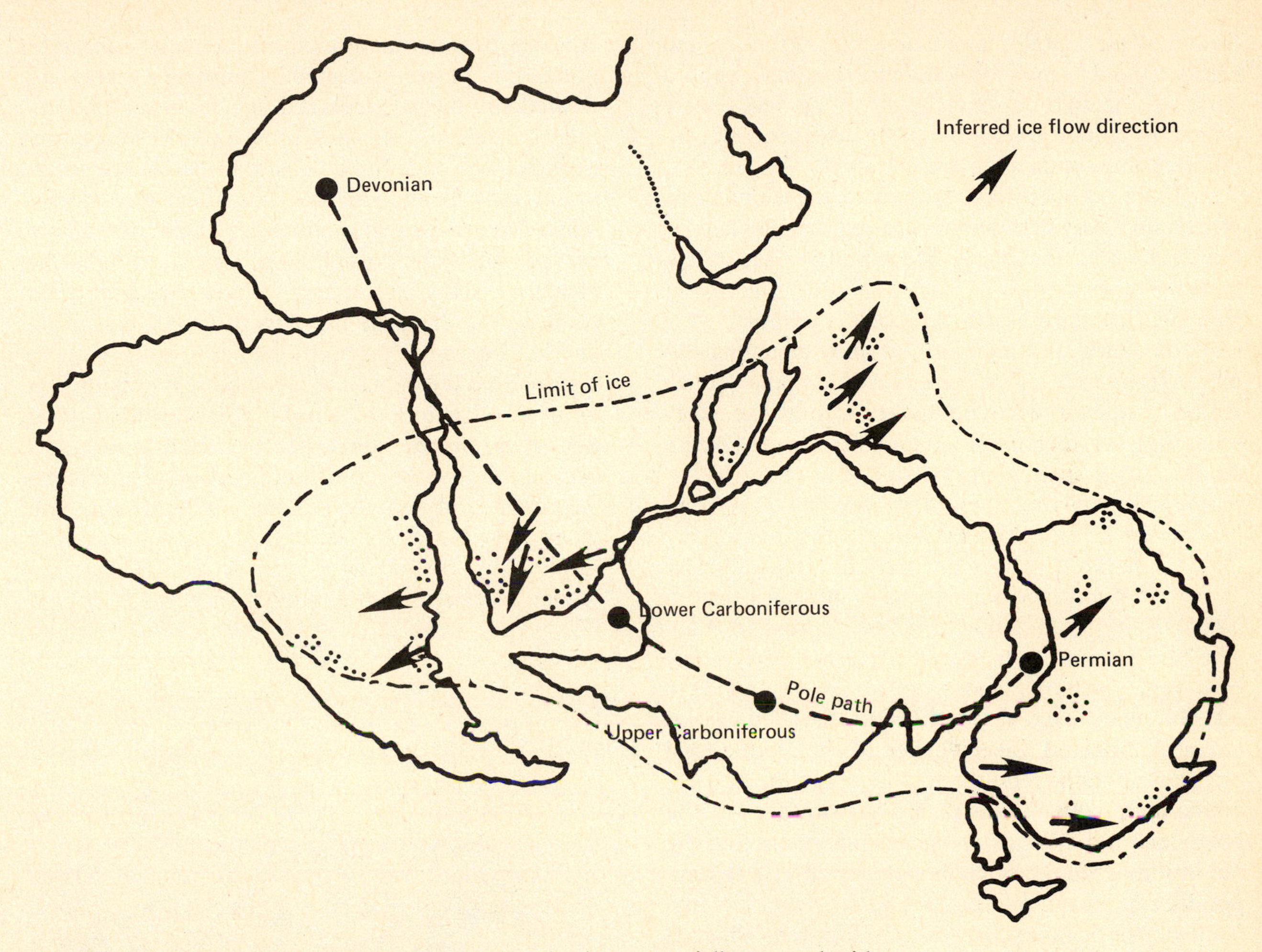

FIGURE 2.2 The nested continents of Gondwanaland were partially covered with glaciers in the late Carboniferous and early Permian, a time when North America and Eurasia lay on the equator. The dashed line circumscribes the limit of the single great ice sheet. Arrows indicate the inferred direction of ice flow. The black dots and associated broken line indicate the path and location of the wandering south pole. In the Cretaceous period, Gondwanaland had broken apart. In the late Carboniferous and early Permian, Gondwanaland was partially covered with glaciers. Later, as Gondwanaland moved south, the southern limit of the land mass again became buried under a permanent layer of ice. (From A. N. Strahler, 1971.)

all other land masses, became a final refuge for marsupial life. Free from competing placental forms, marsupials were able to radiate into a variety of forms and to fill niches similar to those occupied by placental mammals elsewhere. Meanwhile, marsupial mammals that existed in North America and South America were rapidly replaced by placental mammals, except for the opossum. South America, isolated from North America during the lower Cenozoic (Tertiary), supported diverse and unique forms of placental mammalian life. After a land bridge, the Isthmus of Panama, became exposed between the two continents, these mammals were unable to compete with the more advanced placentals that moved down from the north.

While competition resulted in the elimination of certain kinds of mammals, a major influence on the developing fauna was the rapidly changing climate of the northern land mass. In the lower Eocene, Europe, North America, and Asia were still connected in a manner that allowed animals to move from one to the other. At that time North America appeared to be the center of the early evolution of placental mammals. From North America they spread to Europe, but the cold climate inhibited the movement of mammals from North America to Asia. When North America became separated from Europe, new mammalian groups that evolved in Europe could not cross to North America, but they did have access to Asia. The mid-Eocene saw the return of a warm, moist climate; a semitropical rain forest extended to Alaska, and tropical conditions prevailed as far north as England. The climate of the Bering land bridge was benign enough to encourage the passage of mammals from Asia to North America. In the Oligocene the climate cooled again, restricting the migration of mammals to those tolerant of cold temperatures and eliminating tropical forms of plants from northern lands. From the Oligocene through the Pliocene, the cooling trend continued. The movement of animals from Eurasia to North America was restricted to such cold-tolerant species as mammoths and humans, and the flora of the continent was basically a modern one.

Meanwhile, the fauna of the Old World tropics remained free of the influences of great temperature changes, although the increasing aridity in Africa that began in the Oligocene and Miocene brought about the replacement of tropical forests by grasslands, on which evolved huge herds of ungulate fauna and eventually humans.

Cooling of the Pliocene continued until the Pleistocene ushered in the Ice Age (Figure 2.3). Uplift of continents, volcanism, and contrasts in climate placed severe stress on flora and fauna, but these factors were nothing compared to the great ice sheets. Northern Europe and northern North America were transformed periodically into arctic regions as glaciers formed. Temperate forests retreated southward, replaced first by boreal forests, then by tundra, and finally by ice. These ice ages occurred not once but at least four times. Ice sheets swept rock from one area and deposited it elsewhere, wore down mountains, filled in valleys, carved out lakes (including the Great Lakes), changed sea levels and water temperatures, changed the drainage patterns of rivers, and in the end left behind relict arctic climates in unglaciated regions that bordered the ice sheets. It was the age of woolly mammoth, woolly rhinocerous, mastodon, royal bison, camel, peccary, saber-toothed tiger, and dire wolf and of the invasion of the North American continent by humans. It was also the time when a number of species of birds, fish, and insects, isolated by the advance and retreat of glaciers, tended to evolve into a number of new species.

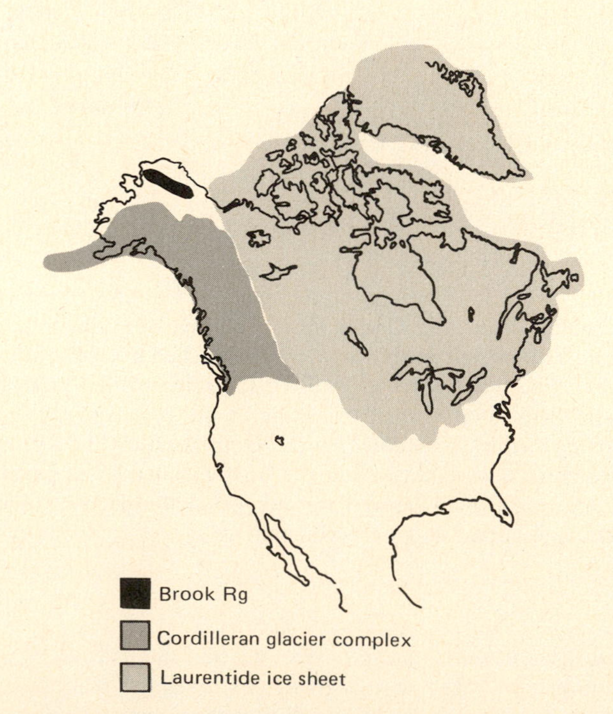

FIGURE 2.3A *Glaciation during the Pleistocene.* Northern North America was covered by four ice sheets, the limits of each usually marked by a terminal moraine. The last, and perhaps the most significant, was the Wisconsin ice sheet. (After Flint, 1970:545)

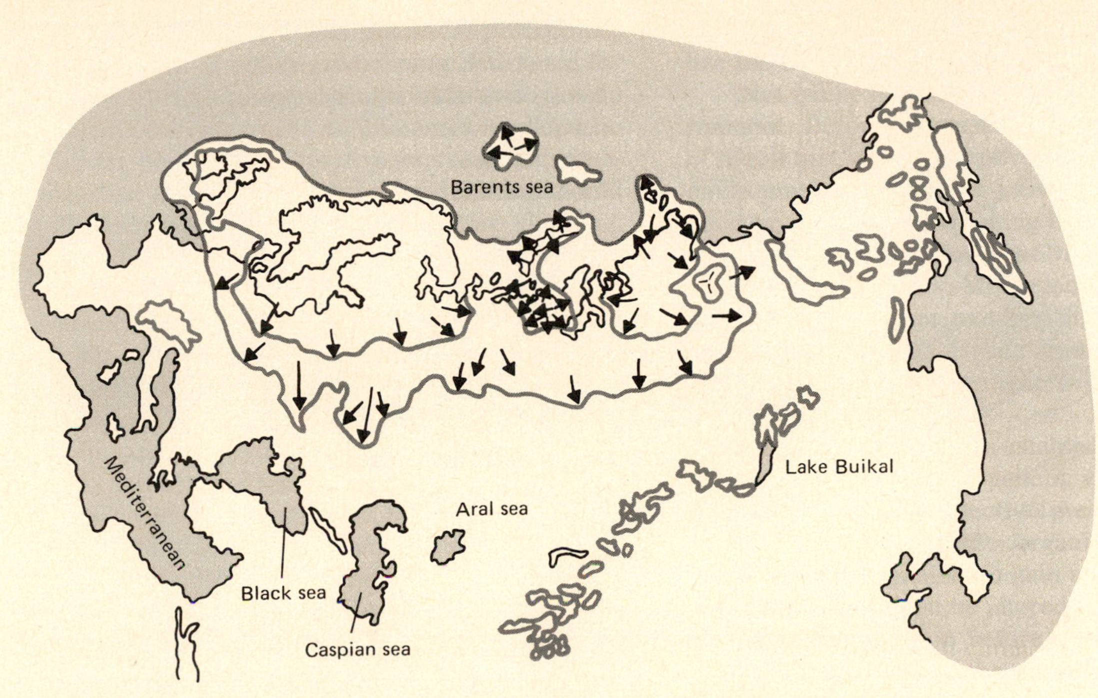

FIGURE 2.3B *Glaciation during the Pleistocene.* Northern Eurasia was covered by ice sheets similar to those covering North America. The most important was the last, known as the Weischselian. Note the disjunct glacier in the region of the Alps. (After Flint, 1970:545)

The Nature of Natural Selection

The preceding cursory review of past changes on Earth serves to emphasize that plants and animals have evolved through time. As environmental conditions slowly changed, flora and fauna responded—either adjusting to a new set of conditions or perishing. Earth is still a dynamic planet. Continents continue to drift, and environments are changing more rapidly than they did in the past because of massive human intervention. To survive as a species, organisms must adapt to the changing environment. The problem is that many conditions are changing faster than organisms can adapt to them.

ADAPTATION

Adaptation implies the ability of an organism to live in harmony or conformity with its environment. Adaptation comes about through the interaction of organisms with their environment. If an organism can tolerate a given set of conditions to such a degree that it can not only survive as an individual but also leave mature reproducing progeny in the population, it contributes its genetic traits to the population's gene pool. It is adapted to its environment. If an organism leaves few or no mature reproducing progeny, it contributes little or nothing to the gene pool. It is poorly adapted to its environment. Those individuals that contribute the most to the gene pool are said to be the most fit, and those that contribute little or nothing to the gene pool are said to be the least fit. The *fitness* of an individual is measured by its reproducing offspring, especially descendants, over several generations. That differential reproductive success of individual organisms is *natural selection.* It is the ability of individuals to leave the most reproducing offspring.

Adaptation is one of those terms in ecology that has been pushed nearly into a state of uselessness because it has been employed in different ways (see Stern, 1970; Harper, 1982). One, and perhaps the most common, meaning of *adaptation* is any form of behavior that is assumed to be the result of natural selection. In other words, natural selection is responsible for most of what we observe in plants and animals. A second meaning is a change in the physical, physiological, or behavioral traits that results from some environmental pressure. The change supposedly improves the ability of one organism to survive and reproduce compared to another that did not change. A third meaning is any physical or physiological feature or form of behavior used to explain the ability of an organism to live where it does. Adaptations are observed because some ancestors left more descendants than others.

A current argument (see, for example, Gould, 1982; Harper, 1982) is that although some features of an organism may be the result of natural selection, others may have different origins. Some features and functions may have been put into current use from a previous use. Or a structure or feature currently in use may have been derived from a structure that had no previous function. Not every feature of an organism is necessarily functional. If neutral in its effects, or if an accidental by-product of other selection changes, a feature may be retained because it is not subject to any selection pressures. As such, it becomes a source of raw material for future evolution (see Box 2.1).

Box 2.1
Adaptation: The Term

Adaptation is one of a number of ecological terms used to cover so many situations that it has become almost meaningless. To bring the term back into a more precise usage, S. J. Gould of Harvard University and Elizabeth Vrba (1982) of the Transvaal Museum in Pretoria, South Africa, have suggested the following, including a couple of new terms using the root word *aptus* (''fit''):

Aptation. The static phenomenon of being fit.

Adaptation. Any structure or behavior that has been pushed toward fitness by natural selection.

Exaptation. Any structure or behavior that has been shifted into current use from a different previous use or co-opted for its current use from a structure that previously had no function.

Preaptation. Potential but unrealized exaptations; corresponds to **preadaptation**, a word that implies evolution of a structure in anticipation of some future function.

These terms have certain advantages over the word *adaptation* as currently used and probably will be accepted and used in ecological literature.

VARIATION: THE BASIS OF NATURAL SELECTION

The community, aquatic or terresterial, consists of many different, locally defined groups of individuals similar in structure and behavior. Individuals within these groups interbreed, oak tree with oak tree, white-footed mouse with white-footed mouse, largemouth bass with largemouth bass. Collectively, individuals within each group make up a genetic population, or *deme*.

Beyond one local population may be other similar demes. They may be separated by distance, or they may be only indistinctly separated—more or less adjacent and continuous over a wide area. Whatever the situation, hereditary material to a greater or lesser degree passes from one population to the other. Some adjacent demes may interbreed so freely that they become essentially one. Others may not interbreed at all. If a local population of a plant or animal dies out, it will, if conditions are favorable, be replaced by individuals from surrounding populations. Individuals that die are replaced by their offspring, so that the population tends to persist through the years, the inheritable features passing from one generation to the next.

Individuals that make up the deme are not identical. Just as wide individual variation exists among human populations, so the same variation exists among individuals of sexually reproducing plants and animals. This variation is the raw material of natural selection.

That variation exists within a population can easily be demonstrated. All one needs to do is to select some 100 or so specimens from a local population and observe and record the variations of a single character: the tail length of some species of mouse, the number of scales on the belly of a snake, the shapes and sizes of sepals and petals on flowers, the rows of kernels on ears of corn. These observations can be tabulated as frequency distributions (Figure 2.4). Many of the specimens will have characteristics with the same value. The value that is most common is called the *mode*. Other values will vary above and below the mode. The frequencies of these values fall off away from the mode; that is, fewer and fewer individuals are in each class.

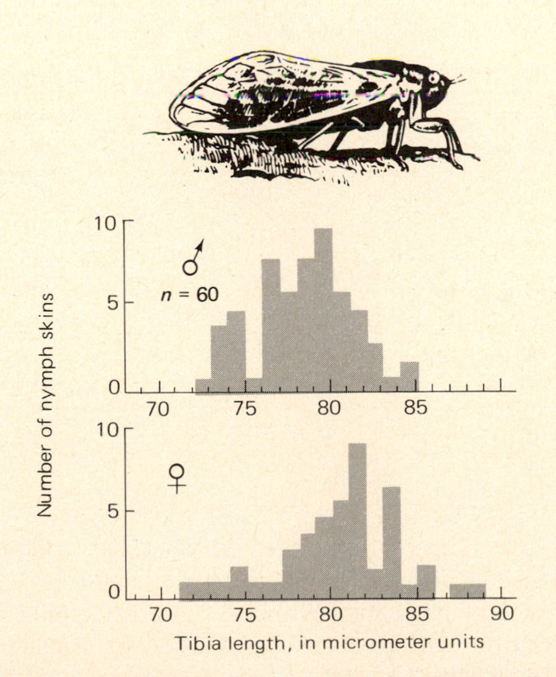

FIGURE 2.4 A histogram showing the frequency distribution of the hind tibia lengths of nymphal exuviae (shed skin) of the periodical cicada, *Magicicada septendecim*. (After Dybas and Lloyd, 1962:447.)

The frequency distribution of these variable characters tends to follow a bell-shaped curve, the normal curve of probability. In some situations the distribution may deviate from the normal bell-shaped curve. These differences may point out some facts about variation within a population. The variations may be due to heredity, to environment, or, more often than not, to an interaction of both.

The characteristics of a species and variations in individuals are transmitted from parent to offspring. The sum of hereditary information carried by the individual is the *genotype*. The genotype directs the development of the individual and produces the characters that make up the individual's morphological, physiological, and behavioral characteristics. The external, or observable, expression of the genotype is the *phenotype*.

Some of the conspicuous individual variants in the phenotype—such as shortened tails, missing appendages, enlarged muscles, or other features that result from disease, injury, or constant use—are not inheritable. Those are acquired characteristics, which the early evolutionist Lamarck hypothesized erroneously were passed down from one generation to another. However, some acquired characteristics that have been environmentally induced are inherited; or to say it more accurately, the ability of an organism to acquire such characteristics is inherited. The ability of a genotype to give rise to a range of phenotypic expressions under different environmental situations is known as *phenotypic plasticity*. Some genotypes have a narrow range of reaction to environmental conditions and therefore give rise to fairly constant phenotypic expression. Some of the best examples of phenotypic plasticity are found among plants. The size of plants, the ratio of reproductive tissue to vegetative tissue, and even the shape of the leaf may vary widely at different levels of nutrition, light, and moisture (Figure 2.5).

Sources of Variation

Of major importance to natural selection and adaptation are genetic, or inherited, variations within a population—variations that arise from mutations and especially from the shuffling of genes and chromosomes in sexual reproduction.

The genetic control mechanisms are in the chromosomes, found within the nucleus of plant

(a) Red oak leaves

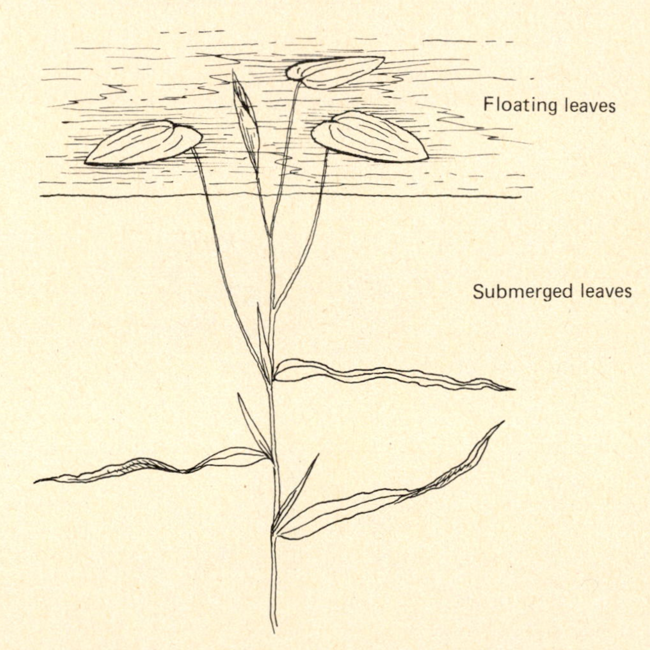

(b) Pondweed

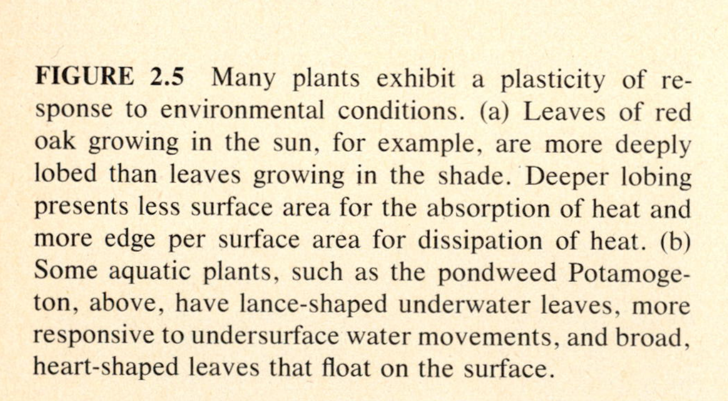

FIGURE 2.5 Many plants exhibit a plasticity of response to environmental conditions. (a) Leaves of red oak growing in the sun, for example, are more deeply lobed than leaves growing in the shade. Deeper lobing presents less surface area for the absorption of heat and more edge per surface area for dissipation of heat. (b) Some aquatic plants, such as the pondweed Potamogeton, above, have lance-shaped underwater leaves, more responsive to undersurface water movements, and broad, heart-shaped leaves that float on the surface.

and animal cells. Each chromosome consists mostly of long, coiled strands of deoxyribonucleic acid, DNA. DNA is the information template from which all cells in the organism are copied. It is a complex molecule in the shape of a double helix, resembling a twisted ladder in its construction. The long strands, comparable to the uprights of a ladder, are formed by an alternating sequence of deoxyribose sugar and phosphate groups. The connections between the strands, or the rungs, consist of pairs of nitrogen bases, adenine, guanine, cytosine, and thymine. In the formation of the rungs, adenine is always paired with thymine, and cytosine is paired with guanine.

The DNA molecule is divided into smaller units, the *nucleotides,* which consist of three elements: phosphate, deoxyribose, and one of the nitrogen bases bonded to a strand at the deoxyribose sugar. The information of heredity is coded in the sequential pattern in which the base pairs occur. According to current theory, each species is unique in that base pairs are arranged in a different order and probably in different proportions from every other species.

Each chromosome carries units of heredity called *genes*, the informational units of the DNA molecule. Genes are composed of several hundred nucleotides. Eukaryotic (eukaryotes are organisms that have cells containing a nucleus surrounded by a membrane) genes are split into a series of alternating peptide coding regions called *exons* and intervening noncoding information *introns* (Figure 2.6). The two are transcribed together as single, high–molecular-weight ribonucleic acid, RNA. For RNA to become a functional messenger, the introns

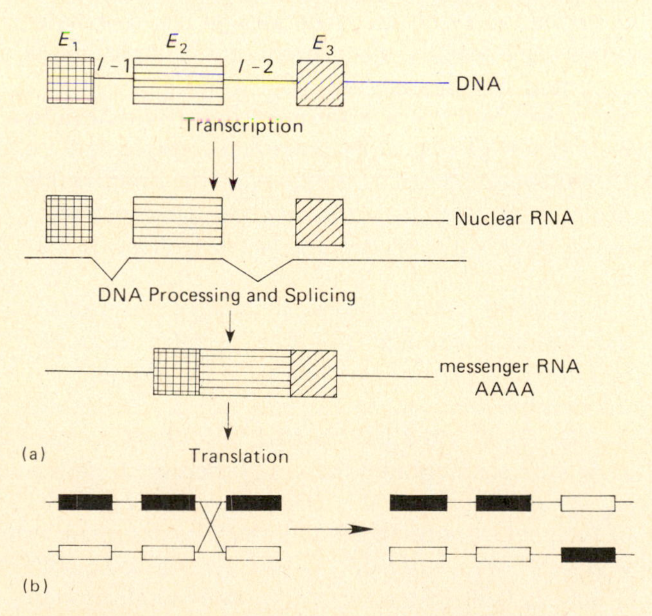

FIGURE 2.6 (a) Eukaryotic genes split into a series of alternating peptide coding regions, *exons,* and intervening sequences, *introns,* transcribed together as a single, high–molecular-weight RNA species. (b) In recombination, exons can get shuffled about, resulting in new exon arrangements with evolutionary implications. (After Hunkapiller et al., 1982:166, 167.)

must be cut away from the exons and the exons spliced to form a continuous translatable RNA. In the process exons can become shuffled. Genes are made up of a number of discrete exons that together encode a specific function. If the exons become shuffled, they can generate a new functional combination. Apparently, introns help isolate exons and maintain the integrity of the genetic information.

Because chromosomes are paired in the body cells, genes are also paired. The position a gene occupies on a chromosome is known as a *locus.* Genes occupying the same locus on a pair of chromosomes are termed *alleles.* If each pair of alleles affects a given trait in the same manner, the two alleles are called *homozygous.* If each pair of alleles affects a given trait in a different manner, the pair is called *heterozygous.*

When cells reproduce (a process of division called *mitosis*), each resulting cell nucleus receives the full complement of chromosomes, or the *diploid* number. In organisms that reproduce sexually, the germ cells, or gametes (egg and sperm), result from a process of cell division, *meiosis*, in which the pairs of chromosomes are split so that the resulting cell nucleus receives only one-half of the full complement, or the *haploid* number. When egg and sperm unite to form a zygote, the diploid number is restored.

Recombination of Genetic Material. When two gametes combine to form a zygote, the gene contents of the chromosomes of the parents are mixed in the offspring. Because the number of possible recombinations is infinitely large, recombination is the immediate and major source of variation. Recombination does not result in any change in genetic information, as mutations do, but it does provide different combinations of genes upon which selection can act. Because some combinations of interacting genes are more adaptive than others, selection determines the variations or new types that will survive in the population. The poorer combinations are eliminated by selection and the better ones retained.

The amount or degree of recombination influencing the amount of variability in a population is limited by a number of characteristics of a species. One limitation is the number of chromosomes and, thus, the number of genes involved. Another is the frequency of crossing over, the exchanging of corresponding segments of homologous chromosomes during meiosis. Others include gene flow between populations, the length of generation time, and the type of breeding—for example, single versus multiple broods in a season in animals and self-pollination versus cross-pollination in plants.

Mutation. A mutation is an inheritable change of genetic material. It may be either a gene mutation (point mutation) or a chromosome mutation.

Gene, or point, mutation is an alteration in the sequence of one or more nucleotides. During meiosis the gene at a given locus is usually copied exactly and eventually becomes part of the egg or sperm. On occasion the precision of this duplication process breaks down, and the offspring DNA is not an exact replication of the parent DNA. The alteration may be a change in the order of nucleotide pairs, a substitution of one nucleotide pair for another, the deletion of a pair, or various kinds of transpositions.

Most gene mutations have very small or no apparent effects. Mutations of single genes that produce larger effects are usually deleterious. Point mutations are important because they restore and maintain variation in the gene pool, especially when selection reduces the frequency of unfavorable alleles or eliminates them, thus reducing the size of the gene pool. Such mutations, however, do not direct evolutionary change.

Chromosomal mutations, or macromutations, may result from a change in the number of chromosomes or a change in the structure of the chromosome. A change in chromosomal number can arise in two ways: (1) the complete or partial duplication of the diploid number rather than the transmission of the haploid number or (2) the deletion of some of the chromosomes.

Polyploidy is the duplication of entire sets of chromosomes. It can arise from an irregularity in meiosis or from the failure of the whole cell to divide at the end of the meiotic division of the nucleus. The individual body cell ordinarily is diploid ($2N$), or twice the haploid number. The union of a diploid gamete with a haploid gamete yields a triploid, usually sterile. Two diploid gametes ($2N$) may unite to form a tetraploid, also usually sterile. Few naturally occurring tetraploids have arisen within a single species (autotetraploid). One example is *Epilobium*, the common fireweed of open fields and thickets. The union of two diploid gametes from different species results in a hybrid tetraploid (allotetraploid). Such tetraploids are fertile because the chromosomes upon division behave as if they were diploids and produce balanced $2N$ chromosomes in the gametes (instead of the normal $1N$).

Allopolyploidy is common in plants. The condition is rare in animals because an increase in sex chromosomes would interfere with the mechanism of sex determination and the animal would be sterile. Polyploidy in plants would be rare, too, if plants depended on cross-fertilization. But plants can be self-fertilized, and they reproduce vegetatively, which gives polyploidy a selective advantage among many species. Polyploid plants differ from the normal diploid individuals of the same species in appearance and are generally larger, more vigorous, and usually more productive. Most of our agricultural plants are polyploids.

Another form of macromutation is duplication or deletion of a part of a normal complement of chromosomes. Such deletions or duplications result in abnormal phenotypic conditions. (One such condition is Down's syndrome, or mongolism, in humans.)

A change in the physical structure of a chromosome may occur in the form of deletion, duplication, translocation, or inversion of segments of the chromosome.

Deletion is the loss of part of a chromosome. A definite segment and the genes thereon are missing in the offspring cell. Occasionally, the functions of the genes at the missing loci will be assumed by genes in some other part of the chromosome, but if the segment lost is large, the individual dies. In heterozygous individuals deletion may permit the manifestation of characters determined by recessive genes. The loss of a short segment results in a marked effect on development.

Duplication involves an addition to the chromosome. In general, duplication is less harmful than deletion and in some cases may have little or no effect on the phenotype. Duplication may increase both the genetic material and the effect of certain genes on development, or it may cause an imbalance in gene activity, reducing the vitality of an organism.

Translocation is the exchange of segments between two nonpaired (nonhomologous) chromosomes. The genes in the translocated segment become linked to those of the recipient chromosome. If the translocation is reciprocal, all of the chromosome material is present, even though it is rearranged. Individuals carrying such translocations are usually normal. If the translocation is not reciprocal, some genes will be transferred to completely different chromosomes, and the linkage relationship becomes altered drastically.

An *inversion* is an alteration of the sequence of genes in a chromosome. It may occur when a chromosome breaks in two places and the segment between the breaks becomes turned around, reversing the order of genes with respect to an unbroken chromosome. When the altered chromosome is paired with a normal chromosome in a heterozygous individual, the alteration interferes with pairings. The crossing over between the inverted and normal chromosomes that is necessary to align each gene with its homologue usually produces abnormal, nonviable gametes.

Mutations may be neutral, beneficial, or disadvantageous, depending upon the environmental circumstances and the genetic background in which they arise. If the effect on the phenotype is deleterious, the mutant allele will be selected against. So will mutations that serve a function already filled by an established gene that appears in an established population adapted to its environment or in one that is highly heterozygous. But if a mutation is advantageous or neutral, it may be retained, especially if it confers some selective advantage in a changing environment or if it appears in a highly inbred, and thus homozygous, population. Individual mutations, however, are rarely agents of recognizable change; that is usually brought about by an accumulation of mutations, each of which alters slightly the appearance or function of an organism. The ultimate outcome of most mutations is the maintenance of variability in the gene pool of a population. Without mutations genetic variations over many generations would be so reduced that a population would not respond any further to natural selection.

Multigene Families. Eukaryotic genes are found in multigene families. They consist of information units that are similar in structure, overlap in their functions, and are linked in a straight line (tandem) (Figure 2.7). Multigene families may range from a few gene copies to thousands of identical to variable copies of ribosomal genes and assorted DNAs.

Such multigene families add considerably to genetic variability. Single genes can encode only a single gene product. Multigene families, however, are heterogeneous collections of closely related proteins, adding to genetic diversity. Members of a multigene family may be expressed simultaneously or individually. Multigene families are able to shield gene copies from natural selection. A deleterious mutation of one gene copy is buffered by other normal gene copies if those gene copies can assume part of the mutated gene's function.

Hardy-Weinberg Equilibrium

Variations found in a population are transmitted to the next generation through sexual reproduction when a sample of the gene pool contained in ga-

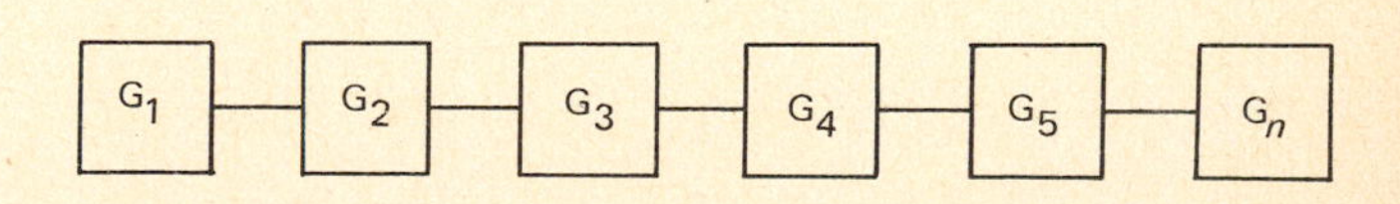

FIGURE 2.7 Model of a multigene family. Multigene families range in size from a few copies of genes to thousands of copies. Gene copies within families can be nearly identical or vary considerably, can vary in size from one species to another, and can encode a heterogeneous collection of closely related protein species. Genes within the same gene family tend to evolve in a similar manner. (After Hood, Campbell, and Elgin, 1975:306.)

metes joins to form zygotes. If genes occur in two forms, *A* and *a*, then any individual carrying both forms can fall into three possible diploid classes: *AA*, *aa*, and *Aa*. Individuals in which the alleles are the same—*AA* or *aa*—are homozygous, and those in which the alleles are different—*Aa*—are heterozygous. Haploid gametes produced by homozygous individuals are either all *A* or all *a*; those by heterozygous, half *A* and half *a*. These recombine in three possible ways in sexual reproduction: *AA*, *Aa*, and *aa* (Figure 2.8). Thus, the proportion of gametes carrying *A* and *a* is determined by the individual genotypes, the genes received from the parents.

Eggs and sperm unite at random. If the mating pattern is also random—that is, if the chance that an individual will mate with another individual having a certain genotype is equal to the frequency of that genotype in the population—then one can predict the genotypes in the next generation.

Assume that a population homozygous for the dominant *AA* is mixed with an equal number from a population homozygous for the recessive *aa*. Their offspring, the F_1 generation, will then consist of 0.25 *AA*, 0.50 *Aa*, and 0.25 *aa* (Figure 2.9). These proportions are called genotypic frequencies. The allele frequency of both *A* and *a* is 0.5, because 25 percent of the population is pure *A* allele, and 25 percent is pure *a* allele. Fifty percent of the population possesses *Aa* alleles. Dividing that by 2, because that portion of the population carries both *A* and *a*, we obtain 25 percent for each.

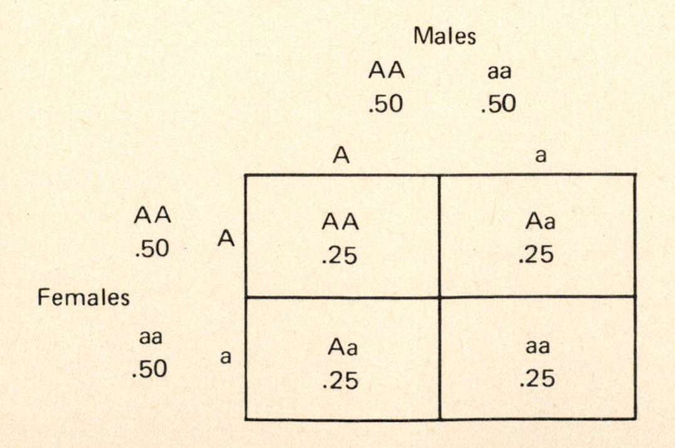

FIGURE 2.8 Mixing two homozygous populations.

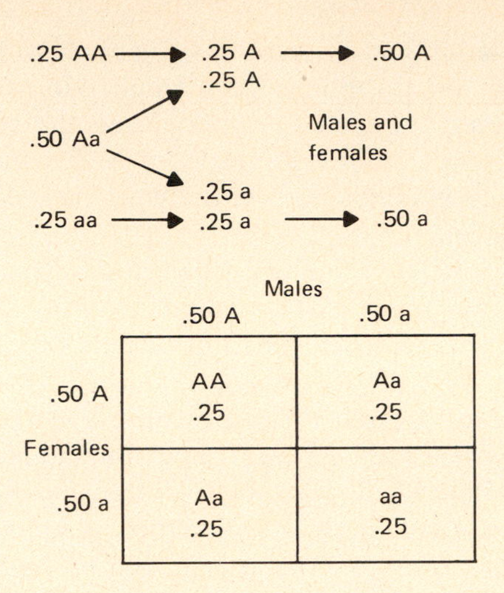

FIGURE 2.9 Proportions in the F_1 generation.

That proportion will be maintained through successive generations of a bisexual population (Figure 2.9) if certain conditions exist: (1) reproduction must be random; (2) mutations must not occur, or else they must be in equilibrium—that is, the rate of mutation from *A* to *a* is the same as from *a* to *A*; (3) the population is large enough so that changes by chance in the frequency of genes are insignificant; (4) migration is negligible; (5) natural selection does not affect the locus; and (6) generations are nonoverlapping.

The equilibrium of these three genotypes can be expressed as the Hardy-Weinberg law. Essentially, it is this: If the frequency of allele *A* is p and the frequency of *a* is q, where $(p + q) = 1.0$, then $(p^2 + 2pq + q^2) = 1.0$, in which p^2 is the genotypic frequency of individuals homozygous for *A*, q^2 is the frequency of individuals homozygous for *a*, and $2pq$ is the frequency of heterozygous individuals, *Aa*. In the hypothetical population above, the proportion of genotypes in the F_1 generation will be $(0.5)^2 + 2(0.5 \times 0.5) + (0.5)^2$.

The same tendency can be demonstrated even if the ratio is not the classic Mendelian 1:2:1. Imagine a population in which the ratio of *A* alleles (p) to *a* alleles (q) is 0.6 to 0.4. The frequency of the genotypes in the F_1 generation will be 0.36 *AA*, 0.48 *Aa*, and 0.16 *aa*, and the gene frequency will be $(0.6)^2 + 2(0.6 \times 0.4) + (0.4)^2$. From this one can

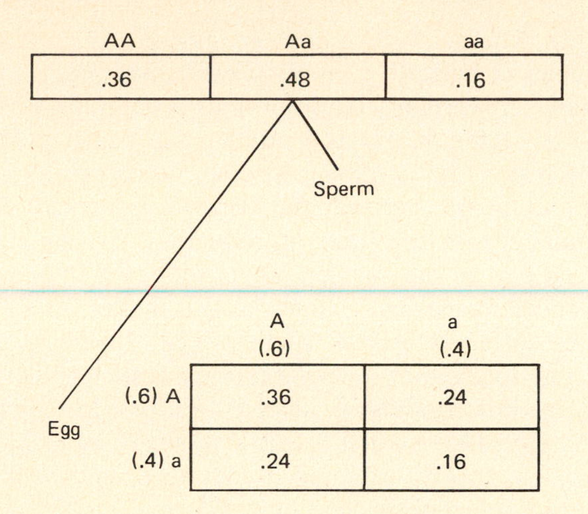

FIGURE 2.10 Illustration of the Hardy-Weinberg law.

conclude that all succeeding generations will carry the same proportions of the three genotypes (Figure 2.10), provided the assumptions mentioned earlier are fulfilled.

The stated assumptions are never perfectly fulfilled in any real population, so the Hardy-Weinberg law must be considered theoretical—a distribution against which actual observations can be compared. Nevertheless, the Hardy-Weinberg law is of fundamental importance in theoretical population genetics. It allows one to state approximately what genotypic frequencies will be from a knowledge of gene frequencies alone, provided that the population has discrete generations. An important implication of this law is that in the absence of specific evolutionary forces to change the frequencies, Mendelian inheritance alone is enough to maintain genetic variability in a population. Hardy-Weinberg equilibrium is attained in one generation of random mating in populations with nonoverlapping generations. In populations with overlapping generations, Hardy-Weinberg equilibrium comes more slowly.

Natural Selection and Evolution

Natural selection results when environmental forces, both abiotic and biotic, favor certain genotypes, as expressed by phenotypes, over others. It functions through nonrandom reproduction within a population. Not every individual in a population is able to contribute its share of genetic characteristics to the next generation or leave surviving offspring. Some fail to survive to reproduce. Others fail to mate or produce offspring. Some individuals leave more reproducing offspring than others. Their contribution to the gene pool or fitness increases at the expense of others.

An outcome of natural selection is a change in gene frequency in a population through time, or *evolution*. Evolution is a process of change in biological systems due mostly to forces in the environment: changes in moisture, temperature, food, habitat, interspecific and intraspecific competition, and predation. Although natural selection is a major force in evolution, natural selection per se is not evolution, nor is evolution natural selection.

TYPES OF NATURAL SELECTION

Natural selection can proceed without any major changes in the phenotype. Random changes in gene composition take place in all populations. If the changes produce individuals that vary widely from the average characteristics of the population, they tend to be less fit and produce fewer offspring. The frequency of their genes is reduced in favor of more normal ones. Such selection is called *stabilizing* because it favors phenotypes near the population mean at the expense of the two extremes (Figure 2.11). Such selection is characteristic of relatively stable environments. However, that does not mean that some changes directed by natural selection are not taking place. Many plants and animals are faced with the problem of maintaining fitness in the face of changing environmental pressures. For these, selection may involve certain behavioral or physiological changes needed just to maintain fitness. Leigh Van Valen (1973) has made an analogy regarding such selection using the Red Queen in Alice's *Adventures through the Looking Glass:* "Now here, you see, it takes all kinds of running you can do to keep in the same place." Although there is variability in the genotype, the population mean does not vary.

In other situations selection may be disruptive

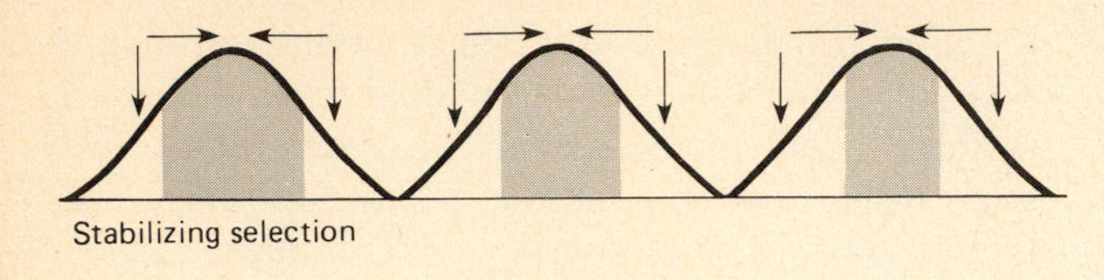

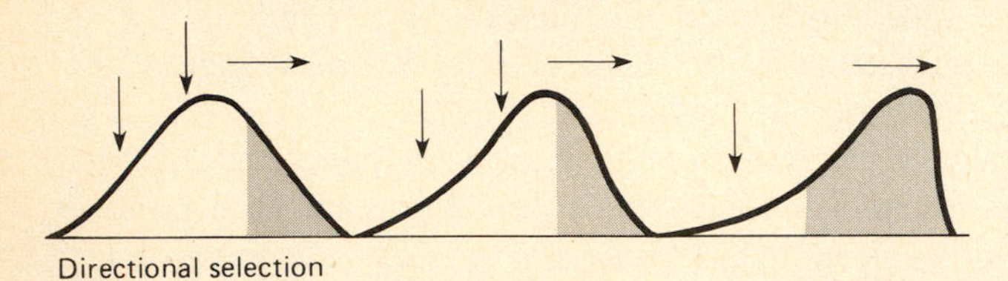

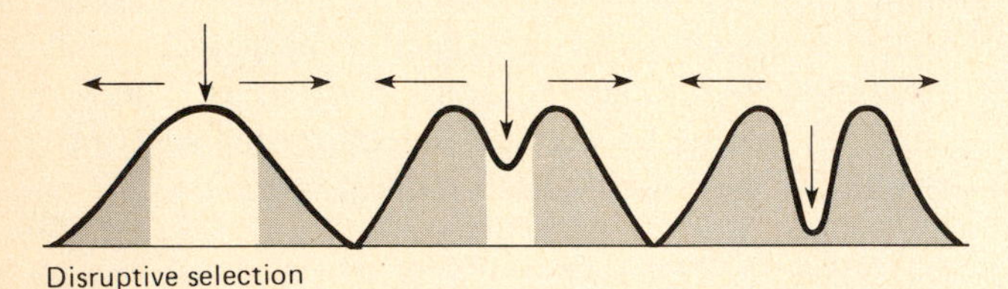

FIGURE 2.11 Three types of selection. Stabilizing selection favors organisms with values close to the population mean. Consequently, little or no change is produced in the population. Directional selection favors one extreme and tends to move the mean of the population toward that extreme. Disruptive selection increases the frequencies of the extremes. The shaded areas represent the phenotypes being selected for. Downward-pointing arrows represent selection pressures; horizontal arrows represent the direction of evolutionary change.

or directional. Both result in a relatively rapid evolutionary change but can revert once selection pressures are relaxed.

In *disruptive* selection (Figure 2.11), both extremes are favored simultaneously, although not necessarily to the same degree. Such selection results when members of a species population are subject to different selection pressures operating in different environments, such as microhabitats. It usually results in *polymorphism,* in which a population contains two or more genotypes.

An example is the swallowtail butterfly *Papilio dardanus,* widely distributed across Africa (Figure 2.12). The females mimic an associated inedible species of butterfly that possesses a warning coloration. (A warning coloration tells a potential predator that the bearer is inedible. See Chapter 17.) The male is not a mimic. Instead, he retains a specific color pattern that the females recognize, essential for mating and reproduction. Within any given part of the species range, the females mimic an inedible species common to that region. At least three different female mimics exist. Intermediate female forms that do not bear any resemblance to the inedible models are selected against. In regions where inedible models are absent, so are mimicking females. In this example predation promotes disruptive selection in the population.

Selection is *directional* when one extreme phenotype is favored at the expense of others (Figure 2.11). In that case the mean phenotype is shifted toward the extreme, provided that heritable variations of an effective kind are present.

An example of directional selection is found in the survival of Darwin's medium ground finch (*Geospiza fortis*) on the 40-ha islet of Daphne Major, the Galapagos, from 1975 through 1978 (Boag and Grant, 1981). During the 1970s, the island received regular rainfall (127–137 mm), resulting in an abundance of seeds and a large finch population (1,500 banded birds). In 1977, however, only 24 mm of rain fell, and the island experienced drought. Plant production, and thus seed production, which furnished the major food in the dry season, declined drastically. Small seeds in particular declined in abundance faster than large seeds, increasing the average size and hardness of available seed. The finches—which, in general, fed on smaller seeds—were forced to turn to the larger ones, usually ignored in normal years. Larger birds in the population ate the larger seeds, while smaller birds apparently had difficulty finding food. Large birds, especially males with large beaks, survived best because they were able to crack large, hard seeds. Females experienced heavy mortality. Overall, the population declined 85 percent from mortality and possibly emigration (Figure 2.13).

Intense selection on Darwin's medium ground finches suggests that natural selection is not a con-

FIGURE 2.12 Disruptive selection resulting in mimicry in the African swallowtail butterfly *Papilo dardanus*. (Mimicry involves the resemblance of one species to another for protective or aggressive purposes. Among butterflies mimicry usually involves the resemblance of a mimic, palatable to predators, to an unpalatable model.) The African swallowtail is widely distributed across the continent as a number of races, including Dardanus in west and southwest Africa, Cenea in southern Africa, and Polytropus and Tibullus in east Africa. Males of all races are nonmimetic. Females of each race have wing patterns and coloration that mimic inedible species in various regions. In west and central Africa, female *dardanus* var. *hippocoon* mimic *Amauris niavius niavius*, a black and white butterfly. On the east coast of Africa, *A. n. niavius* is replaced by *A. n. dominicanus,* which is mimicked by the variety *hippocoonides*. In South Africa the variety *cenea* mimics *Amauris echeria*. It is black with small white to pale yellow spots in the fore wing and white spots on the margin of the hind wings. Throughout much of Africa *P. d. trophonius* females mimic *Danaus chrysippus,* producing a butterfly with the pattern but not the color of *hippocoonides*. This is *trophonius,* with black fore wings carrying white patches and reddish brown hind wings. This form is rather rare. The genetics of these mimics is fairly well understood.

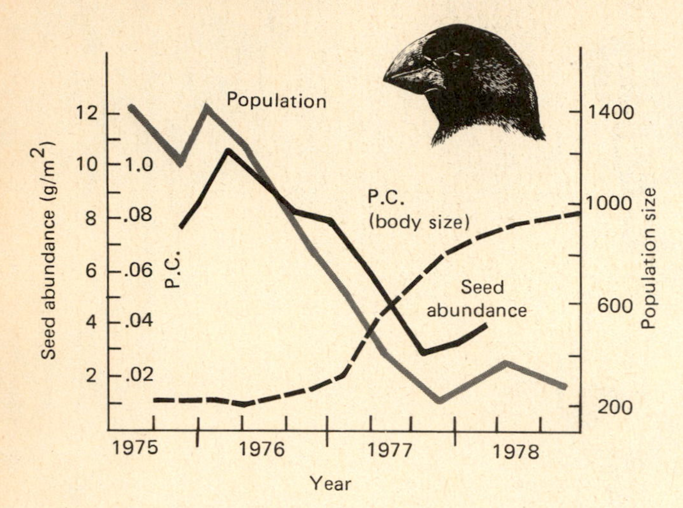

FIGURE 2.13 Evidence of directional selection in Darwin's finch *Geospiza fortis*. One graph presents the population estimate on the island of Daphne Major based on censuses of marked populations (standard deviation lines deleted), and a second graph estimates of seed abundance, excluding two species of seeds never eaten by any Galapagos finches. Populations declined in the face of a seed scarcity brought about by a prolonged drought. A third graph describes changes in principal component (P.C.) score 1, an index of body size. Note how the body size of all birds, especially the males, increased during the drought period. That suggests that smaller-bodied birds were being selected against and large-bodied birds were being favored. Results suggest that the most intense selection in a species occurs during periods of unfavorable environmental conditions. (After Boag and Grant, 1981:83.)

tinuous process but rather exerts its greatest influence during periods of environmental stress or selection "bottlenecks" (Wiens, 1977) during a small portion of an organism's life history. It also suggests how a small, isolated, relatively sedentary, morphologically variable population of a species under selective pressure of a variable environment can experience a rapid evolution of morphological characters.

To cite another example, roadside populations of the common bitterweed *Helenium amarum* appear to be diverging away from field populations of the same species. Bitterweed is common on disturbed sites throughout the southeastern coastal plain and piedmont of the United States. It is a winter annual whose seeds germinate in autumn immediately after seedfall. The plants overwinter as rosettes and in spring send up flower stalks after the day length has exceeded 13 hours. Populations of roadside plants are under a different selection pressure than old field plants—that of roadside mowing. Mowing seems to favor rapidly growing plants that are shorter, branch at a lower level, and have greater reproductive potential than old field plants. These and other differences are retained by progeny grown in a greenhouse environment (Wojcik, 1968).

Regardless of the kind of selection, the characteristics selected are not necessarily the best of all possible traits but rather the most suitable of those available. Characteristics evolved under one set of circumstances may be inappropriate for subsequent environments. Present individuals are products of past selection and are stuck with what their ancestors have provided them. Organisms cannot evolve adaptations for future events that differ from those of the past.

SOME CONSTRAINTS ON NATURAL SELECTION

Natural selection may account for only a part of evolutionary change. Other forces may determine which organisms leave descendants and which do not. In doing so these forces influence evolutionary change, constraining or modifying the effects of natural selection.

One such force is the *founder effect*. It may come about when a fraction of a population is isolated from its parent population or when a patch of new or previously unoccupied habitat is colonized by a few individuals. These individuals carry only a fraction or a small sample of the gene pool of the parent population. If chance is involved in the selection of founding individuals, the new populations will differ from the parent population and from each other. All genes in subsequent populations will stem from the few carried by the founders and from mutations and immigrants. Small size of the population enforces inbreeding among related individuals. Recessive genes, once shielded by the heterozygous condition, become exposed. Some may be deleterious, reducing survival. Other homozygous genes, at a selective disadvantage in the old environment, may enable the carrier to adapt to the new environment. Genes neutral in effect in the parent population may advance the fitness of founder individuals. A small sample size and a limited diversity of genetic material may have a powerful influence on the development of a new species.

Also altering the genetic composition of a small population is *genetic drift,* a random change in gene frequency. The gene pool of each succeeding generation is a sample of the parental population that gave rise to it. As such, it is subject to sampling error. As in any sample, error is greatest in small populations. Eventually, that error in small populations results in a loss of genetic diversity and an increase in homozygosity. If drift lowers the fitness of the population, it will become extinct. But if the genes involved confer a selective advantage in the new environment, they will enhance the fitness of the population. Thus, as noted previously, a small sample size and a limited diversity of genetic material may have a powerful influence on the development of a new species, and chance may play an important role in evolutionary change.

Founder effect and genetic drift may be critical in the reestablishment of new populations of species threatened with extinction. Such populations involved in reintroductions are small samples selected from a parent population. The genetic history of one such small population involves the northern elephant seal (*Mirounga angustirostris*). The elephant seal was reduced by fur hunters to about 20 animals in the 1890s. Once protected, the population has grown to over 30,000 individuals. Bonnell and Selander (1974) examined the genetic variation among the seals by means of electrophoresis. They found no variation in a sample of 24 electrophoretic loci. In contrast, the population of southern elephant seal, never severely reduced, still retains high genetic variability. The lack of genetic variation in the northern elephant seal increases its vulnerability to environmental change.

Another constraint is the nature of the original model and its embryological development. Evolution through natural selection has to act on already developed and more or less complexly organized systems. Evolutionary processes can only tinker with the systems (Jacob, 1977). What changes take place have to be within the limits imposed by the nature of the original. In general, response to selection will follow the line of least resistance, the easiest route, and not necessarily the optimal one (Stebbins, 1970). Further, each population has its own limited genetic resources on which selection can act. The direction evolution takes depends in part on those genes exposed to selection.

Organisms themselves can influence the direction of evolution. That is especially true among interacting pairs of species such as predator and prey, parasites and hosts, and competitors (see Chapters 15 and 16). In those cases one species acts as a selection force. In turn, the other reciprocates. The trait of species A has evolved in response to the specific trait of species B. And the trait of species B has evolved in response to the specific trait of species A. For example, a host species over some period of time may evolve a way to resist attacks of a parasite. The parasite then evolves new ways of breaching the host's defenses. (See, for example, Holmes, 1983; May and Anderson, 1983.)

Such an interaction in which two genetically independent but ecologically related species influence evolutionary changes in each other is *coevolution.* Coevolution may drive a member of a pair to become a specialist, which increases its probability of extinction because its narrow requirements may not be available in the future. Coevolution as defined involves an interacting pair, which often can be difficult to prove as competition in the field. In other instances coevolutionary response may be more general. For example, plants may have evolved chemical defenses against herbivorous insects as far back as the Cretaceous period and transmitted those defenses to modern-day descendants. These chemical defenses now act to protect plants against modern-day insects. This evolution of a particular trait in a number of plants was guided by insect herbivory in the past, and the evolution of insects was guided by some prior evolution of plant defenses. The process of traits' evolving in a number of species in response to traits in several other species is known as *diffuse coevolution*. The genetic changes in interacting species are not highly coupled.

Although coevolution is not a new idea per se—it was hinted at by Darwin—the concept of this ecological and evolutionary interaction has only recently become an area of study. The term *coevolution* was coined by Paul Erlich and Peter Raven in 1964 and appeared in a paper on the evolutionary influences plants and herbivorous insects had on each other. The idea of coevolution has been expanded further in books edited by Gilbert and Raven (1975) and Futuyma and Slatkin (1983). Some of the theory has been developed by Janzen (1978, 1980) and Roughgarden (1979, 1983).

GROUP SELECTION

The general consensus is that natural selection acts on the individual (phenotype) of a species. However, it is difficult to reconcile such traits as warning coloration or alarm calls or any act that benefits a population at the expense of the individual with natural selection. Such acts and traits are called *altruistic*. Altruism, strictly defined, is the sacrifice of one's own well-being in the service of another. To state it in more genetic terms, an altruist has a beneficial effect on the genotypic fitness of others, while experiencing a net decrease in fitness relative to the recipients of the altruistic act (Michod, 1982).

In attempting to explain such traits, some population ecologists have suggested the concept of *group selection*. Group selection results from differences among populations of a species. The characteristic selected evolves because it increases the fitness of the group, though it may decrease the fitness of any individual within the group. Thus, group selection operates in opposition to individual natural selection.

The idea of group selection was suggested by Sewell Wright (1931, 1935), when he proposed *interdemic selection*. It might result among species with small local populations that are isolated enough to allow some differentiation in gene frequencies. But sufficient contact does occur between demes to allow the gradual spread of an advantageous genetic complex to other demes. Consider a deme possessing a high frequency of a gene that decreases mortality or increases reproduction. And suppose it is surrounded by demes possessing the gene at lower frequencies. The deme with the adaptive gene will develop some selective advantage over neighboring demes. It will produce a greater surplus population that will emigrate to surrounding demes or into empty habitat patches. In such a manner a gene that confers some benefit on the group enables the subpopulation possessing it to succeed to such a point that eventually the entire population will consist of individuals possessing the gene. Group selection can then be defined as changes in gene frequency resulting from differential extinction or productivity of groups. Because the groups themselves are short-lived, extinction of groups is not as relevant as differential productivity of groups or the contribution of groups to the mating pool (Michod, 1982).

V. C. Wynne-Edwards (1962) proposed that animal species could limit their population densities by altruistic behavioral acts. He suggested that if a population is to persist, it must possess some homeostatic process that will dampen fluctuations. If a population grows too large, it will run out of food or other resources necessary for survival. Through group selection individuals acquire certain behavioral traits that allow members of the population to assess the density of the local population. That stimulates members of the population to reduce reproductive efforts when their numbers become too great. Such a population not only would survive but would colonize areas once occupied by fast-growing but now extinct populations.

Group selection can operate only under certain conditions. Populations of selfish members must become extinct, and their place must be occupied by populations of altruists. But altruistic populations are unstable. If one is invaded by selfish genes, either by mutation or by immigration, selfish genes would rapidly replace altruistic genes. The selfish genes would have the selective advantage in individual selection. The altruists could come out ahead only if the selfish populations became extinct at a very high rate. Thus, group selection within family-structured populations appears to be necessary for the evolution of altruism. Unfavorable altruistic selection within family groups must be balanced by favorable selection between groups.

The concept of group selection is highly controversial and, although rejected by many evolutionists, does not seem to want to die. Group selection, for example, is being considered in a different context as selection between species (Alexander and Borgia, 1978; Fowler and MacMahon, 1982). Natural selection removes individuals from a population. Species selection removes species from an ecosystem. Species may be established through individual or gene selection, but they succeed or fail as a result of competition with other species. This concept is explored further in Chapter 3.

KIN SELECTION

Kin selection is based on differences among collections of related individuals. It involves the evolution of a genetic trait expressed by one individual (termed the actor) that affects the genotypic fitness of one or more other individuals related to the actor in a nonrandom way (Michod, 1982). If no fitness interactions exist, or if individuals interacting are only randomly related (possess some similar alleles but are not direct descendants), kin selection is not involved.

Kin selection acts in small groups such as parents and offspring or a group of very closely related individuals. Theoretically, kin selection increases the average genetic fitness of the group at the expense of the fitness of some individuals. Those individuals may help in the rearing of related young but do not reproduce themselves (see Hamilton, 1972; E. Wilson, 1975).

Kin selection is favored when an increase in fitness of closely related individuals is great enough to compensate for the loss of fitness in the altruistic individual. If the breeders and helpers are very close kin, as they usually are, the helpers serve to increase the fitness of the genetic traits they hold in common. The genes are copies of the same DNA sequence acquired from a common ancestor. The closer the relative, the closer the replication. Although they do not leave behind offspring of their own (at least for one breeding season), helpers improve both their own fitness and that of their kin because they have a genetic relationship with the offspring. Although altruism is not necessary for kin selection, it can increase under kin selection.

Two aspects of fitness are involved in kin selection. One is *direct fitness,* which comes by the genes in the individual's own offspring. The other is *indirect fitness,* obtained by copies of samples of the individual's genes in the offspring of other related individuals (Brown and Brown, 1981). An individual can achieve indirect fitness by improving the direct fitness of very close relatives such as a parent or sibling. It can do that by helping relatives rear their young. Through a lifetime an individual can acquire direct fitness through its own offspring and indirect fitness by aiding closely related offspring. Direct fitness added to indirect fitness results in *inclusive fitness,* a concept used to explain many aspects of social behavior (Hamilton, 1964).

An example of helping behavior can be found in the Florida scrub jay (*Aphelocoma coerulescens*). Pairs of jays may be assisted by nonbreeding helpers who aid their parents or close kin in rearing their young (Woolfenden, 1975). Florida scrub jays rarely breed before 2 years of age and may not breed for several years after that. Many unmated males and females remain in the local area on familiar ground among familiar birds. Female helpers rarely remain for more than two years because they join the breeding population. Unmated males may be around longer. Because of the males' assistance, the family size of the pairs they help increases. As the family groups expand, their territories expand. Eventually, a dominant male helper may claim a part of the enlarged territory as his own or inherit it and become a breeder (Woolfenden and Fitzpatrick, 1977). By helping, the male has come out ahead. He has increased his inclusive fitness and improved his own direct fitness by increasing the opportunity to become a breeder. In helping, the male may not be altruistic at all. He may be practicing a form of selfish behavior, increasing his own opportunities by cooperating with breeding pairs (Woolfenden, 1971; Ligon, 1981). But his actions still fall under the concept of kin selection, as defined.

Like group selection, kin selection is open to

some questions. Is it an integral part of the evolution of sibling cooperation and helping at the nest? Is it an important cause and part of group living? Or is kin selection a form of selfish individual selection achieved under the guise of cooperation? These and many other questions need to be answered before kin and group selection can be understood (see Alexander and Tinkle, 1981).

Summary

Since its inception, life on Earth has changed through geological time. Land masses moved apart, seas drowned or retreated from continental land masses, mountains rose, climates changed. In such a changing environmental background, life evolved from very simple forms to highly complex plants and animals. Certain groups flourished and disappeared, replaced by other forms more successful in new environments. These changes through time were the outcome of evolutionary processes.

The basis of evolution is genetic variation found in organisms as reflected in the phenotype. Theoretically, in biparental populations variations in gene frequencies and gene ratios remain in equilibrium under certain conditions. These include random reproduction, equilibrium in mutation, no gene flow from one population to another, and a relatively large population size. In nature such conditions do not occur. Populations are often local units or demes that vary in size and are subjected to their own set of selection pressures. Mutations do occur, although they change gene composition slowly. In sexual reproduction, genes are scrambled and recombined in an immense number of variations. Mating and survival are nonrandom. All add up to a marked departure from genetic equilibrium, a change in the frequency of the gene pool. That change over time is evolution.

The direction evolution takes depends upon a number of conditions and events. One is natural selection—nonrandom reproduction. The successful are not just those that survive but rather those that leave behind reproducing offspring. Another is a small population size with limited genetic variation on which selection can work. The nature of the organism, its pattern of development, and its structure place limitations on evolutionary change.

Natural selection tends to eliminate the less fit alleles and favor the most fit. If it favors those close to the population mean, selection is stabilizing and produces little change in the population. It is characteristic of more or less stable environments. If selection favors the extremes, it is disruptive or diversifying. Such selection may occur when segments of a population are subjected to different selection pressures environmentally induced. Or selection may favor one extreme and move the population mean in that direction. Characteristic of a changing environment, directional selection produces the most rapid phenotypic and genotypic changes.

There is some evidence, highly controversial, that selection may act at a level higher than the individual. One such form of higher-level selection, group selection, may involve entire local populations or demes. Another, kin selection, may involve groups of closely related individuals.

Although natural selection may seem to imply a steady improvement or perfection of a species through time, that is not the case. Selection pressure may only enable individuals to maintain their fitness over time. Fitness is a response to current environmental conditions and not to some future condition. Individuals with the greatest fitness in some future situation happen to have the genetic variability that enables them to survive in a changed environment. Nor does natural selection work for the long-term good of the species. While it may improve fitness over the short term, natural selection may actually work against the long-term survival of a species, especially in situations where natural selection favors specialization.

Review and Study Questions

1. Associate each of Earth's geological eras with the rise of life and the appearance and disappearance of the various groups of plants and animals.

2. Distinguish between natural selection, adaptation, and evolution. How do they relate?
3. What is meant by fitness? Direct fitness? Indirect fitness? Inclusive fitness?
4. What is a deme? What is the significance of the deme in natural selection and evolution?
5. What is the mean? Mode? A frequency distribution? How do they relate to variation in the physical characteristics of an organism?
6. Briefly explain the structure of a gene. What is DNA? RNA? Exon? Intron? How do all of these function?
7. Distinguish between a genotype and a phenotype. On what does selection work? What is phenotypic plasticity?
8. Define *allele, locus, homozygous, heterozygous, diploid, haploid, meiosis, mitosis, gene pool.*
9. What are the major sources of variation in the gene pool?
10. What is a mutation? Distinguish between a point mutation and a chromosomal mutation.
11. What is a multigene family? What is its significance in genetic variation?
12. Contrast stabilizing, directional, and disruptive selection.
13. What is the founder effect? Genetic drift? What is their significance? How do they act as constraints on natural selection?
14. What are some other constraints on natural selection?
15. What is coevolution?
16. What is group selection? What is the relationship between group selection and altruism?
17. What is kin selection? How does it contrast with group selection?
18. Are there any strong arguments for selection above the level of the individual?

Suggested Readings

Alexander, R. D., and D. W. Tinkle, eds. 1981. *Natural selection and social behavior: Recent research and new theory.* New York: Chiron.

Brown, J. H., and A. C. Gibson. 1983. *Biogeography.* St. Louis: Mosby.

Futuyma, D. J. 1979. *Evolutionary biology.* Sunderland, Mass.: Sinauer Associates.

Milkman, R., ed. 1982. *Perspectives on evolution.* Sunderland, Mass.: Sinauer Associates.

Solbrig, O. T., and D. J. Solbrig. 1979. *An introduction to population biology and evolution.* Reading, Mass.: Addison-Wesley.

Stebbins, G. L. 1982. *Darwin to DNA: Molecules to humanity.* San Francisco: Freeman.

Wilson E. O., and W. H. Bossert. 1971. *A primer of population biology.* Sunderland, Mass.: Sinauer Associates.

CHAPTER 3

Species and Speciation

Outline

Objectives

Upon completion of this chapter, you should be able to:

1. Define a species and distinguish between the concept of a morphological species and a biological species.
2. Explain what is meant by a cline, an ecotype, and a geographic isolate.
3. Define an isolating mechanism and describe several types.
4. Define speciation and distinguish between allopatric, parapatric, and sympatric speciation.
5. Explain how polyploidy can produce an instant species.
6. Define character displacement.
7. Explain adaptive radiation and convergence.
8. Compare gradualism and punctuated equilibrium as processes of speciation.
9. Explain the concept of species selection.
10. Discuss the rate of evolution.

The discussion up to this point has concerned changes within a single species resulting from various evolutionary forces—genetic variation, mutation, and environmental changes (*anagenesis*). Of greater ecological interest is the splitting of one species into two or more genetic lines (*cladogenesis*). Each species so formed might diverge still further into more new species. Such diversification eventually leads to the formation of new taxa, such as genera and families.

What Is a Species?

One has little difficulty distinguishing a robin from a wood thrush or a white oak from a red oak. Each has certain morphological characteristics—most useful in field guides—that set them apart from other organisms. Each is an entity, a discrete unit to which a name has been given. That is the way Carl von Linné, who gave us our system of classification, saw a great number of plants and animals. He, like others of his day, regarded the many organisms as fixed and unchanging units, the products of special creation. Differences and similarities were based on color pattern, structure, proportion, and other characteristics. From these criteria species were described, separated, and arranged into groups. Each species was monotypic; that is, it contained only those individuals that fairly well approximated the norm or the type for the species, the specimens from which the species was described. Some variation was permissible, but these variants were considered accidental, although some slight changes within the species were admitted to be possible. This is the *morphological species,* a classical concept still alive, useful, and necessary for classifying the vast number of plants and animals (Box 3.1).

Later, the studies of Darwin on variation, of Wallace on geographical distribution, and of Mendel on genetics emphasized that variation within a species was the rule. Because of sexual dimorphism, male and female of the same species looked like separate species. Many closely related forms replaced each other geographically and, in doing so, intergraded with each other. So smooth and gradual was the transition that it was difficult to separate precisely one from another. What was needed was a better definition of a species—one that emphasized interbreeding rather than exact morphological differences.

Ernst Mayr (1942:120) defined the species as "a group of actually or potentially interbreeding populations that are reproductively isolated from other such groups." This concept of the *biological species* embodies a group of interbreeding individuals living together in a similar environment in a given region and under similar ecological relationships. The individuals recognize each other as potential mates. They are a genetic unit, in which each individual holds for a short period of time a portion of the contents of an intercommunicating gene pool.

Such a definition has its limitations. It applies only to bisexual organisms, excluding asexual ones. The concept of biological species arose mainly in vertebrate systematics, where it has the widest application. The biological species is less generally accepted by botanists, for among the plants as well as among many invertebrate animals, the biological species, as defined, is inadequate. Among them asexual or vegetative reproduction is common, and many plants rely upon it. As one might well imagine, such organisms possess very little genetic variability and, for the most part, have lost their capacity to adapt to environmental change. Variability has been sacrified to take maximum advantage of a given environmental situation, and here they are preeminently successful.

Species may be sympatric, allopatric, or parapatric. *Sympatric species* are those that occupy the same area at the same time and thus have the opportunity to interbreed. They are, in effect, "good" species, because they are reproductively isolated. *Allopatric species* are those occupying areas separated by time and space. Because they do not have the opportunity to meet similar species, there is no indication of whether they are potentially capable of interbreeding with them or not. Only if barriers separating two allopatric species are broken, allowing the two to come together, can one test repro-

Box 3.1
Naming and Classifying Organisms

Although humans had been classifying plants and animals for centuries, there was no uniformity about any classification scheme. Botanists and naturalists had been trying since Aristotle's day (384–322 B.C.) to come up with a useful method. No system devised was satisfactory until a young Swedish naturalist and medical doctor, Carl von Linné, who Latinized his name to Carolus Linnaeus (1707–1778), developed one that is still in use today.

Linnaeus, who taught medicine and botany at the University of Uppsala, had a passion for classifying things—plants, animals, diseases, minerals, and many other items that came to his attention. He introduced a binomial, or two-name, system of naming plants and animals—a system that finally brought order out of chaos and led to a standardized method of naming all living organisms in Latin. Latin was then the universal language of science. Because it was a dead language, it would never change in grammar, syntax, or words. That gave it a great degree of uniformity and stability.

Linnaeus explored in Lapland and northern Europe; he had collectors send him plants from other parts of the world. Out of this came a book in 1735, *Species Plantarum*. In it he listed and named all plants then known to science, giving each plant two Latin names—a generic name and a species name. Later, in 1758, he published a tenth edition, *Systema Naturae*, in which he consistently applied the same idea to animals.

Before Linnaeus could assign names to each organism, he had to sort out his plants and animals into some broad general groups. First, all living things, he reasoned, belonged to two great groups, or kingdoms: plant and animal. The animal kingdom he subdivided into six parts: mammals, birds, reptiles, fishes, insects, and a catchall group, vermes (or worms). Among the mammals, for example, he noted that some organisms were similar. It was not too difficult to identify a cat as such, or a dog, although there are many different kinds of each. All have certain identifying characteristics. Cats have retractile claws, rounded heads, and well-developed but relatively small, rounded ears. Some have long tails; other have short tails, ear tufts, and a ruff on the cheeks and throat. All the long-tailed cats, then, were placed in one group, or genus, to which Linnaeus gave the Latin name *Felis*, for cat; and to the short-tailed cats he gave the name *Lynx*. Doglike mammals, too, are readily distinguishable, and he could break them down into groups of animals similar in appearance: the sharp-nosed little foxes, the very doglike wolves, and their descendants, the true dogs. To the wolves and dogs, Linnaeus gave the name *Canis*, Latin for dog. Each genus then contained a number of different animals, all with some broad structural characteristics in common, yet with differences in the finer details: color, size, thickness of hair, and the like. Thus, the genus could be broken down into still finer groups, the individual species. To name this, Linnaeus gave another, second name to the species. The wolf became *Canis lupus*, the mountain lion *Felis concolor*, and the African lion *Felis leo*.

All catlike animals, according to the Linnaean system, are grouped together into a still larger classification, the family, which takes its name from one of the genera it embraces. In the case of cats, the genus *Felis* was used. By adding the standardized suffix *idae* onto the root word *Felis*, the cat family became Felidae; and by a similar process the dog family became Canidae. Dogs and cats, skunks and weasels (Mustelidae), raccoons (Procyonidae), bears (Ursidae), seals (Phocidae), and so on were classified into families. All have several general features in common: five toes with claws, shearing teeth, well-developed canine teeth or fangs, and other skeletal structures. These animals were then grouped together and placed in an even higher category, the order (in this case the order Carnivora). All orders of animals giving birth to living young, possessing body hair, and nourishing young with milk were placed in an even higher category, the class Mammalia. All animals having a backbone were grouped in a phylum (in this instance Chordata), belonging to the animal kingdom.

The final result was a hierarchical scheme of classification in which similar species were grouped together and placed in a genus, similar genera were grouped into families, similar families into orders. The complete hierarchical classification of a mountain lion looks like this:

Kingdom: Animal
Phylum: Chordata
Class: Mammalia
Order: Carnivora
Family: Felidae
Genus: *Felis*
Species: *concolor*

As time went on, the number of known animals swelled, and the need arose to refine this hierarchy even further. Phyla were divided into subphyla, classes into subclasses and infraclasses, orders into suborders, families into subfamilies (and even enlarged into superfamilies), and species into subspecies.

To take another example, the white-footed mouse would be classified as follows:

Kingdom: Animal
Phylum: Chordata
Class: Mammalia
Subclass: Theria
Infraclass: Metheria
Order: Rodentia
Suborder: Myomorpha
Superfamily: Muroidea
Family: Cricetidae
Subfamily: Cricetinae
Genus: *Peromyscus*
Species: *leucopus*

Note that the superfamily ends in *oidea,* the family in *idae,* and the subfamily in *inae*.

The principal categories used in plant classification are somewhat different:

Kingdom: Plant
Division: Trachaeophyta
Subdivision: Pteropsida
Class: Angiospermae
Subclass: Dicotyledoneae
Order: Fagales
Genus: *Fagus*
Species: *grandifolia*

The plant classified here is the American beech.

ductive isolation. Often two allopatric species are not reproductively isolated, and when barriers are eliminated, the two species behave as one. Such has been the case, for example, with the red-shafted flicker and the yellow-shafted flicker. The two species, once allopatric, interbred when they became sympatric and are now considered as one species, *Colaptes auratus*. *Parapatric species* are those that meet only along the borders of their ranges. There is minimal genetic exchange between the two populations, and such hybrids are at a strong selective disadvantage.

Some sympatric species may be *sibling species*—ones that are similar in their appearance or morphology but do not interbreed. They may differ in behavior, in ecology, or in physiology and chro-

TABLE 3.1 SOME DIFFERENCES IN THE SIBLING SPECIES OF THE 17-YEAR AND 13-YEAR CICADAS *(HOMOPTERA, CICADIDAE, MAGICICADA)*

	M. septendecim	*M. tredecim*	*M. cassini*[a]	*M. tredecassini*	*M. septendecula*[a]	*M. tredecula*
Size	NO STATISTICAL DIFFERENCE IN SIZE BETWEEN ANY OF THESE					
Body color	Black above, reddish below; appendages reddish; pronotum reddish yellow; prothoracic pleura reddish yellow	Same as *M. septendecim,* except radial *W* in forewing heavily clouded	Black above, almost black below; appendages reddish; pronotum black; prothoracic pleura black		Pronotum black; prothoracic pleura black; tibia reddish or with narrow black apical markings	Same as *septendecula,* except apical tarsal segments reddish all the way to tip; abdominal sternites with prominent reddish bands
Brood	17 years	13 years	17 years	13 years	17 years	13 years
Call	Low-pitched buzzing phrases; fairly even in intensity, ending with a drop in pitch; phaaaaaaaraoh: 1–3 sec		Rapidly delivered tick series, alternated with high-pitched, sibilant buzzes; noticeable rise and fall in pitch and intensity; ticks 2–3 sec, buzz 1–3 sec		High-pitched, brief phrases in series of 20–40 at rate of 3–5 per sec; entire call 7–10 sec	
Chorus	Even, monotonous roaring or buzzing; no regular fluctuations in intensity or pitch; individual males not synchronized; most intense in morning		Shrill, sibilant buzzing; rise and fall in intensity due to synchronization among individual males; most intense in afternoon		More or less continuous repeating of short, separated buzzes; no regular fluctuations in pitch or intensity; individual males not synchronized; most intense around midday	

[a] The only reliable way to distinguish *septendecula* from *cassini,* other than by song, is by morphometric characters. These color characteristics are very inconsistent. Where *cassini* occurs on the edge of its range in the absence of *septendecula,* some character displacement seems to take place. The abdominal sternites of *cassini* have reddish bands as prominent as those of *septendecula.*

SOURCE: Based on data from Alexander and Moore, 1962; M. Lloyd, personal communication.

mosomal structure. To the human eye the species may be virtually indistinguishable, but the differences are apparent to the animal. Examples of such sibling species are the three species of 17-year periodical cicadas, *Magicicada septendecim, M. cassini,* and *M. septendecula,* and the three species of 13-year periodical cicadas, *M. tredecim, M. tredecassini,* and *M. tredecula.* The members of each group emerge in synchrony with each other (Table 3.1).

Geographic Variation in a Species

Because of the widespread variation of many morphological, physiological, and behavioral characteristics in a widely distributed species, significant differences often exist among populations of different regions. One local group may differ, more or less, from other local populations. The greater the distance between populations, the more pronounced the differences become. Geographic variants reflect the selective environmental forces acting on various genotypes, adapting each population to the locality it inhabits. Geographic variation shows up as clines, ecotypes, and geographic isolates.

CLINE

A *cline* is a gradual geographic variation in some phenotypic character such as size or coloration. Clines are usually associated with an ecological gradient such as temperature, moisture, light, or altitude. Continuous variation results from an uninterrupted gene flow from one population to another along the gradient. Because of that, it is impossible to group demes into separate entities, even though the two extremes may behave as two distinct species.

An example of a cline is the meadow frog, *Rana pipiens,* one of the most familiar of all North American amphibians. It has the largest range, occupies the widest array of habitats, and possesses the greatest amount of morphological variability of any North American ranid, and it is the only one successfully established throughout the prairie country. But the variability and adaptability of the meadow frog are orderly, not haphazard. The species embraces a number of temperature-adapted races on a north-south gradient (Moore, 1949a, 1949b). When the populations of the north and south extremes are compared, the differences are pronounced; yet between the two extremes, no break in variations occurs. Embryos of southern meadow frogs have an upper-limit temperature tolerance of 4°C above that of the northern embryos, although both survive equally well at low temperatures. Southern forms have smaller eggs and a slower rate of development at low temperatures. In fact, so wide are these physiological differences that crossing the northern and southern forms results in defective individuals, although normal hybrids are produced when the meadow frog is crossed with either of the two gopher frogs, *Rana areolata areolata* and *Rana a. capito,* and the pickerel frog, *R. palustris.*

Many species of plants exhibit clinal gradation, some in size and other structural characteristics, others in time of flowering, growth, or other physiological responses to the environment. Clinal differences in plants can be demonstrated by transplant studies in which a series of populations from different environments is grown together under one uniform environment in field and greenhouse. Further comparisons of differences can be obtained by growing such a series under several environmental conditions. Such studies have revealed that a number of prairie grasses—among them blue grama *(Bouteloua gracilis),* side-oats grama *(B. curtipendula),* big bluestem *(Andropogon gerardi),* and switchgrass *(Panicum virgatum)*—flower earlier in northern and western regions of their ranges and progressively later toward the south and east (McMillan, 1959). The goldenrod *Solidago sempervirens* flowers progressively later in the season from north to south along the Atlantic coast.

ECOTYPE

Clinal variations across very short transects, in contrast to a continental one, are *ecotypes.* An ecotype is a genetic strain of a population that is adapted to the unique local environmental conditions in which it is found. For example, the phenotype of a plant species may vary on an altitudinal gradient; or a population of a species inhabiting a mountaintop may differ from a population at the bottom of the slope. Often the distribution of ecological variants of a species will be mosaic rather than clinal. That frequently is the situation where several habitats to which the species is adapted reoccur throughout the range of the species. Some of these ecotypes may have evolved independently from different local populations. (These are called polytypic ecotypes.)

The yarrow *Achillea* blankets the temperate and subarctic Northern Hemisphere with an exceptional number of ecological races. One species, *A. lanulosa,* intensively studied by Jens Clausen and his associates (Clausen, Keck, and Hiesey, 1948), occurs at all altitudes in the Sierra Nevada of California. It exhibits considerable variation, an adaptive response to different climatic environments at various altitudes. Populations at lower altitudes are taller and have higher seed production than those at higher altitudes. Montane populations have a distinctive small size and low seed production.

GEOGRAPHIC ISOLATE

The geographic isolate is a population or a group of populations that is prevented by some extrinsic barrier from effecting a free flow of genes with others of the same species (Mayr, 1963) (Figure 3.1). The degree of isolation depends upon the efficiency of the extrinsic barrier, but rarely is the isolation complete. These geographic isolates, or races (Figure 3.2), and to some extent the clinal variants, taxonomically make up the subspecies as "an aggregate of local populations of a species, and differing taxonomically from other populations of the species" (Mayr, 1942:106). The geographical races of a continental species are, more often than not, connected by intermediate forms, or intergrades, so that it is virtually impossible to draw a line that will separate them all.

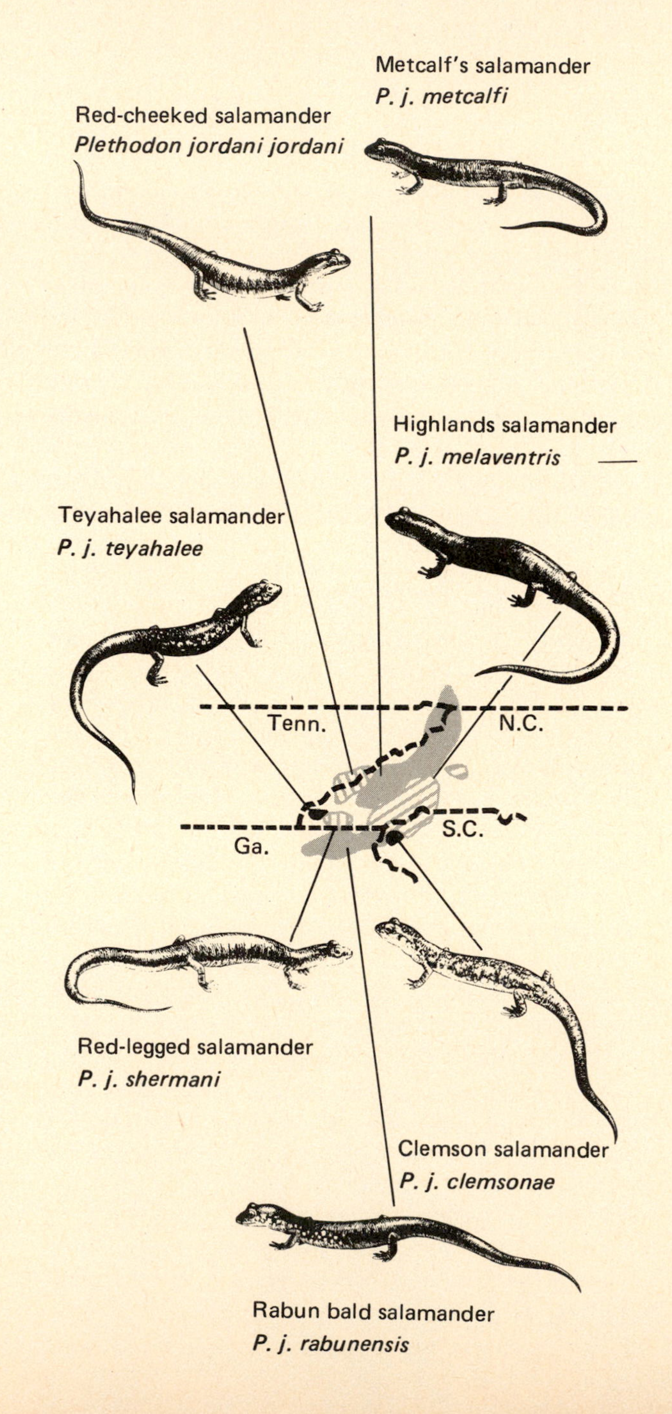

FIGURE 3.1 Geographic races and subspeciation in the *Plethodon* salamanders of the Appalachian highlands. These salamanders of the *jordani* group originated when the population of the salamander *P. yonahlossee* became separated by the French Broad valley. In the eastern part the separated population developed into Metcalf's salamander, which spread northeastward, being the only direction in which any group member could find suitable ecological conditions. South, southwest, and northwest the mountains end abruptly, limiting the remaining *jordani*. Metcalf's salamander is the most specialized and ecologically divergent and the least competitive. Following separation of Metcalf's salamander, the isolation of the red-checked salamander from the red-legged and the rest of the group resulted from the deepening of the Little Tennessee River. Remaining members are still somewhat connected, especially around the headwaters of the Little Tennessee. (Based on data from Hairston and Pope, 1948, and K. Schmidt, 1960, *Checklist of Amphibians and Reptiles.)*

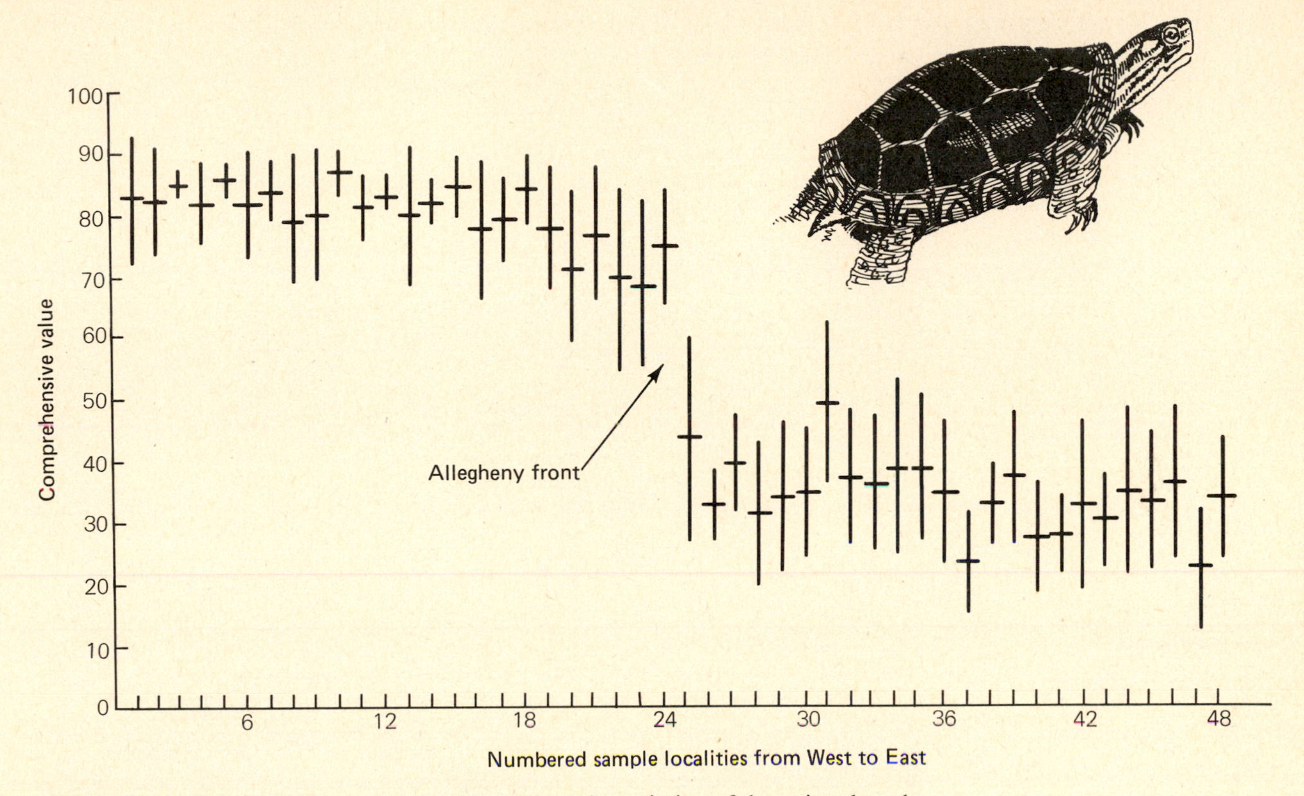

FIGURE 3.2 Effect of a mountain barrier on the subspeciation of the painted turtle. (*Chrysemys picta*). The painted turtle consists of two subspecies, the eastern painted turtle (*C. p. picta*) and the midland painted turtle (*C. p. marginata*). North of the high Appalachians, the two subspecies intergrade. In West Virginia the high Appalachians form an ecological barrier that effectively separates the two subspecies. This separation is illustrated in the plots of composite values of such measurable characteristics as seam alignment, width of plastron, plastron figure, and others. The horizontal bar is the mean value; the vertical bar is the standard deviation. (Data from L. Clack, 1975.)

ECOGEOGRAPHIC RULES

Clines are usually associated with an ecological gradient such as temperature, moisture, altitude, or light. Changes may take place over a relatively small area in response to some varying environmental condition, or they may take place over a much larger area. This phenomenon has resulted in a number of ecological rules, summarized in Table 3.2. These rules may hold in a general sort of way, but underlying physiological assumptions do not hold up. Bergmann's rule, for example, assumes that large animals expend less energy to maintain body heat because of a small surface-area–volume ratio and thus are better adapted to live in colder climates. But the patterns invoked by the rules may result from natural selection for other advantages of size, limb proportions, and the like. For this reason ecogeographic rules, which are useful as general observations, are more of historical than current interest.

TABLE 3.2 RULES CORRELATING VARIATIONS WITH ENVIRONMENTAL GRADIENTS (SUBJECT TO FREQUENT EXCEPTIONS)

Rule	Statement
Bergmann's rule	Geographic races possessing a smaller body size are found in the warmer parts of the range; races of larger body size are found in the cooler climate.
Allen's rule	The extremities of animals—the ears, tail, bill, etc.—are shorter in the cooler part of the range than in the warmer part.
Gloger's rule	Among warm-blooded animals, black pigments are most prevalent in the warm and humid areas, reds and yellow in the arid areas, and reduced pigmentation in the cool areas.
Jordan's rule	Fish living in warm waters tend to have fewer vertebrae than those living in cool waters.
—	Races of birds living in the warmer part of the range lay fewer eggs per clutch than those races living in the cooler part of the range.

Polymorphism

The occurrence of several distinct forms of a species in the same habitat at the same time is polymorphism. It may involve differences in morphological characters and in physiology. The important feature about polymorphism is that the forms are distinct and the characteristic involved is discontinuous. There are no intermediates. Polymorphism may arise from disruptive or divergent selection and appears to be environmentally induced.

A classic example of genetic polymorphism as a result of changing environmental selection is industrial melanism in the peppered moth *Biston betularia* in England. Before the middle of the nineteenth century, the moth, as far as is known, was always white with black speckling on the wings and body (Figure 3.3). But in 1850, near the manufacturing center of Manchester, a black form of the species was caught for the first time. The black form, *carbonaria,* increased steadily through the years until it became extremely common, often reaching a frequency of 95 percent or more in Manchester and other industrial areas. From these places *carbonaria* spread into rural areas far from industrial cities. The form came about through the spread of dominant and semidominant mutant genes, none of which is recessive. The increased frequency and spread has been brought about through the selective pressure of predation. The typical form of the peppered moth has a light color

FIGURE 3.3 Normal and melanistic forms of the polymorphic moth *Biston betularia* at rest on a lichen-covered tree. The spread of the melanistic form, *carbonaria,* in industrial areas is associated with improved concealment of black individuals on soot darkened, lichen-free tree trunks. Away from the industrial areas, the normal color is most frequent because black individuals resting on lichen-covered trunks are subject to heavy predation by birds.

pattern that makes it inconspicuous when it rests on lichen-covered tree trunks. The dark form is very conspicuous on the lichen-covered trunks and is subject to heavy predation. But the grime and soot of the industrial areas killed or reduced the lichen on the trees and turned the bark a nearly uniform color of black. On such trees the black form had the selective advantage. In the polluted woods the light form bore the brunt of predation. (For an excellent review of this topic, see the book by the biologist who did most of the experimental work on the moth, H. B. D. Kettlewell, 1973.)

In the case of the peppered moth, environmental changes converted a disadvantageous allele or a mutant gene into an advantageous one, permitting the latter to spread through the population. Such a polymorphism will exist as long as environmental conditions favor the dark form. Or in other situations such polymorphisms will exist until the new advantageous form has completely replaced the original form or has so swamped it that the original can be maintained only by recurrent mutation. Such a situation is known as a *transient polymorphism*.

Polymorphism may also result from environmental modification of gene action. That is possible only when the two environments—for example, background color—are present at the same time in the same place. Environmentally controlled polymorphism, favoring two or more forms, is the optimal expression of the characters concerned. All intermediates are at a disadvantage and are usually eliminated. The North American black swallowtail *(Papilio polyxenes)* and the European swallowtail *(P. machaon)* are good examples. Both swallowtails pupate either on green leaves and stems or on brown ones. Through selection both have acquired a genetic constitution that produces green-colored pupae in green environments and brown in brown environments. Green pupae would be quite conspicuous in winter, but butterflies emerge from the green in late summer. Those in brown pupae do not emerge until the following spring (Sheppard, 1959).

The swallowtails are examples of the most common type of polymorphism—*stable* or *balanced*. That is a condition in which an apparent optimum proportion of two forms exists in the same habitat and is maintained by environmental heterogeneity. Any deviation in one direction or the other is a disadvantage.

There exists a good deal of circumstantial evidence that genetic polymorphism is related to a patchy environment. A minimum amount of experimental evidence exists, however, to support the hypothesis that a heterogeneous environment has a major role in the maintenance of the polymorphism or to explain how environmental differences can result in polymorphism (Hedrick, Ginevan, and Ewing, 1976).

How Species Stay Apart: Isolating Mechanisms

Each spring there is a rush of courtship and mating activity in woods and fields, lakes and streams. Fish move into their spawning grounds, amphibians migrate to breeding pools, birds are singing. During this frenzy of activity each species remains distinct. Song sparrows mate with song sparrows, trout with trout, wood frogs with wood frogs, and few mistakes are made, even between species similar in appearance.

The means by which the many diverse species remain distinct are *isolating mechanisms*. These include any morphological characters, behavioral traits, habitat selection, or genetic incompatibility which enable different species to remain apart. Isolating mechanisms may be premating or postmating. Premating mechanisms—those that prevent interspecific crosses—include ecological ones: habitat and seasonal isolation; behavior; and mechanical or structural incompatibility. Postmating mechanisms reduce the full success of interspecific crosses.

If two potential mates in breeding condition have little opportunity to meet, they are not likely to interbreed. Habitat selection, even on a local basis, can effectively reinforce this isolation, a type that is important among frogs and toads (see Bogert, 1960). Different calling and mating sites among concurrently breeding frogs and toads tend to keep the species separated. The upland chorus frog and the closely related southern chorus frog breed in the same pools, but ecological preferences tend to separate—partially, at least—the calling aggregations of the species. The southern chorus frog calls from concealed positions at the base of grass clumps or among vegetational debris, while the upland chorus frog calls from more open locations.

Temporal isolation (differences in the timing of the breeding and flowering seasons) effectively separates some sympatric species. The American toad, for example, breeds early in the season, while the Fowler's toad breeds a few weeks later (Blair, 1942). Fluctuations in environmental stimuli can time mating seasons. Among the narrow-mouthed toads, *Microhyla olivacea* breeds only after rains, while *M. carolinensis* is little influenced by rain (Bragg, 1950). Since temporal isolation is incomplete, call discrimination is also involved (Blair, 1955); nevertheless, some hybridization does occur.

Ethological barriers (differences in courtship and mating behavior) are the most important isolating mechanisms in animals. The males of animals have specific courtship displays, to which, in most instances, only females of the same species respond. These displays involve visual, auditory, and chemical stimuli. Some insects, such as certain species of butterflies and fruit flies, and some mammals possess species-specific scents. Birds, frogs and toads, some fish, and such "singing" insects as the crickets, grasshoppers, and cicadas have specific calls that attract the "correct" mates. Visual signals are highly developed in birds and some fish. Species-specific color patterns, structures, and display, which give rise to a high degree of sexual dimorphism among such bird families as the hummingbirds and ducks, have apparently evolved under sexual selection (Sibley, 1957). Among the insects, the flight paths and flash patterns of fireflies on a summer night are the most unusual visual stimuli. The light signals emitted by various species differ in timing, brightness, and color, which may range from white through blues, greens, yellows, orange, and red (Barber, 1951; Lloyd, 1966).

Mechanical isolating mechanisms involve structural differences that make copulation or pollination between closely related species impossible. Although evidence for such mechanical isolation among animals is scarce, differences in floral structures and intricate mechanisms for cross-pollination commonly exist among plants (see Grant, 1963). If hybrids should occur, they could possess such unharmonious combinations of floral structures that they would be unable to function together, either to attract insects to them or to permit insects to enter the flowers.

These three types of isolating mechanisms—ecological, behavioral, and mechanical—prevent the wastage of gametes, diminish the appearance of hybrids, and permit populations of incipient species to become wholly or partly sympatric.

A fourth type of isolating mechanism, the reduction of mating success, does not prevent the wastage of gametes, but it is highly effective in preventing cross-breeding. If hybrids do result, they may be sterile or at a selective disadvantage and thus be eliminated from the breeding population.

How Species Arise

The great diversity of plants and animals in the world causes one to wonder how all these species arose. Each is adapted to an ecological niche in the ecosystem to which it belongs, and each is genetically independent. The process by which this has come about, by which one form becomes genetically isolated from the others, is *speciation*—the multiplication of species. Speciation is accomplished in most plants and animals by an interaction of heritable variation, natural selection, and barriers to gene flow.

ALLOPATRIC SPECIATION

The most usual type of speciation is *allopatric*, or *geographic*, speciation. The first step in geographic speciation is the splitting up of a single interbreeding population into two spatially isolated populations. Imagine, for a while, a piece of land, warm and dry, occupied by species A. Then at some point in geologic time, mountains uplift, land sinks and becomes flooded with water, or some great vegetational catastrophe occurs, which splits the piece of land and separates a segment of species A from the rest of the population. The newly isolated segment will now become species A′. It now occupies an area of cool, moist climate in our imaginary land.

Because it represents only a random sample of the population of species A, A′ will possess a slightly different ratio of genetic combinations. The climatic conditions are different; the selective forces are different. Natural selection will favor any mutation or recombination of existing genes that will result in better adaptation to a cool, moist climate. Similar selection for a warm, dry climate will continue in population A on the original land mass. With different selective forces acting on them, the two populations will diverge. Accompanying genetic divergence will be changes in physiology, morphology, color, and behavior, resulting in ever-increasing external differences, until A′ becomes a geographic race. A′, however, is still capable of interbreeding with species A if given the opportunity.

If geographic barriers break down before isolating mechanisms are effective, then interbreeding takes place and the hybrid individuals produced are fully fertile and viable (Figure 3.4). If they are not at a selective disadvantage in competition with the parent populations, the genes of one race will be incorporated with the gene complex of the other. There will be a period of increased variability in the rejoined populations, and new adaptive forms will be established. Eventually, the variability will decrease to a normal amount, and once again there will be a single, freely interbreeding population.

If the barrier remains, however, further evolutionary diversification occurs. The two populations become increasingly different, and isolating mechanisms become more fully established. The two populations have now become *semispecies,* but genetic differences are not sufficient to prevent interbreeding if the two should come in contact. If the barrier falls at this stage, individuals of the two populations may interbreed and produce hybrid offspring. These offspring form a hybrid zone between the two semispecies populations.

Individuals in the hybrid zone may range from character combinations of species A to those of emerging species. Some hybrids may be indistinguishable from one parent stock or the other. Others may show a high degree of divergence. In some situations the hybrid zone may be relatively narrow. In others the hybrids may form a cline between the two parent populations. One or more characters may filter further than others into the range of the semispecies. Some genes acquired through hybridization are retained because they apparently confer some selective advantage, while disadvantageous genes are eliminated. This incorporation of genes from hybrids is *introgressive hybridization.* In other cases hybrids produced among semispecies are less fertile and viable than the parent stock because they contain discordant gene patterns. Their reproductive potential, if they are fertile at all, is low; they produce few offspring. Hybrids are at a selective disadvantage, while any color pattern, voice, behavior, or the like in parent stock that reinforces reproductive isolation will be favored. This selection against hybrids and reinforcement of isolating mechanisms continues until gene flow between the two populations has ceased. Then species A and new species A′ can invade each other's territory, occupy suitable niches—in our example a warm, dry environment and a cool, moist environment—and become wholly or partly sympatric.

A second, and probably more common, form of geographic speciation results from the establishment of a new colony by one founder (a gravid female) or a small number of founders (see the discussion of the founder principle, page 38). Founder individuals may be "surplus" from a rapid population increase in which selection pressures have been relaxed (Carson, 1968). They may be small populations living on the periphery of the species range that have become isolated from the parent population. Or they may be small populations that have found unexploited areas suitable for invasion. In all cases these small populations are away from genetically stable conditions in the center of the population, where variation in the gene pool is great and selection is for the heterozygotic condition. Founder populations have less genetic variation and experience greater selection for homozygosity. Because of small population size and inbreeding, any adaptive mutation carried by the population, especially by a dominant male, becomes fixed rapidly

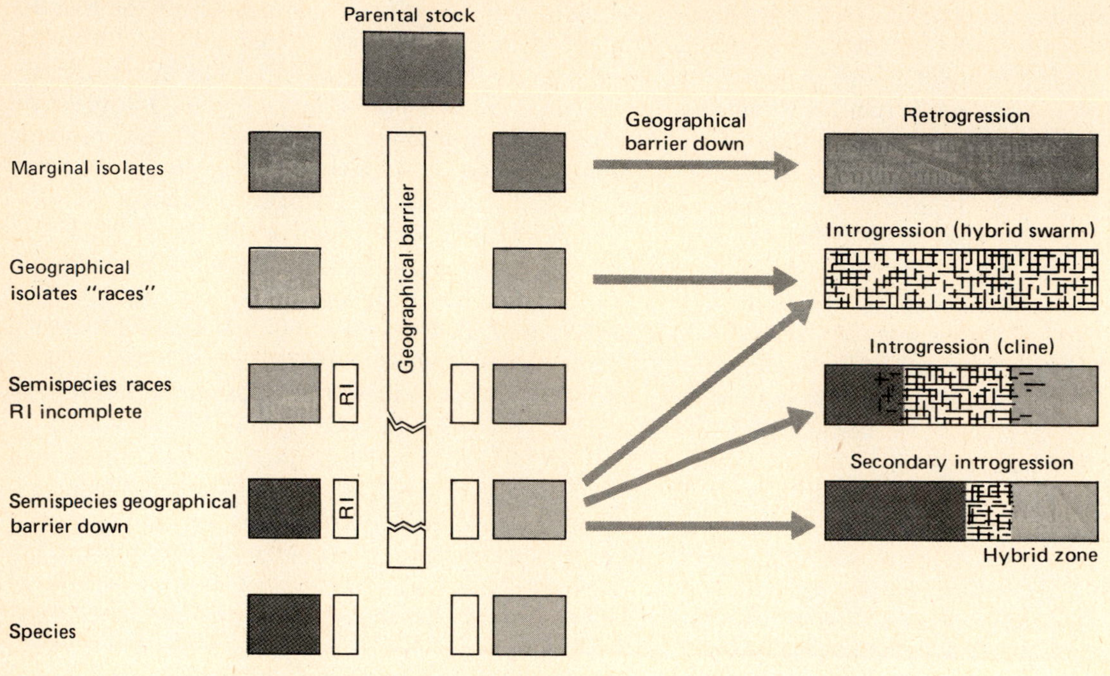

FIGURE 3.4 Geographic, or allopatric, speciation. One species may become two as a result of isolation for a long period of time by a geographical barrier. If the barrier is removed before the separated groups have diverged from the parent stock, they simply rejoin (retrogression). If speciation is incomplete, the two races can interbreed (introgression). If genetic divergence is not great, potential hybrids are not at a disadvantage, and the gene pools of the two races will mesh. If potential hybrids are at a disadvantage, isolating mechanisms are effective, even though some interbreeding does occur (secondary introgression). If genetic divergence is complete, the two species can exist sympatrically after the barrier is removed without interbreeding. (After Mettler and Gregg, 1969:184.)

in the population. Such genetic changes may allow populations to exploit new habitats (Bush, 1975). Once a founder population is well established in a new habitat, reproductive isolation arises by chance. Postmating reproductive isolating mechanisms develop with the fixation of genetic changes. Premating reproductive isolation develops only if the new populations come in contact with the parent population. The numerous species of the fruit fly *Drosophila* in the Hawaiian Islands may have arisen in this fashion (Carson and Kaneshiro, 1976).

PARAPATRIC SPECIATION

Although the idea is prevalent that speciation can take place only with geographic isolation, there is a rapidly growing body of evidence that speciation can occur without geographic isolation (White, 1973, 1974; Bush, 1975). One such type of speciation not related to geographic isolation is parapatric speciation, the evolution of a species as a contiguous (adjacent or nearby) population in a continuous cline. The differentiated populations may meet in a narrow zone of overlap, which may be only a few hundred meters wide. It may take place among organisms that have low vagility (move very little or not at all), are capable of exploiting a somewhat different physical environment adjacent to the species's normal habitat, and become reproductively isolated from the rest of the population. Although suggestive of founders' effect, parapatric speciation differs in that it requires no spatial isolation and reproductive isolating mechanisms arise by selection at the same time genetically unique individuals exploit or colonize a new environment.

Parapatric speciation is characteristic of short-lived, rapidly reproducing species. It occurs in small to medium-sized populations living on the periphery of the main population. Chromosome macromutations frequently initiate speciation with little or no genic differentiation even after speciation is complete. Such populations are characterized by high levels of homozygosity and the lack of long-range dispersal. Diverging populations are in close contact without the usual geographic isolation. Although gene flow does occur between populations, it rarely extends very far into each population because individuals in each experience strong directional selection in two different habitats or microhabitats (Mayr, 1974).

Examples among animals are the flightless morabine grasshoppers endemic to Australia (White, 1973, 1974; Key, 1974), the sessile marine snail *Partula* (Clarke and Murray, 1969), mole crickets (Nevo and Blondheim, 1972), mole rats *(Spalax)* (Nevo, 1969; Nevo and Shaw, 1972), and pocket gophers *(Thomomys)* (Nevo et al., 1974; Thailer, 1974)—all fossorial animals that remain in burrows for most of their lives. Other examples are endemic species of cichlid fish in East Africa (Greenwood, 1974).

Parapatric speciation may be relatively widespread among plants. Plants are nonvagile, so that even with seed dispersal by wind and animals, the level of gene flow between plant populations is so low that effective population size can be measured in meters (Bradshaw, 1972). As a result, plant populations consist of isolated breeding units of various sizes, each more or less adapted to narrow local environmental conditions. It may be particularly prevalent among plants that colonize new habitats, such as soils heavily contaminated by heavy metals (mine spoils, for example) such as nickel and cadmium (Antonovics, 1971).

SYMPATRIC SPECIATION

Sympatric speciation results when premating reproductive isolating mechanisms arise before the populations diverge. It takes place within a population or within the dispersal range of a population. It occurs in the center of a population in a patchy environment rather than on the periphery of a population. Chromosome macromutations probably are not involved. Sympatric speciation probably is a rare event and may be difficult to separate from parapatric speciation.

Sympatric speciation is most apt to occur among insects parasitic on plants or on animals, many of which are host-specific. Such insects are small in size, have short life spans, possess high reproductive rates, have low competitive ability, readily adapt to new conditions, and experience sibmating and other forms of inbreeding.

The first step in sympatric speciation is the formation of a stable polymorphism (Bush, 1975). Such genetic variation as is needed to establish a new host race has to be present in a population before the new host appears. Within the population some individuals must carry two types of genes—one that enables the carrier to recognize and select a new host and one that enables the parasite to counteract any chemical defense of the new host. These genes permit an insect parasite to recognize, move to, and survive on a new host. In a host-specific parasite, mutations at one or two loci would be enough to ensure survival on a host closely related to the original one.

Examples are the new host races of the fruit-eating tephritid fly, the apple maggot or hawthorn fly, *Rhagoletis pomonella*. The fly was once restricted to the hawthorn *(Crataegus)*, but a new race that feeds on apples appeared in 1864 in the Hudson Valley. It spread rapidly through the apple-growing regions of North America. In 1970 in Door County, Wisconsin, a new host race that feeds on cherries was described (Shervis, Bush, and Koval,

1970). The new cherry race apparently developed from the apple race. Each race times its emergence, mating, and oviposition to coincide with the maximum amount of fruit available for oviposition. Thus, the establishment of a population on a new host plant required some genetic alteration of emergence time to coincide with the availability of the appropriate fruit. Among *R. pomonella* the three host races have three emergence times. The cherry race emerges early, to take advantage of the fruiting of cherry in June. The original hawthorn race emerges late, to coincide with the late summer and fall fruiting of hawthorn. The apple race fills the gap between the two. Dispersal of the insect depends upon the presence or absence of host fruit. If the fruit crop is heavy, the insect does not leave the host tree. If fruit is scarce, *R. pomonella* seeks out new host plants for distances up to a mile.

The new cherry race apparently developed when, because of changing market conditions, sour cherries were not picked. Fruit hanging on trees in abandoned and unsprayed orchards provided new egg-laying sites for the insect. The fly was able to exploit this new resource in part because its emergence time extended from late June (cherry time) to October (hawthorn time), although the life span of the individual fly is only 20 to 30 days. The apple and cherry races of *R. pomonella* probably would not have developed without the unintentional help of humans. By devoting large acreages to monocultural fruit production, providing an abundance of food and short dispersal distances, humans created ideal conditions for sympatric speciation. Such speciation might not have occurred in a natural situation of a patchy environment. It is likely that this type of speciation has been repeated a number of times among many of the crop pests.

POLYPLOIDY

Allopatric, parapatric, and sympatric speciation are more or less gradual processes. In sharp contrast is *abrupt speciation,* in which a new species arises spontaneously. The most common method by which abrupt speciation takes place is doubling the number of chromosomes, or *polyploidy* (discussed in Chapter 2), most common in plants. An organism that has double the number of chromosomes in its gametes cannot produce fertile offspring with a diploid member of its ancestral population, but it can do so by mating with another polyploid. Thus, a polyploid has already achieved reproductive isolation. If the polyploid can spread into or exploit a new environment, a new species has been formed. Because plants favor asexual over sexual reproduction in unfavorable conditions, polyploid plants are not at a disadvantage, particularly among perennial herbs. In fact, polyploidy often enables plants to colonize and to tolerate more severe environments. Thus, the availability of new ecological niches favors the establishment of polyploid species.

Many of our common cultivated plants—potatoes, wheat, alfalfa, coffee, and grasses, to mention a few—are polyploids. Polyploidy is rather widespread among native plants, in which it produces a complex of species, as in the case of blackberries *(Rubus),* willows, and birches. The common blue flag of northern North America is a polyploid that probably originated from two other species, *Iris virginica* and *I. setosa,* when the two, once wide-ranging, met during the retreat of the Wisconsin ice sheet. The sequoia is a relict polyploid, its diploid ancestors having become extinct (see Stebbins, 1950).

Character Displacement

In areas where two species overlap their ranges (sympatry), there is a tendency for the differences between them to be accentuated, while outside this area of overlap (allopatry), the differences are weakened or lost entirely (Brown and Wilson, 1956). Such divergence, known as *character displacement,* may be morphological, ecological, behavioral, or physiological. Differences in feeding habits, in anatomical structures that assist in food gathering, and in nesting sites would reduce competition; differences in reproductive behavior would prevent interspecific hybridization.

Although the concept of character displacement has been widely and uncritically accepted, evidence for it is not overwhelming. Evidence for character displacement requires that two closely related species be similar in character in allopatric situations and more distinctive in sympatric situations. The characters under consideration should be displaced in relation to one another.

One among a number of potential examples is character displacement in the bill length of two *Dendroica* warblers that inhabit pine forests, the pine *(D. pinus)* and the yellow-throated *(D. dominica)*. Allopatric populations of the two species have approximately the same bill lengths (Figure 3.5). But strongly sympatric populations on the Delmarva Peninsula (Delaware, Maryland, and Virginia) of the eastern United States differ in the length of bill. The yellow-throated warbler, which probes deeply into pine cones for insects, has a longer bill than the pine warbler. Differences in bill length help to reduce competition for food between the two species. Character displacement, however, is not universal and perhaps should be considered no more than a "weak rule" (MacArthur and Wilson, 1967).

Adaptive Radiation

Character displacement may be one beginning of a divergence of a species into different niches (see Chapter 15), a pathway toward adaptive radiation. The direction and degree to which a population of a species diversifies are influenced by the response of a population to the selective pressures of a new situation and to the ecological opportunities available.

All species of organisms are adapted to some particular environment, but because the environment is limited, overpopulation can result. That in itself is a selective force, for the time finally comes when those individuals able to utilize some unexploited environment and resource are at an advantage. Under reduced competition, those individuals have some opportunity to leave progeny behind. By eliminating disadvantageous genes, selection will strengthen the ability of the population to utilize the new niche it has occupied.

Not every organism can adapt to a new environment. Before a species can enter a new mode of life, it first must have physical access to the new environment. And having arrived there, the species must be capable of exploiting the niches available. Here animals, particularly the vertebrates, have an advantage over plants. Most plants, although easily dispersed, are rather exacting in their habitat requirements. Animals, while finding it difficult to cross barriers, are better able to cope with a new environment.

Once an organism has established a beachhead, it must possess sufficient genetic variation to establish itself under the selective pressures of climate and competition from other organisms. Adaptations that permit an organism to gain a foothold are only temporary; they must be altered, strengthened, and improved by selection before an organism can efficiently utilize the niche. Finally, an ecological niche must be available for exploitation. Competition in a new habitat must be either absent or slight enough to allow the new invader to survive in its initial colonization. Such niches have been available to colonists of some remote islands, such as the Galapagos, the Hawaiian Islands, and the archipelagos of the South Pacific. The abundant empty niches available in these diversified islands when the first colonists arrived encouraged the rapid evolution of species.

Darwin's finches are a classic example of colonization and diversification in an unexploited environment. The original finch population probably consisted of chance migrants from South America. Because of the paucity of invaders, inhibited by the great distance from the mainland, the successful immigrants were able to spread out in a number of evolutionary directions to exploit the islands' resources. David Lack (1947a) suggested a four-stage model to explain the speciation process in Darwin's finches. His model is supported in part by the field studies of P. R. Grant and his associates (1976). In the *first stage*, individuals of a mainland finch species arrived on the Galapagos and established a new population on an island. In the *second stage*, indi-

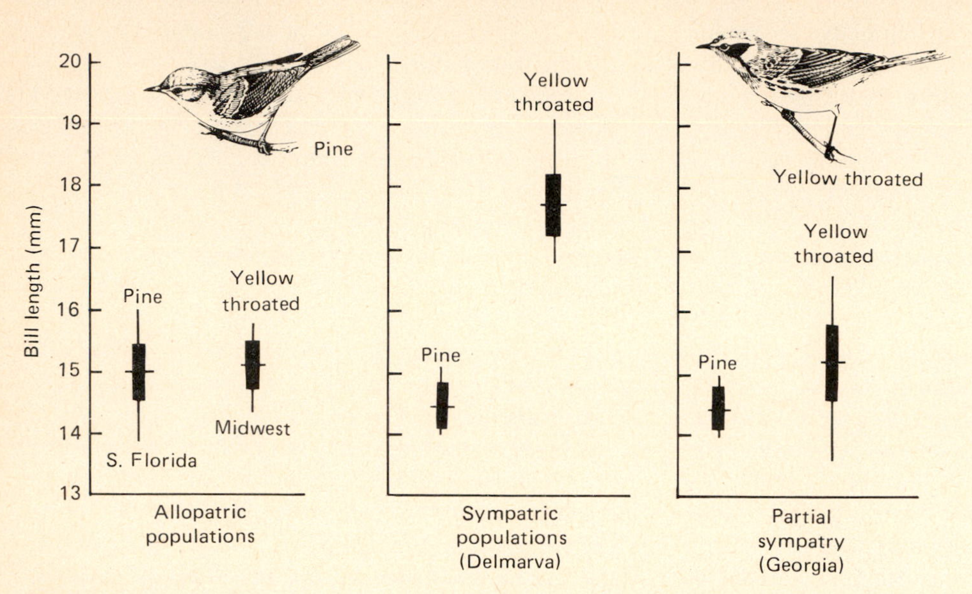

FIGURE 3.5 Possible character displacement in the bill lengths of the yellow-throated warbler (*Dendroica dominica*) and the pine warbler (*D. pinus*). Where the species are allopatric, the bill lengths are similar. Where the species are broadly sympatric, some differences in bill lengths exist. In areas where the birds are weakly sympatric, little difference exists. The long bill of the yellow-throated warbler in the Delmarva region (Delaware, Maryland, Virginia) is associated with its habit of probing into pine cones for insects, a food source not exploited by the pine warbler. Vertical lines represent range of variations; crossbars, the mean; and black rectangles, one standard deviation on either side of the mean. (After Ficken, Ficken, and Morse, 1968).

viduals of the original island population spread to different islands and established new populations. Some evolutionary changes may have taken place. In the *third stage,* individuals from a derived population returned to the island holding the original population, with two possible outcomes. Individuals could interbreed and fuse with the original population. Or if the individuals were moderately different, the birds would tend to discriminate and mate only with their own kind, reinforcing the differences between the two populations. If the two did interbreed, hybrid offspring would be selected against. The *fourth stage* involved multiple repeats of stages 2 and 3, culminating in the formation of 14 different species of finches. (For further information and discussion, see Lack, 1947a; Grant, 1981; P. R. Grant and B. R. Grant, 1980; P. R. Grant and N. Grant, 1983; Grant et al., 1976). Such divergence of one species into a number of different forms—each adapted to different ecological niches, able to exploit new environments or to tap a new source of food—is called *adaptive radiation*.

A similar development took place in the Hawaiian Islands among the honeycreepers, Drepanididae, which evolved into finchlike, creeperlike, and woodpeckerlike forms (see Figure 3.6). They so completely occupied the diverse niches on the island that they prevented adaptive radiation by later colonies of thrushes, flycatchers, and honey eaters. The ancestor of the honeycreepers was probably a nectar-feeding coerebidlike bird with an insectivorous diet somewhat similar to *Himatione* (see Figure 3.6k). After colonizing one or two islands, stragglers undoubtedly invaded other sur-

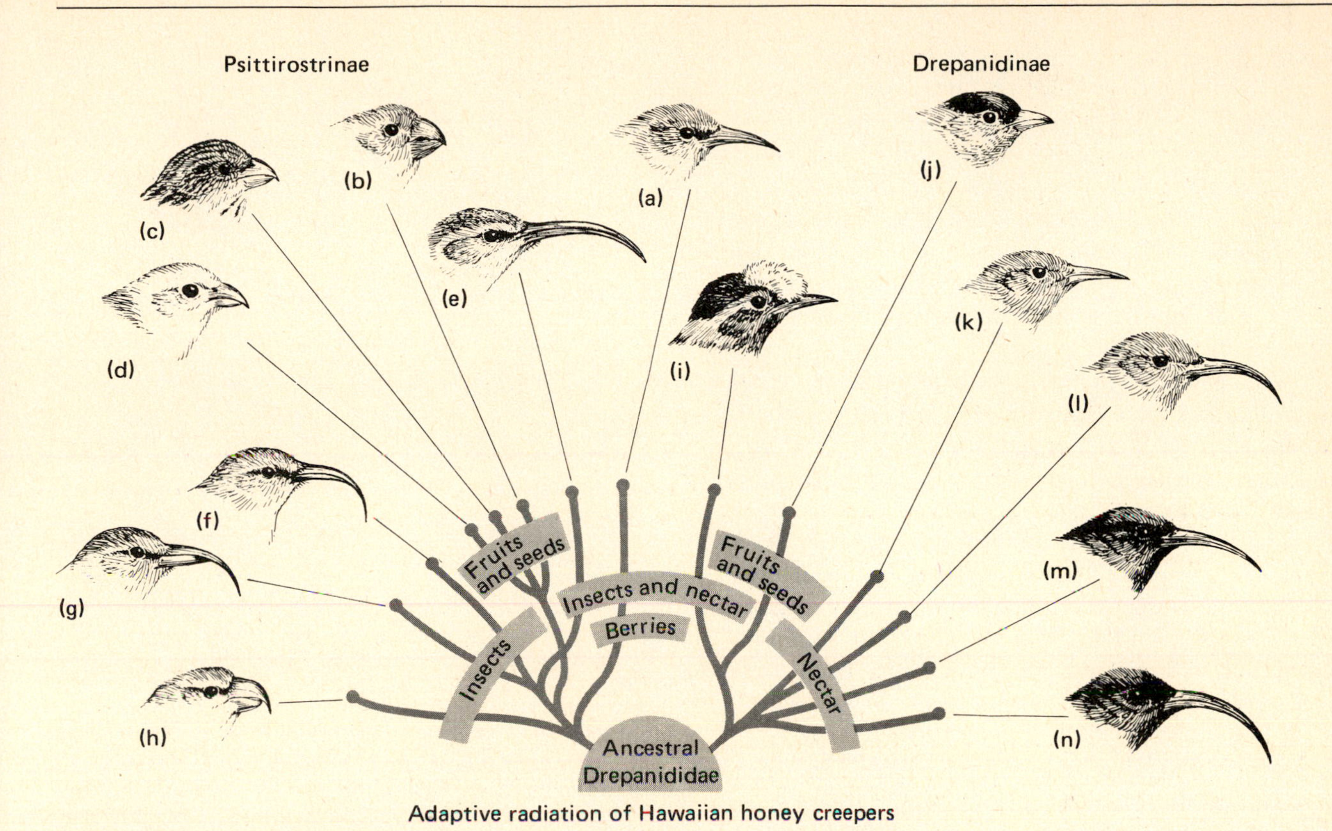

FIGURE 3.6 Adaptive radiation in the Hawaiian honeycreepers, Drepanididae. Selected representatives of two subfamilies, Drepanidinae and Psittirostrinae, illustrate how the family evolved through adaptive radiation from a common ancestral stock. Both subfamilies show a certain degree of parallel evolution based on diet. (a) *Loxops virens,* probes for insects in the crevices of bark and folds of leaves and feeds on nectar and berries; (b) *Psittirostra kona,* a seedeater, extinct; (c) *Psittirostra cantans,* feeds on a wide diet of seeds, insects, insect larvae, and fruit; (d) *Psittirostra psittacea,* a seedeater, feeds especially on the climbing screw pine (Freycinetia arborea); (e) *Hemignathus obscurus,* an insect- and nectar-feeder; (f) *Hemignathus ludidus,* an insect-feeder; (g) *Hemignathus wilsoni,* an insect-feeder; (h) *Pseudonestor xanthophrys,* feeds on larvae, pupae, and beetles of native Cerambycidae and grips branches with curved upper beak; (i) *Palmeria dolei,* feeds on insects and the nectar of ohio (*Metrosideros*); (j) *Ciridops ana,* a fruit-eater and seedeater, extinct; (k) *Himatione sanguinea,* feeds on caterpillars and the nectar of ohio: (l) *Vestiaria coccinae,* feeds on loop caterpillars and the nectar from a variety of flowers; (m) *Drepanis funerea,* a nectar-feeder, extinct; (n) *Drepanis pacifica,* a nectar-feeder, extinct. (Drawings based on Amadon, 1947, 1950, and other sources; evolutionary data from Amadon, 1947, 1950.) See also Bock (1970) and Raikow (1976).

rounding islands. Because each group was under somewhat different selective pressure, the geographically isolated populations gradually diverged. By colonizing one island after another and, after reaching species level, recolonizing the islands from which they came (double invasion), the immigrants enriched the avifauna, especially on the larger islands with more varied habitats. At the same time, competition among sympatric forms placed a selective premium on divergence.

This principle is nicely illustrated by the genus *Hemignathus* (see Figure 3.6e, f, g), all members of which are primarily insectivorous. *Hemignathus obscurus,* whose lower mandible is about the same length as its upper mandible, uses its decurved bill like forceps to pick insects from crevices as it hops along the trunks and limbs of trees. The bill of *H. lucidus* is also decurved, but the lower mandible is much shorter and thickened. The bird uses the lower bill to chip and pry away loose bark as it seeks insects on the trunks of trees. In *H. wilsoni* the modification is carried even further. The lower mandible is straight and heavy. Holding its bill open to keep the slender upper mandible out of the way, the bird uses the lower mandible to pound, woodpeckerlike, into soft wood to expose insects. The bill of the genus *Hemignathus* was specialized at the start to feed on insects and nectar.

The honeycreepers also exhibit *parallel evolution,* or adaptive changes in different organisms with a common evolutionary heritage in response to similar environmental demands. The long, thin, decurved bill of *Hemignathus obscurus* is adapted for a diet of insects and nectar; so, too, are the bills of several members of the subfamily Drepanidinae (see Figure 3.6l, m, n).

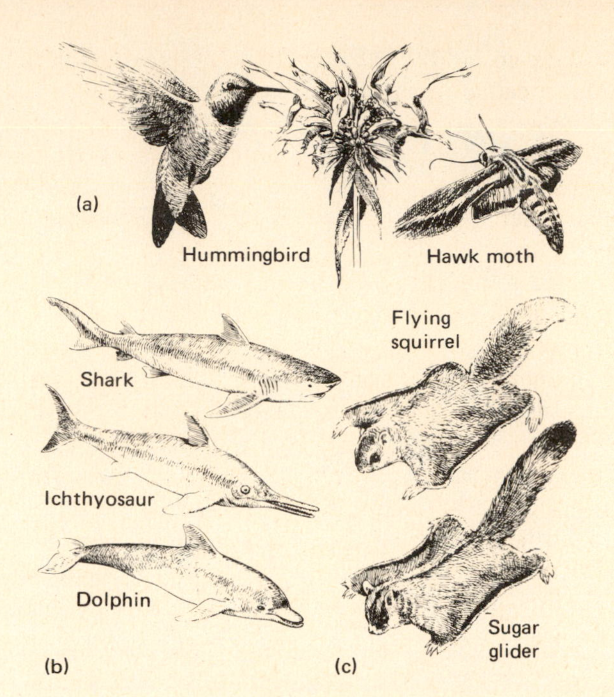

FIGURE 3.7 Convergent evolution in dissimilar organisms. The hummingbird and the hawk moth both feed on nectar, the bird by day, and the moth by night. Both have adaptations of bill and mouth parts for probing flowers; both have the same rapid, hovering mode of flight. The prehistoric marine reptile, the ichthyosaur, the modern shark, a marine mammal, the dolphin, all have the same streamlined shape for fast movement through the water. A North American rodent, the flying squirrel, and an Australian marsupial, the sugar glider, both have a flat, bushy tail and an extension of skin between the foreleg and the hindleg that enable them to glide down from one tree limb to another.

Convergence

Instead of accentuating their differences in zones of overlap, some species will tend to reduce their differences, and their characters will converge (Figure 3.7). Instead of becoming less similar, they become more similar. This apparently happens among species in which reproductive isolation is complete. Selective pressures may tend to favor an increasing resemblance among species as long as reproductive isolation is not upset and the similarities are advantageous to the species (Moynihan, 1968; Cody, 1969). Such convergence can result because sympatric species evolve similar adaptations to the same environment and because it facilitates social reactions among species.

There are a number of possible situations. Sympatric species may become more or less cryptic against background color. Selection may favor dull or dark-colored forms, which might be a reason for the so-called sibling species. Or the animals may become more conspicuous against background color, a trait that can facilitate flocking among individuals of the same and very similar species such as the herons. Sympatric species may become conspicuous, as Batesian or Müllerian mimics. By sharing the same predators, they reduce predation on each other. Convergence in color pattern may be selected for, if it is advantageous for a species to be a social mimic. Such mimicry to facilitate mixed flocking is common among birds of the mountains of neotropical regions (Moynihan, 1968). For example, in parts of the northern Andes, most of the more common and conspicuous species in the mixed flocks of the humid temperate zone are predominantly brilliant blue or blue and yellow.

Finally, some species may evolve striking similarities in voice and coloration where they are sympatric and reduced similarity in areas of allopatry. Such convergence probably arises in response to interspecific territoriality. The more nearly alike they are in color and voice, the more they will behave aggressively as one species. Hybridization is prevented by subtle, species-specific recognition signals. For example, in southern Mexico there are three finches: the collared towhee, *Pipilo ocai;* the rufous-sided towhee, *P. erythrophthalmus;* and the chestnut-capped brown finch, *Atlapetes brunneinucha*. The three form two pairs that are interspecifically territorial. The collared towhee and the brush finch—both colored green, chestnut, black, and white—are so similar in appearance that they can easily be confused in the field. The rufous-sided towhee and the collared towhee have songs that are similar. Because the brush finch and the rufous-sided towhee differ in both plumage and song, their territories overlap (Cody and Brown, 1970).

Similar convergence exists among plants. Within the chaparral vegetation of California, for example, the dominant plants belong to such diverse families as Ericaceae, Rhamaceae, and Rosaceae, yet all are deep-rooted evergreen sclerophyllous shrubs (Mooney and Dunn, 1970a). In fact, throughout all areas with a Mediterranean type of climate, the vegetation has a similar appearance and is dominated by woody, evergreen sclerophyllous species. Even though widely separated geographically and possessing different evolutionary histories, the vegetation has converged in both form and function. This convergence has been in response to similar selective forces, including fire, drought, high temperatures, and low rainfall.

Gradualism Versus Punctuated Equilibrium

A central idea in Darwinian evolution is that populations are always evolving, a step-by-step process of selection from a pool of random variants. Descendants are linked to ancestors by a chain of intermediate genotypes (Figure 3.8). Some lines may be split from ancestral stock by geographic and other forms of isolation, gradually giving rise to new species. Other lines may slowly change into a new species, with the ancestral type becoming extinct. In any situation speciation involves the repetition of favorable adaptations in several lines. Most morphological change takes place along with speciation. Refinement of phenotypic changes results from a gradual or anagenic transformation brought about by natural selection. According to gradualism most species arise by a smooth transformation of an ancestral species. It then changes continuously toward a descendant form. Such a continuous change results in an instability in the species.

A recent, alternative view is punctuated equilibrium. It involves instantaneous speciation followed by a long period of species stabilty or stasis (Stanley, 1979; Eldredge and Gould, 1972). Adaptation is not necessary for evolutionary change; rather, new species arise by the total transformation of a lineage. Punctuated equilibrium is advanced by paleontologists, those who study the fossil record over geological time. Years of study have convinced some of them that species either remain the same for millions of years, changing little, such as the horseshoe crab; end their existence abruptly, geologically speaking, like the dinosaurs; or arise instantaneously. That interpretation appears to be

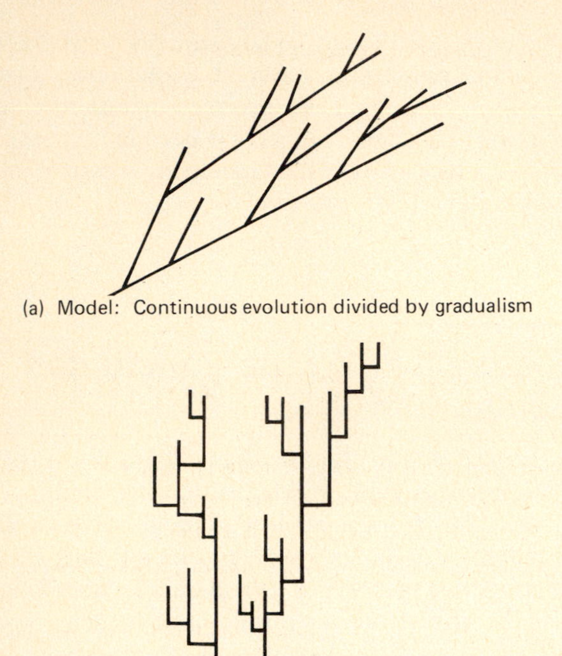

FIGURE 3.8 The classical view of evolution is speciation through gradual change, as illustrated in (a). However, a major process may be abrupt speciation, as diagrammed in (b).

the most reasonable explanation for the geologically short time gaps in the fossil record from one species to another. The fossil record for many lines (such as horses and elephants) is too complete and well-known for any intermediates (as postulated by gradualism) to exist.

A barrier to the acceptance of punctuated equilibrium is the idea of instantaneous speciation. The term *instantaneous* is a relative one. What is an instant to a geologist or paleontologist is a very long time to an evolutionary biologist. An instant in geological time is considered about 1 percent of the lifetime of a species. Most fossil species existed 5 million to 11 million years, which allowed 50,000 to 100,000 years for speciation, a long time for species evolution. Species, according to the theory of punctuated equilibrium, are rapidly established during periods of environmental instability. Old species become extinct, and new evolutionary lines appear. Once established, new species are relatively stable and resist essential change. Differential long-term evolutionary success, however, can still occasionally arise from natural selection acting on the phenotype.

The two approaches may not be that far apart. Instantaneous speciation may require 10,000 to 100,000 years, a sufficient amount of time for natural selection to refine the products of speciation. Local populations would adjust to local environmental conditions and account for any fluctuations within a species during the period of stasis. Most species, once established, rarely evolve beyond differentiation into races or closely related species. All tend to be conservative and retain old habits and habitats. Species evolve new characteristics only when environmental changes force them to do so. Most species, when faced with that situation, become extinct.

Species Selection

Natural selection functions at the level of the individual within a continuous genetic line and brings about changes in both phenotype and genotype. Although natural selection is usually viewed as selecting the most advantageous phenotypes, it acts more to eliminate the disadvantageous, the unfit.

At a higher level, changes in the diversity of species at any given period of time, geological or otherwise, result from species selection, the outcome of differential extinction. All species of life that are presently on Earth and that have ever appeared on Earth have a certain probability of extinction, which is always above zero. That probability varies with different species. Some are more extinction-prone than others (Fowler and MacMahon, 1982).

Highly vulnerable to extinction are those species occupying a small or restricted geographic range. Some, such as island species, have always existed in a restricted range; others have been crowded into parcels of remaining habitat. Such species are more susceptible to the vagaries of natural catastrophe or sudden changes in interspecific relationships, such as the appearance of a new predator or parasite. A single hurricane or fire could wipe out some species. Such was the fate of the remaining population of the heath hen, or eastern prairie chicken, on the island of Martha's Vineyard off the Massachusetts coast.

Although natural selection seems to favor specialists, trophic specialists are more apt to become extinct then generalists. Herbivore species that depend on one species of plant or on particular parts of plants have a much higher probability of extinction than mixed feeders. Members of a straight-line food chain (see Chapter 20) become even more vulnerable to extinction than the species below on which they depend. And species occupying a high trophic level possess a very high probability of extinction.

Large, long-lived species are more vulnerable than small, short-lived species. Natural selection would seem to favor large-bodied, long-lived species. Species selection, however, favors short-lived, smaller species and operates against species with slow rates of evolution. Species with a short generation time have a higher rate of reproduction and a more frequent production of offspring on which natural selection can operate. Short generation time also involves a more frequent shuffling and recombination of genes, providing more genetic diversity over a short period of time and thus greater genetic plasticity. Such species are able to respond more rapidly to environmental changes than long-lived species.

Selective extinction embodies the concept that species are subject to differential or selective failure. The probability of extinction differs among species and depends more on the characteristics of a species than on random processes. Natural selection operates through the selective removal of individuals from a population. That results in a shift in the frequency distribution of certain phenotypic characteristics and changes in gene frequencies relating to those characters. These ultimately result in evolutionary change. Selective extinction, by contrast, removes species rather than genes within a species. Any combination of such qualities as trophic position, size, and generation time that produces high risks increases a species's chance of extinction. Species selection influences the distribution of species within an ecosystem, causing it to differ from the original distribution. Such changes can affect the structure and function of ecosystems.

How Fast Do Species Evolve?

What are the rates of evolution? How fast do species arise? Those are common questions that have no definitive answers. If one keeps in mind that evolution is a change in gene frequency through time, and thus involves some change in phenotype, and that speciation is a multiplication of species, then the two related concepts can be kept apart.

Evolutionary changes involve some adaptations or changes in response to a changing environment. Such changes within a species can occur rapidly. (Consider how fast plant and animal breeders can change the phenotypes and growth efficiencies of domestic plants and animals through selective breeding, a form of goal-directed evolution brought about by human intervention.) In 50 years a single allele controlling industrial melanism in the peppered moth changed in frequency from 0 to 98 percent in populations exposed to industrial pollution. The house sparrow *(Passer domesticus),* introduced into the United States about 100 years ago, spread across North America and differentiated into a number of races (Johnston and Selander, 1964). The continental population differs in size, color, and other morphological characteristics—differences that represent adaptations to different North American environments, from the eastern deciduous forest regions to the desert Southwest. Such changes involve only some features; other features may not change at all. Thus, each species is a mosaic of characteristics that have been passed on unchanged from ancestors and characteristics that have recently evolved.

Rates at which species evolve are difficult to determine. If most species arise by punctuated equilibria or quantum bursts of speciation during periods of environmental change, then determination of rates has little meaning. If some species result from isolation of small populations from parent populations, then rates vary. The numerous species of Hawaiian fruit flies *(Drosophila)* diverged in a few thousand years. Five endemic species of cichlid fish in the African Lake Nabugabo apparently developed in less than 4,000 years, the length of time the lake was isolated from Lake Victoria (Fryer and Iles, 1972). Banana-feeding species of the Hawaiian moth genus *Hedylepta* diverged from a palm-feeding species in about 1,000 years, the time since bananas were introduced to the islands by Polynesians (Zimmerman, 1960). Other species change little. The American and Mediterranean species of the sycamore have evolved little since their separation at least 20 million years ago. When the two species were brought together in English gardens in the 1700s, they hybridized. The hybrid line, known as the London plane tree, is vigorous and fertile. Planted as ornamental, the trees are escaping in places and becoming naturalized.

Summary

For the purposes of description and classification, taxonomists treat plants and animals as a morphological species, a discrete entity that exhibits little variation from the type of the species, the original specimen or specimens on which the description is based. But organisms are highly variable over space and time. That fact has given rise to the concept of the biological species—a group of interbreeding individuals living together in a similar environment in a given region and in similar ecological relationships. This concept, limited as it is to bisexual organisms, is also filled with difficulties, such as the inability to distinguish, among some groups, just where one species begins and another ends. To get around that problem, evolutionary biologists suggest that the species be retained to identify kinds of organisms but that as a concept the species should be regarded as an evolutionary unit.

Species arise by the interaction of heritable variation; natural selection; and some sort of barrier to gene flow between populations, or isolating mechanisms. Isolating mechanisms include any morphological character, behavioral trait, habitat selection, or genetic incompatibility that is species-specific. If isolating mechanisms break down, hybridization results.

The most widely accepted mechanism of speciation involves geographic isolation of one part of a population from another. Each part experiences different selection pressures. The two populations diverge and may ultimately become two distinct species. If geographic barriers break down, the two populations may hybridize, or they may evolve strong—usually premating—isolating mechanisms that reinforce their apartness. That is known as allopatric speciation. Another type of speciation mechanism among populations of organisms that have low vagility and that inhabit contiguous but different and spatially separated habitats is parapatric speciation. It is probably most common among plants. In parapatric speciation both reproductive isolation and speciation take place at the same time. When reproductive isolation precedes differentiation and the process takes places within a population, the outcome is sympatric speciation.

Two views of speciation and evolutionary change are gradualism and punctuated equilibrium. Gradualism says that species arise and differentiate by slow, continuous change within a species line. One species may slowly transform into another. Another species may arise by budding off from a parent population, forming an isolated population, and undergoing a series of changes largely brought about by natural selection. Punctuated equilibrium argues that species arise instantaneously as distinct entities in a geological time span of 10,000 to 100,000 years during periods of great environmental instability.

How rapidly evolution and speciation take place is a topic of debate. Changes within a population at the level of a single allele can take place rapidly over a period of a number of generations. Truly instantaneous speciation, especially among plants, results from polymorphism, an alteration of the number of chromosomes. Some species apparently have never changed for millions of years. Fossil records suggest that new species arise in bursts of speciation, which would make a generalized rate of evolution meaningless. However species arise and regardless of the time span involved, one condition seems certain: Evolution among some populations is always taking place somewhere on Earth, even though evolution may not be a continuous process among all populations.

Review and Study Questions

1. What is a species? A morphological species? A biological species?
2. Why is it difficult to define a species?
3. Distinguish between an allopatric species and a sympatric species; between those and a parapatric species.
4. What is a sibling species? Can you come up with examples other than those given in the text?
5. What is a cline? How does it relate to the species problem?
6. What is an ecotype? A geographical isolate?
7. Define polymorphism. What is its ecological and evolutionary significance?
8. What is an isolating mechanism? Name and describe four classes of isolating mechanisms.
9. What happens if isolating mechanisms break down?
10. What is speciation? Distinguish between allopatric speciation, parapatric speciation, and sympatric speciation.
11. What is polyploidy? What is its evolutionary significance?
12. What property of plants makes them more likely than animals to form hybrid polyploid species?
13. What are character displacement, adaptive radiation, and convergence? How might they interrelate? Can you give some examples?
14. Contrast the concept of gradualism in speciation with that of punctuated equilibrium. Why should punctuated equilibrium be controversial?
15. What is meant by species selection? What is the role of differential extinction?
16. Using the major characteristics affecting the probability of extinction, make a list of some extinction-prone species of plants and animals. Defend your inclusion of each one.
17. Why is it difficult, if not impossible, to establish any definitive rate of evolution or speciation?

Suggested Readings

Eisenberg, J. F. 1981. *The mammalian radiations*. Chicago: University of Chicago Press.

Futuyma, D. J. 1979. *Evolutionary biology*. Sunderland, Mass.: Sinauer Associates.

Gould, S. J. 1977. *Ontogeny and phylogeny*. Cambridge, Mass.: Harvard University Press.

Grant, V. 1971. *Plant speciation*. New York: Columbia University Press.

Lewontin, R. C. 1974. *The genetic basis of evolutionary change*. New York: Columbia University Press.

Simpson, G. G. 1953. *The major features of evolution*. New York: Columbia University Press.

Stanley, S. M. 1979. *Macroevolution: Pattern and process*. San Francisco: Freeman.

Stebbins, G. L. 1974. *Flowering plants: Evolution above the species level*. Cambridge, Mass.: Harvard University Press.

———. 1982. *Darwin to DNA: Molecules to humanity*. San Francisco: Freeman.

White, M. J. D. 1978. *Modes of speciation*. San Francisco: Freeman.

Wilson, D. S. 1980. *The natural selection of populations and communities*. Menlo Park, Calif.: Benjamin/Cummings.

ECOLOGICAL EXCURSION

Natural Selection

Charles Darwin

How will the struggle for existence . . . act in regard to variation? Can the principle of selection, which we have seen is so potent in the hands of man, apply under nature? I think we shall see that it can act most efficiently. Let the endless number of slight variations and individual differences occurring in our domestic productions, and, in a lesser degree, in those under nature, be borne in mind; as well as the strength of the hereditary tendency. Under domestication, it may be truly said that the whole organisation becomes in some degree plastic. But the variability, which we almost universally meet with in our domestic productions, is not directly produced, as Hooker and Asa Gray have well remarked, by man; he can neither originate varieties, nor prevent their occurrence; he can preserve and accumulate such as do occur. Unintentionally he exposes organic beings to new and changing conditions of life, and variability ensues; but similar changes of conditions might and do occur under nature. Let it also be borne in mind how infinitely complex and close-fitting are the mutual relations of all organic beings to each other and to their physical conditions of life, and consequently what infinitely varied diversities of structure might be of use to each being under changing conditions of life. Can it, then, be thought improbable, seeing that variations useful to man have undoubtedly occurred, that other variations useful in some way to each being in the great and complex battle of life, should occur in the course of many successive generations. If such do occur, can we doubt (remembering that many more individuals are born than can possibly survive) that individuals having any advantage, however slight, over others, would have the best chance of surviving and of procreating their kind? On the other hand, we may feel sure that any variation in the least degree injurious would be rigidly destroyed. This preservation of favourable individual differences and variations, and the destruction of those which are injurious, I have called Natural Selection, or the Survival of the Fittest. Variations neither useful nor injurious would not be affected by natural selection, and would be left either a fluctuating element, as perhaps we see in certain polymorphic species, or would ultimately become fixed, owing to the nature of the organism and the nature of the conditions.

Several writers have misapprehended or objected to the term Natural Selection. Some have even imagined that natural selection induces variability, whereas it implies only the preservation of such variations as arise and are beneficial to the being under its conditions of life. No one objects to agriculturists speaking of the potent effects of man's selection; and in this case the individual differences given by nature, which man for some object selects, must of necessity first occur. Others have objected that the term selection implies conscious choice in the animals which become modified; and it had even been urged that, as plants have no volition, natural selection is not applicable to them! In the literal sense of the word, no doubt, natural selection is a false term; but who ever objected to chemists speaking of the elective affinities of the various elements?—and yet an acid cannot strictly be said to elect the base with which it in preference combines. It has been said that I speak of natural selection as an active power or Deity, but who objects to an author speaking of the attraction of gravity as ruling the movements of the planets? Every one knows what is meant and is implied by such metaphorical expressions; and they are almost necessary for brevity. So again it is difficult to avoid personifying the word Nature; but I mean by Nature, only the aggregate action and product of many natural laws, and by laws the sequence of events as ascertained by us. With a little familiarity such superficial objections will be forgotten.

SOURCE: Charles Darwin, *The Origin of Species by Means of Natural Selection* (1859). (New York: Random House Modern Library Edition, 1936), pp. 63–64.

PART III

THE ORGANISM AND THE ENVIRONMENT

INTRODUCTION

Homeostasis and Systems

Laws of Minimum and Tolerance

In 1840 the foremost German organic chemist of his day, Justus von Liebig, published a book, *Organic Chemistry and Its Application to Agriculture and Physiology*. In it he described his analyses of surface soil and plants, and he set forth this simple statement, revolutionary for his day: "The crops on a field diminish or increase in exact proportion to the diminution or increase of the mineral substances conveyed to it in manure." Essentially, what he said was that each plant requires certain kinds and quantities of nutrients, or food materials. If one of these food substances is absent, the plant dies. And if it is present in minimal quantities only, the growth of the plant will be minimal. This became known as the *law of the minimum*.

Continued studies through the years disclosed that not only nutrients but other environmental conditions, such as moisture and temperature, also affected the growth of plants. Later, animals were found to be limited by food, water, temperature, humidity, and other environmental conditions. Eventually, the concept of the law of the minimum was extended to cover all environmental requirements of both plants and animals. Later studies of environmental influences on plants and animals showed that not only does too little of a substance or condition limit the presence or fitness of an organism, but so does too much. Organisms, then, live within a range of too much and too little, the limits of tolerance (Figure III.1). In 1913 this concept of maximum and minimum substances or conditions limiting the presence or success of organisms was incorporated by V. E. Shelford into the *law of tolerance*.

Organisms are limited by a number of conditions and often by an interaction among them. An organism may have a wide range of tolerance for one condition—say, salinity. It is termed euryhaline, from the word *eurys* (meaning "wide") and *haline* (referring to salt). Others may have a narrow range of tolerance for salinity and are called stenohaline, from *stenos* (meaning "narrow"). To further complicate environmental interactions, an organism may have a wide range of tolerance for many conditions but a narrow range for only one. That one narrow range can limit the distribution and fitness of the organism. Further, the maximum and minimum levels of tolerance for a species vary seasonally, geographically, and with the stage of the life cycle. For example, temperatures that are optimal for growth may not be optimal for reproduction. It follows, then, that an organism that exhibits a wide range of tolerance for all environmental influences would be widely distributed.

In some cases when one condition is not optimum for a species, the limits of tolerance are reduced in others. On the other hand, some organisms may utilize an item in surplus to substitute for another that is deficient. Some plants, for instance, respond to sodium when the potassium supply is inadequate (Reitemeier, 1957).

Homeostasis

Most organisms live in a variable environment. Within that environment they have to maintain a

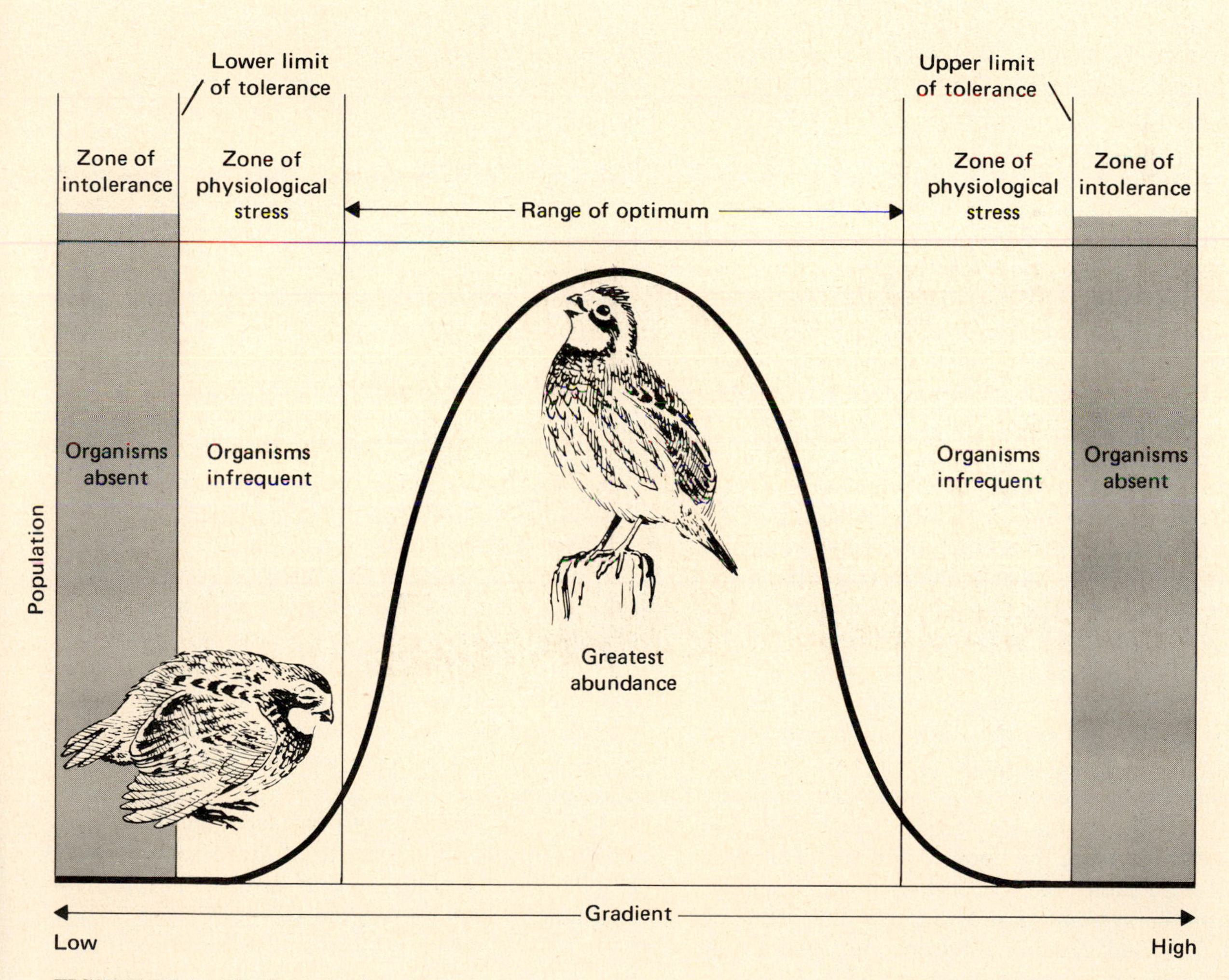

FIGURE III.1 The law of tolerance.

relatively constant internal environment within the narrow limits required by cells. That calls for some means of regulating the internal environment relative to the external one. Organisms have to regulate their body temperature, pH, water, and amount of salts in fluids and tissues, to mention a few factors. Because they take in substances from the environment and use them in cellular chemical reactions, they also have to discharge both excessive intake and waste products of metabolism to the environment to maintain a fairly constant internal environment. The maintenance of these conditions within the tolerance limits of the cells is called *homeostasis*.

Homeostasis involves the feeding of environmental information into a system, which then responds to effects of the input from or changes in external conditions. An example is temperature regulation in humans. The normal temperature for humans is 37°C (98.6°F). When the temperature of the environment rises, sensory mechanisms in the skin detect it and send a message to the brain, which acts (involuntarily) on the information and relays the message to the effector mechanisms that increase blood flow to the skin and induce sweating (Figure III.2). Water excreted through the skin evaporates, cooling the body. If the environmental temperature falls below a certain point, a similar action in the system takes place, this time reducing blood flow and causing shivering, an involuntary muscular exercise producing more heat. This type of reaction, which halts or reverses a movement away from a set point, is called *negative feedback*.

If the environmental temperature becomes extreme, the homeostatic system breaks down. If the environmental temperature becomes too warm, the body is unable to lose heat fast enough to hold the temperature at normal. Body metabolism speeds up, further increasing body temperature, eventually ending in heatstroke or death. If the environmental temperature drops too low, metabolic processes slow down, further decreasing body temperature, eventurally resulting in death by freezing. Such a situation in which feedback reinforces change, driving the system to higher and higher or lower and lower values, is called *positive feedback*.

The idea of homeostasis at the level of the individual can be extended to higher levels: the population, involving intrinsic regulation of size (see Chapter 13), and the ecosystem, encompassing such functions as nutrient cycling (see Chapter 20). All involve the concept of a system.

What is a system? A system is a collection of interdependent parts or events that make up a whole. For example, a radio consists of various transistors, transductors, wires, a speaker, and control knobs, among other things. Each part has a specific function, yet the expression of the role of each depends upon the proper functioning of all the other parts. The whole system fails to function unless there is some kind of input from the outside on which the system can act to produce some kind of output. For the radio the outside input is electrical energy, on which the system acts to pick up certain radio waves, which are transmitted as an output—sound. Thus, all the parts of the radio function as a total system.

There are two basic types of systems: closed and open. A closed system is one in which energy but not matter is exchanged between the system

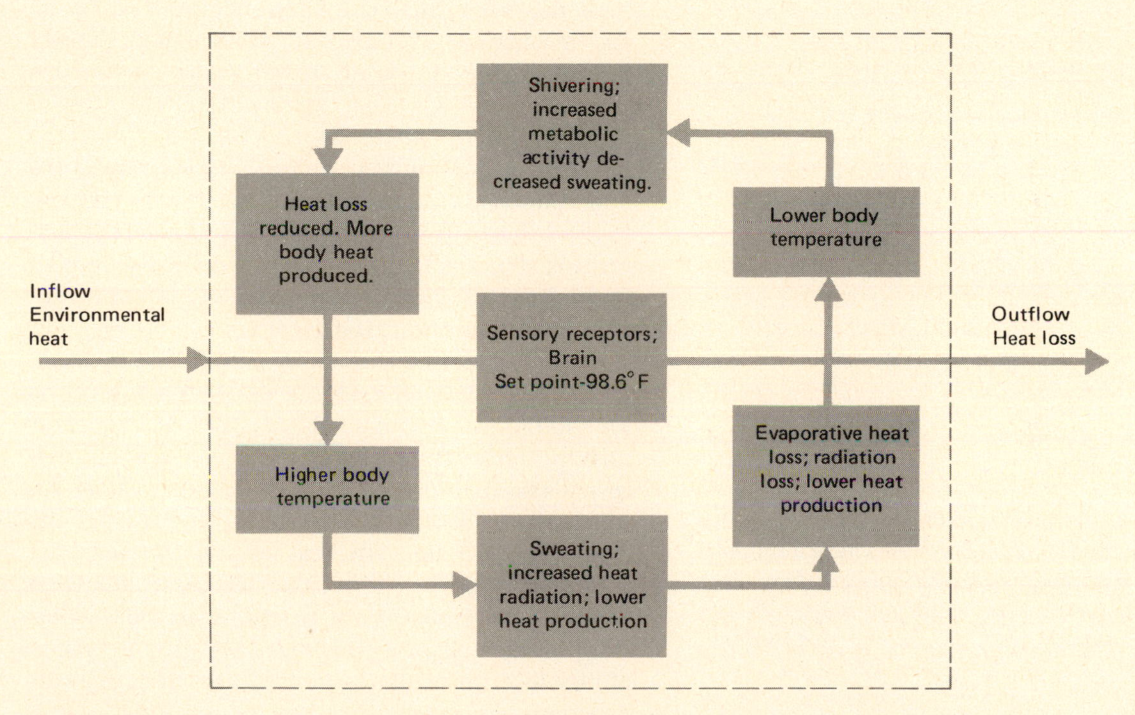

FIGURE III.2 A model of temperature control through feedback mechanisms in humans.

and the environment. The radio is a closed system, and so is Earth. Its only input is energy from the sun. An open system is one in which both matter and energy are exchanged between it and the environment. Living organisms are open systems, and so are ecosystems on Earth. Systems are maintained by balanced exchanges of matter and energy with the environment.

Open systems can be *cybernetic systems,* ones that have a feedback system to make them self-regulating (Figure III.3). To function in such a manner, the cybernetic system has an ideal state, or *set point,* about which it operates. In a purely mechanical system, the set point can be fixed specifically. Consider a dehumidifier set for a humidity level of 50 percent. When the humidity of the air in a room exceeds 50 percent, the switch on the dehumidifier turns on and a fan starts to pull air over the refrigerated coils on which the water condenses to be carried away through a hose or pipe. When sufficient water has been wrung out of the air, the dehumidifier shuts off. The feedback of information

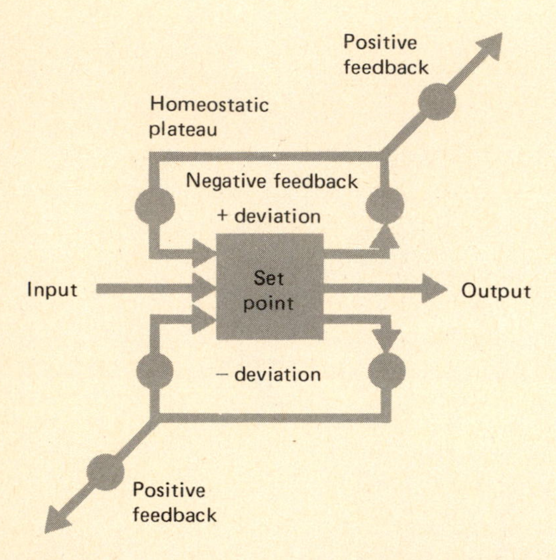

FIGURE III.3 An open system with feedback loops that makes it cybernetic. Cybernetic systems have a set point that the system itself maintains. The homeostatic plateau represents the maximum and minimum values or area within which the system can function by using feedback to regulate itself about the set point. When the system deviates from the set point, a control mechanism brings it back toward the set point. This readjustment is termed negative feedback because it inhibits any strong movement away from the set point. If the system exhibits a continually increasing tendency away from the set point, it is controlled by positive feedback, which can ultimately destroy the system.

on humidity causes the dehumidifier to turn off—a negative feedback mechanism.

Living systems are cybernetic systems that can function at various levels but are always regulated by living organisms. The difference between living and mechanical systems is that in living systems the set point is not firmly fixed. Rather, organisms have a limited range of tolerances, called *homeostatic plateaus,* within which conditions must be maintained. If environmental conditions exceed the operating limits of the system, it goes out of control. Instead of negative feedback governing the system, positive feedback takes over, with a movement away from the homeostatic plateau that can ultimately destroy the system.

The systems approach is especially important to ecology, particularly to an understanding of the function and structure of ecosystems. This approach utilizes the construction of models that represent the real system or parts of the system for the purpose of experimentation. To be valid the model has to mimic the real system, at least over some restricted range, include the most important variables, and be subject to mathematical expression. Models may be constructed to provide a simplified description of a system or to predict changes over time.

One of the most commonly used models in ecology is the compartment model, which is broken down into subcompartments representing subprocesses. The compartment model consists of at least three components: (1) the state variables, selected from many variables involved; (2) the outside forces that drive the system; (3) pathways, flows, or vectors connecting subcompartments with each other. An example is Figure III.4, a highly simplified model of the movement of radioactive material through the food chain to humans. The variables are vegetation, soil, grazing animals, and humans. The source of the input, or driving force, is radio-

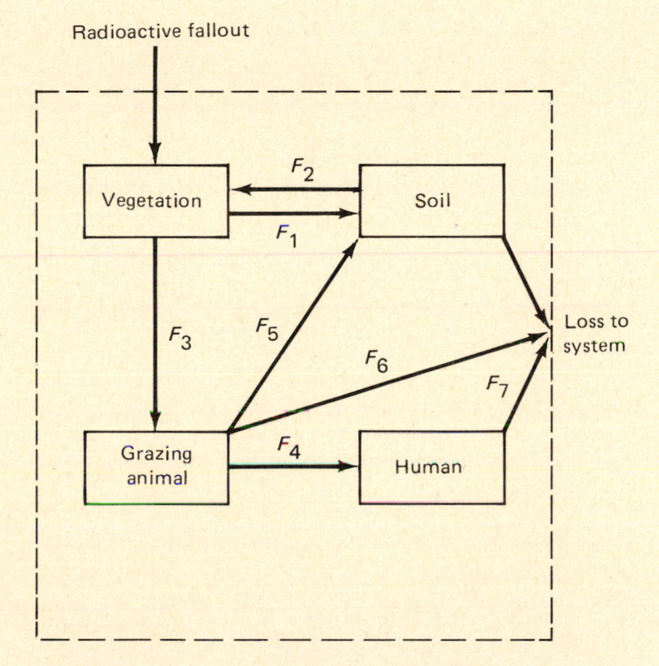

FIGURE III.4 A compartment model of the flow of radioactive material through a food chain to humans. The boxes represent the compartments, or variables; the fallout is the source of input, or driving force. The arrows represent the direction of the flow, or driving force.

active fallout. Of course, the sun, the energy force for photosynthesis, is also involved; otherwise, there would be no uptake of radioactive material. Radioactive fallout settles on soil and vegetation. Vegetation picks up the contaminant through the leaves and through the soil by way of the roots. This material in the vegetation is passed along to grazing animals that consume the plants. In the animals the radioactive material becomes incorporated in meat and milk and is transferred to humans. Some of the radioactive material passes out of the system through wastes from humans and animals and by leaching from the soil.

CHAPTER 4

The Energy Environment

Outline

Objectives

Upon completion of this chapter, you should be able to:

1. Describe the fate of solar energy reaching Earth.
2. Describe how the atmosphere is heated and how atmospheric circulation comes about.
3. Explain adiabatic heating and cooling and how it influences atmospheric circulation and movements.
4. Explain the Coriolis effect and describe how it influences atmospheric circulation and ocean currents.
5. Describe stable and unstable air masses and inversions.
6. Explain the origin of local winds and describe their effects.
7. Describe microclimates and explain their ecological effects.
8. Describe the differences in the microclimates of north-facing and south-facing slopes and in the microclimates of urban and rural areas.
9. Tell why microclimatic differences occur.

Earth in the Sun's Rays

Life on Earth exists in an energy environment. The planet immersed in sunlight intercepts the full spectrum of incoming solar radiation on the outer edge of the atmosphere. The sun's energy comes to Earth at the rate of 2.0 $cal/cm^2/min$ at the outer limits of Earth's atmosphere. That is known as the solar constant. Of the energy that is intercepted by Earth's atmosphere, approximately 30 percent is reflected directly back to space (Figure 4.1). About 11 percent is absorbed by clouds and dust, 8 percent by the atmosphere, and 51 percent by land and oceans. Solar energy that penetrates the atmosphere and reaches Earth's surface is approximately 4 percent ultraviolet light, 44 percent visible light, and 52 percent infrared or long-wave radiation.

Solar energy absorbed at the surface is transformed into chemical energy in very small quantities by photosynthesis; into thermal energy, or heat; and eventually into kinetic energy, or wind. The magnitude of these transformations varies at any one time because Earth is rapidly spinning, only one-half of the planet is sunlit, solar radiation and atmospheric absorption vary, and the surface differs in the ability to absorb solar radiation.

A portion of the solar radiation absorbed by Earth is radiated back to the atmosphere as infrared radiation. Some of the energy heats the atmosphere through turbulent diffusion from land and oceans. Added to that is the liberation of latent heat, energy released when water vapor in the atmosphere condenses and becomes precipitation. These radiation exchanges within the Earth-atmosphere system reduce temperature extremes, making life possible. This is the so-called greenhouse effect, which is

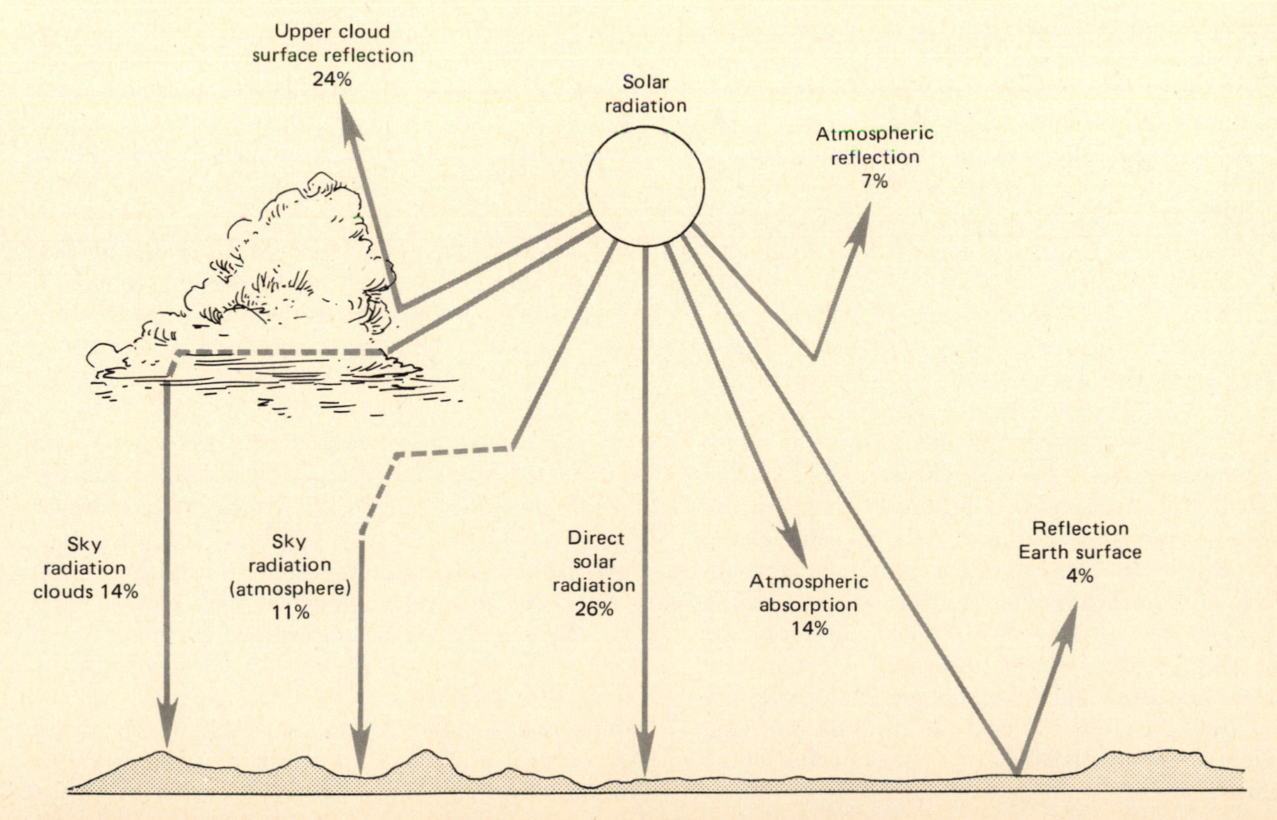

FIGURE 4.1 Disposition of solar radiation upon reaching Earth's atmosphere.

really an atmospheric effect. Shortwave radiation passes through the atmosphere, but returning long-wave radiation cannot readily escape. The enveloping blanket of CO_2 and water vapor reflects much of the long-wave radiation back to Earth's surface. That is not exactly what takes place beneath the greenhouse glass. Greenhouse glass permits a relatively normal radiant energy exchange, but it traps a small volume of air beneath the glass, protecting that air against turbulent heat losses. Unlike the atmospheric blanket, the greenhouse does not trap radiant energy (Munn, 1966; Lee, 1978). Overall, Earth maintains a fairly constant temperature by losing to outer space the same amount of energy it receives.

The sun's energy does not reach Earth uniformly. The shape of Earth and its 23.5° tilt on its axis influence the amount of energy reaching any particular place on Earth. At the equator the rays of the sun strike Earth vertically. There solar energy is more concentrated, and temperatures are higher. At the poles rays of the sun intercept Earth's atmosphere at an angle and have to penetrate a deep blanket of air. More energy is scattered in the atmosphere and reflected back to space. Because energy that reaches the surface is distributed over a wider area, less solar energy per unit of surface is absorbed and temperatures are lower (Figure 4.2).

Because of Earth's ellipical orbit about the sun—which, in effect, tilts Earth toward and away from the sun—the concentration of solar energy, and thus temperature, changes during the 365-day passage, giving us our seasons. When the north pole is tilted away from the sun, no sunlight reaches that part of the globe, more sunlight reaches the south pole, and the sun's rays are concentrated on the equator. These conditions produce winter in the Northern Hemisphere (Figure 4.2). In summer the north pole is tilted toward the sun and the south pole away from the sun. The vertical rays of the sun are shifted northward from the equator, and the temperature rises in the Northern Hemisphere. During the spring and fall equinoxes, both poles are pointed neither toward nor away from the sun, and sunlight is evenly distributed north and south of the equator.

The summer hemisphere has more energy available than the winter hemisphere, and over the year there is an excess of radiation at low latitudes and a deficit at higher latitudes. An outcome of excessive heating of the tropics is a transfer of energy from low to high latitudes. Because of this transfer and the small variation in the amount of solar radiation at low latitudes, the average air temperature there remains rather constant from year to year. Higher latitudes experience greater variations in temperature because of greater variations in the amount of incoming solar radiation.

Atmospheric Circulation

Because the sun's energy is most concentrated at the equator, Earth and the air above it are heated rapidly. As a parcel of lighter warm air rises, it encounters lower atmospheric pressure and expands, expending some of its energy. That expenditure results in a decrease in the internal energy of the air parcel and subsequent cooling. Because no exchange of energy takes place between the air mass and its environment, the process is called *adiabatic*. The rate of decrease in air temperature with height is called the *lapse rate*. For dry air, that is 10°C/km. If the air is moist, water vapor condenses as it cools, and the heat released (latent heat) partially counteracts the cooling process in the air mass. The adiabatic lapse rate for moist air is 6°C/km.

Air heated at the equatorial regions rises until it eventually reaches the stratosphere, where the temperature no longer decreases with altitude. There the air, whose temperature is the same as or lower than that of the stratosphere, is blocked from any further upward movement. With more air rising, the air mass is forced to spread north and south toward the poles. As the air masses approach the poles, they cool, become heavier, and sink over the arctic regions. The heavier air then flows toward the equator, replacing the warm air rising over the tropics (Figure 4.3).

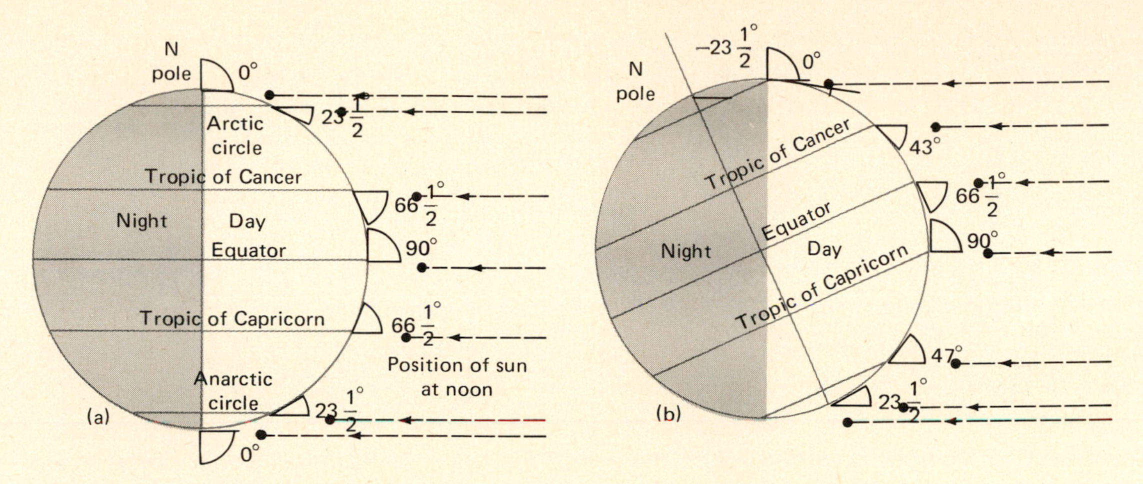

FIGURE 4.2 Altitude of the sun at the equinoxes and solstices. Latitude, the inclination of Earth's axis at an angle of 23.5°, and the rotation of Earth about the sun determine the amount of solar radiation reaching any point on Earth at any time. Earth's surface at all times lies half in the sun's rays and half in the shadow marked by the dividing line, the circle of illumination. The circle of illumination is bisected by the equator and always lies at right angles to the sun. Two times a year—at the vernal and autumnal equinoxes (March 20 or 21 and September 22 or 23)— the circle of illumination passes through the poles. At the time of the summer and winter solstices (June 21 or 22 and December 22 or 23), the circle of illumination is tangent to the Arctic and Antarctic circles. Thus, at the time of the fall and spring equinoxes, the sun's rays fall directly on the equator, so that at noon the altitude of the sun at the equator is 90°. At the north pole the sun is at the horizon and keeps that position as Earth rotates. At the time of winter solstice, the noon altitude of the sun is 66.5° at the equator and 90° at the tropic of Capricorn. The sun remains below the horizon at the north pole and above the horizon at an altitude of 23.5° at the south pole. At the summer solstice, the position of the sun is 90° at the tropic of Cancer. To visualize this, turn the diagram upside down and change the "North" to read "South." At that time of year, the sun remains above the horizon at the north pole for the entire 24 hours. (From Strahler, 1971:52.)

If Earth were stationary and without irregular land masses, the air movement shown in the figure would be the unmodified circulatory pattern of the atmosphere. Earth, however, spins on its axis from west to east. The linear velocity of Earth is greatest at the equator—1,041 miles/hour, or 465 m/second. Any object on the surface there is moving at the same rate. The distance around Earth decreases toward the poles, decreasing linear velocity. At 30° latitude an object on the surface would be moving 403m/second; and at 60° latitude, 233 m/second; and at the poles, 0 m/second. However, Earth still keeps its west-east momentum. Thus, an object moving north from the equator will deflect to the right and will do the same on its return. If the object moves southward from the equator, it will deflect to the

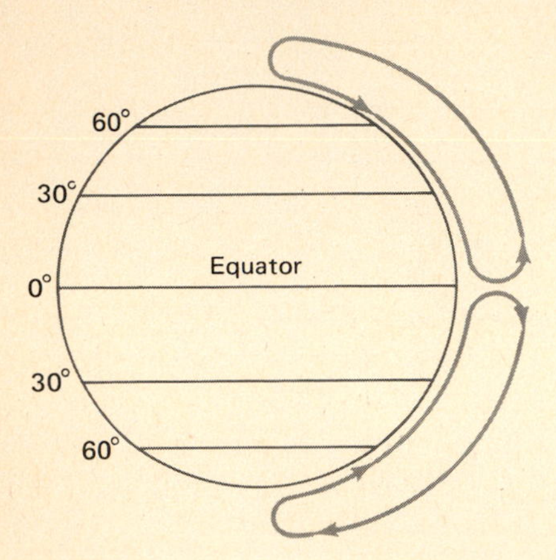

FIGURE 4.3 Circulation of air cells and prevailing winds on an imaginary, nonrotating Earth. Air heated at the equator would rise; spread north and south; and upon cooling at the two poles, descend and move back to the equator.

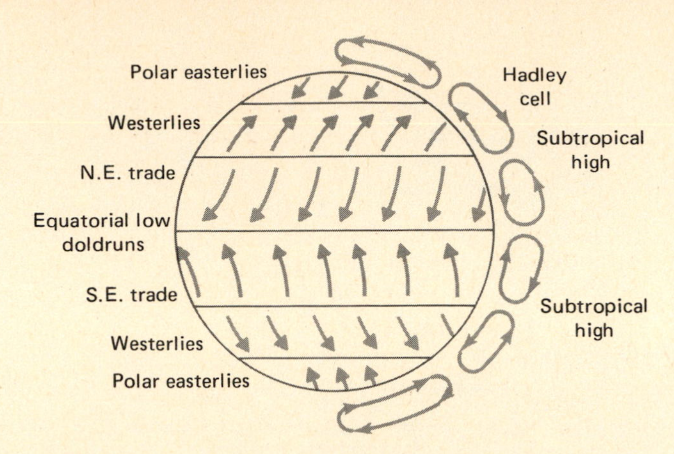

FIGURE 4.4 On a rotating Earth the air current in the Northern Hemisphere starts as a south wind that is moving north but is deflected to the right by Earth's spin and becomes a southwest or west wind. A north wind is deflected to the right and becomes a northeast or east wind. Because the northward air flow aloft just north of the equatorial regions becomes nearly a true westerly flow, northward movement is slowed, with air piling up at about 30° north latitude and losing considerable heat by radiation. Because of the piling up and heat loss, some of the air descends, producing a surface high-pressure belt. Air that has descended flows both northward to the pole and southward to the equator at the surface. The northward-flowing air current turns right to become the prevailing westerlies; the southward-flowing air, also deflected to the right, becomes the northeast trades of low latitudes. The air aloft gradually moves northward, continues to lose heat, descends at the polar region, gives up additional heat at the surface, and flows southward. This flow of air deflected to the right becomes the polar easterlies. Similar flows take place in the Southern Hemisphere, but airflow is deflected to the left by the rotating Earth. The pattern of rising and descending air forms tubes around the Earth called Hadley cells, after the English meteorologist who described them.

left. That is what happens to parcels of air. That phemonenon is known as the Coriolis effect, named after the nineteenth-century French mathematician G. C. Coriolis, who first analyzed it. It is not a simple centripetal force but an apparent effect of a number of forces that act upon objects set in motion on the Earth's spherical surface.

The Coriolis effect, then, deflects air movements and prevents a direct, simple flow from the equator to the poles. In the Northern Hemisphere airflow is deflected to the right and in the Southern Hemisphere to the left. The result is a series of belts of prevailing east winds in the polar regions, the polar easterlies, and, near the equator, the easterly trade winds. In the middle latitudes is a region of west winds known as the westerlies. These belts break the simple flow of air toward the equator and the flow aloft toward the poles into a series of cells (Figure 4.4).

The flow is divided into three cells in each hemisphere. The air that flows up from the equator forms an equatorial zone of low pressure. The equatorial air cools, loses its moisture, and descends. By the time the air has reached 30° latitude, it has lost enought heat to sink, forming a cell of semipermanent high pressure encircling the earth and a region of light winds known as the horse latitudes. The air, warmed again, picks up moisture and rises once more. Some of it flows toward the poles, some toward the equator. Meanwhile, at each pole another cell builds up and flows outward in a southerly and northerly direction to meet the rising warm air flowing toward the poles. This produces a semipermanent low-pressure area at about 60° north and south latitude.

Land heats and cools more rapidly than the oceans. Because the masses of land and water on the earth are not uniform, the surface of the earth experiences uneven heating and cooling. At any given time the temperature changes are much greater over continental areas than over oceans. Oceans act as heat reservoirs; the continents affect the circulation. In winter the west coasts of the continents are warmer than the east coasts because the air reaching the west coasts has traveled over warmer ocean areas.

The interaction of winds and heating produces more or less permanent high-pressure cells known as the subtropical highs in the Atlantic and Pacific oceans; winds and cooling produce low-pressure cells such as the Aleutian and Icelandic lows. The highs are more pronounced during the summer months, the lows during the winter months. Also produced are the monsoon winds, dry winds that blow from continental interiors to the oceans in summer, and winds heavy with moisture that blow from the oceans to the interiors in winter. Last, there are moving air masses with their cyclonic and anticyclonic frontal systems. These major air circulations are responsible for the changing swirls or cloud patterns seen over the Earth from space.

These circulatory patterns and the moisture regimes they influence are responsible for the forests, grasslands, and deserts, which, in turn, reflect the climatic pattern of the planet.

Ocean Currents

The movements of Earth and air and solar energy produce ocean currents, the horizontal movements of water around the planet. In the absence of any land masses, oceanic waters could circulate unimpeded around the globe, as the flow of water does around the Antarctic continent. But land masses divide the ocean into two main bodies, the Atlantic and the Pacific. Both oceans are unbroken from high latitudes, north and south, to the equator; and both are bounded by land masses on either side.

Each ocean is dominated by two great circular water motions, or *gyres,* each centered on a subtropical high-pressure area. Within each gyre the current moves clockwise in the Northern Hemisphere and counterclockwise in the Southern Hemisphere (Figure 4.5). The movements of the currents are caused partly by the prevailing winds, the trades or tropical easterlies on the equator side and the prevailing westerlies on the pole side. The two gyres, north and south, are separated in both oceans by an equatorial countercurrent that flows eastward. That current results from the return of the lighter (less dense) surface water piled up on the western side of the ocean basin by the equatorial current.

As the currents flow westward, they become narrower and increase their speed. Deflected by the continental basin, they turn poleward, carrying cold water with them. The two major currents in the Northern Hemisphere are the Gulf Stream in the Atlantic and Kuroshio Current in the Pacific. Their counterparts in the Southern Hemisphere are the Brazil Current in the Atlantic and the Australian Current in the Pacific.

These ocean currents influence the climates of continental land masses, especially in the coastal regions. The Gulf Stream, for example, carries warm water up along the Atlantic Coast, across to England, then south along the Canary Islands. That warm current moderates the climate of northern North America and England, which otherwise would be much colder. Currents moving toward the equator along the west side of the continents are warmer than adjacent land masses. Air over the water warms, picks up moisture, cools over land, and drops heavy rains, as it does along the coast of British Columbia and the northwestern United States.

Atmospheric Movements

An important aspect of regional and local climates is the daily heating and cooling of air masses. The principal cause of important and sustained decrease and increase in the temperature of masses of air is the adiabatic process in the sinking and rising air mass.

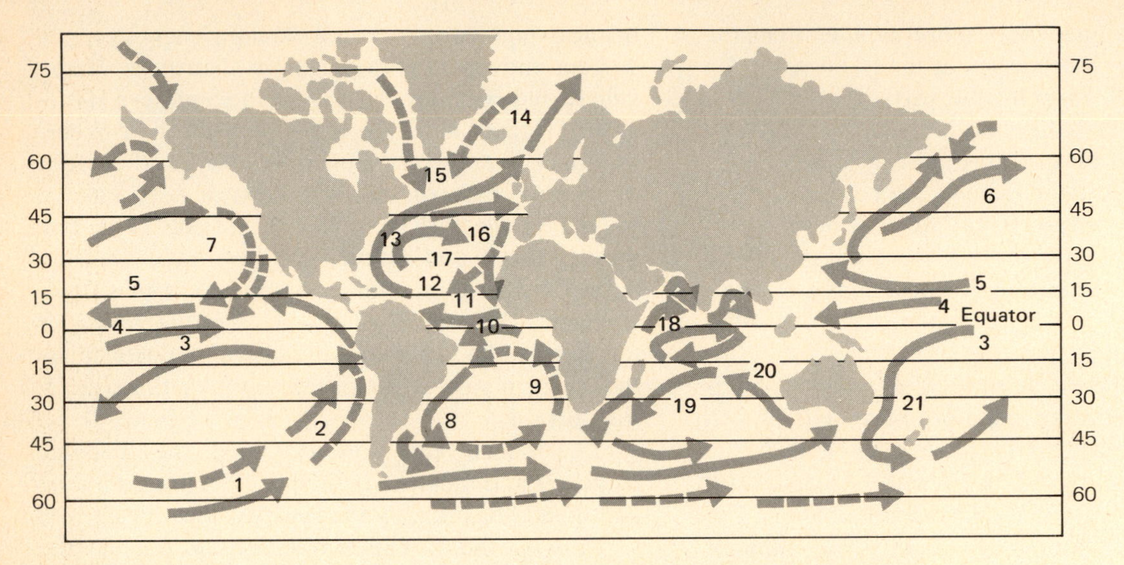

FIGURE 4.5 Ocean currents of the world. Note how the circulation is influenced by continental land masses and how oceans are interconnected by currents. (1) Antarctic West Wind Drift; (2) Peru Current; (3) South Equatorial Current; (4) Counter Equatorial Current; (5) North Equatorial Current; (6) Kuroshio Current; (7) California Current; (8) Brazil Current; (9) Benguela Current; (10) South Equatorial Current; (11) Guinea Current; (12) North Equatorial Current; (13) Gulf Stream; (14) Norwegian Current; (15) North Atlantic Current; (16) Canaries Drift; (17) Sargasso Sea; (18) Monsoon Drift (summer: east; winter: west); (19) Mozambique Current; (20) West Australian Current; (21) East Australian Current. (Dashed arrows represent cool water.) (From Coker, 1947:121.)

If a rising parcel of air is cooled at a dry adiabatic rate that is greater than the lapse rate, it is cooler and denser than the surrounding air. If it becomes immersed in warmer air, it will fall to a level where the surrounding air has the same temperature. When that condition prevails, the atmosphere is said to be *stable*. In a layer of air in which the temperature increases with height—that is, the air is warmer above than it is below—atmospheric conditions are very stable, and the condition is called an *inversion*.

If a parcel of rising air cools at a dry adiabatic rate less than the lapse rate, it is warmer than the surrounding still air and continues to rise, pulling up air from the ground behind it. That action creates eddies, turbulences, and upward currents, which ultimately lead to thunderstorms. Such a condition is termed *unstable*.

Instability of the atmosphere is caused by differential heating of the Earth and its lower atmosphere. Earth by day is heated by shortwave solar radiation (Figure 4.6), which is absorbed in different amounts over the land, depending upon vegetation, slope, soil, season, and so on. Lower layers of the atmosphere—heated by Earth's surface by radiation, conduction, and convection—rise in small volumes; colder air falls. That turnover of the atmosphere produces a turbulence that is increased by winds.

After the sun goes down, Earth begins to lose heat (Figure 4.6). The layer of surface air is cooled, while the air aloft remains near its warmer daytime temperature. In mountainous or hilly country, cold, dense air flows down slopes and gathers in the valleys. The cold air is then trapped beneath a layer of warm air (Figure 4.7). Such radiational inver-

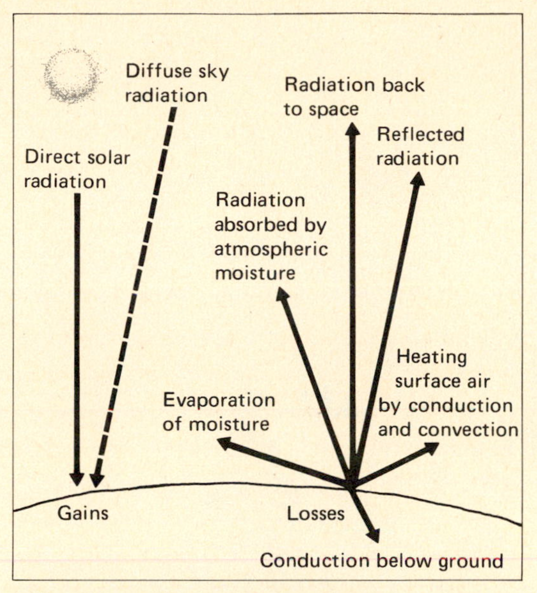

(a) Daytime surface heat exchange

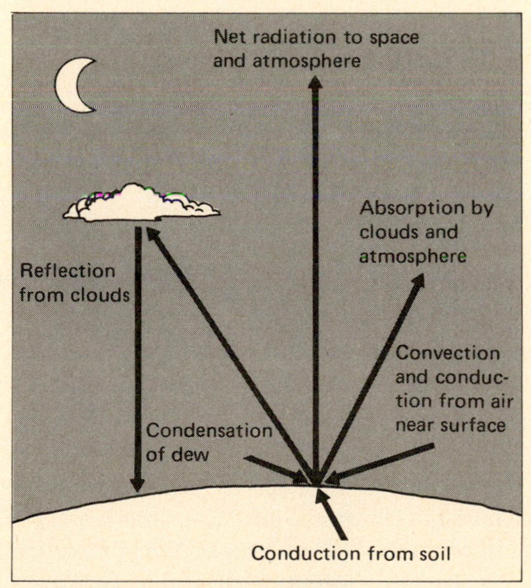

(b) Nighttime surface heat exchange

FIGURE 4.6 Radiant heating of Earth. Solar radiation that reaches Earth's surface in the daytime is dissipated in several ways, but heat gains exceed heat losses. At night there is a net cooling of the surface, although some heat is returned by various processes. (Adapted from Schroeder and Buck, 1970:14.)

sions trap impurities and other air pollutants. Smoke from industry and other heated pollutants rise until their temperature matches the surrounding air. Then they flatten out and spread horizontally. As pollutants continue to accumulate, they may fill the entire area with smog. Such inversions are most intense if the atmosphere is stable.

Similar but more widespread inversions occur when a high-pressure area stagnates over a region. In a high-pressure area, airflow is clockwise and spreads outward. The air flowing away from the high must be replaced, and the only source of replacement air is from above. Thus, surface high-pressure areas are regions of sinking air movement from aloft, called *subsidence*. When high-level winds slow down, cold air at high levels in the atmosphere tends to sink. The sinking air becomes compressed as it moves downward, and as it warms, it becomes drier. As a result, a layer of warm air develops at a higher level in the atmosphere (Figure 4.8). Rarely reaching the ground, it hangs several hundred to several thousand feet above the ground, forming a subsidence inversion. Such inversions tend to prolong the period of stagnation and increase the intensity of pollution. The subsidence inversion that brings our highest concentrations of pollution is often accompanied by lower-level radiation inversions.

Often along the West Coast of the United States, and occasionally along the East Coast, the warm seasons produce a coastal or marine inversion. In this case cool, moist air from the ocean spreads over low land. This layer of cool air, which may vary in depth from a few hundred to several hundred thousand feet, is topped by warmer, drier air, which also traps pollution in the lower layers.

Inversions break up when air close to the ground is heated, causing it to circulate up through the inversion layer, or when a new air mass moves into the area.

Local Winds

Because the atmosphere is constantly in motion, winds are a continual but highly variable influence

FIGURE 4.7 Topography plays an important role in the formation and intensity of nighttime inversions. At night air cools next to the ground, forming a weak surface inversion in which the temperature increases with height. As cooling continues during the night, the layer of cool air gradually deepens. At the same time, cool air moves downslope. Both cause the inversion to become deeper and stronger. In mountain areas the top of the night inversion is usually below the main ridge. If air is sufficiently cool and moist, fog may form in the valley. Smoke released in such an inversion will rise only until its temperature equals that of the surrounding air. Then the smoke flattens out and spreads horizontally just below the thermal belt. (Adapted from Schroeder and Buck, 1970:29.)

on the environment. Winds, especially those near the Earth's surface, are strongly affected by the shape of the topography and local heating and cooling.

Winds that move with the leading edge of the air mass, or frontal system, or that are carried by the general circulation winds aloft are *general winds*. These contrast with local convective winds caused by temperature differences within a locality. Among the most familiar of these are the land and sea breezes experienced along ocean shores and large inland lakes and bays.

In mountainous topography local winds can be exceedingly complex. Differences in the heating of air over mountain slopes and canyon and valley bottoms result in several wind systems. Because of the larger heating surface, air in mountain valleys and canyons becomes warmer during the day. Similarly, the larger cooling area of the valleys causes a reversal of this situation at night. The resulting pressures cause a flow of air upslope by day and downslope by night. Combined valley and upslope winds exit at the ridge tops by day. As the slopes go into the shadows of late afternoon, the slope and valley winds shift direction and become downslope winds.

Local winds affect soil moisture and humidity and have an important bearing on forest fire con-

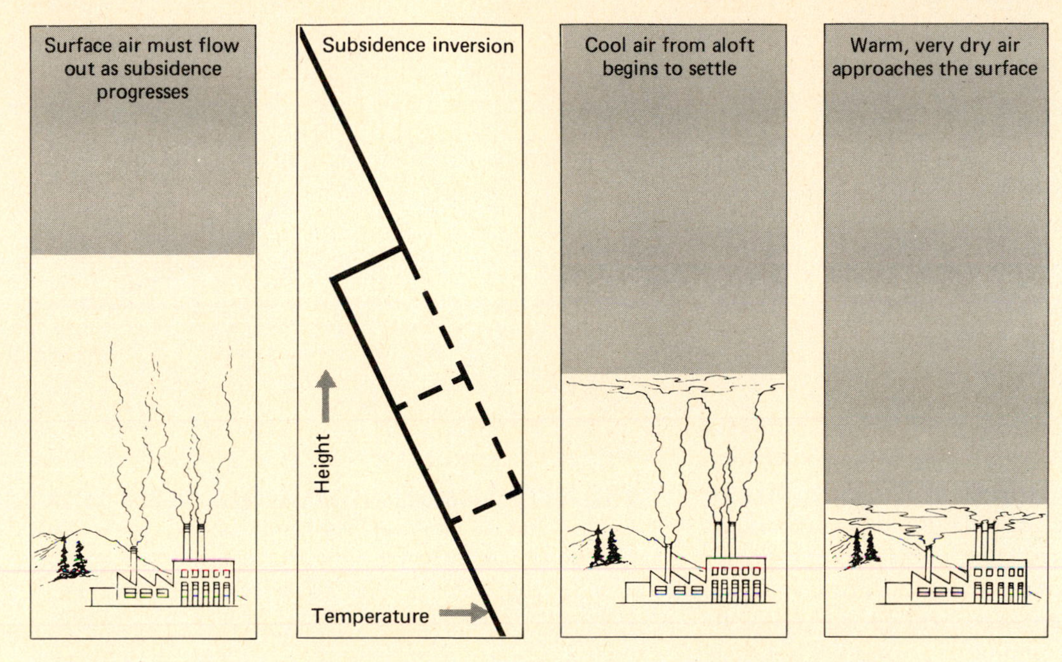

FIGURE 4.8 Descent of a subsidence inversion. The movement of the inversion is traced by successive temperature measurements, as shown by a dashed line. The nearly horizontal dashed lines indicate the position of the descending base of the inversion. The solid line indicates temperature. The temperature lapse rate in the descending layer is nearly dry adiabatic. The bottom surface is marked by a temperature inversion. Two features—temperature inversion and a marked decrease in moisture—identify the base of the descending layer. Below the inversion is an abrupt rise in the moisture content of the air. (Adapted from Schroeder and Buck, 1970:62.)

ditions and local drought situations. The drying action of high, warm winds in winter, when soil moisture is low or unavailable, causes physiological drought. The wind removes humid air about the leaves and increases transpiration. Losing more water than they are able to absorb, evergreens, in particular, dry out, and their foliage turns brown.

Plants that normally grow tall become low and spreading when high winds are frequent and regular, a situation characteristic of the timberline. On high, windswept ridges, cushion plants with small, uniform, crowded branches are most common. Because of constant desiccation, cells of plants growing in these places never expand to normal size, and all organs are dwarfed. Terminal branch shoots are killed back by desiccation; by blasting of ice particles; and along the ocean by the effects of salt spray. As a result, the terminals are replaced by strong laterals, which form a mat close to the ground.

Shallow-rooted trees and trees with brittle woods—such as the willows, cottonwoods, and maples—are thrown or broken by strong winds. Windthrow is most prevalent among trees growing in dense stands that, through logging or natural damage, are suddenly exposed to the full force of the wind.

Hurricanes and other violent windstorms sweeping over forested areas may uproot and break trees across a considerable expanse of land. Hurricanes and strong storms also cause deaths among animals and often carry individuals far from their normal environment and set them down elsewhere.

Winds are an important means of dispersal for

seeds and small animals such as spiders, mites, and even snails (Andrewartha and Birch, 1954; Darlington, 1957). Wind may also play a secondary role in the distribution of small mammals. The deeper accumulation of litter and snow in areas sheltered from the wind supports more small mammals than do more exposed areas (Vose and Dunlap, 1968).

Microclimates

When the weather report states that the temperature is 75°F and the sky is clear, the information may reflect the general weather conditions for the day. But on the surface of the ground in and beneath the vegetation, on slopes and cliff tops, in crannies and pockets, the climate is quite different. Heat, moisture, air movement, and light all vary radically from one part of the community to another to create a whole range of "little," or "micro," climates.

On a summer afternoon the temperature under calm, clear skies may be 28°C at 6 feet (1.83 m), the standard level of temperature recording. But on or near the ground—at the 2-inch (50.8-mm) level—the temperature may be 10° higher; and at sunrise, when the temperature for the 24-hour period is the lowest, the temperature may be 5° lower than at the standard level (Biel, 1961). Thus, in the middle eastern part of the United States, the temperature near the ground may correspond to the temperature at the 1.83-m level in Florida, 700 miles to the south; and at sunrise the temperature may correspond to the 1.83-m level temperature in southern Canada. Even greater extremes occur above and below the ground surface. In New Jersey, March temperatures about the stolons of clover plants 0.5-inch (12.7 mm) above the surface of the ground may be 21°C, while 3 inches (76.2 mm) below the surface, the temperature about the roots is – 1°C (Biel, 1961). The temperature range for a vertical distance of 3.5 inches (88.9 mm) is 20°C. Under such climatic extremes, most organisms exist.

The chief reason for the great differences between the ground and the 1.83-m level is solar radiation. During the day the soil, the active surface, absorbs solar radiation, which comes in short

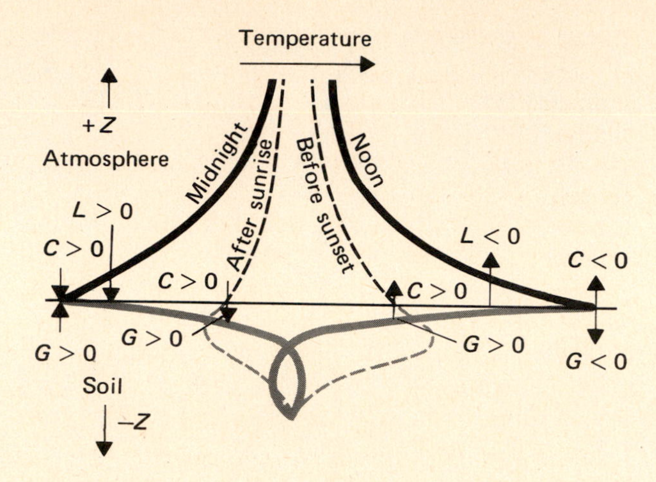

FIGURE 4.9 Idealized temperature profiles in the ground and air for various times of day. C = heat transport by convection; G = heat transport by conduction; L = heat transport by latent heat of evaporation or consideration. (From Gates, 1962:11.)

waves as light, and radiates it back as long waves to heat a thin layer of air above (see Figure 4.9). Since airflow at ground level is almost nonexistent, the heat radiated from the surface remains close to the ground. Temperatures decrease sharply in the air above this layer and in the soil below. Thus, on a sunny but chilly spring or late winter day, you can walk on muddy ground while the air about you is cold.

Heat absorbed by the ground during the day is reradiated by the ground at night. This heat is partly absorbed by water vapor in the air above. The drier the air, the greater the outgoing heat and the stronger the cooling of the surface of the ground and the vegetation. Eventually, ground and vegetation are cooled to the dew point, and water vapor in the air may condense as dew. After a heavy dew a thin layer of chilled air lies over the surface, the result of rapid absorption of heat in the evaporation of dew.

By altering wind movement, evaporation, moisture, and soil temperatures, vegetation influences or moderates the microclimate of an area, especially near the ground. Temperatures at the ground level under the shade are lower than in places exposed to sun and wind. On fair summer days a dense forest cover can reduce the daily range of temperatures at 1 inch (25 mm) by 7° to 12°C, compared with the temperature in the soils of bare fields.

Vegetation also reduces the steepness of the temperature gradient and influences the height of the active surface, the area that intercepts the maximum quantity of solar insolation. In the absence of vegetation or in the presence of very thin vegetation, temperature increases sharply near the soil; but as the plant cover increases in height and density, the leaves of the plants intercept more solar radiation (see Figure 4.10). The plant crowns then become the active surface and raise it above the ground. As a result, daytime temperatures are highest just above the dense crown surface and lowest at the surface of the ground.

Within dense vegetation air movements are reduced to convection and diffusion. In dense grass and low plant cover, complete calm exists at ground level. This calm is an outstanding feature of the microclimate near the ground, since it influences both temperature and humidity and creates a favorable environment for insects and other animals.

Humidity differs greatly from the ground up. Since evaporation takes place at the surface of the soil or at the active surface of plant cover, the vapor content (absolute humidity) decreases rapidly from a maximum at the bottom to atmospheric equilibrium above. Relative humidity increases above the surface, since actual vapor content increases only slowly during the day, while the capacity of the heated air over the surface to hold moisture increases rather rapidly.

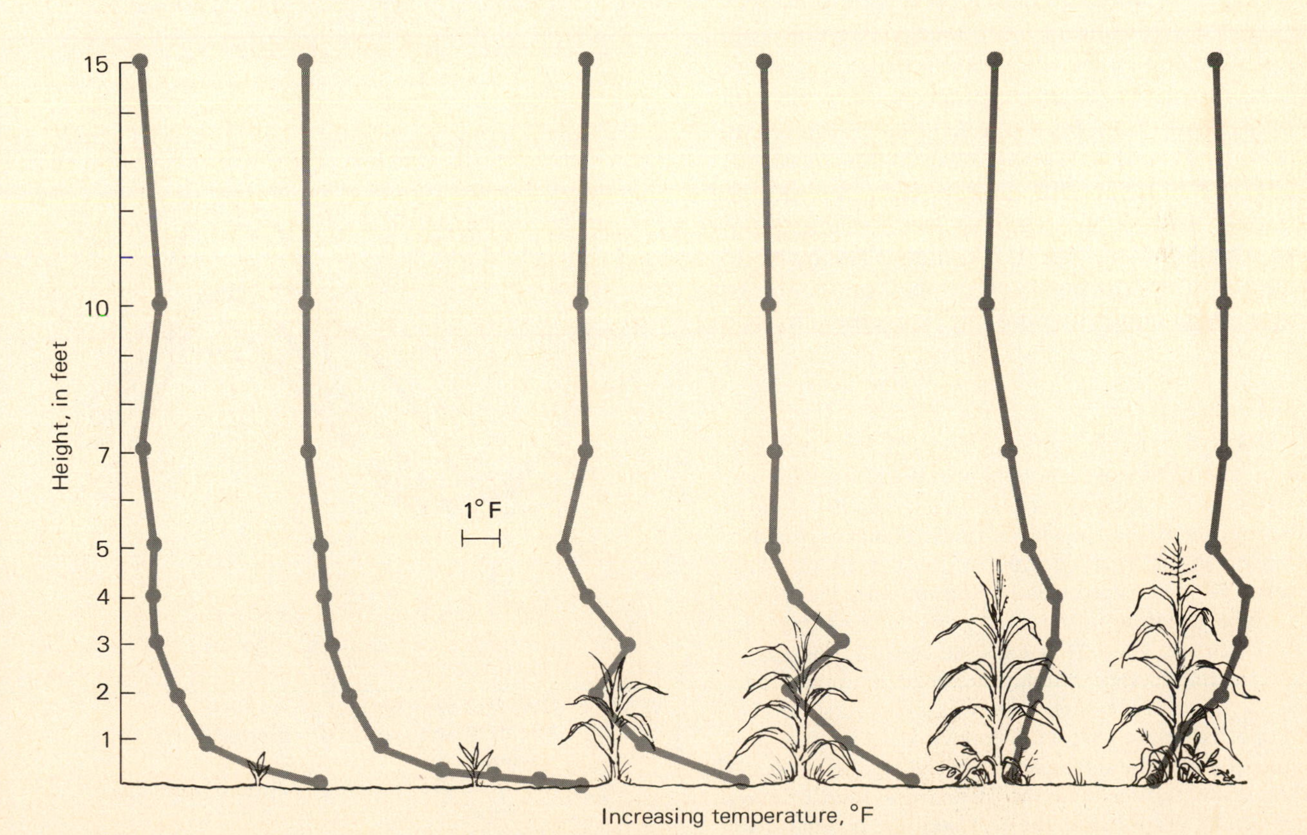

FIGURE 4.10 Vertical temperature gradients at midday in a cornfield from the time of seeding to the time of harvest. Note the increasing height of the active surface. (Adapted from Wolfe, Wareham, and Scofield, 1949:100.)

NORTH-FACING AND SOUTH-FACING SLOPES

Microclimatic variations throughout a given area, then, result from differences in slope, soil, and vegetation. The greatest microclimatic differences exist between north-facing and south-facing slopes. South-facing slopes receive the most solar energy, which is maximal when the slope grade equals the sun's angle from the zenith point. North-facing slopes receive the least energy, especially when the slope grade equals or exceeds the angle of sun ray inflection.

At latitude north 41° (about central New Jersey and southern Pennsylvania), midday insolation on a 20° slope is, on the average, 40 percent greater on south-facing slopes than on north-facing slopes during all seasons. This has a marked effect on the moisture and heat budget of the two sites. High temperatures and associated low vapor pressures induce evapotranspiration of moisture from the soil and plants. The evaporation rate is often 50 percent higher, the average temperature higher, the soil moisture lower, and the extremes more variable on south-facing slopes. Thus, the microclimate ranges from warm, xeric conditions with wide extremes on the south-facing slope to cooler, less variable, more mesic conditions on the north-facing slope. Xeric conditions are most highly developed on the top of south-facing slopes, where air movement is the

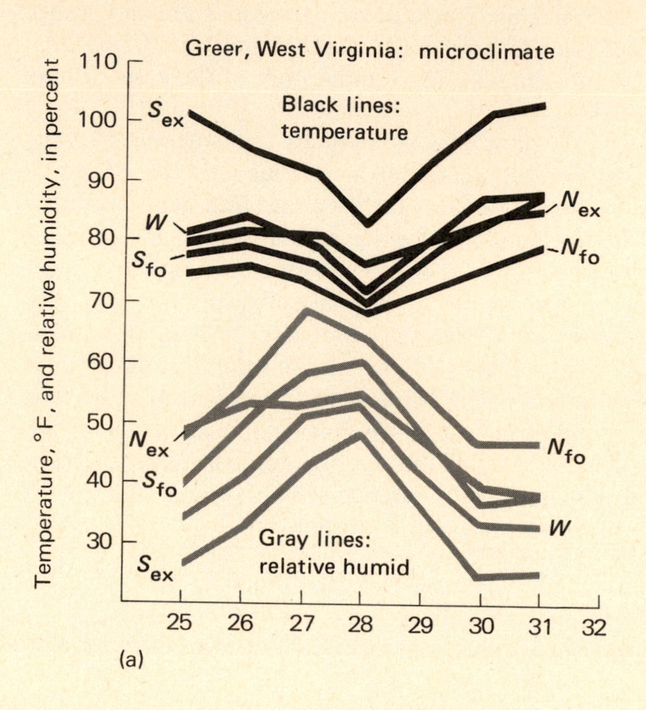

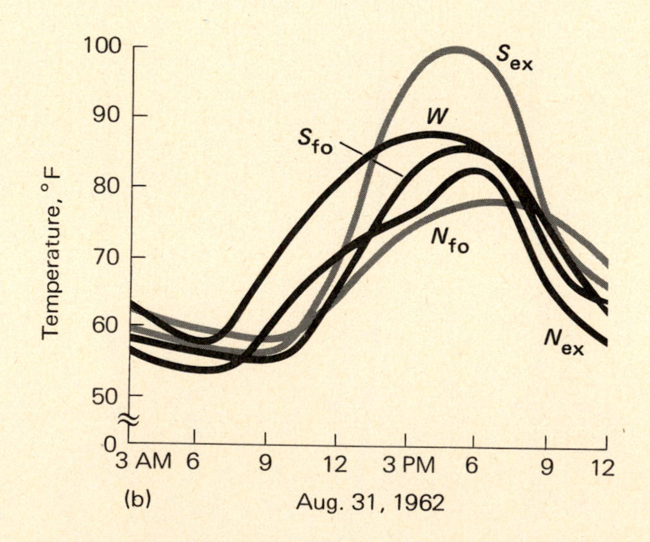

W : Standard weather station on ridge
N_{fo} : Microclimate station on forested north-facing slope
N_{ex} : Microclimate station on exposed north-facing slope
S_{fo} : Microclimate station on forested south-facing slope
S_{ex} : Microclimate station on exposed south-facing slope

FIGURE 4.11 Microclimates of north-facing and south-facing slopes. Daily maximum temperature and minimum relative humidity were recorded by four weather stations measuring microclimate and a single standard weather station during a week in August 1962 at Greer, West Virginia. W = standard weather station on ridge; N_{fo} = microclimate station on forested north-facing slope; N_{ex} = microclimate station on exposed north-facing slope; S_{fo} = microclimate station on forested south-facing slope; S_{ex} = microclimate station on exposed south-facing slope. Note the extremes recorded at exposed sites S_{ex} and N_{ex} and the differences in readings between the microclimate stations and the standard station. Among temperatures recorded at the five stations in August on a sunny day, those recorded at the exposed sites showed the greatest variation. Contrast these with the forested north-facing slope, N_{fo}. (Data from Dr. W. A. van Eck.)

greatest, while the most mesic conditions are at the bottom of the north-facing slopes (see Figure 4.11). In the central and southern Appalachians, north-facing slopes are steeper and include many minor *microreliefs*—small depressions and benches created largely by the upheaved roots of thrown trees. South-facing slopes are longer and less steep because of long-term downward movements of the soil.

The whole north-south slope complex is the result of a long chain of interactions: The solar radiation influences the mositure regime; the moisiture regime influences the species of trees and other plants occupying the slopes (see Figure 4.12); the species of trees, in turn, influence mineral recycling, which is reflected in the nature and chemistry of the surface soil and the nature of the herbaceous ground cover.

The widest climatic extremes occur in valleys and pockets, areas of convex slopes, and low concave surfaces. These places have much lower temperatures at night, especially in winter; much higher temperatures during the day, especially in summer; and a higher relative humidity. Protected from the circulating influences of the wind, the air becomes stagnant. It is heated by insolation and cooled by terrestrial radiation, in sharp contrast to the wind-exposed, well-mixed air layers of the upper slopes. In the evening cool air from the uplands flows down the slope into the pockets and valleys to form a lake of cool air. Often when the warm air in the valley comes in contact with the inflowing cold air, the moisture in the warm air may condense as valley fog.

THE MICROCLIMATE OF THE CITY

Ever since humans became urban dwellers they have not only altered the natural environment; they have also altered the atmosphere, creating a distinctive urban microclimate. The urban microclimate is a product of the morphology of the city and the density and activity of its occupants. In the urban complex, stone, asphalt, and concrete pavement and buildings, with their high capacity for absorbing and reradiating heat, replace natural vegetation, with its low heat conductivity. Rainfall on impervious surfaces is drained away as fast as possible, reducing evaporation. Metabolic heat from masses of people and waste heat from buildings, industrial combustion, and vehicles raise the temperature of the surrounding air. Industrial activities, power production, and vehicles pour water vapor, gases, and particulate matter into the atmosphere in great quantities.

The effect of this storage and reradiation of heat is the formation of a heat island about cities, large and small (see Figure 4.13), in which the temperature may be 6° to 8°C higher than in the surrounding countryside (see Landsberg, 1970). Heat islands are characterized by high temperature gradients about the city. The highest temperatures are associated with areas of highest density and activity, while temperatures decline markedly toward the periphery of the city (see Figure 4.14). Although they are detectable throughout the year, heat islands are most pronounced during the summer and early winter and are more noticeable at night than during the day, when heat stored by pavements and buildings is reradiated to the air. The magnitude of the heat island is influenced strongly by local climatic conditions such as wind and cloud cover. If the wind speed, for example, is above some varying critical value, a heat island cannot be detected.

During the summer the buildings and pavement of the inner city absorb and store considerably more heat than the vegetation of the countryside. In cities with narrow streets and tall buildings, the walls radiate heat toward each other instead of toward the sky. At night these structures slowly give off heat stored during the day. While daytime differences in temperature between the city and the country may not be noticeable, nighttime differences become pronounced shortly after sunset and persist through the night. The nighttime heating of the air from below counteracts radiative cooling and produces a positive temperature lapse rate, while an inversion is forming over the countryside. This, along with the surface temperature gradient, sets the air in motion, producing "country breezes" that flow into the city.

In the winter solar radiation is considerably less because of the low angle of the sun, but heat accumulates from human and animal metabolism,

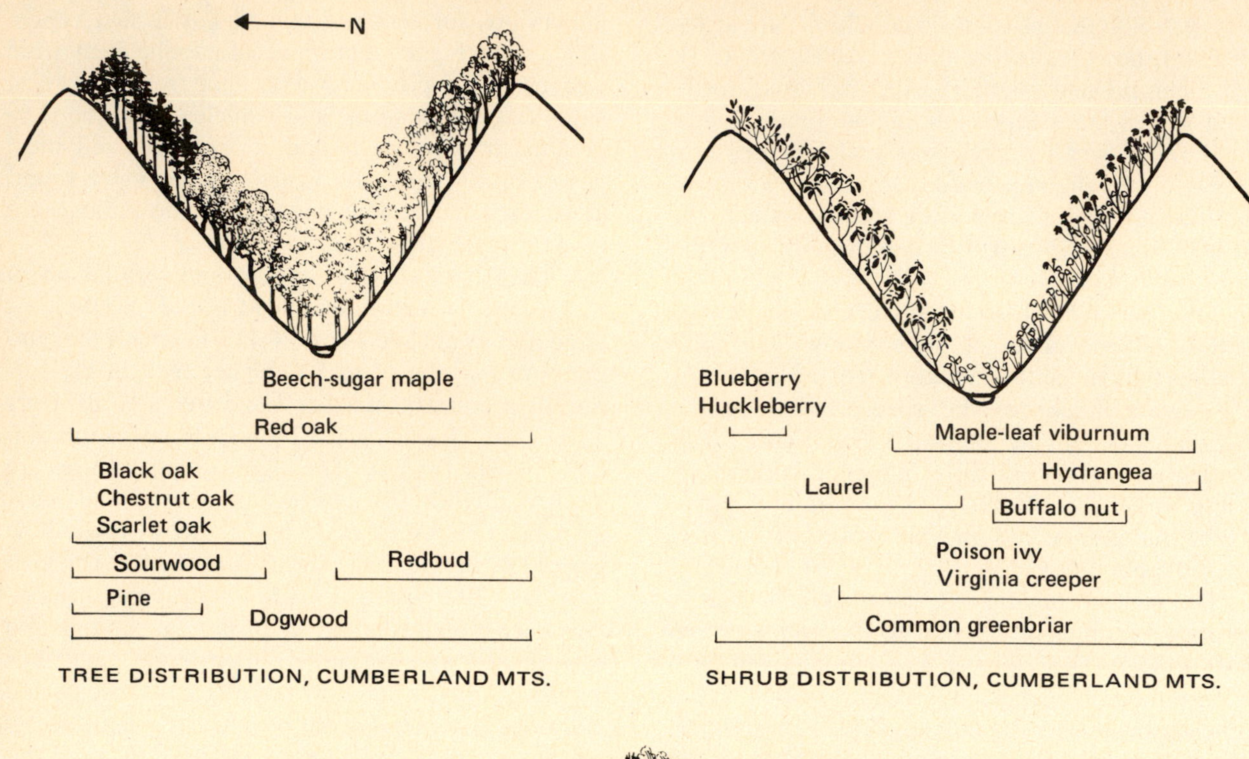

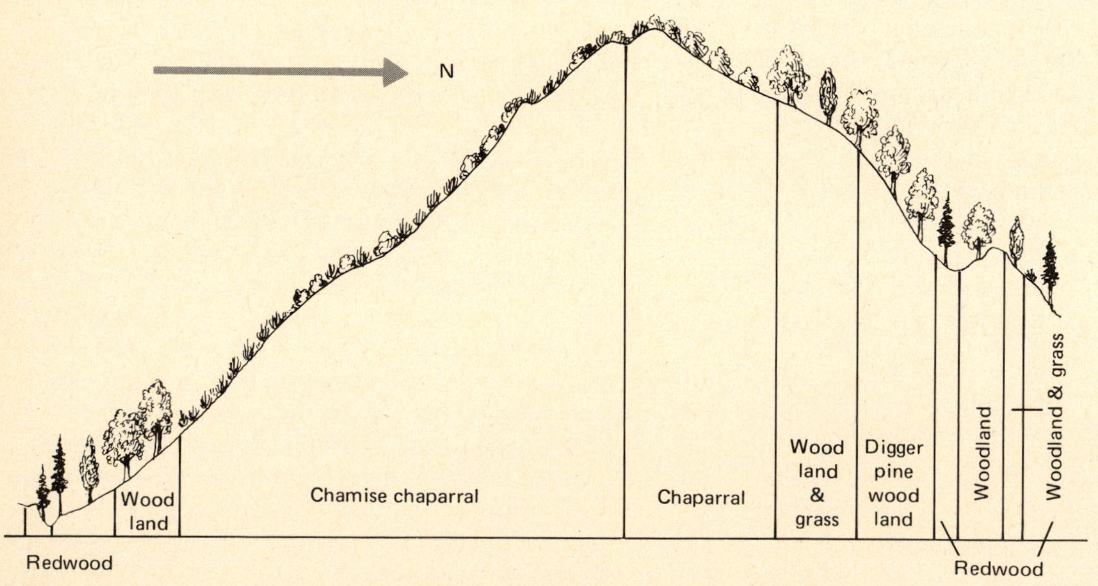

FIGURE 4.12 Influence of microclimate on the distribution of vegetation on north-facing and south-facing slopes. Diagrams shows the distribution of trees and shrubs in the hill country of southwestern West Virginia and the vegetation of the Point Sur area of California. In each case note the similarity of the vegetation on the lower parts of the north-facing and south-facing slopes.

home heating, power generation, industry, and transportation. In fact, the heat contributed by these sources is 2½ times that contributed by solar radiation. This energy reaches and warms the atmosphere directly or indirectly, moderating the winter climate of the city over that of the countryside.

Urban centers influence the flow of wind. Buildings act as obstacles, reducing the velocity of the wind to as low as 20 percent of that of the surrounding countryside, increasing its turbulence, robbing the urban area of the ventilation it needs, and inhibiting the movement of cool air in from the outside. Strong regional winds, however, can produce thermal and pollution plumes, transporting both heat and particulate matter out of the city and modifying the rural radiation balance a few miles downwind (Clarke, 1969; Oke and East, 1971).

Throughout the year urban areas are blanketed with particulate matter, CO_2, and water vapor. The haze reduces solar radiation reaching the city, which may receive 10 to 20 percent less than the surrounding countryside. At the same time, the blanket of haze absorbs part of the heat radiating upward and reflects it back; part of this heat warms the air, and part warms the ground. The higher the concentration of pollutants, the more intense is the heat island. The particulate matter has other micro-

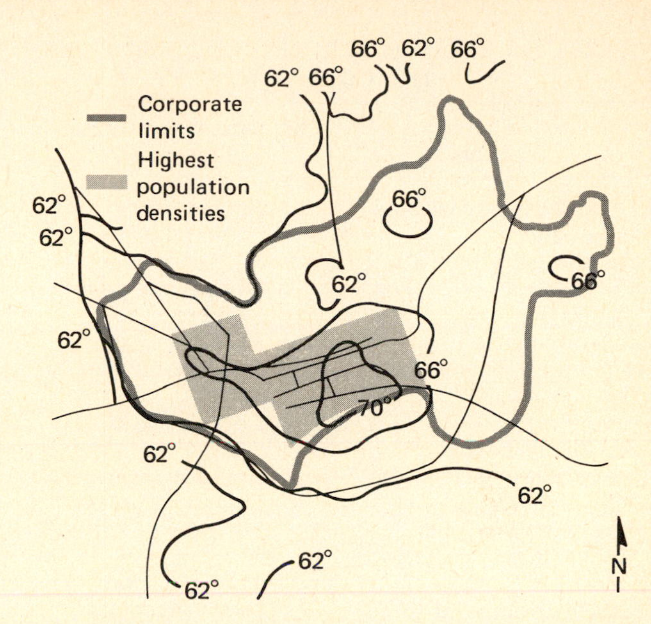

FIGURE 4.14 Thermal pattern of night air in a small city—Chapel Hill, North Carolina. Note that the highest temperatures are inside the corporate limits, where the population and activity are the greatest. (From Kopec, 1970:604.)

climatic effects. Because of the city's low evaporation rate and the lack of vegetation, relative humidity is lower in the city than in surrounding rural areas. But the particulate matter acts as condensation nuclei for water vapor in the air, producing fog and haze. Fogs are much more frequent in urban areas than in the country, especially in the winter (see Table 4.1).

Another consequence of the heat island is increased convection over the city. Updrafts, together with particulate matter and large amounts of water vapor from combustion processes and steam power, lead to increased cloudiness over cities and increased local rainfall, both over cities and over regions downwind. Evidence of weather modification by pollution is the increase in precipitation and stormy weather about La Porte, Indiana, downwind from the heavily polluted areas of Chicago, Illinois, and Gary, Indiana, and close to moisture-laden air over Lake Michigan. Since 1925 there has been a 31 percent increase in precipitation, a 34 percent increase in thunderstorms, and a 240 percent increase in the occurrence of hail (Changnon, 1968).

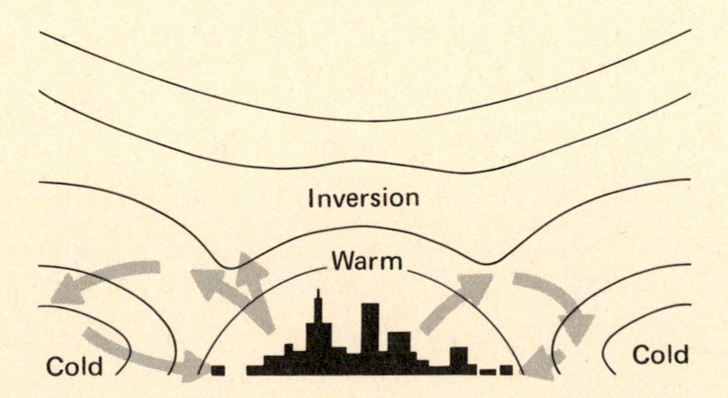

FIGURE 4.13 Idealized scheme of nighttime air circulation above the city in clear, calm weather. A heat island develops over the city. At the same time, a surface inversion develops in the country. This results in the flow of cool air toward the city, producing a country breeze in the city at night. Lines are temperature isotherms; arrows represent wind. (From Landsberg, 1970:1721.)

TABLE 4.1 CLIMATE OF THE CITY COMPARED TO THE COUNTRY

Elements	Comparison with Rural Environment
Condensation nuclei and particles	10 times more
Gaseous admixtures	5-25 times more
Cloud cover	5-10 percent more
Winter fog	100 percent more
Summer fog	30 percent more
Total precipitation	5-10 percent more
Relative humidity, winter	2 percent less
Relative humidity, summer	8 percent less
Radiation, global	15-20 percent less
Duration of sunshine	5-15 percent less
Annual mean temperature	0.5°-1.0°C more
Annual mean wind speed	20-30 percent less
Calms	5-20 percent more

SOURCE: Adapted from data in Landsberg, 1970.

Summary

Solar radiation—transformed into chemical, thermal, and kinetic energy—is the basis of life on earth. Variations in heat budgets and Earth's daily rotation produce the prevailing winds, move ocean currents, and influence rainfall patterns over Earth.

Daily heating and cooling of air masses cause them to rise and sink above Earth. Under certain conditions the temperature of air masses increases with height rather than decreases. Such an air mass is very stable, creating an inversion, which can trap atmospheric pollutants and hold them close to the ground.

Local winds are produced by temperature changes in air masses along coastal areas and in mountainous regions. Differences in the heating of air over mountain slopes and valley bottoms produce upslope breezes by day and downslope breezes by night. Such local winds can influence microclimatic conditions.

The actual climatic conditions under which organisms live vary considerably from one local area to another. These variations, or microclimates, are influenced by topographic differences, height above the ground, vegetative cover, exposure, and other factors. Most pronounced are environmental differences between ground level and upper strata and between cool north-facing and warm south-facing slopes.

Other microclimates exist over urban areas. A city is characterized by the presence of a heat island. Compared to surrounding rural areas, a city has a higher average temperature, particularly at night, more cloudy days, more fog, more precipitation, a lower rate of evaporation, and lower humidity.

Review and Study Questions

1. What is the solar constant? What is the fate of the sun's energy when it reaches Earth's atmosphere?
2. Why does less solar energy reach the poles than the equator?
3. How does Earth's interception of the sun's rays through the year influence the seasons?
4. How does Earth maintain a thermal balance?
5. What is adiabatic heating and cooling? What is the lapse rate?
6. What is the Coriolis effect, and how does it affect atmospheric circulation?
7. What causes ocean currents? What are gyres?
8. When is an air mass stable? Unstable? What is an inversion? A radiational inversion? A subsidence inversion?
9. What are local winds? How are they caused, and what is their effect?
10. What is a microclimate?

11. Explain how temperatures change above and below vegetation during the growing season. How might that affect animal life?

12. Explain why the microclimate of a north-facing slope is different from that of a south-facing slope. What are the differences?

13. What are the microclimatic differences between urban and rural areas? Explain why these differences occur.

Suggested Readings

Gates, D. M. 1962. *Energy exchange in the biosphere.* New York: Harper & Row.

Geiger, R. 1965. *Climate near the ground.* Cambridge, Mass.: Harvard University Press.

Lee, R. 1978. *Forest microclimatology.* New York: Columbia University Press.

Lowery, W. P. 1969. *Weather and life.* New York: Academic Press.

Strahler, A. 1971. *The earth sciences.* New York: Harper & Row.

CHAPTER 5

Life in the Thermal Environment

Outline

Objectives

Upon completion of this chapter, you should be able to:

1. Describe the thermal balance of organisms.
2. Distinguish between poikilothermy and homoiothermy.
3. Discuss ectothermy and endothermy, including the selective advantages of each.
4. Describe some behavioral mechanisms used by animals to maintain thermal balance.
5. Explain how fur, feathers, and fat aid animals in the maintenance of a constant body temperature.
6. Discuss how animals employ changes in metabolic rate and evaporative cooling to maintain body temperature.
7. Explain how hyperthemia can help an animal tolerate high environmental temperatures.
8. Describe the physiology of cold tolerance in some animals.
9. Explain the mechanisms of countercurrent circulation and its adaptive value.
10. Distinguish among hibernation, aestivation, torpor, and diapause.
11. Discuss the nature of hibernation.
12. Describe the metabolic mechanisms employed by plants to meet environmental extremes of heat and cold.
13. Discuss endothermism in plants.
14. Explain the role of temperature in the distribution of plants and animals.

Heat Balance

All organisms live in a thermal environment. The relationship between the organism and its thermal environment involves an exchange of energy between the two. The major source of absorbed energy is solar radiation, which may be direct, diffused from the sky, or reflected from the ground (see Figure 4.1). In addition, shortwave radiation of the sun is transmitted back as infrared thermal radiation from rocks, soil, vegetation, and atmosphere. Organisms gain heat by absorbing it from the environment and by producing metabolic heat. In turn, they lose heat to the environment.

The transfer of heat between organisms and environment involves conduction, convection, and evaporation. Heat moving by diffusion from a warm solid object to a cool object is *conduction*. How fast heat is moved depends upon the degree of contact between the two objects and their temperature difference. The greater the temperature difference and the thinner the conducting layer, the faster heat moves. *Convection* takes place when air or water moves over an object (fluid-solid or fluid-fluid). Heat transfer by convection is much more rapid than by conduction for a given temperature difference. Another source of heat loss is *evaporation*, which depends upon the difference in vapor pressure between the air and the object as well as the resistance of the surface to the loss of moisture. If the humidity of the air is high, little evaporative loss occurs.

Another important type of heat transfer is infrared or *thermal* radiation. Radiant energy impinging on an object is absorbed and then converted into heat at the surface of the absorbing object. You have experienced that type of heat transfer when you stand in front of a fire. Thermal radiation provides a route of heat transfer among organisms and objects in the environment. Radiant exchange of energy integrated with conduction, convection, and evaporation determines the thermal flux of an organism and its environment.

In a very general way, when the surrounding ambient temperature is lower than the temperature of the organism, heat is lost to the environment. When the surrounding ambient temperature is higher than the temperature of the organism, heat moves from the environment to the plant or animal. The problem, then, is for the organism to balance heat gains with heat losses.

The heat balance of an organism may be summarized by the following expression:

Heat gain (solar radiation + thermal radiation + food energy storage)
= heat loss (thermal radiation + conduction + convection + evaporation)

(See Figure 5.1.)

Poikilothermy Versus Homoiothermy

Physiologically, organisms may be divided into two broad groups: poikilotherms and homoiotherms. Popularly, poikilothermic animals are called "cold-blooded," even though their body temperature may be higher than some homoiotherms. And homoiotherms are known as "warm-blooded." Actually, poikilotherms are characterized by body temperatures that vary through the day and homoiotherms by body temperatures that remain rather constant through the day.

The same organisms can be divided into two other broad groups: ectotherms and endotherms. *Ectotherms* depend upon the environment as a source of heat to maintain body temperature. *Endotherms* depend upon internally produced metabolic heat to maintain body temperature. Ectotherms are characterized by a low metabolic rate and high thermal conductance between body and environment. Most of their energy production—50 to 98 percent—is from anaerobic metabolism, which depletes cellular energy stores and accumulates lactic acid in the tissues. That results in rapid physical exhaustion, often within 3 to 5 minutes. Endotherms are characterized by a high metabolic rate and low thermal conductance. Because of efficient cardiovascular and respiratory systems that bring oxygen to the tissues, endotherms can main-

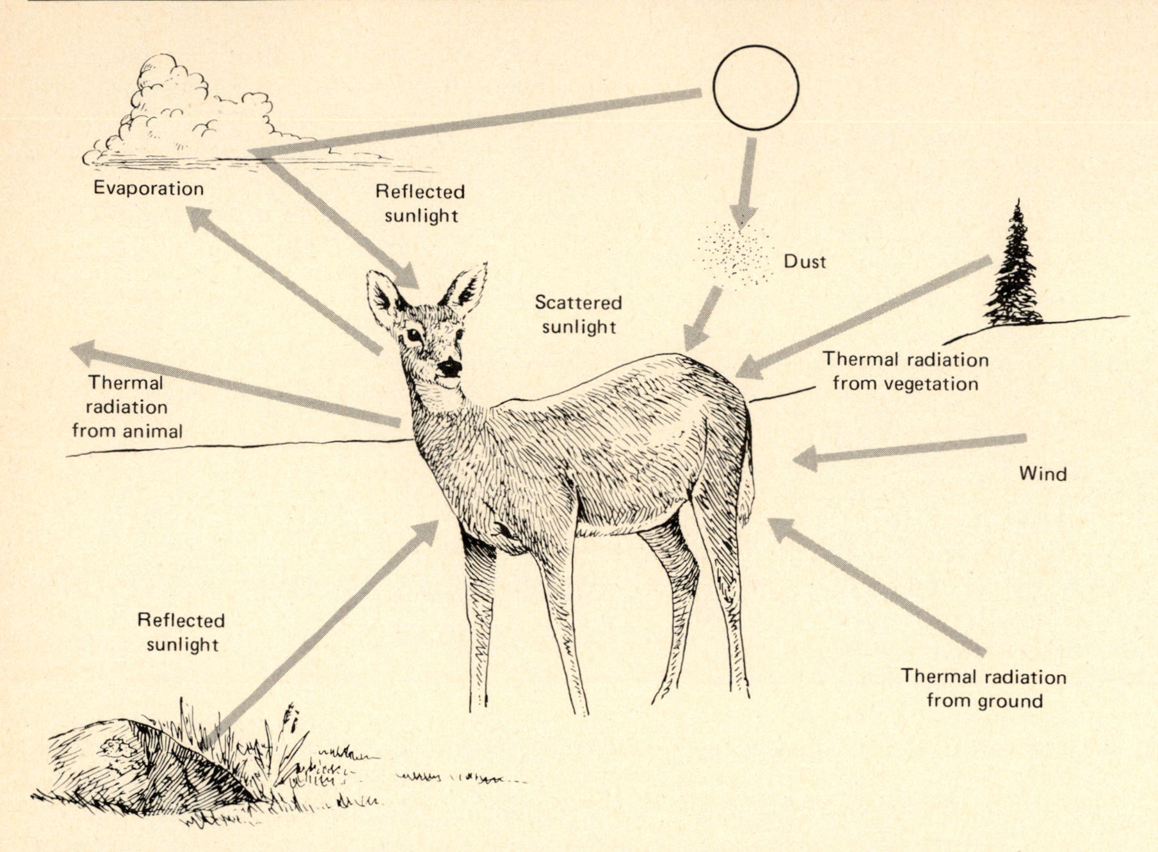

FIGURE 5.1A Exchange of energy between a mammal and its environment. The flow of energy includes (1) shortwave radiation from the sun and solar radiation scattered from clouds and other compartments of the environment and (2) long-wave thermal radiation emitted from the substrate, vegetation, and atmosphere. The mammal (and other animals) loses energy by (1) emitting long-wave infrared radiation; (2) exchanging energy to and from ambient air; (3) convection, conduction, and evaporation. (After Gates, 1962:13.)

tain a high level of aerobic energy production. Thus, they are able to sustain a high level of physical activity without exhaustion of energy. What causes physical exhaustion in endotherms is mostly the heat generated by the activity.

Poikilothermy and homoiothermy as well as ectothermy and endothermy represent extremes in a gradient of physiological responses, and the two groups of terms are not synonomous. Most homoiotherms are endotherms, but not all of them are. Nor are all poikilotherms necessarily ectotherms. Some poikilotherms exhibit a great degree of endothermy, just as numerous homoiotherms exhibit a degree of ectothermy.

ADVANTAGES AND DISADVANTAGES OF ECTOTHERMY AND ENDOTHERMY

Both ectothermy and endothermy have their advantages and disadvantages. Endotherms are able to function independently of external temperatures and to a point are able to exploit a wider range of thermal environments. They can generate energy rapidly when the situation demands, such as escape from predators and pursuit of prey. The major disadvantage of endothermy is the high energetic cost of aerobic metabolism, which leaves a minimum amount of energy available for biomass accumulation. A high resting metabolism is not compatible

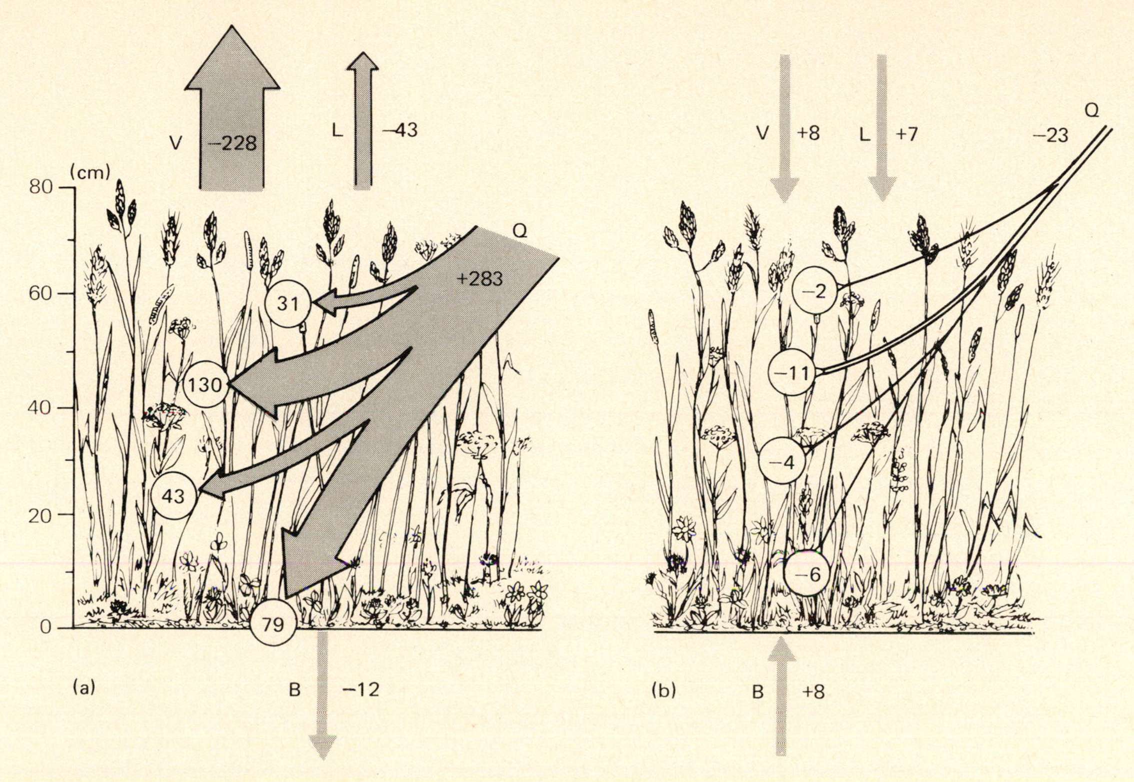

FIGURE 5.1B Energy exchange—absorption and emission—in a meadow on a sunny day (a). At night (b) net radiation as well as heat exchange is reversed. Q = net radiation; V = evaporation; L = sensible heat convection; B = soil heat flux. Figures are cal/cm^2. The active layer in the day lies between 30 and 55 cm; it absorbs 45 percent of net radiation. The second most active layer, the lowermost, absorbs 28 percent. During input 80 percent of the radiant energy is used for evaporation of water, 15 percent for sensible heat convection, and 5 percent to raise soil temperature. (After Cernusca, 1976:148.)

with a low-energy way of life. As a result, metabolic costs weigh heavily against smaller endotherms, which places a lower limit on body size.

Ectotherms are able to allocate more of their energy intake to biomass rather than to metabolic needs. Because they do not depend upon internally generated body heat, ectotherms can curtail metabolic activity in times of food and water shortage or temperature extremes. Their low energy demands enable ectotherms to colonize areas of limited food and water. Nor are they restricted to a minimal body size because of metabolic heat loss. That permits ectotherms to exploit resources and habitats unavailable to endotherms.

BODY SIZE

A close relation exists between body size and basal metabolic rate. Basal rate of metabolism is proportional to body mass raised to the ¾ power. Thus, as a unit of body weight increases, the weight-specific metabolic rate decreases. A doubling of body mass, for example, would increase the basal rate of metabolism by 75 percent. Conversely, as body mass decreases, basal metabolism increases, exponentially with very small body sizes (Figure 5.2). With any given taxonomic group, small animals have a higher metabolism and require more food per unit of body weight than large ones. Small

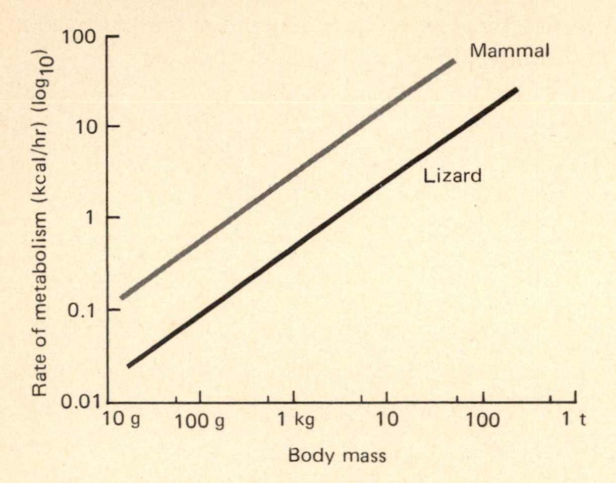

FIGURE 5.2 Basal rate of metabolism in mammals (endotherms) and standard rate of metabolism (30°C) in lizards (ectotherms) as a function of body mass. Rate of metabolism is greater in mammals than in lizards of corresponding size. The slope is close to 0.75 for both ectotherms and endotherms. The metabolic rate does not increase in direct proportion to body mass. (After McNab, 1978:9.)

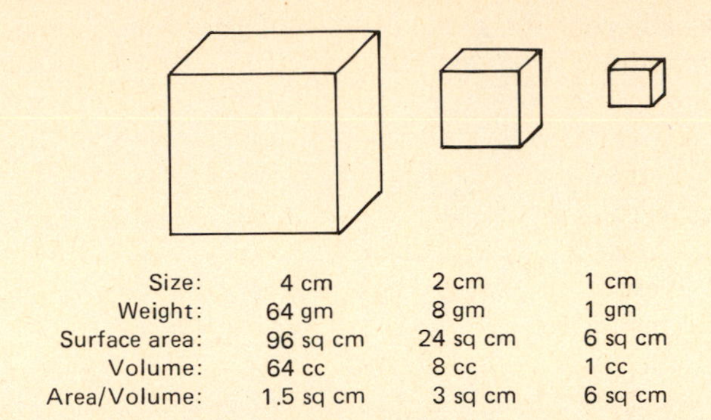

Size:	4 cm	2 cm	1 cm
Weight:	64 gm	8 gm	1 gm
Surface area:	96 sq cm	24 sq cm	6 sq cm
Volume:	64 cc	8 cc	1 cc
Area/Volume:	1.5 sq cm	3 sq cm	6 sq cm

FIGURE 5.3 These three cubes—4, 2, and 1 cm—point out the relationship between surface area and volume. A small object has more surface in proportion to its volume than a large object of similar shape.

shrews (*Sorex* spp.), for example, require daily an amount of food (wet weight) equivalent to their own body weight. It is as if a 150-pound human were to need 150 pounds of food daily to stay alive. Thus, small animals are forced to spend most of their time seeking and consuming food.

Part of the reason lies in the ratio of surface area to body mass (Figure 5.3). Heat is lost to the environment in proportion to surface area exposed. All other things being equal, large animals lose less heat to the environment than small ones. To maintain homeostasis, small animals have to burn energy rapidly. In fact, the weight-specific rates of small endotherms rise so rapidly that below a certain size they could not meet their energy demands. Five grams is about as small as an endotherm can be and still maintain a metabolic heat balance. Few endotherms ever get that small. Because of that problem, most young birds and mammals are born in an altricial state and begin life as ectotherms, depending upon body heat of the parents to maintain their body temperatures. That, of course, allows young animals to allocate most of their energy to growth.

Among ectotherms standard metabolic rates are a function of prevailing environmental temperatures. Because they do not depend on internally generated body heat to maintain tissue temperature, such ectotherms as reptiles and amphibians have a resting metabolic rate only 10 to 20 percent of that of birds. During periods of inactivity their temperatures drop to ambient, with a reduction in metabolic rate. Ectotherms, then, are able to exploit a world of small body size.

Large ectotherms, especially large reptiles, retain some of their body heat because of a small body surface to volume ratio. That, possession of scaled skin, and a greater resistance to changes in environmental temperature allow such ectotherms to acquire a degree of endothermy. From such large ectotherms in the past, endothermism in birds and mammals, both reptilian descendants, probably evolved. It seems likely that mammals evolved from Theraspids, partially committed to endothermy because of their large body size and modified skin. Once mammals acquired endothermism, it was retained by smaller types (see McNab, 1980).

POIKILOTHERMY

Rates of metabolism among poikilotherms are controlled by environmental temperatures. Rising temperatures increase the rates of enzymatic activity, which controls metabolism and oxidation of carbohydrate reserves. For every 10°C rise in temperature, the rate of metabolism in poikilotherms doubles. Obviously, poikilotherms have an upper limit that they can tolerate. Conversely, when ambient temperatures fall, metabolic activity declines. However, most poikilotherms are able to maintain a somewhat constant body temperature by behavioral means during periods of activity. Lizards, for example, may vary their body temperatures no more than 4° to 5°C when active, and amphibians 10°C (Figure 5.4). The range of body temperatures at which poikilotherms carry out their daily activities is called the active temperature range (ACT). By limiting their ACT, poikilotherms can adapt their

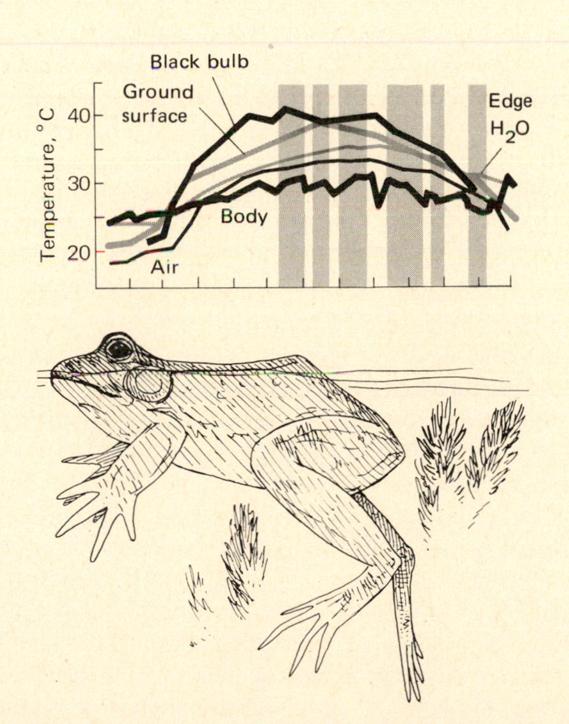

FIGURE 5.4 The body temperature of a bullfrog measured telemetrically. Dips in the black bulb temperature indicated the effects of cloud cover, convection, or both. Water temperature around the pond's edge varies from one location to another as much as 2° to 3°C. Thus, while in shallow water a frog may show a higher body temperature than that recorded for edge water. Note the relative uniformity of temperature the bullfrog maintains by moving in and out of the water. (From Lillywhite, 1970:164.)

physiological and developmental processes to a limited range, still at a low metabolic cost.

Plants and microorganisms are definitive poikilotherms. Their tissue temperature closely tracks that of the environment, and they do not have to respond instantly to temperature changes with a high rate of energy expenditure; nor do they need to maintain a definite minimal metabolism to ensure survival. They are able to exist at a relatively low and variable metabolic rate maintained over a long period of time.

Some aquatic poikilotherms, like most fish, are trivial homoiotherms. Completely immersed in a watery environment, they do not maintain any appreciable difference between their body temperature and that of the environment. Any heat produced by metabolism is lost largely through respiration. When warm venous blood reaches the gills, its heat is quickly lost to water flowing over them, and its temperature drops to that of the surrounding water. If the water is warmer than blood coming to the gills, heat is transferred and carried back to the deep body tissues. Because the temperature of the aquatic environment is relatively stable seasonally, fish and aquatic invertebrates maintain a fairly constant seasonal body temperature. Seasonally they become temperature specialists and exist with a very low range of temperature variation. They are in effect homoiotherms.

Fish and aquatic invertebrates adjust seasonally to changing water temperatures by *acclimatization*. They are able to adjust their activity to changing temperatures by adjusting their physiology, which involves cellular and enzymatic changes over a period of time. Poikilotherms have an upper and lower limit of tolerances to temperature. If they live at the upper end of their tolerable range, poikilotherms will adjust their physiology to that thermal environment, but they do so at the expense of being able to tolerate the lower range of the thermal environment. Similarly, during cold temperature periods, the animals are able to adjust to a lower temperature that would have been lethal when they were acclimatized to a higher temperature range. Because water temperatures change slowly through the year, aquatic poikilotherms are able to make the adjustment slowly. But if they are suddenly subjected to a temperature higher or lower than the one to which they are acclimated, they will experience thermal shock and die.

HOMOIOTHERMY

By means of internal heat production, homoiotherms maintain a body temperature higher than ambient. A high body temperature enables homoiotherms to remain active regardless of environmental temperatures. It is associated with specific enzyme systems designed to operate within a narrow temperature range. A high body temperature may be necessary to maintain the intricate nervous coordination associated with highly developed behavioral patterns.

High body temperatures may have evolved from an inability of large animals to dissipate rapidly the heat produced during periods of high activity (Heinrich, 1976, 1979). That situation would favor enzyme systems that function at a high temperature, with a set point around 40°C. Otherwise, the animal would have to expend an enormous amount of water for evaporative cooling to ambient temperature. The ability to operate at a high temperature provides homoiotherms with greater endurance and the means of remaining active at low environmental temperatures.

Strategies for Thermal Regulation in Animals

All living organisms need to maintain some thermal balance with the environment. They have to sustain a sufficiently high temperature to carry on metabolic activities, avoid gaining and losing excessive amounts of heat from and to the environment, and find ways of avoiding the physiological stresses of heat and cold. To accomplish that, animals, both poikilothermic and homoiothermic, have evolved certain strategies involving ectothermism and endothermism.

BEHAVIORAL MECHANISMS

To maintain a tolerable and fairly constant body temperature, terrestrial and amphibious poikilotherms resort to behavioral mechanisms. These animals—including many insects such as butterflies, dragonflies, damselflies, and others—bask in the sun to raise their body temperatures to the level necessary to become highly active (heliothermism). When they become too warm, these animals seek the shade. Amphibians can maintain fairly uniform body temperatures in summer by moving between different thermal environments. A frog can raise its internal body temperature 10°C above ambient by basking in the sun. When it becomes too warm, the frog moves into the water to cool. By moving in and out of the water and sunlight, a frog can maintain a fairly constant body temperature in relation to air temperature (Figure 5.4). Reptiles raise or lower their bodies relative to the ground and change body shape to increase or decrease the conduction of heat between themselves and the rocks or soil on which they rest. They may also seek shade or sunlight or burrow into the soil to adjust their temperature. Thus, the body temperature of poikilotherms does not necessarily equal or follow air or ambient temperature.

Behavioral means of reducing heat gains and losses are also practiced by homoiotherms. In the heat of a summer's day, birds and mammals seek shady places to escape solar radiation. Desert mammals go underground by day and emerge to become active by night. In winter some mammals, such as rabbits, go underground during periods of inactivity. Larger mammals, such as deer, seek the thermal cover of conifers and rhododendron thickets. On sunny days they seek open places, especially south-facing slopes, to soak up the sun's heat. Mammals such as flying squirrels and birds such as penguins and bobwhite quail huddle together during periods of cold, reducing individual surface area and conserving body heat.

INSULATION

To regulate the exchange of heat between the body and the environment, homoiotherms and certain poikilotherms utilize some form of insulation—a covering of fur, feathers, or layers of body fat. Such insulation reduces conductance. The higher the value of the insulation, the lower the conductance. For mammals, fur is a major barrier to heat flow, but its insulation value varies with thickness, which is greater on large mammals than small ones. Small mammals are limited in the amount of fur they can carry, because a thick coat could reduce their ability to move around. Mammals change the thickness of their fur with the season. A heavy coat of winter fur is shed for a much thinner one in summer. Aquatic mammals, especially those of arctic regions, and such arctic and antarctic birds as auklets and penguins have a heavy layer of fat or blubber beneath the skin. While the temperature of the skin's surface may be near that of the surrounding water, the temperature beneath the fat is the same as that of the deep body core. Because provision has to be made for some thermal windows through which excess body heat can escape, birds and mammals do not have all parts of the body insulated equally. The trunk, especially the thoracic section, is more heavily insulated than the extremities. Birds reduce conductance by fluffing the feathers and drawing the feet into them, making the body a round feather ball. Some arctic birds, such as ptarmigan, have the tarsi feathered; among most, the tarsi are scaled and act as thermal windows.

Although the major function of insulation is to retain body heat, it may also serve to block absorption of heat by the body. In a hot thermal environment, an animal either has to rid itself of excess body heat (see "Physiological Mechanisms") or prevent heat from being absorbed by the body. One means is to reflect solar radiation by light-colored fur or feathers. Another is to grow a heavy coat of fur to insulate the body against environmental heat, a method employed by large mammals of the desert, notably the camel. Heat is absorbed by outer layers of hair and lost to the environment.

Insulation is not a property of homoiotherms alone. It is employed by a number of poikilotherms as well. Moths and bumblebees have a dense, furlike coat over the thoracic region, which serves to retain the high temperature of flight muscles while the insects are in flight (Figure 5.5). The long, soft setae of caterpillars, together with changes in body posture, act as selective insulation by reducing convective heat exchange (Casey and Hegel, 1981). Although caterpillars produce no internal heat of any consequence, the setae act to reduce the rate of convective heat exchange without affecting the uptake of radiant heat.

PHYSIOLOGICAL MECHANISMS

To maintain a constant body temperature, homoiotherms have to employ some physiological means of reducing body heat or increasing it. To become active at relatively low temperatures, some poikilotherms become endothermic, raising body temperature far above ambient.

Metabolism

When ambient temperatures fall to a point beyond which body insulation is no longer effective in reducing heat loss, homoiotherms increase basal metabolism. As a last resort the animal may revert to shivering, a form of involuntary muscular activity that increases heat production.

A number of small mammals can increase heat production by nonshivering thermogenesis. That involves the metabolic burning of highly vascular brown fatty tissue capable of a high rate of oxygen consumption. Brown fat is found about the head, neck, thorax, and major blood vessels. It receives a rich supply of blood vessels and is well innervated by the sympathetic nervous system. The numerous blood vessels and yellow cytochrome pigment give the fat its brownish color (hence, its name). Brown fat is prominent in three groups of mammals: cold-acclimated adults, hibernators, and newborn young.

To become active, some poikilotherms become endothermic and raise their body temperatures above ambient by physiological means. Certain insects, such as the sphinx moth and bumblebee, raise their body temperatures by increasing their rate of metabolism through rapid contraction of the wing muscles. For example, to fly, the bumblebee has to raise its muscle temperature considerably to

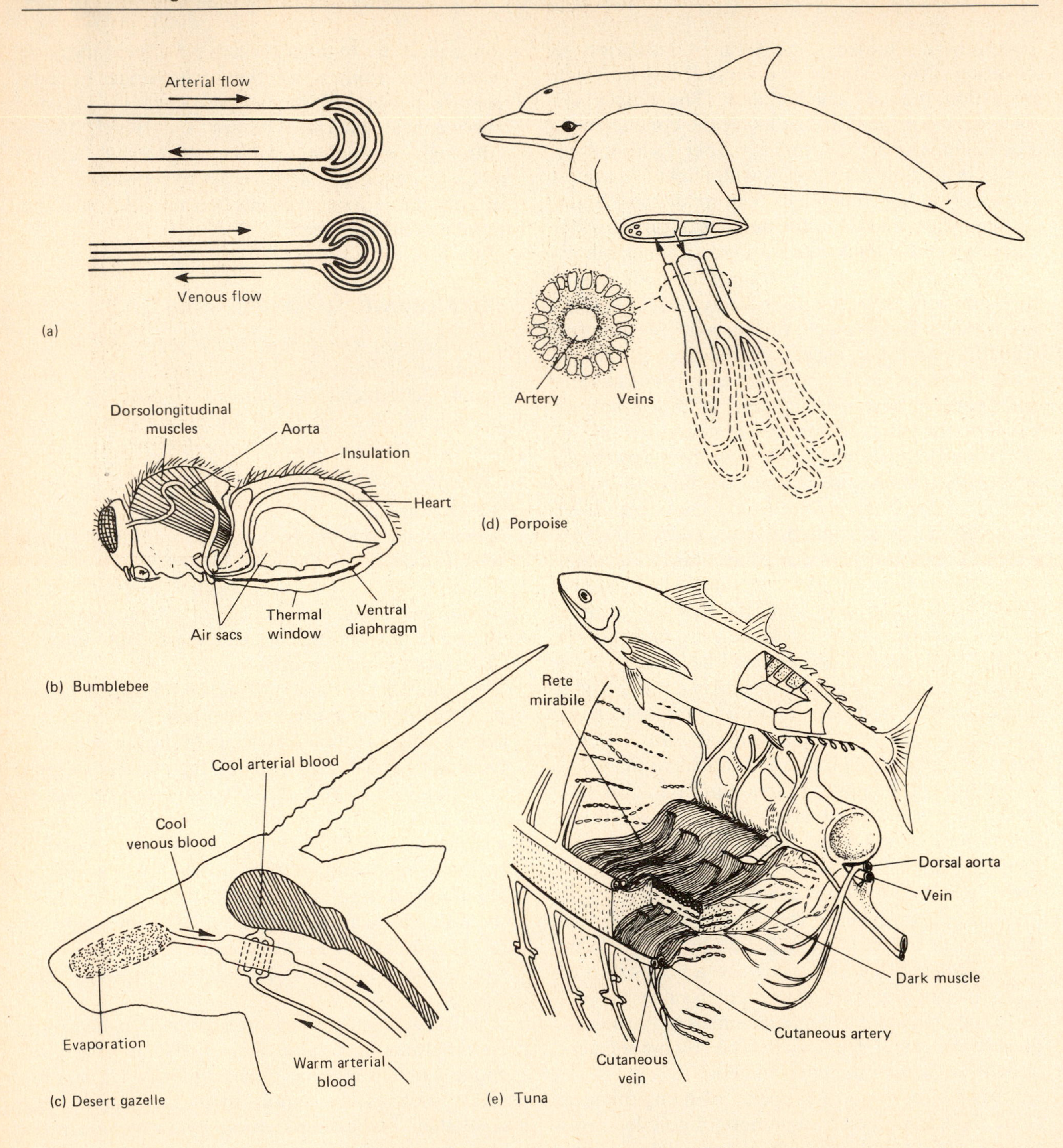
Arterial flow
Venous flow
(a)
Artery
Veins
(d) Porpoise
Dorsolongitudinal muscles
Aorta
Insulation
Heart
Air sacs
Thermal window
Ventral diaphragm
(b) Bumblebee
Rete mirabile
Cool arterial blood
Cool venous blood
Evaporation
Warm arterial blood
(c) Desert gazelle
Dorsal aorta
Vein
Dark muscle
Cutaneous artery
Cutaneous vein
(e) Tuna

generate sufficient energy to move the wings rapidly enough for flight (Heinrich, 1976). To do that the bumblebee contracts its flight muscles, producing metabolic heat by shivering. The bee is able to do this without moving the wings. Once it has achieved a muscle temperature of about 35° to 40°C, the bee is able to fly. It regulates its muscle temperature through the day by means of insulating pile on its thorax and upper abdomen and by the thermal window on the ventral side of the abdomen through which excess heat can pass (see Figure 5.5b). Some species of large dragonflies regulate body temperature by controlling metabolic heat production. When perched they elevate their body temperature by whirring their wings. During flight they may regulate body temperature by alternating flight, which produces a great deal of heat, with gliding, which increases heat loss by forced convection.

FIGURE 5.5 Countercurrent circulation illustrated. (a) A generalized model of countercurrent circulation in an animal. Arteries close to the veins allow an exchange of heat from the warm arterial blood moving from the body core to warm the returning cool venous blood. (b) Temperature regulation in the bumblebee involves a heavy insulation of pile on the thorax and dorsal side of the abdomen and a lack of pile on the ventral side of the abdomen, which acts as a thermal window. A narrow petiole between the thorax and abdomen and air sacs in the anterior part of the abdomen retard heat flow from the thorax to the abdomen. The heart pumps cool blood anteriorly to the thorax. As blood passes in the aorta through the petiole, it is heated by warm blood flowing back from the thorax. (After Heinrich, 1976:564.) (c) The desert gazelle can keep a cool head in spite of a high body core temperature by means of a rete. Arterial blood passes in small arteries through a pool of venous blood cooled by an evaporation process as it drains from the nasal region. (After Taylor, 1972.) (d) The porpoise and its relatives, the whales, use flippers and flukes as a temperature-regulating device. Arteries in the appendages are surrounded by several veins. Venous blood returning to the body core is warmed through heat transfer from arterial blood. In turn, arterial blood flowing to the appendages is cooled before reaching the flippers, thus retaining body heat. (After Schmidt-Nielsen, 1970:272.) (e) Tuna and sharks possess a heat exchanger, or rete, a vascular structure in which roughly equal numbers of very small arteries (about 0.04 mm in diameter) are interspersed with very small veins (about 0.08 mm in diameter). The rete maintains a high temperature in the swimming muscles independent of the temperature of the water in which the fish swims. The swimming muscles of most fish are supplied with blood from a large dorsal aorta that runs along the vertebral column and sends branches out to the periphery. But in tuna and sharks, the blood vessels that supply the dark red swimming muscles run along the sides of the fish just beneath the skin. From these arise many parallel fine blood vessels that eventually make up the rete. The cold end of the heat exchange is at the surface of the fish; the warm end is located in the swimming muscles. Arterial blood the same temperature as the water flows from the gills in fine arteries into the rete. Arterial blood gains heat from the venous blood coming from the muscles. When venous blood has reached the large veins beneath the skin, it has lost its heat, which has been returned to the muscles by way of the arterial blood. Such a heat exchanger allows tuna and sharks to maintain a muscle temperature as much as 14°C warmer than the water in which the fish swim. (Based on Carey and Teal, 1966:1468; adapted from Schmidt-Nielsen, 1979:272.)

Evaporative Cooling

Many birds and mammals employ evaporative cooling to reduce the body heat load. They lose some heat by evaporation of moisture diffused through the skin. Birds and mammals can accelerate evaporative cooling by sweating and panting. Sweating is restricted to certain groups of mammals, particularly horses, pigs, humans, and others that possess sweat glands in the skin. Panting in mammals and gular fluttering in birds increase the movement of air over moist surfaces in the mouth and esophagus. In gular fluttering a bird vibrates the floor of the mouth rapidly while holding the mouth open. In addition to panting and gular fluttering, doves (Columbidae) increase evaporative cooling by inflating the esophagus. The inflated esophagus comes in contact with a subcutaneous vascular plexus that forms an imcomplete collar about the neck. Heat transfers from the warm blood of the plexus to the cooled esophageal membranes (Gaunt, 1980).

Hyperthermia

Storing body heat is not exactly a sound option to avoid high thermal environments because of an animal's limited tolerance for high body temperatures. But certain animals—such as the camel, oryx, and some gazelles—do just that. The camel, for example, stores up body heat by day and unloads it by night by conduction and radiation, especially when available water is limited. Its temperature can fluctuate from 34°C in the morning to 41°C by late afternoon. By storing body heat these animals reduce the need for evaporative cooling and thus reduce water loss. They decrease the thermal gradient between the environment and the body, thereby decreasing the heat flow from the environment. And finally—in the camel, especially—the fur presents a barrier to heat gain from the environment.

Cold Tolerance

Many poikilothermic animals of temperate and arctic regions withstand long periods of below-freezing temperatures in winter. They escape the cold through supercooling and resistance to freezing.

Supercooling takes place when the body temperature falls below freezing without freezing body fluids. The amount of supercooling that can take place is influenced by the presence of certain solutes in the body. Supercooling is employed by some arctic marine fish, certain insects of temperate and cold climates, and reptiles exposed to occasional cold nights.

Some intertidal invertebrates of high latitudes and certain aquatic insects actually survive the cold by freezing and then thawing out when the temperature moderates. In some, more than 90 percent of the body fluids may become frozen, and the remaining fluids contain highly concentrated solutes. Ice forms outside the shrunken cells, and muscles and organs are distorted. After thawing they quickly resume normal shape (Kanwisher, 1959).

Other animals, particularly arctic and antarctic fish and many insects, resist freezing because of the presence of glycerol (antifreeze) in body fluids. Glycerol protects against freezing damage and lowers the freezing point, increasing the degree of supercooling. Wood frogs (*Rana sylvatica*), spring peepers (*Hyla crucifer*), and gray tree frogs (*H. versicolor*) can hibernate just beneath the leaf litter because they accumulate glycerol in their body fluids. That compound is associated with frost tolerance in these frogs and enables them to resist freezing in winter (Schmid, 1982).

COUNTERCURRENT CIRCULATION

To conserve heat in a cold environment and to cool vital parts of the body in a hot environment, a number of animals have evolved countercurrent heat exchangers (Figure 5.5a). For example, the porpoise, swimming in cold arctic waters, is well insulated with blubber, but it could experience an excessive loss of body heat through its uninsulated flukes and flippers. It maintains its body core temperature by exchanging heat between arterial and venous blood in these structures. Arteries carrying warm blood from the heart to the extremities are completely surrounded by veins (Figure 5.5d). Warm arterial blood loses its heat to the cool venous blood returning to the body core. As a result, blood entering the flippers is cool, so that little body heat is lost to the environment, while blood returning to the deep body is warmed. In warm waters, where the animals need to get rid of excessive body heat, blood bypasses the heat exchanger. Venous blood returns unwarmed through the veins close to the skin's surface to cool the body core. Such vascular arrangements are common in the legs of mammals and birds and the tails of rodents, especially the beaver.

Many animals have arteries and veins divided into a large number of small, parallel, intermingling vessels that form a discrete vascular bundle or net known as a *rete*. In a rete the principle is the same as in the blood vessels of the whale's flippers. Blood flows in opposite directions, and a heat exchange takes place.

A countercurrent heat exchange also functions to dissipate heat. The oryx, an African desert antelope exposed to high daytime temperatures, and running African gazelles can experience elevated body temperatures yet keep the highly heat-sensitive brain cool by a rete in the head (Figure 5.5c). The external carotid artery passes through a cavernous sinus filled with venous blood cooled by evaporation from the moist mucous membranes of the nasal passages. Arterial blood passing through the cavernous sinus is cooled on the way to the brain, reducing the temperature of the brain to 2° to 3°C lower than the body core. Such a mechanism is also found in dogs and other animals that pant and maintain a high skin temperature. Animals that sweat, and thus keep the skin temperature cool, have to cool the blood to the temperature at which they must maintain the brain.

Countercurrent heat exchangers are not restricted to homoiotherms. Certain poikilotherms that assume some degree of endothermism also employ the same mechanism. The swift, highly predaceous tuna and mackerel sharks possess a rete in the band of dark muscle tissue used for sustained swimming effort (Figure 5.5). Metabolic heat produced in the muscle warms up the venous blood, which gives up the heat to the adjoining newly oxygenated blood returning from the gills. Such a countercurrent exchange increases the temperature and power of the muscles because warm muscles are able to contract and relax more rapidly.

The swordfish *Xiphias gladius*, like most fish, maintains a body temperature essentially that of the surrounding water. But it protects its eyes and brain from rapid cooling during its excursions into deep cold water with a heat exchanger. The heat exchanger is a rete, associated with the eye muscle, that arises from the carotid artery. The brain heater, rich in mitochondria and cytochrome c, keeps the brain and the eye significantly warmer than the water. Such brain heaters also occur in the white marlin, sailfish, and other billfish.

The bumblebee, too, uses a countercurrent heat exchanger to maintain the high thoracic temperature necessary for flight and foraging. The heart of the bumblebee is a tube lying close to the dorsal surface of the abdomen. It extends in a ventral loop beneath a large insulating air sac in the anterior abdominal wall (Figure 5.5b). The loop passes through the thin waist and next to the ventral diaphram to the thorax. The loop brings cool blood from the heart into close proximity to warm blood flowing through the ventral diaphram and in spaces surrounding the loop back to the abdomen. Heat from the warm blood flowing back to the abdomen is transferred to cool blood entering the thorax through the loop. Without such heat exchange the bumblebee would lose heat to the environment so that flight and heat production could no longer continue.

HIBERNATION, AESTIVATION, AND TORPOR

To escape the stress of heat and cold, a number of animals become dormant or inactive, emerging when environmental conditions are more amenable to living. Because the body temperature of inactive ectotherms tracks that of the environment, these animals have no choice but to become dormant during cold weather.

In addition to supercooling, many insects exhibiting frost hardiness enter a resting stage called *diapause*, characterized by a cessation of feeding, growth, mobility, and reproduction. Among many insects diapause is a genetically determined, obligatory resting stage before development can proceed. It is timed mostly through photoperiod (see Chapter 17) and is associated with falling temperatures. Diapause prevents the appearance of a sensitive stage of development at a time when low temperatures would kill the individuals. Diapause ends with the lengthening of photoperiod and the return of warm temperatures.

Diapause is often called hibernation, which it is not. *Hibernation* is the term given to the state of winter inactivity in many poikilotherms and small homoiotherms. Hibernation is a state of torpor characterized by the cessation of coordinated locomotory movements; a reduction of heart rate, respiration, and total metabolism; and a body temperature below 10°C. Hibernating homoiotherms, however, maintain the ability to rewarm spontaneously from the hibernating state using only endogenously generated heat.

Hibernating poikilotherms experience such physiological changes as decreased blood sugar, increased liver glycogen, altered concentration of blood hemoglobin, altered carbon dioxide and oxygen content in the blood, altered muscle tone, and darkened skin. A few species of frogs that overwinter in the forest litter have a high concentration of glycol in the body and can resist the freezing of body tissues.

Entrance into hibernation by homoiotherms is a controlled physiological process difficult to explain and difficult to generalize from one species to another. Some hibernators, such as the marmot, feed heavily in late summer to lay on large fat reserves, from which they will draw energy during hibernation. Others, like the chipmunk, lay up a store of food instead. All hibernators, however, have to acquire a metabolic regulatory mechanism different from that of the active state. Tissues of deeply hibernating mammals must be able to function adequately at both of the two temperatures at which they exist.

During hibernation the lowered heartbeat may be very even, or at times no heartbeat can be detected for as long as a minute, followed by a period of relatively rapid beats. Greatly reduced respiration may take place at evenly spaced intervals, or it may involve long periods of no breathing followed by several deep respirations. Some hibernators respond to subzero (0°C) ambient temperature by increasing both heart and respiratory rates to speed up the metabolic rate. All seem to awaken spontaneously from time to time. Those hibernators that do not build up fat reserves feed on stored food or, if the weather is mild, emerge briefly above ground.

Associated with entrance into hibernation are a number of other physiological changes (see Lyman et al., 1982). One of the most important involves high carbon dioxide (CO_2) levels in the body and a change in the blood acid level. As the animal enters hibernation, it accumulates metabolically produced CO_2 in the body by reducing CO_2 output in breathing (ventilation) below that produced by metabolism. The increase in CO_2 builds up respiratory acidosis. Acidosis affects cellular processes. It inhibits glycolysis, lowers the threshold for shivering, and reduces the metabolic rate.

As the animal arouses from hibernation, it hyperventilates without any change in the breathing intervals. CO_2 stores in the body decrease, and blood pH rises. These changes are followed by shivering in the muscles, resulting in high lactic acid production and a rise in metabolism.

Some hibernators, such as marmots and woodchucks, prepare for entrance into hibernation by changing the burrow atmosphere. They plug up the entrance to the burrow with soil and nesting material, which greatly reduces air circulation and allows the CO_2 level to build up inside the burrow. That facilitates the buildup of CO_2 stores in the body. Once in hibernation, the animal has a metabolic rate low enough to permit the oxygen (O_2) level to rise and the CO_2 level to drop by diffusion through the ground and burrow wall. When the animal arouses, the O_2 limits in the burrow are near normal, providing favorable conditions for a rapid return to normal metabolic rates. The high metabolic rate of arousal then increases the CO_2 in the burrow atmosphere, allowing the animal to enter hibernation again.

Hibernation provides certain selective advantages to small homoiotherms. Maintaining a high body temperature during periods of cold is energetically too costly because of a high rate of body heat loss and a scarcity of food. It is easier and far less expensive to allow the body temperature to drop. That eliminates the need to keep warm and to seek scarce food, and it reduces the metabolic costs of maintaining body tissues. Even with periodic arousal, which is metabolically costly, the animal expends less energy coming out of torpor than it would have used had it remained homoiothermic and inactive during the same period.

In contrast, grizzly and black bears do not hibernate but instead enter a winter sleep. A bear's metabolism during dormancy is 50 to 60 percent of its normal euthermic state. Its heartbeat drops from 40 to 10 beats per minute. It metabolizes fat exclusively as a source of energy. Through its winter sleep the bear is capable of coordinated movements if aroused, and the female gives birth to her cubs. Bears tranquilized and relocated during the early part of the winter sleep may arouse, move out of their human-selected den, and choose another before retiring again.

A dormancy similar to that of hibernation is experienced by some desert mammals, birds, and amphibians during the hottest and driest parts of the year. Such summer dormancy is called *estivation*. It is less easily detected, and thus less studied, than hibernation.

Closely related to hibernation is the daily torpor experienced by a number of birds, such as hummingbirds and poorwills, and small mammals, such as bats, pocket mice, kangaroo mice, and even white-footed mice. Such daily torpor is not necessarily associated with any scarcity. Rather, it seems to have evolved as a means of reducing energy demands over that part of the day in which the animals are inactive. Nocturnal mammals, such as bats, go into torpor by day, and hummingbirds go into torpor by night. As the animal goes into torpor, its body temperature falls steeply. As the homoiothermic response is relaxed, the temperature declines to within a few degrees of ambient. Arousal returns body temperature rapidly to normal as the animal renews its metabolic heat.

Adaptation to Temperature in Plants

Plants are fixed in place in their environment. Unlike animals, they cannot move to a more favorable situation. Thus, at any given time an individual plant may experience a range of variable temperatures (Figure 5.6). For example, in early spring in the northeastern United States, temperatures about the stolons of clover plants ½ inch above the ground may be 21°C, while below the surface the temperature of the soil about the roots may be −1°C. Over a distance range of only 9 cm, the temperature range is 22°C.

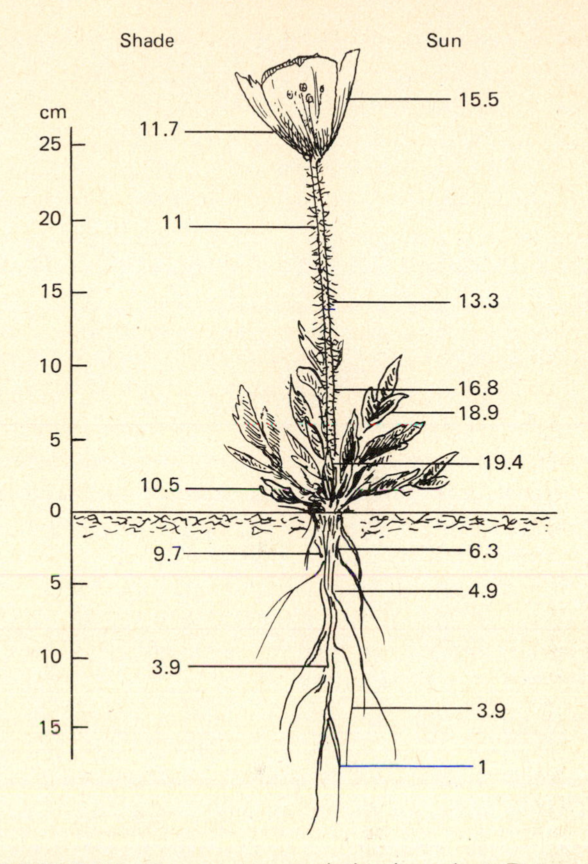

FIGURE 5.6 Temperature variation in a plant. Data are for the arctic plant *Novosieversia glacialis* on a sunny day with an air temperature of 11.7°C. Note the wide range in temperature experienced by various parts of the plant above and below the ground. (Adapted from H. Walter, 1973:190.)

To complicate the picture, an individual plant consists of subpopulations of leaves, buds, and twigs. Some of these plant parts may be fully exposed to the insolation of the sun, while others are shaded by twigs and leaves above. Because of different exposures to the thermal environment and different tolerances to heat and cold, some leaves, buds, and twigs may succumb to environmental extremes while the whole plant still lives.

Plant metabolism contributes little to the internal temperature of the plant, but strong insolation from the sun can increase internal leaf temperatures 10° to 20°C above ambient temperature by day. How much heat a plant gains depends upon the reflectivity of leaves and bark, the orientation of the leaves in relation to incoming radiation, and the size and shape of the leaves. Surfaces perpendicular to the sun's rays absorb more heat than those lying at some other angle. Leaves sharply angled to 70° intercept little midday sun, instead intercepting most of the solar radiation during the morning and evening hours. Because plants lose heat by convection and evapotranspiration, the size and shape of leaves are important. Deeply lobed leaves, like those of some oaks, and small, finely cut leaves, like those of semitropical thorn acacias, lose heat more effectively than broad, unlobed leaves. They expose more surface area per volume of leaf to the air than large leaves. Temperature within an individual leaf may vary because the edges of leaves can cool faster than the central part of the blade. As a result, leaf margins may collect dew or experience frost damage while the midportion of the leaf is unaffected.

METABOLIC ADAPTATIONS

Extremes of both heat and cold kill plants by damaging membranes, inactivating enzymes, and denaturing proteins. Plants have evolved some mechanisms that enable them to deal with both heat and cold. One is the resistance of plant protein to freezing. Plants can tolerate subzero (0°C) temperatures if the temperature decreases slowly, allowing ice to form outside the cell walls. The effect is that of dehydration, which can be reversed when temperatures rise. If the temperature falls too rapidly to permit the transfer of water to outside the cell, intracellular freezing and frost damage to cell structure and function occur.

Tolerance by plant protoplasm is mostly genetic, varying among species and among geographically separated populations of the same species. But the level of tolerance depends upon the degree of hardening and environmental conditions. This is most evident in seasonally changing environments in which plants develop tolerance to cold winter temperatures. Plants develop frost hardening through the fall and achieve maximum hardening in winter. Plants acquire frost hardness—the turning of sensitive cells into hardy ones—through the formation or addition of protective compounds in the cell. That involves the synthesis and proper distribution of antifreeze substances in the cell, such as the accumulation of certain sugars, amino acids, and nontoxic protective compounds. Plants retain their tolerance to cold until growth starts in the spring.

Death of a plant by heat results because of a disturbance to nucleic acid and protein metabolism. Paradoxically, heat resistance is highest when resistance to cold is also highest and is lowest during main periods of growth. When exposed to heat, plants can initiate adaptive processes in cells, especially in the midday of summer, when a short-term increase in heat resistance takes place. This resistance is lost by evening.

During the growing season some plants avoid frost damage by lowering the freezing point of cell fluids, which is accomplished by increasing certain dissolved substances such as sugars and sugar alcohols. That results in the cell sap's supercooling for short periods of time. Supercooling takes place when cell sap is lowered to a temperature somewhat below freezing without its freezing immediately. Further resistance to chilling is obtained by insulation. Some species of arctic and alpine plants and very early blooming species of temperate regions possess hairs that may act as heat traps and prevent cold injury. The interior temperature of cushion-type or ground-hugging leaves may be 20°C higher than the surrounding air.

ENDOTHERMISM IN PLANTS

A very few plants are endothermic to a degree. Exhibiting some endothermism are certain members of the largely tropical arum lily family, which includes philodendrons as well as the skunk cabbage (*Symplocarpus foetidus*) of eastern North America. The arum lily family is characterized by a fleshy stem of flowers or inflorescense, called a spadix, enclosed by a leafy hood, or spathe. Some tropical species, notably philodendron, can temporarily maintain a core temperature in the spadix of 38° to 46°C in an air temperature of 4° to 39°C (Nagy, Odell, and Seymour, 1972). That is accomplished by sterile male flowers, which produce heat by a regulated rate of oxidative metabolism. The flowers consume oxygen at a rate approaching that of the sphinx moth and flying hummingbirds. Apparently, the flower head volatizes chemicals that mimic decomposing flesh and attract certain pollinating insects.

The most endothermic, however, is the skunk cabbage which pushes its green and purple spathe and spadix up through the the cold mud and snow of early spring. R. M. Knutson (1974) measured the temperatures of skunk cabbage spadices over a 3-year period. He found that the spadix maintained an internal temperature 15° to 30°C above ambient air temperatures of −15° to +15°C. In doing so, skunk cabbage consumed oxygen at a rate comparable to that of a homoiothermic animal of the size of a shrew or a hummingbird. Skunk cabbage remained endothermic over the 14-day period of flowering. Throughout that period the temperature of the spadices remained about the same. On colder days and when the temperature dropped by night, skunk cabbage generated more heat. As long as the temperatures remained above freezing, spadix temperatures remained at a nearly constant 21° to 22°C. Skunk cabbage obtains its energy from the large quantity of respiratory tissue in the spadix as well as a nearly inexhaustible supply of energy stored in the deep, fleshy root. Attracted to the warmth of the spadix are ectothermic insects of early spring.

Temperature and Distribution

Most living organisms apparently have a temperature range outside of which they fail to grow or reproduce. Within the favorable range, organisms have an optimum or preferred temperature at which they best maintain themselves. The optimum temperature may vary within the life cycle or with the particular process involved. The optimum temperature for photosynthesis is lower than that for respiration. The seeds of many plants will not germinate and the eggs and pupae of some insects will not hatch or develop normally until chilled. Although the Atlantic lobster will live in water with a temperature range of 0° to 17°C, it will breed only in water warmer than 11°C. Brook trout grow best at 13° to 17°C, but the eggs develop best at 8°C.

Because the optimum temperature for the completion of the several stages of the life cycle of many organisms varies, temperatures for some species are so different from those for other species that the animals cannot inhabit the same area. Some organisms, particularly plants growing under suboptimal temperatures, cannot compete with the surrounding growth, a situation that would not exist under optimal conditions.

Generally, the range of many species is limited by the lowest critical temperature in the most vulnerable stage of its life cycle, usually the reproductive stage. A classic example of temperature limitation on animal distribution is found among four species of ranid frogs (J. A. Moore, 1949a). The wood frog breeds in late March, when water temperature is about 10°C. Its eggs can develop at temperatures as low as 2.5°C. Larval stages transform in about 60 days. This frog ranges into Alaska and Labrador, further north than any other North American amphibian or reptile. The meadow frog breeds in late April, when water temperatures are about 15°C, and the larvae require around 90 days to develop. As a result, the northern limit of its range is southern Canada. The southernmost species of the three, the green frog, does not breed until the water is about 25°C, and the eggs will develop at 33°C, a lethal temperature for the others. Its eggs, however, will not develop at all until the temperature exceeds 11°C. The range of the green frog extends only slightly above the northern boundary of the United States.

A number of examples of temperature limitation can be found among plants. The northern limit of the sugar maple closely parallels the 1.6°C (35°F) mean annual isotherm. The paper birch, a cold-climate species, is found as far north as the 35°C July isotherm and seldom grows naturally where the average July temperature exceeds 52°C. The distribution of the black spruce follows a similar pattern. Its southern distribution is approximately the same as that of the paper birch, and its northernmost outliers are seldom far from the mean July isotherm of 33°C. In southeastern North America the northern limit of the loblolly pine is set by winter temperature and rainfall (see Fowells, 1965).

Summary

All life lives in a thermal environment. Organisms must balance heat inputs and heat outputs between themselves and their environment. Animal life is either poikilothermic or homoiothermic. Body temperatures of the former fluctuate with the environment, while body temperatures of the latter are fairly constant, regardless of environmental temperatures. Animals are also either ectothermic, depending upon the environment as a source of heat, or endothermic, depending upon internally produced metabolic heat to maintain body temperature. All four related types represent extremes in a gradient of physiological response.

To maintain a thermal balance between the organism and the environment, animals employ a variety of approaches. They may resort to behavioral activities, such as seeking the sun or shade, or they may employ insulating layers of fur, feathers, or fat to keep body heat in or environmental heat out. Homoiotherms and some poikilotherms may increase metabolic rates to increase body heat. Endotherms may resort to evaporative cooling or even to hyperthermia to handle excessive heat loads. Some cold-tolerant poikilotherms utilize supercooling, involving the synthesis of glycerol in body fluids to resist freezing in winter. Other animals possess a well-developed countercurrent circulation, which involves exchange of body heat between arterial and venous blood. Such a mechanism allows the animals to retain body heat by reducing heat loss through the extremities or to cool blood flowing to such vital organs as the brain. Some animals enter a state of winter dormancy or hibernation to reduce the high energy costs of staying warm. Hibernation involves a whole rearrangement of metabolic activity to run at a very low level. Heartbeat, breathing, and body temperature are all greatly reduced. A similar slowdown is involved in aestivation, or summer dormancy. Torpor involves a much shorter period of metabolic slowdown. Daily torpor does not involve the extensive metabolic changes characteristic of deep hibernation. Often confused with hibernation is diapause, which is a resting stage in development or growth associated with falling temperatures.

Plant metabolism contributes little to internal plant temperature, but plants have evolved certain mechanisms to resist extremes of heat and cold. These include reflectivity of leaves and bark, leaf size and shape, orientation of leaves toward the sun, and frost hardening. The latter involves the synthesis of certain protective antifreeze substances in the cells. Plants are ectothermic, but a few members of the arum family are seasonally endothermic. That feature in skunk cabbage appears to attract pollinating insects.

Temperature regimes set upper and lower limits within which organisms can grow and reproduce. These optimum temperature ranges influence the spatial distribution of many plants and animals.

Review and Study Questions

1. What is involved in thermal balance of organisms? What physical processes are involved?
2. Distinguish between poikilothermy and homoiothermy. Between ectothermy and endothermy. What are the differences between the two sets of terms? What are the selective advantages and disadvantages of each?
3. What is the relationship among body size, metabolic rates, and temperature regulation?
4. What is the active temperature range of poikilotherms?

5. What is acclimation? Why can fish be considered "trivial homoiotherms"?

6. Speculate on how homoiothermy may have evolved.

7. What are some behavioral mechanisms employed by poikilotherms to maintain fairly constant body temperatures? By homoiotherms?

8. How can insulation of fur, feathers, and fat aid animals in maintaining their thermal integrity?

9. How do homoiotherms respond when ambient temperatures fall below body temperature? Poikilotherms?

10. How do homoiotherms utilize evaporative cooling?

11. Explain how some animals can use hyperthermia, or buildup of body heat, to reduce the ultimate heat load imposed by the environment.

12. Both plants and animals employ supercooling as a means of tolerating low environmental temperatures. What is involved in each? How are they similar?

13. What is countercurrent circulation? Give some examples. Explain how such circulation enables animals to maintain a fairly constant body temperature.

14. Distinguish among hibernation, aestivation, torpor, and diapause.

15. What metabolic changes are involved in entrance to hibernation? What distinguishes deep hibernation from other forms, including torpor? Why is the black bear not a true hibernator?

16. How do plants build up tolerance to or adjust to cold? To heat?

17. Speculate on how endothermy might have developed in the temperate zone member of the arum family, the skunk cabbage.

18. Obtain some good range maps of certain selected trees and compare the ranges with mean annual and summer isotherms. Point out any relationships. See Fowells (1965) for detailed climatological requirements of various species of North American trees.

Suggested Readings

Bannister, P. 1977. *Introduction to physiological plant ecology*. New York: Halsted Press.

Blight, J. 1973. *Temperature regulation in mammals and other vertebrates*. New York: Elsevier.

Etherington, J. R. 1982. *Environment and plant ecology*. 2d ed. New York: Wiley.

Folk, G. E. 1974. *Textbook of environmental physiology*. 2d ed. Philadelphia: Lea & Febiger.

Heinrich, B. 1979. *Bumblebee economics*. Cambridge, Mass.: Harvard University Press.

Heinrich, B., ed. 1981. *Insect thermoregulation*. New York: Wiley.

Hill, R. W. 1976. *Comparative physiology of animals: An environmental approach*. New York: Harper & Row.

Larcher, W. 1975. *Physiological plant ecology*. New York: Springer-Verlag.

Lyman, C. P., A. Malan, J. S. Willis, and L. C. H. Wang. 1982. *Hibernation and torpor in mammals and birds*. New York: Academic Press.

Schmidt-Neilsen, K. 1979. *Animal physiology: Adaptation and environment*. New York: Cambridge University Press.

Strain, B. R., and W. B. Billings, eds. 1975. *Vegetation and the environment*. Vol. 6, *Handbook of vegetation science*. The Hague: W. Junk.

Turner, N. C., and P. J. Kramer, eds. 1980. *Adaptation of plants to water and high temperature stress*. New York: Wiley.

CHAPTER 6

Life in the Moisture Environment

Outline

Objectives

Upon completion of this chapter, you should be able to:

1. Explain how the structure of water affects its behavior.
2. Discuss the ecologically important physical properties of water.
3. Describe the global distribution of water.
4. Describe the water cycle and global water balance.
5. Explain relative humidity, saturation, pressure deficit, and water potential.
6. Distinguish between poikilohydric and homoiohydric plants.
7. Discuss the various mechanisms employed by plants to reduce the impact of water stress.
8. Describe ways in which animals respond to moisture stress.
9. Describe some means by which plants and animals cope with a saline environment.
10. Discuss the relationship between moisture gradients and plant distribution.
11. Tell how organisms respond to or are affected by excessive moisture.

Water is essential to all life. Means of obtaining and conserving it have shaped the nature of terrestrial life. Means of living within it have been the overwhelming influence on aquatic life. Because of its enormous importance, water and its properties need some further discussion.

The Structure of Water

Because of the physical arrangement of its hydrogen atoms and hydrogen bonds, liquid water consists of branching chains of oxygen tetrahedra. The physical state of water—whether liquid, gas, or solid—is determined by the speed at which hydrogen bonds are being formed and broken. Heat increases that speed; hence, weak hydrogen bonds cannot hold molecules together as they move faster. The thermal status of water in a liquid state is such that hydrogen bonds are being broken as fast as they form.

At low temperatures the tetrahedral arrangement is nearly perfect; when water freezes, the arrangement is an almost perfect lattice with considerable open space between ice crystals and, thus, a decrease in density. This is the reason why ice floats. As the temperature of frozen water is increased, this molecular arrangement becomes looser and more diffuse, resulting in random packing (because of the continuous breaking and reforming of hydrogen bonds) and contraction of molecules. The higher the temperature rises, the more diffuse the pattern becomes, until the whole structure (and the hydrogen bonds) breaks down and water melts.

Upon melting, water contracts and its density increases up to a temperature of 3.98°C. Beyond this point the loose arrangement of the molecules means a reduction in density again. The existence of this point of maximum density at approximately 4°C is of fundamental importance to aquatic life.

A change in temperature, and thus a change in the density of water, produces movements of water within ponds and lakes. The heaviest water, at 4°C, moves to the bottom of a body of water, forcing less dense water upward. During summer a layer of warm, light water floats on top of heavier, cool water, resulting in a temperature stratification of water in ponds and lakes. In winter, water near freezing tends to be on the surface, while warmer, denser water is at the bottom, resulting in a winter stratification. When the surface is covered by an even less dense layer of ice, the water beneath is insulated from freezing. During spring and fall—when surface waters warm and cool, respectively—water becomes more dense and sinks. This downward movement, accompanied by the stirring action of the wind, results in a semiannual transport and circulation of nutrients and oxygen throughout the body of water (see Chapter 25).

Seawater behaves somewhat differently. The density of seawater (salinity of 24.7 percent and higher)—or rather, its specific gravity relative to that of an equal volume of pure water (specific gravity equals 1) at atmospheric pressure—is correlated with salinity. At 0°C the density of seawater with a salinity of 35 percent is 1.028. The lower its temperature, the greater the density of seawater; the higher the temperature, the lower the density. No definite freezing point exists for seawater. Ice crystals begin to form at a point on the temperature scale that varies with salinity. As pure water freezes out, the remaining unfrozen water becomes higher in salinity and lower in its freezing point until finally a solid block of ice crystals and salt is formed.

The Physical Properties of Water

SPECIFIC HEAT

Water is capable of storing tremendous quantities of heat energy with a relatively small rise in temperature. It is exceeded in this only by ammonia, liquid hydrogen, and lithium. Thus, water is described as having a high *specific heat,* the number of calories necessary to raise one gram of water one degree centigrade. The specific heat of water is given the value of 1.

Since such great quantities of heat must be absorbed before the temperature of natural waters such as ponds, lakes, and seas can be raised 1°C, they warm up slowly in the spring and cool off just as slowly in the fall. This prevents wide seasonal fluctuations in the temperature of aquatic habitats and moderates the temperatures of local or worldwide environments.

LATENT HEAT

Not only does water have a high specific heat; it also possesses the highest heat of fusion and heat of evaporation, collectively called *latent heat,* of all known substances that are liquid at ordinary temperatures. Large quantities of heat energy must be removed before water can change from a liquid to a solid, and conversely, it must absorb considerable heat before ice can be converted to a liquid. It takes approximately 80 calories of heat to convert 1 g of ice to a liquid state when both are at 0°C. This is equivalent to the amount of heat needed to raise the same quantity of water from 0° to 80°C.

Evaporation occurs at the interface between air and water at all ranges of temperature. Here, again, considerable amounts of heat are involved; 536 calories are needed to overcome the attraction between molecules and convert 1 g of water at 100°C into vapor. This is as much heat as is needed to raise 536 g of water 1°C. When evaporation occurs, the source of thermal energy may be the sun, the water itself, or objects in or around it. Rendered latent at the place of evaporation, the heat involved is returned to actual heat at the point of condensation. Such phenomena play a major role in worldwide meteorological cycles.

VISCOSITY

The viscosity of water is also high because of the energy contained in the hydrogen bonds. Viscosity can be visualized best if one imagines or observes liquid flowing through a glass tube or a clear plastic hose. The liquid moving through the tube behaves as if it consisted of a series of parallel concentric layers flowing one over another. The rate of flow is greatest at the center; but because of the amount of internal friction between layers, the flow decreases toward the sides of the tube. This same phenomenon can be observed along the side of any stream or river with uniform banks. The water along the banks is nearly still, while the current in the center may be swift. This resistance between the layers is called *viscosity.*

This lateral or laminar viscosity is complicated by another type, eddy viscosity, in which water masses pass from one layer to another. This creates turbulence both horizontally and vertically. Biologically important, eddy viscosity is many times greater than laminar viscosity.

Viscosity is the source of frictional resistance to objects moving through the water. Since this resistance is 100 times that of air, animals must expend considerable muscular energy to move through the water. A mucous coating on fish reduces surface resistance. Streamlining does likewise; in fact, the body forms of some aquatic organisms have evolved under the stresses of viscosity. The faster an aquatic organism moves through the water, the greater the stress placed on the surface and the greater the volume of water that must be displaced in a given time. Replacement of water in the space left behind by the moving animal adds additional drag on the body. An animal streamlined in reverse, with a short, rounded front and a rapidly tapering body, meets the least resistance in the water. The acme of such streamlining is the sperm whale.

SURFACE TENSION

Within all substances particles of the same matter are attracted to one another. Water is no exception. Molecules of water below the surface are symmetrically surrounded by other molecules. The forces of attraction are the same on one side of the molecules as on the other. But at the water's surface, the molecules exist under a different set of conditions. Below is a hemisphere of strongly attractive similar water molecules; above is the much smaller attractive force of the air. Since the molecules on the surface are drawn into the liquid, the liquid surface tends to be as small as possible, taut like the rubber of an inflated balloon. This is *surface tension,* which is important in the lives of aquatic organisms.

This skin of water is able to support small objects and animals, such as the water striders and water spiders that run across the pond surface. To other organisms surface tension is a barrier, whether they wish to penetrate into the water below or to escape into the air above. For some the surface tension is too great to break; for others it is a trap to avoid while skimming the surface to feed or to lay eggs. If caught in the surface tension, the insect may flounder on the surface. The imagoes of mayflies and caddisflies find surface tension a handicap in their efforts to emerge from the water. Slowed down at the surface, these insects become easy prey for trout.

Surface tension is important in other ways to all life. It is the force that draws liquids through the pores of the soil and the conducting networks of plants. To overcome this force, aquatic insects and plants have evolved structural adaptations that prevent the penetration of water into the the tracheal systems of the former and the stomata and internal air spaces of the latter.

The Distribution of Water

Although one views water as something of a local phenomenon, such as a stream or autumn rains, it forms a single worldwide resource distributed in land, sea, and the atmosphere and unified by the hydrological cycle. It is influenced by solar energy, by the currents of the air and oceans, by heat budgets, and by the water balances of land and sea.

Oceans cover 71 percent of the Earth's surface (see Table 6.1). With a mean depth of 3.8 km (2.36 miles), they hold 93 to 97 percent of all the Earth's waters (depending on the estimate used) (Kalinin and Bykov, 1969). Thus, fresh water usable by humans represents only 3 percent of the planet's water

TABLE 6.1 WORLD'S WATER RESOURCES

Resource	Volume (W), Thousands of km^3	Annual Rate of Removal (Q) (in Thousands of km^3)	and Process of Removal	Renewal Period (T) ($T = W/Q$)
Total water on Earth	1,460,000	520	Evaporation	2,800 years
Total water in the oceans	1,370,000	449	Evaporation	3,100 years
		37	Difference between precipitation and evaporation	37,000 years
Free gravitational waters in the Earth's crust (to a depth of 5 km)	60,000	13	Underground runoff	4,600 years
Amount of which is present in the zone of active water exchange	4,000	13	Underground runoff	300 years
Lakes	750		—	
Glaciers and permanent snow	29,000	1.8	Runoff	16,000 years
Soil and subsoil moisture	65	85	Evaporation and underground runoff	280 days
Atmospheric moisture	14	520	Precipitation	9 days
River waters	1.2	36.3[a]	Runoff	12(20) days[b]

[a] Not counting the melting of Antarctic and Arctic glaciers.
[b] 12 days for small river systems. 20 days for major rivers draining into the sea.
NOTE: Average error is probably 10–15 percent.
SOURCE: G. P. Kalinin and V. D. Bykov, *Impact of Science on Society* 19(2):139. © Unesco, 1969.

supply. Of the total fresh water on Earth, 75 percent is locked up in glaciers and ice sheets—enough to maintain all the rivers of the world at their present rate of flow for the next 900 years. If oceans contain 97 percent of the world's water, then nearly 2 percent of the remainder is tied up in ice. This leaves less than 1 percent of the world's water available fresh. Freshwater lakes contain 0.3 percent of the freshwater supply, and at any one time rivers and streams contain only 0.005 percent of that supply. Soil moisture accounts for approximately 0.3 percent, and another very small portion of the Earth's water is tied up in living material.

More stable is the groundwater supply, which accounts for 25 percent of our fresh water. Groundwater fills the pores and hollows within Earth just as water fills pockets and depressions on the surface. Estimates—necessarily rough and inaccurate—place renewable and cyclic groundwater at 7×10^6 km^3 (Nace, 1969), or approximately 11 percent of the freshwater supply. Some of the groundwater is "inherited," as in aquifers in desert regions, where the water is thousands of years old. Because inherited water is not rechargeable, heavy use of these aquifers for irrigation and other purposes is mining the supply. In the foreseeable future, the supply could be exhausted. A portion of the groundwater, approximately 14 percent, lies below 1,000 m. Known as fossil water, it is often saline and does not participate in the hydrological cycle.

The atmosphere, for all its clouds and obvious close association with the water cycle, contains only 0.035 percent fresh water. Yet it is the atmosphere and its relation to land and oceans that keeps the water circulating over the Earth.

The Water Cycle

Leonardo da Vinci wrote that "water is the driver of nature." Perceptive as he was, even he could not have appreciated the full meaning of his statement based on the scientific knowledge of his time. Without the cycling of water, biogeochemical cycles could not exist, ecosystems could not function, and life could not be maintained. Water is the medium by which materials make their never-ending odyssey through the ecosystem.

Precipitation is the driving force of the water cycle (see Figure 6.1). Beginning as water vapor in the atmosphere, the moisture coalesces into droplets and ice crystals, which eventually fall as some form of precipitation. As the precipitation reaches the Earth, some of the water falls directly on the ground and some falls on vegetation, on litter on the ground, and on urban structures and streets. It may be stored, hurried off, or in time, it may infiltrate the soil.

Because of interception, which can be considerable, various amounts of water never reach the ground but evaporate into the atmosphere. In urban areas a great portion of the rain falls on roofs and sidewalks, which are impervious to water. The water runs down gutters and drains to be hurried off to rivers.

The precipitation that reaches the soil moves into the ground by infiltration, the rate of which is governed by soil, slope, type of vegetation, and the characteristics of the precipitation itself. In general, the more intense the rain, the greater the rate of infiltration, until the infiltration capacity of the soil, determined by soil porosity, is reached.

During long wet spells and heavy storms, the soil may become saturated; or intense rainfall or rapid melting of snow can exceed the infiltration capacity of the soil. At this point water becomes overland flow. In places it becomes concentrated into depressions and rills, and the flow changes from sheet flow to channelized flow, a process that can be observed even on city streets as water moves in sheets over the pavement and becomes concentrated into streetside gutters. Again, the amount of runoff that affects the erosion of soil depends on slope, soil texture, soil moisture conditions, and the type and condition of vegetation.

In the undisturbed forest, infiltration rates usually are greater than intensity of rainfall, and surface runoff does not occur. In urban areas infiltration rates may range from zero to a value exceeding the intensity of rainfall on certain areas where soil is open and uncompacted. Because of low infiltration, runoff from urban areas might be as much as

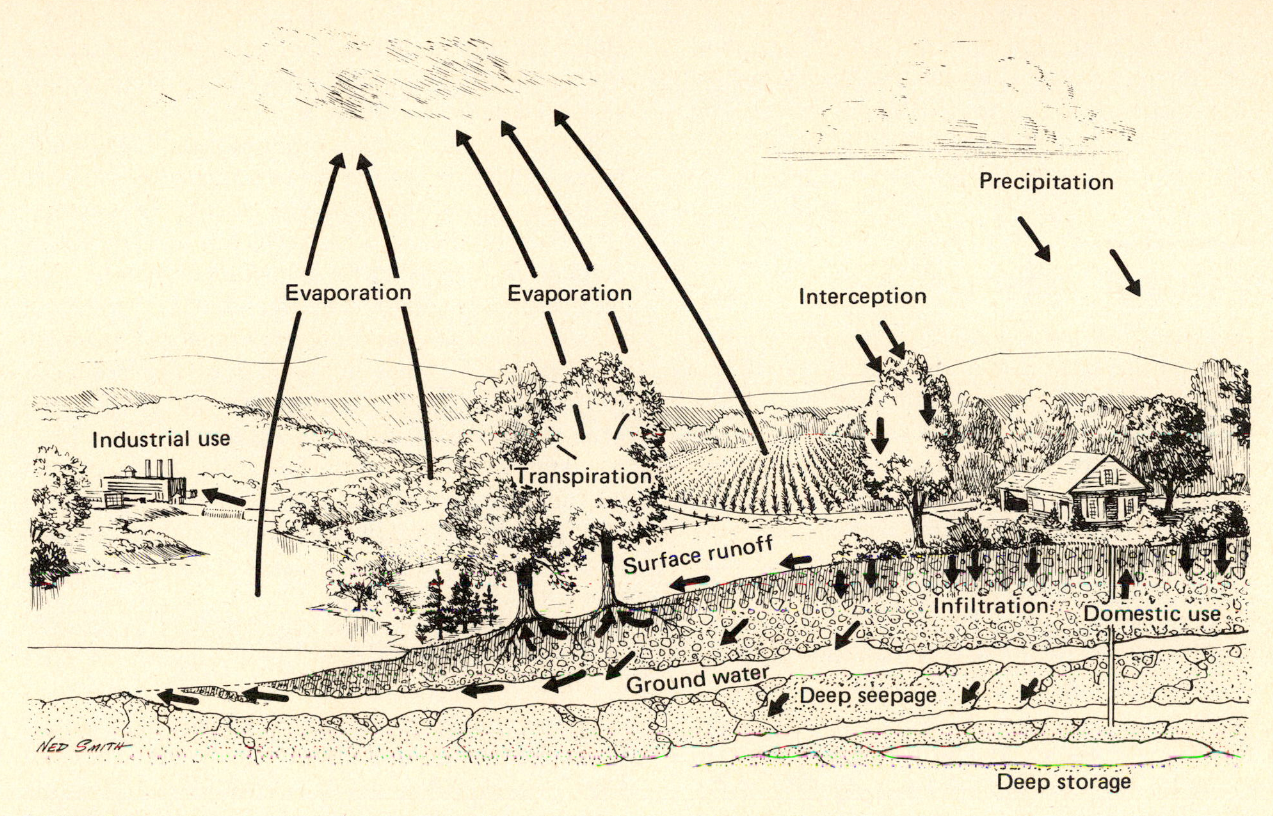

FIGURE 6.1 The water cycle, showing the major pathways of water through the ecosystem.

85 percent of the precipitation (Lull and Sopper, 1969).

Water entering the soil will percolate or seep down to an impervious layer of clay or rock to collect as groundwater. From here the water finds its way into springs, streams, and eventually rivers and seas. A great portion of this water is utilized by humans for domestic and industrial purposes, after which it reenters the water cycle through discharge into streams or into the atmosphere.

Part of the water is retained in the soil. The portion held between soil particles by capillary forces is *capillary water.* Another portion adheres to soil particles as a thin film. This is *hydroscopic water* and is unavailable to plants. The maximum amount of water that a soil can hold at one-third atmosphere after gravitational water is drained away is called *field capacity,* which varies considerably with the texture of the soil. Highly porous sandy soils have low field capacity, while fine-textured clay and humic soils have a high field capacity.

The water remaining on the surface of the ground and on vegetation, as well as the water in the surface layers of streams, lakes, and oceans, *evaporates*—a process by which more water molecules leave a surface than enter it. The rate at which water moves back into the atmosphere is governed by the vapor-pressure deficit of the atmosphere.

As surface soils dry out, evaporation from them ceases because there exists a dry barrier through which little soil water moves. At this point major water losses from the soil take place through the leaves of plants. Plants take in water through the roots, and they lose it through the leaves. As long as sufficient water is available, the leaves remain turgid, and the openings of the stomata are maximal, which permits an easy inflow of carbon dioxide to the leaf but at the same time permits a large loss of water. This loss will continue as long as moisture is available for roots in the soil.

The Global Water Cycle

The molecules of water that fall in a spring shower might well have been part of the Gulf Stream a few weeks before and perhaps spent some time in the Amazon tropical rain forest before that. The local storm is simply part of the mass movement and circulation of water about Earth, a movement suggested by the changing cloud patterns over the face of the planet. The atmosphere, oceans, and land masses form a single gigantic water system, which is driven by solar energy. The presence and movement of water in any one part of the system affects the presence and movement in all other parts.

The atmosphere is one key element in the world's water system. At any one time the atmosphere holds no more than a 10- to 11-day supply of rainfall in the form of vapor, clouds, and ice crystals. Thus, the turnover of water molecules is rapid. Because the source of water in the atmosphere is evaporation from land and sea, there are global differences in the amount of evaporation and the amount of moisture in the atmosphere at any given point. Evaporation at lower latitudes is considerably greater than evaporation at higher latitudes, reflecting the greater heat budgets produced by the direct rays of the sun. Evaporation is greater over oceans than over land. Oceans account for 84 percent of evaporation—considerably more than they receive in return from precipitation (see Figure 6.2). Land areas contribute 16 percent of the annual evaporation, yet they intercept a greater amount.

Moisture in the atmosphere moves with the general circulation of the air. Air currents, hundreds of kilometers wide, are in fact giant, unseen rivers moving in great swirls above the Earth, only a part of whose moisture falls as precipitation in any one place. In the equatorial areas the trade winds move moisture-laden air toward the equator, where it is warmed. Hot air over the equator rises, cools, and drops its moisture as rain. Thus, equatorial regions are areas of maximum precipitation. Air that rises over the equator descends earthward in two subtropical zones around 30° north and 30° south latitude. As the air descends, it warms and picks up moisture from land and sea. The highest annual losses to evaporation occur in the subtropics of the western North Atlantic and Northern Pacific, or the Gulf Stream and the Kinshio Current. North of this

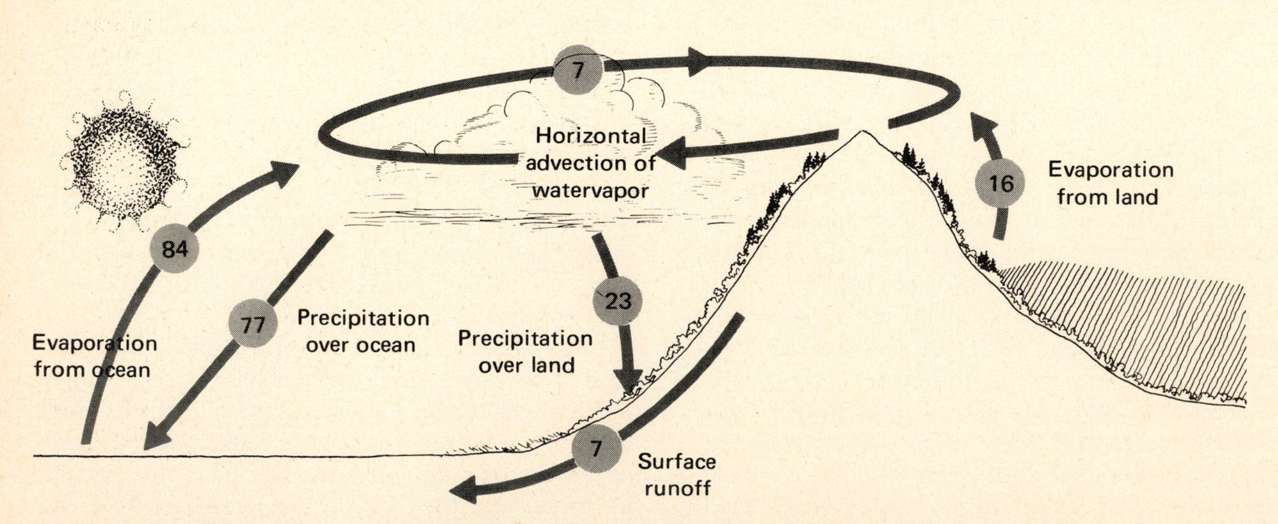

FIGURE 6.2 The global budget of water. The mean annual precipitation of 83.6 cm has been converted into 100 units.

are two more zones of ascending air and low pressure that produce the west coast areas of maximum rainfall. In high latitudes the air descends again in the polar regions, where it remains dry.

The excess of precipitation over evaporation is eventually carried to the sea by rivers. Rivers are the primary movers of water over the globe and carry many more times the amount of water their channels hold. By returning water to the sea, they tend to balance the evaporation deficit of the oceans. Sixteen major rivers discharge 13,600 km^3 annually, or 45 percent of all the water carried by rivers. Adding the next 50 largest rivers brings the total to 17,600 km^3, or 60 percent of all the water discharged to the sea.

Evaporation, precipitation, detention, and transportation maintain a stable water balance on the Earth. Consider the amount of water that falls on the Earth in terms of 100 units (Figure 6.2). On the average, 84 units are lost from the ocean by evaporation, while 77 units are gained from precipitation. Land areas lose 16 units by evaporation and gain 23 units from precipitation. Runoff from land to the ocean makes up 7 units, which balances the evaporative deficit of the ocean. The remaining 7 units are circulating as atmospheric moisture.

In its global circulation, water also influences the heat budgets of the Earth. As already suggested, the highest heat budgets are in the low latitudes, the lowest in the polar regions, and a balance between incoming and outgoing cold and heat is achieved at 38° to 39° latitude. Excessive cooling of higher latitudes is prevented by the north and south transfer of heat by the atmosphere in the form of sensible and latent heat in water vapor and by warm ocean currents.

Examined from a global point of view, the water cycle emphasizes the close interaction between the physical and geographic environments of the Earth. Thus, the water problem, often considered in local terms, is actually a global problem, and local water management schemes can affect the planet as a whole. Problems result not because an inadequate amount of water reaches Earth but because it is unevenly distributed, especially relative to human population centers. Because humans have strongly interjected themselves into the water cycle, the natural usable water resources have decreased, and water quality has declined. The natural water cycle has not been able to compensate for the detrimental effects of humans on water resources.

Rainfall and Humidity

The moisture content of the air is usually expressed as *relative humidity,* the percentage of moisture in the air relative to the amount of water the air could hold at saturation at the existing temperature. If air is warmed while its moisture content remains constant, the relative humidity drops because warm air can hold more moisture than cool air. As the relative humidity drops, the *saturation pressure deficit* (the difference between the partial pressure of water at saturation and the prevailing vapor pressure of the air at the same temperature) increases, and increased evaporation takes place.

Relative humidity varies during a 24-hour period. Generally, it is lower by day and higher by night. Relative humidity under a closed forest canopy is higher than on the outside during the day, and it is lower than on the outside during the night.

Over normal surfaces relative humidity during the day usually increases with height because of the decrease in temperature with height. That contrasts with absolute humidity, which decreases with height. But if a temperature inversion occurs (see Chapter 4), especially at night, the relative humidity decreases upward to the top of the inversion, then changes little or increases only slightly.

In any one area relative humidity varies widely from one spot to another, depending upon topography. Variations in humidity are most pronounced in mountain country. Low elevations warm up and dry out earlier in the spring than high elevations. Because daytime temperatures decrease with altitude, as does the dew point (the temperature at which atmospheric water condenses), relative humidities are greater on the tops of mountains than in the valleys. As nighttime cooling begins, temperature change with altitude is reversed. Cold air rushes downslope and accumulates at the bottom (see Chapter 4). Through the night, if additional cooling occurs, the air becomes saturated with moisture, and fog or dew forms by morning. Dew is the condensation of moisture on ground surfaces, vegetation, and other objects because the surface is cooler than the air. Fog is a cloud hanging close to the ground. Differences in humidity and in the resulting formation of fog and dew can produce vegetative differences on mountain slopes. They are most pronounced on slopes of mountains along the Pacific Coast of North America.

Temperature and wind both exert a considerable influence on evaporation and relative humidity. An increase in air temperature causes convection currents. That sets up an air turbulence, which mixes surface layers with drier air above. Wind movements associated with cyclonic disturbances also mix moisture-laden air with drier air above. As a result, the vapor pressure of the air is lowered, and evaporation from the surface increases.

Moisture relationships closely relate to the distribution of rainfall. Seasonal distribution of rainfall is more important than average annual precipitation. A great difference exists between a region receiving 120 cm of rain rather evenly distributed throughout the year and a region in which nearly all of the 120 cm of rain falls during a several-month period. In the latter situation, typical of tropical and subtropical climates, organisms must face a long period of drought. An alternation of wet and dry seasons influences the reproductive and activity cycles of organisms as much as light and temperature in the temperate regions.

Response to Moisture

Just as organisms need to maintain a thermal balance with their environment, so they also need to maintain a water balance. Water taken up by an organism must equal water lost; absorption must equal evaporation, transpiration, and excretion. It is difficult, however, to separate responses of organisms to variations of moisture in the environment from responses to variations of temperature because of the close relationship between the two. Evaporative water loss, for example, is strongly influenced by high temperatures in both plants and animals, especially those that use evaporation as a way of staying cool.

Organisms live in a moisture environment, ranging from environments entirely aquatic to those deficient in moisture, either physically (as in arid regions) or physiologically (as in saline habitats). At one extreme, organisms must conserve water by preventing its loss to the environment. At the other extreme, organisms must get rid of excess water absorbed or prevent an excessive intake of water. Most organisms fall somewhere between these two extremes.

PLANT RESPONSE TO MOISTURE

Lack of water is a major selective force in the evolution of plants' ability to cope with moisture stress. A large group—including algae, fungi, lichens, and most mosses—is *poikilohydric*. Its water status always tends to match atmospheric moisture conditions, because it possesses no protection against water loss. As water becomes less available, the plants dry out. Their cells shrink without disturbing the fine protoplasmic structures within them, and their vital processes gradually become suppressed. When moisture conditions improve, the plants imbibe water, and the cells fill and resume normal functioning. Such plants restrict their growth to moist periods, and their biomass is always small.

Other plants, mostly ferns and seed plants, are *homoiohydric*. Within limits they are able to maintain a stable water balance independent of fluctuations of atmospheric moisture conditions. That is made possible by water stored in a vacuole within the cell, the evolution of a protective cuticle that slows down evaporation, stomata that regulate transpiration, and an extensive root system to draw water from the soil.

Homoiohydric plants lose water to the atmosphere by an evaporative process, transpiration. *Evaporation* is the physical process of converting liquid water to vapor. *Transpiration* is the evaporation of water from internal surfaces of living plants and its subsequent diffusion from the plant.

As long as the vapor pressure inside the leaf is greater than the vapor pressure of the atmosphere, a plant loses water to the environment. Plants maintain a water balance by drawing water from the soil with their roots and conducting it to stems and leaves. Evaporation takes place at the interface between atmosphere and water within the leaf.

Homoiohydric plants don't give up water easily. They hold it back by a combination of osmotic pressure and turgor pressure of water in the cells pushing outward against the cell walls. The two make up *water potential,* the measure of energy in water (see Box 6.1). Water tends to move in the direction of low water potential. Thus, water moves from the soil into roots and from leaf to atmosphere. Plants are able to control up to a point the movement of water from leaf to atmosphere. They balance that loss by removing water from the soil as long as it is available. In effect, transpiration dries out the soil.

Response to Drought

The most common form of water stress in plants is water deficiency. It may be a physical drought brought about by the lack of soil water and excessive transpiration, or it may be a physiological drought of winter. In cold and windy weather, plants may be unable to replenish their water supply because of dry or deeply frozen soil. In winter the water ducts of plants are filled with ice. If the sun warms the twigs but the soil is not thawed sufficiently to allow the plants to replace water lost through transpiration, twigs and even the whole plant experience dehydration.

The ability of a plant to withstand water stress depends upon its *drought resistance,* the sum of its drought tolerance and drought avoidance. *Drought tolerance* is the ability of a plant to maintain its physiological activity in spite of a lack of water or to survive drying of its tissues. Relatively few species possess any great degree of drought tolerance during the growing season, mainly because the water potential decreases during drought. Instead, most plants depend upon some type of *drought avoidance,* which involves some interaction between their internal water status and that of the environment. Predictably, drought avoidance is most effective among plants of xeric habitats and less effective among species of mesic sites. Drought avoidance may involve phenological and physiological mechanisms.

Phenological avoidance is most common among desert ephemeral plants. Mainly annuals, they germinate, grow, flower, and go to seed only when rains come and wet the surface soil. Seeds of these plants may lie dormant for years waiting for the right moisture conditions.

The most common approach to drought avoidance among plants is physiological. A plant attempts to maintain its internal water potential as close as possible to soil water potential. It has to balance the rate of withdrawing water from the soil with transpirational losses from the leaves. To do that the plant has to exercise some control over transpiration. The most usual way is to regulate the stomatal openings in the leaf through which water escapes. A plant responds to drought by partially closing its stomata and opening them for shorter periods of time. In the early period of stress, the plant closes its stomata and reduces transpiration during the hottest part of the day and resumes normal activity in the cooler hours. As the water balance worsens, a plant opens its stomata only in the morning. As drought continues, the plant may keep its stomata closed, and it may fold or roll the leaves to reduce surface area. The plant now transpires through the cuticle, which gives off only a fraction of the water lost through the stomata.

Closing the stomata cuts down on transpiration, but at the same time, it reduces the carbon dioxide (CO_2) intake necessary for photosynthesis (see Chapter 18). As a result, the rate of photosynthesis declines. That, in turn, affects carbon metabolism, with some wide-ranging effects over time. The ability of the leaf to store carbon in a usable form is reduced. The lack of carbon can affect not only the metabolism of the leaf but other organs as well. Carbon reduction is often accompanied by a lowering of the nitrate concentration in the sap, which affects nitrogen metabolism in the plant.

Xeric species are able to continue photosynthesis at a more negative water potential over an extended period of drought than mesic species. Mesic species close their stomata at a much higher

Box 6.1
A Word about Water

When water evaporates, more molecules leave the surface of water (or any wet object) than enter it. Water molecules continue to leave the surface as long as the air above can accommodate them. When the number of molecules leaving the water equals the number returning, the air is saturated.

The amount of water that air can hold varies with its temperature. Warm air can accommodate more moisture than cool air. For a given temperature there is a fixed quantity of water that air can hold in a given space (see figure). Cold air becomes saturated with less water vapor per unit volume than warm air. The concentration of water vapor in air is measured in terms of mass/volume, such as g/m^3, or as weight of water vapor/unit of dry air.

At saturation, air has a certain saturation vapor pressure. Vapor pressure is that part of the barometric pressure attributed to the presence of water vapor in the air. At sea level the standard barometric pressure is 1,013 millibars (29.92 inches, 76 cm). The vapor pressure of cold, dry air at sea level is 2 millibars (0.06 inches, 0.15 cm) and of warm, humid air

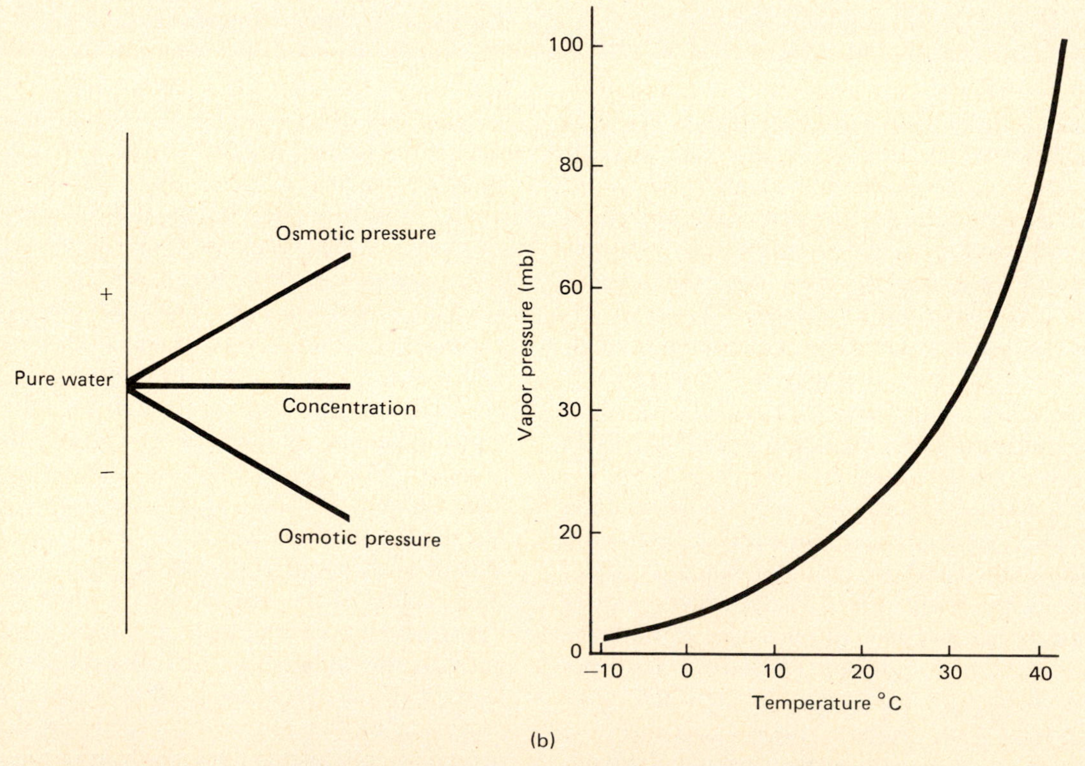

(a) Graph depicting that osmotic potential equals the negative of osmotic pressure. As the concentration of a particular substance increases, osmotic pressure increases and osmotic potential decreases. Osmotic potential is a measure of the free energy of water molecules in the solution compared to the free energy of water molecules in pure water. (b) Vapor pressure of saturated air.

30 millibars (0.9 inches, 2.3 cm), or about 1/35 of the mercury column. This relationship is the *partial pressure of water vapor.*

Saturation vapor pressure is a function of temperature. The difference between the amount of water vapor held at saturation vapor pressure and the actual vapor pressure at any given temperature represents the *vapor-pressure deficit*. Water vapor in the atmosphere moves in response to differences in vapor pressure between any two levels, such as the surface of a pond and the air above. The rate of movement is proportional to vapor-pressure differences, or the vapor-pressure gradient.

Another feature is *water potential.* Water potential is the measure of the energy of water as determined by two opposing forces, osmotic potential and turgor pressure. *Osmotic potential,* the negative of osmotic pressure, is the tendency of a solution to gain water when separated from pure water by an ideal, selectively permeable membrane. *Osmotic pressure* is the minimum pressure that must be applied to a solution to keep it from gaining water when separated from pure water by an ideal, selectively permeable membrane. *Turgor pressure* is the pressure of a solution within a cell pushing out against a cell wall. When the magnitude of osmotic potential and turgor pressure is equal, the water potential is zero. There is no net movement of water into or out of a cell. Osmotic potential, and thus water potential, is lowered by the addition of a solute or the removal of a solvent. Water moves into an area of lower water potential.

level of leaf water potential. Such action works for a short-term drought, but unless the drought ends quickly, mesic plants may experience metabolic damage (see Chabot and Bunce, 1979).

There are other physiological measures a plant can take. One is to reduce transpiration across the cuticle by changing its thickness, structure, or the chemical composition of the lipids it contains. A plant can increase its water uptake by extending its root system into areas of the soil still untapped. And a plant can increase water storage in its cells, or its succulence. That is a common mechanism among some desert plants. Desert succulents are exemplified by the cacti, which usually lack leaves and carry on photosynthesis in the stem. Succulents obtain their water from the upper layers of soil after it rains and store that water in their cells. They conserve the water by separating the light and CO_2 fixation reactions of photosynthesis (see Chapter 18). That allows the plants to go for a long period without opening the stomata. When succulents do open their stomata, it is at night, when water losses are minimal.

Lacking succulence, some plants of arid and semiarid regions become drought-deciduous, shedding their leaves during the dry season. Some desert shrubs, such as ocotillo (*Fouquieria splendens*), are ephemerally deciduous, shedding their leaves four or five times a year and renewing them with each new rain. Other plants possess extremely small leaves or lack leaves altogether. Those plants, such as paloverde (*Cercidium floridum*), carry on photosynthesis through green stem tissue rich in chlorophyll. Such plants have the advantage of being able to start photosynthesis immediately when the rains come, without the time lag otherwise necessary for the formation of new photosynthetic tissue.

Plants of riverine and streamide habitats and deep depressions in arid and semiarid regions escape drought by tapping a deep, permanent or semipermanent water supply. Such trees and shrubs, called *phreatophytes* (well plants), may send roots down to a depth of 60 or 80 m to reach water, although 10 to 30 m is more typical. Examples are cottonwood (*Populus fremontii*), willows (*Salix* spp.), salt cedar (*Tamarix* spp.), and mesquite (*Prosopis glandulosa*).

Long periods of dry weather that result in soil drought reduce plant growth and cause the dieback of plants or outright death. Drought-injured plants are vulnerable to outbreaks of insects and are highly susceptible to fire. Drought can also influence the composition of plant communities. Weaver and Albertson (1956), for example, found that during the drought of 1933 to 1939 on the plains, buffalo grass either disappeared entirely from some ranges or was reduced to small, scattered patches. Its more resistant associate, blue grama, was never killed uniformly and persisted. When the drought ended, in 1940 to 1942, buffalo grass responded rapidly to moisture and blue grama declined.

ANIMAL RESPONSE TO MOISTURE

Animals, too, need to maintain a water balance with their environment. Their adaptations to moisture are more complex than those of plants. Involved is a more or less universal mechanism, the excretory system, to rid the body of excess water or to conserve it.

Aquatic organisms living in fresh water have a higher osmotic pressure in their bodies than the surrounding water. Their problem is to rid themselves of excess water taken in through permeable membranes. Protozoans accomplish that by means of contractile vacuoles. Freshwater fish maintain osmotic balance by absorbing and retaining salts in special cells in the body and by producing copious amounts of watery urine. Other animals maintain dilute urine in comparison to body fluids. Aquatic amphibians balance the loss of salts through the skin by absorbing ions directly from the water and by active transport across the skin and gill membranes. Aquatic animals do not drink.

Terrestrial animals have three major means of gaining water and solutes: through drinking, through food, and through production of metabolic water from food. They lose water and solutes through urine, feces, evaporation over the skin, and respiration.

Animals have a variety of strategies for coping with water balance. Amphibians store water from the kidneys in the bladder and, if circumstances demand it, conserve water for metabolism by reabsorbing it through the bladder wall. They can also gain water directly across the skin. Birds and reptiles employ gut, kidney, and a salt gland. Both birds and reptiles have similar types of kidneys and possess a cloaca, a common receptacle for the digestive, urinary, and reproductive tracts. Urine leaves the kidney and collects in the cloaca, where it is modified in volume and composition and converted into a semisolid paste. Some (or most) water is reabsorbed back into the body. Mammals possess kidneys capable of producing urine with high osmotic pressure and inorganic ion concentration. They lose water through the kidneys, respiratory system, and sweat glands.

Like plants, animals of arid environments are faced with a severe problem of water balance. They can solve the problem in one of two ways—either by evading the drought or by avoiding its effects. Animals of semiarid and desert regions may evade drought by leaving the area during the dry season. That is the strategy employed by many of the large African ungulates. Some small animals, such as the spadefoot toad of the southwestern United States, aestivate below ground and emerge when the rains return. Some invertebrates, such as the flatworm *Phagocytes vernalis,* which occupies ponds that dry up during the summer, encyst. Other aquatic or semiaquatic animals retreat deep into the soil until they reach groundwater level. Many insects undergo diapause, just as they do when confronted with unfavorable temperatures.

Other animals remain active during the dry season but conserve moisture by keeping as much water as possible in the body. One way is to reduce respiratory water loss. Some small desert rodents reduce the temperature of respired air by a countercurrent heat exchange in respiratory passages. Moisture-laden warm air from the lungs passes over the cooled nasal membranes, leaving behind condensed water on the walls of the nasal passages. Dry, warm air inhaled by the rodent is humidified by the evaporation of water on the surface membranes of the nasal passages. An African desert ungulate, the oryx, reduces daytime losses of moisture by becoming hyperthermic (see Chapter 5). A substantial rise in its daytime body temperature reduces evaporative losses. The oryx further reduces water loss by suppressing sweating and by panting only at very high temperatures. Further, the oryx reduces its metabolic rate; by lowering the internal production of calories, the animal reduces the need for evaporative cooling. By night the oryx reduces its nonsweating evaporation across the skin by reducing its metabolic rate below that of the daytime. With a lowered nighttime body temperature, the saturation level for water vapor in exhaled air is lower.

There are other approaches to the problem. Some small desert mammals reduce water loss by remaining in burrows by day and emerging by night. Many desert mammals, from kangaroo rat to camel, produce highly concentrated urine and dry feces and extract water metabolically from the food they eat. In addition, some desert mammals can tolerate a certain degree of dehydration. Desert rabbits may withstand water losses up to 50 percent of their body weight; camels can tolerate a 27 percent loss.

Water balance can have other effects. Moisture influences the speed of development and even the fecundity of some insects. If the air is too dry, the eggs of some locusts and other insects may become quiescent. There is an optimum humidity at which nymphs develop the fastest. Some insects lay more eggs at certain relative humidities than above or below that point. Heavy rains and prolonged wet spells cause widespread death among mammals and birds, especially the young, from drowning, exposure, and chilling. Excessive moisture and cloudy weather kill insect nymphs, inhibit insect pollination of plants, and spread parasitic fungi, bacteria, and viruses among both plants and animals.

Problems of Saline Environments

Saline environments—oceans, salt marshes, estuaries, and alkaline deserts—present their organisms with their own sets of water balance problems. The salt water environment is a physiological desert, where the concentration of salts outside the bodies of organisms can dehydrate them osmotically. In marine and brackish environments, the problem is one of inhibiting the loss of water through the body wall by osmosis and preventing an accumulation of salts in the system. Algae and invertebrates get around that problem by possessing body fluids that have the same osmotic pressure as seawater. In a way, that adds to the problem of marine invertebrates because they cannot obtain fresh water through their food. Marine teleost fish absorb water with the salt into the gut. They secrete magnesium and calcium through the kidneys and pass these ions off as a partially crystalline paste. The fish excrete sodium and chlorine by pumping the ions across membranes of special cells in the gills. This pumping process is one type of *active transport.* It involves the movement of salts against a concentration gradient at the cost of metabolic energy. Sharks and rays retain urea to maintain a slightly higher concentration of salt in the body than in surrounding seawater. Birds of the open sea are able to utilize seawater, for they possess special salt-secreting glands located on the surface of the cranium (Schmidt-Nielsen, 1960). Gulls, petrels, and other seabirds excrete from these glands fluids in excess of 5 percent salt. Petrels and tube-nosed swimmers forcibly eject the fluids through the nostrils; other species drip the fluids out of the internal or external nares.

Among marine mammals the kidney is the main route for the elimination of salt. Porpoises have highly developed renal capacities to eliminate salt loads rapidly (Malvin and Rayner, 1968). In marine mammals the urine has a greater osmotic pressure than blood and seawater (hyperosmotic); the physiology is poorly understood.

Vertebrates in the Arctic and Antarctic have special problems. As seawater freezes, it becomes colder and more salty. The only alternative for most organisms is to increase the solute concentration in the body fluids to lower the body temperature. Some species of fish in the Antarctic possess in their blood a glycoprotein antifreeze substance that enables the animals to exist in temperatures below the freezing point of blood.

Plants inhabiting saline environments are known as *halophytes,* able to survive and to complete their life cycle at high salinities (Flowers, Troke, and Yeo, 1977). These salt-tolerant plants are capable of growing in soils with a salt concentration of 0.2 percent or higher. Characteristically, plants of saline environments accumulate high levels of ions within the cells, especially in the leaves. That concentration, which may equal or exceed that of seawater, allows halophytes to maintain a high cell water content in the face of a low external water potential. Some halophytes compensate for an increase in sodium and chlorine uptake with a corresponding dilution of internal solutions with water stored in tissues. In addition, some plants exhibit high internal osmotic pressure many times that of freshwater and terrestrial plants; possess salt-secreting glands, usually on the leaves, from which the excess salt can be washed away by rainwater; secrete heavy cutin on the leaves; and are succulent (Chapman, 1960). A number of desert plants allow only certain ions to pass across cellular membranes and keep others out. That allows the plants to absorb essential nutrients from the soil and helps maintain internal osmotic pressures higher than those of the surroundings.

Halophytes vary in their degree of tolerance. Some, such as salt marsh hay grass (*Spartina patens*), are intolerant and do best at low salinities. Others, such as salt marsh cord grass (*S. alternifolia*), do best at moderate levels of salinity. A few, such as glassworts (*Salicornia* spp.), are tolerant of high salinities. All halophytes appear to be able to grow in nonsaline environments but exist where they do because they are unable to compete with nontolerant species.

Response to Excessive Moisture

Plants usually experience a deficit of water rather than an excess. But at times of heavy, prolonged rainfall and flooding, terrestrial plants face an excessive amount of water about their roots. Water displaces air in the soil, reducing the oxygen supply and limiting the metabolism and growth of roots.

Some plants of floodplains and swamps have evolved anatomical structures that enable them to endure flooding. Mangroves have pneumatophores, and cypress have root knees that protrude above the water and carry oxygen to the roots. Some grasses and herbaceous plants have hollow tubes leading from the leaves to the roots through which oxygen diffuses. Plants intolerant of flooding may undergo metabolic disturbances, including an excessive accumulation of ethanol brought about by anaerobic conditions. Plants tolerant of flooding avoid that problem by switching over to the production of malic acid under anaerobic conditions (Crawford, 1972). Malic acid is utilized in photosynthesis (see Chapter 18). Trees growing on poorly drained soils may develop shallow, spreading root systems. Their roots are unable to grow downward because oxygen levels below the ground line are too low for root growth. Such trees are subject to windthrow and other storm damage and are sensitive to drought and frost.

Moisture and Plant Distribution

Moisture or the lack of it can have a major influence on the distribution of plants on both a geographic and a local basis. For example, western redcedar (*Thuja plicata*) and western hemlock (*Tsuga heterophylla*) grow where the average annual rainfall in western North America is around 32 inches (Little, 1971).

The influence of moisture on the local distribution of plant communities is well illustrated by the grassland vegetation of the central plains. In a study of the prairie, meadow, and marsh vegetation of Nelson County, North Dakota, Dix and Smeins (1967) divided the soils into 10 drainage classes, ranging from excessively drained to permanent standing water. They determined the indicator species for each drainage class and then divided the vegetational display into six units, corresponding to the drainage pattern (see Figure 6.3). The uplands fell into high prairie, midprairie, and low prairie and the lowlands into meadow, marsh, and cultivated depression. High prairies dominated the excessively drained areas and were characterized by stands of needle-and-thread grass (*Stipa comata*), western wheatgrass (*Agropyron*), and prairie sandweed (*Calamovilfa longifolia*). The midprairie—

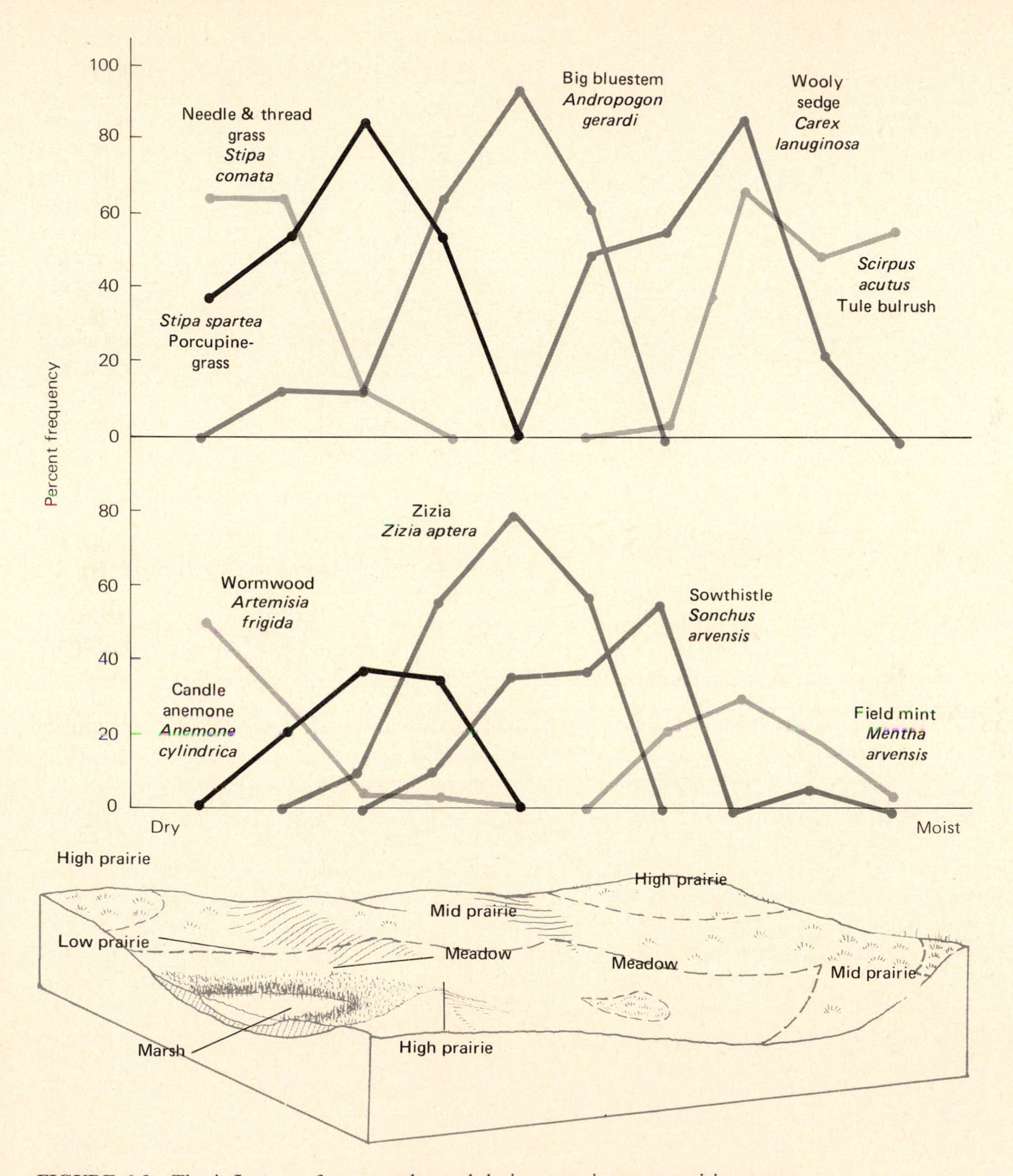

FIGURE 6.3 The influence of topography and drainage regimes on prairie vegetation. The hypothetical block diagram of North Dakota landscape shows the relative positions of vegetational units. The distributional curves of selected species are based on a drainage gradient. The high prairie is the vegetational unit of excessive drainage. (From Dix and Smeins, 1967:34.)

considered to be the climax, or true prairie—was dominated by big bluestem (*Andropagon gerardi*) and little bluestem (*A. scoparius*), porcupine grass, and prairie dropseed (*Sporobolis* spp.). Low prairie on soils of moderate moisture was characterized by big bluestem, little bluestem, yellow Indian grass (*Sorghastrum nutans*), and muhly (*Muhlenbergia* spp.). Lowlands that occupied soils in which the drainage was sluggish and the water table was within the rooting depth of most plants were characterized by canary grass (*Phalaris* spp.), sedge (*Carex*), and *Scolochloa festucacea*. Meadows on even wetter soils were dominated by northern reedgrass (*Calamagrostris inexpansa*), wooly sedge (*Carex lanuginosa*), and spikerush (*Eleocharis* spp.). Marshes covered by permanently standing water contained stands of reed (*Phagramites*), cattails (*Typha* spp.), and tule bulrush (*Scirpus acutus*). Cultivated depressions were usually colonized by spikerush and water plantain (*Alisma* spp.).

Although each drainage class supported a characteristic stand of vegetation, no one species was associated solely with another species. Each species had its own set of optimal drainage requirements and behaved independently of other species. At any particular site, the drainage conditions influenced the combination of dominant plants (see Chapter 8). Because of this, each community blended into the others. The only sharp breaks came where the changes in drainage situations were sharp and severe.

In a similar manner soil moisture influences the distribution of woody plants in the northern Rocky Mountains. In northern Idaho warm, dry lowland soils, droughty in summer, support Idaho fesque (*Festuca idahoensis*) and snowberry (*Symphoricarpos* spp.). As soil moisture improves upslope, ponderosa pine (*Pinus ponderosa*) and Douglas-fir (*Pseudotsuga menziesii*) replace the lowland plants; on midslope, grand fir appears and, still higher, western red cedar, western hemlock (*Tsuga heterophylla*), and subalpine fir (*Abies lasiocarpa*). On the upper parts of south-facing slopes, where the soil may become as droughty as that of the lowlands, wheat grass and Idaho fesque replace woody plants (Daubenmire, 1968a).

Interaction of Temperature and Moisture

A close interaction exists between temperature and moisture in terrestrial environments; and the two determine in large measure the climate of a region and the distribution of vegetation. Low-moisture conditions are more extreme when temperatures are high or low. Moisture, in turn, influences the effects of temperature, a fact observable to everyone. Cold is more penetrating when the air is moist, and high temperatures are more noticeable when relative humidity is high.

Mean monthly temperatures and relative humidities or precipitation can be plotted on a graph to form a climograph, which gives a composite picture of the climate of an area (see Figure 6.4). The mean monthly temperature and the mean monthly rainfall or relative humidity are plotted on the vertical and horizontal axes, respectively, as a single dot. Twelve dots for the year connected together form an irregular polygon, which can be compared with another for similarity or difference in fit. In this way climates can be compared much more easily than by tables. Such climographs are useful to contrast or compare one region or one year with another. Often this is done to determine the suitability of an area for the introduction of exotic animals, particularly game birds.

East-west zonation of vegetation follows a pattern of moisture distribution more than one of temperature. If temperature alone controlled plant distribution in North America, the vegetation zones would be in broad belts running east and west. Only in the far north do the vegetation zones (tundra and the coniferous forest) stretch in these directions. Below, vegetation is controlled by precipitation and evaporation, the latter influenced considerably by temperature. Since available moisture becomes less from east to west, vegetation follows a similar pattern, with belts running north and south. Humid regions along both coasts support natural forest vegetation. This zone is broadest in the east. West of this eastern forest region is a subhumid zone, where precipitation is lower and evaporation

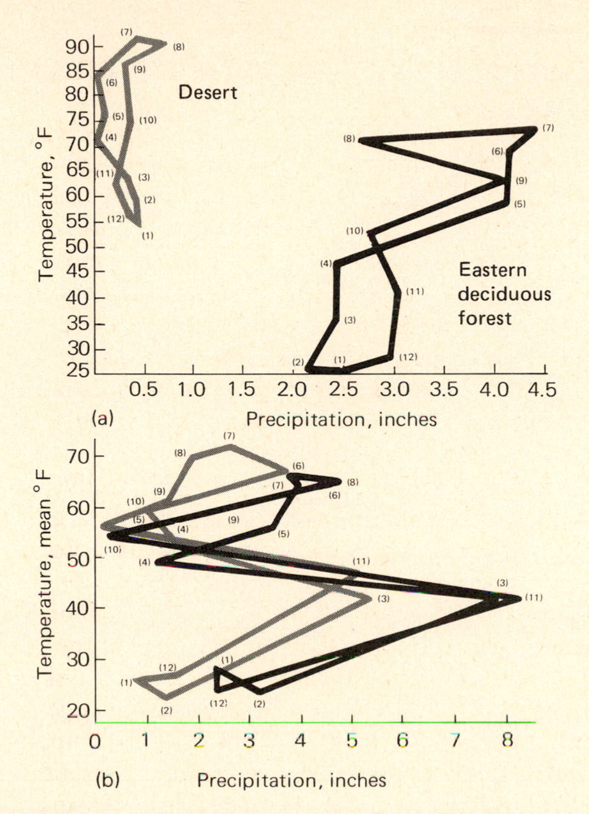

FIGURE 6.4 Temperature-moisture climograph. (a) A climograph comparing two very different regions, the desert and the eastern deciduous forest of North America. Note how the hot, dry climate of the West differs graphically from the cool, temperate, moist climate of the East. These 12-sided polygons give a picture of moisture conditions and permit comparison of one set of moisture conditions with another. (Data: mean temperature and precipitation 1941–1950 for Yuma, Arizona, and Albany, New York.) (b) A climograph comparing conditions on the rain-shadow side and the high-rainfall side in the Appalachian Mountains in West Virginia.

higher. Here the ratio of precipitation to evaporation is about 60 to 80 percent, and the land supports a tall-grass prairie. Beyond this is semiarid country, where the precipitation-evaporation ratio is 20 to 40 percent; it supports a short-grass prairie. To the west of this and on the lee of the mountains is the desert.

In mountainous country, both east and west, vegetation zones reflect climatic changes on an altitudinal gradient (see Figure 6.5). These belts often duplicate the pattern of latitudinal vegetation distribution. In general, the belts include the land about the mountain base, which has a climate characteristic of the region. Next is a higher montane level, which has greater humidity and temperatures that decrease as altitude increases. Here the forest vegetation changes from deciduous to coniferous. Beyond this is a subalpine zone, which includes coniferous trees adapted to a more rigorous climate than the montane species. Above this is the alpine or tundra zone, where the climate is cold and cloudy. Here trees are replaced by grasses, sedges, and small, tufted plants. Between the alpine and the subalpine lies the krummholz, a land of stunted trees. On the very top of the highest mountains is a land of perpetual ice and snow.

Summary

The physical properties of water make it an ideal medium for life. It is most dense at 4°C, a feature of considerable importance to aquatic ecosystems. The result is a warm upper layer of water in lakes and ponds in summer and floating ice in winter. It has a high specific heat and a high latent heat, which moderate the climate near bodies of water and aid in the heat balance of organisms. Its high viscosity and surface tension are important in the capillary flow of water, especially in the soil and the conducting vessels of plants.

Most of Earth's water is in the oceans. Less than 1 percent is available as free fresh water. Water moves through the water cycle, involving precipitation, interception, infiltration, surface flow, and evaporation. The key to the water cycle is the atmosphere. Evaporation into and precipitation from the atmosphere, detention in oceans and land, and atmospheric transport maintain the global water balance. Oceans lose more water by evaporation and gain less from precipitation than land areas. The loss is made up by runoff from land.

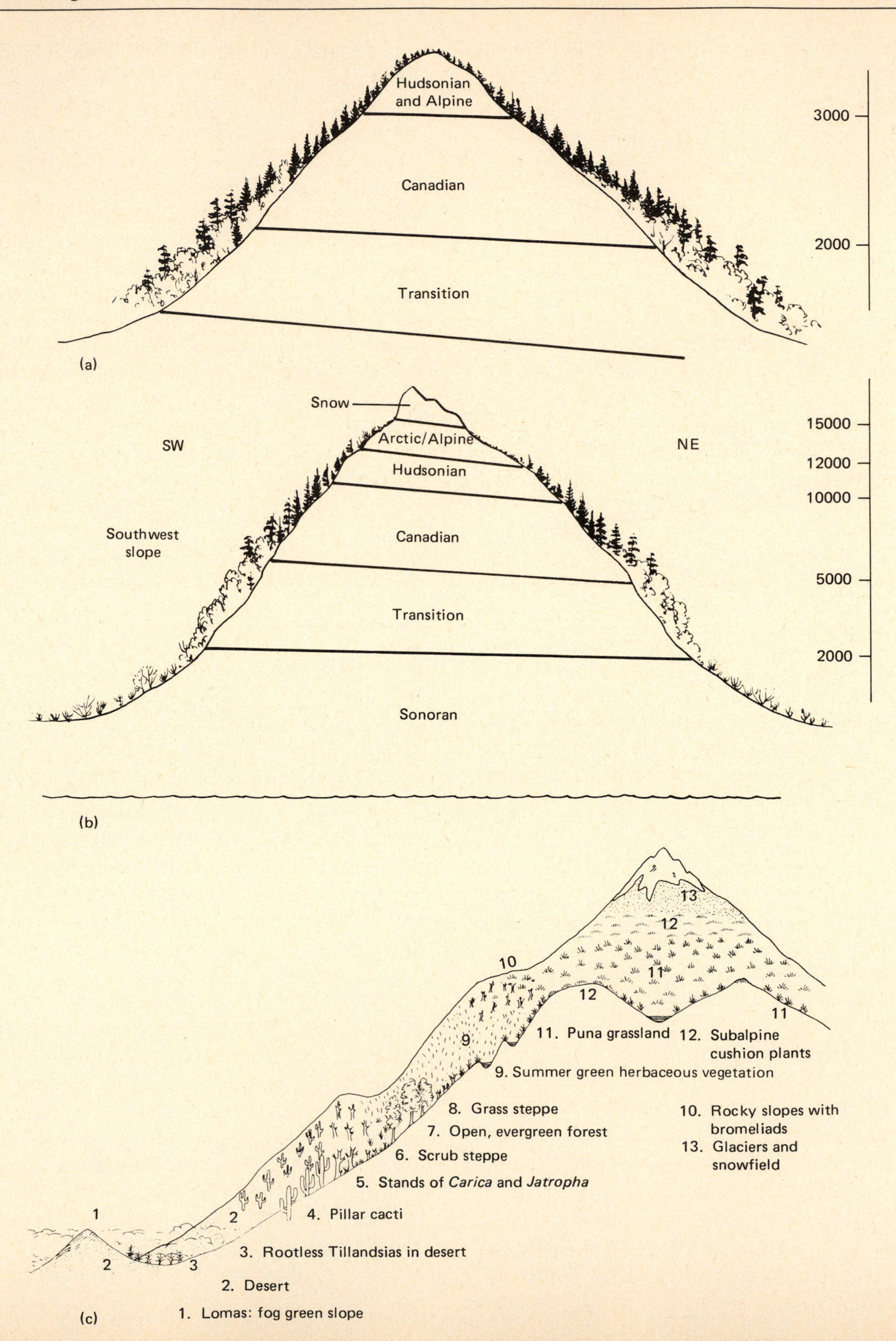
Hudsonian
and Alpine
Canadian
Transition
3000
2000
(a)
Snow
Arctic/Alpine
Hudsonian
Canadian
Transition
Sonoran
SW
NE
Southwest
slope
15000
12000
10000
5000
2000
(b)
1. Lomas: fog green slope
2. Desert
3. Rootless Tillandsias in desert
4. Pillar cacti
5. Stands of Carica and Jatropha
6. Scrub steppe
7. Open, evergreen forest
8. Grass steppe
9. Summer green herbaceous vegetation
10. Rocky slopes with bromeliads
11. Puna grassland
12. Subalpine cushion plants
13. Glaciers and snowfield
(c)

Atmospheric moisture is expressed in terms of relative humidity. Relative humidity is higher by night than by day and is greater at high elevations than at low, but it varies widely with topography, vegetation, wind, and temperature.

All life responds to a moisture regime. Water stress is a major selective force in the evolution of plants and animals. Some plants are poikilohydric, with a moisture content matching that of the environment. Others are homoiohydric and maintain a stable water balance independent of environmental fluctuations. They accomplish that by a combination of osmotic pressure and turgor pressure, collectively known as water potential, which decreases from the soil to the leaf. Because of water potential, water is able to move from the soil through the plant to the atmosphere.

To maintain a water balance during long periods of dry weather, plants depend upon drought resistance—a combination of drought avoidance and drought tolerance. Plants avoid drought by being ephemeral, such as the desert annuals that germinate and bloom only when rains are sufficient. Or plants may reduce transpiration water loss by closing their stomata, increasing leaf thickness, possessing a waxy cuticle, storing water in cells (succulence), or shedding leaves during the dry season. Other plants extend roots into areas of soil where the water has not been tapped. Some species, called phreatophytes, of riverine habitats, have deep roots that reach a permanent water supply.

Animals likewise need to maintain a water balance. It is usually accomplished by some sort of excretory system, ranging from contractile vacuoles in protozoans to kidneys in birds and mammals. Some animals avoid drought by encysting; others go into diapause or aestivation. Desert animals reduce water loss by becoming nocturnal, thus avoiding the heat of day; by producing highly concentrated urine; by using only metabolic water; and by tolerating a certain degree of dehydration.

Organisms inhabiting saline environments live in a physiologically arid region. Their problem is to maintain osmotic balance with the environment. They accomplish this by possessing some means of getting rid of excess salt. Plants may accumulate and excrete salt through their leaves or store water in them. Many animals have salt-excreting glands.

Moisture or the lack of it has a major influence on the distribution of plants, both geographically and locally. Some are restricted to moist (or mesic) sites, others to dry (or xeric) sites. Temperature and moisture also interact to further influence the nature of plant distribution and plant communities.

Review and Study Questions

1. Why is water most dense at 4°C? What is the ecological significance of this fact?
2. What is the ecological importance of the fol-

FIGURE 6.5 Altitudinal zonation in mountains. (a) Mount Marcy in the Adirondack mountains of New York state. The Transition forest on the lower slope is northern hardwoods. The mid-slope is Canadian with paper birch, red spruce, and balsam fir. The upper slopes, Hudsonian and Alpine, support dwarf spruce, willow, and heaths. (b) A composite Rocky Mountain situation in which the lower slope supports oak, which gives way to chaparral and junipers (Sonoran), and to oak and lodgepole pine on the mid-slope (Transition). Higher up, these trees are replaced by ponderosa pine and Douglas-fir, then spruce and fir (Hudsonian), and finally tundra (Alpine). (C) Zonation of vegetation in the Andes (latitude of Lima, Peru) is more complex, ranging from desert and lomas on the lower slopes through zones of desert, shrubs, grasslands, and forests to snowfields. (After Walter, 1971.)

lowing characteristics of water: (1) specific heat, (2) latent heat, (3) viscosity, (4) surface tension?

3. Why is fresh water limited?
4. Describe the water cycle.
5. Define precipitation, infiltration, capillary water, field capacity.
6. Compare precipitation and evaporation over oceans with that over land.
7. How does the global water cycle affect Earth's heat budget?
8. What is relative humidity? Saturation pressure deficit?
9. Why is relative humidity higher at night than by day? At higher elevations than low?
10. How do wind and temperature affect evaporation and relative humidity?
11. Distinguish between a poikilohydric plant and a homoiohydric plant. Give some examples.
12. What is transpiration? Water potential?
13. Distinguish between drought resistance and drought tolerance.
14. In what ways do plants cope with drought or water stress?
15. What are the advantages and disadvantages for a plant in reducing transpiration by closing the stomata?
16. What is succulence? How does that condition enable a plant to withstand aridity?
17. What are phreatophytes?
18. In what ways do animals maintain a moisture balance?
19. How do animals of arid regions avoid dehydration?
20. What are some problems of living in a saline environment? How do plants and animals of a salt marsh get around the problems?
21. In what way does moisture influence local plant distribution?
22. What problems are created for plants by flooding?
23. How do temperature and moisture interact to influence plant distribution?

Suggested Readings

Daubenmire, R. F. 1974. *Plants and environment*. 3d ed. New York: Wiley.

Kramer, P. J. 1969. *Plant and soil water relationships*. New York: McGraw-Hill.

Lange, O. L., L. Kappen, and E. D. Schulze, eds. 1976. *Water and plant life: Problems and modern approaches*. Ecological Studies no. 19. New York: Springer-Verlag.

Lee, R. 1980. *Forest hydrology*. New York: Columbia University Press.

Levitt, J. 1972. *Responses of plants to environmental stress*. New York: Academic Press.

ECOLOGICAL EXCURSION

Water
General Considerations

Lawrence J. Henderson

It was assuredly not chance that led Thales[1] to found philosophy and science with the assertion that water is the origin of all things. Whether his belief was most influenced by the wetness of animal tissues and fluids, or by early poetic cosmogonies, or by the ever present importance of the sea to the Ionians, however vague his conception of water may, indeed must, have been, he at least expressed a conclusion which proceeded from experience and serious reflection. Later, when positive knowledge had already grown to be a substantial basis for speculation, both meteorological and chemical views contributed to the decision of Empedocles and Aristotle to include water among the elements. And it is especially worthy of note that of earth, air, fire, and water the last is the only one which happens to be an individual chemical compound. From that day to this the unique position of water has never been shaken. It remains the most familiar and the most important of all things.

Within a comparatively recent time, to be sure, it has definitely lost its claim to be a true element, in the modern sense, but meanwhile almost every great development of science has but contributed to make its importance more clear. In physics, in chemistry, in geology, in meteorology, and in biology nothing else threatens its preëminence. The physicist has perforce chosen it to define his standards of density, of heat capacity, etc., and as a means to obtained fixed points in thermometry. The chemist has often been almost exclusively concerned with reactions which take place in aqueous solution, and the unique chemical properties of water are of fundamental significance in most of the departments of his science. In geology neptunism has at length won a certain though incomplete triumph over plutonism, and the action of water now appears to be far the most momentous factor in geological evolution. The meteorologist perceives that the incomparable mobility of water, which depends upon its peculiar physical properties and upon its existence in vast quantities in all three states of solid, liquid, and gas, is the chief factor among the properties of matter to determine the nature of the phenomena which he studies; and the physiologist has found that water is invariably the principal constituent of active living organisms. Water is ingested in greater amounts than all other substances combined, and it is no less the chief excretion. It is the vehicle of the principal foods and excretory products, for most of these are dissolved as they enter or leave the body. Indeed, as clearer ideas of the physico-chemical organization of protoplasm have developed it has become evident that the organism itself is essentially an aqueous solution in which are spread out colloidal substances of vast complexity. As a result of these conditions there is hardly a physiological process in which water is not of fundamental importance.

All of these circumstances, which completely justify the interest in water which Thales and Aristotle, and nearly all later students of nature have manifested, depend in great part upon the quantity of water which is present outside the earth's crust, and upon its often unique physical and chemical properties. Such properties are our present concern. Doubtless if it were not for the enormous quantity of water which exists upon our planet, all its physical properties would be of little avail to bring about its universal importance in nature. This, however, as has been above explained, appears to be neither an accidental nor an uncommon phenomenon.

Of the total extent of the earth's surface the oceans make up about three fourths, and they con-

[1] Thales (640–550 B.C.) was one of the seven wise men of ancient Greece. He tried to explain matter in terms of earth, fire, water, and air. He was also an early mathematician and astronomer who successfully predicted an eclipse of the sun. (R. L. S.)

SOURCE: Lawrence J. Henderson, *The Fitness of the Environment* (New York: Macmillan, 1913), 72–79.

tain an amount of water sufficient, if the earth were a perfect sphere, to cover the whole area to a depth of between two and three miles. This corresponds to about 0.2 percent of the volume of the globe. The occurrence of water is, moreover, not less important and hardly less general upon the land. In addition to lakes and streams, water is almost everywhere present in large quantities in the soil, retained there mainly by capillary action, and often at greater depths. The atmosphere also contains an abundance of water as aqueous vapor and as clouds. Now the very occurrence of water upon the earth, and especially its permanent presence, is due in no small degree to its chemical stability in the existing physical and chemical conditions. This stability is of great moment in the various inorganic and organic processes in which water plays so large a part. In the first place the chemical reactions in which it is concerned during the process of geological evolution, though they are no doubt in the total of great magnitude, are both slow and far from violent. Long since any very active changes of this sort, so far as the superficial part of the crust is concerned, have run their course. In the second place water is really, at the temperature of the earth and in comparison with most other chemical substances, an extremely inert body, for the union of hydrogen with oxygen is so firm that it is not readily dissolved.

Thus water exists as a singularly inert constituent of the atmosphere, as a liquid nearly inactive in chemical processes on the surface and in the soil, and everywhere as a mild solvent which does not easily attack the substances which in great variety dissolve in it. The chemical changes which do follow upon solution are not such as to produce substantial chemical transformations, and most substances can pass through water unscathed. The nature of water, then, is a great factor in the chemical stability, which, no less than the physical stability of the environment, is essential to the living mechanism.

CHAPTER 7

Life in the Light Environment

Outline

Objectives

Upon completion of this chapter, you should be able to:

1. Describe the nature of light as it reaches Earth.
2. Explain the fate of visible light in the plant canopy and in the aquatic environment.
3. Contrast shade-tolerant and shade-intolerant plants.
4. Discuss the role of light in the daily and seasonal cycles of plants and animals.
5. Discuss circadian rhythms and their relation to the biological clock.
6. Explain several basic models of the biological clock.
7. Discuss the relationship between biological clocks and critical daylength and their role in the annual rhythms in the lives of animals and plants.
8. Describe circadian rhythms, circannual rhythms, phase shift, free-running phenomena, *Zeitgebers,* entrainment, critical daylength, and short-day and long-day organisms.

Nature of Light

Light is that part of solar radiation (see Chapter 4) in the visible range embracing wavelengths of 0.40 microns to 0.70 microns. This segment of the solar spectrum is known as *photosynthetically active radiation* (PAR), because it is utilized in photosynthesis (discussed in Chapter 18). But light is an important environmental influence in other ways. The ability or inability of plants to grow in the shade of others affects the nature and structure of plant communities. The structure of plants influences the kind and number of animals in a given area (see Chapter 22). Light influences the daily and seasonal activities of animals and the seasonal cycles of plants.

Light that arrives at Earth's atmosphere is not quite the same light that arrives on Earth (Figure 7.1). The high-level ozone layer absorbs nearly all wavelengths, but especially the violets and blues. Molecules of atmospheric gases scatter the shorter wavelengths, giving a bluish color to the sky and causing Earth to shine in space. Water vapor scatters all wavelengths so that an atmosphere with much water vapor is whitish—thus, the grayish appearance of a cloudy day. Dust scatters long wavelengths to produce reds and yellows in the sky. Because of the scattering of solar radiation by dust and water vapor, part of it reaches Earth as diffuse light from the sky known as *skylight*.

Light intercepted by Earth is either reflected, absorbed, or transmitted through objects. Of greatest ecological interest is light reaching vegetation. About 6 to 12 percent of photosynthetically active light striking a leaf is reflected. The degree of reflection varies with the nature of the leaf surface. Because green light is most strongly reflected, leaves appear green. Most of the red light is absorbed by chloroplasts in the mesophyll of the leaf and used in photosynthesis. A remaining fraction of light is transmitted through the leaf. How much depends upon its thickness and structure. A leaf may transmit up to 40 percent of the light it receives, but 10 to 20 percent is more usual. Transmitted light is primarily green and far red. Light

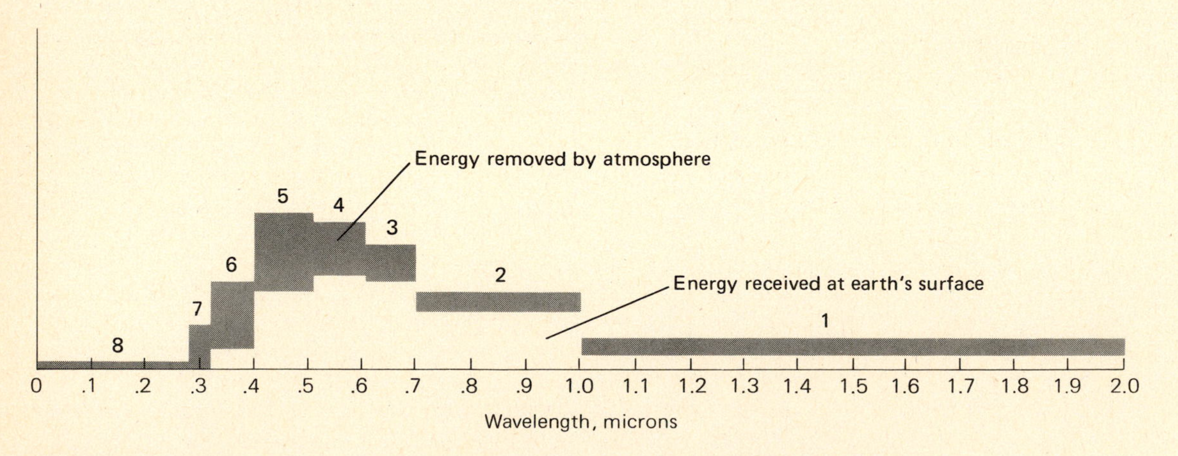

FIGURE 7.1 Energy in the solar spectrum before and after depletion by the atmosphere from a solar altitude of 30°. Figures above the bars indicate: (1) near infrared, with wavelengths over 1 micron; (2) near infrared, 0.7–1.0 microns; (3–5) visible light [(3) red; (4) green, yellow, and orange; (5) violet and blue]; (6–7) ultraviolet. Note the strong reduction in ultraviolet. Nearly all the wavelengths are absorbed by ozone at high levels. The region of peak energy is shifted toward the red end of the spectrum. Visible light in blue wavelengths is scattered rather than absorbed, producing the blue light of the sky. (From Reifsnyder and Lull, 1965:21.)

filtering through a forest is mostly in wavelengths over 0.50 microns. In the depths of a forest, even green light may be extinguished.

Light and Plant Activity

RADIATION IN THE PLANT CANOPY

When you walk into a forest in summer, one of the most obvious environmental changes is a decrease in light. If you were able to view the understory of a grassland, you would observe much the same thing. Most of the sunlight that floods open spaces is intercepted by a leafy canopy. The amount of light that penetrates a stand varies with the nature and position of the leaves. Horizontal leaves arranged in layers intercept more light than leaves arranged in an upright position. As the layers of leaves increase, less light reaches the ground. By contrast, leaves arranged at a 45° angle to the perpendicular allow more light to penetrate the canopy.

Only about 1 to 5 percent of the light striking the canopy of a typical temperate hardwood forest reaches the forest floor. In a tropical rain forest, only 0.25 to 2 percent gets through. More light travels through pine stands—about 10 to 15 percent—but densely crowned Norway spruce allows only 2.5 percent of the light in the open to reach the forest floor. Woodlands comprised of trees with relatively open crowns, such as oaks and birches, allow light to filter through. There light dims out gradually, as it does in grasslands. In grasslands most of the light is intercepted by the middle and lower layers (Figure 7.2).

Foliage density, the key to light interception, is expressed as *leaf area index* (LAI). LAI is the percentage of ground area covered by leaves. It is expressed as:

$$\frac{\text{Total leaf area}}{\text{Area of ground}}$$

An LAI of 3, for example, means that a given area of ground is covered by 3 times that area of leaves arranged in layers.

Individual trees themselves, especially those growing in the open, are affected by the thinning of light through the canopy. Trees with relatively open crowns have leaves distributed within the crown. Light intensity is great enough for interior leaves to survive and carry on photosynthesis. Trees such as cedars and spruce have canopies so dense that

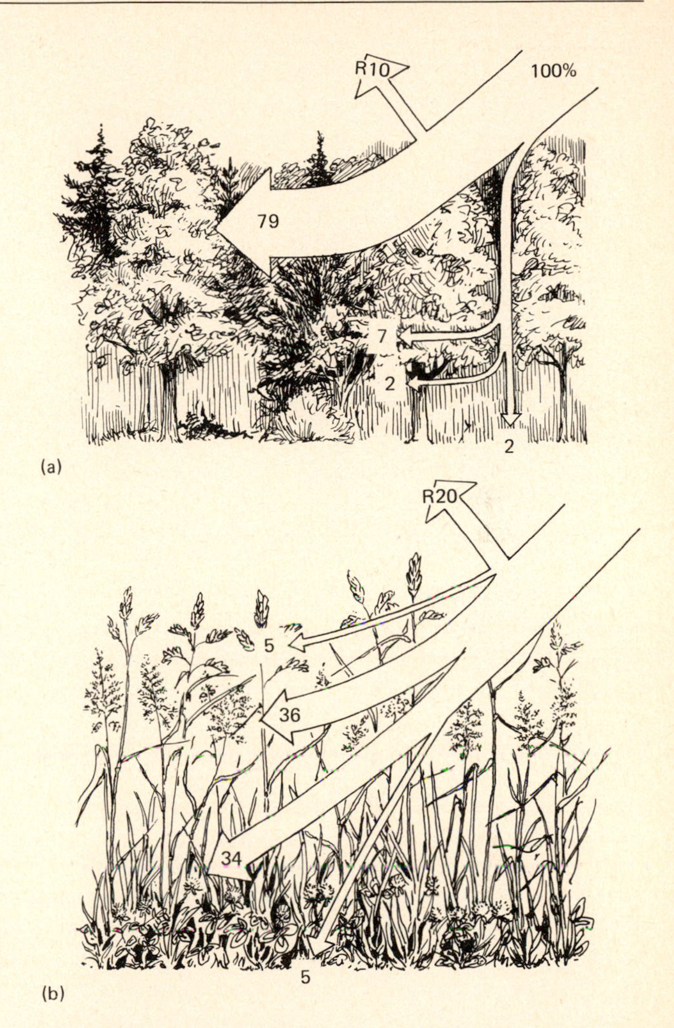

FIGURE 7.2 (a) Attenuation of radiation in a boreal mixed forest. Ten percent of the incident photosynthetically active radiation is reflected from the upper crown, and the greatest absorption occurs within the crown. In the meadow (b) 20 percent of photosynthetically active radiation is reflected from the upper surface, while the greatest absorption occurs in the middle and lower regions, where the leaves are most dense. Only 2 to 5 percent of the incident photosynthetically active radiation reaches the ground. (Adapted from Larcher, 1975:15.)

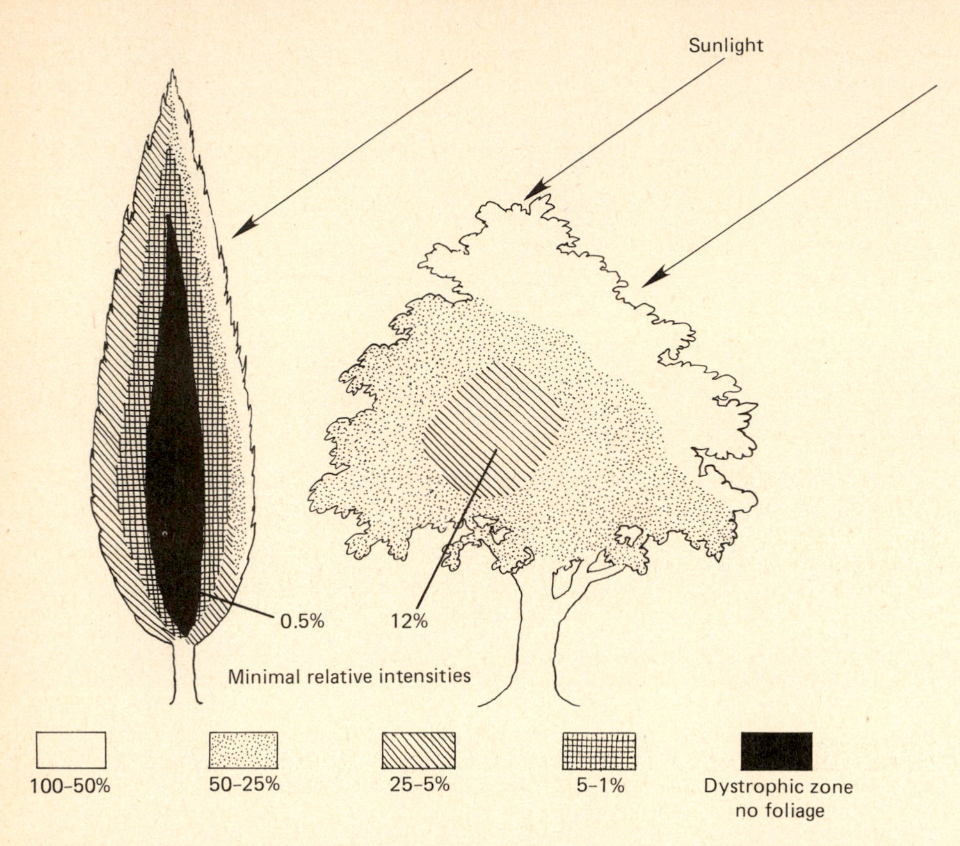

FIGURE 7.3 Reduction of light in a dense crown of cypress and an open crown of olive on a clear summer day. A dense growth of needles in the cypress quickly depletes light to 0.5 percent of illumination in the open. The interior of the tree is too dark to support any foliage. For this reason dense conifers lack interior foliage. The more open crown of the olive, with small reflecting leaves, receives diffuse light strong enough to support leafy branches even within the darkest part of the crown. (From Larcher, 1975:14.)

no leaves can grow in the deepest part of the crown (Figure 7.3).

Although light in the forest dims down through the canopy, some sunlight does penetrate gaps in the crown and reaches the forest floor at open sky intensity as sunflecks.

SHADE TOLERANCE

How a plant responds to shade depends in part on its tolerance to shade. In a general way plants are either shade-tolerant or shade-intolerant (shade avoiders). Shade avoiders are plants of open sites receiving full sunlight. They establish themselves rapidly on disturbed sites; can tolerate extreme site conditions; achieve fast growth in the open; and in the case of trees and shrubs, produce seed at an early age (see Chapter 22) and disperse their seeds widely. They carry on photosynthesis more efficiently in full sunlight and have both a high rate of photosynthesis and a high rate of respiration. They rapidly convert photosynthate into growth.

By contrast, shade-tolerant plants are not competitive with shade-intolerant plants in full sun. They have a lower rate of photosynthesis and, perhaps more important, a lower rate of respiration than shade avoiders. For those reasons they grow more slowly in all environments. Many have the ability to carry on photosynthesis at low light intensities as well as the ability not only to open their stomata in dim light but to open them rapidly. That enables the plants to take advantage of increased sunlight from sunflecks over a short period of time. They may live suppressed in a shaded understory for many years, but when released from overhead shade, they respond immediately with rapid growth.

When forced to grow in the shade, shade avoiders favor stem growth to leaf growth (an effort to reach light), and their leaves are widely spaced, reducing shade on individual leaves. At the same time, these plants experience a proportional reduction in photosynthesis while still undergoing a high rate of respiration.

Shade tolerance or intolerance, however, may involve more than light. Moisture, too, may play a role. Plants of the understory may experience a shortage of soil moisture resulting from competition from canopy vegetation. Intolerant plants might be able to carry on photosynthesis at a rate higher than required to balance respiration (*light compensation point*), but the rate is not enough to allow roots to expand into deeper soil where moisture is available.

The story of a plant's reaction to shade may be told in its leaves. Plants that show little plasticity in leaf shape in either sun or shade seem to function only under one set of environmental conditions. Other plants are more flexible. They change leaf shape and size in response to light conditions. Sun leaves are deeply lobed and relatively thick, with well-developed support and conduction systems. Shade leaves are either weakly lobed or not lobed, have a large surface area per unit of weight, a thinner epidermis, fewer stomata, and less support and conduction tissue. In part, their structure reflects less heat, moisture, and physical stress. Suddenly exposed to the sun, shade leaves lose an excessive amount of moisture and intercept too much light, damaging the chloroplasts. The shock results in the death of leaves. When shaded bark on trunks of trees is exposed to direct sunlight, the light stimulates adventious buds in the bark to develop into *epicormic branches*.

The amount of light that penetrates a terrestrial stand of vegetation varies with the season (Figure 7.4). In early spring in temperate regions, when leaves are just expanding, 20 to 50 percent of the light may reach the forest floor. Spring flowering plants make use of this flood of light by completing the reproductive phase of their life cycle before the canopy closes. When less than 10 percent of the light reaches the forest floor, flowering is over. In fall, when leaves begin to drop, increased light again reaches the forest floor, and another minor surge of flowering, involving certain goldenrods and asters, takes place. In dense coniferous forests, where light is low throughout the year, only ferns and mosses grow in the dim light.

LIGHT IN THE AQUATIC ENVIRONMENT

Aquatic environments also experience a reduction in light. As light, particularly sunlight, strikes the surface of water, as much as 10 percent of it at midday is scattered or reflected. In early morning and late afternoon, when the angle of the sun is low, reflection is greatly increased, and little light penetrates the water. That, in effect, shortens the daylength beneath the surface relative to actual daylength.

Light that penetrates the water drops off exponentially with increasing depth. Some of it is absorbed and scattered by water molecules as well as by dissolved substances and suspended sediments. In turbid waters or waters supporting a heavy growth of phytoplankton, light may penetrate only a few centimeters below the surface. As water depth increases from 0.1 to 100 m, visible light becomes limited more and more to a narrow band of blue light at wavelengths of about 0.50 microns. That is part of the reason why water of deep, clear lakes looks blue. Eventually, blue light is filtered out, and the remaining green light is poorly absorbed by chlorophyll. Depths at which green light occurs are occupied by red algae, which possess the supplementary pigments that enable them to use the energy of green light. In effect—except in the clearest, shallow water—aquatic habitats are essentially shade habitats, and the plants are adapted to grow under reduced light.

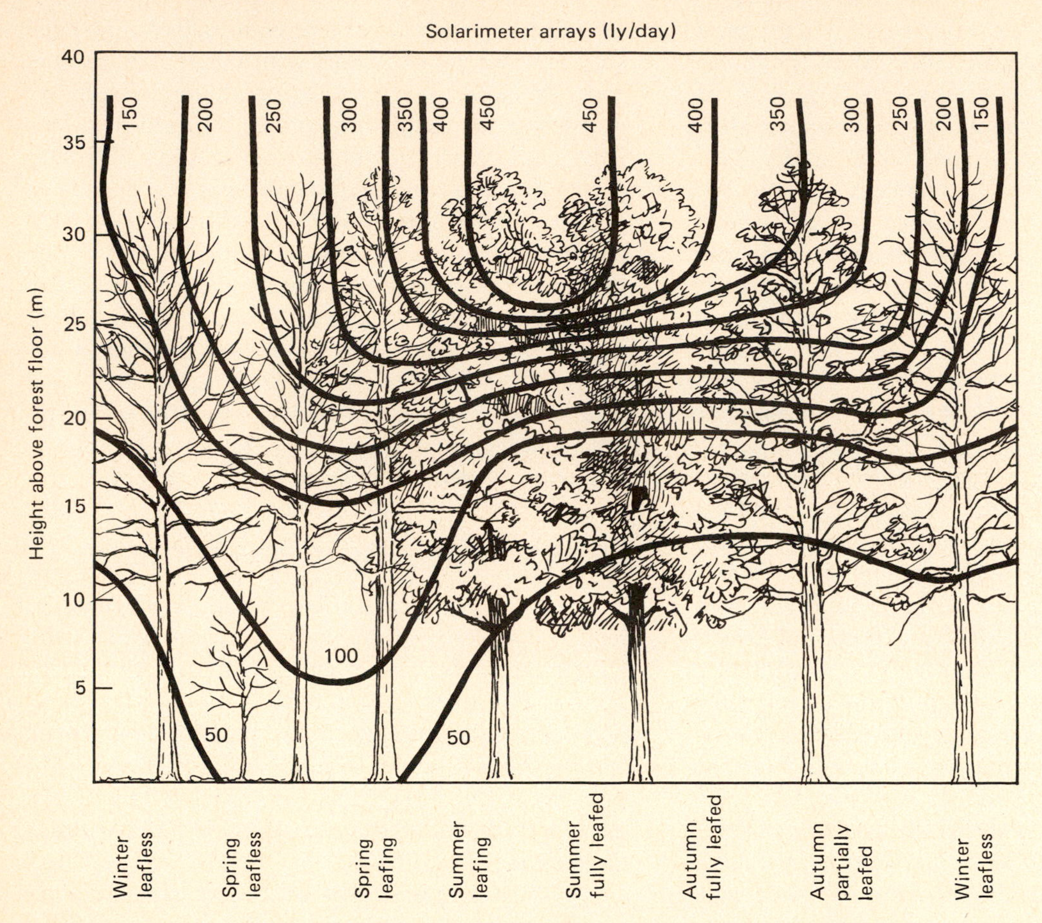

FIGURE 7.4 Synthesized annual course of average daily total solar radiation received within and above a tulip-poplar forest measured as langleys per day. The greatest intensity of solar radiation occurs in midsummer, but the canopy attenuates most of the light so that little more than 10 percent of open light reaches the forest floor. Most illumination reaches the forest floor in spring, when trees are still leafless, the time of spring flowers. The least radiation is received in winter, with lower solar elevations and shorter day lengths. As a result, the amount of solar radiation reaching the forest floor is little more than that of midsummer. (After Hutchinson and Matt, 1977:205.)

Photoperiodism

One aspect of communities with which everyone is familiar is rhythmicity, the recurrence of daily and seasonal changes. Dawn ends the darkness, and bird song signals its arrival. Butterflies, dragonflies, and bees become conspicuous, hawks seek out prey, and chipmunks and tree squirrels become active. At dusk, light fades and daytime animals retire, the blooms of water lilies and other flowers fold, and animals of the night appear. Foxes, raccoons, flying squirrels, owls, and moths take over niches occupied by others during the day. As the seasons progress, daylength changes and, with it, other conspicuous activities. Spring brings migrant birds and initiates the reproductive cycles of many animals and plants. In fall the trees of temperate regions become dormant, insects and herbaceous plants disappear, summer-resident birds return south, and winter visitors arrive. Underlying these rhythmicities is the movement of the Earth relative to the sun. The Earth's rotation on its axis results in the alternation of night and day. The tilt of the Earth's axis along with its annual revolution around the sun produces the seasons.

DAILY PERIODICITY: CIRCADIAN RHYTHMS

Because life evolved under the influences of daily and seasonal environmental changes, it is natural that plants and animals would have some rhythm or pattern to their lives that would synchronize them with fluctuations in the environment. For years biologists have been intrigued by the means by which organisms keep their activities in rhythm with the 24-hour day, including such phenomena as the daily pattern of leaf and petal movements in plants, emergence of insects from pupal cases, and sleep and wakefulness of animals (see Figure 7.5). At one time biologists thought that these rhythmicities were entirely exogenous—that is, that the organisms responded only to external stimuli such as light intensity, humidity, temperature, and tides. Laboratory investigations, however, indicate that this is not the complete answer.

At dusk in the forests of North America, a small squirrel with silky fur and large, black eyes emerges from a tree hole. With a leap the squirrel sails downward in a long, sloping glide, maintaining itself in flight with broad membranes stretched between its outspread legs. Using its tail as rudder and brake, it makes a short, graceful, upward swoop that lands it on the trunk of another tree. This is the flying squirrel, *Glaucomy volans,* perhaps the commonest of all our tree squirrels. But because of its nocturnal habits, this mammal is seldom seen by most people. Unless it is disturbed, the flying squirrel does not come out by day. It emerges into the forest world with the coming of darkness; it returns to its nest with the first light of dawn.

If the flying squirrel is brought indoors and kept under artificial conditions of night and day, the animal will confine its periods of activity to darkness, its periods of inactivity to light. Whether the conditions under which the animal lives are 12 hours of darkness and 12 hours of light or 8 hours of darkness and 16 hours of light, the onset of activity is always shortly after dark. The squirrel's day-to-day activity forms a 24-hour period. This correlation of the onset of activity with the time of sunset suggests that light has a regulatory effect on the activity of the squirrel.

But the *photoperiodism* (response to changing light and darkness) exhibited by the squirrel is not quite so simple. There is more to it than the animal's becoming active because darkness has come. If the squirrel is kept in constant darkness, it still maintains a relatively constant rhythm of activity from day to day (DeCoursey, 1961). But in the absence of any external time cues, the squirrel's activity rhythm deviates from the 24-hour periodicity exhibited under light-and-dark conditions. The daily cycle under constant darkness varies from 22 hours, 58 minutes to 24 hours, 21 minutes, the average being less than 24 hours (most frequent: 23 hours, 50 minutes and 23 hours, 59 minutes) (DeCoursey, 1961). The length of the period maintained under a given set of conditions is an individual characteristic. Because of the deviation of the average cycle length from 24 hours, each individual squirrel gradually drifts out of phase with the day-night changes of the external world (see Figure 7.6). If the same animals are held under continuous light, a very abnormal condition for a nocturnal animal, the activity cycle is lengthened, probably because the animals, attempting to avoid running in the light, delay the beginning of their activity as much as they can.

Circadian Rhythms

The flying squirrel and many other forms of life studied to date, including humans, all possess a rhythm of activity that under field conditions exhibits a periodicity of 24 hours. Moreover, when these organisms are brought into the laboratory and held under constant conditions of light, darkness, and temperature, away from any external time

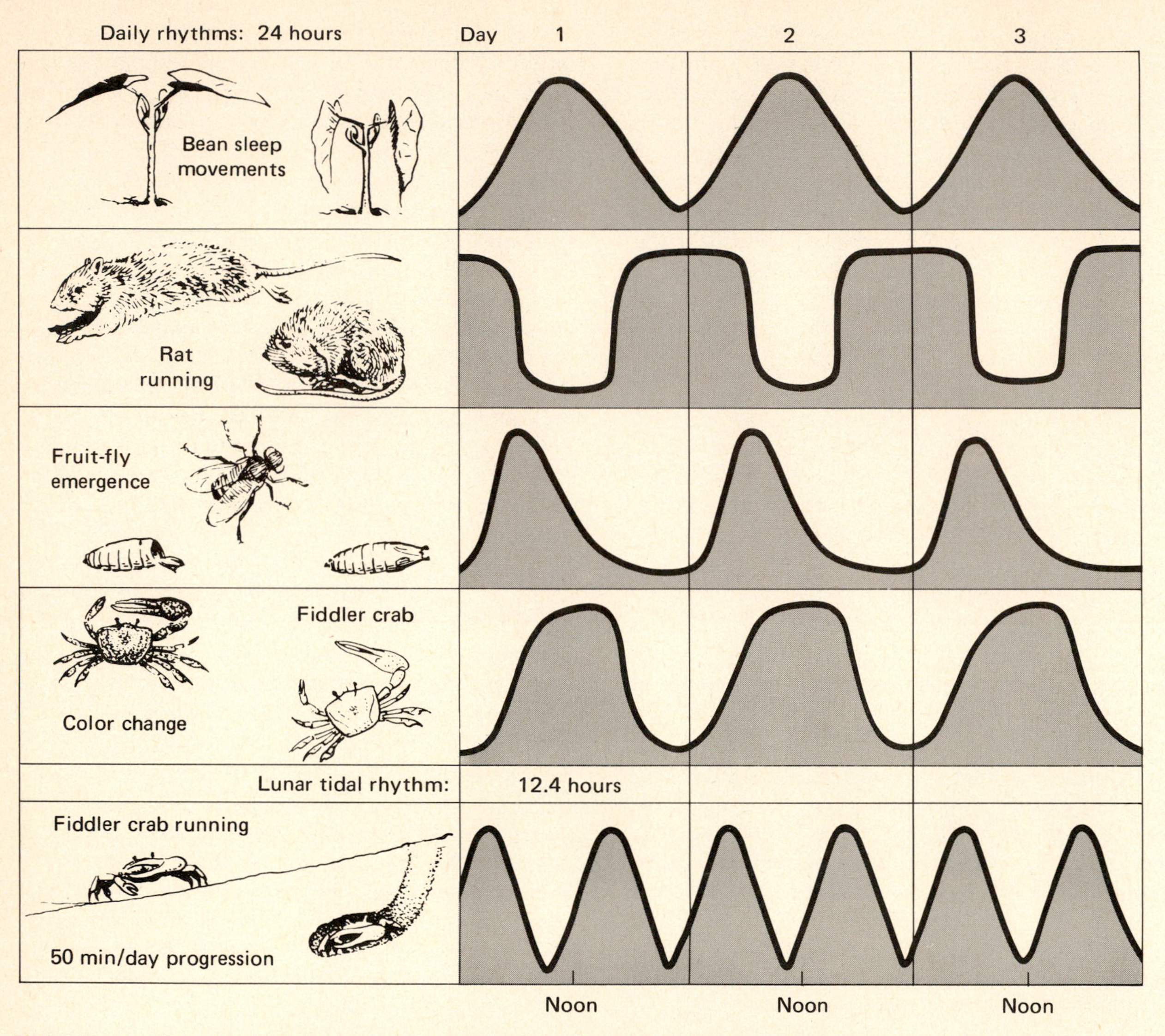

FIGURE 7.5 Examples of rhythmic phenomena experimentally demonstrated to persist under constant conditions in the laboratory, illustrating diagrammatically the natural phase relationships to external physical cycles. (Redrawn by permission from F. A. Brown, *Science* 130 [1959]:1537.)

cues, they still exhibit a rhythm of activity of approximately 24 hours. Because these rhythms approximate but seldom match the periods of the Earth's rotation, they are called *circadian* (from the Latin, *circa,* "about," and *dies,* "day"). The period of the circadian rhythm, the number of hours from the beginning of activity on one day to the beginning of activity on the next, is referred to as *free-running*. In other words, it exhibits a self-sustained oscillation under constant conditions.

Zeitgebers

Thus, many plants and animals are influenced by two periodicities: the internal circadian rhythm of

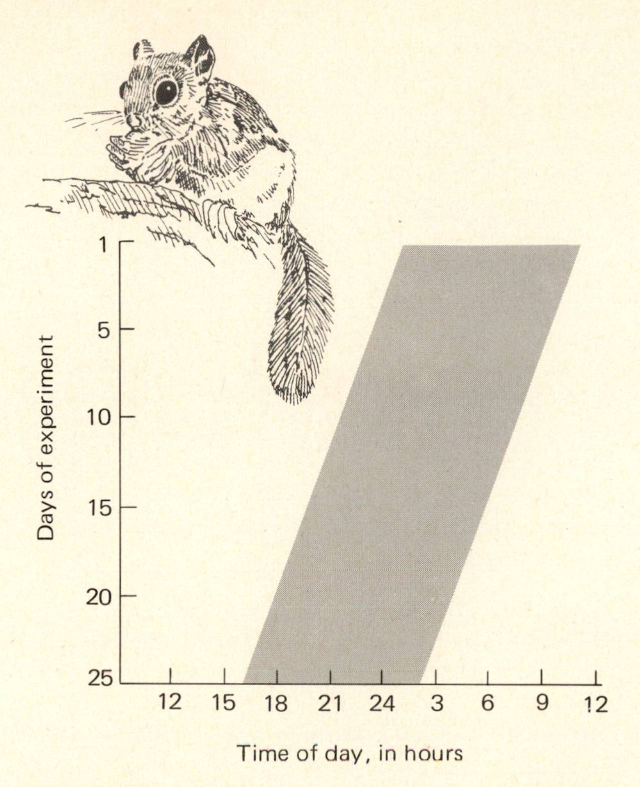

FIGURE 7.6 Drift in phase of the activity rhythm of a flying squirrel held in continuous darkness at 20°C for 25 days. (Adapted by permission from P. J. DeCoursey, *Cold Spring Harbor Symposia on Quantitative Biology* 25 [1960]:51.)

approximately 24 hours and the external environmental rhythm, usually precisely 24 hours. If the activity rhythm of the organism is to be brought into phase, or synchrony, with the external one, then some environmental "timesetter" must adjust the endogenous rhythm to match that of the outside world. The most obvious timekeepers, cues, synchronizers, or *Zeitgebers* (Aschoff, 1958) are temperature and light. Of the two, light is the master *Zeitgeber.* It brings the circadian rhythm of many organisms into phase with the 24-hour photoperiod of the external environment.

Entrainment

The activity rhythm of organisms shows an entrainment to light-dark cycles. The flying squirrel, both in its natural environment and in artificial day-night schedules, synchronizes its daily cycle of activity to a specific phase of the light-dark cycle. This was demonstrated in a series of experiments by DeCoursey (1960a,b, 1961). Flying squirrels were held in constant darkness until their circadian rhythms of activity were no longer in phase with the natural environment. Then they were subjected to a light-dark cycle that was out of phase with their free-running period. If the light period fell in the animals' subjective night, it caused a delay in the subsequent onset of activity. Synchronization took place in a series of stepwise delays until the animals' rhythms were stabilized with the light-dark change (see Figure 7.7). If the light period fell at the subjective dawn or at the end of the dark period (when the animals' activity period was about to end), it caused an advance of activity toward the dusk period. And if the light fell in the animals' inactive day phase, it had no effect.

The flying squirrels do not need to be exposed to a whole light-dark cycle to bring about a shift in the phase of the activity rhythm. A single 10-minute light period is sufficient to cause a *phase shift* in the locomotory activity, provided that it is given during the squirrel's light-sensitive period (DeCoursey, 1960a,b).

THE BIOLOGICAL CLOCK

The idea that plants and animals have an internal timekeeping mechanism, or a biological clock, has been suspected for a long time. The relative preciseness of the timekeeper has been demonstrated for years. A major question, however, has been the location of the clock in living organisms.

Location

Basically, the biological clock is cellular. In one-celled organisms and plants, the clock seems to be located in individual cells. But in multicellular animals the clock is associated with the brain.

Skillful surgical procedures have allowed circadian physiologists to discover the location of the clock in some insects, birds, and mammals. In most insects studied, the clock, including the photoreceptors, is located either in the optic lobes or in tissue between the optic lobes and the brain. In the cockroach and cricket, however, the photorecep-

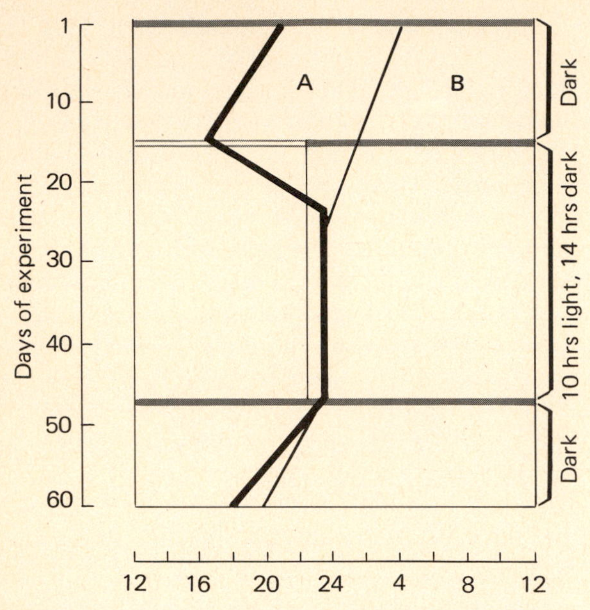

FIGURE 7.7 Diagramatic representation of the synchronization of flying squirrels with circadian rhythm of less than 24 hours in constant darkness to a cycle of 10 hours light and 14 hours darkness. In squirrel A the rephrasing light fell during its subjective night; synchronization was accomplished by a stepwise delay. The onset of activity was stabilized shortly after light-dark change. In squirrel B light fell in the subjective day, and the free-running period continued unchanged until the onset drifted up against the "dusk" light change. This prevented it from drifting forward by a delaying action of light. When it was returned to constant darkness, the onset of activity continued the forward drift. (Redrawn from P. J. DeCoursey, *Cold Spring Harbor Symposia on Quantitative Biology* 25 [1960]:52.)

tors for the entrainment of circadian rhythm of locomotion and stridulation are located in the compound eye, but the controlling clock is in the brain (Saunders, 1982; Beck, 1980). In birds the clock evidently is located in the pineal gland, deep in the lower central part of the brain (Farner and Lewis, 1974; Gwinner, 1978). In mammals it appears to be located in a number of specialized cells (suprachiasmatic nuclei) just above the optic chiasm, the place where the optic nerves from the eyes intersect.

Mechanism

To function as a timekeeper, the biological clock has to fulfill certain conditions. It has to be an internal mechanism with a natural rhythm of about 24 hours. If it weren't, it could not be set to the natural rhythm of 24 hours. Its rhythm must have the capability of being changed, or reset. The biological clock is reset by reoccurring environmental signals, such as changes in the time of dawn and dusk. The clock has to be able to run continuously in the absence of any environmental timesetter. And it has to be able to run the same at all temperatures (that is, it has to be temperature-compensated). Cold temperatures can't slow it down, nor can warm temperatures speed it up. If that happened, the clock could not keep in phase with environmental time.

Models

Two basic models of biological clocks have been proposed. One is an oscillating circadian rhythm sensitive to light (Bunning, 1960). The cycle or time-measuring process begins with the onset of light or dawn. The first half (12 hours) is light-requiring (*photophil*), and the second half is dark-requiring (*scotophil*). Short-day effects are produced when light does not extend into the dark period. Long-day effects are produced when it does (Figure 7.8). Because this simple model does not explain all photoperiodic responses, a variation is a two-oscillator model in which one oscillation is regulated by dawn and the other by dusk.

A second model is the hourglass (Saunders, 1982). It does not involve circadian components, nor is it free-running. It is started by a light-on, light-off stimulus. The timing process begins with darkness, and it shuts off in prolonged darkness or at the beginning of a light period. A light period then prepares it for resetting by the darkness.

A third model combines the hourglass and the single-oscillator model (Pittendrigh, 1966). The light-sensitive phase of the oscillator is more a resetting mechanism. The oscillation is reset to a particular phase at the end of the prolonged light period, producing something of an hourglass effect (Figure 7.9). There are other, more complex varia-

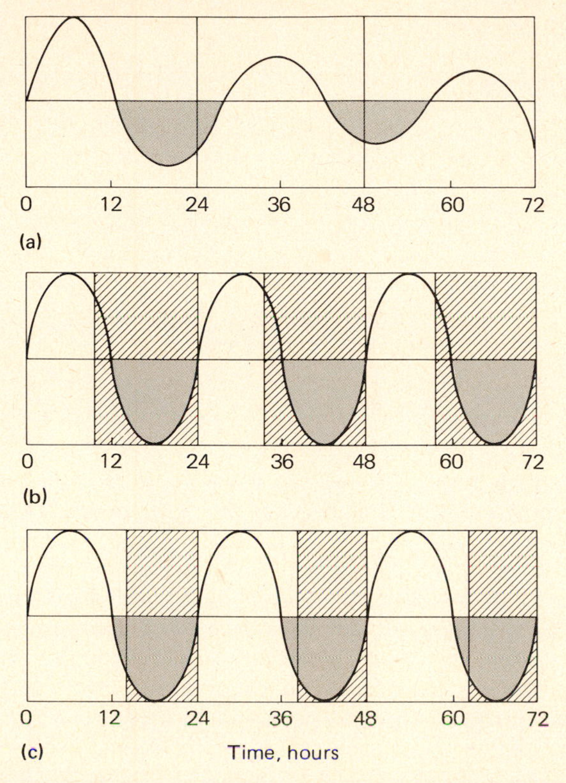

FIGURE 7.8 The Bunning model. Oscillations of the clock cause an alternation of half-cycles with quantitatively different sensitivities to light (white versus black). The free-running clock in continuous light or continuous darkness tends to drift out of phase with the 24-hour photoperiod. Short-day conditions allow the dark to fall within the white half-cycle; in the long day, light falls in the black half-cycle. (From Bunning, 1960:253.)

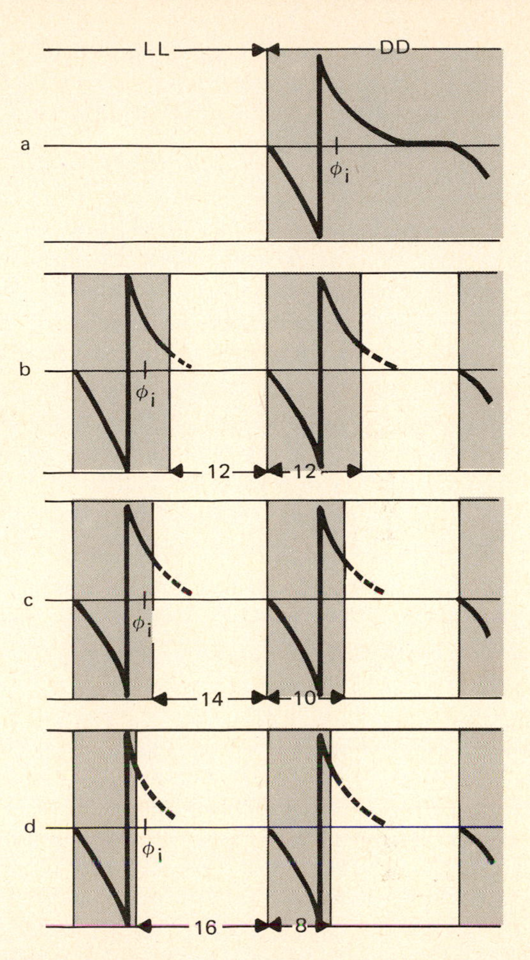

FIGURE 7.9 Hourglass oscillation in pupal eclosion (emergence) in the fruit fly *Drosophila pseudoobscura*. It involves a circadian oscillator or pacemaker as part of the total circadian system of the fly. (a) Constant light (LL) supresses the oscillation; on transfer to constant darkness (DD), it resumes its motion, starting from the circadian time scale (Ct) of 12 hours. (b, c, d) Photoperiods of 12 hours or more also dampen the oscillation, which starts up again from Ct 12 with the onset of darkness each night. As the photoperiod increases, light at dawn falls further back into late subjective night. The photoinducible phase at the end of critical night is not illuminated by photoperiods of 12 and 14 hours, but it is illuminated by a 16-hour photoperiod. (After Pittendrigh, 1966:297.)

tions on these models described in detail by Beck (1980) and D. S. Saunders (1982).

A fundamental question about biological clocks is the nature of the system. Is there one master clock driving all other rhythms? Or is a population of clocks involved? A general consensus is that among higher, multicellular organisms a number of oscillators or pacemakers are involved, each associated with different rhythms influencing various physiological and behavioral phenomena. These pacemakers may be integrated groups of cells within organs such as the central nervous system, where they have specific timekeeping functions. In some cases these oscillators may be coupled to a master or to other pacemakers; or they may be individually set or entrained.

Function

The biological clock provides plants and animals with a sense of time, keeping them in phase with their changing daily environment. But the biological clock does more. It provides some animals, such as the honeybee, with a time memory. It enables the insect to return to a known source of nectar and pollen when it is most available (Renner, 1960). Because its memory involves a circadian rhythm, the bee can remain in a hive for a couple of days during bad weather and still remember the time of day when a particular group of plants is producing nectar. The biological clock also serves as a time-compensation mechanism. A number of animals orient themselves to the sun and use that as a compass direction. These animals are able to maintain direction because their orientation mechanism involves a circadian clock that compensates for the sun's apparent movement during the day. Such time compensation has been demonstrated in bees (von Frisch, 1954; Renner, 1960), starlings (Kramer, 1950; Hoffman, 1965), fish (Hasler, 1960), turtles (Carr, 1962), and others.

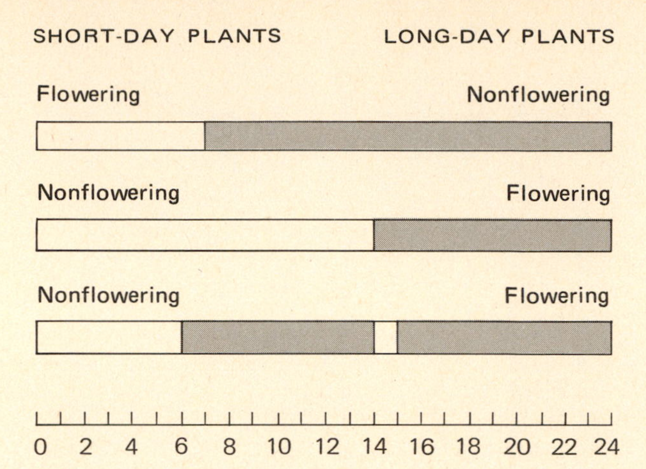

FIGURE 7.10 The time of flowering in long-day and short-day plants is influenced by photoperiod. When plants are held under short-day and long-night conditions, short-day plants are stimulated to flower and long-day plants are inhibited. When daylength is increased, flowering is inhibited in the short-day plants and stimulated in the long-day plants. If the dark period of the short-day and the long-day plant is interrupted, each reacts as if it had been exposed to a long day: the long-day plant flowers; the short-day plant does not. In reality, short-day and long-day plants respond not to the length of light but to the length of darkness. The two might more accurately be called long-night and short-night plants.

ANNUAL PERIODICITY

The biological clock acts as a time-measuring device, able to track changing daylengths or nightlengths through the year. It signals the beginning or end of seasonal changes in behavioral or physiological activity.

Critical Daylength

The signal is *critical daylength* (Figure 7.10). When a period of light reaches a certain portion of the 24-hour day, it inhibits or promotes a photoperiodic response. Critical daylength varies among organisms, but it usually falls somewhere between 10 and 14 hours. Through the year plants and animals compare that time scale with the actual length of day or night. As soon as the actual daylength or nightlength is greater or lesser than the critical daylength, the organism responds appropriately. Some organisms are *day neutral*—not affected by daylength but rather controlled by some other influence, such as rainfall or temperature. Others may be *short-day* or *long-day* organisms. Short-day plants, for example, are those whose flowering is stimulated by daylengths shorter than the critical daylength. Long-day plants are those whose flowering is stimulated by daylengths longer than a particular value. The latter usually bloom in late spring and summer.

Because the same period of dark and light occurs two times a year as the days lengthen in spring or grow shorter in fall, the organism could get its signals mixed. For some insects that would be impossible because the sensitive period occurs only once in their developmental phase. But in other organisms the situation is different. For them the distinguishing characteristic is the direction from which the critical daylength is approached. In one situation the critical daylength is reached as long days move into short and at another time as short days move into long.

Diapause

In the cotton fields of the southern United States lives the pink cotton bollworm, the larva of a tiny moth. Except for a few hours directly after hatching, the larva spends its life in the flower buds or bolls of cotton. At the fourth larval instar stage, the insect goes into *diapause,* a stage of arrested growth, over winter. The onset of diapause comes in late August; but not until near the autumnal equinox, September 21, when the night becomes equal to or longer than the day, does the number of diapausing larvae sharply increase. In late winter, as the days begin to lengthen, the insect comes out of diapause and continues its growth. The emergence from diapause reaches its maximum right after the spring equinox, when the days are just slightly longer than those that induced diapause.

When the larvae of the pink bollworm were exposed to regimes of light and dark in the laboratory, the insect would go into diapause only when the light phase of the 24-hour day was 13 hours or less (Adkisson, 1964). If the larvae were exposed to a light period of 13.25 hours, the insect was prevented from going into diapause. So precise is the time measurement in the insect that a quarter-hour difference in the light period determines whether or not the insect goes into diapause. Diapause terminated most rapidly under photoperiods of 14 hours, less rapidly at 16 and 12. Thus, to the pink bollworm, the shortening days of late summer and fall forecast the coming of winter and call for diapause; and the lengthening days of late winter and early spring are the signals for the insect to resume development, pupate, emerge as an adult, and reproduce.

Reproductive Cycles

Experimental work with a number of species of birds has shown that the reproductive cycle is under the control of an exogenous seasonal rhythm of changing daylengths and an endogenous physiological response timed by a circadian rhythm. After the breeding season the gonads of birds studied to date regress spontaneously. This is the *refractory* period, a time when light cannot induce gonadal activity, the duration of which is regulated by daylength (see Farner, 1959, 1964a; Wolfson, 1959, 1960). Short days hasten the termination of the refractory period; long days prolong it. After the refractory phase is completed, the *progressive* phase begins in fall and winter. During this period the birds fatten, they migrate, and their reproductive organs increase in size. This process can be speeded up by exposing the birds to a long-day photoperiod. Completion of the progressive period brings the birds into the *reproductive* stage. A similar photoperiodic response exists in the cyprinid fish, the minnows (Harrington, 1959).

Seasonal cycles of photoperiodism influence the breeding cycles of many mammals (see Figure 7.11). The flying squirrel has two peaks of litter production, the first in early spring, usually April, in the northeastern United States, and the second in late summer, usually August. To produce litters in April, the flying squirrel must be in breeding condition in January and February.

Muul (1965) investigated the responses of gonadal development to changing photoperiod under laboratory conditions. He found that in the flying squirrel the descent of the testes into the scrotum (in nonbreeding condition, the testes of the squirrel are held in the body cavity) occurred in January under short-day and long-night conditions. An accelerated increase in daylength hastened the descent. The experimental animals held under a natural sequence of photoperiod came into re-reproductive condition and produced litters at the same time as squirrels in the wild in the absence of temperature cues. After the animal's maximum photoperiod in the summer, the testes regressed and remained in that condition while the photoperiod was decreasing. If the decrease in photoperiod was unseasonably accelerated, so that the animal's minimal photoperiod came two months early, followed by increased photoperiod, the testes descended two months early. Thus, in the flying squirrel the testes of the male descend and ovulation in females takes place when the photoperiod increases from 11 to 15 hours, and ovulation ceases and the testes regress when the photoperiod decreases.

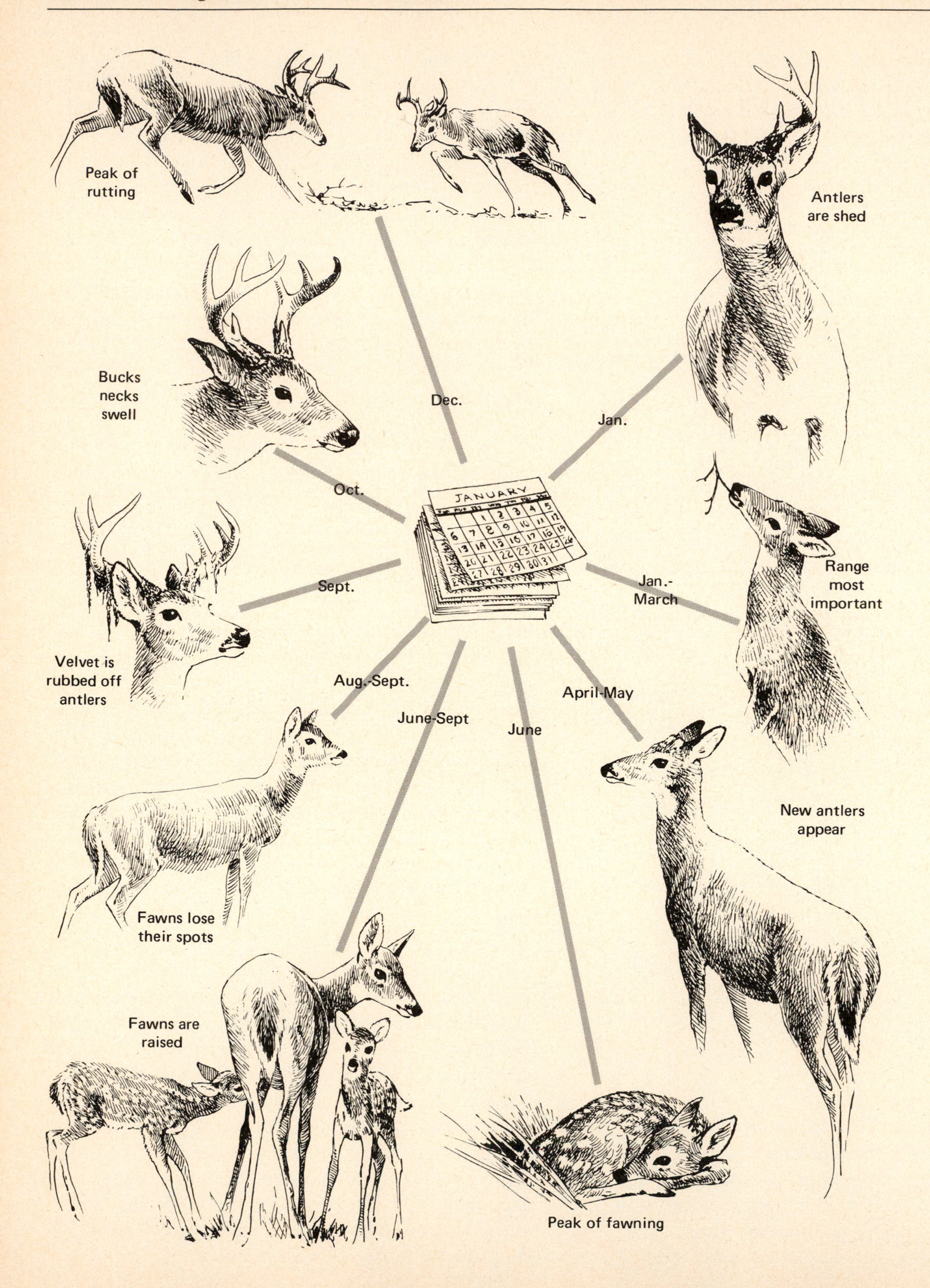
Peak of rutting
Dec.
Antlers are shed
Jan.
Bucks necks swell
Oct.
JANUARY
Range most important
Jan.-March
Sept.
Velvet is rubbed off antlers
Aug.-Sept.
June-Sept
June
April-May
New antlers appear
Fawns lose their spots
Fawns are raised
Peak of fawning

Other Seasonal Patterns

Other seasonal patterns are also influenced by photoperiod. One example is the food-storage behavior of the flying squirrel (Muul, 1965). Squirrels held in the laboratory under seasonal photoperiods and controlled constant temperatures exhibited an intensity of food storage similar to that of animals held under natural conditions. Squirrels exposed to seasonal temperatures and a controlled photoperiod of 15 hours of daylight, equivalent to summer conditions, showed no intense food-storage activity through the winter. When the light was reduced to 12 hours in March, simulating early fall conditions, the squirrels exhibited a marked rise in food-storage activity. Another group was held under constant temperature and a controlled photoperiod of 15 hours of daylight, which was reduced to 13 hours in mid-December. Within a week this group increased food-storing activity sharply and continued it from January through March. These and other experiments demonstrate that food-storing behavior is triggered when the critical daylength reaches about 12 hours in early fall. Such control synchronizes the exploratory and storing behavior of the squirrels with the ripening of the mast crop—nuts and acorns—and prevents a premature harvest.

Antler growth of male deer is also a photoperiodic response. Growth is triggered by the lengthening days of spring, and the velvet is shed in the shortening days of fall (Goss et al., 1974). Normally, the deer is in velvet about one-third of the year. When the duration of that year is changed by altering the frequency of daylength, antlers may be replaced as often as two, three, or four times a year, or only once every other year, depending on how much the light cycle has been increased or decreased (Figure 7.12). If deer are held under constant photoperiod, they will still develop antlers, but only if the daylength ratio of light to dark exceeds 12¾L to 11¼D. If held under a constant daylength of 12L/12D, the annual cycle of antler growth cannot be expressed, and antlers may not be replaced for several years.

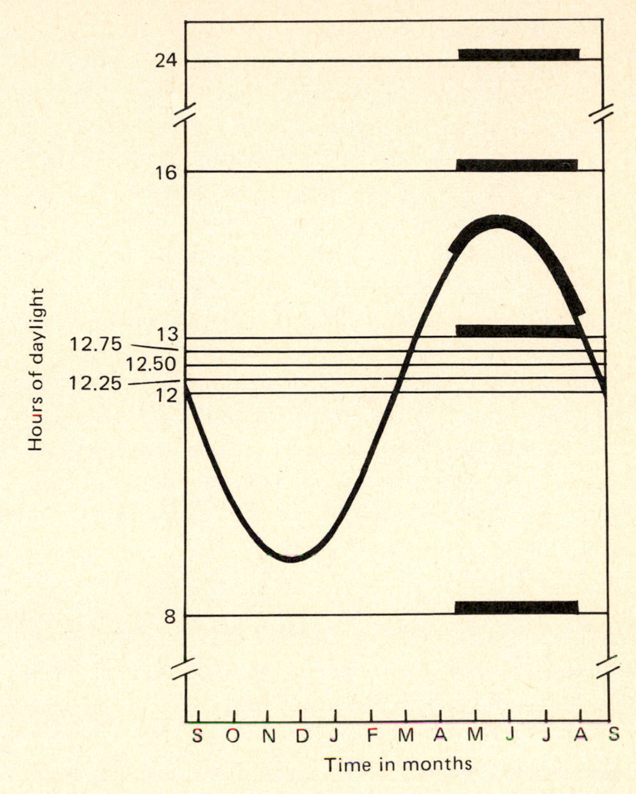

FIGURE 7.12 Response of antler growth in Sika deer to constant daylengths of different duration. The curve represents the natural course of daylength during a year at 42° north latitude. Under natural conditions, antlers are in velvet from early May to early September. When yearling Sika deer are held under light/dark conditions of 8L/16D, 13L/11D, 16L/8D, and 24L/0D starting with the autumnal equinox, all shed and regrew antlers in spring in synchrony with the outdoors. Deer exposed to light periods of 12, 12.25, 12.50, and 12.75 hours failed to replace their antlers the following year. (After Goss et al. 1974:407.)

FIGURE 7.11 Seasonal cycle of the white-tail deer. The annual cycle is attuned to the decreasing daylength of fall, during which the breeding season begins, and to the lengthening days of spring, when antler growth begins.

Seasonal Morphs

Photoperiod can also result in seasonal morphs among a number of insects. One example is the Pierid butterfly, the veined, white *Pieris napa,* in western North America. Populations of this butterfly along the California coast have two generations a year and occur in two phenotypes. One is a dark spring form with heavy black scaling on the veins of the hind wings and the other a lighter summer form almost devoid of black scales. Inland populations have only one generation a year, and all adults are the dark form. Both populations, however, are capable of producing both phenotypes. The dark forms are induced by the short days of autumn and emerge from overwinter diapause pupae. The light summer forms are induced by the long days of summer and emerge from nondiapause pupae (Shapiro, 1977). Such photoperiodic responses may be important in the reproductive isolation of closely related populations and the evolutionary process of speciation (see Beck, 1980; Tauber and Tauber, 1976a, 1976b).

Circannual Clocks

All of the activities discussed above exhibit a circannual periodicity, suggesting the existence of a circannual clock. To prove its existence, one needs to demonstrate conditions similar to those required of a circadian clock: (1) a free-running rhythm in the absence of environmental cues, (2) a period that approximates but deviates from a 365-day cycle, and (3) temperature independence. Considerable evidence exists that a circannual clock controlling body weight and hibernation exists in the golden-manteled ground squirrel (*Citellus lateralis*) of western North America. When held for several years under constant temperature regimes and constant artificial days of 12L/12D, the squirrels exhibited circannual rhythms of about a year (Pengelley and Asmundson, 1974). Such a rhythm has certain advantages. It enables the animals to prepare well in advance for winter and to arouse in spring without the need of environmental cues, since the animals are inaccessible in a deep burrow. It also brings the animals into breeding condition almost immediately after hibernation, so reproduction can take place in the spring.

Most birds show strong circannular rhythms in molt, body weight (fat deposition), testis growth, and nocturnal restlessness (Berthold, 1974). But whether one needs to invoke or demonstrate a circannual clock is questionable. The cycles can be explained by seasonal photoperiod response and a circadian clock.

ADAPTIVE VALUE OF PHOTOPERIODICITY

Light and dark may be the *Zeitgebers* that time the phase of an organism's circadian activity rhythm, but the rhythms may relate more directly to other aspects of the environment, which, ecologically, are more significant to the organism than light and dark per se. The transition from day to night, for example, is accompanied by such environmental changes as a rise in humidity and a drop in temperature. Wood lice, centipedes, and millipedes, which lose water rapidly when exposed to dry air, spend the day in a fairly constant environment of darkness and dampness under stones, logs, and leaves. At dusk, when the humidity of the air is more favorable, they emerge. In general, these animals show an increased tendency to escape from light as the length of time they spend in darkness increases. On the other hand, their intensity of response to low humidity decreases with darkness. Thus, they come out at night into places too dry for them during the day; and as light comes, they quickly retreat to their dark hiding places (Cloudsley-Thompson, 1956, 1960).

Among some animals the biotic rather than the physical aspects of the environment may relate to the activity rhythm. Deer undisturbed by humans may be active by day, but when they are hunted and disturbed, they become strongly nocturnal. Predators must relate their feeding activity to the activity rhythms of their prey. Moths and bees must visit flowers when they are open and provide a source of food. And the flowers must have a rhythm of opening and closing that coincides with the time when the insects that pollinate them are flying.

The entrainment of the phase of its activity rhythm to a natural light-dark cycle means more to an organism than simply an adjustment to a precise 24-hour period. More important, the entrainment serves to time the activities of plants and animals to a day-night cycle in a manner that is appropriate to the ecology of the species. Over the year photoperiod enables organisms to adjust to seasonal changes in the environment. It keeps the organisms' seasonal activities—time of flowering, seeding, courtship, mating, reproduction, migration, hibernation—in tune with the rest of the population and the world about them.

Summary

Light, like moisture and temperature, is an important aspect in the lives of plants and animals, which goes beyond its major role in photosynthesis. As visible light passes through Earth's atmosphere, certain wavelengths, especially violets and blues, are reduced more than others. It is even further reduced in wavelength and intensity as light is reflected, absorbed, or transmitted by layers of vegetation or by the depth of water in aquatic communities. Based on their response to the intensity of light reaching them, plants are either shade-tolerant or shade avoiders. Shade-tolerant plants have lower rates of photosynthesis, lower rates of respiration, and physiological characteristics that enable them to grow and survive under conditions of low light. Shade-intolerant plants, the shade avoiders, have higher rates of photosynthesis and higher rates of respiration, which result in an inability to function efficiently under low light.

Light also influences the reproductive and activity cycles of plants and animals. An almost universal feature of life is an internal biological clock, physiological in function. Its basic structure is probably chemical and is involved in the makeup of the cell. It is free-running under constant conditions, with an oscillation or fluctuation that has its own inherent frequency. For most organisms this fluctuation deviates more or less from 24 hours; for that reason it is called a circadian rhythm. Under natural conditions the circadian rhythm is set or entrained to the 24-hour day by external time cues, or *Zeitgebers,* which synchronize the activity of plants and animals with the environment. Because the most dependable external timesetter is light and dark—day and night—most of the selected species studied so far are entrained to a 24-hour photoperiod. The onset and cessation of activity are usually synchronized with dark and dawn, the response depending upon whether the organisms are diurnal (light-active) or nocturnal (dark-active).

The biological clock is useful not only to synchronize the daily activities of plants and animals with night and day, consistent with the ecology of the species, but also to time the activities with the seasons of the year. The possession of a self-sustained rhythm with approximately the same frequency as that of the environment enables organisms to predict such situations as the coming of spring. It brings plants and animals into a reproductive state at a time of year when the probability of survival of offspring is the highest. It synchronizes within a population such activities as mating and migration, dormancy and flowering. The acquisition and refinement of a physiological timekeeper, geared to *Zeitgebers,* that provides organisms with distinct and species-specific or population-specific synchronization with the environment is a result of natural selection.

Review and Study Questions

1. What is photosynthetically active radiation?
2. What part of the light that arrives on Earth's surface is reduced as it travels through Earth's atmosphere?
3. What is the fate of visible light that is intercepted by Earth's surface, and more specifically, by the leaf?
4. What is leaf area index? How does it relate to light penetration of the vegetative canopy?
5. Contrast a shade-tolerant plant with shade avoiders.
6. What is the light compensation point? Epicormic branches in trees?
7. What is photoperiodism?

8. What is a circadian rhythm? How does it relate to the 24-hour day? What is meant by the term *free-running?*
9. What is a *Zeitgeber?* Phase shift? How do the two relate?
10. What is a biological clock? What does it consist of in some animals, and where is its possible location?
11. What conditions must a biological clock fulfill to function as a timekeeper?
12. Discuss four ways in which a biological clock can function.
13. What is critical daylength? A long-day organism? A short-day organism?
14. How is diapause influenced by changing daylength?
15. How does photoperiodicity influence reproductive cycles? Food-storage behavior? Antler development in deer?
16. How does photoperiodism influence polymorphism in some species of insects?
17. Argue for and against the existence of a circannual clock.
18. What is the adaptive value of photoperiodism?
19. Speculate on how photoperiodism could have a role in speciation.
20. Nothing was said in the chapter about other types of periodicities, especially lunar and tidal rhythms in some organisms. Look up these rhythms in such references as J. Palmer, *An Introduction to Biological Rhythms* (New York: Academic Press, 1976) and P. J. DeCoursey, *Biological Rhythms in the Marine Environment* (Columbia: University of South Carolina Press, 1976).
21. Also ignored in the text is any discussion of circadian rhythms in humans, a very important aspect of physiology that affects all of us. The jet lag effect involves a phase shift plus a resetting of the biological clock. See M. C. Moore-Ede, F. S. Sulzman, and C. A. Fuller, *Clocks that time us. Physiology of the circadian timing system.* (Cambridge, MA: Harvard University Press, 1982).

Suggested Readings

Beck, S. D. 1980. *Insect photoperiodism*. 2d ed. New York: Academic Press.

Bunning, E. 1973. *The physiological clock*. 3d ed. New York: Academic Press.

Pengelley, E. T., ed. 1974. *Circannual clocks: Annual biological rhythms*. New York: Academic Press.

Saunders, D. S. 1982. *Insect clocks*. 2d ed. Elmsford, N.Y.: Pergamon Press.

CHAPTER 8

The Soil Environment

Outline

Objectives

Upon completion of this chapter, you should be able to:

1. Provide a general description of soil and its features.
2. Name some of the essential biological elements and indicate their importance.
3. Explain how physical and chemical weathering and the action of plants and animals interact in the process of soil development.
4. Describe the soil profile and distinguish between the several soil horizons.
5. Distinguish between mull and mor humus.
6. Contrast calcification, podzolization, laterization, and gleization.
7. Explain the importance of cation exchange in the soil and the role of soil acidity.
8. Define soil texture and describe the sizes and combinations of the various particles that make up soil.
9. Explain the significance of soil color.
10. Point out the features of the soil as an environment for life.
11. Discuss the selective pressures, especially as they relate to acidity, of soils on plants.

Soil is the foundation of terrestrial communities. It is the site of decomposition of organic matter and of the return of mineral elements to the nutrient cycle (see Chapter 21). Roots occupy a considerable portion of the soil, to which they tie the vegetation and from which they pump water and minerals in solution needed by plants for photosynthesis and other biogeochemical processes. Vegetation, in turn, influences the development of soil, its chemical and physical properties, and its organic matter content. Thus, soils act as a pathway between the organic and mineral worlds.

A Definition of Soil

As familiar as it is, soil is difficult to define. One definition has soil as a natural product formed from weathered rock by the action of climate and living organisms. Another states that soil is a collection of natural bodies of Earth that is composed of mineral and organic matter and is capable of supporting plant growth. Such definitions seem inadequate or stilted. Indeed, one eminent soil scientist, a pioneer of modern soil studies, Hans Jenny, will not give an exact definition of soil. In his book *The Soil Resource* (1980: 364), he writes:

> Popularly, soil is the stratum below the vegetation and above hard rock, but questions come quickly to mind. Many soils are bare of plants, temporarily or permanently; or they may be at the bottom of a pond growing cattails. Soil may be shallow or deep, but how deep? Soil may be stony, but surveyors [soil] exclude the larger stones. Most analyses pertain to fine earth only. Some pretend that soil in a flower pot is not soil, but soil material. It is embarrassing not to be able to agree on what soil is. In this the pedologists are not alone. Biologists cannot agreed on a definition of life and philosophers on philosophy.

But of one fact we are sure. Soil is not just an abiotic environment of plants. It is teeming with life—billions of minute animals, bacteria, and fungi. The interaction between the abiotic and the biotic make soil a living system.

Biological Elements

Most of the nutrients required by life exist in mineral form in Earth's crust (see Chapter 21). They are made available by weathering and chemical processes and are taken up by plants. Soil is the major reservoir of most nutrients that support life.

Living organisms require at least 30 to 40 elements for growth, development, and reproduction. The bulk of living matter consists of hydrogen, carbon, oxygen, nitrogen, and sulfur. These *bulk elements* are concentrated in living tissue in grams per kilogram and are needed in gram amounts daily.

Carbon is the key atom of all forms of life. Along with oxygen and hydrogen, it makes up the bulk of organic matter. In the ecosystem, carbon occurs as carbon dioxide (CO_2), carbonates, and fossil fuel as well. Carbon dioxide makes up only 0.03 percent of the atmosphere. The amount of CO_2 in natural waters is variable, for it occurs in free and combined states. The pH of aquatic media and of the soil has a pronounced influence on the proportions of the two types. Carbon dioxide combines with water to form weak carbonic acid, H_2CO_3, which dissociates as follows:

$$CO_2 + H_2O \rightleftharpoons H_2CO_3 \rightleftharpoons H^+ + HCO_3$$
$$\rightleftharpoons H^+ + CO_3^{++}$$

Carbon dioxide in solution and carbonic acid make up free carbon dioxide; the bicarbonate (HCO_3) and carbonate (CO_3^{++}) ions are the combined forms. The presence of bicarbonate and carbonate ions in soil or water helps buffer or maintain a certain pH of that medium.

Nitrogen is a component of all proteins, amino acids, and nucleic acids. It makes up 78 percent of the atmosphere as molecular nitrogen (N_2), but most plants can utilize it only in a fixed form, such as nitrates and nitrites (see Chapter 21). Most nitrogen in the soil is found in organic matter. Nitrates leached from the soil and transported by water are an important source of nitrogen for aquatic communities.

Oxygen (O_2), which makes up 21 percent of Earth's atmosphere, is a by-product of photosynthesis. Three major sources of oxygen are carbon dioxide, water, and molecular oxygen, all of which interchange atoms with one another. Other sources include nitrate and sulfate ions, which release oxygen upon decomposition and supply it to a number of living organisms.

Sulfur is a basic constituent of protein. Sulfur is supplied by rainwater and by organic matter in the soil. Excessive sulfur can be toxic to plants. High concentrations limit the uptake of calcium.

Another group of elements is needed in concentrations lower than the bulk elements. These include phosphorus, potassium, calcium, magnesium, and sodium. They are concentrated in grams per kilogram but are required only in fractions of a gram per day. These are known as *macroelements* or *macronutrients*.

Calcium is necessary in animals for the proper acid-base relationship, for the clotting of blood, for the contraction and relaxation of the heart muscles, and for the control of fluid passage through the cells. It gives rigidity to the skeleton of vertebrates and is the principal component of the exoskeletons of insects and the shells of mollusks and arthropods. A number of bivalves and other mollusks are restricted to hard water because of insufficient calcium in soft water to harden their shells. In plants calcium combines with pectin to form calcium pectate, a cementing material between cells. Plant roots need a supply of calcium at the growing tips to develop normally.

Phosphorus is a component of proteins and of many enzymes. It plays a major role in energy transfer at the cellular level in plants and animals.

Magnesium is an integral part of chlorophyll and is active in the enzyme systems of plants and animals. Low intake of magnesium by grazing ruminants causes a serious disease, grass tetany, which may result in death.

Potassium is involved in the formation of sugars and starches in plants and operates in the stomatal openings of leaves. It is also important in the synthesis of proteins, normal cell division and growth, and carbohydrate metabolism in animals.

Sodium and *chlorine*, used in minute quantities by plants, are indispensable to vertebrate animals. These elements are important in the maintenance of acid-base balance, the total osmotic pressure of extracellular fluids, and the formation and flow of gastric and intestinal secretions.

Numerous other elements are needed in much lower concentrations, measured in milligrams or micrograms per kilogram of tissue. Because they were not easily quantified by analytical methods in the past, they came to be known as trace elements. Today we know them as *micronutrients*. Included among these is *iron*, active in photosynthesis and nitrogen fixation and part of the complex proteins that serve as activators and carriers of oxygen in the blood and as transporters of electrons. *Zinc* is needed in the formation of auxins in plant growth substances and is a component of enzyme systems in plants and animals. *Copper* in plants is concentrated in the chloroplasts, where it affects the photosynthetic rate, is involved in oxidation-reduction reactions, and acts as an enzyme activator. *Molybdenum* acts as a catalyst in the conversion of gaseous nitrogen into usable forms by free-living nitrogen-fixing bacteria and blue-green algae. *Iodine* is essential for thyroid metabolism. *Boron* is essential to some 15 functions of plants, including carbohydrate metabolism, water metabolism, and translocation of sugars. *Silicon* is needed by diatoms and by grasses and cereals, in which it conveys resistance to pathogenic bacteria and fungi. Ruminant animals need *selenium*, which is tied to vitamin E activity. All of the micronutrients can be toxic if available or supplied in quantities far greater than needed.

Soil Development

PHYSICAL WEATHERING

Soils begin with the weathering of rocks and their minerals. Exposed to the combined action of water, wind, and temperature, rock surfaces peel and flake away. Water seeps into crevices, freezes, expands, and cracks the rock into smaller pieces. Accompanying this disintegration and continuing long afterward is the decomposition of the minerals themselves. Water and carbon dioxide combine to form carbonic acid, which reacts with calcium and magnesium in the rock to form carbonates. These either accumulate deeper in the rock material or are carried away, depending on the amount of water passing through. Primary minerals that contain aluminum and silicon, such as feldspar, are converted to secondary minerals such as clay. As iron is especially reactive with water and oxygen, iron-bearing minerals are prone to rapid decomposition. Iron remains oxidized in the red ferric state or may be reduced to the gray ferrous state. Fine particles, especially clays, are shifted or rearranged within the mass by percolating water and on the surface by runoff, wind, or ice.

Eventually, the rock is broken down into loose material; this may remain in place, but more often than not, much of it is lifted, sorted, and carried away. Material transported from one area to another by wind is known as *loess;* that transported by water as *alluvial, lacustrine* (or lake), and *marine* deposits; and that transported by glacial ice as *till.* In a few places soil materials come from accumulated organic matter such as peat. Materials remaining in place are called *residual.*

This mantle of unconsolidated material is called the *regolith.* It may consist of slightly weathered material with fresh primary minerals, or it may be intensely weathered and consist of highly resistant minerals such as quartz. Because of variations in slope, climate, and native vegetation, many different soils can develop in the same regolith. The thickness of the regolith, the kind of rock from which it was formed, and the degree of weathering affect the fertility and water relations of the soil.

ACTION OF PLANTS AND ANIMALS

Eventually, plants root on this weathered material. More often than not, intense weathering goes on under some plant cover, particularly in glacial till and water-deposited materials, because they are already favorable sites for some plant growth. Thus, soil development often begins under some influence of plants. They root, draw nutrients from mineral matter, reproduce, and die. Their roots penetrate and further break down the regolith. The plants pump up nutrients from its depths and add them to the surface and, in doing so, recapture minerals carried deep into the material by weathering processes. Through photosynthesis, plants capture the sun's energy and add a portion of it in the form of organic carbon—approximately 18 billion metric tons, or 1.7×10^{17} kilocalories—to the soil each year. This energy source, the plant debris, enables bacteria, fungi, earthworms, and other soil organisms to colonize the area.

The breakdown of organic debris into humus is accomplished by decomposition and, finally, mineralization. Higher organisms in the soil—millipedes, centipedes, earthworms, mites, springtails, grasshoppers, and others—consume fresh material and leave partially decomposed products in their excreta. This is further decomposed by microorganisms, the bacteria and fungi, into various compounds of carbohydrates, proteins, lignins, fats, waxes, resins, and ash. These compounds are then broken down into simpler products, such as carbon dioxide, water, minerals, and salts. The latter process is called *mineralization.*

The fraction of organic matter that remains is called *humus.* It is not stable, as it represents a stage in the decomposition of soil organic matter. New humus is being formed as old humus is being destroyed by mineralization. The equilibrium set up between the formation of the new humus and the destruction of the old determines the amount of humus in the soil.

CHEMICAL WEATHERING

The activities of soil organisms, the acids produced by them, and the continual addition of organic matter to mineral matter produce profound changes in the weathered material. Rain falling upon and filtering through the accumulating organic matter picks up acids and minerals in solution, reaches mineral soil, and sets up a chain of complex chemical reactions. This continues in the regolith. Calcium, potassium, sodium, and other mineral elements, soluble salts, and carbonates are carried in solution by percolating water deeper into the soil or are washed away into streams, rivers, and eventually the sea. The greater the rainfall, the more water moves down through the soil and the less moves upward. Thus, high precipitation results in heavy leaching and chemical weathering, particularly in regions of high temperatures. These chemical reactions tend to be localized within the regolith. Organic carbon, for instance, is oxidized near the surface, while free carbonates precipitate deeper in the rock material. Fine particles, especially clays, also move downward.

These localized chemical and physical processes in the parent material result in the development of layers in the soil, called *horizons*, which impart to the soil a distinctive *profile*. Within a horizon a particular property of the soil reaches its maximum intensity, and away from this level it decreases gradually in both directions. Thus, each horizon varies in thickness, color, texture, structure, consistency, porosity, acidity, and composition.

SOIL HORIZONS

In general, soils have four major horizons: an organic, or O, horizon and three mineral horizons—the A, characterized by major organic matter accumulation, by the loss of clay, iron, and aluminum, and by the development of a granular, crumb, or platy structure; the B, characterized by an illuvial concentration of all or any of the silicates, clay, iron, aluminum, and humus, alone or in combination, and by the development of a blocky, prismatic, or columnar structure; and the C, material underlying the two horizons that is either like or unlike the material from which the soil is presumed to have developed. Below all this may lie the R horizon, the consolidated bedrock.

Because the soil profile is essentially a continuum, often there is no clear-cut distinction between one horizon and another. Horizon subdivisions (see Figure 8.1) are indicated by arabic numbers—for example, O_1, O_2, A_1, A_2, and so forth; lowercase letters are used to indicate significant qualitative

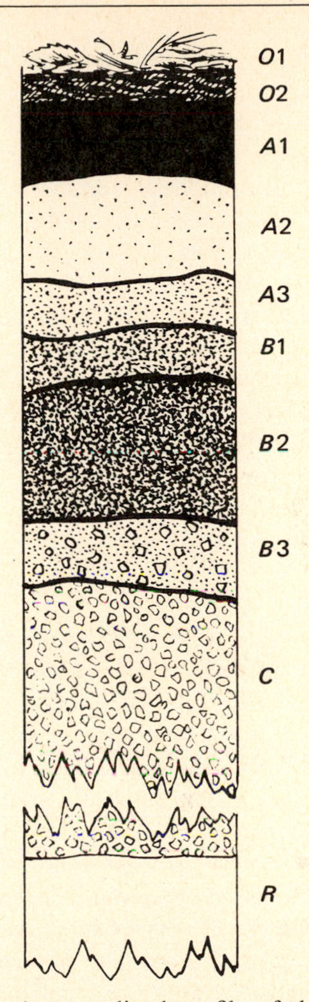

FIGURE 8.1 A generalized profile of the soil. Rarely does any one soil possess all of the horizons shown. O_1—Loose leaves and organic debris. O_2—Organic debris partly decomposed or matted. A_1—A dark-colored horizon with high content of organic matter mixed with mineral matter. The A horizon is the zone of maximum biological activity. A_2—A light-colored horizon of maximum leaching. Prominent in podzolic soils; faintly developed or absent in chernozemic soils. A_3—Transitional to B, but more like A than B. Sometimes absent. B_1—Transitional to B, but more like B than A. Sometimes absent. B_2—A deeper-colored horizon of maximum accumulation of clay minerals or of iron and organic matter; maximum development of blocky or prismatic structure or both. B_3—Transitional to C. C—The weathered material, either like or unlike the material from which the soil presumably formed. A gley layer may occur, as well as layers of calcium carbonate, especially in grasslands. R—Consolidated bedrock.

departures from the central concept of each horizon—for example, A_{2g} or B_t.

The O horizon—once designated as L, F, H, or A_0 and A_{00}—is the surface layer, formed or forming above the mineral layer and composed of fresh or partially decomposed organic material, such as that found in temperate forest soils. It is usually absent in cultivated soils. This, with the upper part of the A horizon, is the region where life is most abundant. It is subject to the greatest changes in soil temperatures and moisture conditions and contains the most organic carbon. And it is the site where most or all decomposition by organisms takes place.

THE ORGANIC HORIZON, OR FOREST FLOOR

Of all the horizons of the soil, none is more important or ecologically more interesting than the forest floor, or the organic horizon. A close relationship exists between litter and humus, and the environmental conditions in the forest community—the internal microclimate of the soil, the moisture regime, its chemical composition, and its biological activity. The forest floor plays a dominant role in the life and distribution of many forest plants and animals, in the maintenance of soil fertility, and in many of the soil-forming processes. The nature and quality of the forest organic layer depend in part on the kind and quality of forest litter. And the fate of that litter and the development of a horizon are conditioned by the activity of microflora and soil animals. In fact, many humus forms undergo initial breakdown in the bodies of animal organisms. To complete the circle, the composition and density of the soil fauna are influenced by the litter. Thus, as Bernier (1961) put it, the forest humus is both "a consequence and a cause" of local ecological conditions.

The importance of the organic layer was stressed early in the history of ecology. Darwin, in his famous work "The Formation of Vegetable Mould through the Action of Worms, with Observations on Their Habits" (1881), pointed out the influence of these animals on the soil. At about the same time, in 1879 and 1884, the Danish forester P. E. Müller described the existence of two types of humus formation in the temperate forest soil; these he called *mull* and *mor.* Not only did he observe differences in vegetation, soil structure, and chemical composition, but he discovered differences in their fauna also. Müller considered mull and mor as biological, rather than purely physiochemical, systems and regarded the fauna present as aiding in their formation. Others have regarded mull and mor from the physical and chemical point of view, with little regard for the biological mechanisms involved. Actually, both are a result of the interaction of all three.

Mor

Mor, characteristic of dry or moist acid habitats, especially heathland and coniferous forest, has a well-defined, unincorporated, and matted or compacted organic deposit resting on mineral soil. It results from an accumulation of litter that is slowly mineralized and remains unmixed with mineral soil. Thus, a sharp distinction or break exists between the O and A horizons.

Slow though mineralization may be, it is the manner in which this process proceeds that distinguishes mor from other humus types. The main decomposing agents are fungi, both free-living and mycorrhizal, which tend to depress soil animal activity and produce acids; nitrifying bacteria may be absent. The vascular cells of leaves disappear first, leaving behind a residue of mesophyll tissue. Proteins within the leaf litter are stabilized by protein-precipitating material, making them, in some cases, resistant to decomposition. Because of limited volume, pore space, acidity, the type of litter involved, and the nature of its breakdown, mor is inhabited by a small biomass of the smaller soil animals. These organisms have little mechanical influence on the soil; instead, they live in an environment of organic material cut off from the mineral soil beneath (see Figure 8.2).

Mull

Mull, on the other hand, results from a different process. Characteristic of mixed and deciduous woods on fresh or moist soils with a reasonable

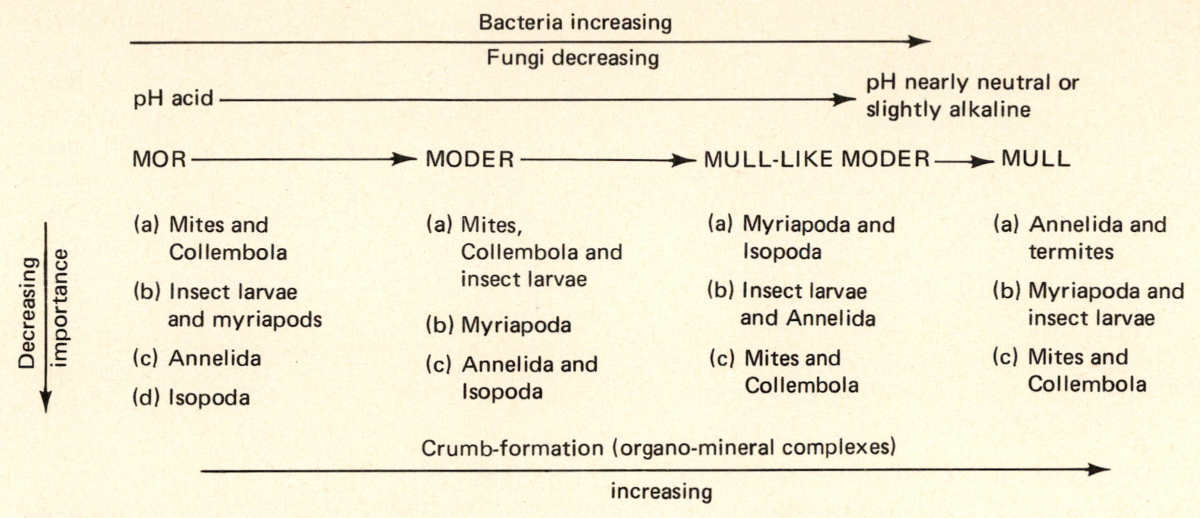

FIGURE 8.2 The sequence of humus types and related processes. Note the inverse relationship between bacteria and fungi as the humus sequence goes from mor to mull, as well as the pronounced changes in invertebrate life. (From J. A. Wallwork, *Ecology of Soil Animals* [New York: McGraw-Hill, 1973:35].)

supply of calcium, mull possesses only a thin scattering of litter on the surface, and the mineral soil is high in organic content. All organic materials are converted into true humic substances, and because of animal activity these are inseparably bound to the mineral fraction, which absorbs them like a dye. There is no sharp break between the O and A horizons.

Because of less acidity and a more equitable base status, bacteria tend to replace fungi as the chief decomposers, and nitrification is rapid. Soil animals are more diverse and possess a greater biomass, reflecting a more equitable distribution of living space, oxygen, food, and moisture and a smaller fungal component. This faunal diversity is one of mull's greatest assets, because the humification process flows through a wide variety of organisms with differing metabolisms. Not only do these soil animals fragment vegetable debris and mix it with mineral particles, thus enhancing microbial and fungal activity, but they also incorporate the humified material with mineral soil. This constant interchange of material takes place from the surface to the soil and back again. Plants extract nutrients from the soil and deposit them on the surface. Then the soil plants and animals reverse the process.

Moder

Between the two extremes, mull and mor, is *moder,* the insect mull of Müller. In this humus type, plant residues are transformed into the droppings of small arthropods, particularly Collembola and mites. Residues not consumed by the fauna are reduced to small fragments, little humified and still showing cell structure. The droppings, plant fragments, and mineral particles all form a loose, netlike structure held together by chains of small droppings. In acid moder the shape of the droppings is destroyed by the washing action of rainwater; and under more extreme conditions humus leached from the droppings acts as a binding substance to form a dense, matted litter approaching a mor. At the other end of the spectrum, the borderline between moder and mull, the droppings of large arthropods, which are capable of taking in considerable quantities of mineral matter with food, are common. However, moder differs from mull in its higher organic matter content, restricted nitrification, and a more or less mechanical mixture of the organic components with the mineral, the two being held together by humic substances, yet separable. In other words, the organic crumbs are deficient in mineral matter, in contrast to mull, in which the mineral and organic parts are inseparably bound together.

Profile Differentiation

The differentiation of the soil profile into horizons and the nature of the soil material—its content and distribution of organic matter, its color, and its chemical and physical characteristics—are influenced over large areas by the combined action of vegetation and its prime determinant, climate (see Figure 8.3). Thus, the soil beneath native grassland differs from that beneath native forest.

CALCIFICATION

Grassland vegetation developed in the subhumid-to-arid and temperate-to-tropical climates of the world—the plains and prairies of North America, the steppes of Russia, and the veldt and savannah of Africa. Dense grass-root systems may extend many feet below the surface. Each year nearly all of the vegetative material above the ground and part of the root system are turned back to the soil as organic residue. Although this material decomposes rapidly the following spring, it is not completely gone before the next cycle of death and decay begins. The humus then becomes mixed with mineral soil by the action of the soil inhabitants, developing a soil high in organic matter. The humus content is greatest at the surface and declines gradually with depth.

Because the amount of rainfall in grassland regions is generally insufficient to remove calcium and magnesium carbonates, these are removed only down to the average depth reached by the percolating waters. The high calcium content of the surface soil is maintained by grass, which absorbs large quantities from lower horizons and redeposits them on the surface. Likewise, there is little loss of clay from the surface layer. This process of soil development has been called *calcification,* and the soil used to be called *pedocal.*

Soils developed by calcification have a distinct A horizon of great thickness and an indistinct B horizon characterized by an accumulation of calcium carbonate. The A horizon is high in organic matter and nitrogen, even in tropical and subtropical regions.

PODZOLIZATION

Forests are the dominant vegetation in the humid regions. Here the cycle of organic matter accumulation differs from that of the grassland. Only part of the organic matter—leaves, twigs, and some trunks—is turned over annually. Leaves, which are the largest source of organic matter, and the vegetation of the ground layer remain on the surface. Dead roots add little to soil organic matter, because they die over an irregular period and are not concentrated near the surface. Because only the leaves are returned regularly to the soil and much of the mineral matter and energy is tied up in trunks and branches, most of the currently available nutrients turned back to the soil come from annual leaf fall. The amount of nutrient return varies with the species composition of the forest, since trees differ in the nutrient content of their leaves. For example, basswood, quaking aspen, hickories, American elm, and flowering dogwood contain more calcium in their leaves and return more to the soil than sugar maple, red maple, yellow birch, and red oak (Lutz and Chandler, 1946). And the latter return more than beech, red pine, white pine, and hemlock.

Rainfall in forested regions is sufficient to leach away many basic elements, especially calcium, magnesium, potassium, iron, and aluminum. Because trees take up fewer bases than grasses, they generally return an insufficient amount to the surface soil to prevent it from becoming acid. The degree of acidity, however, will vary depending on the forest composition and its site. Some forests in the southern Appalachians, particularly those containing yellow poplar and basswood and growing on north and northeast slopes, have a rather high, often neutral, pH in their surface horizons, even though they grow on soils weathered from acid sandstone (unpublished data from van Eck and Smith). Increased acidity may cause the dispersion and downward movement of organic and clay colloids.

A soil developed by this process is called *podzolic* and the process *podzolization,* from the Russian, meaning "ash beneath," referring to the leached horizon of strongly podzolized soils. The latter are characterized by a white A_2 horizon and a brilliant yellow-brown B horizon, the result of accumulations of iron and aluminum compounds and humus (see Figure 8.3). Iron accumulations in some podzol soils may act as a cement, creating a hardpan layer in the B horizon. This layer, called *ortstein,* impedes the free circulation of air and water. Soils developed by this process have been referred to as *pedalfer.*

LATERIZATION

In the humid subtropical and tropical forested regions of the world, where rainfall is heavy and temperatures high, the soil-forming process is much more intense. Because temperatures are uniformly high, the weathering process in these regions is almost entirely chemical, brought about by water and its dissolved substances. The residues from this weathering—bases, silica, alumina, hydrated aluminosilicates, and iron oxides—are freed. Since precipitation usually exceeds evaporation, the water movement is almost continuously downward. With only a small quantity of electrolytes present in the soil water because of continual leaching, silica and aluminum silicates are carried downward, while sesquioxides of aluminum and iron remain behind. The reason for this is that these sesquioxides are relatively insoluble in pure rainwater, while the silicates tend to be precipitated as a gel in solutions containing humic substances and electrolytes. If humic substances are present, they act as protective colloids around iron and aluminum oxides and prevent their precipitation by electrolytes.

The end product of such a process is a soil composed of silicate and hydrous oxides, clays, and residual quartz, deficient in bases, low in plant nutrients, and intensely weathered to great depths. Because of the large amount of iron oxides left, these soils show a variety of reddish colors and generally lack distinct horizons. Below, the profile is unchanged for many feet. The clay has a stable structure, and unless precipitated, the iron is hardened into a cemented laterite. The soil is very pervious to water and is easily penetrated by plant roots. This soil-forming process is termed *laterization* or *latosolization.*

CALICHE

Arid and semiarid regions have relatively sparse vegetation. Because of the lack of plant growth, which is limited by low rainfall, there is very little organic matter and nitrogen in the soil. Light precipitation results in slightly weathered and slightly leached soils high in plant nutrients. Their horizons are usually faint and thin. In these regions occur areas where soils contain excessive amounts of soluble salts, either from parent material or from the evaporation of water draining in from adjoining land. The infrequent rainwater penetrates the soil, but soon afterward evaporation at the surface draws the salt-laden water upward. The water evaporates, leaving saline and alkaline salts at or near the surface to form a crust, or *caliche.*

GLEIZATION

Calcification, podzolization, and laterization are all processes that take place in well-drained soil. Under poorer drainage conditions a different soil-development process is at work. The slope of the land determines to a considerable extent the amount of rainfall that will enter and pass through the soil, the concentration of erosion materials, the amount of soil moisture, and the height at which the water will stand in the soil (see Figure 8.4). The amount of water that passes through or remains in the soil determines the degree of oxidation and breakdown

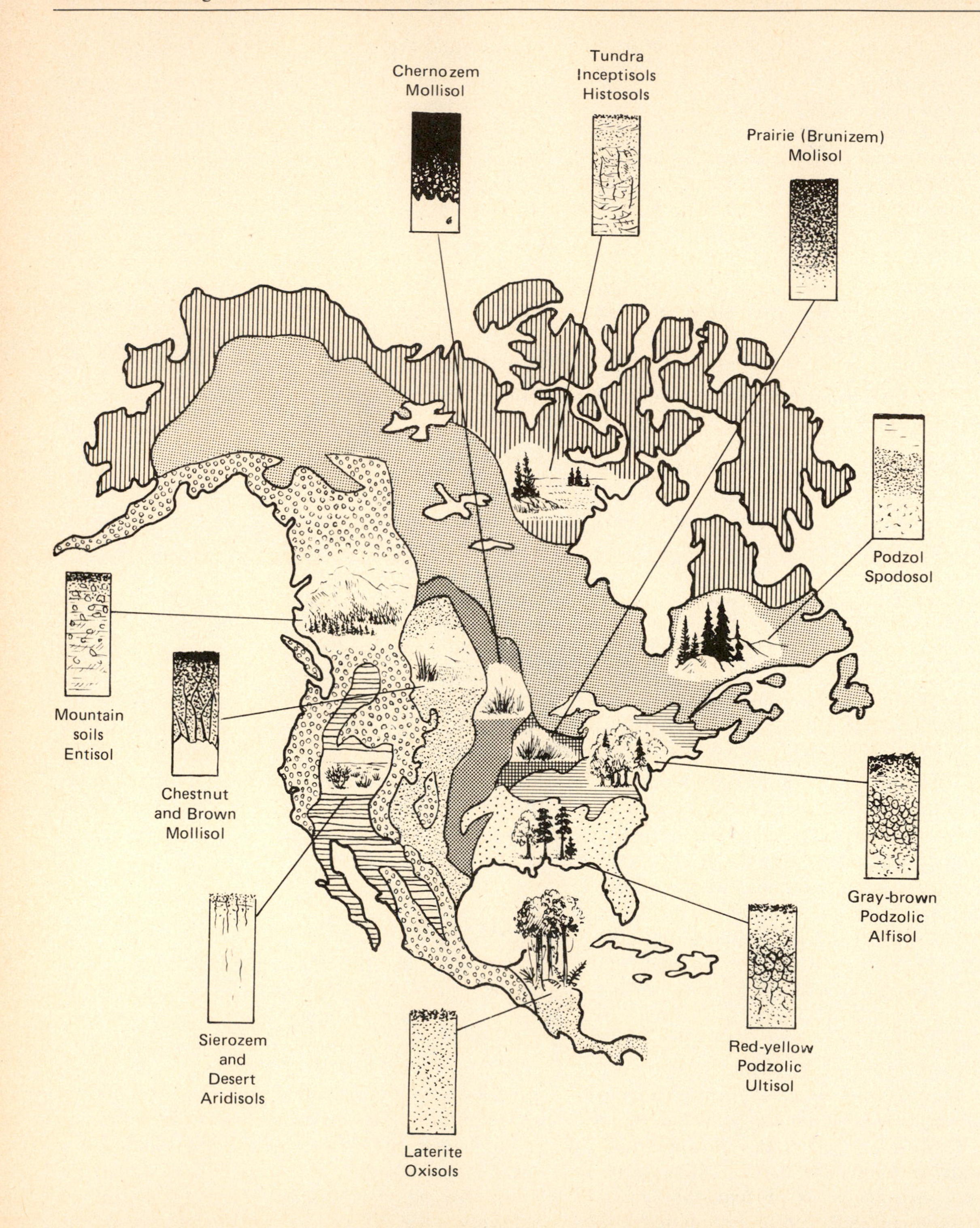
Chernozem
Mollisol
Tundra
Inceptisols
Histosols
Prairie (Brunizem)
Molisol
Podzol
Spodosol
Mountain
soils
Entisol
Chestnut
and Brown
Mollisol
Gray-brown
Podzolic
Alfisol
Sierozem
and
Desert
Aridisols
Red-yellow
Podzolic
Ultisol
Laterite
Oxisols

of soil minerals. In areas where water stays near or at the surface most of the time, iron, because of an inadequate supply of oxygen, is reduced to ferrous compounds; these give a dull gray or bluish color to the horizons.

This process is called *gleization* and may result in compact, structureless horizons. Gley soils are high in organic matter because more is produced than can be broken down by humification, which is greatly reduced because of the absence of soil microorganisms. On gentle to moderate slopes, where drainage conditions are improved, gleization is reduced and occurs deeper in the profile. As a result, the subsoil will show varying degrees of mottling of grays and browns. On hilltops, ridges, and steep slopes where the water table is deep and the soil well drained, the subsoil is reddish to yellowish brown because of the presence of oxidized iron compounds.

DRAINAGE CLASSES

In all, five drainage classes are recognized (Figure 8.4). Well-drained soils are those in which plant roots can grow to a depth of 91 cm without restriction due to excess water. On moderately well-drained soils, plant roots can grow to a depth of 45 to 50 cm without restriction. In somewhat poorly

FIGURE 8.3 The great soil groups. This map of North America shows the general distribution of the important zonal, or great soil, groups of the continent and points out the general relation of soils to vegetation and climate. The majority of the soils illustrated here (the exception is the tundra) are those that develop on well-drained sites. In the humid regions, bases do not accumulate in the soils because of the leaching processes associated with high rainfall. Podzol soils, or spondosols—characterized by a very thin organic layer on top of a gray, leached soil lying over a dark-brown horizon—generally develop in cool, moist climates under coniferous forests. Under deciduous forests gray-brown podzolic soils, or alfisols, develop. These differ from podzols in that the leaching is not so excessive and there is a horizon of grayish brown leached soil beneath the organic layer. The red and yellow soils, or ultisols, occur in the warm, temperate, moist climate of southeastern North America. Developed through podzolization with some laterization, yellow soils are characterized by a grayish yellow leached horizon over a yellow one; the red soils, by a yellowish brown leached soil over a deep red horizon. In tall-grass country with temperate, moist climates are prairie or brunizem soils, or mollisols, the result of calcification. This soil is very dark brown in color, grading through lighter brown with depth. West of this lies the chernozen, another mollisol, a black soil high in organic matter, some 3 to 4 feet deep, that grades into lime accumulations. It develops under tall- and mixed-grass prairie. Closely related are the chestnut and brown soils, also mollisols, dark brown and grading into lime accumulations at 1 to 4 feet. These soils develop under mixed- and short-grass prairie. In desert regions are siernozem and desert soils, or aridisols. They are grayish in color, low in organic matter, and closely underlain with calcareous material. Lateritic soils, or oxisols, are typical of tropical rain forests, where decomposition is rapid, and have a thin organic layer over a reddish leached soil. In high mountains are a variety of soils, here vaguely classified as mountain soils, or entisols. Many of them are stony and lack any well-developed horizons. Tundra soils are variable, but the common one is a gley; they are subject to considerable disturbance from frost action and are underlain with a permanently frozen substrate.

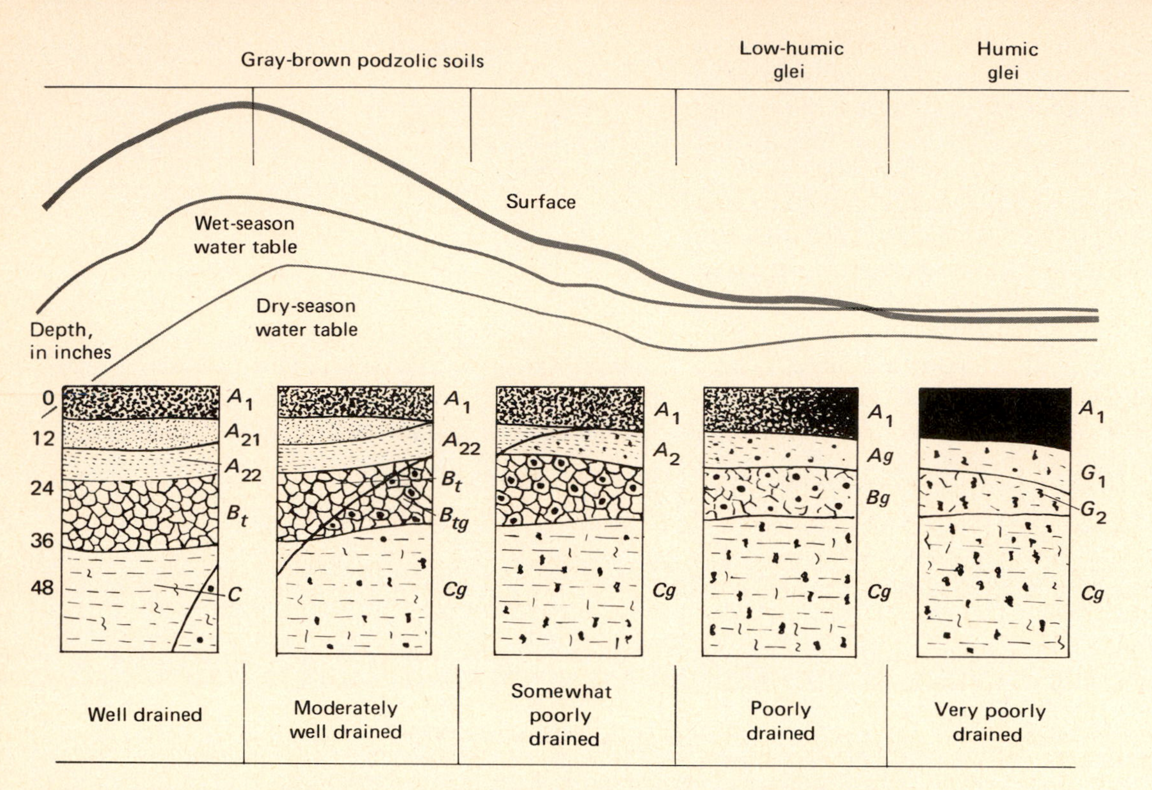

FIGURE 8.4 Effect of drainage on the development of alfisol soils. Wetness increases from left to right. This diagram represents the topographic position the profiles might occupy. Note that the strongest soil development takes place on well-drained sites where weathering is maximum. The least amount of weathering takes place on very poorly drained soils where the wet-season water table lies above the surface of the soil. *G* or *g* indicates mottling; *t* indicates translocated silicate clays. (After Knox, 1952:1.)

drained soils, plant roots cannot grow beyond a depth of 30 or 35 cm. Poorly drained soils are wet most of the time. They are usually characterized by the growth of alders, willows, and sedges. On very poorly drained soils, water stands on or near the surface most of the year.

Soils and Time

Weathering of rock material, accumulation, decomposition, and mineralization of organic material, loss of minerals from the upper surface, gains in minerals and clay in lower horizons, and horizon differentiation all require considerable time. Well-developed soils in equilibrium with weathering and erosion may require 2,000 to 20,000 years for their formation. But soil differentiation from parent material may take place in as short a time as 30 years. Certain acid soils in humid regions develop in 100 years because the leaching process is speeded by acidic materials. Parent materials heavy in texture require much longer to develop into "climax" soils because they impede the downward flow of water. Soils develop more slowly in dry regions than in humid ones. Soils on steep slopes often remain young regardless of geological age because rapid erosion removes the soil nearly as fast as it is formed. Floodplain soils age little through time because of the continuous accumulation of new materials.

Young soils are not as deeply weathered as old soils and are more fertile, since they have not been exposed to the leaching process as long. The latter tend to be infertile because they have long been subjected to leaching of nutrients without replacement from fresh material.

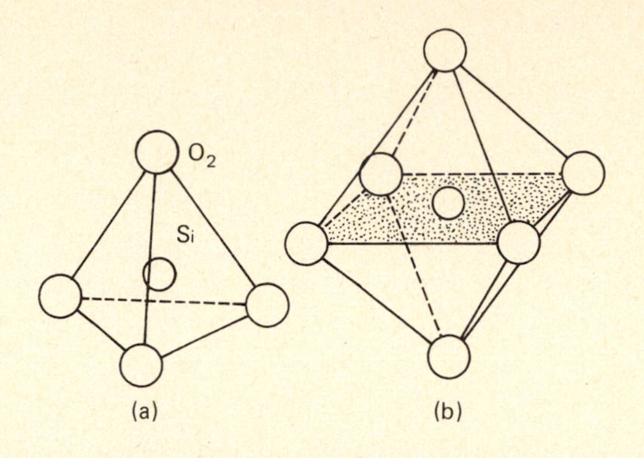

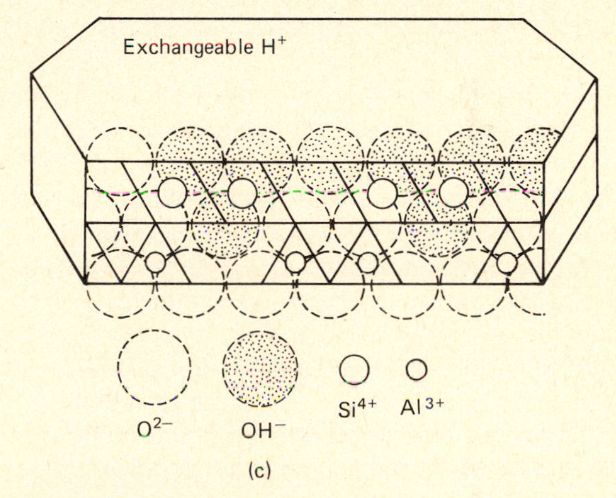

FIGURE 8.5 Structure of clay colloids. Basic building units of aluminosilicate crystals are (a) the four-sided tetrahedron with a small Si^{4+} in the central cavity formed by $4O^{2-}$; (b) eight-sided octahedron with Al^{3+} or Mg^{2+} in the interstices of $6O^{2-}$ and OH^-. (Spheres not drawn to size.) (c) Cross-section of one layer of mineral pyrophyllite shows arrangement of sheets of tetrahedrons (black, T) and octahedrons (shaded, O). Dotted half-circles represent oxygen planes. (After Jenny, 1980:51, 54.)

Soil Chemistry

Chemical elements in the soil are adsorbed on soil particles and dissolved in soil solution and are a constituent of mineral and organic matter. These ions move from soil to plant, from plant to animal, and into the biogeochemical cycle (see Chapter 21). In aquatic systems the ions are dissolved and obey the laws of diffusion and dilute solutions. In soils, ions are limited in their mobility because they are closely held to solid particles of clay and humus.

The key to the availability of nutrients in the soil is the nature of the clay-humus complex. It is made up of platelike particles in the soil called *micelles* (Figure 8.5). Micelles consist of sheets of tetrahedron and octahedron aluminosilicates (silicate, aluminum, iron combined with oxygen, and hydroxyl ions). The interior of the plates is electrically balanced, but the edges and sides are negatively charged. They attract positive ions, water molecules, and organic substances. The number of negatively charged sites on soil particles that attract positively charged cations is called the *cation exchange capacity* (Figure 8.6). These positively charged ions can be replaced by still other ions in the soil solution.

Cation exchange capacity varies among soils, depending upon the structures of clays. Some are made up of octahedron and tetrahedron sheets arranged in large lattices that swell and expand when moist. These possess an extensive surface area, externally and internally, which allows ion exchanges not only between micelles but within the micelles as well. Others have octahedrons and tetrahedrons arranged randomly. They are amorphous clays that do not expand when moist, and cation exchange is of a lesser magnitude.

Negatively charged ions attract cations of Ca^+, Na^+, Mg^+, K^+, and H^+, among others. Some cling to the micelles more strongly than others. H^+ ions are especially tenacious. Hydrogen ions added by rainwater, cationic acids from organic matter, and metabolic acids from roots displace other cations, such as Ca^+. If not taken up by plants, these cat-

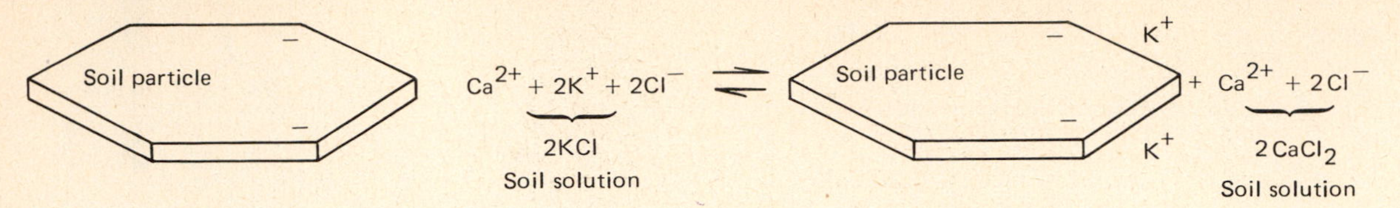

FIGURE 8.6 An illustration of cation exchange on a soil particle. Calcium adsorbed on the soil particle is replaced by two potassium ions. Calcium goes into solution combining with Cl^{-} to form calcium chloride.

ions are carried deeper into the soil or are removed altogether through the groundwater and frequently move into aquatic systems. The cation exchange capacity has a pronounced effect on soil fertility and availability of nutrients to plants.

Ions are available to plants only when dissolved in soil solution. Ions in soil solution maintain an equilibrium with ions adsorbed on micelles, and these, in turn, maintain an equilibrium with absorbed ions in the micelles. As plants remove ions from the soil solution in the vicinity of the roots, other ions diffuse to the region. That, in turn, enhances the release of ions from the micelles.

Acidity is one of the most familiar of all chemical conditions in the soil. If you raise plants, grow a garden, or maintain a lawn, you are aware of the implications of soil acidity. And of major concern are the effects of acid rain on terrestrial and aquatic ecosystems (see Chapter 21). Soil acidification results from the removal of bases by the leaching effects of water moving through the soil profile, the withdrawal of exchangeable ions by plants, the release of organic acids by roots and microorganisms, and the dissociation of $CaCO_3$. If the soil is poorly buffered against acidic inputs, then soil acidity increases.

Typically, soils range between a very acid pH of 3 and a pH of 8, strongly alkaline. Soils just over a pH of 7 (neutral) are considered basic and those of 6.6 or less acid.

Soil acidity has a pronounced effect on nutrient availability. As soil acidity increases, the proportion of exchangeable Al^{+} increases, and Ca^{+}, K^{+}, Na^{+}, and others decrease. Such changes bring about not only nutrient deprivation but also aluminum toxicity. Harmful effects of low pH in both soil and aquatic environments are due not so much to the acid as to the toxic Al^{+} and Fe^{+} ions released in soil and water.

Soil Color

Color has little direct influence on the function of a soil, but considered with other properties, it can tell a good deal about the soil. In fact, it is one of the most useful and important characters for the identification of soil. In temperate regions dark-colored soils generally are higher in organic matter than light-colored ones. Well-drained soil may range anywhere from very pale brown to dark brown and black, depending upon organic matter content. But it does not always follow that dark-colored soils are high in organic matter. Soils of volcanic origin, for example, are dark in color. In warm temperate and tropical regions, dark clays may have less than 3 percent organic matter. Red and yellow soils result from the presence of iron oxides, the bright colors indicating good drainage and good aeration. Other red soils obtain their color from parent material and not from soil-forming processes. Well-drained yellowish sands are white sands containing a small amount of organic matter and such coloring material as iron oxide. Red and yellow colors increase from cool regions to the equator. Quartz, kaolin, carbonates of lime and magnesium, gypsum, and various compounds of ferrous iron give whitish and grayish colors to the soil. The grayest are permanently saturated soils in which iron is in the ferrous form. Imperfectly and poorly drained soils are mottled with various shades of yellow-brown and gray. The colors of soils are determined by the use of standardized color charts.

Soil Texture

Differences among soils and among horizons within a soil are primarily reflected by variations in texture, arrangement, and color. The *texture* of a soil is determined by the proportion of different-size soil particles (Figure 8.7). Texture is partly inherited from parent material and partly a result of the soil-forming process. Particles are classified on the basis of size into gravel, sand, silt, and clay. Gravel consists of particles larger than 2.0 mm. Sand ranges from 0.05 to 2.0 mm, is easily seen, and feels gritty. Silt consists of particles from 0.002 to 0.05 mm in diameter, which can scarcely be seen by the naked eye and feel and look like flour. Clay particles too small to be seen under an ordinary microscope are colloidal in nature. Clay controls the most important properties of soils, including plasticity and exchange of ions between soil particles and soil solution. Most soils are a mixture of these various particles.

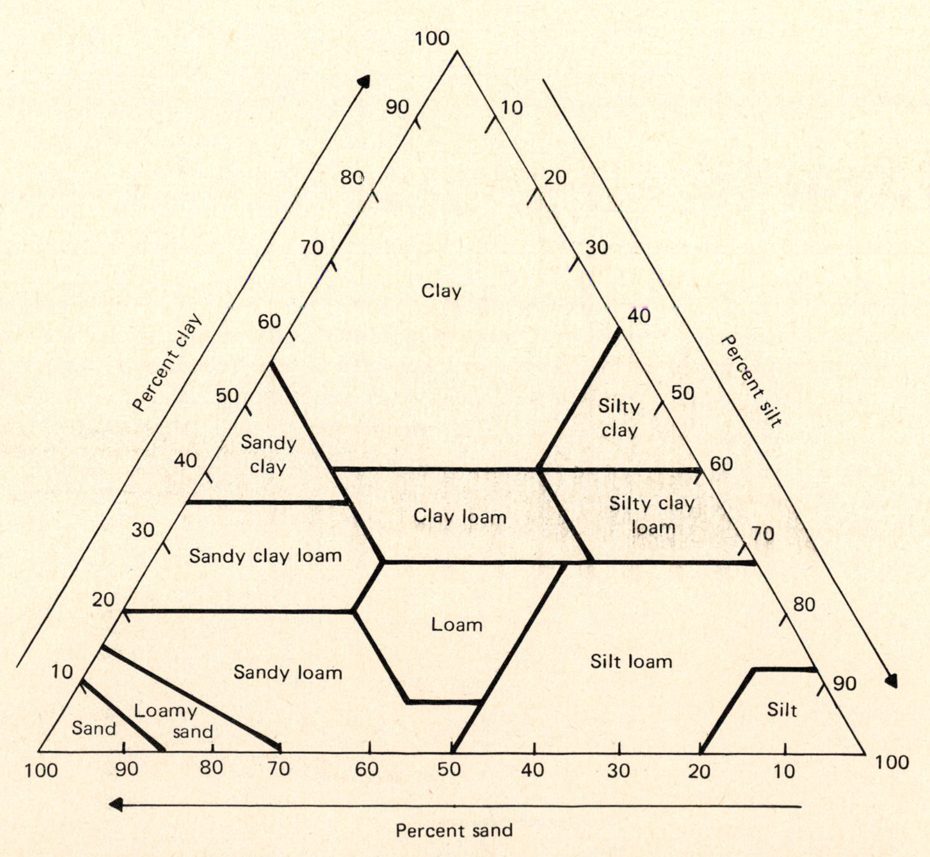

FIGURE 8.7 A soil texture chart showing the percentages of clay (below 0.002 mm), silt (0.002 to 0.05 mm), and sand (0.05 to 2.0 mm) in the basic soil textural classes.

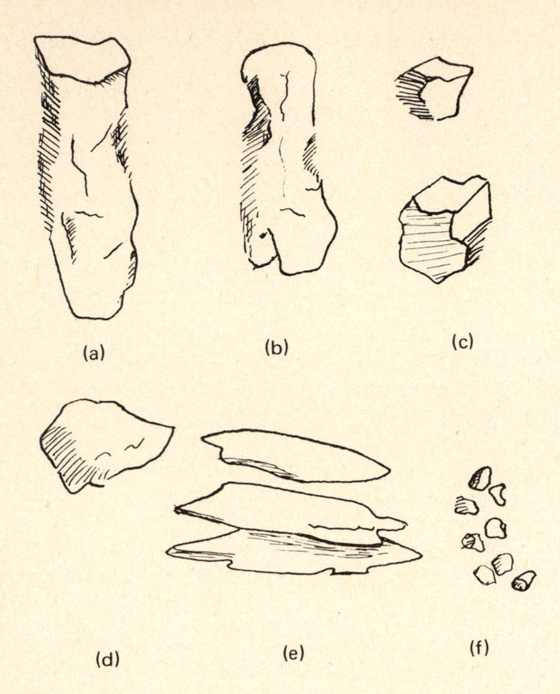

FIGURE 8.8 Some types of soil structure. (a) Prismatic, (b) columnar, (c) angular blocky, (d) subangular blocky, (e) platelike, (f) granular.

Soil particles are held together in clusters or shapes of various sizes, called *aggregates* or *peds*. The arrangement of these aggregates is called *soil structure*. As with texture, there are many types of soil structure. Soil aggregates may be classified as granular, crumblike, platelike, blocky, subangular, prismatic, and columnar (Figure 8.8). Structureless soil can be either single-grained or massive. Soil aggregates tend to become larger with increasing depth. Structure is influenced by texture, plants growing on the soil, other soil organisms, and the soil's chemical status.

Soil Classification

Each combination of climate, vegetation, soil material, slope, and time results in a unique soil, the smallest repetitive unit of which is called a *pedon*. Soils may vary considerably even within a small area. Changes in slope, drainage, and soil material account for local differences between soil individuals. These soil individuals are roughly equivalent to the lowest category in the soil classification system—the *soil series*.

The present soil taxonomic system consists of orders (see Table 8.1), suborders, great groups, families, and series. In the new classification system, the names of lower classification units always end on the formative syllable of their respective order, preceded by other syllables connotative of various soil properties. At the highest taxonomic level, emphasis is placed on the presence or absence of certain diagnostic soil horizons that result from the interaction of soil-forming factors, primarily climate and vegetation. Soil series are named after the locality in which they were first described. For example, the Ovid series in New York was named after the town of Ovid, the Miami after the Miami River in western Ohio. Like the species among plants and animals, soil series are defined in terms of the largest number of differentiating characteristics and occur in fairly limited areas. Higher categories of taxonomy combine series into larger groupings distinguished by even fewer differentiating properties. At the highest level of classification, one recognizes only classes that correspond roughly with broad climatic zones.

Every soil series has neighboring soil series with unlike properties into which it grades abruptly or gradually. If these several soils found side by side have developed from the same soil material but differ mainly in natural drainage and slope, they are said to form a *catena* (see Figure 8.9). When soils are mapped, unlike soils may be grouped together into *associations* for reasons of scale or practical use. When occurring in inseparable patterns, soils are mapped as *complexes*. In detailed mapping, a soil series can be subdivided into *types, phases,* and *variants*.

The Soil as an Environment

The soil is a radically different environment for life than the one above the surface, yet the essential requirements do not differ. Like animals that live

TABLE 8.1 NEW SOIL ORDERS, DESCRIPTION, AND APPROXIMATE EQUIVALENTS IN OLD CLASSIFICATIONS

Order	Derivation and Meaning	Description	Approximate Equivalents
Entisol	Coined from *recent*	Dominance of mineral soil materials; absence of distinct horizons; found on floodplains	Alluvial soils, azonal soils, regosol, lithosol
Vertisol	L., *verto,* for "inverted"	Dark clay soils that exhibit wide, deep cracks when dry	Grumusols
Inceptisol	L., *inceptum,* for "beginning"	Texture finer than loamy sand; little translocation of clay; often shallow; moderate development of horizons	Brown forest soil, sol brun acide, acide, humic gley, weak podzols
Aridisol	L., *aridus,* for "arid"	Dry for extended periods; low in humus, high in base content; may have carbonate, gypsum, and clay horizons	Sierozems, red desert soils, solonchak
Mollisol	L., *mollis,* for "soft"	Surface horizons dark brown to black with soft consistency; rich in bases; soils of semihumid regions	Chestnut, chernozem, prairie; some brown and brown forest and associated humic gleys
Spodosol	Gr., *spodos,* for "ashy"	Light gray, whitish A_2 horizon on top of a black and reddish B horizon high in extractable iron and aluminum	Podzol, brown podzolic soils
Alfisol	Coined from *Al* and *Fe*	Shallow penetration of humus, translocation of clay; well-developed horizons	Gray-brown podzolic, gray wooded soils, noncalcic brown soils, some planisols
Ultisol	L., *ultimus,* for "last"	Intensely leached; strong clay translocation, low base content; humid, warm climate	Red-yellow podzolic, red-brown laterite, some latisols
Oxisol	Fr., *oxide,* for "oxidized"	Highly weathered soils; red, yellow, or gray; rich in kalolinite, iron oxides, and often humus; in tropics and subtropics	Laterites, latosols
Histosol	Gr., *histos,* for "organic"	High content of organic matter	Bog soils, muck

outside the soil, soil fauna require living space, oxygen, food, and water.

The soil in general possesses several outstanding characteristics as a medium for life. It is relatively stable, both chemically and structurally. Any variability in the soil climate is greatly reduced compared to conditions above the surface. The atmosphere remains saturated, or nearly so, until soil moisture drops below a critical point. The soil affords a refuge from high and low extremes in temperature, wind, evaporation, light, and dryness, permitting soil fauna to make relatively easy adjustments to unfavorable conditions.

On the other hand, soil has low penetrability. Movement is greatly hampered. Except to such channeling species as earthworms, soil pore space

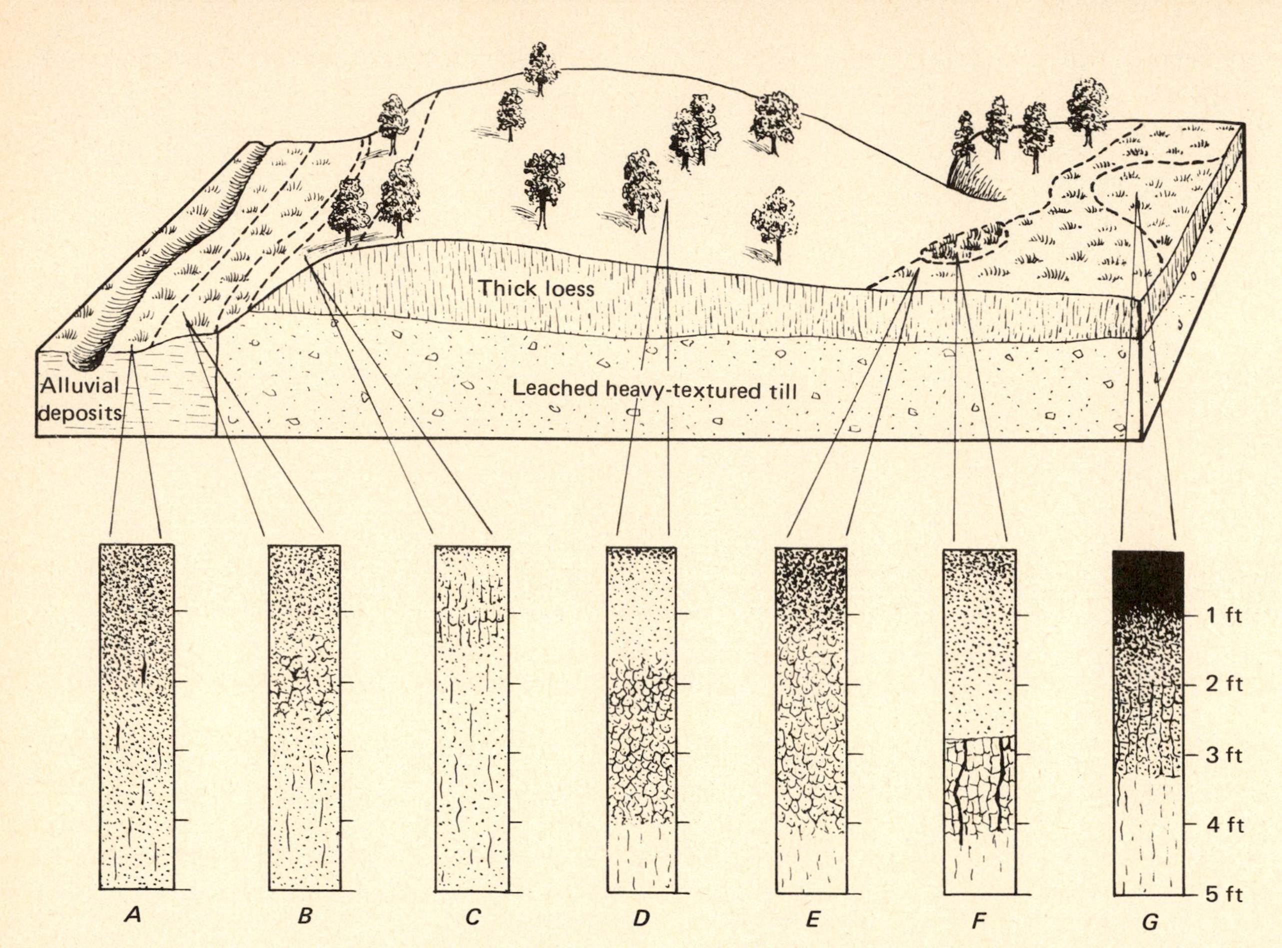

FIGURE 8.9 Effect of topography and native vegetation on the soil. This diagram shows the normal sequence of eight representative soil types from the Mississippi to the upland of Illinois. The drawing also illustrates how bodies of soil types fit together in the landscape. Boundaries between adjacent bodies are gradations or continua rather than sharp lines.

The lower part of the diagram pictures the profiles of seven of the soils, showing the color and thickness of the surface horizon and the structure of the subsoil. Note how the natural vegetation that once covered the land (trees for forest, grass clumps for grass) influenced surface color. The diagram also shows how topographic position and distance from the bluff influence subsoil development.

Profile a is a bottomland soil (Sawmill) formed from recent sediments and has not been subjected to much weathering. Profile b (Worthen), on the foot slope, also developed from recent alluvial material and shows little structure. Profile c (Hooper), on the slope break, developed from a thick loess on top of leached till, while the soil on the bottom of the slope developed directly from the till. Profile d is an upland soil (Seaton) formerly covered with timber. It has a light surface color and lacks structure; this resulted from a rapid deposition of loess during early soil formation that held soil weathering to a minimum. Profile e represents an upland soil (Joy) developed under grass. Note the dark surface and the lack of structure, again a result of rapid deposition of loess. Profile f (Edgington) is a depressional wet spot. Extra water flowing from adjacent fields increased the rate of weathering, resulting in light-colored grayish surface and subsurface and a blocky structure to the subsoil. This indicates a strongly developed soil. The depth of subsoil suggests that considerable sediment has been washed in from the surrounding area. Profile g (Sable) represents a depressional upland prairie soil. Note the deep, dark surface and coarse, blocky structure. Abundant grass growth produced the dark color. (After Veale and Wascher, 1956:57.)

is very important, for this determines the nature of the living space, the humidity, and the gaseous condition of the environment.

The variability of these conditions creates a diversity of habitats, reflected in the diversity of species found in the soil (Birch and Clark, 1953) (Figure 8.10). The number of different species found in the soil is enormous, representing practically every invertebrate phylum. There are 250 species of Protozoa alone in English soils (Sandon, 1927). The number of species of soil animals exclusive of Protozoa found in a variety of habitats in Germany varies from 68 to 203 (Frenzell, 1936). In the soil of a beech woods in Austria live at least 110 species of beetles, 229 species of mites, and 46 species of snails and slugs (Franz, 1950). E. C. Williams (1941) counted 294 species of soil animals, exclusive of Protozoa, in the Panama rain forest.

Only a part of the soil litter is available to most soil animals as living space. Spaces between the surface litter, cavities walled off by soil aggregates, pore spaces between individual soil particles, and root channels and fissures are all potential habitats. Most of the soil fauna are limited to pores and cavities larger than themselves. The distribution of these forms in different soils is often determined in part by the structure of the soil (Weis-Fogh, 1948), for there is a relationship between the average size of soil spaces and the fauna inhabiting them (Kuhnelt, 1950). Large species of mites inhabit loose soils with a crumb structure, in contrast to smaller forms inhabiting compact soils. Larger soil species are confined to upper layers, where the soil interstices are largest (Haarløv, 1960).

Water in the spaces is essential, because the majority of soil fauna is active only in water. Soil

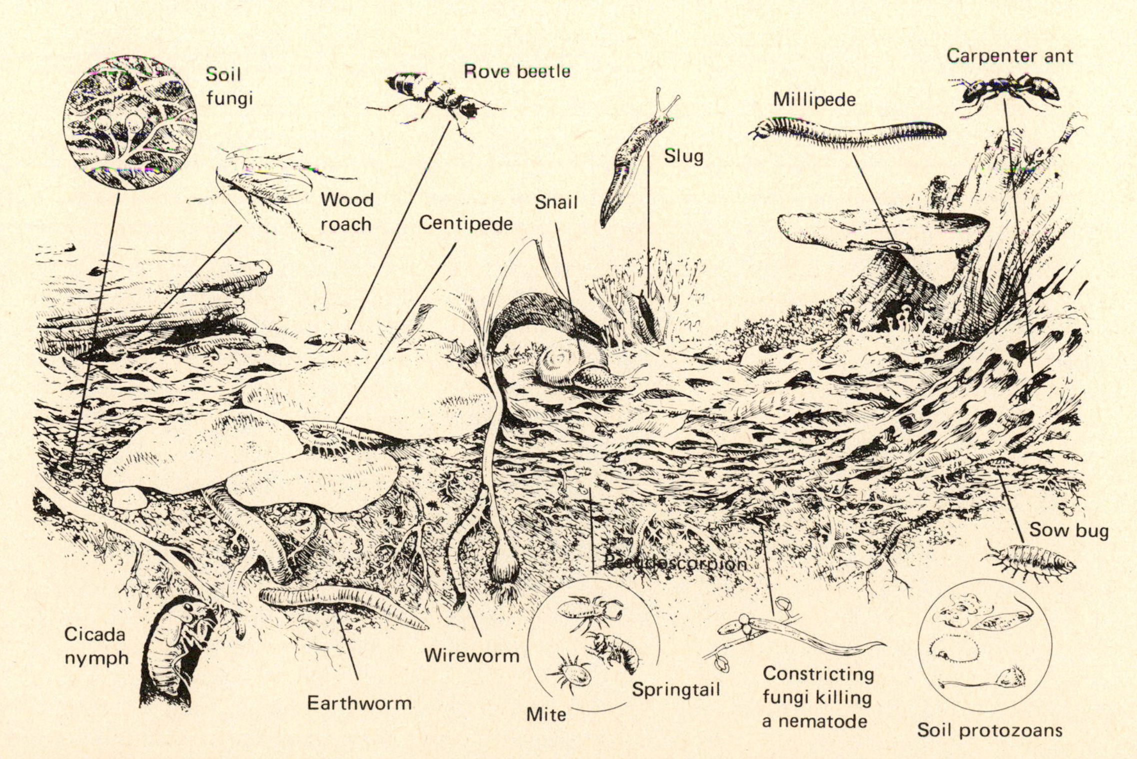

FIGURE 8.10 Life in the soil. This drawing shows only a tiny fraction of the organisms that inhabit soil and litter. Note the fruiting bodies of fungi, which in turn, furnish food for soil-associated organisms.

water is usually present as a thin film coating the surface of soil particles. This film contains, among other things, bacteria, unicellular algae, protozoa, rotifers, and nematodes. Most of these are restricted in their movements by the thickness and shape of the water film in which they live. Nematodes are less restricted, for they can distort the water film by means of muscular movements and thus bridge the intervening air spaces. If the water film dries up, these species encyst or enter a dormant state. Millipedes and centipedes, on the other hand, are highly susceptible to desiccation and avoid it by burrowing deeper into the soil (Kevan, 1962).

Excess water and lack of aeration are detrimental to many soil animals. Excessive moisture, which typically occurs after heavy rains, is often disastrous to soil inhabitants. Air spaces become flooded with deoxygenated water, producing a zone of oxygen shortage for soil inhabitants. If earthworms cannot evade this zone by digging deeper, they are forced to the surface, where they die from excessive ultraviolet radiation. The snowflea (a collembola) comes to the surface in the spring to avoid excess soil water from melting snow (Kuhnelt, 1950). Many small species and immature stages of larger species of centipedes and millipedes may be completely immobilized by a film of water and unable to overcome the surface tension imprisoning them. Adults of many species of these organisms possess a waterproof cuticle that enables them to survive some temporary flooding.

Vegetation and Soils

The fact that a close relationship exists between vegetation and soils is most apparent when parent material or strong biological activity has a marked effect on soil development and characteristics. These, in turn, determine plant communities. The vegetation reflects the soil and further influences soil development.

Two contrasting vegetational communities exist on high-lime (or limestone) and low-lime (or acid) soils. In general, limestone soils support plants that have a high uptake of calcium, need an adequate supply of phosphorus and magnesium, and are intolerant of aluminum ions, which inhibit the uptake of phosphorus. Plants of acid soils cannot tolerate a high level of calcium, which inhibits their uptake of potassium and iron. The vegetation of acid soils includes such plants as heaths, laurel (*Kalmia*), blueberries (*Vaccinium*), oaks (*Quercus*), and chestnut. Limestone soils in eastern North America support such plants as hickory, red cedar (*Juniperus virginiana*), maples, walking ferns (*Camptorsorus*), spleenwort (*Asplenium*), and blazing star (*Liatris*). Limestone soils, especially where outcrops are common and the soils shallow, are xeric sites because crevices and channels in the rocks carry the water far below the surface.

A strong division between the plant communities of limestone and dolomitic soils and those of sandstone or noncarbonate soils exists in the semiarid, subalpine White Mountains of eastern California (Marchand, 1973). This discontinuity is most pronounced along the boundaries of the two geological contacts. The noncarbonate soils are dominated by sagebrush (*Artemisia tridentata*), silvery lupine (*Lupinus argenteus*), and *Koeleria cristata*. The carbonate soils are dominated by bristlecone pine (*Pinus aristata*), limber pine (*Pinus flexilis*), phlox (*Phlox covillei*), and locoweed (*Astragalus kentrophyta*).

In the Pacific Northwest and northern California are large areas of serpentine soils derived from green-colored rocks such as igneous peridotite and metamorphic serpentine rocks. Such soils are low in calcium and high in nickel, magnesium, and chromium, in that order. The effects of these elements—modified by topography, elevation, and water availability—are most conspicuous where plants on serpentine soils grow next to plants on adjacent nonserpentine soils. Serpentine areas are characterized by stunted vegetation and an unusual flora, which includes endemics. A typical serpentine community is Jeffrey pine (*Pinus jeffreyi*)—grass woodland, which contrasts strongly with adjacent stands of Douglas fir. At high elevations the serpentine community is dominated by grasses and at lower elevations, where the moisture conditions are better, by a greater diversity of species. Other areas of serpentine soils occur in Quebec, where tundra contrasts with tiaga on nonserpentine soils; in northern California, where chaparral on serpentine contrasts with oak woodland; in Appalachian regions; and in Cuba, where savanna and shrub contrast with tropical forests (for details see Whittaker, Walker, and Kruckeberg, 1954; Whittaker, 1960; Franklin and Dyrness, 1973).

Summary

All soils are the base for terrestrial ecosystems. Soil is the site of the decomposition of organic matter and of the return of mineral elements to the nutrient cycle. It is the home of animal life, the anchoring medium for plants, and their source of water and nutrients. Soil begins with the weathering of rocks and minerals, which involves the leaching out and carrying away of mineral matter. Its development is guided by slope, climate, original material, and native vegetation. Plants rooted in the weathering material further break down the substratum, pump up nutrients from its depths, and add all-important organic material. This material, through decomposition and mineralization, is converted into humus, an unstable product that is continuously being formed and destroyed by mineralization.

As a result of the weathering process, accumulation and breakdown of organic matter, and the leaching of mineral matter, horizons or layers are formed in the soil. Of these there are four: the O, or organic, layer; the A horizon, characterized by accumulation of organic matter and loss of clay and mineral matter; the B horizon, in which mineral matter accumulates; and C, the underlying material. These horizons may be further divided into subhorizons.

Of all the horizons none is more important than the humus layer, which plays a dominant role in the life and distribution of plants and animals, in the maintenance of soil fertility, and in much of the soil-forming process. Humus is usually grouped into three types: mor, characteristic of acid habitats, whose chief decomposing agents are fungi; mull, characteristic of deciduous and mixed woodlands, whose chief decomposing agents are bacteria; and finally, moder, which is highly modified by the action of soil animals.

Profile development is influenced over large areas by vegetation and climate. In grassland regions the chief soil-forming process is calcification, in which calcium accumulates at the average depth reached by percolating water. In forest regions podzolization—involving the leaching of calcium, magnesium, iron, and aluminum from the upper horizon and the retention of silica—takes place. In tropical regions laterization, in which silica is leached and iron and aluminum oxides are retained in the upper horizon, is the major soil-forming process. Gleization takes place in poorly drained soils. Organic matter decomposes slowly, and iron is reduced to the ferrous state.

Differences between soils and between horizons within soils are reflected by variations in texture, structure, and color. Each combination of climate, vegetation, soil material, slope, and time results in a unique soil, of which the smallest repetitive unit is the pedon. Soil individuals equivalent to the lowest category in the soil classification system are the soil series. These may be further categorized into families, great groups, suborders, and orders.

Review and Study Questions

1. Why is it difficult to define soil?
2. What are some of the elements essential for life? What are the bulk elements? Macroelements? Microelements? What is the relationship of these elements to soil?
3. What role does physical and chemical weathering play in soil development?
4. What is the role of plant and animal life?
5. What is mineralization? Humus? Do the two relate?
6. Distinguish between mull and mor humus. What is their ecological significance?

7. Compare calcification, podzolization, and laterization.
8. What is gleization? Under what conditions does it take place?
9. What are the five major drainage classes? What is their ecological significance?
10. What is a micelle? How do micelles relate to cation exchange capacity?
11. What is cation exchange capacity? How does it relate to nutrient uptake? To soil acidification?
12. What is soil texture? How does it relate to soil particle size?
13. How do soil particles relate to soil structure?
14. Why develop a soil classification scheme? What is a soil catena? A soil series?
15. What characterizes the soil as an environment for life?
16. In what way is soil, especially its acidity and alkalinity, a selective force in the evolution of plants?
17. Modern soil surveys describe soils on a local basis. They also relate local soils and soil characteristics to agriculture, forestry, wildlife, suburban construction, and engineering. Check with your local or state soil conservation office to determine whether a soil survey is available for your area. If so, obtain a copy and study it. Note the pattern of soil distribution. What are the local soil series? Relate local soils to agricultural development and to forest distribution.

Suggested Readings

Brady, N. C. 1974. *The nature and properties of soils.* 8th ed. New York: Macmillan.

Jenny, H. 1980. *The soil resource.* New York: Springer-Verlag.

Lutz, H. J., and R. F. Chandler. 1946. *Forest soils.* New York: Wiley. (Out of print and old, but excellent.)

UNESCO. 1969. *Soil Biology.* Paris: UNESCO.

Wallwork, J. A. 1973. *Ecology of soil animals.* New York: McGraw-Hill.

ECOLOGICAL EXCURSION

Formation of Soil

Peter Farb

The soil lies upon the planet much as the rind upon an orange. But the rind of the soil is not uniform everywhere: it is twenty feet and more deep in places like the Florida Everglades, less than an inch thick in sections of Canada where the glaciers have scraped the rocks bare. It is usually thickest at the Equator, and thins out toward the poles. Its color may be red, as in western Oklahoma, or black, as in Kansas. It may be sand or shale or clay, or fit any one of hundreds of other soil classifications. But always, it is the sole link between the lifeless earth and the teeming world on its surface.

The mountains appear to loom large and the depths of the sea to be unfathomed. Yet, for a ball the size of the earth, these surface irregularities are considerably less, comparatively, than on the rind of the orange. Even these slight configurations on the earthen ball are continually being worn down: the highest peaks lose tiny grains that start on an endless journey. "All soil," once wrote the Harvard geologist Nathaniel Southgate Shaler, "is rock material on its ways to the deep."

Landslides may launch the grain of soil, the grinding of rocks and glaciers propel it, winds catch and sweep it for miles. The soil that now lies upon the continents is merely at its resting stage, momentarily pausing and giving sustenance before taking up its long journey once again. A mountainside is a busy place, soil particles coming and going, grains dislodging each other, rubbing together like sandpaper. With time, there is no substance on earth too hard to be broken down into what may ultimately become soil.

John Burroughs, the Hudson River Valley naturalist, in his native Catskills watched the mountains of rock being created into soil:

> The mountain melts down into a harvest field; vocanic scoria changes into a garden mold; where towered a cliff now basks a green slope; where the strata yawned now bubbles a fountain; where the earth trembled, verdure now undulates. Your lawns and your meadows are built upon the ruins of the foreworld. The leanness of granite and gneiss has become the fat of the land.
>
> —from *Leaf and Tendril,* 1908

The rocks must endure all the ordeals of nature. The sun heats them to tremendous temperatures during the summer day, expanding the minerals lodged there; at night, the coolness contracts them. Repeatedly carried on over decades and centuries and millennia, the strain of expansion and contraction splits off pieces of rock; even the bulkiest must soon crumble.

Although the outside of a rock may be hot to the touch, a few inches inside toward the core it is still cold. That is because rocks are poor conductors of heat. Thus the outside layers expand and contract more than the core, eventually flaking off. Geologists call this "onion weathering," and it is very apparent in the deserts, where the relentless sun expands the rock skin to the limit, and the sudden temperature drop at night contracts it so quickly as to snap off the upper layers. At night in the desert mountains may be heard what sound like rifle shots. They are nature's artillery, shooting off rock that will someday become soil.

Larger fragments of rock, broken off by physical weathering, are further fractured. When they reach creeks and rivers they are rolled along by the waters, ground into ever finer particles. A bend in a river is nearly always sand or gravel, because a stream slows down there and drops its load of fine soil particles. During floodtime, when the stream leaves its course, it spreads the deposited grains thinner and further inland. On the deltas and flood plains of most rivers are built up masses of particles

---

SOURCE: Peter Farb, *Living Earth* (New York: Harper & Row, 1959), 24–28.

drawn from a drainage area that reaches out thousands of miles into the very heart of continents. Each year the Mississippi River dumps 525 million tons of soil and rock fragments into the Gulf of Mexico. This means that about every 7500 years, a foot of soil and rock is removed over the entire drainage basin of the Mississippi, yet it is constantly renewed as more parent material turns into soil.

The soils deposited by river systems are known as "alluvial;" much of the state of Mississippi, for example, has been formed in this way. Alluvial soils are among the richest on the planet, and a large percentage of the world's population is fed from those formed by the Mississippi, Yangtze, Ganges, Amazon, Nile. The high fertility of these soils is due to the fact that the particles were drawn from diverse areas, with a whole range of rocks and minerals to provide copious amounts of every essential for plant growth.

Not only must the rocks submit to physical forces, but there are chemical ones as well. The rainwater that falls upon the rocks and flows over them is by no means pure. As it falls, it picks up some carbon dioxide from the air which dissolves and forms carbonic acid. It is not rain alone that spatters upon the rocks, but acids powerful enough to etch them away and even form limestone caves.

Finally, the forces of life itself create soil. Rocks are broken apart by the expansion of roots in crevices and are also dissolved by acids secreted from the roots. The roots continue to perpetuate the life of their piece of earth by bringing up water from the lower horizons of the soil, water containing dissolved minerals which eventually lodge in the leaves. When the leaves fall and decay, the minerals from the depths are then deposited on the surface of the soil. The giant oak in a meadow, which appears to be the very symbol of rest, is working actively underground to renew its own soil world.

The vegetation of this planet is not merely living off the stored-up fertility of the soil; it is constantly adding to the soil. Scientists have filled a tub with a weighted amount of earth and planted a tree seedling in it. After five years they have found that the tree is perhaps a couple of hundred pounds heavier than when planted, but the soil has lost only a mere pound or two. Where, then, did the tree gain its bulk? From the sun, air, and water used during its growth. So when the tree dies, it returns to the earth not only the elements it obtained directly from the soil, but also the new supplies it manufactured. Vegetation adds to soil fertility, and there is no way to husband a piece of land so well as to let something grow on it and return its residue to the soil.

Thus, the soil is formed from the wearing away of the rocks, and because there are many kinds of rocks, with many possible combinations, countless sorts of soils will arise. Soils formed by limestones, for example, differ from every other kind because limestone rocks were created almost wholly through the action of living things under the oceans, later to be left by receding seas. Limestones consist mostly of corals, shells, and lime filtered out of the sea water by minute animals. Chalk, such as we see in the white cliffs of Dover, where it lies a few thousand feet thick, is the remains of little sea animals called Foraminifera. Coquina rock, used in the construction of St. Augustine and many old buildings on the Florida coast, is a warm-water limestone composed mostly of shell fragments. The peat of the Florida Everglades is underlaid by thick limestone, manufactured by sea animals; whereas peat is often too acid, except for specialized plant growth, the limestone layer has neutralized it and made the Everglades one of the richest agricultural areas in the world.

On many mountainsides, the past and future of soil can be seen on rock outcroppings at the summit. One end, embedded in the mountain, may be covered with a thin layer of soil; the other end of the rock is bare and crumbling. At the base is scarcely an inch of soil, deep enough to hold some annual weeds and a venturesome tree seedling; then comes a zone of crumbling mosses, as the soil mantle grows thinner. The mosses melt into a spreading film of crusty green plants—the lichens—then scattered lichen colonies, finally bare rock eaten away by weathering.

The advancing wave of the lichens is silently at work, dissolving the skin of the rock with acids. Its hairs are wedged between the rock particles, growing and swelling, giving off harsh secretions, dislodging minute bits. Dust from the air settles upon the lichens which, when they die, become a nursery bed of organic and mineral matter, suitable for the mosses.

A mass of splintered rock particles, worn to tiny fragments, some of them microscopic, and deposited in the lowlands, is still not soil. And it will not become soil until life has been added to it. There is no soil without life; they are inseparable. The living things make the difference between a mere mass of mineral particles and a mineral *soil.*

Soils can be made, ultimately, only by the addition to the mineral particles of living and once-living matter, called organic matter. It is stuff that has seen the flame of life—the litter of dead leaves and twigs, husks of sucked-out insects, rotting plant roots, even the pinpoint bodies of microbes. The content of organic matter varies: in some peaty soils of Florida and Michigan, it is nearly wholly organic matter; on the other hand, some desert soils of the Southwest contain but a trace of 1 percent of organic matter. A complex association of life is required in the skeleton of rock fragments and, as a general rule, the greater the complexity of living things that have inhabited the soil, the more productive it is. Soil must have a history, and be able to boast of a pomp of living creatures, each of which has left its mark.

CHAPTER 9

Fire in the Environment

Outline

Objectives

Upon completion of this chapter, you should be able to:

1. Discuss the ecological differences between fires set by lightning and those set by humans.
2. Describe the three types of fire.
3. Discuss the importance of fire frequency to natural ecosystems.
4. Describe some of the effects of fire on soil.
5. Discuss the role of fire as a selective agent in evolution.
6. Explain ways in which fire-dependent plants respond to burning.
7. Discuss how vegetation in fire-dependent ecosystems may have inherently increased its flammability.
8. Describe some fire-adapted species of plants and explain why they require periodic burning.
9. Discuss the effects of fire on vertebrates, especially fire-dependent species.

Fire. The word to us is ambivalent. It connotes warmth and cheer and also fear and destruction. In nature, fire has the same ambivalence. It is both a destroyer and a regenerator of life.

For centuries Western culture has considered fire in forests and grasslands as destructive. European culture did not accept fire as part of the natural order. Europeans regarded burning of vegetation by native peoples as a reflection of their ignorance and as highly destructive. When European settlers arrived in North America, they became concerned about fire suppression. The great forest fires of the Miramichi in New Brunswick in 1825, the Peshtigo fire in Wisconsin in 1871, the Hinkley, Minnesota, fire in 1896, and the Tillamook fire in Oregon in 1933, all fed by great piles of slash left after logging, further emphasized the destructiveness of fire and the need for suppression.

The idea that fires were only destructive affected ecological thinking. Ecologists tended to ignore the importance of fire as an evolutionary force in natural communities, although both F. C. Clements and J. A. Weaver alluded to it in their writings. Early research on fire was carried out by foresters, wildlife biologists, and range managers, beginning mostly in the 1930s. They were interested in the applied side of fire—its use as a tool to obtain maximum production of wood, forage, and wildlife. The ecological relation of fire to plants and the evolution of adaptations to fire received little attention until about 30 years ago. Since then research on the ecology of fire has been extensive, and fire is now regarded as one of the major environmental influences, along with light, moisture, nutrients, and other variables. Much of this change in attitude and thinking about fire must be credited to E. V. Komarek, Sr., of Tall Timbers Research Station, Tallahassee, Florida. For a period of 15 years, from 1962 to 1976, his annual Tall Timbers Fire Ecology Conference and the resulting proceedings focused attention on the role, problems, and uses of fire in world ecosystems. Since then a number of symposia on the role and use of fire have been held worldwide.

Conditions and Causes of Fire

Three conditions are necessary for fire to assume ecological importance: (1) an accumulation of organic matter sufficient to burn, (2) dry weather conditions to render the material combustible, and (3) a source of ignition. The only two important sources of ignition are lightning and humans.

Globally, certain regions possess conditions conducive for burning and the spread of fires set by lightning. One condition is a fire climate, a dry period in which fuel built up in wetter times can burn. The wetter the growing season and the longer and hotter the dry season, the more chances there are for lightning-set fires. Africa is ideally located for the development and occurrence of thunderstorms and lightning (Phillips, 1965; Batchelder, 1967). North America is literally swept with waves of electricity (Komarek, 1964, 1966). Southern and western Australia experience hot summers with low humidity and drying westerly winds (Cochrane, 1968). Long before the emergence of humans, fires periodically swept these regions (Harris, 1958). Thus, each region is characterized by vegetation that evolved under the influence of fire: the grasslands of North America, the chaparral of the southwestern United States, the maquis of the Mediterranean, the South African fynbos (Phillips, 1965), the African grasslands and savannas (Batchelder, 1967), the southern pinelands of the United States, and the even-aged stands of coniferous forests of western North America.

Since the beginning of the Cenozoic era, lightning has been a primary source of ignition. Annually, 16 million thunderstorms occur on Earth (McCann, 1942), causing an average of 100 lightning strokes to the ground per second, 24 hours a day, 365 days a year (Komarek, 1968). Lightning storms are not universally accompanied by precipitation. In temperate regions weather patterns at the end of a drought are often characterized by cloud thunderstorms with no rain. The same is true at the beginning and end of dry periods in tropical regions. When lightning strokes hit the ground, the dry material is kindled, and fires are set off. In the western United States, 70 percent of the forest fires are caused by dry lightning during the summer. Because of the seasonal nature of lightning, fires so caused are most numerous during the growing season, from April through August. At that time fires are the least severe but have the greatest impact as a selective force.

When humans appeared on the scene, fire became an even more powerful influence on vegetation, especially in grassland ecosystems, for they added a new dimension to it. While lightning fires are often random and periodic, human-caused fires are often deliberately set in an attempt to modify or change the environment. Fires became more numerous through the years, and their pattern was and is adjusted to the season, to agricultural calendars, and to religious beliefs. Fires were set to clear ground for agriculture, to improve conditions for hunting, to develop grass and shrubby vegetation attractive to game, to improve forage for grazing, to open up the countryside and reduce ambush cover for enemies, to develop areas for wild fruits and berries, and to make travel easier. Some fires were set for excitement or revenge. Others escaped from campfires and trash burning. Whatever the reason, most fires set by humans burned in the nongrowing seasons of early spring and late fall, when the fires were more intense and the damage more severe.

As humans spread from fire-evolved grasslands and savannas to more humid forested areas, they introduced fire into vegetation types such as hardwood forests, fire-resistant during the lightning season but highly inflammable during a dry spring and fall. Indian fires probably produced some of the open heaths of the northeastern United States (see Russell, 1983) and the glades of Kentucky, helped to maintain oak stands in the central hardwoods region, and were an important ecological influence in the Rocky Mountains (Arno, 1976). Lumbering operations left massive piles of debris that fed extremely hot fires that swept across much of the logged-over country (see Holbrook, 1943). In many places fire burned into deep layers of organic matter and peat to rock and mineral soil, destroying any opportunity for former forest types to return.

The apparent destructiveness of fire stimulated intensive suppression programs, especially in North America. Fire, once a relatively prevalent and selective force in shaping the nature and diversity of vegetation, was eliminated as a part of the natural environment. The results of suppression were just as detrimental to vegetation as too frequent fires. Forest communities became monocultural and stagnated in growth. Accumulation of debris and litter became excessive, and once ignited, supported intensive, forest-destroying fires.

Today ecologists, foresters, range managers, and wildlife biologists recognize fire as a natural force. But rather than rely on random and unpredictable natural fires (except in wilderness areas), they turn to an increased use of fire under controlled conditions of weather, moisture, and season to reduce fuel accumulation; to prepare seedbeds for forest regeneration; to control disease; to maintain certain forest, grassland, and wetland types; and to improve wildlife habitats.

Types of Fire

Fires may be surface, ground or crown. Type and behavior depends upon the kind and amount of fuel, moisture, wind and other meteorological conditions, season, and the nature of the vegetation involved.

SURFACE FIRE

Surface fire is the most common type and feeds on the litter layer. In grasslands it consumes dead grass and mulch, converting organic matter to ash. Usually, it does not harm the basal portions of root stalks, tubers, and underground buds, but it does kill most invading woody vegetation. In the forest, surface fires consume leaves, needles, woody debris, and humus. They kill herbaceous plants and seedlings and scorch the bases and occasionally the crowns of trees. Damage to trees depends upon the intensity of the fire and the susceptibility of the trees to heat. Surface fires may kill thin-barked trees by heating the cambium layer. Thick-barked trees are better protected, but they can be scarred, allowing infection by fungi of decay. Shallow-rooted trees are more vulnerable to surface fires than deep-rooted ones such as oaks and hickories.

CROWN FIRE

The intensity of surface fires depends in part on the amount of debris on the forest floor. If the fuel load is high and the wind strong, surface fires may leap up into the forest canopy, causing a crown fire. Crown fires are most prevalent in coniferous stands because of the flammability of their foliage. If the canopy is unbroken, as is often the case in even-aged stands of pine, spruce, and fir, the fire travels from crown to crown. Tops and branches topple to the ground to further fuel the fire. Crown fires kill all above-ground vegetation, including mature trees, yet certain forest types, such as jack pine, require all-consuming fires to regenerate the stand.

GROUND FIRE

The most destructive fires are ground fires that consume organic matter down to the mineral substrate or bare rock. They are most prevalent in areas of deep, dried-out peat and extremely dry, light organic matter. Such fires are flameless and extremely hot, and persist until all available fuel is consumed. Even if the organic mat is somewhat moist, a hot fire dries out the adjacent fuel and kindles it. In peat, ground fires can eliminate organic matter and create depressions that fill with water to form ponds. In spruce and pine forests, with their heavy accumulation of fine litter, fire can burn down to expose rocks and mineral soil, eliminating any opportunity for that vegetation type to return.

Frequency of Fire

Fire is a semirandom, reoccurring event in ecosystems. How frequently a fire burns over an area, its return rate, is influenced by the occurrence of droughts, the accumulation and inflammability of the fuel, the resulting intensity of the burn, and human interference.

In the grasslands of presettlement North America, fires occurred about every 2 to 3 years. That is time enough to allow dead stems, leaves, and mulch to accumulate. Usually set by lightning, fires raced across the expanse of grassland, the flames carried largely by the dead tops. Temperatures were the hottest at the top of the flames and the lowest near the surface of the soil. Fire quickly converted dead organic matter to ash.

In forest ecosystems the frequency of fires varied greatly, depending upon the type of stand. M. L. Heinselman (1981b) describes six different types of frequencies, each with its own ecological importance:

1. Infrequent light surface fires with more than a 25-year return interval.
2. Frequent light surface fires with a return interval of 1 to 25 years.
3. Infrequent severe surface fires with more than a 25-year return interval.
4. Crown fires and severe surface fires with a short return interval of 25 to 100 years.
5. Crown fires and severe surface fires in combination, with a long return interval of 100 to 300 years.
6. Crown and severe surface fires with a very long return interval of over 300 years.

These types are not exclusive, and in fact occur in combination in many forest ecosystems. For example, a red pine forest may experience a light to moderate surface fire every 25 or 30 years, followed by a crown fire once every 100 to 300 years.

Various forest ecosystems appear to burn and develop under certain fire frequencies. Frequent low-intensity surface fires every 5 to 20 years were typical in presettlement forests of ponderosa pine (*Pinus ponderosa*) in western North America. Such fires reduced the needle layer periodically, preventing the buildup of a heavy fuel load; thinned the stand; eliminated incoming shade-tolerant conifers, such as fir, that could act as a fire ladder to the canopy; and encouraged an open, grassy understory.

Infrequent crown and severe surface fires in combination appeared to be the fate of aspen (*Populus tremuloides*) and jack pine (*P. banksiana*) forests of northern North America. Regionally, light surface fires burned across such forests every 15 to 20 years, with killing fires at 60 to 80 years. Fires at such intervals were necessary to regenerate the stand. Aspen renewed itself by sprouting from roots and jack pine by releasing seed on a newly prepared seedbed of ash-rich soil.

Red pine and white pine forests in the Great Lakes region experienced infrequent surface fires of low intensity with a return interval of about 20 to 38 years that scarred individual trees but killed few. These were punctuated at much longer intervals of 150 to 300 years by severe surface or crown fires that destroyed the stand (Heinselman, 1981a).

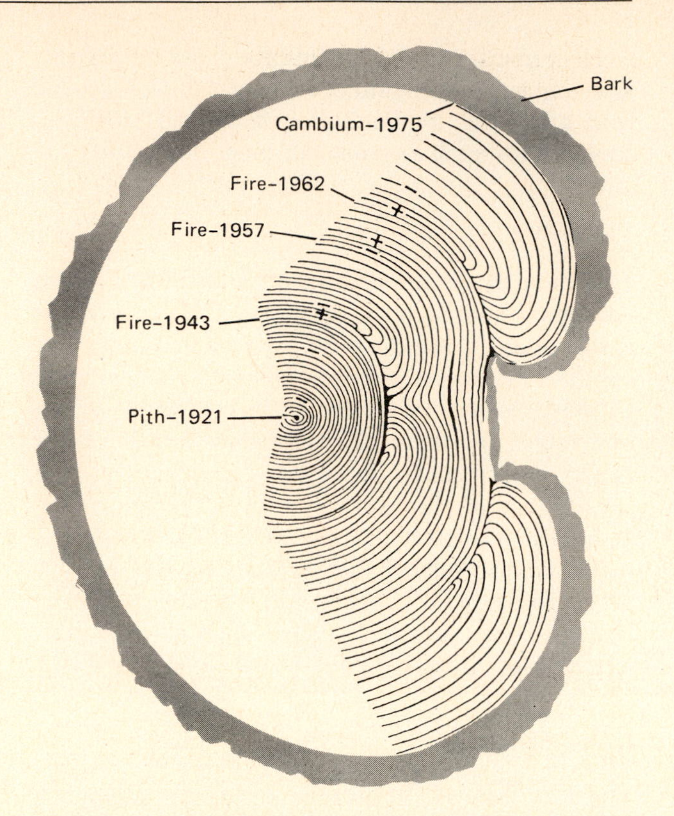

FIGURE 9.1 Dating fires by tree scars. Fire years in a forest stand can be determined in part by counting rings on fire-scarred logs or stumps or by taking increment borings on living trees. In this example every tenth year in from the cambium or inner bark is marked with a −, and scar rings are marked with a +. (From Arno and Sneck, 1977:16.)

Fire History

The frequency of fires in an area is recorded in fire-scarred boles of old trees. Individual trees of most species record more than one fire injury in their trunks and still remain alive. The age of fire scars on living trees is determined from annual growth rings (Figure 9.1). By sampling a number of trees and correlating their fire scars with those of nearby individuals, one can obtain a chronology of fire in a forest (Arno and Sneck, 1977). Old-age stands of ponderosa pine, Douglas-fir (*Pseudotsuga menziesii*), and western larch (*Larix occidentalis*) record fire sequences dating back 300 to 400 years; lodgepole pine (*Pinus contorta*) and western red cedar (*Thuja plicata*), back 200 to 250 years; and giant sequoia (*Sequoiadendron giganteum*), 100 years and more. Further information on fire history can be obtained from the diversity of species in the stand and the age structure of the trees on the area.

A number of fire histories have been developed for various stands. An example is the fire history of the Bitterroot National Forest in Montana (Arno, 1976). One study area was the Tolan Creek watershed (Figure 9.2). Between 1700 and 1900, 30 fires burned through the watershed. The shortest interval between fires was 4 years, the longest 29 years. Because of the short interval between them, the fires apparently were not intense. In some years fires were extensive. In other years some areas burned while other areas experienced small fires or none at all. Based on observations of contemporary fires that were not suppressed on similar forests, low- and medium-intensity fires cover large areas by advancing slowly and sporadically until extinguished by heavy rains of fall and winter.

Since the early 1900s fire suppression in the United States has greatly reduced the frequency and increased the return intervals of fire to the point where return intervals are measured in 1,000+ years (Heinselman, 1981b). Such exclusion has greatly altered the species composition and structure of fire-dependent ecosystems.

Effects of Fire

When you look at a fire-blackened grassland or forest understory, you undoubtedly wonder what effect burning has on the system. There are no good general answers because the effect of fire varies widely, depending upon the amount and flammability of the fuel, the intensity of the fire, the season, moisture, and the nature of the vegetation.

EFFECTS ON SOIL

Fire not only affects vegetation directly. It also impacts the soil, which in turn, further influences plant and animal life.

Temperature

Fire raises the soil's temperature, but the variable amounts of soil moisture present provide a varying insulating effect. Soil temperature does not rise above 100°C until all moisture is evaporated. Because of water's high latent heat of evaporation, considerable heat energy is consumed in drying out the soil. That in itself slows down heat transfer. Measurements of soil temperatures were taken in chaparral under intense, moderate, and light burns (Figure 9.3). Even under hot fires, temperatures did not exceed 200°C at depths of 2.5 cm and cooled off rapidly after fire passed. Postfire soil temperatures increased over prefire temperatures because of the blackened surface and exposure to the sun. Temperatures 3° to 16°C above prefire conditions hastened sprouting and development of roots and shoots, increased decomposition, and often caused seedling mortality.

Organic Matter

More important than fire's effect on soil temperature is its effect on soil organic matter, which fire reduces to ash. Temperatures of 200° to 300°C destroy 85 percent of organic matter; release CO_2, nitrogen, and ash to the atmosphere; and deposit minerals in the form of ash on the soil. Volatilized nitrogen and potassium are released and lost by distillation at temperatures over 200°C, but little is lost under 200°C (Knight, 1966). Most nitrogen and potassium is lost in intense fires—at least 97 percent of the original amount in the forest floor and 67 percent of that in the upper soil horizon. Much of the nitrogen lost, however, is in a form unavailable to plants and is replaced by nitrogen-fixing legumes, whose growth is stimulated by fire; by increased activity of soil microorganisms, including free-living nitrogen-fixing bacteria favored by the increased pH following fire; and by weathering of soil material.

Soil Moisture

As heat destroys organic matter, it breaks down soil aggregates. The larger pores that improve water infiltration and aeration are lost. Bulk density of the soil increases, and permeability decreases. That reduces infiltration of water into the soil, increases surface runoff, and promotes erosion on steep slopes.

Fire can also affect the wettability of certain soils. Decomposing plant material in ponderosa pine and white fir/sequoia forests and chaparral release certain hydrophobic substances that adhere to soil particles and accumulate on the surface and subsurface soil column between burns, creating a nonwettable layer. During a fire the temperature gradient stimulates the volatilization and downward diffusion of these substances to lower soil layers. As a result, the nonwettable layer moves down into the soil (Figure 9.4). How far it moves is influenced by the intensity of the fire. The nonwettable layer now rests between a wettable upper and lower layer (Figure 9.4). As water infiltrates the soil, it reaches the nonwettable layer, where its downward movement is impeded. As the soil above the layer becomes saturated, it slips on the nonwettable layer and slides downslope.

Soil Biota

Fire has both a debilitating and a beneficial effect on soil biota. It reduces fungal populations, in particular those pathogens living in the litter that cause damping off of seedlings. Fire causes a great mortality of nitrifying bacteria, even at a soil temperature of 100°C. It reduces populations of millipedes, spiders, and earthworms, the latter more from the lack of moisture than from the fire. Ants destroyed by fire are quickly replaced by colonists attracted by xeric conditions. On the other hand, fire increases the sporulation and growth of ascomycetes, increases herbaceous legumes, and stimulates the population growth of soil bacteria. Burned areas provide more favorable growing conditions for them because of higher soil temperatures, more favorable pH, a greater amount of available nutrients in ionic form, and greater availability of water.

OTHER EFFECTS

In addition to its impact on the soil, fire has other influences in the ecosystem. It sets the process of stand regeneration in motion by stimulating sprouting from roots and germination of seeds. Fire prepares the seedbed for some species of trees by exposing mineral soil, eliminating competition from fire-sensitive and shade-tolerant species for soil moisture and nutrients. Periodic surface fires thin some coniferous stands such as ponderosa and longleaf pine. Importantly, fire acts as a sanitizer, terminating outbreaks of insects and such parasites as mistletoe by destroying senescent stands and deadwoods and providing conditions for the regeneration of vigorous young trees.

Fire as a Selective Agent

The adaptive traits of plants in fire-prone ecosystems reflect the selective pressure of fire over millions of years of evolution. Such plants respond to fire in three very general ways. They can survive as mature plants with little or no damage. Or the mature plants may die, and their place may be taken by a new generation of young plants of the same species. Or the above-ground parts of the plants may be destroyed, followed by a rapid regeneration from roots and associated structures.

SURVIVAL OF MATURE PLANTS

One way to cope with fire is to evolve defenses against it. That is the evolutionary outcome with

FIGURE 9.2 A series of selected maps showing areas covered by individual fires on the Tolan Creek watershed in the Bitterroot National Forest, Montana, between 1734 and 1900. Only 12 of the 23 fires are illustrated in this series. These are enough to show repeated fires on the area. No fires burned after 1900 because of fire suppression. The watershed supports ponderosa pine, Douglas fir, western larch, lodgepole pine, and white-barked pine.

The general pattern of fires in the watershed was one of frequent fires leaving substantial remnants of earlier age classes of trees. Most pre-1900 fires burned lightly on the drier slopes at lower elevations, perpetuating ponderosa pine as the dominant species in open stands. These fires killed much of the ponderosa pine regeneration and most of the invading Douglas fir. Fires burned with greater intensity on north-facing slopes, where dense, young growth of Douglas fir provided more opportunity for crown fires. Even here, however, much of the old growth—ponderosa pine, Douglas fir, western larch, and some lodgepole pine—survived. In the lower subalpine forest, dominated by lodgepole pine, fires often were of low or medium intensity, spreading mostly on the forest floor. There were occasional hot spots where a stand was killed, largely through bole heating or a "run" through the overstory. (From Arno, 1976:23–29.)

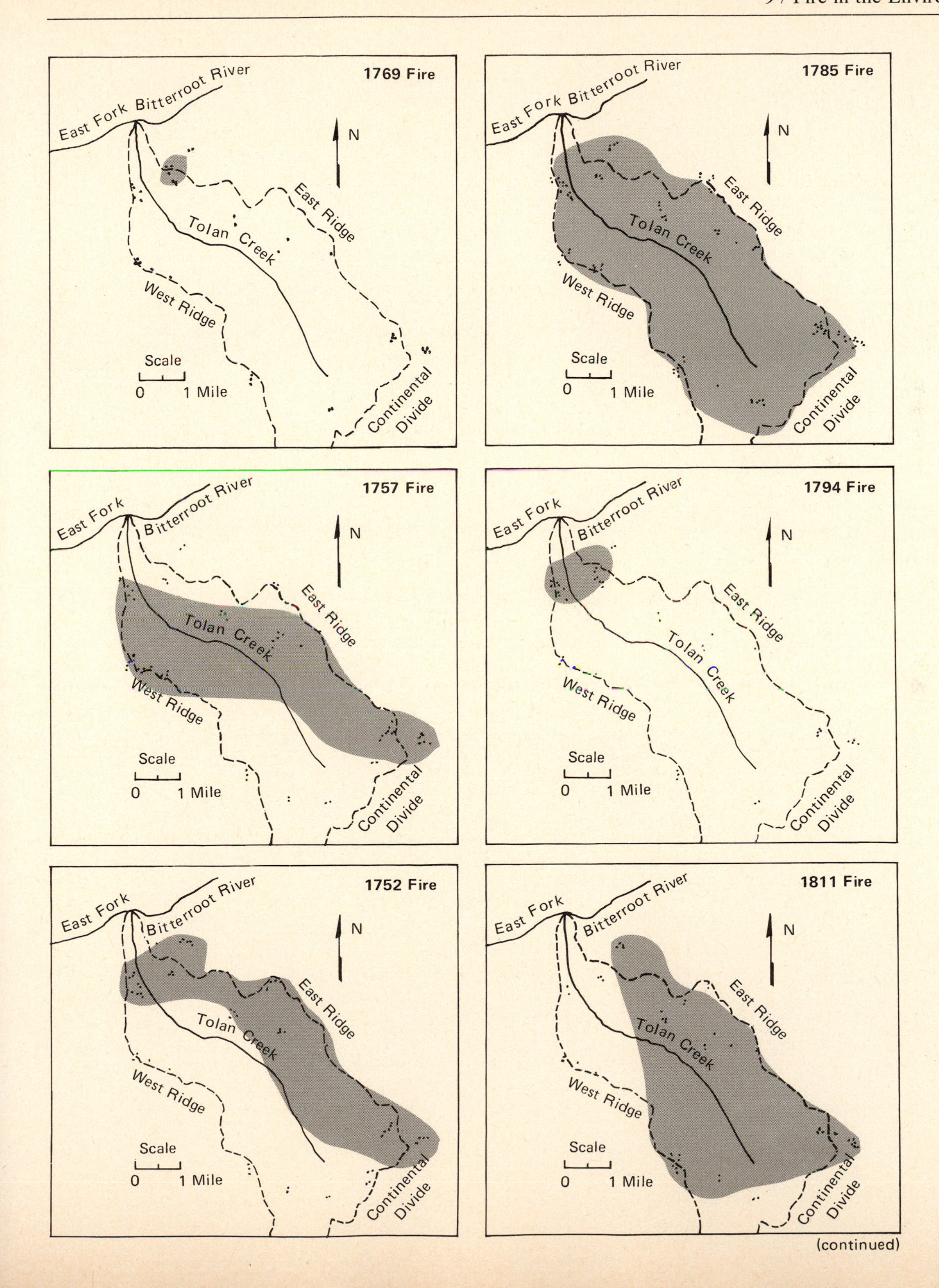
1769 Fire
East Fork Bitterroot River
N
East Ridge
Tolan Creek
West Ridge
Scale
0 1 Mile
Continental Divide
1785 Fire
East Fork Bitterroot River
N
East Ridge
Tolan Creek
West Ridge
Scale
0 1 Mile
Continental Divide
1757 Fire
East Fork
Bitterroot River
N
East Ridge
Tolan Creek
West Ridge
Scale
0 1 Mile
Continental Divide
1794 Fire
East Fork
Bitterroot River
N
East Ridge
Tolan Creek
West Ridge
Scale
0 1 Mile
Continental Divide
1752 Fire
East Fork
Bitterroot River
N
East Ridge
Tolan Creek
West Ridge
Scale
0 1 Mile
Continental Divide
1811 Fire
East Fork
Bitterroot River
N
East Ridge
Tolan Creek
West Ridge
Scale
0 1 Mile
Continental Divide

(continued)

FIGURE 9.2 *continued*

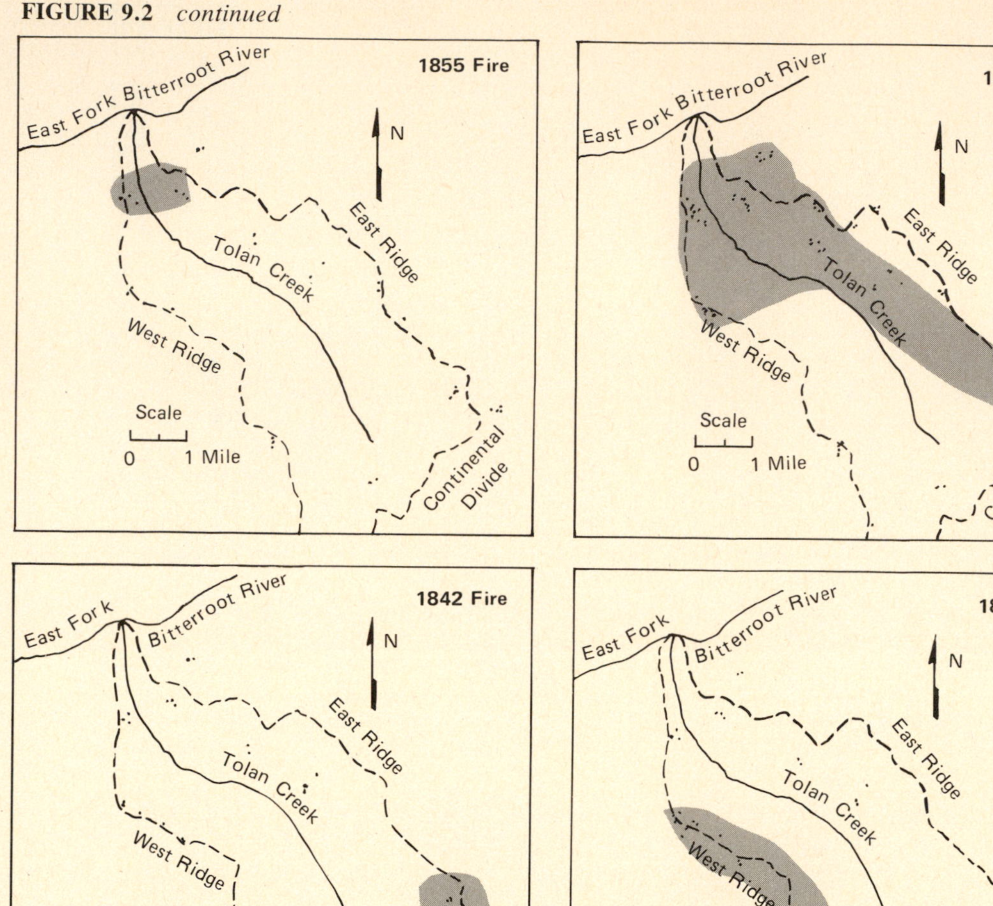

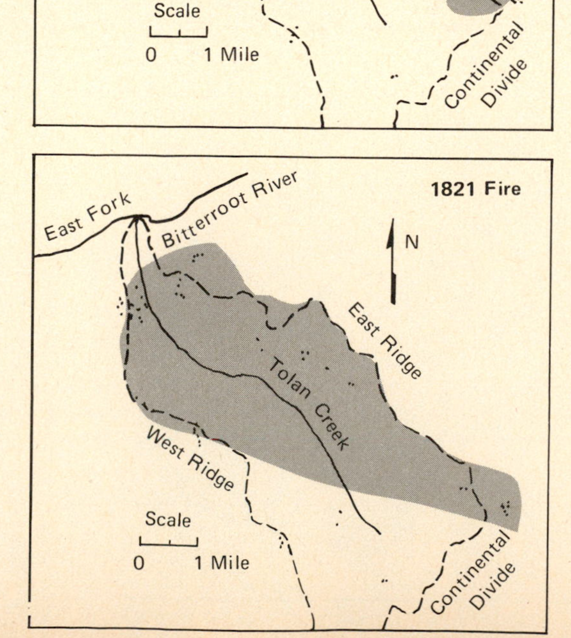

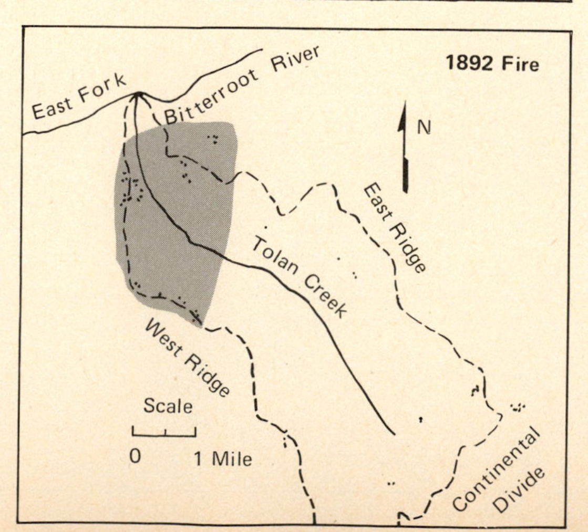

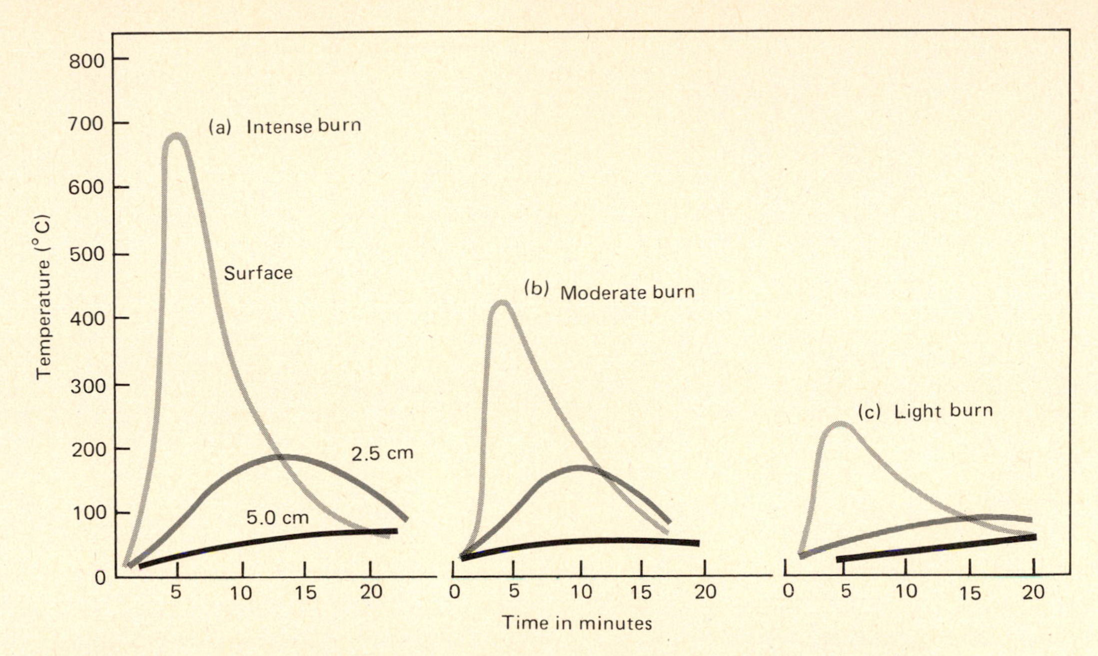

FIGURE 9.3 Typical temperatures at the surface and downward in the soil during an (a) intense, (b) moderate, and (c) light burn in chaparral. Note that the most intense temperatures are on the surface and that little temperature differences exists at the 5.0-cm depth among all three intensities of fire. (After DeBano, Dunn, and Conrad, 1977:69.)

such trees as ponderosa pine, longleaf pine, giant sequoia, oaks, and hickories. One defense is bark thick enough to insulate the cambium from the heat of surface fires. Such protection is not 100 percent effective. Because the heat of fire is rarely uniform about the base of a tree, one part of the trunk may be cool, while another part may burn to the cambium. The injury creates a fire scar that is marked on the annual rings of the trunk (see Figure 9.1).

A second defense is rapid growth, raising the crown high enough above the ground to escape damage from surface fire and to reduce the danger of a surface fire's leaping up into the canopy. Such trees are usually self-pruning, losing their lower branches because of low light intensity. Some trees have fire-resistant foliages, such as summer deciduous trees, whose moist leaves may wilt but will usually not burn.

Some species, particularly long-needled conifers, allow the buildup of a dry, well-aerated mat of litter that supports frequent, periodic, low-intensity surface fires. Such fires prevent the buildup of a heavy fuel load that, once ignited, could destroy the plants. At the same time, such fires release nutrients upon which the plants depend. Similar accumulation of uncompacted litter in grasslands produces much the same results.

DEATH AND REGENERATION

For some plants the death of a mature stand from fire is a means of regenerating and perpetuating the stand. Such destruction results in even-aged stands that arise and become senescent simultaneously, promoting severe but infrequent fires. Such a response works only if the frequency between fires is long enough to allow the plants to mature to produce seed. The seed may be stored either on the plant or in the soil, awaiting the fire to release the seeds or to stimulate germination. By such means,

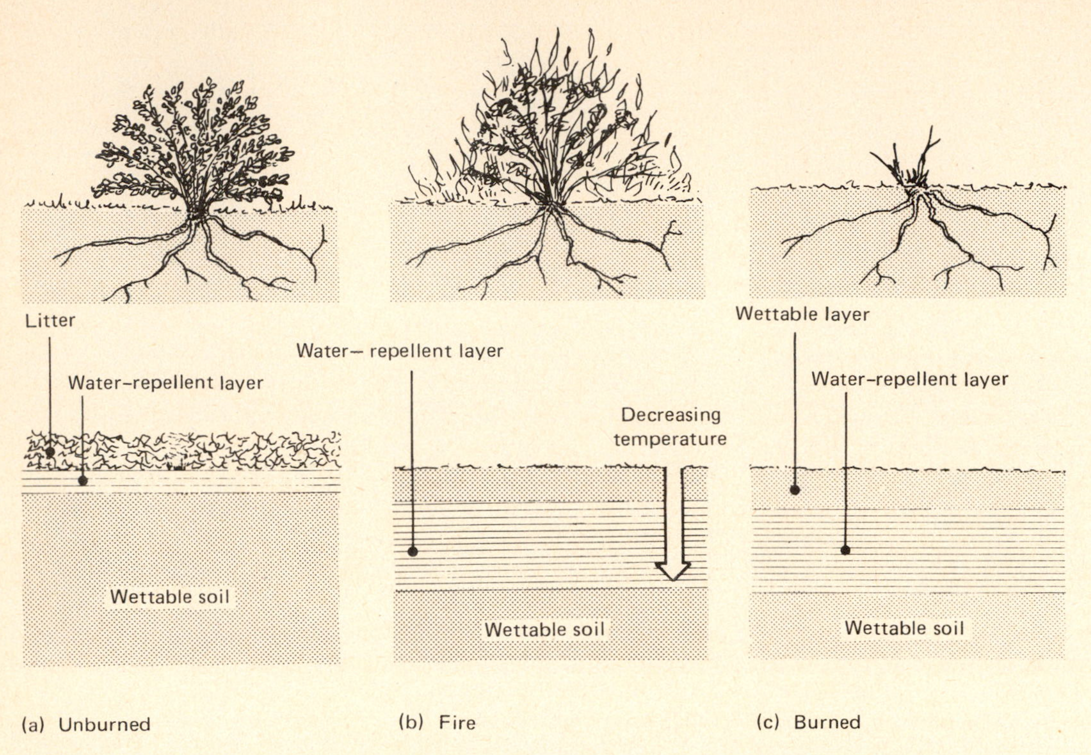

FIGURE 9.4 Water repellency of soil in chaparral before, during, and after fire. (a) Before fire, hydrophobic substances accumulate in the litter and the mineral soil immediately beneath it. (b) Fire burns vegetation and litter, causing hydrophobic substances to move downward along temperature gradients. (c) After a fire a water-repellent layer is located below and parallel to the soil surface on the burned area. This can cause the soil to slide downslope. (After DeBano, Dunn, and Conrad, 1977:71.)

some shade-intolerant species can maintain themselves. Otherwise, they may give way to shade-tolerant species.

Jack pine and lodgepole pine are two coniferous species that retain unripened cones for many years on the trees. Seeds within the cones remain viable until a crown fire destroys the stand, the heat opening the cones, releasing the seeds (*serotiny*), and preparing an open seedbed well fertilized with ash. Jack pine starts bearing cones at the age of 15 to 20 years and becomes senescent at 80. Perpetuation of the stand depends upon a fire interval within that range of years. Thus, serotiny is selected for in those species where fire is spaced with a frequency that allows for maturation but does not exceed it unreasonably. Too frequent fires eliminate the species. Too long a period between fires allows the invasion and ultimate dominance of the stand by shade-tolerant trees. Serotiny in jack pine is most highly developed in populations in the northern part of its range. In southern parts of its range, where jack pine appears in mixed stands, serotiny is not well developed, and the cones will open in the strong heat of a summer sun.

Other species rely on stand destruction and fire-stimulated germination of seeds stored in the soil. Involved is *hardseededness*. The seed coat is impervious to water and other softening agents. The increased temperature of fire-heated soil cracks the hard seed coat or releases the seed from soil-stored chemical inhibitors imposed by overhead living vegetation. Examples of seeds opened by heat are such grassland legumes as beggar's-tick (*Trifolium* spp.). Certain species of annuals and biennials become abundant 1 to 2 years after a chaparral fire, apparently because chemical inhibitors introduced into the soil by the shrub canopy have been destroyed by the heat of the fire (Muller et al., 1968).

Some of the most abundant shrubs in the chaparral of California are *obligate seeders*—plants that regenerate by seeds only rather than by root or stem sprouting. Two-thirds of the species of ceanothus (*Ceanothus*) and manzanita (*Arctostaphylos*) are obligate seeders and depend on locally distributed seeds stored in the soil for reproduction. These obligate seeders at maturity and senescence possess an abundance of dead stems, which feed intense fires, and few potential resprouters. They experience high selection for reproduction by seeds and favor allocation of energy to them. Few sprouts mean less competition for seedlings, for which fire creates large openings (Keeley, 1981).

RESPROUTING

A widespread response to fire is resprouting. Although fire kills the tops and foliage, new growth appears as bud sprouts and root sprouts. Certain trees, particularly a number of *Eucalyptus* species in Australia, possess buds protected beneath the thick bark of larger branches. These buds survive crown fires and break out to develop new foliage. Other plants sprout from buds on roots, rhizomes, root collars, and specialized structures called lignotubers—all protected from fire by the soil. Ferns have subterranean buds on rhizomes that respond to the loss of above-ground foliage. Shrubs such as blackberries and blueberries and trees such as aspens (*Populus*) sprout vigorously from roots. Trees such as oaks and hickories sprout from buds that develop at the root collar just below the ground. Certain species of Mediterranean-type shrubs, including the North American chamise (*Adenostoma fasciculatum*) as well as species in at least seven genera in Australia, have *lignotubers*. These are basal burls from which latent axillary buds are released from inhibition when fire removes the region of growing cells at their tips (apical meristem). Such sprouting is a very effective means of survival or increasing a plant population.

The passage of fire may stimulate some plants to produce flowers and set seed. A well-studied example is the Australian endemic yacca (*Xanthorrhoea australis*). It is a thick-stemmed, sparsely branched plant up to 2.5 m tall. Its stems are well protected from fire by a densely packed mass of fire-resistant leaf bases. The needlelike leaves produced in the crown turn downward against the stem to accumulate as dry thatch. The thatch burns readily. After a fire the plant, which flowers rarely beforehand, flowers immediately (Gill, 1977). Other plants with fire-induced flowering include the South African fire lilies, such as *Haemanthes canaliculatus* (Levyns, 1966), and the North American members of the lily family, death camus (*Zygadenus fremontii*) and soap plant (*Chlorogatum pimerideanum*) (Muller et al., 1968), as well as tussock grasses in New Zealand (Gill, 1981) and prairie grasses of North America. Such flowering responses may have certain selective advantages. Cyclic flowering stimulated by fire would tend to reduce predation on flowers and seeds by inhibiting a population increase of dependent predatory species (see Chapter 17) and would increase germination and growth by preparing a seedbed well fertilized with ash, by reducing competition, and by increasing light.

Although these adaptive traits appear to be a direct response to selection by fire, especially to a wide variety of fire regimes, they may better represent a selection for a particular life cycle. The traits that enhance survival during fire may also enhance survival in other types of environmental stress such as drought, overgrazing and overbrowsing, or defoliation by insects. Fire-stimulated flowering may be a mechanism to produce an abundance of flowers and the subsequent seed crop to satiate predators or to exploit a fire-prepared seedbed. A fire-triggered event could result in an even-aged stand and subsequent accumulation of seeds in the soil. Simply because a trait appears to be fire-selected does not mean that it is only fire-selected.

Fire-Adapted Vegetation

Some vegetation—such as Australian eucalyptus, California chaparral, South African fynbos, jack pine, and lodgepole pine—requires the rejuvenating effects of periodic fires. Such vegetation evolved in a fire environment, according to a hypothesis advanced by Mutch (1970), and acquired certain inherent characteristics that enhance fire spread and increase flammability. These characteristics may be chemical, physical, or physiological (Philpot, 1977).

Highly flammable vegetation, such as chamise, is characterized by foliage with a low mineral content and a high burning rate. High mineral content, especially silica, retards burning and in grass is concentrated in the flower stalks. Such plants, particularly trees and shrubs, also have a high content of flammable resins, waxes, terpenes, and other volatile products that are almost explosive in high-temperature fires. Chamise, for example, has 34 percent of its potential heat contained in volatile products in the foliage.

Plants that burn easily also have a growth form with a high surface area to volume. They are finely branched, have small or needlelike leaves that dry out quickly, burn rapidly, and carry fire from part to part and plant to plant. They usually grow in dense stands with an unbroken canopy that carries fire well. Another trait is an increase in standing dead material as the plants age, adding to the fuel load.

SOME FIRE-ADAPTED SPECIES

Mediterranean Climate Vegetation

The Mediterranean climate is found in the region about the Mediterranean Sea (the origin of its name), on the west coasts of continents, and in general between 30° and 45° north and south latitudes. Mediterranean climate is characterized by hot summers and mild winters, moderate precipitation in winter, summer drought, and short periods in summer of sunny skies. Two Mediterranean-type ecosystems are California chaparral and Australian eucalyptus.

Chaparral. The name is derived from the Spanish word *chaparro,* a thicket of shrubby evergreen oaks. Chaparral is a catchall name for a group of sclerophyllous or hard-leaf plants characteristic of the region. It consists of shrubs belonging to a diversity of families—notably the rose family, including chamise (*Adenostoma fasciculatum*); the buckthorn family, including many species of *Ceanothus*; the heather family, represented by a number of species of manzanita (*Arctostaphylos*); and the oak family, with one prominent member scrub oak (*Quercus dumosa*). Generally, chaparral occupies steep slopes and rocky soils low in organic matter. Some species, such as chamise and ceanothus, are shade-intolerant and grow on xeric sites. Others, such as scrub oak, grow on mesic sites. Fuel buildup during the growing season, slow decomposition, accumulation of dead material, and the volatile nature of many of the shrubs make chaparral highly vulnerable to fire during the summer months.

The most widespread and abundant of the chaparral species is chamise, which grows in pure stands over extensive areas. It grows 2 to 11 m tall and has small, fascicled leaves and fibrous gray bark. The plant is somewhat resinous, a feature that gives it the name *greasewood.* It has a deep taproot as well as lateral, spreading roots. Growth begins in January, increases until April or May, and ceases in June. Periodic fires that burn the chaparral destroy the tops of the plant. Chamise responds by resprouting from a below-ground burl (lignotuber). Sprouts originate from a narrow band 0.5 to 2.5 cm below the surface. With deep roots already in place, the sprouts can draw on already available water and nutrients and start to grow even before the onset of winter rains. Chamise is also a relatively prolific seeder, and seedlings may be abundant after a fire. Few survive, however, for they cannot compete for light with the annual and biennial growth that springs up from seeds after a fire. For chamise, resprouting is the means of survival and rejuvenation after a fire. Seeds are a means of dispersal into new habitat.

Eucalyptus. In Australia, Mediterranean-type vegetation is dominated by many species of eucalyptus, with short, medium, and long boles and a flattish, interlacing, nearly continuous crown of leathery, evergreen leaves. Some species, such as mountain ash (*Eucalyptus regnans*), grow on moist sites and possess a shrubby understory. Others, such as stringybark (*E. obliqua*) and narrowleaf peppermint (*E. radiata*), grow on moist sites and possess a grass understory. All are subject to periodic fires, which may range from tree-killing, very severe bushfires to moderate to light bushfires. In these the trees are rarely burned, but the bark may be scorched and charred, and the trees are defoliated by radiant heat from the ground.

Eucalpytus responds by resprouting, not from the roots but from the trunk and branches. The timing, form, vigor, rate, density of sprouts, and transition into a mature canopy differ among the various species. Basically, new foliage regenerates from dormant shoots, either along the trunk, on main branches, only in the upper branches, or some combination of all three, depending upon the species. Foliage regeneration in stringybark, for example, is confined to upper limbs. In others, such as narrowleaf peppermint, death of the canopy stimulates vigorous sprouting of dormant buds from the trunk in cracks and fissures of the bark (Cochrane, 1968). Short lateral branches cloak the trunk in erect green plumes, which range from 20 to 90 cm, depending upon the species. Blackened trunks may be masked by green in 3 to 7 months. The trees retain their lateral foliage until it is ultimately shaded out as trees slowly regain their prefire form.

One eucalyptus, mountain ash, is killed by fire and regenerates only by seeds. Too frequent fires can eliminate this species from a locality. Mountain ash depends upon seeds stored in the canopy in capsules. In the absence of fire, most of the seeds are lost to predators. Any seedlings produced on unburned sites rarely survive because of the lack of light and fungal infections. But during a fire when the crown is scorched, seeds are released and fall to the ground on a nutrient-rich seedbed free of competing plant growth and pathogens (Gill, 1981).

Longleaf Pine

Most species of pine are dependent to some extent on fire for regeneration and maintenance of the stand, but none is so closely related to fire as longleaf pine (*Pinus palustris*) of the coastal southeastern United States. It is so dependent on fire in its normal life cycle that it cannot continue without it. Without fire longleaf pine would give way to hardwood forest, and the seedlings could not become established. Creeping surface fires started by lightning and Indians in presettlement times burned through longleaf pine stands every 2 to 3 years, clearing the forest of litter and incoming hardwood growth. Occasionally, severe fires destroyed a stand and prepared a seedbed for a new generation of pine.

Longleaf pine requires a burned area free of litter for a seedbed. Seeds shed from September to December germinate within 1 week of seedfall. Little of the seedling appears above ground. The cotyledons are at ground level, and the hypocotyl, the growing shoot, is just below the surface. The root, however, extends 6 cm into the ground. As the longleaf pine seedling develops, it grows long needles that look like grass. This is the so-called grass stage. For 2 to 10 years, the seedling remains at this stage. While the shoots grow little, a strong taproot is pushing deeper into the soil, and the seedling is gaining in diameter and developing a thick bark. Seedlings in the grass stage are resistant to fire. Even if young needles are burned, they are quickly replaced because of food reserves in the taproot. Surface fires at the grass stage not only eliminate competing growth but also destroy both the fungal pathogen brown spot and infected needles. Once established, seedlings grow rapidly, putting on as much as a meter of top growth a year. Fast growth puts the crown well above the ground and out of reach of surface fires.

So important is fire to the longleaf pine that management of commercial stands requires periodic prescribed burns. Although after the sapling and pole stages, longleaf pine can thrive without fire, periodic burning is necessary to inhibit incoming hardwood growth and reduce the buildup of fuel.

Giant Sequoia

The giant sequoia, one of the oldest trees on Earth, exists today because of the role fire plays in its regeneration. "Fire," wrote John Muir (1878:814), "the great destroyer of Sequoia, also furnishes bare, virgin ground, one of the conditions essential for its growth from the seed."

Giant sequoia produces cones and seeds at an early age of 100 to 200 years. Seeds remain viable in green photosynthetic cones for 20 years or more. Over time some cones are always opening, releasing their seed to the wind, or seeds are freed when chickarees (*Tamiasciurus douglasii*) feed on the cone scales. Most seeds perish unless they land on a disturbed site.

Sequoia requires a fire-prepared seedbed of soft, friable soil onto which the extremely small seeds can drop and in which they can become lightly buried. Light necessary for seedling growth is lethal to seeds resting on top of the soil, exposed to direct sunlight. Fire also destroys fungal pathogens and eliminates competition. Once established, shade-intolerant seedlings grow rapidly if soil moisture is adequate and if they grow fast enough to overtop competition and keep their foliage in the sunlight. As the tree ages, it develops a thick bark highly resistant to fire. Over the centuries frequent surface fires burned through the understory (see the "Ecological Excursion"), eliminating debris and shade-tolerant plants. With fire exclusion over the past 80 years, a dense understory of white fir (*Abies concolor*) and incense cedar (*Libocedrus decurrens*) has developed. Such an understory creates a fire ladder, which could carry a devastating fire into the crowns of sequoia.

Grassland

Both grassland climate and grassland vegetation favor fire. Prolonged droughts associated with high temperatures and drying winds, characteristic of grassland regions, provide environmental conditions favorable for fire. Annually the above-ground portion of the plants dies, while roots, shoots, buds, corms, and rhizomes remain active below ground. Dead tops accumulate in a mass of loose, well-oxygenated, easily combustible material.

In presettlement times fires set both by lightning and by aboriginal humans swept across prairies and savannas, consuming dead material in their path. In places the material burned to the ground; in other more moist areas, dead matter only partly burned, producing a mosaic of burned and partly burned patches. Fires occurred mostly in summer, when the fuel ignited easily. They produced stress during the reproductive season, when selection pressures function best. Most fire-dependent grasslands were those in more humid regions where large amounts of above-ground biomass accumulated. Unless burned periodically, these grasslands were soon dominated by woody vegetation.

Grasses such as *Aristida, Andropogon,* and *Setaria* evolved several fire-related traits. One is the ability to take up silica and incorporate it in grass tissue. (Silica oxides are responsible for the glossy appearance and stiffness of grass stems.) Silica increases both stiffness and fire resistance of flowering stalks, enabling grasses to hold their seed heads above the hottest part of the flames. Below-ground roots, buds, and crowns protected from the heat of fire send up new growth, stimulated by nutrients released by fire from dead plant material. At the same time, fire scarifies hard-coated seeds, especially those of legumes, stimulating them to germinate. Fire reduces fungal diseases and eliminates incoming woody growth.

Fire and Vertebrates

Fire has consistently been pictured as an agent destructive to animal life. In severe fires a number of animals may be lost directly to fire and indirectly to predators who take advantage of prey suddenly driven from or deprived of cover. Many animals, however, especially those that live in burrows, survive fires. Studies show that most birds and mammals on areas swept by surface fires remain on or return to the areas. Some species may decrease, but ground-foraging birds tend to increase (Wirtz, 1977; Wright and Bailey, 1982).

Those are short-term effects. Of greater consequence are the long-term effects. Destruction of cover by severe fires alters the habitat and eliminates for a time those species dependent on it. At the same time, fire creates new habitat for a different set of species, especially those that favor shrubby and open land. Many animals tolerate pre- and postfire conditions and in fact are dependent upon such fluctuations in habitat. Fire produces a mosaic of shrubs, young timber, and open land. The interspersion of open land with different kinds of cover is essential for such vertebrates as snowshoe hare, moose, black bear, white-tailed and mule deer, ruffed grouse, and sharp-tailed grouse.

A few species are wholly dependent on fire. One was the eastern prairie chicken, or heath hen (*Tympanuchus cupido cupido*). Its extinction in part can be attributed to the suppression of fires necessary to maintain large shrubby openings dominated by blueberries and other heaths (Thompson and Smith, 1970). The endangered Kirtland's warbler (*Dendroica kirtlandii*) is wholly fire-dependent. Restricted to jack pine forests in the lower peninsula of Michigan, the warbler requires large blocks (40+ ha) of even-aged stands of pine 1.5 to 4.5 meters tall with branches close to the ground. Smaller or larger trees are unacceptable. Intervals of fire sufficient to maintain blocks of these young jack pines are necessary to maintain the habitat of the species.

Another type of fire-dependent birds is a group of *Sylvia* warblers that occupy the Mediterranean shrublands of Sardinia (Walter, 1977). Of the five species, two are fire-dependent. One species, the Dartford warbler (*S. sarda*), can occupy a habitat patch that has been burned within the last 6 years. Another species, Marmora's warbler (*S. undata*), occupies only 18- to 24-year-old tall, shrubby growth. Two of the remaining species are fire-adapted, and one is fire-tolerant.

Summary

Fire is a major force in the evolution of plants and the development of vegetation. Globally, certain regions of Earth, especially those characterized by a Mediterranean-type climate, are more conducive to fires than others. There, fire-adapted vegetation has evolved under periodic burning. The only two sources of periodic fires are lightning and humans.

Fire may be surface, crown, or ground. Surface fires are mostly low-intensity. They do not destroy overhead vegetation, but they do reduce fuel on the ground and understory woody growth. Crown fires, the most spectacular, destroy the overstory. Ground fires burn deeply into the organic layer and have a catastrophic effect on the ecosystem.

Of great ecological importance is the frequency, or return interval, of fire. Too frequent fires can destroy fire-dependent vegetation by destroying the plants before they have had time to mature and seed. Too long a time between fires can result in the elimination of fire-tolerant vegetation by shade-tolerant species. The latter has been the result of years of fire exclusion by humans.

Fire has both beneficial and adverse effects, depending on the nature of the fire, characteristics of the fuel, moisture, wind, temperature, and other meteorological conditions. Fire results in the loss of some soil nutrients by volatilization, but it also makes nutrients locked in undecomposed plant tissues available to plants. It reduces the water-holding capacity of the soil by reducing soil organic matter and increases erosion on steep slopes. Fire reduces fungal populations, many of them pathogens, and stimulates the growth of soil bacteria by increasing available nutrients, soil pH, and soil temperature. Because soil insulates against the heat of fire, buds, roots, corms, and other underground biomass are protected and respond with new growth.

Fire is a strong selective agent in the evolution of plants. Plants have evolved defenses against fire, such as thick bark and rapid growth, which enable them to survive as mature plants. Others may die from fire, but they regenerate a new stand from seeds stored on the plant in cones opened by the heat of fire (serotiny) or from seeds stored in the soil. The underground parts of many plants survive while their foliage is destroyed. They resprout from roots or from dormant buds protected by thick bark on the trunk. Other plants respond to fire by fire-stimulated flowering and seeding. Although many traits appear to be adaptations to fire, they may represent responses to a broader group of stresses, including drought, overgrazing, and overbrowsing.

Fire dependence appears to be an evolved trait that makes for flammability. Such plants have volatile resins and terpenes in their foliage, are finely branched, accumulate dead material, and have a low mineral content in their foliage, which increases their ability to burn.

There are a number of fire-dependent plants and plant communities. These include chamise, one of the most abundant species in California chaparral. It renews itself after fire by vigorous resprouting. Australian eucalyptus resprouts vigorously from dormant buds on limbs and trunk. Longleaf pine, a most fire-dependent species, reseeds on a fire-prepared seedbed. It has a fire-resistant seedling grass stage that requires fire to eliminate competition and a fungal disease, brown spot. Periodic surface fires in young and mature stands reduce fuel buildup and eliminate incoming hardwoods in the understory. Giant sequoia—with its serotinous cones, small seeds, and fire-resistant bark—requires fire for reestablishment and for the elimination of a shade-tolerant understory of white fir and incense cedar, which increases the danger of crown fire and poses the potential of replacing sequoia.

Although fire over the short term may cause the death of vertebrates and destroy the habitat for a number of species, many survive a surface fire and return or recolonize the area. Many species depend upon periodic fires to maintain their habitat and to provide a mosaic of vegetation types required in their life cycle. A few species—such as the Kirtland warbler, the Dartford warbler, and Marmora's warbler—are strongly fire-dependent. Without periodic fires to maintain their restricted habitat requirements, the birds face extinction.

Review and Study Questions

1. Distinguish among surface fire, crown fire, and ground fire.
2. What conditions are necessary for fire to assume ecological importance?
3. Why are lightning and humans the only two possible causes of periodic fires?
4. Contrast the effects of short-interval surface fires in a fire-dependent ecosystem with long-interval fires with few or no surface fires in between.
5. How can fire exclusion bring about an extinction of fire-dependent species?
6. How might the following responses be adaptations to fire: (1) resprouting from roots, (2) serotinous cones, (3) thick bark, (4) dormant buds protected by bark?
7. What characteristics of certain vegetation, such as chamise, encourage flammability?
8. How are the following adapted to a fire regime: (1) longleaf pine, (2) giant sequoia, (3) eucalyptus, (4) chamise, (5) grassland?
9. Find out how other plants are fire-adapted—for example, jack pine, redwood, lodgepole pine, Douglas-fir.

Suggested Readings

(*Note:* Literature is abundant.)

Fischer, W. C. 1980. *Index to proceedings: Tall Timbers Fire Ecology Conferences*. Gen. Tech. Rept. RM-81. Fort Collins, Colo.: USDA Forest Service.

Komarek, E. V., Sr., ed. 1963–1976. *Proceedings: Tall Timbers Fire Ecology Conferences*. Vols. 1–15. Tallahassee, Fla.: Tall Timbers Research Station.

Kozlowski, T. F., and C. E. Ahlgren, eds. 1974. *Fire and ecosystems*. New York: Academic Press.

Mooney, H. A., and C. E. Conrad, eds. 1977. *Proceedings: Symposium on the Environmental Consequences of Fire and Fuel Management in Mediterranean Ecosystems*. Gen. Tech. Rept. WO-3. Washington, D.C.: USDA Forest Service.

Mooney, H. A., et al., eds. 1981. *Proceedings: Symposium on Fire Regimes and Ecosystem Properties*. Gen. Tech. Rept. WO-26. Washington, D.C.: USDA Forest Service.

Pyne, S. J. 1982. *Fire in America*. Princeton, N.J.: Princeton University Press.

Slaughter, C. W., R. J. Barney, and G. M. Hansen, eds. 1971. *Fire in the northern environment: A symposium*. Portland, Oreg.: U.S. Pacific Northwest Forest and Range Experiment Station.

Stokes, M. A., and D. H. Dieterich. 1980. *Proceedings: Fire History Workshop*. Gen. Tech. Rept. RM-81. Fort Collins, Colo.: USDA Forest Service.

ECOLOGICAL EXCURSION

Fire in the Sequoias

John Muir

In the forest between the Middle and East forks of the Kaweah, I met a great fire, and as fire is the master scourge and controller of the distribution of trees, I stopped to watch it and learn what I could of its works and ways with the giants. It came racing up the steep chaparral-covered slopes of the East Fork cañon with passionate enthusiasm in a broad cataract of flames, now bending down low to feed on the green bushes, devouring acres of them at a breath, now towering high in the air as if looking abroad to choose a way, then stooping to feed again, the lurid flapping surges and the smoke and terrible rushing and roaring hiding all that is gentle and orderly in the work. But as soon as the deep forest was reached the ungovernable flood became calm like a torrent entering a lake, creeping and spreading beneath the trees where the ground was level or sloped gently, slowly nibbling the cake of compressed needles and scales with flames an inch high, rising here and there to a foot or two on dry twigs and clumps of small bushes and brome grass. Only at considerable intervals were fierce bonfires lighted, where heavy branches broken off by snow had accumulated, or around some venerable giant whose head had been stricken off by lightning.

I tethered Brownie on the edge of a little meadow beside a stream a good safe way off, and then cautiously chose a camp for myself in a big stout hollow trunk not likely to be crushed by the fall of burning trees, and made a bed of ferns and boughs in it. The night, however, and the strange wild fireworks were too beautiful and exciting to allow much sleep. There was no danger of being chased and hemmed in, for in the main forest belt of the Sierra, even when swift winds are blowing, fires seldom or never sweep over the trees in broad all-embracing sheets as they do in the dense Rocky Mountain woods and in those of the Cascade Mountains of Oregon and Washington. Here they creep from tree to tree with tranquil deliberation, allowing close observation, though caution is required in venturing around the burning giants to avoid falling limbs and knots and fragments from dead shattered tops. Though the day was best for study, I sauntered about night after night, learning what I could and admiring the wonderful show vividly displayed in the lonely darkness, the ground-fire advancing in long crooked lines gently grazing and smoking on the close-pressed leaves, springing up in thousands of little jets of pure flame on dry tassels and twigs, and tall spires and flat sheets with jagged flapping edges dancing here and there on grass tufts and bushes, big bonfires blazing in perfect storms of energy where heavy branches mixed with small ones lay smashed together in hundred cord piles, big red arches between spreading root-swells and trees growing close together, huge fire-mantled trunks on the hill slopes glowing like bars of hot iron, violet-colored fire running up the tall trees, tracing the furrows of the bark in quick quivering rills, and lighting magnificent torches on dry shattered tops, and ever and anon, with a tremendous roar and burst of light, young trees clad in low-descending feathery branches vanishing in one flame two or three hundred feet high.

One of the most impressive and beautiful sights was made by the great fallen trunks lying on the hillsides all red and glowing like colossal iron bars fresh from a furnace, two hundred feet long some of them, and ten to twenty feet thick. After repeated burnings have consumed the bark and sapwood, the sound charred surface, being full of cracks and sprinkled with leaves, is quickly overspread with a pure, rich, furred, ruby glow almost flameless and smokeless, producing a marvelous effect in the night. Another grand and interesting sight are the fires on the tops of the largest living trees flaming above the green branches at a height of perhaps two hundred feet, entirely cut off from the ground-fires, and looking like signal beacons on watch towers. From one standpoint I sometimes saw a dozen

SOURCE: John Muir, *Our National Parks* (Boston: Houghton Mifflin, 1904), 307–314.

or more, those in the distance looking like great stars above the forest roof. At first I could not imagine how these Sequoia lamps were lighted, but the very first night, strolling about waiting and watching, I saw the thing done again and again. The thick, fibrous bark of old trees is divided by deep, nearly continuous furrows, the sides of which are bearded with the bristling ends of fibres broken by the growth swelling of the trunk, and when the fire comes creeping around the feet of the trees, it runs up these bristly furrows in lovely pale blue quivering, bickering rills of flame with a low, earnest whispering sound to the lightning-shattered top of the trunk, which, in the dry Indian summer, with perhaps leaves and twigs and squirrel-gnawed cone-scales and seed-wings lodged in it, is readily ignited. These lamp-lighting rills, the most beautiful fire streams I ever saw, last only a minute or two, but the big lamps burn with varying brightness for days and weeks, throwing off sparks like the spray of a fountain, while ever and anon a shower of red coals comes sifting down through the branches, followed at times with startling effect by a big burned-off chunk weighing perhaps half a ton.

The immense bonfires where fifty or a hundred cords of peeled, split, smashed wood has been piled around some old giant by a single stroke of lightning is another grand sight in the night. The light is so great I found I could read common print three hundred yards from them, and the illumination of the circle of onlooking trees is indescribably impressive. Other big fires, roaring and booming like waterfalls, were blazing on the upper sides of trees on hillslopes, against which limbs broken off by heavy snow had rolled, while branches high overhead, tossed and shaken by the ascending air current, seemed to be writhing in pain. Perhaps the most startling phenomenon of all was the quick death of childlike Sequoias only a century or two of age. In the midst of the other comparatively slow and steady fire work one of these tall, beautiful saplings, leafy and branchy, would be seen blazing up suddenly, all in one heaving, booming, passionate flame reaching from the ground to the top of the tree and fifty to a hundred feet or more above it, with a smoke column bending forward and streaming away on the upper, free-flowing wind. To burn these green trees a strong fire of dry wood beneath them is required, to send up a current of air hot enough to distill inflammable gases from the leaves and sprays; then instead of the lower limbs gradually catching fire and igniting the next and next in succession, the whole tree seems to explode almost simultaneously, and with awful roaring and throbbing a round, tapering flame shoots up two or three hundred feet, and in a second or two is quenched, leaving the green spire a black, dead mast, bristled and roughened with down-curling boughs. Nearly all the trees that have been burned down are lying with their heads uphill, because they are burned far more deeply on the upper side, on account of broken limbs rolling down against them to make hot fires, while only leaves and twigs accumulate on the lower side and are quickly consumed without injury to the tree. But green, resinless Sequoia wood burns very slowly, and many successive fires are required to burn down a large tree. Fires can run only at intervals of several years, and when the ordinary amount of firewood that has rolled against the gigantic trunk is consumed, only a shallow scar is made, which is slowly deepened by recurring fires until far beyond the centre of gravity, and when at last the tree falls, it of course falls uphill. The healing folds of wood layers on some of the deeply burned trees show that centuries have elapsed since the last wounds were made.

When a great Sequoia falls, its head is smashed into fragments about as small as those made by lightning, which are mostly devoured by the first running, hunting fire that finds them, while the trunk is slowly wasted away by centuries of fire and weather. One of the most interesting fire actions on the trunk is the boring of those great tunnel-like hollows through which horesemen may gallop. All of these famous hollows are burned out of the solid wood, for no Sequoia is ever hollowed by decay. When the tree falls the brash trunk is often broken straight across into sections as if sawed; into these joints the fire creeps, and, on account of the great size of the broken ends, burns for weeks or even months without being much influenced by the weather. After the great glowing ends fronting each other have burned so far apart that their rims cease to burn, the fire continues to work on in the centres, and the ends become deeply concave. Then heat being radiated from side to side, the burning goes on in each section of the trunk independent of the other, until the diameter of the bore is so great that the heat radiated across from side to side is not sufficient to keep them burning. It appears, therefore, that only very large trees can receive the fire-auger and have any shell rim left.

Fire attacks the large trees only at the ground, consuming the fallen leaves and humus at their feet, doing them but little harm unless considerable quantities of fallen limbs happen to be piled about them, their thick mail of spongy, unpitchy, almost unburnable bark affording strong protection. Therefore, the oldest and most perfect unscarred trees are found on ground that is nearly level, while those growing on hillsides, against which falling branches roll, are always deeply scarred on the upper side, and as we have seen are sometimes burned down. The saddest thing of all was to see the hopeful seedlings, many of them crinkled and bent with the pressure of winter snow, yet bravely aspiring at the top, helplessly perishing, and young trees, perfect spires of verdure and naturally immortal, suddenly changed to dead masts. Yet the sun looked cheerily down the openings in the forest roof, turning the black smoke to a beautiful brown, as if all was for the best.

PART IV
POPULATION ECOLOGY

INTRODUCTION

Populations

The individual organism does not exist alone. In order to survive and carry out its life cycle, even the most solitary of animals, such as the wolverine, has to encounter and interact with others of its own kind living somewhere in the same area. In doing so, it becomes part of the population. A *population* is a group of similar individuals of the same kind living in the same place at the same time. Members of a population interact with each other. They may cooperate in hunting and in nest building; they mate; they rear young; they compete for food, space, and mates when those resources are in short supply. What happens to individuals in a population has an effect on the lives of others within the population.

Populations have certain unique features. They have an age structure, density, and distribution in space and time. They exhibit a birthrate, a mortality rate, and a growth rate. But like the individuals that comprise them, populations occupy the same space at the same time as do populations of other species. They may compete for the same food or nest sites. Individuals of one population may feed on individuals of another. If food is scarce or low in quality, individuals may fail to reproduce. In preying on a species, the predator necessarily eliminates individuals of the prey population, influencing its mortality rate and growth rate. The interrelations of one population with another can influence the whole web of interactions that makes up the structure and function of an ecosystem.

Population ecology developed around animals. Only recently have plants received comparable treatment. Although some traditional approaches used in the study of animal populations can be applied to them, plants do not lend themselves to quite the same demographic analyses (nor do some of the invertebrates, such as corals and hydroids that grow in colonies). The reasons lie in the nature of the plants themselves. Plants exist as *genets,* genetic individuals that reproduce sexually. But the genet itself is a collection of subunits: buds, twigs, shoots, and leaves above ground and roots, rhizomes, and their extensions below ground. These subunits, termed *metapopulations* (J. White, 1979), have their own demography: birthrates, death rates, and growth rates. Within its metapopulation a genet can acquire new subunits and lose subunits through mortality while the genet itself remains alive. These subunits may remain connected physically or organically for years, or they may separate and live individually (a characteristic exploited in plant propagation).

Consider a tree—a maple, for example—or a shrub. The plants established by seed are individuals or genets. You can consider each plant as a population of buds that develop into new growth that in turn produces more buds. The birth and death of these individual subunits determine not only the rate of growth but also the form the plant will take (see J. White, 1979; Halle, Oldeman, and Tomlinson, 1978; Harper and Bell, 1979). If the tree or shrub is cut, new buds develop along the root collar and send up sprouts or coppice growth. Although the original individual is dead, the genet lives on through new vegetative growth. Some trees, such as black locust and aspen, grow underground by adding root extensions that in turn send up new shoots or suckers. New metapopulations, known as *ramets,* develop in an area surrounding the original genet. These clones may cover considerable area and be of different ages. Because they are a part of the metapopulation from the original growth reiterations of the parent plant, the genet age may be very old, although the individual trees may be very young. As a result, life spans of a genet are indeterminate for many plants, especially trees, shrubs, and a number of perennial herbs. Such characteristics of plants complicate studies of their demography.

CHAPTER 10

Density, Distribution, and Age

Outline

Objectives

Upon completion of this chapter, you should be able to:

1. Define population density, crude density, and ecological density.
2. Describe the types of population distribution.
3. Distinguish between fine-grained and coarse-grained distribution.
4. Discuss problems and methods of determining the density of a population.
5. Describe the age structure of a population and define stable age distribution.
6. Discuss the significance of different types of age pyramids.
7. Discuss ways in which the ages of plants and animals can be determined.
8. Contrast age structure of plants with that of animals.

Three attributes of a population most obvious even to a very casual observer are density, distribution, and age structure. You are very aware of the change in density and distribution of human populations from the concentrated masses of the city to the sparse populations of the countryside. You may notice any increase or decrease of starlings, Japanese beetles, or gypsy moths from one year to another. And you observe young rabbits, speckled-breasted young robins, deer fawns, and the full array of age classes in human populations, which emphasize age structure.

Density

When you get caught in traffic going home, make your way through crowded streets, or enjoy the freedom of lightly traveled roads, you are aware of population density. You respond to that density in some way. You may seek out the crowds, or you may escape to the peaceful countryside or the quiet of your house or apartment. Then again, you may become frustrated at the delays high densities impose on you.

Individuals in natural populations are also affected by density in some way. Trees in crowded stands may grow more slowly, and some may succumb to a lack of water, nutrients, and light, unequally shared. Scarce food may be denied to smaller or less aggressive individuals in a mammalian population; and access to nest sites may be denied to some birds because not enough sites exist to meet the demand. Having too few individuals in a population may reduce their chance of finding a mate or inhibit the performance of behavioral activities essential to the welfare of the population. Density of a population affects the spread of diseases and parasites and influences the risks of individuals' succumbing to predation. In affecting the welfare of individuals in a population, density in part controls the birthrates, mortality rates, and growth rates of a population.

However important it may be, density is an elusive characteristic, difficult to define and hard to determine. Density can be characterized as the number of individuals per unit of space—as so many per square mile, per hectare, or per square meter. That is *crude density*. But individuals in a population do not occupy all the space within a unit, because not all of it is suitable habitat. A biologist might estimate the number of deer living in a square-mile area. The deer, however, will not utilize all the land within the area because of human habitation, land use practices, and the lack of cover and food. Goldenrods inhabiting old fields grow in scattered groups or clumps because of soil conditions and competition from other old field plants.

No matter how uniform a habitat may appear, it is usually patchy, because of microdifferences in light, moisture, temperature, or exposure, to mention a few physical conditions, or because of the lack of sites available for colonization. Each organism occupies only those areas that can adequately meet its basic requirements.

To account for those conditions or effects, the density of organisms should refer to the amount of area available as living space. That would be *ecological density*. Ecological densities are rarely estimated because what portions of the habitat represent living space are difficult to determine. In a Wisconsin study, the densities of bobwhite quail were expressed as the number of birds per mile of hedgerow rather than as birds per acre (Kabat and Thompson, 1963). That is ecological density.

Distribution

SPATIAL DISTRIBUTION

How organisms are distributed over space has an important bearing on density. Individuals of a population may be distributed randomly, uniformly, or clumped (Figure 10.1). Individuals of a population are distributed randomly if the position of each is independent of the other. Forest trees of the canopy layer are frequently spaced at random. So are some invertebrates of the forest floor, particularly spiders (Cole, 1946; Kuenzler, 1958), and the clam *Mulinia lateralis* of the intertidal mud flats of the northeastern coast of North America.

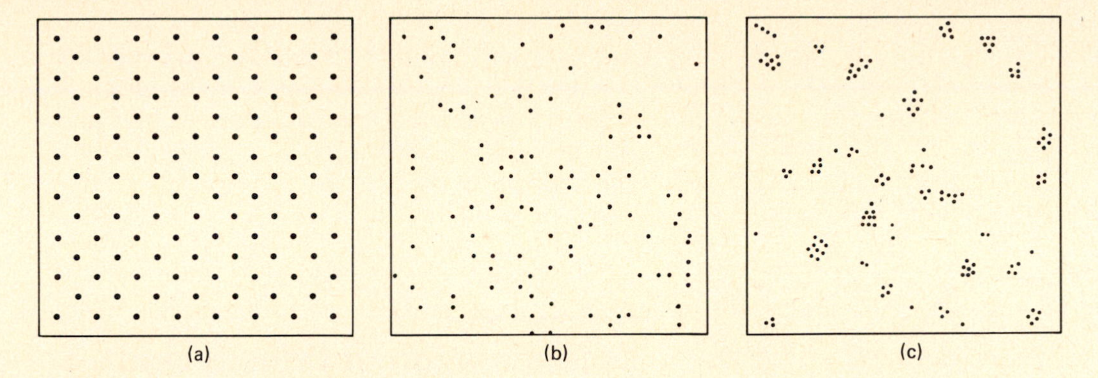

FIGURE 10.1 Patterns of distribution. Note the difference among (a) uniform, (b) random, and (c) clumped.

By contrast, individuals may be distributed uniformly—that is, more or less evenly spaced (Figure 10.1). In the animal world uniform distribution usually results from some form of intraspecific competition, such as territoriality. Uniform distribution happens among plants when severe competition exists for crown or root space, as among some forest trees, or for moisture, as among desert plants (Beals, 1968; Phillips and MacMahon, 1981).

The most common type is clumped (sometimes called contagious) distribution, which consists of patches or groups of organisms scattered over a given area (Figure 10.1). Clumping produces aggregations, the result of responses by plants and animals to habitat differences, daily and seasonal weather changes, reproductive patterns, and social behavior. The distribution of human populations is clumped or aggregated because of social behavior, economics, and geography. There are various degrees and types of clumping. Groups of varying sizes may be randomly or nonrandomly distributed over an area. Aggregations may range from small ones to a single centralized aggregation. If environmental conditions encourage it, populations may be concentrated in long bands or strips along some feature of the landscape such as a river, leaving the rest of the area unoccupied.

Another way of looking at the distribution of individuals is to consider their pattern of dispersion. A species' habitat is usually patchy, consisting of several types of vegetation or microenvironments. If environmental patches are clumped and the areas between them are large, the environment is considered coarse-grained. If areas between patches are small (approaching homogeneity or sameness), the environment is considered fine-grained. Individuals that utilize only one type of habitat, or environmental patch, are considered coarse-grained species and their distribution is highly clumped (Figure 10.2). Individuals that utilize all patches in proportion to their occurrence (in other words occupy the entire area) are considered fine-grained species. As the population of a coarse-grained species increases, the density within the preferred patches increases. At very high densities, surplus individuals in the patches are forced into different but marginal patches. Thus as a population of a coarse-grained species increases, individuals may spread randomly over the area, approaching a fine-grained distribution (see MacArthur and Levins, 1964; Pielou, 1974; Wiens, 1976).

Populations are not uniformly distributed over a region. While the local distribution of populations may be defined by the spacing of individuals, species are distributed on a larger scale as discrete populations. In turn, individual populations may be concentrated in clusters within a given region. Regional distributions of populations make up the range of a species (Figure 10.3). The boundaries of a species' range are not fixed, but rather fluctuate. Habitat changes, competition, predation, climatic changes—all can influence the extent of a species' range. It may expand one year, contract another. The many variables that influence the distribution

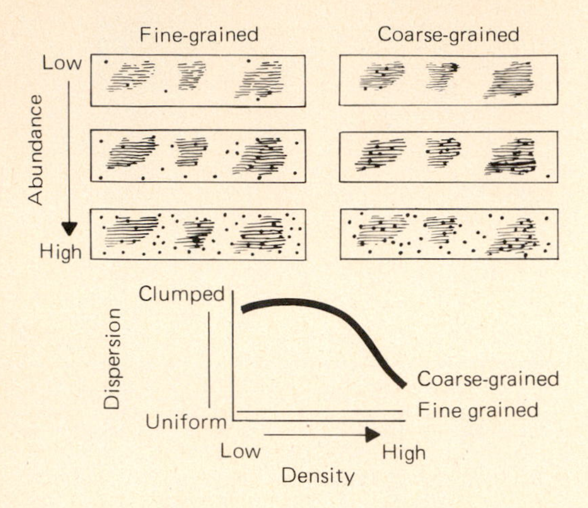

FIGURE 10.2 Habitat or patch occupancy patterns for a fine-grained and a coarse-grained population as density increases from low to high levels. The fine-grained population may retain a uniform distribution pattern over a wide range of densities. The dispersion of a coarse-grained population is initially clumped because individuals settle in preferred patches but tends toward greater uniformity as density increases and individuals are forced to settle in less preferred patches. (From Wiens, 1976:94.)

of a species over its range and the regional adaptations of a species are the subject matter of biogeography.

TEMPORAL DISTRIBUTION

Organisms in populations are not only dispersed in space; they are also distributed in time. Temporal distribution may be circadian (related to daily changes of daylight and dark), lunar, tidal, and seasonal. Distribution may also involve longer periods, including annual fluctuations, successional changes, and evolutionary changes. The environmental rhythm of daylight and dark is responsible for flowers' opening their petals at a specific time during the day, attracting nectar-feeding insects, for the daily migration of plankton from deeper to upper layers of water, for the emergence of nocturnal animals and the retirement of diurnal animals. Some organisms that inhabit intertidal zones show rhythm in their behavior, including presence and absence coinciding with cycles of high and low tide. Seasonal changes result in changes in the populations of wildflowers in forest and field, in the return and departure of migrant animals. Most natural populations show some difference in spring, summer, fall, and winter.

Determination of Density and Dispersion

Density and dispersion of individuals in a population are almost inseparable. Density per unit area provides only minimal information about a population and is less important than how the individuals in a population are arranged spatially. Crude density does not tell us about the unevenness of concentration and dispersion of individuals. A change of boundaries can change the level of density (Figure 10.4).

Fortunately, ecologists rarely need to know the exact abundance or density of a population. (*Abundance* is the number of individuals in a given area, in contrast to *density,* which is the number expressed per unit area.) That can be determined only by a direct count of all individuals in a population. Except for unusual habitat situations, such as antelope living on an open plain or waterfowl concentrated in marshes or some other restricted habitat, direct counts are either extremely difficult or impossible to obtain. Even if possible, direct counts may be too time-consuming or too expensive to conduct for the information gained.

For these reasons ecologists use other means of obtaining information on population density. One is sampling, in which the area of study is divided into subunits in which animals or plants are counted in a prescribed manner. From the sample data is determined the mean density of the unit sampled. The mean is then multiplied by the total number of sampled and unsampled areas to arrive at the estimated population. Confidence limits can be calculated on the estimate. (For details see Brower and Zar, 1977; Cox, 1976; Southwood, 1978; Greig-Smith, 1964; Mueller-Dombois and Ellenberg, 1974;

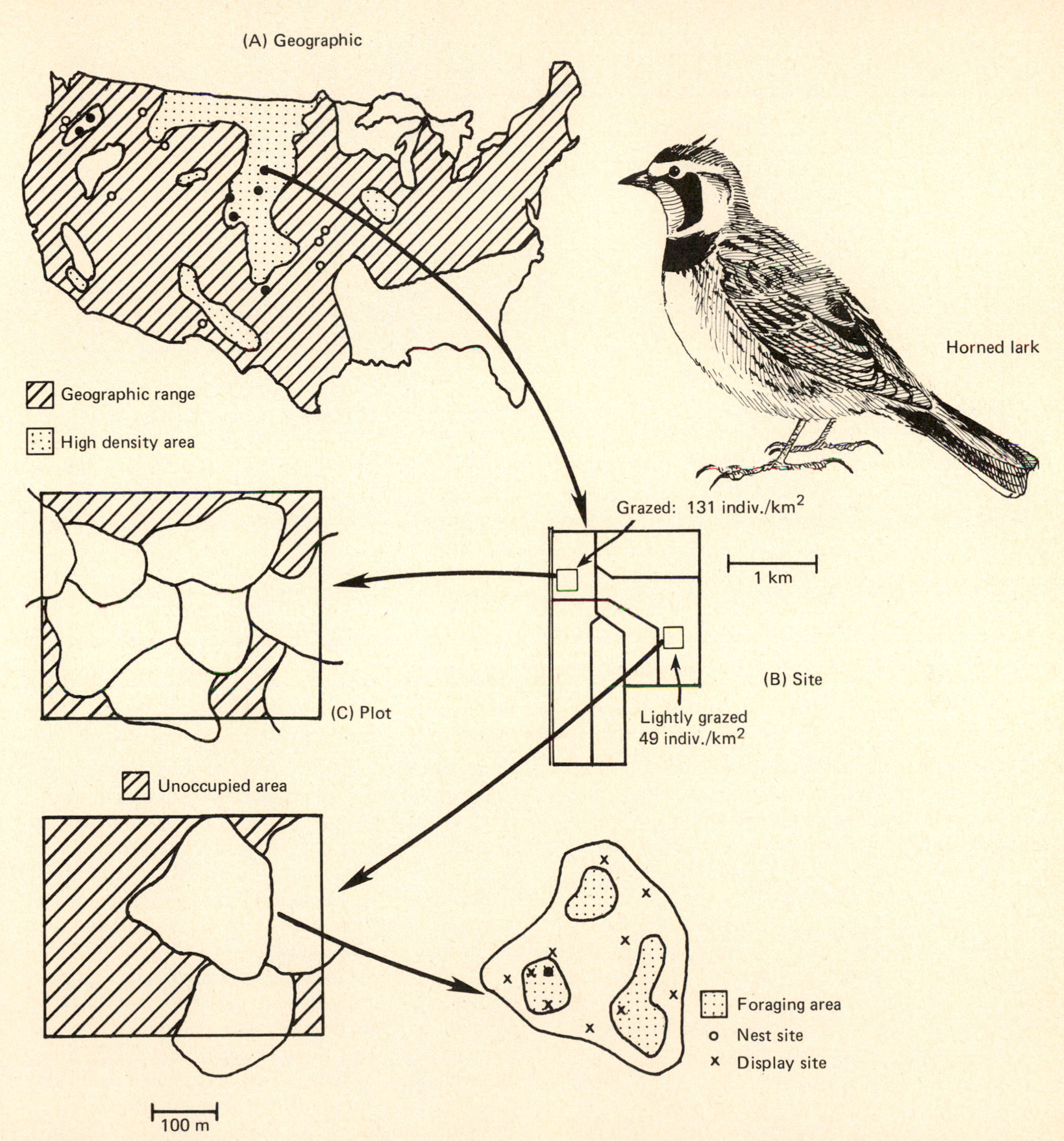

FIGURE 10.3 Populations of an organism are not equally distributed over its range. Some areas have higher densities than others, as indicated by the range map of the horned lark (*Eromophila alpestris*) (a). Within a region the distribution of the bird is influenced by the availability of habitats (b). Within a given habitat the bird's distribution is influenced by territorial behavior (c). Within each territory certain parts may be used for different purposes, such as singing and nesting. (From Wiens, 1973.)

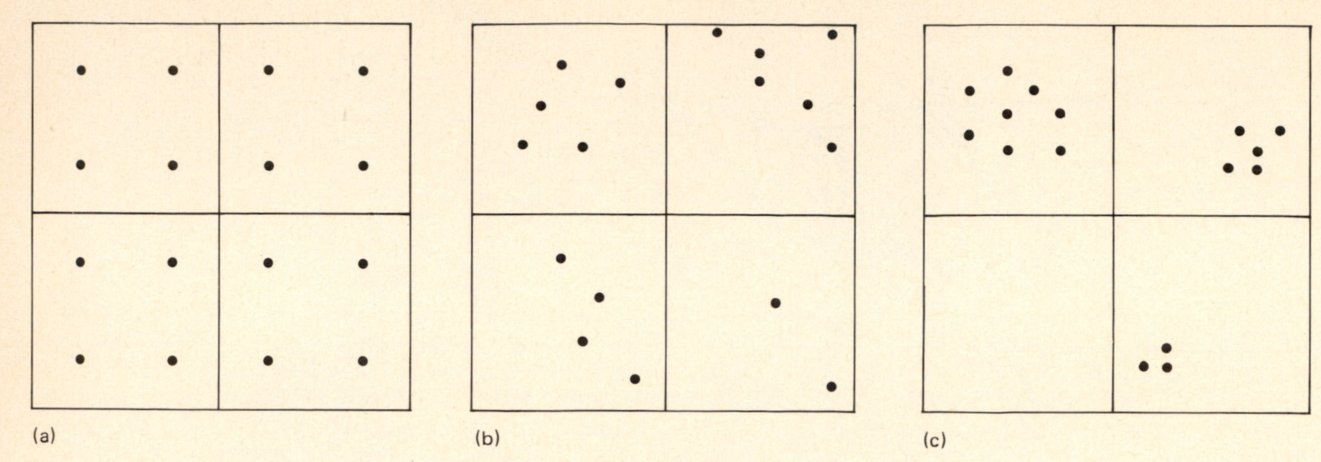

FIGURE 10.4 Crude density for each of the subpopulations enclosed within their respective squares is 16. If the area is divided into four sampling units and one randomly chosen for a sample of the population, the estimates would be quite different, depending upon which unit was selected. In the uniform population any one sampling unit would give a correct estimate of the population; 4×4 (to explode the population to the full area) = 16. For the random population the estimates would be 20, 20, 16, and 8. In the clumped population the estimates would be 32, 20, 12, 0. Thus, estimates of crude density should indicate how individuals of a population are distributed. This little example also points out the problems ecologists encounter when sampling populations.

Poole, 1974; Tanner, 1978.) Sampling is used most widely in the study of populations of plants and sessile animals because no movement in and out of the area is involved. Sampling is also important in the study of enclosed fish populations, such as those found in lakes and ponds.

For most ecological work with animals, indices of relative abundance may be sufficient (Figure 10.5). You might count the number of mourning doves seen along a prescribed length of road traveled year after year, or you might note the number of drumming ruffed grouse or the number of singing birds of one or several species of birds heard along a standardized route, or you might record the number of tracks crossing dusty roads. The results can be converted to numbers of individuals seen or heard singing per mile or per hour. Such counts by themselves are meaningless; they cannot stand alone. But if you have a series of such index figures collected from the same area over a period of years, you can follow trends in density or abundance. Or you can obtain counts from different areas during the same year to compare numbers from one habitat to another. Most population data on birds and mammals are based on indices of relative abundance rather than on direct counts.

Getting some idea of how individuals of a population are dispersed also involves sampling the population. One method uses sample plots and compares the data to some mathematical distribution such as a Poisson, variance-to-mean ratio, and chi-square goodness-of-fit. Other methods involve measuring distances between individuals or distances to individuals from random points located in the habitat and treating the data to appropriate analysis. (See Brower and Zar, 1977.)

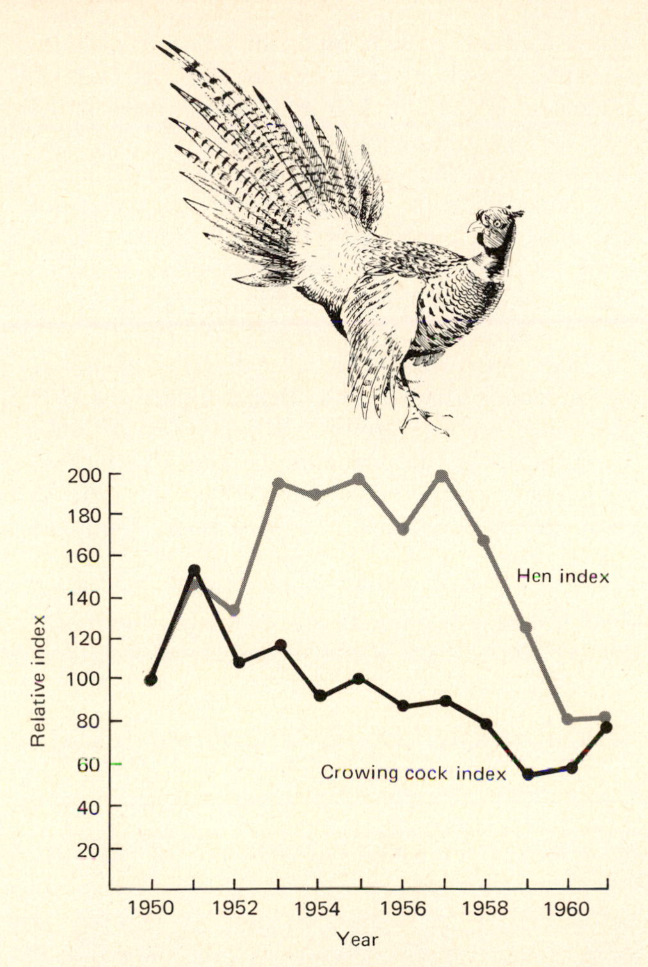

FIGURE 10.5 An index of population abundance. Density trends in a ring-necked pheasant population in Wisconsin were followed with a crowing cock index, obtained by running spring crowing counts over a prescribed route. The hen index is obtained by multiplying the average number of calls per stop in the crowing cock counts and the number of hens observed per cock in the preceding winter. (After Wagner, Besadny, and Kabat, 1965:14.)

Age Structure

Unless a population consists of seasonal breeders with nonoverlapping generations, such as annual plants and animals, it will be characterized by a certain age structure that influences considerably the birthrates and death rates of a population.

Populations can be divided into three ecological periods: prereproductive, reproductive, and postreproductive. In plants the prereproductive period is usually termed the juvenile period. The relative length of each period depends largely on the life history of the organism. Among annual species the length of the prereproductive period has little significant effect on the potential rate of population growth. In longer-lived plants and animals, the length of the prereproductive period has a pronounced effect on the population's rate of growth. Organisms with a short prereproductive period often increase rapidly with a short span between generations. Organisms with a long prereproductive period generally increase more slowly and have a long span of time between generations.

Theoretically, all continuously breeding populations tend toward a *stable age distribution;* that is, the ratio of each age group in a population remains the same if the age-specific birthrates and the age-specific survival rates are the same. When the proportion of age classes in a closed population tends to become constant, mortality equals natality. If the stable situation is disrupted for any reason—such as a natural or human-caused catastrophe, disease, famine, or emigration—the age composition will tend to restore itself upon the return of normal conditions. Changes in age class distribution reflect changes in the production of young, their survival to maturity, and the period of life when most mortality occurs.

Any influence that causes age ratios to shift because of changes in age-specific death rates affects the population birthrate. In populations in which life expectancy for the oldest age classes is reduced, a higher portion falls into the reproductive class, automatically increasing the birthrate. Conversely, if life is extended, a greater portion of the population falls into the postreproductive class, thus reducing the birthrate. Rapidly growing populations are usually characterized by declining death rates, especially in the very young age classes, which inflates the younger age groups. Declining or stabilized populations are characterized by lower birthrates with fewer young to rise into the reproductive age classes and by a larger portion in the older age classes.

AGE STRUCTURE OF ANIMAL POPULATIONS

The age structure of a population represents the ratio of the various age classes in a population to each other at a given time. Determination of age structure requires some means of obtaining the ages of members of a population. For humans this is not a problem, but it is with wild populations. Age data for wild animals can be obtained in a number of ways, the methods varying with the species concerned. The most accurate, but most difficult, method is to mark young individuals in a population and to follow their survival through time. Such a method, however, requires both a large number of marked individuals and, of course, time. For this reason other less accurate methods are employed. These methods include an examination of a representative sample of carcasses for the wear and replacement of teeth in deer and other ungulates; the growth rings in the cementum of teeth of carnivores and ungulates; the annual growth rings in the horns of mountain sheep; and, in rabbits, the weight of eye lens which increases with age. Among birds aging methods involve observations of plumage changes and wear which can only separate juveniles from subadults (in some species) and adults.

Age structure is visualized best by means of age pyramids (Figure 10.6) that compare the percentage of one age group in a population to other age groups. As the population changes with time, the number of individuals and, thus, the ratio in each age class changes. Growing populations are characterized by a large number of young which expands the base of the age pyramid. This large class of young eventually moves up into the reproductive age classes. If the young are as prolific as

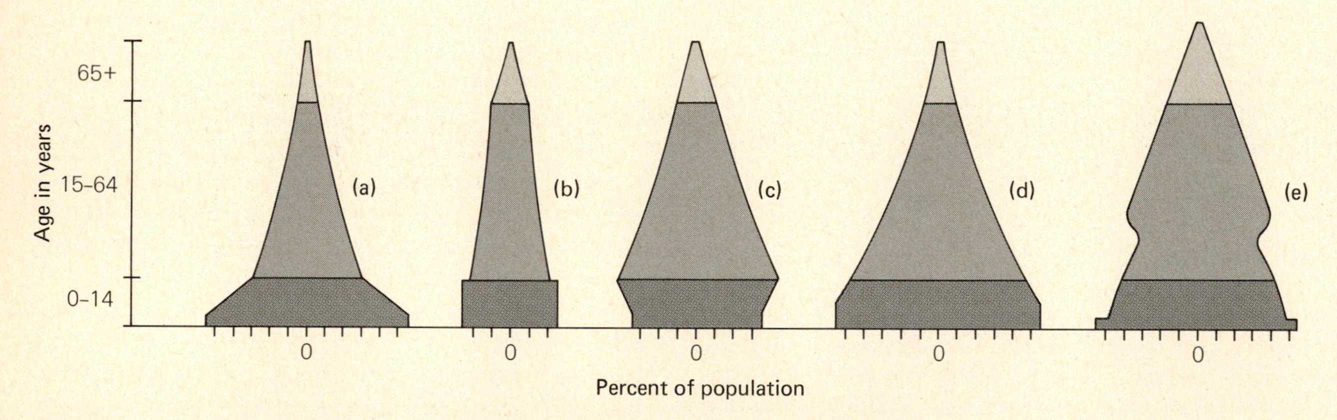

FIGURE 10.6 Diagrams of different types of age structures in populations. The diagrams are divided into three major parts based on an economic viewpoint: the young dependent ages, 0 to 14; the economically active ages, 15 to 64; and the elderly dependents, 65 years plus. (a) The broad base and pinched top suggest youthful population and foretell a rapid population growth. This is typical of certain South American and Asian countries. (b) This age structure suggests minimal population growth and perhaps an aging population. It is characteristic of the United States and Western European nations. (c) The narrow base reflects a declining birthrate. In time, if the trend continues, the structure will approach that of pyramid (e). (d) The age structure of a population in which death rates are declining. The profile will approach that of (a) as more young enter the reproductive age classes. (e) This age structure suggests a decline in the birthrate at an earlier time. The pyramid could change as the young enter the reproductive period.

the parents, the young age class will expand further. Declining populations are characterized by fewer individuals being added to the population and a higher proportion of the individuals in the population moving into the older age classes. With fewer young, fewer individuals will enter the reproductive age classes, further depressing the population. In a way the age structure changes as a population increases, decreases, or remains the same.

Although such generalizations are most applicable to human populations because of their long life span and low juvenile mortality, age structure is much less predictive of population growth in wild animal populations. Population growth for wild animals also depends on the close relationship between animals and their resources, both habitat and food, and the competitive relationships among individuals for those resources. For example, a normally increasing population of a wild species should have an increasing number of young; a decreasing population, a decreasing number of young. But within this framework there can be a number of variations. A population may be decreasing, yet show an increasing percentage of young. Or a population might remain static with an increasing percentage of young, suggesting a high mortality of older individuals. Only by correlating such information as density, mortality, and reproduction with changing age distribution can age structure provide some insight into the dynamics of wild populations.

Because age structure of human populations more accurately reflects population dynamics than those of wild animals, age structure is best illustrated by age pyramids of human populations. Figure 10.6 presents five generalized age pyramids of human populations. Pyramid A has a broad base, reflecting inflated young age groups, and a very narrow top, reflecting very few individuals in the old age classes. Such a pyramid indicates a youthful population in which increasing numbers will enter the reproductive period. Pyramid B is narrow. The ratio between one age class and another is about the same. It suggests that the population is neither growing nor declining. It has reached zero population growth. Pyramid C illustrates a population with a large reproductive age group but declining young. This reflects a decreased birthrate brought about by some form of birth control and predicts a smaller reproductive age class later on. The rate of growth in such a population is being checked. Pyramid D approaches the situation in A. It represents a population in which the death rate is declining and the population is headed for rapid expansion. In contrast, pyramid E has a relatively large portion of its population in the pre- and postreproductive age periods and a marked indentation in the reproductive classes. This suggests at least two situations. Either the population has declined markedly in its recent history, as illustrated in pyramid C, or it has experienced a heavy emigration of members of the younger reproductive age classes when they were entering the working age period. Or the situation could be a combination of both.

The age pyramids of a few selected countries provide some examples of the situations depicted in the general age pyramids. The age pyramid of Sweden (see Figure 10.7), based on the 1970 census

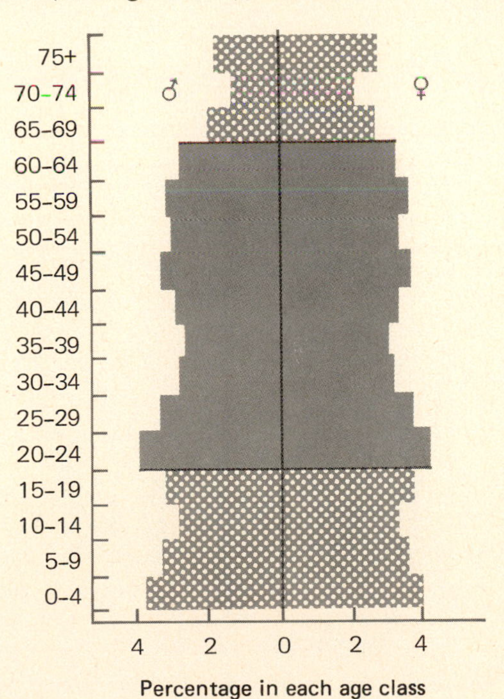

FIGURE 10.7 The age pyramid for Sweden (1970). It is characteristic of a country that is approaching zero population growth. The pyramid is broken into dependency ratio components. (Data from U.N. Demographic Yearbook, 1982:946.)

figures, is one of a population with zero population growth and a stable age distribution. The age pyramid of the United States for 1970 (see Figure 10.8) is similar to that in Figure 10.6(c). The number of young is declining, the population growth is slowing, and the population is aging. The bulge in the 5-to-25-year age classes foretells an increasing portion of the population moving into the reproductive age classes. Even if the reproductive rate were only at the replacement level of 2.1, the population would continue to increase for some time because of the larger number of reproductive individuals. At the replacement rate, it would be 70 years before the United States could reach zero growth. The age pyramid of India (see Figure 10.9) is a broad-based one characteristic of an expanding population with high fertility.

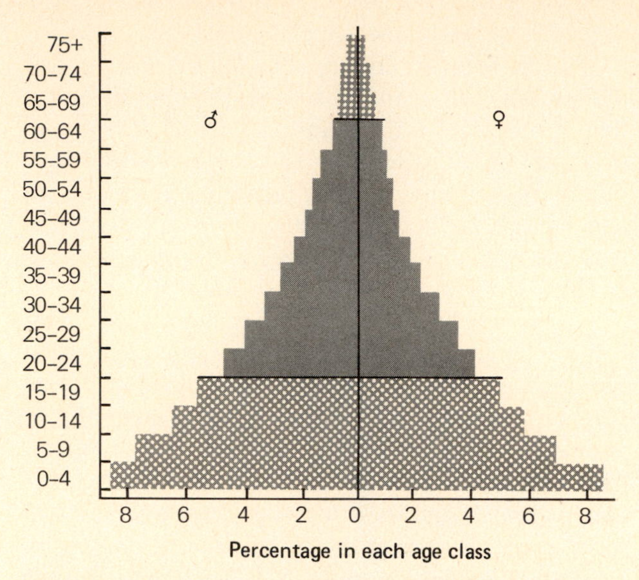

FIGURE 10.9 Age pyramid for India (1970) shows a rapidly expanding population. Note the broad base of young who will enter the reproductive age classes. (Data from *U.N. Demographic Yearbook,* 1970:860.)

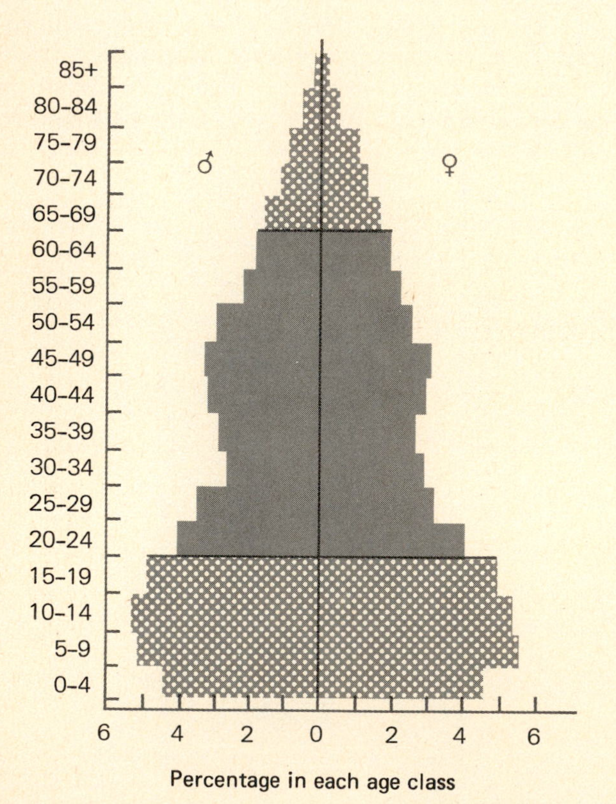

FIGURE 10.8 The age pyramid for the United States (1970), divided into dependency ratio components. The pyramid shows a constrictive shape. The youngest age class is no longer numerically the largest. This type of age structure reflects declining fertility, which in time will further distort the age distribution. (Data from U.S. Bureau of Census, 1970.)

From the age structure one can determine another useful statistic, the dependency ratio. This ratio relates the load placed on the productive portion of the population by the young and dependent old. The ratio is determined by dividing the sum of the number of people less than 20 years old and the number of people 65 years and over by the number of people 20 to 64 years old (see Figure 10.7). The dependency ratio provides an assessment of the costs of education, health care, social security, and old-age assistance that must be provided by the productive age classes. One outcome of a stable age distribution and zero population growth is the large burden of social services the working group is forced to bear. This is especially true in a population that is moving from a decreasing number of young to an increasing number of old. Also bearing a heavy dependency burden are rapidly expanding populations, with their large number of young.

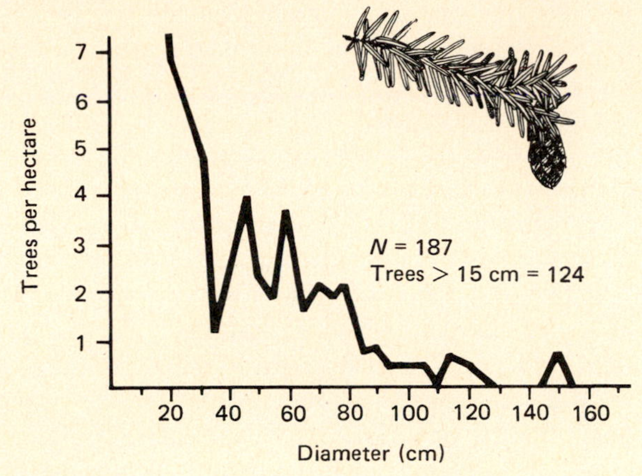

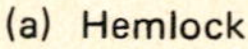

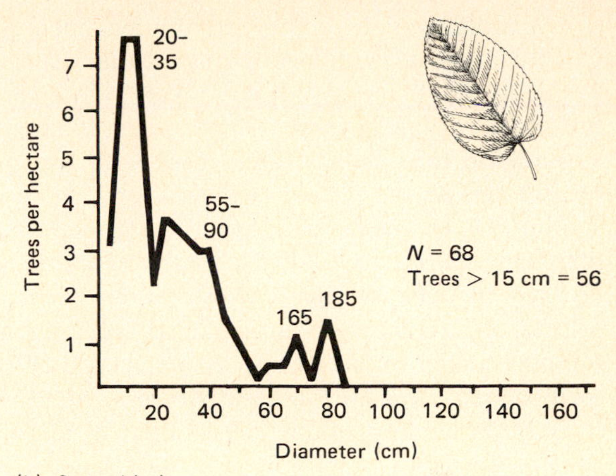

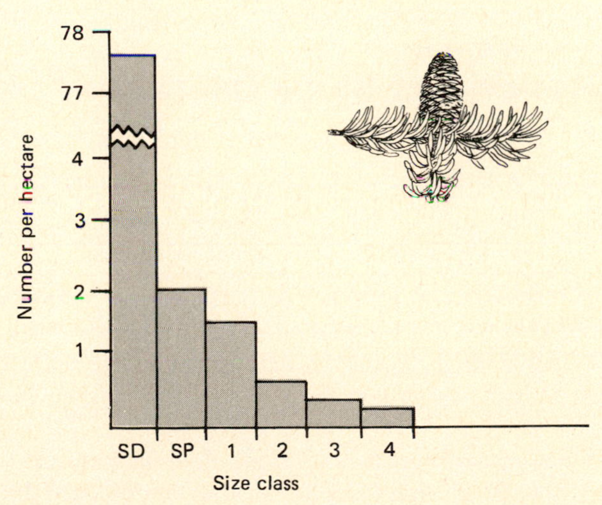

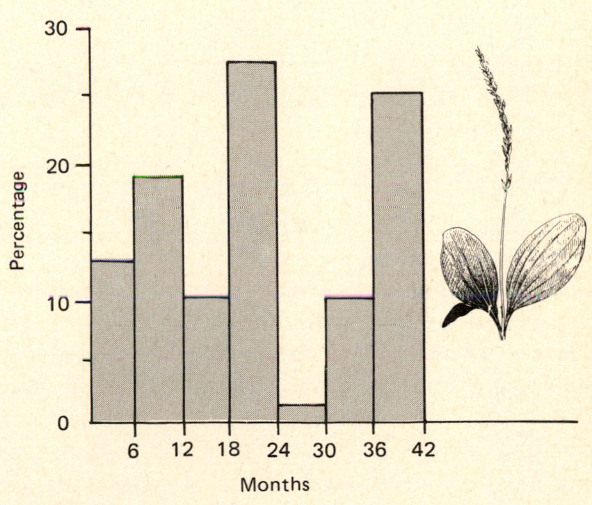

FIGURE 10.10 Age structure of some selected plant populations. (a) Hemlock (*Tsuga canadensis*) stand and (b) sweet birch (*Betula lenta*) stand. The graphs illustrate the distribution of diameter frequency (size) of trees of a sample size greater than 60 for which ages are known. The numbers above the peaks are predicted average age in years for class midpoint. When more than one diameter class is incorporated in the figure, a range of ages is given. (From Lorimer, 1980:1174.) (c) Balsam fir (*Abies balsamifera*) stand on Lake Superior in Ontario. Size classes are seedlings (sd), 1 to 3 cm diameter at breast height (dbh); saplings (sp) greater than 3 cm and less than or equal to 8 to 9 cm, and trees broken down into 7.62-cm diameter classes. (From Hett and Loucks, 1976). (d) Colony of common plantain (*Plantago rugelii*). The age structure is based on marked individuals of known ages (From Hawthorn and Cavers, 1976:516).

AGE STRUCTURE OF PLANT POPULATIONS

Studies of the age structure of plant populations are few (Figure 10.10). The reasons are in part because demographic techniques have been late in coming to the study of plant populations and in part because of the difficulty of determining age since some plants reproduce both sexually and asexually.

A modification of age structure has been used by foresters for years as one guide to timber management. They employed size (DBH, diameter at breast height) as an indicator of age on the logical assumption that diameter increases with age. The greater the diameter, the older the tree. Such assumptions, foresters discovered, were valid for dominant canopy trees. But with their growth suppressed by lack of light, moisture, or nutrients, understory trees and even so-called seedlings and saplings added little to their diameter size. While their smaller diameter suggested youth, the small trees often were the same age as their large dominant associates.

Trees can be aged approximately by counting annual growth rings. Attempts to age nonwoody plants, especially those with corms that seem to exhibit rings of annual growth, have not been very successful (P. A. Werner, 1978). The most accurate method of determining the age structure of relatively short-lived herbaceous plants is to mark individual seedlings and follow them through their lifetimes (Cook, 1979; Hawthorn and Cavers, 1978).

Because the growth and survival of juvenile trees and seedling reproduction are often inhibited by the competitive effects of the dominant overstory trees, the distribution of age classes in many forests is nonstable. One or two age classes dominate and hold onto the site until they die or are removed, allowing new young age classes to develop. There is a disproportionately higher number of old individuals, and when they succumb, there is a high local recruitment of young trees that in turn will dominate the site for years.

Summary

Groups of plants and animals of the same kind living in a particular area are considered populations. The size of a population in relation to the area it occupies is its density. Populations and individuals within a population are distributed in some kind of pattern over the landscape. Some are uniformly distributed, some are randomly distributed, but most are clumped, resulting in aggregations.

The pattern of clumped distribution is called coarse-grained when the clumps and areas between them are large and fine-grained when the areas between them are small.

Individuals comprising the population may be divided into three ecological periods: prereproductive, reproductive, and postreproductive. Visually these periods may be represented as age pyramids.

The proportion of each of the major age groups influences considerably the birthrate, the mortality rate, and population growth. Populations tend toward a stable age distribution at which the proportion of each age class stays the same as long as births and deaths remain constant. When the proportion of age classes becomes constant, deaths equal births.

Review and Study Questions

1. Distinguish between ecological density and crude density.
2. What are the three ways in which populations may be dispersed across the landscape?
3. What are the three age periods into which individuals in a population may be divided? Discuss their significance to a population.
4. What is meant by stable age distribution?
5. Contrast age structure in an animal population with age structure in a plant population. Explain the reason for the difference.
6. Age structure as depicted by age pyramids represents the past history of a population. Yet often age structure is used to predict future population trends. Think of some drawbacks to using present age structure to forecast future population developments. What can age pyramids tell us?

Suggested Readings

See the "Suggested Readings" section at the end of Chapter 14.

ECOLOGICAL EXCURSION

Age Pyramids and Population History

Robert Leo Smith

The history of a population shows in its age distribution. An example is two contrasting areas in Appalachia. One is a county characterized by family-owned agricultural enterprises, small industry, and a stable population. The second is a county characterized by an unstable, exploitative coal industry and explosive population growth.

In the agricultural county with a relatively low population per square mile, the age pyramids for 1930 and 1940 have an expanded base of young, while the percentages shrink in successive age classes (see Figure 10.11). Members of the young age classes obviously left the area as they entered the reproductive age but not in numbers so great as to distort the 15-to-34-year-old age classes in the pyramid. The excess young were drained away, leaving a residual to maintain a fairly constant population level. Since 1950 the age pyramid has suggested a stable population. The ratio of young is low enough to produce a narrow base for the pyramid, and the sizes of the age classes from 15 up are similar. The dependency ratio is fairly high. In 1930 the ratio was 73:20; in 1970, 63:29. The drop in the dependency ratio reflects a drop in the birthrate rather than an increase in the death rate. The fertility ratio, the number of births per 1,000 women aged 20 to 44, was 550:1,000 in 1930; in 1970 it was 327:1,000. In 1930, 35 percent of the population was under 15; in 1970 the percentage declined to 24.2. The percentage of those 65 years and older increased from 7.25 in 1950 to 14.66 in 1970, and the working age classes increased from 57.70 to 61.2 percent. The decline in birthrate did not result from any real decline in the reproductive age classes.

Changes in crude birth and death rates suggest further stability in the population. In 1930 the crude birthrate was 22.6 and the crude death rate 8.1. By 1970 the crude birthrate had dropped to 15.5, and the crude death rate had increased to 13.6. By 1972 the population had overachieved zero population growth. The crude birthrate was 13.4 and the crude death rate 14.7.

Age pyramids for the coal-mining county, by contrast, reflect an exploding population (see Figure 10.12). The age pyramid for 1930 reflects the heavy immigration of the previous decade. The pyramid has disproportionately large 25-to-44-year-old age classes. The profile of the pyramid for 1940 is equilateral, characteristic of a population experiencing a high birthrate and a high death rate. Since the region was not experiencing a high death rate, the profile resulted from the movement of one age class into another over time. The small old-age classes reflect a young population. In 1950 the profile is still that of an equilateral triangle but with an increase in the age classes over 65.

The great migration of the 1950s shows up in the pyramid for 1960. The sharply pinched middle emphasizes a large decrease in the 15-to-34-year-old age classes. The old age classes show a marked increase, while the 0-to-15-year-old age classes reflect only a moderate decline. For example, the percentage of the population under 15 in 1940 was 40.81; in 1960, 39.94 percent; while the percentage of those 65 years and over increased from 1.99 percent to 8.99 percent. The age classes from 15 to 64 declined from 62.30 to 54.42 percent.

The pyramid for 1970 shows a further aging and decline of the population. The older age classes (45 to 65+) increased, the middle of the pyramid is still pinched, and the number of young is declining. The dependency ratio in 1960 was 83.77; by 1970 this ratio had declined to 68.49.

In spite of a declining population, the birthrate in the coal-mining county is very high. In 1940 the crude birthrate was 35.6; by 1970 it had dropped to 20.2, but it increased to 21.2 in 1972. The crude death rate increased from 9.8 in 1940 to 12.4 in 1972. By subtracting the crude death rate from the crude birthrate, one arrives at a crude rate of increase of 0.8 percent a year compared to 2.58 percent in 1940. The fertility rate in 1970 was still a very high 522:1,000, but lower than the figures of 651 for 1930 and 674 for 1960. Thus, the region still

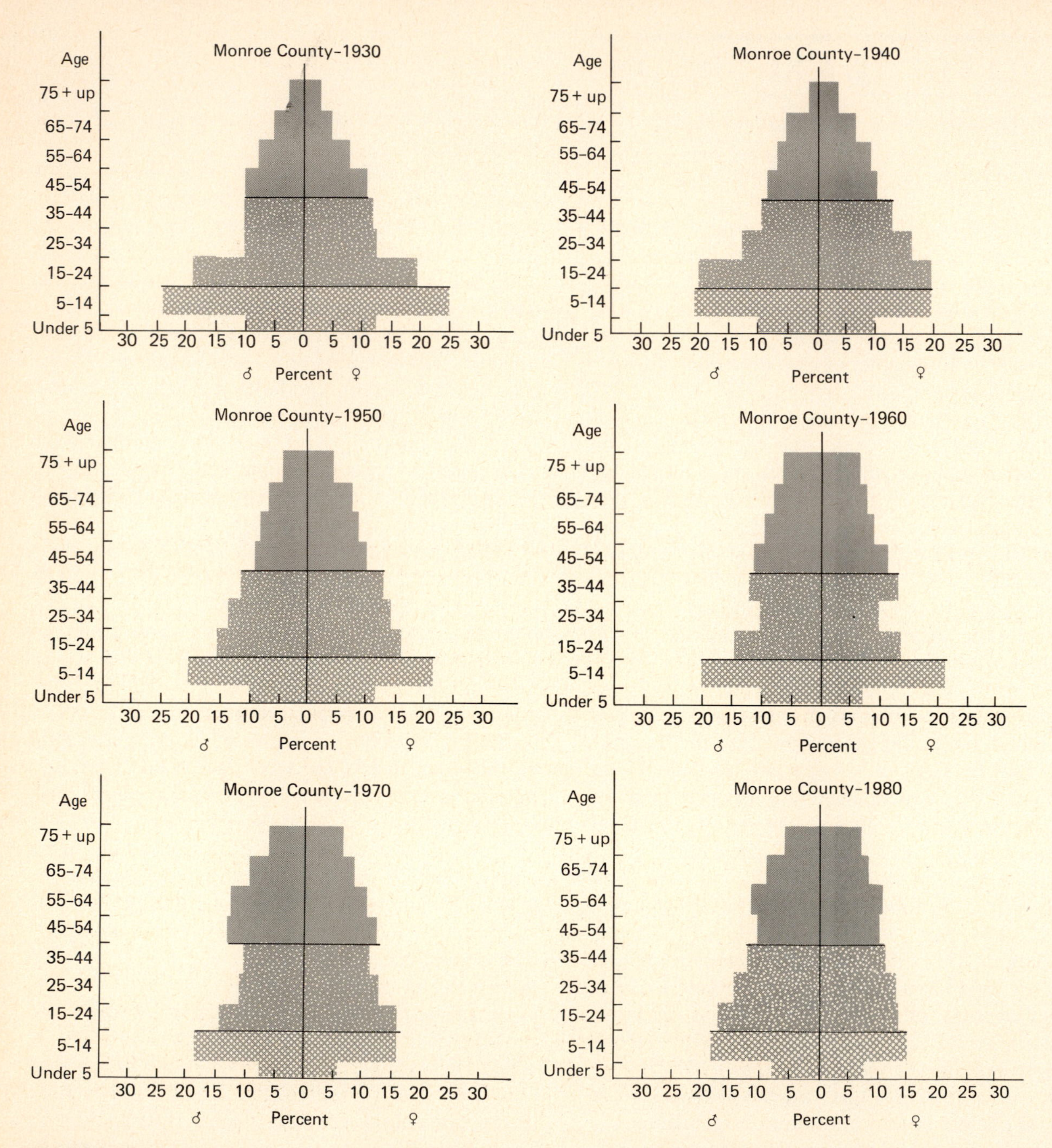

FIGURE 10.11

supplies considerable numbers of potential immigrants to other areas.

The demographic outlook in the coal-mining county is one of a continuing population decline in spite of a high birthrate. As the old individuals die, they will not be replaced by young people growing up, for upon reaching the working age, most of the young people leave the region. Two alternatives are

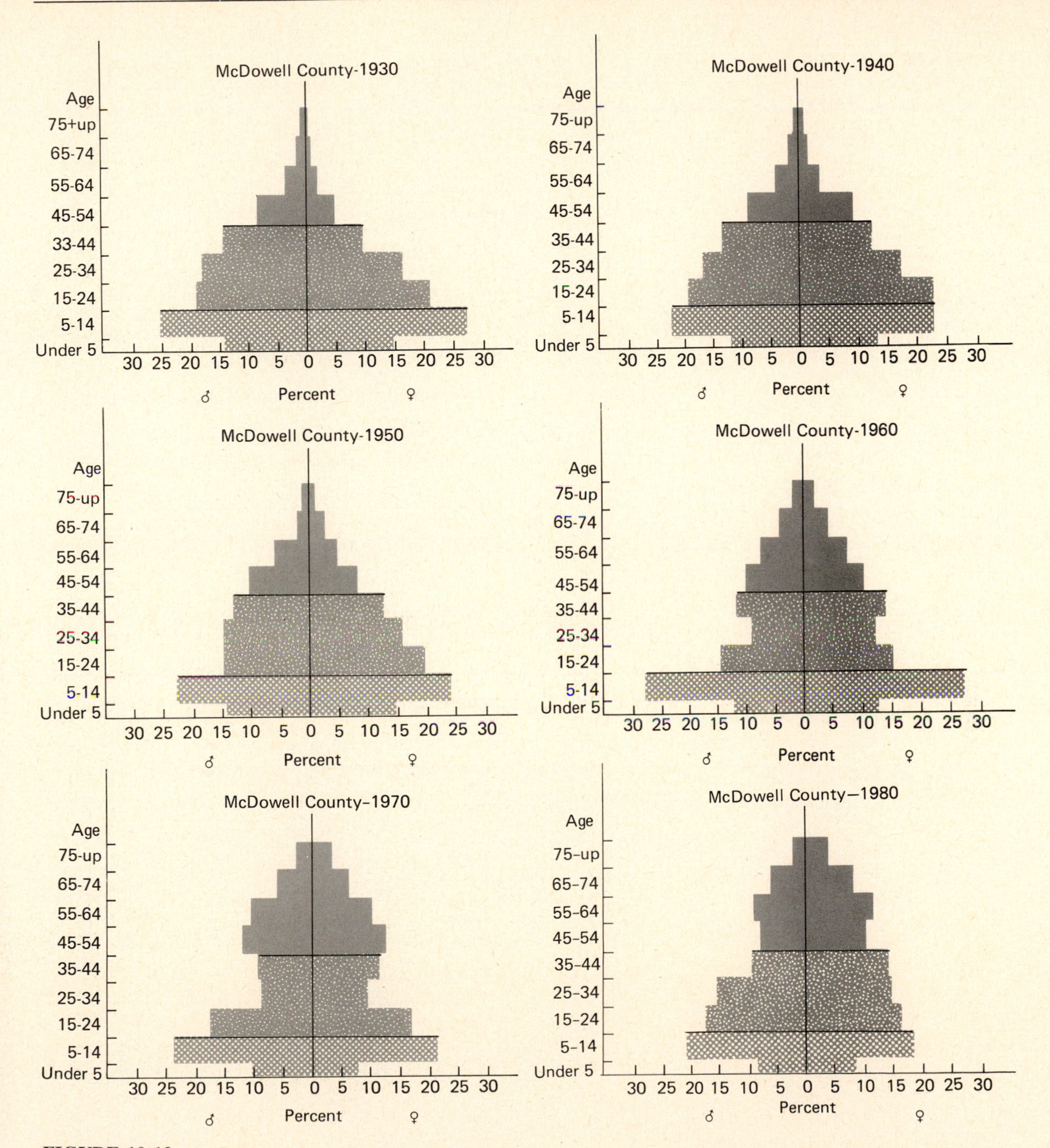

FIGURE 10.12

possible. Either the population can level off at some much lower density and maintain some sort of stability, as it is doing now with increased demands for coal; or the population can become extinct. The exploitative pattern and history of the region, especially with the destruction of the mountains by surface mining, point to the latter eventuality.

CHAPTER 11

Mortality, Natality, and Survivorship

Outline

Objectives

Upon completion of this chapter, you should be able to:

1. Express mortality as the probability of dying and the probability of survival.
2. Understand the construction of a life table and the derivation and meaning of life expectancy.
3. Distinguish among several kinds of life tables.
4. Plot mortality and survivorship curves.
5. Distinguish among different types of survivorship curves.
6. Explain how net reproductve rates are determined.
7. Explain the construction of fecundity tables and their usefulness in the study of population dynamics.
8. Construct a life table diagram.

The demography of a population mostly involves a statistical review of the arrivals and departures of individual members of the population. They may arrive by way of immigration and depart by emigration. But mostly the comings and goings involve the births of new individuals (natality) and the deaths of others (mortality). The difference between the two processes determines the growth and decline of a population.

Mortality

Mortality, which begins even in the uterus and the egg, is usually expressed as the probability of dying, or the mortality rate. It is obtained by dividing the number of individuals that died during a given interval of time by the number alive at the beginning of the period: $m = d/N_t$. The complement of the probability of dying is the probability of survival, the number of survivors at the end of a given time interval divided by the number alive at the beginning of the interval. Because the number of survivors is more important to a population than the number dying, mortality is better expressed either as the probability of surviving or as life expectancy, the average number of years to be lived in the future by members of the population.

To obtain a clear and systematic picture of mortality and survival, a life table can be constructed. The life table, first developed by students of human populations and widely used by actuaries of life insurance companies, is simply an account book of deaths. It consists of a series of columns headed by standard notations, each of which describes mortality relations within a population when age is considered. It begins with a group or cohort of a certain size, usually 1,000 at birth or hatching (see Tables 11.1 through 11.5). The cohort of 1,000 is obtained by converting data collected in the field to the equivalent numbers had the starting density of the cohort been 1,000.

The columns include x, the units of age; l_x, the number of organisms in a cohort that survive to age x, $x + 1$, and so on; and d_x, the number or fraction of a cohort that dies during the age interval x, $x + 1$. The column d_x can be summed to give the number dying over a particular period of time. If l_x and d_x are converted into proportions—that is, if the number of organisms that died during the interval x, $x + 1$ is divided by the number of organisms alive at the beginning of age x, the result is q, the age-specific mortality rate. The q_x column, however, cannot be summed to give the overall mortality rate at any specific age.

Two other columns are L_x, the average years lived, and T_x, the total years lived. These two columns are used to calculate e_x, the life expectancy at the end of each interval. The values for L_x are obtained by summing the number alive at the age intervals x and $x + 1$ and dividing the sum by 2. T_x is calculated by summing all the values of L_x from the bottom of the table to the top. Life expectancy, e_x, is obtained by dividing T_x for a particular age class x by the l_x value for that age class.

VERTEBRATE LIFE TABLES

At one time data for life tables could be obtained only for laboratory animals and humans (Table 11.1). As census methods and age determination techniques became more and more refined, sufficient data for at least an approximate life table could be acquired for some species (Deevey, 1947; Hickey, 1952; Caughley, 1966). It is difficult to obtain information on mortality and survival in wild animals. Mortality can be estimated by determining the ages at death of a large number of animals born at the same time (see Table 11.2). This procedure could involve the marking or banding of a considerable number of animals. Such a method provides information for the d_x column. One can record the ages at death of animals marked at birth but not necessarily born during the same season or year. Data from several years and several cohorts are pooled to provide the information for the d_x column. Another approach is to determine the age of death of a representative sample of carcasses of the species concerned (see Chapter 10). Such information also goes into the d_x column. Recording the ages of death of a sample of a population wiped out by some catastrophe could provide data for an l_x series. Life tables derived from the aging of animals taken during a hunting season provide information

TABLE 11.1 LIFE TABLES FOR WHITE MALE AND FEMALE HUMANS IN THE UNITED STATES, AS OF 1972

x, Age (Years)	q_x, Mortality Rate	l_x, Survivorship	d_x, Mortality	e_x, Life Expectation
White, Male				
0-1	0.0182	100,000	1,824	68.3
1-5	.0033	98,176	323	68.5
5-10	.0023	97,853	220	64.7
10-15	.0024	97,633	238	59.9
15-20	.0075	97,395	735	55.0
20-25	.0096	96,660	927	50.4
25-30	.0083	95,733	799	45.9
30-35	.0089	94,934	845	41.3
35-40	.0123	94,089	1,154	36.6
40-45	.0198	92,935	1,840	32.0
45-50	.0331	91,095	3,013	27.6
50-55	.0516	88,082	4,542	23.5
55-60	.0827	83,540	6,913	19.6
60-65	.1255	76,627	9,618	16.1
65-70	.1793	67,009	12,014	13.1
70-75	.2575	54,995	14,160	10.4
75-80	.3624	40,835	14,798	8.1
80-85	.4745	26,037	12,354	6.3
85 and over	1.0000	13,683	13,683	4.7
White, Female				
0-1	0.0137	100,000	1,370	75.9
1-5	.0025	98,630	252	75.9
5-10	.0017	98,378	164	72.1
10-15	.0014	98,214	141	67.2
15-20	.0029	98,073	285	62.3
20-25	.0032	97,788	314	57.5
25-30	.0035	97,474	339	52.7
30-35	.0047	97,135	461	47.9
35-40	.0071	96,674	687	43.1
40-45	.0114	95,987	1,093	38.4
45-50	.0180	94,894	1,710	33.8
50-55	.0260	93,184	2,425	29.3
55-60	.0402	90,759	3,649	25.1
60-65	.0585	87,110	5,096	21.0
65-70	.0899	82,014	7,370	17.1
70-75	.1445	74,644	10,784	13.6
75-80	.2361	63,860	15,078	10.4
80-85	.3534	48,782	17,237	7.8
85 and over	1.0000	31,545	31,545	5.7

SOURCE: *Vital Statistics of the United States 1972*, vol. 2, part A, Table 5–1 (Washington, D.C.: U.S. Department of Health, Education, and Welfare, Public Health Service).

TABLE 11.2 DYNAMIC LIFE TABLE FOR RED DEER HINDS ON THE ISLE OF RHUM, 1957

x, Age (Years)	l_x, Survivors at Beginning of Age Class x	d_x, Deaths	1,000 q_x, Mortality Rate/1,000	e_x, Further Expectation of Life (Years)
1	1,000	0	0	4.35
2	1,000	61	61.0	3.35
3	939	185	197.0	2.53
4	754	249	330.2	2.03
5	505	200	396.0	1.79
6	305	119	390.1	1.63
7	186	54	290.3	1.35
8	132	107	810.5	0.70
9	25	25	1,000.0	0.50

SOURCE: Lowe, 1969: 435.

TABLE 11.3 TIME-SPECIFIC LIFE TABLE FOR RED DEER HINDS ON THE ISLE OF RHUM

x	l_x	d_x	1,000 q_x	L_x	T_x	e_x
1	1,000	137	137.0	931.5	5,188.0	5.19
2	863	85	97.3	820.5	4,256.5	4.94
3	778	84	107.8	736.0	3,436.0	4.42
4	694	84	120.8	652.0	2,700.0	3.89
5	610	84	137.4	568.0	2,048.0	3.36
6	526	84	159.3	484.0	1,480.0	2.82
7	442	85	189.5	399.5	996.0	2.26
8	357	176	501.6	269.0	596.5	1.67
9	181	122	672.7	120.0	327.5	1.82
10	59	8	141.2	55.0	207.5	3.54
11	51	9	164.6	46.5	152.5	3.00
12	42	8	197.5	38.0	106.0	2.55
13	34	9	246.8	29.5	68.0	2.03
14	25	8	328.8	21.0	38.5	1.56
15	17	8	492.4	13.0	17.5	1.06
16	9	9	1,000.0	4.5	4.5	0.50

SOURCE: Lowe, 1969: 435.

There are two kinds of life tables. One is a cohort or dynamic life table, recording the fate of a group of animals all born at the same time. An example of a dynamic life table is the one for red deer on the Isle of Rhum off Scotland (see Table 11.2). It is based on the number of deer born in 1957, obtained from census data (the young are distinguishable from adults), and from samples of deer that died from 1957 to 1966.

The other type is the time-specific life table, in which the mortality of each age class in a given population is recorded over a year. It is constructed from a sample of animals of each class taken in proportion to their numbers in a population. It involves the assumption that the birth and death rates are constant and that the population is stationary. An example of a time-specific life table is the one for the red deer (see Table 11.3), which is constructed from data on the age distribution of a population sampled in one year, 1957. The l_x schedule shows survival from year 1 rather than from birth. for the l_x column because the sample came from a living population. But the data are biased in favor of older age classes, especially if the data are collected between breeding seasons.

INSECT LIFE TABLES

The life tables described are typical of long-lived animals in which generations overlap and in which different ages are alive at the same time. However, a tremendous number of animals are annual species. They have one breeding season, and generations do not overlap, so that all individuals belong to the same age class. The l_x values can be obtained by observing a natural population over its annual season and estimating the size of the population at each time of observation. For many insects the l_x value can be obtained by estimating the size of the surviving population from egg to adult. If records are kept of weather, abundance of predators and parasites, and the occurrence of disease, deaths from various causes can also be estimated.

The life table (Table 11.4) represents the fate of a cohort from a single egg mass. The age interval, or x column, indicates life history stages, which are of unequal duration. The l_x column indicates the number of survivors at each stage. The d_x column gives a breakdown of deaths by causes in each stage. In the particular population summarized in Table 11.4, dispersion and predation account for most of the losses. Note that no life expectancy is calculated because there is none. All the adult population will die in late summer.

PLANT LIFE TABLES

Mortality in plants is beginning to receive considerable conceptual treatment. But mortality and survivorship in plants are not easily condensed to the summaries recorded in life tables. For one, age is difficult to determine. Mortality of individuals usually stimulates increased growth of the survivors, resulting in an increase in biomass and in the size of the metapopulation of buds, leaves, and stems of the individuals. Seedlings make up a large numerical proportion of individual genets but an extremely small proportion of the biomass and the metapopulation. In a plant population, size rather than age may be more important. Further, it is difficult to separate and even identify individual genets. Are the plants that appear to be individual genets really ramets, a part of the metapopulation? The "parent" plant may die, yet the genet lives on in the sprouts and root suckers. Thus, the plant demographer has to deal with mortality (and natality) on two levels, the genet and the metapopulation (Table 11.5, Figure 11.1).

TABLE 11.4 LIFE TABLE OF A SPARSE GYPSY MOTH POPULATION IN SOUTHEASTERN NEW YORK

x	l_x	d_{xf}*	d_x	$100\ q_x$
Eggs	450	Parasites	67.5	15
		Other	67.5	15
		Total	135.0	30
Instars I-III	315	Dispersion, etc.	157.5	50
Instars IV-VI	157.5	Parasites	7.9	5
		Disease	7.9	5
		Other	118.1	75
		Total	133.9	85
Pre-pupae	23.6	Desiccation, etc.	0.7	3
Pupae	22.9	Vertebrate predators	4.6	20
		Other	2.3	10
		Total	6.9	30
Adults	16.0		5.6	35
Generation	—	—	439.6	97.69

* d_{xf} is the factor responsible for d_x.

SOURCE: R. W. Campbell, "Studies on Gypsy Moth Population Dynamics," in *Forest Insect Population Dynamics: Proceedings of Workshop*, ed. W. E. Waters, Res. Paper NE-125 (Washington, D.C.: USDA Forest Service, 1969), 29–51.

TABLE 11.5 LIFE TABLE FOR A NATURAL POPULATION OF *SEDUM SMALLII*

x	D_x	A_x	A'_x	l_x	d_x	$1{,}000q_x$	L_x	T_x	e_x
Seed produced	4	0–4	−100	1,000	160	160	920	4,436	4.4
Available	1	4–5	− 10	840	630	750	525	756	0.9
Germinated	1	5–6	+ 13	210	177	843	122	230	1.1
Established	2	6–8	+ 35	33	9	273	28	109	3.3
Rosettes	2	8–10	+ 81	24	10	417	19	52	2.2
Mature plants	2	10–12	+126	14	14	1,000	7	14	1.0

SOURCE: Sharitz and McCormick, 1973: 730.

The life table approach in the study of plant demography is most useful in studying (1) seedling mortality and survival, (2) population dynamics of perennial plants whose individuals can be marked as seedlings, and (3) life cycles of annual plants. An excellent example of the life table approach to plant mortality in an annual plant is the one developed by R. R. Sharitz and J. F. McCormick for *Sedum smallii* (Table 11.5). The time of seed formation is considered the initial point in the life cycle. The l_x column indicates the number of plants alive at the beginning of each stage and the d_x column the number dying during the seed and seedling stages. The L_x column gives the mean number of plants alive during the life cycle and the T_x column the total number of plants remaining to the members of the population at the beginning of each life cycle stage. Life expectancy of these annual plants dropped rapidly in the seed stages and returned to a high level after seedling establishment. Although individuals that became established had a good chance of surviving, the high early mortality resulted in a low mean life expectancy.

Another approach to the life table of plants is the yield table developed by foresters (Table 11.6). Like the life table for vertebrates, the yield table considers age classes and number of trees in each class, with additional columns giving diameter, basal area, and volume. Yield tables chart the mortality of trees by the reduced number of individuals in each age class, but as the numbers decline, basal area and biomass increase. Mortality does not necessarily reflect a declining population but rather a maturing one. Like life tables, yield tables are not constant for a species; they are constructed for different site classes that take into consideration different environmental conditions under which the species grows.

TABLE 11.6 YIELD TABLES FOR DOUGLAS-FIR ON FULLY STOCKED ACRE, TOTAL STAND

	SITE INDEX 200		
Age, Years	Trees, Per Acre Number	Av. DBH, in.	Basal Area, ft^2
20	571	5.7	101
30	350	9.0	154
40	240	12.2	195
50	176	15.3	224
60	138	18.2	248
70	113	20.9	268
80	97	23.3	285
90	84	25.6	299
100	75	27.6	312
110	69	29.4	323
120	63	31.1	332
130	59	32.7	341
140	55	34.3	350
150	51	35.8	357
160	48	37.2	364

Note: DBH is the diameter at breast height.
SOURCE: McArdle and Meyer, 1949:14.

Mortality and Survivorship Curves

From the life table two kinds of curves can be plotted: mortality curves, based on the q_x column, and survivorship curves, based on the l_x column.

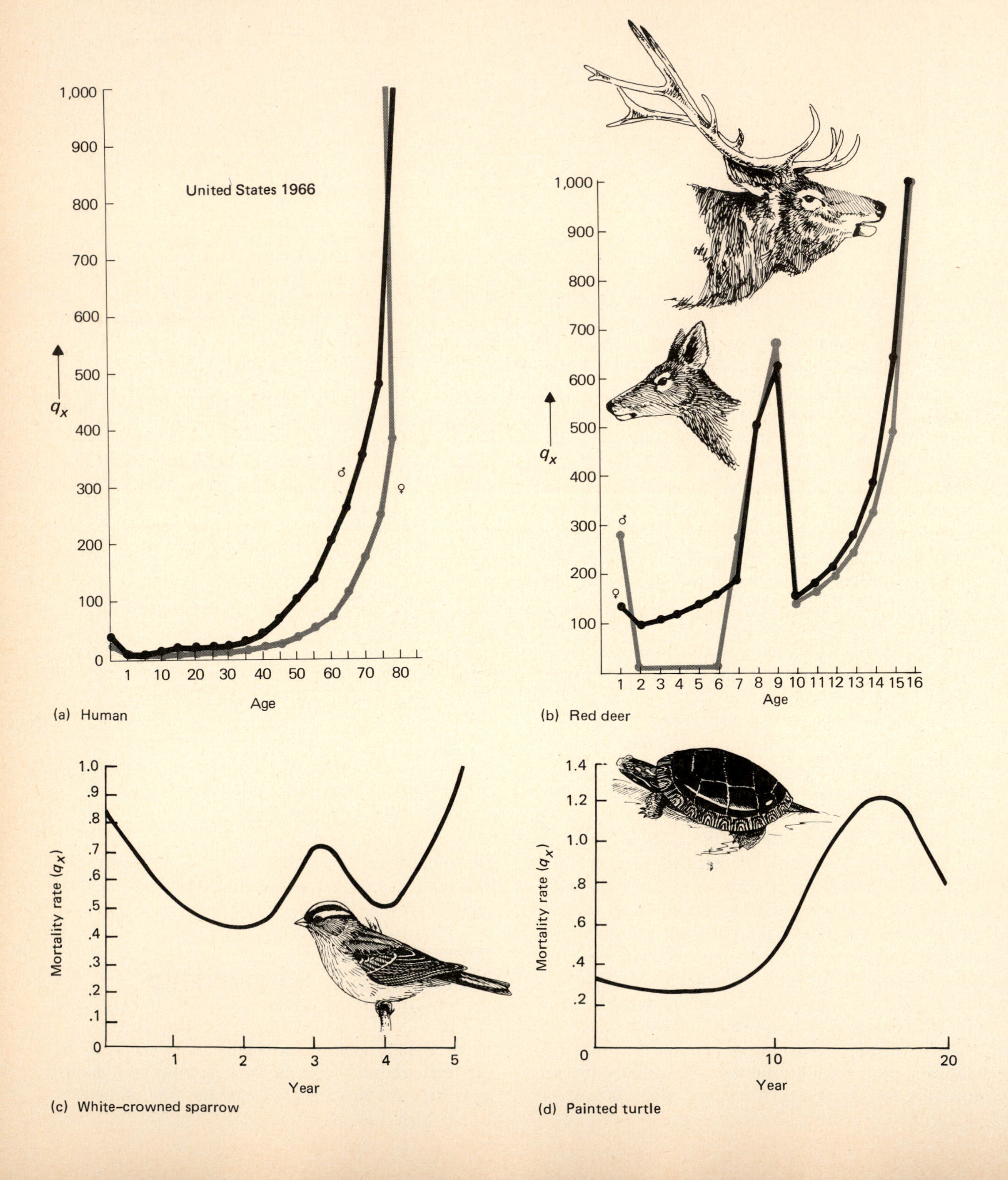
1,000
900
800
700
600
500
400
300
200
100
0
q_x
United States 1966
♂
♀
1
10
20
30
40
50
60
70
80
Age
(a) Human
1,000
900
800
700
600
500
400
300
200
100
q_x
♂
♀
1
2
3
4
5
6
7
8
9
10
11
12
13
14
15
16
Age
(b) Red deer
1.0
.9
.8
.7
.6
.5
.4
.3
.2
.1
0
Mortality rate (q_x)
1
2
3
4
5
Year
(c) White-crowned sparrow
1.4
1.2
1.0
.8
.6
.4
.2
0
Mortality rate (q_x)
10
20
Year
(d) Painted turtle

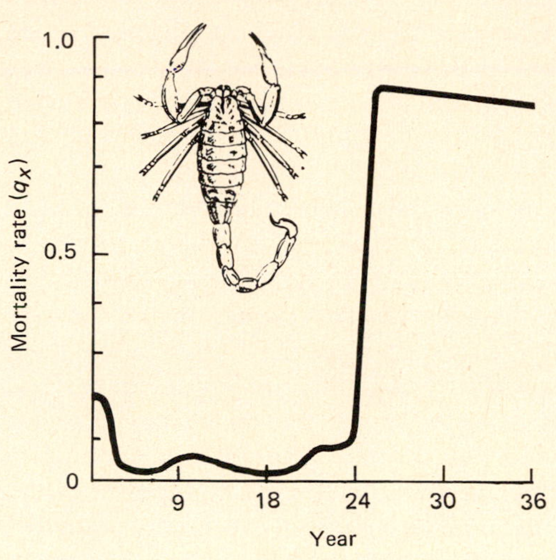

(e) Scorpion

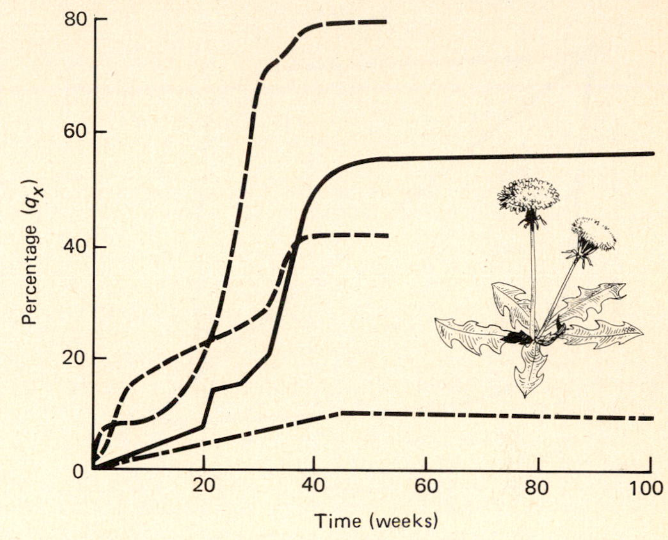

(f) Dandelion

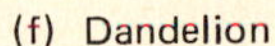

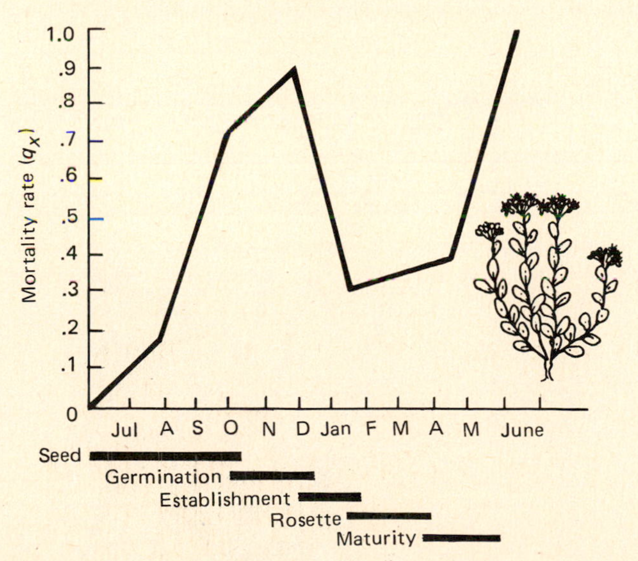

(g) Sedum

FIGURE 11.1 An array of mortality curves for some vertebrates and invertebrates and plants. The mortality curves are roughly J shaped, best exemplified by curves (a), (d), (e), and (f). The break in the J-shape in the others results first from a sharp rise in mortality in midlife, followed by high survival of the remaining individuals for a few years, before the mortality rate increases again. [Based on data for (b) from Lowe, 1969; for (c) from Baker, Mewaldt, and Stewart, 1981; for (d) from Tinkle, Condon, and Rosen, 1981; for (e) from Polis and Farley, 1980; graph for (f) from Solbrig and Simpson, 1974:4479; and data for (g) from Scharitz and McCormick, 1973.]

MORTALITY CURVES

Mortality curves plot mortality rates directly in terms of $1{,}000q_x$ against age. They consist of two parts: (1) the juvenile phase, in which the rate of mortality is high; and (2) the postjuvenile phase, in which the rate decreases as age increases until mortality reaches some low point, after which it increases again (see Figure 11.1 A to G). For mammals a roughly J-shaped curve results. For plants the mortality curve may assume a number of patterns, depending upon the type of plant, annual or perennial, and the method used to plot the data (see Figure 11.1 F and G).

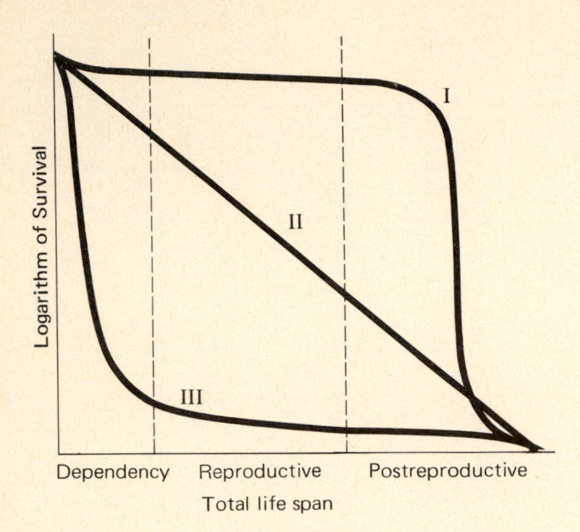

FIGURE 11.2 Three basic types of survivorship curves. The vertical scale may be graduated arithmetically or logarithmically. If plotted logarithmically, the slopes of the line will show the following rates of change: Type I, curve for animals living out the full physiological life span of the species; Type II, curve for organisms in which the rate of change in mortality is fairly constant at all age levels, a more or less uniform percentage decrease in the number that survive; Type III, curve for organisms with a high mortality early in life.

SURVIVORSHIP CURVES

Survivorship curves may be plotted in a number of ways. The usual method is to plot the logarithmic number of survivors, the l_x column against time, with the time interval on the horizontal coordinate and survivorship on the vertical coordinate (see Figure 11.2). Another method is to plot survivorship against time intervals scaled as percentage deviations from mean length of life. That allows the direct comparison of survivorship curves of organisms that have very different life spans.

The validity of survivorship curves depends upon the validity of the life table. Life tables, and thus survivorship curves, are not typical of some standard population; instead, they depict the nature of a population at different places at different times under different environmental conditions. For this reason survivorship curves are useful for comparing the population of one time, area, or sex with the population of another.

Survivorship curves fall into three general types (Figure 11.2) (Deevey, 1947). If mortality rates are extremely high in early life—as in oysters, fish, many invertebrates, and some plants (see Figures 11.3 E and 11.4 A)—the curve is concave, called Type III. If mortality rates are constant at all ages, the survivorship curve will be linear, or Type II (see Figures 11.3 B and 11.4 B). Such a curve is characteristic of the adult stages of birds, rodents, and reptiles as well as many perennial plants. When individuals tend to live out their phys-

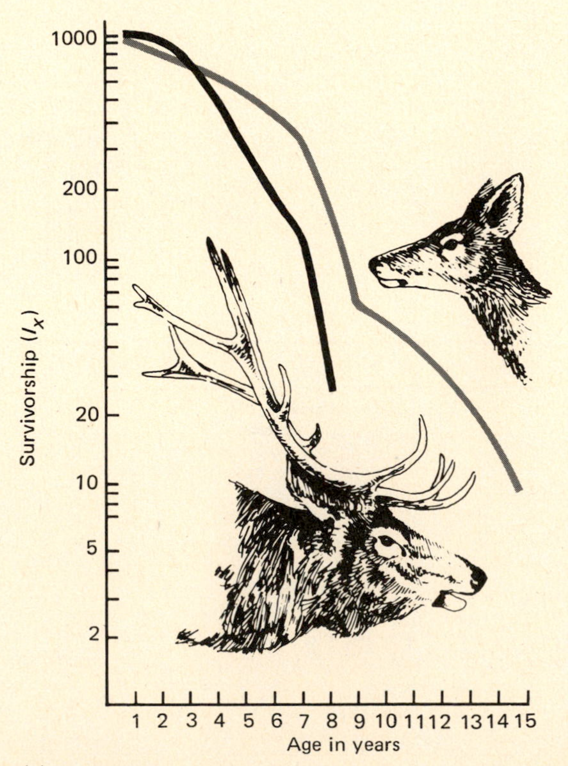

(a) Red deer

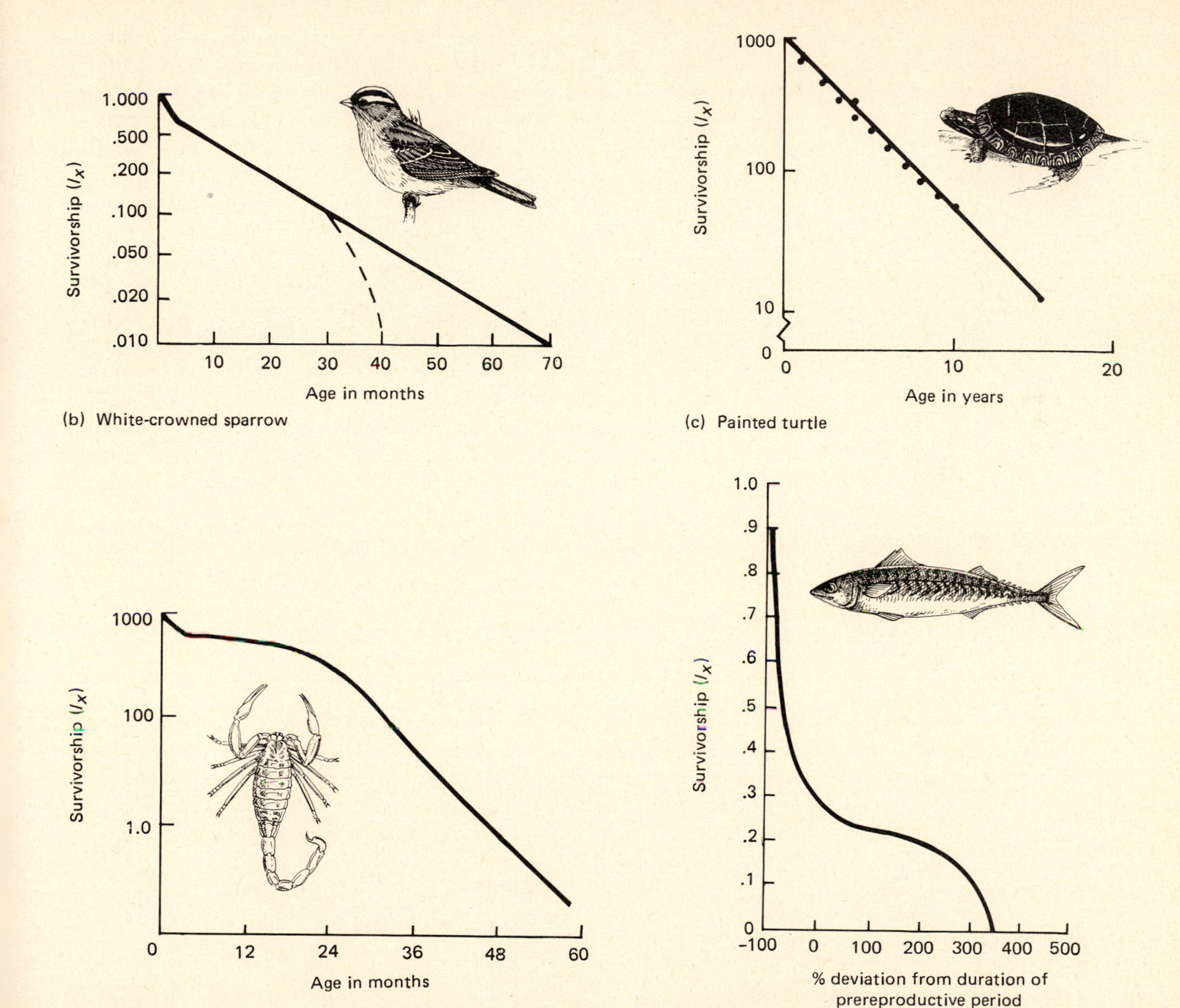

FIGURE 11.3 An array of survivorship curves for some vertebrates and invertebrates. The Atlantic mackerel exhibits a Type-III curve; the white-crowned sparrow, Type II, which appears to be typical for birds; and the red deer, Type I. [Data for (a) from Sette, 1943; for (b) from Baker, Mewaldt, and Stewart, 1981; for (c) from Tinkle, Condon, and Rosen, 1981; for (d) from Polis and Farley, 1980; and for (e) from Lowe, 1969.]

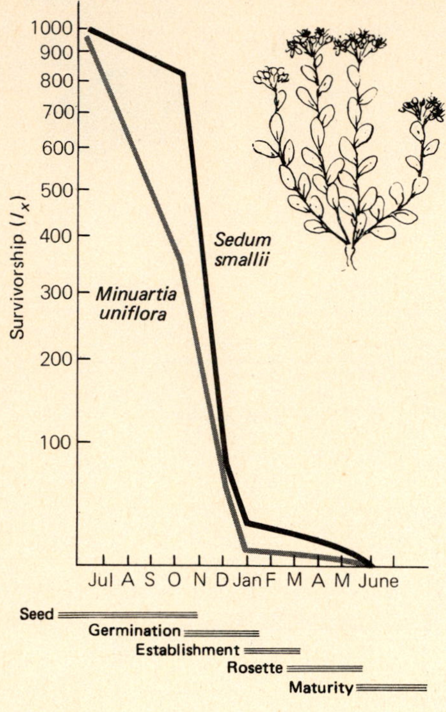

(a) Sedum

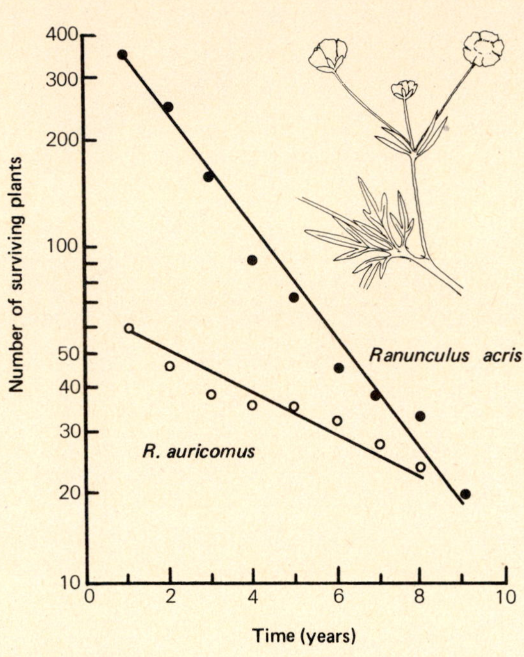

(b) Ranuculus

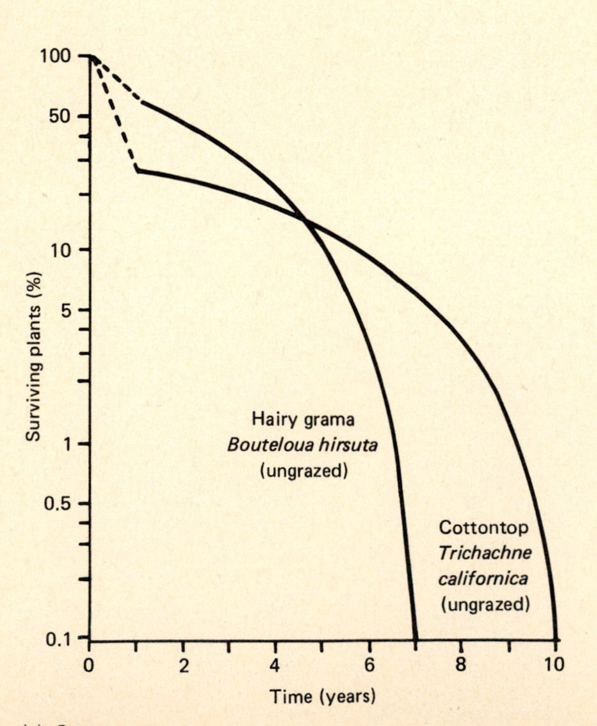

(c) Grass

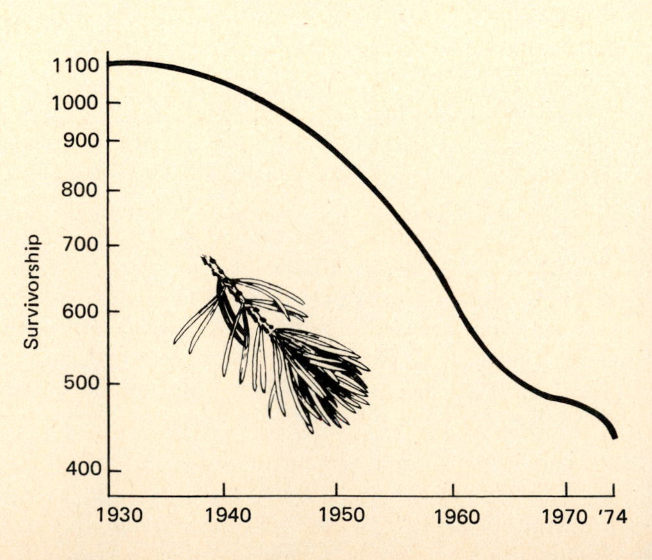

(d) Jack pine

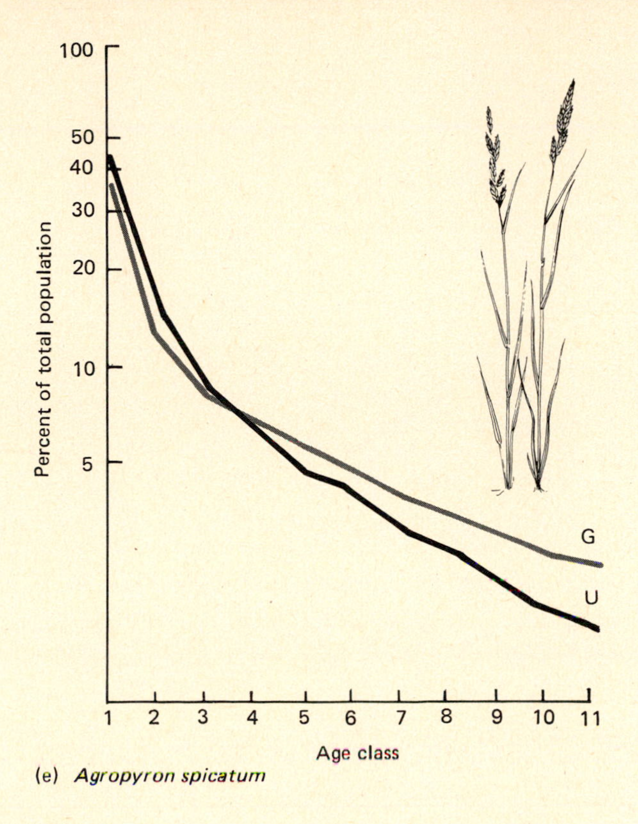

FIGURE 11.4 An array of survivorship curves for plants. *Sedum* suggests a Type-III survivorship curve; jack pine, a Type I; and Ranunculus, a Type II. [Data for (a) from Scharitz and McCormick, 1973; for (b) from Sarukhan and Harper, 1974; for (c) from West, Rea, and Harniss, for 1979:382; (d) from Yarranton and Yarranton, 1975:311; and for (e) from Sarukhan and Harper, 1974.]

iological life span and when there is a high degree of survival throughout life followed by heavy mortality at the end of a species' life span, the curve is strongly convex, or Type I. Such a curve is typical of humans and other mammals and some plants (see Figures 11.3 C and 11.4 D).

The generalized survivorship curves are conceptual models only, but they serve as models to which survivorship of a species can be compared. Most survivorship curves are intermediate between two of the models (Figures 11.3 D and 11.4 C). Consider, for example, the series of survivorship curves of the population of Sweden from the middle of the eighteenth century to 1959 (Figure 11.5). During the early eighteenth century, mortality of young was high, and life expectancy was relatively low. With advances in medicine and modifications of the environment, the survivorship curves reflect a shift in mean life expectancy toward the maximum.

Mortality in subunits or metapopulations of plants—the buds, leaves, shoots, twigs, and roots—

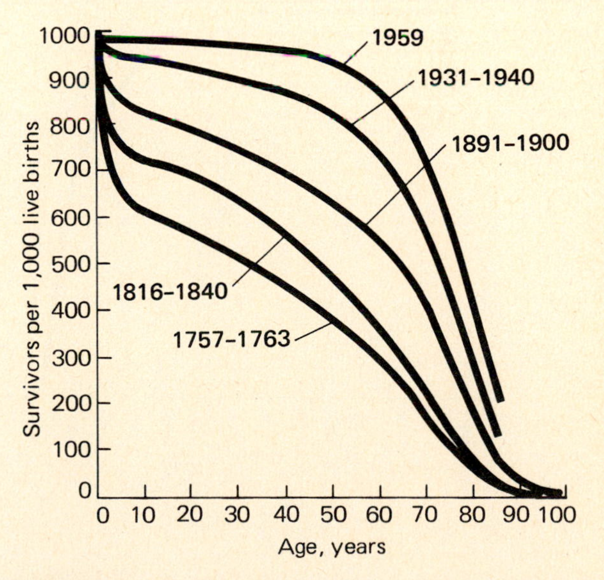

FIGURE 11.5 Survivorship curves for the population of Sweden over several centuries. Note that as health conditions improved and the standard of living increased, survivorship curves began to approach the physiological life span, changing from convex to concave. (From C. Clark, 1967.)

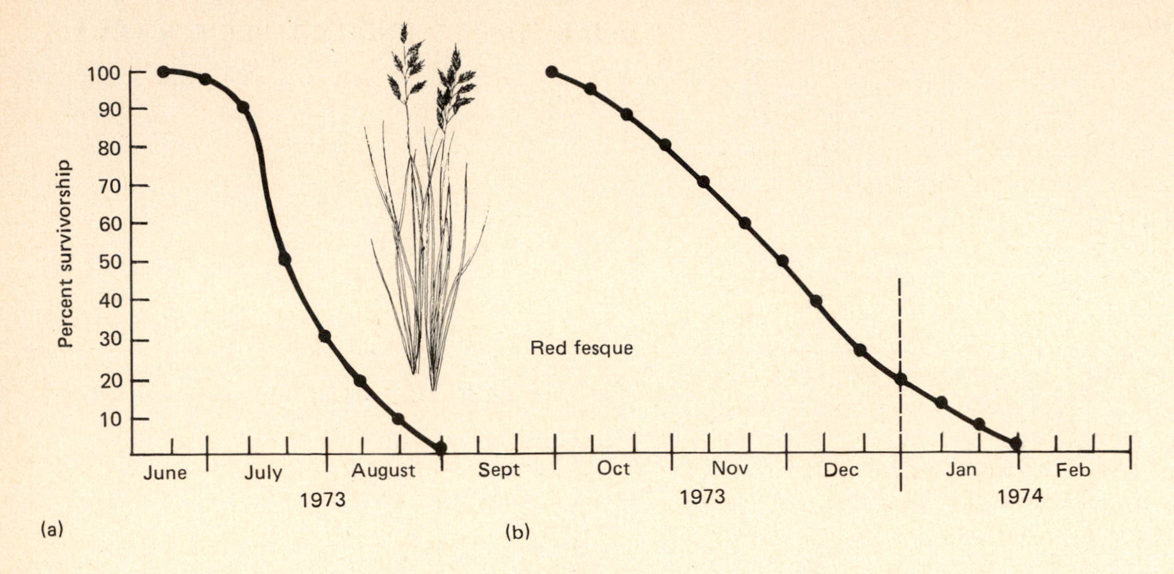

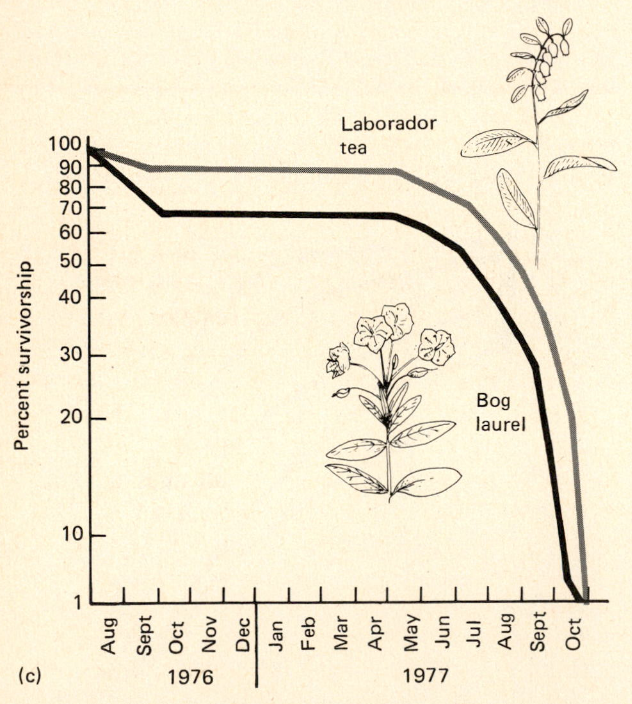

FIGURE 11.6 An array of survivorship curves for metapopulations of plants. (a and b) Marked leaves of red fesque *(Festuca rubra)* at two different times of the year; (a) one summer; (b) late summer extending into winter (after P. Williamson, 1976:1063). (c) Two broad-leaf evergreen shrubs of the bog Labrador tea (*Ledum groenlandicum)* and bog laurel *(Kalmia polifolia)*. (After Reader, 1978:1251.)

may also be analyzed by the use of survivorship curves (Figure 11.6). Knowledge of leaf mortality is important to estimate correctly primary production by plants using the harvest method (see Chapter 19). Failure to allow for the "birth" of new leaves and the "death" or turnover of mature leaves can result in lower estimates of primary production (see P. Williamson, 1976).

Natality

The greatest influence on population increase is usually natality, the production of new individuals in a population. Natality, measured as a rate, may be expressed as either a crude birthrate or a specific birthrate. If the number of births in a given period, such as a year, is divided by the estimated population at the midpoint of the period and the result multiplied by 1,000, the figure represents the *crude birthrate,* expressed as births per 1,000 population per unit of time. A more precise way of expressing the birthrate is the number of births per female of age x per unit of time, because reproductive success varies with age. If females of reproductive age are divided arbitrarily into age classes and the number of births for each age class is tabulated from this, an *age-specific schedule of births* can be constructed (see Table 11.7). Because population increases are a function of the female, the age-specific birth schedule can be modified further by determining only the mean number of females born in each female age group. This information is known as the *gross reproductive rate*. This contrasts with the *net reproductive rate, R,* the number of females left during a lifetime by a newborn female or the mean number of females born in each female age group. Because it is calculated by multiplying the gross reproductive rate, m_x, by the survival of each class, l_x, it includes adjustments for mortality in females in each age group.

How net reproductive rates are determined can be demonstrated in the fecundity table for white-crowned sparrows (Table 11.7). The fecundity table uses the survivorship column, l_x, from the life table and an m_x column, the mean number of females born to females in each age group. To calculate the net reproductive rate, the l_x values are converted to a proportionality, in this case by dividing each age value by 1,000 females of the birth year. Age 0 (x column) females lay no eggs; therefore, their m_x value is 0. The m_x value for a female aged 1 year is 3.142, and the m_x values increase with age. To adjust for mortality, the m_x values are multiplied by the corresponding l_x, or survivorship, values. The resulting value, $l_x m_x$, gives the mean number of females born in each age group adjusted for survivorship.

TABLE 11.7 FECUNDITY TABLE FOR FEMALE WHITE-CROWNED SPARROWS

x*	l_x	m_x†	$l_x m_x$	v_x‡
0	1.000	0	0	1.042
1	0.167	3.142	0.525	6.234
2	.083	3.333	.277	6.221
3	.048	3.556	.171	4.994
4	.012	3.750	.045	5.750
5	.006	4.000	.024	4.000
		R_0 =	1.042	28.241

* *Note:* Age 0 is taken from the egg stage; subsequent birthdays are July 1.
† m_x is the clutch size.
‡ v_x (reproductive value) is the age-specific expectation of future offspring $\sum_{t=x}^{\infty} (l_t/l_x)m_x$.

SOURCE: M. C. Baker, L. R. Medwaldt, and R. M. Stewart, 1981: 642.

Thus, for age class 1 the m_x value is 3.142, but when adjusted for survival, the value drops to 0.525, and for age 5 the m_x value of 4.0 drops to 0.024, reflecting poor survival of the adult females. If you multiply the proportion of females living from age x to age t, l_t/l_x, by m_x, you obtain the reproductive value of each age group, v_x. Note that in spite of their lower m_x value, females age 1 have the highest reproductive value. When the adjusted m_x values, $l_x m_x$, are summed over all ages at which reproduction occurs, the sum represents the number of females that will be left during a lifetime by a newborn female, or R_0. If the R_0 value is 1, the female has replaced herself. If the value is less than 1, the female is not replacing herself, and the population may be declining. If the value is much over 1, the population is increasing.

Examples of fecundity tables appear in Tables 11.7 through 11.10. Reproduction in the red deer (Table 11.8) begins with the 3-year-olds, and fecundity increases until age 7 years, when a decline in mean number of births begins. For a red deer population with the survivorship indicated, the re-

TABLE 11.8 FECUNDITY TABLE FOR RED DEER

x	l_x	m_x	l_xm_x
1	1.000	0	0
2	.863	0	0
3	.778	0.311	0.242
4	.694	.278	.193
5	.610	.308	.134
6	.526	.400	.210
7	.442	.476	.210
8	.357	.358	.128
9	.181	.447	.081
10	.059	.289	.017
11	.051	.283	.014
12	.042	.285	.012
13	.034	.283	.010
14	.025	.282	.007
15	.017	.285	.005
16	.009	.284	.003
		R_0 =	1.316

SOURCE: Lowe, 1969: 444.

TABLE 11.9 FECUNDITY TABLE FOR PAINTED TURTLES

x	l_x	m_x	l_xm_x
0	1.0000	0	0
1	.6700	0	0
2	.5092	0	0
3	.3870	0	0
4	.2941	0	0
5	.2235	0	0
6	.1169	2.8	0.475
7	.1229	2.8	.361
8	.0981	2.8	.274
9	.0746	2.8	.208
10	.0567	2.8	.159
15	.0144	2.8	.040
20	.0009	2.8	.002
		R_0 =	1.518

SOURCE: Data from D. W. Tinkle, J. D. Condon, and P. C. Rosen, 1981: 1430.

productive rate is 1.316. The females are more than replacing themselves. Fecundity in the white-crowned sparrows (Table 11.7) increases with age; the reproductive rate is 1.025. The females are just replacing themselves. Painted turtles (Table 11.9), a very long-lived species, do not begin to reproduce until 6 years of age, and the fecundity is stable at 2.8 throughout reproductive life. Their reproductive value of 1.51 suggests a growing population. A scorpion population (Table 11.10) studied did not fare as well. Reproduction begins at 3 years; fecundity is the same throughout the female's reproductive lifetime. R_0 was 0.887, indicating that females were not replacing themselves, and the population must be declining.

Natality in plants, like mortality, has its conceptual problems because plants reproduce both sexually and asexually. If you consider only the genet, or genetic individual, then natality is restricted to sexual reproduction. Involved are two separate populations, seeds and seedlings, and two separate processes, the production of seeds and the germination of seeds. Except for annuals and biennials, which have one reproductive effort resulting in the death of the parent plant, seed production by individual plants is hard to estimate. Woody plants and other perennials, even within a population, vary in longevity and in seed production over a period of years as well as in the ability of the seeds to germinate.

The formal equivalent of births in plants is germination. Before "birth" seeds usually undergo a varying period of dormancy, often necessary before they can sprout. The seeds of some plants remain dormant for years, buried in soil or mud as a seed

TABLE 11.10 FECUNDITY TABLE FOR SCORPIONS

x (Months)	l_x	m_x	l_xm_x
22	0.439	0	0
23	.410	0	0
24	.382	0	0
36	.072	9.95	0.716
48	.014	9.95	.139
60	.003	9.0	.027
72	.0005	9.0	.005
		R_0 =	0.887

SOURCE: Data from G. A. Polis and R. D. Farley, 1980: 624.

bank, until they are exposed to conditions conducive to germination. Once the seed has germinated, the seedling is subject to mortality. Thus, the plant population at all times consists of two parts, one growing and producing seeds and the other stored as seeds in a dormant state.

But among many plants genetic individuals make up only a fraction of the plant population. Also involved are ecologic individuals, the ramets, and populations of buds, leaves, shoots, and sprouts, some connected to the "parent" plant and others living independently. Do asexually reproduced plants represent natality? Genetically they are a part of the original genetic individuals, even though they eventually may be involved in seed production. Can you count them as new individuals? And does the production of new growth by a plant—the development of buds, the growth of new shoots, the unfolding of new leaves—represent natality at the level of the metapopulation, just as the death of buds and leaves represents mortality? Both have a pronounced influence on the plant itself. Such questions, however, are yet to be resolved.

Summary

Population size is influenced by the number of individuals added to the group by births and immigration and by the number leaving by death and emigration. The difference between the two determines the growth and decline of populations.

Mortality, concentrated in the young and the old, is the greatest reducer of populations. Mortality and its complement, survivorship, are best analyzed by means of a life table, an age-specific summary of mortality operating on a population.

From the life table one can derive both mortality curves and survivorship curves. They are useful for comparing demographic trends within a population and among populations living under different environmental conditions and for comparing differences in survivorship among various species. In general, mortality curves assume a J shape. Survivorship curves fall into one of three major types: Type I, in which survival of young is very low; Type II, in which mortality, and thus survivorship, is constant through all ages; and Type III, in which individuals tend to live out their physiological life spans. Survivorship curves follow similar patterns in both plants and animals.

Survivorship and mortality in plants are complicated by the survivorship and mortality of parts of plants. The metapopulation of leaves, twigs, and clones also exhibits its own demographic characteristics.

Birth has the greatest influence on population increase. Like deaths, births are age-specific. Certain age classes contribute more to the population than others. Reproductive values of various age groups as related to their survival can be determined from fecundity tables derived from the life table.

As with mortality, natality in plants is complicated by vegetative reproduction and sexual reproduction through seed production and germination. In general, the latter are considered true natality because they involve the production of genets, or genetic individuals. Again, clones, leaves, buds, and twigs exhibit their own "birth" rates.

Review and Study Questions

1. What are mortality, natality, survivorship, and fecundity? How are they interrelated?
2. Although they are not specifically stated in the text, what life history characteristics influence birthrates?
3. What is a life table? What information is needed for its construction?
4. What are the advantages and weaknesses of a life table in the study of population dynamics?
5. What are the differences between a mortality curve and a survivorship curve? From which columns of the life table are they derived?
6. Why are birth and natality more difficult to

study in plants than in animals? Consider the concepts of natality, mortality, and survivorship as they relate to plants.

7. The text contains graphs of survivorship and mortality, but none for fecundity. That is left to you. For each of the mortality curves given, construct a fecundity curve from data in the fecundity tables. Plot both the fecundity curve and the mortality curve on the same graph. What can you conclude from a study of the two graphs concerning the populations of each of the species?

Suggested Readings

See the "Suggested Readings" section at the end of Chapter 13.

CHAPTER 12

Population Growth

Outline

Objectives

Upon completion of this chapter, you should be able to:

1. Distinguish between exponential and logistic growth.
2. Understand the relationship between net reproductive rate and the rate of increase.
3. Explain what is meant by carrying capacity and its relationship to the logistic growth curve.
4. Explain the influence of time lags on population growth.
5. Discuss cycles and irregular fluctuations in populations.
6. Explain the difference between population growth in animals and population growth in plants.
7. Discuss reasons why populations become extinct.

Mortality and natality are the two major forces influencing population growth. If births exceed deaths, the population increases. If births equal deaths, the population remains the same. And if deaths exceed births, the population is headed for extinction. Two additional influences on population growth are immigration, an influx of new individuals into a population, and emigration, the dispersal of individuals from a population.

Exponential Growth

The rate at which populations change can be estimated from R_0, the net reproductive rate as determined from the fecundity table (see Table 11.7). Consider the populations of four species of animals: the red deer, white-crowned sparrow, painted turtle, and scorpion. According to the fecundity tables presented in Chapter 11 for a local population of each of the four species, the net reproductive rate for the painted turtle was 1.51; for the red deer, 1.32; for the white-crowned sparrow, 1.04; and for the scorpion, 0.887. When $R_0 = 1$, females are replacing themselves, and the population remains the same. When R_0 is greater than 1, the population increases. How fast it increases is revealed by how much R_0 exceeds 1. When R_0 is less than 1, the population is declining.

The general equation for population growth is $N_t = N_0R_0$, in which N_t is the population size at some given time in the future, N_0 is the initial population, and R_0 is the net reproductive rate. To determine the population growth of the painted turtle, as an example, you might start with an initial population of $N_0 = 100$. R_0 is 1.51. In the following generation, time 1, the population would be:

$$N_1 = N_0R_0 = (100)(1.51) = 151$$

Growth to the next generation, time 2, would be:

$$N_2 = N_0R_0R_0 = (100)(1.51)(1.51) = 228$$

or

$$N_2 = N_1R_0 = (151)(1.51) = 228$$

The equation can be stated as $N_{t+1} = N_tR_0$ or, expressed more concisely, $N_t = N_0R_0^t$, in which R is raised to the power of the appropriate generation. For example, $N_2 = N_0R_0^2$. Growth rates for the four species are given in Table 12.1.

The equation can also be written as

$$R_0 = N_{t+1}/N_t$$

If you lack sufficient life table data to construct a fecundity table, you can estimate R_0 from the ratio of numbers at successive time intervals. For example, using data for the painted turtle in Table 12.1:

$$R_0 = N_4/N_3 = 519/344 = 1.51$$

The equation $N_t = N_0R_0^t$ describes a population that grows exponentially, like compound interest. Such growth can occur when R_0 is greater than 1, the environment remains constant, and resources are excessive. Exponential growth of populations of the four species is shown in Table 12.1 and Figure 12.1. Note how the shape of the curve varies with the value of R_0. The closer R_0 comes to unity, the slower the growth. In fact, the white-crowned sparrow population is barely replacing itself; and the scorpion population, with an R_0 value of less than 1, is declining rather than increasing exponentially.

These growth curves suggest several features of population growth. It is influenced by heredity and by life history features, such as age at the beginning of reproduction, the number of young produced, survival of the young, and the length of the reproductive period.

A population may increase at an exponential rate until it overshoots the ability of the environment to support it. Then the population declines sharply from starvation, disease, or emigration. From the low point the population may recover to undergo another phase of exponential growth; it may decline to extinction; or it may recover and fluctuate about some level far below the high level once reached.

The J-shaped, or exponential, curve is characteristic of some insects and of vertebrates introduced into a new and unfilled environment. An example of an exponential growth curve is the rise

TABLE 12.1 EXPONENTIAL GROWTH OF LOCAL POPULATIONS OF FOUR SPECIES OF ANIMALS: PAINTED TURTLE, RED DEER, WHITE-CROWNED SPARROW, AND SCORPION ($N_t = N_0R_0^t$; N = 100)

Year	Painted Turtle R = 1.51	Red Deer R = 1.32	White-crowned Sparrow R = 1.04	Scorpion R = 0.887
0	100	100	100	100
1	151	132	104	89
2	228	174	108	79
3	344	230	112	70
4	519	304	117	62
5	785	400	122	55
6	1,185	529	126	48
7	1,789	689	132	43
8	2,702	922	136	38
9	4,081	1,216	142	34
10	6,162	1.606	148	30

(and decline) of a human population in southwestern West Virginia (Figure 12.2a). Growth between 1830 and 1950 was exponential. Between 1950 and 1960 the population decreased dramatically. The decline produced a curve typical of a population unresponsive to the carrying capacity of the environment. Growth stops abruptly and declines sharply in the face of environmental deterioration. Another example is the reindeer herd on St. Paul Island (Figure 12.2b). Introduced on St. Paul in 1910, the

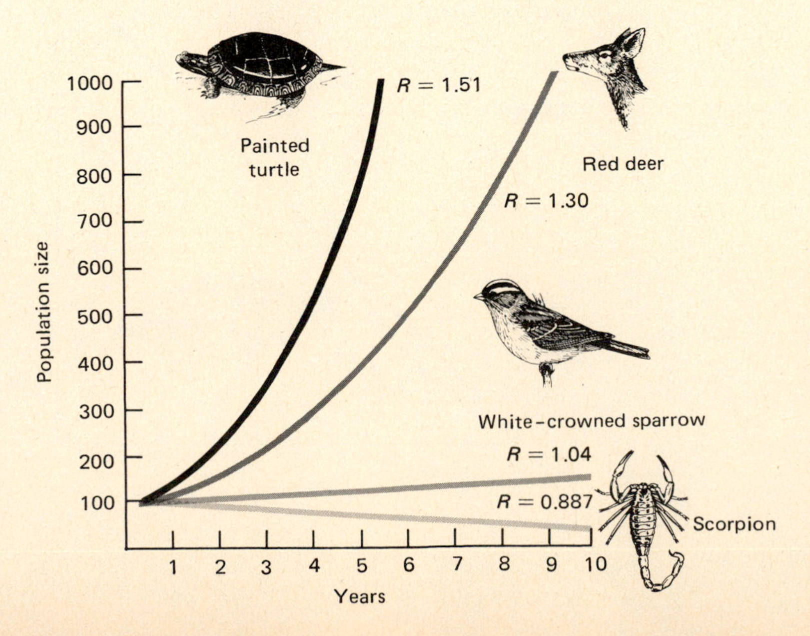

FIGURE 12.1 Exponential growth for populations of four species, each with a different value of R_o. The number 100 represents the initial populations. Note that the scorpion, with an R_o value of 0.887 (with a value of R_o = 1 needed for replacement), is headed for extinction at the local level (see Table 12.1). (Data for graphs based on Tinkle, Condon, and Rosen, 1981, for the painted turtle; Lowe, 1969, for the red deer; Baker, Mewaldt, and Stewart, 1981, for the white-crowned sparrow; and Polis and Farley, 1980, for the scorpion.)

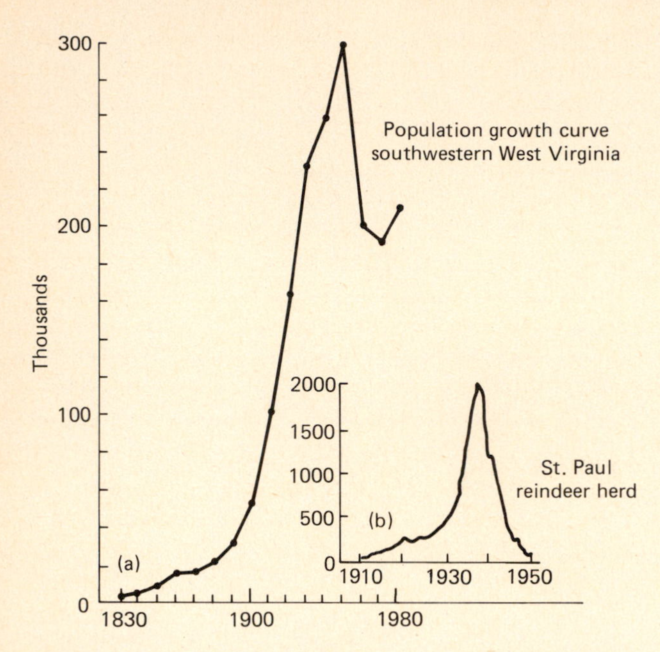

FIGURE 12.2 (a) Population growth curve for five southwestern West Virginia coal counties. Note the exponential growth followed by a sharp population decline. (Prepared by R. L. Smith from U.S. Census Bureau data; from R. L. Smith, *The Ecology of Man: An Ecosystem Approach,* [New York: Harper and Row, 1976].) (b) Exponential growth of the St. Paul reindeer herd and its subsequent decline. (From Scheffer, 1951.)

reindeer expanded rapidly from 4 males and 21 females to a herd of about 2,000. So severely did the reindeer overgraze their range that the herd plummeted to 8 animals by 1950.

RATE OF INCREASE

R_0, the net reproductive rate, is usually designated by the Greek letter lambda and, in turn, is often replaced by the term e^r, where e is a constant, the base of natural logarithm 2.71828, and r is the power to which e is raised. The term r, the rate of increase, is obtained by taking the natural log of R_0. Thus, for $R_0 = 1.51$, $\log_n R_0 = 0.412$. Thus,

$$e^r = 2.71828^{0.412} = 1.51$$

Values for r for $\log_n R_0$ for each of the four populations appear in Table 12.2.

TABLE 12.2 VALUES FOR R_0 (NET REPRODUCTIVE RATE) AND r (RATE OF INCREASE) FOR FOUR SPECIES OF ANIMALS

Species	$R_0 = e^r$	r
Painted turtle	1.51	0.412
Red deer	1.32	0.277
White-crowned sparrow	1.04	0.039
Scorpion	0.887	−0.120

The rate of increase is the slope of the population growth curve when the graph is plotted as the logarithms of the numbers. Such a graph assumes a straight line (Figure 12.3).

Determination of r by such methods results only in an approximation of the actual rate of increase. To determine the value of r_m, the intrinsic rate of increase, requires a more complex mathematical calculation, which is beyond the scope of this book.

The rate of increase depends upon the exponential rate at which a population grows if it has a stable age distribution appropriate to current life

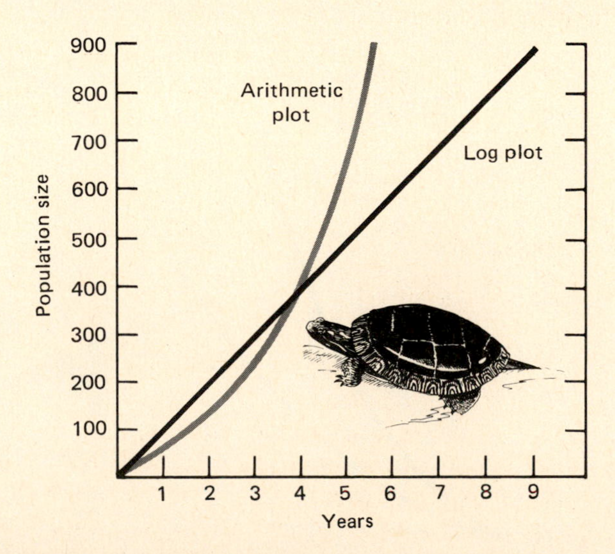

FIGURE 12.3 Exponential growth of a painted turtle population plotted arithmetically and logarithmically. (Based on data in Tinkle, Condon, and Rosen, 1981.)

table parameters of age, survival, and fecundity. It also depends upon mean fecundity and mean survival at each age in the population. Because age structure is seldom stable and survival and fecundity vary over time, r, like R_0, reflects the past and not the present.

Nevertheless, the use of r, the rate of increase, has certain advantages over R_0, the net reproductive rate. The latter is expressed as unity, + or −, but when population growth is measured as r, it has the same value as an equivalent rate of population decrease. Thus, if the scorpion population was increasing at the same rate as it was decreasing, $r = +0.120$, the value of R_0 would be 1.127. The fact that a population with an R_0 value of 0.887 is declining at the same rate, -0.120, as a population with an R_0 of 1.127 is increasing is not apparent when R_0 values are compared. Thus, r allows a direct comparison of rates and converts easily from one form to another. In addition, r allows the doubling time of population growth to be calculated as follows: $0.6931/r$. The doubling time for the turtle population is $0.6931/0.412 = 1.68$ years.[1]

Now, with some of the mystery removed from r, the term will be used from this point on. To present the exponential growth curve in terms of r, you would use the equation $n_t = N_0 e^{rt}$. Thus, for the painted turtle at time 4,

$$N_4 = 100\,(2.71828^{0.412 \times 4}) = 100\,(2.71828^{1.648}) = 520$$

(see Figure 12.3).

Logistic Growth

For populations in the real world, the environment (as assumed by the exponential growth rate) is not constant, the resources are not unlimited. As the density of a population increases, competition among its members for available resources also increases. With fewer resources to share and with an unequal distribution of those resources available, mortality increases, fecundity decreases, or both occur. As a result, population growth declines with increasing density, eventually reaching a level at which population growth ceases. That level is called carrying capacity and is expressed as K. The population theoretically is in equilibrium with its resources or environment. In other words, population growth is density-dependent, in contrast to exponential growth, which is independent of population growth.

Inhibitions on the growth of a population by competition among its members can be described mathematically by taking the exponential equation $N_t = N_0 e^{rt}$ and adding to it some variables to account for the effects of density. If you use the equation involving R_0, then the brake on population growth is described by

$$\frac{R_0 - 1}{K}$$

The equation reads

$$N_{t+1} = \frac{N_t R_0}{1 + \dfrac{(R_0 - 1)N}{K}}$$

The term $\frac{R_0 - 1}{K}$ is usually replaced by a, so the equation reads

$$N_{t+1} = \frac{N_t R_0}{1 + aN_t}$$

The same equation using r can be expressed as

$$N_t = \frac{K}{1 + e^{a - rt}}$$

where $a = r/K$. This equation assumes discrete periods of reproduction. The more usual equation is a differential one that assumes continuous breeding, and reproduction is expressed in terms of the rate of growth per individual. The equation is:

$$\frac{dN}{dt} = rN\left(\frac{K-N}{K}\right) = rN\left(1 - \frac{N}{K}\right)$$

in which dN/dt represents the instantaneous rate of change in population density N. The rate of growth

[1] $N_t/N_0 = 2$; $e^{rt} = 2$; $\therefore rt = \log_n 2 = 0.6931$; $\therefore t = 0.6931/r$.

is unaffected by competition when N approaches 0 and slows when N approaches K.

That equation describes the logistic, sigmoidal, or S-shaped, growth curve (see Figure 12.4). It is called logistic because it calls attention to the logistical problems of allocating scarce resources to an expanding population. Population increase is slow at first, then accelerates until it reaches maximum. As density increases, growth slows, marked by the inflection point in the curve. As the population approaches carrying capacity, the curve flattens. Or to state it more precisely, as $K - N$ approaches 0, dN/dt also approaches 0. When $N = K = 0$, the population has reached equilibrium.

The logistic equation suggests that populations function as systems, regulated by positive feedback. Growth results from positive feedback (as illustrated by the exponential curve), ultimately slowed by the negative feedback of competition and resource availability. As the population approaches the upper limit of the environment, it theoretically responds instantaneously as density-dependent reactions set in. Rarely does such feedback work as smoothly in practice as the equation suggests. Often adjustments lag, and available resources may be sufficient to allow the population to overshoot equilibrium. Unable to sustain itself on the available resources, the population then drops to some point below carrying capacity, but not before it has altered resource availability to future generations. Its recovery as determined by the reproduction rate is influenced by the density of the previous generation and the recovery of the resources, especially the food supply. These build a time lag into population recovery.

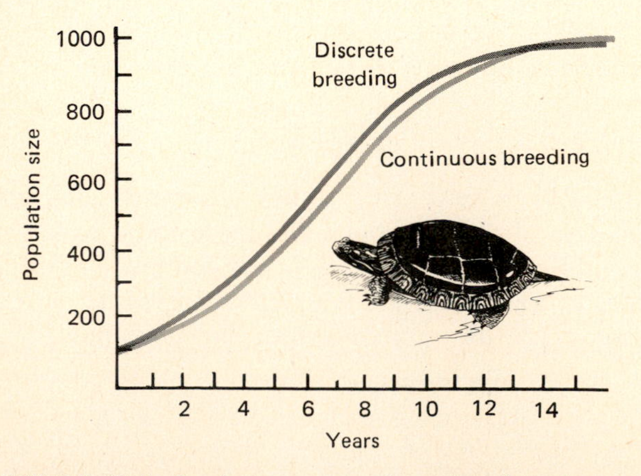

FIGURE 12.4 Logistic growth of a painted turtle population plotted as discrete and continuous breeding.

Time lags result in fluctuations in the population. The population may fluctuate widely without any reference to equilibrium size. Such populations may be influenced by some powerful outside, or extrinsic, force such as weather or by some chaotic changes inherent in the population (see May, 1976). A population may fluctuate about the equilibrium level, K, rising and falling between some upper and lower limits. Such fluctuations are called stable limit cycles. Some populations oscillate between high and low points in a manner more regular than one would expect to occur by chance. These are commonly called cycles.

The two most common oscillation intervals in animal populations are 9 to 10 years, typified by the lynx and snowshoe hare, and 3 to 4 years, typified by lemmings (Figure 12.5). These cyclic fluctuations are confined largely to simpler ecosystems, such as the northern coniferous forest and tundra. Cycles in the snowshoe hare, in part, involve an interaction between the hare and its overwinter food supply, mostly small aspen twigs. Overutilization of the food supply by a growing population of hares reduces the ability of the plants to recover from excessive pruning. Ultimately, the decreased plant growth triggers an overwinter food shortage, resulting in a heavy mortality of hares. While the hare population is low, the vegetation recovers, stimulating a resurgence of the hare population and initiating another cycle.

Fluctuations in populations are characterized by certain demographic parameters. L. Keith and his associates (Meslow and Keith, 1968; Keith and Windberg, 1978) followed a snowshoe hare population through two periods of decline and one period of increase (15 years, 1.5 cycle periods) in the Rochester district of central Alberta, Canada. The decline, which began before the peak winter population, was characterized by a high winter-to-spring weight loss, a decrease in juvenile growth rate, a decrease in juvenile overwinter survival, a reduction in adult survival beginning 1 year after the

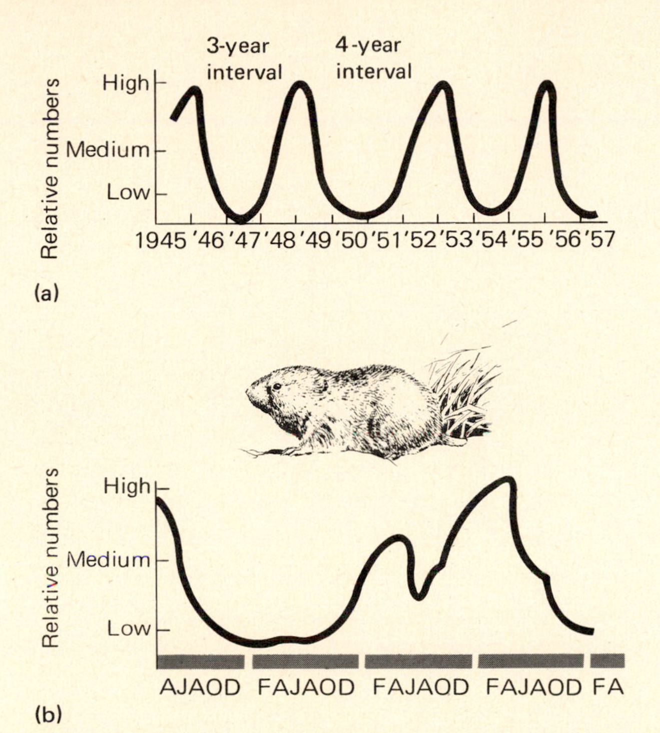

FIGURE 12.5 (a) Generalized curve of the 4-year cycle of the brown lemming near Barrow, Alaska. (b) Generalized curve of a single oscillation in a short-term cycle showing subordinate fluctuations; D, December; O, October; J, July; A, April. (After F. Pitelka, 1957:79, 80.)

population peak and continuing to the low point, and a decrease in reproduction. The upswing of the cycle, which set in about 3 years after the peak winter and slowed the rate of decline, was characterized by a lower winter-to-spring weight loss, increased juvenile growth rate, increased overwinter juvenile survival, and increased reproduction. Potential natality during the upswing was at least double that of the worst decrease years immediately preceding the low. Similar parameters are common to feral house mouse populations (DeLong, 1966) and microtine rodents (Krebs and Myers, 1974).

The logistic growth curve with all its fluctuations is common to many populations liberated in a new environment. Left undisturbed, they usually exhibit a single overshoot. The curve involves four stages (Caughley, 1970) (see Figure 12.6): (1) the initial increase stage, the period between the establishment of the population and the attainment of the initial peak; (2) the initial stabilization period at peak population; (3) the decline; and (4) the post-decline stage, in which the population adjusts to a lower level. As the population approaches K, either upward or downward, growth departs from the logistic and fluctuates between some lower and upper levels.

Human population may also follow the logistic model in a similar fashion. The growth curve for an agricultural county in southeastern West Virginia is sigmoid or S-shaped (see Figure 12.7). The population reached an asymptote, or K, of about 13,000 in 1900 and has fluctuated about that level since. One dip in the population occurred in 1930; a recovery followed. A second decline took place between 1950 and 1960.

The population growth rate of plants is necessarily different from that of animals. The rate of increase, r, is bipartite: one part due to seed production and the other due to growth of individual plants as expressed by clonal spread. The growth rate of plants, like that of most animals, is rarely continuous. Examples of growth rates that could be described by the differential equation:

$$\frac{dN}{dt} = rN\left(1 - \frac{N}{K}\right)$$

are algae and duckweeds (*Lemna* spp.). Other plant populations grow mostly in pulses, controlled in part by the periodic seed production that occurs within the life cycle of the plant. One generation may dominate a site until some event such as a fire or a timber harvest eliminates that population and permits another to grow.

Harper and White (1974) point out the close relationship that exists between N, the number of plants in a population, and w, the weight of the plants, including all their subunits. When N, density, is high, a relationship exists between the growth of survivors and the mortality of individual plants. As mortality increases, the growth of individual plants increases. This observation gives rise to several generalizations about the growth of plant populations: (1) The rate of mortality of individuals is related to the stress of population density. (2) The growth of survivors is related to the rate of mortality of individual plants in the population. (3) In

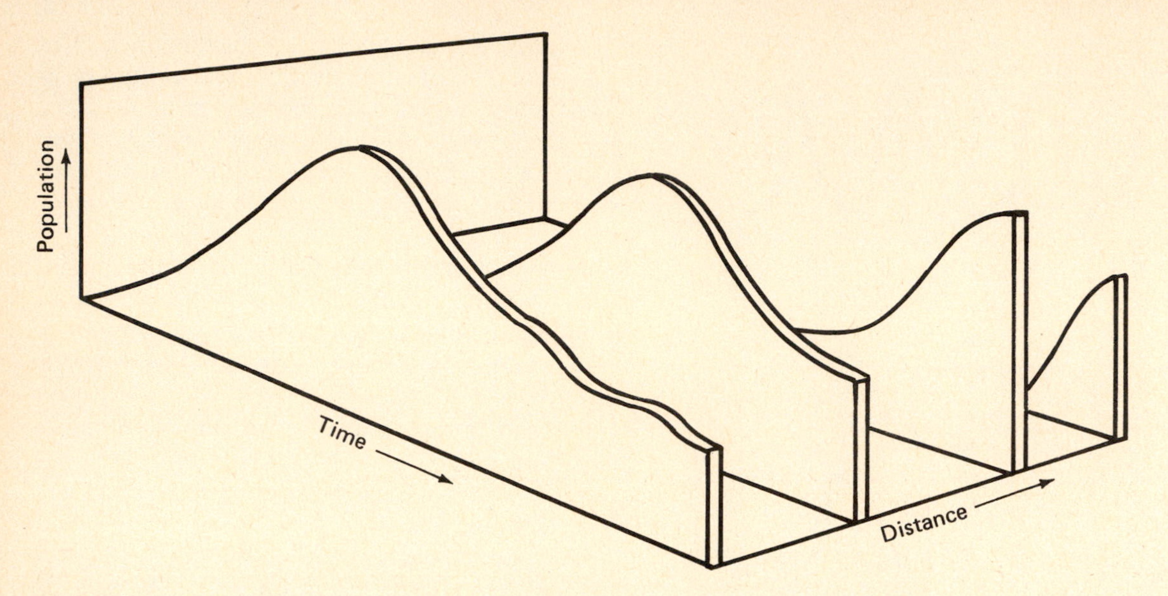

FIGURE 12.6 A model of the population increase of an ungulate herbivore released in a vacant habitat. The curve on the extreme left shows population growth, irruptive fluctuation, decline, and postdecline stability at the point of initial colonization. The curve on the extreme right shows the stage of population growth on two successive fronts. The population at the first front is at peak density; the population at the next advancing front is in the exponential growth phase; and the population on the advancing front is becoming established. (Adapted from Caughley, 1970:57.)

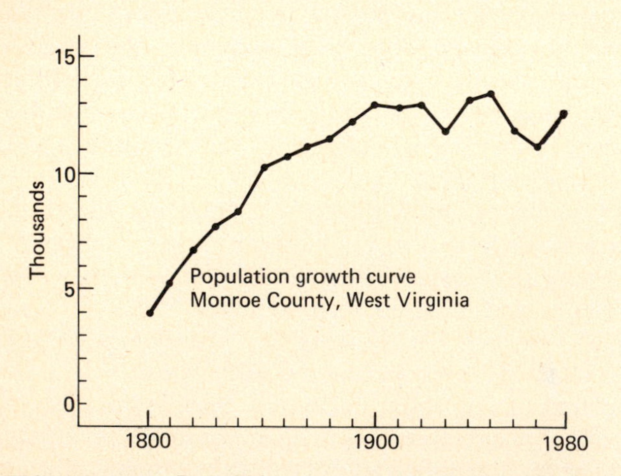

FIGURE 12.7 Population growth for Monroe County, West Virginia. Monroe County is agricultural with most land in local ownership. The growth curve is logistic. (From R. L. Smith, *The Ecology of Man: An Ecosystem Approach,* 2d ed. [New York: Harper & Row, 1976].)

populations characterized by stress of density, more growth and weight become incorporated into a few individuals. As a result, in a population of even-aged individuals, a hierarchy of size in plants develops, with relatively few dominants and a large class of suppressed individuals. (4) The rate of mortality is constant with time. (5) Density-dependent mortality of individual plants is preceded by the death of subpopulations such as lower limbs and leaves.

Extinction

As the growth curves suggest, population growth is maximum when there are neither too many nor too few individuals (the Allee effect, Figure 12.8). When the population falls below or exceeds these points, the rate of increase declines. Increasing

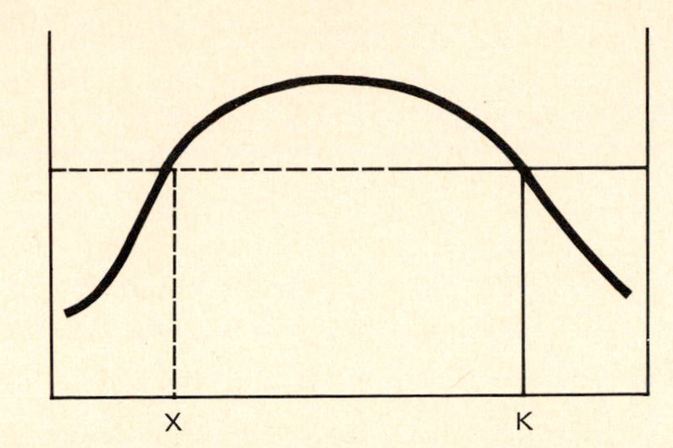

FIGURE 12.8 The Allee effect. Population growth is highest when populations are at moderate densities. Populations decline when the population size exceeds some upper-limit K or falls below some lower limit. In both situations *r* is negative.

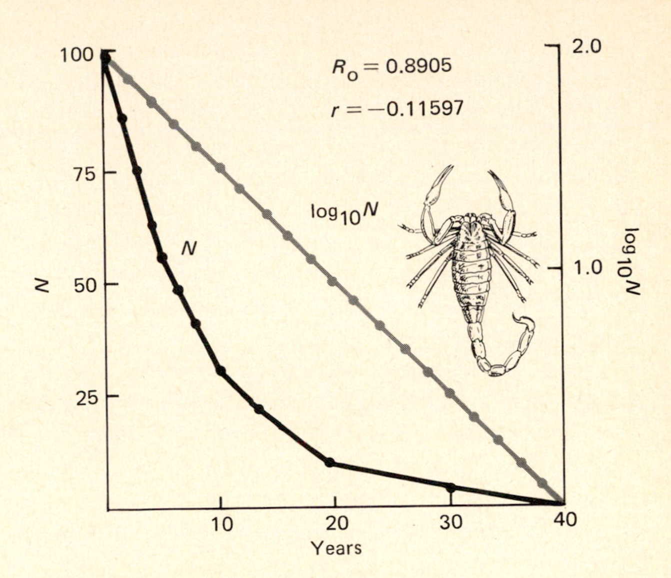

FIGURE 12.9 Exponential decline of a population becoming extinct, based on data for a scorpion population. (Based on data from Polis and Farley, 1980.)

sparseness is associated with a reduction in the rate of increase. The population may become so low that *r* becomes negative and the population declines toward extinction (see Figure 12.9).

The heath hen is a classic example. Formerly abundant in New England, this eastern form of prairie chicken was driven eastward by excessive hunting to Martha's Vineyard, off the Massachusetts coast, and to the pine barrens of New Jersey. By 1880 it was restricted to Martha's Vineyard. Two hundred birds made up the total population in 1890. Conservation measures increased the population to 2,000 by 1917. But that winter a fire, gales, cold weather, and excessive predation by goshawks reduced the population to 50. The number of birds rose slightly by 1920, then declined to extinction in 1925. The last bird died in 1932.

There are several causes for the decline of sparse populations. When just a few animals are present, the females of reproductive age may have only a small chance of meeting a male in the same reproductive condition. Many females remain unfertilized, reducing average fecundity. A small population faces the prospect of an increased death rate. The fewer the animals, the greater may be the individual's chances of succumbing to predation. And small populations may not be large enough to stimulate social behavior necessary for successful reproductive activity.

Extinction is a natural process, albeit a selective one (Stanley, 1979; Fowler and MacMahon, 1982). Species differ in their probability of extinction, a probability dependent in part on the characteristics of the species rather than wholly on random factors. Some of the qualities of a species that favor a high rate of extinction are large body size, small or restricted geographic range, habitat or food specialization, lack of genetic plasticity, and loss of alternate prey species among omnivores.

Extinctions are not spread evenly over Earth's history. Most extinctions occurred in a geologically brief period of time (less than several tens of millions of years). One occurred in the late Permian 225 million years ago when 90 percent of the shallow-water marine invertebrates disappeared (Raup and Sepkoski, 1982). Another occurred in the Cretaceous 65 to 125 million years ago when the dinosaurs vanished. That was perhaps brought about by some extraterrestrial influence such as asteroids striking Earth, interrupting oceanic circulations, influencing climatic conditions, and accompanying turbulent volcanic and mountain-building activity.

One of the great extinctions of mammalian life took place during the Pleistocene when such species as the woolly mammoth, giant deer, mastodon, and giant sloth vanished from Earth. Some students of the Quaternary believe that climatological changes brought about by the advance and retreat of the ice sheet caused the extinctions. Others argue that the disappearance of certain large mammals was caused by overkill by Pleistocene hunters, especially in North America, as human populations swept through North and South America between 11,550 and 10,000 years ago (Martin, 1973). Or perhaps the large grazing herbivores could not withstand the combined predatory pressure of both the humans and the large carnivores.

One usually thinks of modern-day extinction as taking place simultaneously over the full range of a species. But the process of extinction does not work in that fashion. It begins with isolated local extinctions caused when conditions so deteriorate in a given area that species of plants and animals disappear. Eventually, one local extinction added to another sums to total extinction.

The most important cause of extinction today is habitat alteration, which is a local phenomenon. Cutting and clearing away of forest, drainage and filling of wetlands, conversion of grasslands to croplands, construction of highways and industrial complexes, and urbanization and suburbanization greatly reduce available habitat for many species. When a habitat is destroyed, its unique plant life is eliminated, and the animals must either adapt to changed conditions or leave the area and seek a new place to settle.

Because of the rapidity of habitat destruction, no evolutionary time exists for a species to adapt to changed conditions. Forced to leave, the dispossessed usually find the remaining habitats filled and face competition from others of their own kind or from different species. Restricted to marginal habitats, the animals may persist for a while as nonreproducing members of a population or succumb to predation or starvation. As the habitat becomes more and more fragmented, the animals are broken down into small, isolated, or "island," populations, out of contact with other populations of their species. As a result, genetic variations in the isolated populations are reduced, making members of those populations less adaptable to environmental changes. The maintenance of local populations often depends heavily on the immigration of new individuals. As distance between local populations, or "islands," increases and as the size of the local populations declines, the continued existence of each local population becomes more precarious. As the local population falls below some minimum level, it may become extinct simply through random fluctuations. The current rate of massive habitat changes brought about by humans may result in mass extinctions in the next century.

Summary

Population growth results from differences between additions through births and immigration and removals through death and emigration. When additions exceed removals, the population increases. The difference between the two (when measured as an instantaneous rate) is r, the rate of increase. It is derived from the net reproductive rate, R_0, determined from a fecundity table or from the change in population size from one period to another. Both take into account age and survivorship.

In an unlimited environment, populations expand geometrically or exponentially, described by a J-shaped curve. Such growth may occur when a population is introduced into an unfilled habitat.

Because resources are limited, geometric growth cannot be sustained indefinitely. Population growth eventually slows and arrives at some point of equilibrium with the environment, the carrying capacity (termed K).

However, natural populations rarely achieve a stable level, but fluctuate about some mean. Fluctuations that tend to range between some upper and lower limit are termed stable limit cycles. Some fluctuations have peaks and lows that occur more regularly than one would expect by chance. The two most common intervals are 3 to 4 years, as in lemmings, and 9 to 10 years, as in snowshoe hare.

When removals from a population exceed additions, populations may decline to extinction. Extinction is a natural process taking place over long periods of time. Old species disappear and new species arise. When populations are very small, chance events alone can lead to extinction. But at the present time, extinctions have accelerated as human populations have expanded, causing rapid destruction of habitats and contaminating the environment.

Review and Study Questions

1. Contrast exponential and logistic growth of a population.
2. Contrast R_0, net reproductive rate, with *r*, rate of increase. How does *r* relate to R_0?
3. How fast would the painted turtle population grow if the initial population were 50 instead of 100? Demonstrate by drawing a growth curve with a population of 50 instead of 100 and draw your own conclusions.
4. Under what conditions would you expect a population to exhibit exponential growth? How might you measure *r*, rate of increase per individual, in an unlimited environment?
5. How might you determine the carrying capacity of a given habitat if the population tended to fluctuate rather than arrive at and remain at some upper level as the logistic equation assumes?
6. Discuss extinction as a process. Relate that process to some of our current endangered species, such as the California condor, whooping crane, and blue whale.

Suggested Readings

See the "Suggested Readings" section at the end of Chapter 13.

CHAPTER 13

Population Regulation

Outline

Objectives

Upon completion of this chapter, you should be able to:

1. Define competition and distinguish between scramble and contest competition.
2. Discuss the effects of intraspecific competition on growth and reproduction, on density and increase in biomass, and on density-dependent mortality.
3. Explain the significance of the $-3/2$ law in plant population dynamics.
4. Discuss the role of stress in population regulation.
5. Distinguish between presaturation dispersal and saturation dispersal.
6. Discuss the significance of each type of dispersal in population regulation.
7. Discuss the influence of genetic changes in population growth.
8. Describe social hierarchy and explain how it may function in population regulation.
9. Discuss territoriality and explain the advantages of owning a territory and the costs of territorial defense.
10. Explain the role of territoriality in population regulation.
11. Explain the significance of space capture in plants.

No population continues to grow indefinitely. Even those exhibiting exponential growth, like the St. Paul reindeer herd, ultimately confront the limits of the environment. Most populations, however, do not behave in an exponential fashion. As the density of a population increases, interactions among members of the population and the availability of resources result in increased mortality, reduced natality, or both. If the population drops below the density the environment is able to support, mortality decreases, natality increases, and the population grows.

Involved are positive feedback of population growth when conditions are favorable and negative feedback of population decline when conditions are unfavorable. Throughout most of the sigmoid curve, both types of feedback operate, with a change in the relative importance of each. In the early stages of population growth, positive feedback dominates. In the later stages of population growth, negative feedback slows down positive feedback and dominates when K is reached. The speed of the response involves the population's impact on the resources. If individuals have removed a resource, such as food, faster than it can be replaced, then the present population has impoverished the environment for the next generation. That slows positive feedback and introduces a time lag in population response. Through such positive and negative feedback, a population arrives at some form of regulation (Figure 13.1).

Implicit in the concept of population regulation is density dependence. Density-dependent effects influence a population in proportion to its size. At some low density there is no interaction. Above that point the larger the population becomes, the greater is the proportion of individuals affected. Density-dependent mechanisms act largely through environmental shortages and competitive interactions among members of the population to obtain these resources. If the effects of a particular influence do not change proportionately with population density, or if the proportion of individuals affected is the same at any density, then the influence is density-independent. Most often, density-independent effects are related to weather.

At one time ecologists diverged widely in their views on the relative importance of density-dependent and density-independent influences on populations. The arguments were largely semantic, stemming from different points of view. Ecologists now recognize that the numbers of organisms are determined by an interaction between biotic regulating mechanisms and the environment, a point that will become evident later.

Intraspecific Competition

Population regulation, in part, involves competition among individuals of the same species for environmental resources. Competition results only when a needed resource is in short supply relative to the number seeking it. As long as resources are abundant enough to allow each individual a sufficient amount for survival and reproduction, no competition exists. When resources are insufficient to satisfy adequately the needs of all individuals, the means by which they are allocated has a marked influence on the welfare of the population.

When resources are limited, a population may exhibit one of two responses. All the individuals can share the resources equally with none of them obtaining enough for growth or reproduction, provided the population density remains high or competition remains intense. Such competition is called *scramble*. Or some individuals can claim enough resources for maintenance and reproduction while denying others shares of the resources. That type of competition is called *contest*. Generally under the stress of limited resources, a species population will exhibit only one or the other type of competition. Some are scramble species and others are contest species, but the nature of intraspecific competition within a species may vary according to the stage in the life cycle. For example, the larval stages of some insects may experience scramble competition and the adult stages face contest competition.

The outcome of scramble and contest competition differs. Scramble competition can produce chaotic oscillations in the population over time, and it limits the average density of the population below that which the resources could support if an adequate amount of resources were supplied only to part of the population. For that reason, scramble competition can result in a wastage of resources relative to population growth. In contest competition the deleterious effects of limited resources are confined to a fraction of the population because unsuccessful individuals are denied access to the resources by successful competitors. That eliminates or greatly reduces the wastage of resources, permits the maintenance of a relatively high population density, and maintains some numerical constancy.

Both scramble and contest types of competition are similar in that they have a threshold level below which no competition takes place and all individuals survive or maintain fitness. Above that threshold, scramble competition results in little production of offspring at best and, ultimately, could reduce the population to zero. Once a population experiencing contest-type competition passes that threshold, a fraction of the individuals will get all the resources they need to survive and produce offspring. The remaining individuals will get less than they need and produce no offspring or die. Thus, in scramble competition most individuals get less than they need, while in contest-type competition only a portion of the population experiences that fate.

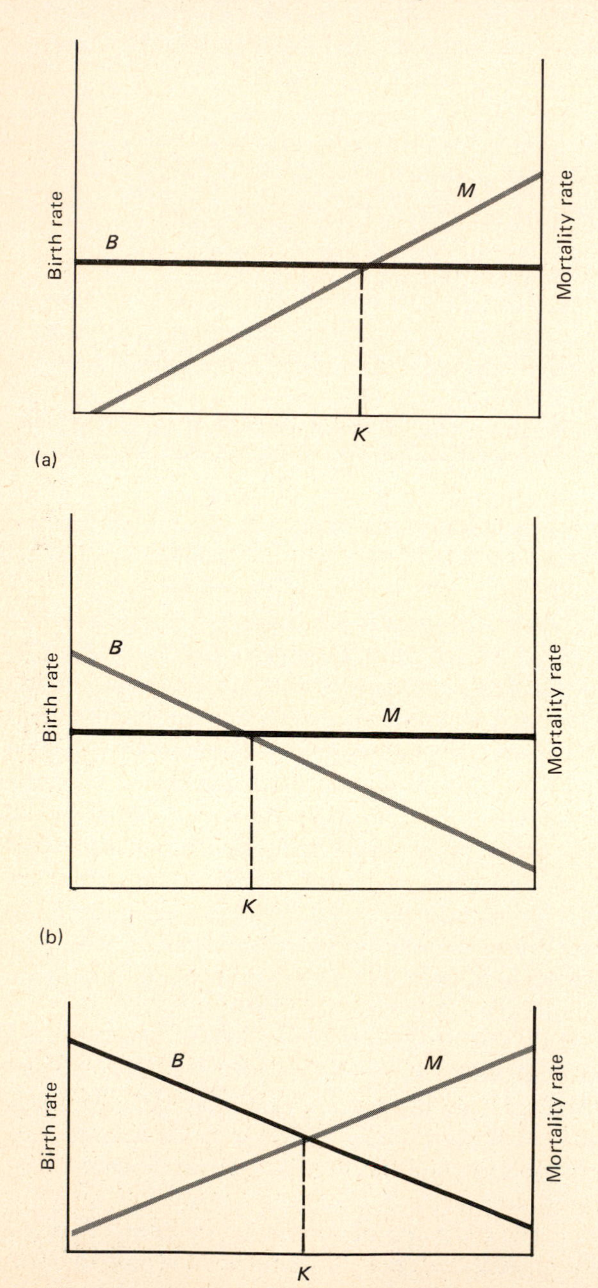

FIGURE 13.1 Population regulation, if it is to function, requires density-dependent birthrates and density-dependent mortality rates. In (a) the birthrate is independent of population density, as indicated by the straight line. It remains unchanged as the population increases. But in (a) the death rate increases as population increases. As long as the birthrate exceeds the death rate, the population increases toward *K*. At *K* the population reaches equilibrium, maintained by increasing mortality. In (b) the situation is reversed. The mortality rate is independent of population density, but birthrate declines as density increases until the population reaches *K*. At that point equilibrium is maintained by a decreased birthrate. In (c) both birthrates and death rates are density-dependent, and the population reaches equilibrium when the birthrate exceeds the death rate. Fluctuations in each will tend to hold the population at or near the equilibrium point and influence population density. If the birthrate increases, then the mortality rate increases.

SOME CONSEQUENCES OF INTRASPECIFIC COMPETITION

Because the intensity of intraspecific competition is density-dependent or density-proportional, evidence of its happening comes slowly. It involves no sudden thresholds; it increases gradually, affecting largely the quality of life rather than the survival of individuals, although through time its effects become accentuated, ultimately affecting individual survival and reproduction.

Retarded Growth and Delayed Reproduction

As population density increases toward a point at which resources are insufficient to meet needs, individuals in those populations characterized by scramble competition reduce their intake of food. That slows the rate of growth and inhibits reproduction. Examples of this inverse relationship between density and rate of body growth may be found among populations of poikilothermic vertebrates. Dash and Hota (1980) discovered that frog larvae reared experimentally at high densities failed to grow normally (Figure 13.2a). They experienced slower growth, required a longer time to reach the size at which transformation from the tadpole stage takes place, and had a lower probability of completing metamorphosis. Those that did reach threshold size were smaller than those living in less dense populations. Fish living in overstocked ponds exhibit a similar response to density (Figure 13.2b). Bluegills *(Lepomis macrochirus)*, for example, normally grow to the size of a dessert plate; but in overstocked and underharvested farm ponds, they rarely grow beyond the size of a half dollar and never reproduce.

Like fish and tadpoles, plants, too, may respond to density in a scramble fashion through reduction in individual growth. Botanists call it *phenotypic plasticity*. Individuals adjust their growth form, size, shape, number of leaves, flowers, and production of seeds in a scramble fashion to the limited resources available. In some plants, reproduction by seed at high densities is almost nonexistent (Putwain, Machin, and Harper, 1968); it is solely by vegetative means. As the density of plants increases, the number of vegetative offspring also declines. For example, at high densities the genets of perennial rye grass *(Lolium perenne)* produce few tillers, and their weight is low. At low densities the genets produce many tillers, and their weight is high (Kays and Harper, 1974).

Decrease in Density and Increase in Biomass

A close relationship exists between plant density and growth of individual plants as measured by biomass accumulation. Up to a certain point, all plant seedlings exhibit an increase in biomass. But

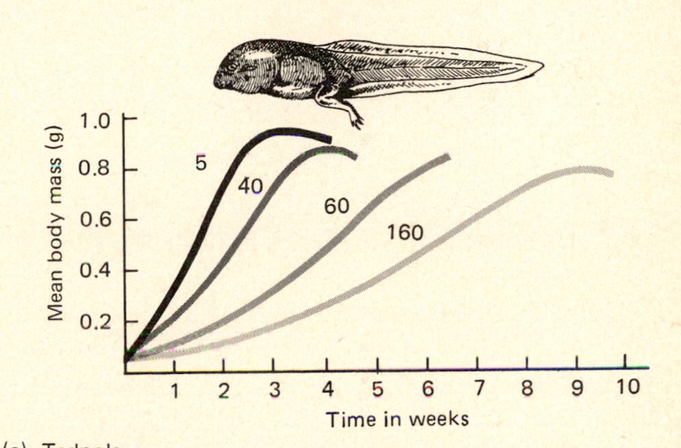

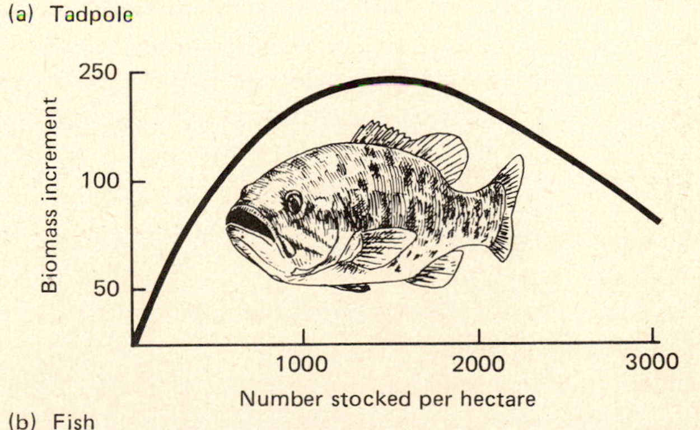

FIGURE 13.2 Two examples of the effect of population density on the growth of individuals in a population. (a) Influence of density on the growth rate of the tadpole *Rana tigrina*. Note how rapidly individual growth rates decline as density increases from 5 to 160 individuals confined in the same space. (After Dash and Hota, 1980:1027.) (b) Growth of fish as influenced by population density. Note again how rapidly growth declines with density. (After Bachael and LeCren, 1967.) Both graphs reflect the scramble type of competition.

as their size increases, plants interfere with each other, competing for the same resources—light, moisture, nutrients, space. Plants initially respond to competition and increasing density in a plastic manner, modifying their form and size. Up to a point, plants are successful. But when their capability to respond in a plastic manner is exceeded, the population begins to experience density-dependent mortality. That usually begins when the percentage of plant cover per unit area exceeds 100 percent. Response takes place the earliest in dense seedling populations, but in time it also occurs in less dense populations as plant size increases. As biomass of individual plants increases, fewer plants per unit area are needed for competition to develop. As a result, plants starting at different densities converge toward some common value, with decreases through time.

If one plots the logarithm of the mean weight of plants against the logarithm of mean density over time, the slope of the line is close to −3/2 (Figure 13.3). The line represents the limit in the amount of biomass that can be supported for a given plant density. A high plant density means low mean weight, and conversely, a high mean weight means low plant density, a point emphasized in yield tables for forest trees. Such an inverse relation between plant weight and density through time is common among most plants. It is called the self-thinning, or −3/2, law (Yoda et al., 1963; White and Harper, 1970).

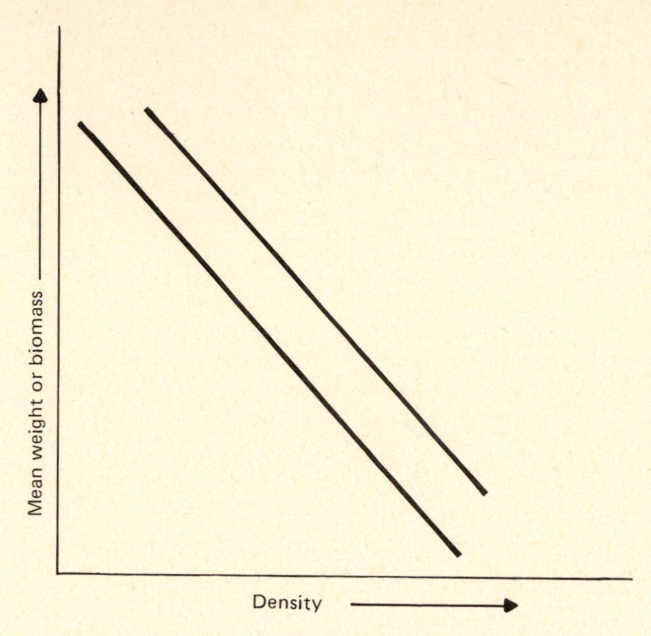

FIGURE 13.3 The −3/2 power law. As the density of a plant population decreases, the mean weight of surviving plants increases. This results from progressive thinning within a population of plants, an outcome of intraspecific competition. (After Harper, 1977.)

Density-Dependent Mortality—Case of the African Buffalo

Among some animals, large mammals in particular, the effects of density become most prominent as the population approaches carrying capacity. But to demonstrate density dependence in such a population you need to know something about the level of the population. Is it still expanding? Or is it at some equilibrium level? One way to determine that is to reduce a population and allow it to expand again.

That is what happened to the African buffalo *(Syncerus caffer),* studied in detail on the Serengeti by A. R. E. Sinclair (1977). The African buffalo (not to be confused with a bison) is a large bovine ungulate of the African savanna. Its food is grass, preferably the protein-rich leaves. In 1895, rinderpest, a measlelike disease of cattle, spread from domestic cattle to the African buffalo and wildebeest *(Connochaetes taurinus).* Herds were decimated in the early 1900s. Veterinarians eventually eliminated the disease, which caused a high mortality among juveniles as late as 1964. Released from heavy juvenile mortality, the African buffalo expanded dramatically, reaching an apparent equilibrium density in the 1970s.

As the herd approaches equilibrium, the population also approaches stationary age distribution, and adult mortality increases proportionately with density. The critical time of year for the buffalo population is the dry season, when food may become scarce. Rainfall determines the productivity of grass. The greater the rainfall, the more vigorously the grass grows, increasing the amount of forage available in the dry season. Equilibrium den-

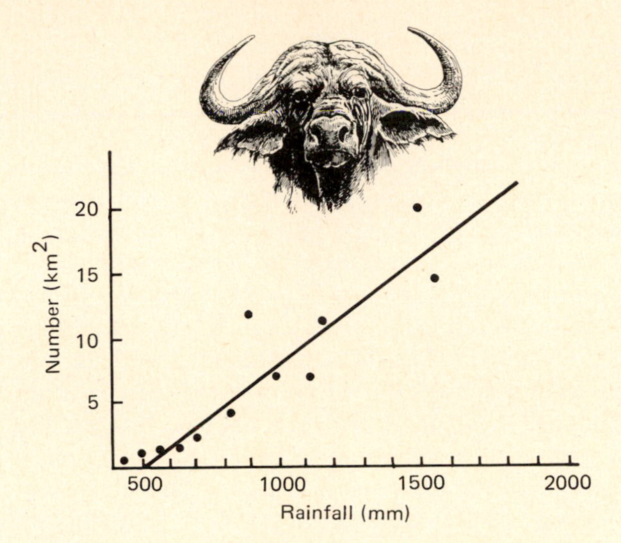

FIGURE 13.4 A model diagram showing the relationship between rainfall, forage production (grass), density of the population of African buffalo, mortality, natality, and individual response of buffalo to environmental stress. The dry season causes a decline in quantity and quality of forage, resulting in malnutrition, lowered resistance to disease, and increased probability of mortality. The relationship becomes accentuated during periods of drought that reduce rainfall during the rainy season. Such adult mortality results in population decline. (Based on data from Sinclair, 1977:214.)

sity varies with mean annual rainfall; the greater the rainfall, the greater the density of buffalo (Figure 13.4).

During the wet season on the African savanna, food is abundant, but during the dry season, the quality of food declines as the grasses dry. The buffalo become more selective, seeking green leaves, moving to the moist riverine habitat, breaking up into smaller units, and utilizing different areas. As the dry season progresses, buffalo become less selective, consuming dry leaves and stems that they would otherwise have rejected. As food quantity and quality decline, competition in a scramble fashion becomes keener. Individual buffalo feeding in any one area reduce the food available to neighboring animals. The more buffalo present, the less food there is available for each individual. Eventually, the quality of food as measured by its available protein drops below maintenance level, and the animals use up their fat reserves. Undernourished and lacking the protein intake necessary to maintain immunity to disease and parasites they normally harbor, adults become susceptible to any disease or parasite they host. Older animals are the most vulnerable. The number dying depends upon the rapidity with which the adults use up their energy reserves before the coming of the rainy season and new growth. If the next season sees more rainfall and that rainfall extends sporadically into the dry season, the mortality of adults the following year is reduced.

Thus, mortality of adults, as influenced by rainfall and poor resources, determines equilibrium density and regulates the population. By contrast, juvenile mortality appears to be density-independent. It results from one or several randomly fluctuating environmental variables and causes fluctuations in the population. In turn, density-dependent adult mortality compensates for the disturbances and dampens fluctuations.

The examples suggest that populations do respond to increasing numbers in a density-dependent fashion through intraspecific competition. The timing of the response depends upon the nature of the population itself (Fowler, 1981). Among larger mammals, such as the African buffalo, with a long life span and relatively low reproduction, regulating mechanisms do not function until the population nears carrying capacity (Figure 13.5). Among organisms characterized by high reproduction and short life spans, response to density may occur much earlier, as evidenced by the slowing of the growth rate of frogs and tadpoles.

Some Mechanisms of Population Regulation

What form does intraspecific competition take, especially in situations in which resources such as food, space, and mates are not equally shared? How are resources allocated? What happens to

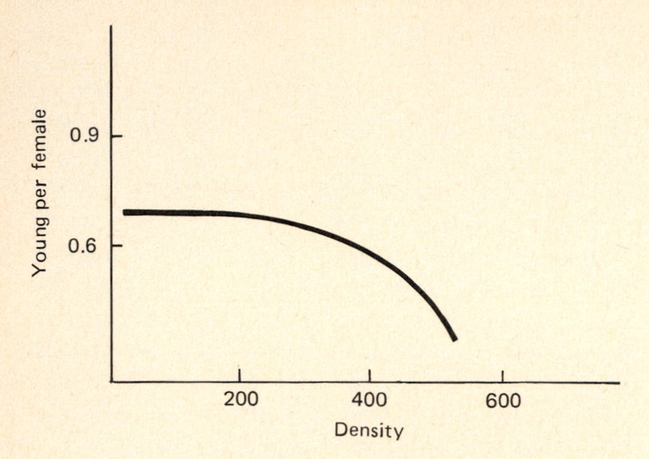

FIGURE 13.5 An example of nonlinear density-dependent change in a large mammal population, specifically the American bison. The birthrate of the bison, expressed as young per female, declines after the population has reached a certain density and then declines sharply as the density increases. (From Fowler, 1981:607, from data of J. E. Gross et al.)

those individuals who do not receive their share? Intraspecific competition can operate in subtle and not so subtle ways. Its outcome affects the fitness of the individuals involved and may influence population density.

STRESS

As a population reaches a high density, individual living space becomes restricted and often aggressive contacts among individuals increase. One hypothesis of population regulation is that increased crowding and social contact result in stress (Christian, 1963, 1978; Christian and Davis, 1964; Davis, 1978). Such stress triggers hyperactivation of the hypothalamus-pituitary-adrenocorticular system which in turn alters gonadotrophic secretions. Such profound hormonal changes result in a suppression of growth, a curtailment of reproductive functions, and delayed sexual activity among individuals in the population. These hormonal changes may suppress the immune system and cause further breakdown in white blood cells, increasing an individual's vulnerability to disease. Social stress among pregnant females may increase intrauterine mortality and cause inadequate lactation and subsequent stunting of nurslings. Thus, stress results in decreased births and increased mortality.

Such population-regulating effects have been confirmed in confined laboratory populations of several species of mice and to a lesser degree in enclosed wild populations of woodchucks *(Marmota monax)* (Lloyd et al., 1964) and Old World rabbits *(Oryctolagus cuniculus)*. K. Myers and his associates (1967, 1971) experimented with rabbits at several densities in different living spaces within confined areas of their natural habitat. Those living in the smallest space, in spite of a decline in numbers, suffered the most debilitating effects. Sexual and aggressive behavior increased, reproduction declined, fat about the kidneys decreased, and kidneys showed inflammation and pitting on the surface. Weights of liver and spleen decreased, and adrenal size increased.

Young rabbits born to stressed mothers were stunted in all body proportions and in organs. As adults they exhibited such behavioral and physiological aberrations as a high rate of aggressive and sexual activity and large adrenal glands relative to body weight. Low body weight, the lack of lipids in the adrenals, abnormal adrenals, and poor survival indicated a lack of fitness in such rabbits. In contrast, rabbits from low- to medium-density populations showed excellent health and survival.

Pheromones or chemical releasers present in the urine of adult rodents inhibit reproduction among members of a population. The function is suggested in a study involving wild female house mice *(Mus musculus)* living in high-density and low-density populations confined to grassy areas within a highway cloverleaf. Urine from females of a high-density population was absorbed onto filter paper. The paper was placed with juvenile wild female mice held individually in laboratory cages. Similarly, urine from females in low-density and sparse populations was placed with other juvenile test females. Juvenile females exposed to urine from high-density populations experienced delayed puberty, while females exposed to urine from low-density populations did not. The results suggest that pheromones present in the urine of adult females in high-density populations may delay puberty and help to slow further population growth (Massey and Vandenbergh, 1980). Juvenile female house mice exposed to urine of dominant adult males accelerate the onset of puberty (Lombardi and Vandenbergh, 1977).

The response of plants to stress is influenced by their adaptiveness to stress situations. For example, some plants are adapted to shade. Under the light-intercepting canopy of other plants or similar low-light conditions, shade-adapted plants will respond to that stress by growing slowly, conserving energy by reducing the rate of photosynthesis, and forgoing the production of flowers and seed. In effect, they wait for a time when sufficient light will stimulate rapid growth. Plants adapted to other forms of stress follow a similar pattern (Grime, 1977). Plants adapted to open light respond to low light intensities by growing rapidly in height (which under some conditions will get them up into the light). However, they develop thin cell walls, which reduce the supporting ability of the stems, resulting in weak, spindly individuals. And they are highly vulnerable to fungal infections, which can be fatal.

Under conditions of low nutrients, low moisture, or other environmental stress, some groups of plants will use up their nutrient reserves and fail to produce seeds. That limits their ability to respond to improving environmental conditions. In contrast, ruderal plants, those adapted to persistent and severe disturbance, respond to stress by producing seeds at the expense of vegetative development. Individual plants are small and poorly developed, yet the number of seeds relative to individual plant biomass is high. Seeds of such plants can survive buried in the soil for long periods of time. They are able to germinate rapidly when a disturbance exposes the seeds to light and fluctuating daily temperatures. Examples of such plants are annual weeds.

While such responses to stress do influence individual fitness among plants and the maintenance and expansion of populations, there is little evidence that stress in plants acts in any sort of a regulatory manner.

DISPERSAL

Instead of coping with stress (to state the situation somewhat anthropomorphically), some animals run away from a bad situation. They disperse into submarginal and unoccupied habitat. Although dispersal is most apparent when population density is high, it is a constant phenomenon. Some individuals leave the parent population whether it is crowded or not. Those individuals leaving the population are called *emigrants*. When these same individuals move into a new area or join another population, they become *immigrants*.

When a lack of resources forces individuals to leave the population, the result is saturation dispersal (Lidicker, 1975). The dispersers, who can be considered surplus to the population, appear to be a random sample of the genetic constitution, age structure, and sex ratio of the parent population (Keith and Tamarin, 1981). Ultimately, many of these will die from disease, parasites, predators, and exposure (see Errington, 1963). Because such dispersal results from overpopulation, it has little influence on population regulation.

More important to population regulation is presaturation dispersal, which takes place when the population is increasing, but well before the population reaches a density at which food and cover are overexploited. Individuals who participate are not a random selection of the population. The dispersers consist mostly of females and subadult males and are genetically predisposed to emigrate (see Krebs et al., 1976; Tamarin, 1978; and Keith and Tamarin, 1981).

For presaturation dispersal to occur, the dispersing individuals must be motivated to move, be physically able to do so (not be restrained by some barrier to their movement), and have available a dispersal sink. A dispersal sink is an unfilled habitat, even if marginal or submarginal, into which the animals can settle for at least a short while. These animals must be removed permanently from the original population.

W. D. Lidicker (1975) has hypothesized that presaturation dispersal could play a key role in regulating population growth of voles. Some support for that hypothesis comes from the studies of an island population of beach voles *(Microtus brewerii)* and a mainland population of meadow voles *(M. pennsylvanicus)* by R. Tamarin (1978) and T. Keith and R. Tamarin (1981). A population of meadow voles on the mainland near Manomet, Massachusetts, had a dispersal sink in the surrounding countryside. Individuals—mostly females, who were the best colonizers in the population, and subadult males—left the population before it became excessively high. Their departure changed the age structure, sex ratio, and genetic composition of the nondispersing population (see Figure 13.6). The population of beach voles on small Muskeget Island was noncylic. Effectively confined by the ocean and lacking a dispersal sink, the population over time stabilized near carrying capacity. Would-be dispersers that moved into suboptimal habitat on the island represented a random selection of the parent population.

Most dispersal studies concentrate on voles because the populations are localized and the mammals are numerous, with a high reproductive rate and a short generation time. But evidence that presaturation dispersal takes place in other mammals, especially those with expanding populations, also exists.

An example is the wolf population in Minnesota (Fritts and Mech, 1981). As the size of the wolf population expands, the size of the packs remains stable and surprisingly small. Packs consist of only two to four animals in spite of high production of cubs and low mortality. Tracking of radio-collared wolves indicated that nonbreeding yearling animals left the home pack in autumn to form their own packs. Rather than attempting to breed in their home pack, which would have been impossible (see page 256), most young wolves dispersed early and formed their own social units. By breaking away from the parental pack and establishing themselves

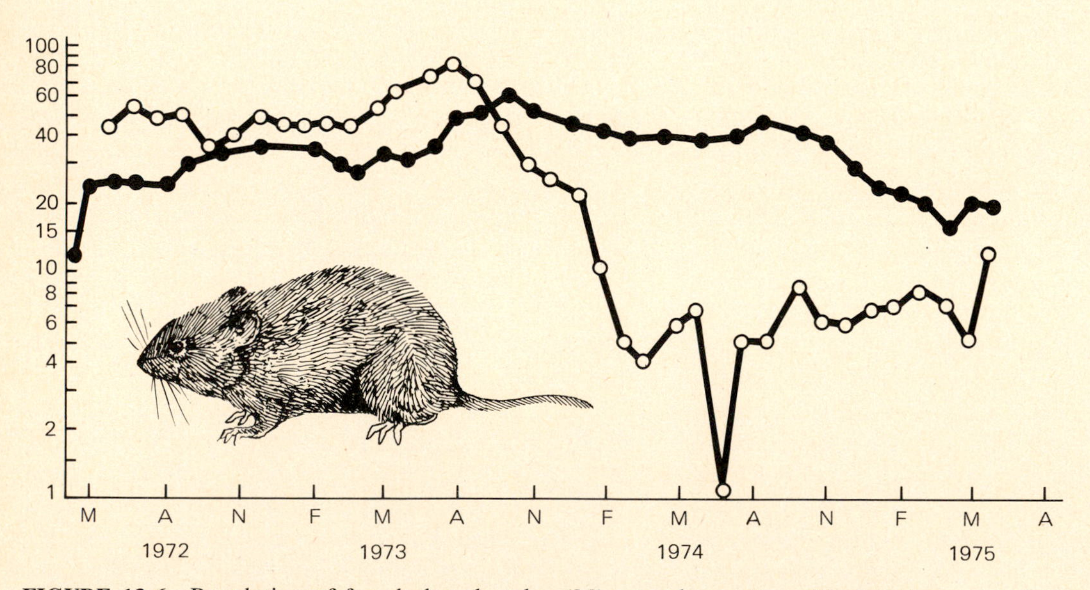

FIGURE 13.6 Population of female beach voles *(Microtus breweri)* on Muskeget Island compared to populations of female meadow voles *(M. pennsylvanicus)* on the mainland (Barnstable, Massachusetts). Note the relative stability of the island population. It has no dispersal sink, and the individuals comprising it seem to be selected for their ability to tolerate high densities. The mainland population appears to be cyclic, reflecting the tendency of individuals to disperse before the population reaches a high density. (After Tamarin, 1978:547.)

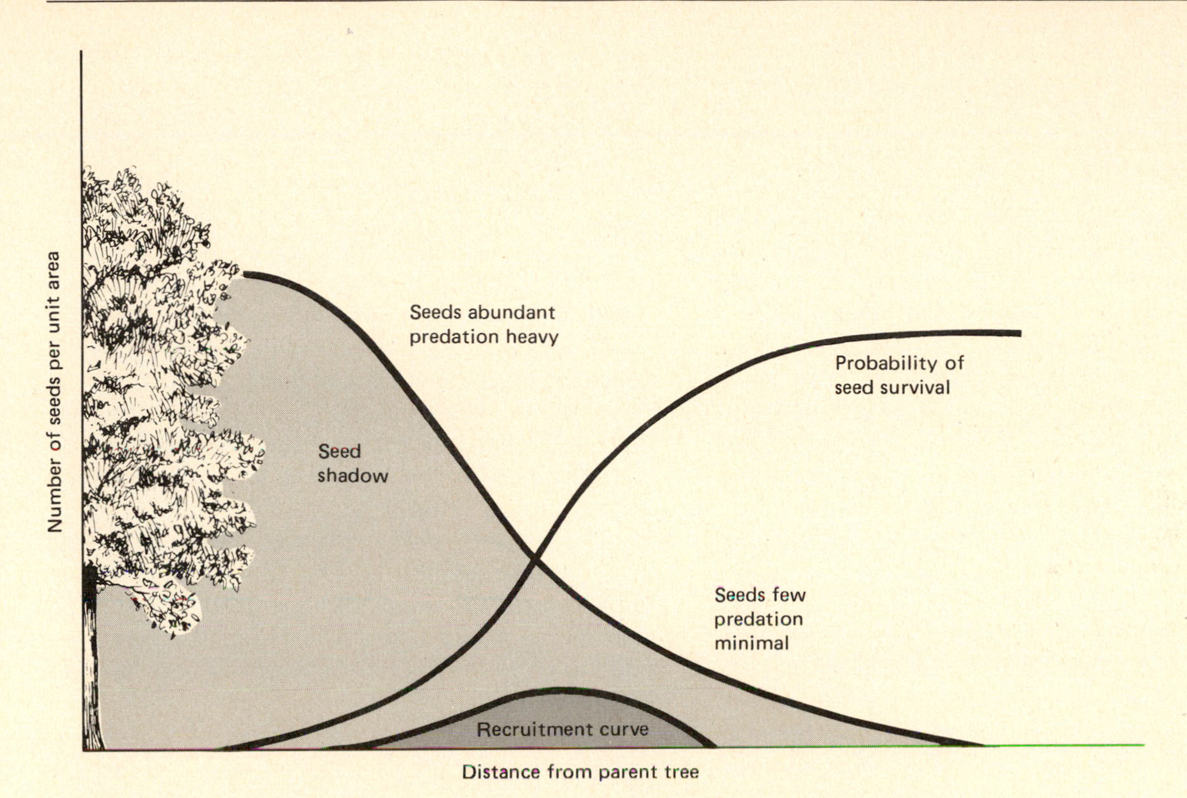

FIGURE 13.7 Model of seed dispersal in relation to distance from the parent plant. All the seeds close to the tree are eaten by seed predators. As the distance from the parent plant increases, the probability of seed survival increases because the density of both seeds and predators decreases. Although the number of seeds decreases rapidly with distance, recruitment or seed survival is higher well away from the plant. (After Janzen, 1970:502.)

in unfilled habitat, young wolves increase the probability of reproductive success and their own fitness.

Dispersal can not only regulate populations and change genetic composition in one area; it can also contribute to population growth and increase the gene flow, genetic variation, and reproductive potential of the receiving population. Through such dispersal, colonizing populations are augmented, and a species is able to expand its range.

Plants are not mobile. The only means they have of expanding their populations into new areas is by some type of seed dispersal. Most plants depend upon currents of air or water, gravity, or transport in the gut, feathers, or fur of animals for the exodus of their seeds.

Successful dispersal of plants may depend upon the frequency of dispersal. The more often the plant has a heavy seed crop, the more often it will have new propagules available for colonization, even though seed losses will be enormous. Regardless of the means of dispersal, most of the seeds are dropped near the parent plant. If one plots the number of seeds against distance, the curve approximates exponential decay (Figure 13.7). Seeds near the parent plant usually are subject to more intense predation by seed-eating animals than are seeds scattered some distance away. Thus, the probability of seed survival increases with the distance away from the parent plant.

GENETIC CHANGE

D. Chitty (1960) and D. Pimentel (1961, 1968) have advanced hypotheses that genetic changes in a population influenced by natural selection at high and low densities might be involved in population regulation. Chitty and Phipps (1967) suggested that observed declines in vole populations might have resulted from selective pressures of mutual interference as measured by levels of aggressiveness. When population densities are low, mutually tolerant individuals adapted to crowding would be selected over aggressive individuals. As density increases, however, the competition for available resources intensifies, and aggressive individuals have the selective advantage. They claim more space and a disproportionate amount of the resources and force others to leave the population. That results in a reduced population, in which aggressive individuals no longer have the selective advantage. As the number of tolerant individuals in the population increases, the population builds up again.

There is little evidence to support genetic change involving social tolerance. In fact, aggressive animals rather than tolerant ones may be the individuals that leave the population. Krebs et al. (1973) observed that the dispersing males of the meadow vole were more aggressive than the males that remained. As density in their control areas increased, the less tolerant individuals were the ones that moved into less densely populated areas.

But there is some evidence that genetic changes are involved in the affairs of populations reaching a high density. Change in the genetic quality of a population was demonstrated in a laboratory population of *Drosophila* by H. Carson (1968). He introduced a wild-type male into a population of an inbred line of females. That small change in the gene pool stimulated a population flush, a rapid increase in population size even though there were no changes in food and space. The population increased three times over the original number within nine generations, when the population abruptly crashed. Carson hypothesized that because other variables apparently were constant, the changes in the gene pool could be considered a stimulus sufficient to trigger a population flush and subsequent crash.

Carson recognized four stages to the flush. In phase 1, injection of new genetic material into a population promotes low mortality and high fecundity. In phase 2, a period of population increase beyond previous levels, selection is relaxed, and all types of individuals survive. Because selection does not act to eliminate the less desirable new genetic combinations produced by high genetic variability, average fitness of the population declines. In phase 3 the population peaks, but the decline in fitness makes the population extremely vulnerable to environmental conditions, with increased pressure on available food and space. The population crashes—often to an extremely low level, lower perhaps than before the flush.

SOCIAL HIERARCHY

Many species of animals that live in groups have some form of social organization. Usually, that social organization is based on dominance of one individual over another, enforced by intraspecific aggressiveness involving threats, fighting, and intolerance. Two opposing forces are at work. One is the need for the presence of others of the same kind. The other is a negative reaction against crowding, the need for personal space.

Social dominance takes many forms. The simplest is a straight-line interaction. An alpha individual is dominant over all others, a beta individual is dominant over all but the alpha, and so on. The social, more complex interactions involve triangular relations throughout the population (Figure 13.8). Such fixed dominance patterns are termed *peck orders*. In some groups peck order is replaced by *peck dominance,* in which social rank is not absolutely fixed. The position of an individual in the hierarchy depends upon environmental conditions, past social rank, close associates, size, weight, hormonal levels, and the like.

A → B → C → D

Straight-line peck order

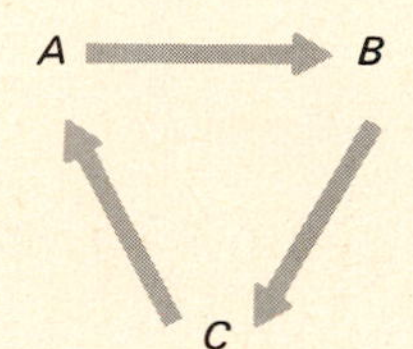

Triangular peck order

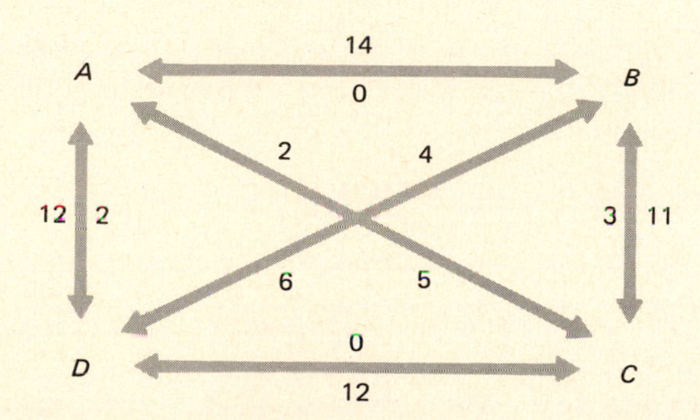

Triangular peck order

FIGURE 13.8 Types of social hierarchy. (a) straight-line peck order, in which one animal is dominant over the one below it. A is the alpha organism; D is the omega. (b) Triangular peck order, in which A is dominant over B, B is dominant over C, and C is dominant over A. (c) A more complex example of a triangular peck order. The double arrows indicate encounters between individuals. The numbers represent the number of wins of one individual over another. For example, A is clearly dominant over B, with 14 wins. B never dominated A. A is also dominant over D, with 12 wins out of 14 encounters. C is also strongly dominant over D, with 12 wins to 0 for D. B is also dominant over D, with 6 wins to D's 4 wins. D is clearly the omega individual. Other interactions can be read in a similar manner.

To the dominant individuals in such a social structure go most of the resources. Among many species the dominant males secure the most mates, thereby ensuring greater fitness at the expense of little or no fitness for the subdominant males. In times of severe shortages of such resources as food, subdominant animals suffer first and often succumb to starvation.

Among wintering white-tailed deer, adult bucks are dominant over adult does, and adult does are dominant over young does and bucks. Even fawns have their own social hierarchy. When food shortages develop in northern wintering congregation areas called yards, subdominant individuals either are excluded from the food or are able to secure only a minimal amount of it. In severe winters these animals either starve or survive in such poor physical condition that they fail to reproduce. Such periods of adverse weather conditions coupled with food shortages can result in a sharp decline in the population (Figure 13.9).

Such declines cannot be called regulatory. Rather, the outcome of that type of contest competition simply ensures that the dominant animals in the population will continue to reproduce successfully. The fitness of such individuals is secured at the expense of subdominant individuals.

Social dominance, however, may play a role in population regulation if it affects reproduction and survival in a density-dependent manner. An example is the wolf. Wolves live in small groups of 6 to 12 or more individuals called packs. The pack is an extended kin group consisting of a mated pair, one or more juveniles from the previous year who do not become sexually mature until the second year, and several related, nonbreeding adults.

The pack has two social hierarchies, one headed by an alpha female and the other headed by an alpha male, the leader of the pack to whom all other members defer. Below the alpha male is the beta male, closely related, often a full brother, who has to defend his position against pressure from males below (see Figure 13.10).

Mating within the pack is rigidly controlled. The alpha male (occasionally the beta male) mates with the alpha female. She prevents lower-ranking

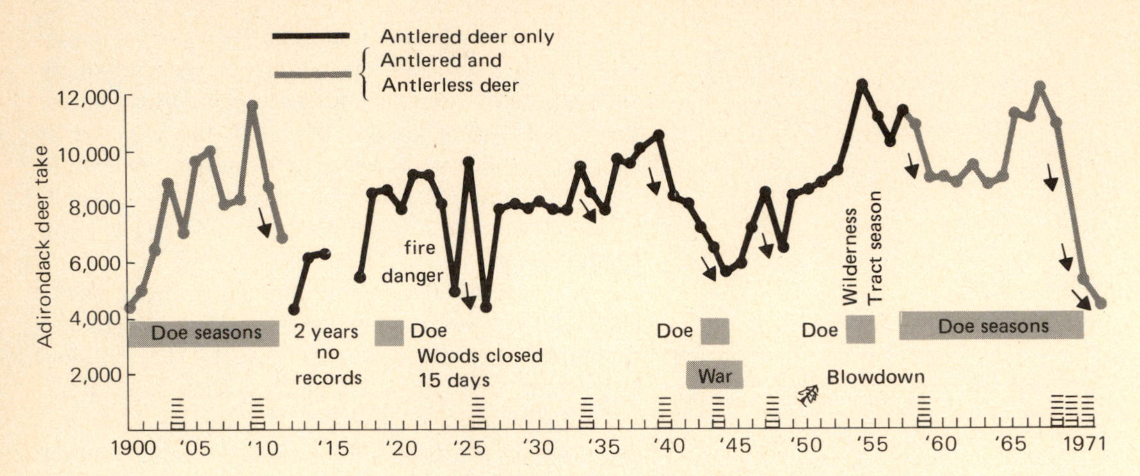

FIGURE 13.9 Fluctuations in a deer population as influenced by environmental conditions. Changes in population density are influenced in part by environmental changes, which, in turn, influence food supply and behavioral interactions. The deer population is subject to hunting for bucks and periodically for does. The bars on the date line on the abscissa indicate years of severe winters and deep snows. Note that bad winters with heavy mortality are followed by sharp declines in the population. The declines are the result of heavy overwinter mortality from starvation and from reduced natality the following spring. A blowdown in 1950 apparently increased the availability of forage in a region characterized by a minimal winter food supply. A bad winter in 1959 caused another decline, but the deer population recovered in spite of an open season on does. Then three severe winters in a row, characterized by prolonged deep snows and cold temperatures, resulted in the death of five out of every six fawns and a precipitous decline in the deer herd. Although weather initiated the declines, the ultimate cause was a lack of suitable food for the deer, which, in turn, resulted in malnutrition and the inability of the deer to maintain body homeostasis. (After Severinghaus, 1972.)

females from mating with the alpha and other males, while the alpha male inhibits mating attempts by other males. Thus, each pack has one reproducing pair and one litter of pups each year. These are reared cooperatively by all the members of the pack.

The level of the wolf population in a region is governed by the size of the packs, which hold exclusive areas (see "Territoriality"). Regulation of the size of the pack is achieved by events within the pack that influence the amount of food available to each wolf. The food supply itself does not affect births and deaths, but the social structure that leads to an unequal distribution of food does. The reproducing pair, the alpha female and the alpha male, has priority for food; they, in effect, are independent of the food supply. The subdominant animals, male and female, with little reproductive potential, are affected most seriously. At high densities the alpha female will expel other adult females from the pack. Other individuals may leave voluntarily. Unless these animals have the opportunity to settle successfully in new territory and form a new pack, they fail to survive (see "Dispersal").

The social pack, then, becomes important in population regulation. As the number of wolves

FIGURE 13.10 Social hierarchy in a small wolf pack. Note the two separate hierarchies, one male and one female, but with one individual occupying the omega position. The alpha male is the dominant individual in the pack. (After Fox, 1980:61.)

increases, the size of the pack increases. Individuals are expelled or leave, and the birthrate relative to the population declines because most sexually mature females do not reproduce. Overall, the percentage of reproducing females declines. When the population of wolves is low, sexually mature females and males leave the pack, settle in unoccupied habitat, and establish their own packs (see "Dispersal") with one reproducing female. As a result, nearly every sexually mature female reproduces, and the wolf population increases. But at very low densities, females may have difficulty locating males to establish a pack and so fail to reproduce or even survive (see Zimen, 1978).

TERRITORIALITY

Social hierarchy is one expression of contest-type competition. Another is territoriality. Often it is difficult to draw a sharp line between the two. Depending upon the season, the degree of crowding, and how the resources are distributed, territoriality can grade into social hierarchy and vice versa.

You see evidence of territoriality as spring comes to the countryside. Songbirds settle on their claims. Male bullfrogs and green frogs defend a piece of wetland during the mating period. Male dragonflies chase intruders away from their perches around the edge of a pond or along a streamside. Bass defend their nests on the pond bottom. Even white-tailed deer does for a few weeks will defend the places where their fawns are hidden.

Such defended areas are called territories. A *territory* may be defined more formally as a defended, more or less fixed and exclusive area maintained by an individual or by a social group such as a wolf pack. Some would give the term an even broader definition and say that a territory exists when individuals or groups are spaced out more than you would expect from a random occupation of suitable habitat (Figure 13.11) (Davies, 1978).

Once an animal has established a territory, the owner must defend it against intruders. In the early period of territorial establishment, conflicts may be numerous. Territorial holders among birds, frogs, and insects usually defend their claims vocally by singing from some conspicuous spot. Songs and calls advertise the fact that the area is already occupied. They are a sort of long-distance warning that tells potential trespassers they should not waste their energies trying to settle there. Birds may shift their song perches throughout the terri-

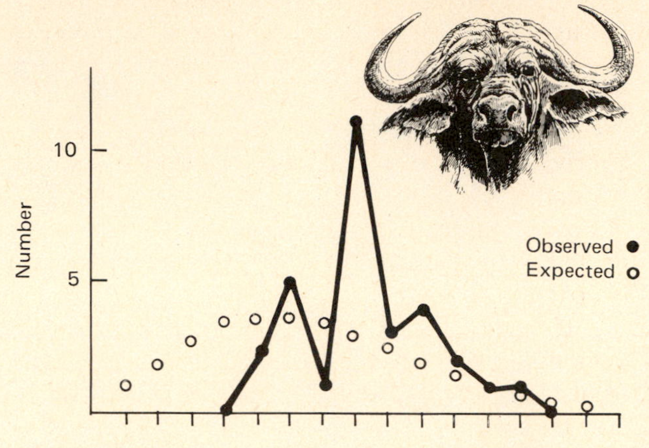

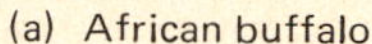

(a) African buffalo

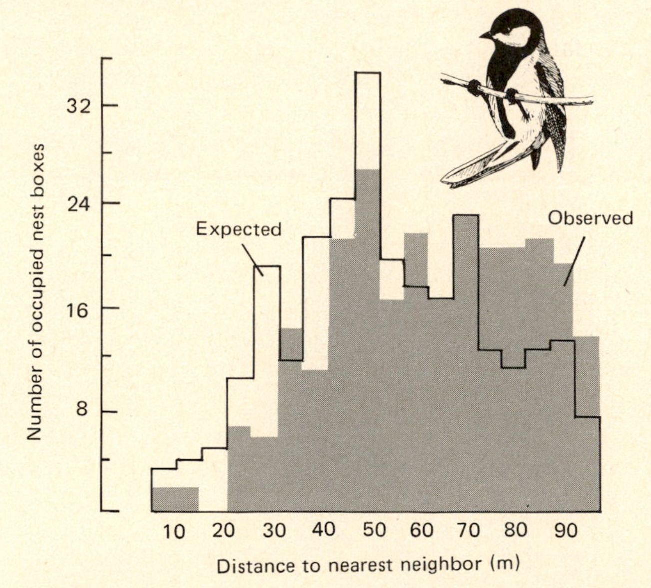

(b) Great tit

FIGURE 13.11 Two examples of spacing in which the animals are spaced more than would be expected from a random distribution over the available suitable habitat. (a) Buffalo herds on the Serengeti (Sinclair, 1977:145); (b) great tit in Wytham Woods (J. R. Krebs, 1971:4).

tory or vary their song patterns, perhaps in an effort to suggest that more than one male is in the area. Song can be very effective at maintaining space between individuals. If a bird is removed from its territory, the space is quickly claimed by another deprived of a territory outright or forced to settle in some suboptimal area. If the territorial male is removed but its song has been recorded and is played in the territory, other males will stay away (Carrick, 1963).

If songs and calls fail and another individual does move into the territory, the owner may confront the trespasser with a visual display. That display may involve raising the crest, fluffing the body feathers, spreading the wings and tail, and waving the wings among birds, erecting the ears and baring the fangs among mammals, and other such movements. Such displays usually intimidate the invader, encouraging it to leave.

If intimidation displays fail, then the territorial owner is forced to attack and chase the intruder, an activity easily observed among many birds in spring. Some birds carry such fighting to an extreme when they constantly fight their own reflection in a windowpane. Inevitably, the territorial owner wins, although he may have to make some adjustments in the territorial boundary to accommodate other neighbors.

Some animals may defend a territory in a more subtle manner by the use of scent markers. A wolf pack, for example, marks the boundary of its territory by well-placed scent posts, frequently renewed (Peters and Mech, 1975). Scent marks are most highly concentrated about the edges of the territory. These scent marks warn neighboring wolf packs about boundary rights. Just as important, these scent posts tell members of the pack that they are within their own territory and prevent accidental straying into hostile territory.

The use of scent or chemical releasers, or pheromones, is widespread among animals, especially mammals and insects. They are secreted from exocrine glands, transmitted as a liquid or a gas, and smelled and tasted by others. They are important not only to mark territory and trails but also to convey such information as the identity of an individual, its sex and social rank, the location of food, and the presence of danger and to attract mates (see E. O. Wilson, 1971; Whittaker and Feeney, 1971).

Types of Territories

Types of territories vary depending upon the animals' needs. Some territories are general-purpose, within the boundaries of which the owner performs all activities: feeding, mating, nesting. Some animals, like certain hawks, defend mating and nesting territories and feed elsewhere. Others, like some frogs and grouse, defend mating territories only. Some, like the barn swallow, defend a territory only about the nest. Hummingbirds may defend only the food supply—patches of nectar-producing flowers.

Why Defend a Territory?

Why should a cardinal or a wolf pack defend a territory? The reasons will vary among animals. For some it is the acquisition and protection of a needed resource, like food, or a reduction in the risk of predation. For others it may be the attraction of a mate. Whatever the apparent reason for defending a territory might be, the basic reason is the benefits derived, which ultimately are an increased probability of survival and improved reproduction success, or, in short, increased fitness. By defending a territory, the individual forces others into suboptimal habitat, reducing their fitness and at the same time increasing the proportion of its own offspring in the population.

Defending a territory is costly business. Territorial defense uses energy and consumes time, interferes with courtship and mating and with the rearing of young. Like all economic endeavors, territorial ownership has its costs and benefits, and the territorial owner has to balance the two (Figure 13.12). There exist situations in which territories are economically defendable and situations in which they are not. A general prerequisite is a predictable resource somewhat dispersed. Acquisition of an area ensures the owner of its resource needs, reducing foraging costs, and allows time for other activities. If resources are unpredictable and very patchy, it may be advantageous for individuals to belong to a group and cooperatively seek needed

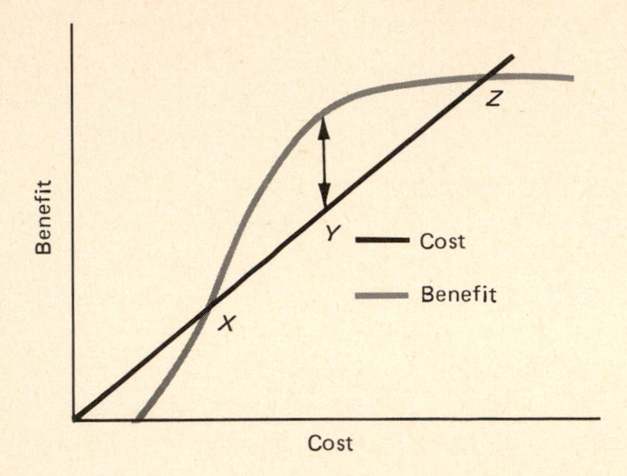

FIGURE 13.12 A graphic model of cost-benefit ratio curves as they might apply to territorial defense. Costs of territorial defense increase as the size of the territory increases. Benefits derived from territorial defense increase in a logistic manner, leveling off after territorial size or resources become more than sufficient to meet the animal's need. The animal can economically defend an area between *X* and *Z*. A maximum cost-benefit ratio occurs at point *Y*.

resources without being restricted to any one area. Spotted hyenas *(Crocuta crocuta)*, for example, live in clan territories in the Ngorongora Crater of Kenya, where resources are predictable, while on the Serengeti Plains, where food is seasonal, they range over wide areas and do not defend a territory (Kruuk, 1972).

Territory Size

Closely associated with the cost-benefit ratio of territorial ownership is territorial size. As the size of a territory increases, the cost of territorial defense increases. Many male birds in spring attempt to claim more ground than they can economically defend. They are forced to draw in the boundaries to make the area more manageable to hold. But there is a minimum size below which the territorial owner cannot go, because it will be too small the meet the animal's needs. The number of territorial owners an area can hold is determined by the total area available divided by the minimum size of the territory. When the territories are filled, the excess are ejected or denied access. These individuals make up the floating population. Somewhere along the gradient of too small and too large is an optimal size for the territory, one from which the owner gains maximum benefits for the costs incurred.

Some animals, notably certain spiders, have fixed territories (Riechert, 1981). The size of the territories remains the same, regardless of spider number or resource availability. Such fixed sizes may be transmitted as part of inherited behavior, or they may be transmitted culturally, from one owner to the next, and between generations.

Other animals are more flexible. Optimum size will vary from year to year and from locality to locality. If a resource such as food is abundant, the territory may be small, and if resources are less abundant, it may be larger. In general, territory size tends to be no larger than required. For example, the territories of the goldenwinged sunbird *(Nectarina reichenowi)* vary greatly in size and in floral composition, but each territory contains just enough of a nectar supply to meet an individual's daily energy requirement (Gill and Wolf, 1975). Such flexibility in the size of territories has been likened by Julian Huxley (1945) to an elastic disk compressible to a certain size. Territory size decreases as density increases, but when the territory compresses to a certain size, the resident resists further compression and denies access to additional settlers. Because aggressive behavior varies among individuals, the most aggressive have the advantage, and the less aggressive are forced to settle elsewhere.

For some animals—birds, in particular—it is not the size of the territory that counts, but its quality (Figure 13.13a). Some males are successful in claiming the best territories, usually measured by some features of the vegetation that make them superior nesting sites (Figure 13.13b). Less successful males occupy suboptimal territories, and many males secure no territory at all.

To successful males come the females, and the "wealthiest" males always obtain a mate. In fact, among some species of birds, such as the dicksissel (Zimmerman, 1971) of the midwestern grasslands and some of the marshland blackbirds (Orions, 1969b), females will accept a male already mated if he holds an outstanding territory. Apparently, the quality of the territory is such that the odds for a

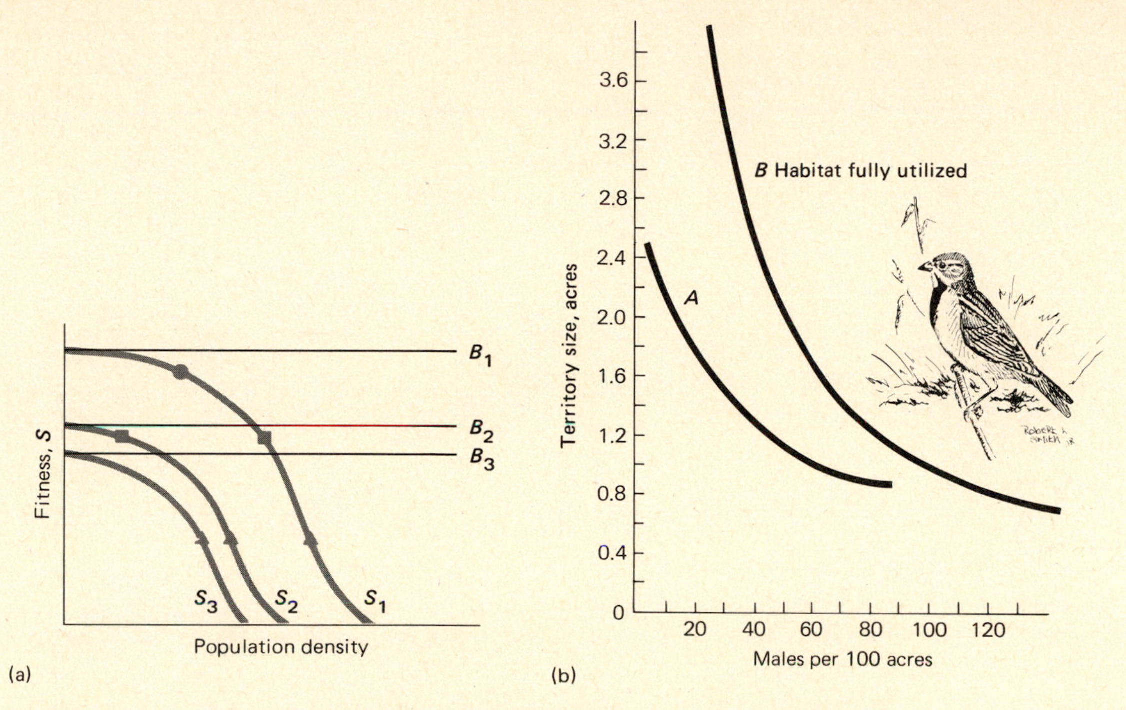

FIGURE 13.13 The relationship between habitat suitability, population density, and settling of habitats of different qualities. (a) A theoretical model of habitat suitability as measured by individual fitness. The suitability of habitat is ranked from B_1, the best, to B_3, the poorest. The curves S_1, S_2, and S_3 show the actual fitness of individuals. As the density in each habitat increases, fitness decreases. Note that even at high densities, fitness in the best habitat exceeds fitness in the poorest habitat at much lower densities. Thus, at low population densities, all individuals should settle in the best habitat, B_1. At intermediate densities some individuals will settle in habitat B_2. At high densities all habitats will be settled. (After Fretwell and Lucas, 1969:29.) (b) A field example of habitat suitability and fitness. As vegetative quality increases for the dicksissel, as measured by litter depth and vegetation density, the number of mates a male acquires increases. (After Zimmerman, 1971:591.)

successful nesting by the female are greater there even though the female may have to share the area with another. At the other extreme are male birds who own such poor territories that they are unable to attract a mate.

Outcome of Territoriality

As a result of contest competition among males for space, some individuals secure optimal territories, while at the other extreme some individuals are denied territory. Thus, a portion of the population does not reproduce because they are excluded from suitable breeding sites by territorial individuals. They make up a surplus population, a floating reserve that would be able to reproduce if a territory became available to them.

The existence of such a floating reserve of potentially breeding adults has been described for a number of species, including the red grouse *(La-*

gopus lagopus scoticus) in Scotland (Watson and Moss, 1971), Australian magpie *(Gymnorhina tibicen)* (Carrick, 1963), Cassin's auklet *(Ptychoramphus aleuticus)* on the northwestern coast of North America (Manuwal, 1974), and the white-crowned sparrow *(Zonotrichia leucophrys nuttalli)* of California (Petrinovich and Patterson, 1982). Studies of a banded white-crowned sparrow population indicated a nonbreeding surplus of potentially breeding birds. In fact, 24 percent of the territorial holders entered the breeding population 2 to 5 years after banding, and 25 percent of the nestlings that acquired territories did so 2 to 5 years after their birth. Territory holders that disappeared during the breeding season were replaced almost immediately.

Although the existence of floating reserve populations is acknowledged, few data exist on the social organization and behavior of surplus birds. Floaters may form flocks with a dominance hierarchy on areas not occupied by territory holders, as do the red grouse and Australian magpies. Or they may live singly off the territories as the white-crowned sparrows do; or they may spend much time on the breeding territories of others.

An example of the latter strategy is provided by the detailed studies of the rufous-colored sparrow *(Zonotrichia capensis)* in Costa Rica by S. Smith (1978). By observing banded birds, both territorial and nonterritorial, and by selectively removing certain individuals, Smith was able to determine the role of the floater, or "underworld" bird. Territorial sparrows in her study area occupied small territories, ranging from 0.05 to 0.40 ha, and made up 50 percent of the total population. The other 50 percent—underworld birds, consisting of both males and females—lived in well-defined, restricted home ranges within other birds' territories (Figure 13.14). Male home ranges, often disjoined, embraced three or four territories. Female home ranges were usually restricted to a single territory. Because home range boundaries of both sexes coincided with territorial boundaries, each territory held two single-sex dominance hierarchies of floaters, one male and one female. When a territorial owner, male or female, disappeared, it was quickly replaced by a local underworld bird of appropriate sex on the territory. These floaters usually entered the territories as young birds hatched some distance away. Territorial owners tolerated them because the cost in energy to exclude them would probably have been greater than the benefits accrued. In addition, because the floaters regulated the number of indi-

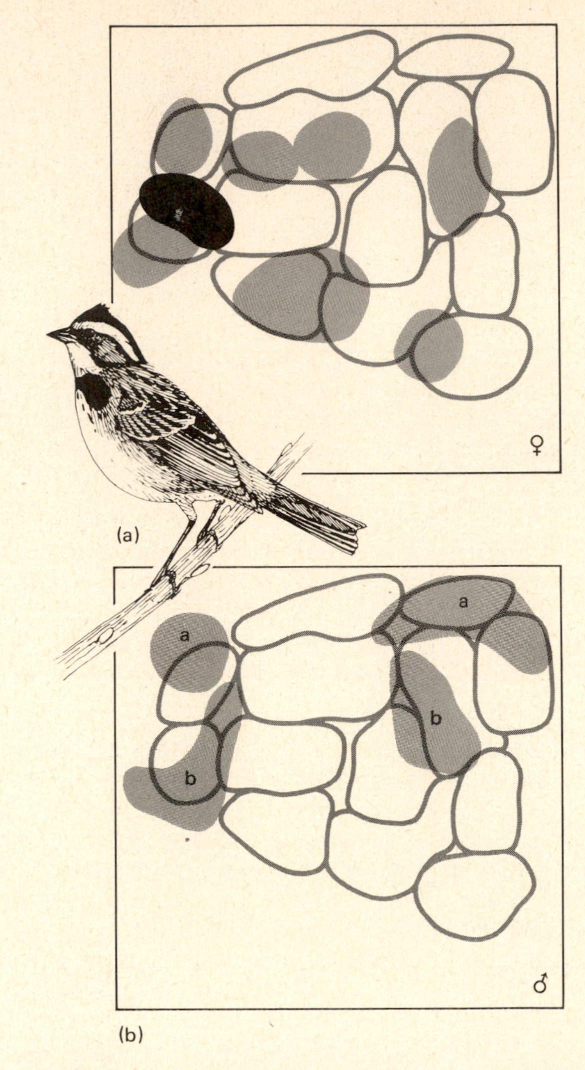

FIGURE 13.14 Territorial boundaries of the rufous-collared sparrow. Home ranges of two "underworld" males and females are superimposed on occupied territories. Eventually, some of the females occupied a territory by replacing a missing bird. The male home ranges *a* and *b* are disjoined, and each includes several territories. (Adapted from S. Smith, 1978:573, 576.)

viduals in a hierarchy, they could reduce territorial defense costs by discouraging other intruders.

A similar situation exists among some territorial spiders (Riechert, 1981). They, too, have a floating reserve of individuals who will quickly claim vacated sites. These floaters live in cracks and crevices within occupied territories. Periodically, the floaters will attempt unsuccessfully to take over an occupied habitat.

Consequences of Territoriality

A consequence of territoriality can be population regulation. If no limit to territorial size exists and all pairs that settle on an area get a territory, then territoriality results only in spacing out the population. No regulation of the population results. But if territories have a lower limit, then the number of pairs that can settle on an area are limited. Those that fail to get a territory have to leave. Thus, territoriality might regulate population density, but only under certain conditions. For example, in the arctic ground squirrel *(Spermophilus undulatus)*, all females are allowed to nest, but territorial polygamous males drive excess males from the colony into submarginal habitats, where they exist as a nonbreeding floating population. The number of breeding males remains constant because losses are continuously replaced from the floating population. In this species territoriality stabilizes the number of breeding males but does not regulate the population, because only the males are surplus.

J. Krebs (1971) removed breeding pairs of great tits *(Parus major)* from their territories in an English oak woodland. The pairs were replaced by new birds, largely first-year individuals, that moved in from territories in hedgerows considered suboptimal habitats. The vacated hedgerow habitats, however, were not filled, suggesting that a floating reserve of nonterritorial birds did not exist. In that case, territoriality limited the number of birds in optimal habitat but did not regulate the population because all the birds were breeding in some habitat.

If an excess of nonterritorial males and females of reproductive age exists in the area—as was the case with the white-crowned sparrows, rufous-collared sparrows, and red grouse—then reproduction is limited by territoriality, and density-dependent population regulation results. Surplus individuals are limited or excluded from breeding, and the population will not increase beyond some upper limit set by the number of territories available.

SPACE CAPTURE BY PLANTS

Plants obviously are not territorial in the same sense that animals are territorial. But plants can capture and hold onto space for a period of time (see Chapter 23). This phenomenon in the plant world can be considered analogous to territoriality in animals, especially if you accept the alternative definition of territoriality—individual organisms spaced out more than one would expect from a random occupancy of suitable habitat.

Plants from dandelions to trees do capture a certain amount of space and defend it over time from intraspecific and interspecific competitors. They exclude other individuals from the same and smaller size classes. When a dandelion plant spreads its rosette of leaves on the ground, it effectively eliminates all other plants from the area covered by its leaves. The faster-growing trees in a forest achieve a height dominance over others, allowing them to expand their canopies over the tops of others. Plants with expansive crowns or rosettes of leaves intercept light, effectively eliminating competition and restricting occupancy of the ground below to shade-tolerant species. Because the root systems of trees more or less mirror in the ground the expanse of the canopy, dominant trees are also in a superior competitive position for nutrients and moisture. Because of their longevity, some plants, especially trees, occupy space over a very long period of time, preventing invasion by other individuals of the same or other species. Plants successful in capturing space increase their fitness at the expense of others. The survival rate of a few adults is high; others are eliminated.

Plants may also capture space by the release of organic toxins that reduce competition for light, nutrients, and space. A variety of phenolic compounds released by roots and by leaves due to litterfall and rainwater throughfall accumulate in the soil and inhibit the germination of seeds and the growth of plants, herbaceous and woody.

Summary

Populations do not increase indefinitely. As resources become less available to an increasing number of individuals, birthrates decrease, mortality increases, and population growth slows. If population declines, mortality decreases, births increase, and population growth speeds up. Between positive and negative feedback, the population arrives at some form of regulation.

Regulation involves intraspecific competition. Competition occurs when resources are in short supply relative to the number seeking them. Competition can take two forms: scramble and contest. In scramble competition, resources are shared equally by all individuals in the population, none of which receives a sufficient amount for growth and reproduction. In contest competition, sufficient resources for growth and reproduction are claimed by dominant individuals while denying them to others. The latter produce no offspring or perish.

Intraspecific competition can result in decreased or retarded growth of individuals, decreased density and increased biomass in individual plants, delayed reproduction, and density-dependent mortality in animals brought about by malnutrition, decline in immunity to diseases, and parasites.

Responses to intraspecific competition may take a number of forms. One is increased stress brought about by crowding, fighting, lack of food or space, or attempts to acquire mates. Stress may result in delayed reproduction, abnormal behavior, increased adrenal activity, reduced growth and production of flowers and seeds, and reduced ability to resist disease and parasitic infections.

Increased stress, however, might lead to dispersal. Dispersal is a constant phenomenon in populations at presaturation levels. Such dispersal acts as a regulatory mechanism. It apparently is not a random selection of individuals; rather, the individuals seem to be genetically programmed to disperse. Many end up in dispersal sinks, such as submarginal habitats, and some successfully occupy new or unfilled habitat. At saturation levels dispersal is a response to overcrowding, and the dispersers are surplus to the population.

Another response to increasing density may involve genetic change. Under favorable environmental conditions and relaxed selective pressures, an increasing population exhibits increasing genetic variability and declining average fitness. When environmental conditions change and selective pressures increase, individuals with less desirable genetic combinations are eliminated, decreasing the population but increasing average fitness.

Intraspecific competition may be expressed as social behavior, a social organization based on the dominance of one individual over another. Dominant individuals secure most of the resources, and in times of resource shortages, the effects are borne by subdominant individuals.

One form of social hierarchy is territoriality, in which a defended or exclusive area is maintained by an individual or a group. Animals defend territories by songs, calls, displays, chemical scents, and fighting. Because of the costs incurred in territorial defense, owners can afford the benefits derived only if their increased fitness exceeds costs. Territoriality is a form of contest competition in which a certain portion of the population is excluded from reproduction. These nonreproducing individuals act as a floating reserve of potential breeders, available to replace losses of territorial holders. In such a manner, territoriality can act as a population regulating mechanism.

Review and Study Questions

1. Explain how density-dependent effects might influence population growth and size.
2. What is competition? Distinguish between scramble and contest competition and discuss the importance of each in population regulation.
3. How might stress influence population growth in mammals? In plants? How do plants respond to stress?

4. Distinguish between saturation dispersal and presaturation dispersal. Why is presaturation dispersal the only one effective in population regulation? What is the importance of a dispersal sink?

5. How does an increasing population influence genetic variability?

6. How is dominance expressed in social hierarchy? In what way can it function as a population regulating mechanism?

7. What is territoriality, and how does it relate to social hierarchy?

8. Discuss the economics of territoriality. Name some conditions under which territoriality might be too expensive.

9. How can territorial behavior increase the fitness of the male holder of the territory? The female? Speculate on why the female may choose the territory rather than the male owning it.

10. How can territoriality function in population regulation? What is the role of the floaters?

11. Speculate on territoriality in plants. How might you study such a phenomenon?

Suggested Readings

Andrewartha, H. G. 1971. *Introduction to the study of animal populations*. Chicago: University of Chicago Press.

Begon, M., and M. Mortimer. 1981. *Population ecology*. Sunderland, Mass.: Sinauer Associates and Oxford, England: Blackwell Scientific Publications.

Grime, J. P. 1979. *Plant strategies and vegetative processes*. New York: Wiley.

Harper, J. L. 1977. *Population biology of plants*. New York: Academic Press.

Hutchinson, G. E. 1978. *An introduction to population ecology*. New Haven, Conn.: Yale University Press.

Ito, Y. 1980. *Comparative ecology*. Cambridge, England: Cambridge University Press.

Krebs, J., and N. B. Davies, eds. 1984. *Behavioral ecology: An evolutionary approach*. 2d ed. Oxford, England: Blackwell Scientific Publications.

Lack, D. 1954. *The natural regulation of animal numbers*. Oxford, England: Oxford University Press.

Solbrig, O. T., ed. 1980. *Demography and evolution in plant populations*. Berkeley: University of California Press.

Tamarin, R. H., ed. 1978. *Population regulation*. Benchmark Papers. Stroudsburg, Pa.: Dowden, Hutchinson, and Ross.

Tanner, J. T. 1978. *Guide to the study of animal populations*. Knoxville: University of Tennessee Press.

Wilson, E. O., and W. H. Bossert. 1971. *A primer of population biology*. Sunderland, Mass.: Sinauer Associates.

ECOLOGICAL EXCURSION

The Mast Feeders

Durward L. Allen

In the fall of 1834, Bartholomew County in south-central Indiana was the scene of a squirrel hunt not unique in its day, but the like of which will not be seen again.

The floor of cool forests in the rich flat-lands was deep with the mast of beech, white oak, and shagbark hickory that crunched into the cushiony duff at every step. Ridges were strewn with the bounty of chestnut oaks, and everywhere timber was a-rustle with swarms of fattening gray squirrels.

More than a century of development had brought the Kentucky rifle (née Pennsylvania) to a high state of perfection, and barking a squirrel at 60 yards was largely a matter of the clear eye and steady hand.

The good men and true of Sand Creek Township considered themselves the elite of the squirrel-hunting world and did not hesitate to advertise it. The riflemen of Wayne Township denied this presumption and challenged them to a contest. The two townships agreed that each would select 50 hunters who would shoot squirrels for three days, at the end of which a grand barbeque would be held, with the losing side footing the bill.

The full details of this shooting fest are not available, but enough was recorded to give some idea of the abundance of the gray squirrel in some of its best native range. The winner of the hunt presented *900 squirrels* at the end of the three days and the runner-up had 783!

Such abundance was not just a local occurrence. Squirrels took heavy toll of cornfields in forest openings, and in 1749 the colony of Pennsylvania paid bounties on 640,000 of them. There was a time in Ohio when county taxes were payable in squirrel scalps, and an early record mentions a gunner who killed 160 in a day—at a time when they were not especially plentiful.

Among frontier hunters the gray squirrel was known as the migratory squirrel, and with good reason. Periodically its numbers built up to a density that became intolerable to the species itself. Probably it is safe to say that the climax in this increase always occurred during a year when mast production was heavy. September and October are the season when young animals are on the move to find comfortable sites for future living, but that is different from the pervasive restlessness that seized entire populations. They would begin to travel, evidently in one direction, not stopping for lakes, streams, or anything else.

During such a movement the residents of Saginaw, Michigan, would awake, mornings, to find bedraggled black squirrels (the commonest color phase in the North) perched on every piling in the river. In a southward migration near Racine, Wisconsin, the animals were passing for two weeks, and it was a month before all had disappeared.

After a migration the woods would seem deserted of squirrels, but it did not take long for their numbers to become conspicuous again. The best evidence indicates that northern gray squirrels could be expected to migrate about every five years.

They were going nowhere, and they got nowhere. On these marches mortality was heavy from all causes, and the numbers were worn down and dissipated. Over large areas gray squirrels disappeared when dense forests were cut, but here and there the animals have again become sufficiently prosperous so that a reversion to the old migratory habit has been observed.

The migrations, it seems, were a device for clearing the land of a too-numerous population and converting a million animals back into the humus of the earth. A part of the time, at least, the gray squirrel is sufficiently immune to the effects of external controls (disease, predators, etc.) so that in the course of the ages it has developed a *sociological* means of getting rid of excess numbers.

In contrast, the fox squirrel is not habitually migratory. Probably because it inhabited forest *edges* rather than extensive continuous stands of

---

SOURCE: Durward L. Allen, *Our Wildlife Legacy* (New York: Funk & Wagnalls, 1954), 89-92.

woodland, it did not build up, in depth, to mass hordes like the gray squirrel. Nevertheless it did, and does, become highly abundant periodically, and studies in the Midwest have shown what happens:

There comes a spring when a late frost, or some other climatic condition, nips the mast crop before it gets started. A food scarcity develops, and young animals are brought into competition with the abundant adults for what acorns and other nuts are still buried in the ground. Ordinarily, from early summer on they would find the oaks and hickories weighted with green mast to be had for the cutting, but now this basic supply is missing. . . . A healthy fall population of fox squirrels should contain about two-thirds young of the year. But when numbers are high and food short the mortality of young is far greater than usual. By autumn they may compose only a quarter of the population.

The real crisis comes when a poor mast year is followed by an old howler of a winter. Ordinarily the squirrels would be well layered with belly fat, and in periods of deep snow, ice, and blizzards they would roll up in the nest and snooze it out. This winter they are thin and must have food regardless of weather. It is hard to find and hard to dig up. A diet of buds will not sustain them for long. The scabies mite (mange) is ever-present and on weakened animals it takes over. Part or all of the hair falls out and open lesions develop in the skin. Undernourishment appears also to induce a deficiency of blood sugar that gives rise to fatal seizures of shock when the animal is excited or over-exerts. Many squirrels become so weak they no longer can climb. They starve, freeze, and are easily taken by predators.

But the end is not yet. Fox squirrels usually breed in midwinter, and in this year of sorrows bare survival is the best they can do. Only those animals come through that are in the most favorable locations. The new crop of young may be drastically reduced or entirely missing in the spring that follows.

Although this situation was first observed in Michigan, it probably is a frequent occurrence in northern squirrel ranges, and something similar has been recorded in Europe. Squirrels are the foremost game animals in Finland, and extensive field research shows that they fluctuate radically in numbers about every five years, usually with ten years between major peaks. In a large area where they were studied, the animals died off within a couple of months in the fall of 1943. It was estimated that the drop in numbers was such that there was one animal where there had been 450 before.

Food failures have been a cause of squirrel declines in that region also, but probably of greater importance are the epizoötics* that go through the population when numbers are at their maximum. Coccidiosis has been of particular importance. Shortages of fir cones, a primary food supply, appear to help induce diseases in the same manner as in Michigan.

It is a recurring picture of numbers building up to a point of instability, the downfall being brought about by one cause or another, or by a combination of causes. A dense population of animals is a precarious structure. It becomes weaker as it builds, and what triggers the collapse is, perhaps, not too important. It is highly important, however, . . . to realize that a reduction from abundance is almost always inevitable.

* The same as an epidemic among humans.

CHAPTER 14

Life-History Patterns

Outline

Objectives

Upon completion of this chapter, you should be able to:

1. Discuss the importance of energy allocation to reproduction.
2. Compare semelparity with iteroparity.
3. Discuss the relationship between the number of young and parental investment in them, and the ecological significance of the relationship.
4. Discuss the relationship among the number of young, fitness of the parents, and fitness of the young.
5. Show the relationship among size, age, and fecundity and discuss possible reasons for the relationship.
6. Compare *r*-selection with *K*-selection and discuss the value and shortcomings of the concept.
7. Discuss the various modes of sexual reproduction.
8. Discuss the nature of sexual selection in animals and plants.
9. Define a mating system and distinguish among monogamy, polygamy, polygyny, and polyandry.

Ecology in recent years has acquired certain anthropomorphic terms such as *strategy* (life-history strategy, reproductive strategy, optimal foraging strategy, mating strategy) and *tactics* (reproductive tactics, foraging tactics). The term *strategy* has several related meanings. One is the science or art of planning and directing military movements and operations. Another is plans, methods, or series of maneuvers for obtaining a specific goal. Tactics are plans or procedures for obtaining a desired end or result. The use of these terms in ecology is unfortunate in some ways, because they imply conscious, detailed planning on the part of living organisms toward a desired end or goal—that oak trees and warblers consciously set out a course in their life cycles. Needless to say, nothing could be further from the real situation.

The term *life-history strategy* in ecology means the selective processes involved in achieving fitness by living organisms. Such processes involve, among other things, fecundity and survivorship; physiological adaptations; modes of reproduction; age at reproduction; number of eggs, young, or seeds produced; parental care; means of avoiding environmental extremes; size; and time to maturity. Success among individuals and species is measured only in terms of successful offspring. How this is achieved becomes the organism's life-history strategy or pattern.

An ideal situation for any organism is to reproduce shortly after birth, produce a large number of highly adapted offspring, live for a long period of time, reproduce frequently, lavish parental care on the offspring, and devote the maximum amount of energy to reproduction. Realistically, any organism has to compromise considerably with the ideal to achieve optimal fitness. The various means by which that goal is achieved are the subject of this chapter.

Reproductive Effort

ALLOCATION OF ENERGY

Energy captured by an organism (see Chapter 19), like currency used to run a household or business, must be allocated to certain essential uses. Some must go to growth, to maintenance, to the acquisition of food, to defending territory and escaping predators; and some has to go to reproduction. To achieve optimal fitness, an organism has to budget its energy carefully. How it budgets that energy is part of its reproductive strategy.

The more energy an organism spends on reproduction, the less it has to budget for growth and maintenance. As a result, the individual may grow more slowly to the next age or fail to survive. For example, Lawlor (1976) found that reproductive females of the terrestrial isopod *Armadillidium vulgare* had a lower rate of growth than nonreproductive females. Nonreproductive females devoted as much energy to growth as reproductive females devoted to both growth and reproduction.

How much energy organisms invest in reproduction varies. Energy costs involve not only the weight of the progeny but also the amount of energy expended in rearing the young. Herbaceous perennials may invest between 15 and 25 percent of annual net production in new plants, including vegetative propagation. Wild annuals (single-stage reproducers) expend 15 to 30 percent; most grain crops, 25 to 36 percent; and corn and barley, 35 to 40 percent. The lizard *Lacerta vivipara* invests 7 to 9 percent of its annual energy assimilation in reproduction (Avery, 1975). The female Allegheny Mountain salamander, *Desmognathus ochrophaeus*, spends 48 percent of its annual energy flow on reproduction, including energy stored in eggs and energy costs of brooding (Fitzpatrick, 1973).

TIMING OF REPRODUCTION

If an organism is to contribute its maximum to future generations, it has to balance the profits of immediate reproductive investment against the costs of future prospects, including fecundity and its own survival, a trade-off between present progeny and future offspring.

One strategy is to invest all energy into growth, development, and storage and then expend the energy in one massive reproductive effort and die. Such a reproductive strategy is employed by most insects and other vertebrates, some species of salmon, annual plants, and some bamboos. Certain bamboos delay flowering for 100 to 120 years, produce one massive crop of seeds, and die (Janzen, 1976). These organisms sacrifice future prospects by expending all their energy in one suicidal act of reproduction. That mode of reproduction is *semelparity.*

Another strategy is to produce fewer young at one time but to repeat reproductive efforts throughout a lifetime. That is called *iteroparity.* For an iteroparous organism the problem becomes one of timing reproduction—early in life or at a later age. Early reproduction reduces survivorship and the potential for later reproduction. Later reproduction increases growth of organisms and improves survivorship but reduces fecundity. Energy expended in repeated reproduction weighs against future prospects as measured by declining fecundity and reduced potential for survival. Mammals, for example, experience increased mortality during and after each reproduction (Figure 14.1).

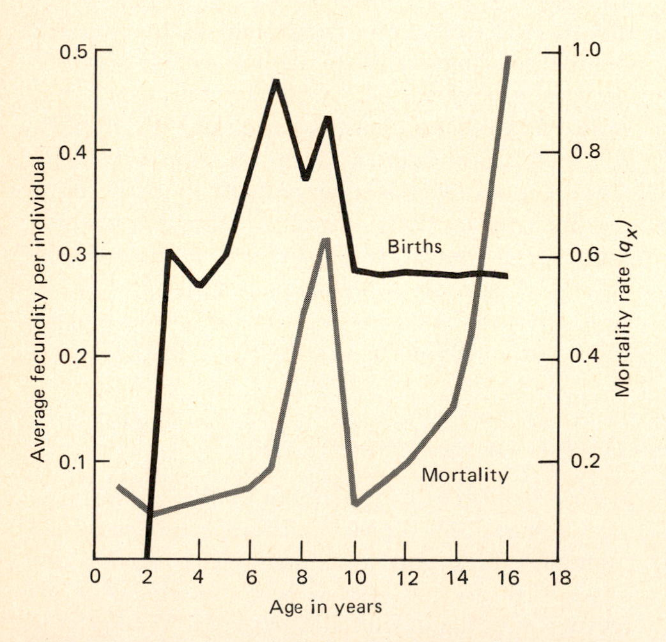

FIGURE 14.1 As an individual increases its reproductive effort, it is likely to reduce its survivorship. In red deer, for example, an increased rate of birth results in an increased rate of mortality. Energy expended in repeated reproduction works against the potential for survival. (Based on data in Lowe, 1969.)

Most organisms make a trade-off between present progeny and future offspring and survival. The nature of these trade-offs varies from species to species and within species. If an individual grows rapidly while investing no energy in reproduction and is short-lived as an adult, its reproductive strategy will probably be a single reproductive effort. Any delays in the production of offspring reduce the probability that the individual will survive to reproduce. But if an organism is longer-lived and its chances for survival as an adult are high, it can reproduce more than once throughout a lifetime. Under these conditions the strategy is usually delayed reproduction. However, some semelparous organisms exhibit delayed reproduction. Mayflies may spend several years as larvae before emerging to the surface of the water for an adult life of several days devoted to reproduction. Certain Pacific salmon do not reproduce until they are 3 to 4 years old. And periodical cicadas spend 13 to 17 years below ground developing before they emerge as adults to stage an outstanding exhibition of a single-term reproductive effort.

PARENTAL INVESTMENT AND NUMBER OF YOUNG

The number of offspring produced in each reproductive effort may range from many small ones to a single large individual. The number produced relates to the amount of parental investment per individual offspring. If a parent produces a large number of young, it can afford only minimal investment per individual. In such cases, animals provide no parental care, and plants store a minimal amount of energy in seeds. Such organisms usually inhabit disturbed sites or unpredictable environments, or places such as the open ocean, where opportunities for parental care are minimal or difficult at best. By dividing energy allocated for reproduction among as many young as possible, parents increase the chances that some young will successfully settle somewhere. By such a strategy parents increase their own fitness but decrease the fitness of the young.

Parents that produce few young are able to expend more energy on each individual. The amount of energy will vary, of course, depending upon the number of young produced and the stage of development of the young, their size, and their stage of maturity when born. Some organisms expend less energy during incubation or gestation and invest more energy after birth of the young. The young are born or hatched in a helpless condition and require considerable parental care. Such animals are called *altricial*. Other animals have longer incubation periods or longer gestation periods, so the young are born in an advanced stage of development and are able to forage for themselves shortly after birth. Such young are called *precocial*. Examples are gallinaceous birds and ungulate mammals. Plants may produce relatively few large seeds with large amounts of stored energy provided for the germinating seedling. Examples are acorns, walnuts, hickory nuts, and coconuts. In such cases, parents decrease their own fitness but increase the fitness of their young.

The number and size of young depend upon the selective pressures under which the organism evolved or to which it must adjust. Among plants, short-lived annuals have numerous small seeds. High production of numerous offspring ensures that some will survive and germinate the next growing season. Optimum seed size and number for perennial plants relate to dispersal ability, colonizing ability, and the need to escape predation. Plants that colonize disturbed or unpredictable environments may have small, windblown seeds that can be carried great distances. Because a minimal amount of energy is invested in each seed, these plants can afford heavy losses of seeds to ensure that some successfully survive and reproduce. Plants subject to heavy predation may have small seeds that provide less attractive food packages. Again, an abundance of seeds is an insurance that some will escape predation (see Chapter 17).

Plants associated with more stable environments may produce fewer and larger seeds, with a large store of energy that the seedling can use to get established. Such plants may invest a considerable amount of energy in the production of toxins or heavy seed coats to reduce predation. Some plants may adjust their production of seeds to meet current environmental conditions. Such adjustments may involve a genetic polymorphism or a plastic response. For example, the common weed purslane speedwell, *Veronica peregrina*, a plant of moist soils, growing in an environment where moisture conditions are optimal produces few but heavy seeds. The same species growing where moisture is limited and competition from grass is keen grows taller and produces lighter seeds that are better able to disperse (Linhart, 1974). Or the adjustment may involve a plastic response, as with the annual *Polygonum cascadense*. Plants growing on harsh, open environments allocate proportionately more of their energy to reproduction than those growing in more moderate habitats (Hickman, 1975).

Animals exhibit similar adjustments. "Annual" species, such as insects that overwinter in the egg stage, produce enormous numbers of small eggs. Birds such as quail and mammals such as rabbits and mice that are subject to heavy predation produce large numbers of young.

One way by which animals can apportion energy to reproduction is through the adjustment of clutch size and litter size. David Lack (1954) proposed that clutch size in birds evolved to equal the average largest number of young the parents can feed. Thus, clutch size is an adaptation to food supply. Temperate species, he argued, have larger clutches because increasing daylength provides the parents a longer time to forage for food to support large broods. In the tropics, where daylength does not change, food is a limited resource.

M. Cody (1966) modified Lack's ideas by proposing that clutch size results from different allocations of energy to egg production, avoidance of predators, and competitive ability. In temperate regions periodic local climatic catastrophes can hold a population below carrying capacity. Natural selection then favors a high rate of increase and thus larger clutches on the average than in tropical regions. The predictability of the tropical climate makes the maintenance of carrying capacity more important than increased production of young. More energy is expended in avoiding predation and in meeting competition.

These theories are supported by field examples. Birds in temperate regions do have larger clutch sizes than those in the tropics (Figure 14.2).

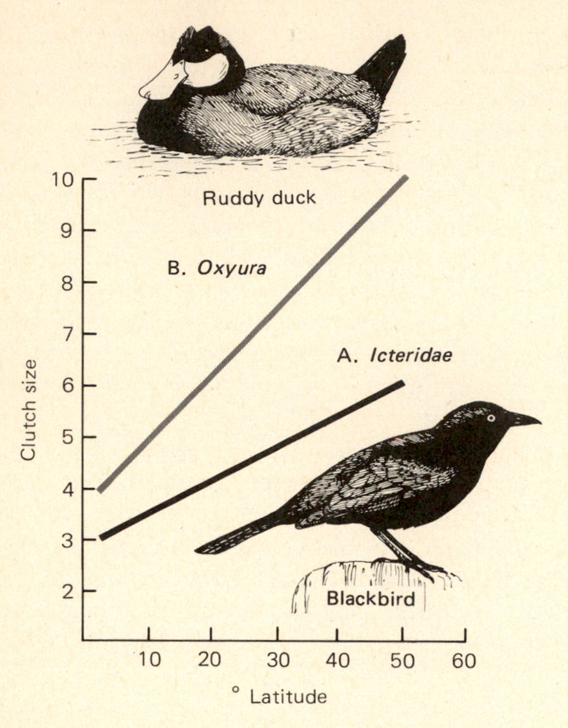

FIGURE 14.2 Birds of the same family tend to have smaller clutches at low latitudes than at high latitudes. Clutch sizes for the family Icteridae (blackbirds) and for the Genus *Oxyura* (ruddy ducks) of the family Anatidae illustrate this trend. (Adapted from Cody, 1966:179.)

Mammals at higher latitudes have larger litters than those at lower latitudes (Lord, 1960). Lizards living at lower latitudes have smaller clutches and higher reproductive success than those living at higher latitudes. Even among insects, tropical species produce fewer eggs and fewer clutches than their temperate-region counterparts (Landahl and Root, 1969).

Within a given region, production of young may reflect abundance of food, and the optimal clutch size may not be the average for the region. The optimal size may be more flexible, a response to available resources. Spight and Emlen (1976) provided two marine snails, *Thias lamellosa* and *T. emarginata,* with an increased food supply. Adult *T. lamellosa* increased their average egg clutch sizes from 930 to 1,428, and *T. emarginata* spawned more times during the year. G. Hogstedt (1980) discovered that the size of clutches varied among local populations of the European magpie *(Pica pica)* during the same year. Large clutches and high adult survival were associated with high territorial quality. Average clutch size, optimal for the birds, went with average territories. Birds able to acquire the highest-quality territories had the largest clutches. Hogstedt also noted that colonial birds and other species that lack individual foraging territories have smaller clutches of similar size.

AGE, SIZE, AND FECUNDITY

A relationship exists between fecundity and size and age. Among plants and poikilothermic animals, in particular, fecundity increases with size and age. Among homoiotherms the relationship is not so clear-cut. Age at maturity and seed production varies among trees. Some, like Virginia pine *(Pinus virginiana),* will start to produce seed at 5 to 8 years when growing in open stands. In dense stands Virginia pine may delay seed production for 50 years. Quaking aspen *(Populus tremuloides)* produces seeds at the age of 20 years and white oak not until the age of 50 years. Acorn production varies with size. Trees 16 inches in diameter at breast height produce around 700 acorns; those 24 to 26 inches produce 2,000 or more acorns (Downs and McQuilkin, 1944).

Perennial herbaceous plants and even annuals exhibit size and fecundity relationships. Among the *Plantago* species, for example, annuals have a higher reproductive effort and seed output per unit of leaf area than perennial species; but perennials with a larger average leaf area produce a greater weight of seeds per plant (Primack, 1979). Perennials delay flowering until they have attained a sufficiently large leaf area. Flowering plants are larger than nonflowering ones. Annuals show no such relationship between leaf area and reproductive effort, which seems to be independent of plant size once reproduction starts. But size differences among annuals do result in differences in seed production. Small plants produce few seeds, even though the plants themselves may be contributing the same proportionate share to reproductive efforts. Annual jewelweeds (*Impatiens* spp.) growing in optimal environments produce both outcrossed and self-fertilized flowers, the latter produced late in the season (see "Mating Strategies"). Jewelweeds growing in stressed environments produce only a few self-fertilized flowers and seed early in the season (Waller, 1982).

Similar patterns exist among poikilothermic animals. Fecundity in fish increases with size, which, in turn, increases with age. Because early fecundity reduces both growth and later reproductive success, there is a selective advantage to delaying sexual maturation until more body growth is achieved. Gizzard shad *(Dorosoma cepedianum)* reproducing at 2 years of age produce 59,000 eggs. Those delaying reproduction until the third year produce 379,000 eggs. Egg production then declines to about 215,000 eggs in later years. Among these fish only about 15 percent spawn at 2 years of age; 80 percent breed at 3 years.

Delayed maturation is characteristic of fish and other poikilotherms. Among 104 species of freshwater fish in the United States, only 15 percent mature the first year, and 21 percent the second year. Sixty-four percent of all species mature at 3 years of age or older. By contrast, 55 percent of 171 species of placental mammals reach reproductive maturity at 1 year of age, 20 percent the second year. Only 25 percent mature at the age of 3 or older (Bell, 1980).

An increase in fecundity is associated with a decrease in body mass and an increase in the rate of metabolism. Fenchel (1974) pointed out that such relationships exist over a wide range of unicellular and multicellular organisms, including poikilotherms and homoiotherms. Brian McNab (1980) has called attention to a close correlation among fecundity, metabolism, and food among mammals. Mammals with a high rate of metabolism have higher fecundity than mammals with a low rate of metabolism. Because basal metabolism in mammals increases with a decrease in size, small mammals have a higher metabolic rate than large ones.

Metabolic activity apparently is also related to quantity and quality of food in space and time. Mammals that feed on tree leaves, dried fruits and seeds, and insects have lower rates of metabolism than grazing ungulates, most rabbits and hares, microtine rodents, and carnivores. A seed-eating rodent with the same body size as a grazing rodent would have a lower rate of metabolism and lower fecundity. Unlike some poikilotherms, mammals cannot divert energy from maintenance to reproduction. They all have certain basic metabolic costs needed to maintain homeostasis, which cannot be changed. Energy above that is available for reproduction. All mammals, then, probably have as high a rate of increase as their metabolisms will support.

r-SELECTION AND *K*-SELECTION

The review of reproductive strategies suggests two points: (1) Species living in harsh or unpredictable environments in which mortality is largely independent of population density allocate more energy to reproduction and less to growth, maintenance, and adjustments to the environment. (2) Species living in a stable or more predictable environment in which mortality results from density-related factors and competition is keen allocate more energy to nonreproductive activities. The former are known as *r-strategists* because environmental conditions keep growth of such populations on the rising part of the logistic curve. In such situations, genotypes with a high r are favored. The latter are called *K-strategists* because they are able to maintain their densest populations at equilibrium (asymptote) or carrying capacity (MacArthur and Wilson, 1967).

According to this concept, K-strategists and r-strategists are under different selection pressures. r-selection favors those genotypes that confer the highest possible intrinsic rate of increase; rapid development; early maturity; small body size; short life span; early, even single reproductive effort; a large number of offspring; and minimal parental care. As a result, r-strategists typically are opportunists; they have the ability to colonize temporary or disturbed habitats where competition is minimal. They have means of wide dispersal and respond quickly to disturbance to the population. K-selection favors genotypes that confer the ability to cope with physical and biotic pressures, to tolerate relatively high population densities, delayed reproduction, slower development, larger body size, relatively long life span, repeated reproduction, and few offspring. Animals care for young; plants produce seed with stored food. *K*-strategists are resource-limited, and the population remains at or near carrying capacity. These qualities, along with the lack of a means of wide dispersal, make *K*-strategists poor colonists.

However useful the terms *r-selection* and *K-selection* may be conceptually, they create their own problems. One rests with *K*. *K* is a parameter and not a function of life-history traits. Individuals cannot be under direct selective pressure to increase *K*. Selection can only act on *r*, which is a function of life-history traits: age, survivorship, and fecundity. Thus, *r* and *K* are not equivalent terms.

The concept of both *r*-selection and *K*-selection assumes deterministic environments: one that is unpredictable, another that is predictable or stable. But environments are stochastic, fluctuating. If in a fluctuating environment adult mortality fluctuates and juvenile mortality does not, then selection should favor early maturity, larger reproductive effort, and more young. And under the same conditions if juvenile mortality or birthrate fluctuates and adult mortality does not, then selection should favor later maturity. Thus, one can arrive at the same strategies by considering the response of individuals to selective pressure only in terms of *r* (Sterns, 1976).

r-selection and *K*-selection can be considered as polar entities of a continuum from *r* to *K* (Pianka, 1970). Such a view tempts a classification of species as either *r*-selected, *K*-selected, or somewhere in between. But you can't very well compare elephants with mice. The concept of *r*-selection and *K*-selection is most useful when applied to individuals within a population or to populations within a species. Under certain conditions individuals or populations will exhibit *r*-selected traits or *K*-selected traits. Meadow mice living in situations where dispersal can take place easily exhibit characteristics of *r*-selection, while those living under conditions in which there is no dispersal sink assume *K*-selected characteristics (see Chapter 13) (Tamarin, 1978).

Because of the difficulty of forcing plants into *r* and *K* categories, J. Grime (1977, 1979) has proposed the idea of ruderal, or *R*-strategists; competitive, or *C*-strategists; and stress-tolerant, or *S*-strategists. *R*-strategists reproduce early in life, possess high fecundity, experience lethal reproduction (semelparity), occupy uncertain or disturbed habitats, and have well-dispersed seeds. *C*-strategists and *S*-strategists occupy more stable environments and are relatively long-lived, often drastically reducing the opportunity for seedling establishment, resulting in high juvenile survival. Beyond these characteristics, the two have evolved quite different life-history strategies. *C*-strategists reproduce early and repeatedly, utilizing an annual expenditure of energy stored prior to seed production. *S*-strategists have delayed maturity, intermittent reproductive activity, and long-term energy storage.

Mating Strategies

Sexual reproduction is common to most multicellular organisms. Even those that rely primarily on asexual or vegetative reproduction will revert, even if infrequently, to sexual reproduction. Sexual reproduction allows the gene pool to become mixed, increasing the genetic variability necessary to meet changing selective pressures and to prevent an accumulation of harmful mutations. But for the individual, sexual reproduction is expensive. Each individual can contribute only one-half of its genes to the next generation. The success of that contribution depends upon a member of the opposite sex. For that reason each individual must acquire the best possible mate.

TYPES OF SEXUAL REPRODUCTION

Sexual reproduction can take a variety of forms. The most familiar is separate male and female individuals. Plants with that characteristic are called *dioecious*, and examples are holly trees and stinging nettle *(Urtica)*. The equivalent term for animals is *gonochorustic*. Some organisms may possess both male and female sex organs. They may be *monoecious* or *hermaphroditic*. There is a difference. Hermaphrodites have both male and female organs in the same individual. Among plants that means flowers contain both stamens and ovules. Among animals individuals possess both testes and ovaries, a condition common to a number of invertebrates such as earthworms. Monoecy is restricted to plants, in which the individual possesses separate male and female flowers, such as birch *(Betula)* and aspen *(Populus)* trees. Some hermaphrodites are simultaneous or sequential. The latter type are one sex when young and develop into the opposite sex when mature.

Such sex reversal seems to be stimulated by a social change involving sex ratio in the population. Sex reversal among several species of marine fish can be initiated by the removal of one or more individuals of the other sex. Among some coral fish, removal of females from a social group stimulates an equal number of males to change sex and become females (Friche and Friche, 1977). Among other species, removal of males stimulates a one-to-one replacement of males by sex-reversing females (D. Y. Shapiro, 1980).

Plants also exhibit a gender change (Freeman, Harper, and Charnov, 1980). One such plant is jack-in-the-pulpit, *Arisaema triphyllum* (Doust and Cavers, 1982), a clonal woodland herb whose genet (genetic individual) is a perennial corm and whose ramet is a single annual shoot. Jack-in-the pulpit produces staminate (male) flowers one year, an asexual vegetative shoot the next, and a carpellate (female) or monoecious shoot the next. Over its life span a jack-in-the pulpit may produce both genders as well as a nonsexual vegetative shoot, but in no particular sequence. Usually, an asexual stage follows a gender change. Gender changes in some plants are normally stimulated by some environmental changes in moisture and light. But gender change in jack-in-the-pulpit appears to be triggered by an excessive drain on its photosynthate by female flowers. If the plant is to survive, one carpellate flowering could not follow another. To avoid death, the plant reduces its reproductive effort the next year by changing its gender or becoming vegetative.

Mating in hermaphroditic and monoecious species can involve either outcrossing, in which genetic variability is assured, or self-fertilization, which involves no genetic change. Apparently, most hermaphroditic animals are not self-fertilized. Earthworms mate with other individual earthworms, for example. Some animal hermaphrodites are completely self-sterile. A few, like certain land snails, are self-fertilized. Many hermaphroditic plants are self-compatible but have evolved different means to prevent self-fertilization. Anthers and pistils may mature at different times or the pistil may extend well above the stamens. Other plants have evolved more effective means. One is a genetic mechanism that prevents the growth of a pollen tube down the style of the same individual. Other hermaphroditic species may be divided into two or three morphologically different types. Pollination can take place only between members of different types.

Although many hermaphroditic organisms possess mechanisms to reduce self-fertilization, the capability of self-fertilization in hermaphrodites does carry certain advantages, especially among plants. A single self-fertilized individual is able to colonize a new habitat and then reproduce itself, establishing a new population. Other hermaphrodites produce self-fertilized flowers under stressed conditions, ensuring a new generation. Jewelweed, for example, under normal and optimal environmental conditions produces cross-pollinated *(chistogamous)* flowers. But under adverse environmental conditions or after the chistogamous flowers have set seed, jewelweed (as well as violets and other species) produces tiny, self-fertilized *(cleistogamous)* flowers that never open. They have vestigial petals, no nectar, and few pollen grains and remain in a green, budlike stage. Thus, if outcrossing fails or never develops, the plants have ensured a next generation by self-fertilization.

SEXUAL SELECTION

Possessing an optimal reproductive strategy is of little consequence unless the organism has some means not only of bringing the sexes together but also of enabling individuals to choose a partner with the greatest possible fitness. That involves some form of *sexual selection*. Sexual selection consists of members of one sex competing among themselves to mate with members of the other sex (intrasex competition) and members of the other sex (usually female) showing preference for those that win. Supposedly, males (and in some situations females) that win the intrasex competition are the fittest, and by selecting a mate from among those males, females ensure their own fitness.

The problem is, how does a female select a mate with the greatest fitness? That is still a largely unanswered question. Actually, the female has little to go on. She might select a winner from among males that bested others in combat—as in bighorn sheep, elk, and seals—or ritualized display. Or females may select mates based on intensity of courtship display. Whatever the situation, the selection process comes down to salesmanship on the part of the male and sales resistance on the part of the female (G. C. Williams, 1975). The female may attempt to elicit as much courtship behavior as possible from a potential mate. Females, for example, may force males to display in groups, as in prairie chickens and in some frogs and toads (see K. D. Wells, 1977; Howard, 1978a). They usually choose the larger and probably older males. In those species in which males monopolize a group of females such as elk and seals, the females accept the dominant male. But even in that situation the females are not completely placid and do have some choice. Protestations by female elephant seals over the attention of a dominant male may attract other large males nearby who may attempt to dislodge the male from the group. Such behavior ensures that the females will mate only with the highest-ranking male (Cox and Le Boeuf, 1977).

Often females choose males who offer the highest-quality territories (for example, see Howard, 1978b; Searcy, 1979; J. L. Zimmerman, 1971). In that situation the question is whether the female selects the male and accepts the territory that goes with him or whether she selects the territory and accepts the male that goes with it. By choosing a male with a high-quality territory, the female can best assure her own fitness.

Whatever the case, the ultimate strategy for both male and female is to assure their own maximum fitness. But what increases male fitness is not necessarily what improves female fitness. Sperm are cheap. With little investment involved, males should mate with as many females as possible to achieve maximum fitness. Because females invest considerably more in reproduction, it is to their advantage to be more selective in choosing a mate. In general, they should refuse to mate with a promiscuous male or one with several mates. Instead, they should favor one-to-one relationships in which the male helps to care for the young. For males such a relationship carries certain risks. The male must guard the female from other males to ensure that the offspring he helps to raise are really his own. Females are always sure that the offspring they raise are their own.

Sexual selection is considered mostly an attribute of animals. Although Darwin recognized such selection functioning in plants, one might have difficulty imagining how plants would engage in sexual selection. Mostly, sexual selection in plants is associated with the evolution of dioecy from largely hermaphroditic plants (see Willson, 1979; Bawa, 1980; Ross, 1982; and especially Willson and Burley, 1983). It may involve largely intrasexual competition among hermaphroditic flowers for the dispersal of pollen. Selective advantage would go to those with the larger flowers that provided the most pollen and nectar. The more energy these flowers put into pollen, the less energy is available for seed production by the female component. If some hermaphrodites carry a mutant gene for partial female sterility, all their energy could be directed toward pollen production. Such flowers would have a selective advantage over normal hermaphrodites. At the same time, male-sterile flowers would not waste energy on pollen production. They would have a selective advantage over normal and female-sterile hermaphrodites. They could channel most of their resources to seed production and would not have to contend with their own pollen. Ultimately, that type of sexual selection could lead to the evolution of dioecy.

Sexual selection in both plants and animals involves major selective processes acting on both sexes during the whole reproductive process. The role of intrasexual competition among males, however, is more easily observed and better understood than female sexual selection. We still do not know what criteria females use to make the very important choice of the fittest male. And we still have difficulty separating intramale competition for mates from female selection of males.

MATING SYSTEMS

The behavioral mechanisms and the nature of the social organization involved in an organism's obtaining a mate are called a *mating system*. A mating system includes such aspects as the number of mates acquired, the manner in which they are acquired, the nature of the pair bond, and the pattern of parental care provided by each sex. The various mating systems result in a different fraction of the male population acquiring mates. Mating, as already described, becomes competitive, and selection works against those males deprived of mates.

The structure of mating systems runs from monogamy through many variations of polygamy. *Monogamy* is the formation of a pair bond between one male and one female, and both parents typically care for the young. Monogamy is prevalent among insectivorous birds; carnivorous birds and mammals; certain herbivorous mammals, such as beaver and muskrats; gibbons; and humans. *Polygamy* is the acquisition by an individual of two or more mates, none of which is mated to other individuals. A pair bond exists between the individual and each mate. There are two kinds of polygamy. One is

TABLE 14.1 AN ECOLOGICAL CLASSIFICATION OF MATING SYSTEMS

Monogamy	Neither sex has opportunity of monopolizing additional members of the opposite sex. Fitness often maximized through shared parental care.
Polygyny	Individual males frequently control or gain access to multiple females.
Resource defense polygyny	Males control access to females *indirectly*, by monopolizing critical resources.
Female (or harem) defense polygyny	Males control access to females *directly*, usually by virtue of female gregariousness.
Male dominance polygyny	Mates or critical resources are *not economically monopolizable*. Males aggregate during the breeding season and *females select mates* from these aggregations.
Explosive breeding assemblages	Both sexes converge for a short-lived, highly synchronized mating period. The operational sex ratio is close to unity and sexual selection is minimal.
Leks	Females are less synchronized, and males remain sexually active for the duration of the females' breeding period. Males compete directly for dominant status or position within stable assemblages. Variance in reproductive success and skew in operational sex ratio reach extremes.
Rapid multiple-clutch polygamy	Both sexes have substantial but relatively *equal* opportunity for increasing fitness through multiple breedings in rapid succession. Males and females each incubate separate clutches of eggs.
Polyandry	Individual females frequently control or gain access to multiple males.
Resource defense polyandry	Females control access to males *indirectly*, by monopolizing critical resources.
Female access polyandry	Females do not defend resources essential to males but, through interactions among themselves, may limit access to males. Among phalaropes, both sexes converge repeatedly at ephemeral feeding areas where courtship and mating occur. The mating system most closely resembles an explosive breeding assemblage in which the operational sex ratio may become skewed with an excess of females.

SOURCE: Emlen and Oring, 1977: 217.

polygyny, in which an individual male gains control of or access to two or more females. The other is the less common *polyandry,* in which a female gains control of or access to two or more males. Mating systems in which one male has several short pair bonds with different females in sequence, as in grouse and white-tailed deer, are called *polybrachygyny.* Finally, there is *promiscuity,* in which males and females copulate with one or many of the opposite sex but form no pair bonds. True promiscuity is rare or nonexistent in the natural world. The various types of polygamous relationships, of which there are a number, have been classified by Emlen and Oring (1977) according to the means by which individuals gain access to the limited sex, male or female (Table 14.1).

Summary

Life-history strategy includes all the selective processes used by an organism to achieve optimal fitness. These processes include, among others, fecundity, survival, mode of reproduction, age at reproduction, parental care, number of young produced, and time to maturity—all related to the amount of energy allocated to reproduction.

To achieve optimal fitness, an organism has to balance immediate reproductive efforts against the costs of future prospects. At one extreme is semelparity, in which organisms expend all their energy in one lethal reproductive effort. At the other end is iteroparity, in which organisms produce fewer young at one time but have repeated reproduction through a lifetime.

The number of young produced relates to the amount of parental investment. Organisms that produce a large number of offspring have a minimal investment in each offspring. They can afford to send a large number into the world with a chance that a few will survive. By so doing, they increase parental fitness but decrease the fitness of the young. Organisms that produce few young invest considerably more in each individual, providing them with a greater ability to survive. Such organisms increase the fitness of the young at the expense of the fitness of the parents.

A large number of young is characteristic of annual plants, short-lived mammals, insects, and semelparous species. Few young are characteristic of long-lived species. Iteroparous species may adjust the number of young in response to environmental conditions and the availability of resources. In general, clutch and litter sizes increase from the tropics toward the arctic.

A direct relationship between size and fecundity exists among plants and poikilotherms. The larger the size, the more young are produced. Mammals exhibit a relationship between fecundity and metabolism. Mammals with a high rate of metabolism have high fecundity. Because basal metabolism decreases with size, small mammals have both high metabolism and a higher fecundity.

A review of reproductive strategies suggests two types of selection: r-selection and K-selection. r-selection favors genotypes with a high rate of increase, rapid development, early maturity, small body size, short life span, large numbers of offspring, minimal parental investment in individual young, and minimal parental care. K-selection favors genotypes that tolerate high population densities, delayed reproduction, slow development, large body size, relatively long life, fewer offspring, large parental investment in individual young, parental care in animals, and large seeds with stored food in plants.

Life-history strategies also involve mating strategies, associated with sexual reproduction. Sexual reproduction commonly involves individual males and individual females. Plants with separate males and females are called dioecious. Organisms possessing both male and female sex organs may be monoecious or hermaphroditic. Hermaphrodites have both male and female organs in the same individual or flower. Monoecy, restricted to plants, involves separate male and female flowers on the same individual.

An important component of mating strategy is sexual selection. In general, males compete among themselves for the opportunity to mate with the other sex, and females show a strong preference for the winner. Supposedly, males that win intrasexual competition are the most fit. By choosing among those males, females ensure their own fitness. There is some evidence for sexual selection among plants. It appears to be associated with the evolution of dioecy from largely male-dominated hermaphroditic flowers for the dispersal of pollen.

Review and Study Questions

1. What is meant by energy allocation? How does it affect reproduction and growth?
2. Compare semelparity with iteroparity. What environmental conditions favor semelparity?
3. Discuss the range of iteroparity from a few young at a late age to many small litters at an early age. What are the advantages and disadvantages of reproduction at an early age?
4. Contrast parental investment in many young and few young. Under what conditions should each of the investments be made?
5. Compare altricial and precocial young. How do parental investments in each differ?
6. How might plants and animals apportion energy to reproduction according to resource availability?
7. What is the relationship between size and fecundity in plants and poikilothermic animals? Does the size relationship hold among homoiotherms, especially mammals? Explain.
8. What is the relationship between fecundity and metabolism?
9. What is the difference between r-selected and K-selected organisms?
10. What are some of the major weaknesses of the above concept? In what situations is the concept most useful?
11. What is dioecy? What is the difference between a monoecious and a hermaphroditic organism?
12. What are some selective advantages of hermaphroditism?
13. What is sexual selection? In what ways does selection differ between males and females?
14. How might sexual selection evolve in plants? Expand the topic beyond what is presented in the chapter. See Willson, 1983; Willson and Burley, 1983.
15. Define a mating system and distinguish among monogamy, polygamy, polygyny, and polyandry.

Suggested Readings

Bajema, C. J., ed. 1984. *Evolution by sexual selection theory.* Benchmark Papers in Systemic and Evolutionary Biology. New York: Scientific and Academic Editions.

Gubernich, D. J., and P. H. Klopfer, eds. 1981. *Parental care in mammals.* New York: Plenum.

Hrdy, S. B. 1981. *The woman that never evolved.* Cambridge, Mass.: Harvard University Press.

Krebs, J. R., and N. B. Davies, eds. 1978. *Behavioral ecology: An evolutionary approach.* Oxford, England: Blackwell Scientific Publications.

Thornhill, R., and J. Alcock. 1983. *The evolution of insect mating systems.* Cambridge, Mass.: Harvard University Press.

Wasser, S. K., ed. 1983. *Social behavior of female vertebrates.* New York: Academic Press.

Willson, M. F. 1983. *Plant reproductive ecology.* New York: Wiley.

Willson, M. F., and N. Burley. 1983. *Mate choice in plants: Tactics, mechanisms, and consequences.* Princeton, N.J.: Princeton University Press.

Wilson, E. O. 1980. *Sociobiology: The abridged edition.* Cambridge, Mass.: Harvard University Press.

CHAPTER 15

Interspecific Competition

Outline

Objectives

Upon completion of this chapter, you should be able to:

1. Describe the various types of interrelationships among populations.
2. Define commensalism, amensalism, allelopathy, protocooperation, mutualism, symbiosis, predation, parasitism, and interspecific competition.
3. Contrast interference and exploitative types of competition.
4. Describe four theoretical outcomes of interspecific competition.
5. Explain competitive exclusion.
6. Explain how potentially competing species may coexist.
7. Explain resource partitioning.
8. Define the niche and distinguish between a fundamental niche and a realized niche.
9. Explain niche overlap, niche width, niche compression, niche shift, and ecological release.

Although the most intense relationships exist between individuals of the same species, they do not live apart from individuals of other species. Living in close association with each other, different species experience some sort of interspecific relationship. They may compete with one another for some shared resource such as food, light, space, and moisture. One may depend upon the other as a source of food. Others may in some way mutually aid each other. Such interrelations can be beneficial or detrimental to either or both species involved, or they may have no effect at all.

Types of Interrelations

If one designates a positive effect of one species on another as a +, a detrimental effect as a −, and no effect as a 0, one can arrive at six different ways in which a population may interact (Table 15.1). When neither of the two populations affects the other, the relationship is (00),or neutral. If the two populations mutually benefit each other, the interaction is (++), or positive. And if the relationship is mutually detrimental, it is negative (− −). The remaining three interactions are (0+), (−0), and (− +).

When one species maintains or provides a condition necessary for the welfare of another but does not affect its own well-being by doing so, the interaction (0+) is called *commensalism*. An example is an epiphytic plant growing on the trunk or limb of a tree. The tree provides support only, and the epiphyte gets its nutrients through its aerial roots. The interaction (0−), in which one species reduces or adversely affects the population of another but remains unaffected itself, is called *amensalism*. An example may be the release of a toxic substance by one organism that inhibits the growth or survival of another. The effects of juglone released into the soil by black walnut suppress the growth of heaths and other plants beneath the tree. The release of such chemical substances is called *allelopathy*. Amensalism is difficult to demonstrate in the laboratory, and its role in nature is questioned. It may be considered a form of competition.

TABLE 15.1 INTERSPECIFIC RELATIONSHIPS

	+	0	−
+	+ +	+ 0	− +
0	0 +	0 0	0 −
−	− +	− 0	− −

\+ + Both increases from the interrelationship: *mutualism* and *protocooperation*.

\+ 0 Fitness of one increases; the other remains neutral: *commensalism*.

− + Fitness of one is reduced; fitness of the other is increased: *predation* and *parasitism*.

0 − Fitness of one declines; fitness of the other is unchanged: *amensalism*.

− − Fitness of both declines: *competition*.

0 0 No influence on each other: *neutralism*.

Perhaps ecologically more important than amensalism and commensalism is that relationship which benefits both populations (++). If the relationship is not essential for the survival of the population, then the relationship is nonobligatory and is called *protocooperation*. If the relationship is obligatory, then both populations are dependent in some way on the other, and the association is called *mutualism*.

Mutualism and protocooperation and sometimes commensalism are collectively termed *symbiosis,* a word of Greek origin, meaning simply "living together." The term was introduced by the German botanist A. de Bary in 1897 as the German word *symbiose*. He used the word to refer to both beneficial and adverse effects resulting from such a relationship between different organisms, including parasites and their hosts. However, symbiosis can be equated with relationships resulting in mutual benefits to both parties involved. Equating symbiosis with mutualism probably resulted in part from de Bary's use of lichen-algae partnerships as examples of symbiosis. Today the use of the word *symbiosis* is confused (Lewin, 1982; Goff, 1982). Its uses range from its historical meaning (embracing all situations in which different organisms live together in close association, including parasitism) to situations best described as obligatory mutualism.

Strong suggestions have been made to use the term in its original historical meaning to avoid confusion. Part of the argument for such broad usage is the lack of any clear-cut distinction on the continuum of relationships from parasitism to mutualism. In fact, the lichen-algae associations, once the standard example of mutualism, turn out to be controlled parasitism, in which the fungus is actually an obligatory parasite of the alga (Ahmadjian and Jacobs, 1982).

While it is easier to cite examples of commensalism and protocooperation, proving an association to be mutualistic is much more difficult. An observed relationship, as the fungal-algal association demonstrates, may not be what it seems. Rather than accept the nature of symbiotic relationships on observational evidence, suggestions have been made to explore the relationships between different associated organisms both qualitatively and quantitatively, including spatial and temporal relationships, the degree of dependence, the mode of dependence (physical, nutritional, physiological, and so on), and the effect of one organism on another (Lewin, 1982; Goff, 1982; Whitfield, 1979).

Two other relationships that have been of greatest interest and importance to population ecologists are competition (− −), which is detrimental to populations of both species—each inhibits the other—and predation and parasitism (− +), in which the population of one species benefits at the expense of another. Predation involves the killing and consumption of prey. Parasitism involves one organism's feeding on another, where the prey or host is rarely killed outright. The two, parasite and host, live together for some time. The host survives, although its fitness is reduced. If it does succumb, the host does so because in a weakened condition it is unable to resist other infections. It is not to a parasite's selective advantage to kill its host, for by doing so, it eliminates its own habitat and energy source. Parasites that eventually cause the death of a host are *parasitoids*. Examples are certain wasps and flies that lay their eggs in or on the host. Their larvae develop inside the host and eventually kill it. Their effect is much like that of a predator that needs to kill only a single prey item in its life cycle.

Interspecific Competition

The relationship in which the populations of both associated species are affected adversely (− −) is *interspecific competition*. Interspecific competition, like intraspecific competition, involves the seeking of a resource in short supply, but two or more species are involved. Both types of competition may take place simultaneously. Gray squirrels, for example, may be competing among themselves for acorns during a poor crop year. At the same time, white-footed mice, white-tailed deer, wild turkeys, and blue jays may be vying for the same crop. Because of competition, individuals within a species may be forced to broaden the base of their foraging efforts, and populations of various species may be forced to turn away from acorns to reduce their food base relative to other species and concentrate on food more efficiently obtained. Thus, intraspecific competition selects for a broadening of the resource base, or generalization, while interspecific competition favors a reduction of the overall resource base and more specialization—points to be explored later.

Like intraspecific competition, interspecific competition comes in two forms, *interference* and *exploitation*. Interference competition involves direct or aggressive interactions between competitors. Exploitative competition depletes the resource to a level where it is of little value to each population.

The concept of interspecific competition is one of the cornerstones of evolutionary ecology. Competition, the struggle to survive, the survival of the fittest, was one of the foundations on which Darwin based his idea of natural selection. Because it is advantageous for a species to avoid it, competition has been regarded as the major force behind species divergence and specialization. Although the concept had a strong influence on ecological thinking, it is one of the least known and more controversial areas of ecology.

In the early part of the twentieth century, two mathematicians, Alfred Lotka (1925) and Vittora Volterra (1926, 1931), independently arrived at mathematical expressions to describe the relationship between two species utilizing the same resource. The equations are modifications of the logistic growth equation (Chapter 12), one for each species. To these pairs of logistic equations Lotka and Volterra added competition coefficients, constants that describe the degree to which species 1 inhibits the growth of species 2 and species 2 inhibits the growth of species 1. These equations consider both intra- and interspecific competition:

$$\frac{dN_1}{dt} = r_1N_1\left(\frac{K_1 - N_1 - \alpha_{12}N_2}{K_1}\right)$$

$$\frac{dN_2}{dt} = r_2N_2\left(\frac{K_2 - N_2 - \alpha_{21}N_1}{K_2}\right)$$

where K_1 and K_2 are the carrying capacities for species 1 and species 2, respectively; α_{12} is a constant representing the inhibitory effect of species 2 on species 1; and α_{21} is the inhibitory effect of species 1 on species 2. Initially, each species grows at its own intrinsic rate of increase, and each experiences intraspecific competition. In the presence of interspecific competition, the outcome is influenced by the relative values of K_1, K_2, α_{12}, and α_{21}. In the absence of either species, no interspecific competition exists, and population growth is regulated by intraspecific competition. But in the presence of competing species, species 1 decreases the carrying capacity, K, for species 2 at a certain rate, and species 2 decreases the carrying capacity for species 1 at a certain rate, because each has to share limited resources with the other.

OUTCOMES OF COMPETITION

Depending upon the combination of values for the Ks and the *alpha*s, the Lotka-Volterra equations predict four different potential outcomes. In two situations one species wins out over the other. In one case species 1 inhibits further increase in species 2 while it can still increase itself, and species 2 dies out. In the other case species 2 inhibits further increase in species 1 while still continuing to increase itself, and species 1 eventually disappears. In the third situation each species when abundant inhibits the growth of other species more than it inhibits its own growth. The outcome depends upon which species initially is numerically the most abundant. The two species may coexist for a while in an unstable equilibrium. In real-life situations the outcome may depend upon which of the two species has the competitive advantage in the face of environmental change over time. In the fourth situation the two species achieve a stable equilibrium and coexist. In that situation neither population reaches carrying capacity, K, and individuals of each species inhibit their own population growth more than they inhibit that of the other species.

EVIDENCE FROM THE LABORATORY

The theoretical Lotka-Volterra equations stimulated studies of competition in the laboratory, where under controlled conditions the outcome is more easily determined. One of the first to study competition experimentally was the Russian biologist G. F. Gause (1934). In one experiment he introduced two species, *Paramecium aurelia* and *P. caudatum,* together in a tube containing a fixed amount of bacterial food. *P. caudatum* died out. The success of *P. aurelia* resulted from its higher rate of increase and its ability to survive at a high density (Figure 15.1).

In another experiment Gause introduced *P. caudatum* in the same solution with another species, *P. bursaria.* In that experiment both species were able to reach stability because *P. bursaria* confined its feeding to bacteria on the bottom, and the other organism fed on bacteria suspended in solution. Although the two species used the same food supply, they occupied different parts of the culture, with each species thus utilizing food that essentially was unavailable to the other.

Thomas Park (1948, 1954) studied the competitive relationships between two laboratory populations of flour beetles, *Tribolium castaneum* and *T. confusum.* They found that the outcome of competition depended considerably on environmental conditions and humidity and on fluctuations in the total number of eggs, larvae, pupae, and adults. Often the final outcome of competition was not determined for generations. In general, however, *T. castaneum* usually won out over *T. confusum.* But if the two competitors were exposed to a sporozoan parasite, *Adelina tribolium, T. confusum* usually won.

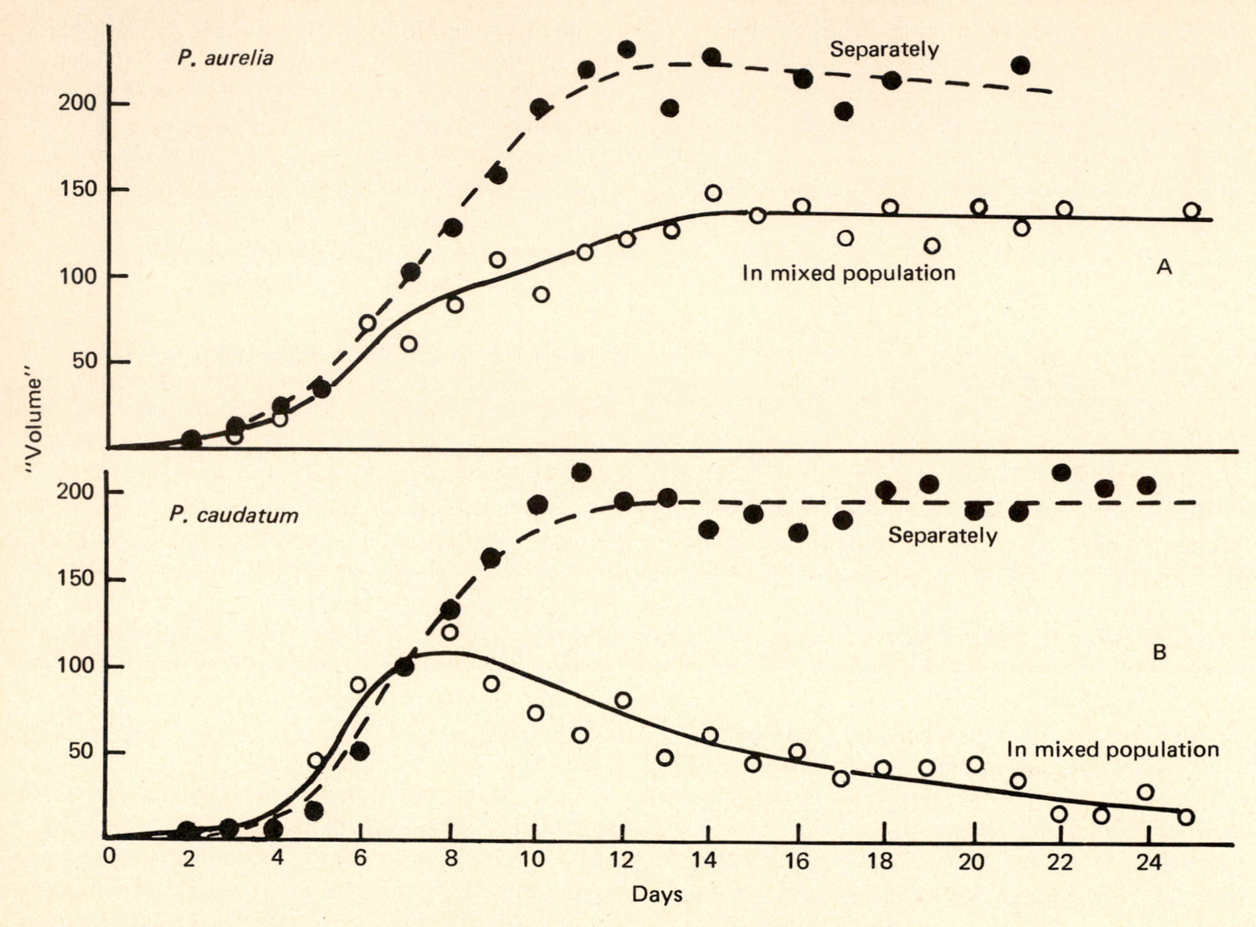

FIGURE 15.1 Competition experiments with two species of *Paramecium*. The graphs show the growth of two related ciliated protozoans, *P. aurelia* and *P. caudatum,* when grown separately and when grown in a mixed culture. In the mixed culture *P. aurelia* outcompetes *P. caudatum,* and the result is competitive exclusion. (From Gause, 1934:98.)

The British plant ecologist J. L. Harper (1961) performed similar experiments with four species of duckweed *(Lemna)*. When grown alone, each of the four species had different growth rates under crowded and uncrowded conditions. For example, when grown alone, *L. polyrrhiza* had a higher rate of population growth than *L. gibba*. When the two species were grown together, *L. gibba* excluded *L. polyrrhiza* (Figure 15.2).

In these laboratory experiments the patterns were similar. The two species involved grew exponentially at low densities. As their densities increased, their actual rates of increase, and thus their population growth, declined, and each began to have an influence on the other. Under these conditions one species had a greater fitness than the other. A point was reached where the increase of one species became zero, while the other population continued to grow. As it grew, its competitive influence increased, and the rate of increase for the other species became negative.

Competitive Exclusion

In three of the four situations predicted by the Lotka-Volterra equations, one species wins and the other loses, eventually driven to extinction, pro-

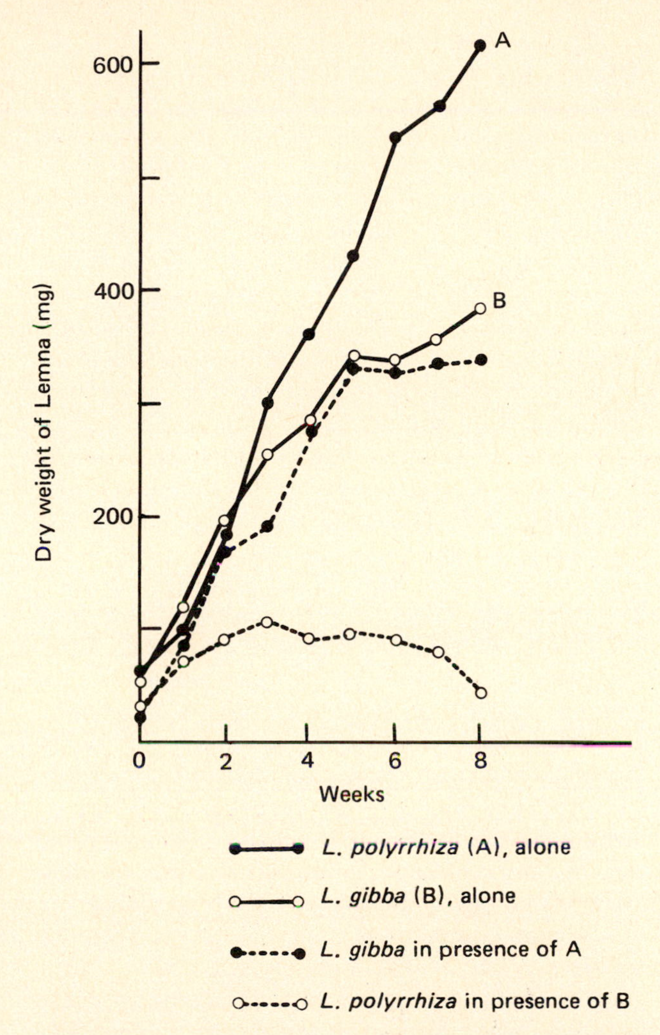

FIGURE 15.2 Competition experiments with species of duckweed *(Lemna)*. In this experiment, populations of two species were grown by J. L. Harper. When grown alone, *L. polyrrhiza* and *L. gibba* exhibit somewhat similar rates of growth, although *L. polyrrhiza* attains greater population levels. When grown in a mixed culture, *L. gibba* shows little difference in growth rates, but *L. polyrrhiza* in the presence of *L. gibba* is eliminated. (After Harper, 1961.)

vided the interaction proceeds to saturation level, or *K*. Evidence provided by laboratory experiments tends to support the mathematical models. In competitive situations if one species produces enough individuals to prevent the population increase of another, it can reduce that population to extinction or exclude it from the area.

These observations led to the concept called Gause's principle, although the idea was far from original with him. More recently, the concept has been called the *competitive exclusion principle* (Hardin, 1960), which can be stated as "Complete competitors cannot coexist." Basically, that principle means that if two noninterbreeding populations possess exactly the same ecological requirements, and if they live in exactly the same place, and if population A increases the least bit faster than population B, then A will eventually occupy the area completely and B will become extinct.

The competitive exclusion principle is little more than an ecological definition of a species (see Chapter 3). A corollary of the principle is that if two species coexist, they must have different ecological requirements. Obviously, two separate species cannot have identical requirements; being different species, they use the environment differently. Two or more species, however, can compete for essential resources without being complete competitors. The competitive exclusion principle has other problems. It assumes that competitors remain genetically unchanged long enough to exclude the other species, that immigrants from other areas with different conditions do not move into the population of the losing species, and that environmental conditions remain constant (Pielou, 1974). The idea, however, has stimulated a more critical look at competitive relationships in natural situations, including studies to determine what ecological conditions are necessary for coexistence among species.

EVIDENCE FROM THE FIELD

Demonstrating interspecific competition in laboratory populations is one thing; demonstrating competition under natural conditions is another. In the field one has no control over the environment; one has difficulty knowing whether the populations involved are at or below carrying capacity (or saturation); one lacks full knowledge of the life-history requirements or the subtle differences between the species involved. Further, the competitive outcome among associated species may have been settled through evolutionary time. What one observes in the field may be the results of competition in the past rather than competition itself.

Competitive Exclusion

Few clear-cut examples of competitive exclusion have been experimentally demonstrated. One of the best is the timeworn example of the two species of barnacles *Chthamalus stellatus* and *Balanus balanoides* on the Scottish coast, studied by Joseph Connell (1961). *Chthamalus* typically lives on the higher part of the intertidal zone, and *Balanus* lives on the lower part. By clearing experimental patches of rock of all barnacles, Connell was able to demonstrate that free-swimming planktonic larvae of both forms were able to attach themselves to rocks at any point in the intertidal zone and develop into sessile adults. In the lower zone the faster-growing *Balanus* crowds off the rock any *Chthamalus* growing there or else grows over them. *Balanus,* however, is not as resistant to dessication as *Chthamalus,* which is tolerant of long periods of exposure to air and sun. Because the upper intertidal zone is exposed the longest to air and sun, *Balanus* is excluded from the upper zone by physical dessication. *Chthamalus* is able to exist there free from competition. Thus, *Chthamalus* is excluded from the lower zone by competition from *Balanus,* and *Balanus* is excluded from the upper zone by its own physiological intolerance of environmental conditions.

Another example of exclusion brought about by physiological tolerance, aggressive behavior, and restriction to habitats in which one organism has the competitive advantage over the other involves small mammals. On the eastern slope of the Sierra Nevadas live four species of chipmunks: the alpine chipmunk *(Eutamias alpinus),* the lodgepole pine chipmunk *(E. speciosus),* the yellow pine chipmunk *(E. amoenus),* and the least chipmunk *(E. minimus),* at least three of which have strongly overlapping food requirements.

Each species occupies a different altitudinal zone (Figure 15.3). The line of contact is determined partly by interspecific aggression (Heller, 1971; Sheppard, 1971; Chappel, 1978). The upper range of the least chipmunk is determined by aggressive exclusion from the dominant yellow pine chipmunk. Although the least chipmunk is capable of occupying a full range of habitats from sagebrush desert to alpine fell fields, it is restricted in the Sierras to sagebrush habitat. It is physiologically more capable of handling heat stress than the others, enabling it to inhabit extremely hot, dry sagebrush. If the yellow pine chipmunk is removed from its habitat, the least chipmunk moves into the vacated open pine woods; but if the least chipmunk is removed from the sagebrush, the yellow pine chipmunk does not invade the habitat. The aggressive behavior of the lodgepole pine chipmunk in turn determines the upper limit of the yellow pine chipmunk. The lodgepole pine chipmunk is restricted to shaded forest habitat because it is the most vulnerable to heat stress. Most aggressive of the three, the lodgepole pine chipmunk may limit the downslope range of the alpine chipmunk. Thus, the range of one species, the least chipmunk (and possibly two if you include the alpine chipmunk), like that of the barnacles *Chthamalus,* is determined through aggressive exclusion by another species and its ability to survive and reproduce on habitat hostile to the other species. The other two species are restricted to certain habitats by their own physiology and by aggressive exclusion.

Under competitive situations in the past, aggressive behavior was probably selected for in the lodgepole pine and yellow pine chipmunks because of a seasonally limited, often patchy food supply that can be cached for winter use and economically defended. Aggressiveness was probably not selected for in the least chipmunk because such activity would not be metabolically feasible in the hot sagebrush desert outside the physiological range of the other species. Instead, selection was for tolerance of heat stress.

In both of these examples, competitive exclusion does not fit the theoretical framework described by the Lotka-Volterra equations. The competing species do not occupy the same habitat and increase simultaneously in it, so that exclusion results from superior fitness of one over the other, as demonstrated in laboratory populations. Instead, exclusion is determined quickly by physiological

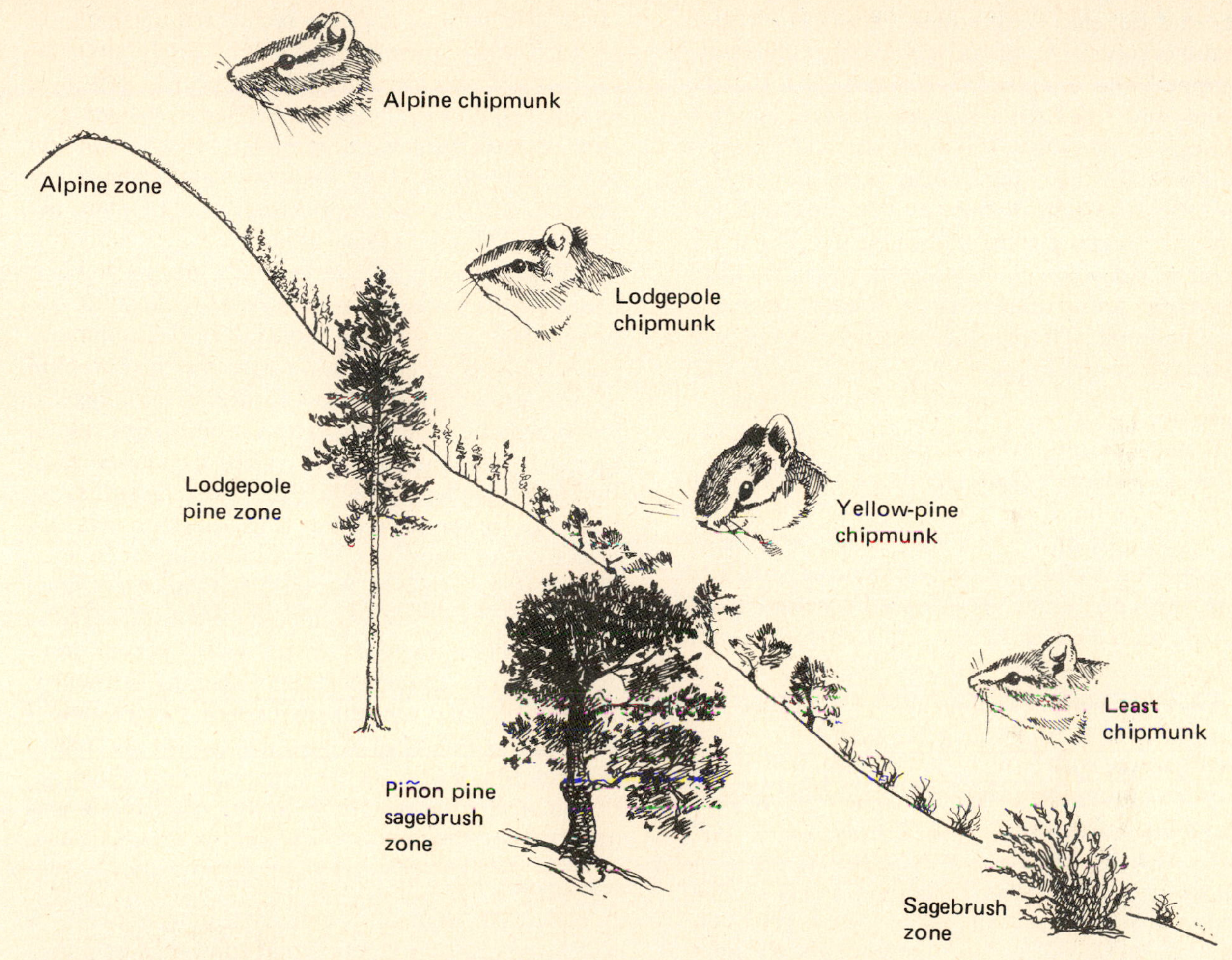

FIGURE 15.3 A transect of the Sierra Nevadas in California, latitude 38° north, showing vegetational zonation and altitudinal range of four species of chipmunks *(Eutamias)* that inhabit the east slope. (Transect after Heller and Gates, 1971:424.)

tolerances and by aggressive behavior. Both the barnacles and chipmunks are good examples of interference competition.

Unstable Equilibrium

An unstable equilibrium usually results in one species' winning over another. Under natural conditions the species that wins more often than not is determined by the environmental conditions under which the two exist at any one time.

In some ways, unstable equilibrium can be observed more easily among plants than among animals. From the time seedlings germinate and develop, demand for growing space, light, moisture, and nutrients increases. Plants that most effectively utilize resources in short supply have the best chance for survival. Species of plants whose roots are in the same horizon have to compete for limited moisture and nutrients. In western North America the shallow-rooted annual cheat grass *Bromus tectorum* grows early in spring and often reduces moisture to the point where slow-growing annuals, perennials, and even shrubs are unable to withstand the competition (Holmgren, 1956). In drier regions plants that develop roots rapidly after germination have the competitive advantage. Weaker plants are overtopped and eventually crowded out by more vigorous and aggressive species. Under optimum conditions some plants will dominate and suppress their associates. During moist years the prairie grass little bluestem dominates the grassland, but during dry years it almost disappears. Then two associates, side-oats grama and blue grama, grasses of the same genus associated with each other, become dominant. These situations are examples of exploitative competition.

Coexistence

Coexistence results when competing species are able to share resources among them, although such competition reduces the fitness of both. Among animals, especially birds, coexistence may involve interspecific territoriality, in which the area is divided among the competing species, each behaving toward the other as if both or all were the same species (see Orions and Willson, 1964).

Although coexistence resulting from interspecific competition is usually associated with closely related species with similar habitat requirements, it also exists among very different taxonomic groups. An example is the competitive relationship between rodents and ants in the desert scrub of Arizona. Both groups feed on seeds of desert plants; both overlap considerably in the size of seeds they consume. Production of those seeds is strongly influenced by desert rainfall. A strong correlation between the number of species of rodents and ants and mean annual rainfall suggests that ants and rodents may compete for the limited amount of seeds produced.

J. Brown and D. W. Davidson (1977) and D. W. Davidson (1977a, 1977b) undertook field experiments to determine to what extent rodents and ants competed for a limited seed supply. They established six experimental plots 36 m in diameter. From two of the plots they removed rodents by trapping and excluded immigrants by fencing. From another two plots they eliminated the ants by repeated applications of an insecticide. The remaining two plots, from which both ants and rodents were removed, were the controls. They discovered that ants and rodents did indeed compete with each other. Removal of rodents resulted in an increase of ants, who consumed as many seeds alone as rodents and ants combined did previously. Removal of ants had a similar effect on rodents. Where both ants and rodents were removed, the amount of seeds increased. Thus, both ants and rodents depressed each other's population and perhaps the fitness of the seed plants as well.

Observations of coexistence between plant competitors have been largely experimental. H. Leith (1960) showed that mixtures of perennial ryegrass *(Lolium perenne)* and white clover *(Trifolium repens)* growing in pastures form moving mosaics. Patches dominated by ryegrass tend to be invaded by clover and patches of clover by grass. Van Den Bergh and DeWit (1960) seeded sweet vernal grass *(Anthoxanthum odoratum)* and timothy *(Phleum pratense)* together in field plots in different proportions. They compared the ratio of tillers of the two species after the first winter with the ratio after the second winter. In plots where sweet vernal grass was in excess, the proportion of timothy increased, and where timothy was in excess, sweet vernal grass increased. That experiment points out that in some mixes of species, the species involved experience more intraspecific than interspecific competition.

Allelopathy

A particular form of competition among plants is *allelopathy,* the production and release of chemical substances by one species that inhibit the growth of other species. These substances may range from acids and bases to relatively simple organic compounds that reduce competition for nutrients, light, and space. Produced in profusion in natural communities as secondary substances, most compounds remain innocuous, but a few may influence community structure. For example, broomsedge *(Andropogon virginicus)* produces chemicals that inhibit the invasion of old fields by shrubs and thus maintains its dominance (Rice, 1972). And the allelopathic effects of goldenrods *(Solidago),* asters *(Aster),* and certain grasses prevent tree regeneration in glades (Horsley, 1977).

However, as in all competitive interactions, toxins may not be involved in plant competition to the extent hypothesized. J. L. Harper (1977) suggests that toxic interactions in higher plants may not be as common as supposed for two reasons: (1) Higher plants rapidly evolve tolerances to such environmental toxins as zinc, nickel, and copper and have developed tolerances to herbicides. (2) Complex organic molecules are broken down by the action of soil microbes, and plant toxins probably experience the same fate.

RESOURCE PARTITIONING

Observations of a number of species sharing the same habitat suggest that they coexist by partitioning available resources among them. Animals may utilize different kinds and sizes of food, feed at different times, or in different areas. Plants may occupy a different position on a soil moisture gradient, require different proportions of nutrients, have different tolerances for light and shade. Each species exploits a portion of the resources that in one way or another becomes unavailable or is unavailable to others. Such partitioning is regarded as an outcome of interspecific competition.

A Theoretical Consideration

Consider species A, which, in the absence of any competitor, utilizes a range of different-sized food items (Figure 15.4). One can picture that utilization as a bell-shaped curve on a graph, with food as the ordinate and fitness as the abscissa. Most individuals feed about the optimum. Individuals at either tail feed on larger or smaller food items. As population size increases, the range of food taken may increase as intraspecific competition forces some individuals to seek food at the two extremes. Such intraspecific competition fosters increasing genetic variability in the population.

Now allow a second species, B, to enter the area. When its resource use curve is superimposed on the curve of species A, B shows considerable overlap. Selective pressure from interspecific competition forces both species A and species B to narrow their range of resource use. Natural selection will favor those individuals living in areas of minimal or no overlap. Ultimately, the two species will narrow their ranges of resource use. They will diverge, moving to the left and the right on the graph. Direct interspecific competition will be reduced, and the two species will coexist. Thus, while intraspecific competition favors expansion of the resource base, interspecific competition narrows the range. The populations involved have to arrive at some balance between the two.

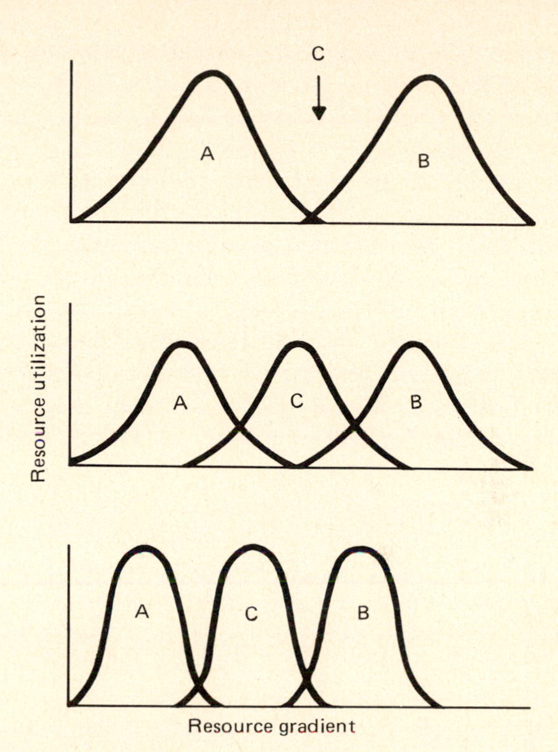

FIGURE 15.4 Theoretical resource gradient utilized by three competing species, A, B, and C. A and B share the resource gradient with minimal overlap. A third species, C, whose optimal resource utilization lies between A and B, competes with A and B. In response to selection pressures, A and B narrow their range of resource utilization to the optimum, and C utilizes that portion of the resource used at less than an optimal level by A and B.

Enter the third species, C, which utilizes a food resource in the area of overlap between A and B. Because it can make optimal use of resources in that part of the gradient, it will force A and B to further restrict the range of resources available to them. The result is a partitioning of resources along the resource gradient, which may be food size, location of food, temporal availability, and the like. Such a group of functionally similar species whose members interact strongly with one another but weakly with the remainder of the community is called a *guild* (Root, 1967).

Field Examples

Any number of intensive field studies will turn up examples of presumed resource partitioning. David Lack (1971) noted that three species of the genus *Parus* living in the broadleaf woods of Europe and Great Britain feed in different parts of the tree canopy and consume different food throughout most of the year. The blue tit *(P. caeruleus),* small in size, feeds high in the oak trees, where it gleans insects, mostly under 2 mm long, from twigs, buds, leaves, and galls. The marsh tit *(P. palustris)* feeds either in the shrubby understory or on twigs and branches below 20 feet. It consumes mostly insects about 3 to 4 mm long, as well as seeds and fruits. The great tit *(P. major)* feeds mostly on the ground, where it concentrates on insects over 6 mm long and eats acorns, hazelnuts, and seeds of wood sorrel. During the summer, when they rear their young, all three species feed heavily on caterpillars, which are in abundant supply.

Robert MacArthur (1958) observed a similar partitioning among five species of warblers inhabiting the spruce forests of the northeastern United States. Each fed in a different part of the canopy, and each was specialized behaviorally to forage in a somewhat different manner.

A similar partitioning of resources exists among plants. One example involves three species of annuals growing together in a field on prairie soil abandoned 1 year after plowing. Each plant exploits a different part of the soil resource (Figure 15.5). Bristly foxtail *(Setaria faberi)* has a fibrous, shallow root system that exploits a variable supply of moisture. It possesses the ability to recover rapidly from water stress, to take up water rapidly after a rain, and to carry on a high rate of photosynthesis even when partially wilted. Indian mallow *(Abutilon theophrasri)* has a sparsely branched taproot extending to intermediate depths where moisture is adequate during the early part of the growing season but is less available later on. That plant is able to carry on photosynthesis at a low water potential. The third species, smartweed *(Polygonium pensylvanicum),* possesses a taproot that is moderately branched in the upper soil layer and develops mostly below the rooting zone of other species, where it has a continuous supply of moisture (Wieland and Bazzaz, 1975).

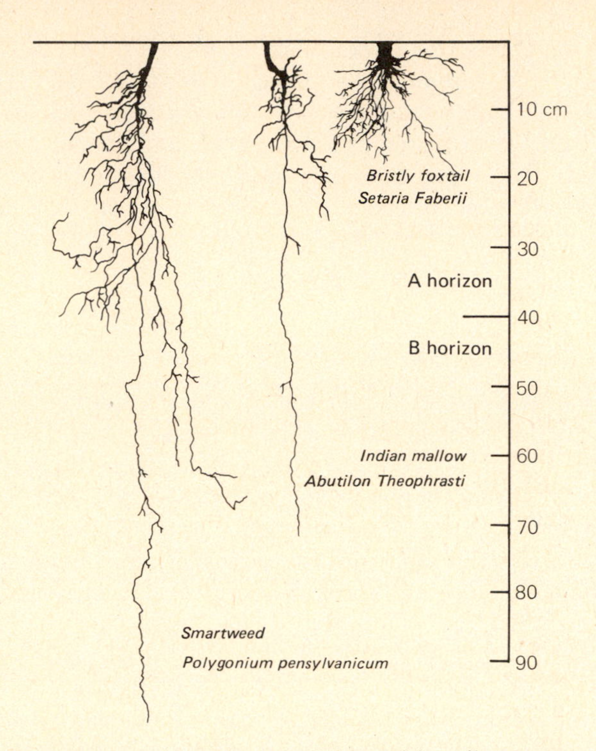

FIGURE 15.5 Partitioning of the prairie soil resource at different levels by three species of annuals in a field 1 year after disturbance. (From Wieland and Bazzaz, 1975:686.)

Such patterns of resource use are examples of interspecific competition in action. By dividing the resources in some manner, each species avoids direct competition with the other(s). Such competition may be taking place currently, or it may have occurred in the past. Such competition is also regarded as instrumental in directing the divergence of species.

Although the observed relationship among the several warblers, tits, and plants seems to fit the concept of resource partitioning and they would appear to serve as examples of interspecific competition and coexistence in action, the evidence is circumstantial. Few experiments have been carried out in the field to test interspecific competition and resource partitioning, although relationships have been demonstrated in laboratory situations (see "The Niche"). The relationship among the species observed may be the result of current competition or the outcome of competition in the past. Or the species could have adapted to the environment in different ways that have no relationship to competition.

One field example in which circumstantial evidence strongly supports interspecific competition and resource partitioning involves guilds of seed-eating rodents on three major deserts of western North America (Great Basin, Mojave, and Sonoran) (Brown, Reichman, and Davidson, 1979; Bowers and Brown, 1982). The guild in each of these deserts includes pocket mice and kangaroo rats of different body sizes (Figure 15.6). Species of similar body size occur together less frequently than would be expected on the basis of chance, and rarely do species of similar body size coexist locally. Differences in body size among four commonly associated seed-eating rodents allow each to utilize different seed sizes and to forage in different microhabitats. Behavioral differences, too, are involved. The quadrapedal pocket mouse *Perognathus penicillatus* feeds in dense vegetation and under shrub canopy, where it is protected from predators. The much swifter bipedal kangaroo rat *Dipodomys merriami* feeds on larger seeds in more open areas.

Other field experiments designed to confirm interspecific competition produce different results. W. B. Kincaid and G. N. Cameron (1982) conducted removal experiments to determine the degree of resource partitioning and interspecific competition between the cotton rat *Sigmodon hispidus* and the harvest mouse *Reithrodontomys fulvescens* in a Texas prairie during a population high. They measured food and habitat use in control areas and in experimental areas in which one or the other species was removed. The cotton rat fed mostly on grasses, the harvest mouse on insects. There was a 32 percent overlap in food utilization and an 82 percent overlap in habitat. Removal of each species resulted in changes in food utilization and habitat use no greater than would be expected by chance alone. Thus, removal of one species had no pronounced effect on the other, suggesting that interspecific competition was minimal in determining diet and habitat use by the two rodents.

The Niche

Closely associated with (in fact, almost inseparable from) the concept of interspecific competition is the concept of the niche. The *niche* is one of those terms in ecology that defies a rigorous definition. Ecologists seem to know what it means, but attempts to put that meaning down on paper satisfactorily are elusive.

The word *niche* in everyday terms means a recess in a wall where you place something, usually an ornamental object; or it is a place or position in life suitable or appropriate for a person. In ecology it means an organism's place and function in the environment. Or does it?

One of the first to propose the idea of the niche in ecology was Joseph Grinnell, a California ornithologist (1917, 1924, 1928). He suggested that the niche be regarded as a subdivision of an environment occupied by a species. Essentially, Grinnell was describing the habitat of a species. Charles Elton (1927), in his classic book *Animal Ecology,* considered the niche as the fundamental role of the organism in the community—what it does, its relationship to its food and its enemies. Basically, this concept stresses the occupational status of the species. Other definitions are variations on the same theme. One considers the habitat as the animal's address and the niche as its occupation (Odum, 1959). Another considers the niche only as a functional position (Whittaker, Levin, and Root, 1973), while the habitat and the niche combined comprise the ecotope, "the ultimate evolutionary context of the species." Another definition regards the niche as embracing all the ways in which a given individual, population, or species conforms to its environment (Pianka, 1978).

The definition that most closely links the niche to competition is the one proposed by G. E. Hutchinson (1957). It reflects and is based on the competitive exclusion principle. According to the

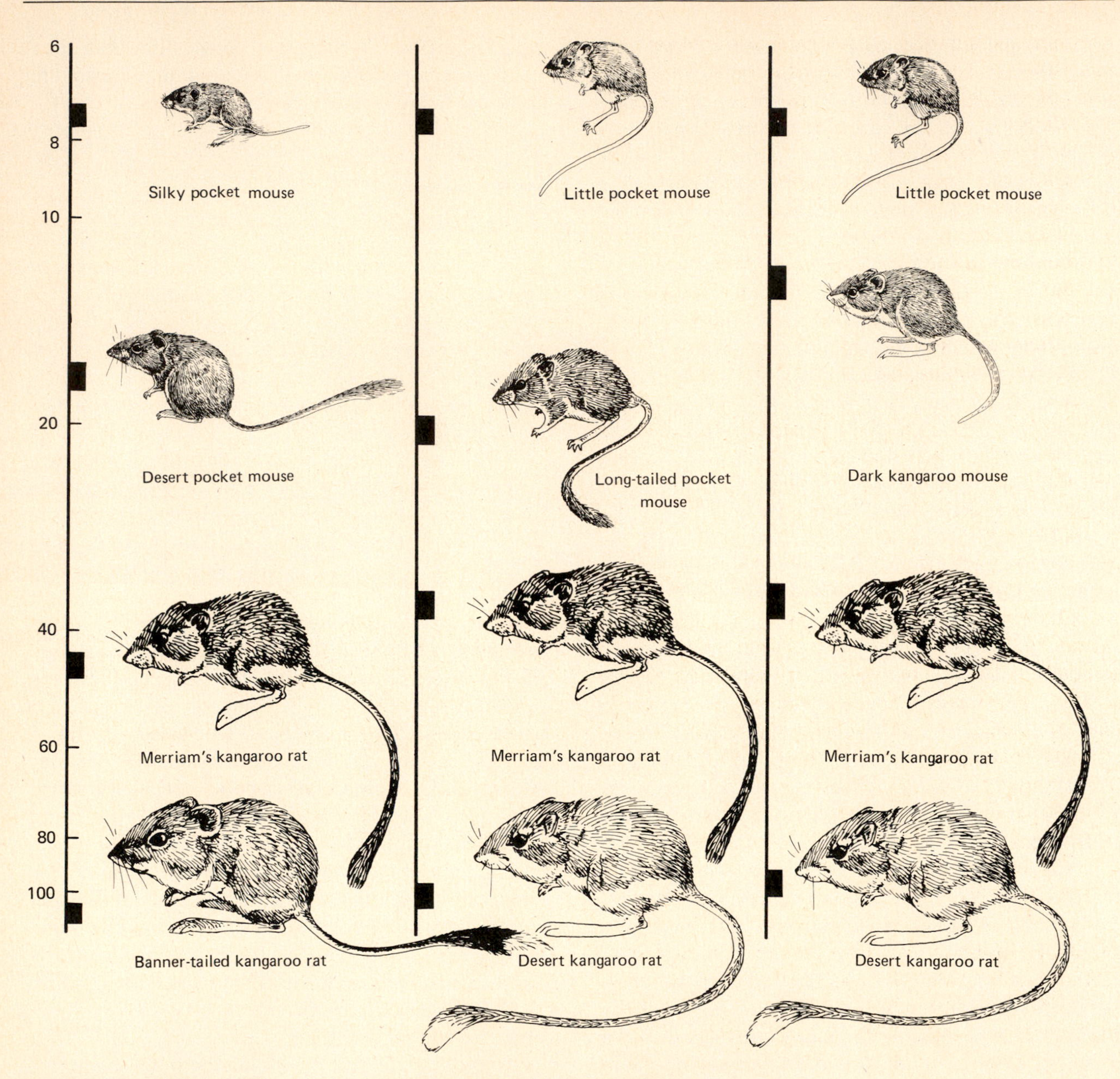

FIGURE 15.6 Niche differentiation and resource partitioning, perhaps as an outcome of interspecific competition, is illustrated by guilds of rodents found in the major desert regions of North America: the Mohave, the Sonoran, and the Great Basin. Size distribution of the four species involved is very similar, even though the identity of the species differs among the deserts. Sizes are plotted on a logarithmic scale so that equal spacing indicates equal ratios. Body mass in each community ranges from 7 to 100 g, and the minimum body size ratio between species exceeded 1.75. (Data from Bowers and Brown, 1982:396.)

Hutchinsonian concept, an organism's environment consists of many physical and biological variables, each of which can be considered as a point in a multidimensional space. Hutchinson called that space the *hypervolume*.

One can visualize the multidimensional niche to a certain extent by creating a three-dimensional one. Consider three niche-related variables for a hypothetical organism: food size, foraging height, and humidity (Figure 15.7). Suppose the animal can consume only a certain range of food size. Food size, then, is one dimension of its niche. Add to this foraging height, the area to which it is limited for seeking food. If one graphs that on a second axis and encloses the space, one has a rectangle, representing a two-dimensional niche. Suppose, too, that the animal can survive and reproduce only within a certain range of humidity. Humidity can be plotted on a third axis. By enclosing that space, one comes up with a volume, a three-dimensional niche. Of course, a number of variables, both biotic and abiotic, influence a species' or an individual's fitness. A number of these niche dimensions, *n*—difficult to visualize and impossible to graph—make up the *n*-dimensional hypervolume that would be a species' niche. An individual or a species free from the interference of another could occupy the full hypervolume or range of variables to which it is adapted. That is the idealized *fundamental niche* of a species.

The fundamental niche assumes the absence of competitors, but rarely is this the case. Competitive relationships force the species to constrict a portion of the fundamental niche it could potentially occupy. In those parts its fitness might be reduced to zero. The conditions under which an organism actually exists in any given situation are its *realized niche* (Figure 15.8), which may be further restricted by the absence of certain features of the niche at any given point in time and space. Like the fundamental niche, the realized niche is an abstraction. In their studies ecologists usually confine themselves to one or two niche dimensions, such as a feeding niche, a space niche, or a tolerance niche.

Putwain and Harper (1970) studied the population dynamics of two species of dock, *Rumex acetosa* and *R. acetosella,* each growing in hill

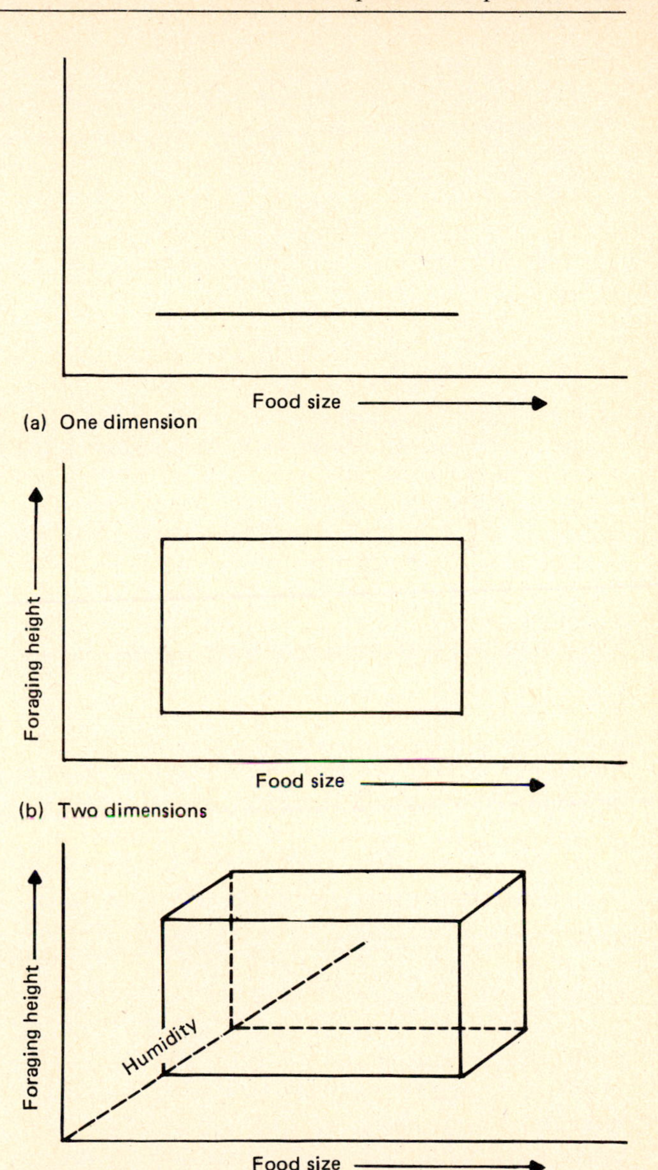

FIGURE 15.7 An illustration of niche dimension. Assume three elements comprising the hypothetical organism's niche: food size, foraging height, and humidity. Graph (a) represents a one-dimensional niche involving food size. In graph (b) a second dimension has been added, foraging height. By enclosing that space, one obtains a two-dimensional niche. Now suppose the organism can survive and reproduce only within a certain range of humidity, graphed as a third axis. By enclosing all those points, one arrives at (c) a three-dimensional niche space, or volume, for the organism.

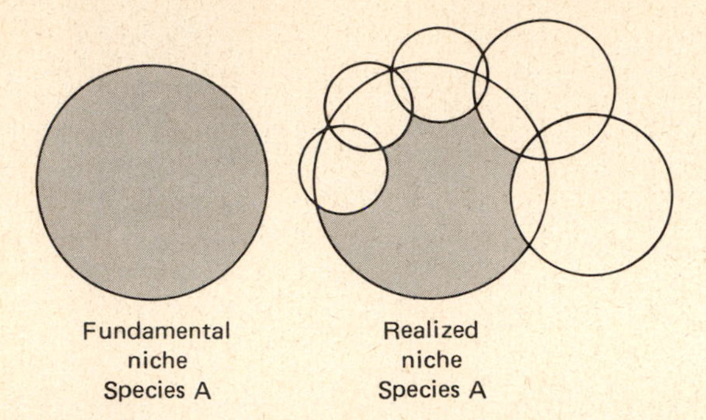

FIGURE 15.8 Fundamental and realized niches. The fundamental niche of a species represents the full range of environmental conditions, biological and physical, under which the species can successfully exist. However, under pressure of superior competitors, the species may be completely displaced from part of its fundamental niche and forced to retreat to that part of the fundamental niche hypervolume to which it is most highly adapted. The portion that it occupies is its realized niche.

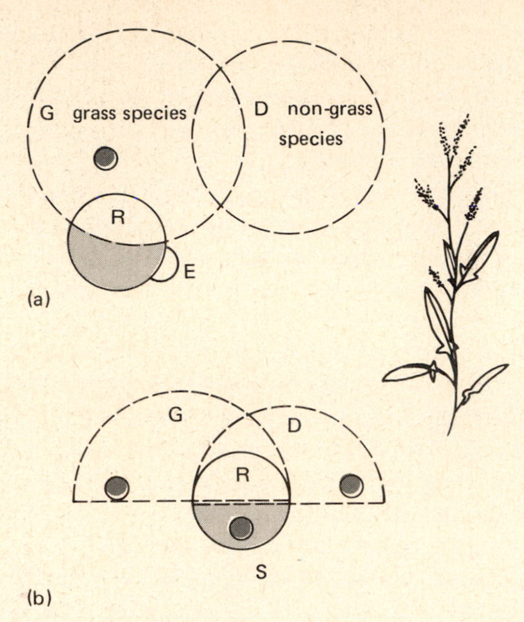

FIGURE 15.9 Diagrammatic representation of niche relationship of *Rumex acetosa* and *R. acetosella* in mixed grassland swards. In each diagram the fundamental niches of grass species (G) and nongrass species (D) overlap. The fundamental niche of *Rumex* species (R) is shown as a continuous line, and the realized niche is shaded. E is that part of the fundamental niche of *R. acetosa* which is expressed in the presence of nongrass species only and does not overlap the fundamental niches of G and D. The fundamental niches of seedlings—shown by the small, dark-colored circles—are contained within the fundamental niches of grasses, nongrasses, and *Rumex*. (From Putwain and Harper, 1970.)

grasslands in North Wales. *R. acetosa* grew in a grassland community dominated by velvet grass *(Holcus lanatus)* and red and sheep fesques *(Festuca rubra* and *F. ovina). R acetosella* grew in a community dominated by sheep fesque and bedstraw *(Galium saxatile)*. To determine interference and niches of the two docks, Putwain and Harper treated the flora with specific herbicides to remove selectively in different plots (1) grasses and (2) forbs, except *Rumex* species. All species except *R. acetosella* spread rapidly after the grasses were removed. *R acetosella* increased only after both grasses and nongrasses were removed. The niches of the two plants are diagrammed and explained in Figure 15.9. The presumed fundamental niche of *R. acetosella* (R) overlaps the fundamental niches of both grasses and other forbs (D). Only when these competitors are eliminated does *R. acetosella* realize its fundamental niche. *R. acetosa,* however, overlaps only with the grasses, and only their removal is necessary to permit that dock to expand.

Note that the niches of seedlings differ from those of the mature plant. The fundamental and realized niches of an organism can change with its growth and development. Insects with a complex life cycle may occupy one niche as a larva and an entirely different niche as an adult. In other organisms niche space can change as the organism matures because food and cover requirements change as the organism grows larger.

NICHE OVERLAP

Niche overlap occurs when two or more organisms use a portion of the same resource, such as food, simultaneously. The amount of niche overlap is assumed to be proportional to the degree of competition for that resource. With little or no competition, niches may be adjacent to one another with no overlap, or they may be disjunct (Figure 15.10). At the other extreme, according to competition theory, under intense competition, the fundamental niche of one species may be completely within or correspond exactly to another, as in the case of the *Rumex* seedling. In such instances there can be two outcomes. If the niche of species 1 contains the niche of species 2 and species 1 is competitively superior, species 2 will be eliminated entirely. If species 2 is competitively superior, it will eliminate species 1 from that part of the niche space that species 2 occupies. The two species then coexist within the same fundamental niche.

When fundamental niches overlap, some niche space is shared and some is exclusive, enabling the two species to coexist (Figure 15.10). However, niche overlap does not mean high competitive interaction. In fact, the reverse may be true. Competition involves a resource in short supply. Extensive niche overlap may indicate that little competition exists and resources are abundant. Pianka (1972, 1975) has suggested that the maximum tolerable overlap in niches should be lower in intensively competitive situations than in environments where competition is minimal.

For simplicity, niche overlap is usually considered as one- or two-dimensional. In reality, of course, a niche involves the utilization of many types of resources: food, a place to feed, cover, space, and so on. Rarely do two or more species possess exactly the same niche involving all requirements. While species may show overlap on one gradient, they may not on another. The total of all competitive interactions may be less than the competition or niche overlap suggested on one gradient alone (Figure 15.11).

Differences in niches are not solely a characteristic of various species. Division of feeding space, food size, and other niche components may exist between sexes of the same species. The male red-eyed vireo, for example, obtains its insect food in the upper canopy, the female in the lower canopy and near the ground. The feeding area between the two overlaps only about 35 percent (P. Williamson, 1971). Although they may utilize similar foods, each secures insects from different levels (Figure 15.12). A pronounced difference in bill size allows the male Arizona woodpecker *(Dendrocopus arizonae)* to forage on the trunk and the female to hunt on the branches (Ligon, 1968).

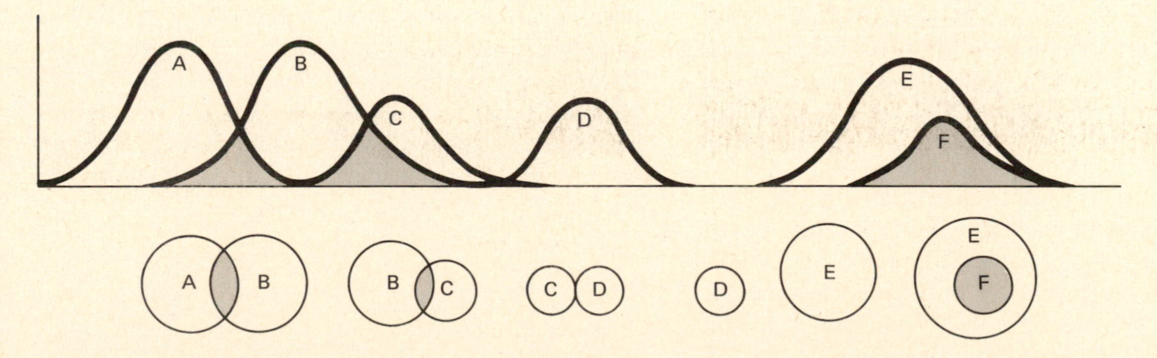

FIGURE 15.10 Different types of niche relationships visualized as graphs on a resource gradient and as Venn diagrams. Species A and B have overlapping niches of equal breadth but are competitive at opposite ends of the resource gradient. B and C have overlapping niches of unequal breadth. Species C shares a greater proportion of its niche with B than B does with C. (In this example, however, B shares its niche also with A at the other end of its resource niche.) C and D occupy adjacent niches with little possibility of competition. D and E occupy disjunct niches, and no competition exists. Species F has a niche contained within the niche of E. If F is superior to E competitively, it persists, and E shares that part of its niche with F. (Compare this with Figure 15.9.) (Adapted from Pianka, 1983.)

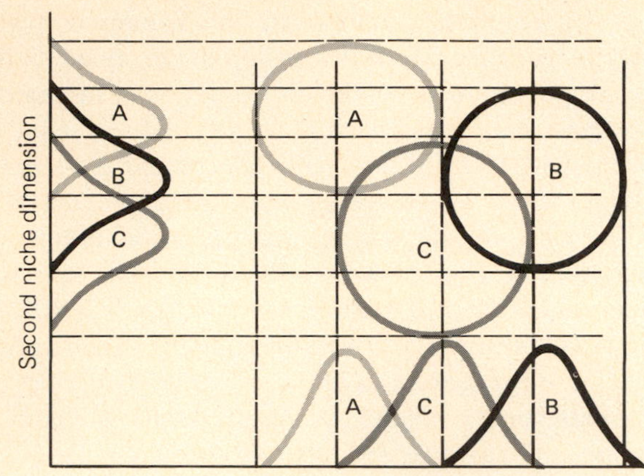

FIGURE 15.11 Niche relationship based on two gradients. Models of a niche as a single gradient do not indicate the degree of true niche overlap where other gradients are involved. Two species may exhibit considerable overlap on one gradient and little or none on another. When several niche dimensions are considered, niche overlap may be reduced considerably, as illustrated here. On resource gradient 1, A and B exhibit no overlap, and on resource gradient 2, they overlap equally and in an opposite way. When both niches are considered (circles), A and B do not overlap. C on resource gradient 2 overlaps equally with B and very little with A. On resource gradient 2, C overlaps with both A and B. When both gradients are considered, C overlaps mostly with B and very little with A.

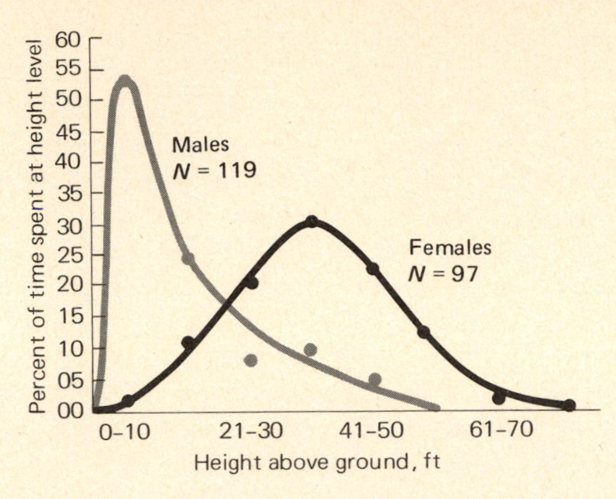

FIGURE 15.12 Separation of male and female red-eyed vireos *(Vireo olivaeus)* by height of foraging. Mean height for males is 37.1 feet; standard deviation, 12; standard error (S.E.), 1.0; range, 9 to 75. Mean height for females is 14.2 feet; standard deviation, 10.8; S.E., 1.1; range, 2 to 50. (From P. Williamson, 1971.)

NICHE WIDTH

If one plotted the range of resources—for example, food size—utilized by an animal or the range of soil moisture conditions occupied by plants, the length of the axis intercepted by the curve would represent niche width (Figure 15.13). Theoretically, *niche width* (also called niche breadth and niche size) is the extent of the hypervolume occupied by the realized niche. A more practical definition is the sum total of the variety of different resources exploited by an organism (Pianka, 1975). Measurement of a niche usually involve the measure of some ecological variable such as food size or habitat space (Figure 15.14).

Niche widths are usually described as narrow or broad. The wider the niche, the more generalized the species is considered to be. The narrower the niche, the more specialized is the species. Generalist species have broad niches and sacrifice efficiency in the use of a narrow range of resources for the ability to use a wide range of resources. As competitors they are superior to specialists if resources are somewhat undependable. Specialists equipped to exploit a specific set of resources occupy narrow niches. As competitors they are superior to generalists if resources are dependable and renewable. A dependable resource supply is closely partitioned among specialists with low interspecific

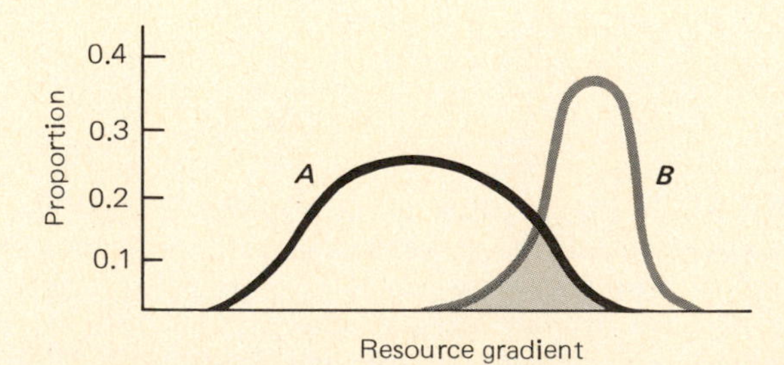

FIGURE 15.13 Hypothetical distribution of a species with a broad niche (A) and a species with a narrow niche (B) on a resource gradient. The niches overlap (shaded area). Species A overlaps species B more than species B overlaps species A.

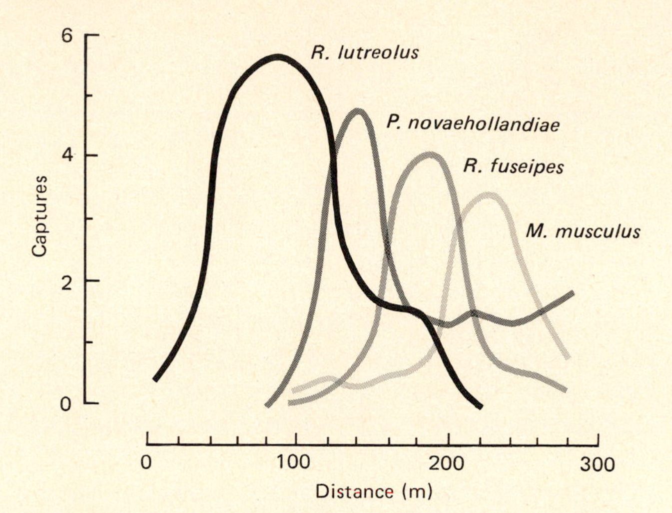

FIGURE 15.14 Partitioning of habitat space by four species of murid rodents in Australia. The numbers of individuals captured are plotted against distance—a transect across a study area with increasing elevation. The transect followed topography and thus a moisture gradient. Both, in turn, affected structure and composition of vegetation, creating habitat differences. Note that in spite of niche overlap, the abundance of each species is highest at some point on the gradient where the populations of other are lower. (After B. J. Fox, 1981:1418.)

overlap (Roughgarden, 1974). If resource availability is variable, generalist species are subject to invasion and close packing with other species during periods of resource abundance.

NICHE COMPRESSION AND ECOLOGICAL RELEASE

Niche width provides some indication of resource utilization by a species. If a community consisting of a number of species with broad niches is invaded by competitors, intense competition along points on the resource gradient may force the original occupants to restrict or compress their use of space, range of foods, or other activities to patches of habitat providing optimal resources. Competition that results in contraction of habitat rather than a change in the type of food or resource utilized is called *niche compression* (MacArthur and Wilson, 1967).

Conversely, if interspecific competition is reduced, a species may expand its niche by utilizing space previously unavailable to it. Niche expansion in response to reduced interspecific competition is called *ecological release*. Ecological release may occur when a species invades an island that is free of potential competitors, moves into habitats it never occupied before on the mainland, and increases its abundance (Cox and Ricklefs, 1977). Such expansion may also follow when a competing species is removed from a community, allowing remaining species to move into microhabitats they previously could not occupy.

NICHE SHIFT

Associated with compression and release is another response, *niche shift*. Niche shift is the adoption of changing behavioral and feeding patterns by two or more competing populations to reduce interspecific competition. The shift may be a short-term ecological response or a long-term evolutionary response involving some change in basic behavioral or morphological traits.

Werner and Hall (1976, 1979) demonstrated niche shift in three competing species of sunfish (Centrarchidae): the bluegill *(Lepomis maerochirus)*, the pumpkinseed *(L. gibbosus)*, and the green sunfish *(L. cynellus)*. When the three species were stocked in separate replicated experimental ponds, the food and habitat preferences were similar, and their average growth rate increased. The three preferred the emergent and submerged vegetation about the edges of the ponds and the large items of food they found there. When the three species were stocked together in equal densities, they occupied the vegetated zone about the edges. As food resources declined, the bluegill and pumpkinseed left, leaving behind the green sunfish, a more efficient forager, in the vegetation. Bluegill and pumpkinseeds concentrated their foraging efforts on bottom invertebrates, mainly Chironominae. But the pumpkinseed—with its short, widely spaced gill rakers that do not become fouled when the fish sorts through the bottom sediments—was better able to exploit that food supply. The bluegill—with long, fine gill rakers that retain small prey—foraged in the open-water column.

The fish also exhibited shifts in habitat utilization. When bluegill and green sunfish were confined together in equal densities in a pond, the green sunfish eventually caused the bluegill to shift to open water; but in the absence of green sunfish, bluegills invaded the dense vegetation. Bluegill and pumpkinseed have a dietary overlap of 50 to 55 percent. When the two are stocked together in a pond, the bluegill again is ultimately forced to the open-water habitat. In the end, however, the bluegill may be the superior competitor. Young sunfish of all three species feed on zooplankton in open water. This places the young of green sunfish and pumpkinseeds in direct competition for small-sized food with both adult and young bluegill. A more efficient open-water forager, the bluegill probably affects the recruitment of the other two species. Possibly its generalized diet and its ability to shift its habitat as the situation requires account for the bluegill's being the most common sunfish in ponds.

Summary

Relations among species may be positive (+), or beneficial; negative (−), or detrimental; or neutral (0). There are six possible interactions: (00), neutral (of little ecological interest); (++), in which both populations mutually benefit each other (mutualism and protocooperation); (− −) in which both populations are affected adversely (competition); (0+), in which one population benefits and the other is unaffected (commensalism); (0−), in which one population is harmed and the other is unaffected (amensalism); and (+ −), in which one population benefits and the other is harmed (predation, parasitism).

Interspecific competition—the seeking of a resource in short supply by individuals of two or more species, reducing the fitness of both—may be one of two kinds: interference and exploitative. Exploitative competition depletes resources to a level of little value to either population. Interference involves aggressive interactions. A particular form of interference competition is allelopathy, the secretion of chemical substances that inhibit the growth of other organisms.

As described by the Lotka-Volterra equations, four outcomes of interspecific competition are possible. Species 1 may win over species 2; species 2 may win over species 1. Both of these represent competitive exclusion. A third possibility is unstable equilibrium, in which the potential winner is determined by the one most abundant at the outset. And a final possible outcome is stable equilibrium, in which the two species coexist, but at lower population levels than if each existed in the absence of the other.

The competitive exclusion principle—two species with exactly the same ecological requirements cannot coexist—has conceptual difficulties. It has, however, stimulated critical examinations of other competitive relationships, especially how species coexist and how resources are partitioned. Groups of functionally similar species that share a spectrum of resources and interact strongly with each other are termed guilds.

Closely associated with the concept of interspecific competition is the concept of the niche. Basically, a niche is the functional role of an organism in the community. What that functional role is might be constrained by interspecific competition.

In the absence of any competition, an organism occupies its fundamental niche. In the presence of interspecific competition, the fundamental niche is reduced to a realized niche, the conditions under which an organism actually exists. When two different organisms use a portion of the same resource, such as food, these niches are said to overlap. Overlap may or may not indicate competitive interaction.

The range of resources used by an organism suggests its niche width. Species with broad niches are considered generalist species, while those with narrow niches are considered specialists. Niche compression results when competition forces an organism to restrict its type of food or constrict its habitat. In the absence of competition, the organism may expand its niche and experience ecological release. Organisms may also undergo niche shift by changing their behavioral or feeding patterns to reduce interspecific competition.

Review and Study Questions

1. Describe various relationships among individuals of different species involving +, increase in fitness; −, decrease in fitness; and 0, no effect.
2. Define commensalism, amensalism, predation, parasitism, and parasitoidism.
3. Distinguish among symbiosis, mutualism, and protocooperation.
4. Define interspecific competition and distinguish between interference and exploitative competition.
5. Based on the Lotka-Volterra competition equations, describe four outcomes of competition.
6. What is the competitive exclusion principle? What is the problem with the concept?
7. What conditions often determine the outcome of unstable equilibrium under natural situations?
8. What permits coexistence among competing species?
9. What is allelopathy? How might it be considered a form of competition?
10. What is resource partitioning? What are some weaknesses of the concept?
11. Define niche.
12. Distinguish between a fundamental niche and a realized niche.
13. What is niche overlap? Niche width? How is niche width usually measured?
14. What is meant by niche compression, ecological release, and niche shift?

Suggested Readings

Cody, M. L., and J. M. Diamond, eds. 1975. *Ecology and evolution of communities.* Cambridge, Mass.: Harvard University Press.

May, R. M., ed. 1981. *Theoretical ecology: Principles and applications.* 2d ed. Sunderland, Mass.: Sinauer Associates.

Pianka, E. R. 1983. *Evolutionary ecology.* 3d ed. New York: Harper & Row.

Pontin, A. J. 1983. *Competition and coexistence of species.* Marshfield, Mass. and London: Pitman Publishers.

Price, P. W., C. N. Slobodchikoff, and W. S. Gaud, eds. 1984. *A new ecology: Novel approaches to interactive systems.* New York: Wiley.

Strong, D. R., Jr., D. Simberloff, L. G. Abele, and A. B. Thistle, eds. 1983. *Ecological communities: Conceptual issues and the evidence.* Princeton, N.J.: Princeton University Press.

Note: See also *American Naturalist* 122 (no. 5, November 1983) for a group of controversial articles on interspecific competition, as well as Grime (1979), Hutchinson (1978), and Krebs and Davies (1978), cited in the "Suggested Readings" section at the end of Chapter 13.

ECOLOGICAL EXCURSION

Niches

Charles Elton

. . . It should be pretty clear by now that although the actual species of animals are different in different habitats, the ground plan of every animal community is much the same. In every community we should find herbivorous and carnivorous and scavenging animals. We can go further than this, however: in every kind of wood in England we should find some species of aphid, preyed upon by some species of ladybird. Many of the latter live exclusively on aphids. That is why they make such good controllers of aphid plagues in orchards. When they have eaten all the pest insects they just die of starvation, instead of turning their attention to some other species of animal, as so many carnivores do under similar circumstances. There are many animals which have equally well-defined food habits. A fox carries on the very definite business of killing and eating rabbits and mice and some kinds of birds. The beetles of the genus *Stenus* pursue and catch springtails *(Collembola)* by means of their extensile tongues. Lions feed on large ungulates—in many places almost entirely zebras. Instances could be multiplied indefinitely. It is therefore convenient to have some term to describe the status of an animal in its community, to indicate what it is *doing* and not merely what it looks like, and the term used is "niche." Animals have all manner of external factors acting upon them—chemical, physical, and biotic—and the "niche" of an animal means its place in the biotic environment, *its relations to food and enemies*. The ecologist should cultivate the habit of looking at animals from this point of view as well as from the ordinary standpoints of appearance, names, affinities, and past history. When an ecologist says "there goes a badger" he should include in his thoughts some definite idea of the animal's place in the community to which it belongs, just as if he had said "there goes the vicar."

The niche of an animal can be defined to a large extent by its size and food habits. We have already referred to the various key-industry animals which exist, and we have used the term to denote herbivorous animals which are sufficiently numerous to support a series of carnivores. There is in every typical community a series of herbivores ranging from small ones (*e.g.* aphids) to large ones (*e.g.* deer). Within the herbivores of any one size there may be further differentiation according to food habits. Special niches are more easily distinguished among carnivores, and some instances have already been given.

The importance of studying niches is partly that it enables us to see how very different animal communities may resemble each other in the essentials of organisation. For instance, there is the niche which is filled by birds of prey which eat small mammals such as shrews and mice. In an oak wood this niche is filled by tawny owls, while in the open grassland it is occupied by kestrels. The existence of this carnivore niche is dependent on the further fact that mice form a definite herbivore niche in many different associations, although the actual species of mice may be quite different. Or we might take as a niche all the carnivores which prey upon small mammals, and distinguish them from those which prey upon insects. When we do this it is immediately seen that the niches about which we have been speaking are only smaller subdivisions of the old conceptions of carnivore, herbivore, insectivore, etc., and that we are only attempting to give more accurate and detailed definitions of the food habits of animals.

SOURCE: Charles Elton, *Animal Ecology* (London: Sidgwick & Jackson, Ltd., 1927), 63–64.

CHAPTER 16

Predation

Outline

Objectives

Upon completion of this chapter, you should be able to:

1. Define predation and distinguish among parasitism, parasitoidism, herbivory, and cannibalism.
2. Describe in a general way the Lotka-Volterra model of predation and its theoretical and experimental outcomes.
3. Define functional response and describe the three types.
4. Define search image and switching and relate their importance to predation.
5. Explain numerical response and total response.
6. Discuss the significance of cannibalism in population regulation.
7. Explain the concept of optimal foraging.

Among the relationships between organisms, one in which nature is "red in tooth and claw" is predation. Most of us associate predation with a hawk taking a mouse, a wolf killing a deer. That is a narrow view of predation. It involves much more than what is implied by the common notion of the word. A fly laying its eggs on a caterpillar to develop there at the lethal expense of the victim is exhibiting a form of predation called *parasitoidism.* The parasitoid attacks the host (the prey) indirectly by laying its eggs on the host's body. When the eggs hatch, the larvae feed on the host, ultimately killing it. True *parasitism* involves the predator's living on or within the host. The parasite is physiologically dependent upon the host and seldom kills the host outright (which, in effect, would destroy both its habitat and its food). A deer feeding on woody shrubs and grass and a mouse eating a seed are forms of predation called *herbivory.* Seed consumption is a form of outright predation. Grazing on plants without killing them is a form of parasitism. A special form of predation is *cannibalism,* in which the predator and the prey are the same species. Thus, predation in its broadest sense can be defined as one living organism feeding on another living organism.

Predation in its actions and reactions is more than just a transfer of energy. It represents a direct and often complex interaction of two or more species, of the eaters and the eaten. The numbers of some predators may depend upon the abundance of their prey, and the population of the prey may be controlled by its predators. Each can influence the fitness of the other and favor new adaptations.

Theory of Predation

The two mathematicians Lotka (1925) and Volterra (1926) extended their models of population growth and competition to predation. Independently they proposed mathematical statements to express the relationship between predator and prey populations. They provided one equation for the prey population and another for the predator population.

The growth equation for the prey population involving the maximum rate of increase per individual and the removal of the prey from the population by the predator is:

$$\frac{dN_1}{dt} = r_1N_1 - PN_1N_2$$

where N_1 = density of the prey population
r_1 = intrinsic rate of increase in the absence of predation
N_2 = density of the predatory population
P = coefficient of predation
N_1N_2 = assumption that removal of prey from the population is proportional to the chance encounter between predator and prey

The growth equation for the predator is influenced by the density of the prey population:

$$\frac{dN_2}{dt} = P_2N_1N_2 - d_2N_2$$

where P_2 = coefficient expressing effectiveness of the predator
d_2 = density-independent mortality rate of the predator

The underlying assumptions of the Lotka-Volterra equations are that the prey population grows exponentially and that reproduction in the predator population is a function of the number of prey consumed.

These equations show that as a single predator population increases, the single prey population decreases to a point at which the trend is reversed. The prey increases, followed by an increase in the predator population. The rise and fall of the two populations result in oscillations in each (Figure 16.1).

About a decade later A. H. Nicholson, an ecologist, and W. Bailey, an engineer and mathematician, developed a mathematical model for a host-parasitoid relationship. That model predicted increasingly violent oscillations in single-predator–single-prey populations living together in a limited area with all the external conditions constant.

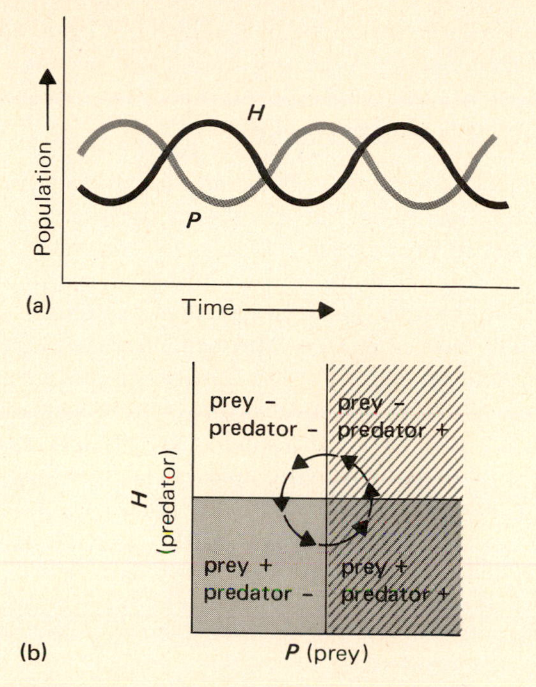

FIGURE 16.1 The Lotka-Volterra model of predator-prey interactions. The Lotka-Volterra equations can be visualized with the two graphs. (a) The abundance of each population is plotted as a function of time. The model shows the joint abundance of species. An increased abundance of prey is followed by an increase of predators. (b) The zero growth curves of both predator and prey are straight and intersect at right angles. Responses of each population are indicated by a minus sign for population decline and a plus sign for population increase. Predators increase to the right of the vertical line; prey increase below the vertical line. The Lotka-Volterra model assumes that reproduction of predators is a function of the number of prey consumed.

Both models state that as the predator population increases, it will consume a progressively larger number of prey until the prey population begins to decline. In turn, the abundance of prey influences the reproductive rate of the predator population. In time, the number of predators overshoots the availability of prey, and the predator population declines to a point where the reproduction of prey more than balances its losses through predation. The prey then increases, followed by an increase in the population of predators. The cycle or oscillation may continue indefinitely. The prey is never quite destroyed by the predator; the predator never completely dies out.

Both equations are too simple and unrealistic. They overemphasize the influence of predators on prey populations. They ignore genetic changes, stress, emigration, aggression, the availability of cover and hiding places, the difficulty of predators in locating prey as prey numbers become scarcer, and other parameters that influence fluctuations in predator-prey populations. The continuing appeal of the Lotka-Volterra equations to population ecologists lies in the straightforward mathematical descriptions of cyclic behavior in predator and prey populations.

EXPERIMENTAL SYSTEMS

The biologist G. F. Gause (1934) attempted to demonstrate the Lotka-Volterra predation model experimentally, just as he did the competition models. He reared together under constant environmental conditions a prey population, *Paramecium caudatum,* whose populations grow exponentially, and a predator, the ciliate *Didinium nasutum,* whose reproduction rate is independent of food intake, a feature not considered by Gause. Thus, Gause's experiments did not test the Lotka-Volterra model of predation. However, Gause did demonstrate that in a laboratory habitat the predator *Didinium* exterminated its prey, regardless of the density of the two populations. After the prey was destroyed, the predators died of starvation. Only by introducing prey periodically to the medium was Gause able to maintain the predator population and prevent it from dying out. In that manner, he was able to maintain the two populations together and produce regular fluctuations in both. The predator-prey relations were ones of overexploitation and annihilation unless immigration took place from other prey populations.

In another experiment Gause introduced sediment in the floor of the tube habitat. There the prey could escape from the predator. When predators eliminated the prey from the clear medium, they died from lack of food. The *Paramecium* that took refuge in the sediment continued to multiply and eventually took over the medium.

Over 20 years later, in a different type of experiment, C. Huffaker (1958) attempted to learn if an adequately large and complex laboratory environment could be established in which a predator-prey system would not be self-exterminating. Involved were the six-spotted mite *Eotetranychus sexmaculatus* and a predatory mite, *Typhlodromus occidentalis*. Whole oranges, placed on a tray among a number of rubber balls the same size, provided food and cover for the spotted mite. Such an arrangement permitted the experimenter to control both the total food resource available and the pattern of dispersion by covering the oranges with paper and sealing wax to whatever degree desired and by changing the general distribution of oranges among the rubber balls. Huffaker could manipulate conditions to simulate a simple environment in which the food of the herbivore was concentrated or a complex environment in which the food was widely dispersed, partially blocked by barriers, and in which refuges were lacking. In both situations the two species at first found plenty of food available for population growth. Density of predators increased as the prey population increased. In the environment where the food was concentrated and dispersion of the prey population was minimal, predators readily found prey, quickly responded to changes in prey density, and were able to destroy the prey rapidly. In fact, the system was self-annihilative. In the environment where the primary food supply and prey were dispersed, predator and prey went through two oscillations before the predators died out. The prey recovered slowly.

Several important conclusions resulted from the study. First, predators cannot survive where their prey population is low. Second, a self-sustaining predator-prey population cannot be maintained without immigration of prey. Third, the complexity of prey dispersal and predator searching relationships, combined with a period of time for the prey population to recover from the effects of predation and to repopulate the areas, has more influence on the period of oscillation than the intensity of predation.

FUNCTIONAL RESPONSE

Models of interactions between predator and prey suggest two distinct responses of the predator to changes in prey density. One is for the individual predator to eat more prey as the prey population increases or take them sooner. Or the predators can become more numerous through increased reproduction or immigration. The former is called a *functional response* and the latter a *numerical response*.

The idea of a functional response was introduced by the English entomologist M. E. Solomon (1949) and was explored in detail by C. S. Holling (1959, 1961, 1966). The basis of a functional response is that a predator (or parasite) will take or affect more prey as the density of prey increases. Holling classified functional responses into three types (Figure 16.2).

In a Type I response the number of prey affected per predator increases in a linear fashion to a maximum as prey density increases. Predators of any given abundance take a fixed number of prey during the time they are in contact, usually enough to satiate themselves. Trout feeding on an evening hatch of mayflies is an example. Type I produces density-independent mortality up to satiation.

In a Type II response the number of prey affected rises at a decreasing rate toward a maximum value. A dominant component of a Type II response is handling time by the predator. Handling time includes time spent pursuing the prey, subduing, eating, and digesting it. Because handling time limits the amount of prey that the predator can process per unit of time, the number of prey taken per unit of time slows to a plateau while the number of prey is still increasing. For that reason a Type II functional response rarely acts as a stabilizing force on a prey population, unless the prey occurs in patches.

The plateau in Type II may be influenced by an aggregative response of predators to patches of high density. Patches of profitable foraging may attract other predators to the area (Figure 16.3).

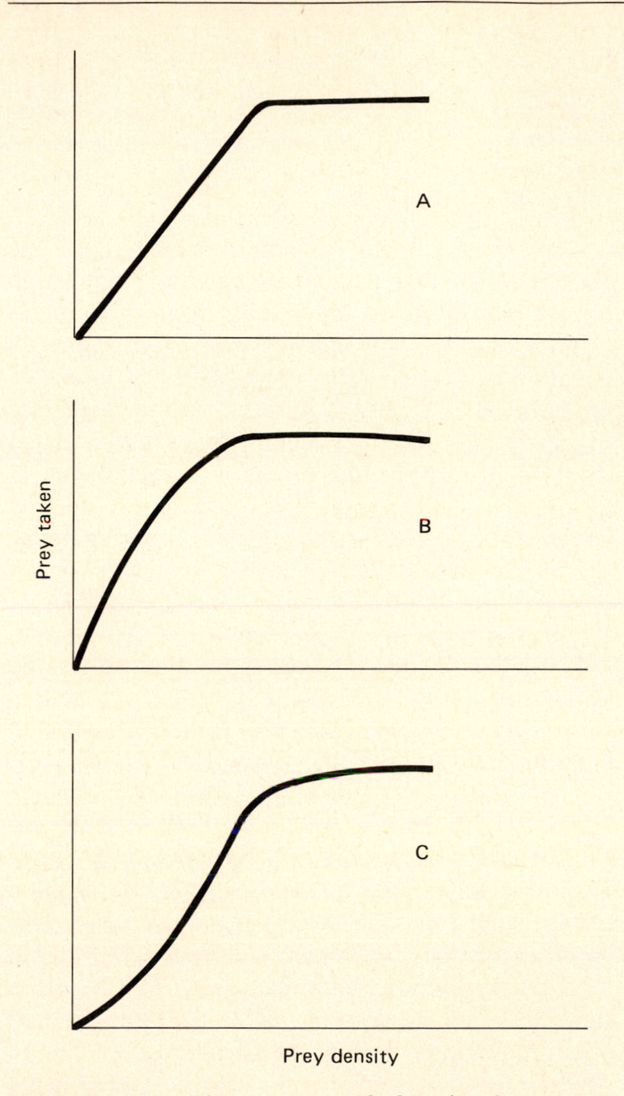

FIGURE 16.2 Three types of functional response curves. (A) Type I, in which the number of prey eaten per predator increases linearly to a maximum as prey density increases. (B) Type II, in which the number of prey eaten rises at a decreasing rate to a maximum value. (c) Type III, in which the number of prey taken is low at first, then increases in a sigmoid fashion approaching an upper asympote. (From Holling, 1959:300.)

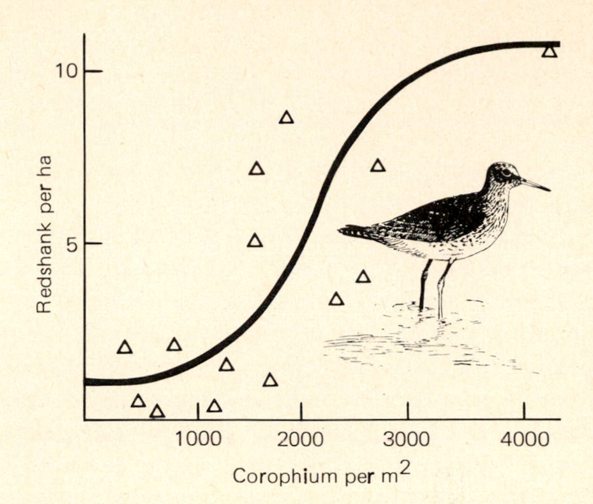

FIGURE 16.3 Aggregative response in the redshank *(Tringa totanus)*. The curve plots the density of the predator (the redshank) in relation to the average density of arthropod prey *(Corophium)*. (After Hassel and May, 1974:576.)

When one or two members of a predator species discovers and begins to feed on the prey item, other members of the population observe the congregating animals and join them (Curio, 1976). As the numbers of predators increase on the patch, they interfere with each other. That increases each predator's search time, reduces the efficiency of predation, and encourages a number of predators to leave the crowded area.

Type III functional response is more complex than Type II. In Type III the number of prey taken is low at first, then increases in a sigmoid fashion approaching a plateau at which the attack rate remains constant. Type III functional response can potentially stabilize a prey population, because the attack rate varies with prey density. At low densities the attack rate is negligible, but as prey density increases (as indicated by the upward sweep of the curve), predatory pressure increases in a density-dependent fashion.

Type II responses involve situations of varying densities of a single prey species. Type III responses invariably involve two or more prey species. Predators take most or all of the individuals that are in excess of a certain minimum number as determined, perhaps, by the availability of prey cover and the prey's social behavior. The population level at which the predator no longer finds it profitable to hunt the prey species has been called the *threshold of security* (Figure 16.4) by P. Errington (1946).

Type III responses have been called compensatory because as prey numbers increase above the threshold, surplus animals become vulnerable to predation through intraspecific competition. Below the threshold of security, the prey species compensates for its losses through increased reproduction and greater survival of young. Below the threshold of security, functional response of the predator is very low; above the threshold, functional response is marked. An example of that type of predation is detailed by Errington (1963) in his notable long-term study of the muskrat. Adult muskrat *(Ondatra zibethicus)* well established on breeding territories and occupying an optimal habitat were largely free from predation by mink *(Mustela vison)* and had the greatest fitness. Animals excluded from the territory or forced to occupy submarginal habitats were highly vulnerable to predation or, if they were not, usually failed to reproduce. In either case, these animals could be considered surplus. Their demise did not affect the overall fitness of the prey population.

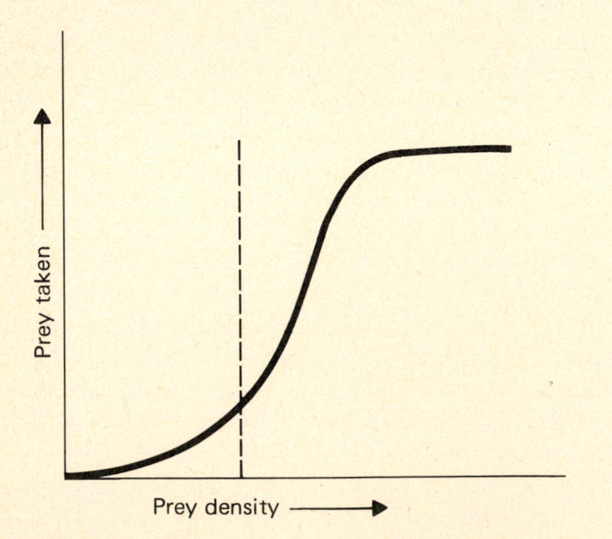

FIGURE 16.4 Compensatory predation as illustrated by a functional response curve. There is no response to the left of the vertical line, which represents the threshold of security for the prey.

SEARCH IMAGE

Another reason for the sigmoidal shape of the Type III functional response curve may be the search image. According to the search image hypothesis (Tinbergen, 1960), when a new prey species appears in an area, its risk of becoming prey is low. The predator has not yet acquired a search image for the species. Once the predator has secured a palatable item of prey, the predator finds it easier to locate others of the same kind. The more adept and successful the predator becomes at securing a particular prey item, the longer and more intensely it concentrates on that item. In time the number of the prey species becomes so reduced or its population so dispersed that encounters between it and the predator lessen. The search image for that species begins to wane, and the predator reacts to another species.

Although the search image hypothesis has been criticized (Murdoch and Oaten, 1975), it is an important and integral part of predation. Such a conclusion is supported by studies of food selection and foraging behavior of red-winged blackbirds (Alcock, 1973), white-crowned sparrows *(Zonotrichia leucophrys)* (Simons and Alcock, 1973), European thrushes *(Turdus)* (Smith, 1974a, 1974b), and titmice (*Parus* spp.) (Smith and Sweatman, 1974). In general, these birds develop a search image for the type of food fed upon previously and learn where they are most likely to find food (see "Optimal Foraging"). Search image and hunting profitability are the complementary components of the Type III response curve.

SWITCHING

The Type III functional response curve is characteristic of situations embracing more than one prey species. As such, it involves a facultative predator and an alternate prey. Although a predator may have a strong preference for a certain prey, it can turn to an alternate, more abundant prey species that provides more profitable hunting. If rodents, for example, are more abundant than rabbits and quail, foxes and hawks will concentrate on rodents. That idea was advanced early by Aldo Leopold in his book *Game Management* (1933), in which he describes alternate prey species as buffer species because they stand between the predator on the one hand and game species on the other. If the population of the buffer prey is low, the predators will turn to game species; foxes and hawks will concentrate on rabbits and quail.

A predator's turning to an alternate, relatively more abundant prey has been termed *switching* (Murdoch, 1969). In switching, the predator concentrates a disproportionate amount of feeding on the more abundant species and pays little attention to the rarer species. As the relative abundance of the two prey species changes, the predator turns its attention to the more common prey (Figure 16.5).

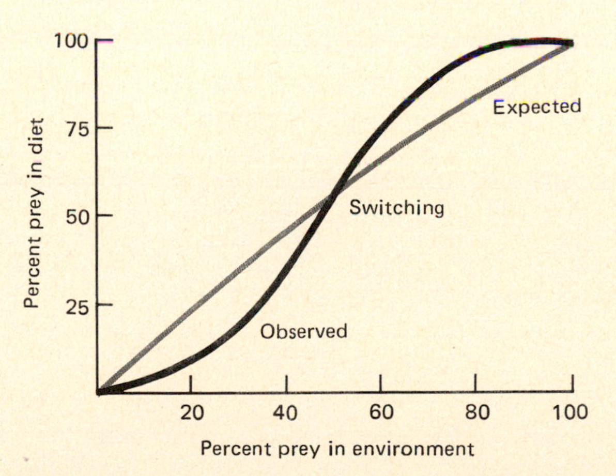

FIGURE 16.5 A model of switching. The straight line represents a situation in which the proportion of a given prey is the same in both the environment and the diet of the predator. It represents a constant preference with no switch. The curved line represents the proportion of prey species actually taken. At one point the number of prey taken is proportionately less than what occurs in the environment. Switching occurs at the point where the curved line crosses the straight. At that point the number of prey taken is disproportionate to the number in the environment.

At what point in prey abundance a predator switches depends considerably on the threshold of security for the prey species involved. The threshold of security may be much lower for a highly palatable prey. A predator may hunt longer and harder for a palatable species before it turns to a more abundant, less palatable alternate prey. Conversely, the predator may turn from the less palatable species at a much higher level of abundance than it would if a more palatable species were involved. For the palatable prey that situation is maladaptive. It is at a selective disadvantage when associated with less palatable prey. Such selective pressure favors temporal or spatial isolation of the more palatable from the less palatable species, or it favors mimicry of a less palatable species by a palatable species (see Chapter 17).

In spite of switching, some predators deliberately seek out certain prey, no matter how scarce. Predators in a California grassland exhibited a distinct preference for meadow voles over harvest mice and other rodents even through alternate prey was more abundant (Pearson, 1966). In fact, the abundance of alternate prey apparently enables carnivores to maintain a sufficiently high population and to obtain sufficient energy to permit continued predatory pressure on the preferred species. Among herbivores, deer exhibit a pronounced preference for certain species of browse plants (Klein, 1970). Meadow mice often concentrate on seeds of preferred grasses, even though the seeds of other species are abundant and available (Batzli and Pitelka, 1970).

NUMERICAL RESPONSE

As the density of prey increases, the number of predators may also increase. Numerical response takes three basic forms (Figure 16.6): (1) direct response, in which the number of predators in a given area increases as the prey density increases; (2) no response, in which the predator population remains proportionately the same; and (3) inverse response. An example of the latter is the response of parasites to an increasing density of budworms in Canada. Parasites increased sharply at first, then declined

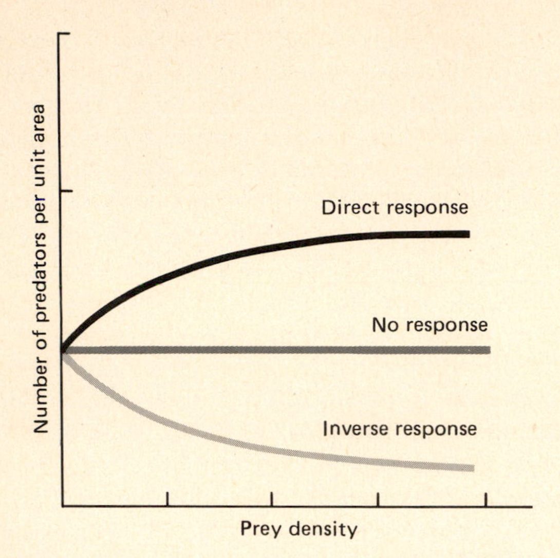

FIGURE 16.6 Basic forms of numerical response. (From Hassell, 1966:576.)

rapidly as the pupal density of the budworm increased (Morris, 1963).

Most numerical responses involve an increase by reproductive effort. Because reproduction does require a certain minimal time, a lag exists between an increase of a prey population and a numerical response by a predator population. For example, the population of great horned owl *(Bubo bubo)* in a 62-square-mile area in Alberta, Canada, increased over a three-year period, 1966 to 1969, from 10 birds to 18 as the population of its prey, the snowshoe hare, increased sevenfold (Rusch et al., 1972). The portion of owls nesting increased from 20 to 100 percent as the biomass of snowshoe hare in the owls' diets increased from 23 to 50 percent. In that situation, nutrition, controlled by the amount of prey available, influenced fitness. If food is limited because of low prey density, fitness is necessarily low (Figure 16.7), and positive numerical response is low. With increasing prey density, fitness increases, and the numerical response is proportionately higher.

Numerical response may involve both an aggregative response and an increase in fitness. An example can be found among the "fugitive" warblers of northern forests, especially the Tennessee *(Vermivora peregrina),* Cape May *(Dendroica tigrina),* and bay-breasted *(D. castanea)* warblers, whose abundance is dictated by an outbreak of spruce budworm. During such periods, populations of the bay-breasted warblers increased from 10 to 120 pairs per 100 acres (Mook, 1963; Morris et al., 1958), and Cape May and bay-breasted warblers have larger clutches than associated warbler species (MacArthur, 1958). In fact, Cape May and possibly bay-breasted warblers apparently depend upon occasional outbreaks of spruce budworm for their continued existence. At those times, these two species are able to increase more rapidly than other

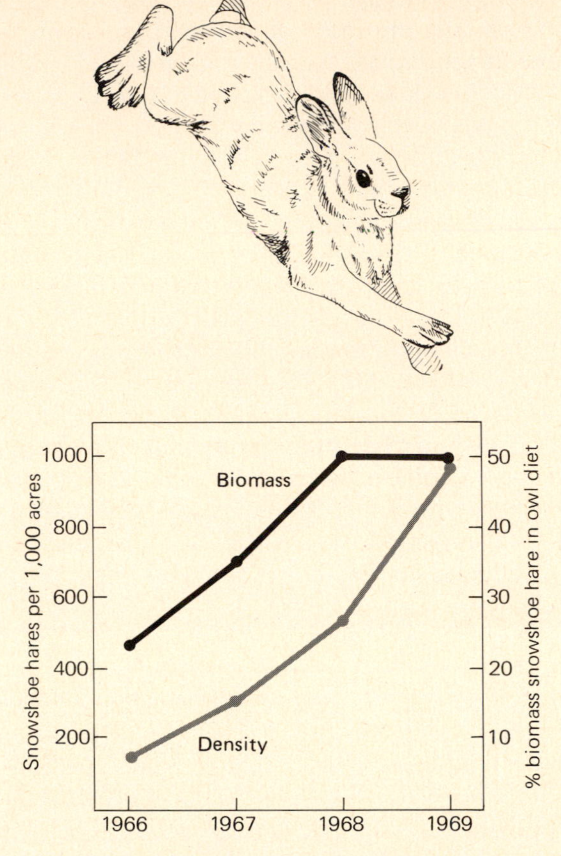

FIGURE 16.7 Numerical response in the snowshoe hare–horned owl system. Density of the snowshoe hare in spring in its habitat is plotted with the percentage of biomass of the hare in the diet of the great horned owl. (From Rusch et al., 1972:291.)

warblers because of extra-large clutches. But during years between outbreaks, they are reduced in numbers and even become extinct locally.

TOTAL RESPONSE

Functional and numerical responses may be combined to give a total response, and predation may be plotted as a percentage (Figure 16.8). If that is done, predation falls into two types: (1) the percentage of predation declines continuously as prey density increases, and (2) the percentage of predation rises initially and then declines. The latter results in a dome-shaped curve produced by the sigmoid (Type III) functional response curve to prey density and by direct numerical response.

Cannibalism

A special form of predation is cannibalism, more euphemistically called intraspecific predation. Cannibalism involves the killing and eating of an individual of the same species. Cannibalism is more widespread and important in the animal kingdom than many ecologists may wish to admit, probably because of the moral significance attached to it. In reality, cannibalism is common to a wide range of animals, both aquatic and terrestrial, from protozoans and rotifers through centipedes, mites, and insects to frogs, toads, fish, birds, and mammals, including humans. Interestingly, about 50 percent of terrestrial cannibals, mostly insects, are normally herbivorous species, the ones most apt to encounter a shortage of protein. In freshwater habitats the bulk of cannibalistic species are predaceous, as they all are in marine ecosystems (Fox, 1975a, 1975b, 1975c).

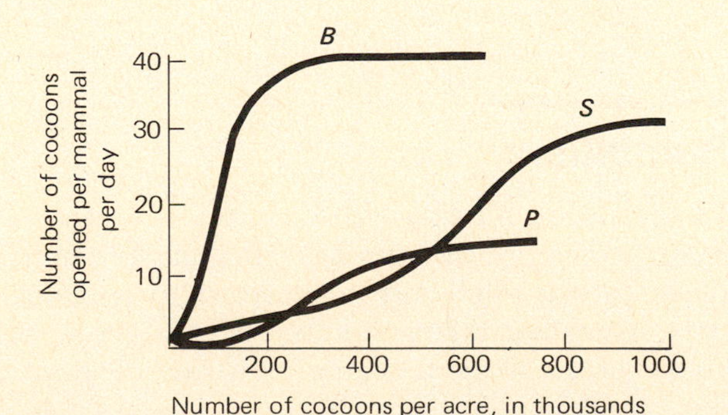

FIGURE 16.8 Total response of predators to prey density expressed as a percentage of predation to prey density. Total response includes both functional response and numerical response. S = *Sorex* shrew; B = *Blarina* shrew; C = *Peromyscus* mouse. (From Holling, 1959:304.)

Cannibalism has been associated with stressed populations, particularly those facing starvation. While some animals do not become cannibalistic until other food runs out, others do so when the relative availability of alternative foods declines and individuals in the population are disadvantaged nutritionally (Alm, 1952). It is probably initiated when hunger triggers search behavior, lowers the threshold of attack, increases the time spent foraging, and expands the foraging area. It is consummated when the individual encounters vulnerable prey of the same species (see Polis, 1981). Other conditions that may promote cannibalism are (1) crowded conditions or dense populations, even when food is adequate; (2) stress, especially when induced by low social rank; and (3) the presence of vulnerable individuals—such as nestlings, eggs, or runty individuals—even though food resources are adequate.

Whatever the cause, the intensity of cannibalism is influenced by local conditions and by the nature of local populations. In general, cannibalism fluctuates greatly over both long and short periods of time. Among some predaceous fish, such as walleye *(Stizostedium vitreum)* (Fortney, 1974), and insects such as freshwater backswimmers *(Notonecta hoffmanni)* (Fox, 1975b, 1975c), cannibalism is most prevalent in summer, which coincides with a sudden decrease in normal prey and a reduction of spatial refuges for the young.

Not all individuals in a population become cannibals (see Polis, 1981). Intraspecific predation is usually confined to older and larger individuals. Those receiving the brunt of cannibalism are the small and the young. But not always. In some situations, groups of smaller individuals will attack and devour larger individuals.

Demographic consequences of cannibalism depend upon the age structure of the population and the feeding rates of various age classes. Even at very low rates, cannibalism can produce demographic effects. Three percent cannibalism in the diet of walleyes could account for 88 percent of mortality among the young (Chevalier, 1973). Cannibalism can account for 23 to 46 percent of the mortality among eggs and chicks of herring gulls (Parsons, 1971), 8 percent of young Belding ground squirrels (Sherman, 1981), and 25 percent of lion cubs (Bertram, 1975). If a large proportion of either an entire population or a vulnerable age class is eaten, it can result in violent fluctuations in recruitment.

Cannibalism can become a mechanism of population control that rapidly decreases the number of intraspecific competitors as food becomes scarce. It is unlikely to bring about local population extinctions because it is short term. It decreases as resources become relatively more available to survivors and as vulnerable individuals are scarcer or harder to find. By reducing intraspecific competition at times of resource shortages, cannibalism may actually reduce the probability of local extinction of a population. In the long run, however, cannibalism could be self-defeating because it runs counter to the second law of thermodynamics and trophic level dynamics (see Chapter 19). The exceptions would be among those animals whose young feed on different trophic levels than the adults. In effect, cannibalism would be harvesting young grazers.

Cannibalism can be a selective advantage to survivors. Survivors gain a meal and eliminate both a potential competitor for food and at the same time a potential conspecific (of the same species) predator. With the population reduced, the survivor has more food, enhancing its chances for further survival, rapid growth, and increased fecundity.

Cannibals can also increase their own fitness by reducing the fitness of competitors. They can do that in several ways. Individuals can kill and eat other individuals of the same sex and reduce competition for mates. Or they can eat the offspring of a competitor, as adults of Belding ground squirrels do. Among some animals—insects, in particular—the females will kill and eat the male after mating, reducing the probability that other females will encounter a mate.

Cannibalism can be a selective disadvantage if individual survivors become too aggressive and destroy their own progeny or genotype completely, reduce their genotype faster than genotypes of conspecific competitors, or reduce their chances of successful reproduction by eliminating suitable mates.

However, selection can balance advantages against disadvantages. In some situations the disadvantages of cannibalism are less severe than starvation and reproductive failure caused by inadequate nutrition. For example, parents cannibalizing some of their own offspring can increase the probability of survival and fitness of either parents or surviving offspring or both (see Polis, 1981; Rohwer, 1978) and utilize rather than waste energy already invested in them. If a population is reduced by starvation, the survivors may be nutritionally stressed. If individuals are removed by cannibalism, density is reduced early, and per capita food supply remains high. Survivors have improved their fitness because, being well-fed as juveniles, they grow faster, survive better, and produce more young.

Parasitism

Parasitism is a relationship in which two organisms live together, one deriving its nourishment from the other without killing it. Parasites, strictly speaking, draw nourishment from the tissues of larger hosts, a case of the weak attacking the strong. Parasites are physiologically dependent on the host, live much shorter lives, and have a high reproductive potential. Heavy infestations of parasites may cause hosts to die from secondary infections or to experience stunted growth, emaciation, or sterility.

Parasites exhibit a tremendous diversity in ways to exploit their hosts and adaptations for doing so. Parasites may be plants or animals, bacteria, viruses, or protozoans, and they may parasitize plants or animals or both. They may live on the outside of the host (ectoparasites) or within its body (endoparasites). Some are full-time parasites; others are only part-time. Part-time parasites may be parasitic as adults and free-living as larvae or the reverse. Parasites have developed numerous ways to gain entrance to their hosts, even to the point of using several hosts as dispersal agents. They have evolved various means and degrees of mobility, ranging from free-swimming ciliated forms to forms totally dependent on other organisms for transport. They have developed diverse ways of securing themselves to the host to maintain their positions. Some, such as the tapeworm, have become so adapted to the host that they no longer require a digestive system. They absorb food directly through their body walls.

Parasites may be restricted to one host. Some parasites of birds—especially certain tapeworms—can live only on one particular order or genera (see Baer, 1951). Some parasites may live their entire life cycles on one host, while others require more than one host (Figure 16.9).

Successful parasitism represents something of a compromise between two living populations. Parasites and hosts that evolved together have developed a sort of mutual toleration, with a low-grade, widespread infection that remains as long as conditions are favorable for the host.

SOCIAL PARASITISM

Another form of parasitic relationship is social, in which one organism is dependent on the social structure of another. Social parasitism may be temporary or permanent, facultative or obligatory, within a species or between species. Four forms of social parasitism can be defined in terms of those types of relationships.

The first is a temporary facultative parasitism within a species. That type is rather well developed among ants and wasps. For example, a newly mated queen of the wasp genus *Polistes* or *Vespa* will attack an established colony of her own species and displace the resident egg-carrying queen (E. O. Wilson, 1975). Intraspecific nest parasitism is another example. Parasitic females lay eggs in nests of host females of the same species. It is well developed among ants and wasps and among certain groups of birds, especially waterfowl (Anatidae). Among these waterfowl are black-bellied tree ducks *(Dendrocygna autumnalis),* goldeneyes *(Bucephala clangula),* and wood ducks *(Aix sponsa),* hole-nesting species. The host female responds to parasitism by reducing the size of her clutch by a number that will adjust the final clutch size to a given number of eggs. The earlier the parasitic female lays her eggs in her host's nest, the greater will be her share of the clutch. Anderson and Eriksson (1982) suggest that such brood parasitism evolved among those waterfowl because suitable nest sites are scarce, nests are easy to locate, and ducks do not defend their nests. Just as important, female waterfowl have a strong affinity for their places of birth. That makes it likely that the host and the parasite are related genetically.

A second type of social parasitism is temporary facultative parasitism between species. An example can be found among the formicine ant genus *Lasius*. A newly mated queen of the species *L. reginae* will enter the nest of a host species, *L. alienus,* and kill its queen. The *alienus* workers will care for the *reginae* queen and her brood. In time the *alienus* workers, deprived of their own queen and thus replacements, die out, and the colony then consists of *reginae* workers.

A third type of social parasitism is temporary obligatory parasitism between species. Although common in ants, the most outstanding examples are obligatory egg or brood parasitism in birds. Brood parasitism has been carried to the ultimate by the cowbirds and Old World cuckoos, both of which have lost the art of nest building, incubation, and caring for the young. They pass off these duties to the host species by laying eggs in their nests. The brown-headed cowbird of North America removes one egg from the nest of the intended victim, usually the day she is to lay, and the next day lays one of her own as a replacement. Some host birds counter by ejecting the egg from the nest. Others hatch the egg and rear the young cowbird, usually to the detriment of their own offspring. The host's young may be pushed from the nest or die from lack of

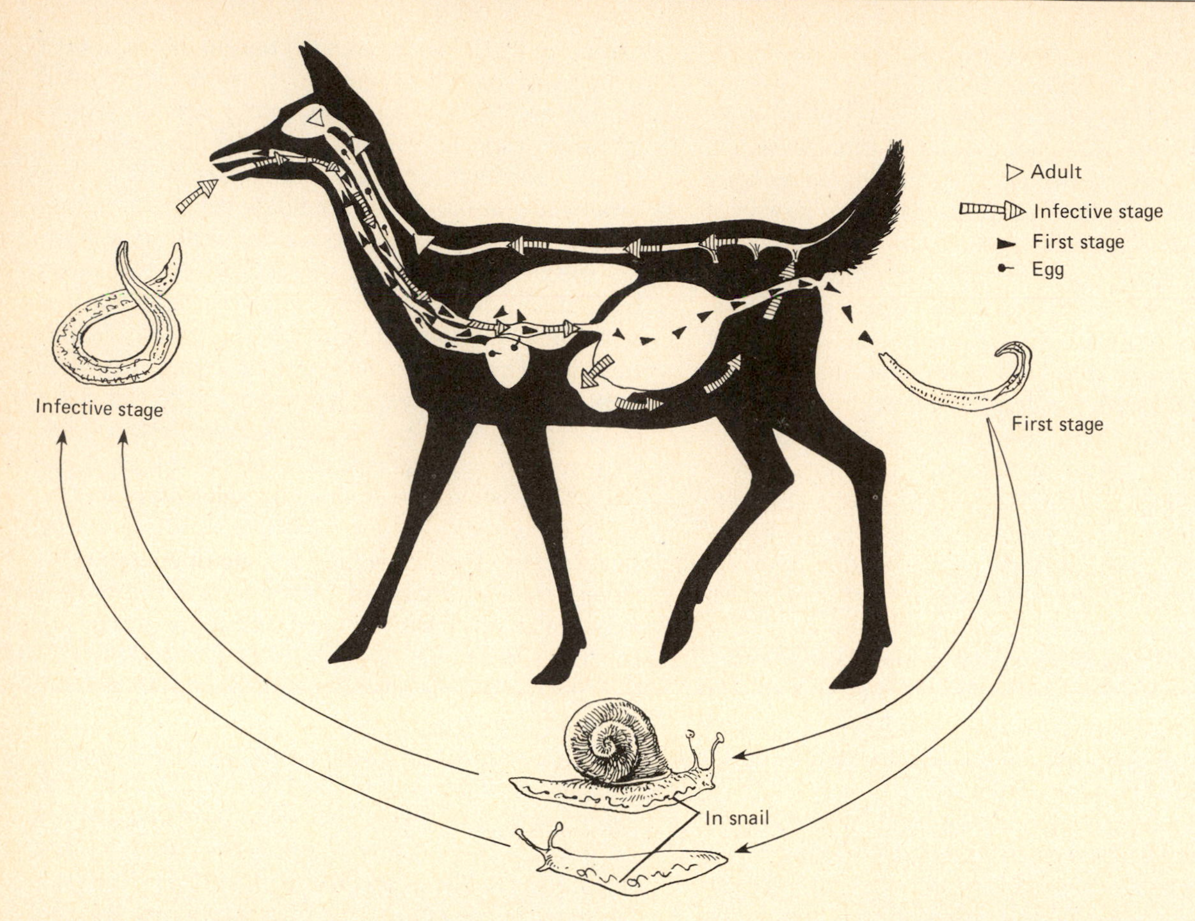

FIGURE 16.9 The life cycle of a parasite requiring an alternate host, the brainworm (actually a lungworm) *Parelaphostrongylus tenuis* in white-tailed deer (but transmissible to moose and elk). Adult worms in the host, the deer, release eggs into the venous circulation. The eggs reach the lung as emboli (undissolved material). In the lungs the eggs develop into first-stage larvae and hatch. The larvae move into air spaces in the lungs and pass up the bronchial tubes. Subsequently, the larvae are swallowed and pass in the feces. In the external environment the first-stage larvae invade the foot of terrestrial snails as they move across feces. There the larvae grow, molt twice, and give rise to the third and infective stage. Deer acquire the third stage by ingesting infected snails on vegetation. In the alimentary canal the larvae leave the tissues of the snail and penetrate the gastrointestinal wall, cross the peritoneal cavity, and follow the lumbar and other nerves to the vertebral canal, a journey of about 10 days. The migrating third-stage larvae enter the dorsal horns of the gray matter of the brain, where they develop for about a month. Forty days after infection, the subadult worms enter the spinal subdural space. Here they mature and migrate anteriorly to the cranium, where they deposit eggs in the venous circulatory system. Some females may lay eggs in the brain, and the eggs develop to the first stage in the meninges. (After U. R. Strelive in Davidson et al., 1981:141.)

food because of the more aggressive nature and larger size of the young cowbird.

A fourth type of social parasitism is permanent obligatory parasitism between species. The parasitic form spends its entire life cycle in the nest of the host (E. O. Wilson, 1975). That type of social parasitism is common among ants and wasps. In most cases the species are workerless, and queens have lost the ability to build nests and care for the young. The queen gains entrance to the nest of the host and either dominates the host queen or kills her outright and takes over the colony.

Optimal Foraging

Have you ever watched for some length of time a robin hopping across a lawn? It flies in, lands, hops along for a short distance as if sizing up the situation, and stops. Then it moves on deliberately in a series of irregular paths across the grass. The bird pauses every few feet, stares ahead as if in deep concentration or cocks its head toward the ground. It either moves on or crouches low as if to brace itself. It pecks quickly in the ground and pulls out an earthworm. On occasions the earthworm pulls back, and in the tug of war the worm wins. The robin does not push the action. It lets go and hops to another spot to repeat the food-seeking activity.

If you analyze the behavior of the robin, it can be divided into four parts. The robin had to decide where to hunt for food. Once on the area, it had to search for palatable items of food. Having located some potential food (in our situation the earthworm just below the surface of the ground), the robin had to decide whether to pursue it or not. If it began the pursuit, by pecking in the ground, then the robin attempted to follow through with a capture, in which it might or might not be successful. By capturing the earthworm, the robin earned some units of energy. Hopefully, the robin gained more energy than it expended in its round of foraging activity.

The problem facing the robin, and all other animals, is securing sufficient energy to maintain itself, feed its young, and lay up fat for migration or winter, yet not spend too much time doing so. Time, in the parlance of the robin, is energy. To spend too much time in securing energy without a sufficient return for the effort results in a type of personal bankruptcy—a loss of fitness. If the robin fails to find suitable and sufficient food in the area of lawn it is searching, it leaves to search elsewhere. But if successful, the bird probably will return until the spot is no longer an economical place to feed.

The means animals employ to secure food are termed their *foraging strategy.* Of particular theoretical interest to ecologists is the optimal foraging strategy, against which actual foraging strategy can be measured. An optimal foraging strategy is one that provides the maximum net rate of energy gain, thus endowing the animal with the greatest fitness. A foraging strategy involves two separate but related components. One is optimal diet; the other is optimal foraging efficiency. Ecologists have come up with certain rules (hypotheses) for optimal foraging.

OPTIMAL DIET

When the robin searches the lawn for food, it should, by optimal foraging theory, select only those items that provide the greatest energy return for the energy expended. That places an upper and lower limit on the size of food items accepted as well as the degree of palatability. If a food item is too large, it requires too much time to handle; if too small, it does not deliver enough energy to cover the costs of capture. Our robin should reject or ignore less valuable items such as small beetles and caterpillars and give preference to smaller and medium-sized earthworms. These more valuable food items would be classified as preferred food, ones taken out of proportion to their availability relative to all other food items. Our robin, according to optimality theory, should take the most valuable food items first. When those items have been depleted, the robin should turn its attention to the next most valuable food items. Eventually, the animal should expand its diet to include the poorer foods when the discovery of high-value foods falls below a certain rate.

Although such a theory makes practical sense, it is difficult to test in the field. Not only would one have to know exactly what items the animals were consuming; one would also have to know the relative availability of the food items in the habitat. Some studies have been done under controlled conditions involving birds (great tit, *Parus major;* Krebs, Kacelnick, and Taylor, 1978), fish (brown trout, *Salmo trutta;* Ringler, 1979), and invertebrates (shore crabs, *Carcinus maenas;* Elner and Hughes, 1978). Another involving bluegill sunfish *(Lepomis macrochirus)* was done by Werner and Hall (1974). They presented a group of 10 bluegills with three sizes of *Daphnia* in a large aquarium. After a period of foraging, the fish were killed and their stomachs examined to determine the number and size of *Daphnia* taken. When the density of the prey presented was low, the fish consumed the three sizes according to the frequency encountered. They showed no preference for size (Figure 16.10). When the prey population was dense, the fish consumed the largest prey items. When presented with an intermediate number, the fish consumed the two largest size classes. The results of these feeding trials support optimal foraging theory.

Feeding trials involving the shore crab and its prey, the blue mussel *(Mytilus edulis),* gave much the same results (Elner and Hughes, 1978). The crabs' diet extended to smaller mussels as the preferred size became scarce. But at no time did crabs exclude small mussels completely from their diet, even when larger mussels were more than abundant enough to fill them. The crabs did not pass up good food when encountered, even though it did come in smaller packages. That is probably the way animals in the wild respond.

Two field studies provide some insight into how animals forage under natural conditions. J. Goss-Custard (1977a, 1977b) studied food selection by redshanks *(Tringa totanus)* on mudflats containing different sizes of polychaete worms. He found

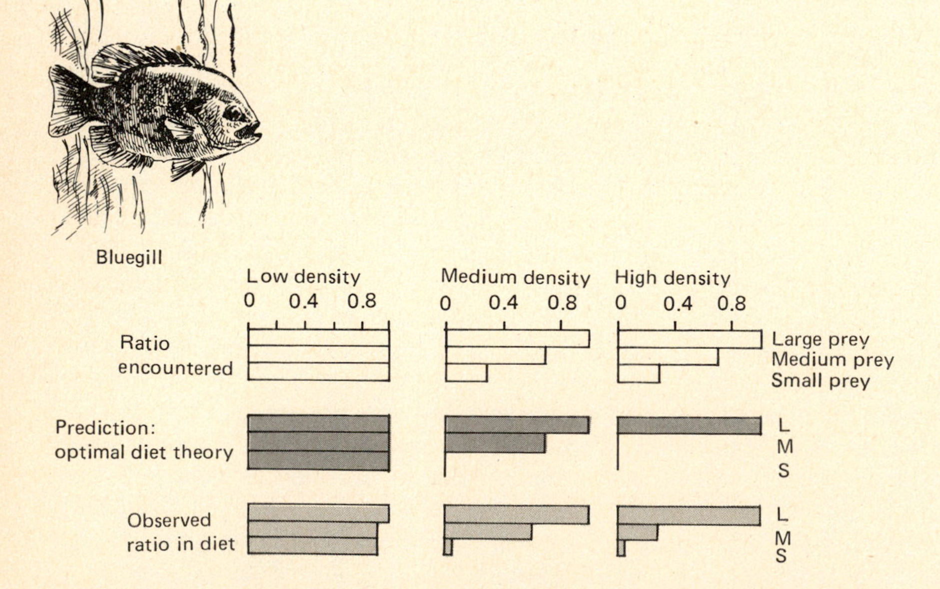

FIGURE 16.10 Optimal choice of diet in the bluegill sunfish preying on different sizes of *Daphnia*. The histograms show the ratio of encounter rates with each size class at three different densities, the prediction of optimal ratios in the diet, and the observed ratio in the diet. Note the preference of the bluegill for large prey. (After Werner and Hall, 1974:1048.)

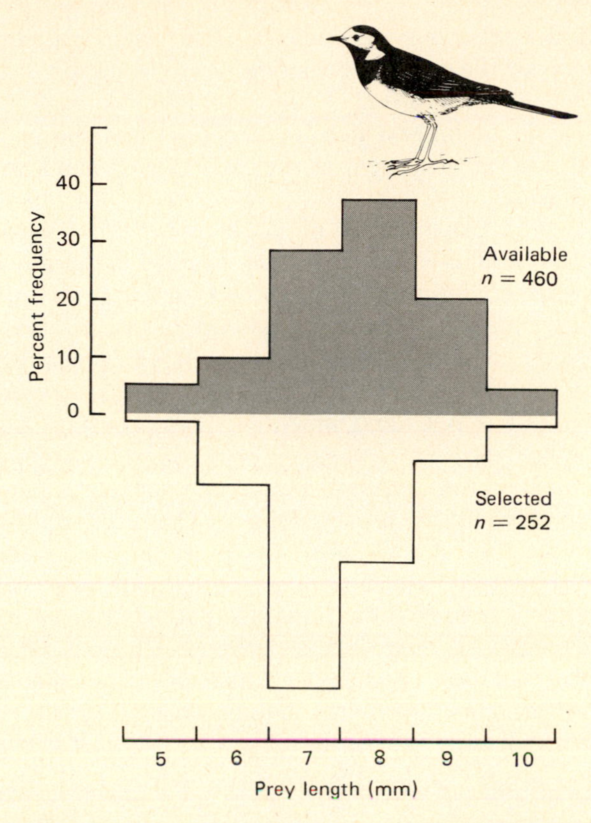

FIGURE 16.11 Pied wagtails show a definite preference for medium-sized prey, which are taken in disproportionate amounts compared to sizes of prey available in the environment. (Davies, 1977:48.)

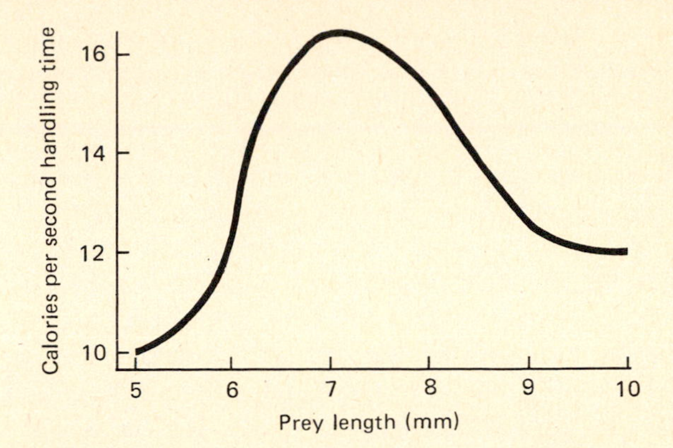

FIGURE 16.12 The prey size chosen by pied wagtails is the optimal size for providing a maximum amount of energy per handling time. Small sizes provide too few calories. Large sizes require too much handling time. (Davies, 1977:48.)

that as the number of large worms increased, redshanks became more selective and tended to ignore small worms, regardless of how common they were, as long as the density of large worms remained high. These field studies were supported by laboratory studies.

N. Davies (1977) studied the feeding behavior of the pied wagtail *(Motacilla alba)* and the yellow wagtail *(M. flava)* in a pasture field near Oxford, England. The birds fed on various dung flies and beetles attracted to cattle droppings. Prey of several sizes were available: large, medium, and small flies and small beetles. The wagtails showed a decided preference for medium-sized prey (Figure 16.11). The size of prey selected corresponded to the optimum-size prey the birds could handle profitably (Figure 16.12). The birds ignored small sizes. Although easy-to-handle small prey did not return sufficient energy, large sizes required too much time and effort to handle.

OPTIMAL FORAGING EFFICIENCY

The robin lives in a heterogeneous, or patchy, environment. It is made up of clumps of trees, shrubby thickets or plantings, and open lawn. The robin could forage in all these patches of vegetation, utilizing all food items it came across, but it doesn't. Rather, the bird concentrates its foraging activities in patches of open lawn. Moreover, the robin may restrict its searching to certain parts of the lawns where the environment is more favorable for earthworms. The bird may concentrate its attention on the most productive patches until the earthworms are depleted or have moved deeper in the soil in response to a drying surface. Then the robin is forced to turn its attention to other patches and other foods, such as ground beetles, flies, sowbugs, snails, and millipedes.

By its behavior the robin has been following the rules of optimal foraging: (1) concentrate foraging activity in the most productive patches; (2) stay with those patches until their profitability falls to a level equal to the average for the foraging area as a whole (Figure 16.13); (3) leave the patch once it has been reduced to the level of average productivity; (4) ignore patches of low productivity.

How well animals go by these rules has also been the object of experimentation, both in the laboratory and in the field. S. F. Hubbard and R. M. Cook (1978) studied the foraging behavior of a parasitoid, the ichneumon wasp *Memeritis canescens*, in a laboratory arena containing patches of the host, larvae of *Ephestia cantella*. Hosts in densities of 64, 32, 16, 8, and 4 were placed in petri dishes filled to the brim with plaster of paris, the hosts' substrate. Space between the larvae was filled with wheat bran. One result of the experiment showed that all patches were depleted of larvae to a common level of host abundance. The richest patches suffered the greatest depletion. As exploitation proceeded, the amount of time spent by the wasp in patches of highest density declined, and the proportion of time spent in the next richest patch increased.

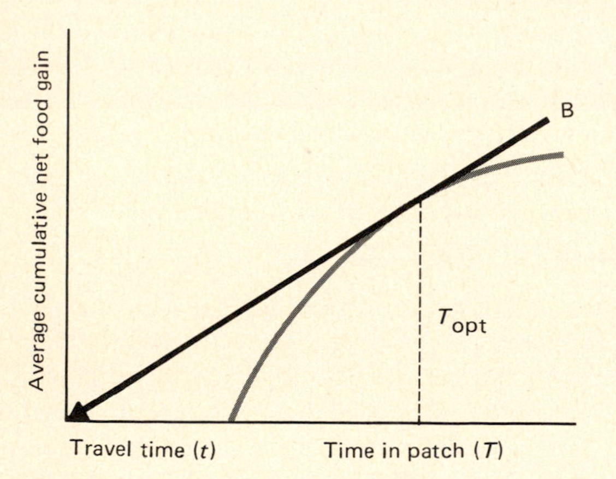

FIGURE 16.13 How long should a predator remain in a habitat patch seeking food? This graph provides a theoretical answer. Time spent in travel and time spent in a habitat patch are plotted against food depletion (net accumulation of gain in food, which declines as available food is depleted). The curve represents the cumulative amount of food harvested relative to time in the patch (which is high early in time). The straight line represents average food intake per unit of time for the habitat as a whole. Where line AB touches the curve represents the point at which the predator has reached average cumulative net food gain for the habitat as a whole. Beyond this point net food gain declines below average. It is no longer profitable for the predator to remain. Thus, the point represents the optimal time (T_{opi}) for the predator to leave and seek a more profitable food patch. (J. R. Krebs, 1978:42.)

The experiments of Zack and Falls (1976a, 1976b, 1976c) studied foraging strategy under more natural conditions. They exposed captive ovenbirds individually to a patchy food supply (mealworms) presented in natural outdoor pens in typical habitat.

In one experiment Zack and Falls presented four patch locations, which they held constant, although they interchanged prey densities. They found that the ovenbirds increased their search path exponentially with prey density. The birds rapidly shifted their search efforts as prey densities were interchanged. Because less search path was required per prey found in dense patches, the birds concentrated their efforts in areas of high profitability and took a higher percentage of prey available in these sites. Ovenbirds did not always visit every patch location during the observation periods, but they always visited the densest patches. That suggests the bird's discovery of one or more profitable feeding areas discourages the sampling of other sites to assess their profitability. The ovenbird's tendency to quit searching after encountering one or more profitable patches and its ability to learn the location of and return to patches of high prey density may limit the number of patches the bird will exploit. The ovenbird may utilize other patches only if it discovers them by chance.

In another set of experiments Zack and Falls (1976b) exposed ovenbirds to various sets of patches. When the birds were exposed to a single patch, they quickly concentrated foraging there. The birds took equal amounts of food at different prey densities and did not vary the amount of search path per visit. However, as patch density increased, ovenbirds decreased the number of visits to the patch and the amount of search path per prey located. Also, the birds reduced exploratory search outside the patch at high prey densities. That suggests that with low depletion and high renewal rates, a single profitable feeding site might be enough.

In a second experiment Zack and Falls presented two prey patches successively. Prey in patch 1, presented on day 1, was not renewed on day 2. Instead, a new patch with prey was presented. After an initial visit to patch 1, the birds quickly abandoned the first patch location and, after some exploratory behavior, rapidly concentrated their search efforts in the new location.

Zack and Falls (1976c) also investigated when ovenbirds would give up the search in one area and move on to the next. They discovered that the birds did not take some optimal number based on training and previous experience. Instead, they learned rapidly to find patches of prey; they chose feeding sites nonrandomly and avoided areas of no food and patches visited previously. Ovenbirds improved their foraging efficiency by searching nonrandomly within the patches, avoiding areas already exploited. By doing so, ovenbirds were less likely to deplete their prey. If the birds followed a systematic search pattern, they gave up and left when a patch was completely covered. The time at which they gave up the search was unrelated to prey density.

Observation and experimental studies support the hypothesis of optimal foraging up to a point. It is not surprising that they should. Much optimal foraging theory concerns actions you would expect any mobile animal to take: forage in areas where food is abundant, leave when searching is no longer rewarding, select the most palatable and larger items of food, leave the poor items until last, and don't travel any further than necessary to feed. Where the theory breaks down is in the expectation that the animals will select patches in order of their profits or take only optimal food items first and ignore the rest. Such choices may be characteristic of animals foraging in a stable laboratory environment. It is not necessarily the way animals behave in the wild.

Being opportunists, animals will take some less than optimal food items (by ecologists' standards) upon discovery, and they may quit before food items are reduced to some minimal level. Nor will they pass up certain profitable patches because they do not meet some theoretical expectation. Animals quickly learn where food is and where food is not, and they don't waste much time on a patch after it is depleted. Foragers, however, will stay with a patch as long as the rate of replenishment exceeds the rate of depletion. Some animals are highly restricted in their choices. Sedentary animals such as corals, barnacles, and black fly larvae, filter feeders all, have to take what food flows past them. Others have severely restricted foraging patterns that limit their choices of food.

In spite of its weaknesses, the concept of optimal foraging is valuable. It provides a basis or scale against which to measure or compare how animals do forage. By doing so, the theory provides a better insight into how animals select and harvest their food.

Summary

Predation is the consumption of one living organism by another, a relationship in which one organism benefits at the other's expense. However, there is a close interaction between predator and prey. Each can influence the fitness of the other. In its broadest sense predation includes herbivory, parasitism, and cannibalism.

The interaction between predator and prey involves a functional response and a numerical response. A functional response is one in which the predator takes more prey as the density of its prey increases. A numerical response is one in which the number of predators increases as the prey increases.

There are three types of functional responses: Type I, in which the number of prey affected increases in a linear fashion to a maximum as the prey increases; Type II, in which the number of prey affected increases at a decreasing rate toward a maximum value; and Type III, in which the number of prey taken increases in a sigmoidal fashion as the density of prey increases.

Type III response may involve both a search image and switching. Predators develop a search image for a particular prey as that item becomes more abundant. Switching takes place when a predator turns to an alternate, relatively more abundant prey and takes that prey in a disproportionate amount relative to other prey species. Of the three types of functional response, only Type III is important as a population-regulation mechanism.

A particular form of predation is cannibalism, or intraspecific predation. It often but not necessarily occurs when a population is under stress. Cannibalism is particularly common among insects and fish, in which predatory adults feed on grazing young. Cannibalism can become an important mechanism of population control.

Another specialized form of predation is parasitism, in which small organisms live on or within their hosts and draw nourishment from them without necessarily killing them. Parasites whose larval forms draw sustenance from the hosts, ultimately killing them, function much like true predators. They are called parasitoids.

Social, or nest, parasitism occurs when a parasite lays its eggs in the nest of another. Social parasitism may be nonobligatory and intraspecific. The female lays part of her clutch in the nest of another female of the same species. That type of parasitism is widespread among waterfowl. Interspecific nest parasitism is obligatory because the female builds no nest of her own. In both cases the parasites increase their own fitness at the expense of the fitness of the host.

Central to the study of predation is the concept of optimal foraging, a strategy that obtains for the predator a maximum rate of net energy gain. There is a break-even point above which foraging is profitable and below which it is not. Optimal foraging involves an optimal diet, one that includes the most efficient size of prey for both handling and net energy return. Optimal foraging efficiency involves the concentration of activity in the most profitable patches of prey and the abandonment of those patches when they are reduced to the average profitability of the area as a whole.

Review and Study Questions

1. Define parasitism, parasite, herbivory, and cannibalism.
2. What are the basic features and underlying assumptions of the Lotka-Volterra equations for predation?
3. What is a functional response to predation? A numerical response?
4. Distinguish among Type I, Type II, and Type III functional responses. Why are Type I and Type II not important in population regulation?
5. What is meant by search image? Switching? How do they relate to Type III responses?
6. Speculate about why cannibalism is as widespread as it appears to be among animals. What triggers cannibalism, and what are its demographic effects?
7. What are some special adaptations of parasites? Why is it disadvantageous for parasites to kill their hosts?
8. What is social parasitism? Speculate about why nest parasitism might have evolved and why some individuals engage in intraspecific nest parasitism.
9. What is optimal foraging strategy?
10. According to optimal foraging theory, what constitutes an optimal diet?
11. How is foraging influenced by a patchy environment?

Suggested Readings

Curio, E. 1976. *The ethology of predation*. New York: Springer-Verlag.

Krebs, J. R., and N. B. Davies, eds. 1978. *Behavioral ecology: An evolutionary approach*. Oxford, England: Blackwell Scientific Publications.

Kruuk, H. 1972. *The spotted hyena*. Chicago: University of Chicago Press.

Taylor, R. J. 1984. *Predation*. New York: Chapman and Hall.

Zaret, T. M. 1980. *Predation and freshwater communities*. New Haven, Conn.: Yale University Press.

Note: See also the "Suggested Readings" section at the end of Chapter 13.

ECOLOGICAL EXCURSION

Winter Crisis at Goose Lake Marsh

Paul L. Errington

Of all the "Ecological Excursions," this one requires some comment. The selection is taken from Paul Errington's Muskrat Populations. *The book is a summary of Errington's 25 years and 30,000 hours of field study of the muskrat* (Ondatra zibethicus) *and its major predator, the mink* (Mustela vison). *Errington, professor of zoology at the University of Iowa until his death in 1962, earned an international reputation as an authority on vertebrate predation. He studied a number of predator-prey relationships, but his chief subjects were muskrats and mink. Out of his enormous amount of field work, Errington developed a theoretical base for understanding the interaction between vertebrate predators and prey, which he summarized in a classic paper, "Predation and Vertebrate Populations"* (Quarterly Review of Biology *21:14–177, 221–245). He demonstrated that vertebrate predators rarely regulated prey populations, that predators lived mostly on the victims of intraspecific strife and deteriorating habitat. In addition to being an outstanding field scientist, he was also able to present his findings in fluently written prose. Consider this account of an incident in the history of the muskrat population on Goose Lake Marsh in central Iowa. He tells of the plight of muskrats facing a major crisis during a period of drought that carried on into the winter. What was a time of crisis for the muskrat was a time of plenty for the mink. But, as Errington shows, predation had little effect on the muskrat population. It was environmental conditions that were the ultimate causes of the devastation of the muskrat population, with predation by mink being the proximate cause. Predation or not, the muskrat population was doomed.*

SOURCE: Paul Errington, *Muskrat Populations* (Ames: Iowa State University Press, 1963), 265–267.

Until late January, the population of the wetter north half was visibly secure. It was not known to have suffered any losses whatever since fall, despite intrusions into lodges by probably three minks. The minks fed chiefly on fishes congregated in the muskrat channels beneath the lodges, on remains of dead ducks and coots, and to some extent on mice and rabbits of the surrounding land.

But, early in February, the wet marsh of the north half was freezing to the bottom. Muskrats were traveling in increasing numbers on top of the ice from one lodge to another in search of better quarters. Mortality began with a single mink victim about the first of the month. On February 6, an animal, blind and with flesh of tail frozen and chewed away, died from cold on the ice. Four muskrats (including one marked by toe-clipping) were known to have abandoned the marsh as the crisis became worse, and others doubtless did also.

The night of February 8 (the second of two cold nights during which temperatures fell to about 15 degrees below zero Fahrenheit) virtually completed the transition from tolerable to intolerable living conditions for the remaining muskrats of Goose Lake, although a few of the better lodges retained some degree of habitability for another week or longer. In less than three weeks, Goose Lake was devoid of living muskrats, very spectacularly as a result of their increasing vulnerability and through the agency of minks.

Not all of the February mink killings could be figured out from signs, but a relatively complete record of the exploitation by the minks was obtained from visits to the marsh on 15 of 18 days during the period of greatest mortality and from occasional visits thereafter. It is believed that the fates of very nearly all of the muskrats dying at Goose Lake after February 8 were accounted for—20 in all.

Prior to the latter date, most of the muskrat activity on top of the ice had been in the southwest and southeast corners of the wetter north half where restless animals had explored unoccupied lodges and the burrows and vegetation of the shore. On the night of February 8, minks were known to have killed a muskrat in a land hole southwest of the north half and another one on or near the shore of the southeast corner. The final places of deposition of the carcasses could not be ascertained because of poor tracking conditions. Little could be learned of the circumstances of death of a third victim except that it occurred. On February 9, a mink killed three muskrats in the southeast corner of the wetter north half—apparently all there were—and left them piled uneaten inside of a lodge.

In late afternoon of February 11, five muskrats left the main lodge in the southwest corner for a smaller unoccupied lodge 45 yards away, where two of them remained. The other three continued to another unoccupied lodge 80 yards farther on. In the evening or night, a mink made three round-trip visits from shore to the latter lodge, dragging away two muskrats (to cache them in different holes in the upper chambers of abandoned muskrat burrows 55 to 75 yards away) and returning unsuccessful from the third visit. The third muskrat here was alive the next day, but the mink returned in the evening (February 12) to kill it and drag it 265 yards across the snow to another land hole. The mink later visited the first-mentioned lodge having the two muskrats, killed one and dragged it 190 yards to the same land hole in which it had deposited its first kill of the night. It returned for the last muskrat of the original five that had journeyed forth from the main lodge in the southwest corner, injured it but did not kill it. This muskrat returned to the main lodge, staggering in the snow and leaving blood as it rested; an animal corresponding to it died in the lodge, to be eaten by its fellow muskrats.

No sign of mink was seen for the night of February 13 in the southwest corner of the north half of Goose Lake. More muskrats left the main lodge to station themselves in the smaller lodge 45 yards away. During the night of February 14, the mink came back and killed three muskrats in and about the smaller lodge. It left two carcasses cached where killed and dragged the third to another small lodge near by.

During the night of February 15, one of the two dead muskrats cached as above was carried away by a mink, and a mink visited a place in the east central part of the north half where hitherto secure muskrats were coming out on the ice. Here, the mink killed two and left them uneaten (except for a few bites taken out of the neck of one) in a feed house. One of the victims had liver lesions of the hemorrhagic disease. The next night, the mink returned to kill and eat a third muskrat.

By February 16, only one place on the marsh was known to have living muskrats, and one muskrat was killed and largely eaten by a mink during the night. A second was killed and left uneaten on the night of February 20. A third—evidently the very last one—was killed on February 27. On March 1 and 5, checkups were made under favorable field conditions without finding any sign of muskrats remaining alive, in lodges, land holes, or anywhere else about Goose Lake.

Of 40 mink scats deposited from February 6 to February 26, 38 consisted of muskrat remains. After annihilation of the local muskrats, the minks (still about 3 individuals distinguishable on the basis of tracks) fed upon muskrat carcasses cached in various land holes and lodges for perhaps another week and even continued visiting the fragments of some carcasses past the middle of March. Only 2 of 9 mink scats found for this period contained muskrat remains, nevertheless, and it could be seen that this source of food soon became exhausted. Apart from 3 carcasses cached in the bottom of a lodge that filled with water during a thaw and froze over, nearly all of the muskrats killed by the minks in February were sooner or later eaten by minks, though often some days or weeks after the actual killing.

Altogether, mortality of 74 muskrats was detected at Goose Lake from the beginning to the end (late March) of the 1947–48 drought emergency.

Compared with other winter drought crises observed, the one at Goose Lake in February, 1948, was singularly severe in view of the depth of the water over the muskrat-occupied parts at freeze-up. In the winter of 1945–46, muskrats had, in fact, lived with considerable security in the rush-grown south half of the marsh after freezing of the shallow surface water and much of the mud underneath; but they still had had access to rootstocks and other rich sources of foods imbedded deep in unfrozen mud. In 1947–48, except for the vegetation making up the lodges, the entire food supply of the muskrats of the north half became encased in ice as cold weather continued—the coontail and pondweed growths, the rootstocks of yellow water lily, and the fishes that found their last refuge in the muskrat channels—and the muskrats simply could not gnaw it out in satisfactory quantities. For a time, certain individuals seemed to specialize on frozen fishes, and there was feeding by muskrats on remains of some of the muskrats killed by minks; yet, as the water of different parts of the north half froze to the bottom, the muskrats broke out of their lodges virtually in groups, to search diligently about the vicinity for more livable quarters. As a rule, the muskrats were in fair to good flesh and uninjured by cold at the times of their death, but their alimentary tracts tended to be rather empty or filled with harsh material. It was apparent that they suddenly had become hungry and desperate.

Another distinction having a bearing on this winter crisis should be pointed out. Instead of the large, insulating lodges of cattail and bulrush marshes, the lodges of the north half of Goose Lake were small shells of ice-lined coontail and algal masses, heaped around holes or cracks in the ice. Even the largest were eaten-out shells, having disproportionately large basins within, and very limited ramifications of burrows and channels underneath. Inferior for protection during cold snaps, caving in during thaws, requiring frequent repairs whatever the weather, these flimsy structures doubtless did not impose the handicap on their occupants as did the sealing of the food supply by the ice, but they surely were not much of an asset when the crisis came. Indeed, Iowa muskrats may sometimes remain alive for weeks in midwinter in drought-exposed marshes—the necessity for outside foraging and the presence of minks notwithstanding—if only they can withdraw into the better types of lodges.

On the whole, the February situation exemplified a combination of adverse factors beyond muskrat limits of toleration. The mink predation followed standard patterns and was appraised as being symptomatic of vulnerability rather than a true cause of population decline of the muskrats. Except possibly for a group of three muskrats living partly in a land burrow and partly in adjacent lodges, few of the animals killed by minks would have stood, under the existing circumstances, much chance of survival even in the absence of the minks.

CHAPTER 17

Coevolution

Outline

Objectives

Upon completion of this chapter, you should be able to:

1. Define coevolution.
2. Relate coevolution to predator-prey systems, seed dispersal, plant pollination, and mutualism.
3. Explain plant-herbivore systems and herbivore-carnivore systems.
4. Discuss the relationship of herbivory to plant fitness and herbivore fitness.
5. Explain the relationship of plants, herbivores, and predators to predator-prey cycles.
6. Discuss some mechanisms of prey defense.
7. Discuss mechanisms predators have evolved to counteract prey defenses.
8. Discuss coevolution of seed-dispersal systems.
9. Explain mutualism and discuss how mutualism might have evolved.

I hesitate to title this chapter "Coevolution" because it may imply that all pairs of predators and prey, parasites and hosts, seeds and their dispersers, plants and pollinators are the outcome of coevolution. *Coevolution* is the joint evolution of two interacting populations in which the selection pressures are reciprocal. The trait of one species evolves in response to the trait of another, and that trait has evolved in response to the trait of the first (Janzen, 1980). Any evolutionary change in one member changes the selection forces acting on the other. It is basically a game of adaptation and counteradaptation to changing selection pressures imposed on one taxon by the other.

Such a restricted definition implies that interacting species "grew up" together over evolutionary time. There is no proof that contemporary examples of coevolution did so (Janzen, 1980). The chances are that many coevolutionary pairs that are now the objects of study are not inhabiting the environments in which they evolved. Instead, the organisms probably evolved in a mix of different habitats through time, each with somewhat different selective pressures. When these plants or animals invaded or found themselves in a different habitat, they adjusted to the plants or animals at hand. If the traits already acquired fit the needs of the plant and herbivore or herbivore and carnivore, then the two interact in a manner that suggests long-term convergent evolution. If the fit is good, then further evolutionary changes are minor. In such a manner a highly coevolved system can result with no evolutionary change in either partner. The relationship would then continue to be selected for by current interaction.

One example, described by D. Janzen and P. Martin (1982), is the relationship between some New World flora in Costa Rica and two domesticated large herbivores, horses and cattle, which replaced the large herbivores that became extinct 10,000 to 20,000 years ago. In the lowland forests of Costa Rica, two trees, among others, *Crescentra alata* and *Pithecellobium saman,* have hard, ripe fruits that are ignored by native dispersal agents. But the fruits of *Crescentra* are eaten by range horses and those of *Pithecellobium* by cattle. The seeds pass through the guts of both unharmed, while the fruits of many other trees go uneaten by any herbivore and rot on the ground. Undoubtedly, these fruits and seeds coevolved with Pleistocene herbivores. With their demise, one member of the pair was lost until domesticated herbivores were introduced into the area. If it were not for the fact that the fruit predators were domesticated, one could easily interpret the relationship as coevolution. A similar type of relationship probably exists among many plant-herbivore and predator-prey pairs.

Another approach to coevolution is to consider it as a less restrictive, more general response of groups of species to one another. A particular trait evolves in one or more species in response to a trait or suite of traits in several other species. Plants could have evolved chemical and physical defenses against a diverse array of herbivorous insects. In turn, many insects evolved the ability to detoxify a wide range of plant chemicals. Similarly, animals might have evolved a generalized immune system in response to a wide range of parasites. Plants adapt to nonspecific pollinators and insects to the flowers they pollinate. Such interactions are termed *diffuse coevolution.* This contrasts with a pairwise response between one species and another, such as a specialized parasite and its host.

The concept of diffuse coevolution more closely matches the original concept of coevolution proposed by P. Ehrlich and P. Raven (1964). In a paper on butterfly habits and plant chemistry, they introduced but did not define the term. However, they viewed coevolution as a stepwise response of an organism to another resulting in a series of adaptive radiations (see Chapter 3). Plants acquired a diverse spectrum of secondary compounds early in their history in response to herbivory. Evolution of chemical defenses enabled plants to colonize new habitats. An array of insects in those new habitats evolved in response to plant defenses. Their idea of coevolution did not refer to reciprocal adaptation. Instead, insects evolved adaptations to plant defenses, and plants evolved defenses against herbivory, but seldom was genetic change in either highly coupled.

While most coevolved traits appear to be more diffuse than pairwise, pairwise interactions do seem to have evolved in some instances, especially between parasites and their hosts.

Predator-Prey Systems

The usual approach in ecology is to consider predation on plants and predation on animals as two separate entities. But they are not separate. Predator-prey interactions at one trophic level influence predator-prey interactions at other trophic levels. We cannot really understand a herbivore-carnivore system without understanding plants and their herbivores; nor can we understand plant-herbivore relations without understanding predator-herbivore relationships. All three trophic-level interactions are related, but our understanding of these interrelations is meager. Mentally we still operate on two trophic levels.

PLANT-HERBIVORE SYSTEMS

A deer pulling tender young twigs from a sapling, horses clipping grass in a pasture, caterpillars voraciously chewing tree leaves, bark beetles feeding on cambium, chipmunks eating acorns—all are instances of predation on plants. Evidence of such herbivory occurs to some extent on the leaves and fruit of almost any plant you examine.

If you measure the amount of biomass actually consumed by herbivores, it may be small, perhaps 6 to 10 percent relative to biomass available, except in years of some major insect outbreaks or in the presence of an overabundance of deer. But consumption is not a good measure of the importance of herbivory to a plant. True, the tree lives, and the grass is still green. But grazing on plants has a more subtle impact—that of affecting the fitness of both plants and herbivores.

Effects on Plant Fitness

Removal of plant tissue—leaf, bark, stems, roots, sap—affects a plant's fitness and its ability to survive, even though it may not be killed outright. Loss of foliage and subsequent loss of roots decrease plant biomass, reduce the vigor of the plant, place it at a competitive disadvantage with respect to surrounding vegetation, and lower its reproductive effort or fitness. That is especially true in the juvenile stage, when the plant is most vulnerable to increased mortality and reduced competitive ability.

Although a plant may be able to compensate for the loss of leaves by increasing photosynthetic assimilation in the remaining leaves, it may be adversely affected by the loss of nutrients, depending upon the age of the tissues removed. Young leaves are dependent structures, importers and consumers of nutrients drawn from reserves in roots and other plant tissues. As the leaf matures, it becomes a net exporter of nutrients, reaching its peak before senescence sets in. Grazing herbivores such as sawfly and gypsy moth larvae, deer, and rabbits concentrate on more palatable, more nutritious leaves. They tend to reject older leaves because they are less palatable, are high in lignin, and often contain secondary compounds (tannin, for example; see page 329). If grazers concentrate on young leaves, they remove considerable quantities of nutrients.

Plants respond to defoliation with a flush of new growth that drains nutrients from reserves that otherwise would have gone into growth and reproduction. Defoliation also draws on the plants' chemical defenses, a costly response. Often the withdrawal of nutrients and phenols from roots exposes the roots to attack by root fungi while the plant marshals its defenses in the canopy (Parker, 1981). If defoliation of trees is complete, as often happens during an outbreak of gypsy moths or fall cankerworms (*Alsophila pometaria*), replacement growth differs from the primary canopy removed. The leaves are smaller, and the total canopy area may be reduced by as much as 30 to 60 percent (Heichel and Turner, 1976). Defoliation in a subsequent year may cause an even further reduction in leaf size and number. Some trees may end up with only 29 to 40 percent of the original leaf area to produce food in a shortened growing season.

Severe defoliation and subsequent regrowth alter the tree physiologically. Growth regulators controlling bud dormancy are changed when leaves are removed. The plant uses up reserve food to maintain living tissues until new leaves are formed. Buds for the next year's growth are late in forming. The refoliated tree is out of phase with the season, and the drain on nutrient reserves adversely affects the tree over winter because twigs and tissues are immature at the onset of cold weather. Such weakened trees are more vulnerable to insect and disease attack the next year. And because nutrient reserves are used for regrowth and maintenance, no resources are available for reproduction. Defoliation of coniferous species results in death.

Some plant predators, such as aphids, do not consume tissue directly but tap plant juices instead, especially on new growth and young leaves. Sap suckers can decrease growth rates and biomass of woody plants by 25 percent.

Damage to the cambium and the growing tip (apical meristem) may be more important in some plants. Deer, rabbits, mice, and bark-burrowing insects feed on those parts, often killing the plant or changing its growth form.

Moderate grazing, even in a forest canopy, can have a stimulating effect, increasing biomass production, but at some cost to vigor and at the expense of nutrients stored in roots. The degree of stimulation depends upon the nature of the plant, nutrient supply, and moisture. In general, the biomass of grass is increased by grazing up to a point. Then biomass production declines. Adverse effects are greatest when new growth is developing. Defoliation then, as in deciduous trees, results in a loss of biomass, decreased growth, and delayed maturity (Andrzejewska and Gyllenberg, 1980). Grasses, however, are well adapted to grazing, and most benefit from it. Because the meristem is placed close to the ground, the older rather than the more expensive young tissue is consumed first. Grazing stimulates production by removing older tissue functioning at a lower rate of photosynthesis, reducing the rate of leaf aging, thus prolonging active photosynthetic production and increasing the light intensity on underlying young leaves, among other things. Some grasses can maintain their fitness only under the pressures of grazing, even though defoliation reduces sexual reproduction (McNaughton, 1979; Owen, 1980). Exclusion studies show that in the absence of grazing, some dominant grasses disappear. There is even some evidence that saliva produced by grazers stimulates the growth of grass (Dyer, 1980).

Effects on Herbivore Fitness

Herbivory is a two-way street. Plants, although seemingly passive in the process, have a pronounced effect on the fitness of herbivores. For herbivores it is not the quantity of food that is critical—usually there is enough biomass—but the quality. Because of the complex digestive process needed to break down plant cellulose and convert plant tissue into animal flesh, high-quality forage rich in nitrogen is essential. Without that, herbivores can starve to death on a full stomach. Low-quality foods are tough, woody, fibrous, undigestible. High-quality foods are young, soft, and green or are storage organs such as roots, tubers, and seeds. Most food is low-quality, and herbivores forced to live on such resources experience high mortality or reproductive failure (Sinclair, 1977). Added to the problem of quality is the chore of overcoming the various defenses of plants that make the food unavailable, hardly digestible, unpalatable, or even toxic.

Secondary plant substances can affect the reproductive performance of some mammals. Isoflavonoids in plants—usually found concentrated in legumes, particularly alfalfa and ladino clover—mimic estrogenic hormones, especially progesterone. When consumed, these isoflavonoids exert an estrogenic effect and induce a hormonal imbalance in grazing herbivores that results in infertility, difficult labor, and reduced lactation. Secondary compounds also serve as reproductive cues for some voles (Berger et al., 1981). A particular compound, 6-methoxybenzoxaxolinone (6-MBOA), rapidly stimulates reproductive effort in montane voles (*Microtus montanus*). When voles feed on grass, they stimulate the injured plant tissue to release an enzyme that converts a precursor compound abundant in young growing tissue to 6-MBOA. The ingested chemical serves as a cue that the vegetative growing season has begun. Such a chemical cue allows the voles to produce offspring when food resources will be available to them. These chemical signals are important to the voles because they live in an environment where food resources are unpredictable and depend upon the timing of snowmelt and other environmental conditions. Yearly differences in the appearance of new vegetative growth and of 6-MBOA may influence population fluctuations in this vole (Negus, Berger, and Forslund, 1977).

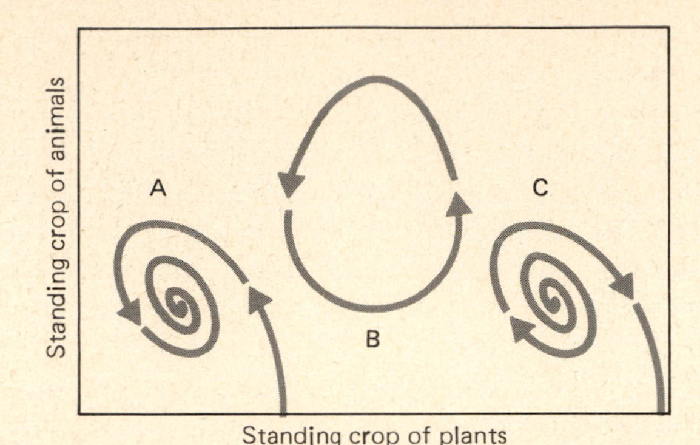

FIGURE 17.1 Diagram of possible outcomes of a herbivore-vegetation system or a predator-prey system: (A) stable equilibrium, (B) stable limit cycle, and (C) unstable equilibrium. (*After Caughley, 1976a.*)

HERBIVORE-CARNIVORE AND PARASITE-HOST SYSTEMS

Herbivory supports carnivory. Unlike herbivores, carnivores are not faced with a lack of quality in their food. The quality is there—all highly proteinaceous and easily digestible. It is quantity that frequently is lacking. That dictates a somewhat different relationship between eater and eaten. Numbers of prey become important. Fitness of the predator depends upon its ability to capture prey; and fitness of herbivores depends upon their ability to elude predation and at the same time overcome plant defenses. That sort of a combination puts a squeeze on herbivores.

A general observation is that some sort of stability exists between predator and prey. Rarely do predators exterminate their prey. On occasions the prey population may outstrip the ability of the predator to contain it. At times predators can reduce or depress prey populations. There are three possible outcomes for such interactions: stable oscillations, stable limit cycles, and unstable oscillations (Figure 17.1). The unstable oscillation is more of a laboratory than a field phenomenon.

A stable oscillation results when a prey population is self-limiting (see Chapter 13) and the predator has little impact on the prey population. Some prey species possess strong limiting mechanisms, such as territoriality. Under such conditions, as Tanner (1975) has demonstrated, the prey has a higher intrinsic rate of increase than the predator. An example is muskrats and their major predator, the mink (Errington, 1963). Muskrat populations are limited by territorial behavior. According to Errington the predator takes only the surplus prey—those that contribute nothing to the population. However, Tanner suggests from his simulation model involving intrinsic growth rates of muskrats and mink that the predator is capable of holding the prey at a density lower than the habitat could support in the absence of predation.

Some stability in predator-prey interactions may be achieved with prey populations that are not self-limiting if the rate of increase of the prey species is less than that of the predator. In such cases the predator must be self-limiting through competitive and behavioral interactions. Examples are ungulates and hares, populations of which appear to be predator-controlled. Removal of the predator from the system results in a marked increase in the prey population, especially ungulates, with a resulting strong impact on vegetation.

A stable limit cycle results when the rates of increase of the predator and the prey are approximately the same and the prey exhibits only a weak interspecific population regulation, as exemplified by the snowshoe hare and lynx. In such cases population numbers undergo well-defined cyclic changes over time (May, 1973). The amplitude of the cycle (the upper and lower numbers the population reaches during the cycle) is determined by such variables as birthrate, death rate, predation rate, and so on. The period (time to complete one cycle) is also fixed. If the system is disturbed, a stable limit cycle will return to equilibrium.

Occupancy of two habitats by the prey can dampen a stable limit cycle if the second habitat provides poor protection from predators and is inhabited by overflow or surplus animals from the optimal habitat. In some ways that is a refuge situation and is typical of snowshoe hare populations in the Rocky Mountains and central Appalachians (Tanner, 1975; Wolff, 1980).

Relationships between predator and prey and parasites and host can be manipulated easily enough on paper. But how do these interactions work out in the field under natural conditions? Several predator-prey systems have been studied in some detail in the field. One is the three-trophic-level interaction of vegetation, snowshoe hare, and lynx. The other is the European rabbit and its viral parasite myxomatosis.

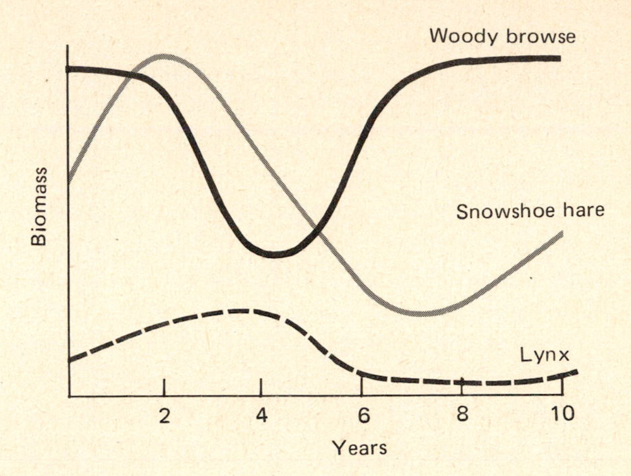

FIGURE 17.2 A graphic model of the vegetation-herbivore-predator cycle involving woody vegetation, snowshoe hare, and lynx. Note the time lag between the cycle of vegetation recovery, the growth and decline of the snowshoe hare population, and the rise and fall of the predator population. (*Adapted from Keith, 1974.*)

An Example of Predator-Prey Interaction

The basic demographic features of the snowshoe hare cycle have been outlined earlier. The pattern of the cycle is shown in Figure 17.2. The cycle is generated by a hare-vegetation interaction. A decline in peak densities of snowshoe hares is initiated by an overwinter (September to May) shortage of food. The hares' winter food is mostly woody browse. Essential browse consists largely of stems less than 3 mm in diameter, young growth with a concentration of nutrients (Keith, 1974; Pease, Vowles, and Keith, 1979; Wolff, 1980). The hare-vegetation interaction becomes critical when essential browse falls below that needed to support the population over winter, approximately 300 g per individual per day of stems 3 to 4 mm in diameter. Excessive browsing and girdling during population increase reduce subsequent increases of woody growth, bringing about a food shortage, which causes a high winter mortality of juvenile hares and lowered reproduction the following summer. In addition, plant defensive mechanisms, particularly secondary metabolites (see page 329), further reduce food availability. For example, a toxic phenol (pinosylvin methyl ether) in highly digestible winter buds and catkins of green alder (*Alnus crispa*) greatly reduces the palatability of those items, usually rejected by the hares (Bryant et al., 1983).

The decline in hares elevates the predator-hare ratio, intensifying predation on the hare that extends the period of decline and drives the population to a level below that which the habitat could support. The low hare population allows vegetation to recover and accumulate biomass. It also causes a sharp decline in predator populations (mostly lynx) because of reduced reproductive success brought about largely by the early loss of young. With the decline of predators and a growing abundance of winter foods, the hare population begins to rise sharply, starting another cycle.

The periodicity of this 10-year cycle is determined mainly by the vegetation-hare and predator-hare interactions. The length of the increase phase depends upon (1) the average rate of population growth from low to high densities and (2) the amount of biomass of essential woody browse present. The length of the decrease phase depends upon

(1) the response of woody plants to overutilization by hares, including a delayed density-dependent reduction of new growth, and (2) the intensity and duration of predation after the decline of hares from a food shortage.

Ten-year cycles of the snowshoe hare are characteristic of the boreal region. South of the boreal region, some disjunct populations of snowshoe hares exhibit cyclic fluctuations, and some do not. Why such is the case has been studied by Dolbeer and Clark (1975), Tanner (1975), and Wolff (1980). In coniferous forests and associated regions south of the boreal forests, snowshoe hares exhibit cyclic fluctuation only in a uniform environment of spruce and fir (which, of course, is also characteristic of the boreal forest). But in regions where the environment is very patchy, where many kinds of vegetation patterns exist, cycles do not occur. Hares occupying high-quality habitats are protected from predation, while hares living in areas of poor cover are subject to predation. In effect, patches of high-quality habitat act as refuges from which surplus animals are able to repopulate poorer habitat patches. These hares, in turn, are eliminated by predators. Such dispersal and predation tend to hold the population of hares in better habitats at a level at which they do not overutilize their food supply, thus damping cyclic behavior.

An Example of Parasite-Host Interaction

The parasite-host interaction between the European rabbit and the viral infection myxomatosis is an example of coevolutionary forces at work, the adaptation and counteradaptation of parasites and hosts (see Holmes, 1983; May and Anderson, 1983).

To control the introduced European rabbit, the Australian government introduced its viral parasite in the population. The first epidemic of myxomatosis was fatal to 97 to 99 percent of the rabbits. The second resulted in a mortality of 85 to 95 percent; the third, 40 to 60 percent (Fenner and Ratcliffe, 1965). The effect on the rabbit population was less severe with each succeeding epizootic, suggesting that the two populations were becoming integrated and adjusted to each other.

In this adjustment, attenuated genetic strains of virus, intermediate in virulence, tended to replace highly virulent strains. Too high a virulence killed off the host; too low a virulence allowed the rabbits to recover before the virus could be transmitted to another host. Also involved was a passive immunity to myxomatosis conferred upon the young born to immune does. Finally, a genetic strain arose in the rabbit population providing an intrinsic resistance to the disease.

The transmission of myxomatosis virus is dependent upon *Aedes* and *Anopheles* mosquitoes, which feed only on living animals. Rabbits infected with the more virulent strain live for a shorter period than those infected with a less virulent strain. Because the latter live for a longer period, the mosquitoes have access to that virus for a longer time. That gives the less virulent strain a competitive advantage over the more virulent.

In those regions where the less virulent strains have a competitive advantage, the rabbits are more abundant because fewer die. That means more total virus is present in those regions than in comparable areas where more virulent strains exist. Thus, the virus with the greatest rate of increase and density within the rabbit population is not the one with the selective advantage. Instead, the virus whose demands are balanced against supply has the greatest survival value.

Prey Defenses

In an evolutionary context predator and prey play a sort of game. Prey evolve often elaborate means of defense. To survive under selective pressures, the predator must come up with ways to breach the defenses. The relationship between the two involves a flux of adaptive genetic change in each. Predators, however, do not seem to closely track changes in their prey. Over evolutionary time predators appear to have experienced an adaptive gap between themselves and the prey (Bakker, 1983). Predators, as suggested by fossil records, did not evolve rapidly enough to track escape adaptations in their prey. Thus, they possess a suboptimal and barely adequate efficiency in predation.

Every structure involved in defense or capture may not have evolved under selective pressures. The mechanisms might be borrowed from some other use. Lignins in plants, for example, decrease digestibility of the plants for the herbivore but at the same time provide a major support for the plants. Venom of snakes may protect the animals from enemies, yet it is also the means of capturing and subduing prey.

CHEMICAL DEFENSES

Chemical defenses are widespread among both plants and animals. They serve to warn potential predators and to repel or inhibit would-be attackers.

Plant Defenses

Chemical defenses involving secondary metabolic products are a first line of defense by plants against herbivores. The basis of chemical defense is the accumulation of an array of toxic proteins—lectins and protease inhibitors, alkaloids, cyanogenic glycosides, cyanolipids, digestive-inhibiting polyphenols, terpenes, and tannins. These secondary products, poisonous to the plant itself, may be stored in vacuoles within cells and released only when the cells are broken, or they may be stored and secreted by epidermal glands to function as a contact poison or volatile inhibitor (see Rosenthal and Janzen, 1979).

Production and storage of such metabolites are expensive to the plant and require a trade-off between defense and reproductive effort. Over evolutionary time the processes involve the relationship between plant availability as prey or host and the presence of herbivores. The mode of defense varies with the nature of the plants, which can be divided into two groups, apparent and unapparent (Feeny, 1975b) or available and predictable and unavailable and unpredictable (Rhoades and Cates, 1976). The two sets of terms are synonymous—the former from the viewpoint of the herbivore, the latter from the viewpoint of the plant.

Apparent plants are large, easy to locate, available to herbivores, usually long-lived, and woody. They possess the most expensive type of defense—quantitative. Such a defense, not easily mobilized, is most effective against herbivore specialists. Involved are mostly tannins and resins concentrated near the surface tissues of leaves, in bark, and in seeds. They form indigestible complexes with leaf proteins, reduce the rate of assimilation of dietary nitrogen, reduce the ability of microorganisms to break down leaf proteins in herbivore digestive systems, and lower palatability. The problem with such a defense is slow response. After defoliation by gypsy moths, oaks a year later increased tannin and phenolic content of their leaves and increased their toughness (Schultz and Baldwin, 1982).

Unapparent plants are short-lived, mostly annuals and perennials, and scattered in space and time. These plants employ a qualitative defense involving plant secondary substances such as cyanogenic compounds and alkaloids that interfere with metabolism. These substances can be synthesized quickly at little cost, are effective at low concentration, and are readily transported to the site of attack. They can be shuttled about the plants from growing tips to leaves to stems, roots, and seeds. And they can be transferred from seed to seedling. These substances protect mostly against generalist types of herbivores.

Animal Defenses

Chemical defenses are widespread among many groups of animals. Some species of fish release pheromones from the skin into the water, which act as alarm substances and induce a fright reaction in other members of the same or related species (Pfeiffer, 1962). The pheromone is produced in specialized cells in the skin that do not open to the surface, so that it is released only when the skin is broken. Fish in the vicinity receive the stimulus through the olfactory organs. Such alarm substances are most common among fish that are social, nonpredaceous, and lack defensive structures.

Arthropods, amphibians, and snakes employ secretions to repel predators. Strongly odorous, easily detected defensive substances are produced in often copious quantities by arthropods (Eisner and Meinwald, 1966; Eisner, 1970). The secretions are produced by glands containing large saclike reservoirs that are essentially infoldings of the body wall and are discharged through small openings.

They may ooze onto the animal's body surface, as in millipedes; be aired by evagination of the gland, as in beetles; or be sprayed out for distances up to several feet, as in grasshoppers, earwigs, and stink bugs. These secretions effectively repel birds, mammals, and insects alike by their effect on the predator's face and mouth. Active components in the defensive secretions of many arthropods include toxic secondary substances such as saponins, gossypol, and cyanogenic glycosides that they have taken from plants and sequestered in their own bodies. Some arthropods, however, synthesize their own poisons, as do venomous snakes, frogs, and toads.

CRYPTIC COLORATION

One approach to defense is to make the predator's task of finding prey items more difficult. Certain color patterns and behaviors evolved by prey function to hide them from predators. Such cryptic colorations involve patterns, shapes, postures, movements, and behaviors that tend to hide the organisms. Some animals are protectively colored, blending them into the background of their normal environment. Such protective coloration is common among fish, reptiles, and many ground-nesting birds. Countershading or obliterative coloration, in which the lower part of the body is light and the upper part is dark, reduces the contrast between the unshaded and shaded areas of the animal in bright sunlight. Object resemblance is common among insects. For example, the walking stick resembles a twig, and some insects resemble leaves. Some animals possess eyespot markings, which intimidate potential predators, attract their attention away from the animal, or delude them into attacking a nonvulnerable part of the body. Eyespot patterns in Lepidoptera seem to intimidate predators by imitating the eyes of a large avian predator that attacks small, insectivorous passerine birds.

Associated with cryptic coloration is flashing coloration. Certain butterflies, grasshoppers, birds, and ungulates, such as white-tailed deer, display extremely visible color patches when disturbed and put to flight. The flashing coloration may distract and disorient predators. When the animal comes to rest, the bright or white colors vanish, and the animal disappears into its surroundings (see Harvey and Greenwood, 1978, for review).

WARNING COLORATION AND MIMICRY

Animals that possess pronounced toxicity and other chemical defenses often possess warning coloration, bold colors with patterns that serve as warnings to would-be predators. The black-and-white stripes of the skunk, the bright orange of the monarch butterfly, and the yellow and black coloration of many bees and wasps serve notice of danger to their predators. All their predators, however, must have some unpleasant experience with the prey before they learn to associate the color pattern with unpalatability or pain.

Similar animals associated with inedible species sometimes evolve a similar mimetic or false warning coloration. That phenomenon was described some 100 years ago by the English naturalist H. W. Bates, in his observations of tropical butterflies. The type of mimicry he described, now called Batesian, is the resemblance of an edible species, the mimic, to an inedible one, the model. Once the predator has learned to avoid the model, it avoids the mimic also. Batesian mimicry is disadvantageous to the model because the predator will encounter a number of tasty mimics and thus take longer to avoid the model, which, in turn, will suffer greater losses in the learning process. The greater the proportion of mimics to the model, the longer the learning time of the predator. Usually, the number of mimics is fewer than the model.

Among North American butterflies, the palatable viceroy butterfly mimics the monarch, most of which are distasteful to birds (Brower, 1958) (Figure 17.3). Both model and mimic are orange in ground color with white and black markings; they are remarkably alike. Yet the viceroy's nonmimetic relatives are largely blue-black in color.

A less common type of mimicry, called Mullerian, involves both unpalatable models and unpalatable mimics. Such mimicry is advantageous to both. The pooling of numbers between the model and the mimic reduces the losses of each, because the predator associates distastefulness with the pat-

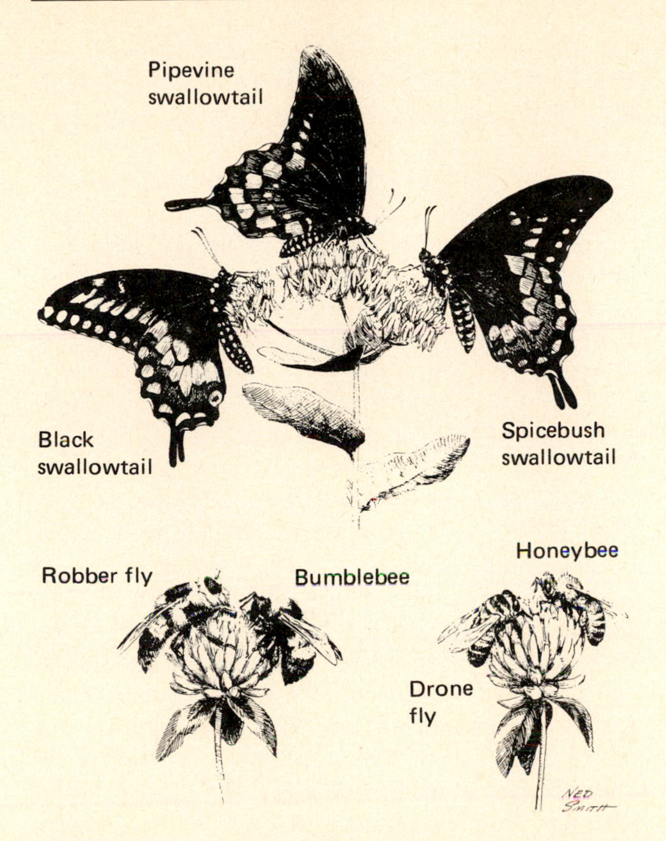

FIGURE 17.3 Mimicry in insects. The model, the distasteful pipevine swallowtail, has as its mimics the black swallowtail and the spicebush swallowtail. The black female tiger swallowtail is a third mimic. All these butterflies are found in the same habitat. The robber fly illustrates aggressive mimicry. It is a mimic of the bumblebee, on which it preys. The drone fly is a mimic of the honeybee.

tern without having to handle both species. Mullerian mimicry differs from Batesian in that feedback from handling either species is negative, reinforcing the learning process on the part of the predator. Batesian mimics and models are not of the same phylogenetic line, while Mullerian mimics include members of the same genus and family, probably because their ability to utilize and store poisons from plants has become fixed in an evolutionary line.

Mimicry is usually considered an evolutionary response in animals, but animals in search of food may have stimulated mimicry in the plant kingdom also. Many examples probably go undetected. Gilbert (1975), in his study of the passionflower butterfly (*Heliconius*) and its food plant, the passionflower (*Passiflora*), found evidence of plant mimicry. *Passiflora,* a tropical vine of the New World comprising around 350 species, has a wide range of intraspecific and interspecific leaf and stipule shapes. The number of *Passiflora* species found in any one area is about 2 to 5 percent of the 350 species. *Passiflora* species are used as an egg-laying site and source of larval food by some 45 highly host-specific species of *Heliconius,* each of which utilizes a limited group of the plants. The visually sophisticated butterflies learn the position of the vines and return to them on repeated visits. Within a habitat the leaf shapes of passionflowers vary among the species. Under visual selection of butterflies, passionflowers apparently evolved leaf forms that make them more difficult to locate. Because the larval food niche is larger than that of egg-laying females, there has been selective pressure for divergence among *Passiflora* species. Probably as a result of those selection pressures, *Passiflora* leaf shapes converge on those of associated tropical plants that *Heliconius* find inedible. So close are the convergences that plant taxonomists have named some *Passiflora* species after the genus they resemble.

ARMOR AND WEAPONS

Some of the least costly defenses available, at least to plants, are structures that make penetration by predators difficult, if not impossible. These include tough leaves, hairy leaves, thorns, and spines—structures that have evolved early in the history of the plants when they may have been subject to even greater predatory pressure. Because they represent little investment, plants can still retain them. Thick, hard seed coats provide protection from seed-eating animals. The problem with such a defense is that the seeds need to be scarified—the hard seed coat softened—so the seedling itself can escape. If scarification is not achieved, the seedling is sealed in, never to germinate.

Animals resort to similar armor. Clams, armadillos, turtles, and numerous beetles all withdraw into their armor coats or shells when danger approaches. Their problem is an inability to assess the outside environment. Is the predator large or small, still present or departed? How much foraging time should the animal sacrifice before daring to open up its defenses? Another type of animal defense is exhibited by porcupines and echidnas, whose quills (modified hairs) are effective in discouraging predators.

ALARMS, MOBBING, AND DISTRACTION DISPLAYS

Some animal defenses can be behavioral. One such technique is the alarm call, which is given at the moment of potential danger when a predator is sighted. High-pitched alarm calls are not species-specific. They are recognized by a wide range of animals close by. An unsolved problem about alarm calls is to whom they are directed—the predator or conspecific prey. If directed toward potential prey, the alarm call could be either altruistic or selfish. If the alarm exposed the caller's position to the predator, the caller could draw the predator's attention away from other conspecifics, including kin. Or it could attract more conspecifics for cooperative defense and lower its risk of being taken. Alarm calls do function to warn close relatives, as in the case of Belding ground squirrels (Sherman, 1977). Highly sedentary, closely related females live in close proximity to each other. Adult and 1-year-old females living with relatives respond quickly to danger and give most of the alarm calls, which function to warn offspring and other relatives. Beyond that there are few conclusive studies on the evolution and function of alarm calls.

Alarm calls often bring in numbers of potential prey that respond to the situation by mobbing or hustling the predator. An example is the harassment of an owl perched in a tree by many small birds attracted to the scene by general alarm calls. Mobbing may involve harassment at a safe distance or direct attack. The ultimate outcome for the prey is to reduce the risk of predation to themselves and their offspring. But like alarm calls, the evolution and adaptive significance of mobbing are still obscure.

Distraction display is any type of behavior used to attract the attention of a predator away from eggs or young. Distraction displays are most common among birds. Birds with precocious young, such as the killdeer, usually exhibit the most vehement distraction displays at the time the eggs hatch and altricial birds, such as vesper sparrows, at the time the young fledge. Because the beneficiaries of distraction display are the immediate offspring, the behavior probably evolved through kin selection.

GROUP DEFENSE

For some prey, living in groups may be the simplest form of defense. Groups, especially in mobbing situations, would probably deter a predator that would not be so inhibited when facing only one or two prey individuals. Sudden explosive group flight can confuse a predator, unable to decide which individual to follow. The more prey are congregated, the less is the chance of any one individual's being taken. By keeping close together, individuals reduce their chances of being the one closest to the predator. By maintaining a tight or cohesive group, prey make it difficult for any predator to obtain a victim.

PREDATOR SATIATION

A more subtle defense is the timing of reproduction so that most of the offspring are produced in a very short period of time. Then food is so abundant that the predator becomes satiated, allowing a percentage of the reproduction to escape. That is a strategy employed by such ungulates as the wildebeest and caribou. It involves synchronized birth of young, with the result that young are so abundant in one short period of time that the predators become satiated, and the remaining young quickly grow beyond a size easily handled by the predators (Schaller, 1972; Bergerud, 1971). The 13- and 17-year appearances of periodical cicadas (*Magicicada* spp.) function in much the same manner. By appearing suddenly in enormous numbers, they quickly satiate predators and do not need to evolve costly defensive mechanisms. While huge numbers of adults do succumb to predators, the losses hardly dent the total population. In other years the predators must seek alternative prey. Thus, the cicadas' major defense is to prevent predators from ever evolving any dependence upon them.

Predator satiation is a major strategy in plants and is most prevalent in those species lacking chemical defenses. It involves four different approaches (Janzen, 1971).

The first is to distribute seeds so that all of a seed crop is not equally available to all members of a seed predator population. Seeds of most trees are concentrated near the parent plant, and the number of seeds declines rapidly as the distance from the tree increases. Predators are attracted to the vicinity of the parent tree, while seeds scattered some distance away are missed by predators. These survivors produce most of the recruitment.

A second approach is to reduce the time of seed availability. If all the seed matures at one time, seed predators will be unable to utilize the entire crop before germination.

The third approach is the evolution of a protective covering or mild toxicity, so that the seed can be exploited only by predators that are partially or wholly host-specific. That reduces the number of predators that can use the seed for food.

A fourth approach is to produce a seed crop periodically rather than annually. The longer the time interval between seed crops, the less opportunity dependent seed predators have to maintain a large population. Seed predators often experience local increases in density after good seed years but decline rapidly when the food supply is depleted. That reduces the number of predators available to exploit the next seed crop. The production of a periodic seed crop depends upon the synchronization of seed production among individuals. That synchrony is usually achieved by a weather event, such as late frosts or protracted dry spells. As individuals of a tree species in a community become synchronized, strong selection pressures build against nonsynchronized individuals because they experience heavy seed predation between peak seed years. Such individuals either drop out of the community over evolutionary time or become synchronized.

Predator Offense

As prey evolved ways of avoiding predators, predators necessarily had to evolve better ways of hunting and capturing prey (Bakker, 1983).

HUNTING TACTICS

Predators have three general methods of hunting: ambush, stalking, and pursuit. Ambush hunting involves lying in wait for prey to come along. The method is typical among some frogs, certain insects like the mantid, lizards, and alligators. Ambush hunting has a low frequency of success, but it requires a minimal amount of energy. Stalking, typical of herons and some cats, is a deliberate form of hunting with quick attack. The predator's search time may be great, but pursuit time is miminal. Thus, it can afford to take smaller prey. Pursuit hunting, typical of many hawks, involves minimal search time, because the predator usually knows the location of prey. But pursuit time is relatively great, and in general, the predator must secure relatively large prey. Searchers spend more time and energy to encounter prey. Pursuers, theoretically, spend more time to capture and handle prey once it is noticed.

SIZE OF PREY

The various species of predators have certain energy requirements that can be met only by profitable foraging. Predators cannot afford prey that is too small to meet their energy requirements unless that prey can be captured quickly and is in abundance. Predators also have an upper size limitation. The prey may be too big to consume or too difficult or dangerous to handle. In fact, some prey become invulnerable to predation through body growth.

Some predators have evolved methods for killing prey much larger than themselves. Just as prey may live in groups to lessen the risk of predation, so predators may hunt in groups and by cooperative effort take down large prey. Wolves, African hunting dogs, jackals, and African lions are examples of predators that take large prey by hunting cooperatively. Certain snakes and arthropods may use venom to kill large prey.

Conversely, predators much larger than their prey have evolved ways of filtering organisms from their environment, particularly in aquatic communities. Examples are the net-spinning caddisflies and baleen whales that feed on krill. Intermediate-sized predators are usually hunters, while very small predators are usually parasitoids. The boundaries between the various predator-prey sizes are usually set by some energy cost-benefit ratio.

CRYPTIC COLORATION AND MIMICRY

If prey can use cryptic coloration, concealing coloration, and mimicry to their advantage, predators can do the same. Cryptic and concealing coloration enable them to blend into the background or break up their outlines. Predators can deceive their prey by resembling the host or prey. An example is the model bumblebee and the mimic robber fly (*Mallophora bomboides*). Not only does the robber fly benefit from reduced predation, but it also exploits the model for food. The robber fly preys on Hymenoptera by preference, and its resemblance to its prey allows it to escape notice until it is too late for the bee to flee or defend itself (Brower and Brower, 1962). The female of certain species of fireflies imitates the mating flashes of other species, attracting to her males of those species, which she promptly kills and eats (Lloyd, 1980). That type of mimicry is called *aggressive*.

BREACHING CHEMICAL DEFENSES

Although plants do possess powerful chemical defenses that work well against generalist herbivores, specialists manage to breach them. Many, if not most, toxic plants have specialist herbivores that can exploit their foliage or seeds. These specialists in some way absorb or metabolically detoxify foreign substances. Detoxification may be achieved by oxidation and conjugation reactions that yield inactive excretable products or by the action of gastrointestinal flora that degrade secondary metabolic compounds. Some herbivores are able to store plant poisons and use them in their own defense or in the production of pheromones.

HUNTING ABILITY

Predators have acquired various adaptations that improve their hunting ability in addition to such weapons as fangs and claws. Bats, for example, are able to produce ultrasonic pulses through the nose or mouth that enable them to detect prey by echolocation (for a good summary see Vaughan, 1978). Night-hunting owls can locate prey by hearing rather than by sight. Their feathered facial disks reflect the sounds of prey and direct them to the ears (Konishi, 1973). The owl's large ear openings are positioned asymmetrically, enabling it to detect differences in the elevation of prey. The owl's ability to fly noiselessly allows it to come upon the prey without alerting the victim. Day-hunting harriers flying over densely grown grass fields have similar facial disks and placement of ears that enable the hawks to locate by sound voles hidden in the grass (W. R. Rice, 1982).

Seed Dispersal

Some plants turn the predator-prey relationship around and attract predators to aid in the dispersal of seed. The interaction might best be described as protocooperation.

To utilize animals as agents of dispersal, plants must evolve types of fruits or seeds that will attract fruit-eating, or *frugivorous,* animals and at the same time discourage consumption of unripe fruit. Plants accomplish that by cryptic coloration, such as green, unripened fruit among green leaves; by unpalatable texture; repellent substances; and hard outer coats. When seeds mature, plants can attract frugivores to ripened fruit by presenting attractive odors, altering the texture of fruits and seeds, improving succulence, and acquiring a high content of sugars and oils.

If ripe fruits are attractive to frugivores, they are also attractive to other organisms that take no part in seed dispersal and that damage the fruit or cause it to rot or spoil (Janzen, 1977). These include bacteria, fungi, and invertebrates such as fruit flies, bees, and wasps. Plants can counter by timing the ripening of fruit when nondispersal predators are the least abundant (fall in the temperate regions), by shortening the time fruits are available, by reducing nutritional balance, by lowering the quality of food, and by using secondary compounds to restrict predation to those frugivores able to handle the toxic substances. The latter two are maladaptive and extreme measures because they reduce the number of frugivores attracted to the fruit (Herrera, 1982).

Plants have two alternative approaches to the dispersion of seeds by animals. One is to become opportunistic and evolve fruits that can be exploited by a large number of dispersal agents. Such plants opt for quantity dispersal, the scattering of a high number of seeds at a sustained rate with the chance that a diversity of consumers will drop some seeds in a favorable site.

Such a strategy is typical of plants of temperate regions, where most fruit-eating birds and mammals are opportunistic consumers, which rarely specialize in any one kind of fruit. Because these opportunistic fruit-eaters do not depend on the fruit for their basic sustenance, there is no need for plants to provide a balanced diet. Their fruits are usually succulent and rich in sugars and organic acids and contain small seeds that pass through the digestive tract unharmed. To accomplish such a passage, plants must evolve seeds with hard coats resistant to digestive juices. Seeds of such plants often will not germinate unless they have been conditioned or scarified by passage through the digestive tract. Large numbers of small seeds may be so dispersed, but relatively few are deposited on suitable sites.

In temperate regions fruits ripen in early and late summer, when the young of the year are no longer dependent on highly proteinaceous food and both adults and young can turn their attention to a growing abundance of fruits. The fruits may be small-seeded or large-seeded, high or low in nutrients, sweet or otherwise. The fruits of early summer ripen before migratory birds pass through. Small-seeded fruits—blackberries, blueberries, mulberries—are sweet, fragrant, easily invaded by fungi and yeasts, and highly attractive to birds and small mammals. The seeds easily pass through the gut, but dispersal distance is restricted. Large-seeded fruits, mostly cherries (*Prunus*), hang high, where they are taken by raccoon, opossum, turkey, and robins, and the fruits that drop are eaten by foxes and skunks. These animals distribute the seeds more widely, often into such suitable habitats as old fields.

Fruits of late summer and fall ripen when migrant birds come through. They congregate in flowering dogwoods, spicebush, autumn olive, and wild grape to feed on high-quality fruits with large seeds but also nutrient-rich flesh. Such fruits do not last long on the trees and shrubs, and their seeds are scattered widely by birds. Fruits of lower quality have less fats and sugars, are unlikely to be invaded by microorganisms, and hang on well into winter; hence, they are available to birds over a longer period of time and are sure to be consumed once more palatable, short-lived succulent fruits are gone. In such a manner those fruits avoid competition for dispersers in early fall.

A second alternative available to plants is to evolve fruits adapted to a relatively small group of highly specialized frugivores and sacrifice a high rate of dispersal for quality of dispersal (McKey, 1975). That involves a higher cost per seed because more energy must be expended in production of nutritional fruit and a large seed containing high energy reserves for a germinating seedling. If a plant depends upon a small group of highly specialized frugivores, then the animals would in turn have to depend upon the fruit crop for their sustenance to be effective dispersers. The plant or combination of plants would have to provide a balanced diet, requiring expensively produced fruits rich in proteins and oils. Such a dependence on fruits by birds would result in repeated visits to the plant, ensuring that seeds will be removed quickly after maturation and reducing chances that seeds will be left to rot on the tree.

Although there is a strong tendency to equate the dependence of fruiting plants on frugivores with the mutualistic relationship of plants and animal pollinators, the obligatory dependency is questioned by some (Wheelwright and Orians, 1982). Unlike specialized plant-pollinator systems, most fruiting plants, even in the tropics, are visited by a diversity of frugivores, and the fruit-eating animals consume a large number of different fruits. They are specialists only in that they live largely on fruits. Unlike pollinators, frugivores are not speciality feeders on one or several plants only. Their possession of soft-lined stomachs is not necessarily to treat seeds gently (McKey, 1975) but perhaps to avoid digestion of toxins from thin-coated seeds by regurgitating them. Problems restricting mutualistic relationships in seed dispersal include the inability of plants to provide a reward for quality seed dispersal, because the frugivores are paid in advance; the inability of dispersers to recognize quality habitats for the seeds and deposit the seeds therein; and the advantages accruing to the plant for having a large number of dispersers.

Mutualism

Mutualism is a positive, obligatory relationship between two species, each of which has its survival, growth, and reproductive ability enhanced by the presence of the other. Evidence suggests that the relationship is reciprocal exploitation rather than a cooperative effort acting for the benefit of each (Barrett, 1983). Facultative or nonobligatory relationships that are beneficial but not necessary for the existence of both species are more accurately but anthropomorphically considered protocooperation.

Some forms of mutualism are so permanent and obligatory that the distinction between the two interacting populations becomes blurred. A good example is mycorrhiza, a mutualistic relation of plant roots with the mycelium of fungi. So important is this mutualism to the growth of forest trees and to the functioning of forest ecosystems that it is the object of expanding research activity. Common to many trees of temperate and tropical forests, one form, *ectomycorrhizae,* produces shortened and thickened roots that suggest coral. The hyphae work between the root cells and extend into a network on the outside of the root, which acts as a substitute for root hairs. Mycorrhizae aid in the decomposition of litter and the translocation of nutrients, especially nitrogen and phosphorus, from the soil into the root tissue (Zak, 1964; Marx, 1971). Mycorrhizae increase the capability of roots to absorb nutrients (Voigt, 1971), provide selective ion absorption and accumulation, mobilize nutrients in infertile soil, and render unavailable substances available (especially those bound up in silicate materials). In addition, mycorrhizae reduce susceptibility of the host to invasion of pathogens by utilizing root carbohydrates and other chemicals attractive to pathogens. They provide a physical barrier to pathogens, secrete antibiotics, and stimulate the roots to elaborate chemical inhibitory substances (Marx, 1971). In return, the roots provide support and a constant supply of carbohydrates.

A balanced association exists between the two. Any alteration in light or nutrient availability creates a deficiency of carbohydrates and thiamine for the fungi; and interruption of photosynthesis causes a cessation of fruiting of the mycorrhizae. Thus, any interruption or imbalance in the continuous supply of metabolites impairs or destroys the association.

Mycorrhizal interactions extend beyond the fungi and roots. While some mycorrhizae have above-ground fruiting bodies—the familiar mushrooms whose spores are released to the air—others have below-ground fruiting bodies known as truffles. These mycorrhizae depend on small mammals, especially voles, to disperse their spores. Small mammals are able to detect underground truffles by smell. They dig up the truffles, eat them, and defecate, spreading the mycorrhizal spores across the forest floor. Mycorrhizal spores cannot germinate until they come in contact with tree roots. Thus, a three-way obligatory relationship exists (Figure 17.4). The tree depends upon mycorrhizae for nutrient uptake from the soil. The mycorrhizae depend upon tree roots for an energy source and upon small mammals for dispersal of spores. And small mammals obtain a significant portion of their food from truffles.

Some of the most important and complex mutualistic relationships are between plants and their pollinators—insects, birds, and bats. For example, Gilbert (1975) describes a coevolved mutualism between *Heliconius* butterflies, *Anguria* concurbit vines, which the butterflies pollinate, *Passiflora,* and their larval host plants. *Passiflora,* in turn, has coevolved mutualistic defenses against its heliconine herbivores: predaceous ants and vespid wasps are attracted to the plant by nectar glands located on leaves and stems. (For a review of such extrafloral nectar production and mutualistic interactions, see Bentley, 1977.)

Although there are a number of types of interactions between plants and animals and between animals, the most common mutualisms involve plant-pollinator systems, for good reasons (see Feinsinger, 1983). There is an advantage for plants in having a select group of dependable pollinators visit their flowers and carry pollen directly from plant to plant. In turn, the pollinators are assured of a high-quality reward, nectar, paid upon delivery of the pollen. It is advantageous for the pollinators to have a dependable source of food to which they have more or less restricted access.

To ensure that only certain pollinating insects, birds, and bats secure nectar, some plants have evolved ways of restricting visitors. These include asymmetrical flowers, which require special behavior or structures on the part of the visitors to enter; inaccessible placement of nectar at the end of a long corolla tube; hairs and other devices that make entry difficult; and toxic constitutents in the reward that deter general visitors. To attract such pollinators, plants offer a highly defendable reward, a large volume of nectar per flower, high in both quality and nutrients and available over a long flowering season. In turn, the pollinators have evolved specialized morphological structures such as long bills and long tongues, specialized behavior, and metabolic changes allowing them to utilize nectar as a major food.

Examples of mutualism seem to support the suggestion that mutualistic relationships evolved from predator-prey or parasite-host relationships (N. Smith, 1968). Initially, one member of the interaction increases the stability of the resource level for the second. In time, energy benefits accrue to the second member, and perhaps its activities begin to improve the fitness of the first. Selection then favors mutual interaction. In a plant-pollinator relationship, the pollinators may have first exploited the nectar of flowers. As a result of the exploitation, some flowers were pollinated by the animals. The improved efficiency and success of pollination through the activity of pollinators improved the fitness of the plants. Selection then favored the development of mechanisms to maintain the interactions, such as sugar-rich nectar to keep the pollinators coming.

In spite of its importance and presumed relationship to other two-species interactions, mutualism is just beginning to receive theoretical consideration. May (1976), Christiansen and Fenchel (1977), and Post, Travis, and DeAngelis (1980) have suggested mathematical models involving Lotka-Volterra equations that might serve as a beginning. Because each species benefits the other, an increase in one population directly influences or is influenced by an increase in the other. However, because there is an upper limit to population growth based on carrying capacity, the influence of each has to have a saturation point. The environment may impose an upper limit on population growth. In a plant-pollinator system, a certain population level of both interacting species is necessary before any equilibrium is possible and both populations reach maximum stable densities. If plant density is too low and pollinators have difficulty finding plants, the number of pollinators will decline below replacement level.

If such a mutualistic system is in a predictable environment, it will persist. If it is in an unpredictable environment subject to disturbances, the population of one or the other is apt to fall below replacement levels, with probable extinction. That suggests why mutualism is much more apparent in tropical than in temperate regions. Another reason may be that in tropical regions insects and certain birds have a high and continual impact on plants throughout the year. That results in a stronger coevolutionary response between plant and herbivore.

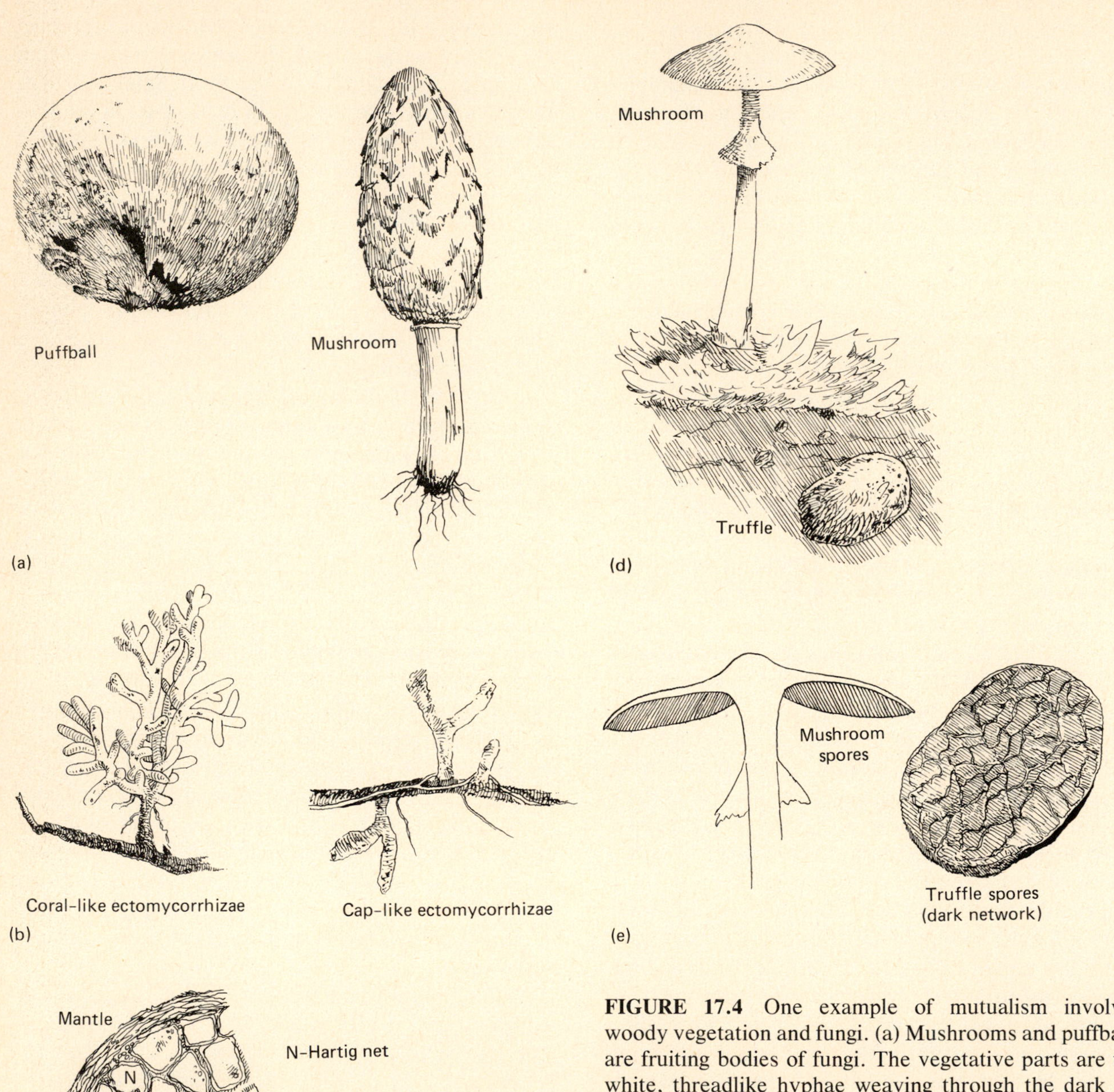

FIGURE 17.4 One example of mutualism involves woody vegetation and fungi. (a) Mushrooms and puffballs are fruiting bodies of fungi. The vegetative parts are the white, threadlike hyphae weaving through the dark organic litter. Follow the hyphae carefully, and you may find them threading among the rootlets of trees.

(b) Some rootlets may support caplike, nodular, or corallike growths of various colors, such as brown, black, white, red, and yellow. They are known as mycorrhizae or fungus roots. These particular ones are ectomycorrhizae, so named because most of their growth is outside the root.

(c) Ectomycorrhizae form a mantle of fungi about the tips of the rootlets. Hyphae invade the tissues of rootlets

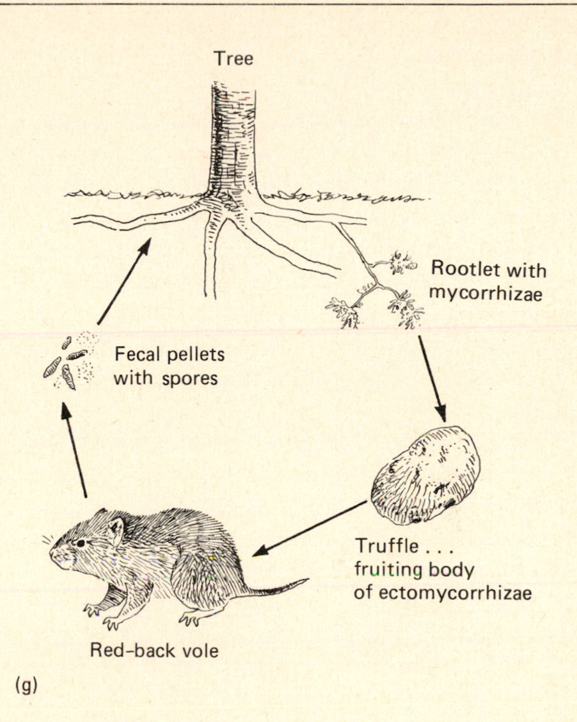

and then grow between the cells. This growth is known as the Hartig net. By growing in such close association with rootlet cells, fungi obtain from their host simple carbohydrates, amino acids, and other nutrients necessary for the completion of their life cycle.

The tree gains much more in return. Because the hyphae radiate from the fungus root into the soil, ectomycorrhizae act as extensions of the tree rootlets. Mycorrhizae enable the tree to increase the uptake of nutrients from the soil and convert complex organic substances and minerals into nutrients the plant can use. They produce growth regulators that increase the life of rootlets and protect the rootlet from disease. Without ectomycorrhizae certain groups of trees, especially conifers, oaks, and birches, could not become established, survive, and grow. In fact many mycorrhizae are specific to certain trees such as pines and Douglas fir.

(d) There are two general types of mycorrhizae: those whose fruiting bodies are among the familiar mushrooms and puffballs and those whose fruiting bodies, called truffles, are below ground. But not all fungi that produce mushrooms and puffballs are mycorrhizal.

(e) Ectomycorrhizal fungi spread by spores produced by fruiting bodies. Above ground mushrooms discharge their spores to the air. Some are carried long distances by the wind. Other spores drop nearby. These fruiting bodies, appear, mature, and die quickly. Some are eaten by mammals of the forest–deer, squirrels, and mice.

(f) Below-ground truffles produce spores within the fleshy structure and depend upon mammals to disperse them. Woodland mice and voles, chipmunks, flying squirrels, and pocket gophers seek out truffles by their smell, which may be cheesy, spicy, or fruity. Each species of truffle has its own distinctive odor, which grows stronger as the spores mature. These mammals digest the fleshy part of the truffle, but the spores become concentrated in fecal pellets. These mammals carry spores from forested to nonforested areas and thus inoculate the open soil with mycorrhizae.

If a rootlet of a seedling tree is close to spores, the fungus and the root develop a mutualistic relationship, and the seedling is off to a good start. Lacking this association, the seedling might never become established or may grow poorly. The lack of ectomycorrhizae is one reason why forest trees often fail to invade open areas.

(g) Thus, a three-way relationship exists among forest trees, ectomycorrhizae, and small mammals. Each in a way depends upon the other. The ectomycorrhizae need the tree for energy. The tree needs the ectomycorrhizae for the uptake of nutrients from the soil. One group, the truffles, depends upon small mammals to disperse the spores, and small mammals feed heavily on the truffles. By scattering the spores, small mammals aid in the continuation of their food supply and the forest.

Summary

In exploitative or feeding situations, interacting species often adjust their relationship to one another. Prey under the selective pressures of predation evolve some means of defense to counter the effects of predation. Predators improve their efficiency in exploiting prey in order to survive. Each, in effect, has a reciprocal influence on the evolution of the other. The reciprocal influence is called coevolution.

Two coevolved relationships are plant-herbivore systems and predator-prey systems. Each of these two systems in turn is affected by the other. Herbivory affects plant fitness by reducing the amount of photosynthate and the ability to produce more. Plants respond by trying to replace losses. Failure to do so means a loss of fitness or even death. At the same time, plants affect herbivore fitness by denying herbivores palatable or digestible food or by producing secondary compounds that interfere with reproduction. Herbivore-carnivore systems involve relations that influence the fitness of each and introduce stable or unstable oscillations into the system.

In response to selective pressures of predation, prey species, both plant and animal, have evolved measures of defense. One is the development of color patterns or behavior to escape detection or inhibit predators. Warning coloration in prey tells a predator that a prey item is distasteful or disagreeable in some manner. Usually, the predator has to experience at least one encounter with such a prey to learn the significance of a color pattern.

Often palatable species will mimic unpalatable species, thus acquiring some protection from predators. Batesian mimicry is the development of the appearance of a distasteful or dangerous model by an edible species. Mullerian mimicry is the development of a common warning coloration by different species, all of which are distasteful. Although mimicry is usually associated with animals, mostly insects, there is increasing evidence of mimicry among plants to hide from specialized herbivores or to attract seed dispersers.

Cryptic coloration serves to blend the organism with its surroundings, making it hard to detect. Although warning and cryptic coloration and mimicry are usually associated with prey species, such mechanisms are also employed by predators to increase hunting efficiency.

Chemical defense, widespread among plants and animals, usually involves a distasteful or toxic secretion that repels, warns, or inhibits a would-be attacker. Plants employ secondary metabolic products such as alkaloids, phenolics, and cytogenic glycosides. Chemical defense is most successful against generalist herbivores. Certain specialists breach the chemical defense and detoxify the secretions or sequester the toxins in their own tissues as a defense against predators.

Another form of defense is predator satiation. Reproduction is timed such that offspring are so abundant that predators can take only a fraction of the very young, leaving a number to escape by growing to a size too large for the predator to handle.

Some plants have turned the situation around and utilize predators to disperse seeds. By packaging seeds in attractive fruits, plants encourage their consumption by animals that will transport seeds in their digestive tracts.

Mutualism is the beneficial relation between two interacting species that cannot live separately. Like predator-prey interactions, mutualism may be the result of coevolution over a long period of time. Some of the most complex mutualistic relationships evolved between plants and pollinators. These relationships probably developed out of some form of predator-prey relationship.

Review and Study Questions

1. What is coevolution?
2. What effect do herbivores have on plant fitness?
3. How do plants affect the fitness of herbivores?
4. How can predation lead to stable oscillations in predator-prey populations?

5. How do chemical defenses function in plants? In animals?

6. Distinguish between apparent and unapparent plants. How do they differ in defense against herbivory?

7. What is cryptic coloration? How is it utilized by prey species? By predators?

8. What is the difference between Batesian mimicry and Mullerian mimicry? What is aggressive mimicry?

9. How do alarm calls, mobbing, and distraction display aid in animal defense? How might these actions increase the fitness of individual prey? Could they decrease fitness?

10. How can predator satiation function as a prey defense? Discuss such a defense from an individual prey's point of view.

11. Contrast ambush with the pursuit and stalking methods of hunting. Evaluate each from a cost-benefit point of view. Under what conditions might each have evolved?

12. How do predators solve the problem of prey size?

13. What are some ways in which herbivores might get around a plant's chemical defenses?

14. How might seed-dispersal systems have evolved? Is fruit predation a chancy way of distributing seeds? Why? What about the role of seed predators?

15. What is mutualism? Look up some examples of mutualism and examine them critically. Are they really mutualistic? Could mutualism act to stabilize interacting populations?

Suggested Readings

See the "Suggested Readings" section at the end of Chapter 16, especially those works dealing with the big predators.

Futuyma, D. J., and M. Slatkin, eds. 1983. *Coevolution.* Sunderland, Mass.: Sinauer Associates.

Gilbert, L. E., and P. H. Raven. 1975. *Coevolution of animals and plants.* Austin: University of Texas Press.

Nitechi, M. H., ed. 1983. *Coevolution.* Chicago: University of Chicago Press.

Price, P. N. 1980. *Evolutionary biology of parasites.* Princeton, N.J.: Princeton University Press.

Rosenthal, G. A., and D. H. Janzen, eds. 1979. *Herbivores: Their interactions with secondary plant metabolites.* New York: Academic Press.

PART V

THE ECOSYSTEM

INTRODUCTION

The Ecosystem Concept

Some ecological terms have become commonplace. One is *ecosystem*. Like other well-used terms, *ecosystem* has a range of definitions. It has been defined as a partially or completely self-contained mass of organisms, as all organisms present in an area together with their physical environment, and as all the energetic interactions and material cycling that link organisms in a community with one another and with their environment. Definitions along the latter line come closer to the real meaning of an ecosystem.

As a biological term, *ecosystem* has a recent origin. It was coined by the English ecologist A. G. Tansley in 1935. He presented the term in an article in the journal *Ecology* in which he took issue with the use and abuse of certain vegetational concepts and terms. He was critical of the concept of the biotic community because "animals and plants are too different in nature to be considered as members of the same community. The whole complex of organisms present in an ecological unit may be called the *biome*." He argued (1935:229):

> The fundamental concept appropriate to the biome considered together with all the effective inorganic factors of its environment is the *ecosystem,* which is a particular category among the physical systems that make up the universe. In an ecosystem the organisms and the inorganic factors alike are *components* which are in relatively stable dynamic equilibrium. . . .
>
> It is the systems so formed which, from the point of view of the ecologist, are the basic units of nature on the face of the earth. Our natural human prejudices force us to consider the organisms (in the sense of the biologist) as the most important parts of these systems, but certainly the inorganic factors are also parts—there could be no systems without them, and there is constant interchange of the most various kinds within each system, not only between the organisms but between the organic and the inorganic.

Thus, the prefix *eco-* indicates environment, and *-system* refers to a complex of coordinated units.

An ecosystem is basically an energy-processing system whose components have evolved together over a long period of time. The boundaries of the system are determined by the environment—that is, by what forms of life can be sustained by the environmental conditions of a particular region. Plant and animal populations within the system represent the objects through which the system functions.

Inputs into the system are both biotic and abiotic. The abiotic inputs are energy and inorganic matter. Radiant energy, both heat and light, imposes restraints on the system (1) by influencing temperature and moisture regimes, which determine what organisms can live in a system under a given set of environmental extremes, and (2) by affecting the productive capability of the system. Inorganic matter consists of all nutrients, water, carbon dioxide, oxygen, and so forth that affect the growth, reproduction, and replacement of biotic material and the maintenance of energy flow. Some of the materials in this chemical environment are necessary for the maintenance of the system, while others may be toxic or detrimental to its functioning.

The biotic inputs include other organisms that move into the ecosystem as well as influences imposed by other ecosystems in the landscape. No ecosystem stands alone. One is influenced by others. A stream ecosystem, for example, is strongly influenced by the terrestrial ecosystem through which it flows. Island ecosystems are influenced by their distance from continental land masses.

In simplest terms, all ecosystems, aquatic and

terrestrial, consist of three basic components—the producers, the consumers, and abiotic matter. Traditionally, the abiotic inputs—CO_2, O_2, H_2O, nutrients derived from weathering of materials and from precipitation, and so on—are considered as abiotic components of the system. Here they are considered as inputs to the system rather than as a part of the ecosystem (Figure V.1).

The *producers,* or *autotrophs,* the energy-capturing base of the system, are largely green plants. They fix the energy of the sun and manufacture food from simple organic and inorganic substances. Autotrophic metabolism is greatest in the upper layers of the ecosystem—the canopy of the forest and the surface water of lakes and oceans.

The *consumers,* or *heterotrophs,* utilize the food stored by autotrophs, rearrange it, and finally decompose the complex materials into simple inorganic compounds again. In this role their function is to regulate the rate of energy flow and nutrient cycling and to stabilize the system. The heterotrophic component is often subdivided into two subsystems, consumers and decomposers. The consumers feed largely on living tissue, and the

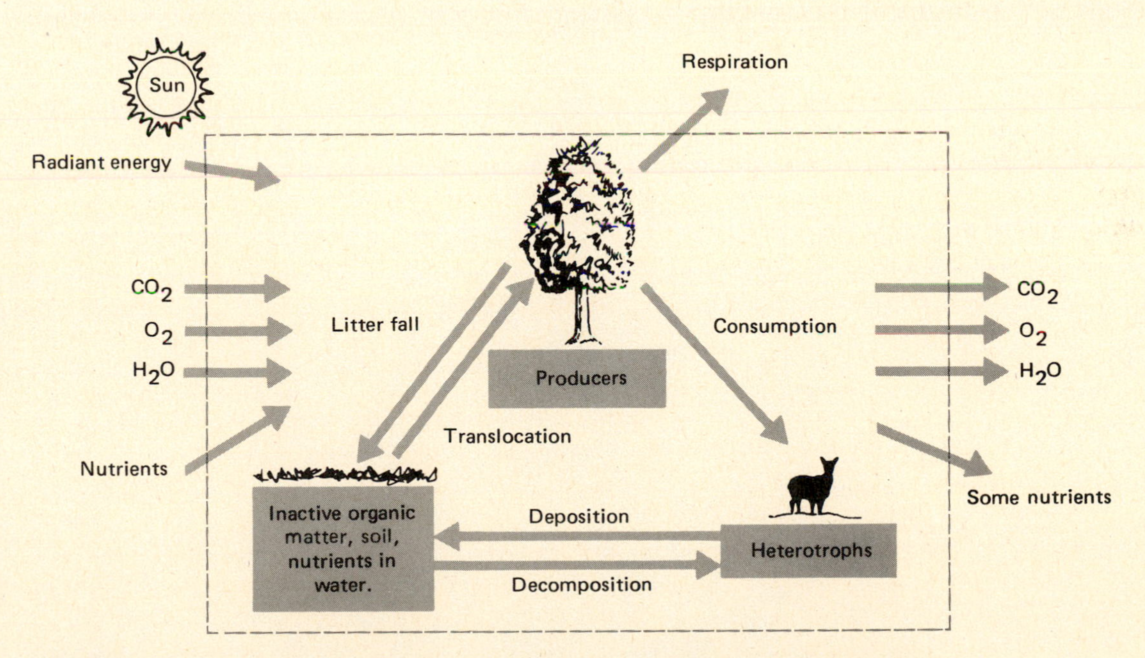

FIGURE V.1 Schematic diagram of an ecosystem. The dashed lines represent the boundary of the system. The three major components are the producers, the consumers, and the abiotic elements—inactive organic matter, the soil matrix, nutrients in solution in aquatic systems, sediments, and so on. The lines indicate interactions within the system. Inputs to the ecosystem from the environment include CO_2, O_2, H_2O, nutrients, and radiant energy. Outputs to the environment include CO_2, O_2, H_2O, some nutrients, and heat of respiration. (*Adapted in part from O'Neill, 1976:1245.*)

decomposers break down dead matter into inorganic substances. But no matter how they may be classified, all heterotrophic organisms are consumers, and all in some way act as decomposers. Heterotrophic activity in the ecosystem is most intense where organic matter accumulates—in the upper layer of the soil and the litter of terrestrial ecosystems and in the sediments of aquatic ecosystems.

The third, or *abiotic,* component consists of the soil matrix, sediments, particulate matter, dissolved organic matter in aquatic ecosystems, and litter in terrestrial ecosystems. All the dead or inactive organic matter is derived from plant and consumer remains and is acted upon by the decomposer subsystem of the heterotrophs. Such organic matter is critical to the internal cycling of nutrients in the ecosystem.

The driving force of the system is the energy of the sun, which causes all other inputs to circulate through the system. The various outputs—or more correctly, outflows—from one subsystem become inflows to another. While energy is utilized and dissipated as heat of respiration, the chemical elements from the environment are recycled by organisms within the system. How fast nutrients turn over in the system is influenced by these consumers, which function as rate regulators in the ecosystem.

CHAPTER 18

Photosynthesis and Decomposition

Outline

Objectives

Upon completion of this chapter, you should be able to:

1. Distinguish among the C_3, C_4, and Crassulacean acid cycles in photosynthesis.
2. Describe the anatomical and physiological differences between C_3, C_4, and CAM plants.
3. Explain the ecological importance of the three photosynthetic pathways.
4. Discuss some of the limits on the rate of photosynthesis.
5. Explain the significance of leaf area index to photosynthesis.
6. Describe photosynthetic efficiency.
7. Define decomposition and discuss its importance.
8. Describe the pathway of decomposition.
9. Discuss the role of saprophages, microbial organisms, and macroorganisms in the decomposition process.
10. Describe nutrient immobilization and mineralization and tell how they relate.
11. Name some influences on decomposition.
12. Compare rates of decomposition within and among ecosystems.

At the end of winter in the temperate regions and at the end of the dry season in the semitropical regions, the buds of trees and shrubs swell and new grass sprouts, enveloping the countryside with a haze of green. As the season progresses, the fresh green of spring expands into the full greenness of summer growth. With the approach of late summer, the greenness dulls. In temperate regions the leaves acquire some color of autumn; in semitropical regions they turn dull yellows and dried browns. In both regions the leaves fall to the ground, accumulating as litter beneath the trees and drifting into streams, ponds, and lakes. There the leaves are softened by water and rain, fragmented by litter-feeding organisms, and attacked by fungi and bacteria.

The cycle of the year has witnessed another time of growth, during which green plants utilized the energy of the sun to convert CO_2 and H_2O into new plant tissue, and a time of decay, during which material drawn from Earth was returned and made available for the renewal of growth. Thus, Earth functions on two major processes without which life could not exist: photosynthesis and decomposition.

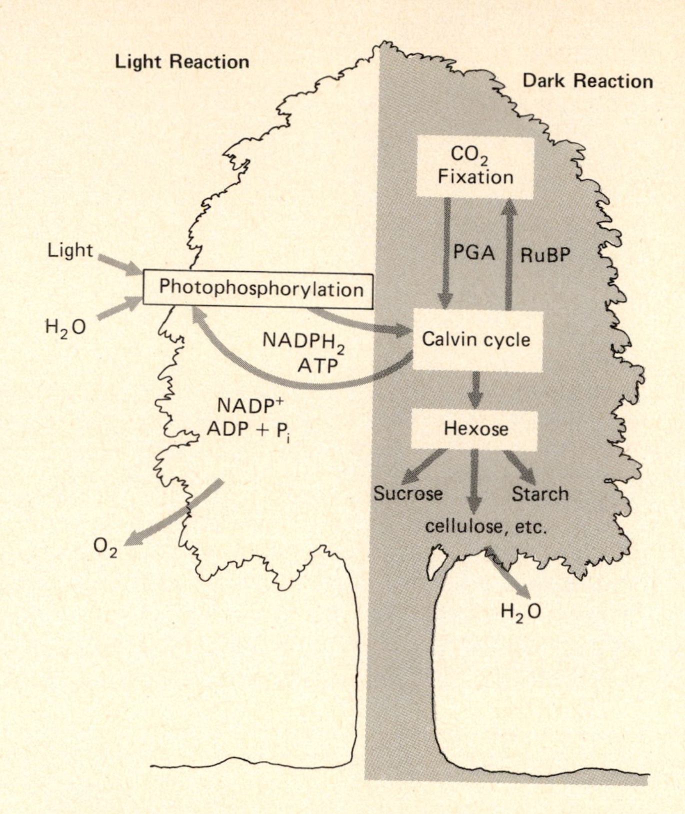

FIGURE 18.1 Basic processes of photosynthesis, involving both light reactions (on left) and dark reactions (on right).

Photosynthesis

Photosynthesis is the conversion of CO_2 and H_2O into carbohydrates using the energy of light captured by chlorophyll. It is carried out primarily by chlorophyll-bearing plants, aquatic and terrestrial. Minor contributions are made by photosynthetic bacteria that use hydrogen and various organic compounds as electron donors.

Although photosynthesis is commonly expressed as

$$6CO_2 + 12H_2O + \text{light energy} \rightarrow C_6H_{12}O_6 + O_2$$

it is a complex reaction involving a sequence of reactions that ultimately arrive at glucose (Figure 18.1). Glucose is only an intermediate compound. The final products are more complex sugars, free amino acids, proteins, fats and fatty acids, vitamins, pigments, and coenzymes. All of these substances probably are synthesized in chloroplasts by reactions involving photoelectron transport and photophosphorylation (the addition of one or more phosphate groups to a molecule by light energy). Synthesis of various products may take place in different parts of plants or under different environmental conditions. Mature leaves of some species of plants may produce only simple sugars, while young shoots and rapidly developing leaves may produce fats, proteins, and other products.

This is not an appropriate place to describe in detail the photosynthetic process. You can find complete reviews in various texts on plant physiology and good basic summaries in several outstanding general biology texts. What is of interest here are those features of photosynthesis that relate to ecological functions, especially energy transfer and adaptation of plants to various environmental situations.

CALVIN CYCLE

Photosynthesis consists of two parts: a light reaction and a dark reaction. The light reaction converts light energy into chemical energy and, as the term implies, takes place in light. The dark reaction converts CO_2 into sugars and starches. In spite of its name, the reaction does not require darkness but rather takes place independent of light, although this, too, is challenged.

Involved in the light reaction are two pigments or photosystems that absorb light energy (photosystem I, active at 700 nanometers far red wavelength, and photosystem II, active at 680 nanometers); water; and a hydrogen carrier, NADP (nicotinamide adenine dinucleotide phosphate), in the chlorophyll. Ignoring the details of the interactions of the two photosystems and the exchange of energy, what happens is this: the light reaction traps light energy and stores it for a short time in a high-energy phosphate bond in ATP (adenosine triphosphate). Hydrogen, obtained from water, is used to reduce $NADP^+$ to NADPH, and O_2 is given off as a gas:

$$2H_2O + 2NADP^+ \xrightarrow[\text{chlorophyll}]{\text{light}} 2NADPH + 2H^+ + O_2$$

In the dark reaction, CO_2 enters the process. In the presence of ATP, cells of chloroplasts fix CO_2 into carbohydrates:

$$\begin{gathered} ATP + NADPH + H^+ + CO_2 \rightarrow \\ ADP + P_i{}^* + NADP^+ + \text{carbohydrate} \end{gathered}$$

The reaction, so described, leaves much of the story missing. Involved is a five-carbon compound, ribulose biphosphate, RuBP. It combines with CO_2 to produce a six-carbon molecule that soon breaks into two three-carbon molecules of phosphoglyceric acid, PGA (Figure 18.2). Each PGA molecule next

* P_i is an inorganic phosphate; an abbreviation for the free phosphate group $HPO_4{}^{2-}$ added directly to ADP.

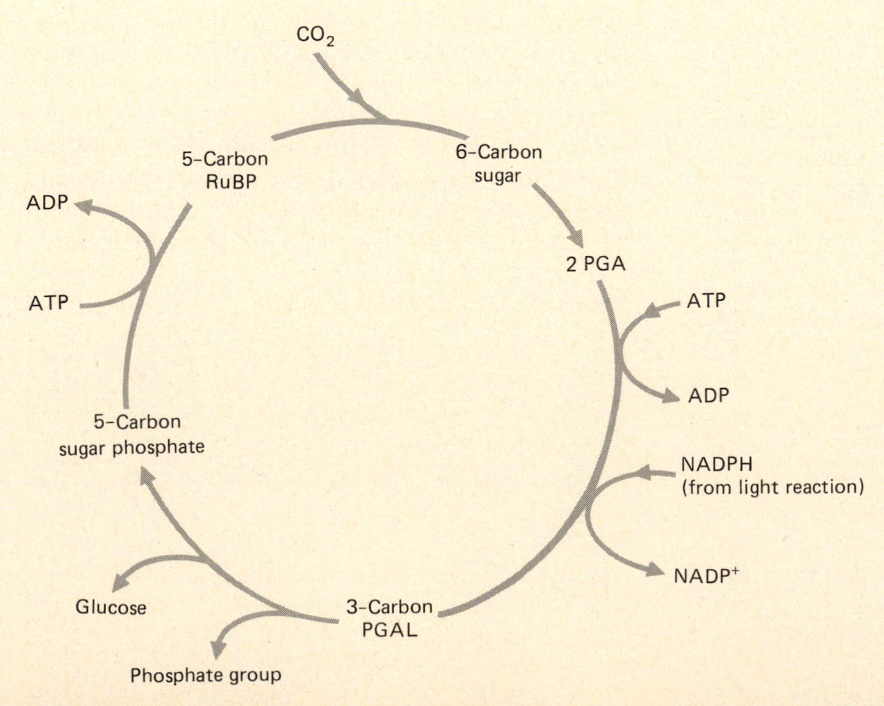

FIGURE 18.2 The Calvin cycle of photosynthesis.

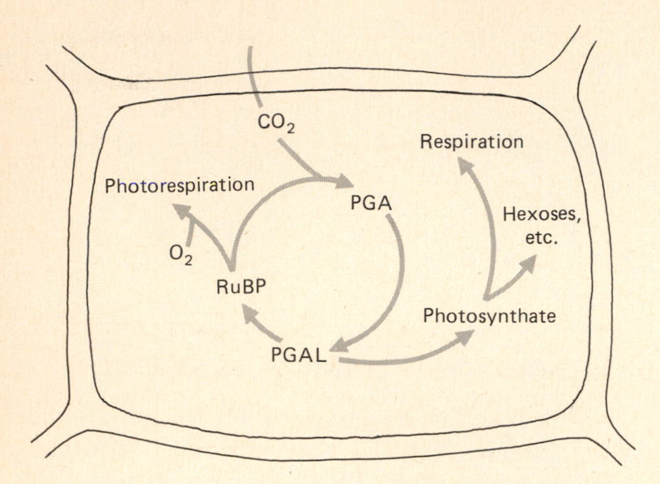

FIGURE 18.3 Basic dark reaction of the C_3 pathway of photosynthesis. (*After Barbour, Burk, and Pitts, 1980:310.*)

receives a second phosphate with a high-energy bond from a molecule of ATP. This new high-energy bond is then broken to supply the energy needed to reduce PGA to PGAL, phosphoglyceraldehyde. The reaction involves an addition of H^+ from NADPH. Some of the three-carbon PGALs are joined to form hexoses, six-carbon sugars. Other PGALs in combination with another ATP are used to reform new five-carbon RuBP, returned to the system to react with CO_2.

This sequence of reactions is known as the Calvin cycle, named for Melvin Calvin and his associates, who described it in a series of publications, the first of which appeared in 1948. The Calvin cycle, the most common form of photosynthetic reaction, is referred to as the C_3 pathway (Figure 18.3), and plants that possess it as their sole photosynthetic reaction are called C_3 plants.

PHOTORESPIRATION

Associated with the C_3 pathway is photorespiration, not to be confused with respiration. Green plants, like animals, undergo typical cellular respiration and generate energy with the breakdown of carbohydrates and the subsequent release of CO_2. But most plants also undergo another form, photorespiration, which produces CO_2 from ribulose biphosphate.

Studies of photosynthesis led to the discovery that the evolution of CO_2 in light is greater than in darkness, when the opposite should be the case. Investigations showed that RuBP, the CO_2 acceptor molecule in the C_3 pathway, also accepts O_2 in competition with its own carboxylation reaction. What apparently happens is that some RuBP and an enzyme, oxygenase, react with oxygen to yield one molecule of PGA and one molecule of phosphoglycollate. The PGA is returned to the reduction pentose phosphate cycle to form hexose sugars, while phosphoglycollate gives rise to CO_2 and some glucose, but with a considerable expenditure of energy.

The reaction is apparently a wasteful process, tying up RuBP that could be used for fixing CO_2. When O_2 is fixed, energy is lost, and only half as much PGA is produced. On the average, half of the CO_2 fixed in photosynthesis is lost in photorespiration. A rise in temperature increases that loss and decreases the yield of photosynthesis. Why plants should engage in such an energy-wasting process is not clear. One hypothesis is that photorespiration may be a safety valve used for the dissipation of energy in case CO_2 becomes limiting. Under that condition, pools of ATP and $NADP^+$ may become fully charged. To utilize that energy, RuBP fixes oxygen and prevents injury to chlorophyll and a breakdown of the photosynthetic system. When CO_2 is not limiting, the "escape" system still functions to a certain extent.

HATCH-SLACK CYCLE

In 1965 plant physiologists—especially H. P. Kortschak and his co-workers in Hawaii—discovered that the first detectable products of photosynthesis in sugarcane leaves were malic and aspartic as well as phosphoglyceric acids and not the latter alone. Subsequently, the problem was studied by M. D. Hatch and C. R. Slack and their co-workers in Australia. They determined that a different pathway was involved in photosynthesis in sugarcane and some other plants. Carbon dioxide was first fixed by acceptor molecules of phosphoenolpyruvate (PEP), a three-carbon compound, to form the four-carbon acids malate and aspartate. This reaction takes place in chlorophyll-containing mesophyll cells. The malic and aspartic acids are transported to the cells of bundle sheaths (Figure 18.4), where the acids are broken down by enzymes to release the CO_2. The CO_2 is then used to form the three-carbon PGL as in the Calvin cycle. The three-carbon carrier molecule returns to the mesophyll cell, where it is converted to PEP to receive more CO_2 (Figure 18.4). Because these plants utilize a four-carbon molecule in photosynthesis, they are called C_4 plants, and the photosynthetic pathway a C_4 system.

The major difference between the C_3 and C_4 pathways is the initial fixation of CO_2. The C_4 pathway involves an extra step in the process, which has certain advantages. By fixing CO_2 in the mesophyll and transporting it to the cells of the bundle sheath, the plants can reduce the concentration of CO_2 in the mesophyll to near zero while maintaining a high concentration of CO_2 in the cells of the bundle sheath. PEP involved in the transport is more reactive with CO_2 than RuBP and thus fixes it more efficiently at low concentrations. As a result, oxygen has no opportunity to inhibit the photosynthetic process. The Hatch-Slack cycle, however, requires two additional ATP molecules to fix each molecule of CO_2, but the rate of photosynthesis is twice as fast.

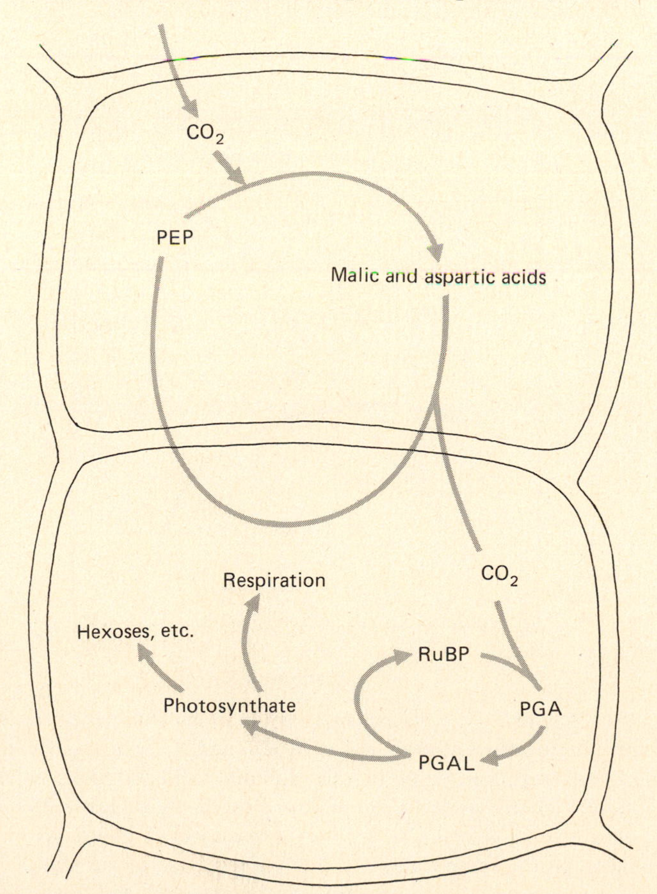

FIGURE 18.4 Basic dark reaction of the C_4 pathway of photosynthesis. Note the relationship of the Calvin cycle to the C_4 pathway. (*After Barbour, Burk, and Pitts, 1980:310.*)

CRASSULACEAN ACID METABOLISM (CAM)

A third type of photosynthetic process is the Crassulacean acid metabolism (CAM) (Figure 18.5). It is common to a small group of desert plants, mostly desert succulents, in the families Cactaceae, Euphorbiaceae, and Crassulaceae. To reduce the loss of water from transpiration, these plants close the stomata by day and open them by night to take in CO_2, the reverse of other plants (Figure 18.6). CO_2 is fixed by PEP at night and stored as malate in the vacuoles of cells. By day the stomata close, and malate moves from the vacuole into the cell. CO_2 is removed from malate and, in the presence of light

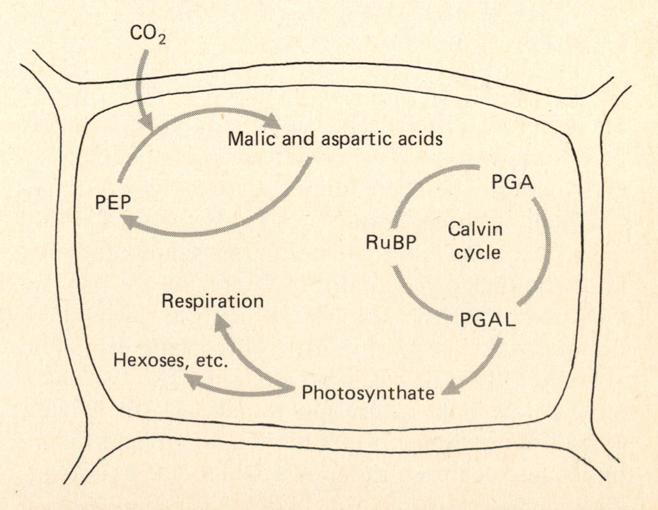

FIGURE 18.5 Basic dark reaction in the CAM cycle. Compare this with the C_4 cycle (Figure 18.4). (*After Barbour, Burk, and Pitts, 1980:310.*)

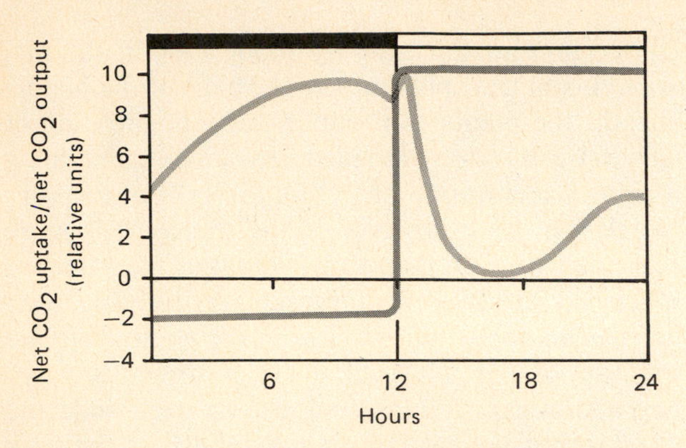

FIGURE 18.6 A model of CO_2 exchange in a CAM plant and a non-CAM plant in a photoperiod of 12:12 hours dark:light rhythm. Note the CAM intake of CO_2 during the night, while the non-CAM plant exhibits a loss of CO_2. The non-CAM plant has a high intake of CO_2 during the day. The CAM plant, however, does not experience a negative CO_2 balance. (*After Kluge, 1972.*)

energy, is used to refix CO_2 by the way of the Calvin cycle. CAM is similar to the C_4 pathway except there is no division of the process between different cells.

DIFFERENCES AMONG C_3, C_4, AND CAM PLANTS

Anatomically and physiologically, C_3 and C_4 plants are different (Table 18.1). C_3 plants have vascular bundles surrounded by a layer or sheath of colorless cells, and the cells containing chlorophyll are irregularly distributed throughout the mesophyll of the leaf. C_3 plants have an optimum temperature for photosynthesis of 16° to 25°C (Figure 18.7). The CO_2 compensation point (at which the CO_2 uptake is equal to CO_2 evolution by the plant) is in the range of 30 to 70 parts per million CO_2. C_3 plants have a low light-saturation threshold also (Figure 18.8). Photosynthesis is inhibited when light falling on leaves reaches a range of 1,000 to 4,000 footcandles.

Plants exhibiting the C_4 cycle have vascular bundles surrounded by distinctive bundle sheath cells rich in chlorophyll, mitochondria, and starches. Mesophyll cells are arranged laterally about the bundle sheaths and have few chloroplasts. Compared to C_3 plants, C_4 plants have a higher optimum temperature for photosynthesis, 30° to 45°C. They have a lower CO_2 compensation point, 0 to 10 parts per million, and exhibit no photorespiration, thus saving energy for other metabolic processes. They have a high light-saturation threshold; in fact, it is difficult for C_4 plants to reach light saturation, even in full sunlight (10,000 to 12,000 footcandles). The net rate of photosynthesis for C_4 plants in full sunlight is 40 to 80 mg CO_2/dm^2 leaf area/hour, while the net rate for C_3 plants is 15 to 35 mg CO_2/dm^2 leaf area/hour (see Black, 1973).

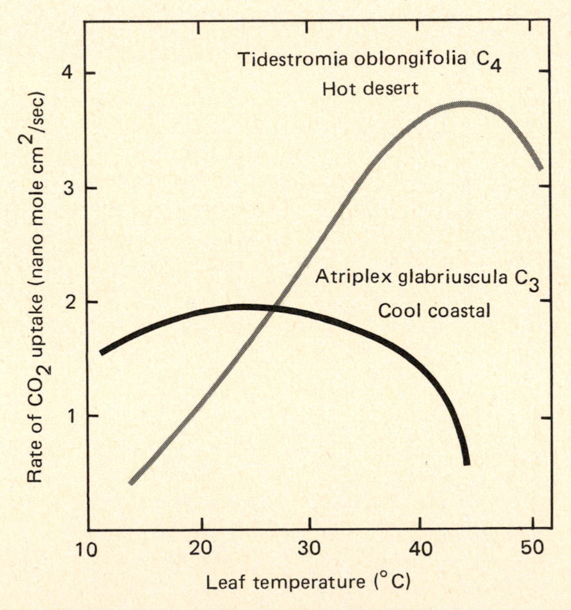

FIGURE 18.7 The effect of changes in leaf temperature on the photosynthetic rates of C_3 and C_4 plants. C_3 plants—represented by orache, a plant of the cool coastal habitat grown under a temperature regime simulating its natural one—exhibit a decline in the rate of photosynthesis as the temperature of the leaf increases. The plant also reaches a maximum photosynthetic output at a lower temperature than the C_4 plant, represented by Arizona honeysweet. This desert species increases its rate of photosynthesis as the temperature of the leaf increases, up to about 50°C. (*After Bjorkman, 1973:56.*)

TABLE 18.1 SOME DISTINGUISHING CHARACTERISTICS OF C_3, C_4, AND CAM PLANTS

Characteristic	C_3 Plants	C_4 Plants	CAM Plants
Leaf anatomy	Palisade, spongy parenchyma	No mesophyll differentiation, large bundle sheath	No mesophyll differentiation, big cells with large vacuoles
Light-saturation point	3,000–6,000 footcandles	8,000–10,000 + footcandles	?
Optimum temperature	15°–30°C	30°–45°C	30°–35°C
CO_2 compensation point	30–70 ppm	0–10 ppm	0–4 ppm
Maximum photosynthetic rate (mg CO_2/dm^2 leaf area/hr)	15–35	30–45	3–13
Maximum growth rate (g/dm^2/day)	1	4	0.02
Photorespiration	High	Low	Low
Stomata behavior	Open day, closed night	Open day, closed night	Closed day, open night

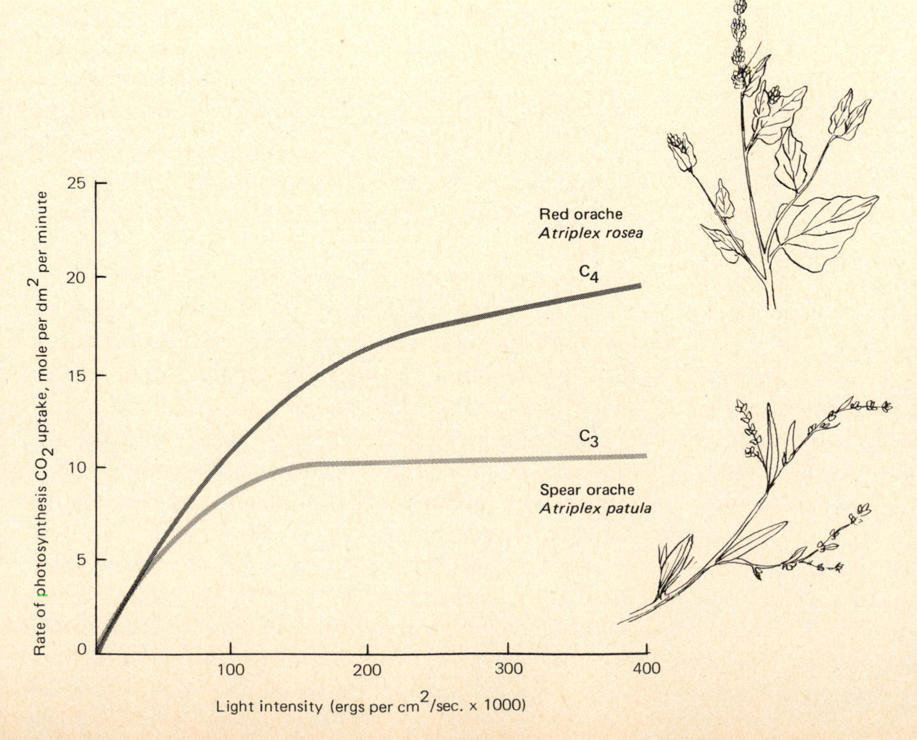

FIGURE 18.8 Effect of changes in light intensity on the photosynthetic rates of C_3 and C_4 plants grown under identical conditions. In this case, the conditions are a 16-hour day, 25°C by day, 20°C at night, with ample water and nutrients. The C_3 species, spear orache, exhibits a decline in the rate of photosynthesis as measured by CO_2 uptakes as light intensity increases. The C_4 species, red orache, shows no such inhibition. (*After Bjorkman, 1973:53.*)

In spite of the differences, plants cannot be divided arbitrarily into C_3 and C_4 groups. Some are difficult to place on the basis of either CO_2 compensation or primary products of photosynthesis. Some C_3 species do not become light-saturated at low light levels (see Singh, Singh, and Yadava, 1980).

Differences between C_3 and C_4 plants are ecologically important. Because they are more efficient photosynthetically under conditions of high light and temperature, C_4 plants may be more competitive and adjust more easily to environmental stress than C_3 plants. C_4 plants are mostly native to warm and tropical regions, where both light intensity and water are limiting. At low temperatures, the photosynthetic yield of C_3 plants exceeds that of C_4 plants; it matches that of C_4 plants at 25° to 30°C and is decidedly inferior to that of C_4 plants at temperatures over 30°C. (For a review see Berry and Bjorkman, 1980.)

Although C_4 plants are sensitive to low temperatures, no step in the C_4 pathway appears to be unusually sensitive to low temperatures. Features that provide C_4 plants with their advantage at high temperatures (such as stability of proteins and enzymes and transport of photoelectrons) are not necessarily related to the C_4 pathway. The C_4 pathway probably evolved among plants with a tolerance for higher temperatures in various water-limited environments. There selection advantage would favor those plants with an internal mechanism able to concentrate CO_2. Such characteristics, advantageous in warm climates, would be disadvantageous in cool climates. For that reason, the C_4 pathway probably became linked genetically to those characteristics that favor heat tolerance.

C_4 plants, which undoubtedly evolved in tropical climates and then migrated to temperate climates, have the more advanced photosynthetic process. The C_3 pathway is common to all plants, including those with a C_4 pathway. The latter is not found in algae, mosses and lichens, ferns, and more primitive angiosperms. It is known to exist in hundreds of monocotyledonous and dicotyledonous species comprising 100 genera and 10 plant families. Grasses comprise about half of the known C_4 species. In North America most of these are subtropical in distribution. From Florida to Texas 65 to 82 percent of the grass species are C_4 plants; in the Central Plains, 31 to 61 percent; and in the northern part of the continent, 0 to 23 percent (Terri and Stowe, 1976). Those with the C_4 pathway are warm-season grasses; the rest are cool-season grasses, putting on growth and flowering in spring and early summer. Apparently, no C_4 plants grow on the tundra.

CAM plants, also inhabitants of hot climates, not only fix CO_2 from malate, like C_4 plants; they fix CO_2 from the atmosphere as well, depending on environmental conditions. Because CAM plants live in areas where water is deficient, they have to maintain a positive carbon balance, or at least prevent a negative balance, during long dry periods and still maintain a positive water balance. These plants evolved an adaptive variant of the C_4 photosynthetic pathway to achieve the maximum possible carbon gain with a minimum loss of water. They achieve that at the expense of rapid growth.

LIMITS ON RATES OF PHOTOSYNTHESIS

Photosynthesis is limited by both environmental conditions and the nature of community structure. These limits include such environmental variables as light intensity, temperature, moisture, atmospheric gases, soil nutrients, and competition and such plant variables as leaf area, geometry of vegetational canopy, and photosynthetic capacity. To further complicate the picture, all of these variables interact.

One important variable requires some comment—leaf area index (LAI) (see Chapter 7), which involves the relationship of leaves to light interception. Some plants have an optimal LAI at which there is minimal shading of one leaf by another (Pearce, Brown, and Blaser, 1965). As LAI increases beyond that point, photosynthesis decreases because of increased shading of one leaf by another. Other plants do not show this response (Loomis, 1967), primarily because of the angle of the leaf, which influences light interception. Horizontal leaves perpendicular to incoming light intercept the most light, but a number of layers of horizontal leaves reduces the amount of light reaching lower leaves. Leaves growing at an angle require a much higher LAI to intercept the same quantity of light as horizontal leaves. As their LAI increases above a certain value, angled leaves carry on photosynthesis at a faster rate. Such an adaptation in leaf position to obtain maximum photosynthetic rates and still possess a high LAI is characteristic of corn, grasses, beets, turnips, and other row crops.

EFFICIENCY OF PHOTOSYNTHESIS

Photosynthetic efficiency in converting the energy of the sun to organic matter can be assessed from two viewpoints: (1) the amount of energy required for the evolution of a molecule of oxygen and (2) the ratio of calories per unit area of harvested vegetation to total solar radiation intercepted. In terms of energy input, photosynthesis is a rather inefficient process. To release 1 mole of oxygen and to fix 1 mole of carbon dioxide, a green plant needs an estimated 320 kilocalories of light energy. For each mole of oxygen evolved, approximately 120 kilocalories of energy are fixed. That is an efficiency of approximately 38 percent. Efficiencies calculated for isolated chloroplasts and for some algae amount to 21 to 33 percent (Kok, 1965; Wassink, 1969; Bassam, 1965).

When calories stored relative to light energy available are considered, efficiency is considerably less. The usable spectrum, 0.4 to 0.7 wavelengths, is only half the total energy incident on vegetation. Highest short-term efficiency measured over a period of weeks of active growth may amount to 12 to 19 percent (Wassink, 1969). In most instances, however, photosynthetic efficiency is computed on either an annual or a growing season basis.

A sampling of ecological efficiencies is given in Table 18.2. These figures cannot be accepted with a great deal of confidence, nor are they really comparable. All of them involve a number of assumptions in their determination. There is no standard method for determining photosynthetic efficiency. In many cases radiant energy incident upon the vegetation studied was estimated from tables or measurements taken in the region but not in the study area. Temperature, moisture, nutrient status of the soil, and other variables are not held constant. Estimates are often based on total energy available for the year rather than during the growing season alone, which lowers estimates considerably. If based on the growing season, estimates might be as high as 10 percent instead of 1 to 3 percent. Error is also introduced if energy fixation is based on estimates of community peak standing crops rather than on the determination of the peak standing crop of individual species within the community. Because different species reach peak standing crop at different periods of the growing season, estimates made only at one period seriously underestimate photosynthetic efficiency.

Decomposition

Photosynthesis is a packaging process. Energy of light, carbon dioxide, and water are utilized to form simple carbohydrates. Reactions from that point produce more complex carbohydrates and incorporate mineral nutrients from soil and water into plant proteins, fats, and carbohydrates.

Decomposition is the opposite process—the ultimate breakdown of organic matter into its basic constituents of carbon dioxide, water, inorganic nutrients, and release of energy. Compared to photosynthesis—which is a relatively straightforward, however complex, photoreaction—decomposition is a messy affair. It involves a number of organisms, innumerable pathways, and the influence of a wide range of environmental variables. For that reason decomposition is difficult to study and has not received the same intensive investigation as photosynthesis. In many ecological texts decomposition tends to be ignored or given only minimal attention. Yet without decomposition, life on Earth could not exist.

Decomposition essentially is the reduction of organic matter of higher molecular weight to that of lower molecular weight, leading ultimately to the transformation of organic matter to inorganic matter. It is not, as you are frequently led to believe, the final stop in the food web where plant and animal remains are processed by decomposer organ-

TABLE 18.2 COMPARATIVE YIELDS AND PHOTOSYNTHETIC EFFICIENCIES OF SOME PLANT COMMUNITIES

Community	Location	Growing Season, Days	Dry Matter, $g/m^2/day$	Efficiency, %	Reference
Corn	Minnesota	92	11.00	2.10	Ovington and Lawrence (1967)
1-year weed field	New Jersey	120	19.50	3.80	Botkin and Malone (1968)
Meadow steppe	West Siberia	100	15.00	3.40	Rodin and Bazilevic (1965)
Short-grass steppe	Colorado	115	22.00	3.40	Moir (1969)
Tall-grass prairie	Colorado	140	8.00	1.20	Moir (1969)
Tall-grass prairie	Colorado	218	3.85	0.77	Sims and Singh (1971)
Short-grass prairie	Colorado	266	5.38	0.80	Sims and Singh (1971)
High mountain grassland	Colorado	92	10.74	1.20	Sims and Singh (1971)
Desert grassland	Colorado	267	1.58	0.16	Sims and Singh (1971)
Oak-pine forest	Long Island	250		0.90	Whittaker and Woodwell (1969)
Polar desert	Devon Island	50–60	0.59	0.03	Bliss (1975)
Sedge-moss meadow	Devon Island	50–60	12.36	0.79	Bliss (1975)
Lichen heath	Norway	105	10.80	0.70	Wielgolaski and Kjelvik (1975)
Wet meadow	Norway	120	36.30	2.30	Wielgolaski and Kjelvik (1975)
Birch forest	Norway	150	28.20	1.70	Wielgolaski and Kjelvik (1975)

isms. Actually, the greater part of the decomposition process is the conversion of organic matter into animal and microbial tissue (Figure 18.9), which moves into new food chains. Ultimately, over varying periods of time, organic matter does end up inorganic, but by very indirect routes.

PATHWAYS OF DECOMPOSITION

Decomposition involves a number of processes, including the leaching of soluble compounds, consumption by animals, fragmentation, bacterial and fungal breakdown, consumption of bacterial and fungal organisms, excretion of organic and inorganic compounds by organisms, and flocculation of colloidal organic matter into larger particles. Organic matter acted upon may range in size from whole organisms to small fractions as coarse particulate organic matter, fine particulate organic matter, and dissolved organic matter.

One pathway of decomposition begins when herbivores eat plants and carnivores consume prey. Not only does the animal extract minerals and nutrients from the food it eats for its own nutrition, but it also deposits a substantial portion as partially decomposed material—feces. The amount and nature of fecal material deposited, especially by herbivores, depends in part on the digestibility of the

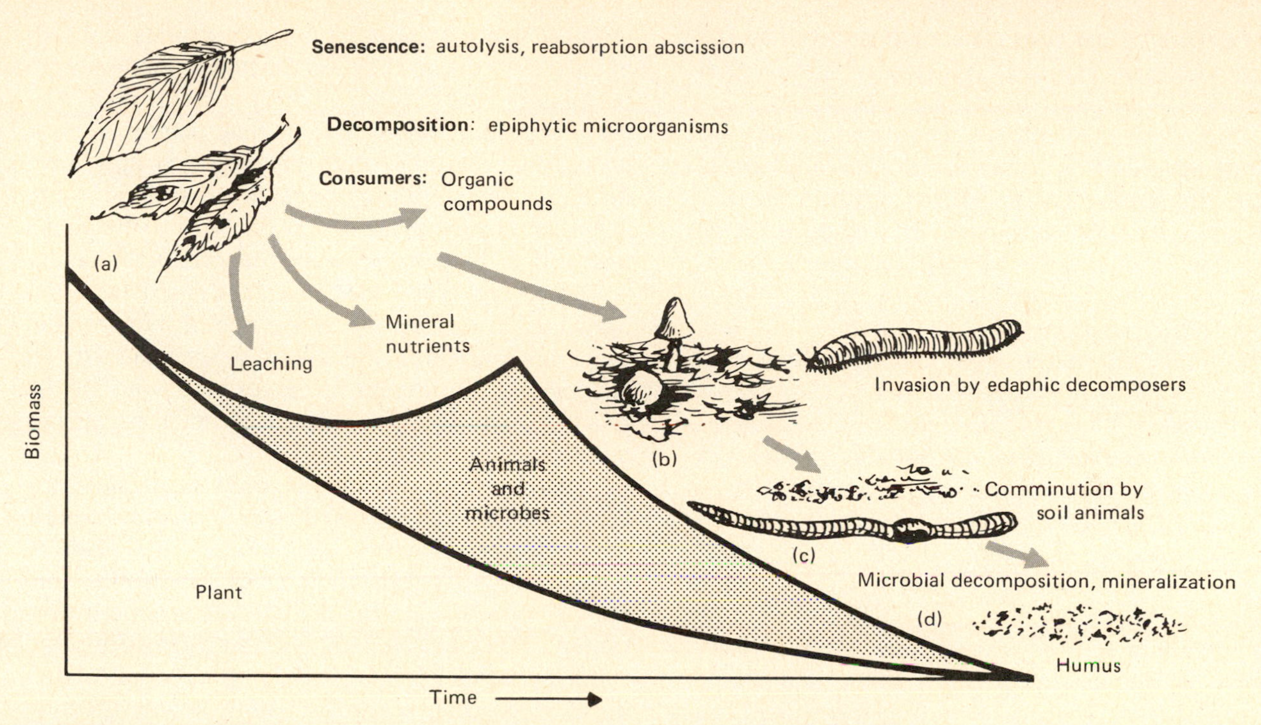

FIGURE 18.9 A diagrammatic representation of a terrestrial decomposition cycle. (a) Initial stage of decomposition, involving senescence of the leaf, during which some nutrients are lost by leaching, autolysis, and reabsorption by the plant. Some nutrients in the leaf are lost to primary consumers such as caterpillars and aphids. The senescent leaf is also colonized by epiphytic saprophytes. (b) Leaves fall to the ground to accumulate as leaf litter. Dead leaves are colonized by fungi and bacteria, which utilize soluble nutrients first. The leaves are then colonized by a succession of litter-feeders—reducer-decomposers—and by an array of microbes. (c) Action of reducer-decomposers and microbes reduces litter to humic substances. Further breakdown of organic matter continues, resulting in mineralization. (d) Finally, the humus becomes mixed with mineral soil. Throughout the decomposition process, plant, animal, and microbial biomass has also been changing. Plant biomass, at its maximum at the time of initiation of decomposition, declines with time until it is incorporated into soil. Animal and microbial biomass increases as those organisms transform plant tissue into heterotroph tissue. Populations of these organisms reach their maximum at the leaf litter stage and then decline as less plant biomass is available to support them. (*After Stout, Tate, and Molloy, 1976:98.*)

plant and the ability of the animal's digestive system to handle plant material. Herbivores generally select vegetation low in lignin and digest most of the easily soluble compounds. Partially digested lignins are eliminated and are available for microbial colonization.

Some coprophagous herbivores digest little of the material they consume but rather deposit it as largely fragmented material conditioned for colonization by bacteria. These herbivores, such as rabbits, reingest their own feces and utilize the gut flora as well as the microflora that develops on the feces outside of the body. The microbes convert the indigestible carbohydrates into digestible microbial tissue.

Decomposition of leaves begins while they are still on the plant. As the leaves approach senescence, the plant itself reabsorbs much of the nutrients into its permanent parts. During the growing season, plants produce varying quantities of exudates that support an abundance of surface microflora. These organisms feed on exudates and any cellular material that sloughs off. The same exudates account for the nutrients leached from leaves during a rain.

While some microbes are utilizing exudates of leaves, others are utilizing organic material from the roots of living plants. The soil region immediately surrounding the roots, known as the rhizosphere, and the root surface itself, known as the rhizoplane, support a host of microbial-feeders on root litter and root exudates. Root exudates may consist of simple sugars, fatty acids, and amino acids. The kind of exudates in the rhizosphere can influence quantitative and qualitative differences in the microflora of rhizospheres of different plant species (F. E. Clark, 1969a, 1969b).

DECOMPOSER ORGANISMS

Organic detritus is attacked by a variety of *saprophages,* organisms that live on dead matter. Saprophages can be divided into two groups: the microscopic and the macroscopic. Together there may be over one million of them in a square meter of the first 7 to 10 cm of a temperate hardwood forest soil. Approximately 40 percent of these organisms are bacteria; about 50 percent are microscopic fungi; 5 to 9 percent may be protozoans; and 0.05 percent may be true fungi (mushrooms). Animals that can be seen by the naked eye, the macrofauna, account for only 0.04 percent of the total population.

The dominant detrital-feeding organisms are heterotrophic bacteria and fungi. Bacteria may be *aerobic,* requiring oxygen as an electron acceptor; or they may be *anaerobic,* able to carry on their metabolic functions without oxygen by using some inorganic compound as the oxidant. Anaerobic bacteria are commonly found in aquatic muds and sediments. As a group they are the major decomposers of animal matter. Major decomposers of plant material are the fungi, whose hyphae penetrate plant and animal material. Both bacteria and fungi produce enzymes necessary to carry out specific chemical reactions. Enzymes are secreted onto the plants and animal matter. Some of the products are absorbed as food, and the remainder is left for other organisms to utilize. Once a group of bacteria and fungi has exploited the material according to its capabilities, another group moves in to continue the process. Thus, a succession of microorganisms takes place in the detritus.

Macroorganisms include such small, detritus-feeding animals as collembolas or springtails, mites, millipedes, earthworms, nematodes, and slugs in terrestrial ecosystems and crabs, mollusks, and mayfly, stonefly, and caddisfly larvae in aquatic ecosystems. Detritus-feeders, such as earthworms and caddisfly larvae, fragment organic material, breaking it into smaller pieces by both mechanical action and digestion, mixing it with soil in the case of the earthworm, excreting it, and even adding substances to the material that stimulate microbial growth. These same organisms also consume bacteria and fungi associated with the detritus, as well as small invertebrates and protozoans clinging to the material. Others graze on the microbes. They feed on the detrital particles to remove bacteria and fungi and prepare the surface for recolonization by microbes. These grazers may reduce bacterial and fungal populations, inhibiting the effects of increased population density, accelerating the multiplication of soil microbes, and thus speeding up microbial activity.

A sort of complementary relationship exists between the macrodecomposers and the microdecomposers. The macroorganisms fragment the detrital material, making it more available to smaller detrital feeders and to bacteria and fungi. Ultimately, the material is reduced in size to a point where even microbial activity cannot continue. At this point bacteria assimilate organic compounds, in effect concentrating them into larger particles that in turn are made even larger by bacterial aggregation. This material is again available to macrodecomposers. They, in turn, may produce fecal pellets larger than the material digested, providing surfaces available for microbial colonization. In such a manner decomposing organic matter may be passed among feeding groups until it finally reaches an inorganic state.

PROCESS OF DECOMPOSITION

Leaves fallen on the ground, in streams, and at the edges of ponds are colonized by a growing population of mostly fungi and some bacteria. These microorganisms quickly extract the most soluble substances and soften the leaves, making them available for detrital-feeding organisms. Among the first to invade the material are sugar-consuming fungi and bacteria. Once the glucose is utilized, the detritus is invaded by other bacteria and fungi feeding on cellulose.

As the microorganisms work on the plant debris, they assimilate nutrients and incorporate them in their tissues. As long as these nutrients are a part of the living microbial biomass, they are unavailable for recycling. That is known as *nutrient immobilization*. The amount of mineral matter that can be tied up by microbes varies greatly. Some exhibit luxury consumption, ingesting more than they need for maintenance and growth. Bacteria and fungi are very short-lived. They die or are consumed by litter invertebrates. Death and consumption, as well as the leaching of soluble nutrients from the decomposing substrate, release minerals contained in the microbial and detrital biomass. This process, known as *mineralization*, makes nutrients available for use by plants and microbes.

Thus, a cycle of immobilization and mineralization takes place within the soil. Nutrients are temporarily immobilized in microbial tissue. As microbes die, the nutrients are released or mineralized and become available for uptake again. Microbial uptake occurs simultaneously with mineralization. The amount of nutrients available for plants depends in part on the magnitude of uptake by microbial decomposers.

The process of decomposition is aided by the fragmentation of detritus by litter-feeding invertebrates. They consume parts of leaves, opening them up to microbial invaders. The action of such litter-feeders as millipedes and earthworms may increase exposed leaf area to 15 times its original size (Ghilarov, 1970). Because the net assimilation (see Chapter 19) of plant detritus by litter-feeders is less than 10 percent, a great deal of the material passes through the gut of these organisms. They utilize only the easily digested proteins and carbohydrates. Mineral matter in fecal material is readily attacked by microbes. Some litter-feeders, such as earthworms, enrich the soil with vitamin B_{12}. In addition, they mix organic matter with soil, thus bringing the material into contact with other microbes.

Evidence suggests that in aquatic environments bacteria and, to a limited extent, fungi act more as converters than as regenerators of nutrients, whereas phytoplankton and zooplankton play a major part in the cycling of nutrients. Phytoplankton, macroalgae, and zooplankton furnish dissolved organic matter, with algae being the main contributor. Phytoplankton and other algae excrete quantities of organic matter at certain stages of their life cycle, particularly during rapid growth and reproduction. Twenty-five to 75 percent of the regeneration of nitrogen in the marine environment takes place by autolysis (enzymatic breakdown of plant and animal tissue upon death) of phytoplankton and zooplankton rather than by bacterial decomposition (Johannes, 1968). In fact, 30 percent of the nitrogen contained in the bodies of zooplankton is lost by autolysis within 15 to 30 minutes after death, too rapidly for any significant bacterial action to occur.

Bacteria, phytoplankton, and zooplankton utilize inorganic nutrients as well as such organic nutrients as vitamin B_1 (necessary for the growth of both phytoplankton and zooplankton) and organic sources of nitrogen and phosphorus. In effect, they tend to concentrate these nutrients by incorporating them into their own biomass.

Important in the concentration of nutrients are the bacteria, which use dissolved organic matter as a substrate for growth. Both dissolved and colloidal matter condense on the surface of air bubbles in the water, forming organic particles on which bacteria flourish (R. T. Wright, 1970; Riley, 1973). Fragments of cellulose supply another substrate for bacteria. Bits of plant detritus, bacteria, and phytoplankton are consumed by both bacteria and planktonic animals (Strickland, 1965a).

As in terrestrial situations, the utilization of these organic nutrients by bacteria results in both the increase and the immobilization of nutrients. Bacteria can use nutrients such as phosphorus in excess of their need. Such luxury consumption can reduce the available supply of nutrients to phytoplankton.

Bacteria are consumed by ciliates and zooplankton, which, in turn, excrete nutrients in the form of exudates and fecal pellets in the water. Zooplankton, too, in the presence of an abundance of food consumes more than it needs and can reduce microbial populations. Zooplankton, then, will excrete half or more of the ingested material as fecal pellets, which make up a significant fraction of suspended material. These pellets are attacked by bacteria that utilize the nutrients and growth substances they contain. Thus, the cycle starts over again (Figure 18.10).

Aquatic muds are largely anaerobic habitats. Fungi are absent, and the decomposer bacteria are largely facultative anaerobes. Incomplete decomposition often results in the accumulation of peat and organic muck. Nevertheless, particulate matter supports a rich bacterial population. For example, the snail *Hydrobia* feeds on detritus found in mud flats (Newell, 1965). Its fecal pellets are devoid of nitrogen but rich in carbohydrates, suggesting that the snail cannot digest cellulose. If these pellets are held in filtered seawater, the nitrogen content quickly rises. The rise in nitrogen is accomplished by the growth of marine bacteria that colonize the fecal material and utilize the nitrogen dissolved in seawater. The fecal pellets, enriched with nitrogen in the form of bacterial protein, are reingested by the snail, a form of coprophagy. The snail digests the bacterial bodies, and the resultant fecal pellets are again devoid of nitrogen. Colonization of fecal pellets is repeated.

Thus, bacteria function primarily to concentrate nutrients rather than to release them to the environment by decomposition. That task is accomplished largely by algae, zooplankton, and detritus-feeding animals that release certain nutrients and metabolites into the water by physical and chemical breakdown of plant and animal material, releasing cellular contents and excreting organic matter into the water. (For some details see Anderson and MacFadyen, 1976; Saunders et al., 1980.)

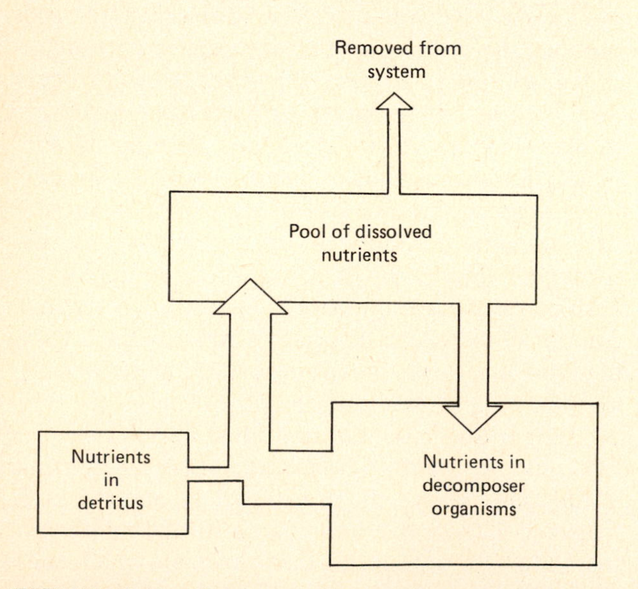

FIGURE 18.10 Nutrient cycling in microorganisms, emphasizing their role in nutrient immobilization. (*After Fenchel and Harrison, 1976:296.*)

INFLUENCES ON DECOMPOSITION

Decomposition is influenced by a number of environmental and biotic variables. Among them are moisture, temperature, exposure, type of microbial substrate, and vegetation. Both temperature and moisture greatly influence microbial activity by affecting metabolic rates. Decomposing litter in wetlands experiences conditions ranging from complete submergence to complete exposure to the atmosphere, from anaerobic to aerobic. Lower rates of decomposition take place under anaerobic conditions, while decomposition proceeds most rapidly under moist aerobic conditions. Alternate wetting and drying results in an increase in microbial growth and respiration during the rewetting period. In wetlands, decomposition proceeds most rapidly under alternate wetting and drying. Continuous dry spells and arid conditions reduce both activity and growth of microbial populations, inhibiting decomposition.

Slope exposure, especially as it relates to temperature and moisture, and the type of vegetation can increase or decrease decomposition. Witkamp (1963) found that bacterial counts from north-facing slopes were nine times higher in hardwoods than in coniferous stands, but counts from hardwood and coniferous stands on south-facing slopes did not differ. That undoubtedly was due to drier conditions on south-facing slopes.

Nutritional composition of leaves in litter has a powerful influence on decomposition. Easily decomposable and highly palatable leaves from such species as redbud, mulberry, and aspen support initially higher populations of decomposers than litter from oaks and from pine, which is high in lignin. Earthworms have a pronounced preference for such species as aspen, white ash, and basswood; take with less relish and do not entirely consume sugar maple and red maple; and do not eat red oak at all (Johnston, 1936). In a European study (Lindquist, 1942), earthworms preferred the dead leaves of elm, ash, and birch; ate sparingly of oak and beech; and did not touch pine or spruce needles. Millipedes likewise show a species preference (van der Drift, 1951). Thus, decomposition of litter from certain species proceeds more slowly than decomposition of litter from other species. On easily decomposable material, initially high populations of microbes decline with time as energy is depleted. But on more resistant oak and pine litter, initially low population densities of microbes increase as decomposition proceeds (Witkamp, 1963).

RATE OF DECOMPOSITION

How fast does organic matter decompose? There are several ways to find out, all based on the loss of biomass in one form or another.

One method is the use of paired plots on a given area (the Wiegart-Evans method). Dead material from one plot is removed and weighed at time t_0. At a later date, time t_1, dead material is removed from the other plot and weighed. From that data the rate of decomposition can be calculated.

A second method is the use of nylon mesh bags of varying mesh sizes. Litter is put in the bag and the bag is placed on top of the litter, within the litter, or in the upper layer of soil. Loss in weight over time is recorded. By placing litter in bags with mesh too fine to allow macrofauna to enter, one can estimate the comparative role of litter fauna in decomposition. By adding a chemical to inhibit the activity of microfauna and flora, one can determine the role of leaching only in decomposition. A similar method is used in stream ecosystems. There packets of leaves are placed in streams.

A third, more artificial method involves the use of strands of cellulose material, such as fiber paper or cotton-wool strips, placed in mesh bags instead of natural litter. The method has the advantage of standardizing litter material, allowing a comparison among several communities.

The problem with methods involving biomass loss is the inability to separate the loss of weight in litter and the gain in weight by a growing microbial biomass.

A fourth method involves radiocarbon measurements. Wheat straw or natural litter is labeled with ^{14}C. As the material decomposes, the radioisotope carbon becomes distributed in various fractions of soil organic matter.

The evolution of CO_2 from soil and litter is another measure of decomposition. CO_2 production declines as decomposition advances and energy is consumed—provided, of course, that no new litter is added to the decomposing material. Needless to say, that is not the situation under natural conditions. Another drawback is the near impossibility of separating respiration of roots from respiration of microbes.

In spite of limitations, some estimates of decomposition rates have been determined (Table 18.3). All communities studied, aquatic and terrestrial, share one feature. Initial decomposition rates are high because of leaching of soluble compounds and consumption of highly palatable tissue by microbes. As decomposition proceeds, the rate slows because only less decomposable material remains (Figure 18.11).

Leaf litter confined in mesh bags in a hard beech *(Nothofagus)* forest in New Zealand lost 20 percent of its weight in 6 months, but only one-third of it had decomposed at the end of 4 years (Stout, Tate, and Molloy, 1976). However, a total

TABLE 18.3 RATES OF DECOMPOSITION OF SELECTED ORGANIC MATERIAL IN FRESH WATER

Species	%/day
Particulate leaf organic matter	
Elm, *Ulmus americana*	0.05–0.15
Beech, *Fagus grandiflora*	0.25
Red oak, *Quercus rubra*	0.27
Sugar maple, *Acer saccharum*	0.33–1.07
Yellow birch, *Betula lutea*	0.5–1.2
Walnut, *Juglans nigra*	0.70
White ash, *Fraxinus americana*	1.20
Basswood, *Tilia americana*	1.75
Fine particulate matter	
Phytoplankton	14.0
Detritus	8.0
Zooplankton	1.0–33.0
Dissolved organic matter	
Dissolved carbohydrates	0.38–0.45
Dissolved nitrogen	0.50
Dissolved organic carbon	2.7–3.4

SOURCE: Leaf data: Petersen and Cummins, 1974; fine and dissolved organic matter: Saunders, 1974.

litter volume of 1,400 g/m^2 on the forest floor and an annual litterfall of 500 g/m^2 suggested a turnover time of 3 years under natural conditions in that forest stand. That rate compares with a mean resident time of carbon in a yellow-poplar stand in Tennessee of 3.1 years for rapidly decomposable material. Slowly decomposable material had a resident time of 59 years.

Deciduous leaves placed in litter bags in streams lose 10 to 30 percent of their weight as dissolved organic matter a few days after immersion. The remaining 70 to 90 percent takes much longer to break down and disappear—up to 50 to 400 days (Saunders, 1974).

A common way of expressing decomposition rates in terrestrial communities in which new litter accumulates as old disappears is the index k. It is the average fraction of accumulated organic matter lost in one year. If input (I) equals decomposition (X), then k is the ratio of the two, I/X. The more usual form is $x = x_0e^{-kt}$ (Olson, 1963), which is an exponential equation (see Chapter 12). The index k provides a convenient comparison of decomposition rates of various plant parts, plant species, and plant communities (Table 18.4 and Figure 18.12). Leaves and fine roots decompose most rapidly; tree logs most slowly. Depending on size, logs may require 100 years or more to disappear. Bogs, being cold and wet, retain organic matter 100 times longer or more than tropical rain forests, in which decomposition is so rapid that organic material is lost in less than a year (Figure 18.12).

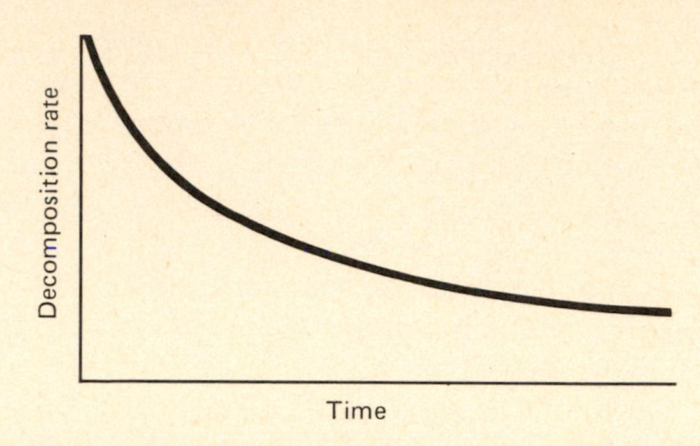

FIGURE 18.11 A generalized curve for decomposition. Decomposition declines exponentially with time. (*After G. W. Saunders, 1976:349.*)

TABLE 18.4 DECOMPOSITION CONSTANTS FOR DOUGLAS-FIR LITTER COMPONENTS

Component	k/yr
Needles	0.22–0.28
Branches	0.06–0.09
Cones	0.06–0.08
Bark	0.01–0.04

SOURCE: D. W. Johnson et al., 1982:205.

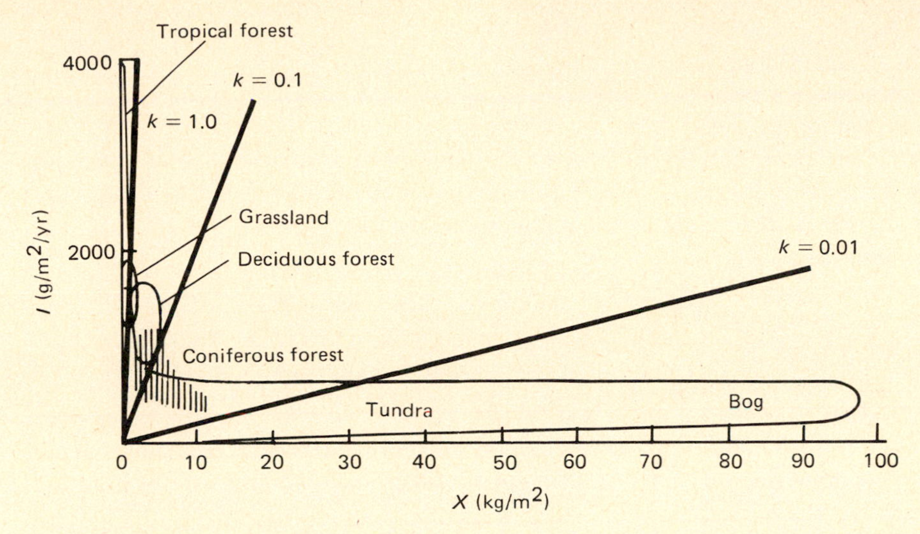

FIGURE 18.12 Input and accumulation of organic matter in a number of terrestrial ecosystems. The ratio of input to decomposition is k. Note the relatively high input and low accumulation of organic matter in tropical forests and grasslands and the high accumulation in the tundra and bog, where cold and moisture inhibit decomposition. (*After Olson, 1963.*)

Summary

The most basic processes in the functioning of ecosystems are photosynthesis and decomposition. Photosynthesis is the process by which green plants utilize the energy of the sun to convert carbon dioxide and and water into carbohydrates. Plants possess one of three different photosynthetic pathways. Most plants use the C_3 pathway, dominated by the Calvin cycle, which involves the formation of a three-carbon phosphoglyceric acid used in subsequent reactions. C_4 plants utilize a four-carbon process, in which the carbon dioxide taken into the leaf reacts to form malic and aspartic acids. The CO_2 is fixed in these compounds and is then released to the Calvin cycle.

These two groups of plants, C_3 and C_4, possess structural and physiological differences that are important ecologically. C_4 plants can carry on photosynthesis at higher leaf temperatures, at higher light intensities, and at lower CO_2 concentrations than C_3 plants. C_3 plants are less efficient because of photorespiration, in which plants substitute O_2 for CO_2 in part of the Calvin cycle. C_3 plants are distributed from the arctic to the tropics. C_4 plants are most successful in a hot, arid environment. Because of their low CO_2 compensation point, C_4 plants are able to fix CO_2 in spite of the closure of stomata to reduce water loss. Most highly adapted to an arid environment are plants with a third type of photosynthetic pathway, CAM. These plants absorb CO_2 by night, when water loss is reduced, store CO_2 as malic acid in cell vacuoles, and release it during the day to the Calvin cycle.

In spite of the various pathways, the overall efficiency of photosynthesis, measured as a ratio of gross or net photosynthesis to solar radiation, is low. During the growing season, plants achieve an efficiency of roughly 1 to 3 percent or lower if net rather than gross photosynthesis is used in the calculation.

Involved in the return of nutrients to the ecosystem and the final dissipation of energy are the decomposer organisms. The true organisms of decay, those responsible for the conversion of organic compounds to inorganic ions, are the heterotrophic bacteria and fungi. But the decomposition process involves a diverse group of organisms, from consumers, which return part of the material ingested in a partially decomposed form to other decomposer organisms, to reducer-decomposers, which fragment detrital material into smaller particles. These are then more accessible to bacteria and fungi. Another group, the microbial grazers, feeds on detrital particles, mainly for the bacteria and fungi growing on them. These organisms reduce microbial populations and thus influence microbial activity. Bacteria and fungi, as well as other decomposers, immobilize nutrients—remove them from circulation—by incorporating them in their body tissues. In terrestrial ecosystems, bacteria and fungi play the major role in decomposition. In aquatic ecosystems, bacteria and fungi act more as converters, while phytoplankton and zooplankton play a major role in the cycling of nutrients.

Review and Study Questions

1. Contrast photosynthesis and decomposition. Why are they the two basic functions in an ecosystem?
2. What characterizes the Calvin cycle, the Hatch-Slack cycle, and the Crassulean acid cycle in photosynthesis?
3. What are the C_3, C_4, and CAM pathways of photosynthesis?
4. What is the ecological significance of each?
5. What distinguishes a C_3, C_4, and CAM plant anatomically and physiologically?
6. Why is the C_4 pathway considered an evolutionary advancement over the C_3 pathway? Under what environmental conditions was it probably selected?
7. How does LAI relate to photosynthetic rates?
8. How efficient is photosynthesis in storing energy?
9. Why is decomposition more than just the traditional end point in a food chain? When does decomposition begin?
10. Characterize the following as they relate to decomposition: saprophage, anaerobic bacteria, aerobic bacteria, fungi, microbial decomposers, macroorganisms, and microbial grazers.
11. How do detrital-feeders relate to bacterial and fungal populations? How does this relationship influence decomposition?
12. What is nutrient immobilization? Mineralization?
13. What is the role of bacteria, phytoplankton, and zooplankton in the decomposition process in the aquatic environment?
14. What are the major environmental influences on decomposition?
15. How do decomposition rates vary among different plant parts?
16. How do temperature and moisture influence decomposition and organic accumulation in various ecosystems?

Suggested Readings

Anderson, J. M., and A. MacFadyen, eds. 1976. *The role of terrestrial and aquatic organisms in the decomposition process.* Seventeenth Symposium, British Ecological Society. Oxford, England: Blackwell Scientific Publications.

Bliss, L. C., O. W. Heal, and J. J. Moore, eds. 1981. Part V of *Tundra ecosystems: A comparative analysis.* New York: Cambridge University Press.

Breymeyer, A. I., and G. M. Van Dyne, eds. 1980. Chapters 2 and 7 in *Grasslands, systems analysis and man*. Cambridge, England: Cambridge University Press.

Burris, R. H., and C. C. Black, eds. 1976. *CO_2 metabolism and plant productivity.* Baltimore: University Park Press.

Cooper, J. P. 1975. *Photosynthesis and productivity in different environments*. New York: Cambridge University Press.

Lange, O. L., L. Kappen, and E. D. Schulze, eds. 1976. Part V of *Water and plant life*. New York: Springer-Verlag.

LeCren, E. D., and R. H. Lowe-McConnell, eds. 1980. Chapter 7 in *The functioning of freshwater ecosystems*. New York: Cambridge University Press.

CHAPTER 19

Primary and Secondary Production

Outline

Objectives

Upon completion of this chapter, you should be able to:

1. Explain briefly the laws of thermodynamics.
2. Define entropy.
3. Point out how the laws of thermodynamics relate to ecology.
4. Define gross primary production, net primary production, and respiration.
5. Discuss how plants allocate net primary production to different parts and functions.
6. Discuss the significance of the ratio of above-ground and below-ground accumulation of biomass.
7. Tell why net primary production declines over time, especially in forest ecosystems.
8. Compare net primary production of various world ecosystems and explain why such variation occurs.
9. Describe secondary production.
10. Compare assimilation efficiencies of poikilotherms and homoiotherms and explain the differences.

Production in ecosystems involves the fixation and transfer of energy, the ultimate source of which is the sun. Fixation of solar energy is accomplished by plants through photosynthesis (Chapter 18). Fixed energy or photosynthate in the form of carbohydrates accumulates as plant biomass and becomes available to nonphotosynthetic organisms who convert it to heterotrophic biomass. This fixation and transfer of energy is governed by the nature and laws of energy.

Nature of Energy

The energy of the sun, already discussed, comes as thermal energy (Chapter 5) and as light energy (Chapter 7). Thermal energy heats Earth; light energy is utilized in photosynthesis. In both instances the transfer of radiation is considered as the movement of particles from one point to another. Energy can be said to flow; but it can flow only if there is an energy source and an energy sink. For Earth the energy source is the sun. The planet absorbs part of the energy and gives up energy to a sink, outer space.

The flow of energy is mediated at the molecular level. Thermal energy is characterized by its rapid distribution among all molecules in the system without any chemical reaction. The effect of thermal energy is to set the molecules into a state of random motion and vibration. The hotter the object, the more thermal energy is being absorbed and the more vibrational and rotational change is taking place. These motions tend to spread from a hot body to a cooler one, transferring energy from one to the other. The energy of light waves, on the other hand, especially the blue and red wavelengths, causes electronic excitations that can lead to photochemical reactions. The energy of light or photons sends one of a pair of electrons to a higher state or orbit. Uncoupled from its partner, it is free to be involved in photochemical reactions.

Energy exists in two forms, potential and kinetic. Potential energy is energy at rest—that is, capable of and available for performing work. Kinetic energy is energy of motion and results in work, performed at the expense of potential energy. Work, which results from the expenditure of potential energy, is of at least two kinds: the storage of energy and the arranging or ordering of matter with no storage.

Laws of Thermodynamics

The expenditure and storage of energy are described by two laws of thermodynamics. The first law of thermodynamics states that energy is neither created nor destroyed. It may change forms, pass from one place to another, or act upon matter in various ways, but regardless of what transfers and transformations take place, no gain or loss in total energy occurs. Energy is simply transferred from one form or place to another. When wood is burned, the potential energy present in the molecules of wood equals the kinetic energy released, and heat is evolved to the surroundings. This is an *exothermic reaction.*

On the other hand, energy from the surroundings may be paid into a reaction. Here, too, the first law holds true. In photosynthesis, for example, the molecules of the products store more energy than the reactants. The extra energy is acquired from the sunlight, but again, there is no gain or loss in total energy. When energy from outside surroundings is put into a system to raise it to a higher energy state, the reaction is *endothermic.*

Although the total amount of energy involved in any reaction, such as burning wood, does not increase or decrease, much of the potential energy involved is degraded into a form incapable of doing any further work. It ends up as heat, disorganized or randomly dispersed molecules in motion, useless for further transfer. The measure of this relative disorder is called *entropy.*

Transfer of energy involves the second law of thermodynamics. It states that when energy is transferred or transformed, part of the energy assumes a form that cannot be passed on any further. When coal is burned in a boiler to produce steam, some of the energy creates steam, and part is dispersed as heat to the surrounding air. The same thing happens to energy in the ecosystem. As energy is transferred from one organism to another in the form of food, a large part of that energy is degraded as heat—hence, no longer transferable—and the remainder is stored as living tissue.

But biological systems seemingly do not conform to the second law, for the tendency of life is to produce order out of disorder, to decrease rather than increase entropy. The second law theoretically applies to the isolated, closed system in which no energy or matter is exchanged between the system and its surroundings. An isolated system approaches thermodynamic equilibrium—that is, a point at which all the energy has assumed a form that cannot do work. A closed, isolated system tends toward a state of minimum free energy (energy available to do work) and maximum entropy. An open system maintains a state of higher free energy and lower entropy. In other words, the closed system tends to run down; the open one does not. As long as there is a constant input of free energy to the system and a constant outflow of energy in the form of heat and waste, the system maintains a steady state. Thus, life is an open system maintained in a steady state.

Primary Production

The flow of energy through the community starts with the fixation of sunlight by plants, a process that in itself demands the expenditure of energy. A plant gets its start by living on the food stored in the seed until its production machinery is working. Once started, the green plant begins to accumulate energy. Energy accumulated by plants is called *production* or, more specifically, *primary production,* because it is the first and basic form of energy storage. The rate at which energy accumulates is known as *primary productivity.* All of the sun's energy that is assimilated—that is, total photosynthesis—is *gross primary production.*

Because plants, like other organisms, must overcome the tendency of energy to disperse, free energy (that available to do work) must be expended for production as well as for reproduction and maintenance. The energy required for this is provided by the reverse of photosynthesis: metabolic respiration, which results in the production of CO_2 and H_2O and the liberation of energy. Most plants have to contend with photorespiration (see Chapter 18), associated with photosynthesis but not with metabolism. That, too, acts as a drain on photosynthesis. Energy remaining after respiration and stored as organic matter is *net primary production,* or plant growth. Net primary production can be described by the following equation:

Net primary production (NPP) = gross primary production (GPP) − autotrophic respiration (R)

Production is usually expressed as kilocalories per square meter per year (kcal/m^2/year). However, production may also be expressed as dry organic matter in grams per square meter per year (g/m^2/year). If either of these two measures is employed to estimate efficiencies and other ratios, the same unit must be used for both the numerator and the denominator of the ratio. Only calories can be compared with calories, dry weight with dry weight.

BIOMASS

Net primary production accumulates over time as plant biomass. Part of this accumulation is turned over seasonally through decomposition. Part is retained over a longer period as living material. The amount of this accumulated organic matter found on a given area at a given time is the *standing crop biomass.* Biomass is usually expressed as grams of organic matter per square meter (g/m^2), calories per square meter (cal/m^2), or some other appropriate measure per unit of area. Thus, biomass differs from production, which is the rate at which organic matter is created by photosynthesis. The biomass present at any given time is not the same as productivity.

Biomass Allocation

Plants budget their production of photosynthate, distributing it in a systematic way to leaves, twigs, stems, bark, roots, flowers, and seeds. What each part receives depends upon its demands for energy, growth, and maintenance. How much is allocated to each component is difficult to determine because of how time-consuming is the chore of cutting trees and clipping herbs, extracting roots, separating each of the components, weighing them, and determining both their energy and their nutrient contents.

The pattern of allocation varies with the plant. Single-celled planktonic algae are well supplied, indeed immersed, in nutrients and light. They accumulate large supplies, growing rapidly and dividing into new individuals. The photosynthate increases population size, which is a measure of their growth.

Annuals begin their life cycle in the spring with the germination of overwintering seeds. With only one growing season in which to complete its life cycle, an annual has to allocate its photosynthate first to leaves, which, in turn, become involved in photosynthesis, which increases vegetation biomass. At the time of flowering, the plant decreases the amount of energy and nutrients allocated to leaves. They are supplied only with enough to maintain themselves; in fact, lower leaves may die. Most of the photosynthate is diverted to reproduction. In the sunflower, for example, the biomass of leaves declines from approximately 60 percent of the plant during growth to 10 to 20 percent by the time the seeds are ripe. When in bloom, the sunflower distributes 90 percent of its photosynthate to the flower head and the rest to leaves, stem, and roots (Eckardt et al., 1971). If the annual grows on a dry or nutrient-poor site, it has to allocate more energy and nutrients to roots and less to leaves and flowers, lowering its competitive ability.

Perennials maintain a vegetative structure over a period of several to many years. They begin their life cycle like an annual, but once established they allocate their photosynthate in a very different manner. Before perennials expend any energy on reproduction, they divert excess photosynthate to the roots. In some species, like the skunk cabbage, the roots develop into massive storage organs. Energy and nutrients stored in the roots make up a reserve upon which the plants draw when they begin early spring growth. When they are ready to flower, perennials divert energy going into storage to the production of flowers and fruit. As the flowers fade and the fruit ripens, the plants once more send photosynthates to the roots to build up the reserves they will need the next spring.

Consider the grass blue grama *(Bouteloua gracilis)*, a C_4 species. Eighty-five percent of the net carbon fixed in photosynthesis is translocated. Of that, approximately 24 percent goes to shoots; 24 percent to the crown, from which new growth develops; and 52 percent to roots (Detling, 1979). Under environmental stress, perennials, including grasses, typically allocate more of their photosynthate to roots than to leaves and shoots.

Trees and woody shrubs live a long time, which greatly influences the manner in which they distribute photosynthate. Early in life leaves may make up more than one-half of their biomass (dry weight), but as trees age, they put down more woody growth. Trunks and stems become thicker and heavier, and the ratio of leaves to woody tissue changes. Eventually, leaves account for only 1 to 5 percent of the total mass of the tree. Thus, the production system that supplies the energy and nutrients is considerably less than the biomass it supports. Much of the photosynthate goes into permanent capital, inaccessible to the plant and ultimately available only to decomposers. Thus, much of the energy of woody plants goes into support and maintenance, which increases as the plants age.

When deciduous trees leaf out in spring, they expend one-third of their reserve energy on expansion and growth of leaves. This expenditure, of course, is repaid as the leaves carry out photosynthesis during the spring and summer. After leaves, trees give preference to flowers; then cambium, new buds, and deposits of starch in roots and bark; and finally, new flower buds. Reproduction and vegetative growth compete for energy allotments. If photosynthate is limited, vegetative growth gets first claim. Because energy demand by reproduction is high—up to 15 percent in pines, 20 percent in beech, and 35 percent or more in fruit trees—trees can afford an abundance of fruit only periodically, once every 2 or 3 years in deciduous trees and 2 to 6 years in conifers. As the growing season ends, photosynthates are withdrawn from leaves and the excess sent to roots, woody tissue, and buds.

Evergreen trees have a somewhat different situation. Because their photosynthetic machinery can function year round when temperature and moisture conditions permit, they do not need to draw on root reserves for new growth in spring. They can afford to wait until later in the growing season to send out new shoots. Then evergreens can draw on new photosynthesis built up earlier in the spring. For the same reason, new growth develops rapidly and matures in a few weeks.

The allocation of biomass has been determined for some forest stands (for a summary of some forest stands in the eastern United States, see Whittaker et al., 1974). For example, in a young oak-pine forest on Long Island, 25 percent of the net primary production was allocated to stem wood and bark, 40 percent to roots, 33 percent to twigs and leaves, and 2 percent to flowers and seeds. Among the shrubs, 54 percent of the net primary production was allocated to roots, 21 percent to stems, and 23 percent to leaves (Whittaker and Woodwell, 1969).

The proportionate allocation of net production to above-ground and below-ground biomass, to shoot and root, tells much about different ecosystems and about different components within ecosystems. A high shoot-root ratio indicates that most net production goes into a plant's supportive functions. Plants with a large root biomass are more effective competitors for water and nutrients and can survive more successfully in harsh environments because most of their active biomass is below ground. Plants with a low shoot-root ratio have most of their biomass above ground and assimilate more light energy, resulting in higher productivity. Tundra communities existing in an environment with long, cold winters and a short growing season have shoot-root ratios that reflect a harsh environment. Wet grass and sedge communities have a shoot-root ratio of 1:2.1; dwarf shrub communities, 1:3.1; low shrub communities, 1:2.0. Even forested tundra has a relatively high ratio for a forest: 1:0.8. Further south, midwest prairie grasses have a shoot-root ratio of 1:3, indicative, perhaps, of cold winters and a limited moisture supply. In forest ecosystems, with their high above-ground biomass, the shoot-root ratio is low. For the Hubbard Brook Forest in New Hampshire, the shoot-root ratio (based on the data of Gosz, Likens, and Bormann, 1976) for trees is 1:0.213; for shrubs, 1:0.5; and for herbs, 1:1. As one could predict, the shoot-root ratio increases through the vertical strata from the canopy to the floor.

Biomass Distribution

Above-ground biomass is distributed vertically in the ecosystem. Vertical distribution of leaf biomass in terrestrial communities and the concentration of plankton and floating and submerged vegetation influence the penetration of light, which, in turn, influences the distribution of production in the ecosystem. The region of maximum productivity in the aquatic ecosystem is not the upper sunlit surface (strong sunlight inhibits photosynthesis) but some depth below, depending upon the clarity of the water and the density of plankton or vegetative growth (Figure 19.1). As depth increases, light intensity decreases until it reaches a point at which the light received by the vegetation is just sufficient to meet respiratory needs and production equals respiration (Figure 19.2). This is known as the *compensation level.* In the forest ecosystem a similar situation exists. The greatest amount of photosynthetic biomass (Figure 19.3) as well as the highest net photosynthesis is not at the top but at some point below maximum light intensity. In spite of wide differences in plant species and in types of ecosystems, the vertical profiles of biomass of the various ecosystems appear to be similar.

Within the vertical profile, biomass varies seasonally and even daily. In grasslands and old field ecosystems, much of the net production is turned over every year. The standing crop of living material in an old field in Michigan amounted to about 4×10^3 kg/ha in late summer, compared to 80 kg/ha in late spring. But at this time the standing crop in dead matter was nearly 3×10^3 kg/ha (Golley, 1960). The above-ground biomass of a tall-grass prairie that included both living and dead material was about twice that of the standing crop, the living

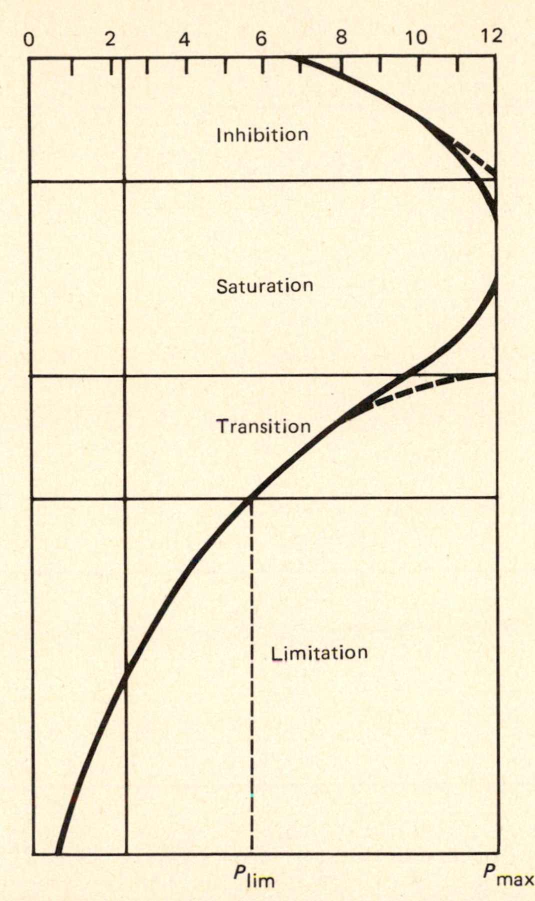

FIGURE 19.1 A generalized diagram of photoassimilation versus depth in a homogenous population of phytoplankton. Note that photosynthesis is inhibited at and near the surface because of high light intensity (irradiance). Maximum photosynthesis takes place somewhat below the surface where light is at saturation. Photosynthesis declines rapidly with depth because of reduction in irradiance and a change in the spectral composition of light. Under actual conditions, the photosynthetic profile would be irregular because of changes in biomass with depth under different environmental regimes. (*Adapted from D. F. Westlake, 1980:166.*)

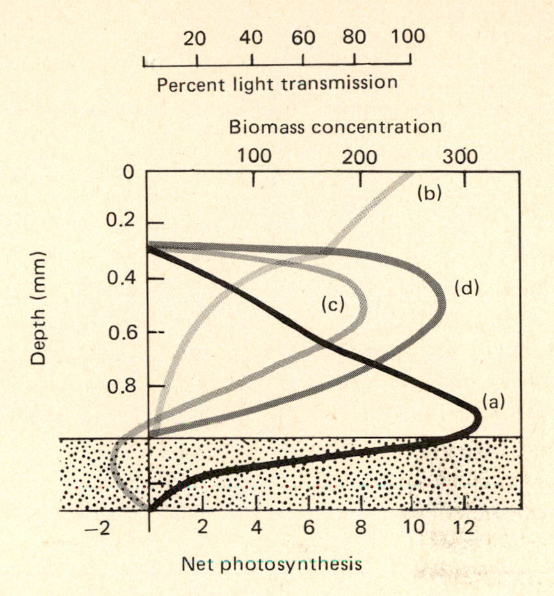

a-Biomass concentration (g dry wt/m^3)
b-Percent light transmission
c-Net photosynthesis (dg O_2m^3/hr)
d-Net photosynthetic capacity (mgO_2 (g dry wt)/hr)

FIGURE 19.2 Diagram of vertical profiles of biomass, light transmission, net photosynthetic capacity, and net photosynthesis of a stand of a submerged macrophyte, *Vallisneria denseserrulata*. The biomass profile has little influence on the profile of net photosynthesis, which is influenced largely by irradiance and photosynthetic capacity. Most of the photosynthetically active biomass is just below the water's surface. Photosynthetic capacity of the vegetation, which decreases within the deeper parts of the stand, is independent of plant biomass because of morphological differences in the plants, decreased irradiance, and shading. Maximum photosynthesis is concentrated near the surface of the stand, whereas maximum biomass is near the base of the growth. (*Adapted from D. F. Westlake, 1980:179.*)

material added during the growing season (Kucera, Dahlman, and Koelling, 1967). The above-ground biomass has a turnover rate of approximately 2 years and the below-ground biomass of roots, 4 years. In a forest ecosystem a considerably greater portion of net production is tied up in wood, as exemplified by an oak-pine forest on Long Island. Leaves, fruit, flowers, deadwood, and bark contributed 4,342 g/m^2/year, for a total of 653 g/m^2, or about 58 percent of net primary production (Woodwell and Marples, 1968).

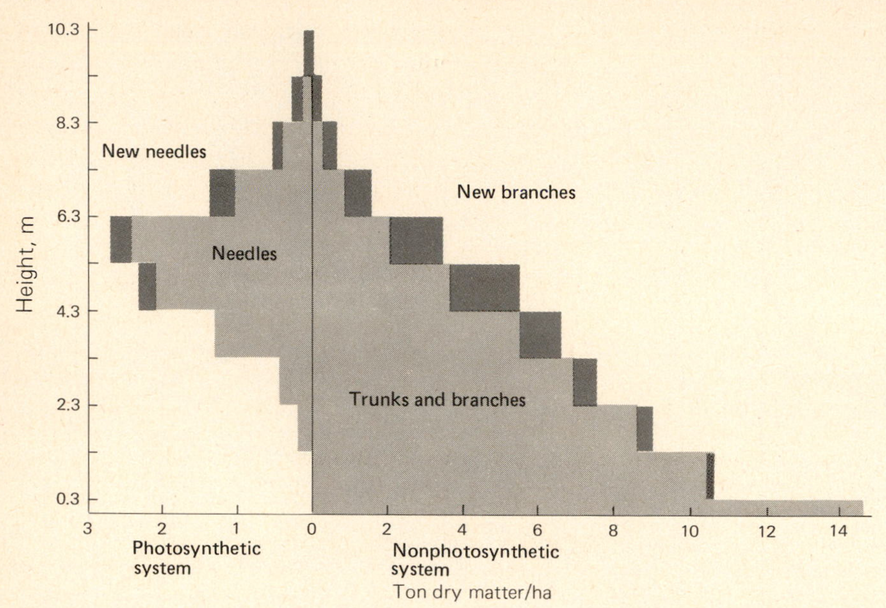

FIGURE 19.3 Structure and productive systems of a pine spruce-fir forest in Japan. The dark gray areas represent new needles and new branches. Note that most of the active photosynthetic area lies below the uppermost parts of the canopy, a pattern similar to that of the phytoplankton community. (*After Monsi, 1968.*)

NET PRODUCTION

Little of the energy assimilated by plants goes into organic production. Most of the light absorbed by plants is converted to heat and lost through convection and radiation. What fraction of light energy is used in photosynthesis goes into gross production, and what is left over after respiration goes into net production. Production efficiency (see Box 19.1), the ratio of net primary production to gross primary production, is on the average rather high. Algae and corn have an efficiency of 80 to 85 percent; eastern deciduous forests, 42 percent; prairie grasses, 66 percent; tundra vegetation, 50 percent; and submerged aquatic plants, 65 to 75 percent. The most efficient plants are those such as grasses that do not maintain a high supporting biomass as woody tissue (see Table 19.1).

Productivity of ecosystems is influenced by temperature and precipitation (Figure 19.4). Such relationships show up in variations in productivity of different global ecosystems (Figure 19.5). The most productive terrestrial ecosystems are tropical rain forests, with high rainfall and warm temperatures; their net productivity ranges between 1,000 and 3,500 g/m^2/year (Figure 19.6). Temperate forests, where rainfall and temperatures are lower, range between 600 and 2,500 g/m^2/year (Whittaker and Likens, 1973). Shrublands, such as heath balds and tall-grass prairie, have net productions of 700 to 1,500 g/m^2/year (Whittaker, 1963; Kucera, Dahlman, and Koelling, 1967). Desert grasslands produce about 200 to 300 g/m^2/year, whereas deserts and tundra range between 100 and 200 g/m^2/year (Rodin and Bazilevic, 1967). Net production of the open sea is generally low. The productivity of the North Sea is about 170 g/m^2/year, and that of the Sargasso Sea 180 g/m^2/year. However, in some areas of upwelling, such as the Peru Current, net production can reach 1,000 g/m^2/year. These differences in net primary productivity from tropic to arctic regions are reflected in litter production. In tropical forests that ranges from 900 to 1,500 g/m^2/year; in temperate forests, 200 to 600 g/m^2/year;

Box 19.1

Production Efficiencies

Production, both primary and secondary, can be expressed in a number of ways. The following are the more important ones, including those used in the text.

Terms

GPP	Gross primary production
NPP	Net primary production
R	Respiration
P	Secondary production: tissue growth, reproduction, exoskeleton growth, biomass change
C	Consumption: plant material ingested
F	Egestion: feces, urine, gas, other products
A	Assimilation: food or energy absorbed

Equations

1. Photosynthetic efficiency: GPP/solar radiation
2. Assimilation efficiency, plants: GPP/light absorbed
3. Production efficiency: R/GPP
4. Maintenance efficiency: R/NPP
5. Effective production: NPP/GPP
6. Assimilation efficiency, animals: A/C
7. Ecological growth efficiency: P/C
8. Growth efficiency: P/A

and in arctic and alpine regions, 0 to 200 g/m^2/year (Bray and Gorham, 1964).

Some ecosystems have consistently high production. Such high productivity usually results from an additional energy subsidy to the system. This subsidy may be a warmer temperature, greater rainfall, or circulating or moving water that carries food or additional nutrients into the community. In agriculture such subsidies include the use of fossil fuel for cultivation, application of fertilizer, and the control of pests. Swamps and marshes, ecosystems at the interface of land and water, may have a net productivity of 3,300 g/m^2/year. Estuaries, because of inputs of nutrients from rivers and tides, and coral reefs, because of inputs from changing tides, may have a net productivity between 1,000 and 2,500 g/m^2/year. Among agricultural ecosystems, sugarcane has a net productivity of 1,700 to 1,800 g/m^2/year; hybrid corn, 1,000 g/m^2/year; and some tropical crops, 3,000 g/m^2/year.

NET PRODUCTION AND TIME

Annual net production changes with time and age. For example, a Scots pine *(Pinus sylvestris)* plantation in England achieved a maximum annual net production of 22 × 10^3 kg/ha at the age of 20 years. It then declined to 12 × 10^3 kg/ha at 30 years of age (Ovington, 1961). Woodlands apparently achieve their maximum annual net production in the pole stage, when the dominance of trees is greatest and the understory is minimal (Figure 19.7). The understory makes its greatest contribution to annual net production during the early and mature stages of the forest. As the age of a forest stand increases, more and more of the production is needed for maintenance, and very little gross production is left for growth (Figure 19.8). The pattern is well illustrated by Douglas fir in the Pacific Northwest. Seventy percent of the net production of a 20- to 40-year-old stand accumulates as stored biomass. In a 450-year-old stand, only 6 to 7 percent of gross photosynthesis is available for net production, and most of that is converted to detritus (Grier and Logan, 1977).

Production varies not only among different types of ecosystems but also among similar systems and within one system from year to year. Production is influenced by such factors as nutrient availability, precipitation, temperature, length of growing season, animal utilization, and fire. For example, herbage yield of a grassland may vary by a factor of 8 between wet and dry years (Weaver

TABLE 19.1 COMPARATIVE ESTIMATES OF PRODUCTIVITY AND RESPIRATION FOR SEVERAL NORTH AMERICAN ECOSYSTEM TYPES

	Old-growth Douglas-fir Forest	Eastern Deciduous Forest	Eastern Oak-pine Forest	Prairie	Tundra	Potato Field	Rye Field
Gross primary production (GPP)	7.72	1.62	1.32	0.64	0.24	1.29	1.00
Autotrophic respiration (R_A)	7.20	0.94	0.68	0.22	0.12	0.43	0.34
Net primary production (NPP)	0.52	0.68	0.60	0.42	0.12	0.85	0.66
Heterotrophic respiration (R_H)	0.36	0.52	0.37	0.27	0.11	0.50	0.31
Net ecosystem production ($GPP - R_E$)	0.16	0.16	0.27	0.15	0.01	0.36	0.35
Ecosystem respiration ($R_E = R_H + R_A$)	7.56	1.47	1.05	0.49	0.23	0.93	0.65
Production efficiency (R_A/GPP)	0.93	0.57	0.52	0.34	0.50	0.34	0.34
Effective production (NPP/GPP)	0.07	0.42	0.45	0.66	0.50	0.66	0.66
Maintenance efficiency (R_A/NPP)	13.80	1.38	1.13	0.51	1.00	0.52	0.52
Respiration allocation (R_H/R_A)	0.05	0.55	0.54	1.26	0.90	0.86	0.91
Ecosystem productivity (NEP/GPP)	0.02	0.10	0.20	0.23	0.05	0.28	0.35

NOTE: Values in kilograms carbon per square meter per year.

SOURCE: Old-growth Douglas-fir forest: Grier and Logan, 1977; eastern deciduous and oak-pine forests, prairie, and tundra: Reichle, 1975:259.

and Albertson, 1956). Overgrazing of grasslands by cattle and sheep or defoliation of forests by such insects as the gypsy moth can seriously reduce net production. Fire in grasslands may result in increased productivity if moisture is normal, but reduced productivity if precipitation is low (Kucera, Dahlman, and Koelling, 1967). An insufficient supply of nutrients, especially nitrogen and phosphorus, can limit net productivity, as can mechanical injury of plants, atmospheric pollution, and the like.

MEASURING PRIMARY PRODUCTION

Estimating the rates of energy fixation and storage is difficult. Several major techniques for determining primary production are in current use. All estimate energy fixation indirectly by the amount of oxygen produced or CO_2 uptake and release and by biomass accumulation.

Infrared Gas Analysis

The flux of CO_2 to and from a plant or part of a plant can be measured with an infrared gas analyzer. The plant or plant part is sealed in a controlled environment chamber. Photosynthesis and dark respiration can be monitored continuously, and changes in photosynthesis under different conditions can be determined by regulating the environment. The drawback to the method is the small number of plants that can be measured and the lack of portability in the field.

The infrared gas analysis technique can be used in the field, however, to measure the uptake of carbon dioxide and its release in respiration. A sample of the community—a twig and its leaves, a segment of a tree stem, ground cover and soil surface, or even a portion of the total community is enclosed in a plastic tent. Air is drawn through the enclosure, and the carbon dioxide concentration of incoming

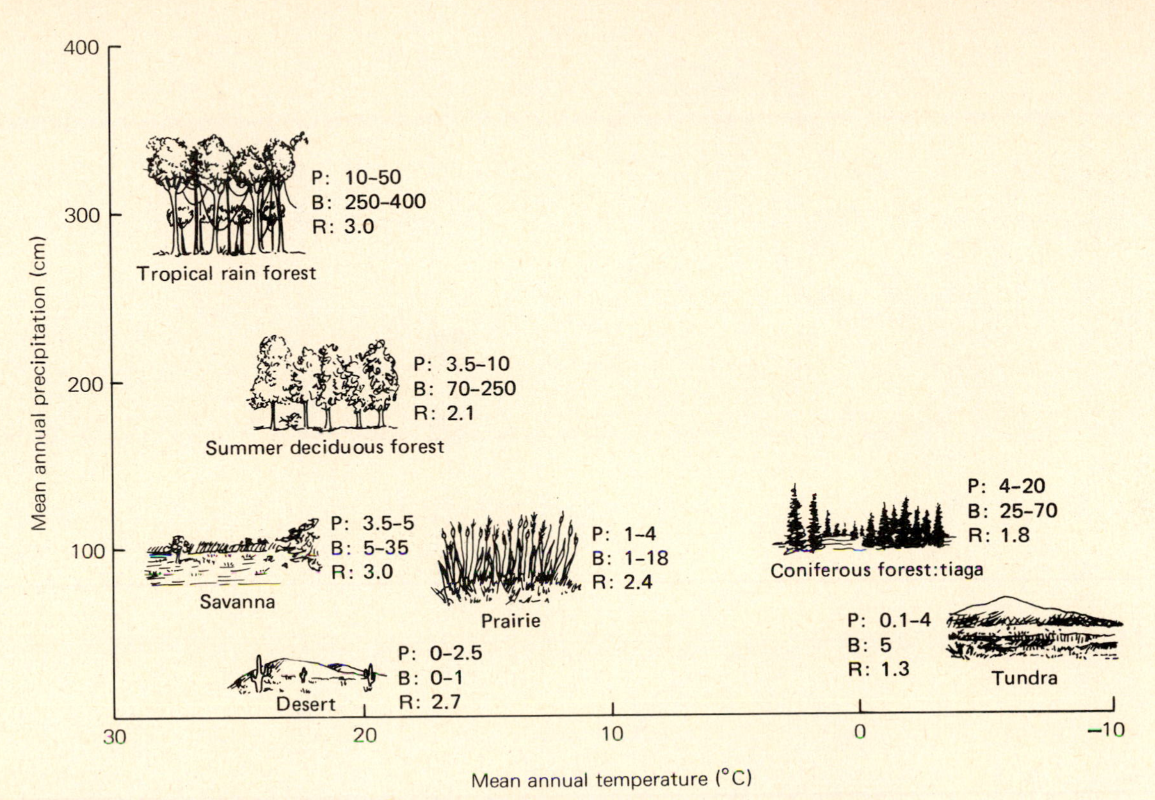

FIGURE 19.4 A distribution of primary production, biomass, and radiation input relative to climate and geographical distribution. P = primary production (ton/ha); B = biomass (ton/ha); R = solar radiant input (kcal/m^2/yr, 0.3–3.0 microns). (*After Etherington, 1982:355.*)

and outgoing air is measured with the infrared gas analyzer. The assumption is that any carbon dioxide removed from incoming air has been incorporated into organic matter. A similar sample may be enclosed in a dark bag. The amount of carbon dioxide produced in the dark bag is a measure of respiration. In the light bag the quantity of carbon dioxide would be equivalent to photosynthesis minus respiration. The two results added together indicate gross production.

Labeled Radioactive $^{14}CO_2$

Measurement of the rate of uptake of radioactive carbon (^{14}C) by plants is one of the most sensitive techniques available to measure net photosynthesis under field conditions. The method involves exposing photosynthetic tissue to an atmosphere containing radioactively labeled $^{14}CO_2$ and measuring the amount of $^{14}CO_2$ fixed per unit of time, usually of short duration. A Plexiglas chamber is sealed on the surface of a leaf for 10 to 20 seconds. Air with $^{14}CO_2$ is passed over the leaf. After exposure to the label, a known portion of the leaf is removed and killed immediately. Leaf disks are returned to the laboratory for analysis of the amount of ^{14}C fixed.

In aquatic communities the method involves the addition of radioactive carbonate ($^{14}CO_3$) to a sample of water containing its natural phytoplankton population. After a short period of time, the plankton material is strained from the water, washed, and dried. Radioactivity counts are taken, and from them calculations estimating the amount of CO_2 fixed in photosynthesis are made. The estimate is based on the assumption that the ratio of

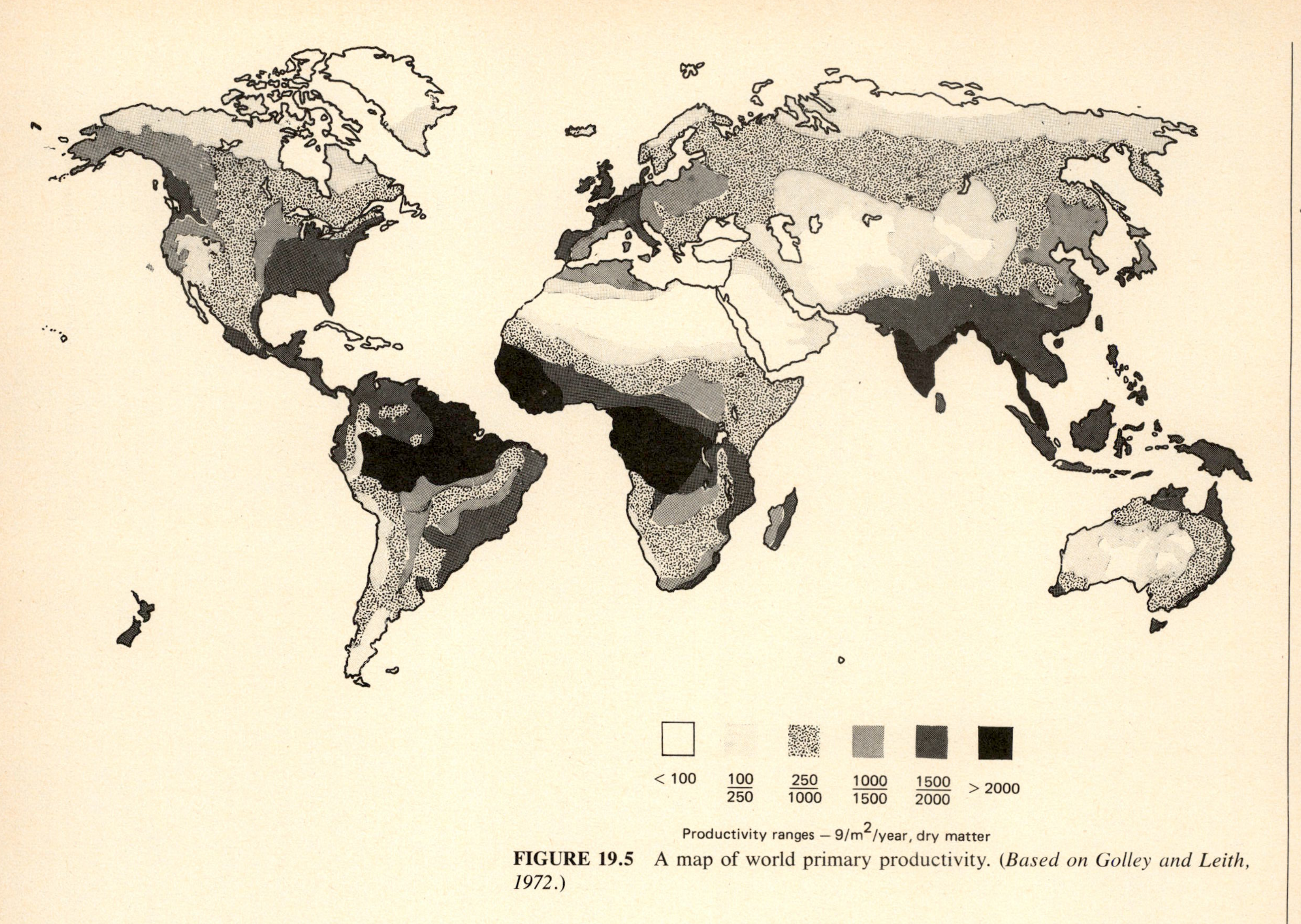

FIGURE 19.5 A map of world primary productivity. (*Based on Golley and Leith, 1972.*)

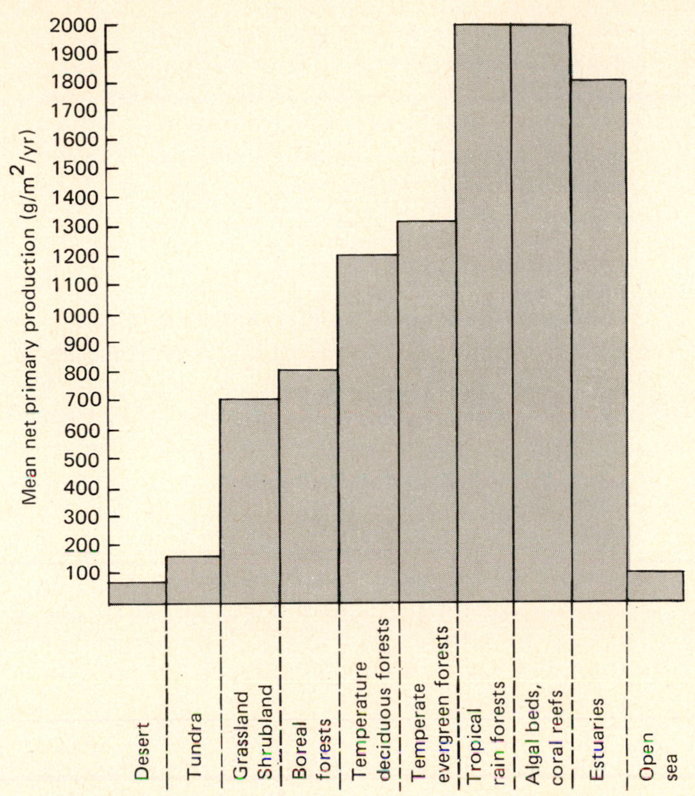

FIGURE 19.6 Comparative productivity of world ecosystems.

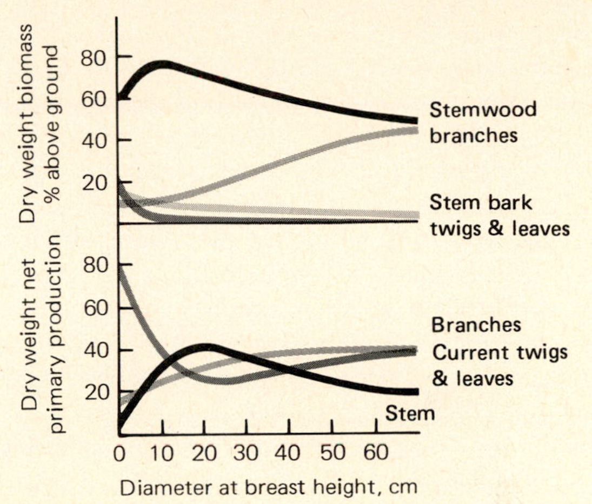

FIGURE 19.7 Relation of above-ground biomass and production to tree size in the Hubbard Brook Forest. Trends in above-ground biomass and production percentages in relation to diameter at breast height (dbh) involve 63 sample trees of three major species: sugar maple, yellow birch, and beech. Note that as the trees increase in size, the ratio of branches to stems increases. In larger trees, branches account for a greater proportion of net primary production than stems. In smaller trees, current leaves and twigs account for the greater percentage of primary production. The percentage declines rather rapidly as trees approach pole stage, then increases as trees mature. (*From Whittaker et al., 1974:239.*)

activity of $^{14}CO_3$ added to the activity of phytoplankton is proportionate to the ratio of total carbon available to that assimilated.

The method has certain deficiencies in both terrestrial and aquatic situations. Some terrestrial plants discriminate against $^{14}CO_2$. One cannot be sure that $^{14}CO_2$ is being absorbed at the same rate as normal CO_2. In addition, $^{14}CO_2$ may be diluted by CO_2 released by respiration. In aquatic communities the method does not measure adequately changes in the oxidative state of the carbon fixed. All of the carbon fixed is not retained by the producers. Some tends to seep out of algal cells as water-soluble organic compounds used by bacteria. The various primary producers have different abilities to utilize available light. And the amount of carbon fixed is influenced by the species composition of the plant community.

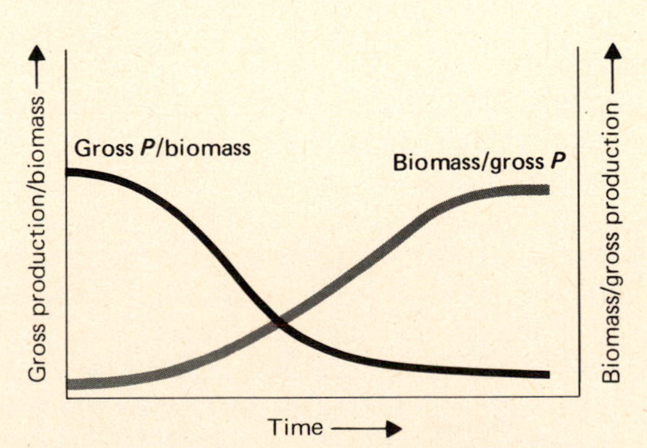

FIGURE 19.8 Model showing the change through time in ratios between gross community primary production and biomass or production efficiency and between biomass and gross community primary production or maintenance efficiency. Note the early high production efficiency and the later accumulation of biomass. (*After Cooke, 1967:719.*)

Light-Dark Bottle

The light-dark bottle method is commonly used in aquatic communities. It is based on the assumption that the amount of oxygen produced is proportional to gross production. Two bottles containing a given concentration of phytoplankton are suspended at the level from which the samples were obtained. One bottle is black to exclude light; the other is clear. In the light bottle a quantity of oxygen proportional to the total organic matter fixed (gross production) is produced by photosynthesis. At the same time, phytoplankton is using some of the oxygen for respiration. Thus, the amount of oxygen left is proportional to the amount of fixed organic matter remaining after respiration, or net production. In the dark bottle, oxygen is being used but is not being produced. The amount of oxygen utilized, obtained by subtracting the amount of oxygen left at the end of the run (usually 24 hours) from the quantity at the start, gives a measure of respiration. The amount of oxygen in the light bottle added to the amount used in the dark bottle provides an estimate of total photosynthesis, or gross production.

This method, too, has shortcomings. It fails to take into account bacterial respiration and growth of phytoplankton in the light bottle. And the procedure is based on the assumption that respiration in the dark is the same as that in the light.

Chlorophyll

An estimate of production of some ecosystems can be obtained from chlorophyll and light data. The technique is based on the close relationship between chlorophyll and photosynthesis at any given light intensity. The relationship remains constant for different species of plants and thus communities, even though the chlorophyll content of organisms varies widely as a result of nutritional status and the duration and intensity of light to which the plant is exposed.

The method, best adapted to aquatic communities, involves the determination of the chlorophyll *a* content of the plant or leaf per gram or per square meter, which under reasonably favorable conditions, remains the same. Because the quantity of chlorophyll *a* tends to increase or decrease with the amount of photosynthesis, the chlorophyll per square meter indicates the food-manufacturing potential at the time. The photosynthesis-chlorophyll ratio remains rather constant even in cells whose chlorophyll content varies widely because of nutrition or the duration and intensity of light to which the plants were previously exposed.

Harvest

The harvest method is widely employed in terrestrial communities. It is most useful for estimating the production of cultivated land and range, where production starts from zero at seeding or planting time, becomes maximum at harvest, and is subject to minimal utilization by consumers.

Briefly, the technique involves the clipping or removal of vegetation at periodic intervals, drying to a constant weight, and expressing that weight as biomass in grams per square meter per year. The biomass can be converted to calories, and the harvested material expressed as kilocalories per year or per growing season.

To estimate community production from a sample taken only at the end of the growing season seriously underestimates production. In order to arrive at accurate estimates, plant material must be sampled throughout the growing season and the contribution of each individual species determined. Different species of plants reach their peak production at different times during the growing season.

Some studies determine only above-ground production and leave the roots unsampled. Other studies attempt to estimate below-ground production. Sampling the root biomass is difficult at best. Although the roots of some annuals and crop plants may be removed from the soil, the problem becomes more difficult with grass, herbaceous plants, and forest trees. Investigators face the almost impossible task of separating new roots from older ones. They have the added problem of estimating the turnover of short-lived small roots and root materials and the variability of sampling.

Dimension Analysis

Because plants of different ages, sizes, and species make up forest and shrub ecosystems, a modified harvest technique known as dimension analysis is used. Dimension analysis involves the measurement of height, diameter at breast height (DBH), and diameter growth rate of trees in a sample plot. A set of sample trees is cut, weighed, and measured, usually at the end of the growing season. Height to the top of the tree, DBH, depth and diameter of the crown, and other parameters are taken. Total weight, both fresh and dry, is determined, as is the weight of trunk and limbs. Roots are excavated and weighed. By various calculations, the net annual production of wood, bark, leaves, twigs, roots, flowers, and fruit is obtained. From that information the biomass and production of trees in the sample are estimated and then summed for the whole forest. The biomass of ground vegetation and litter is also estimated. Because energy utilized by plants and plant material consumed by animals and microorganisms are not accounted for, the harvest method estimates net community rather than net primary production. Because of forestry interest in biomass production, a growing body of data is available for selected trees.

Secondary Production

Net production is the energy available to the heterotrophic components of the ecosystem. Theoretically, at least, all of it is available to the grazers or to the decomposers; but rarely is it all utilized in this manner. The net production of any given ecosystem may be dispersed to another food chain outside of the ecosystem through removal by humans or other agents such as wind or water currents. For example, about 45 percent of the net production of the salt marsh is lost to estuarine water (Teal, 1962). Much of the living material is physically unavailable to the grazers—they cannot reach the plants. Living organic matter, as long as it is alive, is unavailable to decomposers and detritus-feeders, and dead materials may not be relished by grazers. The amount of net production available to herbivores may vary from year to year and from place to place. The quantity consumed will vary with the type of herbivore and the density of the population.

Once consumed, a considerable portion of the plant material, again depending on the kind of plant involved and the digestive efficiency of the herbivore, may pass through the animal's body undigested. A grasshopper assimilates only about 30 percent of the grass it consumes, leaving 70 percent as wastes (Smalley, 1960). Mice, on the other hand, assimilate about 85 to 90 percent of what they consume (Golley, 1960; R. Smith, 1962).

Energy, once consumed, either is diverted to maintenance, growth, and reproduction or is passed from the body as feces and urine (see Figure 19.9). Part of the energy will be lost through urine, and depending on the nature of the organism, the loss of energy can be variable and often quite high (Cook, Stoddard, and Harris, 1954). Another portion is lost as fermentation gases. Of the energy left after losses through feces, urine, and gases, part is utilized as "heat increment," which is the heat

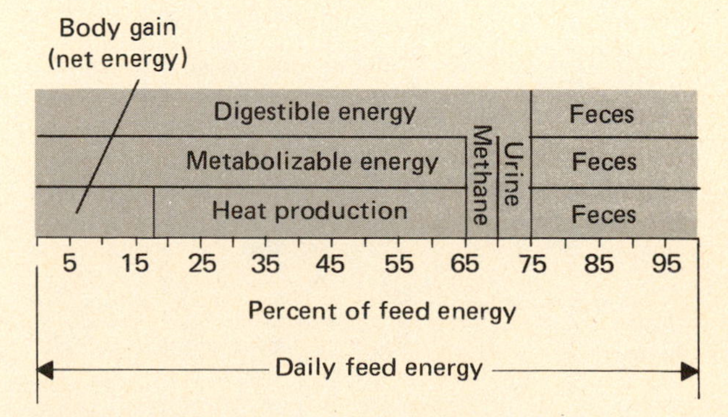

FIGURE 19.9 Relative values of the end products of energy metabolism in the white-tailed deer. Note the small amount of net energy gained (body weight) in relation to that lost as heat, gas, urine, and feces. The deer is a herbivore, a first-level consumer. (*After Cowan, 1962:5.*)

needed for metabolism above that required for basal or resting metabolism. The remainder of the energy is "net" energy, available for maintenance and production. This includes energy involved in capturing or harvesting food, muscular work expended in the animal's daily routine, and energy needed to keep up with the wear and tear on the animal's body. The energy used for maintenance is lost as heat.

Maintenance costs—highest in the small, active, warm-blooded animals—are fixed or irreducible. In small invertebrates energy costs vary with the temperature: a positive energy balance exists only within a fairly narrow range of temperatures. Below 5°C spiders become sluggish and cease feeding, and have to utilize stored energy to meet their metabolic needs. At approximately 5°C assimilated energy approaches energy lost through respiration. From 5° to 20.5°C spiders assimilate more energy than they respire. Above 25°C the ability of the spider to maintain a positive energy balance declines rapidly (Van Hooke, 1971).

The energy left over from maintenance and respiration goes into production, including new tissue, fat tissue, growth, and new individuals. This net energy of production is *secondary production*. Within secondary production there is no portion known as gross production. That which is analogous to gross production is actually assimilation. Secondary production is greatest when the birthrate of the population and the growth rates of the individuals are highest. This usually coincides, for obvious reasons, with the time when net production is also highest.

This scheme is summarized in Figure 19.10. It is applicable to any consumer organism, herbivore or carnivore. The herbivore represents the energy source of the carnivore; and as in the case of the plant food of the herbivore, not all of the energy contained in the body of the herbivore is utilized by the carnivore. Part (such as hide, bones, and internal organs) is unconsumed, and the same metabolic losses can be accounted for. At each transfer considerably less energy is available for the next consumer level.

Just as net primary production is limited by a number of variables, so is secondary production.

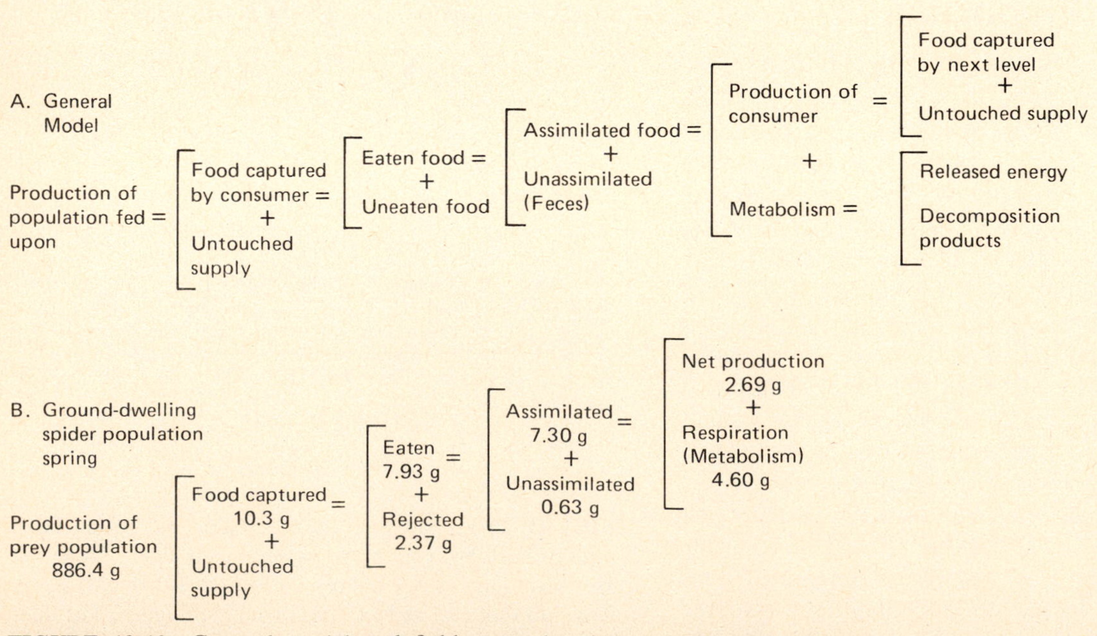

FIGURE 19.10 General model and field example of the components of energy metabolism in secondary production. (*Data from Moulder and Reichle, 1970.*)

The quantity, quality (including the nutrient status and digestibility), and availability of net production are three limitations. So is the degree to which primary and available secondary production are utilized.

The latter can be examined using two different ratios. One is the ratio of assimilation (the energy extracted from the food) to ingestion (the food actually eaten), or A/C. This measures the efficiency with which the consumer extracts energy from the food it consumes. The other is the ratio of productivity to assimilation, or P/A. This measures the efficiency with which the consumer incorporates assimilated energy into new tissue, or secondary production.

The ability of the consumer population to convert the energy it ingests varies with the species and the type of consumer. Insects that feed on plant tissues, such as grasshoppers, are more efficient producers than insects that feed on plant juices (Andrzejewska and Gyllenberg, 1980). Larval stages of insects are more efficient producers than the adult stage. Homoiotherms have a high assimilation, but because they utilize about 98 percent of that energy in metabolism, they have poor production efficiency (Table 19.2). Poikilotherms utilize about 79 percent of their total assimilation in metabolism. They convert a greater portion of their assimilated energy into biomass than do homoiotherms. The difference, however, is balanced by assimilation efficiency. Poikilotherms have an efficiency of around 30 percent in digesting food, while homoiotherms have an efficiency of around 70 percent. The poikilotherm has to consume more calories to obtain sufficient energy to meet the needs of maintenance, growth, and reproduction.

TABLE 19.2 SECONDARY PRODUCTION OF SELECTED CONSUMERS, KCAL/M^2/YR

Species	Ingestion (I)	Assimilation (A)	Respiration (R)	Production (P)	A/I	P/I	R/A	Reference
Harvester ant (h)	34.50	31.00	30.90	0.10	0.1	0.0002	0.99	Odum, Connell, and Davenport (1962)
Plant hopper (h)	41.30	27.50	20.50	7.00	0.67	0.169	0.75	Wiegert (1964)
Salt marsh grasshopper (h)	3.71	1.37	0.86	0.51	0.37	0.137	0.63	Teal (1962); Smalley (1960)
Spider, small < 1 mg (c)	12.60	11.90	10.00	1.90	0.94	0.151	0.84	Moulder and Reichle (1974)
Spider, large > 10 mg (c)	7.40	7.00	7.30	-3.00	0.95	—	1.04	Moulder and Reichle (1972)
Savannah sparrow (o)	4.00	3.60	3.60	0	0.90	0	1.0	Odum, Connell, and Davenport (1962)
Old field mouse (h)	7.40	6.70	6.60	0.1	0.91	0.014	0.98	Odum, Connell, and Davenport (1962)
Ground squirrel (h)	5.60	3.80	3.69	0.11	0.68	0.019	0.97	Wiegert and Evans (1967)
Meadow mouse (h)	21.29	17.50	17.00	—	0.82	—	0.97	Golley (1960)
African elephant (h)	71.60	32.00	32.00	8	0.44		1.0	Petrides and Swank (1966)
Weasel (c)	5.80	5.50	—	—	0.95	—	—	Golley (1960)

NOTE: h = herbivore; o = omnivore; c = carnivore.

Summary

A major function of ecosystems is energy flow, which supports life on Earth. Energy is governed by the laws of thermodynamics. The second law—which states that as energy is transferred or transformed from one state to another, a portion is no longer usable—is particularly applicable to energy flow in ecosystems. As energy moves through an ecosystem from sun to plants to consumers, much of it is lost as heat of respiration. Energy is degraded from a more organized to a less organized state, or entropy. However, a continuous flux of energy from the sun into ecosystems prevents them from running down.

To carry on photosynthesis and fix energy, plants must utilize part of the energy they fix. The total amount of energy fixed by plants is gross primary production. The amount of energy left after plants have met their respiratory needs is net primary production, which shows up as plant biomass. The amount of accumulated biomass on any given area at a particular time is standing crop biomass. Energy fixed by plants is allocated to different parts of the plant and to reproduction. Plants allocate energy first to leaves, then to flowers. Excess production goes to roots and other supporting tissue, where some of the reserves are available for growth the following year.

Efficiency of production varies among plants and among ecosystems. Most efficient are those plants such as grasses and annuals that do not maintain a high supporting biomass. The least productive are old trees, which expend most of their gross production in maintenance. Productivity of ecosystems is influenced by temperature and precipitation. The most productive are tropical rain forests and estuaries; the least productive are tundras and warm deserts.

Net production is available to consumers either directly as plant tissue or indirectly through animal tissue. Once consumed and assimilated, energy is diverted to maintenance, growth, and reproduction and to feces, urine, and gas. Change in biomass, measured as weight change and reproduction, represents secondary production. Efficiency varies. Homoiotherms have a high assimilation efficiency but low production efficiency because they have to expend so much energy in maintenance. Poikilotherms have low assimilation efficiency but high production efficiency. Much of the energy goes into growth rather than maintenance.

Review and Study Questions

1. What conditions are necessary for energy flow?
2. What are the first and second laws of thermodynamics? How do they relate to ecology?
3. What is entropy? How is it related to energy flow?
4. What is primary production? Primary productivity? Gross primary production? Net primary production? Respiration?
5. What is meant by compensation level?
6. What is the significance of shoot-root ratio to production in terrestrial ecosystems?
7. What is energy or biomass allocation in plants? How is energy allocated?
8. How does net production relate to the age of an ecosystem?
9. What influences productivity?
10. What ecosystems have a high net productivity? Low? Why?
11. If you were trying to determine plant biomass in an old field, how would you sample the vegetation to get an accurate estimate?
12. What method might one employ to determine primary production of an old field? A forest? A lake?
13. How can the measurement of oxygen production and utilization of CO_2 provide some estimate of primary production?

14. What is secondary production? How does it differ from primary production?
15. What is the difference in energy allocation and efficiency between homoiotherms and poikilotherms?

Suggested Readings

Bliss, L. C., O. W. Heal, and J. J. Moore, eds. 1981. Chapters 6, 7, 10, 11, and 12 in *Tundra ecosystems: A comparative analysis*. New York: Cambridge University Press.

Breymeyer, A. I., and G. M. Van Dyne, eds. 1980. Chapters 2, 3, 4, 5, and 6 in *Grasslands, systems analysis, and man*. New York: Cambridge University Press.

Cooper, J. P. 1975. *Photosynthesis and productivity in different environments*. New York: Cambridge University Press.

Edmonds, R. L., ed. 1982. Chapters 5 and 6 in *Analysis of coniferous forest ecosystems in the western United States*. US/IBP Synthesis, Series no. 14. Stroudsburg, Pa.: Dowden, Hutchinson & Ross Publishing Company.

Edmondson, W. T., and G. G. Winberg. 1971. *A manual on methods for the assessment of secondary production in fresh waters*. IBP Handbook no. 17. Oxford, England: Blackwell Scientific Publications.

LeCren, E. D., and R. H. Lowe-McConnell, eds. 1980. Chapters 5, 6, and 9 in *The functioning of freshwater ecosystems*. New York: Cambridge University Press.

Leith, H., and R. H. Whittaker, eds. 1975. *Primary productivity in the biosphere*. New York: Springer-Verlag.

Milner, C., and R. E. Hughes. 1968. *Methods for the measurement of the primary production of grassland*. IBP Handbook no. 6. Oxford, England: Blackwell Scientific Publications.

Newbould, P. J. 1967. *Methods in estimating primary production of forests*. IBP Handbook no. 2. Oxford, England: Blackwell Scientific Publications.

Petrusewicz, K., ed. 1967. *Secondary productivity of terrestrial ecosystems*. 2 vols. Warsaw, Poland: Pantsworve Wydawnictwo Naukowe.

Petrusewicz, K., and A. Macfadyen. 1970. *Productivity of terrestrial animals*. IBP Handbook no. 13. Oxford, England: Blackwell Scientific Publications.

Reichle, D. E., ed. 1981. Chapters 7 and 8 in *Dynamic properties of forest ecosystems*. New York: Cambridge University Press.

Vollenweider. R. A. 1969. *A manual on methods for measuring primary production in aquatic environments*. IBP Handbook no. 12. Oxford, England: Blackwell Scientific Publications.

CHAPTER 20

Food Chains, Trophic Levels, and Energy Flow

Outline

Objectives

Upon completion of this chapter, you should be able to:

1. Describe a food chain and a food web.
2. Distinguish among herbivores, carnivores, omnivores, and decomposers.
3. Explain the advantages of considering feeding groups as biophages and saprophages.
4. Distinguish between detrital and grazing food chains.
5. Define a trophic level.
6. Relate food webs to trophic levels.
7. Discuss energy flow through the food chain, with emphasis on efficiency of energy transfer.
8. Compare pyramids of biomass, energy, and numbers.
9. Define net ecosystem production.
10. Discuss energy budgets of a community.

Food Chains

The energy stored by plants is passed along through the ecosystem in a series of steps of eating and being eaten, the *food chain* (see Figure 20.1). Food chains are descriptive. When worked out diagrammatically, they consist of a series of arrows, each pointing from one species to another for which it is a source of food. In Figure 20.1, for example, grass plants are consumed by grasshoppers, grasshoppers become food for meadowlarks, and meadowlarks are preyed upon by marsh hawks or harriers. Thus, we have a relationship that can be written as follows:

Grass → grasshopper → meadowlark → harrier

But as the diagram indicates, no one organism lives wholly on another. Resources are shared, especially at the beginning of the chain. Plants are eaten by a variety of mammals and insects, and some of the animals are consumed by several predators. Thus, food chains become interlinked to form a *food web,* the complexity of which varies within and between ecosystems.

At each step in the food chain, a considerable portion of the potential energy is lost as heat of respiration. As a result, organisms in each trophic level pass on less energy than they receive. That limits the number of steps in any food chain to four or five. The longer the food chain, the less energy is available for the final members.

COMPONENTS OF FOOD CHAINS

Herbivores

Feeding on plant tissues is a host of plant consumers, the *herbivores*. They are capable of converting energy stored in plant tissue into animal tissue. Their role is essential in the community, for without them the higher trophic levels could not exist. The English ecologist Charles Elton, in his classic book *Animal Ecology* (1927), suggested that the term *key industry* be used to denote animals that feed on plants and are so abundant that many other animals depend on them for food.

Only the herbivores are adapted to live on a diet high in cellulose. Modification in the structure of the teeth, complicated stomachs, long intestines, a well-developed caecum, and symbiotic flora and fauna enable these animals to use plant tissues. For example, ruminants, such as deer, have a four-compartment stomach. As they graze, these animals chew their food hurriedly. The material consumed descends to the first and second stomachs (the rumen and reticulum), where it is softened to a pulp by the addition of water, kneaded by muscular action, and fermented by bacteria. They convert part of the celluloses, starches, and sugars to short-chain volatile fatty acids. These are rapidly absorbed into the bloodstream and oxidized to provide the mammal's chief form of energy. At leisure, ruminants regurgitate the undigested portion, chew it more thoroughly, and swallow it again.

The lagomorphs—rabbits, hares, and pikas—have a simple stomach and a large caecum. In the formation of fecal pellets, part of the ingested material is attacked by microorganisms and is expelled into the large intestine as moist, soft pellets surrounded by a proteinaceous membrane. The soft pellets, much higher in protein and lower in crude fiber than the hard fecal pellets, are reingested (coprophagy) by the lagomorphs. The amount of feces recycled by coprophagy may range from 50 to 80 percent. This reingestion is important, for it provides bacterially synthesized B vitamins and ensures more complete digestion of dry material and better utilization of protein (see McBee, 1971).

Carnivores

Herbivores, in turn, are the energy source for *carnivores,* animals that feed on other animals. In a general and popular sense, carnivores are considered to be larger organisms that kill and eat smaller prey. But if one considers carnivory in the broadest sense, then any organism that feeds on another organism or the tissue of an organism is a carnivore, functionally speaking. Thus, carnivory could include *parasites* that live in or on, and draw their nourishment from, their hosts without actually killing them; and it would include the *parasitoids,* which draw nourishment from the host until the

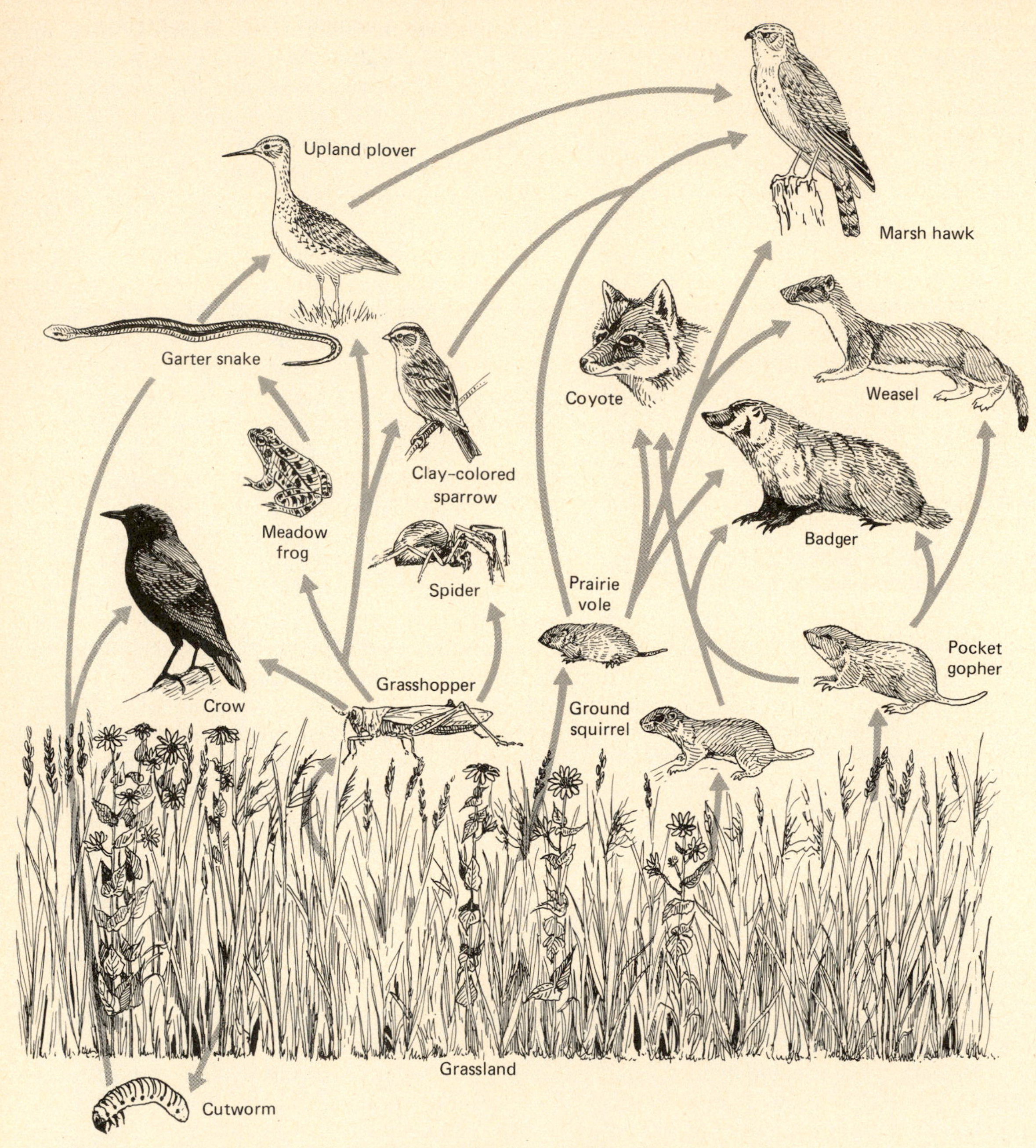

FIGURE 20.1 A food web within a prairie grassland community. Arrows flow from the eaten to the consumer. (*Based on Bird, 1930:383.*)

host dies. At this point the parasitoid transforms into another stage of its life cycle, becoming independent of the host.

Organisms that feed directly on grazing herbivores are termed *first-level carnivores* or *second-level consumers*. First-level carnivores represent an energy source for second-level carnivores. The typical carnivore is well adapted for a diet of flesh. Hawks and owls have sharp talons for holding prey and hooked beaks for tearing flesh. Mammalian carnivores have canine teeth for biting and piercing. Cheek teeth are reduced, but many forms have sharp-crested shearing or carnasial teeth.

Omnivores

Not all consumers can be fitted neatly into a trophic level, for many do not confine their feeding to one level alone. The red fox feeds on berries, small rodents, and even dead animals. Thus, it occupies herbivorous and carnivorous levels, as well as acting as a scavenger. Some fish feed on both plant and animal matter. The basically herbivorous white-footed mouse also feeds on insects, small birds, and bird eggs. Many species, including the white-footed mouse, are cannibalistic. They eat the flesh of their own kind, not only in an extreme food shortage, as one might like to believe, but frequently as a way of supplementing their diet. The food habits of many animals vary with the seasons, with stages in the life cycle, and with the size and growth of organisms. Consumers that feed on both plant and animal matter are termed *omnivores* (Figure 20.2).

Decomposers

Decomposers make up the so-called final feeding group. As discussed in Chapter 18, that is an oversimplified view of a complex functional group of organisms. All consumers to some degree function as decomposers. They either reduce enzymatically part of the material ingested or fragment it into smaller pieces, making it more accessible to other consumers, including bacteria and fungi.

Food chains involving the traditional decomposers, bacteria and fungi, usually reach up into the herbivore-carnivore food chains. In fact, decomposers—so frequently considered as some distant feeding group unclassifiable in the general scheme of food chains—actually function as herbivores and carnivores, depending upon the source of their food: dead plant or animal material. Perhaps only the bacteria and fungi that transform organic compounds into inorganic nutrients usable by photosynthetic plants should be considered outside of the general classification of herbivores and carnivores.

Biophages and Saprophages

Although descriptive, the terms *herbivore, carnivore,* and *omnivore* have probably outlived their usefulness as functional terms in energy transfer through food webs. Instead, heterotrophs should be considered either as biophages, those organisms utilizing living material, or saprophages, those utilizing nonliving matter (Wiegert and Owen, 1971). First-order biophages feed on plants and are the traditional herbivores, while first-order saprophages feed on dead plant material as well as organic material egested by first-order biophages. First-order biophages, in turn, are utilized at death by second-order saprophages, which are really functional carnivores. In turn, first-order saprophages may be utilized by second-order biophages. Such an approach incorporates decomposers into general feeding groups at various levels in the food chain (Figure 20.3).

Other Feeding Groups

Several other feeding groups are also involved in energy transfer. *Parasites* spend a considerable part of their life cycle living on or in, and drawing their nourishment from, their hosts. Most do not kill their hosts, but some, the *parasitoids,* do draw nourishment from their hosts until they die. At that point the parasitoid transforms into another stage of its life cycle, becoming independent of the host. Functionally, parasites are specialized carnivores and herbivores.

Scavengers are animals that eat dead plant and animal material. Among these are termites and various beetles that feed on dead and decaying wood, and crabs and other marine invertebrates that feed on plant particles in water. Bot flies, dermestid beetles, vultures, and gulls are only several of many animals that feed on animal remains. Scavengers

FIGURE 20.2 The red fox is an example of an omnivore, an animal that feeds on a trophic level above and below it. The food habits of the red fox are seasonal. The timing of flowering and the onset of breeding activities of animals influence the availability of food through the year. Note the prominence of fruits and insects in summer and fall and of rodents in spring and winter. (*Based on data from Scott, 1955.*)

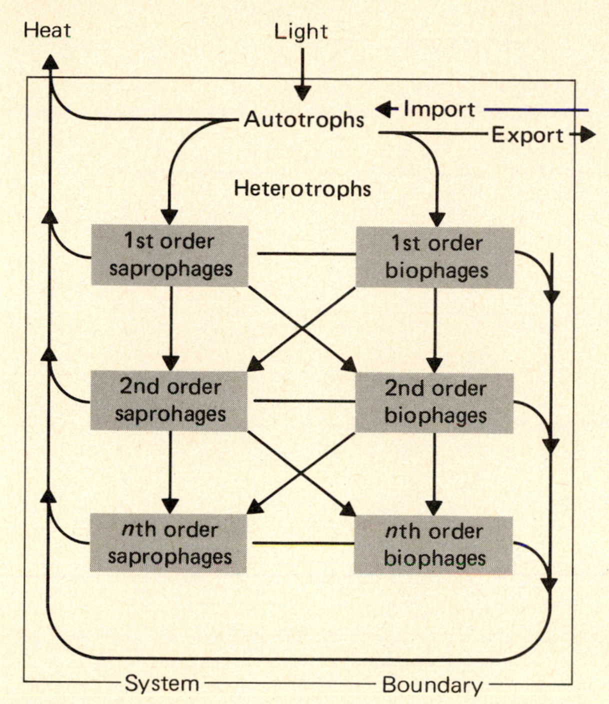

FIGURE 20.3 A model of two-channel energy flow that separates energy flow into two pathways, one utilized by organisms that feed on living matter and the other utilized by organisms that feed on nonliving matter. (*After Wiegert and Owen, 1971:74.*)

may be considered either herbivores or carnivores, or saprophages.

Saprophytes are plant counterparts of scavengers. They draw their nourishment from dead plant and animal material, chiefly the former. Because they do not require sunlight as an energy source, they can live in deep shade or dark caves. Examples of saprophytes are fungi and Indian pipe. The majority are herbivores, but some do feed on animal matter.

MAJOR FOOD CHAINS

Within any ecosystem there are two major food chains, the *grazing food chain* and the *detritus food chain* (see Figure 20.4). Because of the high standing crop and relatively low harvest of primary production, most terrestrial and shallow-water ecosystems are characterized by the detrital food chain. In deep-water aquatic systems—with their low biomass, rapid turnover of organisms, and high rate of harvest—the grazing food chain is the dominant one.

The amount of energy shunted down the two routes varies among communities. In an intertidal salt marsh, less than 10 percent of living plant material is consumed by herbivores, and 90 percent goes the way of the detritus-feeders and decomposers (Teal, 1962). In fact, most of the organisms of the intertidal salt marsh obtain the bulk of their energy from dead plant material. Fifty percent of the energy fixed annually in a Scots pine plantation is utilized by decomposers. The remainder is removed as yield or stored in tree trunks (Ovington, 1961). In some communities, particularly undergrazed grasslands, unconsumed organic matter may accumulate and the materials remain out of circulation for some time, especially when conditions are not favorable for microbial action. The decomposer or detritus food chain receives additional materials from the waste products and dead bodies of both herbivores and carnivores.

Detrital Food Chain

The detrital food chain is common to all ecosystems, but in terrestrial and littoral ecosystems, it is the major pathway of energy flow, because so little of the net production is utilized by grazing herbivores. Of the total amount of energy fixed in a tulip poplar *(Liriodendron)* forest, 50 percent of the gross production goes into maintenance and respiration, 13 percent is accumulated as new tissue, 2 percent is consumed by herbivores, and 35 percent goes to the detrital food chain (Edwards, unpublished, as cited by O'Neill et al., 1975). Two-thirds to three-fourths of the energy stored in a grassland ecosystem undergrazed by domestic animals is returned to the soil as dead plant material, and less than one-fourth is consumed by herbivores (Hyder, 1969). Of the quantity consumed by herbivores, about one-half is returned to the soil as feces (Breymeyer and Kajah, 1976). In the salt marsh ecosystem, the dominant grazing herbivore, the grasshopper, consumes just 2 percent of the net production available to it (Smalley, 1960).

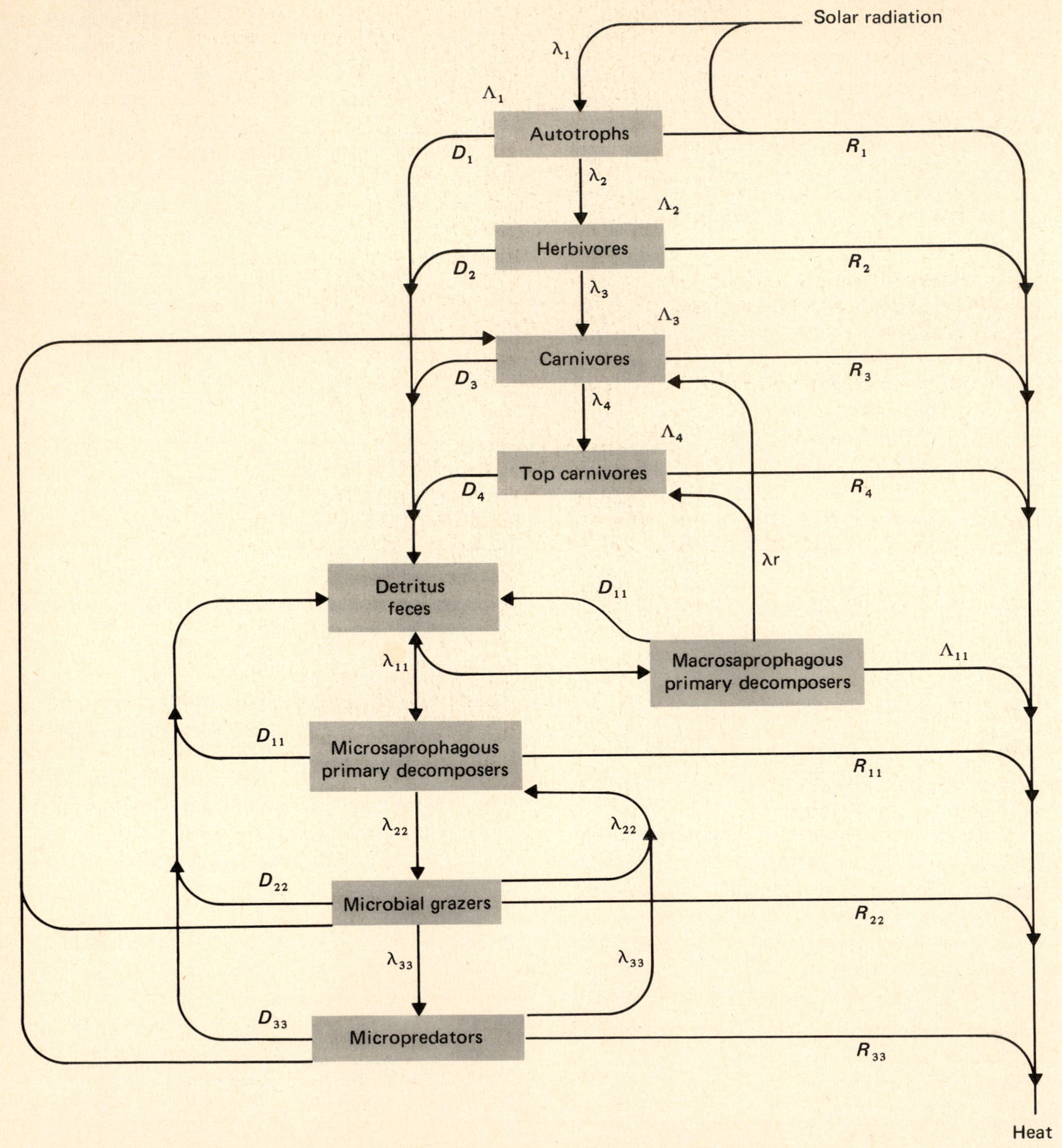

FIGURE 20.4 A model of detritus and grazing food chains. Two pathways lead from the autotrophs, one toward the grazing herbivore, the other toward the detritus-feeders. Note the interrelationships between the two food chains. (*After Paris, 1969.*)

The forest litter, the habitat of a number of detrital-feeding invertebrates, is a good place to seek an example of a detrital food web. Gist and Crossley (1975) provide information on trophic relationships among a group of selected litter animals. Although they collected no data for energy flow per se, the movement of radioactive calcium and phosphorus permitted the construction of a food web (Figure 20.5). The litter is utilized by five groups: millipedes (Diploda), orbatid mites (Cryptostigmata), springtails (Collemba), cave crickets (Orthoptera), and pulmonate snails (Pulmonata). Of these, the mites and springtails are the most important litter-feeders. These herbivores are preyed upon by small spiders (Araneidae) and predatory mites (Mesostigmata). The predatory mites feed on annelids, mollusks, insects, and other arthropods. The spiders feed on predatory mites. Springtails, pulmonate snails, small spiders, and cave crickets are preyed upon by carabid beetles, while medium-sized spiders feed upon cave crickets and other insects. The medium-sized spiders, in turn, become additional items in the diet of beetles. Beetles, spiders, and snails are consumed by birds and small mammals, members of a grazing food chain. Detrital food webs are linked, through predation, to grazing food chains at higher consumer levels.

Food chains involving *saprophages* may take two directions: toward the carnivores or toward microorganisms (Figure 20.4). The role of these

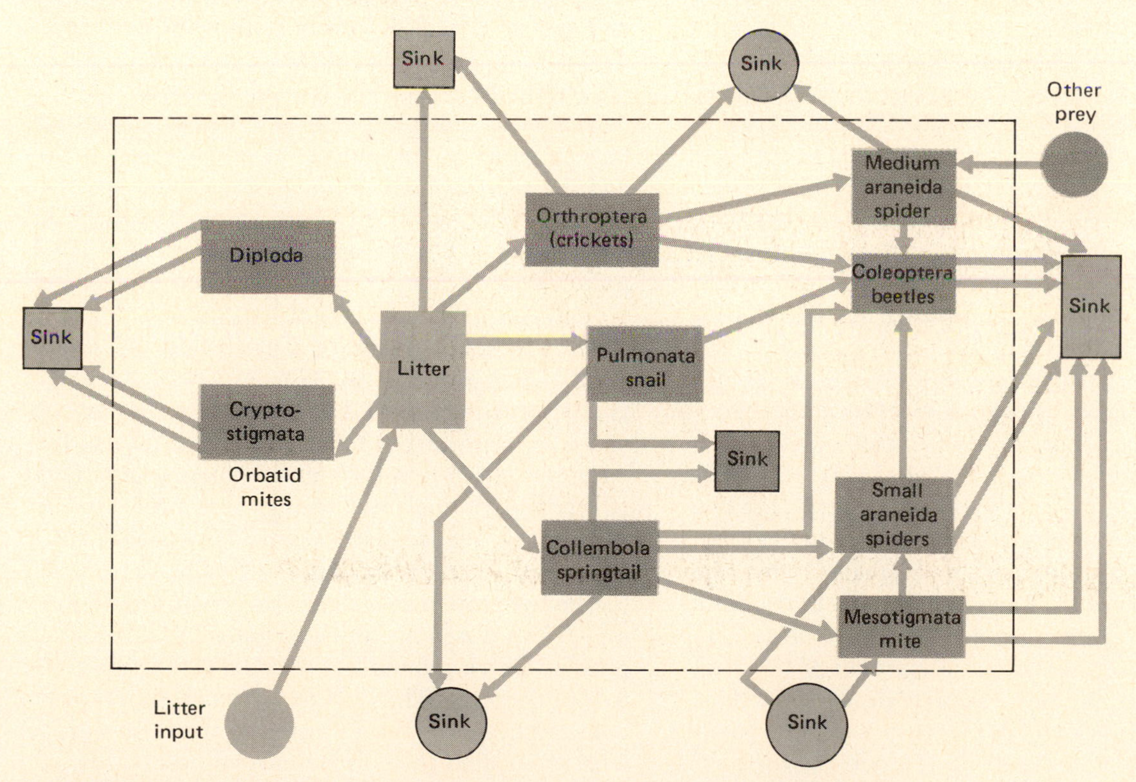

FIGURE 20.5 Model of a detrital food chain involving litter-dwelling invertebrates. The dashed lines represent the boundary of the system. Note that the detrital food chain involves a herbivorous component—millipedes and mites to the left and crickets and springtails to the right. The herbivores support a carnivorous component. The detrital food web, like the grazing food web, can become quite complex. (*After Gist and Crossley, 1975:86.*)

feeding groups in the final dissipation of energy has already been mentioned. But they also serve as food for numerous other animals. Slugs eat the larvae of certain Diptera and Coleptera, which live in the heads of fungi and feed on the soft material. Mammals, particularly red squirrels and chipmunks, eat woodland fungi. Dead plant remains are food sources for springtails and mites, which, in turn, are eaten by carnivorous insects and spiders. These, in turn, are energy sources for insectivorous birds and small mammals. Blowflies lay their eggs in dead animals, and within 24 hours the maggot larvae hatch. Unable to eat solid tissue, they reduce the flesh to a fetid mass by enzymatic action and feed on the proteinaceous material. These insects are food for other organisms.

Grazing Food Chain

The grazing food chain is the most obvious one (Figure 20.1). Cattle grazing on pasture land, deer browsing in the forest, rabbits feeding in old fields, and insect pests feeding on garden crops represent basic consumer groups of the grazing food chain. In spite of its conspicuousness, the grazing food chain is not the major one in terrestrial and many aquatic ecosystems. Only in some open-water aquatic ecosystems do the grazing herbivores play a dominant role in energy flow.

Although voluminous data exist on phytoplankton production, filtration rates by grazing zooplankton, and production efficiencies of zooplankton (for a summary see Wetzel, 1975), relatively few data are available on the flow of energy, grazing rates, biomass turnover rates for phytoplankton, and turnover of zooplankton biomass within the same aquatic system. Carter and Lund (1968) found that certain grazing protozoans feeding on certain planktonic algae consumed 99 percent of the population in 7 to 14 days. In a review of available data, M. Brylinsky (1980) came up with a turnover rate (production/biomass) of 113 percent for phytoplankton, 7.1 percent for herbivorous zooplankton, and 1.2 percent for carnivorous zooplankton. The high turnover rate for phytoplankton reflects the high production rate and low biomass of unicellular algae.

In terrestrial systems a relatively small portion of primary production goes the way of the grazing food chain. Over a three-year period, only 2.6 percent of the net primary production of a tulip-poplar forest was utilized by grazing herbivores, although holes made in the growing leaves resulted in the loss of 7.2 percent of photosynthetic surface. Andrews and his associates (1974) studied energy flow through a short-grass plains ecosystem involving ungrazed, lightly grazed, and heavily grazed plots. Even on the heavily grazed grassland, cattle accounted for only 15 percent of the net above-ground primary production, or 3 percent of total net primary production. On such grasslands, however, cattle grazing may account for the consumption of 30 to 50 percent of above-ground net primary production. About 40 to 50 percent of energy consumed by cattle is then returned to the ecosystem and the detrital food chain as feces (Dean et al., 1975).

Although the above-ground herbivores are the conspicuous feeders, below-ground herbivores can have a pronounced impact on primary production and the grazing food chain. Andrews and his associates (1975) found that the below-ground herbivores—consisting mainly of nematodes (Nematoda), scarab beetles (Scarabaeidae), and adult ground beetles (Carabidae)—accounted for 81.7 percent of total herbivore assimilation on the ungrazed short-grass plains, 49.5 percent on the lightly grazed grassland, and 29.1 percent on the heavily grazed one. Ninety percent of the invertebrate herbivore consumption took place below ground, and 50 percent of the total energy was processed by nematodes. On the lightly grazed grassland, cattle consumed 46 kcal/m^2 during the grazing season, and the below-ground invertebrates consumed 43 kcal/m^2. When a nematicide was added to a midgrass prairie, above-ground net production increased 30 to 60 percent. Thus, below-ground herbivorous consumption can impose a greater stress on a grassland ecosystem than above-ground herbivores.

Trophic Levels

If all organisms that obtain their food in the same number of steps (that is, all those that feed wholly or in part on plants, wholly or in part on herbivores, and so on) are superimposed, the structure can be collapsed into a series of points representing the trophic or feeding levels of the ecosystem. Thus, each step in a food chain represents a trophic level. Animals feeding wholly on plants—for example, the grasshopper and vole—occupy a single trophic level. But most animals at higher levels, such as the red fox, participate simultaneously in several trophic levels because of variation in their diets. Their total food intake has to be apportioned among the several trophic levels involved. The first trophic level belongs to the producers, or plants; the second trophic level to the herbivores, or first-level consumers; the third level to the first-level carnivores, or second-level consumers; and so on.

Although trophic levels typically do not include the decomposers, logically they should. Decomposers should be considered herbivores or carnivores, depending on the nature of their food source. Decomposers feeding on dead plant material, as well as bacteria occupying the rumen of ungulate animals or the guts of termites, should be considered functional herbivores, or first-level consumers; decomposers feeding on the bodies of dead animals should be considered second-level consumers; and so on. In that manner, all the various steps in energy transfer in an ecosystem can be placed in some trophic level. That is the approach used in the construction of pyramids of biomass and energy, considered later.

Energy Flow through a Food Chain

Energy flow through natural food chains is difficult to study, but one energy flow pattern that has been worked out carefully (Golley, 1960) involves old field vegetation, the meadow mouse, and the least weasel (Figure 20.6). The mouse was almost exclusively herbivorous, and the weasel lived mainly on mice. The vegetation converted about 1 percent of the solar energy into net production, or plant tissue. The mice consumed about 2 percent of the plant food available to them and the weasels about 31 percent of the mice. Of the energy assimilated, the plants lost about 15 percent through respiration, the mice 68 percent, and the weasels 93 percent. The weasels used so much of their assimilated energy on maintenance that a carnivore preying on a weasel alone would not be able to survive.

An ecological rule of thumb allows a magnitude of 10 reduction in energy as it passes from one trophic level to another. Thus, if 1,000 kilocalories of plant energy were consumed by herbivores, about 100 kilocalories would be converted into herbivore tissue, 10 kilocalories to the first-level carnivore production, and 1 kilocalories to the second-level carnivore. However, based on data available, a 90 percent loss of energy on the average from one trophic level to another may be high. Certainly, a wide range in the efficiency of conversion exists among various feeding groups (see Table 20.1). Production efficiency (see Box 19.1) in plants (net production/solar radiation) is low, ranging from 0.34 percent in some phytoplankton to 0.8 to 0.9 percent in grassland vegetation. Plant production consumed by herbivores is utilized with a varying degree of efficiency. Herbivores consuming green plants are wasteful feeders, but not nearly as wasteful as those feeding on plant sap (Table 20.1). More energy loss occurs in assimilation. Assimilation efficiencies vary widely among poikilotherms and homoiotherms (Table 20.2). Homoiotherms are much more efficient than poikilotherms (see Chapter 5). However, carnivorous animals, even poikilotherms, have very high assimilation efficiency (see Box 20.1). Predaceous spiders feeding on invertebrates have assimilation efficiencies of over 90 percent (Moulder, Reichle, and Auerbach, 1970; Van Hooke, 1971). Because of high maintenance or respiratory costs, homoiotherms have low production efficiency (production/consumption) compared to poikilotherms (Table 20.2). Only about 2 to 10 percent of the energy consumed by herbivorous homoiotherms goes into biomass production, less than the 10 percent average suggested by the rule of thumb. However, poikilotherms convert 17 percent of their consumption to herbivorous biomass. On midwestern grasslands average herbivore production efficiency, involving mostly poikilotherms,

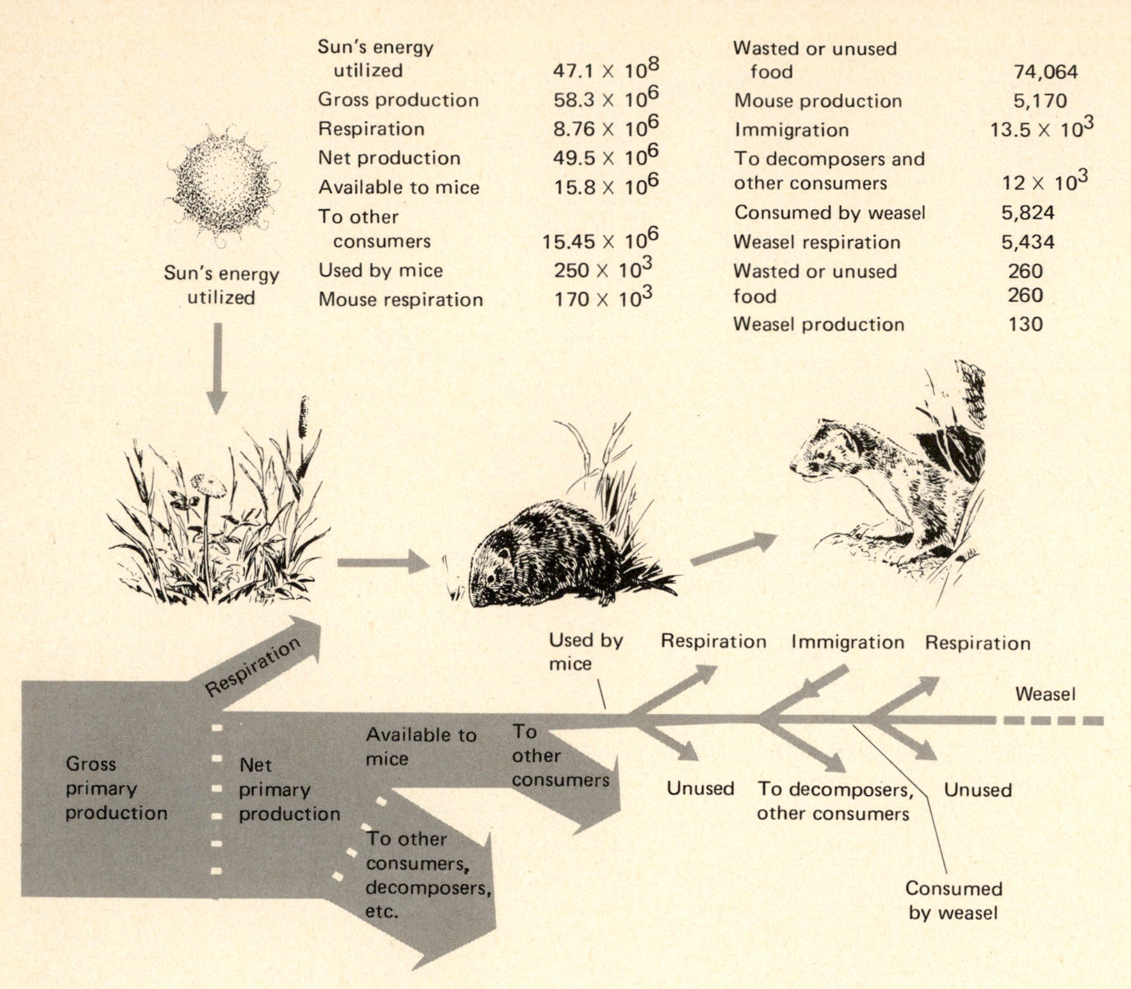

FIGURE 20.6 Energy flow through a food chain in an old field community in southern Michigan. The relative sizes of the blocks suggest the quantity of energy flowing through each channel. Values are in cal/ha/year. (*Based on data from Golley, 1960.*)

ranged from 5.3 to 16.5 percent (Table 20.3). Production efficiency on the secondary consumer, or carnivore, level ranged from 13 to 24 percent.

Transfer of energy from one trophic level to another tells the real story, but such data are hard to collect. The ratio of phytoplankton to secondary zooplankton production in open freshwater ecosystems is 1:7.1 according to Brylinsky (1980), and the ratio of herbivore zooplankton production to carnivorous zooplankton production is 1:2.1. Efficiencies are lower in the benthic community: 2.2 for herbivores and 0.3 for carnivores.

Energy transfer efficiency (consumption at trophic level n/net production at level $n - 1$) among invertebrate consumers on a short-grass plain is 9 percent for herbivores (Gyllenberg, 1980), 10 to 28 percent for saprophages (Kajah et al., 1980), 38 percent for above-ground predators, and 56 percent for below-ground predators. Reichle (1971) calculated the trophic-level production efficiency (assimilation at trophic level n/net production at level $n - 1$) for soil and litter invertebrates in a deciduous forest ecosystem as: saprophages, 0.11 to 0.17; phytophages, 0.02 to 0.07; and predators, 0.02.

TABLE 20.1 CONSUMPTION, WASTE, AND PRODUCTION (KCAL/M²/SEASON) BY CONSUMERS IN A MIXED PRAIRIE GRASSLAND

Consumers	Consumption	Wastage	Production
Above-ground			
Primary consumers			
Plant-tissue–feeders			
Mammals	1.60	.53	.018
Arthropods	6.33	4.35	1.13
Plant-sap–feeding arthropods	24.72	49.44	5.16
Pollen-nectar–feeding arthropods	.98	.20	.26
Seed-feeders			
Birds	.16	.00	$.505 \times 10^{-8}$
Mammals	.20	.00	.002
Arthropods	.17	.03	.04
Dead plant-litter–feeding arthropods	9.88	4.26	1.48
Secondary consumers			
Predators			
Birds	.88	.00	5.66×10^{-8}
Mammals	.20	.00	.003
Arthropods	3.25	.65	.94
Scavenger arthropods	.34	.17	.04
Below-ground invertebrates			
Primary consumers			
Plant-tissue–feeders	244.26	107.05	34.39
Plant-sap–feeders	382.38	240.10	64.74
Fungal-feeders	33.66	18.55	5.51
Bacteria-feeders	66.53	13.31	14.63
Secondary consumers			
Predators	117.83	23.57	27.63
Protozoa-feeders	31.65	6.33	7.92

SOURCE: Scott, French, and Leetham, 1979: 97, 98, 99.

TABLE 20.2 ASSIMILATION EFFICIENCY AND PRODUCTION EFFICIENCIES FOR HOMOIOTHERMS AND POIKILOTHERMS

Efficiency	All Homoiotherms	Grazing Arthropods	Sap-feeding Herbivores	Lepidoptera	All Poikilotherms
Assimilation					
A/C	77.5 ± 6.4	37.7 ± 3.5	48.9 ± 4.5	46.2 ± 4.0	41.9 ± 2.3
Production					
P/C	2.0 ± 0.46	16.6 ± 1.2	13.5 ± 1.8	22.8 ± 1.4	17.7 ± 1
P/A	2.46 ± 0.46	45.0 ± 1.9	29.2 ± 4.8	50.0 ± 3.9	44.6 ± 2.1

SOURCE: Based on data from Andrzejewska and Gyllenberg, 1980.

TABLE 20.3 CONSUMER EFFICIENCY (SECONDARY PRODUCTION/SECONDARY CONSUMPTION)

		Producers		Herbivores		Carnivores	
Habitat	Growing Season, Days	Production	% Efficiency	Production	% Efficiency	Production	% Efficiency
Short-grass plains	206	3,767	0.8	53	11.9	6	13.2
Midgrass prairie	200	3,591	0.9	127	16.5	37	23.7
Tall-grass prairie	275	5,022	0.9	162	5.3	15	13.9

SOURCE: Based on data from Andrzejewska and Gyllenberg, 1980.

Box 20.1

Ecological Efficiencies

Secondary consumption efficiency =

$$\frac{\text{Secondary production}}{\text{Secondary consumption}} \quad \frac{P}{C}$$

Turnover efficiency =

$$\frac{\text{Production at level } n}{\text{Biomass at level } n} \quad \frac{P}{B}$$

Energy transfer efficiency =

$$\frac{\text{Consumption at trophic level } n}{\text{Production at level } n-1} \quad \frac{C_n}{P_{n-1}}$$

Trophic-level efficiency =

$$\frac{\text{Assimilation at level } n}{\text{Production at level } n-1} \quad \frac{A_n}{P_{n-1}}$$

$$\frac{\text{Production at level } n}{\text{Production at level } n-1} \quad \frac{P_n}{P_{n-1}}$$

Ecological Pyramids

If one sums all of the biomass or living tissue contained in each trophic level and all of the energy transferred between, one can construct pyramids of biomass and energy for the ecosystem (see Figure 20.7).

The pyramid of biomass indicates by weight or other means of measuring living material the total bulk of organisms or fixed energy present at any one time—the *standing crop*. Since some energy or material is lost in each successive link, the total mass supported at each level is limited by the rate at which energy is being stored below. In general, the biomass of the producers must be greater than that of the herbivores they support, and the biomass of the herbivores must be greater than that of the carnivores. That usually results in a gradually sloping pyramid for most communities, particularly the terrestrial and shallow-water ones, where the producers are large and characterized by an accumulation of organic matter; life cycles are long; and the harvesting rate is low.

But this does not hold for all ecosystems. In such aquatic ecosystems as lakes and open seas, primary production is concentrated in the microscopic algae, characterized by a short life cycle, rapid multiplication of organisms, little accumulation of organic matter, and heavy grazing by herbivorous zooplankton. At any one point in time, the standing crop is low. As a result, the pyramid of biomass for these aquatic ecosystems is inverted; the base is much smaller than the structure it supports.

When production is considered in terms of energy, the pyramid indicates only the amount of energy flow at each level. The base on which the pyramid of energy is constructed is the quantity of organisms produced per unit of time or, stated dif-

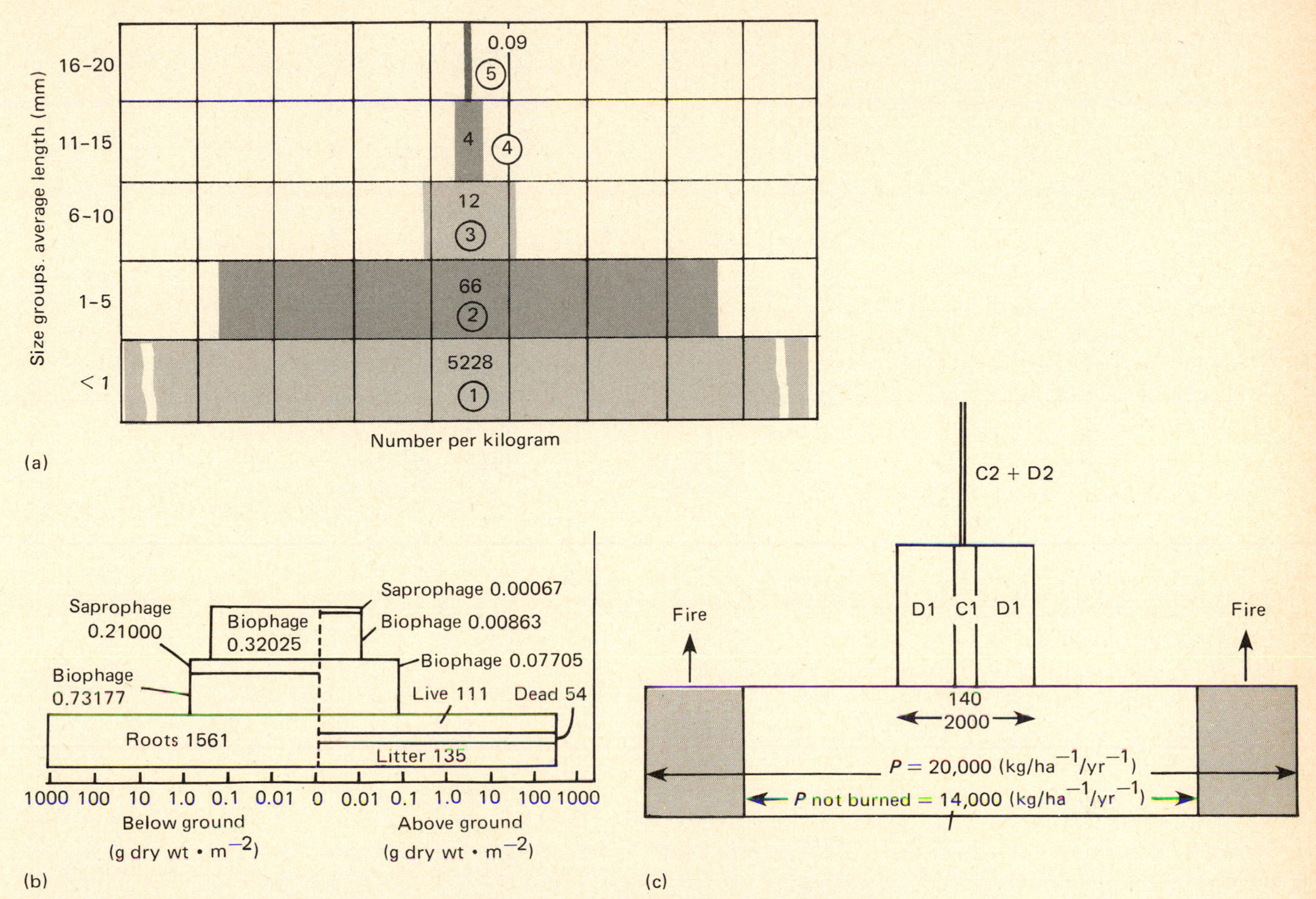

FIGURE 20.7 Examples of ecological pyramids. (a) A pyramid of numbers among the metazoans of the forest floor in a deciduous forest. (*After Park, Allee, and Shelford, 1939.*) (b) A pyramid of biomass for a northern short-grass prairie for July. The base of the pyramid represents biomass (g dry weight/m^2) of producers; the second (middle) level, primary consumers; and the top, secondary consumers. Above-ground biomass (right) and below-ground biomass (left) are separated by a dashed vertical line. The trophic-level magnitudes are plotted on a horizontal logarithmic scale. The compartments are divided on a vertical linear scale according to live, standing dead, and litter biomass or biophagic and saprophagic consumer biomass. Note that unlike the conventional pyramids, this one recognizes the detrital as well as the grazing food chain components on the same trophic levels. (*After French, Steinhorst, and Swift, 1979.*) (c) A pyramid of energy for the Lamto Savanna, Ivory Coast. *P* is primary production; C1, primary consumers; C2, secondary consumers; D1, decomposers of vegetable matter; D2, decomposers of animal material. Again, the decomposer, or detrital, and grazing food chains have been collapsed into the same trophic levels, since they should be functional. (*After Lamotte, 1975:216.*)

ferently, the rate at which food material passes through the food chain. Some organisms may have a small biomass, but the total energy they assimilate and pass on may be considerably greater than that of organisms with a much larger biomass. On a pyramid of biomass these organisms would appear much less important in the community than they really are. Energy pyramids are sloping because less energy is transferred from each level than was paid into it. This is in accordance with the second law of thermodynamics. In instances where the producers have less bulk than the consumers, particularly in open-water communities, the energy they store and pass on must be greater than that of the next level. Otherwise, the biomass that producers support could not be greater than that of the producers themselves. This high energy flow is maintained by a rapid turnover of individual plankton rather than by an increase in total mass.

Another pyramid commonly used in ecological literature is the pyramid of numbers (see Figure 20.7). This pyramid was sugggested by C. Elton (1927), who pointed out the great difference in the number of organisms involved in each step of the food chain. Although the organisms on the lower trophic levels are the most abundant, the pyramid is occasionally inverted. A single tree, for example, would represent only one organism at the producer level yet supports thousands of consumer animals. Successive links of carnivores decrease rapidly in number until there are very few carnivores at the top. The pyramid of numbers ignores the biomass of organisms. Although the numbers of a certain organism may be greater, the total weight, or biomass, of the organisms may not be equal to that of the larger organisms. Neither does the pyramid of numbers indicate the energy transferred or the use of energy by the groups involved. And because the abundance of members varies so widely, it is difficult to show the whole community on the same numerical scale.

The pyramid of numbers is often confused with a similar one in which organisms are grouped into size categories and then arranged in order of abundance. Here the smaller organisms are again the most abundant; but such a pyramid does not indicate the relationship of one group to another.

Food Webs and Trophic Levels—A Second Look

Until recently, the theory of food webs involved two generalizations. One stated that the maximum number of trophic levels in a food web is controlled by the second law of thermodynamics—the inefficiency of energy transfer from one trophic level to another. The second stated that complex food webs are more stable than simple ones. In the 1970s, especially the latter part of the decade, these two generalizations were questioned (May, 1973; Pimm and Lawton, 1977; Pimm, 1982), stimulating renewed interest in both theoretical and field studies of food webs. While neither of the two basic generalizations has been proved wrong, current studies are providing some new insights into food webs and trophic structure.

Of interest are such questions as: What determines the size of food webs and the number of trophic levels involved? How are food webs organized and structured? How are food webs affected by the successful invasion of a new species and the deletion of current ones? Are complex food webs more stable than simple ones? What processes, including population dynamics and energy flow, are involved in the pattern of food webs?

J. E. Cohen (1978) and, more recently, F. Briand (1983a, 1983b) assembled published food webs—terrestrial, marine, and freshwater—in an effort to detect the existence of structure or nonrandomness in food webs. Using detailed mathematical analyses, Cohen demonstrated that most food webs in the observed communities could be collapsed into one-dimensional representations termed interval graphs. These show overlap or the lack thereof.

Briand analyzed 62 food webs by multivariate analysis. He was able to separate food webs into two large groups—those found in fluctuating environments characterized by variations in temperature, salinity, pH, moisture, and other conditions and those found in constant environments. Food webs in constant environments are characterized by greater species richness and more trophic links (connectance) than food chains in fluctuating environments. These tended to be shorter with fewer trophic links, possibly because environmental constraints imposed greater rigidity on the shape of the webs. Organisms in fluctuating environments may be forced to optimize their foraging, restricting their choice of prey (see Chapter 16). The widest food chains, those with the greatest number of herbivorous species, were the shortest. Narrow food webs, by contrast, had the greatest fraction of top carnivores. In general, Briand found that distinctive habitats or ecosystems possessed their own distinctive trophic structure (Figure 20.8). Food webs in benthic regions, for example, are shorter than pelagic food webs, probably because of their detrital base.

How food chains might be assembled is another question. Are they a product of random assemblage or not? To seek some answer to that question, P. Yodzis (1981, 1982, 1983) simulated the development of a food web based on natural parameters. Starting with a number, N, of basal species or producers, Yodzis added additional species, each of which had a certain ecological efficiency, making a fraction of its own consumption available to the next trophic level. Each new species added had to obtain its energy from production available from other species. Each species had to choose a food source already utilized by another. At the same time, a newly arriving species had a total production, a fraction of which had to be made available to a subsequently arriving species. There is, however, an upper and lower limit to the number of species a predator can exploit. There arrives a point at which total production is too low to allow a new species to enter. Thus, the limitation of energy transformation (the second law) forces a pattern to a food web. Yodzis found a certain degree of similarity between his simulated food webs and real-world food webs.

The kinds of species that can enter a food web are also a subject for simulation studies, the results of which conform to what one might expect. Generalist species most easily invade simple food webs, and specialists, capable of exploiting a restricted source of energy, are best able to invade complex webs (Drake, 1983). All these studies suggest that food webs are not random assemblages but rather are regulated by properties of existing food webs and the nature of invading species.

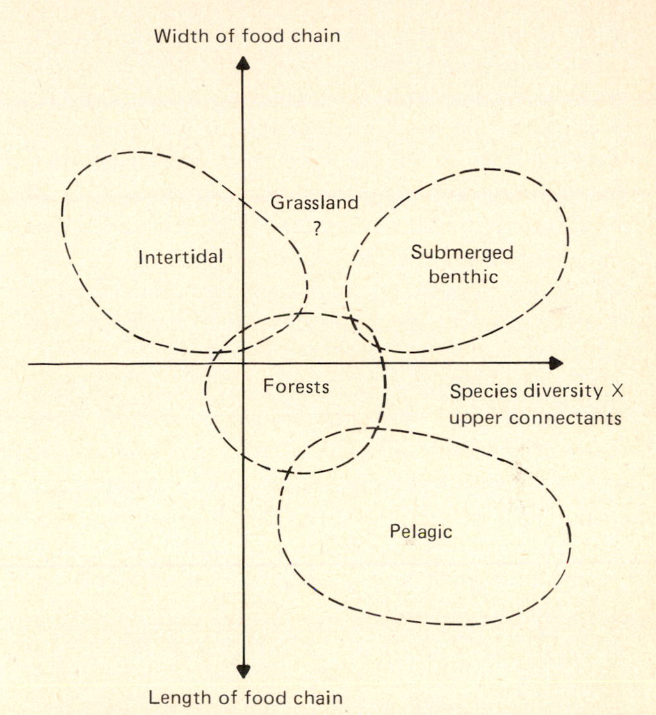

FIGURE 20.8 Food web characteristics of species richness and connectance (linkage) at the upper trophic levels (sc), width of food web, and length of food web for several major ecosystems. The pelagic ecosystem has long food webs and high species richness and connectance. The intertidal ecosystem has wide and short food webs and low species richness and connectance. (*From Briand, 1983b:39.*)

What happens when a species is removed from a food web? Theoretically, there is little effect if the species removed is a generalized prey species or a generalized predator or a member of a simple, straight-line food chain (Pimm, 1980). But if the species removed is a key predator, its removal can have a pronounced effect. The removal of a species causes the greatest loss of species in the trophic level beneath when the predator has a controlling influence on the equilibrium density of the prey and the prey are generalists in their food habits. Removal has the least effect when the predator exerts a controlling influence on the equilibrium density of the prey but the prey are specialized. For example, R. Paine (1974) removed a key predator, the starfish *Piaster,* from a rocky intertidal community dominated by a diverse group of invertebrates. With the loss of the key predator, the community was quickly dominated by a mussel, eliminating or reducing the populations of the subordinate species and changing a complex food web to a more simplified one. Field and computer simulation studies seem to support the hypothesis that complex food webs are less stable than simple ones.

Studies of published food webs and computer simulations have their serious shortcomings. Food webs used in the analyses were not obtained with a food web study in mind but rather to characterize certain environments and relationships among a few species. Published food webs are rather general at the lower trophic levels, especially at the producer level, and most detailed at the upper trophic levels. Often the food relations between eater and eaten are based on general observations rather than on detailed food habit studies of all species involved. Published food chains are also mostly grazing ones, which ignore contributions of the various trophic levels and species involved to the detrital food web, the most important of all. The underlying assumption involved in the simulation studies is that net energy produced on one trophic level is passed on to the next trophic level. Further, food webs vary in any ecosystem seasonally (for example, note the food habits of the red fox, illustrated in Figure 20.2) or diurnally (as food webs in an intertidal zone). Food chains are also subject to the foraging behavior of the herbivores and carnivores involved and the heterogeneity or patchiness of the environment. Such spatial and temporal variations may weaken any possibility of rigorous quantification of linkages in food chains.

Community Energy Budgets

Since energy flow involves both inputs and outputs, the efficiency and production of an ecosystem can be estimated by measuring the quantity of energy entering the community through the various trophic levels and the amount leaving it (Figure 20.9). Such information can give *net ecosystem production* (or net community production, as it is often called), which can be expressed as follows:

Net ecosystem production (biomass accumulation)
= gross primary production − plant respiration
− animal respiration − decomposer respiration

A balance sheet for energy flow and production, with debit and credit sides, can be drawn up for a community (see Table 20.4).

Few communities have been studied intensively enough to present such a broad picture, but some studies are available. One is of a salt marsh, an autotrophic community (Teal, 1962), and the other of a cold spring, a heterotrophic community (Teal, 1957), because the major energy source was plant material that had fallen into the water. Of the energy transformed by organisms in the spring, 76 percent entered as leaves, fruit, and branches of terrestrial vegetation. Photosynthesis accounted for 23 percent, and 1 percent came from immigrating caddisfly larvae. Of this total input, 71 percent was dissipated as heat, 1 percent lost through the emigration of adult insects, and 28 percent deposited in the community. In the salt marsh the producers themselves were the most important consumers, for plant respiration accounts for 70 percent of gross production, an unusually high figure (Scots pine, for example, utilizes only about 10 percent in respiration). Plants are followed by the bacteria, which utilize only one-seventh as much as the producers. Primary and secondary consumers come in a poor third, using only one-seventh as much energy as the bacteria.

In spite of energy budget sheets, these examples point out how incomplete and fragmentary our knowledge of energy flow through ecosystems is. To understand something of energy flow through ecosystems, one has to know something of energy flow through populations within the ecosystem and then relate this information to the flow of energy from one trophic level to another. Herein lies the weakness of the model. Energy budgets of several ecosystems to date are based in part on assumptions rather than on known values of energy flow.

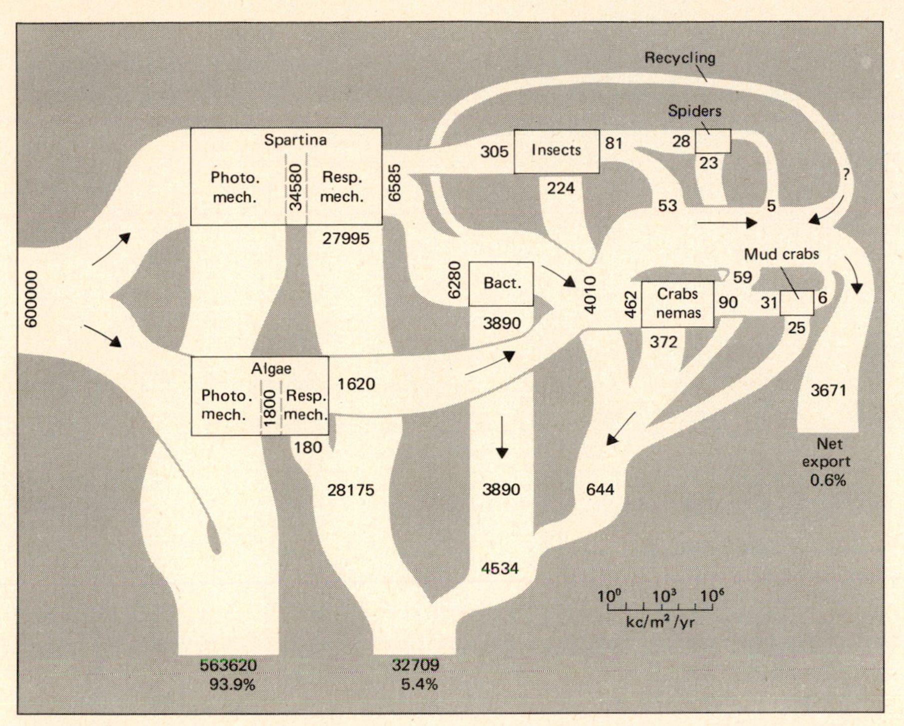

FIGURE 20.9 Energy flow through a Georgia salt marsh. Note the importance of the detrital food chain. (*From Teal, 1962:622.*)

TABLE 20.4 COMMUNITY ENERGY BALANCE SHEETS FOR AN AUTOTROPHIC AND A HETEROTROPHIC COMMUNITY

Autotrophic community: the salt marsh	
Input as light	600,000 kcal/m²/year
Loss in photosynthesis	563,620 kcal/m²/year, or 93.9% of light
Gross production	36,380 kcal/m²/year, or 6.1% of light
Producer respiration	28,175 kcal/m²/year, or 77% of gross production
Net production	8,205 kcal/m²/year
Bacterial respiration	3,890 kcal/m²/year, or 47% of net production
First-level consumer respiration	596 kcal/m²/year, or 7% of net production
Second-level consumer respiration	48 kcal/m²/year, or 0.6% of net production
Total energy dissipation by consumers	4,534 kcal/m²/year, or 55% of net production
Export	3,671 kcal/m²/year, or 45% of net production
Heterotrophic community: temperate cold spring	
Organic debris	2,350 kcal/m²/year, or 76.1% of available energy
Gross photosynthetic production	710 kcal/m²/year, or 23% of available energy
Immigration of caddis larvae	18 kcal/m²/year, or 0.6% of available energy
Decrease in standing crop	8 kcal/m²/year, or 0.3% of available energy
Total energy dissipation to heat	2,185 kcal/m²/year, or 71% of available energy
Deposition	868 kcal/m²/year, or 28% of available energy
Emigration of adult insects	33 kcal/m²/year, or 1% of available energy

SOURCE: Salt marsh: Teal, 1962; cold spring: Teal, 1957.

Too little is known of energy flow through any population to allow a clear picture of energy flow through an ecosystem. And what knowledge we do possess of energy flow through any population is often unreliable. If this knowledge is to be used, one has to assume that energy flow through a population is constant or that, if fluctuating, it is at least predictable. But probably it is not, since energy flow through populations is variable, depending on ecological conditions.

For example, growth, and thus energy storage, by salmonid fish is related to the size of their food (Palohemio and Dickie, 1970). When particle size is small, fish expend more energy obtaining food, and use a smaller portion in growth. Variations in temperature affect the rate of food assimilation, and variations in salinity and nitrogenous wastes in the water affect both energy turnover and efficiency of utilization. Within any ecosystem, animal populations may have a pronounced influence on the rate of energy fixation of plants. Overgrazing and overbrowsing reduce the amount of primary production in grassland and forest ecosystems. Overexploitation of a prey species, either by humans or by predators, can affect the amount of secondary production. The nutrient composition of soils or plants can limit energy fixation and storage in both primary and secondary production. Lack of boron in the soil, for example, can severely depress the growth of alfalfa; and the low nutrient status of a soil, and thus of plants, can affect the production of animals.

We lack enough detailed studies of trophic efficiency to give us any clear picture of the structure of ecosystems. But the concept of energy flow is valuable as a guideline for future studies and as a basis for understanding some of the interrelationships and interactions of humans and their environment.

Summary

A basic function of the ecosystem is the flow of energy. The energy of sunlight fixed by the autotrophic component of the ecosystem is available to the heterotrophic component, of which the herbivores are the primary consumers. Herbivores, in turn, are a source of food for carnivores. At each step or transfer of energy in the food chain, a considerable amount of potential energy is lost as heat, until ultimately, the amount of available energy is so small that few organisms can be supported by that source alone. Animals further up on a food chain often utilize several sources of energy, including plants, and thus, in their feeding habits, become omnivores.

Energy flow in the ecosystem takes two routes: one through the grazing food chain, the other through the detritus food chain. In the latter, the bulk of production is utilized as dead organic matter by saprophages.

The loss of energy at each transfer limits the number of trophic levels, or steps, in the food chain to four or five. At each level, biomass usually declines; so if the total weight of individuals at each successive level is plotted, a sloping pyramid is formed. In certain aquatic situations, however, where there is a rapid turnover of small aquatic producers or phytoplankton, the pyramid of biomass may be inverted. But in either case, energy decreases from one trophic level to another and is pyramidal.

The ratios of energy flow in or between trophic levels, in or between populations of organisms, or in or between individual organisms are ecological efficiencies. Any number of ratios can be determined. Among some of the most useful are assimilation efficiency, consumption efficiency, production efficiency, and energy transfer efficiency.

Although knowledge of energy flow in ecosystems is fragmentary and difficult to come by, the concept of energy flow is a valuable guide to the study and understanding of ecosystem functioning and of the relationship of humans to their environment.

Review and Study Questions

1. What is a food chain? A food web?
2. Define herbivore, carnivore, omnivore, decom-

poser, parasitoid, scavenger, biophage, and saprophage. What is the functional role of each?

3. What are the two major food chains? How are they interrelated?
4. What is a trophic level? Relate the levels to three kinds of ecological pyramids.
5. Under what situation can a pyramid of biomass be inverted?
6. What is ecological efficiency?
7. What are the problems involved in setting up an energy budget for a natural community?
8. What is the difference between an autotrophic and a heterotrophic community?

Suggested Readings

See the "Suggested Readings" section at the end of Chapter 19; also:

Breymeyer, A. I., and G. M. Van Dyne, eds. 1980. Chapter 11 in *Grassland, systems analysis, and man.* New York: Cambridge University Press.

French, N., ed. 1979. Chapter 4 in *Perspectives in grassland ecology.* New York: Springer-Verlag.

CHAPTER 21

Nutrient Cycles

Outline

Objectives

Upon completion of this chapter, you should be able to:

1. Define and distinguish among the types of biogeochemical cycles.
2. Describe the oxygen cycle.
3. Explain the carbon cycle in general and on a global scale.
4. Describe the nitrogen cycle, including the four major processes involved.
5. Describe the phosphorus cycle and compare the terrestrial and aquatic components of it.
6. Explain the sulfur cycle and its relation to the acid rain problem.
7. Describe how chlorinated hydrocarbons follow a pseudonutrient cycle.
8. Define turnover time, turnover rate, and residence time.
9. Develop a model of nutrient flow through the ecosystem.
10. Distinguish between oligotrophy and eutrophy and point out how they relate to nutrient flow.
11. Discuss mechanisms of nutrient conservation in the ecosystem.
12. Describe nutrient budgets, including their major inputs and outputs.

The existence of the living world depends upon the flow of energy and the circulation of materials through the ecosystem. Both influence the abundance of organisms, the metabolic rate at which they live, and the complexity of the ecosystem. Energy and materials flow through the ecosystem together as organic matter; one cannot be separated from the other. The continuous round trip of materials, paid for by the one-way trip of energy, keeps ecosystems functioning.

All the essential nutrients and many others besides, including a number of human-made materials such as chlorinated hydrocarbons, flow from the nonliving to the living and back to the nonliving parts of the ecosystem in a more or less circular path known as the *biogeochemical cycle* (*bio* for living; *geo* for water, rocks, and soil; and *chemical* for the processes involved). Some of the material is returned to the immediate environment almost as rapidly as it is removed; some is stored in short-term nutrient pools such as the tissues of plants and animals or the soil and sediment in lakes and ponds; and some may be tied up chemically or buried deep in the Earth in long-term nutrient storage pools before being released and made available to living organisms. Between the easily accessible and the relatively unavailable, there exists a slow but steady interchange.

The important roles in all nutrient cycles are played by green plants, which organize the nutrients into biologically useful compounds; by the organisms of decomposition, which ultimately return them to their simple elemental state; and by air and water, which transport nutrients between the abiotic and living components of the ecosystem. Without these no cyclic flow of nutrients would exist.

There are two basic types of biogeochemical cycles: *gaseous* and *sedimentary*. In gaseous cycles the main reservoir of nutrients is the atmosphere and ocean. In sedimentary cycles the main reservoir is the soil and the sedimentary rocks and other rocks of Earth's crust. Both involve biological and nonbiological agents, both are driven by the flow of energy, and both are tied to the water cycle.

Gaseous Cycles

Because gaseous cycles are closely linked to the atmosphere and oceans, they are pronouncedly global and involve the compounds or substances of which we are most consciously aware: oxygen, carbon dioxide, and nitrogen.

THE OXYGEN CYCLE

Oxygen, the by-product of photosynthesis, is involved in the oxidation of carbohydrates, with the release of energy, carbon dioxide, and water. Its primary role in biological oxidation is that of a hydrogen acceptor. The breakdown and decomposition of organic molecules proceeds primarily by dehydrogenation. Hydrogen is removed by enzymatic action from organic molecules in a series of reactions and is finally accepted by the oxygen, forming water.

Oxygen is very active chemically. It can combine with a wide range of chemicals in the Earth's crust and is able to react spontaneously with organic compounds and reduced substances.

The major supply of free oxygen that supports life is in the atmosphere. There are two significant sources of atmospheric oxygen. One is the photodisassociation of water vapor, in which most of the hydrogen released escapes into outer space. If the hydrogen did not escape, it would oxidize and recombine with the oxygen. The other source is photosynthesis, active only since life began on Earth. Because photosynthesis and respiration are cyclic, involving both the release and the utilization of oxygen, one would seem to balance the other so that no significant quantity of oxygen would accumulate in the atmosphere. At some time in the Earth's history, the amount of oxygen introduced into the atmosphere had to exceed the amount used in the decay of organic matter and that tied up in the oxidation of sedimentary rocks. Part of the atmospheric oxygen represents the portion remaining from the unoxidized reserves of photosynthesis—coal, oil, gas, and organic carbon in sedimentary rocks. The amount of stored carbon in the Earth suggests that 150×10^{20} g of oxygen has been available to the atmosphere, over 10 times as much as now present, 10×10^{20} g (F. S. Johnson, 1970).

The main nonliving oxygen pool consists of molecular oxygen, water, and carbon dioxide, all intimately linked to each other in photosynthesis and other oxidation-reduction reactions and all exchanging oxygen with each other. Oxygen is also biologically exchangeable in such compounds as nitrates and sulfates, which are utilized by organisms that reduce them to ammonia and hydrogen sulfide.

On the surface the oxygen cycle might appear to be quite simple. But because oxygen is so reactive, its cycling is quite complex. As a constituent of carbon dioxide, it circulates freely throughout the biosphere. Some carbon dioxide combines with calcium to form carbonates. Oxygen combines with nitrogen compounds to form nitrates, with iron to form ferric oxides, and with many other minerals to form various other oxides. In these states oxygen is temporarily withdrawn from circulation. In photosynthesis the oxygen freed is split from the water molecule. This oxygen is then reconstituted into water during plant and animal respiration. Part of the atmospheric oxygen that reaches the higher levels of the troposphere is reduced to ozone (O_3) by high-energy ultraviolet radiation.

THE CARBON CYCLE

Because it is a basic constituent of all organic compounds and a major element involved in the fixation of energy by photosynthesis, carbon is so closely tied to energy flow that the two are inseparable. In fact, the measurement of productivity is commonly expressed in terms of grams of carbon fixed per square meter per year.

The source of all the fixed carbon in both living organisms and fossil deposits is carbon dioxide, CO_2, found in the atmosphere and dissolved in the waters of the Earth. To trace its cycling to the ecosystem is to redescribe photosynthesis and energy flow (see Chapter 18).

The carbon contained in animal wastes and in the protoplasm of plants and animals is eventually released by assorted decomposer organisms. The rate of release depends on environmental conditions such as soil moisture, temperature, and precipitation. In tropical forests most of the carbon in plant remains is quickly recycled, for there is little accumulation in the soil. The turnover rate of atmospheric carbon over a tropical forest is about 0.8 year (Leith, 1963). In drier regions such as grasslands, considerable quantities of carbon are stored as humus. In swamps and marshes, where dead material falls into the water, organic carbon is not completely mineralized and is stored as raw humus or peat and circulated only slowly. The turnover rate of atmospheric carbon over peat bogs is somewhere on the order of 3 to 5 years (Leith, 1963).

Similar cycling takes place in the freshwater and marine environments. Phytoplankton utilizes the carbon dioxide that has been diffused into the upper layers of water or is present as carbonates and converts it into carbohydrates. The carbohydrates so produced pass through the aquatic food chains. The carbon dioxide produced by respiration is reutilized by the phytoplankton in the production of more carbohydrates. Under proper conditions a portion is reintroduced into the atmosphere. Significant portions of carbon bound as carbonates in the bodies of shells, snails, and foraminifers become buried in the bottom mud at varying depths when the organisms die. Isolated from biotic activity, that carbon is removed from cycling and becomes incorporated into bottom sediments, which through geological time may appear on the surface as limestone rocks or coral reefs. Other organic carbon is slowly deposited as gas, petroleum, and coal at an estimated global rate of 10 to 13 g/m^2/year.

The cycling of carbon as carbon dioxide involves its assimilation and respiration by plants, its consumption in the form of plant and animal tissue by animals, its release through their respiration, the mineralization of litter and wood, soil respiration, accumulation of carbon in a standing crop, and its withdrawal into longer-term reserves such as humus and peat fossil deposits (see Figure 21.1).

Diurnal and Seasonal Patterns

If you were to measure the concentration of carbon dioxide in the atmosphere above and within a forest on a summer day, as Woodwell and Dykeman (1966)

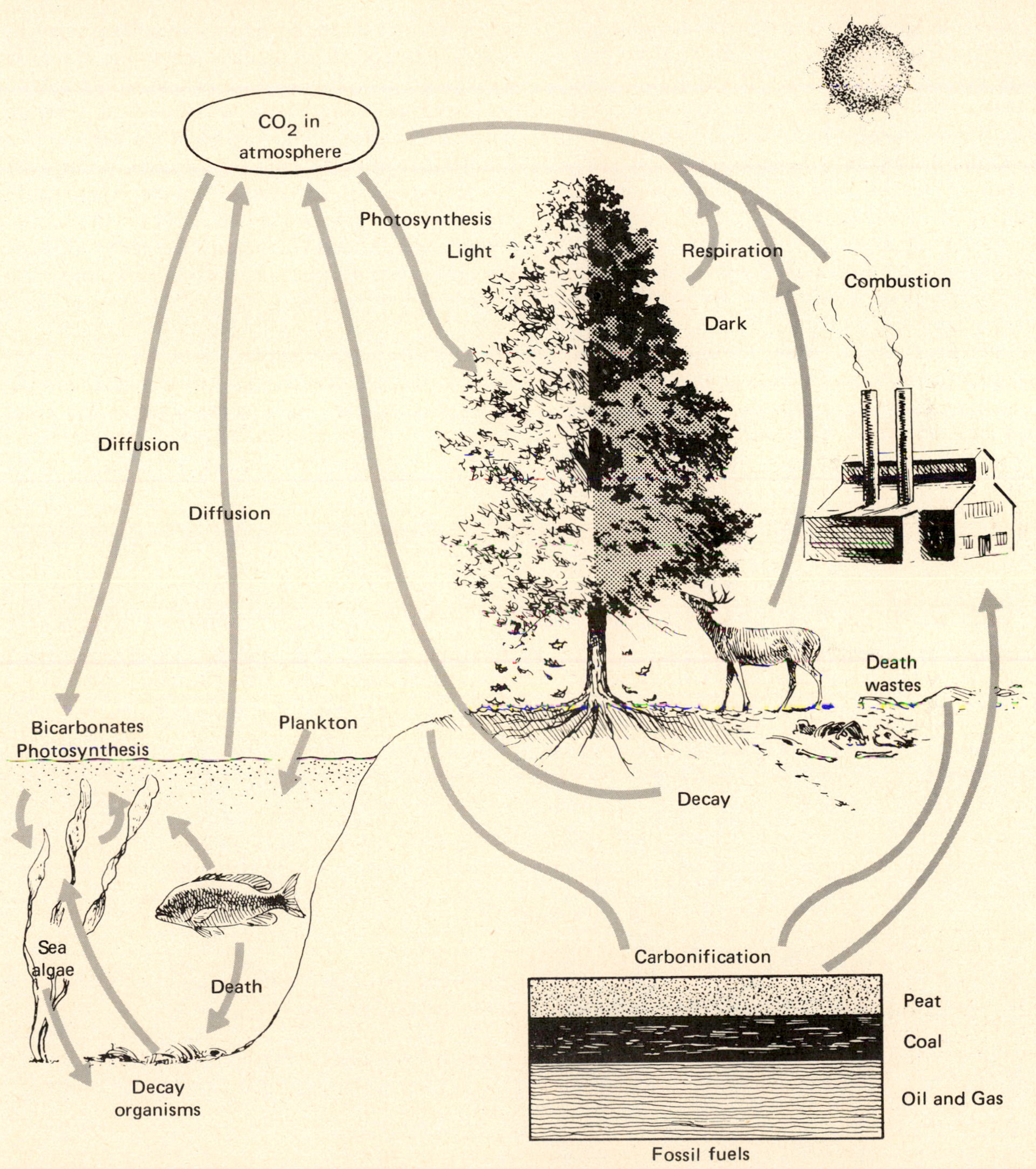

FIGURE 21.1 The carbon cycle in aquatic and terrestrial ecosystems.

did, you would discover that it fluctuates throughout the day (Figure 21.2). At daylight, when photosynthesis begins, plants start to withdraw carbon dioxide from the air, and the concentration declines sharply. By afternoon, when the temperature is increasing and the humidity is decreasing, the respiration rate of plants increases, the assimilation rate of carbon dioxide declines, and the concentration of carbon dioxide in the atmosphere increases. By sunset the light phase of photosynthesis ceases, carbon dioxide is no longer being withdrawn from the atmosphere, and its concentration in the atmosphere increases sharply. A similar diurnal fluctuation takes place in aquatic ecosystems.

Likewise, there is a seasonal course in the production and utilization of carbon dioxide that relates both to temperature and to the dormant and growing seasons. In the spring, when land is greening and phytoplankton is actively growing, the daily production of carbon dioxide is high. As measured by nocturnal accumulation in spring and summer, the rate of carbon dioxide production may be two to three times higher than winter rates at the same temperature. The transition from lower to higher rates increases dramatically at about the time of the opening of buds and falls off just as rapidly at about the time when the leaves of deciduous trees start dropping in the fall.

The Global Carbon Cycle

The carbon budget of Earth is closely linked to the atmosphere, land, and oceans and to the mass movements of air around the planet (Figure 21.3). The carbon pool involved in the global carbon cycle

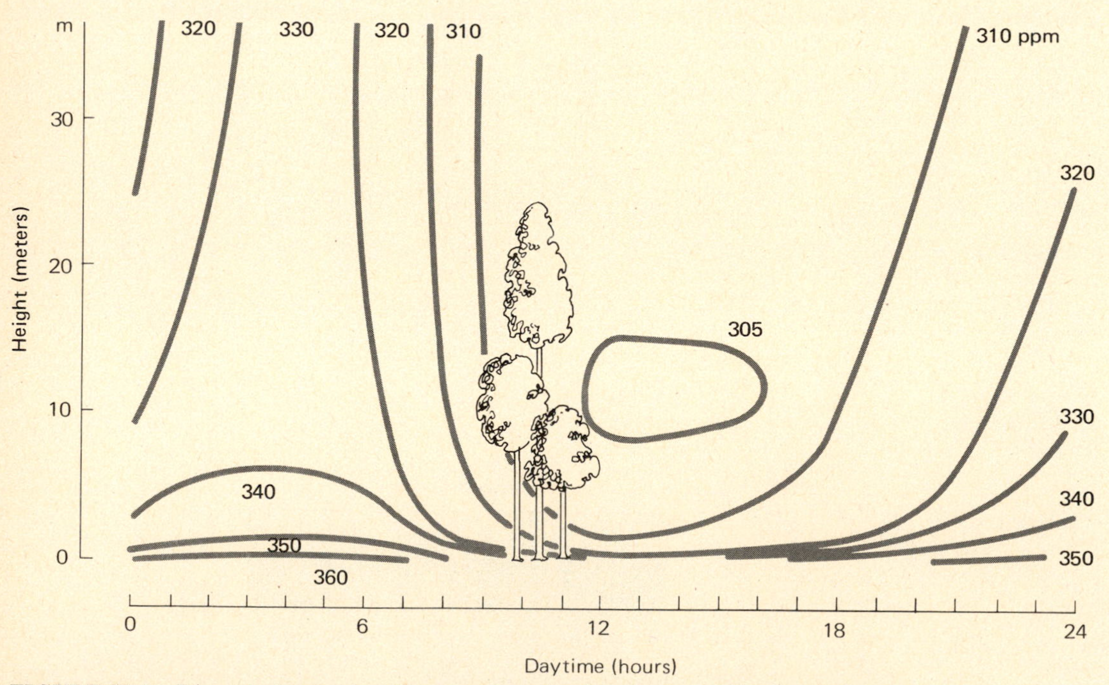

FIGURE 21.2 Diurnal changes in CO_2 concentration in a forest. Note the nighttime increase in CO_2 above and within the forest, especially above the forest floor. This increase results from the respiration of plants and soil microorganisms. During the day, plants actively take in CO_2 in photosynthesis, reducing its concentration in the atmosphere above and within the forest. The concentration of CO_2 begins to increase in late afternoon and evening, when the light phase of photosynthesis winds down. (*After Miller and Rusch, 1960.*)

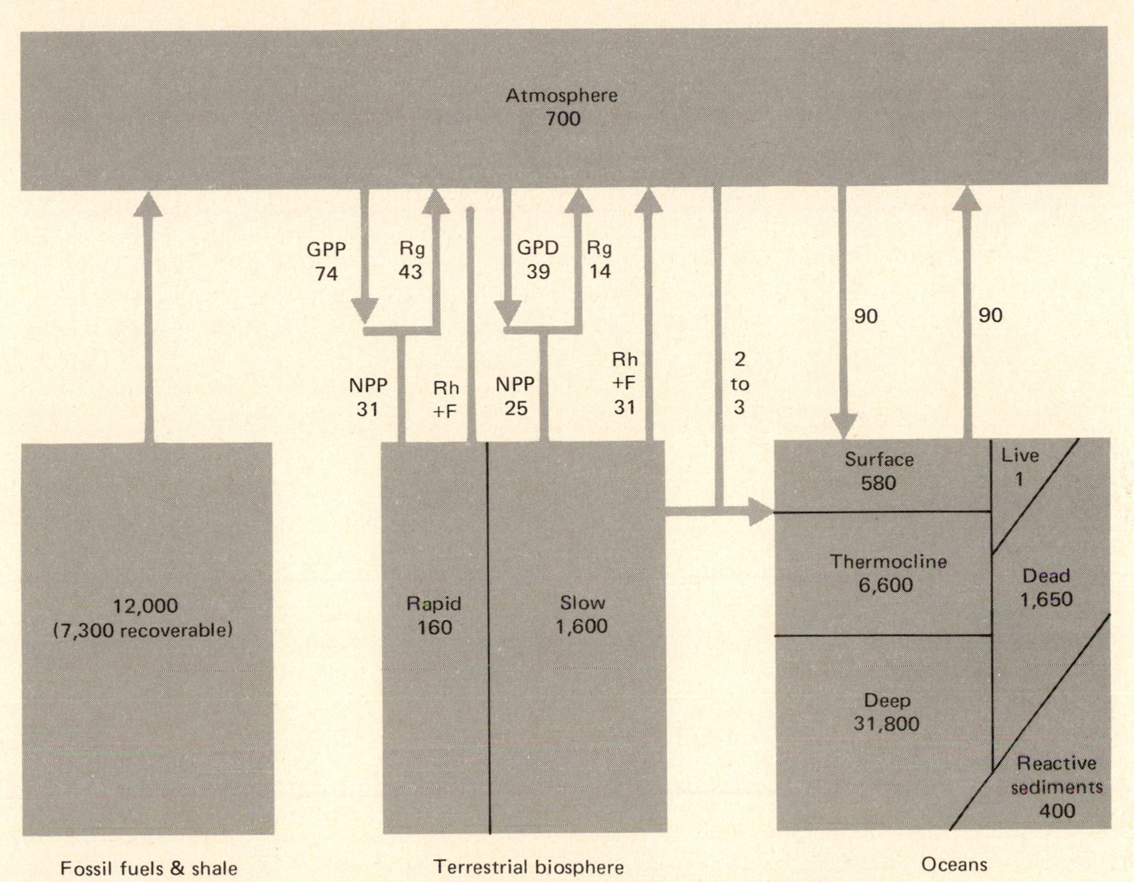

FIGURE 21.3 Global compartments of CO_2. The diagram shows the interconnected subsystem of the atmosphere, ocean, and main continental organic matter pools. Shown are the major fluxes (in gigatons, or Gt, per year) and pool sizes (in Gt) of carbon. Fluxes include gross primary production (GPP), green plant respiration (R_g), net primary production (NPP), respiration by heterotrophs (R_h), and fires (F). (*From Baes et al., 1977:314.*)

amounts to an estimated 55,000 gigatons (a gigaton is equal to 1 billion, or 10^9, metric tons). Fossil fuels and shale account for an estimated 12,000 gigatons. The oceans contain 93 percent of the active carbon pool, over 39,000 gigatons, mostly as carbonate (CO_3^- and bicarbonate ions (HCO_3^-). Dead organic matter in the oceans accounts for 1,650 gigatons of carbon, and living matter, mostly phytoplankton, 1 gigaton. The terrestrial biosphere contains an estimated 1,446 gigatons as dead organic matter and 826 gigatons as living matter. The atmosphere, the major coupling mechanism in the cycling of CO_2, holds about 702 gigatons of carbon.

Carbon cycling in the sea is nearly a closed system. The surface water acts as the site of the main exchange of carbon between atmosphere and ocean. In the surface water, carbon circulates physically by means of currents and biologically through assimilation by phytoplankton and movement through the food chain. Only about 10 percent of the carbon that comes up from the deep in upwellings is used in organic production. The other 90 percent goes back to the deep (Broecker, 1973). Carbon in the sediments is trapped for about 10^8

years, but carbon atoms in the water have a residence time of about 10^5 years, which means that in 100,000 years carbon atoms are completely replaced (see Baes et al., 1977).

Most of the carbon in the land mass is in slowly exchanging matter, about 1,456 gigatons in humus and recent peat and about 600 gigatons in large stems and roots. A small amount, 160 gigatons, is in rapidly exchanging material. Exchanges between the land mass and atmosphere are nearly in equilibrium. Photosynthesis by terrestrial vegetation removes about 113 gigatons. Plant decomposition, plant respiration, and fires return about the same amount. The world output of carbon from soil respiration in terrestrial ecosystems amounts to an estimated 75 gigatons a year. Forests are the main consumers, fixing about 36 gigatons of an estimated total world net production of 65 gigatons per year (Olson, 1970). Forests are also a major reservoir of the terrestrial organic carbon pool, containing 1,485 gigatons of the estimated total 2,216 gigatons.

Affecting the global cycle of CO_2 has been the rapid injection of carbon dioxide into the atmosphere from the burning of fossil fuels and from the clearing of forests and other land use changes. This input, which has been increasing at an exponential rate since the beginning of the industrial revolution some 100 years ago, has increased the CO_2 concentration in the atmosphere from an estimated 260 to 300 parts per million to 330. Up to 1950 two-thirds of the carbon dioxide injected came from biospheric sources and one-third from the burning of fossil fuel. Approximately one-half of the input stays in the atmosphere. The other half is removed by terrestrial plants and the oceans. Oceans are the primary carbon sink (Stuiver, 1978).

A major concern over increased CO_2 in the atmosphere is its possible effect on Earth's temperature. The spectral properties of carbon dioxide in the atmosphere are such that it tends to prevent long-wave radiation reflected from Earth from escaping to outer space and to deflect it back to Earth. Although CO_2 in the atmosphere may tend to raise the temperature of Earth, the effect may be counterbalanced by increased dust in the air, which intercepts shortwave radiation and deflects it back to outer space, thus tending to cool Earth.

THE NITROGEN CYCLE

Nitrogen is an essential constituent of protein, which is a building block of all living material. It is also the major constituent (79 percent) of the atmosphere. The paradox is that in its gaseous state, N_2, abundant though it is, is unavailable to most life. Before it can be utilized, nitrogen must be converted into some chemically usable form. Getting it into that form comprises a major part of the nitrogen cycle.

To be used, free molecular nitrogen has to be fixed. This *fixation* comes about in two ways. One is by high-energy fixation, such as cosmic radiation, meteorite trails, and lightning, which provide the high energy needed to combine nitrogen with the oxygen and hydrogen of water. The resulting ammonia and nitrates are carried to the Earth in rainwater. Estimates (Eriksson, 1952) suggest that less than 8.9 kg N/ha is brought to the Earth annually in this manner. About two-thirds of this comes as ammonia and one-third as nitric acid, H_2NO_3.

The second method of fixation is biological. This amounts to 100 to 200 kg N/ha, or roughly 90 percent of the fixed nitrogen contributed to the Earth each year. This fixation is accomplished by symbiotic bacteria living in association with leguminous and root-noduled nonleguminous plants, by free-living aerobic bacteria, and by blue-green algae.

In agricultural ecosystems the nodulated legumes of approximately 200 species are the preeminent nitrogen fixers. In nonagricultural systems some 12,000 species of plants, from free-living bacteria and blue-green algae to nodule-bearing plants, are responsible for nitrogen fixation. Also contributing to the fixation of nitrogen are free-living soil bacteria. The most prominent of the 15 known genera are the aerobic *Azotobacter* and the anaerobic *Clostridium* (see Mishustin and Shilnikova, 1969). Blue-green algae are another important group of largely nonsymbiotic nitrogen fixers. Of the some 40 known species, the most common are in the genera *Nostoc* and *Calothrix,* which are found both in soil and in aquatic habitats. Recently, certain lichens (*Collema tunaeforme* and *Peltigera rufescens*) were also implicated in nitrogen fixation

(Henriksson and Simu, 1971). Lichens with nitrogen-fixing ability possess nitrogen-fixing blue-green species as their algal component.

Another source of nitrogen is organic matter. The wastes of animals broken down by decomposition release nitrates and ammonia into the ecosystem. All of these nitrogenous products are involved in another phase of the nitrogen cycle: the processes of nitrification, denitrification, and ammonification.

In *ammonification* the amino acids are broken down by decomposer organisms to release energy. It is a one-way reaction. Ammonium, or the ammonia ion, is directly absorbed by plant roots and incorporated into amino acids, which are subsequently passed along through the food chain. Wastes and dead animal and plant tissues are broken down to amino acids by heterotrophic bacteria and fungi in soil and water.

Nitrification is a biological process in which ammonia is oxidized to nitrate and nitrite, yielding energy. Two groups of microorganisms are involved. *Nitrosomonas* utilize the ammonia in the soil as their sole source of energy because they can promote its oxidation to nitrite ions and water. Nitrite ions can be oxidized further to nitrate ions in an energy-releasing reaction. The energy left in the nitrite ion is exploited by another group of bacteria, the *Nitrobacter,* which oxidize the nitrite ion to nitrate.

Nitrates are a necessary substrate for *denitrification,* in which nitrogen in the nitrate form is reduced to the gaseous form by the denitrifiers, represented by fungi and the bacteria *Pseudomonas*. Like nitrification, denitrification takes place under certain conditions: a sufficient supply of organic matter, a limited supply of molecular oxygen, a pH range of 6 to 7, and an optimum temperature of 60°C.

With the basic and necessary processes just described, the nitrogen cycle (Figure 21.4) can be followed briefly. The sources of inputs of nitrogen under natural conditions are the fixation of atmospheric nitrogen; additions of inorganic nitrogen in rain from such sources as lightning fixation and fixed "juvenile" nitrogen from volcanic activity; ammonia absorption from the atmosphere by plants and soil; and nitrogen accretion from windblown aerosols, which contain both organic and inorganic forms of nitrogen.

In terrestrial ecosystems, nitrogen, largely in the form of ammonia or nitrates, depending on a number of variable conditions, is taken up by plants, which convert it into amino acids. The amino acids are transferred to consumers, which convert them into different types of amino acids. Eventually, their wastes (urea and excreta) and the decay of dead plant and animal tissue are broken down by bacteria and fungi into ammonia. Ammonia may be lost as gas to the atmosphere, acted upon by nitrifying bacteria, or taken up directly by plants. Nitrates may be utilized by plants, immobilized by microbes, stored in decomposing humus, or leached away. This material is carried to streams, lakes, and eventually the sea, where it is available for use in aquatic ecosystems.

In aquatic ecosystems, nitrogen is cycled in a similar manner, except that the large reservoirs contained in the soil humus are largely lacking. Life in the water contributes organic matter and dead organisms that undergo decomposition and the subsequent release of ammonia and, ultimately, nitrates.

Under natural conditions nitrogen lost from ecosystems by denitrification, volatilization, leaching, erosion, windblown aerosols, and transportation out of the system is balanced by biological fixation and other sources. Both chemically and biologically, terrestrial and aquatic ecosystems constitute a dynamic equilibrium system, in which a change in one phase affects the other.

Intrusion into the Nitrogen Cycle

Human intrusion into the nitrogen cycle has either upset the steady state or shifted the system into a new steady state. Cultivation of grasslands, for example, has caused a steady decline in the nitrogen content of the soil (Jenny, 1933). The mixing and breaking up of the soil increased the rate of decomposition and decreased new additions of organic matter. Removal of harvested crops caused additional losses. Harvesting of timber results in a heavy outflow of nitrogen from forest ecosystems, not only in the timber removed but also in short-

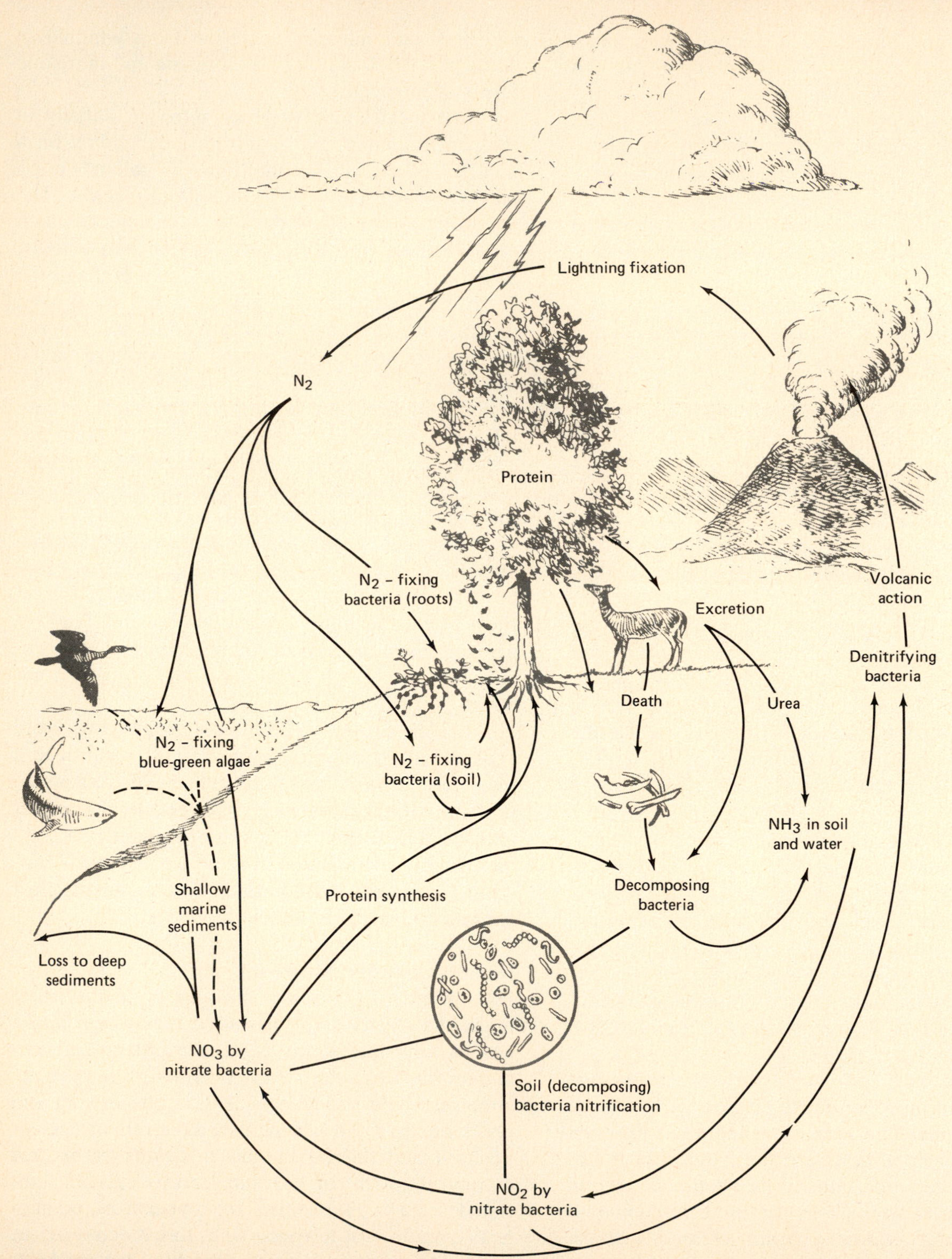

FIGURE 21.4 The nitrogen cycle in terrestrial and aquatic ecosystems.

term nitrate losses from the soil (Bormann et al., 1968). Heavy addition of commercial fertilizer increases the amount of nitrogen in cropland. Unless properly applied, a considerable portion may be lost as nitrate nitrogen to the groundwater. Additional inputs of nitrogen into the natural cycle from industrial nitrification and large-scale cultivation of nitrogen-fixing legumes may be at a higher rate than the total amount denitrified. The difference represents a nitrate buildup in soils and groundwater and the nutrient enrichment of rivers, lakes, and estuaries.

Automobile exhausts and industrial combustion add nitrous oxides, especially nitrogen dioxide (NO_2), to the atmosphere. There nitrogen dioxide is reduced by ultraviolet light to nitrogen monoxide and atomic oxygen (O). Atomic oxygen reacts with oxygen (O_2) to form ozone (O_3). Ozone reacts with nitrogen monoxide to form nitrogen and oxygen, thus closing the cycle. In the presence of sunlight, atomic oxygen from the photochemical reduction of NO_2 also reacts with a number of reactive hydrocarbons to form reactive intermediates called radicals. These radicals then take part in a series of reactions to form still more radicals, which combine with oxygen, hydrocarbons, and NO_2. As a result, nitrogen dioxide is regenerated, nitric oxide disappears, and ozone accumulates, along with a number of secondary pollutants, to produce photochemical smog. In addition, NO_2 reacts with moisture in the atmosphere to form weak nitric acid, which is carried to soil and water in precipitation (see the discussion of acid precipitation on page 417).

Sedimentary Cycles

The mineral elements required by living organisms are obtained initially from inorganic sources. Available forms occur as salts dissolved in soil water or in lakes, streams, and seas. The mineral cycle varies from one element to another, but essentially it consists of two phases: the *salt solution phase* and the *rock phase*. Mineral salts come directly from the Earth's crust through weathering. The soluble salts then enter the water cycle. With water, they move through the soil to streams and lakes and eventually reach the seas, where they remain indefinitely. Other salts are returned to the Earth's crust through sedimentation. They become incorporated into salt beds, silts, and limestones; after weathering, they again enter the cycle.

Plants and many animals satisfy their mineral requirements from mineral solutions in their environments. Other animals acquire the bulk of their minerals from the plants and animals they consume. After the death of living organisms, the minerals are returned to the soil and water through the action of organisms and the processes of decay.

There are many different kinds of sedimentary cycles. Some, such as sulfur, are something of a hybrid between the gaseous and the sedimentary because they have reservoirs not only in the Earth's crust but also in the atmosphere. Others, such as phosphorus, are wholly sedimentary—the element is released from rock and deposited in both shallow and deep sediments of the sea.

THE PHOSPHORUS CYCLE

Because phosphorus is unknown in the atmosphere and none of its known compounds has an appreciable vapor pressure, the phosphorus cycle is closed. It can follow the hydrological cycle only part of the way, from land to sea (Figure 21.5). Under undisturbed natural conditions, phosphorus is in short supply. It is freely soluble only in acid solutions and under reducing conditions. In the soil, phosphorus is held in very slightly soluble minerals that quickly establish an equilibrium between phosphorus in solution and phosphorus adsorbed on surfaces of soil particles. Much of it becomes immobilized as phosphates of either calcium or iron. Even superphosphate applied to cropland may be converted rapidly to unavailable inorganic compounds. Additional phosphorus occurs in stable but biologically inactive organic forms. Phosphorus's natural scarcity in aquatic ecosystems is emphasized by the almost explosive growth of algae in water receiving heavy discharges of phosphorus-rich wastes.

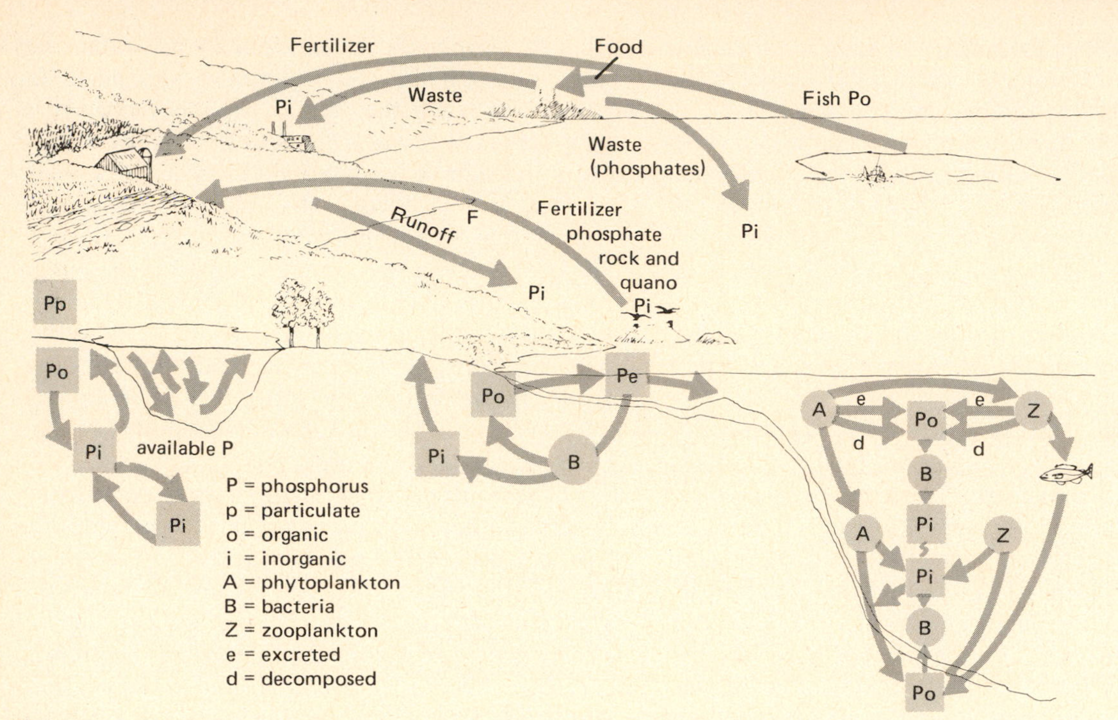

FIGURE 21.5 The phosphorus cycle in aquatic and terrestrial ecosystems.

The main reservoirs of phosphorus are rock and natural phosphate deposits, from which the elements are released by weathering, leaching, erosion, and mining for agricultural use. Some of it passes through terrestrial and aquatic ecosystems by way of plants, grazers, predators, and parasites; and it is returned to the ecosystem by excretion, death, and decay. In terrestrial ecosystems organic phosphates are reduced by bacteria to inorganic phosphates. Some are recycled to plants, some become immobilized as unavailable chemical compounds, and some are immobilized by incorporation into the bodies of microorganisms. Some of the phosphorus of terrestrial ecosystems escapes to lakes and seas, both as organic phosphates and as particulate organic matter.

In marine and freshwater ecosystems, the phosphorus cycle moves through three compartments: particulate organic phosphorus, dissolved organic phosphates, and inorganic phosphates. Inorganic phosphates are taken up rather rapidly by all forms of phytoplankton, which, in turn, may be ingested by zooplankton or detritus-feeding organisms. Zooplankton, in turn, may excrete as much phosphorus daily as is stored in its biomass (Pomeroy, Mathews, and Shik Min, 1963), keeping the cycle going. More than half of the phosphorus zooplankton excretes is inorganic phosphate, which is taken up by phytoplankton. In some instances 80 percent of this excreted phosphorus is sufficient to meet the needs of the phytoplankton population. The remainder of the phosphorus in aquatic ecosystems is in organic compounds that may be utilized by bacteria, which fail to regenerate much dissolved inorganic phosphate. Bacteria are consumed by the microbial grazers, which then excrete the phosphate they ingest (Johannes, 1968). Part of the phosphorus is deposited in shallow sediments and part in deep water. In the ocean some of the latter may be recirculated by upwelling, which brings the phosphates from the unlighted depths to the photosynthetic zones, where they are taken up by phytoplankton. Part of the phosphorus contained in the bodies of plants and animals is deposited in the shallow sediments and part in the deeper ones. As a result, the surface waters may become depleted of phosphorus and the deep waters saturated. Because phosphorus is precipitated largely as calcium compounds, much of it becomes immobilized for long periods in the bottom sediments. Upwelling returns some of it to the photosynthetic zones, where it is available to phytoplankton. The amount available is limited by the insolubility of calcium phosphate.

Intrusion into the Phosphorus Cycle

As with other biogeochemical cycles, human activities have altered the phosphorus cycle. Because the cropping of vegetation depletes the natural supply of phosphorus in the soil, phosphate fertilizers must be added. The source of phosphate fertilizer is phosphate rock. Because of the abundance of calcium, iron, and ammonium in the soil, most of the phosphate applied as fertilizer becomes immobilized as insoluble salts. In 1968, for example, 50 percent more phosphate fertilizer was added to cropland than was lost to the oceans from global runoff from all sources.

Part of the phosphorus used as fertilizer is removed in harvested crops. Transported far from the point of fixation, this phosphorus in vegetables and grain is eventually released as waste when foodstuffs are processed or consumed. Concentration of phosphorus in the wastes of food-processing plants and feedlots adds a quantity of phosphates to natural waters. Greater quantities are supplied by urban areas, where phosphates are concentrated in sewage systems. Sewage treatment is only 30 percent effective in removing phosphorus, so 70 percent remains in the effluent and is added to the waterways. In aquatic ecosystems the phosphorus is taken up rapidly by the vegetation, resulting in a great increase in biomass. Unless new input of phosphorus continues, as it would through sewage effluents, the phosphorus is lost to the sediments. Eventually, all the phosphorus mobilized by humans becomes immobilized in the soil or in the bottom sediments of ponds, lakes, and seas.

Phosphorus requirements for agricultural production depend largely on the utilization of natural deposits of phosphate rock and, to a lesser extent, on the harvest of fish and guano deposits. Because so much of the phosphate applied to soil is eventually immobilized in deep sediments and because the activity of phytoplankton seems inadequate to keep phosphorus in circulation, more of the element is being lost to the depths of the sea than is being added to terrestrial and freshwater aquatic ecosystems (Hutchinson, 1957).

THE SULFUR CYCLE

The sulfur cycle is both sedimentary and gaseous (Figure 21.6). It involves a long-term sedimentary phase in which sulfur is tied up in organic and inorganic deposits, is released by weathering and decomposition, and is carried to terrestrial and aquatic ecosystems in salt solution. The gaseous phase of the cycle permits the circulation of sulfur on a global scale.

Sulfur enters the atmosphere from several sources: the combustion of fossil fuels, volcanic eruptions, the surface of the oceans, and gases released by decomposition. It enters the atmosphere initially as hydrogen sulfide, H_2S, which quickly oxidizes into another volatile form, sulfur dioxide, SO_2. Atmospheric sulfur dioxide, soluble in water, is carried back to Earth in rainwater as weak sulfuric acid, H_2SO_4. Whatever the source, sulfur in a soluble form is taken up by plants and is incorporated through a series of metabolic processes, starting with photosynthesis, into such sulfur-bearing amino acids as cystine. From the producers, sulfur in amino acids is transferred to consumer groups.

Excretions and death carry sulfur in living material back to the soil and to the bottoms of ponds, lakes, and seas, where the organic material is acted upon by bacteria, releasing the sulfur as hydrogen sulfide or sulfate. One group, the colorless sulfur bacteria, both reduces hydrogen sulfide to elemental sulfur and oxidizes it to sulfuric acid. The green and purple bacteria, in the presence of light, utilize hydrogen sulfide as an oxgen acceptor in the photosynthetic reduction of carbon dioxide. Best-known are the purple bacteria found in salt marshes and in the mud flats of estuaries. These organisms are able to carry the oxidation of hydrogen sulfide as far as sulfate, which may be recirculated and

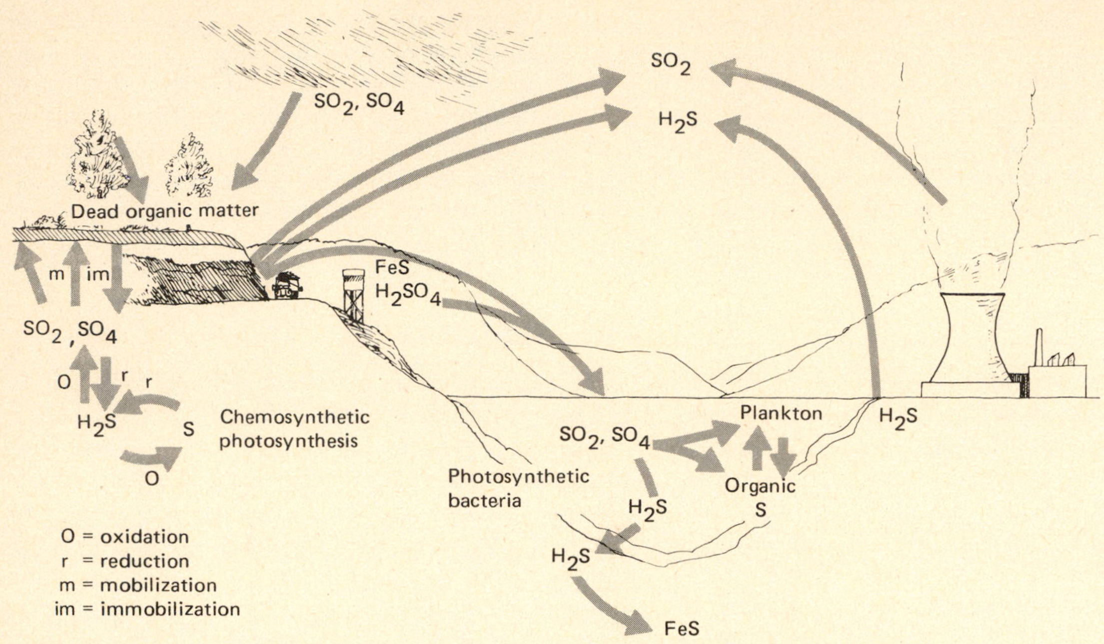

FIGURE 21.6 The sulfur cycle. Major sources are the burning of fossil fuels and acid mine waters from coal mines. O = oxidation; r = reduction; m = mobilization; im = immobilization.

taken up by the producers or used by sulfate-reducing bacteria. The green sulfur bacteria can carry reduction of hydrogen sulfide to elemental sulfur.

Sulfur, in the presence of iron and under anaerobic conditions, will precipitate as ferrous sulfide, FeS_2. This compound is highly insoluble under neutral and alkaline conditions and is firmly held in mud and wet soil. Sedimentary rocks containing ferrous sulfide (called pyritic rocks) may overlie coal deposits. Exposed to air in deep and surface mining for coal, the ferrous sulfide oxidizes and in the presence of water produces ferrous sulfate ($FeSO_4$) and sulfuric acid:

$$2FeS_2 + 7O_2 + 2H_2O \rightarrow 2FeSO_4 + H_2SO_4$$

In other reactions ferric sulfate (Fe_2SO_4) and ferrous hydroxide ($FeOH_3$) are produced:

$$12FeSO_4 + 3O_2 + 6H_2O \rightarrow 4Fe_2(SO_4)_3 + 4Fe(OH_3)$$

In this manner, sulfur in pyritic rocks, suddenly exposed to weathering by human activities, discharges heavy slugs of sulfuric acid, ferric sulfate, and ferrous hydroxide into aquatic ecosystems. These compounds destroy aquatic life and have converted hundreds of miles of streams in the eastern United States to highly acidic water.

The Global Sulfur Cycle

As with nitrogen, oxygen, and other gaseous cycles, the biosphere plays an important role in the sulfur cycle. However, the sedimentary phase makes the total cycle more complex (Figure 21.7). Sources of sulfur include the weathering of rocks, especially pyrites, erosional runoff, industrial production, and decomposition of organic matter. The bulk of sulfur appears first as a volatile gas, hydrogen sulfide. In the hydrosphere, the soil, and the atmosphere, hydrogen sulfide is oxidized to sulfides and sulfates, the forms in which sulfur is most readily circulated. The atmosphere contains sulfate particles, sulfur dioxide, and hydrogen sulfide. The latter is most abundant over continents. The concentration of sulfur as hydrogen sulfide in the un-

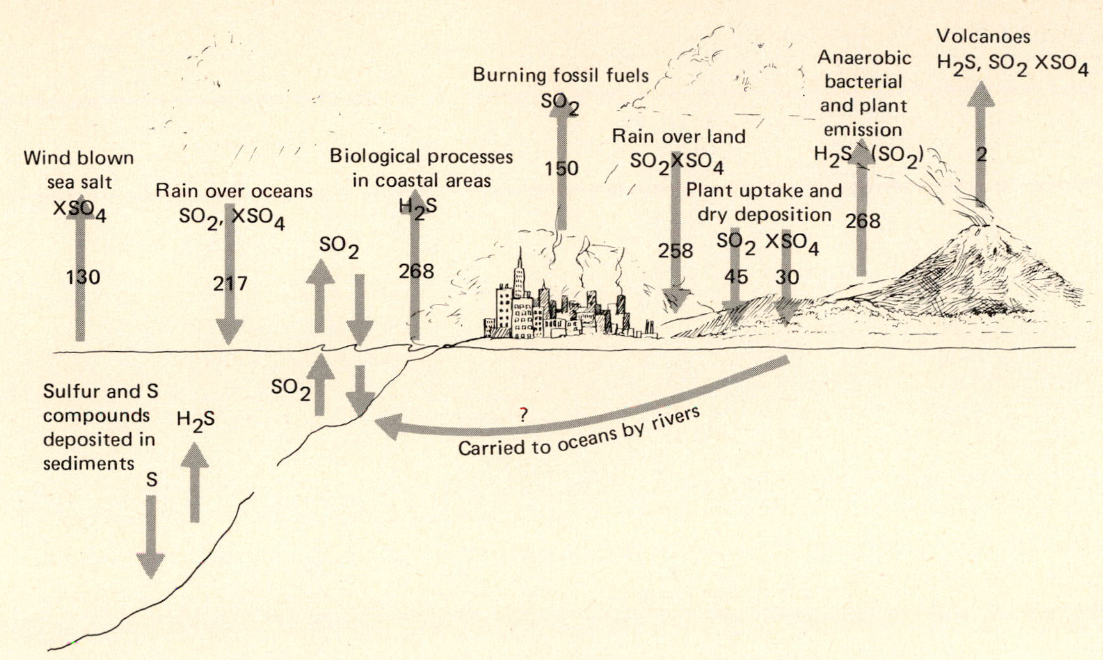

FIGURE 21.7 Sources and sinks of atmospheric sulfur compounds. Units are 10^6 tons per year, calculated as sulphate. (*After Kellogg et al., 1972:594.*)

polluted atmosphere is estimated at 6 g/m^3; as sulfur dioxide, at 1 g/m^3. Part of the sulfur in the atmosphere is recirculated to land and sea by precipitation. The concentration of sulfur dioxide in rain falling over land has been estimated as 0.6 mg/liter and over sea as 0.2 mg/liter, excluding sea spray.

It is almost impossible to estimate the biological turnover of sulfur dioxide because of the complicated cyling within the biosphere. Eriksson (1963) estimates that net annual assimilation of sulfur by marine plants is on the order of 130 million tons. Added to the anaerobic oxidation of organic matter, that brings the total to an estimated 200 million tons. Both industrially emitted sulfur and fertilizer sulfur are eventually carried to the sea. These two sources probably account for the 50-million-ton increase annually of sulfur in the ocean. The balance sheet of sulfur in the global cycle is summarized in Table 21.1.

Intrusion into the Sulfur Cycle

Heavy industry and electrical power generating plants burn tremendous amounts of fossil fuel. A by-product of this combustion, in addition to particulate matter, is sulfur dioxide. Globally, some 147 million tons of sulfur dioxide are poured into the atmosphere each year, but with regional rather than global effects. Once in the atmosphere, sulfur dioxide does not remain in the gaseous state. It reacts with moisture to form sulfuric acid.

Sulfuric acid in the atmosphere has a number of effects. In concentrations of a few parts per million, it is irritating to the respiratory tract. High concentrations of SO_2 have been implicated in such pollution disasters as the one in Donora, Pennsylvania, in 1938 and in New York City in the 1960s. Plants exposed to such atmospheres are killed outright or injured by acidic aerosols during periods of foggy weather, light rain, or high relative humidity.

Emission of sulfur dioxide as well as nitrogen dioxides into the atmosphere is strongly implicated in the *acid rain* problem (Figure 21.8). Acid precipitation involves large regions of the eastern United States and Canada, rural and urban areas of the western United States, Europe, and Japan. The problem results in part from the construction of tall

TABLE 21.1 BUDGET OF SULFUR IN NATURE (10^6 TN/YR)

	LITHOSPHERE							
	ATMOSPHERE		(SEDIMENT ROCKS)		PEDOSPHERE		(OCEANS)	
Item	To	From	To	From	To	From	To	From
River discharge						80	80	
Weathering				15	15			
Fertilizers				10	10			
Precipitation		165			65		100	
Sea spray	45							45
Dry deposition		200			100		100	
Sedimentation			15					15
Industrial	40			40				
Increase in sea			50*					50
Balance								
soils—atmosphere	110					110		
oceans—atmosphere	170							170
Total	365	365	65	65	190	190	280	280
Specification								
as SO_2 sulfur	45	165			90	80	180	95
as SO_2 sulfur	40	200			100		100	
as H_2S sulfur	280					110		170
as other forms of sulfur			65	65				15

* For the balance this has to be treated as an item borrowed by the ocean from the lithosphere.
SOURCE: E. Erikisson, 1963:4005. Copyright by American Geophysical Union.

stacks at power plants and industrial complexes, so built to decrease ground-level concentrations of SO_2. The outcome has been the release of SO_2 into the upper atmosphere, where it is carried by upper-level winds to points far removed from the source. The concentration of SO_2 is further aggravated by the removal of alkaline particulate matter, fly ash, from the stacks by scrubbers to eliminate local particulate fallout.

Precipitation relatively free from contamination, both human and natural, has a pH of around 5.7. But when SO_2 and NO_x are injected into the atmosphere, they are converted into strong acids, resulting in precipitation with a pH of 4.0 or lower. In the western United States, the ratio of SO_2 to NO_x is 1:1; but in the eastern United States, the ratio is 2:1.

The effect of acid rain is a subject of considerable controversy, especially political. Ecologically, the effect is more clear-cut. The impact of acid precipitation—which includes rain, snow, fog, dew, and dry deposition—is greatest on those parts of Earth's surface that are poorly buffered: largely noncalcareous granitic bedrock. Such areas are characteristic of the eastern United States, Canada, and northern Europe. There the acid problem is most pronounced. In those regions terrestrial ecosystems are nutrient-poor and their soils acidic. Aquatic ecosystems, reflecting the terrestrial basins or watersheds in which they lie, likewise are nutrient-poor and slightly acidic. Such lakes and streams have a pH of about 6.0. Already on the border of being too acid for aquatic vertebrate and invertebrate life, such streams are highly sensitive to any flux of acids entering them and quickly become acidic.

The effect of acid precipitation varies with the ecosystem receiving it. Over short time periods terrestrial ecosystems show little effect. Many of the soils on which acid precipitation falls are already acid and the vegetation adapted to acidic conditions. In fact, the addition of S and N to such soil

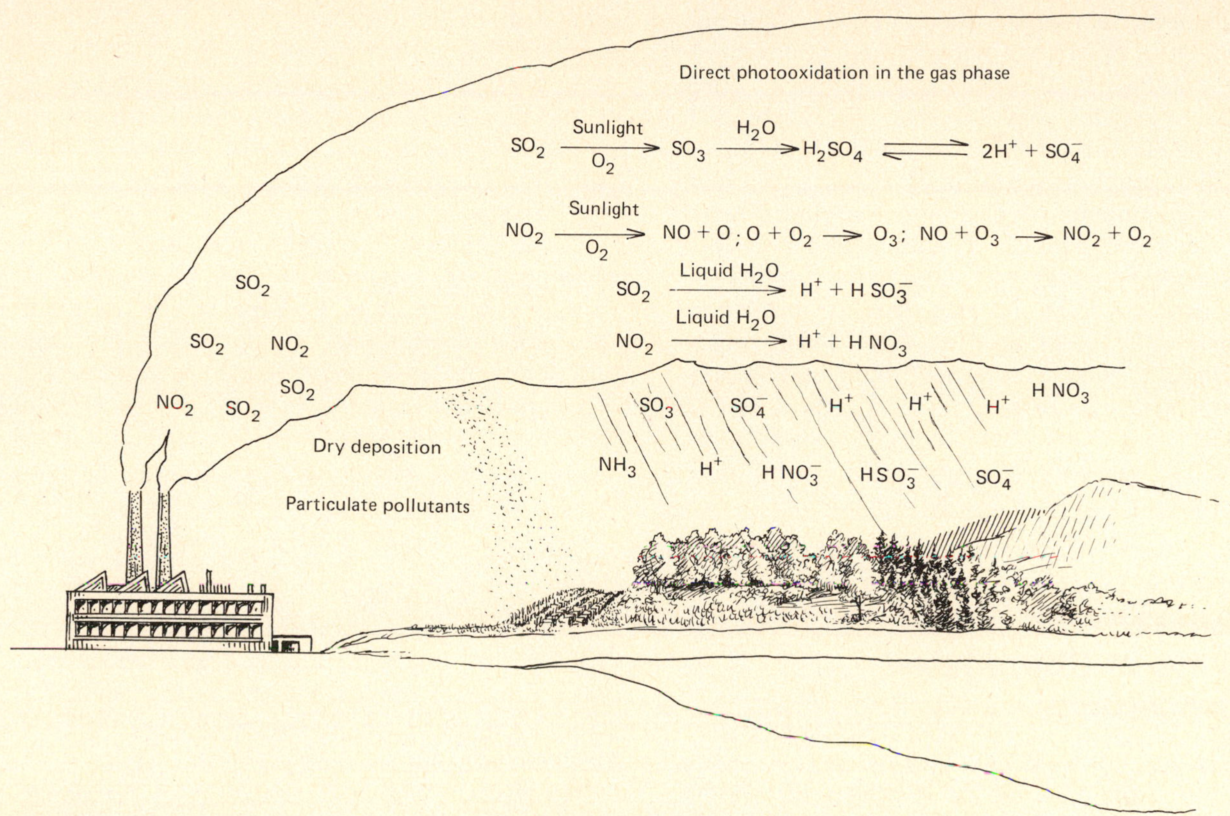

FIGURE 21.8 The formation of sulfuric and sulfurous acids from sulfur oxide pollution, which comes to aquatic and terrestrial ecosystems as dry deposition and acid rain.

may have a beneficial fertilizing effect. On the other hand, such inputs can increase the leaching of Ca, Mg, and K from the soil. The rate of leaching may outstrip replacement by weathering, resulting in nutrient deficiency. Acid precipitation can reduce the availability of P, the rate of nitrogen fixation, and the breakdown of organic matter. At the same time, low pH can mobilize such toxic heavy metals as aluminum and manganese as they are replaced on the soil particles by hydrogen ions.

Acidic leaching of soils can increase the nutrient level of streams and lakes in a watershed, provided the soil has appreciable reserves of calcium. As acid rain percolates through the soil, it becomes neutralized while releasing basic ions. But such enrichment is often canceled by snowmelt and spring rainwater flowing over the surface of frozen ground, following old root channels and animal burrows into the receiving water. Such waters then become acid in spite of the buffering effects of the soil. Through time such soils become more acidic as exchange sites become highly saturated with hydrogen. Substantial amounts of acid will then enter the receiving water.

In the water, sulfate ions replace bicarbonate ions, and solutions once dominated by Ca^{++} and HCO_3^- become dominated by Ca^{++} and $SO_4^=$, pH declines, and the concentration of metallic ions increases. When the pH of the groundwater in the surrounding watershed is 5 or lower, high concentrations of aluminum ions are carried to lakes and streams. In aquatic systems aluminum tends to precipitate the dark humics, increasing transparency of the water, which in itself could stimulate production of phytoplankton. But at low pH, aluminum also tends to precipitate phosphorus, a nutrient already limiting. While adult fish and some other aquatic organisms may be able to tolerate high acidity per se, the combination of Al and H is highly toxic. The combination of high acidity and a high level of aluminum, a typical situation during snow melt, can cause fish kills. The filling of depressions and temporary ponds with such water inhibits the reproduction of frogs and salamanders. Acid waters are toxic to invertebrates also, either killing organisms directly or interfering with their calcium metabolism, causing crustaceans to lose the ability to recalcify their shells after moulting. As acidity increases, desmids and blue-green algae seem to be replaced by green algae.

Cycling of Chlorinated Hydrocarbons

Of all human intrusions into biogeochemical cycles, none has done more to call attention to nutrient cycling than the widespread application of pesticides, especially chlorinated hydrocarbons, and herbicides as well as the leakage of the industrial chemical PCB into the environment. The substance that has been most responsible is DDT.

As early as 1946 Clarence Cottam called attention to the damaging effects of DDT on ecosystems and nontarget species. But the impact of pesticides remained obscure until Rachel Carson exposed the dangers of hydrocarbons in her book *Silent Spring*. The detection of DDT in tissues of animals in the Antarctic, far removed from any applied source of the insecticide, emphasizes the fact that chlorinated hydrocarbons do indeed enter the global biogeochemical cycle and that they become dispersed around the Earth.

DDT (and other chorinated hydrocarbons) has certain characteristics that make this global circulation possible (see Figure 21.9). It is highly soluble in lipids or fats and poorly soluble in water and therefore tends to accumulate in the lipids of plants and animals. It is persistent and stable. It undergoes some degradation, largely from DDT to DDE, and has a half-life of approximately 20 years. It has a vapor pressure high enough to ensure direct losses from plants. It can become adsorbed to particles or remain as a vapor and, in either state, be transported by atmospheric circulation. It can return to land and sea on rainwater.

Insecticides are applied on a large scale by aerial spraying. One-half or more of the toxicant applied in this manner is dispersed to the atmosphere and never reaches the ground. If the vegetative cover is dense, only about 20 percent reaches the ground. For example, in a massive spraying of DDT over 66,000 acres of forest land in eastern Oregon in order to control an outbreak of Douglas fir tussock moth, only about 26 percent of the intended DDT application reached the forest floor. In fact, as little as 1 percent of an aerially applied pesticide may actually hit target organisms.

On the ground or on the water's surface, the pesticide is subject to further dispersion. Apparently, little DDT moves from the surface soil to the subsoil. Pesticides reaching the soil are lost through volatilization, chemical degradation, bacterial decomposition, runoff, and the harvest of organic matter, which can amount to around 1 percent of the total DDT used on the crop.

DDT sprayed on forest and cropland enters streams and lakes, where it is subject to further distribution and dilution as it moves downstream. Insecticides released in oil solutions penetrate to the bottom and cause mortality of fish and aquatic invertebrates (see reviews in Pimentel, 1971; Cope, 1971; Johnson, 1968). Trapped in the bottom rubble and mud, the insecticide may continue to recirculate locally and kill for some days. In lakes and ponds, emulsifiable forms of DDT tend to disperse through the water, but not necessarily in a uniform way. DDT in oil solutions tends to float on the surface and is moved about by the wind. Eventually, pesticides reach the ocean, where they may concentrate in surface slicks. These slicks, which attract plankton, are carried across the seas by ocean currents. In the ocean, part of the DDT residue may circulate in the mixed layer. Some of it may be transferred to deep waters, and more may be lost through the sedimentation of organic matter.

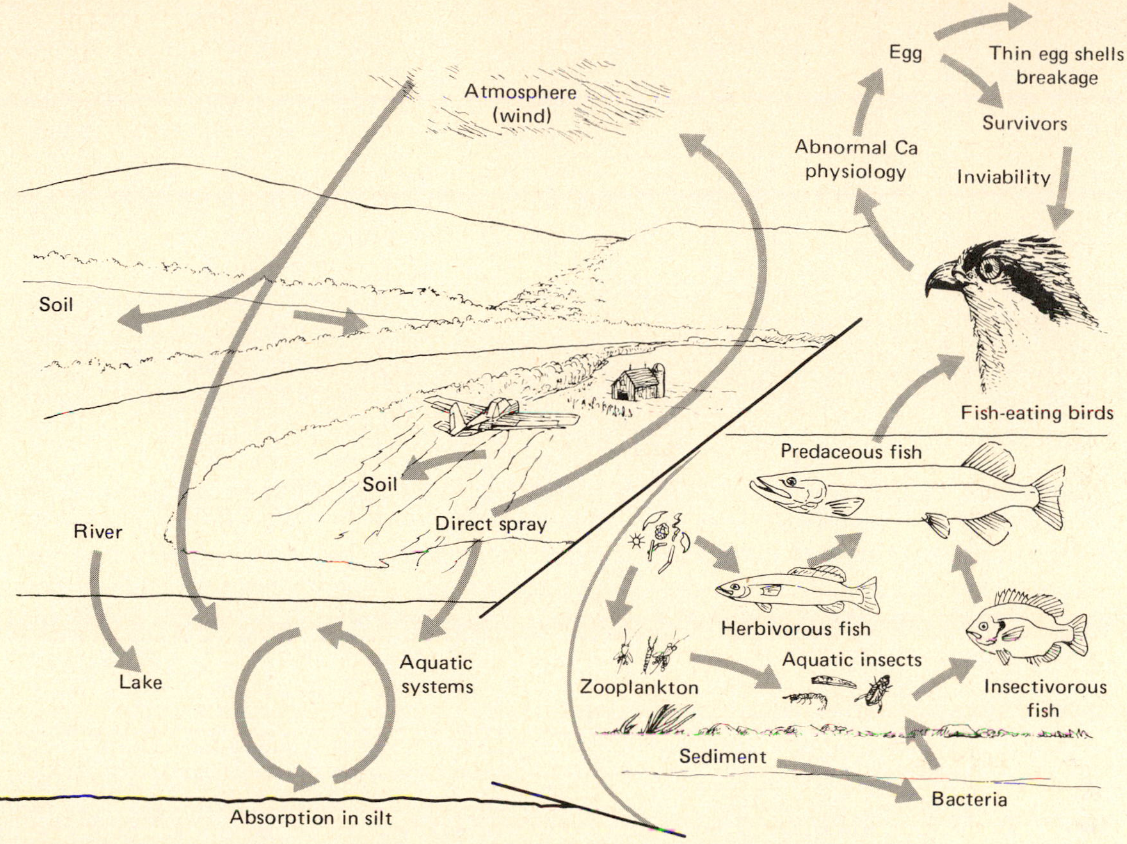

FIGURE 21.9 A chlorinated hydrocarbon cycle. The initial input comes from spraying on croplands and forests. A relatively large portion fails to reach the ground and is carried on water droplets and particulate matter through the atmosphere.

Although considerable amounts of DDT and other pesticides are transported by water, the major movement of pesticide residue takes place in the atmosphere. Not only does the atmosphere receive the bulk of the pesticidal sprays, but it also picks up the fraction volatized from soil, water, and vegetation. If DDT remained as a vapor alone, the saturation capacity of the atmosphere would hold as much DDT as has been produced to date. But the capacity of the atmosphere to hold DDT is increased greatly by the adsorption of residues to particulate matter. Thus, the atmosphere becomes a large circulating reservoir of DDT and other chlorinated hydrocarbons. Residues are removed from the atmosphere by chemical degradation, diffusion across the air-sea interface, and (mostly) by rainfall and dry fallout (Study of Critical Environmental Problems, 1970).

Although the quantity of residues of DDT and other chlorinated hydrocarbons may be relatively small, amounting to one-thirtieth or less of the amount produced each year, the concentrations are still sufficient to have a deleterious impact on marine, terrestrial, and freshwater ecosystems. DDT and related compounds tend to concentrate in the lipids of living organisms, where they undergo little degradation (see Menzie, 1969; Bitman, 1970; Peakall, 1970).

The high solubility of DDT in lipids leads to its concentration through the food chain. While only a portion of the food is ingested by consumer organisms, most of the DDT contained in the food is retained in the fatty tissues of the consumer. Because it breaks down slowly, DDT accumulates to high and even toxic levels. DDT so concentrated is passed on to the next trophic level. The carnivores on the top level of the food chain receive massive amounts of pesticides (Woodwell, Wurster, and Isaacson, 1967; Rudd and Genelly, 1956).

High concentrations of DDT in their tissues often result in death or impaired reproduction and genetic constitution of organisms. A residue level of 5 parts per million in the lipid tissues of the ovaries of freshwater trout causes 100 percent die-off of the fry, which pick up lethal doses as they utilize the yolk sac. High levels of DDT are correlated with the decline of such fish as sea trout and California mackerel. DDT and its degraded products interfere with calcium metabolism in birds. Chlorinated hydrocarbons block ion transport by inhibiting the enzyme ATPase, which makes available the needed energy. This reduces the transport of ionic calcium across membranes and can cause the death of organisms. DDT also inhibits the enzyme carbonic anhydrase (Bitman, 1970; Peakall, 1970), which is essential for the deposition of calcium carbonate in the eggshell and the maintenance of a pH gradient across the membrane of the shell gland.

The prohibition of DDT in the United States and most Western nations in the early 1970s has resulted in the slow recovery of such raptors as bald eagles and ospreys and other fish-eating birds such as pelicans, now that these birds lay normal-shelled eggs again. The ecological problem of pesticides has not lessened, however. More than 500,000 tons of pesticides and herbicides are produced annually and used in agroecosystems to control insect pests and weeds. The impact of these toxicants extends beyond the boundary of agricultural fields.

Most studies have emphasized the effects of pesticides on target and nontarget species and environmental contamination. Largely unknown are the effects of pesticides on whole ecosystems: nutrient cycling, energy flow, food chains, genetics of populations, and species diversity. Pesticides that greatly reduce or eliminate a target species will reduce associated species as well, including predatory populations. With their predators reduced, some species held to a level of relative rarity by predation suddenly increase, becoming pests themselves (Edwards and Thompson, 1973; Edwards et al., 1979). By destroying populations of certain insects or plants, pesticides and herbicides reduce species diversity. Ecosystems respond by replacing one individual species with another. The reduction of grazing zooplankton in certain aquatic ecosystems by pesticides resulted in an increase in phytoplankton populations (Hurlbutt, 1975). The use of herbicides on aquatic floating and emergent vegetation stimulated an increase in mats of blue-green algae. Reduction or loss of insect populations from pesticides also decreased bird and mammal populations dependent upon them (Barrell and Darnell, 1967). Spraying of endrin on crop fields can cause widespread death of passerine birds inhabiting them.

Herbicides and pesticides can interfere with the cycling of nutrients. Chemicals toxic to earthworms and other decomposers can decrease their populations and adversely affect decomposition and in turn reduce soil fertility. Pesticides can further influence nutrient cycling by altering the chemical composition of plants. Certain pesticides can increase the amount of nutrients such as N, P, Ca, and Mg in plants. These changes can influence insects feeding on them. For example, the application of the herbicide 2-4D to a cornfield increased the nitrogen content of the corn. Its improved nutritional quality resulted in a threefold increase in corn leaf aphid populations. Corn borer females were one-third larger than average and laid one-third more eggs (Oka and Pimentel, 1974).

Nutrient Flow

A key element in nutrient cycling is the food base, mostly vegetation, that takes up inorganic nutrients. The amount of nutrients moved through the system is influenced by nutrient availability to plants in soil and water. The quantity of nutrients available depends upon inputs to the detrital pool, the rate at which detritus is decomposed, and the amount of nutrients that go into biomass storage and into soil and humus and detrital sediments and the release of nutrients from that pool (Figure 21.10).

How nutrients are utilized in the ecosystem depends upon the size of the abiotic reserve, the proportion of nutrients stored in the abiotic and biotic components, the movement among them, and the rate of turnover in the recycling pool (see Box 21.1). Turnover depends upon the speed at which nutrients are released from the detrital pool to an available form, leaching, decomposition, and subsequent uptake.

EUTROPHY AND OLIGOTROPHY

Based on their nutrient pools and nutrient availability, ecosystems, both terrestrial and aquatic, can be considered as somewhere on a nutrient gradient extending from oligotrophy to eutrophy. The terms *eutrophy* (from the Greek *eutrophoz,* meaning "well nourished") and *oligotrophy* (meaning "poorly nourished") were introduced by the German limnologist C. A. Weber in 1907. He applied the terms to the development of peat bogs. Later, E. Naumann related the terms to phytoplankton production in lakes. Eutrophic lakes, found in fertile lowlands, held high populations of phytoplankton. Oligotrophic lakes, common to regions of primary rocks, held little phytoplankton.

Until recently, these terms were used only in the context of aquatic ecosystems, and open-water systems at that, even though the historical use of the words implied that eutrophic and oligotrophic bodies of water were found in nutrient-rich and nutrient-poor watersheds.

Oligotrophic ecosystems are characterized by a low content of nutrients, especially nitrogen and phosphorus. If nitrogen is abundant, phosphorus alone may be limiting. Low nutrient availability in terrestrial ecosystems results from low inputs from the external environment through weathering and

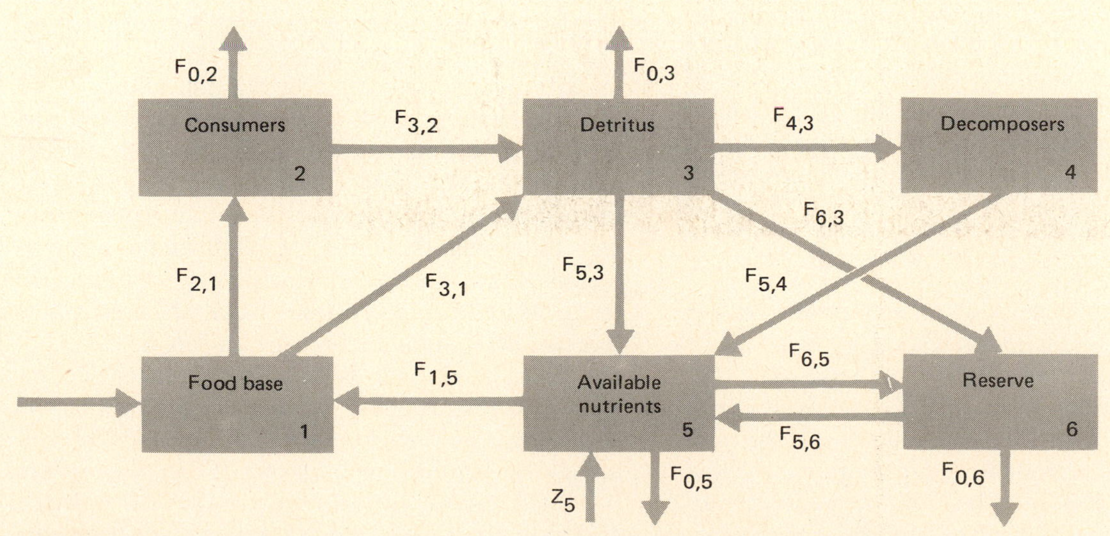

FIGURE 21.10 Model of nutrient flow in an ecosystem. Arrows indicate direction of flow of nutrients between compartments represented by numbered blocks. The subscripts 0, 1, 2, and so on represent compartment numbers. Thus, $F_{3,1}$ indicates flow of nutrients from compartment 1 to compartment 3; F_0 indicates outflow to the environment; and $F_{0,3}$ is outflow to the environment from compartment 3. (*From Webster, Warde, and Patten, 1975:13.*)

Box 21.1
Nutrient Turnover

You will come across several terms in the literature that relate to nutrient flow through ecosystems: *turnover rate, turnover time,* and *residence time*. The terms are related:

Turnover Rate. That fraction of the total amount of a substance in a component which is released or enters an ecosystem in a given length of time. For example, assume that 692 kg N/ha is held in woody and understory vegetation. Annually, 122 kg N/ha is returned as litterfall and throughfall. The turnover rate of N would be 122/692, or 17.6 percent.

Turnover Time. The time required to replace a substance equal to the amount in the component. It is the reciprocal of turnover rate. To give an example using the data from above: 692/122, or 5.7 years, would be required to turn over the amount of N found in woody and herbaceous vegetation.

Residence Time. The time a given amount of a substance remains in a designated compartment of a system. Common to geochemical literature, it is similar to turnover time. For example, carbon atoms in ocean water have an estimated residence time of 10^5 years. That means that in 100,000 years carbon atoms are completely replaced.

precipitation. Such ecosystems are usually on nutrient-poor geological material such as sandstone or granite, or they are on deeply weathered soils that prevent plants from reaching nutrient-rich unweathered parent material. The outcome of low nutrient input is low production of organic matter and a low rate of decomposition. Most of the nutrients are stored above ground in plant biomass and organic litter rather than below ground in the soil. Because of the close relationship between aquatic and terrestrial ecosystems, oligotrophic watersheds tend to possess oligotrophic aquatic ecosystems. Also low in available nutrients, these tend to support low phytoplankton populations but do possess well-developed floating and emergent vegetation rooted in bottom sediments.

Eutrophic ecosystems are characterized by a high nutrient content. Terrestrial systems have a greater proportion of organic matter and nutrients stored in soil. And plants have a rich supply of nutrients available from weathering of soil material. Eutrophic aquatic systems have an abundance of nutrients, especially nitrogen and phosphorus, which stimulate a heavy growth of phytoplankton and blue-green algae and aquatic plants. Increased photosynthetic production leads to an increased regeneration of nutrients and organic compounds through rapid decomposition and further growth.

Visual differences between aquatic eutrophic and oligotrophic ecosystems may be striking. In a very general way, oligotrophic systems tend to be clear and eutrophic systems green with phytoplankton growth. Highly eutrophic systems may have heavy filamentous mats of algae. Differences in net production are also great. On the average, oligotrophic systems have a net productivity of 25 g C/m^2/year and eutrophic ones 75 g C/m^2/year (Westlake, 1980).

No striking differences exist between terrestrial oligotrophic and eutrophic ecosystems. Net primary production, rates of litterfall within broad climatic regions, and biomass are all about the same. In the absence of disturbance, nutrient cycling of the two differs only in the proportion of organic matter and nutrients stored in mineral soil. One difference is humus. In oligotrophic systems humus makes up about 30 percent of total organic matter. In eutrophic systems (such as sugar maple forests and montane tropical rain forests), the humus layer is minimal. But major differences do appear with disturbance. Losses of nutrients to the atmosphere and leaching upon disturbance are relatively low in eutrophic systems. The reasons are the high storage of nutrients in the soil and the ability of the soil to absorb most of the nutrients lost to leaching. Responses to disturbance are mostly influenced by microclimatic changes such as an increase in moisture and light. Disturbance in oligotrophic systems results in leaching and the loss of nutrients to the atmosphere. In response to this tendency for nutrient leakage, species of terrestrial oligotrophic systems have evolved mechanisms to reduce nutrient losses.

MECHANISMS FOR NUTRIENT CONSERVATION

One of the most prominent mechanisms to reduce nutrient losses is a mat of fine roots, humus, and mycorrhizal fungi near the surface. Such mats are best developed in oligotrophic tropical rain forests, but they are also prominent in coniferous forests of temperate regions. These mats physically absorb nutrients entering them almost immediately, with negligible leaching (Stark and Jordan, 1978). Mycorrhizal fungi then take up nutrients from decomposing litter and transfer them to roots. Acidity and a high tannin content of the litter inhibit bacteria that could consume the limited nitrogen available.

Some species of oligotrophic systems withdraw large amounts of nutrients from the soil and accumulate them in biomass. A portion of the nutrients stored in the tree is returned to the forest floor by way of litterfall, stemflow, and leaf wash (Cole, Gessell, and Dice, 1968).

Although regarded as an adaptation to water-stressed environments, evergreenness is another nutrient-conserving mechanism. Thick cuticle, waxy coating, and low water content of leaves make them more resistant to herbivores and parasites, reducing nutrient loss to consumers. The same features also enable them to resist nutrient leaching by rainfall. Year round rather than seasonal leaf fall allows the trees to return nutrients slowly to the soil, in contrast to a rapid return by deciduous trees.

Trees of oligotrophic systems scavenge nutrients from rainfall indirectly through microcommunities of algae and lichens colonizing canopy leaves in tropical rain forests (Jordan et al., 1979), balsam fir forests (Lang, Reiners, and Heier, 1976), and western coniferous forests (Johnson et al., 1982). Algae and lichens adsorb nutrients on their surface. These nutrients become available for uptake when the leaves die and decompose (Figure 21.11).

Such nutrient-conserving mechanisms as evergreenness and root mats are more or less associated with oligotrophic ecosystems. Because of nutrient limitations, they provide some control over the rate of nutrient flow to prevent overutilization. These controls involve at least two mechanisms of nutrient utilization relative to energy flow.

One mechanism is to divert energy production to plant populations with low biomass and with different responses to environmental conditions. Such plant populations reproduce rapidly when conditions are optimal and take up nutrients quickly. A succession of such plant populations through the seasons ensures continuous energy fixation and utilization of nutrients. A second mechanism puts energy conversion in individuals of great bulk and slow reproductive rates. Large individuals are able to survive unfavorable conditions and store large quantities of nutrients in biomass. Although the first mechanism is most typical of aquatic situations and the second of terrestrial ecosystems, especially forests, neither mechanism operates exclusively in only one type of ecosystem. Lakes have rooted aquatics with biomass accumulation, and forests have a seasonal parade of herbaceous understory plants.

Each mechanism has provisions for the conservation of nutrients. One provision is a pool of organic matter. Litter, soil organic matter, and biomass of autotrophs form such a pool in terrestrial ecosystems. In aquatic ecosystems the pool is particulate organic matter.

Turnover of the organic pool is relatively slow, roughly one magnitude lower than the turnover of the vegetative component (O'Neill et al., 1975). Organic matter has a key role in recycling nutrients because it prevents rapid losses from the system. Large quantities of nutrients are bound tightly in organic matter and are not readily available, but their release can be activated by decomposer organisms. If energy, and especially water, are limited, however, as in desert ecosystems, then recycling is minimal because the ecosystem cannot attain a standing crop large enough to deplete the nutrient supply.

Another mechanism is to partition the nutrient reserve between long-term and short-term nutrient

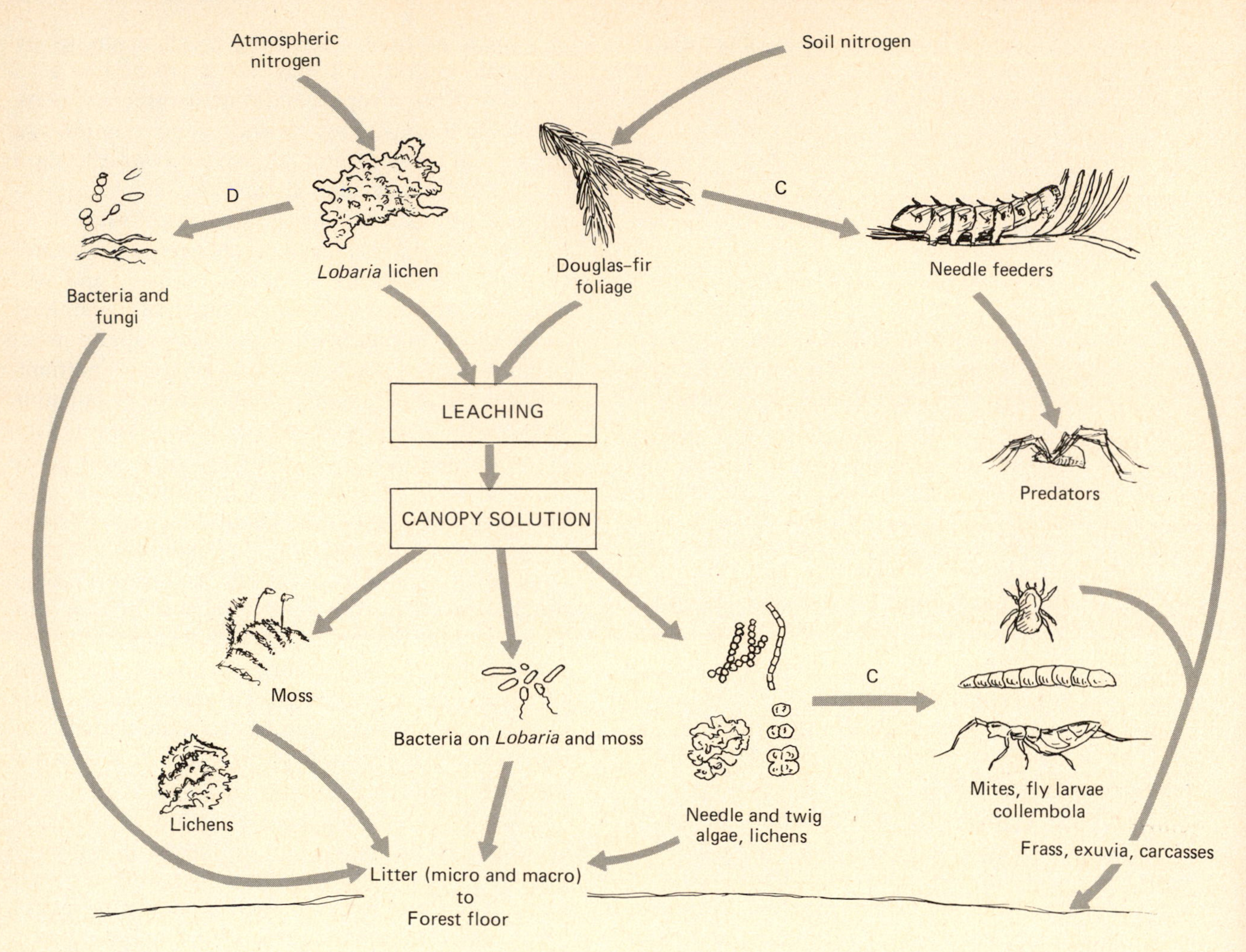

FIGURE 21.11 Nitrogen cycle in the canopy of old-growth Douglas fir. Microecosystems exist in the high canopy of old Douglas fir and probably many other forest trees as well. They consist of primary producers (lichens) and biophage and saprophage consumers. These microecosystems conserve and recycle nutrients such as nitrogen themselves and influence nutrient return to the forest floor by way of leaching and throughfall. (*After Johnson et al., 1982:193.*)

pools. For example, of the structural components of individual plants—wood, bark, twigs, and leaves—leaves are recycled the fastest and wood the slowest. The leaves represent a short-term nutrient pool and wood a long-term reservoir. Thus, in a mature forest, nutrients stored in vegetation are recycled at various time intervals from 1 to 100 or more years. Such partitioning prevents excessive losses of nutrients and releases nutrients slowly to biogeochemical cycles.

Nutrient cycling within forest ecosystems in particular can be influenced by foliage-eating insects such as fall cankerworm *(Alsophila pometaria)* and saddled prominent *(Heterocampa guttivita)* (Swank et al., 1981). Chronic defoliation by such insects results in an increased export of NO_3-nitrogen from the forest to stream ecosystems. Accompanying this export are a number of other changes in ecosystem function. Production of leaves increases and production of wood decreases as energy and nutrients are invested in new foliage. There are large inputs of nutrients to the forest floor through frass (fecal material of caterpillars) and increased litterfall. That results in increased microbial activity, including nitrifying bacteria and associated litter metabolism. The outcome is a significant increase in available nutrients (including mineral N) in soil and litter. Ecosystem responses involve a temporary shift from wood to leaf production, an increased rate of nutrient uptake, and an increased turnover rate of nutrients in litter and soil. In such a manner, forest insects also become a mechanism in the regulation of nutrient cycles.

Nutrient Budgets

Nutrients are constantly being removed or added by natural and artificial processes (see Figure 21.12). In woodland, shrub, and grassland ecosystems, nutrients are returned annually to the soil by leaves, litter, roots, animal excreta, and the bodies of the dead. Released to the soil by decomposition, these nutrients again are taken up, first by plants and then by animals. In freshwater and marine ecosystems, the remains of plants and animals drift to the bottom, where decomposition takes place. The nutrients again are recirculated to the upper layers by the annual overturns and by upwellings from the deep.

The cycle, however, is not a closed circuit within an ecosystem. Nutrients are continuously being imported into as well as carried out of any ecosystem. Appreciable quantities of plant nutrients are carried in by rain and snow (Emanuelsson, Eriksson, and Egner, 1954) and by aerosols (White and Turner, 1970). In Western Europe, at least, the weight of nutrients supplied by these sources is roughly equivalent to the quantity removed in timber harvest (Neuwirth, 1957).

A small quantity of the nutrients carried to the forest by rain and snow is absorbed directly through the leaves, but this hardly offsets the quantity leached out. Rainwater dripping down from the canopy is richer in calcium, sodium, potassium, phosphorus, iron, manganese, and silica than rainwater collected in the open at the same time, although less rainwater reaches the forest floor (Tamm, 1951; Madgwick and Ovington, 1959). Throughfall of rain in an oak woodland accounted for 17 percent of the nitrogen, 37 percent of the phosphorus, 72 percent of the potassium, and 97 percent of the sodium added by the canopy to the soil. The remainder was supplied by fallen leaves (Carlisle, Brown, and White, 1966). The nutrients leached from the foliage are taken up in time by the surface roots and translocated to the canopy. Such localized nutrient cycles may require only a few days for completion.

For some elements the amount carried in by aerosols may exceed that carried in by rain. Estimates of the annual income of nutrients to an English mixed deciduous woodland by airborne particles were 125.2 kg Na/ha, 6.3 kg K/ha, 4.2 kg Ca/ha, 16.2 kg Mg/ha, and 0.34 kg P/ha. The income from aerosols was greater for elements known to occur in droplets, such as sodium, potassium, and magnesium. The income of calcium, associated with terrestrial sources, and phosphorus, associated with biological activity, was greater in rainfall (White and Turner, 1970).

These little nutrient cycles can be followed by means of radioactive tracers. By inoculating white oak trees with 20 microcuries of ^{134}Cs (cesium 134), Witherspoon, Auerbach, and Olson (1962) were able to follow the gains, losses, and transfers of this radioisotope. About 40 percent of the ^{134}Cs inoculated into the oaks in April moved into the leaves by early June (see Figure 21.13). When the first rains fell after inoculation, leaching of radiocesium from the leaves began. By September this loss amounted to 15 percent of the maximum concentration in the leaves. Seventy percent of this rainwater loss reached mineral soil; the remaining 30 percent found its way into the litter and understory. When the leaves fell in autumn, they carried with

Nutrients in
precipitation
Nutrients in
windblown dust
Nutrients in
wood
harvest
Litter fall and
leaching of
nutrients
Nutrients in
wildlife
harvest
Release of nutrients
by weathering and
root decomposition
Nutrient loss through
runoff and erosion

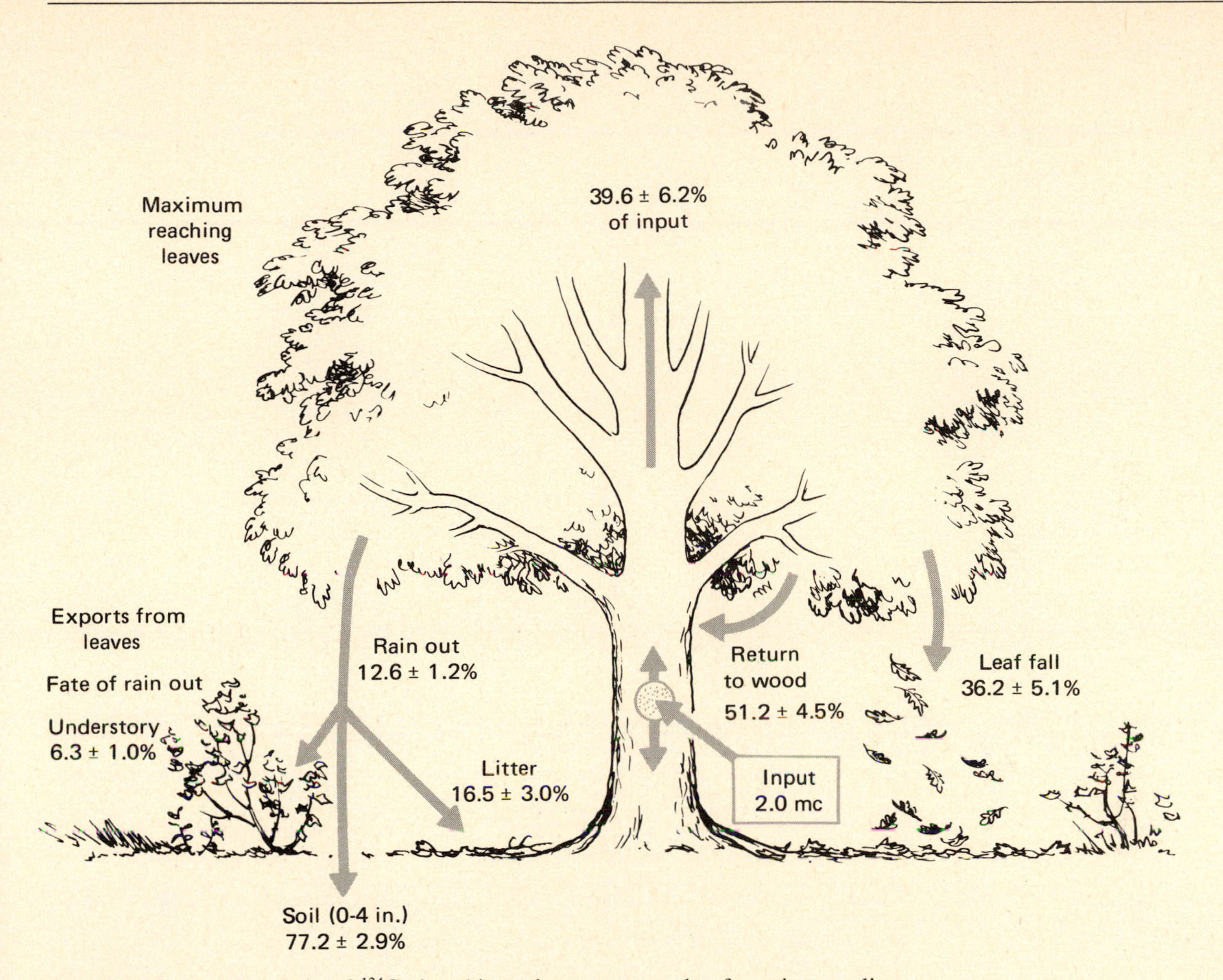

FIGURE 21.13 The cycle of ^{134}Cs in white oak as an example of nutrient cycling through plants. The figures are the average for 12 trees at the end of the 1960 growing season. (*After Witherspoon, Auerbach, and Olson, 1962; courtesy of Oak Ridge National Laboratory.*)

FIGURE 21.12 Generalized nutrient budget in a forest ecosystem. Input of nutrients is through precipitation, windblown dust, litterfall, release of nutrients by weathering, and root decomposition. Outgo is through wood harvest, wildlife harvest, runoff, erosion, and leaching.

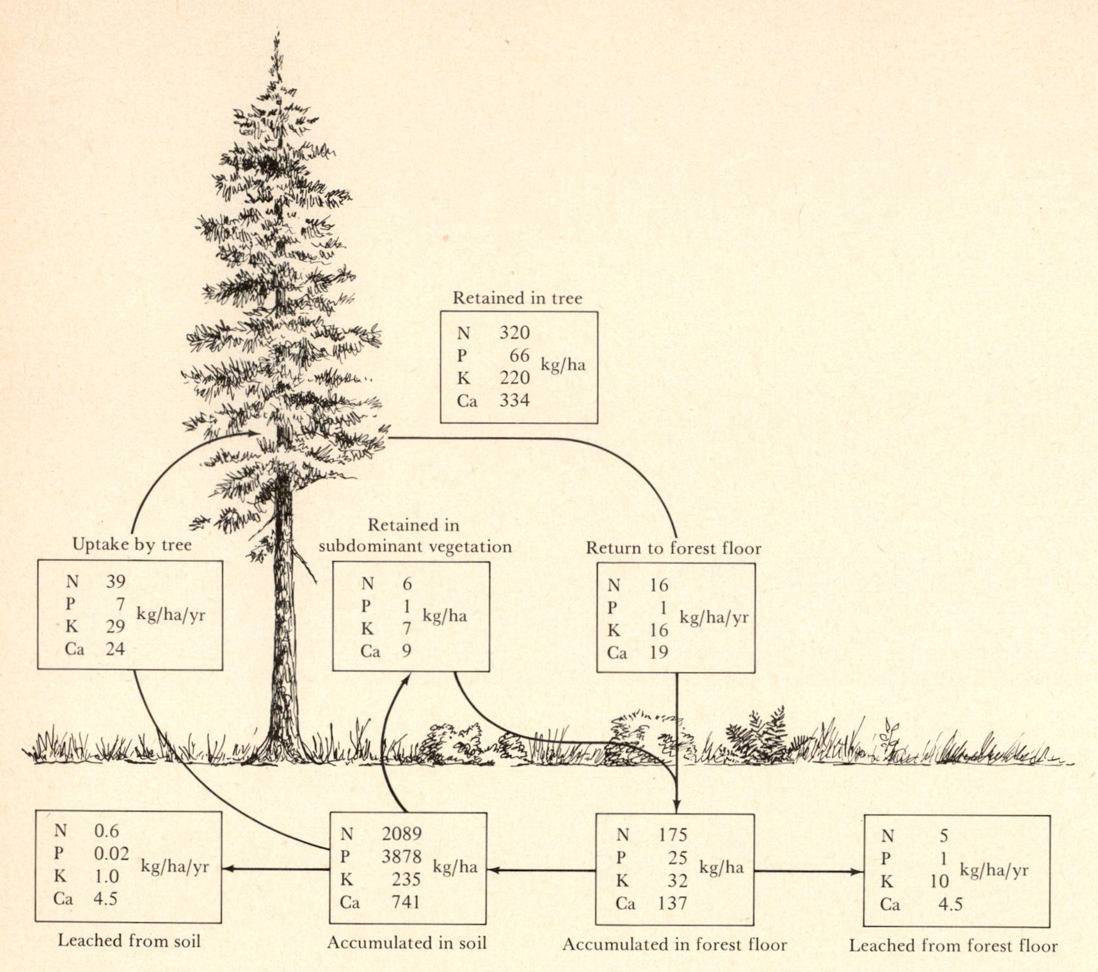

FIGURE 21.14 Distribution and short-term cycling of nitrogen, phosphorus, and calcium in a second-growth Douglas fir forest. (*From R. L. Smith, 1976:12, based on Cole, Gessel, and Dice, 1967.*)

them twice as much radiocesium as was leached from the crown by rain. Over the winter, half of this was leached out to mineral soil. Of the radiocesium in the soil, 92 percent still remained in the upper 4 inches nearly 2 years after the inoculation. Eighty percent of the cesium was confined to an area within the crown perimeter, and 19 percent was located in a small area around the trunk. This suggests that cesium distribution in the soil was greatly influenced by leaching from rainfall and stemflow.

An example of a nutrient budget is one for a Douglas-fir stand, illustrated in Figure 21.14. Ten to 15 percent of the nutrients held by Douglas-fir are located in the root and stump below ground level. Mineral accumulation in the bark is relatively high. The understory plants contain a minimal proportion of the nutrients. The budget also points out that the internal cycle is more important to the nutrient budget than inputs by or losses from precipitation. A portion of the nutrients is stored in the tree, then returned to the forest floor by way of litterfall, stemflow, and leaf wash. Ten percent of the nitrogen, 50 percent of the phosphorus, 71 percent of the potassium, and 22 percent of the calcium are returned by this route. The annual accumulation

TABLE 21.2 UTILIZATION OF NITROGEN, PHOSPHORUS, POTASSIUM, AND CALCIUM FROM SOIL BY SECOND-GROWTH DOUGLAS-FIR

Element	Yearly Uptake (% of Total in System)*	Static Supply (Yr)†	Cyclic Supply (Yrs)
N	1.4	73	125
P	0.2	537	582
K	12.5	8	12
Ca	3.3	30	64

* Includes input by precipitation and return by litter stem flow, and wash.
† Number of years supply will last.
SOURCE: D. W. Cole et al., 1967:220.

of calcium, phosphorus, potassium, and nitrogen is considerable, both in the tree and in the forest floor. The nutrient cycle depends upon the withdrawal of elements from the soil at an annual rate of 34.6 kg N/ha, 6.3 kg P/ha, 19.9 kg K/ha, and 11.5 kg Ca/ha. The time required for total soil depletion at this rate is summarized in Table 21.2. The depletion time is speculative because it does not take into account the change through time in transfer rates between components of the ecosystem or the addition of elements by way of nitrogen fixation, mineral solubility, and accumulation of organic matter (Cole et al., 1968).

Summary

Materials flow from the living to the nonliving and back to the living parts of the ecosystem in a perpetual cycle. By means of these cycles, plants and animals obtain nutrients that are necessary for their well-being.

There are two kinds of cycles: the gaseous (represented by the carbon, oxygen, and nitrogen cycles) and the sedimentary (represented by the phosphorus cycle). The sulfur cycle is a combination of the two.

The oxygen cycle, although deceptively simple, is quite complex. The major sources of atmospheric oxygen are photodissociation of water vapor and the by-product of photosynthesis. Oxygen also circulates freely as a constituent of carbon dioxide. Active chemically, it combines with a wide range of inorganic chemicals, organic substances, and reduced compounds, and it is involved with the oxidation of carbohydrates, releasing energy, carbon dioxide, and water.

The carbon cycle is so closely tied to energy flow that the two are inseparable. It involves the assimilation and respiration of carbon dioxide by plants, its consumption in the form of plant and animal tissue, its release through respiration, the mineralization of litter and wood, soil respiration, and the accumulation of carbon in standing crops and its withdrawal into long-term reserves. The carbon dioxide cycle exhibits both diurnal and annual curves. The equilibrium of carbon dioxide exchange between land, sea, and atmosphere has been disturbed by its rapid injection into the atmosphere by the burning of fossil fuels, but nearly two-thirds is removed from the atmosphere by land vegetation and by the sea.

The nitrogen cycle is characterized by the fixation of atmospheric nitrogen by nitrogen-fixing plants, largely legumes and blue-green algae. Involved in the nitrogen cycle are the processes of ammonification, nitrification, and denitrification.

The sedimentary cycle involves two phases, salt solution and rock. Minerals become available through the weathering of the Earth's crust, enter the water cycle as salt solutions, take diverse pathways through the ecosystem, and ultimately return to the sea or back to the Earth's crust through sedimentation. The phosphorus cycle is wholly sedimentary, with reserves coming largely from phosphate rock. Much of the phosphate used as fertilizer becomes immobilized in the soil, but great quantities are lost in detergents and other wastes carried by sewage effluents.

Sulfur enters both the gaseous and sedimentary phases. A significant portion of sulfur released to the atmosphere is a by-product of the burning of fossil fuel. When such quantities are injected into the atmosphere, SO_2 becomes a major pollutant, affecting or even killing plant growth, causing respiratory afflictions in humans, and producing acid rainfall over parts of Earth.

Of more serious consequence globally are the chlorinated hydrocarbons. Used in insect control, these pesticides have contaminated the global ecosystem and entered food chains. Since they become more concentrated at higher trophic levels, the chlorinated hydrocarbons affect the predacious animals most adversely. Fish-eating birds are endangered because chlorinated hydrocarbons interefere with their reproductive capability.

Important roles in mineral cycling are played at one end by green plants, which take up nutrients, and at the other end by decomposers, which release nutrients for reuse, as well as by air and water, in which return trips are made. In ecosystems more or less in equilibrium, nutrient inputs equal nutrient outputs. Internal cycles through which nutrients move from soil to plant and back to soil assume great importance in the functioning of ecosystems.

Review and Study Questions

1. What are the two types of biogeochemical cycles, and what are their distinguishing characteristics?
2. What are the features of the oxygen cycle?
3. What is the significance and importance of the carbon cycle?
4. What pools of CO_2 are involved in the global carbon cycle? What pool is responsible for the major input?
5. What is the impact on the CO_2 cycle of the burning of fossil fuels?
6. What are the possible consequences of increasing CO_2 input into the atmosphere?
7. What is the role of the ocean in carbon cycling?
8. Characterize the following processes in the nitrogen cycle: fixation, ammonification, nitrification, and denitrification.
9. What biological and nonbiological mechanisms are responsible for nitrogen fixation?
10. What are some sources of input into the nitrogen cycle? Sources of output?
11. What is the source of phosphorus in the phosphorus cycle?
12. Through what three compartments does the phosphorus cycle move in aquatic ecosystems? Describe the circulation through these compartments.
13. What is the source of sulfur in the sulfur cycle? Why does it have characteristics of both the sedimentary and gaseous cycles?
14. What are some sinks in the sulfur cycle?
15. What are the ecological effects of sulfur pollution? Of acid rain?
16. How are DDT and other chlorinated hydrocarbons circulated in ecosystems?
17. What is the impact of chlorinated hydrocarbons in ecosystems? What do they affect and why?
18. How are the cycling of materials and the flow of energy related?
19. Upon what does the food base of the nutrient cycle depend?
20. What influences nutrient utilization in the ecosystem?
21. What mechanisms related to energy flow exist in the ecosystem to ensure conservation of nutrients and control nutrient flow?
22. How does organic matter play a key role in the recycling of nutrients?
23. How do large organic pools and short-term and long-term nutrient pools act to conserve nutrients in an ecosystem?

Suggested Readings

Edwards, C. E. 1974. *Persistent pesticides in the environment*. West Palm Beach, Fla.: CRC Press.

Hasler, A. D., ed. 1975. *Coupling of land and water systems*. New York: Springer-Verlag.

Howell, F. G., J. B. Gentry, and M. H. Smith, eds. 1975. *Mineral cycling in southeastern ecosystems*. Springfield, Va.: National Technical Information Service.

Likens, G. E., F. H. Bormann, R. S. Pierce, J. S. Eaton, and N. M. Johnson. 1977. *Biogeochemistry of a forested ecosystem*. New York: Springer-Verlag.

National Research Council. 1981. *Atmosphere-biosphere interactions: Toward a better understanding of the ecological consequences of fossil fuel combustion*. Washington, D.C.: National Academy Press.

National Research Council. 1983. *Acid deposition: Atmospheric processes in eastern North America*. Washington, D.C.: National Academy Press.

Pomeroy, L. R., ed. 1974. *Cycles of essential elements*. Benchmark Papers in Ecology. Stroudsburg, Pa.: Dowden, Hutchinson & Ross.

Woodwell, G. M., and E. Pecan, eds. 1973. *Carbon and the biosphere*. Springfield, Va.: National Technical Information Service.

ECOLOGICAL EXCURSION

Odyssey

Aldo Leopold

X had marked time in the limestone ledge since the Paleozoic seas covered the land. Time, to an atom locked in a rock, does not pass.

The break came when a bur oak root nosed down a crack and began prying and sucking. In the flash of a century the rock decayed, and X was pulled out and up into the world of living things. He helped build a catkin, which became an acorn, which fattened a deer, which fed an Indian, all in a single year.

From his berth in the Indian's bones, X joined again in chase and flight, feast and famine, hope and fear. He felt these things as changes in the little chemical pushes and pulls which tug timelessly at every atom. When the Indian took his leave of the prairie, X moldered briefly underground, only to embark on a second trip through the bloodstream of the land.

This time it was a rootlet of bluestem which sucked him up and lodged him in a leaf, which rode the green billows of the prairie June, sharing the common task of hoarding sunlight. To this leaf also fell an uncommon task: flicking shadows across a plover's eggs. The ecstatic plover, hovering overhead, poured praises on something perfect; perhaps the eggs, perhaps the shadows, or perhaps the haze of pink phlox which lay on the prairie.

When the departing plovers set wing for the Argentine, all the bluestems waved farewell with tall new tassels. When the first geese came out of the north and all the bluestems glowed wine-red, a forehanded deermouse cut the leaf in which X lay and buried it in an underground nest, as if to hide a bit of Indian summer from the thieving frosts. But a fox detained the mouse, molds and fungi took the nest apart, and X lay in the soil again, foot-loose and fancy-free.

Next he entered a tuft of side-oats grama, a buffalo, a buffalo chip, and again the soil. Next a spiderwort, a rabbit, and an owl. Thence a tuft of sporobolus.

All routines come to an end. This one ended with a prairie fire, which reduced the prairie plants to smoke, gas, and ashes. Phosphorus and potash atoms stayed in the ash, but the nitrogen atoms were gone with the wind. A spectator might, at this point, have predicted an early end of the biotic drama, for with fires exhausting the nitrogen, the soil might well have lost its plants and blown away.

But the prairie had two strings to its bow. Fires thinned its grasses, but they thickened its stand of leguminous herbs: prairie clover, bush clover, wild bean, vetch, lead-plant, trefoil, and baptisia, each carrying its own bacteria housed in nodules on its rootlets. Each nodule pumped nitrogen out of the air into the plant, and then ultimately into the soil. Thus the prairie savings bank took in more nitrogen from its legumes than it paid out to its fires. That the prairie is rich is known to the humblest deer mouse; why the prairie is rich is a question seldom asked in all the still lapse of ages.

Between each of his excursions through the biota, X lay in the soil, and was carried by the rains, inch by inch, downhill. Living plants retarded the wash by impounding atoms, dead ones, by locking them to their decayed tissues. Animals ate the plants and carried them briefly uphill or downhill, depending on whether they died or defecated higher or lower than they fed. No animal was aware that the altitude of his death was more important than his manner of dying. Thus a fox caught a gopher in a meadow, carrying X uphill to his bed on the brow of a ledge, where an eagle laid him low. The dying fox sensed the end of his chapter in foxdom, but not the new beginning in the odyssey of an atom.

An Indian eventually inherited the eagle's plumes, and with them propitiated the Fates, whom he assumed had a special interest in Indians. It did not occur to him that they might be busy casting dice against gravity; that mice and men, soils and songs might be merely ways to retard the march of atoms to the sea.

SOURCE: Aldo Leopold, *A Sand County Almanac with other Essays on Conservation from Round River* (New York: Oxford University Press, 1966), 104–108.

One year, while X lay in a cottonwood by the river, he was eaten by a beaver, an animal which always feeds higher than he dies. The beaver starved when his pond dried up during a bitter frost. X rode the carcass down the spring freshet, losing more altitude each hour than heretofore in a century. He ended up in the silt of a backwater bayou, where he fed a crayfish, a coon, and then an Indian, who laid him down to his last sleep in a mound on the riverbank. One spring an oxbow caved the bank, and after one short week of freshet, X lay again in his ancient prison, the sea.

An atom at large in the biota is too free to know freedom; an atom back in the sea has forgotten it. For every atom lost to the sea, the prairie pulls another out of the decaying rocks. The only certain truth is that its creatures must suck hard, live fast, and die often, lest its losses exceed its gains.

It is the nature of roots to nose into cracks. When Y was thus released from the parent ledge, a new animal had arrived and begun redding up the prairie to fit his own notions of law and order. An oxteam turned the prairie sod, and Y began a succession of dizzy annual trips through a new grass called wheat.

The old prairie lived by the diversity of its plants and animals, all of which were useful because the sum total of their coöperations and competitions achieved continuity. But the wheat farmer was a builder of categories; to him only wheat and oxen were useful. He saw the useless pigeons settle in clouds upon his wheat, and shortly cleared the skies of them. He saw the chinch bugs take over the stealing job, and fumed because here was a useless thing too small to kill. He failed to see the downward wash of over-wheated loam, laid bare in spring against the pelting rains. When soil-wash and chinch bugs finally put an end to wheat farming, Y and his like had already traveled far down the watershed.

When the empire of wheat collapsed, the settler took a leaf from the old prairie book: He impounded his fertility in livestock, he augmented it with nitrogen-pumping alfalfa, and he tapped the lower layers of the loam with deeprooted corn. With these he built the empire of red barns.

But he used his alfalfa, and every other new weapon against wash, not only to hold his old plowings, but also to exploit new ones which, in turn, needed holding.

So, despite alfalfa, the black loam grew gradually thinner. Erosion engineers built dams and terraces to hold it. Army engineers built levees and wing-dams to flush it from the rivers. The rivers would not flush, but raised their beds instead, thus choking navigation. So the engineers built pools like gigantic beaver ponds, and Y landed in one of these, his trip from rock to river completed in one short century.

On first reaching the pool, Y made several trips through water plants, fish, and waterfowl. But engineers build sewers as well as dams, and down them comes the loot of all the far hills and the sea. The atoms which once grew pasqueflowers to greet the returning plovers now lie inert, confused, imprisoned in oil sludge.

Roots still nose among the rocks. Rains still pelt the fields. Deer mice still hide their souvenirs of Indian summer. Old men who helped destroy the pigeons still recount the glory of the fluttering hosts. Black-and-white buffalo pass in and out of red barns, offering free rides to itinerant atoms.

CHAPTER 22

The Community

Outline

Objectives

Upon completion of this chapter, you should be able to:

1. Describe the Raunkiaer life form system and explain its usefulness in describing community structure.
2. Explain how vertical and horizontal structure influence the nature of a community.
3. Distinguish between edge and ecotone and discuss the edge effect.
4. Explain how species dominance influences community structure.
5. Discuss the concept of species diversity.
6. Explain the theory of island biogeography.
7. Discuss the relationship and importance of the theory of island biogeography to habitat fragmentation, species equilibrium, and the management of species populations.

The most visible part of the ecosystem is the biotic: the vegetation and animal life. The assemblage of plants and animals in any given ecosystem or, to be more restrictive, in any given physical environment is a *community*. Thus, a community can be considered not only as the combination of plant and animal populations that comprise the biotic portion of a given ecosystem, such as a forest, but also as the assemblage of organisms that inhabit a fallen log, an acorn, or even a tiny pool of water held in the hollow of a tree. The former is a major, or *autotrophic*, community, independent of others and requiring from the outside only the energy of the sun. The latter is a minor, or *heterotrophic*, community, dependent on the major community for its energy source. Thus, the community is something more than a loose assemblage of independent organisms. The plants and animals that make up the community are interdependent, living together in some orderly fashion and forming a functional unit of the ecosystem.

Although the community usually consists of a certain combination of species, similar communities (an oak-hickory forest, for example) need not consist of exactly the same assemblage of species. This does create some problems in defining a community. Botanists use the term *association* for a plant community possessing a definite floristic composition. They use the word *community* only in a very general sense. Zoologists and many ecologists apply the term equally to specific assemblages and to general groupings. Because of these problems, some ecologists question whether definitive communities actually exist or whether they are simply abstractions resulting from human attempts to arrange communities, like species, into neat categories.

But the general community does exist, and regardless of its species composition, it does possess certain general attributes such as dominance, niche, species diversity, structure, stability, development, and a metabolic role in the functioning of the ecosystem. Both major and minor communities can be looked at in terms of these attributes.

Growth Forms and Life Forms

The form and structure of terrestrial communities can be characterized by the nature of the vegetation. The plants may be tall or short, evergreen or deciduous, herbaceous or woody. Such characteristics can be used to describe growth forms. Thus, one might speak of shrubs, trees, and herbs and further subdivide the categories into needle-leaf evergreens, broadleaf evergreens, evergreen sclerophylls (small, tough, evergreen leaves, as in chamise), broadleaf deciduous, thorn trees and shrubs, dwarf shrubs, ferns, grasses, forbs, and lichens.

Perhaps a more useful system is the one designed in 1903 by the Danish botanist Christen Raunkiaer. Instead of considering the plants' growth form, he classified plant life by the relation of the embryonic or meristemic tissues that remain inactive over winter or a dry period (perennating tissue) to their height above ground. Such perennating tissue includes buds, bulbs, tubers, roots, and seeds. Raunkiaer recognized five principal life forms, which are summarized in Table 22.1 and Figure 22.1. All the species in a region or community can be grouped into the five classes and the ratio between them expressed as a percentage, providing a life form spectrum of the area that reflects the plants' adaptations to the environment, particularly climate (see Table 22.2 and Figure 22.2). A community with a high percentage of perennating tissue well above ground (phanerophytes) would be characteristic of warm climates. A community with most of its plants chamaephytes and hemicryptophytes would be characteristic of cold climates, and a community dominated by therophytes would be characteristic of deserts.

Vertical Stratification

A distinctive feature of a community is vertical stratification (see Figure 22.3), physical and biological. The stratification of a community is determined largely by the life form of the plants—their size, branching, and leaves—which, in turn, influences and is influenced by the vertical gradient of light. The vertical structure of the plant community provides the physical structure in which many forms of animal life are adapted to live. A well-developed forest ecosystem, for example, has several layers of vegetation. From top to bottom, they are the *canopy*, the *understory*, the *shrub*, the *herb* or *ground layer*, and the *forest floor*. And one can continue down into the root layer and soil strata (see Chapter 8).

The canopy, which is the primary site of energy fixation, has a major influence on the rest of the forest. If it is fairly open, considerable sunlight will reach the lower layers and the shrub and the understory tree strata will be well developed. If the canopy is closed, the shrub and the understory trees and even the herbaceous layers will be poorly developed.

The understory consists of tall shrubs such as witch hobble, understory trees such as dogwood and hornbeam, and younger trees; some are the

TABLE 22.1 RAUNKAIER'S LIFE FORMS

Phanerophytes (Gr. *phaneros*, "visible"). Perennial buds carried well up in the air and exposed to varying climatic conditions. Trees and shrubs over 25 cm; typical of moist, warm environments.

Chamaephytes (Gr. *chamai*, "on the ground"). Perennial shoots or buds on the surface of the ground to about 25 cm above the surface. Buds receive protection from fallen leaves and snow cover. Plants typical of cool, dry climates.

Hemicryptophytes (Gr. *krypos*, "hidden"). Perennial buds at the surface of the ground, where they are protected by soil and leaves. Many plants are characterized by rosette leaves and are characteristic of cold, moist climates.

Cryptophytes. Perennial buds buried in the ground on a bulb or rhizome, where they are protected from freezing and drying. Plants are typical of cold, moist climates.

Therophytes (Gr. *theros*, "summer"). Annuals, with complete life cycle from seed to seed in one season. Plants survive unfavorable periods as seeds and are typical of deserts and grasslands.

FIGURE 22.1 Raunkiaer's life forms: (1) phanerophytes, (2) chamaephytes, (3) hemicryptophytes, (4) cryptophytes (geophytes), (5) therophytes. The parts of plants that die back are unshaded; the persistent parts with buds are dark.

TABLE 22.2 LIFE FORM SPECTRA OF MAJOR ECOSYSTEMS

Community	Ph	Ch	He	Cr	Th
Arctic tundra	1	23	61	13	2
Temperate deciduous forest	15	2	49	13	12
Subtropical forest	34	23	10	5	15
Rain forest	54	6	12	3	16
Desert	26	7	18	7	42

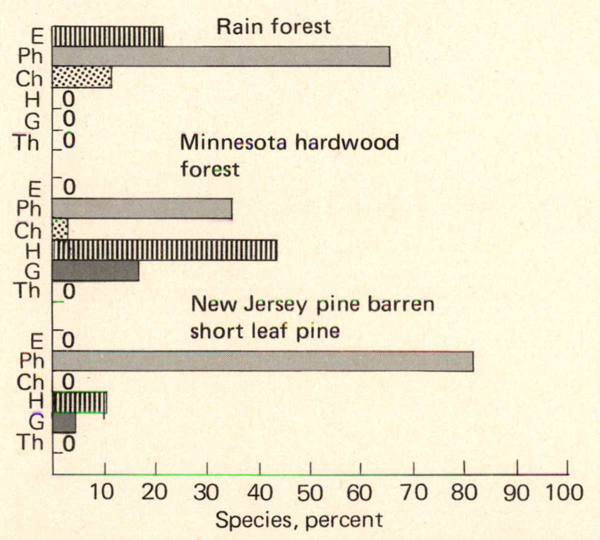

FIGURE 22.2 Life form spectra of a tropical rain forest (*based on data from Richards, 1952*), a Minnesota hardwood deciduous forest (*based on data from Buell and Wilbur, 1948*), and a New Jersey pine barren (*based on data from Stern and Buell, 1951*). Note the importance of epiphytes in the tropical rain forest and their absence in the other two. Also note the conspicuous importance of phanerophytes in the pine forest, the absence of ground plants in the rain forest, and the absence of annuals in all three forests.

same as those in the crown, while others are of different species. Species that are unable to tolerate shade and competition will die; others will eventually reach the canopy after some of the older trees die or are harvested.

The shrub layer differs with the type of forest. In oak forests on south-facing slopes, blueberries are most characteristic; in the moist cove forests grow buffalo nut, hydrangea, and rhododendron. In the northern hardwood forests, witch hobble, maple-leaf viburnum, and striped maple are common.

The nature of the herb layer depends on the soil moisture conditions, the slope position, the density of the overstory, and the aspect of the slope, all of which vary from place to place through the forest.

The final layer, the forest floor, has already been discussed (see Chapter 8) as the site where the important process of decomposition of forest litter takes place and where nutrients are released to the nutrient cycle.

Aquatic ecosystems such as lakes and oceans have strata determined by light penetration, temperature profile, and oxygen profiles (see Figure 25.7). In the summer well-stratified lakes have a layer of freely circulating surface water, the *epilimnion;* a second layer, the *metalimnion,* which is characterized by a *thermocline* (a very steep and rapid decline in temperature); the *hypolimnion,* a deep, cold layer of dense water about 40°C, often low in oxygen; and a layer of bottom mud (see Chapter 25). In addition, two other structural layers are recognized, based on light penetration: an upper zone roughly corresponding to the epilimnion, which is dominated by plant plankton and is the site of photosynthesis, and a lower layer, in which decomposition is most active. The lower layer roughly corresponds to the hypolimnion and the bottom mud.

Ecosystems, terrestrial and aquatic, have similar biological structure. They possess an autotrophic layer concentrated where light is most available, which fixes the energy of the sun and manufactures food from organic substances. In forests this layer is concentrated in the canopy; in grasslands, in the herbaceous layer; and in lakes and seas, in the upper layer of water. Ecosystems also possess a heterotrophic layer that utilizes food stored by autotrophs, transfers energy, and circulates matter by means of herbivory, predation in the broadest sense, and decomposition.

The degree of vertical stratification has a pro-

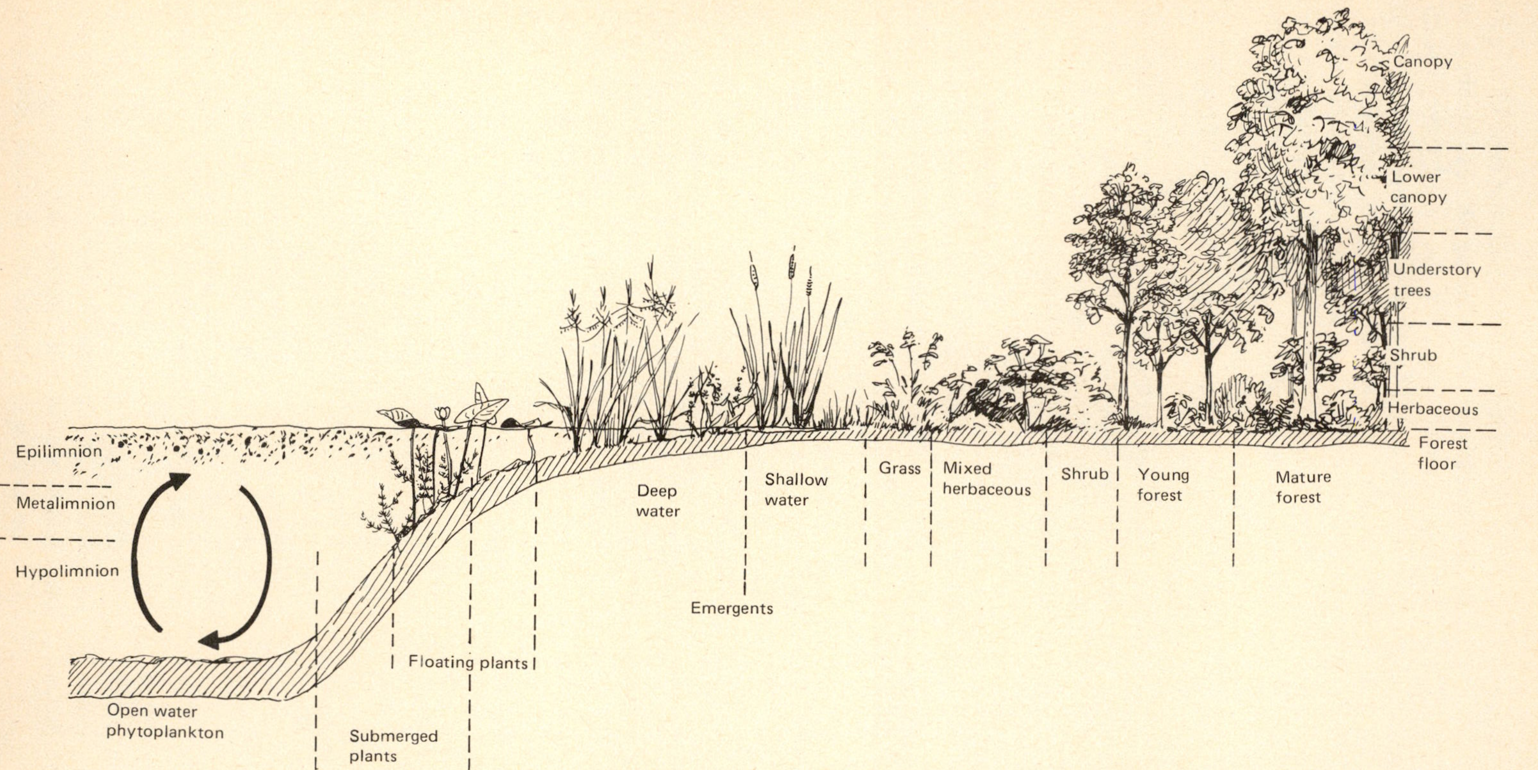

FIGURE 22.3 A vertical section view of communities from aquatic to terrestrial illustrates the general features of each. Both are structurally similar in that the zone of decomposition and regeneration is in the bottom stratum and the zone of energy fixation is in the upper stratum. In succession from aquatic to terrestrial types, stratification and the complexity of the community become greater. Stratification as it influences the distribution of organisms and their niches in aquatic communities is largely physical, influenced by gradients of oxygen, temperature, and light. Stratification in terrestrial communities is largely biological. Dominant vegetation affects the physical structure of the community and the microclimatic conditions of temperature, moisture, and light. Because the forest has four to five strata, it can support a greater diversity of life than a grassland with basically two strata. Floating and emergent aquatic plant communities can support a greater diversity of life than open water.

nounced influence on the diversity of animal life in the community, a relationship demonstrated by R. H. MacArthur and J. W. MacArthur (1961). A strong correlation exists between foliage height diversity and bird species diversity. Increased vertical stratification increases the availability of resources and living space, which favors a certain degree of specialization. Grasslands, with their two strata, hold about 6 to 7 species of birds, all ground nesters. An eastern deciduous forest may support 30 or more species occupying different strata. The scarlet tanager (*Piranga olivacea*) and wood pewee (*Contopus virens*) are canopy species, the Acadian flycatcher (*Empidonax virescens*) is an understory canopy species, and the hooded warbler (*Wilsonia citrina*) is a forest shrub species. Insects show similar stratification. Among the pine bark beetles inhabiting northeastern North America, the large red turpentine beetle (*Dendroctonus valens*) is restricted to the base of trees; the pine engraver beetle (*Ips pini*), to the upper trunk and large branches. A third species, the small and abundant *Pityogenes hopkinsi,* lives on smaller branches in the crown.

FIGURE 22.4 Horizontal stratification of vegetation in an area of Wisconsin countryside. Vegetation patches include fragmented woodland, cropland and pasture, shrub growth, hedgerows, woodland ecotones, and herbaceous growth. This is a good example of a patchy environment.

Horizontal Stratification

Horizontal stratification relates to the distribution of organisms, principally plants, on the ground and across the canopy. Like vertical stratification, it can influence the presence or absence of animal life. The pattern of plant distribution is mostly clumped to varying degrees (see Chapter 10), presenting a mosaic of vegetation and creating a patchy environment (Figure 22.4). The distribution of plants in terrestrial ecosystems is influenced in part by the nature of seed dispersal and vegetative reproduction. Plants with airborne seeds may be distributed widely, while plants with heavy seeds or with pronounced vegetative reproduction will be clumped near the parent plant. The interaction between the physical environment and vegetation is equally important. Herbaceous plants of the forest may be clustered where pools of sunlight reach the forest floor. The nature of the soil, its structure, moisture conditions, and nutrient availability also influence plant distribution, as well as allelopathy (see Chapter 15) and grazing. A pronounced form of horizontal stratification is *zonation,* caused primarily by differences in climatic or edaphic conditions that retard or inhibit the colonization and growth of rooted vegetation. Such stratification is most conspicuous about ponds and bogs (see Chapter 25).

Edge and Ecotone

Closely associated with horizontal stratification are *edge* and *ecotone*. Although the two terms are often used synonymously, they are different. An edge is where two or more different vegetational communities meet. An ecotone is where two or more communities not only meet but intergrade (Figure 22.5).

Edges may result from abrupt changes in soil type, topographic differences, geomorphic differences (such as rock outcrops), and microclimatic changes. Because the adjoining vegetation types are determined by long-term natural features, such edges are usually stable and permanent and are considered *inherent*. Other edges result from such natural disturbances as fire, storms, and floods or from such human-induced disturbances as grazing, timber harvesting, land clearing, and agriculture. The adjoining vegetational types are successional, or developmental (see Chapter 23), and will change or disappear with time. Such edges are termed *induced*. They can be maintained only by periodic disturbance. Induced edges, too, may be abrupt, or they may be transitional, resulting in an ecotone.

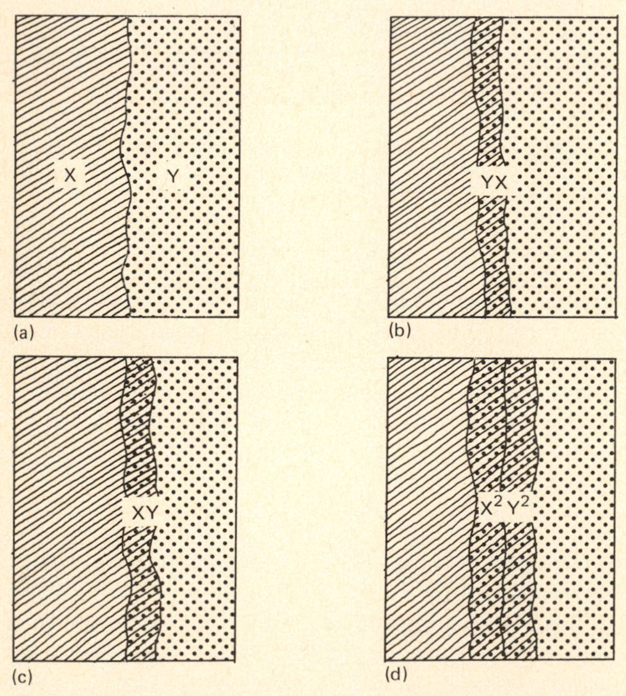

FIGURE 22.5 Edge and types of ecotone that might develop. (a) Abrupt, narrow edge with no development of an ecotone. (b) Narrow ecotone developed by advancement of community Y into community X. (c) Community X advances into community Y to produce ecotone XY. (d) Ideal ecotone development, in which plants from both communities invade each other to create a wide ecotone, X^2Y^2. The latter is the type of ecotone that will serve the most edge species.

Ecotones arise from the blending of two or more vegetational types. Plants competitively superior and adapted to environmental conditions existing in the edge advance as far into either community as their ability to maintain themselves will allow (Figure 22.6). Beyond this point interior plants of adjacent communities maintain themselves. As a result, the ecotone exhibits a shift in dominance of certain species of each community.

It also involves a number of highly adaptable species that tend to colonize such areas. Edge species of plants tend to be opportunistic (see Chapter 14), are shade-intolerant, and are tolerant of a relatively xeric environment, including a high rate of evapotranspiration, reduced soil moisture, and fluctuating temperatures. Animal species of the edge are usually those that require two or more vegetational communities. For example, the ruffed grouse (*Bonasa umbellatus*) requires new forest openings with an abundance of herbaceous plants and low shrubs, dense sapling stands, pole timber for nesting cover, and mature forests for winter food and cover. Because the ruffed grouse spends its life in an area of 10 to 20 acres, this amount of land must provide all of its seasonal requirements. Some species, such as the indigo bunting, are restricted exclusively to the edge situation (Figure 22.7). Because of the species responses, the variety and density of life are often greatest in and about edges and ecotones. This phenomenon has been called the *edge effect* (Leopold, 1933).

Edge effect is influenced by the amount of edge available—its length, width, and degree of contrast between adjoining vegetational communities. The greater the contrast between adjoining plant communities, the greater the species richness should be (Figure 22.8). An edge between forest and grassland should support more species than an edge between a young and a mature forest. The larger the adjoining communities, the more opportunity exists for flora and fauna of adjoining communities as well as species that favor edge situations to occupy the area. If patches of vegetation are too small to sup-

FIGURE 22.6 Stages in edge development. (a) Initial creation of an abrupt edge. Open sides allow light penetration into woodland and development of xeric conditions. (b) Within 10 years or less, vegetation responds to changed conditions. Some trees along the edge adjust to changed conditions and develop epicormic branching and expand their crowns. Other trees, unable to withstand increased light, may die. A dense understory of shade-intolerant woody seedlings and sprouts and herbaceous plants develops. Because of open conditions, edge vegetation invades woodland. (c) Increased expansion of crown, growth of epicormic branching, and height growth of edge understory close the gap between crown and edge vegetation and reduce side lighting. Some species disappear, and edge growth into the woods is reduced. (d) At this stage canopy trees and understory experience increased competition for light. Some species survive; others disappear, depending upon competitive ability. The number of dominant and codominant individuals in the edge declines, and such edge-oriented species as hawthorne, hickory, aspen, and oak replace shrubs such as blackberry. (e) At this stage, the edge is now dominated by a few large trees of the edge with crowns developed low to the ground. Only minimal edge understory remains. (f) In some situations understory vegetation may disappear if the edge is dominated by trees whose crowns reach the ground. (g) When conditions permit, edge species may invade adjacent fields, developing an ecotone. (*After Ranney, Bruner, and Levenson, 1981:83.*)

FIGURE 22.7 Map of territories of a true edge species, the indigo bunting *(Passerina cyanea),* which inhabits woodland edges, large gaps in forests creating edge conditions, hedgerows, and roadside thickets. The male requires tall, open song perches and the female a dense thicket in which to build a nest. (*After Whitcomb et al., 1981:143.*)

port their characteristic species (see page 458), then the area becomes a homogeneous community dominated by edge species.

The edge effect comes about because environmental conditions differ from those of adjacent vegetational communities, especially adjoining forests. Increased solar radiation in the newly created edge, high temperature, and exposure to wind result in a high rate of evaporation. Plants place increased demands on soil moisture. Sudden exposure to sunlight subjects trees to stress from increased heat and light. Some mesic, shade-tolerant trees succumb. Others are injured by sun scald. Light-tolerant species respond by increasing crown growth and epicormic branching (new branching sprouting on the trunk). The effects of microclimatic changes are most pronounced on south-facing and west-facing edges, because they receive the greatest amount of solar radiation. Thus, edge favors xeric, light-demanding species capable of competing successfully for available soil moisture. (These points will be explored further later in the chapter.)

Community Gradients

Although the definition of a community is rather straightforward—an assemblage of species occupying a given area—the nature of the community has been the object of study and dispute for years. Is a community such as an oak-hickory forest a real entity that is definable, describable, and constant from one stand of oak-hickory to another? Or is it something of an abstraction, a collection of different populations that exist together because they have similar environmental requirements? Such questions are still being debated.

The composition of any one community is determined in part by the species that happen to be distributed on the area and can grow and survive under prevailing conditions. Seeds of many plants may be carried by wind and animals, but only those that are adapted to grow in the habitat where they are deposited and are capable of overcoming competition of any species already present will take root and thrive. One adapted species may colonize an area and prevent others equally well adapted from entering. Wind direction and velocity, size of the seed crop, disease, and insect and rodent damage all influence the establishment of vegetation. The exact species that settle an area and the number of individual species that succeed are situations that seldom if ever are repeated in any two places at any two times. Thus, the element of chance is heavily involved. Nevertheless, there is a certain pattern, with more or less similar groups recurring from place to place. Only a relatively small group of species are potential dominants because a limited number are well adapted to the general climate and soils of the region they occupy.

Two opposite views of natural communities exist. One regards communities as distinct natural units or associations. The distribution and abundance of a species in a community are determined by its interaction with other species in the same community. Species making up the community typically are associated with each other and are organized into discrete groups. Groups of stands similar to one another form associations. Stands of one association are clearly distinct from stands of other

FIGURE 22.8 Contrast in edge is important in increasing species richness. A high-contrast edge (a) is more valuable to edge species than a low-contrast edge (b) because two quite different vegetation types adjoin. Low-contrast edges do not provide enough difference between vegetational communities to be of maximum value to edge species. Of greatest value is an advancing edge (c), such as woody vegetation invading an adjoining old field. An advancing edge not only provides variation in height but in effect creates two edges on the site.

associations. The community acts as a unit in seasonal activity, in competition with other communities, in trophic functions, and in succession. This concept of the community, developed by F. E. Clements (1916), in effect regards the community as a sort of supraorganism, the highest stage in the organization of the living world—rising from cell to tissue, organ, organ systems, organism, population, community. The whole is more than the sum of its parts.

For practical reasons of study and description, the idea of distinct, definable communities has advantages. But contrary to that concept, general observations confirm that species comprising a community do not necessarily associate exclusively with one another. Rather, each species appears to be distributed in its own way, according to its own response to varying environmental conditions such as altitude, moisture, temperature, soil, nutrients, light, drainage, and other physical conditions. Some organisms will succeed only in certain environmental situations and tend to be confined to certain habitats. They have a restricted distribution over an environmental gradient (Figure 22.9). Others are more tolerant and occupy a wider distribution on an environmental gradient such as moisture, tem-

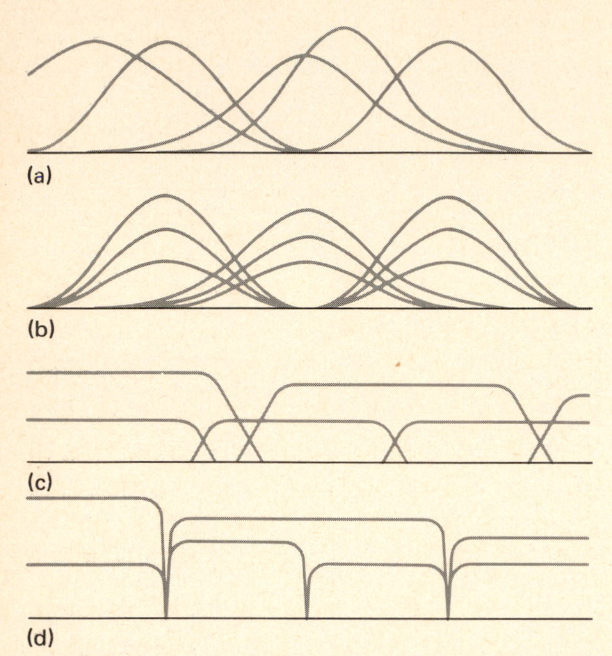

FIGURE 22.9 Four models of species distribution along environmental gradients. (a) The abundance of one species on an environmental gradient is independent of the others. Thus, the association of several species along the gradient changes with the response of the individual species to that gradient. (b) The abundance of one species is associated with that of another. The two or more species are always found in association with each other. (c) The distribution of one species is independent of another on an environmental gradient, but the abundance and distribution of each species is sharply restricted at some point on the gradient by interspecific competition. (d) The distribution of a species is sharply restricted by a change in some environmental variable.

perature, soil, slope position and the like. This sequence of communities showing a gradual change in composition is called a *continuum* (Curtis, 1959). A community on such a gradient can be described as a discrete area in the continuum. Each community is somewhat different from its neighbor, the difference increasing roughly as the distance between them increases.

That is the individualistic approach, advanced independently by H. A. Gleason (1926) in the United States and L. G. Ramensky (1927) in Russia. Interestingly, when these concepts were originally presented, they conflicted with the then-dominant organismic concept in the United States and the association concept in Russia and were ignored, if not rejected, by ecologists. Ultimately, both became accepted in their respective countries during the 1960s. In the United States, Curtis (1959), McIntosh (1958, 1967), and Whittaker (1962, 1965, 1967, 1970) utilized the continuum concept in their gradient analysis approach to vegetation studies over geographical areas.

The gradient approach emphasizes the species rather than the community per se as the essential unit in the analysis of interrelationships and distribution. Species respond independently to the biotic environment according to their own genetic characteristics. They are not bound together into groups of associates that must appear together. Instead, when species populations are plotted along an environmental gradient, the resulting graphs are bell-shaped curves, which overlap in a heterogeneous fashion (Figure 22.10). In this view the community is regarded as a collection of populations of species existing under similar environmental conditions.

The continuum or gradient concept does not negate the basic concept of the community held by the organismic school. Within any one area of the gradient, certain species are going to be dominant, even though associated species will vary. The assemblage of species is not a natural unit in the sense that they are bound together by some obligatory relationship. But the assemblage does operate as a functional unit in the flow of energy, cycling of nutrients, and trophic relationships, just as the organismic advocates propose.

Species Dominance

In a general sort of way, the nature of communities is controlled by either physical or abiotic conditions such as substrate, the lack of moisture, and wave action or by some biological mechanism. Biologically controlled communities are often influenced by a single species or by a group of species that

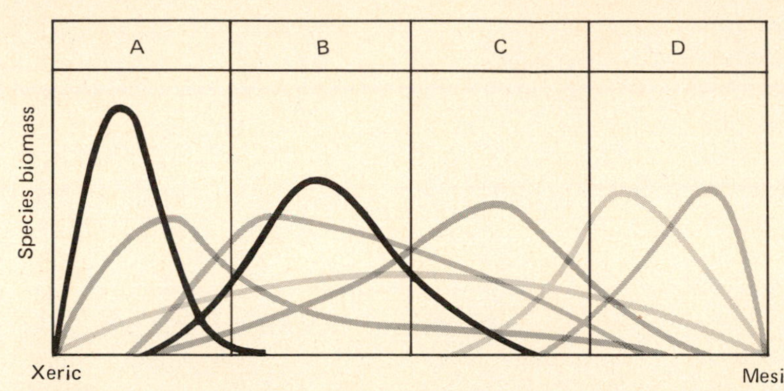

FIGURE 22.10 Vegetation distributed along a moisture gradient from xeric to mesic. Each species responds in its own individual way to moisture, yet sufficient overlap in response exists among species to allow a number of them to associate with each other on the gradient. The nature of the community, its dominants and associated species, depends upon the point at which community boundaries are placed. In the above hypothetical gradient, each demarked community is characterized by its own dominants, although some species are shared with other communities. The communities at either end of the gradient are distinct, although they share one ubiquitous species. Shifting community boundaries would result in changes in community composition and dominants.

modify the environment. These organisms are called *dominants*.

It is not easy to describe a dominant or to determine the dominant species. The dominants in a community may be the most numerous, possess the highest biomass, preempt the most space, make the largest contribution to energy flow or mineral cycling, or by some other means control or influence the rest of the community.

In a practical sense, some ecologists have given the dominant role to those organisms that are numerically superior. But numerical abundance alone is not sufficient. A species of plant, for example, can be widely distributed over the area and yet exert little influence on the community as a whole. In a forest the small or understory trees can be numerically superior, yet the nature of the community is controlled by a few large trees that overshadow the smaller ones. In such a situation the dominant organisms are not those with the greatest numbers but those that have the greatest biomass or that preempt most of the canopy space and thus control the distribution of light. Ecologists measure such dominants by biomass or basal area. Or the dominant organism may be relatively scarce yet by its activity control the nature of the community. The predatory starfish *Piaster,* for example, preys on a number of associated species and reduces competitive interaction among them, so a number of different prey species is able to coexist (Paine, 1966). If the predator is removed, a number of prey species disappears and one becomes dominant. In effect, the predator controls the nature of the community and must be regarded as the dominant.

The concept of dominance involves certain implications. In the first place, the dominant species may not be the most essential species in the community from the standpoint of energy flow and nutrient cycling, although this is often the case. Dominant species achieve their status by occupying niche space that might potentially be occupied by other species in the community. For example, when the American chestnut was eliminated by blight from oak-chestnut forests, the chestnut's position was taken over by other oaks and hickories.

Although dominants frequently shape populations of other trophic levels, dominance necessarily relates to species occupying the same trophic level. If a species or small group of species is to achieve dominance, it must relate to a total population of species, all of which have similar ecological requirements. One or several become dominants because they are able to exploit the range of environmental requirements more efficiently than other species in the same trophic level. The subordinate species exist because they are able to occupy a niche or portions of it that the dominants cannot effectively occupy. Dominant organisms, then, are generalists capable of utilizing a wide range of physiological tolerances. The subdominants tend to be more specialized in their environmental requirements and more limited in their physiological tolerances.

To determine dominance, ecologists have used several approaches. One can measure relative abundance of the species involved, comparing the numerical abundance of one species to the total abundance of all species. Or one can measure relative dominance, which is the ratio of the basal area occupied by one species to total basal area; or one can use relative frequency as a measure (see Box 22.1). Often all three of these measurements are combined to arrive at an *importance value* for each species. This index is based on the fact that most species do not normally reach a high level of importance in the community, but those that do serve as an index, or guiding species. Once importance values have been obtained for species within a stand, the stands can then be grouped by their leading dominants according to importance values. Such techniques are useful in the study and ordination of communities on some environmental gradient.

Box 22.1
Some Measures of Dominance

1. Dominance = $\dfrac{\text{basal area or aerial coverage, species A}}{\text{area sampled}}$
2. Relative dominance = $\dfrac{\text{basal area or coverage, species A}}{\text{total basal area or coverage, all species}}$
3. Relative density = $\dfrac{\text{total individuals, species A}}{\text{total individuals, all species}}$
4. Frequency = $\dfrac{\text{intervals or points where species A occurs}}{\text{total number of sample plots or points}}$
5. Relative frequency = $\dfrac{\text{frequency value, species A}}{\text{total frequency values, all species}}$
6. Importance value = relative frequency + relative dominance + relative density

All the above results may be multiplied by 100.

Simpson's index of dominance:

$$\text{Dominance} = \frac{\Sigma n_i(n_i - 1)}{N(N - 1)}$$

where N = total number of individuals of all species
n_i = total number of individuals of species A

Species Diversity

NATURE OF DIVERSITY

Species dominance implies that some organisms are more abundant than others in a particular community—an almost universal characteristic of ecosystems. Among the array of species that make up the community, relatively few are abundant, and most of them are rare. You can discover this characteristic for yourself by counting all the individuals of different species of plants in a number of sample plots and determining what percentage of each makes up the whole (relative abundance).

Table 22.3 presents the structure of a mature woodland consisting of 24 species of trees over 4 inches DBH (diameter at breast height). Two trees,

TABLE 22.3 STRUCTURE OF VEGETATION OF A MATURE DECIDUOUS FOREST IN WEST VIRGINIA

Species	Number	Percentage of Stand
Yellow-poplar (*Liriodendron tulipifera*)	76	29.7
White oak (*Quercus alba*)	36	14.1
Black oak (*Quercus velutina*)	17	6.6
Sugar maple (*Acer saccarum*)	14	5.4
Red maple (*Acer rubrum*)	14	5.4
American beech (*Fagus grandiflora*)	13	5.1
Sassafras (*Sassafras albidum*)	12	4.7
Red oak (*Quercus rubra*)	12	4.7
Mockernut hickory (*Carya tomentosa*)	11	4.3
Black cherry (*Prunus serotina*)	11	4.3
Slippery elm (*Ulmus rubra*)	10	3.9
Shagbark hickory (*Carya ovata*)	7	2.7
Bitternut hickory (*Carya cordiformis*)	5	2.0
Pignut hickory (*Carya glabra*)	3	1.2
Flowering dogwood (*Cornus florida*)	3	1.2
White ash (*Fraxinus americana*)	2	.8
Hornbeam (*Carpinus caroliniana*)	2	.8
Cucumber magnolia (*Magnolia grandiflora*)	2	.8
American elm (*Ulmus americana*)	1	.39
Black walnut (*Juglans nigra*)	1	.39
Black maple (*Acer nigra*)	1	.39
Black locust (*Robinia pseudaacacia*)	1	.39
Sourwood (*Oxydendrum arboreum*)	1	.39
Tree of heaven (*Ailanthus altissima*)	1	.39
	256	100

yellow-poplar and white oak, made up nearly 44 percent of the stand. The three next abundant trees—sugar maple, red maple, and American beech—each made up a little over 5 percent of the stand. Eight species ranged from 1.2 to 4.6 percent of the stand, while 10 remaining species as a group represented about 5 percent of the stand. Another woodland sample presents a somewhat different picture. The data presented in Table 22.4 show that the community consists of 10 species, of which 2, yellow-poplar and sassafras, make up 84 percent of the stand. Thus, these two forest stands illustrate the pattern of a few common species associated with many rare ones.

These two tables illustrate another characteristic of the distribution of species within a community—species richness or abundance and the evenness of distribution of individuals among the species. The stand described in Table 22.3 is richer in species than the stand in Table 22.4, and the evenness with which the individuals are distributed among the species is greater in the first stand than in the second.

These two parameters, species richness and species evenness, are useful in measuring another attribute of the community: species diversity. A community that contains a few individuals of many species will have a higher diversity than will a community containing the same number of individuals but with most of them confined to a few species. For example, a community with ten species of 10 individuals each has a higher diversity than a community also with ten species but with the 100 individuals apportioned 90, 1, 1, 1, 1, 1, 1, 1, 1, 1.

To quantify species diversity, several indexes have been proposed (Box 22.2). The most widely

TABLE 22.4 STRUCTURE OF VEGETATION OF A DECIDUOUS FOREST IN WEST VIRGINIA

Species	Number	Percentage of Stand
Yellow-poplar (*Liriodendron tulipifera*)	122	44.5
Sassafras (*Sassafras albidum*)	107	39.0
Black cherry (*Prunus serotina*)	12	4.4
Cucumber magnolia (*Magnolia grandiflora*)	11	4.0
Red maple (*Acer rubrum*)	10	3.6
Red oak (*Quercus rubra*)	8	2.9
Butternut (*Juglans cinerea*)	1	.4
Shagbark hickory (*Carya ovata*)	1	.4
American beech (*Fagus grandiflora*)	1	.4
Sugar maple (*Acer saccharum*)	1	.4
	174	100

used is the Shannon-Weiner index, which has been adapted from communication or information theory:

$$H = -\sum_{i=1}^{s}(\rho_i)(\log_2 \rho_i)$$

where H = diversity of species
s = number of species
p_i = proportion of individuals of the total sample belonging to the ith species.

Thus, the index takes into consideration the number as well as the relative abundance of species. When the diversity index, H, is calculated for the two woodlands, the first, as described in Table 22.3, has a diversity index of 3.59, while the woodland described in Table 22.4 has a diversity index of 1.87. This index is obtained using $\log_2$, usually employed in the Shannon-Weiner formula. Diversity calculated using $\log_n$ is 2.49 and 1.30, respectively.

The two components, species richness and evenness, can be separated. The simplest determination of species richness is to count the number of species. In the first woodland that is 24 and in the other, 10. To determine evenness, you first have to calculate H_{maximum}, what H would be if all species in the community had an equal number of individuals. This can be calculated by:

$$H_{\max} = \text{l}_n S$$

where l_n = natural log
S = number of species

For the first woodland $H_{\max}$ is 3.18 and for the second, 2.30.

Evenness (J) is determined by:

$$J = H/H_{\max}$$

Evenness of the first woodland is 0.78 and of the second, 0.57. The first woodland has a more even distribution of species than the second.

Up to this point species diversity has been considered as a measure of diversity within a given community, or *alpha diversity.* Diversity between communities is *beta diversity.* It can be calculated by such techniques as coefficients of community, percent similarity, distance measures, and others. Two examples are given in Box 22.3. (For methods of determining community similarity see, for example, Brower and Zar, 1977.)

DIVERSITY GRADIENTS

Species diversity can be used not only to compare similar communities or habitats within a given region but also to examine global ecosystems. Traveling north from the tropics to the Arctic, one finds the numbers of species of plants and animals de-

Box 22.2

Indexes of Diversity

The Shannon-Wiener index of diversity discussed in the text is only one of a number of diversity indexes. The Shannon-Wiener index is based on information theory. It measures the degree of uncertainty. If diversity is low, then the certainty of picking a particular species at random is high. If diversity is high, then it is difficult to predict the identity of a randomly picked individual. Thus, high diversity means high uncertainty.

Another commonly employed index is Simpson's. It takes a different approach—the number of times one would have to take pairs of individuals at random from the population of individuals of all species to find a pair of the same species. The Simpson index of diversity is the inverse of the dominance index (see Box 22.1):

$$\text{diversity} = \frac{N(N-1)}{\Sigma n_i(n_i-1)}$$

or

$$1 - \frac{\Sigma n_i(n_i-1)}{N(N-1)}$$

Thus, in a collection of species high dominance means low diversity.

The Shannon-Wiener and Simpson indexes take into consideration both the richness and evenness of species. A much simpler index of diversity that does not take evenness into account is Margalef's:

$$\text{Diversity} = (s-1)/\log N$$

where s is the number of species and N is the number of individuals. Such an index does not allow differentiation among diversities of different communities having the same s and N and, therefore, is much less useful.

Box 22.3

Community Similarity

A number of methods are available for comparing similarity or dissimilarity of communities. The one most often recommended is Morisita's index, based on Simpson's index of dominance (see Brower and Zar, 1977; Horn, 1966). However, for illustrative purposes, we will use two simpler approaches: Sorensen's *coefficient of community* and *percent similarity.*

Coefficient of Community

$$CC = \frac{2c}{s_1 + s_2}$$

where c = number of species common to both communities and s_1 and s_2 = number of species in communities 1 and 2.

For the woodland examples:

s_1 = 24 species
s_2 = 10 species
c = 9 species

$$CC = \frac{2(9)}{24+10} = \frac{18}{34} = 52.9$$

Coefficient of community does not consider the relative abundance of the various species. It is most useful when the major interest is the presence or absence of species.

Percent Similarity

To calculate percent similarity (PS), first tabulate species abundance in each community as a percentage. Then add the lowest percentage for each species that the communities have in common. For the two woodlands, 15 species are exclusive to one community or the other. Thus, the lowest percentage for those 15 species is 0, and they need not be added in.

$$\text{PS} = 29.7 + 0.4 + 3.6 + 0.4 + 4.7 + 2.9 + 4.4 + 0.4 + 0.39 = 46.89$$

Percent similarity does consider relative abundance of various species in each community.

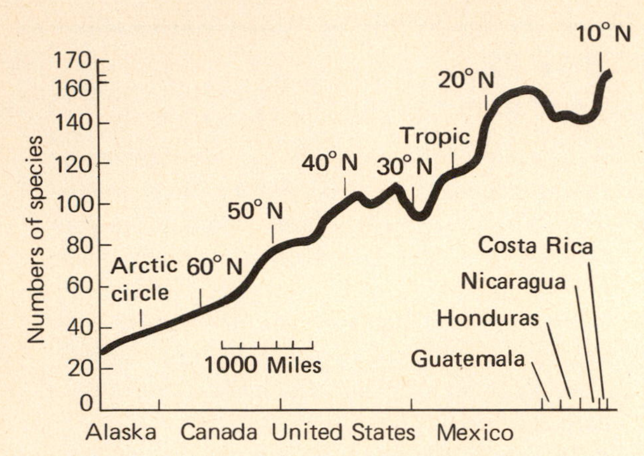

FIGURE 22.11 The diversity of mammals across a continent is influenced by temperature and moisture reflected in latitudinal and altitudinal variation. Species density of North American mammals is depicted graphically from the Arctic through Central America. (*After Simpson, 1964:62.*)

creasing on a latitudinal gradient. Species of nesting birds are much higher in Central America than they are in Newfoundland (Fischer, 1960). The same pattern exists among mammals (Simpson, 1964) (Figure 22.11), fish (Lowe-McConnell, 1969), lizards (Pianka, 1967), and trees (Monk, 1967).

But diversity is not restricted to a latitudinal gradient. In oceans, species diversity increases from the continental shelf, where food is abundant but the environment is changeable, to the deep water, where food is less abundant but the environment is more stable. Mountain areas generally support more species than flat lands, peninsulas have fewer species than adjoining continental areas, and islands, small or remote, have fewer species than large islands or those near continental land masses. From east to west in North America, the number of species of land birds (MacArthur and Wilson, 1967) and mammals (Simpson, 1964) increases. This increased diversity on an east-west gradient relates to an increased diversity of the environment both horizontally and altitudinally. Eastern North America has more uniform topography and climate and, thus, holds fewer species than western North America. However, because of more favorable moisture conditions, amphibians are more abundant and diverse in eastern North America than in the western part of the continent, while reptiles are more diverse in the hot, arid regions of the west (Kiester, 1971) (Figure 22.12).

SPECIES DIVERSITY: SOME HYPOTHESES

Species diversity within and among communities involves three components: spatial, temporal, and trophic. All in some manner relate to niche differentiation.

Differential use of space increases as vertical and horizontal stratification increase. Such stratification provides more microhabitats to exploit and niches to fill. Animals are able to partition habitats among them; the finer the partitioning, the more kinds of organisms can coexist. Plant diversity is influenced by changes in soil type, drainage, nutrient status, elevation, and the like.

Changing temporal use of habitats also increases diversity. Seasonal changes are most pronounced. Spring flowers give way to summer- and then fall-blooming species. Migrant summer nesting birds are replaced by winter migrants. Species active by day are replaced by nocturnal animals as darkness comes. Such temporal changes and replacements greatly increase the total species diversity in a community. Such temporal changes are often overlooked in determining species diversity.

Trophic differences that add to diversity are not so much those involving different trophic levels but, rather, subtle differences in feeding habits among species on the same trophic level. Among groups of ungulates, for example, some are browsers, feeding on woody vegetation, while others feed on herbaceous plants. Among those some feed on young growth; others on more mature growth. Some feed on grasses; others consume herbaceous plants. The degree of specialization is reflected in diversity.

Although temporal, spatial, and trophic components account for differences in diversity, certain questions go unanswered. Why, for example, is species diversity higher in tropical regions than in temperate and arctic regions? That question in particular has intrigued ecologists for years. They have

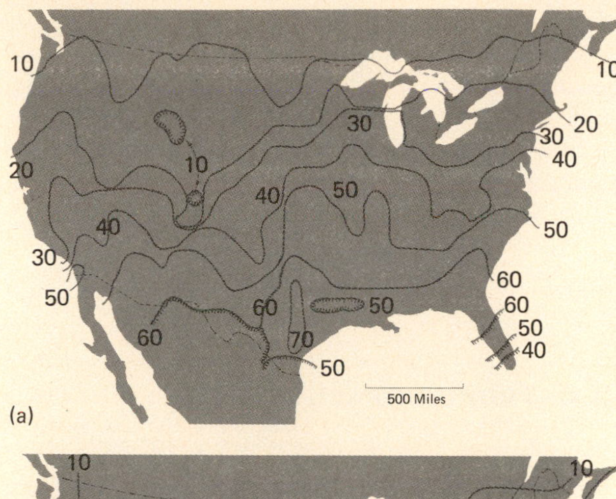

(a)

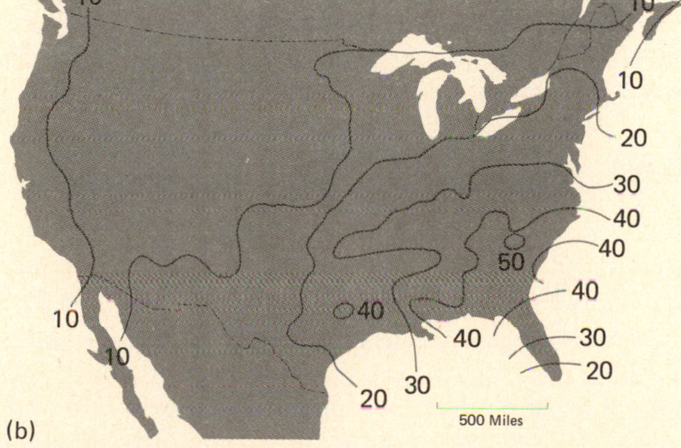

(b)

FIGURE 22.12 (a) Pronounced latitudinal variations occur among reptiles and amphibians. Being poikilothermic and endothermic, reptiles have their greatest density in hot desert regions and lower latitudes of North America. (*From Kiester, 1971.*) (b) Being not only poikilothermic but also highly sensitive to moisture conditions, amphibians reach their greatest diversity in the central Appalachians, then decrease northward and southward as well as westward. Species numbers are lowest in the dry and cold regions of the continent. (*From Kiester, 1971:131, 132.*)

come up with several hypotheses, mostly difficult, if not impossible, to test.

Perhaps the oldest is the *evolutionary time* hypothesis, which dates back to Alfred Wallace (1878). According to the time hypothesis, the tropical regions, in the words of Wallace, are "a more ancient world than that represented by the temperate zone," and, compared to the northern latitudes, relatively undisturbed by glaciation. For that reason, tropical regions have had more time for the evolution of plants and animals. Temperate regions have not experienced sufficient time for species to diverge, adapt to, or occupy completely the changed environment.

Related to the time hypothesis is the *climatic stability* hypothesis. In an unstable climate, species would develop tolerances sufficiently broad to allow them to adjust to and survive in a wide variation in the physical environment, food supply, and the like. Such species, in effect, would occupy broad niches. On the other hand, a stable climatic environment to which they need not constantly respond allows species to adapt to a variety of microclimatic habitats and to specialize in their feeding habits. In effect, they occupy smaller niches.

A climate may be variable over time, but predictable. According to the *climatic predictability* hypothesis, species have evolved ways to take advantage of seasonally predictable variations in climate and depend upon those variations in their life cycle. Plants, for example, put on seasonal growth, flower, and fruit during favorable periods of the year and go dormant during the winter or dry season. Desert annuals germinate and bloom only during periods of adequate rainfall. Animals migrate to a more favorable climate when conditions become severe or food and water become scarce, or they enter hibernation or aestivation and return or become active when environmental conditions become favorable again. This ability to specialize on predictable environmental conditions and temporal changes in food and other resources results in increased species diversity.

The *heterogeneous environment* hypothesis holds that the more complex the structure of the community, the more potential niches it possesses. That allows a greater opportunity for speciation among organisms to exploit those niches. Thus, a tropical rain forest, with its complex vertical structure, provides many more niches and is able to support many more species than a temperate forest, grassland, or arctic tundra. Variations in altitude and topography, locally or regionally, provide additional habitats, adding further to diversity.

The *productivity* hypothesis states that the more resources available in the form of nutrients, plants, or prey species, the more species are able to specialize. The tropical rain forests, with a long growing season and a large variety of plant species, have a high primary production. For that reason they are able to support many more animal species than temperate or arctic regions, with their much lower productivity. The more energy available in a usable form for organisms, the more species the ecosystem can support. Such an argument, however, does not quite hold for plants growing on nutrient-rich sites. Increased nutrient availability results in a reduction in plant species diversity and the dominance of a few species even though primary production is high. This is true both in grasslands (Mahmoud and Grime, 1976) and Costa Rican forests, ranging from dry forests to rain forests (Huston, 1979). However, a high primary production that is both stable and predictable allows the coexistence of more species than would be possible under a less predictable set of conditions. More availability of energy allows a greater specialization across a gradient of resources.

The production hypothesis relates to the *competition* hypothesis. In a more variable environment, the major selection forces come from the physical environment. In a more stable environment such as the tropics, selection forces are largely biotic, especially intraspecific and interspecific competition. Competition favors specialization, resulting in smaller niches.

Another hypothesis involving interpopulation relationships is the *predation* theory. It holds that a random or selective removal of prey species by a predator reduces the level of competition among them. That allows more species to coexist locally than would do so in the absence of predation, because populations of competitors are held low enough to prevent any one from becoming dominant.

Insular Ecology

A general observation holds that as the size of an area increases, species richness also increases up to some maximum point. Such a relationship between area and species richness is obvious on oceanic islands, a point noted by early naturalist explorers. Smaller oceanic islands held fewer species than larger islands, and oceanic islands remote from the mainland and larger islands had fewer species than those near to the mainland. The zoogeographer P. Darlington (1957) suggested a rule of thumb: a tenfold increase in area leads to a doubling of the number of species.

ISLAND BIOGEOGRAPHY THEORY

Such observed relationships were formally presented as a theory of island biogeography by R. MacArthur and E. Wilson (1967). The theory is relatively simple. It states that the number of species on an island is determined by a balance between immigration and extinction (Figure 22.13). *Immigration* will vary with the distance of the island from the mainland or a pool of potentially colonizing species. *Extinction* will vary with the area of the island. The smaller the area, the greater the probability of extinction. Theoretically, at equilibrium each new immigrant species must be matched by a species extinction. The number of species at equilibrium remains stable, but the composition of the species may change. The rate at which one species is lost and a replacement gained is the *turnover rate*.

The concept of island biogeography, as originally developed, related to oceanic islands, but the theory was soon applied to mainland island habitats: pockets of grassland in forests, fragmented woodlands surrounded by seas of agricultural land and urban developments, alpine mountaintops, and the like (MacArthur, 1972). The theory recently has become important in the management of endangered species and the development of nature reserves (for example, see Burgess and Sharpe, 1981; Soule and Wilcox, 1980).

PROBLEMS WITH THE ISLAND BIOGEOGRAPHY THEORY

The theory of island biogeography, which might better be called insular ecology, has been univer-

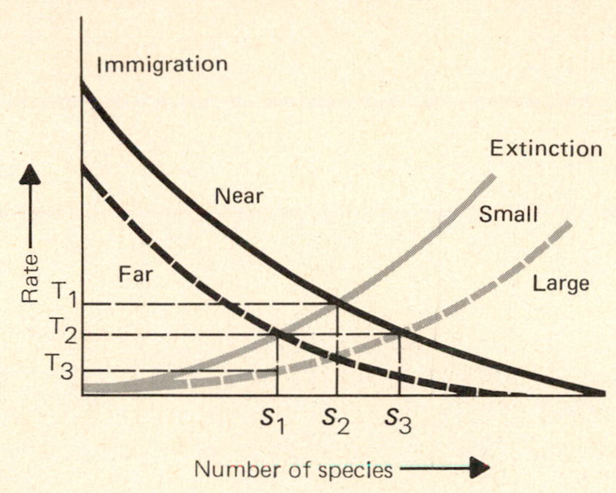

FIGURE 22.13 According to the island equilibrium model, immigration rates balance extinction rates. Immigration rates are distance-related. Islands near a mainland have a higher immigration rate than islands distant from a mainland. Extinction rates relate to area, being higher on small islands than on large ones. As the intersections of immigration and extinction curves indicate, species equilibrium is greater on larger islands than on small ones if both are the same distance from the mainland. The number of species expected on small islands close to the mainland is about the same as expected on large, distant islands. The turnover rate, immigration versus extinctions, is higher on small, near islands. They have small populations, which are more likely to become extinct, but at the same time they receive replacements easily from the nearby mainland. Large, far islands experience a low turnover rate. Their large populations reduce the chance of random extinction, and their distance from the mainland precludes a high immigration rate. (*Based on MacArthur and Wilson, 1967:22.*)

sally accepted, often uncritically. Although its major premise seems to hold, the theory has some weaknesses. One concerns the interpretation of immigration and extinction.

What is an immigrant? Any organism that happens to appear or be on the island or one that has established a reproducing population? As a rule, an immigrant is a species that arrives on and colonizes the island with a minimum number of individuals that breed and increase the population (Simberloff, 1969). Such a definition does not provide any time limit. Is a species that makes an appearance, breeds for a year or two, and disappears a true immigrant or a casual breeding visitor, counted twice in the calculation of a turnover rate? What constitutes hard-core immigrants? We have no satisfactory answer. But to accurately determine species equilibrium, you have to sort out the residents from the transients.

According to the theory, immigration varies only with distance from a resource pool, the size of which is based on the number of species occupying the mainland. It does not consider the ability of the species to emigrate. The presence of a species on the mainland does not necessarily make it a member of the pool. Low populations or poor dispersal ability may eliminate certain species as potential colonists. The presence or absence of other species may influence immigration. Colonization of islands by plants, for example, increases as the number of successful immigrants increases (M. Williamson, 1981). Further, island size may be as important as distance to the immigrants. Bigger islands are easier to find than small ones. Both size and distance may be complicated by the heterogeneity of the habitat. It may be difficult to distinguish that effect from island size, age, and distance from a source of immigrants.

Extinction is a function of rarity or population size. The smaller a population is, the greater the likelihood it will become extinct by chance. According to island biogeography theory, as the number of species on an island increases, the rate of extinction increases. Theoretically, the more species there are on an island, the rarer each one becomes, because the number of individuals of each species declines. Thus, an increased number of species increases the likelihood of any one species' dying out. But when does a species become extinct? When it disappears from an island or when it is reduced to a nonreproducing status?

Island biogeography theory considers only species numbers and not the species themselves. Turnover is measured by immigrations minus extinctions over a period of time. When species A moves onto an island, it is counted as a plus; when it disappears, it is counted again, as a minus. Thus, it is counted twice. But at the same time another species may

FIGURE 22.14 Trends in the numbers of three selected species of birds in Eastern Wood, a woodland island, Bookham Common, Surrey, England. The robin *Erithacus rubecula* is one of the most abundant breeding birds, obviously well adapted to small islands of woodlands. Its population fluctuated only slightly over the 27-year period. The willow warbler *Phylloscopus trochilus,* abundant earlier, became extinct when logging on the tract stopped and woody growth encroached on the area. The willow warbler, a bird of open woods, is dependent upon disturbance for maintenance of its habitat. This species points out the importance of some artificial disturbance in island situations to maintain variation in habitats. At the last census the willow warbler reappeared, thus experiencing both extinction and immigration. The starling *Sturnus vulgaris* immigrated into the area and apparently achieved equilibrium level in the oak wood. Its appearance may have been stimulated by improved nesting sites following cessation of logging. (*Adapted from M. Williamson, 1981:98.*)

be declining slowly and steadily and is counted only when it disappears (Figure 22.14). Island theory, then, virtually ignores population dynamics and life-history requirements of species involved and the recruitment of new immigrants onto an island, which augments or replenishes a species already there.

TESTS OF THE ISLAND BIOGEOGRAPHY THEORY

In contrast to the ready acceptance of island biogeography theory are the relatively few studies designed to test the hypothesis. A major experimental study was the defaunation of minute islands of mangroves in the Florida Keys by D. S. Simberloff (1969). The insect fauna was removed by covering the islands with sheeting and applying insecticides. The sheeting was then removed, and certain groups of arthropods were censused at frequent intervals during the first year and then two and three years later. The islands richest and poorest in species prior to defaunation were also the richest and poorest after defaunation. Islands with the greatest number of species were those closest to the main stands of mangroves. The number of insect species, however, continued to increase beyond the original, and species found frequently were not the same as those present before the experiment. Species turnover was high, in part because of the difficulty of separating transients from breeding colonists.

Few data are available on turnover in more natural situations. Most studies involve surveys on islands that happened to be censused at an earlier time. Species present during the most recent surveys are compared to species lists of previous surveys. The appearance of new species (immigrants) and the disappearance of old ones (extinction) are used to calculate the average turnover rate (for example, see Diamond, 1969, 1971).

Of considerably more interest are long-term annual censuses on a given island. An example is the seasonal census of confirmed nesting birds over a 26-year period in Eastern Wood of Bookham Common, Surrey, England. A 16-ha oak *(Quercus robur)* woods, it was part of the 112-ha woods that covers much of the common. Although small, Eastern Wood is close to a pool of potential immigrants. The woods was under some form of management, including timber cutting, until 1952. After that, encroachment of woody growth changed the habitat. These changes resulted in the only permanent extinction—that of the willow warbler *Phylloscopus trochilus,* associated with open woodland—and the appearance of one new immigrant, the starling *(Sturnus vulgaris).*

Sometime during the 26 years, 44 species of birds appeared in the woods. Of these, 6 apparently did not nest, leaving 38 breeding species. Of these, 4 species had territories extending beyond the woods, and 9 species never had more than two pairs nesting in the woods itself. Eleven more species nested in fewer than 5 years during the 26 years. They had to be considered casual species, counted as immigrants and extinctions in calculating annual turnover. Only 14 species were regular breeders. Thus, much of Eastern Wood's avifauna was subsidized by nearby larger woodland. If the tract were suddenly transformed into a truly isolated island, those 14 species would make up—for a time, at least—the avifauna of the island. Ultimately, only those species with a minimum of 10 breeding pairs would persist. These include the great tit (*Parus major*), blue tit (*Parus caeruleus*), wren (*Troglodytes troglodytes*), robin (*Erithacus rubecula*), and blackbird (*Turdus merula*).

During the 26 years, Eastern Wood experienced considerable turnover of species, with an average of three immigrations and three extinctions a year. Species equilibrium, as determined by immigration and extinction curves, is 32 species, somewhat higher than the 27 species that on the average inhabited Eastern Woods over the years (Figure 22.15).

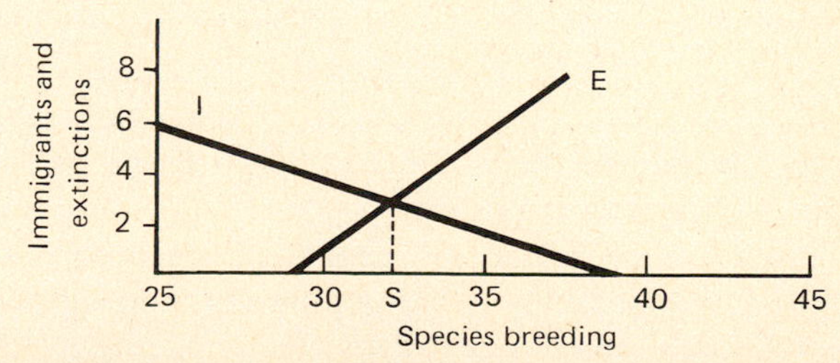

FIGURE 22.15 Immigration and extinction curves for Eastern Wood. The immigration curve is a regression line (the points have been omitted), which cuts the abscissa at 39 species. The maximum number of species that bred at one time or another in Eastern Wood was actually 44. The extinction line is at 45°, indicating one extinction for every species present over 29. The equilibrium point, where immigration intersects extinction, is 32 species. The lines are straight rather than curved because data are discrete, not continuous. (*Adapted from M. Williamson, 1981:101.*)

Data for Eastern Wood suggest that in general the theory of island biogeography is correct—that if the woods held 40 or more species, any new species that might breed in the woods would be a casual one, nesting infrequently at very low density; that the addition of a species would be expected to lead to one additional extinction. The weakness of the theory is its inability to take into account population dynamics and ecology of the individual species involved, which are important in the management of species facing an island existence.

FRAGMENTATION OF HABITATS AND ISLAND DYNAMICS

Eastern Wood is an example of the effects of fragmentation—the breaking up of large tracts of habitat such as woodland and prairie into smaller units separated by seas of urbanization and agricultural crops. In the process of fragmentation, species requiring large parcels of habitat, the interior species, disappear. Some maintain their populations if the population is supplemented by immigrants from a not too distant replacement pool, as in Eastern Wood. Other species, attracted by edge conditions, move in. Thus, some species are lost, some species remain, and some species are gained. The species composition of the habitat island shifts, usually toward edge or generalized species. The ability of interior species to maintain their presence depends upon the size of the fragment and its relationship to a pool of interior species.

What size island of remnant vegetation is needed to maintain regional populations and satisfy habitat requirements of the species concerned? At what island size does species richness reach its maximum? Such questions have stimulated a number of studies of the response of both plants and animals to habitat fragmentation.

There is a point in island size at which no interior species can exist. The size of a fragmented woodland, for example, may be so reduced that the edge merges into the forest interior and the fragment, for all purposes, is forest edge (Figure 22.16). Mesic, shade-tolerant plants are replaced by shade-intolerant, xeric, often opportunistic species over time, and animal species of the interior are replaced by edge species. As the size of the area increases, the ratio of edge to interior decreases, and the number of forest interior species may increase (Figure 22.16).

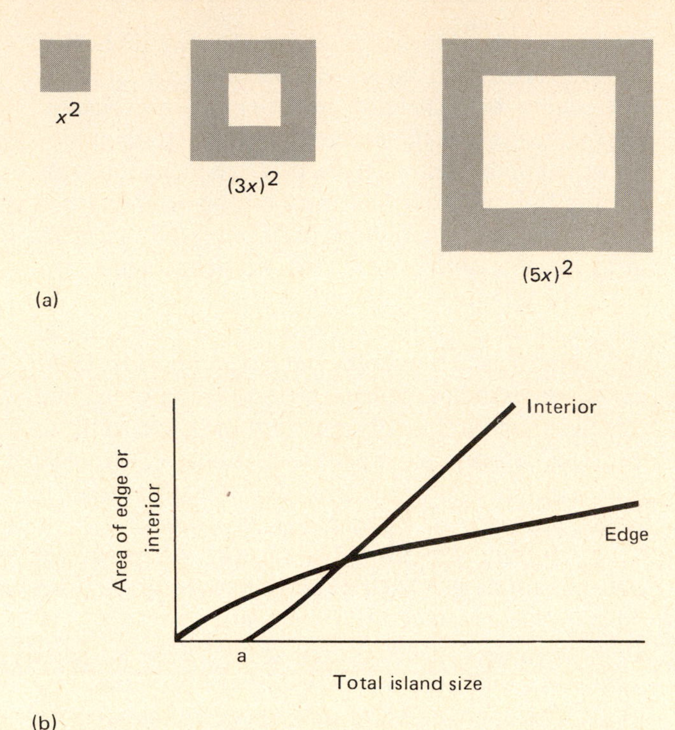

FIGURE 22.16 Relationship of island size to edge and interior conditions. Functionally, all islands are edge. (a) By allowing the depth of the edge to remain constant, the ratio of edge to interior decreases as island size increases. When the size of the island is large enough to maintain mesic conditions, an interior begins to develop. (b) The graph also shows this relationship. Below point a, the size at which interior species can exist, the woodland is edge. As size increases, interior area increases and ratio of edge to interior decreases. This relationship of size to edge holds for square or circular islands. Long, narrow woodland islands whose width does not exceed the depth of the edge would be edge communities, even though their area might be the same as that of square or circular ones. (*After Levenson, 1981:32.*)

The minimum size of forest habitat needed to maintain interior species differs with plants and animals. For forest interior plants the minimum area depends upon the size at which moisture and light conditions become both mesic and shady enough to support shade-tolerant species. Size, however, depends in part on the nature of the edge about the stand (Levenson, 1981; Ranney, Bruner, and Levenson, 1981)—whether it is closed, cutting down on the penetration of light and wind—and on canopy closure. If the stand is too small and too open, the interior environment becomes so xeric that it prevents reproduction by mesic species, both herbaceous and woody. As a result, when mature residual mesic species—such as sugar maple and beech—die, they are replaced by xeric species such as oak.

Several studies provide some insight into the impact of forest fragmentation on flora and fauna. Species richness of plants is greatest in edge situations where xeric species intermingle with some interior species (Figure 22.17). The total number of woody species in Wisconsin woodlots increases with woodlot size up to approximately 2.3 ha. At that size vegetation achieved a maximum balance between edge and residual interior species (Levenson, 1981). Beyond that size species richness declined and finally leveled off at 9.4 ha (23 acres), as mesic conditions returned to the interior and shade-tolerant species persisted. Thus, there is a negative correlation between edge species and the size of forest islands and a positive correlation between interior species and size.

The nature of species replacement, turnover, and immigration in Wisconsin woodlands is influenced by distance of the forest island from seed sources. Seed dispersal to woodlands is aided by

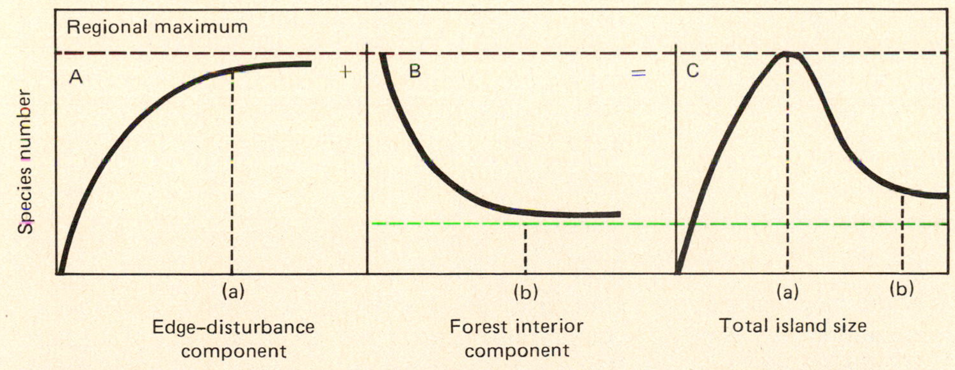

FIGURE 22.17 Species richness of a woodland relates to the ratio of edge to interior, area, and a variety of conditions. (A) Under ideal conditions of both sufficient area and variation in environmental conditions, all woody species of a region could coexist (a). The number would be asymptotic to the number of woody species found in the region. At some point *a* along the curve, interior conditions would prevail over edge. (B) As mesic, low-light conditions develop, light-demanding species will decline and the interior population will stabilize at the number of mesic, shade-tolerant species found in the region (b). (C) Considering edge and interior species together, species richness would follow the curve in C. Smaller islands would hold edge species exclusively up to some point *a*. Then, as interior conditions develop, edge species drop out, and as size increases, mostly interior species remain (b). (*After Levenson, 1981:34.*)

hedgerows and other narrow belts of vegetation linking one forest island with another. Species that are bird-dispersed are more successful immigrants than species dispersed by mammals and the wind. The latter two types tend to become local in distribution, leading to their extinction in many forest islands.

Among birds a similar pattern exists. Small forest islands of 5 ha or less are occupied by edge or ubiquitous species at home in any size forest tract. A New Jersey study (Forman, Galli, and Leck, 1976; Galli, Leck, and Forman, 1976) showed that maximum bird diversity was achieved with woodlands 24 ha in size. However, these woodlands held no true forest interior species such as the worm-eating warbler (*Helmitheros vermivorus*) and the ovenbird (*Seiurus aurocapillus*), which are highly sensitive to forest fragmentation and require extensive areas of woods (Whitcomb et al., 1981; Lynch and Whitcomb, 1977). The presence of forest interior species in smaller woodlands depends upon the nearness of those islands to a pool of replacement individuals. As with forest vegetation, species richness of forest interior species is positively correlated with island size.

ISLAND THEORY AND SPECIES MANAGEMENT

Stimulated though they are by island biogeography theory, studies of forest fragmentation are concerned more with the effect of size, shape, and distribution of forest fragments on the distribution and populations of species than with testing island biogeography theory. Some of their results, however, support island biogeography theory. Large islands of forests do hold more species than small ones; islands more distant from a mainland hold fewer species than near ones; species richness tends to reach some equilibrium related to island size and degree of isolation. But studies also expose some weaknesses of the theory: its treatment of species as equals; its inability to involve population size; and its indifference to the kinds and quality of species involved. Edge and ubiquitous species are favored over rarer, highly size-dependent forest interior species.

But it is not difficult to understand why island biogeography theory, among all ecological theories, relates closely to the management of wild plants and animals. It is relevant to such problems as the size, shape, number, and distribution of reserves and the degree to which fragmentation of habitat leads to species extinction.

Because of the lack of sufficient supporting data, island theory has led to vigorous debate over the very important question: Which is more preferable, one large reserve or two smaller reserves that add up to the same area? One group (Simberloff and Abele, 1976, 1982; Abele and Connor, 1979; Higgs, 1981) argues that two or more smaller reserves will hold more species than one large reserve of the same area. Data on species richness and island number seem to support that argument. Others (Diamond and May, 1976; Wilson and Willis, 1975; Whitcomb et al., 1976) argue that large islands are preferable because the large area will support not only more rare species but large populations of species, making them less vulnerable to random extinction. Populations on small islands may not be large enough to be self-sustaining and need to depend upon a subsidy of immigrants from distant sources, as exemplified by the bird populations on Eastern Wood.

There is more to consider. As large areas are fragmented into smaller ones, the populations they contain are going to decline to a much lower species equilibrium. (Recall the prediction that if Eastern Wood were completely isolated, the species equilibrium would collapse from 32 to 5 species.) For many species it would involve a sharp decline or crash, and the remnant population would be the progenitors of future generations. If the population of any species remains small, it would experience all the problems of small populations: a random sampling of gene frequency, genetic drift, erosion of genetic variation, increased inbreeding and accompanying loss of fitness, all increasing the probability of species extinction.

To avoid a loss of heterozygosity and to reduce inbreeding, a population has to maintain an effective size. The effective population size is not the number of individuals in a population nor the actual number of breeding individuals, but the size that determines the rate of genetic drift. Effective size is very sensitive to an imbalanced sex ratio, especially involving polygamous males and monogamous females. Only a few dominant males make a genetic contribution to the next generation, resulting in a loss of genetic variation. Thus, the future population traces its ancestry to few individuals, increasing the probability of inbreeding and homozygosity. It is almost impossible to estimate an effective population size, but perhaps a population of any less than 500 individuals might be in trouble.

Another problem of small island size is the possible loss of keystone species—those upon which a number of other species depend in one way or another. Often keystone species are plants that provide nectar or fruit. Or they may be species that maintain certain conditions in the habitat upon which other species depend. For example, a sharp reduction in the rabbit population in southern England from myxomatosis resulted in thick growth of meadow grass in fields inhabited by the spectacular large blue butterfly *Maculinea arion*. Heavy grass resulted in the extinction of open-ground ant colonies, the nests of which were utilized by large blue caterpillars. As a result, the large blue is nearly extinct. The loss of one keystone species, the rabbit, a grazing herbivore, resulted in the extinction—locally, at least—of two other species.

Further, small islands may not be large enough to encompass the home range or territory of certain species (see Harris, 1984). An island that may be large enough to support a full complement of both edge and those interior species with relatively small home ranges may not be large enough to support certain wide-ranging species whose area of home range far exceeds that of the island or reserve. Concern should not be over large islands versus small islands but rather over the maintenance of an intact ecosystem. The size of an ecosystem might be dictated by the area needed to support a viable population of animals high on the trophic level or those that require a large area to support them. Thus, ecosystems large enough to support timber wolves, grizzly bear, lions, or elephants undoubtedly would encompass all species concerned. Management should be based not only on island theory and species richness but on a knowledge of species requirements, responses, and population genetics as well.

Summary

A biotic community is a naturally occurring assemblage of plants and animals living in the same environment, mutually sustaining and interdependent, constantly fixing, utilizing, and dissipating energy.

All communities exhibit some form of layering or stratification, which largely reflects the life forms of plants and influences the nature and distribution of animal life in the community. The communities that are most highly stratified offer the richest variety of animal life, for they contain a greater assortment of microhabitats and available niches.

There are two opposing views concerning the nature of the community. According to the organismic school, a community is an integrated unit that has definable boundaries. The individualistic school argues that a community is a collection of populations that require similar environmental conditions. The makeup of any one community is determined in part by the species that happen to be distributed in the area and can grow and survive under prevailing conditions. The exact species that settle in an area and the number that survive are rarely repeated in any two places at the same time, but there is a certain recurring pattern of more or less similar groups. Rarely can different groups of communities be sharply delimited, because they blend together to form a continuum along some environmental gradient.

The place where two different communities meet is an edge. The area where two communities blend is an ecotone. An edge may be inherent, produced by a sharp environmental change such as a topographical feature or soil type. Or an edge may be induced, created by some form of disturbance and changes through time. Because it supports not only selected species of adjoining communities but also a group of opportunistic edge species, an edge has a high species richness.

Communities are organized about dominant species, especially in the temperate regions. The dominants may be the most numerous, possess the highest biomass, preempt the most space, or make the largest contribution to energy flow. But the dominant species may not necessarily be the most important in the community.

Communities may be characterized by species diversity. Species diversity involves two components: species richness, the number of species in a community, and equitability, how evenly the total number of individuals are apportioned among the species. A species diversity index is useful only on a comparative basis, either within a single community over time or among communities. Species diversity within a community is called alpha diversity and that between communities, beta diversity.

A relationship exists between species diversity and area. In a very general way, larger areas support more species than smaller areas. This species-area relationship is involved in the theory of island biogeography, which states that the number of species an island holds represents a balance between immigration and extinction. Immigration rates on an island are influenced by the distance of the island from the mainland or a pool of potential immigrants. Thus, islands the most distant from a mainland would receive the fewest immigrants and the ones closest to the mainland, the most. Extinction rates are influenced by the area of an island. Because small islands hold smaller populations and have less variation in habitat, they experience higher extinction rates than large islands.

The theory of island biogeography has practical applications. The fragmentation of natural habitats such as forests results in islands of remnant habitats in a sea of agricultural or urban lands. The effects of such fragmentation on species are predictable by island theory. It also serves as one guide to the establishment of nature reserves and the management of rare and endangered species.

In such situations, however, more than size and distance of islands are involved. Of critical importance are (1) interrelations of species involved, especially the role of certain keystone species in the maintenance of ecosystem integrity; (2) effective population size necessary to maintain genetic diversity and fitness; and (3) habitat heterogeneity, including early successional or disturbed areas required by some species.

Review and Study Questions

1. Distinguish between a phanerophyte and a therophyte; a chamaephyte and a cryptophyte.
2. Of what value is the Raunkiaer system?
3. Why are epiphytes so few or absent in temperate forests? What are some temperate forest epiphytes?
4. A number of studies of bird species diversity in forests rely on the relationship between foliage height diversity and bird species diversity. What is a weakness of such an approach to the relationship between vegetational diversity and animal diversity?
5. Contrast the stratification of an aquatic community with that of a terrestrial community.
6. What are edge, ecotone, and edge effect?
7. Discuss the reason why, from an ecological and a species point of view, edge should be considered separate from ecotone.
8. Discuss the utilization of a patchy environment by an edge species such as the cottontail rabbit or tiger swallowtail butterfly. What vegetational patches do they need? Why are they found only in edge situations? What is the role of disturbance in the maintenance of edge habitat?
9. How could an abrupt edge be modified to attract more edge species?
10. Species dominance is usually measured by such criteria as biomass, basal area, and number. Can you come up with a situation in which a species with the greatest biomass exerts a controlling influence in a community yet is not a dominant species as they are usually measured?

11. What is the theory of island biogeography?

12. Island biogeography theory considers immigration as a function only of distance from the source and extinction as a function only of population size. What else influences immigration and extinction?

13. What are some problems associated with the measurement of turnover rates? What constitutes a turnover?

14. How does the concept of island biogeography relate to the fragmentation of eastern deciduous woodlands?

15. Contrast interior species with edge species. Are all small island species edge species?

16. The concept of island biogeography is being applied to the establishment and management of nature preserves, wildlife refuges, and so on. There is considerable argument over what is best—many small islands or several large ones. The argument involves some very basic differences in the approach to the study of island biogeography theory. Refer to the literature developing on this topic. Are both sides right?

Suggested Readings

Burgess, R., and D. Sharpe, eds. 1981. *Forest island dynamics in man-dominated landscapes*. New York: Springer-Verlag.

Cody, M., and J. M. Diamond, eds. 1975. *Ecology and evolution of communities*. Cambridge, Mass.: Harvard University Press.

Harris, L. D. 1984. *The fragmented forest*. Chicago: University of Chicago Press.

May, R. M., ed. 1976. *Theoretical ecology: Principles and application*. Philadephia: Saunders.

Pielou, E. C. 1974. *Population and community ecology.* New York: Gordon and Breach.

Williamson, M. 1981. *Island populations*. Oxford, England: Oxford University Press.

CHAPTER 23

Succession

Outline

Objectives

Upon completion of this chapter, you should be able to:

1. Define succession and describe changes that occur in the various seral stages.
2. Discuss the influence F. Clements and H. Gleason had on the development of the concept of succession.
3. Contrast the population approach and the ecosystem approach to succession.
4. Compare the facilitation, tolerance, inhibition, relay floristics, and initial floristic models of succession.
5. Contrast early seral species with late seral species, from both a population and an ecosystem point of view.
6. Distinguish between the effects of large-scale disturbance and small-scale disturbance.
7. Explain gap-phase replacement, cyclic succession, and fluctuations.
8. Compare heterotrophic succession with autotrophic succession.
9. Explain why animal life changes with seral stages.
10. Discuss the concept of the climax and its limitations.
11. Comment on several concepts of stability as they relate to ecosystems.

Abandoned cropland is a common sight in agricultural regions, particularly in areas once covered with forest. No longer tended, the lands grow up in grasses, goldenrod, and other herbaceous species. If you followed the same piece of ground over a long enough period of time, you would observe that the same fields are invaded by blackberries, sumac, and hawthorn, followed by aspen and pine. Many years later, the abandoned cropland will support a forest of maple, hickory, oak, or pine. Thus, over a period of years, one assemblage of plant species has succeeded another until a relatively stable forest occupies the area. This replacement of one community by another over time is called *succession* (Figure 23.1).

Succession is one ecological process that has been described for numerous sites over the years. One of the better examples is the detailed study of old field succession in the Piedmont of North Carolina (Keever, 1950). The year a crop field is abandoned, the ground is claimed by annual crabgrass (*Digitaria sanguinalis*), whose seeds, lying dormant in the soil, respond to light and moisture and germinate. But crabgrass's claim to the ground is short-lived. In late summer the seeds of horseweed, a winter annual, ripen. Carried by the wind, they settle on the old field, germinate, and by early winter have produced rosettes. The following spring, horseweed, off to a head start over crabgrass, quickly claims the field. During the summer, the field is invaded by other plants—white aster (*Aster ericoides*) and ragweed (*Ambrosia artemissifolia*). Competition from the aster and the inhibiting effects of decaying horseweed roots on the horseweed itself allow aster to achieve dominance.

By the third summer, broomsedge (*Andropogon virginicus*), a perennial bunchgrass, invades the field. Abundant organic matter and the ability to exploit soil moisture more efficiently permit broomsedge to dominate the field. About this time, pine seedlings, finding room to grow in open spaces among the clumps of broomsedge, invade the field. Within 5 to 10 years, the pines are tall enough to shade the broomsedge. A layer of poorly decomposed pine needles (duff) that prevents most pine seeds from reaching mineral soil, dense shade, and competition for moisture between successfully germinated seedlings and shallow-rooted parent trees inhibit pines from regenerating themselves on the site. In time, however, shade-tolerant oaks and sweetgum, which have long taproots, exploit a moisture supply unavailable to shallow-rooted pines. The hardwoods grow up through the pines and, as the pines die out (if they are not cut), take over the field. Further development of the hardwood forest continues as shade-tolerant trees and shrubs fill in the understory—dogwood, redbud, sourwood, hydrangea, and others. The sere (sequence of communities) has arrived at a mature or tolerant stage, in which only the dominant species of the crown can reproduce themselves in their own shade.

As the plants change, so do the animals. Early stages are characterized by such arthropods as crickets, grasshoppers, and spiders and by such seed-eating birds as mourning doves. Broomsedge brings in meadowlarks and meadow mice, and the low pine, rabbits. Mature pines shelter pine warblers and sparrows. As pines decline and hardwoods claim the area, downy woodpeckers, flycatchers, and hooded and Kentucky warblers appear.

Similar successional trends take place in such freshwater ecosystems as ponds and lakes (see Chapter 25). Rapidity of succession is determined by depth, by deposition of silt from inflowing waters, and by the density of aquatic vegetation. Succession starts with open water and its planktonic community until enough bottom silt accumulates to support submerged vegetation such as chara or muskgrass and pondweeds. Sediments carried by surface runoff and inflowing streams, augmented by detritus produced within the pond, fill the basin. As depth becomes more shallow, floating plants and emergents colonize the pond. Where sediments build above the water level, alders and willow come in, followed by such trees as silver maples, red maples, and elms. In time, as the site becomes drier, these trees are replaced by oaks and other low-ground or mesic hardwoods.

(a)

(b)

(c)

(d)

(e)

FIGURE 23.1 Successional changes in an old field over 30 years. (a) The field as it appeared in 1942, when it was moderately grazed. (b) The same area in 1963. (c) A close view of the rail fence in the left background of (a). (d) The same area 20 years later. The rail fence has rotted away, and white pine and aspen grow in the area. (e) The same field in 1972. Note how aspen has claimed much of the ground.

Historical Overview of Succession

At one time succession was easily explained. It was unidirectional, predictable change in vegetation on an area over time. One community replaced another in response to changes brought about by the vegetation itself, ending in a self-reproducing terminal community. The entire sequence of communities was a *sere*. Each community was a *seral stage*. The final stage was the *climax*. Succession that started on a site previously uncolonized by plants or animals was *primary*. If primary succession began on a wet substrate, it was called *hydrarch*. If it began on a dry substrate, the succession was called *xerarch*. Ultimately, according to successional theory, both hydrarch and xerarch succession would converge into a relatively mesic regional climatic climax. By contrast, succession that started on a disturbed site that had already supported vegetation was called *secondary*.

The architects of this approach to succession were initially Henry Cowles and then Frederick Clements (see Chapter 1). Clements (1916) viewed succession as a process involving several phases, to each of which he gave his own terminology. Succession began with a bare site, called *nudation*. Nudation was followed by *migration*, the arrival of propagules on the area. Migration was followed by the establishment and growth of vegetation. Clements called this phase *ecesis*. As vegetation became well established, grew, and spread, various species began to compete for space, light, and nutrients. This phase was *competition*, which was followed by *reaction*. That consisted of self-induced (*autogenic*) effects of plants on their habitat. The outcome of this reaction was the replacement of one plant community by another, resulting in the persistence of one species complex or community, or *stabilization*. In spite of vigorous criticism of Clements's concept of succession, the basic processes he outlined are still valid and accepted subconsciously, even by his critics.

The problem with Clements's view of succession lies beyond the outline of the process. To Clements, each stage of succession represented a step in the development of a supraorganism, the climax. The climax was able to reproduce itself, "repeating with essential fidelity the stages of its development." Each seral stage so modified the environment that plants of a particular seral stage could no longer exist there. They were replaced by plants of the next stage until the vegetation arrived at the self-reproducing climax. That marked the end of succession.

Clements had his critics. A contemporary one was H. A. Gleason (see Chapter 1). He viewed succession as a random process in which short-lived colonizing species were eventually replaced by longer-lived species. Plants involved in succession were those that arrived first on the site and were able to establish themselves under the prevailing environmental conditions. The final outcome was determined by competitive and other interactions among the species; by the plants' modification of the microenvironment; and by the ability of the plants to exploit nutrients, moisture, and other environmental inputs on the site.

Later, Frank Egler (1954) recast these views of succession in terms of *relay floristics* and *initial floristic composition*. Relay floristics most nearly corresponds to Clements's idea of succession—groups of associated species marching together and disappearing through time as one group replaces the other. Initial floristics, involved only in secondary succession, proposes that the propagules of most species, both pioneer and late-stage, are initially on the site. Which species becomes dominant is influenced by life-history characteristics and competitive interactions. Short-lived species are eventually replaced by long-lived, but not necessarily climax, species.

Current Views on Succession

The foregoing views on succession are evident in two current divergent theories of succession. One is the holistic ecosystems approach; the other is the reductionist population approach.

The ecosystem approach is a direct descendant of Clements's organismal theory of succession. It considers succession as a developmental process leading to the formation of an emergent entity with its own unique characteristics involving nutrient flow, biomass accumulation, and species diversity (E. P. Odum, 1969, 1983). Succession begins with developmental stages of short-lived intolerant plant species and terminates with a mature stage dominated by long-lived species. The process is community-controlled. Young developmental stages so modify the environment that the existing community is replaced by a subsequent more mature one, better able to exploit the changed environment.

As succession proceeds, ecosystem attributes change. Succession begins, for example, with an unbalanced community metabolism. Gross production in young ecosystems greatly exceeds respiration, and biomass accumulates. As the ecosystem matures, respiration begins to equal production. Ultimately, in the terminal or mature stage of succession, production balances respiration. The ratio of biomass accumulation increases to a point at which maximum biomass is maintained per unit of energy flow. Nutrient cycling, more open in earlier stages, becomes closed and internal, with minimal dependence on external inputs. Storage and turnover time of nutrients within the system increase during the development of the sere, as do control of nutrient retention and conservation. As species composition of the sere changes through time, species diversity increases. In a way, succession is ecosystem evolution.

That concept of succession as an orderly transition of one seral stage to another driven by self-induced (autogenic) modification of the environment is rejected by those who support a reductionist point of view. They argue that succession can best be explained by physiology, life-history strategies, and population dynamics, especially regeneration and mortality (Peet and Christensen, 1980; Noble and Slatyer, 1980). Noble and Slatyer (1980) have proposed a set of vital attributes applicable to secondary plant succession. Such attributes include the method of arrival on a disturbed site, persistence of a species on the site before and after disturbance, the ability of a species to establish itself and grow to maturity in a developing community, and the time a species requires to reach its critical life stage and replacement. What species become established is largely a stochastic process, one subject to chance. Species composition over time is determined by development, longevity, and response to competition. There is no need to invoke emergent properties or to seek such indices of succession as changes in ratio of gross production to biomass. In fact, argue the reductionists, the patterns described by ecosystem holists do not hold. Species diversity may be greater in the earlier stages than in the mature one. Maximum production of biomass precedes the climax; and the tightest nutrient cycling is somewhere partway on the road to the mature stage.

J. Connell and R. Slatyer (1977) proposed three models of succession based largely on a reductionist approach. One is a *facilitation* model, which is basically a Clementsian approach. Early successional species modify the environment so that it becomes more suitable for later successional species to invade and grow to maturity. In effect, early-stage species prepare the way for later-stage species, thus facilitating their success.

The *inhibition* model involves strong competitive interactions. No one species is completely superior to another. The site belongs to those species that, in the colorful words of a Civil War general describing how armies hold ground, are "the firstest with the mostest." The firstcomers hold the site against all invaders. They make the site less suitable for both early and late successional species. As long as they live and reproduce, they maintain their position. These species relinquish it only when they are damaged or die, releasing space to another species. Gradually, however, species composition shifts as short-lived species give way to long-lived species.

A third model, the *tolerance* model, suggests that later successional species are neither inhibited nor aided by species of earlier stages. Later species can invade a site, become established, and grow to maturity in the presence of those that preceded them. They can do so because the later species tolerate a lower level of some resources than the earlier ones. Such interactions lead to communities composed of those species most efficient in exploiting resources, either by interference competition (see Chapter 15) or by utilizing sources unavailable to other species. An example might be a highly shade-tolerant species that could invade, persist, and grow beneath the canopy because it is able to exist at a lower level of one resource—light. Ultimately, through time, that species would prevail.

Current ideas about succession are at best muddled. The process, so obvious on disturbed sites, is so variable from place to place, from ecosystem to ecosystem, that it defies broad generalizations. Problems for ecologists rise in part because no one can follow succession sequentially on a given site for a sufficiently long period of time to develop any definitive inferences about succession. Instead, our ideas about succession are derived from studies of variously aged stages on different sites—studies over space rather than over time.

Mechanisms of Succession

Primary or secondary, the basic processes or mechanisms of succession are similar. Primary succession always begins with a site never before colonized by life. Secondary succession begins with a disturbed area that supports or has supported life. Secondary succession can begin at any point along a gradient of disturbance from a barren area to one supporting some advanced stages of vegetative growth such as a clear-cut forest.

POPULATION ATTRIBUTES

Succession in its very earliest stages begins with an open site (nudation). The barren area is eventually colonized by early successional species, variously called pioneer, opportunistic, or fugitive. The three names are synonymous, although each carries with it a somewhat different yet appropriate connotation. On a primary site these species have to move into the area (migration), carried there by wind or animals. On a secondary site, residual propagules—seeds lying dormant in the soil, roots, and rhizomes—play an important role in colonization. These are joined by migrants from an available seed pool. What species arrive on the site depends upon such variables as distance, vagility of seeds, wind direction, size of seed crop, and other factors. In other words, colonization is largely a matter of chance. Having arrived on the site, the pioneers have to establish themselves and grow (ecesis). Success of the colonists depends upon their ability to cope with a stressful environment and to compete with individuals of their own and other species. Because they are tolerant of and grow faster under initial stressful conditions, annual and biennial species usually appear first.

Plants of early seral stages in general are small in size, low-growing, have short life cycles, and reproduce annually by seeds or send out new growth from buds near the ground (geophytes). They grow rapidly, attain dominance quickly, and put most of their production into photosynthetic tissue. Pioneers maintain their dominance by suppressing for a while the growth of any later-stage plants that might exist as seedlings beneath them. They produce large numbers of easily dispersed small seeds, which can remain dormant for a long time in the soil waiting for favorable environmental conditions to germinate. These plants respond quickly to disturbance, especially exposure of mineral soil. And they are tolerant of fluctuating environmental conditions, particularly a wide range of daily temperatures on the soil's surface, alternate wetting and drying, and intense light.

Because pioneer plants of the initial stage of succession have to renew synthetic structures each year, they eventually lose their temporary dominance. In time, they are suppressed by the taller, more vigorous growth of later-stage plants, which carry over biomass from year to year. These plants gradually assume dominance and are better able to exploit the site. At the same time, they also place a much greater demand on resources such as nutrients, light, and moisture. As the availability of resources decreases and demand increases, competition within and among populations of plants becomes more intense. Mortality, as reflected in self-thinning, increases (Chapter 13). As the population degenerates, resources are released for use by plants of still later stages (Peet, 1981).

Plants of late stages of succession grow more slowly and are relatively long-lived and thus are able to dominate the site over a much longer period of time. Much of their production goes into storage as biomass and into maintenance. They produce relatively few, heavy seeds, which are primarily dispersed by animals or gravity. Their seeds are large, providing an abundance of nutrients to get new seedlings started, but the seeds' vitality and longevity are rather low. The species mostly are specialists, adapted to a narrow range of environmental conditions in which the plants either hoard resources or use them more effectively (see Chapter 21 and the discussion of *K*-selection in Chapter 14).

Thus, succession may be considered as the expression of differences in colonizing ability, growth, and longevity of species adapted to grow along a gradient of changing environmental conditions. As environmental stresses and conditions change, plant species also change gradually. The replacement of one or several species or groups of species by another results in part from interspecific competition, which permits one group of plants to suppress slower-growing species. As slower-growing plants of earlier stages are supplanted by species of later stages, the structure of the community as dictated by growth forms and longevity of the plants also changes. Eventually, succession arrives at a point where long-lived species create a relatively stable community called the climax (see page 483).

Succession as described emphasizes the role of population dynamics and life-history strategy in seral changes. But as seral stages advance, environmental conditions change. In some situations, the control plants have over their environment also has a pronounced influence (facilitation), as suggested by classic successional theory. This is especially true in boreal ecosystems dominated by spruce. Succession proceeds from mesic balsam-poplar and white spruce to a hydric, slow-growing black spruce stage. An organic mat of needles thickens on the forest floor, and deepening shade encourages the growth of a layer of moss. This blanket of organic matter and moss creates a colder soil temperature and a rise and thickening of the permafrost. Cold soil, poor drainage, and an accumulation of nutrients by moss bring about a decline of the spruce. The spruce stage then moves into a climax vegetation of a treeless moss-lichen association (Viereck, 1970).

Similar influences occur elsewhere on primary sites such as sand dunes and glacial deposits, in particular. Through deposition of organic matter and shading of the surface, plants reduce surface evaporation, increase nutrient availability, and otherwise modify the environment enough to permit more demanding plants to invade. Some pioneer species, such as crabgrass and sunflower, may produce chemicals that inhibit their own growth (E. L. Rice, 1972), an allelopathic effect that paves the way for invasion by grasses that are not affected by the toxins of weeds (Parenti and Rice, 1969). Grasses, in turn, may inhibit nitrogen-fixing bacteria, thereby slowing succession to the next stage (E. L. Rice, 1972).

Succession can be considered as an expression of differences in colonizing ability, growth, and survival of organisms adapted to a particular set of conditions on an environmental gradient. The position of a species on the gradient is determined by the length of life cycles, time of reproduction, reproductive output, and other characterisitics (Pickett, 1976). As certain environmental conditions and stresses change, populations of plants and sessile animals, such as barnacles and mussels, also change. The replacement of one or several by others results from interspecific competition and the interactions of herbivores, predators, and disease, which permit one group of organisms to suppress slower-growing or less tolerant species.

ECOSYSTEM ATTRIBUTES

The ecosystem approach to succession views the process as a property of the community driven by changes in attributes between youthful and mature systems. According to the ecosystem model, early or youthful stages of succession are characterized by relatively few species, low biomass, and dependence on an abiotic source of nutrients. Net community production is greater than respiration (see Table 23.1), resulting in an increase in biomass over time. Energy is channeled through relatively few pathways to many individuals of a few species, and

TABLE 23.1 EXPECTED TRENDS IN ECOLOGICAL SUCCESSION FROM DEVELOPMENTAL TO MATURE SYSTEMS

Attribute	Trend	Accept or Reject*
Biomass	Increases	A
GPP/ER	Approaches 1	R (plants)
GPP/B	Decreases	A
B/ER	Increases	A
Net community production	Decreases	R (plants) A (animals)
Total organic matter	Increases	A
Inorganic nutrient input	External to internal	?
Species richness	Increases	A
Species equability	Increases	R
Stratification	Increases	R
Size of organisms	Increases	R
Niche specialization	Broad to narrow	?
Role of detritus	Increases in importance	?
Growth form	r to K	R
Nutrient conservation	Increases	R

Note: GPP = gross primary production; ER = ecosystem respiration; B = standing crop biomass.
*Accept or Reject: If the trends of attributes are considered hypotheses, then they can be tentatively accepted or rejected based on current data. Accept = A; Reject = R.
SOURCE OF ATTRIBUTES: Odum, 1981: 446.

production per unit of biomass is high. Food chains are short, linear, and largely grazing.

The mature stages in succession are characterized by a greater diversity of species, high biomass, a nutrient source largely organic in nature, and gross production that about equals respiration. Food chains are complex and largely detrital. Inorganic nutrients accumulate in soil and vegetation, and considerable quantities are locked or hoarded in plant tissue (see Chapter 21).

As seral stages advance, physical stratification increases, with a corresponding increase in niche space. Fundamental niches (see Chapter 15) shift from broad, general to narrow, specialized ones. Accompanying these changes is an increase in the diversity of species.

How well does succession conform to the ecosystem model? Studies of ecosystem functions provide some insight into the validity of ecosystem theory. Consider species diversity. Diversity does not necessarily increase with advancing successional stages (Figure 23.2). Early stages in old field succession may have a greater diversity of plants than later stages (Tramer, 1975). Some of the later stages may be dominated by plants with strong allelopathic interference, which reduces species diversity (Bazzaz, 1975). Sassafras *(Sassafras albidum),* for example, maintains itself in relatively pure stands by releasing into the soil at different times of year phytotoxins that inhibit the germination of seeds and the growth of other plants (Gant and Clebsch, 1975). Diversity of both bird species and trees increases with succession but reaches a maximum before the climax stage. In earlier forest stages, species both tolerant and intolerant of shade and other environmental stresses share the site (see Chapter 22). When the community reaches late stages, diversity declines.

Other attributes also fail to match the theory. This is suggested by the results obtained from 15 years of data collected from a developing northern hardwoods forest at Hubbard Brook, New Hampshire. Bormann and Likens (1979) divided the seral stages of forest recovery following clear-cutting into a *reorganization,* or young, seral stage; an *aggradation,* or developmental, seral stage, in which the growing forest is passing through the shrub and pole timber stage; a *transition* phase, when growth is beginning to slow and a number of individual trees die from competition and self-thinning; and finally, a mature, or *steady-state,* phase. Data for ecosystem function in the reorganization and early aggradation phase are actual; those for later stages are simulations based on actual data using the JABOWA model.

The ratio of production to respiration (P-R) did not follow a straight route from > 1 to $\simeq 1$. Instead, the P-R ratio for the reorganization stage and the transition stage was > 1, and for the steady state, equilibrium. Only in the aggradation phase was the P-R ratio greater than 1. These changes in the P-R ratio reflected the loss of biomass through decomposition of dead organic matter following clear-cutting in the reorganization phase and the death of trees in the transition stage. Gross production increased through all stages, and net production was highest in the aggradation stage, which led to a rapid accumulation of biomass. Biomass accumulation was highest in the aggradation stage, declined during the transition stage, and leveled off during the steady-state phase (Figure 23.3). Nutrient cycling was most loose during the reorganization phase, when quantities of nutrients were being exported from the system. The reason was rapid decomposition of detrital material and the inability of incoming vegetation to utilize all of the resources. Nutrient cycling was tightest during the aggradation phase, which was the most highly regulated of all stages. Habitat diversity was highest in the steady-state phase, and species richness was highest in the reorganization stage and the early aggradation stage, with its mix of shade-intolerant, shade-tolerant, and intermediate species.

CONTROL BY PHYSICAL CONDITIONS

The idea of succession is most applicable to biologically controlled ecosystems. Some major ecosystems, however, are controlled by physical conditions, such as strong currents in estuaries, wave action on rocky shores, high temperatures and low rainfall in deserts. Because of the harsh environment, relatively few species can live there. Since

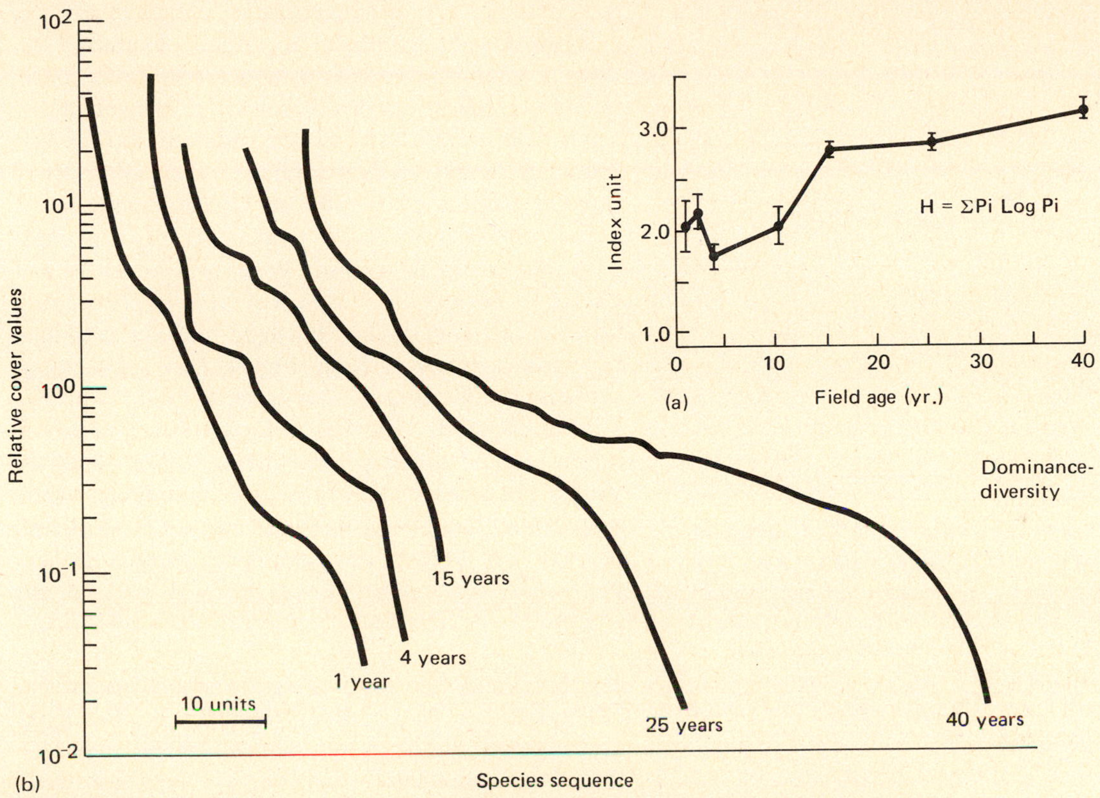

FIGURE 23.2 Relationship between plant species diversity and succession. (a) (Inset) Plant species diversity generally increases with succession and may reach a maximum in the forest stage when shade-tolerant and shade-intolerant species are together in the community. Relatively low species diversity in a successional community (as in the example graphed here) may result from strong dominance by a species with allelopathic interference. High species diversity may result from a high degree of vertical and horizontal microenvironmental heterogeneity. (b) Dominance diversity curves of successional communities are geometric at first. Dominance curves become less steep with time as more species are added, and gradually a log normal distribution with an increase of species possessing intermediate relative importance values develops. (*From Bazzaz, 1975:486, 487.*)

the early-stage species are the same as the climax species, no succession occurs.

Role of Disturbance

Disturbance is an integral component of secondary succession. It initiates succession and maintains some types of ecosystems. A disturbance (often tagged a perturbation) is a disruption of the pattern of an ecosystem by some physical force such as fire, wind, or timber harvesting.

LARGE-SCALE DISTURBANCE

A severe disturbance over a large area results in the colonization of the site by opportunistic species.

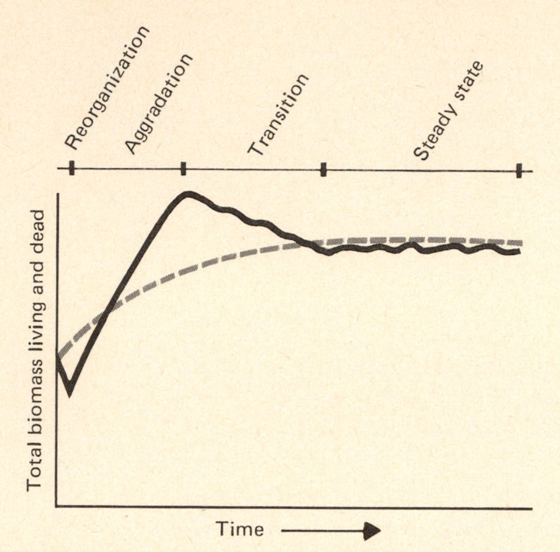

FIGURE 23.3 Model of biomass accumulation during successional development of a forest after clear-cutting. The dashed line is an asymptotic model of net biomass accumulation (living and dead) until a steady state is achieved, as predicted by the ecosystem theory of succession. The solid line is biomass accumulation as observed and predicted by a shifting-mosaic steady-state model. (*From Bormann and Likens, 1979:166.*)

What species will colonize the area depends upon local environmental conditions and dissemination strategies employed. An opportunistic species may become established by a continuous input of seed from the outside or by storage of seed in the soil. Successful colonization of a disturbed site by direct natural seeding depends upon such conditions as distance of seed source, size of seed crop, timing of seed arrival, and favorable microclimate on the site. To take advantage of the disturbed conditions, the opportunist should be nearby; its seeds should reach the site in sufficiently large numbers to ensure some seed survival; and seed should settle on exposed mineral soil to favor rapid germination and successful seedling survival.

Some opportunistic species colonize a disturbed area by means of such propagules as dormant seeds or roots on the area at the time of disturbance. One example is pin cherry *(Prunus pensylvanica),* whose seeds are carried to a forest by birds and small mammals or are deposited on the forest floor by an earlier stand of cherry. Pin cherry seeds can remain dormant for up to 50 years. When the forest canopy is removed and moisture, temperature, and light conditions become favorable, pin cherry seeds germinate, and young trees quickly dominate the site, crowding out the associated blackberry (*Rubus* spp.), which also colonizes the area. If seedling growth is dense, a pin cherry canopy can close in 4 years, eliminating other species except highly shade-tolerant seedlings of sugar maple and beech (Figure 23.4). If seedling growth is moderately dense, species with wind-disseminated seeds, such as yellow birch and paper birch, will also occupy the site. Within 30 to 40 years, pin cherry dies out, allowing birch, sugar maple, and beech to dominate the site. But during its period of tenure, pin cherry contributes numerous seeds to the forest floor, ready to reclaim the site when another disturbance provides the opportunity.

Most existing studies of responses to disturbance relate to tree species. Very few studies have investigated the response of the herbaceous (or field) layer of a forest to disturbance. Ash and Barkham (1976) studied the response of the herbaceous understory of an English coppice forest after cutting. (A coppice forest is one in which stump sprouts or root suckers are maintained as the main source of regeneration, with cutting rotations between 20 and 40 years.) Cutting coppice involves complete canopy removal, resulting in increased surface temperature on the forest floor and full exposure to light. Typically, a number of open-habitat or opportunistic species germinate and become established (Figure 23.5) but are soon excluded by the developing canopy. In spite of the disturbance, characteristic woodland species persist throughout the cycle. Adapted to a high-light regime in spring before the leaves are out, these plants have the ability to tolerate high light intensities. At the same time, they are able to coexist with annuals and open-habitat perennials that cast a shade on the ground like tree cover. Opportunistic species disappear, and woodland herbs again assume dominance, often developing into monospecific stands (see also Zamaro, 1982).

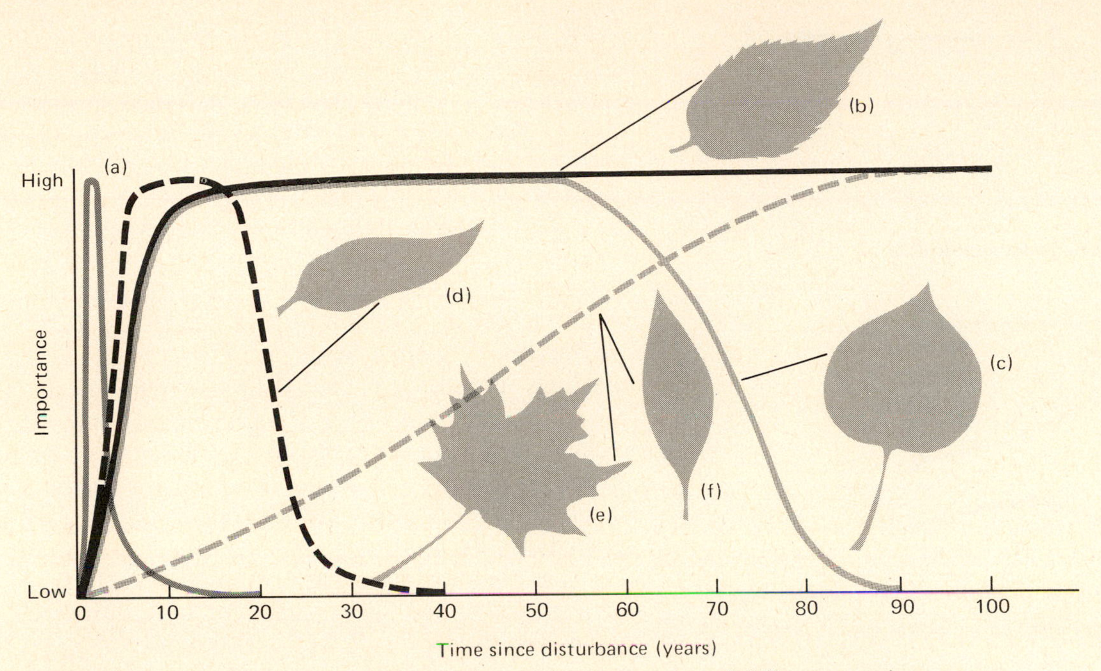

FIGURE 23.4 Diagrammatic representation of the importance of different species along a gradient of time following a disturbance of a typical northern hardwoods forest. Immediately after disturbance, blackberries (a) dominate the site, but they quickly give way to yellow birch (b), quaking aspen (c), and pin cherry (d). Intolerant pin cherry assumes dominance early but within 30 years fades from the forest. Yellow birch, an intermediate species, assumes early dominance, which it retains into the mature or climax stand. Trembling aspen, an intolerant species, begins to drop out after about 50 years. Meanwhile, sugar maple (e) and beech (f), highly tolerant species, slowly gain dominance through time. In about 100 years the mature forest is dominated by beech, maple, and birch. (*After Marks, 1974:75.*)

SMALL-SCALE DISTURBANCE

If the area involved is relatively small or the intensity of disturbance is relatively slight, response to disturbance entails reorganization of the vegetation that occupied the site prior to disturbance. Reorganization may take the form of canopy expansion; growth of new branches on trunks of remaining trees; stump sprout, as in oaks and maples; root suckering, as in aspen; and growth of suppressed advanced regeneration (seedlings) to fill in newly created gaps.

Gap-Phase Replacement

Gaps, large or small, formed in the forest represent sites of temporary reduction in the utilization of light, moisture, and nutrients. Suppressed growth is quickly stimulated by this sudden abundance of resources. If gaps are small, such as those created by the death or removal of individual trees, the response is typically reorganization of vegetation. The canopy expands, and increased light and moisture on the forest floor may stimulate the growth of understory shrubs and advanced regeneration of forest trees. Usually, small gaps favor or stimulate the growth of tolerant species (Trimble, 1973), so that in a terminal or climax community, replacement species are likely to be those of the canopy.

If gaps are large—the kind that would result

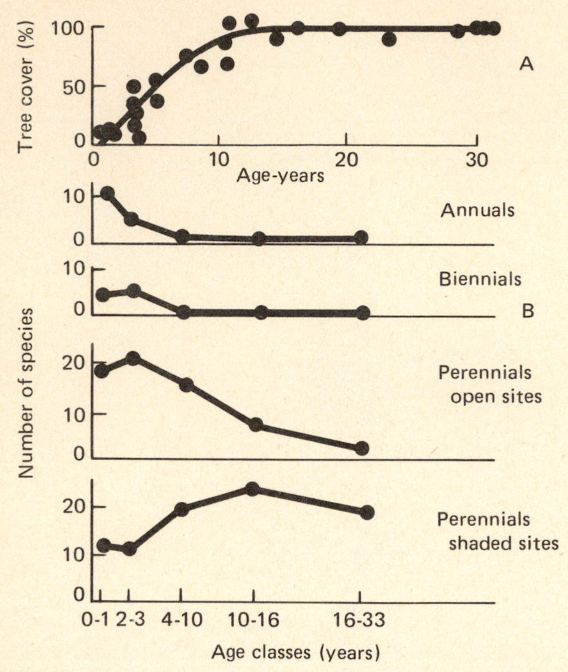

FIGURE 23.5 Response of understory to disturbance in a coppice stand in England. The graphs show changes in the percentage of cover produced by the growth of the canopy and the number of herb and shrub species in the field or ground layer at different times after coppicing. Not all species were present during the 30-year period of growth. Total numbers for each type are indicated. Note the rapid decline of annuals, biennials, and open-site perennials as the canopy closes. Open-site perennials dominate the field layer shortly after coppicing. Perennials of shaded sites show a sigmoidal growth response as the canopy closes. (*From Ash and Barkham, 1976:706.*)

from timber cutting, insect damage, windstorms, and ice storms—the response may involve both vegetational reorganization and invasion by opportunistic species. In the Appalachian hardwood forests, for example, stump sprouts and suppressed seedlings may respond rapidly and fill in the gap in several years (Trimble, 1973). Invasion by opportunistic species is limited, and the future composition of the gap will be determined in part by the competitive interactions of incoming growth. Intolerant species such as yellow poplar and black cherry may outcompete tolerant species such as sugar maple, which will remain in the understory, capable of filling small gaps that might appear later in the established forest. Both yellow poplar and cherry would remain for some time as a component of the mature or equilibrium forest.

Gap formation and gap replacement are important in maintaining diversity in tropical rain forests. Gaps are created by the death of pioneering species or the action of windstorms and landslides. Gaps are filled by a number of small understory species, seedlings and saplings already present in the understory, and by invasion of opportunistic pioneer species (Lang and Knight, 1983). Gap-phase replacement appears to be important in the maintenance of species richness in tropical rain forests (Hartshorn, 1978; Doyle, 1981).

Gap-phase replacement results in patches of different stages of successional or compositional maturity. Chronic small-scale disturbances are important in the maintenance of species richness and structural diversity within a mature forest ecosystem.

Cyclic Replacement

Successional stages that appear to be unidirectional are often in fact phases in a cycle of vegetational development. Such cyclic replacement results from destruction of vegetation by some characteristic of the dominant organism or by periodic disturbances that start regeneration again at some particular stage (Figure 23.6). Such changes are a part of community dynamics, usually occur on a small scale within the community, and are repeated over the whole of the community. Each successive community or phase is related to the others by orderly changes in the upgrade or downgrade series. Such cyclic replacements contribute to community persistence.

These phasic cycles were recognized in the Scottish heaths and described by A. Watt (1947). Scottish heather represents the peak of the upgrade series. After its death a lichen *(Cladonia silvatica)* becomes dominant and covers the dead heather stems. Eventually, the lichen disintegrates to expose bare soil, the last of the downgrade series. The bare soil is colonized by bearberry to initiate

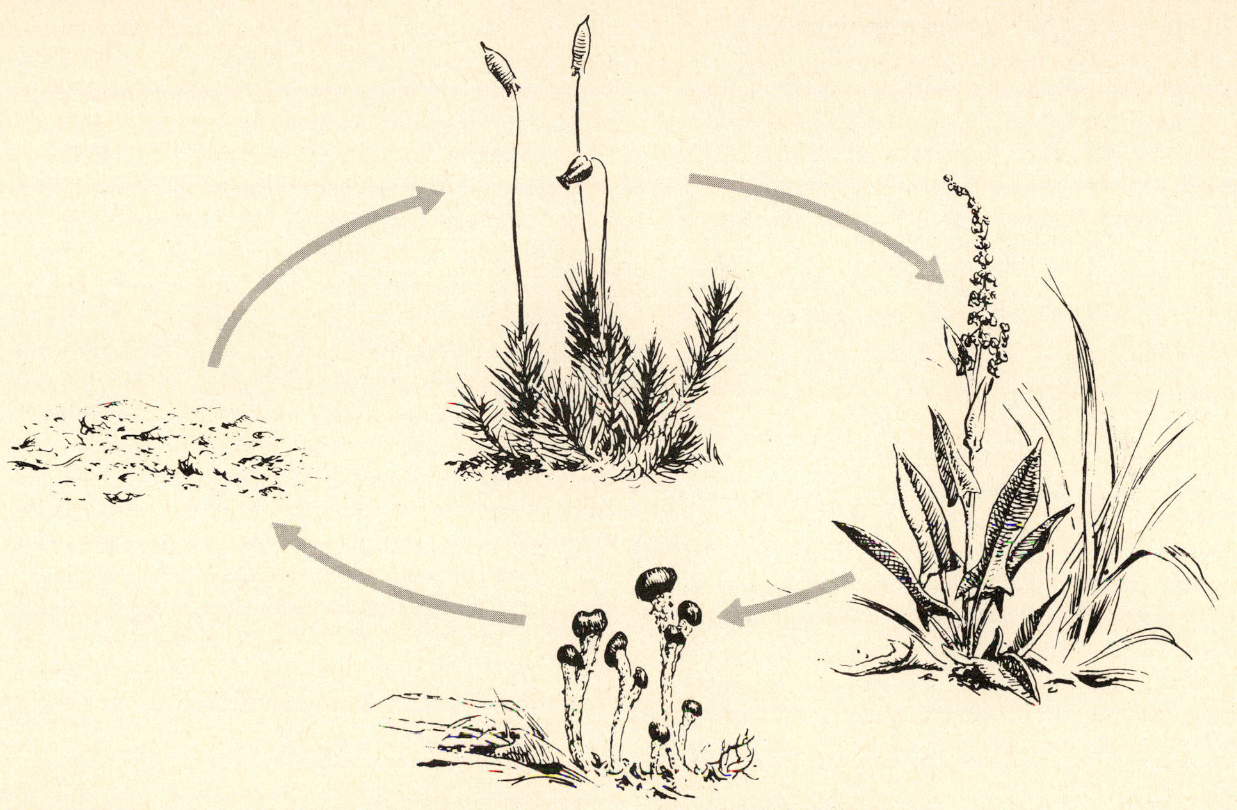

FIGURE 23.6 Cyclic replacement in an old field community; moss, dock, lichen, bare ground, and back to moss.

the upgrade series. Heather then reclaims the area and dominates again. There are other, shorter phasic cycles, one involving heather, lichen, bare soil, and heather again.

Cyclic succession is a relatively common and important phenomenon in a number of ecosystems. Cyclic succession, often initiated by the denudation of ground by ants and ground squirrels, occurs in old field communities of the northern United States. It involves Canada bluegrass, mosses, and lichens (Evans and Cain, 1952). In the desert scrub of Texas, creosote bush *(Larrea tridentata)* exchanges places with Christmas tree cholla *(Opuntia leptocaulis)* with an interval of bare ground between (Yeaton, 1978). The overriding pattern of successional development on the coastal tundra of Alaska is cyclic, controlled primarily by changes in microrelief and drainage regimes (Webber et al., 1980). Cyclic replacement retains the long-term stability of pothole marshes in north central North America. During periods of drought—about every 5 to 20 years—shallow marshes dry. Organic debris accumulated on the bottom decays rapidly, releasing nutrients for recycling and stimulating the germination of seeds. The upgrade of the cycle begins with seed germination on exposed mud. That is followed by a newly flooded stage, with sparse, often well-dispersed vegetation, dominated by annuals and immature perennials; a flooded, dense marsh dominated by perennials; and a deep, open marsh, in part caused by feeding activities of muskrats, rimmed with emergents. The cycle begins anew when the marsh dries (see Weller and Fredrickson, 1973).

Even gap-phase replacement in forest ecosystems may involve cyclic replacement. An example is in the northern hardwood forests of New Hampshire (Forcier, 1975). Involved are sugar maple, beech, and yellow birch. Beech seedlings and saplings are positively associated with yellow birch and sugar maple and negatively with their own canopy. When a gap is created, the first to fill the space is usually the widespread opportunistic yellow birch. It grows rapidly, shading an understory of sugar maple seedlings and saplings. When the short-lived yellow birch dies, its place is taken by sugar maple, beneath which beech seedlings now grow. When sugar maple dies, young beech move into the canopy position.

Fluctuations

Fluctuations are nonsuccessional or short-term changes in ecosystems (Rabatnov, 1974). Fluctuations differ from succession in that floristic composition over time is stable; that is, no new species invade the site in spite of such major disturbances as fire and wind throw. Fluctuations rather than succession appear to be the rule in boreal coniferous ecosystems. Most species reinvade the site directly after a major disturbance without the intervention of earlier successional species. All species capable of taking part in stand development are on the site from the start.

An example is the wave-generation pattern in balsam fir *(Abies balsamae)* forests in the northeastern United States as described by Sprugel (1976). The fluctuation involves a disturbance in which trees continually die off at the edge of a wave and are replaced by vigorous stands of young balsam fir (Figure 23.7). The cycle is initiated when an opening occurs in the forest, exposing trees on the leeward side of the opening to the wind. Dessication of canopy foliage by winter winds, loss of branches and needles in the winter from rime ice forming on the needles, and decreased primary production due to the cooling of needles in the summer cause the death of trees. Their death exposes the trees behind them to the same lethal conditions, and they, in turn, die. This process continues so that a wave of dying trees through the forest is followed by a wave of vigorous reproduction of young balsam fir.

Regeneration waves follow each other at intervals of about 60 years. The regularity of the process is such that all stages of degeneration and regeneration are found in the forest at all times, provided the stand is not cut. This phasic cycle results in a steady state because the degenerative changes in one part of the forest are balanced by regenerative stages in another. The wave-generation process ensures the stability of the forest and prevents its advancement into a hardwoods stage.

Types of Disturbances

FIRE

Fire (see Chapter 9) in climax vegetation or seral stages sets back succession, influences species

FIGURE 23.7 Diagrammatic cross section through a regenerative wave in a balsam fir forest. The wave is initiated at the location of standing dead trees, with mature trees beyond it and with an area of vigorous regeneration below it. In the area where dead trees have fallen, a crop of young fir seedlings is developing. Beyond these is a dense stand of fir saplings, followed by a mature forest and then by a second wave of dying trees. (*After Sprugel, 1976.*)

composition by favoring fire-resistant species and selecting against fire-sensitive ones, and shapes the character of the stand (Curtis, 1959; Lutz, 1956; Heinselman, 1981a, 1981b; for details and examples, see Komarek (1962-1976), *Annual Tall Timbers Fire Ecology Conference*). Fire can inhibit the invasion of shrubs in grassland and stimulate the growth of grass; it can promote the growth of shrubs in forests and stimulate shrubby growth in chaparral. Fire can reduce the numbers of certain species, increase the numbers of others, and eliminate some altogether. Fire can produce successional stability by destroying mature stages and initiating their redevelopment on the same site. In such a manner, fire recycles certain vegetational types such as chaparral and jack pine. Exclusion of fire from fire-dependent ecosystems will alter the type of vegetation that will ultimately claim the site and lead to ecological conditions not experienced by the ecosystem under a natural regime of fire.

Fire is a useful tool for controlling succession and economically maintaining more valuable seral stages. Prescribed burning—fire under control—is used as a management tool to eliminate hardwood understory beneath southern pines and thus to perpetuate these intolerant trees, to prevent the encroachment of woody growth into grassland, to develop openings and browse for wildlife, and to maintain certain shrub communities such as blueberries and emergent marsh vegetation.

TIMBER HARVESTING

Removal of a forest, especially by clear-cutting, returns the land to an earlier stage of succession. Unless followed by fire or badly disturbed by erosion and logging activities, the cutover area rapidly fills in with herbs and shrubs—blackberries and dense thickets of sprout growth and tree seedlings. The area passes quickly through the shrub stage to canopy closure.

Humans can modify the forest to meet their requirements. Early in the life of a new forest, trees economically undesirable and poor in form can be removed. This improves the growing stand economically but not necessarily ecologically. Later, the maximum growth of crop trees can be encouraged by thinning. The increased space between the trees stimulates crown expansion and increases growth.

Many of the most valuable and desirable timber trees exist in the earlier seral stages instead of the climax. Maintaining and reproducing this seral stage is often a problem. Stands of pine, balsam fir, and some spruces, aspen, and yellow poplar are maintained by clear-cutting the mature trees to expose the ground to sunlight. Only under this condition will intolerant seedlings survive. The other extreme in management is selection cutting. In this practice single trees or groups of trees, based on their positions in the stand and their potential for future growth, are cut. Because gap-phase replacement rather than broad successional changes takes place, the forest remains structurally the same.

Mismanagement of the forest—such as clear-cutting excessively large areas, erosion initiated by poor layout of logging roads, and poor slash disposal—can limit the rate of succession and delay the return of the original vegetation. High grading, or taking the best and leaving the poorest, not only eliminates certain tree species from the future forest but also tends to leave genetically inferior trees to supply seed.

GRAZING AND BROWSING

Grazing by domestic and wild animals may arrest succession or even reverse it. Overgrazing of grasslands by domestic stock results in denudation and erosion of the land. In the rangelands of the southwestern United States, overgrazing reduces the organic mat and thus the incidence of fire. Because of this, as well as the reduced competition from grass and dispersal of seeds through cattle droppings, mesquite and other unwanted shrubs rapidly invade the area (Phillips, 1965; Humphrey, 1958; Box, Powell, and Drawe, 1967).

Wild grazing and browsing mammals also influence succession and development. In parts of eastern North America, large populations of white-tailed deer have destroyed reproduction and developed a browse line, the upper limits on a tree at which a deer can reach foliage. The effects of browsing can vary widely. D. A. Marquis (1974), in his exclosure studies on the Allegheny Plateau in

Pennsylvania, found that the white-tailed deer nearly eliminated pin cherry and selectively reduced sugar maple. Beech, birch, and striped maple, not preferred by deer, increased in the stands. Browsing by white-tailed deer resulted in regeneration failures in 25 to 40 percent of the areas he studied. On some cutover areas, forest regeneration was destroyed and the sites dominated by grasses, ferns, goldenrods, and asters.

The African elephant has a pronounced effect on the nature of the ecosystem. When elephant populations are in balance with forage supplies and their movements are not restricted, the elephant plays an important role in creating and maintaining woodland. When elephant numbers exceed their resource base, their feeding habits, combined with fire, can devastate flora, fauna, and soils. Elephant depredation on trees (Figure 23.8) acts as a catalyst to fires, which are the primary cause of converting woodland to grassland (Wing and Buss, 1970).

Insect attacks on vegetation can affect but probably do not alter the course of succession. Outbreaks of spruce budworm result in periodic destruction of monocultural stands of spruce, which regenerates itself. Severe defoliation of spruce in spruce-fir forests of northeastern North America allows the suppressed understory of balsam fir to achieve dominance. Balsam fir, in turn, is subject to spruce budworm attack. Death of fir is followed by regeneration of spruce and fir. In fact, the spruce-fir forest is probably incapable of perpetuating itself without budworm infestation, unless fire intervenes (Ghent, 1958). Outbreaks of gypsy moth in the eastern deciduous forest result in tree mortality and opening of the canopy, but the overall ecological effects are not known.

FIGURE 23.8 Grazing animals can have an important influence on ecosystem succession and stability. The elephant, here uprooting a mopane tree *(Colophospermium)* at the end of the dry season, is important in the life cycle of mopane, which requires the disturbance regime of elephants, and in the maintenance of the savanna ecosystem. Too many elephants results in its destruction; too few elephants results in bush encroachment. (*Photo by T. M. Smith.*)

Insect disturbances are not confined to forests and late seral stages. Grazing (phytophagous) insects may also affect vegetational composition and succession in earlier seral stages. H. McBrien, R. Harmsen, and A. Crowder (1983) reported on a study of an outbreak of the beetle *Trirhabda* in old fields in southeastern Ontario. Three species of these beetles fed preferentially on goldenrod *(Solidago canadensis)*, resulting in a reduction in the percentage cover of that plant from a range of 40 to 70 percent to less than 1 percent. The loss of goldenrod was paralleled by an increase in earlier seral species and thus a temporary reversal in succession.

Herbivores, especially the snail *Littorina littorea*, influence secondary succession in a midzone rocky intertidal community (see Chapter 26) (Lubchenco, 1983). They consume early successional algae such as *Ulva lactuca*. If the algae go ungrazed, they inhibit the appearance of a later successional species, the rockweed *Fucus vesiculosus* (see Chapter 26). Although periwinkles will graze on small *Fucus*, the germlings occupy inaccessible crevices and pits, where they grow until they become too large for periwinkles to consume. In such a manner, a marine herbivore speeds succession on rocky shores.

WIND

A strong element of disturbance in forested ecosystems is wind. Depending on the strength of the wind and its associated storms, disturbance can range from the toppling of a single tree in locally heavy thunderstorms to large-scale windfalls caused by hurricanes (Spurr, 1956). Upturned roots of single and multiple windfalls produce mounds of exposed mineral soil, which are colonized by either tolerant or intolerant species, resulting in gap replacement (Oliver and Stephens, 1977). Large-scale disturbances result in more pronounced changes, as evidenced by the past history of New England forests (Spurr, 1956; Hibbs, 1983). Revegetation following blowdowns in old white pine stands included a mix of short-lived intolerant and tolerant tree species, which arose from sprouts and seedlings already on the site along with wind-dispersed seeds. The effect of wind disturbance was a more rapid return to hardwood forests than would have occurred naturally through long-term mortality of pine.

Succession and Animal Life

As seral stages change, the animal life they support changes with them (Figure 23.9). Because animal life is influenced more by structural characteristics than by species composition, each stage has its own distinctive group of animals. For this reason, the successional stages of animal life might not correspond to the successional stages identified by plant ecologists. Animals would classify a young stand of yellow-poplar or balsam fir under 6 m tall as a shrub stage; a plant ecologist would consider yellow poplar an intolerant tree stage and fir a tolerant tree stage.

Early terrestrial successional stages support animals of grasslands and old fields such as meadowlarks, meadow voles, and grasshoppers. When woody plants invade, a new structural element is added. Meadowlarks decline, and field sparrows and song sparrows appear. When woody vegetation, whether young trees or tall shrubs, claims the area, shrubland animals colonize the area. Field sparrows decline, and thickets are claimed by towhees, catbirds, brown thrashers, and goldfinches. Meadow mice give way to white-footed mice. When woody growth exceeds 6 m and the canopy closes, shrub species decline and are replaced by birds and insects of the forest canopy. As the community matures and more structural elements are added, new species appear, such as tree squirrels, woodpeckers, and birds of the forest understory, such as hooded warblers and wood thrushes.

Diversity of animal life across a range of seral stages varies with the nature of each individual community. In a very general way, shrubland and edge communities and mature stands have a greater diversity of animal life than young forest stands. The key to a diversity of wildlife is a variety of successional stages rather than a homogeneous climax stand and the maintenance of each stage in an area sufficiently large to support its characteristic species.

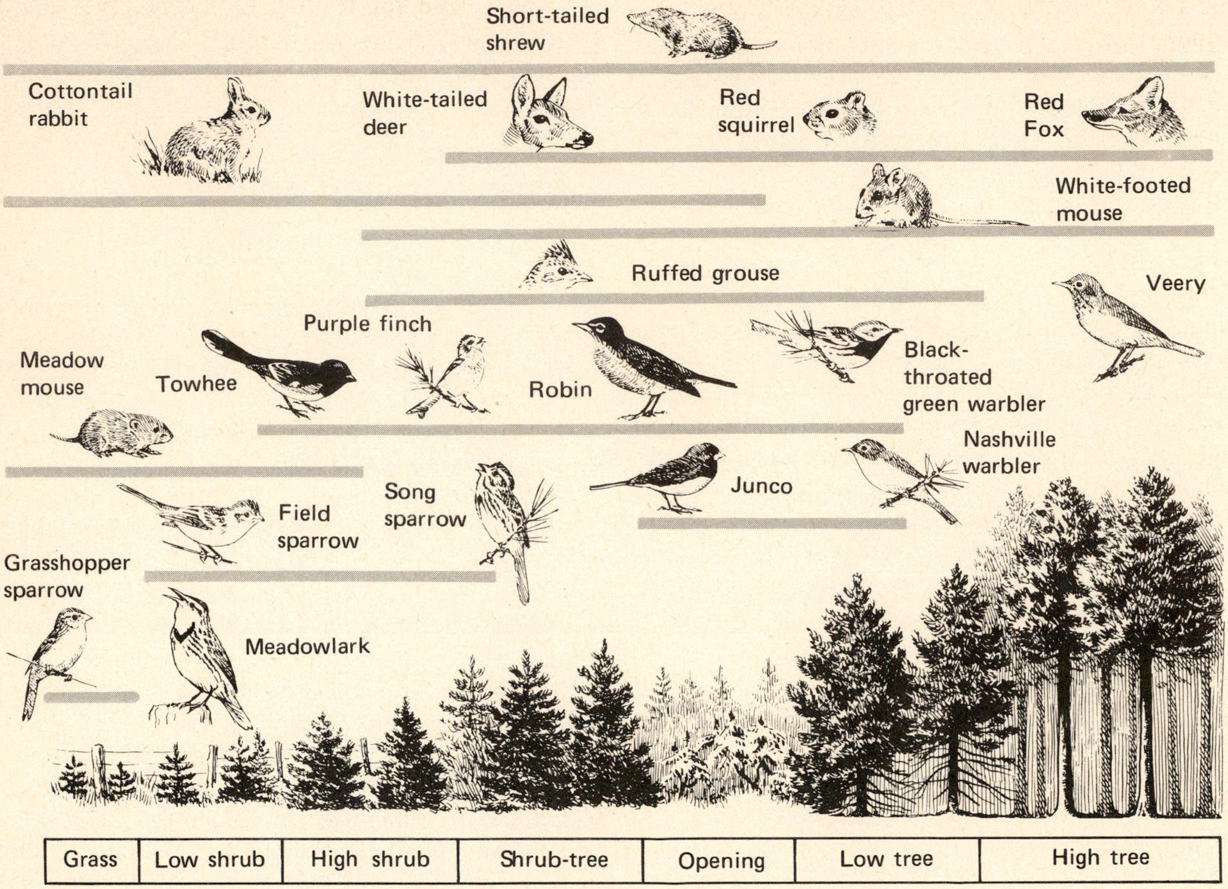

FIGURE 23.9 Wildlife succession in conifer plantations in central New York. Note how some species appear and others disappear as vegetation density and height change. Other species are common to all stages.

Heterotrophic Succession

Within each major community, and dependent on it for an energy source, is a number of microcommunities. Dead trees, animal carcasses and droppings, plant galls, and tree holes all furnish a substrate on which groups of plants and animals live, succeed each other, and eventually disappear, becoming in the final stages a part of the nutrient base of the major community itself. In these instances succession is characterized by early dominance of heterotrophic organisms, maximum energy available at the start, and a steady decline in energy as the succession progresses.

An acorn supports a tiny parade of life from the time it drops from the tree until it becomes part of the humus (Winston, 1956). Succession often begins while the acorn still hangs on the tree. The acorn may be invaded by insects, which carry to the interior pathogenic fungi fatal to the embryo. Most often the insect that invades the acorn is the acorn weevil, *Curculio rectus*. The adult female burrows through the pericarp into the embryo and deposits its eggs. Upon hatching, the larvae tunnel through to the embryo and consume about half of it. If fungi—*Penicillium* and *Fusarium*—invade the acorn, either simultaneously with the weevil or alone, they utilize the material. The embryo then turns brown and leathery, and the weevil larvae become stunted and fail to develop. This represents the pioneer stage.

When the embryo is destroyed, partially or completely, by the pioneering organisms, other animals and fungi enter the acorn. Weevil larvae leave the acorn through an exit hole, which they cut through the outer shell. Through this hole, fungi feeders and scavengers enter. Most important is the moth *Valentinia glandenella,* which lays its eggs on or in the exit hole, mostly during the fall. Upon hatching, the larvae enter the acorn, spin a tough web over the opening, and proceed to feed on the remainder of the embryo and the feces of the previous occupant. At the same time, several species of fungi enter and grow inside the acorn, only to be utilized by another set of occupants, the cheese mites *Tryophagus* and *Rhyzozhyphus.* By the time the remaining embryo tissues are reduced to feces, the acorn is invaded by cellulose-consuming fungi. The fruiting bodies of these fungi, as welll as the surface of the acorn, are eaten by mites and collembola and, if moist, by cheese mites, too. At this time predaceous mites enter the acorn, particularly *Gamasellus,* which is extremely flattened and capable of following smaller mites and collembola into crevices within the acorn. Outside on the acorn, cellulose- and lignin-consuming fungi soften the outer shell and bind the acorn to twigs and leaves on the forest floor.

As the acorn shell becomes more fragile, holes other than the weevils' exits appear. One of the earliest appears at the base of the acorn, where the hilum falls out. Through this hole large animals such as centipedes, millipedes, ants, and collembola enter, although they contribute nothing to the decay of the acorn. The amount of soil in the cavity increases, and the greatly softened shell eventually collapses into a mound and gradually becomes incorporated into the humus.

Thus, microcommunities illustrate one concept of succession: that the change in the substrate may be brought about by the organisms themselves. When organisms exploit an environment, their own life activities make the habitat unfavorable for their own survival and instead create a favorable environment for different groups of organisms. Those responsible for the beginning of succession are quite specialized for feeding in acorns, the later forms are less so, and the final group are generalized soil animals such as earthworms and millipedes.

The Climax

According to theory, as succession slows down, the sere achieves an equilibrium or steady state with the environment. This last seral stage is mature, self-sustaining, self-reproducing through developmental stages, and relatively permanent. The vegetation is tolerant of the environmental conditions it has imposed upon itself. This terminal community is characterized by an equilibrium between gross primary production and total respiration, between the energy captured from sunlight and the energy released by decomposition, between the uptake of nutrients and the return of nutrients by litterfall. It has a wide diversity of species, a well-developed spatial structure, and complex food chains; and its living biomass is in a steady state. The final stable community of the sere is the *climax* community, and the vegetation supporting it is the climax vegetation.

There are three theoretical approaches to the climax. One is the *monoclimax theory,* developed largely by Frederick Clements. This theory recognizes only one climax, determined solely by climate, no matter how great the variety of environmental conditions is at the start. All seral communities in a given region, if allowed sufficient time, would ultimately converge to a single climax. The whole landscape would be clothed with a uniform plant and animal community. All communities other than the climax are related to the climax by successional development and are recognized as subclimax, postclimax, disclimax, and so on.

Another approach is the *polyclimax theory* (Tansley, 1939). The climax vegetation of a region consists of not just one type but a mosaic of vegetational climaxes controlled by soil moisture, soil nutrients, topography, slope exposure, fire, and animal activity (see Daubenmire, 1968).

A third and perhaps more acceptable approach is the *climax pattern hypothesis* (Whittaker, 1953;

MacIntosh, 1958; Selleck, 1960). The composition, species structure, and balance of a climax community are determined by the total environment of the ecosystem and not by one aspect, such as climate alone. Involved are the characteristics of each species population, their biotic interrelationships, availability of flora and fauna to colonize the area, the chance dispersal of seeds and animals, and the soils and climate. The pattern of climax vegetation will change as the environment changes. Thus, the climax community presents a pattern of populations that corresponds to and changes with the pattern of environmental gradients, intergrading to form ecoclines. The central and most widespread community in the pattern is the prevailing or climatic climax, one that most clearly expresses the climate of the area. This central community type, such as an oak-hickory forest or a beech-maple forest, relates the community to the climate in major ecoclines and provides a regional pattern to the vegetation.

Climax is a useful concept for describing and mapping regional vegetation. It is less useful for describing individual stands of mature and old-growth vegetation. To categorize an individual stand as climax, you are faced with such questions as: What is climax vegetation? Do climaxes really exist? Does a sere ever reach equilibrium in species composition and ecosystem attributes? Vegetation is dynamic. Mortality is a feature of the climax. As old trees die, new individuals and species take their places. Thus, species composition changes, even though physiognomy remains the same. Gross production is high, but so is respiration because of the decomposition of woody material. Net production, then, is relatively low, but production never achieves equilibrium with respiration. Biomass continues to accumulate on individual trees. F. Bormann and G. Likens (1979) have suggested that, rather than an equilibrium, a mature stage of forest vegetation achieves a *shifting-mosaic steady state,* in which a standing crop of living and total biomass (living + dead) fluctuates about a mean. Such an ecosystem consists of patches in various stages of seral development, from ones exhibiting high net production and biomass accumulation to ones of senescent and down timber in which respiration exceeds production. The production of these patches remains more or less the same through time.

Studies of old or mature forested ecosystems needed to test the climax theory are rare. Some involve 450- to 1,000-year-old stands of Douglas-fir in the Pacific Northwest (Franklin and Hemstrom, 1981). Technically, these old Douglas-fir forests are not climax, because Douglas-fir is a pioneer species that happens to be very long-lived. The climax species that might replace the Douglas-fir in another 500 years are western hemlock *(Tsuga heterophylla)* and Pacific silver fir *(Abies amabilis).* Yet by the standards of permanency applied to a deciduous forest, these old stands are "climax." These old Douglas-fir stands are uneven-aged when, theoretically, they should be even-aged. Age classes range from 145 years on up within the stand, suggesting occasional disturbances and replacement or perhaps failure of canopy closure over a period of time because of variable seed crops. The variation in age classes within the stand suggests some agreement with the shifting-mosaic steady state. Gross production is high, but so is respiration, and net production is low (see Chapter 24). Ratio of live biomass to dead peaks at 300 to 400 years. Total biomass is the highest at around 750 years, and total dead biomass is greatest at 800 to 1,000 years. The forest is a complex system of many species of saprophytes and epiphytic lichens. Nutrient retention within the system involves complex detrital pathways. Nitrogen is fixed by lichens in tree crowns (see Chapter 21) and by microflora in decomposing logs. Structural diversity is high, with a large range in the size of individual living trees and numerous large, standing dead trees, large downed trees, and decomposing logs.

Succession and Time

Time is an integral component of succession. But time is relative, measured in terms of human experience. Climax vegetation is theoretically permanent, but what is permanent? Vegetation that remains the same over one or several human lifetimes? By that standard, old field vegetation could be climax to ants, birds, or meadow mice.

Nevertheless, successional communities have their time spans, governed in part by the longevity of the plants that make up the seral stages. An old field with annual weeds may exist no more than 1 to 2 years. Pioneer lichens and moss on a granite outcrop may exist for hundreds of years. Seral grass stages may last 10 to 15 or even fewer years before being overtaken by woody growth. Woody growth in a shrub stage—whether true shrubs or incoming tree growth below a height of 6 m—may last an additional 10 to 15 years until pioneering trees take over or the canopy closes. If trees are pioneer, shade-intolerant species such as aspen, the stage may last 25 to 40 years before shade-tolerant species become dominant and hold the site 250 to 500 years.

This time line implies that shade-tolerant species replace shade-intolerant ones and that trees replace shrubs. Not always. A dense growth of shrubs such as meadow sweet, mountain laurel, or Saint-John's-wort may claim a site permanently—50, 60, 70 years without any indication of change. Shade-intolerant species can hold onto a site and maintain their positions for many years. Yellow-poplar, white pine, and red pine, to mention several pioneer species, can colonize an abandoned field and remain dominant for well over 100 years. Douglas-fir, a pioneer species in western North America, may claim a site for over 1,000 years before theoretically giving way to western hemlock, which may need another 500 years to become dominant.

How long a seral stage may last could be simply an academic question, except that time has implications in forestry and wildlife management. Certain types of wildlife habitat are seral and thus ephemeral. Their maintenance may be critical to the welfare of certain species, which requires human interference with the successional process. Certain commercially valuable timber trees are pioneering species that require periodic disturbance to ensure regeneration.

Is Succession Directional?

Classic successional theory holds that succession is directional and therefore predictable. Succession is directional: For succession to operate it has to head somewhere, even if in a circle (cyclic succession). But the concept of succession implies unidirection—a one-way trip. In that case, the answer to the question of whether succession is directional would have to be a qualified no.

One could predict with a high degree of probability that an old field in northeastern North America will, barring further disturbance, return to forest. One would have much more difficulty predicting the kind of forest, even if one had knowledge of previous vegetation. Each successional community, climax and otherwise, is individualistic, a one-time product of abiotic and biotic forces operating during the time of its development. The exact interaction of these forces will not be repeated again. Any new successional community will be molded by current abiotic and biotic inputs. The exact original composition of the species will not be duplicated. Thus, one might predict successfully the physiognomy over a broad region but miss the successional communities on a local level in spite of recent attempts to do so (see Horn, 1975, 1981). There are too many side roads succession can take, influenced by environmental conditions and biotic interactions that do not lead to a climax (Figure 23.10).

Successional direction is implicit in two popular models of forest succession and their variations: JABOWA (Botkin, Janak, and Wallis, 1972; Aber, Botkin, and Melillo, 1978) and FORHAB (Shugart, Crow, and Hett, 1973). These models involve explicit variables, including temperature, moisture, nutrients, light, life-history strategies of species involved, mortality, replacement growth rates, and competition. These models are predictive over large areas as long as conditions conform to the boundaries set by the variables. Such models are useful in planning timber harvests and predicting changes in wildlife habitat and the effects of pollution and fire on forest ecosystems.

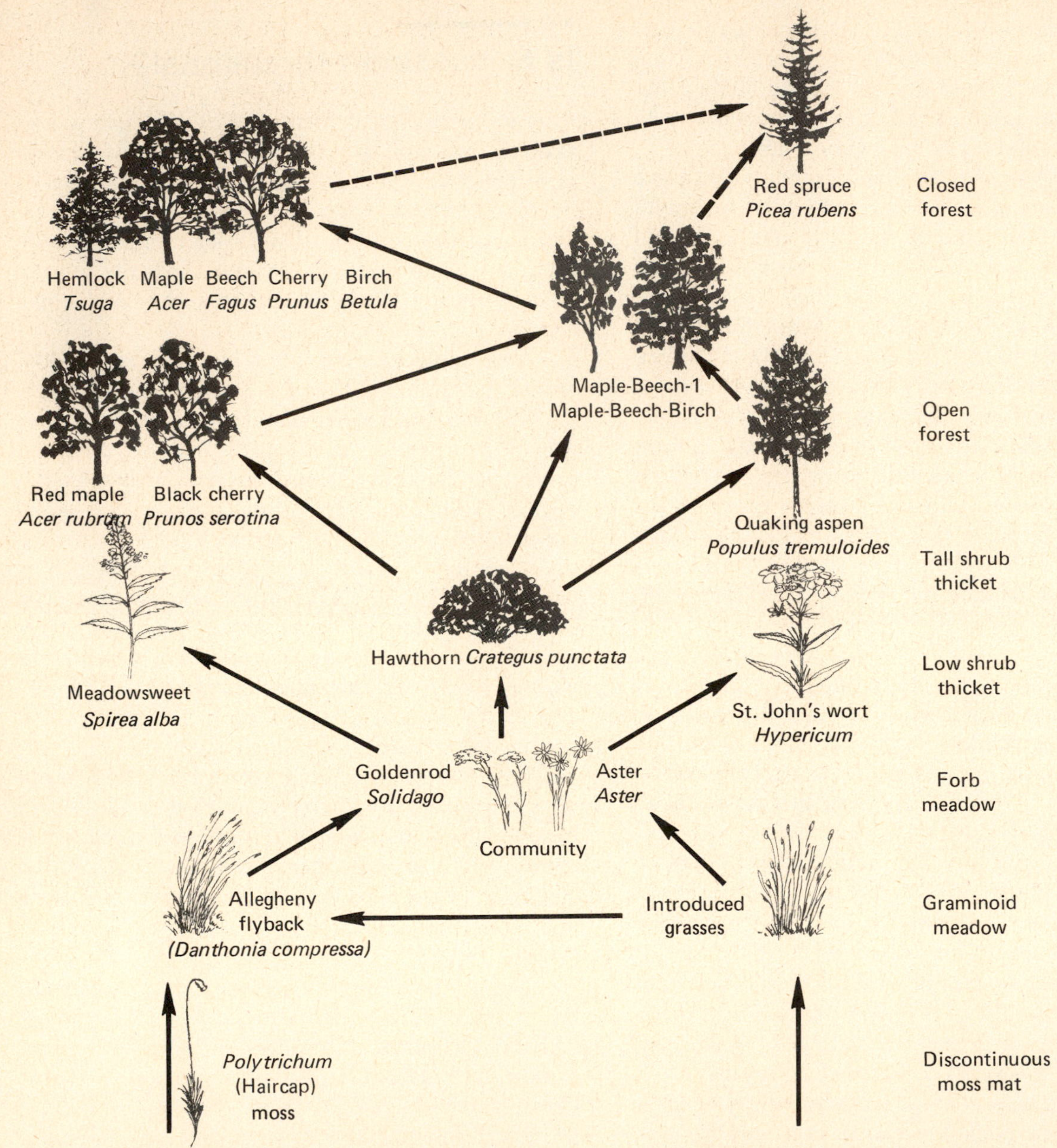

FIGURE 23.10 Flowchart of successional development in Canaan Valley, West Virginia, a high-altitude hanging valley surrounded by still higher mountains. Once covered with spruce forests and open glades, the valley was cut over, burned, and converted into farm and grazing land. Much of this submarginal farmland has been abandoned and is reverting to natural vegetation (which, in turn, is being converted into condominiums). Successional trends in the area vary with the nature of the soil and the depth of the water table. This diagram reflects successional development on moderately well-drained, acidic soil. The starting point is either moss; a native grass, Allegheny flyback; or introduced grasses. Under certain conditions, succession may terminate in shrub communities of either spirea or Saint-John's-wort, stands of which are so dense that vegetative growth beneath them is suppressed. In other instances, succession leads to a hemlock-beech-maple forest with only slight indication that succession might proceed to spruce. (*After R. Fortney, 1975.*)

Succession and Ecosystem Stability

Another theory of succession involves stability. As succession proceeds to a climax, it achieves stability. Stability, in turn, nurtures species diversity, and so on. At one time, symposia were devoted to the subject of biological and ecological stability. One doesn't hear too much about stability anymore, but the idea that a climax community, such as a mature forest, is static and unchanging and that a disturbed forest returns to its original condition is still pervasive.

Ecosystem stability cannot be considered until the word *stability* is defined. Stability carries two connotations. The term can mean the ability to absorb changes, to persist through time. Persistence implies *resilience,* a return to an initial starting point or an equilibrium state like the pendulum of a clock, the ability of something to absorb changes and still persist, and the speed with which it returns to its original condition (Holling, 1973). Resilience does not necessarily imply high stability of individual populations within the system. They may fluctuate widely in response to environmental changes, yet the system itself may be highly resilient. Or stability can mean *resistance,* the ability to absorb disturbance, forces that could dislodge the system from an equilibrium state (see Holling, 1973). Often the resistance of an ecosystem results from its structure. Most resistant systems have large biotic structures such as trees and nutrients and energy stored in standing biomass.

A forest ecosystem, for example, is relatively resistant to disturbance. It can respond to such environmental changes as frost, drought, and insect outbreaks because the system is able to draw on stored reserves of nutrients and energy. But if the forest system is highly disturbed by fire or logging, the system exhibits low resilience. Its return to its original condition as a mature forest is slow. Thus, a sort of inverse relationship exists between resistance and resilience. A disturbance (or perturbation), if strong enough, may result in a different outcome. The system may be so greatly disturbed that it is unable to return to its original state, and a different ecosystem may replace it. When the spruce forests of the central Appalachians were burned, the organic matter was destroyed, and spruce was replaced by blueberry and stunted birch (Figure 23.11).

Based on these two interpretations of stability, are mature ecosystems more stable than earlier stages? If stability implies that organisms are long-lived, that their populations are relatively constant over time and not subject to strong oscillations or rapid replacement, and that the community changes its composition very slowly and resists invasion from other species, then mature systems are more stable. They are highly resistant. But if stability means resilience to disturbance, the ability to return to equilibrium, then early stages of succession are more stable than mature ones. Plow a field of annual weeds and the next spring the field will be claimed by the same weeds again. Trample or tear up a grassy field and, in a year or two, grass will claim

FIGURE 23.11 Forest fires of great intensity can have a profound influence on succession. Intense fires burned over the Allegheny plateau area in West Virginia known as Dolly Sods after the spruce forest was cut. Fire consumed the peatlike ground layer to bedrock and mineral soil. The forest never recovered from the Civil War years' burn. The plateau is now a boulder-strewn landscape with intermittent patches of blueberry, dwarfed birches, mountain ash, and bracken fern.

the field again. But cut a mature forest and 100 years may pass before any semblance of the original forest returns. Thus, as succession proceeds, stability—as defined by constancy and diversity of populations involved—increases. But if stability is defined by the ability of a community to rebound after a disturbance, then it decreases through succession.

The idea of stability is difficult to accept in its original meaning in ecology. Ecosystems are dynamic because they consist of populations of plants and animals that respond to environmental changes. Change is inherent in natural systems. To manage an ecosystem such as a mature forest or wilderness area as an unchanging entity like a painting of a landscape is contrary to the behavior of natural systems. Periodic disturbance is an integral part of the system. To persist, some ecosystems, such as grasslands and pine forests, require periodic disturbance (Figure 23.12). Yet the disturbance must occur within bounds. Outside of these bounds there may be no return (see Figure 23.11). But up to a point, stability of ecosystems requires a certain degree of disturbance (Loucks, 1970; Heinselman, 1981a). Without it some ecosystems cannot be maintained and will change to something different.

Summary

With the passage of time, natural communities change. Today's old fields return to forests tomorrow. Weedy fields in prairie country revert to grass. This gradual directional (but not necessarily unidirectional) change in community composition through time is succession. It is characterized by the replacement of early seral-stage pioneer or opportunistic species by late-stage equilibrium species; a progressive change in species structure, growth form, and tolerance to shading; an increase in biomass and organic accumulation; tighter nutrient cycling; and a gradual balance between community production and community respiration.

Succession may begin with any one or several groups of organisms capable of growing on an open site successfully and arriving there early, as either seeds, spores, or residual propagules. They pre-

(a)

(b)

FIGURE 23.12 An example of changes in vegetation influenced by the exclusion of fire. (a) A photo taken in 1888 shows a north-northwest view of U. S. Army troopers in the vicinity of Fort Maginnis. This fort was located in the Judith Mountains (in the distance) about 20 miles northeast of Lewistown, Montana. Ground cover on the ridge and lower slopes in the distance is apparently dominated by perennial grasses. The distribution of conifers on the far slopes is largely confined to localized sites. The pattern of regenerating conifers suggest that wildfire had burned the stand several decades earlier. (b) The same site on July 18, 1980, 92 years later. The foreground vegetation shows the effects of current livestock grazing. Shrubs, including chokecherry, current, rose, shrubby cinquefoil, and common juniper are more conspicuous in the rocks, as are conifers and shrubs in the left midground. The stream course in the distance at center left of photo, which was formerly treeless, now supports large cottonwoods, Douglas-fir, and ponderosa pine. Absence of fire has allowed profuse growth of Douglas-fir and ponderosa pine on the far slopes. *(Photo (a) by W. H. Culver and photo (b) by W. J. Reich, courtesy of Montana Historical Society. Interpretative text based on a description written by George E. Gruell in* Fire and vegetative trends in the northern Rockies: Interpretations from 1871–1982. *U.S.D.A. Forest Service General Technical Report INT-158, page 54.)*

empt space and continue to exclude or inhibit the growth of others until the former colonists die or are damaged, releasing resources and allowing new, longer-lived species to enter. Or succession may come about because of changes induced by the organisms themselves. As they exploit the environment, their life activities make the habitat unfavorable for their own continued survival and create a favorable environment for different species. Although a conflict exists between these two views of succession, both are involved to a greater or lesser degree. Although any point on the time gradient may give the impression that one discrete community replaces another, seral stages actually represent a pattern of species replacement driven by differential longevity, competition, and chance. Thus, species and not discrete communities are replacing one another, but the end result is seral or community change.

Eventually, communities arrive at some form of steady state with the environment. This stage, usually called the climax, is more or less self-sustaining, largely through small-scale disturbances involving largely gap-phase replacement and cyclic succession. Gap-phase replacement involves both vegetational reorganization and invasion by opportunistic and equilibrium species, depending on the nature of the disturbance and biotic conditions. Cyclic replacement results from destruction of vegetation induced by some periodic biotic or environmental disturbance that starts regeneration again at some particular stage. Each successive phase or community is related to the others by orderly changes in the rise and decline of successive communities. Nonsuccessional or reversible changes or fluctuations in community composition usually result from environmental stresses such as changes in soil moisture, wind, and grazing. As a result of these disturbances, the climax is most often a mosaic of regenerating patches in which the new or replacement species may be the same as the ones being replaced. Although response to disturbance in mature systems may involve replacement of one species by another on a short- or long-term basis, it does result in the maintenance of the community in general. Thus, the stability of a system often requires the introduction of instability to the ecosystem.

Succession in natural communities is often interrupted by major disturbances such as wind, fire, grazing, insect outbreaks, and timber harvesting. Such disturbances can shape and modify the nature of the community and the direction of succession by favoring certain species and eliminating others.

Perhaps the most outstanding characteristic of natural communities is their dynamic nature. They are constantly changing through time—rapidly in early stages of development, more slowly in later stages. Even those communities that are seemingly the most stable slowly change through time.

Review and Study Questions

1. What is succession? A seral stage? A sere?
2. Distinguish between primary and secondary succession.
3. Describe the relay floristics and initial floristic concepts of succession. How would succession differ under the two?
4. Compare the initial floristic model with the inhibition model of succession.
5. Why might the facilitation model of succession be more appropriate to primary succession than to secondary succession?
6. Distinguish among gap-phase replacement, cyclic succession, and fluctuation. In what way may each be associated with the others?
7. What role might allelopathy play in succession?
8. What is the climax? Discuss the validity of the climax concept.
9. How does time relate to the climax concept?
10. Contrast the flow of energy and nutrients in heterotrophic and autotrophic succession. How does heterotrophic succession relate to autotrophic succession?

11. Although little emphasis was placed on *r*- and *K*-selected species and succession, relate these two concepts to species of early- and late-stage succession. How well do the concepts relate? How would you categorize a pine tree versus crabgrass as an invader of an old field?

12. How can disturbance be an important aspect of ecosystem stability?

13. Would the use of the terms *resistant* and *resilient* be preferable to *stability*? Why or why not?

14. An influx of pollutants into an aquatic ecosystem highly disturbs the system. If the source of pollution is removed, the lake or stream gradually returns to its original condition. Is the aquatic ecosystem resistant or resilient? Why?

15. Locate an area with which you have been familiar over time. What vegetational changes have taken place? What disturbances brought them about: land abandonment, logging, grazing, fire? Old aerial photos or a series of them will provide insights.

Suggested Readings

This chapter gives little emphasis to descriptions of actual succession, primary or secondary. However, numerous examples, largely descriptive, exist in the literature. Here are some references to specific types of succession that you may wish to consult (full citations are in the Bibliography):

Primary. Van Cleve and Viereck, 1981: alluvial deposits; Walker et al., 1981; Olson, 1958: sand dunes; Crocker and Major, 1955; Lawrence, 1958: glacial deposits; Oosting and Anderson, 1937, 1939: granite outcrops.

Secondary. Keever, 1950; Van Cleve and Viereck, 1981: temperate forest and coniferous forest regions; Doyle, 1981; Gomez-Pompa and Vazquez-Yanes, 1981; Kowal, 1966: tropical forests.

Bormann, F., and G. Likens. 1979. *Pattern and process in forested ecosystem*. New York: Springer-Verlag.

Cairns, J., Jr., ed. 1980. *The recovery process in damaged ecosystems*. Ann Arbor, Mich.: Ann Arbor Science Publishers, Inc.

Clements, F. 1916. *Plant succession: An analysis of the development of vegetation*. Publication 242. Washington, D.C.: Carnegie Institute of Washington.

Golley, F., ed. 1978. *Ecological succession*. Benchmark Papers. Stroudsburg, Pa.: Dowden, Hutchinson & Ross.

Knapp, R., ed. 1974. *Vegetation dynamics*. The Hague: W. Junk.

Means, J., ed. 1982. *Forest succession and stand development research in the Northwest*. Corvallis: Forest Research Laboratory, Oregon State University Press.

Waring, R., ed. 1979. *Fresh perspectives from ecosystem analysis*. Proc. 40th Annual Biology Colloquium. Corvallis: Oregon State University Press.

West, D. C., H. H. Shugart, and D. D. Botkin, eds. 1981. *Forest succession: Concepts and applications*. New York: Springer-Verlag.

ECOLOGICAL EXCURSION

Graveyard Ecology

Robert Leo Smith

Ecology of a region is told in part by its graveyards—not the neat, trimmed graveyard behind many rural churches, nor the manicured memorial gardens that hide behind a facade, but the graveyards half forgotten and lost back in the woods. There is a great deal more honesty in old graveyards than in modern memorial gardens; they do not hide the fact that death is indeed death, and those that rest there have gone on to their eternal reward.

In old graveyards, too, the headstones are more informative than today's cold markers, and their inscriptions are straightforward. Modern gravestones are impersonal, telling little except that he or she who lies beneath was born in a given year and died in another. Old gravestones are biographical. They usually give the exact date of birth, exact date of death, age to months, relationship, station in life—soldier, farmer, merchant, mother, father—and often an appropriate verse, weathered words which often can be read only with some study and difficulty.

Hunters and woodland hikers come across these old graveyards back in the forest, usually on a ridgetop or an upper slope. They are covered by trees and hidden by shrubs.

Far off the traveled road and reached only by a trail, some old graveyards are still cared for after a fashion. Descendants, perhaps a generation or two removed, still make it their duty to prevent the woods from reclaiming the cemetery. In such graveyards, the informative stones, vegetation and nature of the surrounding area can tell careful observers considerable information about the area's past.

Every graveyard hidden back in the woods reveals one fact, at least: The area thereabout was once occupied by people. Perhaps it was a small farming community, a group of people held together by a common interest, common faith, intermarriage—and a common graveyard. The high tide of prosperity may have been sometime around the latest date inscribed on the tombstone. Because no one was buried in the graveyard after that date, one can surmise that the people moved away. The earliest date on the tombstone indicates roughly when the first people settled there. What happened between the earliest and latest dates on the tombstones and the reason the people left can sometimes be read on the tombstones and in the landscape.

I have come across a number of them and, although each differs in many respects, a similar story seems to fit them all. For the most part, they are family burial grounds, plots removed from a church but close to home. The earliest dates of death are in the early 1800s, often in the 1840s or 1850s. Around that time some families moved in, cleared the land, perhaps finished a task started by lumbering, and built a house, a barn, a shed, probably using lumber from trees they cleared from the hillsides. They built a spring house carefully roofed and walled, or dug a shallow well. They laid the land out into fields separated by fences, created well-worn lanes, planted an orchard, some grapes, and as a final touch, daffodils and a lilac bush by the dooryard.

For a number of years they prospered on subsistence farming. Their demands were frugal and the surplus crops paid for the mortgage. Cows grazed in the fields too far from the barn to crop or too wet to plant. A couple of hogs grew and fattened on acorns and table scraps in a pen down the lane, and a coop full of chickens kept the family in eggs. Potatoes, corn, oats and wheat were planted in spring and hay was cut and stacked in summer. There were apples to be picked in fall and stored in apple cellars and applebutter to be made. There were nuts to be gathered from the walnuts and hickories carefully preserved when the woods were cleared. Corn shocks stood knee-deep in ragweeds and bright orange pumpkins, and covies of quail ran from corn shock to corn shock. When the weather turned cold, usually around Thanksgiving, the hogs

SOURCE: *Panorama Magazine* of *Sunday Dominion Post,* October 26, 1969:4, and *Wonderful West Virginia,* December 1979:27.

were butchered and the meat cured with hickory smoke. Except for flour, sugar and yard goods, there was little need to drive a wagon (later, a Model T) to town 15 or 20 miles away. And when winter settled in, families were pretty well isolated by snow and mud.

So, deep out in the country, rural people lived a self-sufficient and conservative way of life. While children played beneath the shade of spreading oaks, their elders had worries about mortgages, sickness and death. Death often came early, the gravestones tell us, and it claimed the young before it claimed the elders. When death came to the very young it came in twos or threes, usually in late winter or early spring. Some years death among the young was more common than in others, and the graveyard stones hint that some violent form of sickness, such as scarlet fever or diphtheria, swept through the countryside. Other tombstones tell us that young mothers died in childbirth or a father passed on early, leaving a widow and a number of children. The families contributed some sons to the Civil War, Spanish-American War and World War I. That was the last. The boys who made it back from World War I didn't stay on the farm. How could you keep them down on the farm, once they had a taste of seemingly more exciting life in the city and far off parts of the world?

More important, there was no money to be made on field farms. After the West was opened, there was no way to compete with the plains country in growing oats and corn and other cash crops that kept a mortgage from being foreclosed. In fact, through the years the old farm grew poorer and was barely able to support the old folks, let alone a young man with a family. So the old people died in the late 1920s, so the gravestones say, or else they moved away.

Decay crept over the farmsteads. One by one the buildings fell in—the poorly built chicken house first, then the barn and, finally, the house that watched it all, its clapboards blackened by the weather, its windows broken, its doors swung open, its rooms exposed to the wind, rain and snow. Eventually, the house went too.

The fields were claimed even quicker. The first year a family failed to plow the corn stubble or the oats, ragweed and foxtail took over, sumac and blackberry and greenbrier, then locust, yellow poplar and oak. Grassfields stayed open a little longer, but eventually they grew back into woods again. Finally even the dooryard was reclaimed.

The story is all there for hunter and hiker, in the graveyards and open cellars half hidden by grapevines and blackberries. Scratch away the leaf mold of 50 years and you may find that flat stones covered the cellar floor. Look closer and you will find the old spring close to the kitchen door, and the lilac bush half choked by greenbrier and hawthorns. Come back in spring and you will find that the lilac bush still blooms, and the daffodils as well. The trees that shaded the dooryard still spread their branches, vainly trying to keep above the surrounding forest growth. You can still pick out rutted lanes in the woods, lanes that once led from field to farm, and old fencerows marked by unusually straight lines of trees. Curious parallel ridges of ground beneath the trees tell you that the last crop planted here was potatoes, and a lush growth of blackberries may mark a barnyard, a hog yard or a chicken run.

Numerous wet spots and stony soil evident in the woods tell partly that the land was hardly suited for agriculture. Time passed and in the end the trees that once claimed the land have claimed it again. Man was clearly the loser, and the trees and chipmunks, grouse and greenbrier were clearly the winners. The graveyard, the cellar holes, the half-choked lilac bush and the grouse tell it all.

PART VI

A SURVEY OF ECOSYSTEMS

INTRODUCTION

Distribution of Terrestrial Ecosystems

Humans have always been interested in the plants and animals about them. But until very recently in human history, such knowledge was limited to life in the immediate area. As world exploration expanded human horizons, naturalists joined explorers in their global travels, especially in the 1800s. These naturalists included such personages as Alexander von Humboldt, Alfred Wallace, and Charles Darwin. They brought back specimens of new and often strange forms of life, accumulated knowledge on distribution of species, and noted similarities and differences in plants and animals around the world. Botanists, in particular, noted that the world could be divided into great blocks of vegetation—deserts, grasslands, and coniferous, temperate, and tropical forests. These divisions they called *formations,* even though plant geographers had difficulty drawing sharp lines among them. In time, plant geographers, notably J. F. Schouw (1823), attempted to correlate vegetation formations with climatic differences and found that blocks of climate reflected blocks of vegetation and their own life-form spectra.

Zoogeographers lagged behind plant geographers in their study of animal distribution. For one thing, the number of animal species was much greater than the number of plant species, and no clear-cut relation seemed to exist between animal distribution and climate. Both factors complicated distribution studies, but ultimately, zoogeographers accumulated basic information on the global distribution of animals. This was done for birds in 1878 by Philip Sclater, who mapped them into six regions, roughly corresponding to continents. But the master work in zoogeography was done by Alfred Wallace, who is also known for having reached the same general theory of evolution as Darwin. Wallace's realms, with some modification, still stand today.

Biogeographical Realms

There are six biogeographical realms, each more or less embracing a major continental land mass and separated by oceans, mountain ranges, or desert (see Figure VI.1). They are the Palearctic, the Nearctic, the Neotropical, the Ethiopian, the Oriental, and the Australian. Because some zoogeographers consider the Neotropical and the Australian realms to be so different from the rest of the world, these two are often considered as regions or realms equal to the other four combined. They are classified as Neogea (Neotropical), Notogea (Australian), and Metagea (the main part of the world). Each region possesses a certain distinction and uniformity in the taxonomic units it contains, and each to a greater or lesser degree shares some of the families of animals with other regions. Except for Australia, each has at some time in Earth's history had some land connection with another across which animals and plants could pass.

Two regions, the Palearctic and Nearctic, are quite closely related. In fact, the two are often considered as one, the Holarctic. Both are much alike in their faunal composition, and together they share, particularly in the north, such animals as the wolf, hare, moose (called elk in Europe), stag (called elk in North America), caribou, wolverine, and bison.

Below the coniferous forest belt, the two regions become more distinct. The Palearctic is not rich in vertebrate fauna, of which few are endemic. Palearctic reptiles are few and are usually related to those of the African and Oriental tropics. The Nearctic, in contrast, is the home of many reptiles and has more endemic families of vertebrates. The Nearctic fauna is a complex of New World tropical and Old World temperate families. The Palearctic

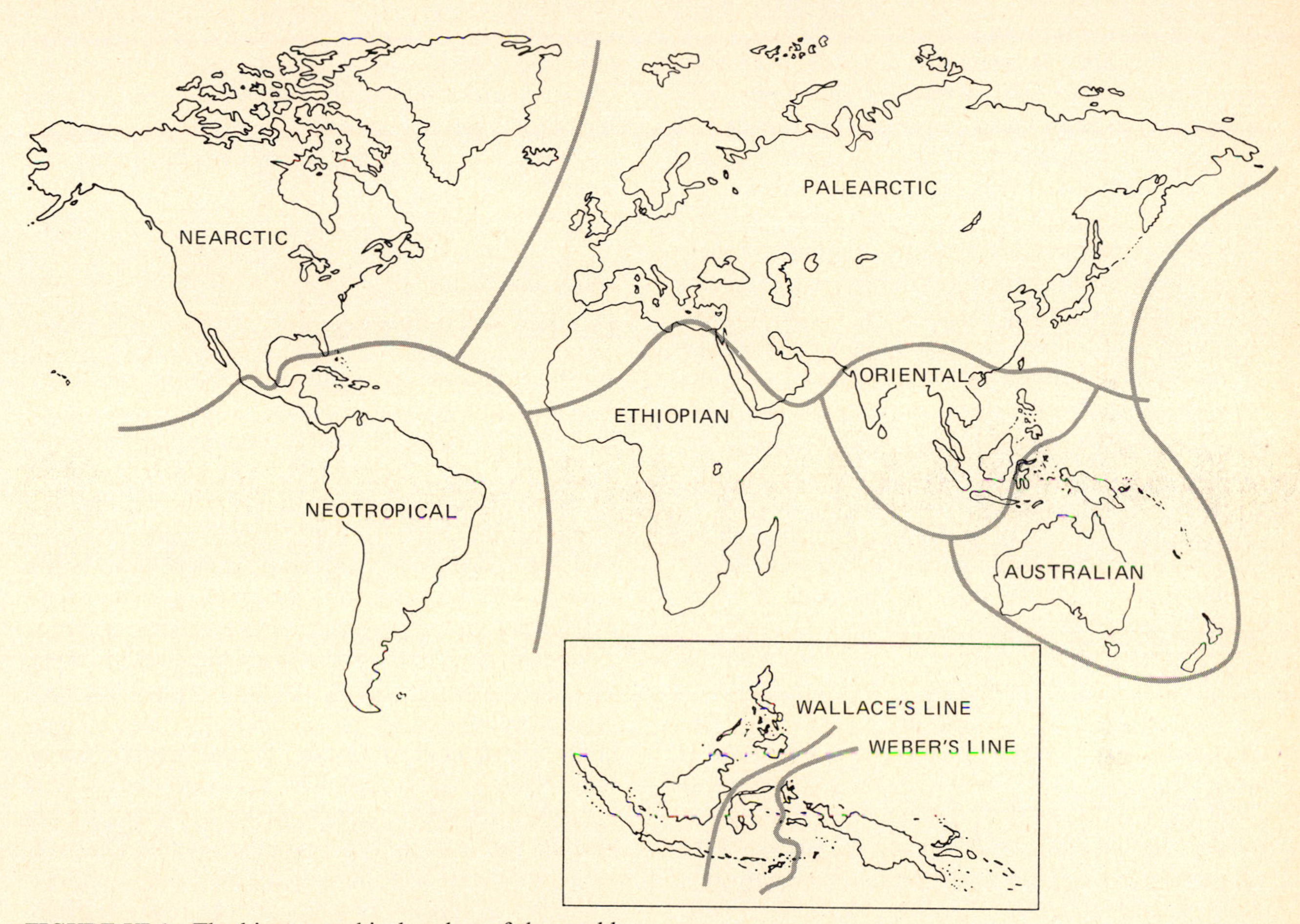

FIGURE VI.1 The biogeographical realms of the world.

is a complex of Old World tropical and New World temperate families.

Isolated until 15 million years ago, the fauna of the Neotropical is most distinctive and varied. In fact, about half of the South American mammals, such as the tapir and llama, are descendants of North American invaders, whereas the only South American mammals to survive in North America are the opossum and the porcupine. Lacking in the Neotropical is a well-developed ungulate fauna of the plains, so characteristic of North America and Africa. However, the Neotropical is rich in endemic families of vertebrates. Of the 32 families of mammals, excluding bats, 16 are restricted to the Neotropical. In addition, 5 families of bats, including the vampire, are endemic.

The Ethiopian realm embraces tropical forests in central Africa and, in the mountains of east Africa, savanna, grasslands, and desert. During the Miocene and Pliocene, Africa, Arabia, and India shared a moist climate and a continuous land bridge, which allowed the animals to move freely between them. That accounts for some similarity in the fauna between the Ethiopian and Oriental regions. Of all the regions, the Ethiopian contains the most varied vertebrate fauna, and in endemic families it is second only to the Neotropical.

Of the tropical realms, the Oriental, once covered with lush forests, possesses the fewest endemic species and lacks a variety of widespread families. It is rich in primate species, including two families confined to the region, the tree shrews and tarsiers.

Perhaps the most interesting and the strangest region, and certainly the most impoverished in vertebrate species, is the Australian. Partly tropical and partly south temperate, this region is noted for its lack of a land connection with other regions; the poverty of freshwater fish, amphibians, and reptiles; the absence of placental mammals and the dominance of marsupials. Included are the monotremes, with two egg-laying families, the duckbilled platypus and the spiny anteaters. The marsupials have become diverse and have evolved ways of life similar to those of the placental animals of other regions.

Life Zones

By the turn of the century, some biologists were attempting to combine plants and animals into one distributional scheme. C. Hart Merriam (1898), founder of the U. S. Bureau of Biological Survey (later to become the Fish and Wildlife Service), proposed the idea of *life zones*. These are transcontinental belts running east and west. The differences among them, expressed by the animals and plants living there, are supposedly controlled by temperature.

Merriam divided the North American continent into three primary transcontinental regions: the Boreal, the Austral, and the Tropical. Each of these regions Merriam further subdivided into life zones. For example, he subdivided the Boreal region into three zones: the Arctic-Alpine zone; the Hudsonian zone, or northern coniferous forest; and the Canadian zone, or southern coniferous forest.

Once widely accepted, life zones are rarely used today, although the terms still creep into the literature on vertebrates. The life zone concept failed because it is not a unit that can be recognized continentwide by a characteristic and uniform faunal or vegetational component. There are wide differences in the various zones between east and west.

Biotic Provinces

Another approach to the subdivision of North America into geographic units of biological significance is the biotic provinces concept, defined and mapped by L. R. Dice in 1943. The biotic provinces concept was an attempt to classify the distribution of plants and animals, especially the latter, on the basis of ranges and centers of distribution of the various species and subspecies. A province embraced a continuous geographic area that contains ecological associations distinguishable from those of adjacent provinces. Each biotic province was further subdivided into ecologically unique subunits, districts, or life belts, based largely on altitude, such as grassland belts and forest belts. The boundaries usually coincided with physiographic barriers rather than with vegetation types. That created problems with the scheme because, although a number of species may be confined to some biotic provinces, the distribution of most is determined by the presence of suitable habitat.

Biomes

Still another approach, pioneered by V. E. Shelford, was simply to accept plant formations as biotic units and to associate animals with plants. Such an approach works fairly well, because animal life does depend upon a plant base. These natural broad biotic units are called *biomes* (Figures VI.2, VI.3). Each biome consists of a distinctive combination of plants and animals in the fully developed or climax community, and each is characterized by a uniform life form of vegetation such as grass or coniferous trees. It also includes the developmental stages, which may be dominated by other life forms. Because the species that dominate the seral stages are more widely distributed than those of the climax, they are of little value in defining the limits of the biome.

On a local and regional scale, communities are considered as gradients, in which the combination of species varies as the individual species respond to environmental gradients. On a larger scale, one can consider the terrestrial and even some of the aquatic ecosystems as gradients of communities and environments on a continental scale. Such gradients of ecosystems are *ecoclines* (Figure VI.4).

In addition to gradual changes in vegetation, there are gradual changes in other ecosystem characteristics. As one goes from highly mesic situations and warm temperatures to xeric situations and cold temperatures, productivity, species diversity, and the amount of organic matter decrease. There is a corresponding decline in the complexity and organization of ecosystems, in the size of plants, and in the number of strata to vegetation. Growth form changes. The tropical rain forest is dominated by phanerophytes and epiphytes, the arctic tundra by hemicryptophytes, geophytes, and therophytes. Wherever similar environments exist on Earth, the same growth forms exist, even though species differences may be great. Thus, different continents tend to have communities of similar physiognomy.

There are six major terrestrial biomes: forests, grasslands, woodlands, shrublands, semidesert shrub, and desert. These six can be further divided into biome types, depending upon climatic conditions and elevation. These various biomes of the world fall into a distinctive pattern when plotted on a gradient of mean annual temperature and mean annual precipitation (Figure VI.5). The plots obviously are rough. Many types intergrade with one another, and adaptations of various growth forms may differ on several continents. Climate alone is not responsible for biome types. Soil and fire can also influence which one of several biomes will occupy a region. Structure of biomes is further influenced by the nature of the climate, whether marine or continental. The same amount of rain, for example, can support either shrubland or grassland.

Ecoregions

A recent attempt at the classification of the biotic world is the ecoregion approach, developed largely by R. W. Bailey (1976) of the U. S. Forest Service. It is based on the ecoregion concept of J. M. Crowley (1967). The classification scheme involves a synthesis of climate types, vegetation associations, and soil types into a single, geographic, hierarchical classification, which reflects both ecological properties and spatial patterns (Tables VI.1, VI.2).

An *ecoregion* is a continuous geographical area across which the interaction of climate, soil, and topography are sufficiently uniform to permit the development of similar types of vegetation (Figure VI.6). Boundaries between ecoregions, however, are arbitrary, drawn where the associations of the two regions cover approximately the same area. Within an ecoregion certain ecological communities similar to those of other regions may exist but nevertheless belong to the ecoregion in which they occur.

The ecoregion classification relates more to management of forest, range, and wildlife resources than to ecology per se. Its purpose is to provide a foundation for ecological management of natural resources. Because each region has its own distinctive flora, fauna, climate, soil, and landform, each requires its own approach to management. The ecoregion approach provides a means of studying the problems of resource management on a regional basis. It provides a framework for the organization and retrieval of data gathered in resource inventories and for the interpretation of inventory data. The concept has been widely accepted in the area of applied ecology.

The classification involves a hierarchy for ecosystems (Table VI.1). The largest category is the domain, a subcontinental area of broad climatic similarity. Domains are divided into divisions, de-

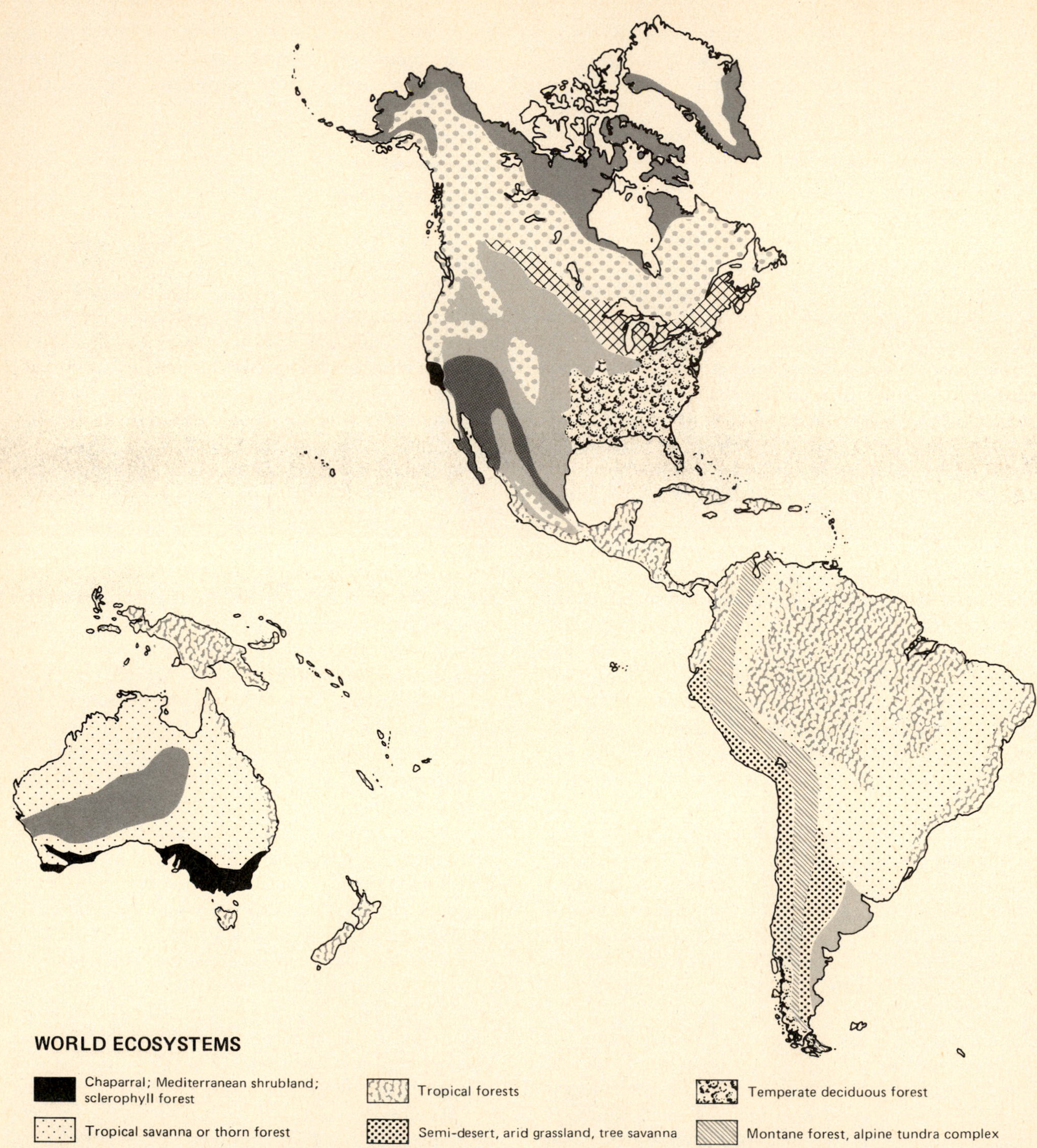

FIGURE VI.2 The major biomes of the world. The arctic and boreal forests are circumpolar and hold many taxonomically related and functionally similar species. The other biomes are similar in type, but most possess different genera and species, which, nevertheless, are similar in function (ecological equivalents).

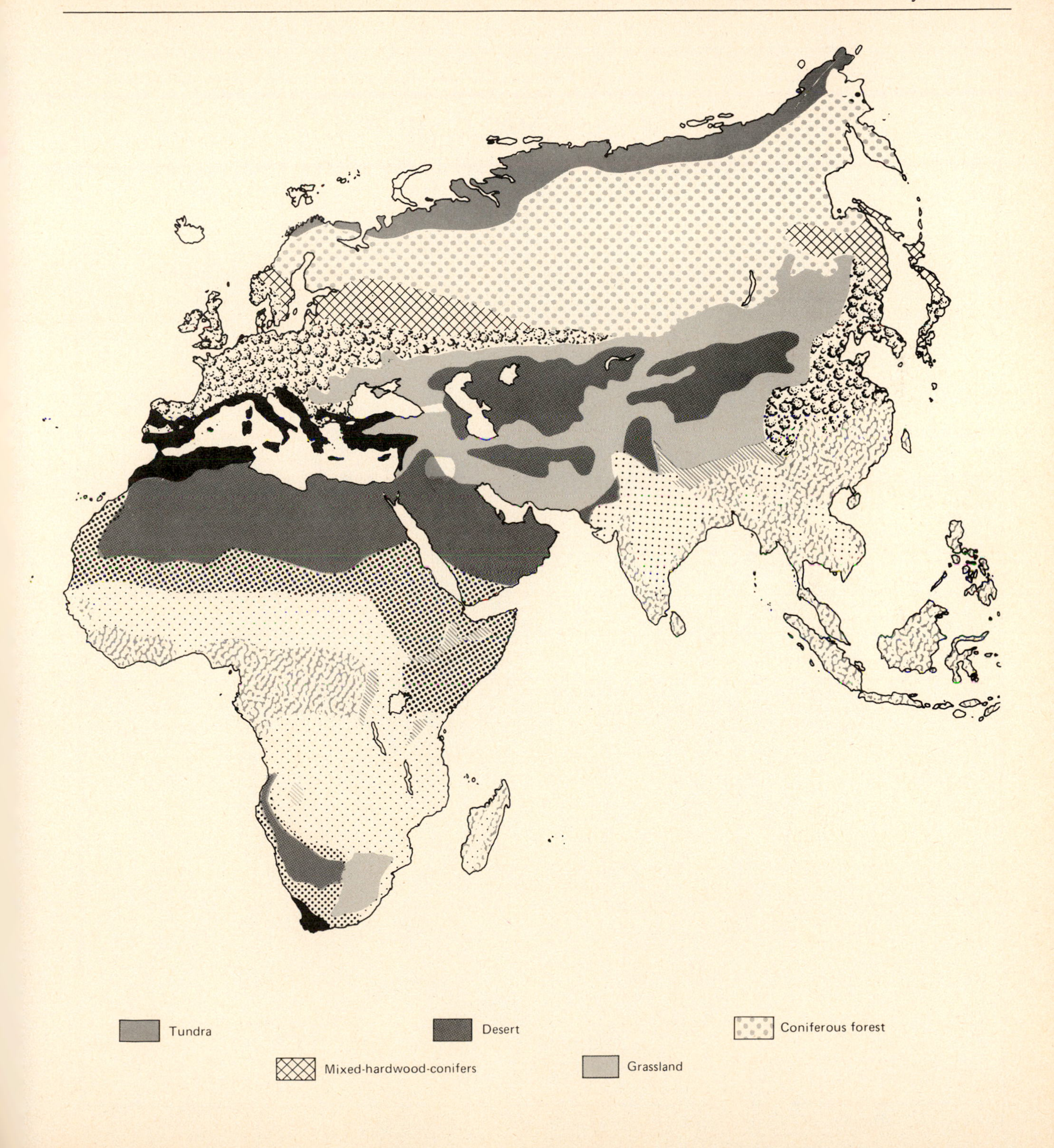
Tundra
Desert
Coniferous forest
Mixed-hardwood-conifers
Grassland

FIGURE VI.3 Biomes of North America. (*Map by John Aldrich, courtesy U.S. Department of the Interior, Fish and Wildlife Service.*)

FIGURE VI.4 Gradients of vegetation in North America from east to west (a) and north to south (b). The east-west gradient runs from the mixed mesophytic forest of the Appalachians through the oak-hickory forests of the central states, the ecotone of bur oak and grasslands, the prairie, the short-grass plains, and the desert. The transect does not cross the Rocky Mountains. The gradient is largely a result of variations in precipitation. The north-south gradient reflects temperature. The transect cuts across the tundra, the boreal coniferous forest, the mixed northern hardwood forest, the mixed mesophytic forests of the Appalachians, the subtropical forests of Florida, and the tropical forests of Mexico.

termined by isolating areas of differing vegetation and regional climates. Divisions are divided into provinces, which correspond to broad vegetational regions having a uniform regional climate and the same type or types of zonal soils (Figure VI.7). Provinces are further subdivided into smaller and smaller categories useful in local areas.

Holdridge Life Zone System

Another useful approach to the classification and study of ecosystems is the Holdridge life zone system (Holdridge, 1967; Holdridge et al., 1971). It is based on the assumption that (1) mature, stable plant formations represent physiognomically discrete vegetation types recognizable throughout the world; and (2) geographical boundaries of vegetation correspond closely to boundaries between climatic zones (Holdridge et al., 1971). Vegetation is determined largely by an interaction of temperature and rainfall.

The Holdridge system divides the world into life zones arranged according to latitudinal regions, altitudinal belts, and humidity provinces (Figure VI.8). The boundaries of each zone are defined by mean annual precipitation and mean annual biotem-

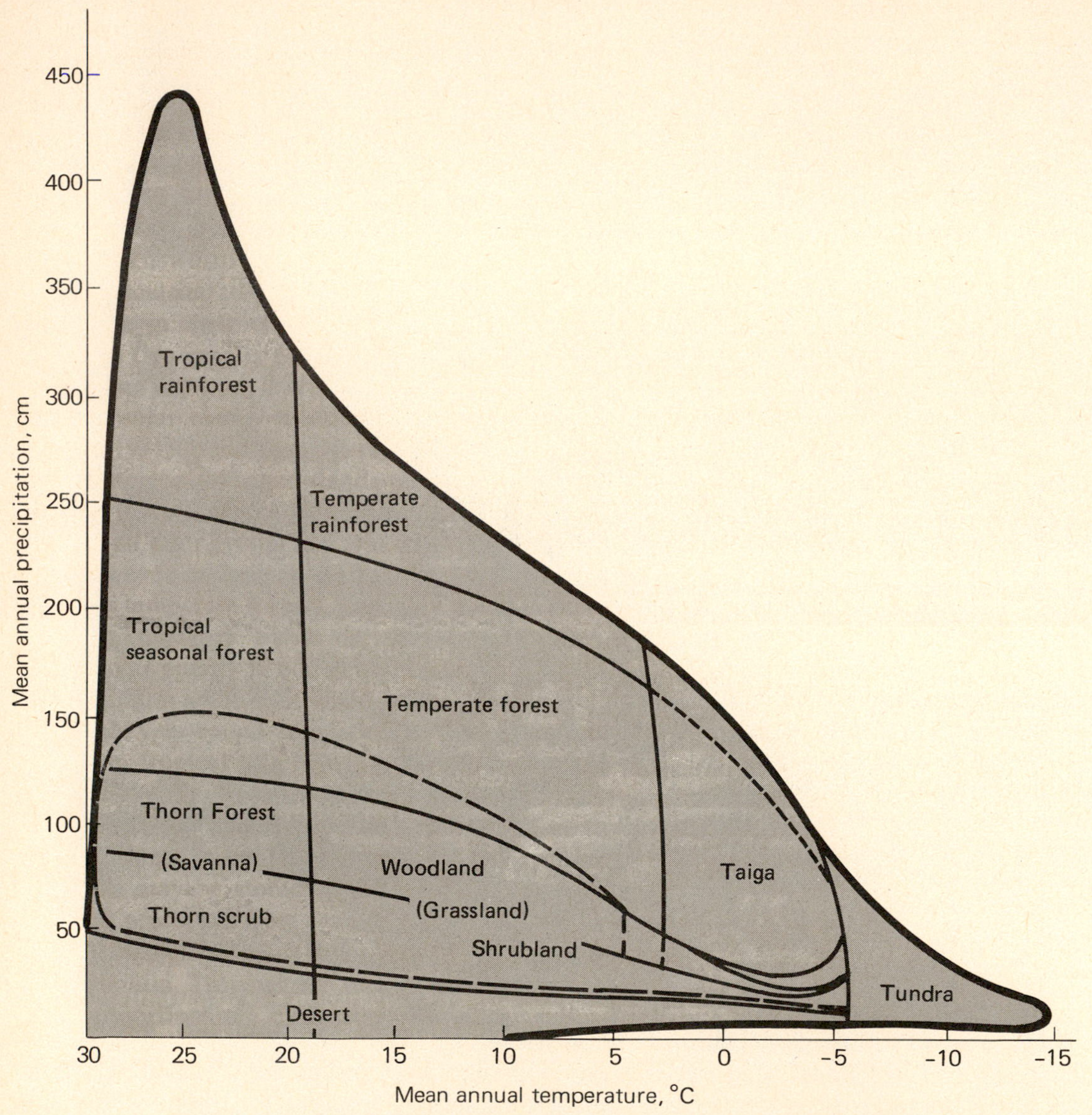

FIGURE VI.5 A pattern of world plant formation types in relation to climatic variables of temperature and moisture. In certain areas where the climate (maritime versus continental) varies, soil can shift the balance between such types as woodland, shrubland, and grass. The dashed line encloses a wide range of environments in which either grassland or one of the types dominated by woody plants may form the prevailing vegetation in different areas. (*After R. Whittaker, 1970:65,* Communities and ecosystems *[New York: Macmillan]*.)

perature. Altitudinal belts and latitudinal regions are defined in terms of mean biotemperatures and not in terms of latitudinal degrees or meters of elevation.

The Holdridge classification is based on three levels: (1) climatically defined life zones; (2) associations, which are subdivisions of life zones based on local environmental conditions; and finally, (3) further local subdivisions based on actual cover or land use. The term *association,* as used by Holdridge, means a unique ecosystem, a distinctive habitat or physical environment and the naturally evolved community of plants and animals.

The subdivision into zones or associations is based on an interaction between environment and vegetation as defined by biotemperature, precipitation, moisture availability, potential evapotranspiration, and potential evapotranspiration ratio.

TABLE VI.1 A HIERARCHY FOR ECOSYSTEMS

Name	Defined as Including
1. Domain	Subcontinental area of related climates
2. Division	Single regional climate at the level of Köppen's types (Trewartha, 1943)
3. Province	Broad vegetation region with the same type or types of zonal soils
4. Section	Climatic climax at the level of Küchler's potential vegetation types (1964)
5. District	Part of a section having uniform geomorphology at the level of Hammond's land-surface form regions (1964)
6. Landtype association	Group of neighboring landtypes with recurring pattern of landforms, lithology, soils, and vegetation associations
7. Landtype	Group of neighboring phases with similar soil series or families with similar plant communities at the level of Daubenmire's habitat types (1968)
8. Landtype phase	Group of neighboring sites belonging to the same soil series with closely related habitat types
9. Site	Single soil type or phase and single habitat type or phase

NOTE: This table is not intended to define the levels precisely, but merely to indicate the general character of the classification.

SOURCE: From R. G. Bailey, 1978, accompanying map; adapted from Crowley (1967) and Wertz and Arnold (1972).

TABLE VI.2 GENERAL ENVIRONMENTAL CHARACTERISTICS OF FIRST- AND SECOND-ORDER ECOREGIONS

Domain	Division	Temperature	Rainfall	Vegetation	Soil
Polar	Tundra	Mean temperature of warmest month <10°C	Water deficient during the cold season	Moss, grasses, and small shrubs	Tundra soils (entisols, inceptisols, and associated histosols)
	Subarctic	Mean temperature of summer is 10°C; of winter, −3°C	Rain even throughout the year	Forest, parklands	Podzols (spodosols and associated histosols)

(continued)

TABLE VI.2 (continued)

Domain	Division	Temperature	Rainfall	Vegetation	Soil
Humid temperature	Warm continental	Coldest month below 0°C, warmest month <22°C	Adequate throughout the year	Seasonal forests, mixed coniferous-deciduous forests	Gray-brown podzolic (spodosols, alfisols)
	Hot continental	Coldest month below 0°C, warmest month >22°C	Summer maximum	Deciduous forests	Gray-brown podzolic (alfisols)
	Subtropical	Coldest month between 18°C and −3°C, warmest month >22°C	Adequate throughout the year	Coniferous and mixed coniferous-deciduous forests	Red and yellow podzolic (ultisols)
	Marine	Coldest month between 18°C and −3°C, warmest month <22°C	Maximum in winter	Coniferous forests	Brown forest and gray-brown podzolic (alfisols)
	Prairie	Variable	Adequate all year, excepting dry years, maximum in summer	Tall grass, parklands	Prairie soils, chernozems (mollisols)
	Mediterranean	Coldest month between 18° C and −3°C, warmest month >22°C	Dry summers, rainy winters	Evergreen woodlands and shrubs	Mostly immature soils
Dry	Steppe	Variable, winters cold	Rain <50 cm/yr	Short grass, shrubs	Chestnut, brown soils, and sierozems (mollisols, aridisols)
	Desert	High summer temperature, mild winters	Very dry in all seasons	Shrubs or sparse grasses	Desert (aridisols)
Humid tropical	Savanna	Coldest month >18°C, annual variation <12°C	Dry season with <6 cm/yr	Open grassland, scattered trees	Latosols (oxisols)
	Rain forest	Coldest month >18°C, annual variation <3°C	Heavy rain, minimum 6 cm/month	Dense forest, heavy undergrowth	Latosols (oxisols)

Note: Names in parentheses in the soil column are soil taxonomy orders (USDA Soil Survey Staff, 1975).

SOURCE: Bailey, 1978, map.

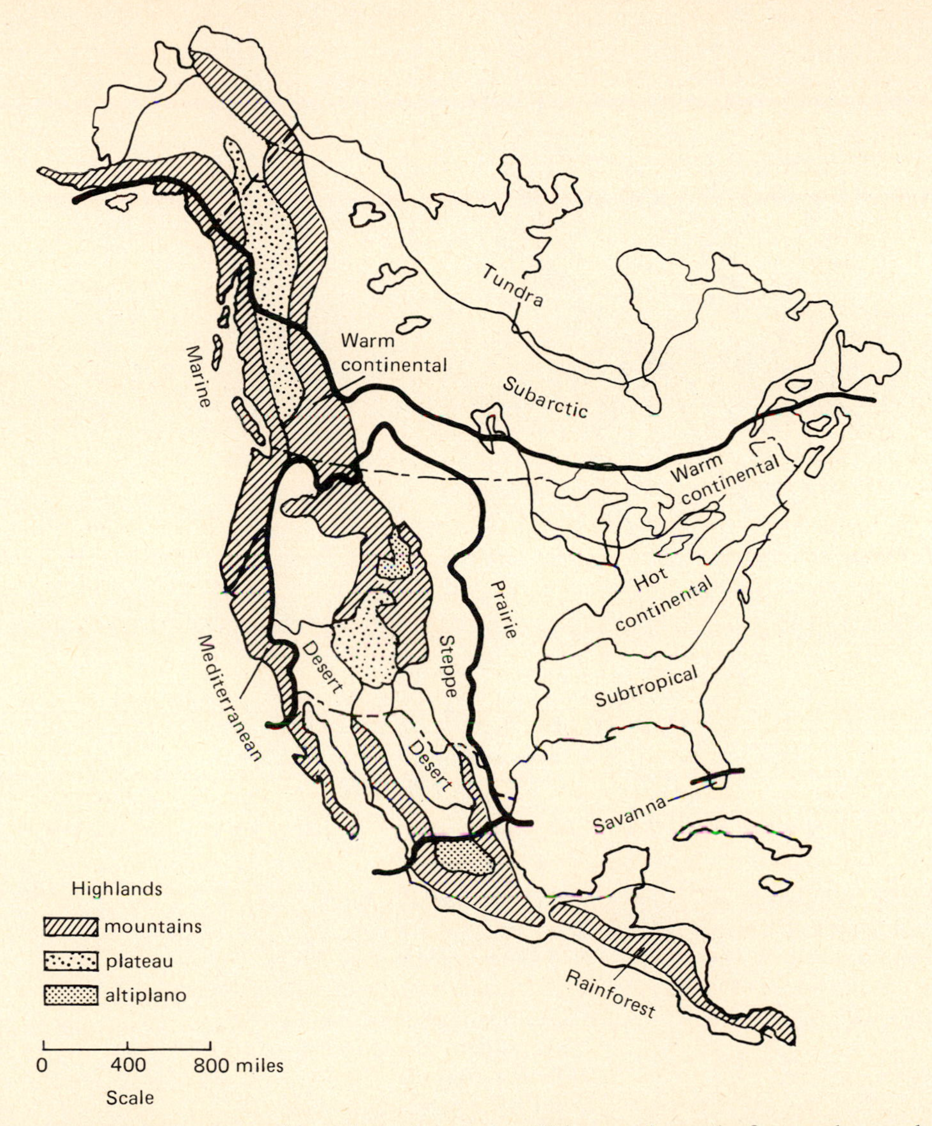

FIGURE VI.6 Ecoregions of North America. This map shows the first- and second-order ecoregions. The heavy lines outline the domains: polar, humid temperate, and humid tropical. The thin lines demark the boundaries of the labeled divisions (see Table VI.1). (*From Bailey, 1978.*)

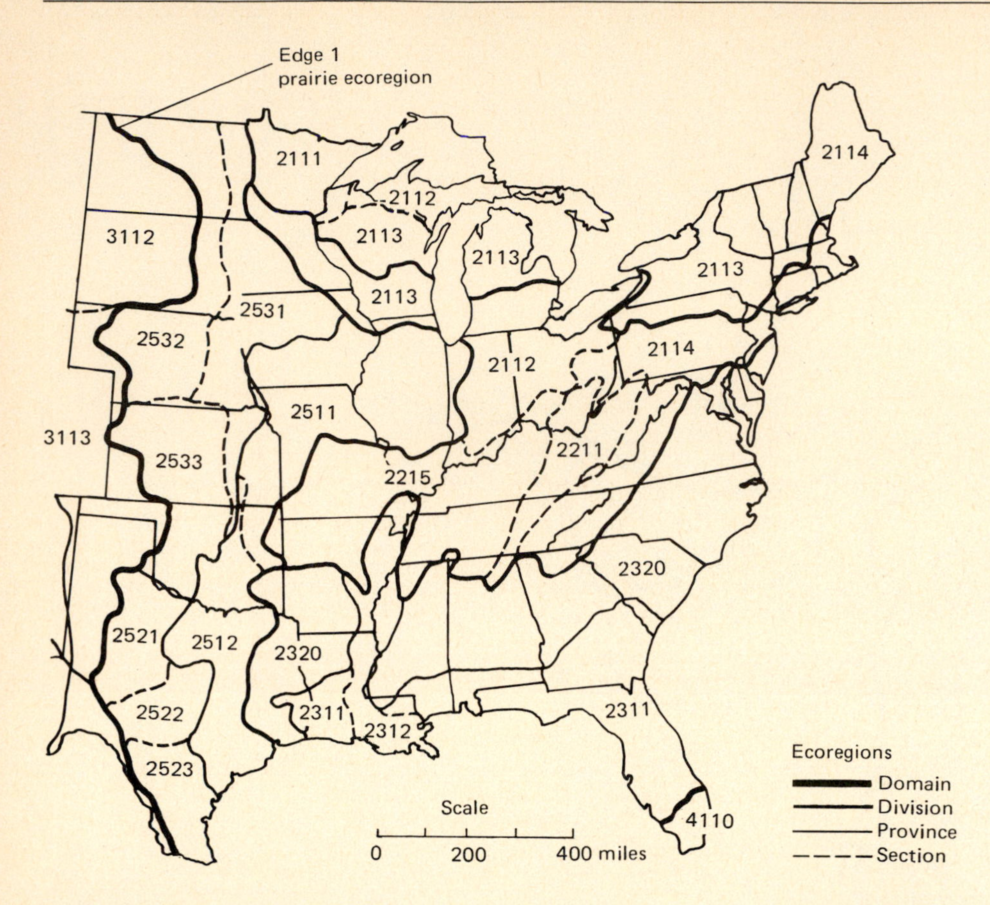

FIGURE VI.7 An example of the further subdivision of ecoregions, showing only the eastern and central United States to the edge of the humid temperate domain. The dark lines indicate the boundaries of provinces, the dashed lines, the sections within. *(For full details, see Bailey, 1978)*

2000 Humid temperate domain
- 2100 Warm continental division
  - 2110 Laurentian mixed forest province
    - 2111 Spruce-fir forest
    - 2112 Northern hardwood-fir forest
    - 2113 Northern hardwood forest
    - 2114 Northern hardwood-spruce forest
- 2200 Hot continental division
  - 2210 Eastern deciduous forest
    - 2211 Mixed mesophytic forest
    - 2212 Beech-maple forest
    - 2213 Maple-basswood forest
    - 2214 Appalachian oak
    - 2215 Oak-hickory forest
- 2300 Subtropical division
  - 2310 Outer coastal plain forest
    - 2311 Beech-sweetgum-magnolia-pine-oak forest
    - 2312 Southern flood plain forest
  - 2320 Southern mixed forest
- 2500 Prairie division
  - 2510 Prairie parkland province
    - 2511 Oak-hickory-bluestem parkland
    - 2512 Oak-bluestem parkland
  - 2520 Prairie brushland
    - 2521 Mesquite-buffalo grass
    - 2522 Juniper-oak-mesquite
    - 2523 Mesquite-acacia
  - 2530 Tall-grass prairie
    - 2531 Bluestem prairie
    - 2532 Wheatgrass-bluestem-needlegrass
    - 2533 Bluestem-grama

4000 Humid tropical domain
- 4100 Savanna division
  - 4110 Everglade province

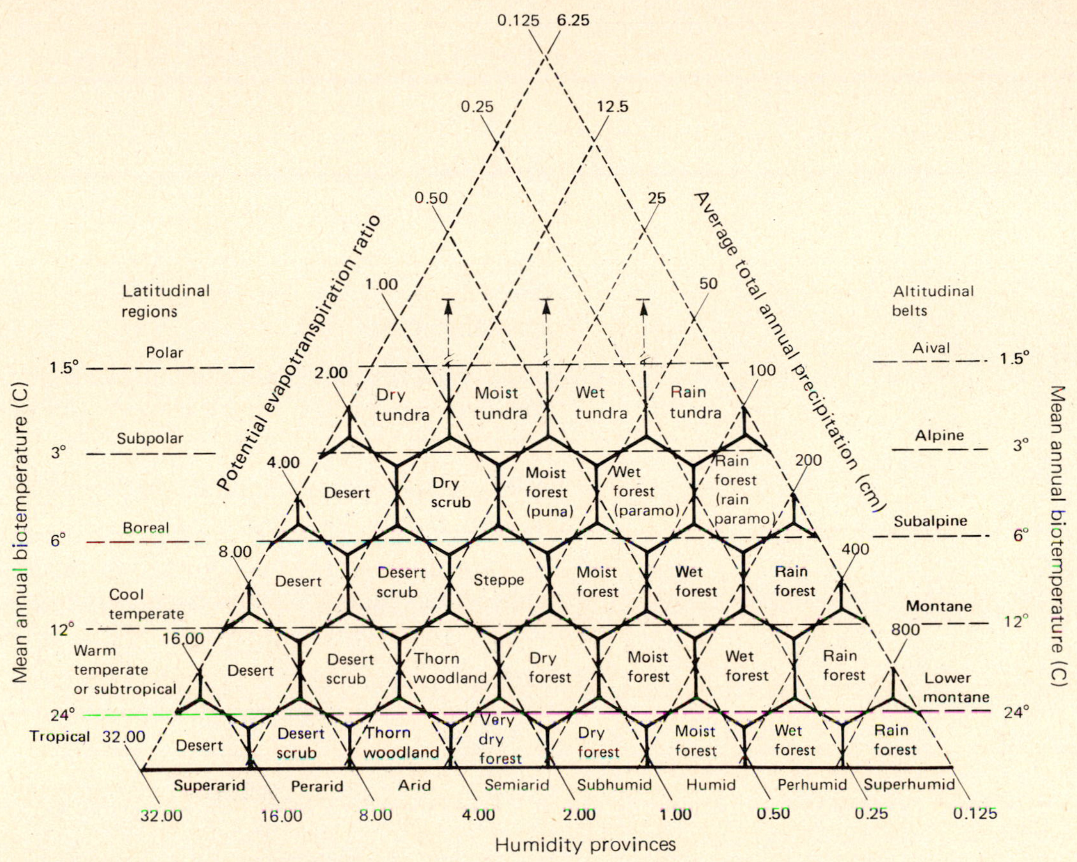

FIGURE VI.8 The Holdridge life zone system for classifying plant formations. (*After Holdridge, 1967.*)

CHAPTER 24

Terrestrial Ecosystems

Outline

Objectives

Upon completion of this chapter, you should be able to:

1. Describe the types and general features of grassland ecosystems.
2. Discuss energy flow and nutrient cycling in grassland ecosystems, emphasizing the role of below-ground biomass and its consumers.
3. Point out the major structural and functional features of savanna ecosystems.
4. Distinguish between climax and seral shrublands and explain why shrublands are so difficult to characterize.
5. Explain the importance of stratification, both biological and environmental, in forest ecosystems.
6. Compare structure and function of coniferous, temperate deciduous, and tropical rain forests and explain the reasons for differences in the functional processes involved in each.
7. Describe the major features of the tundra ecosystem, emphasizing the differences between arctic and alpine tundra.
8. Explain the importance of permafrost to the structure and functioning of tundra ecosystems.
9. Contrast nutrient cycling in tundra ecosystems with that in temperate grassland and forest and tropical ecosystems.
10. Describe the major features of desert ecosystems and explain how aridity influences energy flow and nutrient cycling.

If you were to examine thousands of photos taken from Earth-probing satellites, you would appreciate both the great diversity of land forms and the orderly yet patchy distribution of vegetation around the world. Deserts would appear in certain predictable regions. Earth would be belted with tropical forests. Grasslands and cultivated lands would dominate much of the Northern Hemisphere at lower latitudes, broken by patches of hardwood forests. A great belt of tundra and dark-colored boreal forest would cap the world. These broad vegetational features you would observe represent major terrestrial biomes, preferably termed ecosystems. You could group them in such categories as desert; tundra; grassland; coniferous, deciduous, and tropical rain forest; savanna; and thornbush. That is an incomplete set of categories. To be more precise, you could refine the listing by including more types of tropical vegetation, such as tropical deciduous or seasonal forest. Rain forest, as a category, is rather limited by comparison. You might include such types as temperate broadleaf evergreen forests, alpine shrublands, alpine grasslands, and the like. And you could subdivide the boreal forest into forest and taiga—open, boggy areas with clumps of spruce. But for the purpose of this brief discussion, the categories of ecosystems will be rather broad and inclusive, with an emphasis on structure and function rather than natural history.

Grasslands

At one time grasslands covered about 42 percent of the land surface of the world; today, however, much of it is under cultivation. All grasslands have in common a climate characterized by high rates of evaporation and periodically severe droughts, a rolling to flat terrain, and animal life dominated by grazing and burrowing species. They occur largely where rainfall is between 10 and 30 inches per year, too light to support heavy forest growth and too great to encourage a desert. Grasslands, however, are not exclusively a climatic formation, because most of them require periodic fires for maintenance and renewal and for the elimination of incoming woody growth.

TYPES OF GRASS

Grasses are either sod formers or bunch grasses. The former develop a solid mat of grass over the ground, while the latter grow in bunches (Figure 24.1). Clumps are formed by the erect growth of all shoots, and the grasses spread at the base by means of tillers. Sod-forming grasses, which include such species as Kentucky bluegrass and western wheatgrass, reproduce and spread by means of underground stems. Some grasses may be either sod or bunch, depending on the local environment. Big bluestem, for example, will develop a sod on rich, moist soil and forms bunches on dry soil. Associated with the grasses are a variety of legumes and forbs.

FIGURE 24.1 Growth forms of sod grass (right); crested wheatgrass; and a bunch grass, little bluestem (left). Also shown are root penetration and distribution in the soil (maximum depth: 2 m).

GRASSLAND TYPES

Grasslands are temperate, tropical, or cultivated. Temperate grasslands may be natural; seminatural, including pastures for grazing and meadows for hayland; and arable or tame (Figure 24.2). Tame grasslands may be seeded to perennial forages as rotational grassland or to annual forages. Natural temperate grasslands include the tall-grass, mixed, and desert grasslands and the short-grass plains of North America.

The tall-grass prarie occupies—or rather, occupied—a narrow belt running north and south next to the deciduous forest of eastern North America. In fact, it was well developed within a region that could support forests, but grasslands were maintained by periodic fires. Big bluestem was the dominant grass of moist soils and occupied the valleys of rivers and streams and the lower slopes of hills. Drier uplands in the tall-grass country were dominated by bunch-forming needle grass, side-oats grama, and prairie dropseed.

West of the tall grass is the mixed-grass prairie. Typical of the northern Great Plains, it embraces largely the needlegrass–grama grass community, with needlegrass-wheatgrass dominating gently rolling soils of medium texture (Coupland, 1950). Because the mixed prairie is characterized by great annual extremes in precipitation, its aspect varies widely from year to year. In moist years midgrasses are prevalent, while in dry years short grasses and forbs are dominant.

The short-grass plains reflect a climate where rainfall is light and infrequent (10 to 17 inches in the west, 20 inches to the east), humidity low, winds high, and evaporation rapid. The shallow-rooted short grasses utilize moisture in the upper soil layers beneath which roots do not penetrate. Sod-forming blue grama and buffalo grass dominate the short-grass plains, accompanied by such midgrasses as wheatgrass, side-oats grama, and little bluestem.

From southeastern Texas to southern Arizona and south into Mexico lies the desert grassland, similar in many respects to short-grass plains, except that triple-awn grass replaces buffalo grass. Composed largely of bunch grasses, desert grasslands are widely interspersed with other vegetation types such as oak savanna and mesquite.

Around the world exists a belt of tropical monsoon grasslands extending from western Africa to eastern China and Australia. Within this belt monsoon grasslands fall into an ecoclimatic gradient—arid to semiarid grasslands; medium-rainfall grasslands, found mainly in India, Burma, and northern Australia; the high-rainfall monsoonal or equatorial grasslands of Southeast Asia; and grasslands whose species are adapted to a hot monsoonal summer

(a)

(b)

(c)

FIGURE 24.2 (a) Buffalo, which once roamed the short-grass plains in countless numbers, epitomize the North American grasslands. (*Photo courtesy South Dakota Fish and Game Department.*) (b) Short-grass plains give way abruptly to forested hills in the Black Hills of South Dakota. (c) A cut through a cultivated grassland, a hayfield, showing the several strata in grasslands. Note the tall grass (timothy) and the denser understory (clover and alfalfa) and grass blades.

and cool to cold winters (Whyte, 1968). The tropical grasslands of South America do not fall into this group because geographic conditions do not promote true or false monsoonal conditions. Instead, grasslands of Latin America are largely *steppe,* consisting almost entirely of bunch grass, with no legumes and very few herbs, bushes, or trees (McIlroy, 1972).

Continental tropical grasslands fall into their own ecoclines. In Africa, for example, climatic climax grasslands are confined to desert areas with prolonged drought. Grass cover, dominated by *Aristida,* is low, and soil may be blown away by the wind. With a slight increase in moisture, desert scrub with scattered shrubs of the genera *Commiphora* and *Acacia* becomes the climax type. As rainfall increases, desert scrub gives way to desert grass–*Acacia* savanna; tall-grass–*Acacia* savanna; tall-grass–low-tree savanna; and finally, humid forests. Thus, low-rainfall areas are characterized by quick-growing, tufted types of low ground cover and high-rainfall areas by tall, coarse grasses.

Much tropical grassland exists because of fires that prevent the intrusion of woody vegetation. Grass, in turn, reacts to burning by putting out new shoots, drawing on reserves of moisture and food in rhizomes and roots, which are depleted before the arrival of rain. If the grass is overgrazed, plants weaken and deteriorate and may be replaced by annual or unpalatable species. On the other hand, the elimination of fire is just as disastrous.

STRUCTURE

Grasslands have three basic strata: roots, ground layer, and herbaceous layer (Figure 24.2c). The root layer is more pronounced in grasslands than in any other major community. Half or more of the plant is hidden beneath the soil; in winter that represents almost the total grass plant. The bulk of roots occupies rather uniformly the upper 15 cm or so of the soil profile; they decrease in abundance with depth. The depth to which roots of grass extend is considerable. Little bluestem, for example, reaches 1 to 2 m and forms a dense mat as deep as 0.8 m (Weaver, 1954). In addition, many grasses possess underground stems, or rhizomes, that serve both to propagate plants and to store food.

Roots develop in three or more zones. Some plants are shallow-rooted and seldom extend much below 0.5 m. Others go well below the shallow-rooted species but seldom more than 1.5 m. Deep-rooted plants extend even further into the soil and absorb relatively little moisture from the surface soil (see Chapter 6). Thus, plant roots absorb nutrients from different depths in the soil at different times, depending upon moisture (Weaver, 1954).

The mulch or ground layer is characterized by low light intensity during the growing season, reduced wind flow, and relatively high humidity. Light intensity decreases as the grass grows taller and shades the ground. Temperatures decrease as solar insolation is intercepted by a blanket of vegetation, and wind flow is minimal.

Grasslands that are unmowed, unburned, and ungrazed accumulate a layer of thick mulch on the ground surface. The oldest layer consists of decayed and fragmented remains of fresh mulch. Three or 4 years must pass before natural grassland herbage decomposes completely (Hopkins, 1954). As the mat increases in depth, more moisture is retained, creating very favorable conditions for microbial activity (McCalls, 1943).

Grazing reduces mulch, as do fire and mowing. Light grazing tends to increase the weight of humic mulch at the expense of fresh (Dix, 1960). Moderate grazing results in increased compaction, which favors an increase in microbial activity and a subsequent reduction in both fresh and humic mulch. Heavy grazing severely reduces accumulation of mulch (Zeller, 1961). Burning reduces mulch, but the mulch structure returns 2 or 3 years after a fire on lightly grazed and ungrazed land (Tester and Marshall, 1961; Hadley and Kieckhefer, 1963). Mowing and removal of grass for hay greatly reduces both fresh mulch and humic mulch.

The herbaceous layer consists of three or more strata, more or less variable in height through the year. Low-growing and ground-hugging plants such as dandelion and mosses make up the first stratum. All these become hidden beneath the middle and upper layers as the growing season progresses. The middle layer consists of shorter grasses and such herbs as wild mustard and daisy. The upper layer consists of tall grasses and forbs, conspicuous mostly in autumn.

Grasslands support an array of herbivorous animals, both invertebrate and vertebrate. Natural grasslands are characterized by large ungulate fauna, such as bison of North America, antelopes of African plains, and kangaroos of Australia. Most of these have been replaced by domestic cattle and sheep. In addition, there are numerous herbivorous rodents such as meadow mice and lemmings. But the dominant herbivores are invertebrates, which occupy all strata at some time during the year. The root layer is inhabited by nematodes, the upper soil by earthworms and ants, the ground layer by carabid beetles, and the upper strata by grasshoppers, leafhoppers, and crickets.

FUNCTION

Production and nutrient cycling vary considerably across the array of grassland types, from semiarid to semihumid. Primary production is influenced by both temperature and moisture. It is lowest where precipitation is low and temperatures are high and highest where mean annual precipitation is greater than 800 mm and mean annual temperature is above 15°C (see data in Coupland, 1979; Breymeyer and Van Dyne, 1980). Production, however, is most directly related to precipitation (Figure 24.3). The greater the mean annual precipitation, the greater is the above-ground production. This comes about in part because increased moisture reduces water stress and enhances the uptake of nutrients.

Production is mirrored in biomass accumulation. Mean green shoot above-ground biomass among the array of grasslands studied in the International Biological Program (IBP) ranges from 50 g/m^2 in arid grasslands to 827 g/m^2 in subhumid grasslands. Mean underground biomass ranges from 45 to 4,707 g/m^2. In most tropical grasslands, that biomass is usually less than 1,000 g/m^2 (median: 700 g/m^2). Ratio of root to shoot (or below-ground to above-ground biomass) ranges from 0.2 to 10.3. The ratio is smallest in tropical grasslands, a mean of 0.8. The ratio for seminatural grasslands is a mean of 3.3 and for temperate grasslands, 4.4. Except for a short period of maximum above-ground biomass, below-ground biomass is two to three times that above ground. Primary productivity ranges from 82 g/m^2/year in semiarid regions to 3,396 g/m^2/year in subhumid tropical grasslands. In natural and seminatural temperate grasslands, production ranges from 98 to 2,430 g/m^2/year. Primary

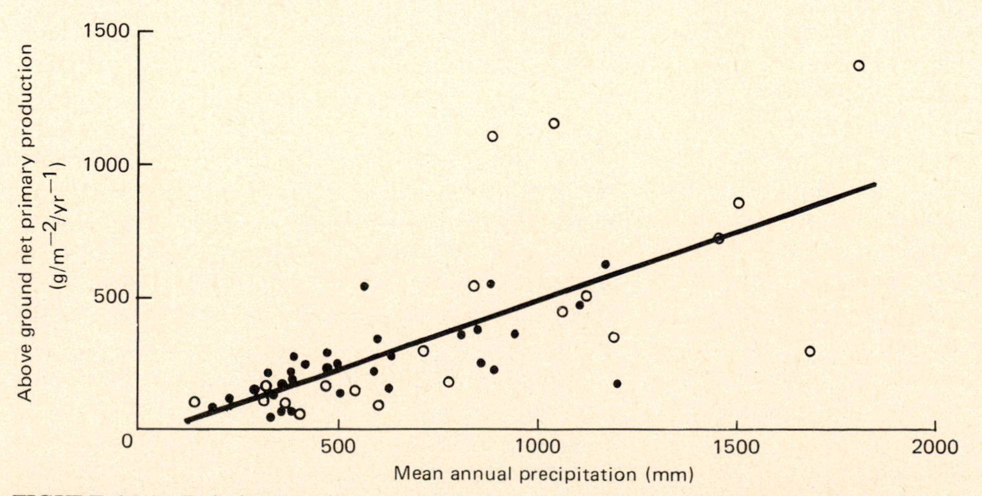

FIGURE 24.3 Relationship between above-ground primary production and mean annual precipitation for 52 grassland sites around the world. North American grasslands are indicated by dots. Annual net production = 0.5 (annual precipitation) − 29; r^2 = 0.51. (*From Lauenroth, 1979:10.*)

productivity in North American natural grasslands ranges from 300 to 1,600 g/m^2/year. Much of the net production of grassland does not appear as aboveground biomass. Seventy-five to 85 percent of the photosynthate is translocated to the roots for storage below ground.

Grasslands have the highest index of productivity per unit of standing crop (20 to 55 percent) among all terrestrial ecosystems. The tundra during its short growing season has an index of 10 to 20 percent. The index for the boreal forest is only 2 to 5 percent and for the temperate deciduous forest, 8 to 10 percent. Of course, a great deal of the energy in woody ecosystems goes into the maintenance of high standing crops, which grasslands do not support. Efficiency of energy capture by grasslands is 0.75 percent for net primary production. Light to moderate grazing increases that efficiency to 0.87 percent.

Many grassland plants are adapted to grazing, having evolved under selective pressure of grazing herbivores, and respond by increasing total primary production up to a point. Although the large herbivores are the most conspicuous consumers, the most important are the invertebrates. The aboveground biomass of invertebrates—including plant consumers, saprovores, and predators—ranges from 1 to 50 g/m^2, and grazing animals amount to about 2 to 5 g/m^2. Below-ground invertebrates exceed 135 g/m^2. The major above-ground consumers are grasshoppers, and the major below-ground consumers are nematodes. Nematodes account for 90, 95, and 93 percent of all below-ground herbivory, carnivory, and saprophagous activity. They account for 46 to 67 percent of root and crown consumption, 23 to 85 percent of fungal consumption, and 43 to 88 percent of below-ground predation. Above-ground herbivores consume 2 to 7 percent of primary production, or 3 to 10 percent if both the amounts consumed and wasted are considered. Below-ground consumption, including wastage, amounts to 13 to 46 percent.

Not only is a large proportion of primary production consumed below ground, but a greater proportion is utilized at each trophic level (Figure 24.4). Efficiency of transfer, however, is about the same. Invertebrates convert about 9 to 25 percent

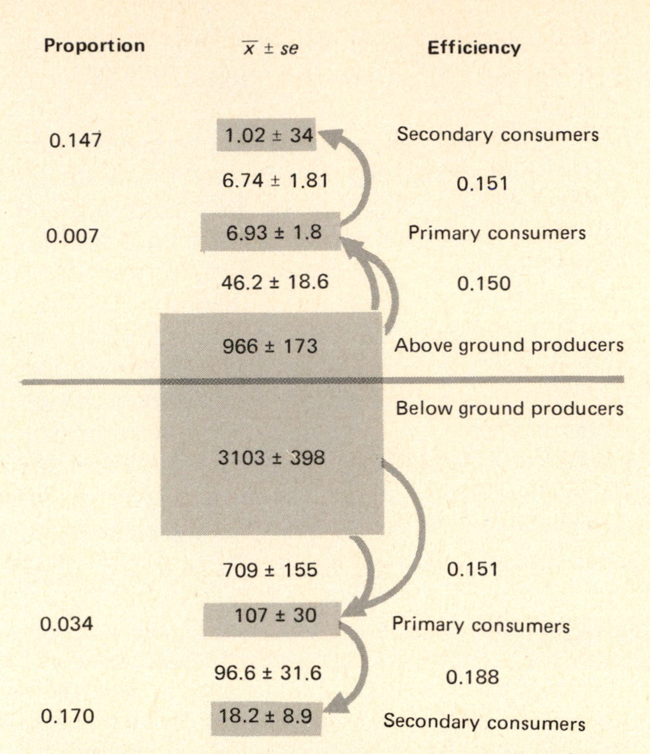

FIGURE 24.4 Grassland production (in boxes) and consumption (between boxes) representing means and standard errors for desert, tall-grass, mixed, and short-grass sites in North American grasslands. Production in one trophic level is represented as the proportion of production in the next trophic level and efficiency as the ratio of production to consumption by each consumer trophic level. (*From French, 1979:185.*)

of ingested energy to animal tissue, while homoiotherm grazers, such as sheep and cattle, convert 3 to 15 percent to animal tissue. Some invertebrate consumers are wasteful. The amount of aboveground vegetation detached or otherwise killed about equals that consumed by vertebrate grazing herbivores.

Most of the primary production, above-ground especially, goes to the decomposers—dominated by fungi, whose biomass is two to seven times that of bacteria. Overall, the decomposer biomass exceeds that of the invertebrates.

Green plant consumers, however, are important in the cycling of nutrients in grassland ecosystems. Invertebrate consumers are highly inefficient in assimilating ingested material. Much of their intake is deposited as feces or frass. Because the nutrients they contain are in a highly soluble form, they are returned rapidly to the system. Large grazing herbivores return a portion of their intake as dung, which becomes a major pathway of nutrient cycling (Figure 24.5). A well-developed coprophagous fauna speeds the decay of manure and accelerates the activity of bacteria in the feces.

Nitrogen provides an example of nutrient cycling in grasslands (Figure 24.6). About 90 percent of nitrogen in grasslands is tied up in soil organic matter, 2 percent in litter, 5 percent in live and dead plant cover, 1 percent in dead shoots, about 0.8 percent in soil microflora, and a very small amount in above- and below-ground invertebrates. When nitrogen is limiting, most of it is shunted to green herbage, where it remains during the growing season. Some of this is consumed by hervibores. Another fraction is translocated below ground to roots. Much of this nitrogen is moved above ground to new growth the following growing season. A fraction is retained in standing dead leaves and returned to the litter and soil surface, where it is acted upon by fungi and bacteria. Part becomes immobilized in microbial biomass but is quickly mineralized upon the death of microbes and reenters the nitrogen cycle. Still another fraction is tied up in humus and resists decomposition. Turnover, however, is fairly rapid, and most of the nitrogen that enters a green plant one growing season will reenter another the following year.

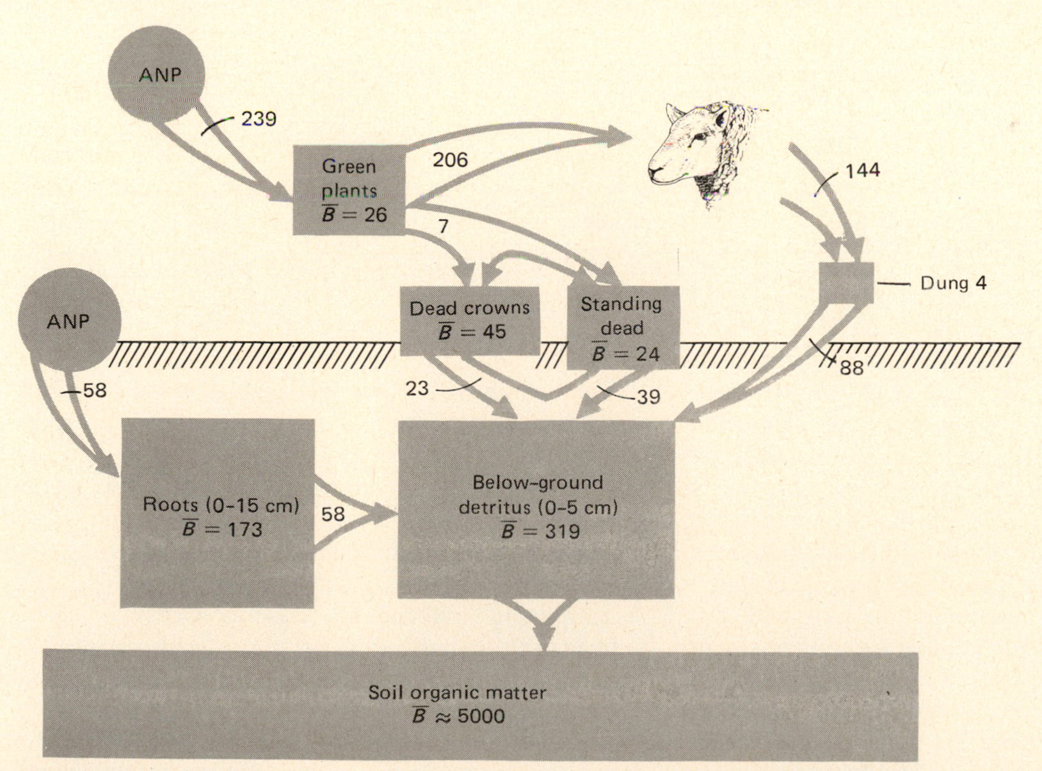

FIGURE 24.5 Carbon (energy) flow in a grazed pasture, indicating the role of grazing herbivores in the functioning of grassland ecosystems. Boxes represent the retention of carbon (g/m^2). Arrows represent the flow of carbon (g/m^2/year). ANP = annual net productivity. (*From Breymeyer, 1980.*)

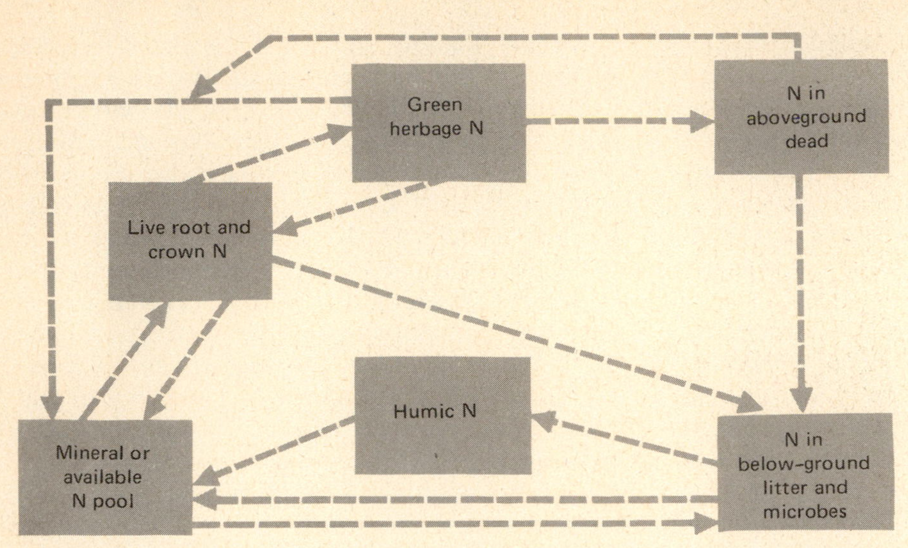

FIGURE 24.6 Diagram of the internal flows of nitrogen in a short-grass prairie. Note the movement (translocation) between the mineral or available nitrogen pool and live crown and root and green herbage, and between below-ground litter and microbes and the available nitrogen pool. These relationships point out the conservative cycling of nitrogen in the grassland ecosystem. (*From Clark, Cole, and Bowman, 1979.*)

Tropical Savannas and Woodlands

Because there is no typical savanna ecosystem, it is difficult to make broad general statements about it. Even the origin of the name, which has its roots in several different languages, is obscure (Bourliere and Hadley, 1983). The word in its several different origins referred to grasslands or plains. Only in the late 1800s was the term applied to the vegetation type it now infers: grassland with scattered trees. As such, a savanna is more of a continuum than a distinct type such as a tundra or a tropical rain forest. Savannas range from grass savanna with very few widely scattered trees to shrub savanna, grassland with scattered shrubs (Figure 24.7a and b), to tree savanna (Figure 24.7c), to savanna woodland or thornbush (Figure 24.7d). Savannas cover much of central and southern Africa, Malaysia, western India, northern Australia, and parts of northern and east-central South America. Some savannas are natural. Others are seminatural or anthropogenic, brought about and still maintained by centuries of human interference.

STRUCTURE

Savannas have several characteristics in common. The major stratum is grass, which is interrupted by trees and shrubs (Figure 24.8). They experience periodic fires and have a pattern of growth controlled by alternating wet and dry seasons. Vegetation exhibits accelerated growth following the first rains of the wet season. The shrub and tree component (except for baobab trees) is short-lived, seldom surviving for more than several decades. Competition may exist between grasses and woody vegetation for topsoil moisture. Similarly, trees compete for soil resources, accounting for the spacing patterns of woody vegetation (Smith and Walker, 1983; Gutierrez and Fuentes, 1979). Grasses for the most part are bunch or tussock. The larger part of the living herbaceous biomass is

(a) (b) (c) (d)

FIGURE 24.7 Savanna ecosystems in southern Africa: (a) grass savanna, (b) shrub savanna, (c) tree savanna with *Acacia,* (d) savanna woodland dominated by *Combreteum.* (*Photos by T. M. Smith.*)

underground, with a highly developed root system throughout the upper 30 cm.

Savannas support or are capable of supporting a large and often varied assemblage of herbivores, both grazing and browsing. This diversity is particularly pronounced in Africa. In spite of their visual dominance, large ungulates consume only about 10 percent of primary production. Dominant herbivores are the invertebrates, including the acridid grasshoppers, seed-eating ants, and detrital-feeding dung beetles and termites (Gillon, 1983).

FUNCTION

Little information exists on the rate of primary production in savannas. Probably a wide range in production exists between grass savanna on the one end and tree savanna and woodlands on the other (see Lamotte and Bourliere, 1983; Rutherford, 1978). Primary production is initiated at the beginning of the wet season. Moisture sends a quick flush of growth into grass and woody plants, supported by translocation of nutrients from roots and by nu-

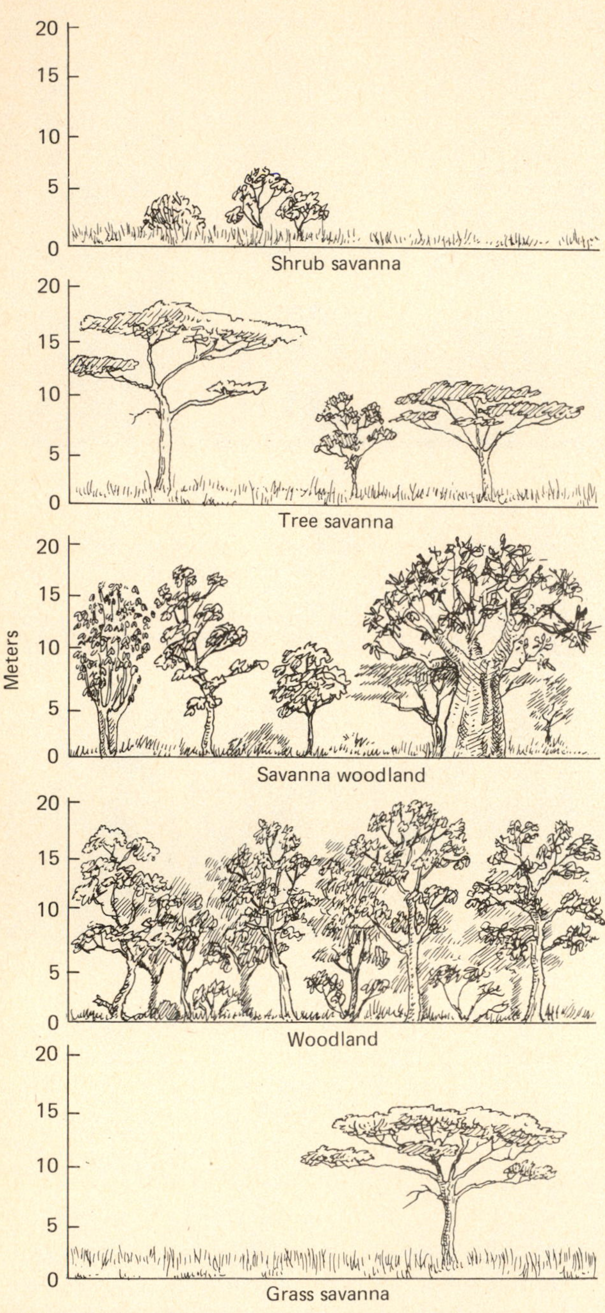

FIGURE 24.8 Profile and structure of a gradient of savanna types: grass, shrub, tree, savanna woodland and woodland. (*Based in part on Menaut, 1983.*)

trients released from materials accumulated in the dry season. Large root systems efficiently transfer nutrients from the soil to above-ground biomass. The nutrient pool, especially N, is low, with much of it tied up in plant and animal biomass. Nutrient turnover, however, is high, with little accumulation of soil organic matter.

Shrublands

Covering large portions of the arid and semiarid world is climax shrubby vegetation. In addition, climax shrubland exists in parts of temperate regions because historical disturbances of landscapes have seriously affected their potential to support forest vegetation (Eyre, 1963). Among such shrub-dominated plagioclimaxes are the *moors* of Scotland and the *macchia* of South America. But outside of these regions, shrublands are seral, a stage in the land's progress back to forest. There they exist as second-class citizens of the plant world (McGinnes, 1972), given little attention by botanists, who tend to emphasize dominant plants. As a result, little work has been done on seral shrub communities (see also Duffey, 1974).

CHARACTERISTICS OF SHRUBLANDS

Shrubs are difficult to characterize. They have, as McGinnes (1972) points out, a "problem in establishing their identity." They constitute neither a taxonomic nor an evolutionary category (Stebbins, 1972). One definition is that a shrub is a plant with woody, persistent stems but no central trunk and a height of up to 15 or 20 feet. But size does not set shrubs apart, because under severe environmental conditions many trees will not exceed that size. Some trees, particularly coppice stands, are multiple-stemmed, and some shrubs have large, single stems. Shrubs may have evolved either from trees or from herbs (for a detailed discussion of the evolution of shrubs see Stebbins, 1972).

Shrub ecosystems, seral or climax, are characterized by woody structure, increased stratification over grasslands, dense branching on a fine scale, and low height. Dense growth of many shrub types, such as hawthorn and alder, develops nearly impenetrable thickets that offer protection for such animals as rabbits and quail and nesting sites for birds. The deep shade beneath discourages any understory growth. Many seral shrubs have flowers that are attractive to insects (and to humans), and their seeds, enclosed in palatable fruits, are easily dispersed by birds and mammals.

The success of shrubs depends largely on their ability to compete for nutrients, energy, and space (West and Tueller, 1972). In certain environments shrubs have many advantages. They have a lower energetic and nutrient investment in above-ground parts than trees. Their structural modifications affect light interception, heat dissipation, and evaporative losses, depending on the species and environments involved. The more arid the site, the more common is drought deciduousness and the less common is evergreenness (Mooney and Dunn, 1970a). The multistemmed forms influence interception and stemflow of moisture, increasing or decreasing infiltration into the soil (Mooney and Dunn, 1970b). Because most shrubs can get their roots down quickly and form extensive root systems, they can utilize soil moisture deep in the profile. This feature gives them a competitive advantage over trees and grasses in regions where the soil moisture recharge comes during the nongrowing season. Because they do not have a high root-shoot ratio, shrubs draw less nutrient input into above-ground biomass and more into roots. Their perennial nature allows immobilization of limiting nutrients and slows the nutrient recycling process, favoring further shrub invasion of grasslands. Subject to strong competition from herbs, some climax shrubs, such as chamise *(Adenostoma fasciculatum),* inhibit the growth of herbs by means of allelopathy (see Chapter 15) (McPherson and Muller, 1969). Only when fire destroys mature shrubs and degrades the toxins do herbs appear in great numbers. As the shrubs recover, herbs decline. The herb species affected have apparently evolved the ability to let their seeds lie dormant in the soil until they are released from suppression by fire.

SHRUBLAND TYPES

Arid Shrublands

In the semiarid country of western North America, in regions bordering the Mediterranean, in Australia, in parts of South America, in western India, and in central Asia are communities of xeric, broadleaf evergreen shrubs and dwarf trees not over 8 feet tall. These sclerophyll communities are characteristic of regions where winters are mild and rainy and summers are long, hot, and dry (for detailed descriptions see McKella, Blaisdell, and Goodin, 1972). Their vegetation varies with geographic location, altitude, and direction of slope.

These shrub communities go by different names in different parts of the world. In the Mediteranean region, shrub vegetation results from forest degradation and falls into three major types. The *garrigue* is low, open shrubland on well-drained to dry calcareous soil and results from the degradation of pine forests. The *maqui,* a higher type of thick shrubland, occurs where there is more rainfall and replaces cork forests. The third type, the *matorral,* combines the other two types and appears to be equivalent to the North American chaparral. In South America, shrublands go by various names, depending on the region. In central Brazil, large areas of low, woody evergreen plants are called *cerrados;* in northeastern Brazil, shrublands, leafless in protracted dry periods, are called *catinga.* Along the Pacific Ocean in Peru and Chile exists a highly diverse shrubland, again suggestive of chaparral (see Soriano, 1972). In southwestern Australia, the shrub country, dominated by low-growing *Eucalyptus,* is known as *mallee.*

In North America the sclerophyllic shrub community is known as *chaparral.* There are three types: (1) the mediterranean-type in California, dominated by shrub oaks and chamise, where winters are wet and summers dry; (2) the inland chaparral of Arizona and New Mexico, dominated by Gambel oak and other species but lacking chamise; and (3) the Great Basin sagebrush (see Figure 24.9a). For the most part, these communities lack an understory, and ground litter, highly flammable, ranges from light to heavy. Heavy seeders, many plant species of the chaparral require the heat and scarring action of fire to induce germination. Periodic fires clear away old growth, making way for the new and recycling nutrients through the ecosystem.

Seral Shrublands
The nature of a shrubland that develops on abandoned land is influenced by previous management of the area, such as the intensity of grazing; the woody plants already in the field (although suppressed); the seeds present in the ground; the source and availability of seed; the dispersal of seed to the field by birds, mammals, and wind; and the colonizing ability of the shrub species.

On drier uplands, invading shrubs rarely exert complete dominance over herbs and grass. Instead, the plants are scattered or clumped in grassy fields (Figure 24.9b). On wet ground the plant community is often dominated by tall shrubs and contains an understory between that of a meadow and a forest (Curtis, 1959). There is some evidence (Egler, 1953; Niering and Egler, 1955; Niering and Goodwin, 1974) that in regions where forest is the normal end of succession, shrubs can form a stable community that will persist for many years.

FUNCTION

The functional aspects of shrubland ecosystems are poorly studied, but data from a few mediterranean-type ecosystems provide some interesting insights into nutrient cycling.

The soils of mediterranean-type ecosystems are low in nutrients and are especially deficient in nitrogen and phosphorus. The cycling of these nutrients appears to be tight and conservative. Nitrates in the soil vary seasonally, reflecting rainfall or the lack of it and microbial activity, which is relatively high during the winter. A flush of microbial activity, which involves decomposition of humus and mineralization of nitrogen and carbon, follows wetting of dry soil. Nitrate accumulation depends upon a progressive drying period after rains. The topsoil gradually dries out to an increasing depth. The resulting improved aeration and increasing soil temperature favor rapid bacterial nitrification and the retention of nitrate ions in the soil. As the dry period continues, nitrates accumulate and remain fixed in the topsoil along with other soluble nutrients (Schaefer, 1973). When the rains arrive and wet the soil, the concentration of nutrients stimulates a flush of growth. If heavy rains suddenly enter dry topsoil, quantities of nutrients may be lost by leaching. In California chaparral much of the nitrogen returned to the soil is lost through erosion (Mooney and Parsons, 1973).

Certain plants of the mediterranean systems exhibit some nutrient-conservation mechanisms. *Ceanothus,* an early successional species in the California chaparral, is a nitrogen fixer (Mooney and Parsons, 1973). In the Australian ecosystem, *Atriplex vesicana,* the dominant plant, lowers the nitrogen content of the surrounding soil and concentrates nitrogen through litterfall in the soil directly beneath the plant. More nitrogen, however, is withdrawn from the soil than is returned by litterfall, which represents about 10 percent of total plant nitrogen. Litter has a lower nitrogen and phosphorus content than fresh leaves, which suggests that the plant withdraws nitrogen and phosphorus from the leaf into the stem before leaffall.

Australian sclerophyllous plants also exhibit phosphorus conservation, involving three mechanisms. First, in aging leaves, phosphorus, like nitrogen, is recirculated through the plant with only minimal losses to litter. Second, a fine mat of mycorrhizal roots penetrates the decomposing litter and takes up phosphorus. Third, polyphosphate and hydrolyzing enzymes are present in the roots of sclerophyllous plants. As orthophosphate is released from decomposing litter in spring, it is stored in roots as long-chain polyphosphate. When growth begins, polyphosphate seems to be hydrolyzed back to orthophosphate and transported to growing shoots (Specht, 1973).

Forests

Of all the vegetational types of the world, probably none is more widespread (or at least was at one time) or more diverse than forests. In spite of the wide range of forest types, all possess certain char-

(a)

(b)

FIGURE 24.9 (a) Shrubland of the cool desert of the Great Basin, here dominated by sagebrush. (b) In eastern North America, shrublands are usually successional communities, but if vegetation is dense, shrub communities may persist for a very long time. This slope is claimed by Saint-John's-wort *(Hypericum virginicum)* and wild indigo *(Baptisia tinctoria)*.

acteristics in common. Dominated by a large aboveground biomass, forests possess a multilayered structure that provides a habitat for a wide assemblage of species.

STRATIFICATION

When you walk in the summer woods, you are observant of at least one fact: you are moving beneath an overhead canopy of leaves. If you observe more closely, you will recognize other layers—a ground or herbaceous layer at your feet, a layer of shrubs, and some understory trees. Then you will be aware of one characteristic of forest ecosystems—their layering or stratification (Figure 24.10).

Stratification results from an interplay of various forces. In part, it reflects species differences in the life form of woody vegetation, such as height, geometry of the canopy, and distribution of foliage. Stratification also reflects the growth patterns of individuals, influenced by the interaction of biological and environmental conditions such as intra- and interspecific competition and light and moisture. More often than not, stratification results from a combination of life form and individual growth patterns, often measured as foliage height diversity. Like other ecological characteristics, stratification is more a continuum than discrete layers or floors, even though field measurements of stratification involve discrete measurements such as intervals of height. In a way, the layering or measurement of it is a human artifact.

Highly developed, uneven-aged deciduous forests usually contain four strata: the upper canopy, consisting of dominant or codominant trees; a lower tree canopy; a shrub layer; and a ground or field layer, consisting of herbs, ferns, and mosses. Its composition varies with the season from the spring, with its trilliums and hepaticas, to the fall, with its woodland asters. The tropical rain forest has an additional stratum—tall trees, whose deep crowns rise above the rest of the forest to form a discontinuous canopy.

Even-aged deciduous forests—a result of fire, clear-cutting, and other disturbances—often have poorly developed strata beneath the canopy because of dense shade. Coniferous forests, too, usu-

Coniferous forest profile

Temperate deciduous forest profile

Tropical rain forest profile

FIGURE 24.10 Profiles showing stratification of three types of forest: coniferous, temperate deciduous, and tropical rain forest. A = upper canopy; B = lower canopy; C = understory tree; D = shrub layer; E = herbaceous or ground layer.

ally have poorly developed lower strata, and the ground layer consists largely of ferns and mosses and very few herbs. Open-canopied pine stands, such as ponderosa and longleaf pines, have a well-developed ground layer of grasses.

ENVIRONMENTAL STRATIFICATION

A forest is like layers of blankets over the ground. The various strata modify the environment from the forest canopy to the forest floor.

Light

Bathed in full sunlight, the uppermost layer of the canopy is the brightest part of the forest. Down through the forest strata, light intensity is progressively reduced to only a fraction of full sunlight. In a summertime oak forest, only about 6 percent of the total midday sunlight reaches the forest floor, and the brightness of the light is about 0.4 percent of that of the upper canopy. In a tropical rain forest, only about 1 percent of midday sun reaches the forest floor. The amount of light reaching the forest floor of coniferous stands varies with the species. Open-crowned pines allow more light to filter through than do closed-crowned pines such as red and white pines. Conditions are somewhat different with spruce and fir. The upper crown, a zone of widely spaced narrow spires, is open and well lit, while the lower crown is densest and intercepts most of the light.

Seasonal differences in light intensity exist within the forest, with the greatest extremes in the deciduous forest. The forest floor receives its maximum illumination during March and April, before the leaves appear. A second, lower peak occurs in fall. The darkest period is in midsummer. Light in the coniferous forest is reduced by approximately the same amount throughout the year because the trees retain their foliage. Illumination is greatest in midsummer, when the sun's rays are direct, and lowest in winter, when the intensity of incident sunlight is lowest.

Humidity

A forest interior is a humid place because of plant transpiration and poor air circulation. Variation of humidity inside the forest is influenced in part by the degree to which lower strata are developed. Within these layers, leaves are adding moisture to the immediately surrounding air, increasing the humidity just above each. Thus, layers of increasing and decreasing humidity may exist from the forest floor to the forest canopy. During the day, when the air warms and its water-holding capacity increases, relative humidity is lowest. At night, when temperature and moisture-holding capacity are low, relative humidity in the forest rises. The lowest humidity in the forest is a few feet above the canopy, where air circulation is best. The highest humidity is near the forest floor, a result of the evaporation of moisture from the ground and the settling of cool air from the strata above.

Temperature

The highest temperatures in deciduous forests are in the upper canopy, because this stratum intercepts solar radiation. Temperatures tend to decrease through the lower strata. The most rapid decline occurs from the leaf litter down through the soil.

The temperature profile of a coniferous forest tends to be somewhat the reverse of that of deciduous forests, particularly in spruce and fir. There the coolest layer is in the upper canopy, perhaps because of greater air circulation, and the temperature increases down through the several strata to the forest floor.

The temperature profile changes through the 24-hour period. At night temperatures are more or less uniform from the canopy to the floor because radiation takes place most rapidly in the canopy. As the air cools, it sinks and becomes slightly heated by the warmer air beneath the canopy. During the day, the air heats, and by midafternoon, temperature stratification becomes most pronounced. On rainy days, temperatures throughout are more or less equalized, because water absorbs heat from warmer surfaces and transfers it to cooler ones.

Temperature stratification varies seasonally. In fall, when the leaves of the deciduous forest drop and the canopy thins, temperatures fluctuate more widely at various levels. Maximum temperatures decrease from the canopy downward but rise again at the litter surface. The soil, no longer shaded by an overhead canopy, absorbs and radiates more heat than in summer. Below the insulating blanket of litter, temperatures decrease through the soil. Thus, two temperature maximums may exist in the profile—one in the canopy, the other on the surface of the litter. Winter temperatures decrease from the canopy down to the understory, where in some forests, temperatures rise and then drop at the litter surface. Temperatures then increase rapidly down through the soil. During spring, conditions are highly variable. Maximum temperatures develop on the leaf litter, which intercepts solar radiation in the leafless woods, and temperatures decrease upward toward the canopy.

Wind

Anyone who has entered a woods on a windy day is well aware how effectively the forest reduces wind velocity. The influence of forest cover on wind velocity varies with the height and density of the stand and the size and density of the crown. Overall, wind velocities inside the forest may be reduced by 90 percent, and velocity near the ground usually ranges from 1 to 2 percent of wind velocity outside. Velocities in open and cutover stands and in wintertime deciduous forests are greater than in dense stands and in coniferous forests; the latter are most effective in reducing the flow of wind.

STRATIFICATION OF ANIMAL LIFE

In general, the diversity of animal life is associated with the growth form and stratification of plants. Some animals are associated with or spend the major part of their lives in a single stratum; others may range over two or more. Arthropods in particular confine their activities to one stratum.

The greatest concentration and diversity of life are found on and just below the ground layer. Many animals, particularly soil invertebrates, remain in the subterranean strata. Some of them, such as millipedes and spiders, move up into the upper strata at night when humidity is favorable, a vertical migration somewhat similar to that of plankton organisms in lakes and seas (Dowdy, 1944). Others—mice, shrews, ground squirrels—burrow into the soil for shelter or food but spend considerable time above ground. The larger mammals live on the ground layer and feed on the vegetation.

Birds move rather freely among several strata (Figure 24.11). Some occupy essentially the ground layer but move up into the trees to feed, to roost, or to advertise territory. Others occupy the shrub and understory layer, and still others are canopy species. Another group, such as woodpeckers, lives in the open space of tree trunks. In forests the diversity of bird species seems to be related more to height, density of foliage, and resulting stratification than to plant species composition.

Stratification of life is most pronounced in tropical rain forests, where six distinct feeding strata are recognized (Harrison, 1962). An insectivorous and carnivorous feeding group, consisting largely of bats and birds, works the *upper air* above the canopy. Within the *canopy* a wide variety of birds, fruit bats, and mammals feeds on leaves, fruit, and nectar. A few are insectivorous and mixed feeders. In the *middle zone of tree trunks* is a world of flying mammals, birds, and insectivorous bats. Ranging up and down the *trunks* are scaunsorial mammals that enter both the canopy and the ground strata to feed on the fruits of epiphytes attached to tree trunks, on insects, and on other animals. Animals on the *ground* include plant-feeders that browse on leaves or eat fallen fruits. The final feeding stratum includes small ground animals—birds and small mammals capable of some climbing—which search the ground litter and lower parts of tree trunks for food.

Invertebrates are common, including highly colored butterflies, beetles, and bees. Among the unseen invertebrates, hidden in loose bark and axils of leaves, are snails, worms, millipedes, centipedes, scorpions, spiders, and land planarians. Termites are abundant in the rain forest and play a vital role in decomposition of woody material. Together with ants, they are the dominant insect life. Ants are found everywhere in the rain forest, from the upper

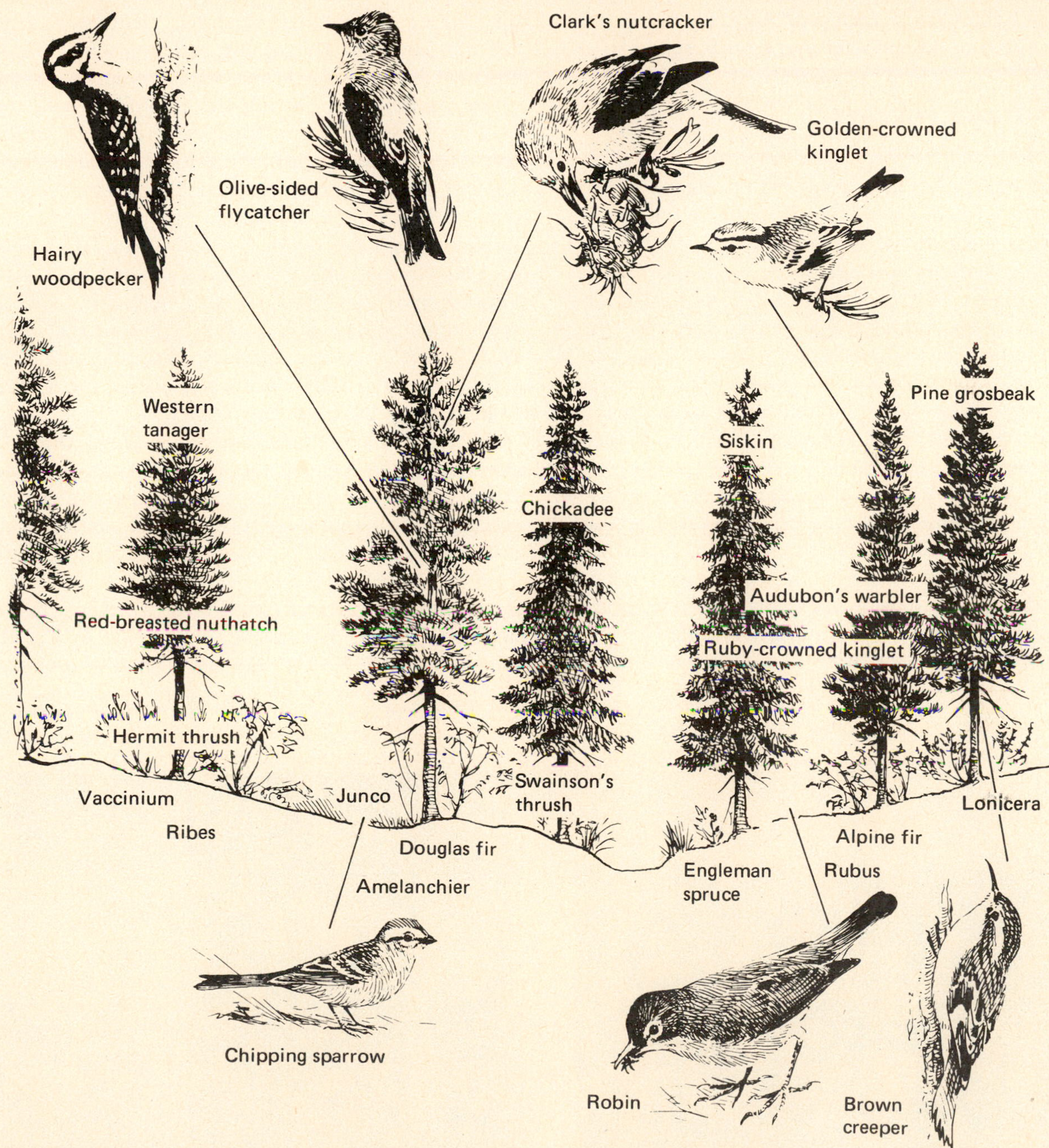

FIGURE 24.11 Foraging niches of some birds in a spruce-fir forest in Wyoming. (*After Salt, 1957:375.*)

canopy to the forest floor, although, like most rain forest life, the majority tend to be arboreal.

CONIFEROUS FORESTS

Types

Extending from northern New York, northern New England, and southern Canada north to the tundra and west through Alaska to the Pacific is the continentwide belt of northern coniferous and *boreal* forest (Figure 24.12). The same type of forest continues its circumpolar range across northern Eurasia. Outliers of the northern coniferous forest extend southward from New England through the high Appalachians. Spruces and balsam fir form the matrix of the forest, with aspen and birch as successional species.

South of Alaska the coniferous forest differs from the boreal forest both floristically and ecologically. The reasons for the change are climatic and topographical. Moisture-laden winds move inland from the Pacific, meet the barrier of the Coast Range, and rise abruptly. Suddenly cooled by this upward thrust into the atmosphere, moisture in the air is released as rain and snow in amounts up to 635 cm per year. During the summer, when winds shift to the northwest, the air is cooled over the chilly north seas. Though the rainfall is low, the cool air brings in heavy fog, which collects on the forest foliage and drips to the ground to add perhaps 127 cm more of moisture. This land of superabundant moisture, high humidity, and warm temperatures supports the *temperate rain forest* of the Pacific Northwest. This is a coniferous forest of lavish vegetation dominated by western hemlock, western red cedar, Sitka spruce, and Douglas-fir (Figure 24.12). Further south, where the precipitation is still high, grows the *redwood forest,* occupying a strip of land about 24 kilometers long.

The air masses that dropped their moisture on the western slopes of the Coast Range descend down the eastern slopes, creating conditions that produced the Great Basin deserts, rise up the western slopes of the Rocky Mountains, cool, and again drop moisture. In the Rocky Mountains grow the *montane coniferous forests* (Figure 24.12), made up of several forest associations influenced to a great extent by elevation. At high elevations, where win-

(a)

(b)

(c)

(d)

(e)

FIGURE 24.12 Some coniferous forest types. (a) A montane coniferous forest in the Rocky Mountains. The dryer lower slopes support ponderosa pine; higher elevations are cloaked with Douglas fir. (b) The temperate rain forest represents the maximal development of the temperate coniferous forest. It is found along the Pacific Coast of North America, where the maritime climate is cool and humid, and is characterized by an abundance of moisture. The temperate rain forest embraces the redwood and sequoia forests of California and the Douglas fir–western hemlock–Sitka spruce forests of the northwest Pacific Coast. Typical is this old Douglas fir stand *(Pseudotsuga menziesii)* with abundant western hemlock *(Tsuga heterophylla)* understory. (c) Subalpine forest dominated by subalpine fir *(Abies lasiocarpa)*. This tree grows with Engelmann spruce and mountain hemlock. (d) It becomes shrubby at the timberline. (e) A northern boreal forest of spruce and balsam fir.

ters are long and snow is heavy, grows the *subalpine* forest (Figure 24.12), characterized by Engelmann spruce and alpine fir. Lower elevations support Douglas-fir and ponderosa pine. Forests similar to those of the Rocky Mountains grow in the Sierras and Cascades.

Other needle-leaf evergreen forests, largely pine, exist in eastern Asia; in western Europe, where the dominant species is Scots pine (*Pinus sylvestris*); and in eastern North America, where most are seral in nature and are often maintained by controlled burning, as in the coastal southern pine region of the United States.

A deciduous seral stage, common to much of the western coniferous forest and the northern coniferous forest as well, is aspen parkland, which supports trembling aspen, the most widespread tree of North America.

Function

Ecosystem function has been studied most extensively in forest ecosystems (see Reichle, 1981). One coniferous forest ecosystem studied in detail is the Douglas-fir. A nutrient budget and cycle for a young stand were presented in Chapter 21. For comparison, consider a Douglas-fir stand 450 years old located in the Andrews Experimental Forest in Oregon. The total organic matter, both living and dead, in the Douglas-fir forest amounted to 1,313,630 kg/ha. Of this, 73 percent was in living organic matter and 23 percent in detrital matter. Most of the living biomass—788,190 kg/ha—was in branches and bole, the accumulation of years; in roots, 172,800 kg/ha; and in foliage, 14,120 kg/ha, about 1.4 percent. Litter accounted for 65 percent of detrital organic matter and the soil, 35 percent. Litter organic matter exceeds soil organic matter because of the accumulation and slow decomposition of needles and the long-term decomposition of large fallen trunks and limbs.

Nutrient budgets, summarized in Tables 24.1 and 24.2, emphasize that cycling within the ecosystem is more important than input by precipitation. Except for calcium, which is not a component of protein and is therefore not dependent upon miner-

TABLE 24.1 NUTRIENT BALANCE SHEET FOR A 450-YEAR-OLD DOUGLAS-FIR STAND, PACIFIC NORTHWEST, KG/HA

	Element				
	N	P	K	Ca	Mg
Input/output					
Atmosphere	2.0	0.3	1.2	3.1	1.2
Leaching to watershed	1.7	0.6	9.7	121.8	10.1
Loss/gain	+0.3	−0.3	−8.5	−118.7	−8.9
Internal cycling					
Litterfall	18.8	4.5	7.3	40.4	3.3
Throughfall	3.2	1.0	13.2	7.0	2.0
Total	22.0	5.5	20.5	47.4	5.3
Live accumulation					
Woody biomass	419	53	108	585	89
Roots	140	12	50	225	52
Foliage	147	33	81	102	16
Total	706	98	239	912	157
Detrital accumulation					
Forest litter	445	62	80	619	160
Soil rooting zone	4,560	34	660	2,040	560
Total	5,005	96	740	2,659	720

SOURCE: Based on data from Cole and Rapp, 1981:382.

TABLE 24.2 ANNUAL ELEMENT BALANCE OF A 450-YEAR-OLD DOUGLAS-FIR FOREST IN THE PACIFIC NORTHWEST, KG/HA/YEAR

	Element				
	N	P	K	Ca	Mg
Process					
Requirement	34.8	7.1	26.7	17.9	3.7
Uptake	23.7	5.8	21.2	53.3	5.6
Internal recycling	11.1	1.3	5.5	(−35.4)	(−1.9)
				0	0

Note: Requirement—annual increment of elements associated with bole and branch wood plus current foliage production.
Uptake—annual increment of elements associated with bole and branch wood plus annual loss through litterfall, leaf wash, and stemflow.
Internal recycling—Requirement − Uptake; deficiency in uptake made up by recycling elements within plant biomass.
SOURCE: Data from Cole and Rapp, 1981:358.

alization for mobility through the ecosystem, only minimal amounts of nutrients are lost from the ecosystem by leaching. Large amounts of nutrients are stored in trees. In the old Douglas-fir stand, 59 percent of the nitrogen found in living biomass, 54 percent of the phosphorus, 64 percent of the calcium, and 57 percent of the magnesium are stored in woody biomass; 21, 34, 34, 11, and 10 percent respectively are stored in the foliage. The remainder is stored in the roots. Some of these nutrients are returned to the forest floor by litterfall and throughfall (stemflow and leaffall) to be taken up by the tree again, and some are recycled internally, including the transfer of nutrients from old to current foliage, to meet deficiencies in uptake by the roots. Uptake of nitrogen, for example, amounts to 23.7 kg/ha/year, 68 percent of requirement (Table 24.2). Uptake of calcium, on the other hand, far exceeds requirement, so that cycling within the tree is not necessary. In fact, Douglas-fir appears to hoard calcium.

Considering the whole forest ecosystem, 12 percent of the nitrogen, 51 percent of the phosphorus, 24 percent of the potassium, 26 percent of the calcium, and 18 percent of the magnesium are stored in vegetation. Considering the detrital component, 91 percent of the nitrogen is found in the soil rooting zone, while 64 percent of the phosphorus is found in the litter layer.

TEMPERATE DECIDUOUS FORESTS

Types

The temperate deciduous forest (Figure 24.13) grows in moderately humid continental climates, characterized by moderate summer rainfall, mild summer temperatures, and cold winters. Dominated by broadleaf deciduous trees, this forest type is found in eastern Asia; in western Europe, where the number of species is low, the result of glaciation; and in eastern and central North America, where the forest types consist of a number of associations. They include the beech-maple and northern hardwood forests (with pine and hemlock) in the northern regions and the oak and oak-hickory forests that cover most of the Appalachian mountains and the central forest region. The temperate deciduous forest reaches its greatest development in the mixed mesophytic forest of the central Appalachians, where the number of temperate tree species is unsurpassed by any other region in the world.

In Europe two major associations are the oak forests and beech forests, the latter shaded, with a sparse understory, and the former more open-crowned, with a developed understory. The Asiatic broadleaf forest—found in eastern China, Japan, Formosa, and Korea—is somewhat similar to the North American temperate forest and contains a

FIGURE 24.13 (a) A virgin stand of mixed mesophytic forest in the central Appalachians, composed of white oak, beech, yellow-poplar, and other species. Note the well-developed understory in this uneven-aged climax stand.

number of plant species in the same genera as those found in North America and western Europe.

Function

There are a number of detailed studies of functions of deciduous forest ecosystems. Among the best known are those of a northern hardwoods forest at Hubbard Brook, New Hampshire (Bormann and Likens, 1979); an oak-pine forest at Brookhaven, Long Island (Whittaker and Woodwell, 1969; Woodwell and Botkin, 1970); and a Belgian oak forest (Duvigneaud and Denaeyer-DeSmet, 1970); as well as others summarized by Reichle (1981).

Consider a 30- to 80-year-old yellow-poplar–oak stand on Walker Branch, Oak Ridge, Tennessee, with an understory of redbud, dogwood, Virginia creeper, and Christmas fern. Net primary production (NPP) was 726 g C/m^2/year. Most of the production was concentrated in leaves, branches, wood, bark, and roots (Figure 24.14). Plant respiration, determined by gas-exchange measurements, was 1,436 g C/m^2/year, or 66 percent of gross primary production of 2,162 g C/m^2/year. Heterotrophic respiration amounted to 670 g C/m^2/year. Subtracting that from NPP gave a net ecosystem production (NEP) of 56 g C/m^2/year.

Total organic matter amounted to 329,100 kg/ha. Ninety percent of the living organic matter—140,100 kg/ha—was in branches, boles, and roots. Foliage renewed annually amounted to 5,100 kg/ha. Less than 1 percent of the total detrital biomass—190,500 kg/ha—was in forest litter. Over 99 percent was in the soil rooting zone. Total detrital biomass exceeds living organic biomass.

Nutrient cycling involves a balance between inputs to the biological system from precipitation

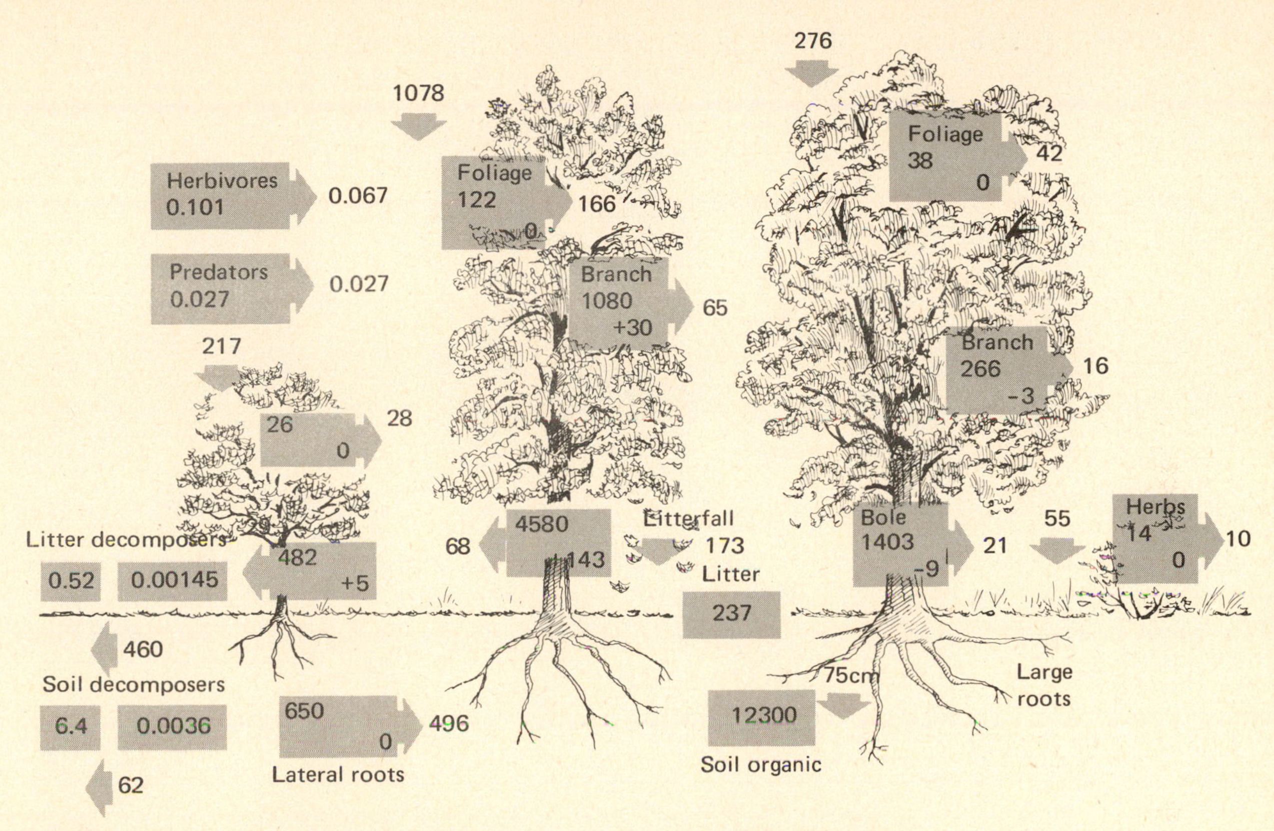

FIGURE 24.14 Annual carbon cycle in a yellow-poplar forest. Units of measure are: carbon transfer—g/m^2/year and pools—g C/m^2. Annual standing crop is given in the center of each box and the net annual increment in the right-hand corner. Two values for NPP and NEP are given. Those in parentheses are based on harvest/allometric methods; the others are based on data from the gas-exchange method. (*After Harris et al., 1975.*)

and outputs or losses from the system through leaching. In the yellow-poplar–oak forest, total input through precipitation for five elements—N, P, K, Ca, and Mg—amounted to 20.44 kg/ha, and losses amounted to 212.78 kg/ha (see Table 24.3). Nitrogen was cycled much more tightly than the other four elements. More nitrogen came into the system through precipitation than was lost through leaching. Greatest losses were with calcium and magnesium. Much of this probably resulted from the weathering of the dolomitic limestone substrate.

Except for calcium, which is taken up in excess of need, uptake of nutrients in the yellow-poplar–oak stand does not meet requirements (Table 24.4). Uptake of nitrogen, for example, amounts to 58.1 kg/ha/year, 66 percent of requirement. Deficiencies have to be met by the cycling of nutrients within the tree biomass. Nutrient accumulations in tree biomass form a considerable pool of nutrients, most of which is unavailable for short-term recycling (Table 24.3). Forty-nine percent of N, 35 percent of P, 42 percent of K, 62 percent of Ca, and 49 percent of Mg are incorporated in woody biomass; 31, 51, 43, 28, and 24 percent respectively in root biomass; and 20, 14, 15, 10, and 27 percent respectively in foliage.

Considering the whole forest ecosystem, 7 percent of N, 3 percent of P, 12 percent of Ca, and less

TABLE 24.3 NUTRIENT BALANCE SHEET FOR 30- TO 80-YEAR-OLD YELLOW-POPLAR–OAK (*LIRIODENDRON-QUERCUS*) FOREST RIDGE, TENNESSEE, KG/HA

	Element				
	N	P	K	Ca	Mg
Input/output					
Atmosphere	8.7	0.54	1.0	9.1	1.1
Leaching to watershed	1.8	0.02	6.8	147.5	77.1
Loss/gain	+6.9	+0.52	−5.8	−138.4	−76
Internal cycling					
Litterfall	36.2	2.7	19.1	58.3	8.3
Throughfall	12.0	0.4	18.4	21.9	3.4
Total	48.2	3.1	37.5	80.2	11.7
Live accumulation					
Woody biomass	189	15	127	462	38
Roots	122	22	132	211	19
Foliage	78	6	45	75	21
Total	389	43	304	748	78
Detrital accumulation					
Forest litter	187	11	14	294	22
Soil rooting zone	7,300	1,400	36,000	6,300	8,700
Total	7,487	1,411	36,014	6,594	8,722

SOURCE: Data from Cole and Rapp, 1981:394.

than 1 percent each of K and Mg are stored in vegetation. The litter layer is the most important nutrient pool, because it is quickly decomposed (average turnover time of 4 years) and recycled, although the bulk of the nutrient pool is in the mineral soil. Nutrients stored in living biomass, especially in the roots, are translocated and recycled through the living biomass, especially the foliage. The foliage, in turn, translocates a considerable portion of its nutrients back to the roots before leaffall. However, mineral cycling can be maintained only if nutrients are pumped from the soil reservoir or are released by weathering of parent material.

TROPICAL RAIN FORESTS

Types

The rain forest or some variation of it—the monsoon forest, the evergreen savanna forest, the mon-

TABLE 24.4 ANNUAL ELEMENT BALANCE OF A 30- TO 80-YEAR-OLD YELLOW-POPLAR–OAK FOREST, OAK RIDGE, TENNESSEE, KG/HA/YEAR

	Element				
	N	P	K	Ca	Mg
Process					
Requirement	87.9	6.3	47.5	82.6	21.7
Uptake	58.1	3.4	40.0	87.8	12.4
Internal recycling	28.9	2.9	7.5	(−5.2) 0	9.3

Note: Requirement—annual increment of elements associated with bole and branch wood plus current foliage production. *Uptake*—annual increment of elements associated with bole and branch wood plus annual loss through litterfall, leaf wash, and stem flow. *Internal recycling*—Requirement − Uptake; deficiency in uptake made up by recycling elements within plant biomass.

SOURCE: Data from Cole and Rapp, 1981:359.

tane rain forest—forms a worldwide belt around the equator. The largest continuous rain forest is in the Amazon Basin of South America. West and central Africa and the Indo-Malayan region are other major areas of tropical rain forest.

The rain forest (Figure 24.15), so called because of its constant high humidity, grows in a land where seasonal changes are minimal. The mean annual temperature is about 26°C; the mean minimum rarely goes below 25°C, and heavy rainfall occurs through much of the year. Under such perpetual midsummer conditions, plant activity continues uninterrupted, resulting in very luxurious growth. Tree species number in the thousands; usually none is dominant, and the majority are represented by a very few individuals. Communities with single dominants are usually limited to certain soils and to particular combinations of soil and topography. The tree trunks are straight, smooth, and slender, often buttressed, and reach 25 to 30 m before expanding into leathery, simple leaves (see Halle, Oldeman, and Tomlinson, 1978). Climbing plants (the lianas), long, thick, and woody, hang from trees like cables, and epiphytes grown on trunks and limbs. The undergrowth of the dark interior is sparse and consists of shrubs, herbs, and ferns. Litter decays so rapidly that the clay soil, more often than not, is bare. The tangled vegetation popularly known as jungle is second-growth forest that develops where the primary forest has been despoiled.

Moonsoon forests and other deciduous and semideciduous forests grow where rainfall diminishes and a pronounced dry season occurs. They differ from rain forests in that they lose their leaves during the dry season. Such forests, most often found in southeastern Asia and India, are commonly known as tropical seasonal forests. Similar stands occur along the Pacific side of Mexico and Central America.

FIGURE 24.15 A view of a tropical rain forest in Puerto Rico. (*Photo by T. M. Smith.*)

Function

Perhaps the most intensely studied tropical forest is the one at the Puerto Rico Nuclear Center. H. T. Odum (1970) and his associates worked up an energy budget for that forest. Incoming solar radiation amounted to 3,830 kcal/m^2/day. Gross production amounted to 131 kcal/m^2/day, of which 116 kcal was used in respiration, leaving a net production of 15.2 kcal/m^2/day, as determined by gas analysis. Roots were responsible for 60 percent of respiration; leaves for 33 percent; and trunks, branches, and fruit for the remainder. Net productivity, as measured by biomass accumulation, was 16.31 kcal/m^2/day. Cumulative biomass addition through wood growth was 0.72 kcal/m^2/day, or 3.8 percent of net production. The remainder of net production passed through the grazing and detrital food chains.

Odum's energy budget is for a specific rain forest, and because site, soil, and other conditions of different tropical forests vary widely, productivity also varies widely. However, a few generalizations can be made about their energy budgets. Tropical forests use about 70 to 80 percent of their energy intake for maintenance and 20 to 30 percent for net production. Average gross primary production is about 67 MT/ha/year, or 28 × 10^3 kcal/m^2/year (Golley, 1972). Mean annual net production is about 21.6 MT/ha. That exceeds temperate forests, averaging 13 MT/ha/year, by a factor of 1.7 and boreal forests, averaging 8 MT/ha/year, by a factor of 2.7 (Farnsworth and Golley, 1973).

Tropical and temperate forests, however, differ somewhat in their efficiency of production. Efficiency in this case is defined as the amount of energy stored in wood, leaves, fruit, and litter divided by total solar energy available to the community. Jordan (1971) found that the rate of wood production in intermediate-aged stands was about the same for both tropical and temperate forests, but the rate of leaf and litter production was higher in the tropics. However, efficiency of wood production was higher in temperate forests, probably because more selective pressure exists in temperate forests, where solar energy is not as abundant, to produce the maximum amount of wood.

High year-round temperatures and abundant rainfall in tropical rain forest areas produce rapid geologic cycling. Because geologic cycling is accelerated, biologic cycles apparently are modified to keep nutrients in the living portion of the system. Nutrients may be stored in living biomass, where they are protected from leaching, or the time the nutrient element remains in the soil may be reduced to a minimum. Tropical rain forests tend to concentrate proportionately more calcium, silica, sulfur, iron, magnesium, and sodium and less potassium and phosphorus than temperate forests. In spite of these differences, rain forests hold larger quantities of nutrients simply because of their much larger biomass.

Much of the mineral recycling takes place through litterfall (Figure 24.16). The ratio of mineral elements held in biomass to the amount returned to the soil by litter provides some estimate of turnover time. Phosphorus, magnesium, calcium, and potassium all appear to have a turnover time of less than 100 years, and most elements are recycled in 20 years.

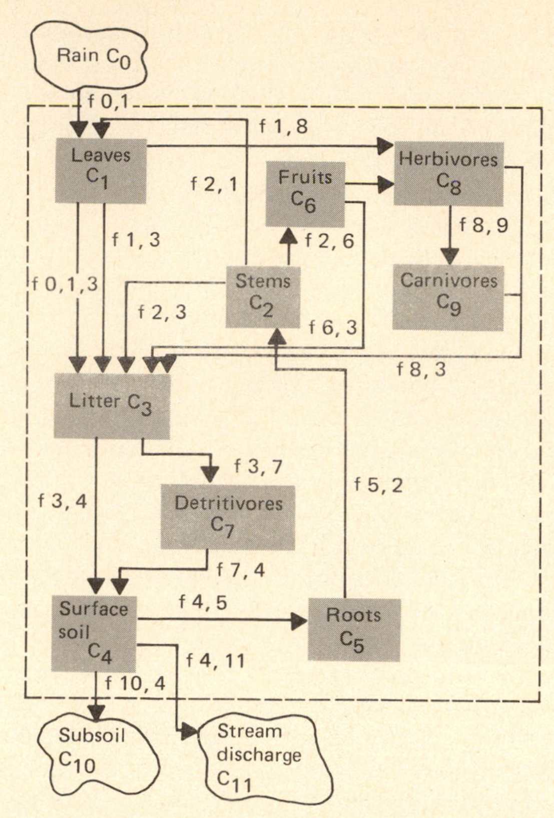

FIGURE 24.16 Diagram of material cycling in a tropical rain forest. Dotted lines indicate system boundary. Boxes identify system components. Arrows indicate transfer functions between components and sources. Note how litter is the crossroads of the system. (*From Golley, 1975.*)

Data for a tropical forest in Panama indicate that inputs of phosphorus and potassium balance outputs (Table 24.5), but more calcium and magnesium are lost from the system than are gained by rainfall input. This suggests considerable input from the soil reservoir. Annual uptake of phosphorus, potassium, and calcium far exceeds stream discharge, while the reverse is true for magnesium. The budget also suggests that phosphorus and potassium might be limiting and that elements are conserved by rapid internal cycling.

Internal cycling may be aided by (1) rapid return of nutrients leached by throughfall, (2) reten-

TABLE 24.5 BIOLOGICAL AND GEOLOGIC CYCLES IN A TROPICAL MOIST FOREST, PANAMA, KG/HA/YEAR

	Element			
	P	K	Ca	Mg
Process				
Input (rain)	1.0	9.3	29.0	5.0
Uptake	11.0	187.0	270.0	30.0
Output	0.7	9.5	163.0	44.0

SOURCE: Based on data from Golley, 1975.

tion of nutrients by fungal hyphae, and (3) uptake by mycorrhizal fungi. Roots and humus of a tropical rain forest form a mat up to 40 cm thick on the surface of the soil. This mat contains up to 58 percent of the feeder roots of the forest trees (Jordan, 1982). The roots, in turn, support an abundance of mycorrhizal fungi, which attach the feeder roots to dead organic matter by hyphae and rhizomorph tissue (see Chapters 17 and 21). Mycorrhizae allow direct physical absorption of nutrients from dead organic matter. Algae in the mat may take up dissolved nutrients from rain and store them in their biomass until released by their death and decomposition. Microorganisms fix nitrogen, and environmental conditions in the mat, especially a low pH and a high concentration of tannin, inhibit the denitrifying bacteria. As a result, few nutrients move down into mineral soil, and most are rapidly recycled back to the living biomass (see Jordan, 1982; Jordan and Herrera, 1981).

TEMPERATE WOODLANDS

In western parts of North America, where the climate is too dry for montane coniferous forests, one finds temperate woodlands (Figure 24.17). These ecosystems are characterized by open-growth small trees beneath which grow well-developed stands of grass or shrubs. There are a number of types of temperate woodlands, which may consist of needle-leafed trees, deciduous broadleaf trees, or sclerophylls or any combination of these. An outstanding example is the pinon-juniper woodland, in which two dominant genera, *Pinus* and *Juniperus*, are always associated. This ecosystem is found on the front range of the Rocky Mountains to the eastern slopes of the Sierra Nevada foothills. In southern Arizona, New Mexico, and northern Mexico occur oak-juniper and oak woodlands; and in the Rocky Mountains, particularly in Utah, one finds oak-sagebrush woodlands. In the Central Valley of California grow evergreen-oak woodlands with a grassy undergrowth. Structurally similar woodland grows around the Mediterranean and in Australia.

Tundras

North of the coniferous forest belt lies a frozen plain clothed in sedges, heaths, and willows, which encircles the top of Earth. This is the arctic tundra (Figure 24.18). The tundra—the word comes from the Finnish *tunturi,* meaning "a treeless plain"—is dotted with lakes and bogs and transected by streams and high ground, is exposed to wind, and is bare and rock-covered. At lower latitudes a similar landscape, the alpine tundra, occurs in mountainous areas. But in the Antarctic a well-developed tundra is lacking. Arctic or alpine, the tundra is characterized by low temperatures, a short growing season, and low precipitation.

GENERAL CHARACTERISTICS

Frost molds the tundra landscape. Alternate freezing and thawing of the upper layer of soil and the presence of a permanent frozen layer in the ground, the *permafrost,* create conditions unique to the tundra. Because of the permafrost, water cannot move into the soil. This reservoir of water lying on the upper surface enables plants to exist in the driest parts of the Arctic.

The symmetrically patterned landforms so typical of the tundra result from frost. Freezing and thawing of soil tend to push stones and other larger material upward and outward from the mass to form the patterned surface. Typical nonsorted patterns

FIGURE 24.17 A temperate woodland of western juniper *(Juniperus occidentalis)* and wheatgrass *(Agropyron)* in western Oregon. This woodland is an ecotone between the montane coniferous forest of ponderosa pine and the sagebrush desert. (*U.S. Forest Service.*)

associated with seasonally high water tables are frost hummocks, frost boils, and earth stripes (Figure 24.19). Sorted patterns are characteristic of better-drained sites. The best known of these are the stone polygons, whose size is related to frost intensity and size of the material (Johnson and Billings, 1962). On sloping ground, creep, frost thrusting, and downward flow of soil change polygons into sorted stripes running downhill. Mass movement of supersaturated soil over the permafrost forms solifluction terraces, or "flowing soil." This gradual downward creep of soils and rocks eventually rounds off ridges and other irregularities in topography. The molding of the landscape by frost action is called *cryoplanation* and is far more important than erosion in wearing down the arctic landscape.

Permafrost in the alpine tundra exists only at very high elevations and in the far north, but frost-induced processes—small solifluction terraces and stone polygons—are still present. The lack of permafrost results in drier soils; only in alpine wet meadows and bogs do soil moisture conditions com-

(a) (b) (c)

FIGURE 24.18 (a) Alpine tundra in the Rocky Mountains in early spring. (b) Stone polygons in a Rocky Mountain alpine tundra. (c) The wide expanse of the arctic tundra. This photograph shows an area near the Sadlerochit River on the Arctic National Wildlife Range, 5 miles from the Arctic Ocean. Note the frost polygons in the foreground. The caribou, a major arctic herbivore, are a part of the Porcupine herd. (*Photo courtesy U.S. Fish and Wildlife Service.*)

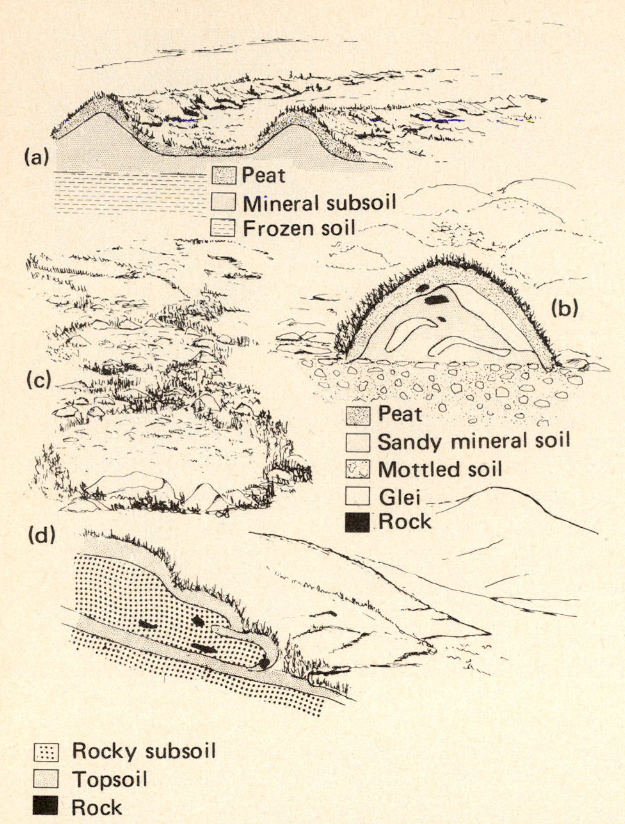

FIGURE 24.19 Patterned ground forms typical of the tundra region: (a) unsorted earth stripes, (b) frost hummocks, (c) sorted stone nets and polygons, (d) solifluction terrace. (*Diagrams adapted from Johnson and Billings, 1962:122.*)

pare with those of the Arctic. Precipitation, especially snowfall and humidity, is higher in the alpine region than in the arctic tundra, but steep topography results in a rapid runoff of water.

The tundra, arctic or alpine, supports a structurally simple vegetation. The number of species tends to be few; the growth is slow; and most of the biomass and functional activity are confined to relatively few groups. Growing and reproductive seasons are short. Most of the vegetation is perennial and reproduces vegetatively rather than by seed. Although it appears homogeneous, the pattern of vegetation is patchy. A combination of microrelief, snowmelt, frost heaving, and aspect, among other conditions, produces an almost endless change in plant associations from place to place (Polunin, 1955).

STRUCTURE

Although the arctic and alpine tundras are somewhat similar, the vegetation of the two differs in species composition and in adaptation to environmental conditions, especially light. Because the atmosphere is thinner in the alpine tundra, light intensity, especially ultraviolet, is high on clear days. Thus, alpine plants reach the saturation point for light at higher intensities than arctic plants, which are adapted to lower light intensities. Arctic plants require longer periods of daylight than alpine plants, a response to 24 hours of summer daylight. Arctic plants propagate themselves almost entirely by vegetative means; alpine plants propagate by means of seeds. The short-lived adventitious roots of arctic plants are short; they are parallel to rhizomes. Those of alpine species are long, are long-lived, and can penetrate to considerable depths because they are not impeded by permafrost (see Billings and Mooney, 1968). Low and hugging the ground, cushion and mat-forming plants, rare in the Arctic, are common to the windy alpine tundra. Over much of the arctic tundra, typical vegetation is a cottongrass—sedge–dwarf heath community.

Structurally, most of the tundra vegetation is underground. Root-shoot ratios of vascular plants range from 3:1 to 10:1. Roots are concentrated in the soil active layer, and above-ground parts seldom exceed 30 cm above the surface. Microflora, the bacteria and fungi, live near the surface and are adapted to the cold. Bacteria are active at $-7.5°C$, and fungal growth continues at 0°C. Fungi, however, can break down plant structural carbohydrates at temperatures below 0°C, while aerobic bacteria at 0°C are restricted to nonstructural carbohydrates and products of fungal decomposition (Flanagan and Bunnell, 1980). Fungi and aerobic bacteria are restricted to approximately the upper 7 to 10 cm of soil. Below that the microflora are restricted to anaerobic bacteria.

Invertebrate fauna are concentrated near the surface, where there are abundant populations of segmented whiteworms (Enchytraeidae), collembolas, and flies (Diptera), chiefly craneflies. Summer brings hordes of blackflies, deer flies, and mosquitoes. Invertebrates are mostly microivores, grazers on bacteria and fungi. Lacking are saprivors. The biomass of invertebrate fauna exceeds the above-ground biomass of vertebrate fauna. Dominant vertebrates are herbivores, including lemmings, caribou, and muskox, the distribution of which varies across the tundra. Although caribou may provide the greatest herbivore biomass, lemmings may consume three to six times as much forage (Batzli et al., 1980). Caribou are extensive grazers, spreading out over the tundra in summer to feed on sedges. Muskox are intensive grazers, restricted to more localized areas, where they feed on grasses, sedges, and dwarf willow. Major arctic carnivores are insect-feeding birds—sandpipers, plovers, longspurs, and waterfowl—snowy owls, hawklike jaegers, wolf, and arctic fox.

FUNCTION

Primary production in arctic ecosystems is low (Figure 24.20). The net annual production of above-ground vegetation ranges from 40 to 110 g/m^2 in sedge, cotton grass, heath, and shrub-heath communities. Net annual below-ground production in sedge and sedge-grass meadows ranges from 130 to 360 g/m^2 (for a summary, see Bliss et al., 1973; Wielgolaski et al., 1981; Miller et al., 1980). Available data suggest that daily primary production rates of above-ground vegetation range from 0.9 to 1.9 g/m^2 and that the efficiency of primary production ranges from 0.20 to 0.5 for the growing season. The rate of production is comparable to that of temperate ecosystems, but the growing season is so short—50 to 75 days—that overall annual production is greatly reduced. Arctic plants make maximum use of the growing season and light by car-

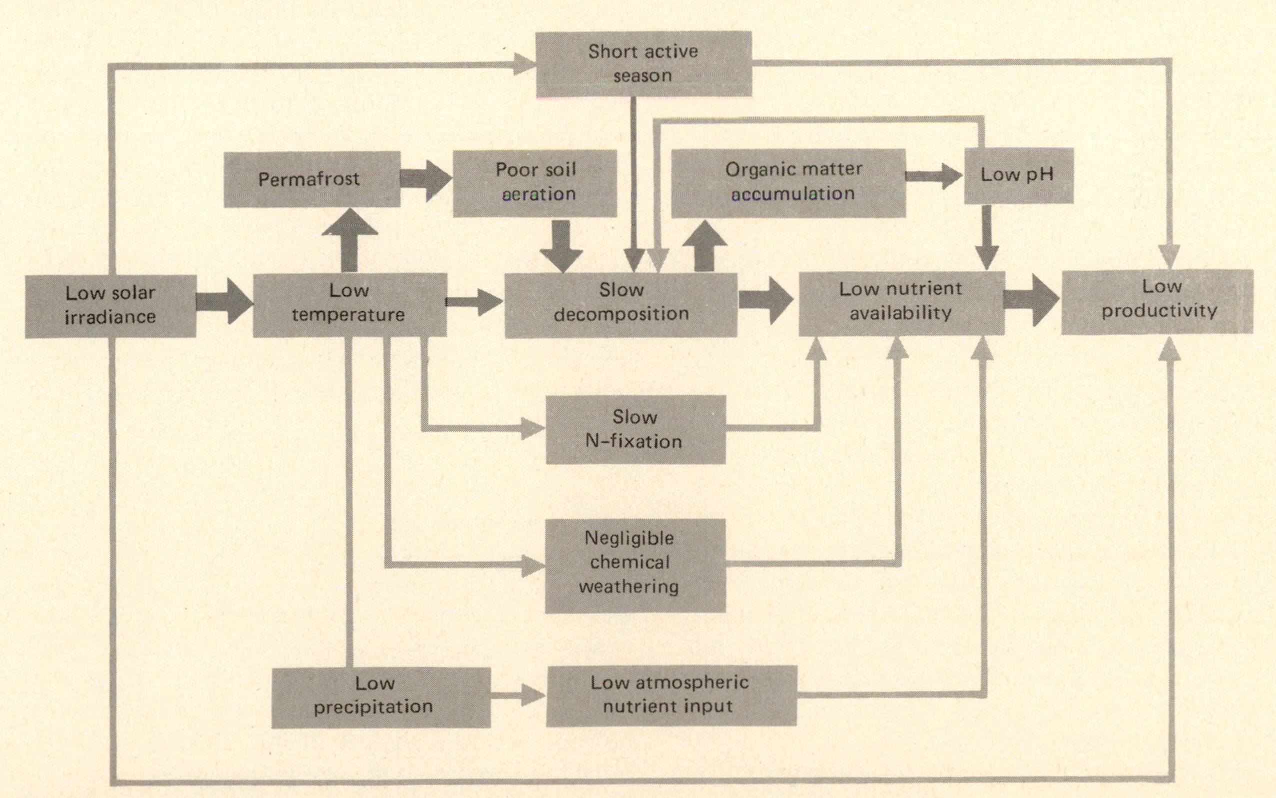

FIGURE 24.20 Causal relationships between solar radiation, low temperatures, and low primary productivity in arctic coastal tundra. Thickness of arrows indicates magnitude of effect. (*From Chapin III, Miller, Billings, and Coyne, 1980:481.*)

rying on photosynthesis at midnight, when the light level is one-tenth of that available at noon. Most of the energy and carbon are stored below ground, 10 times the amount sequestered by temperate grasslands (Chapin, Miller, Billings, and Coyne, 1980).

Nutrient cycling is conservative and tight, with minimal loss outside of the system. Nutrient input is low because of low temperatures and low precipitation that slow weathering and because of a low rate of nitrogen fixation and a minimal input from precipitation. Nitrogen fixation, mostly by blue-green algae, provides about 68 percent of the total nitrogen input and ammonia deposition in precipitation, about 29 percent. Dissolved nitrogen in runoff water and denitrification account for most of the losses. Nearly 81 percent of the total input is stored by the system (Barsdale and Alexander, 1975). Most of the nutrients, especially nitrogen, are bound in soil organic matter (SOM). Soil organic matter replaces plants in temperate ecosystems as the reservoir of organically bound nutrients. Tundra ecosystems depend primarily on the release of nutrients from SOM by decomposition. Pools of soluble nutrients, especially nitrogen and phosphorus, are small relative to the exchangeable pool, and the exchangeable pool is small relative to the nonexchangeable pool (Figure 24.21).

Most of the total annual nutrient flux takes place at snowmelt, when nutrients frozen over winter are released from litter, excreta of animals, and microbes. Most of this nutrient flux, however, is taken up by microbial organisms and moss or byrophytes. Mosses in particular accumulate nutrients and release them slowly or store them in peat. At the outset of the growing season, the uptake of nutrients by microbes exceeds the uptake by plant roots. Vascular plants depend largely on mycorrhizal association for the uptake of nutrients (see Chapter 21). Monocotyledonous plants draw upon nutrients and carbohydrates stored in underground roots. Once above-ground biomass is established, plants have the competitive advantage over microbes for nutrients.

A rapid upward movement of nutrients early in the season at the expense of below-ground biomass supports fast shoot growth when radiation is favorable. Although more nutrients become available as the depth of thaw increases, they usually are in short supply. Tundra plants respond to this inadequate supply of nutrients, especially nitrogen and phosphorus, by reducing their growth to a small number or biomass of leaves and stems that are well supplied with nutrients (Chapin, Tieszen, Lewis, Miller, and McCown, 1980). By contrast temperate plants respond to an inadequate supply by exhibiting symptoms of nutrient deficiency. Within 6 weeks after the start of the growing season, plants begin sending nutrients back underground. As the cold approaches, the plants die back. Dead parts add to the accumulation of organic matter. Nutrients leached from dead leaves are accumulated by mosses or are frozen in place until the following summer's snowmelt.

Animals increase the release of nutrients by either stimulating or short-circuiting the decomposer process. Soil invertebrates consume nearly all the microbial populations near the surface of the soil. Grazing herbivores consume large amounts of above-ground production and distribute nutrients across the tundra through their droppings. They also fell standing litter and live plant biomass, improving both the substrate and environmental conditions for decomposers. Thus, activities of consumers stimulate a continuous turnover in both soluble and exchangeable pools. Nitrogen on the average is turned over 10 or more times and phosphorus 200 times during the growing season. Such efficient short-season cycling allows tundra ecosystems to make maximum use of a limited supply of nutrients.

Deserts

Deserts are defined by geographers as lands where evaporation exceeds rainfall. No specific rainfall can be used as a criterion, but desert conditions may range from extreme aridity to sufficient moisture to support a variety of life. Deserts are positioned on the Earth in two distinct belts, one confined roughly to the area around the tropic of Cancer, the other to the area around the tropic of Capricorn.

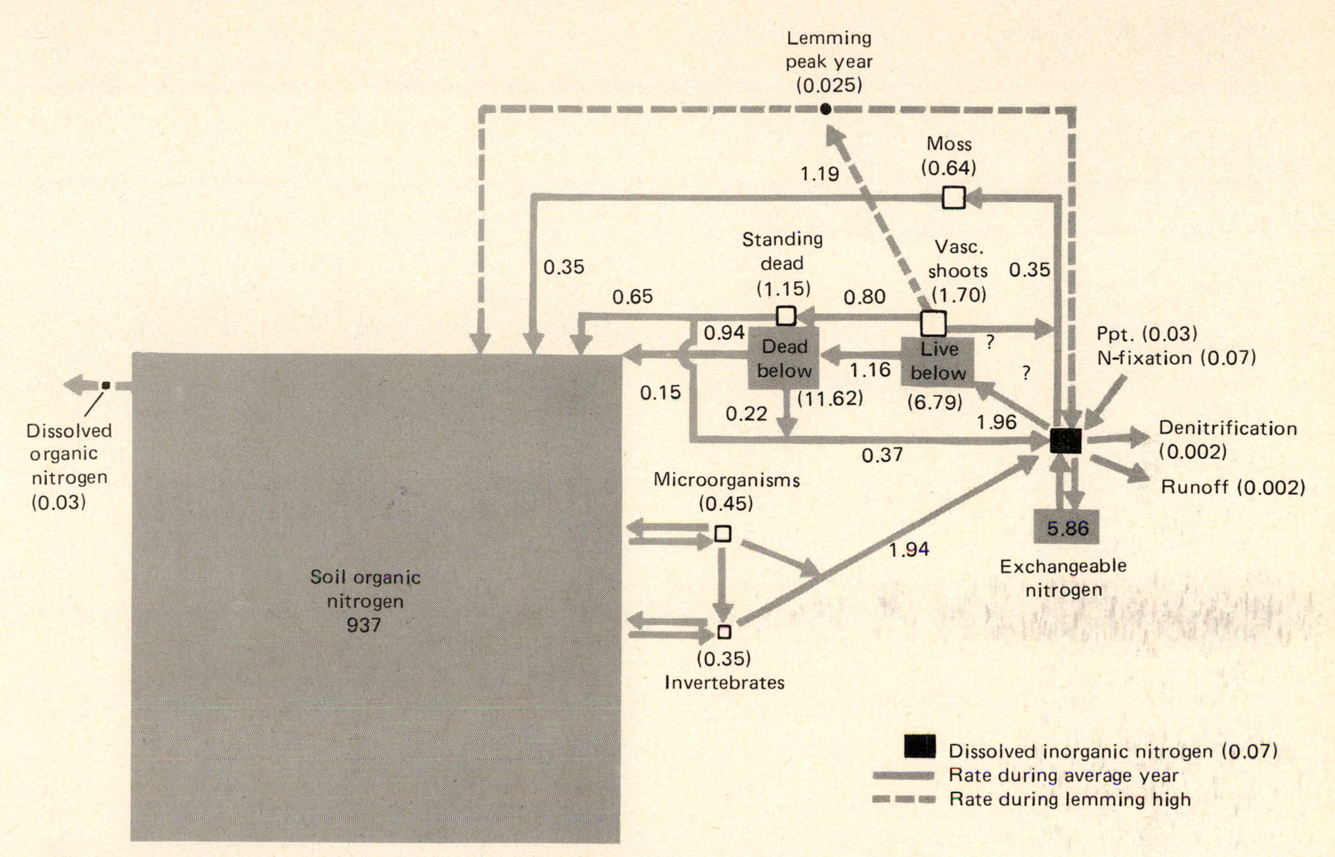

FIGURE 24.21 Annual nitrogen budget in a moist meadow on an arctic coastal tundra to the depth of 20 cm. The area of each box is proportional to its compartment size in g/m^2. Values next to the arrows indicate annual fluxes in g/m^2/year. The budget represents that for years of low lemming populations. Years of high lemming populations are indicated by dashed lines to account for increased flux through the rodents. Amount of leaching from vascular plants is unknown; and mosses are assumed to get all their nutrients from the soil (which they probably do not). (*After Chapin III et al., 1980:471.*) Note the relative amounts of N in the soil organic matter and the small amounts in other compartments.

Deserts result from several forces. High-pressure areas alter the course of rain. The high-pressure cell off the coast of California and Mexico deflects rainstorms moving south from Alaska to the east and prevents moisture from reaching the southwest. In winter the high-pressure areas move southward, allowing winter rains to reach southern California and parts of the North American desert. Winds blowing over cold waters become cold also; they carry very little moisture and produce little rain. Thus, the west coast of California and Baja California, the Namib Desert on coastal southwest Africa, and the coastal edge of the Atacama in Chile may be shrouded in mist yet remain extremely dry.

Mountain ranges also play a role in desert formation by causing a rain shadow on their lee side. The High Sierras and the Cascade Mountains intercept rain from the Pacific and help maintain the arid conditions of the North American desert. The low eastern highlands of Australia effectively block the southeast trade winds from the interior of that continent. Other deserts, such as the Gobi and the interior of the Sahara, are so remote from the source of oceanic moisture that all the water has been wrung from the winds by the time they reach these regions.

CHARACTERISTICS

All deserts have in common low rainfall, high evaporation (from 7 to 50 times as much as precipitation), and a wide daily range in temperature from hot by day to cool by night. Low humidity allows up to 90 percent of solar insolation to penetrate the atmosphere and heat the ground. At night the desert yields the accumulated heat of the day back to the atmosphere. Rain, when it falls, is often heavy and, unable to soak into the dry earth, rushes off in torrents to basins below.

The topography of the desert, unobscured by vegetation, is stark and, paradoxically, partially shaped by water. The unprotected soil erodes easily during violent storms and is further cut away by the wind. Alluvial fans stretch away from eroded, angular peaks of more resistant rocks. They join to form deep expanses of debris, the *bajadas*. Eventually, the slopes level off to low basins, or *playas*, which receive waters that rush down from the hills and water-cut canyons, or *arroyos*. These basins hold temporary lakes after the rains, but water soon evaporates and leaves behind a dry bed of glistening salt.

The desert is not the same everywhere. Differences in moisture, temperature, soil drainage, alkalinity, and salinity result in variations in the amount of vegetation, the dominant plants, and the associated plants. There are hot deserts and cold deserts, extremely dry deserts and ones with sufficient moisture to verge on grasslands or shrublands. The Sahara desert, the world's largest, has minimal vegetation, mostly clustered around oases. The Arabian desert has its tamarisk and the central Asian desert its saltbrush and salsola. The North American desert can be divided into two parts: the northern cool desert—the Great Basin—and the hot desert of the southwest—the Mohave, the Sonoran, and the Chihuahuan. The two, however, grade one into the other.

Sagebrush (*Artemisia* spp.) is the dominant plant of the cool desert (see Figure 24.9a). Hot desert shrubland is dominated largely by creosote bush *(Larrea divaricata)*, accompanied by bur sage *(Hymenoclea)* and burrobush *(Franseria)*. Woody-stemmed and soft, brittle-stemmed shrubs are characteristic desert plants. In the matrix of shrubs grows a wide assortment of other plants: yuccas, cacti, small trees, and ephemerals. In the southwestern American desert, large succulents rise above the shrub level and change the aspect of the desert out of proportion to their numbers (see Figure 24.22). Like forest trees and prairie grasses, most desert species grow best in certain topographical situations. The giant saguaro, the most massive of all cacti, grows on the bajadas of the Sonoran desert with smaller, brilliant-flowered cacti. Other plants—ironwood, smoke tree, palo verde—grow best along the banks of intermittent streams, not so much because of their moisture requirements as because their hard-coated seeds must be scraped and bruised by the grinding action of sand and gravel during flash floods before they can germinate.

Ephemeral plants—those that flower annually—supply seasonal brilliance to the desert. Surviving from one favorable period to another as seeds, they flourish best on sandy soils, for here moisture easily penetrates the soil and the ground warms rapidly, favoring germination of seeds.

In sharp contrast to conditions in the forest, the struggle in the desert is not one of plant against plant for light and space, but one of all plants for moisture. Since little layering exists in the desert plant community, light is not a problem. Because moisture is limited, the plants are mostly low; because of the competition for moisture, they are widely spaced.

FUNCTION

The desert ecosystem differs from other ecosystems in at least one important way. The input of precipitation, somewhat continuous in other eco-

FIGURE 24.22 Saguaro dominates the aspect of this desert in the southwestern United States. Note the water-shaped topography. (*Photo courtesy Arizona Fish and Game Department.*)

systems, is highly discontinuous in the desert ecosystem. It comes in pulses as clusters of rainy days 3 to 15 times a year. Of these, only 1 to 6 may be large enough to stimulate biological activity. Thus, the desert ecosystem experiences periods of inactive steady states broken by periods of production and reproduction. Rain stimulates both processes, and after a short period of adequate moisture, water is scarce again, and both production and biomass decrease to some low steady state.

Primary production in the desert depends on the portion of available water used and the efficiency of plants in using it. We have already discussed the various ways plants have evolved to conserve and efficiently use limited water, such as separating photosynthesis from transpiration by fixing and storing large quantities of CO_2 in the dark.

Data from various deserts in the world (for a summary see Noy-Mier, 1973, 1974) suggest that annual net primary production of above-ground vegetation varies from 30 to 200 g/m^2 in arid zones and from 100 to 600 g/m^2 in semiarid shrublands. Below-ground root production is also low. In arid regions it may amount to 100 to 400 g/m^2 and in semiarid regions 250 to 1,000 g/m^2. The amount of biomass that accumulates and the rate of turnover (the ratio of production to biomass) depend on the dominant type of vegetation. In deserts like the Sonoran, where trees, shrubs, and cacti dominate, annual productivity is about 10 to 20 percent of the above-ground standing crop biomass of 300 to 1,000 g/m^2. In deserts with perennial types of vegetation, annual production is 20 to 40 percent of the biomass of 150 to 600 g/m^2. And of course, annual or ephemeral communities have a 100 percent turnover of both roots and above-ground foliage; annual production is the same as peak biomass. These turnover rates are higher than those of forests and tundra. The ratio of below-ground biomass to above-ground biomass (stems and foliage) for perennial grasses and forbs is between 1 and 20 and for shrubs, between 1 and 3. In general, deserts, unlike the tundra, do not have a high below-ground–above-ground ratio. The root biomass is relatively small, a characteristic of desert plants.

Adding to primary production in the desert are lichens and green and blue-green algae, abundant as soil crusts. These blue-green algal crusts, whose biomass ranges up to 240 kg/ha, are unique because of their unusually high rate of nitrogen fixation: 10 to 20 g/m^2/year. This amount of fixed nitrogen contrasts sharply with the low nitrogen input from rainfall and dry fallout of 2 to 3 g/m^2/year. In spite of high fixation rates, only 5 to 10 g/m^2 of total nitrogen input becomes part of higher plants. Approximately 70 percent of the nitrogen is short-circuited back to the atmosphere as volatilized ammonia and as N_2 from denitrification speeded up by dry alkaline soils (Reichle, 1975; Rixon, 1970) (Figure 24.23).

Grazing herbivores of the desert are opportunists and generalists in their mode of feeding. They consume a wide range of species, plant types, and parts, although the animals do have seasonal preferences. Mule deer, peccary, and desert sheep feed on succulents and ephemerals when available, then switch to woody browse during the dry period. As a last resort, herbivores may consume dead litter and lichens. Small herbivores such as desert rodents and ants are specialists and feed largely on seeds (see Graetz, 1981).

Herbivores in a shrubby desert, under most conditions, consume little more than 2 percent of the above-ground primary production; but seed-eating herbivores can consume most seed production. In one of the few studies of herbivory in the desert, Chew and Chew (1970) found that small grazing herbivores (the jackrabbit and the kangaroo rat) used only about 2 percent of the above-ground net primary production but consumed 87 percent of the seed production, a rate of consumption that could have a pronounced effect on plant composition and plant populations. Fifty-five percent of the energy flow through small mammals in the shrub desert passed through the kangaroo rat, 22 percent through the browsing jackrabbit, and 6.5 percent through the insectivorous grasshopper mouse *(Onychomys torridus)*.

Carnivores, like herbivores, are opportunistic feeders, with few specialists. Lizards feed on ants, and foxes and bobcats on hares and reptiles. But most carnivores, such as foxes and coyotes, have mixed diets that include leaves and fruits; and even insectivorous birds and rodents feed considerably on herbivorous foods. Omnivory rather than car-

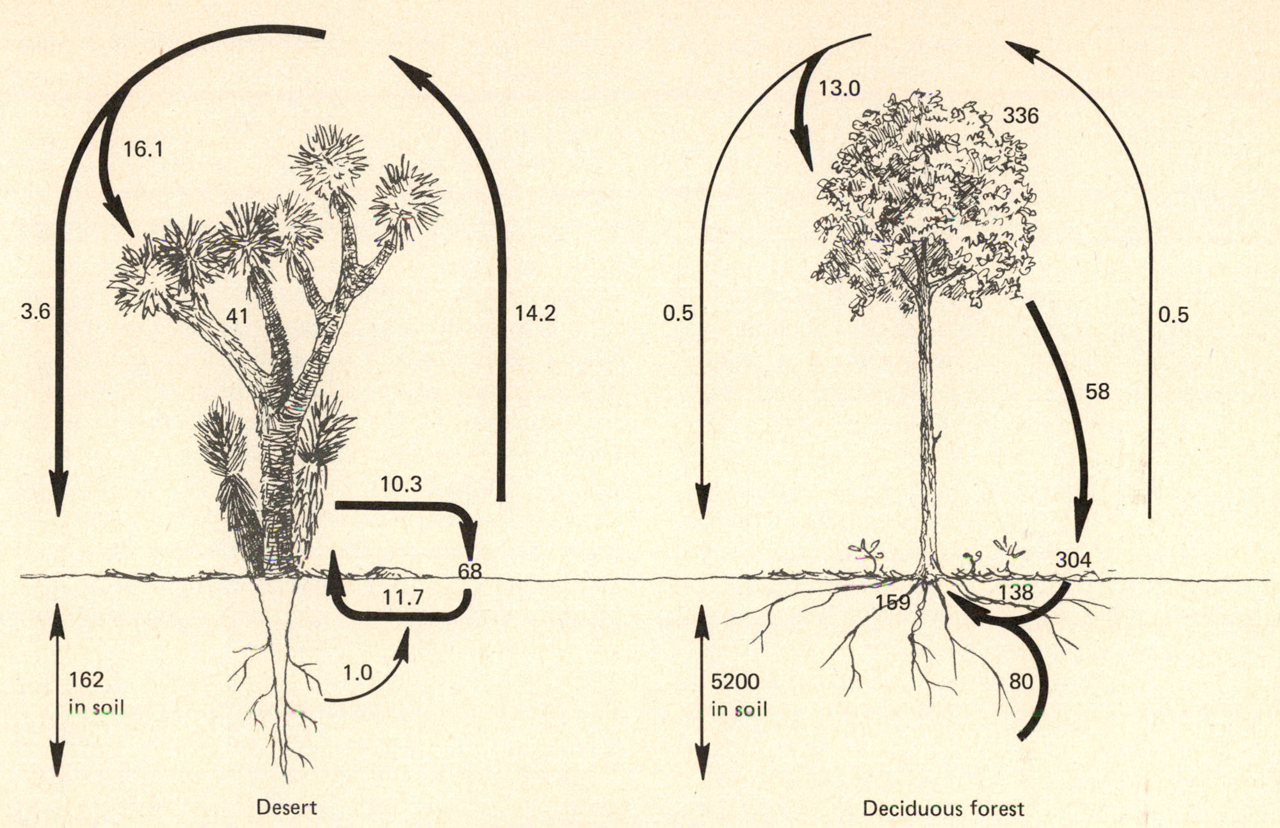

FIGURE 24.23 The nitrogen cycle of the desert differs from that of a deciduous forest. In the forest a considerable portion of the nitrogen is cycled through plants. In the desert considerable quantities of nitrogen are fixed, but most of it is lost through denitrification. (*After Reichle, 1975:260.*)

nivory and complex food webs seem to be the rule in the desert ecosystem.

The detrital food chain seems to be less important in the desert than in other ecosystems. Although most functional and taxonomic groups of soil microorganisms exist in the desert, fungi and actinomycetes are prominent. Microbial decomposition, like the blooming of ephemerals, is limited to short periods when moisture is available. Because of this, dry litter tends to accumulate to a point at which the detrital biomass is greater than the above-ground living biomass. Most of the ephemeral biomass disappears through grazing, weathering, and erosion. Decomposition proceeds mostly through detritus-feeding arthropods such as termites that ingest and "decompose" woody tissue in their guts. In some deserts considerable amounts of nutrients may be locked up in termite structures, to be released when the structures are destroyed. Other important detrital feeders are acarids and various isopods.

Because of limited production and decomposition, nutrients, especially nitrogen, in the desert ecosystem are limiting. The nutrient supply, confined largely to the upper surface of the soil, is vulnerable to erosion and volatilization and can be rapidly depleted by the growth of annual plants. Because tree and shrub growth is sparse in arid regions, nutrient return by litterfall is strongly localized around the plants. This concentrates nutrient recycling and microbial activity around the shrubs.

Summary

The land masses of Earth are clothed in a variety of vegetational types, most of which have been destroyed or drastically changed by human activities over the centuries. Nevertheless, the basic units remain—deserts; tundras; coniferous, temperate deciduous, and tropical rain forests; woodlands; thornbush; savannas; and grasslands. Each has its own distinctive life forms, with their own unique adjustments to energy flow and material cycling.

Natural grasslands occupy regions where rainfall is between 10 and 30 inches a year, but they are not exclusively climatic; many exist through the intervention of fire and human activity. Natural grasslands, once covering extensive areas of Earth, have shrunk to a fraction of their original size. Humans have converted vast areas of grassland to grainland and have converted large areas of forest to seminatural, tame, and seral grasslands for livestock grazing and hay. Seral or natural grasslands consist of bunch or sod grasses, or both, and a herbaceous component. Productivity varies considerably, influenced greatly by precipitation. It ranges from 82 g/m^2/year in semiarid grasslands to 30 times that much in subhumid tropical and tame or cultivated grasslands. The bulk of primary production goes underground to the roots. To a point grazing stimulates primary production. Although the most conspicuous grazing herbivores are large mammals, the major consumers are invertebrates. The heaviest consumption takes place below ground, where nematodes are the dominant herbivores. Most of the primary production goes to the decomposers. Nutrients are recycled rapidly. A significant quantity goes to the roots, to be moved above ground to next year's growth.

Savannas are grasslands with scattered woody vegetation. They are characteristic of regions with alternating wet and dry seasons. Difficult to characterize precisely, savannas range from grass with an occasional tree to shrub and tree savannas. The latter grade into woodland and thornbush with an understory of grass. Much of the nutrient pool is tied up in plant and animal biomass, but nutrient turnover is high, with little accumulation of soil organic matter.

Semiarid regions support climax shrub ecosystems, characterized by densely branched wood structure and low height. The success of shrubs depends upon their ability to compete for nutrients, energy, and space. In semiarid situations shrubs have numerous competitive advantages, including structural modifications that affect light interception, heat losses, and evaporative losses. Nutrient conservation appears to be highly developed, including translocation and concentration of nutrients from surrounding soil to soil beneath the plants by way of litterfall. Shrublands, which go by different names in different regions of the world, are most characteristic of places where winters are mild and rainy and summers are long, hot, and dry. Seral shrublands occupy land going back to forest. Although a successional stage, many seral shrublands remain stable for long periods.

Coniferous, temperate deciduous, and tropical rain forests are the three dominant types of forest cover over the world. Confined to the Northern Hemisphere, the coniferous forest is typical of regions where summers are short and winters long and cold. The deciduous forest—which is richly developed in North America, western Europe, and eastern Asia—grows in a region of moderate precipitation and mild temperatures during the growing season. The tropical rain forest grows in equatorial regions where humidity is high, rainfall is heavy (especially during one season of the year), seasonal changes are minimal, and the annual mean temperature is about 28°C. Of the three, the tropical rain forest is the most highly stratified and contains the greatest diversity of ecological niches. A well-developed deciduous forest is the next most highly stratified.

Accompanying vegetative stratification is stratification of light, temperature, and moisture. The canopy receives the full impact of climate and intercepts light and rainfall; the forest floor is shaded through the year in most coniferous and tropical rain forests and in late spring and summer in deciduous forests. The greatest concentration and diversity of life are on and just below the ground layer.

Most living forest biomass is above ground—concentrated in bole, branches, and foliage—and detrital biomass is concentrated in litter and soil organic matter. The amount of detrital biomass varies with the forest ecosystem. The greatest concentration of litter and soil organic matter is found in coniferous forests, where decomposition is slow. Detrital biomass varies in temperate deciduous forests depending upon species composition. The litter of some trees, such as maple and yellow poplar, decomposes more rapidly than the ligninous, acidic litter of oak. Tropical rain forests experience little accumulation of litter because of rapid decomposition. The major exchangeable nutrient pool for coniferous and temperate deciduous forests is soil organic matter. For tropical rain forests, most of the nutrients are concentrated in a mat of roots, mycorrhizal fungi, and humus on or just below the surface of the soil. Cycling of nutrients, especially nitrogen, is relatively tight, with minimal losses from the ecosystem. Internal cycling involves translocation of nutrients from leaves to roots, litterfall and throughfall, and storage in roots. However, mineral cycling can be maintained over time only by the release of nutrients from parent material by weathering.

The alpine tundra of high mountain ranges in lower latitudes and the arctic tundra that extends beyond the tree line of the far north are at once similar and dissimilar. Both are characterized by low temperature, low precipitation, and a short growing season. Both possess a frost-molded landscape and plant species whose growth forms are low and whose growth rates are slow. The arctic tundra has a permafrost layer; rarely does the alpine tundra. Arctic plants require longer periods of daylight than alpine plants and reproduce vegetatively, while alpine plants propagate mostly by seeds. Over much of the Arctic, the dominant vegetation is cottongrass, sedge, and dwarf heaths. In the alpine tundra, cushion- and mat-forming plants, able to withstand buffeting by the wind, dominate exposed sites. Net primary production is low, and most plant growth occurs underground. In spite of an assemblage of grazing ungulates and rodents, most of the production goes to decomposers. Decomposition, however, is slow, resulting in a great accumulation of peat, which locks up nutrient supplies. Nutrient cycling, especially of N and P, is necessarily conservative and tight, operating on very small pools of soluble and exchangeable nutrients. Most of the cycled nutrients are concentrated in the roots, translocated to growing shoots above ground early in the season, and replaced by production and subsequent return of nutrients to the roots later in the growing season. Animals aid in the cycling of nutrients by consuming some of the primary production and distributing droppings and dung across the tundra.

The desert is a harsh environment whose plants and animals have evolved ways of dealing with aridity and high temperatures. Deserts occupy about one-seventh of the land surface of Earth and are largely confined to a worldwide belt between the tropic of Cancer and the tropic of Capricorn. Deserts result largely from the climatic patterns of Earth as well as the locations of mountain ranges and the remoteness of land areas from sources of oceanic moisture. Functionally, deserts are characterized by low net production, opportunistic grazing herbivores and carnivores, and a detrital food chain that is relatively less important than in other ecosystems. Considerable quantities of nitrogen are fixed by crustlike blue-green algae on the desert floor, but most of the nitrogen is lost back to the atmosphere.

Review and Study Questions

Note: Questions 1 through 7 relate to the Introduction, "Distribution of Terrestrial Ecosystems."

1. Name the six biogeographical regions and briefly indicate their major features.
2. What is the weakness of the life zone concept?
3. What are biotic provinces?
4. What are biomes? Why is the biome concept widely accepted, especially in North America?

5. What are ecoregions? What is their practical application? How do they relate to the biome concept?
6. What is an ecocline?
7. How does the Holdridge life zone system differ from other approaches to classifying terrestrial ecosystems?
8. What is the difference between bunch grass and sod grass?
9. What characteristics do all grasslands have in common?
10. Contrast the production and function of the above-ground and below-ground components of grassland ecosystems. Why does so much of the energy flow and nutrient cycling take place below ground in grasslands?
11. What distinguishes savannas from grasslands, both in structure and in function? Under what climatic conditions do savannas develop?
12. What characteristics do seral and climax shrublands share?
13. What is the major reason for stratification in forest ecosytems? How does structural stratification affect environmental stratification? What is the interaction between them?
14. Contrast nutrient cycling in a coniferous forest with that in a temperate deciduous forest; that in a temperate deciduous forest with that in a tropical rain forest. What makes the tropical rain forest so different from the temperate deciduous forest?
15. What physical and biological features characterize the tundra?
16. How does the alpine tundra differ from the arctic tundra?
17. What is the relationship among permafrost, plant life, and nutrient cycling?
18. Compare primary production of the arctic tundra with that of grassland and desert. Where does the bulk of net primary production of the arctic tundra accumulate and why?
19. What is the role of herbivores in the arctic tundra?
20. What climatic forces lead to deserts?
21. Contrast primary production of the desert with that of a grassland. What is the difference and why?

Suggested Readings

Biogeography

Brown, J. H., and A. C. Gibson. 1983. *Biogeography.* St. Louis: Mosby.

Grasslands

Breymeyer, A., and G. Van Dyne, eds. 1980. *Grasslands, systems analysis and man.* Cambridge, England: Cambridge University Press.

Coupland, R. T., ed. 1979. *Grassland ecosystems of the world: Analysis of grasslands and their uses.* Cambridge, England. Cambridge University Press.

Duffey, E. 1974. *Grassland ecology and wildlife management.* London: Chapman and Hall.

French, N., ed. 1979. *Perspectives on grassland ecology.* New York: Springer-Verlag.

Heal, O. W., and D. F. Perkins, eds. 1978. *Ecology of some British moors and montane grassland.* New York: Springer-Verlag.

Risser, P. G., et al., eds. 1981. *The true prairie ecosystem.* Stroudsburg, Pa.: Dowden, Hutchinson & Ross.

Weaver, J. E. 1954. *North American prairie.* Lincoln, Nebr.: Johnsen.

Savanna

Bourliere, F., ed. 1983. *Tropical savannas.* Ecosystems of the World no. 13. Amsterdam, Holland: Elsevier Scientific Publishing Company.

Sinclair, A. R. E., and M. Norton-Griffiths, eds. 1979. *Serengeti: Dynamics of an ecosystem.* Chicago: University of Chicago Press.

Shrubland

Castri, F. di, and H. A. Mooney, eds. 1973. *Mediterra-*

nean-type ecosystems. Ecol. Studies, vol. 7. New York: Springer-Verlag.

Friedlander, C. P. 1961. *Heathland ecology*. Cambridge, Mass.: Harvard University Press.

McKella, C. M., J. P. Blaisdell, and J. R. Goodin, eds. 1972. *Wildland shrubs: Their biology and utilization*. General Technical Report INT-1. Washington, D.C.: Intermountain Range and Forest Experiment Station, U.S. Department of Agriculture.

Mooney, H. A., ed. 1977. *Convergent evolution in Chile and California Mediterranean climate ecosystems*. New York: Academic Press.

Forest

Bormann, F. H., and G. E. Likens. 1979. *Pattern and process in a forested ecosystem*. New York: Springer-Verlag.

Cousens, J. 1974. *An introduction to woodland ecology*. Edinburgh, Scotland: Oliver Boyd.

Curtis, J. T. 1959. *The vegetation of Wisconsin*. Madison: University of Wisconsin Press.

Edmonds, R. L., ed. 1981. *Analysis of coniferous forest ecosystems in the western United States*. Stroudsburg, Pa.: Dowden, Hutchinson & Ross.

Golley, F. B., ed. 1983. *Tropical rain forest ecosystems: Structure and function*. Ecosystems of the World no. 14A. Amsterdam, Holland: Elsevier Scientific Publishing Company.

Larsen, J. A. 1980. *The boreal ecosystem*. New York: Academic Press.

Reichle, D. E., ed. 1981. *Dynamic properties of forest ecosystems*. Cambridge, England: Cambridge University Press.

Richards, P. W. 1952. *The tropical rain forest*. Cambridge, England: Cambridge University Press.

Whitmore, T. C. 1975. *Tropical rain forests of the Far East*. Oxford, England: Clarendon.

Tundra

Bliss, L. C., O. H. Heal, and J. J. Moore, eds. 1981. *Tundra ecosystems: A comparative analysis*. New York: Cambridge University Press.

Brown, J., P. C. Miller, L. L. Tieszen, and F. L. Bunnell. 1980. *An arctic ecosystem: The coastal tundra at Barrow, Alaska*. Stroudsburg, Pa.: Dowden, Hutchinson & Ross.

Bruemmer, F. 1974. *The Arctic*. New York: Quadrangle.

Rosswall, T., and O. W. Heal, eds. 1975. *Structure and function of tundra ecosystems*. Ecological Bulletin no. 20. Stockholm: Swedish Natural Sciences Research Council.

Wielgolaski, F. E., ed. 1975–1976. *Fennoscandian tundra ecosystems*. Part I, *Plants and microorganisms;* part II, *Animals and systems analysis*. New York: Springer-Verlag.

Zwinger, A. H., and B. E. Willard. 1972. *Land above the trees*. New York: Harper & Row.

Desert

Brown, G. W., Jr., ed. 1976–1977. *Desert biology*. 2 vols. New York: Academic Press.

Goodall, D. W., ed. 1976. *Evolution of desert biota*. Austin: University of Texas Press.

Goodall, D. W., and R. A. Perry, eds. 1981. *Arid-land ecosystems: Structure, functioning and management*. Volumes 1 and 2. Cambridge, England: Cambridge University Press.

Krutch, J. W. 1952. *The desert year*. New York: Slone.

McGinnis, W. G., B. J. Goldman, and P. Paylore, eds. 1969. *Deserts of the world*. Tucson: University of Arizona Press.

Orians, G. H., and O. T. Solbrig, eds. 1977. *Convergent evolution in warm deserts*. Stroudsburg, Pa.: Dowden, Hutchinson & Ross.

West, N. E., and J. Skujins, eds. 1978. *Nitrogen in desert ecoystems*. Stroudsburg, Pa.: Dowden, Hutchinson & Ross.

CHAPTER 25

Freshwater Ecosystems

Outline

Objectives

Upon completion of this chapter, you should be able to:

1. Compare lotic and lentic ecosystems relative to their physical characteristics.
2. Contrast energy flow and nutrient cycling in lotic and lentic ecosystems.
3. Explain why a lotic ecosystem is a continuum.
4. Contrast the functional differences between a headwater stream and a river.
5. Describe the role and the physical position of the various feeding groups in a lotic ecosystem.
6. Compare the role of detritus in lotic and lentic ecosystems.
7. Distinguish between oligotrophy and eutrophy.
8. Describe the various types of wetlands.
9. Contrast rheotrophic with ombrotrophic wetlands.

A very close relationship exists between land and water, between aquatic and terrestrial ecosystems. Primarily through the water cycle, one feeds upon the other. Water that falls on land runs from the surface and moves through the soil to deeper layers to enter groundwater, springs, and streams and eventually ponds, lakes, estuaries, and oceans (see Chapter 6). The nature of any one aquatic ecosystem is influenced heavily by the watershed in which it lies. A *watershed* is a land area drained by a stream or a network of streams. It is characterized by vegetative cover; geology; soils; topography, or steepness of its landscape; and land use. Streams provide the drainage pathway; ponds, lakes, and wetlands act as catch basins. Thus, any watershed may embrace two aquatic ecosystems—a *lentic,* or still water, system and a *lotic,* or flowing water, system.

Flowing Water Ecosystems

Streams may begin as outlets of ponds or lakes, or they may arise from springs or seepage areas as headwater streams high up in the watershed. As water drains away from the source, it flows in a direction dictated by the lay of the land and underlying rock formations. Its course may be determined by the original slope; or water, seeking the least resistant route to lower land, may follow joints and fissures in bedrock near the surface and shallow depressions in the ground. Whatever its direction, water is concentrated in rills that erode small furrows, which soon grow into gullies. Moving downstream, especially where the gradient is steep, the water carries with it a load of debris that cuts the channel wider and deeper and that sooner or later is deposited within or along the stream. At the same time, erosion continues at the head of the gully, cutting backward into the slope and increasing the drainage area.

Just below its source, the stream may be small, straight, and often swift, with waterfalls and rapids. Further downstream, where the gradient is less and velocity decreases, meanders become common, and the stream deposits its load of sediment as silt, sand, or mud. At flood time, material carried by a stream is dropped on level lands over which floodwaters spread to form floodplain deposits. These floodplains are part of a stream or river channel used at the time of high water—a fact few people recognize.

When a stream or river flows into a lake or sea, the velocity of the water is suddenly checked, and the load of sediment is deposited in a fan-shaped area around the mouth to form a delta. Here the course of a stream or river is broken into a number of channels, which are blocked or opened with subsequent deposits. As a result, the delta is characterized by small lakes and swampy or marshy islands.

Streams are classified according to order. A small headwater stream without any tributaries is a first-order stream. When two streams of the same order join, the stream becomes a higher-order stream. Thus, if two first-order streams unite, the resulting stream becomes a second-order one, and when two second-order streams unite, the stream becomes a third-order one. The order of a stream can increase only when a stream of the same order joins it. Its order cannot be increased by the entry of a lower-order stream (for further discussion see Strahler, 1971). In general, headwater streams are orders 1 to 3; medium-sized streams, 4 to 6; and rivers, greater than 6.

PHYSICAL CHARACTERS

The character of a stream is molded by the velocity of the current. Velocity, which varies among streams and within streams, is influenced by the shape and steepness of the stream channel, width, depth, roughness of the bottom, and rainfall. The velocity of flow influences the nature of the bottom and the amount of silt deposition. Fast streams are those whose velocity of flow is 50 cm per second or higher (Nielsen, 1950). At this velocity the current will remove all particles less than 5 mm in diameter and will leave behind a stony bottom. High water increases the velocity; it moves bottom stones and rubble, scours the streambed, and cuts new banks and channels. As the gradient decreases and the width, perhaps the depth, and the volume of water increase, silt and decaying organic matter accumulate on the bottom. The character of the stream changes from one of fast water to slow (see page 559).

Flowing water ecosystems are often a series of two different but interrelated habitats, the turbulent riffle and the quiet pool (Figure 25.1). The waters of the pool are influenced by processes occurring in the rapids above, and the waters of the rapids are influenced by events in the pool.

Riffles are the sites of primary production in the stream (see Nelson and Scott, 1962; Minshall, 1978). Here the periphyton, or *aufwuchs*, organisms that are attached to or move on a submerged substrate but do not penetrate it, assume dominance. Periphyton occupies a position of the same importance as phytoplankton of lakes and ponds. Periphyton consists chiefly of diatoms, blue-green and green algae, and water moss.

Above and below the riffles are the pools. Here the environment differs in chemistry, intensity of current, and depth. Just as the riffles are the sites of organic production, so the pools are the sites of decomposition. They are catch basins of organic materials, for here the velocity of the current is reduced enough to allow part of the load to settle out. Pools are the major sites of CO_2 production during the summer and fall. That is necessary for the maintenance of a constant supply of bicarbonate in solution (Neel, 1951). Without pools, photosynthesis in the riffles would deplete the bicarbonates and result in smaller and smaller quantities of available carbon dioxide downstream.

Free carbon dioxide in rapid water is in equilibrium with that of the atmosphere. The amount of bound carbon dioxide is influenced by the nature of the surrounding terrain and the decomposition taking place in pools of still water. Most of the carbon dioxide in flowing water occurs as carbonate and bicarbonate salts. Streams fed by groundwater from limestone springs receive the greatest amount of carbonates in solution. Because of the coating of algae and ooze on the bottom, little calcium carbonate is added by the action of carbonic acid on a limestone streambed (Neel, 1951).

The degree of acidity or alkalinity, or pH, of the water reflects the CO_2 content as well as the presence of organic acids and pollution. The higher the pH of stream water, the richer the natural waters generally are in carbonates, bicarbonates, and associated salts. Such streams support more abundant aquatic life and larger fish populations than streams with acid waters, which are generally low in nutrients.

The constant swirling and churning of stream water over riffles and falls result in greater contact with the atmosphere. Thus, the oxygen content of the water is high at all levels and is often near the saturation point for existing temperatures. Only in deep holes or in polluted waters does dissolved oxygen show any significant decline.

Temperature of a stream is variable. In general, small, shallow streams tend to follow but lag behind air temperatures, warming and cooling with the seasons but rarely falling below freezing in winter. Streams with large areas exposed to sunlight are warmer than those shaded by trees, shrubs, and high banks. That is ecologically important because temperature affects the stream community.

Overall production of a stream is influenced by the nature of the bottom and the width. Pools with sandy bottoms are the least productive, because they offer little substrate for the aufwuchs. Bedrock, although a solid substrate, is so exposed to currents that only the most tenacious organisms can maintain themselves. Gravel and rubble bottoms support the most abundant life because they have the greatest surface area for aufwuchs, provide many crannies and protected places for insect larvae, and are the most stable (Table 25.1). Food production decreases as the particles become larger or smaller than rubble. Insect larvae, on the other hand, differ in abundance on the several substrates. Bottom production in streams 20 feet wide decreases by one half from sides to center, and in streams 100 feet wide, it decreases one-third (Pate, 1933). Streams 6 feet or less in width are four times as rich in bottom organisms as those 19 to 24 feet wide. That is one reason why headwater streams make such excellent trout nurseries.

STRUCTURE AND FUNCTION

The lotic, or flowing water, system is open and largely heterotrophic (Figure 25.2). A major energy

(a)

(b)

(c)

FIGURE 25.1 (a) A fast mountain stream in a deep woods. The bottom is largely bedrock. (b) Two different but related habitats in a stream, the riffle (foreground) and the pool (background). (c) A slow stream reflects the clouds and sky of a summer afternoon.

source is detrital material carried to it from the outside (*allochthonous*). Much of this organic matter input comes as coarse particulate organic matter (CPOM), leaves and woody debris dropped from streamside vegetation, particles larger than 1 mm in size. Another type of organic input is fine particulate organic matter (FPOM), material less than 1 mm in size, including leaf fragments, invertebrate

TABLE 25.1 BOTTOM TYPES, SHOWING AVERAGE AMOUNTS OF AVAILABLE FISH FOOD PER SQUARE FOOT IN EACH

Number of Determinations	Type of Bottom	AVERAGE WEIGHT, G/FT²		
		1933	1932	1931
32	Silt and mud	3.63	3.07	2.90
100	Rubble	2.96	2.47	0.67
38	Large rubble	2.27	1.55	0.50
31	Small rubble	4.11	3.53	0.85
53	Coarse gravel	1.35	1.51	1.02
30	Fine gravel	1.05	0.93	0.84
6	Sand	0.55	0.10	

SOURCE: Pate, 1933.

feces, and precipitated dissolved organic matter. A third input is dissolved organic matter (DOM), material less than 0.5 micron in solution. One source of DOM is rainwater dripping through overhanging leaves, dissolving the nutrient-rich exudates on them. Other DOM input comes by a geological pathway through subsurface seepage, which brings nutrients leached from adjoining forest, agricultural, and residential lands. Many streams receive inputs from mechanical pathways through the dumping of industrial and residential effluents. Supplementing this detrital input is autotrophic production in streams (*autochthonous*) by diatomaceous algae growing on rocks and by rooted aquatics such as water moss (*Fontinalis*). Energy is lost through two pathways: geological (through streamflow feeding downstream systems) and biological (from respiration).

The processing of this organic matter involves both physical and biological mechanisms. In fall, leaves drift down from overhanging trees, settle on the water, float downstream, and lodge against banks, debris, and stones. Soaked with water, the leaves sink to the bottom, where they quickly lose 5 to 30 percent of their dry matter as water leaches soluble organic matter from their tissues. Much of this DOM is either incorporated onto detrital particles or precipitated to become part of the FPOM. Another part is incorporated into microbial biomass (Figure 25.3).

Within a week or two, depending upon the temperature, the surface of the leaves is colonized by bacteria and fungi, largely aquatic hyphomycetes. Fungi are relatively more important on CPOM because large particles offer more surface for mycelial development (Cummins and Klug, 1979). Bacteria are associated more with FPOM. Microorganisms degrade cellulose and metabolize lignin. Their populations form a layer on the surface of leaves and detrital particles that is much richer nutritionally than the detrital particles themselves (Anderson and Cummins, 1979). Leaves and other detrital particles are attacked by a major feeding group, the *shredders,* insect larvae that feed on coarse particulate organic matter. Among these shredders are craneflies (Tipulidae), caddisflies (Trichoptera), and stoneflies (Plecoptera). They break down the CPOM, feeding on the material not so much for the energy it contains but for the bacteria and fungi growing on it (Cummins, 1974; Cummins and Klug, 1979). Shredders assimilate about 40 percent of the material they ingest and pass off 60 percent as feces.

Broken up by the shredders and partially decomposed by microbes, the leaves, along with invertebrate feces, become part of the fine particulate organic matter, which also includes some precipitated DOM. Drifting downstream and settling on the stream bottom, FPOM is picked up by another feeding group of stream invertebrates, the *filtering* and *gathering collectors*. The filtering collectors include, among others, the larvae of blackflies (Simulidae), with filtering fans, and net-spinning caddisflies, including *Hydropsyche*. Gathering collectors, such as larvae of midges, pick up particles form stream bottom sediments. Collectors obtain much of their nutrition from bacteria associated with the fine detrital particles.

While shredders and collectors feed on detrital material, another group feeds on the algal coating of stones and rubble. These are the *scrapers*, which include the beetle larvae, water penny (*Psephenus* spp.), and a number of mobile caddisfly larvae.

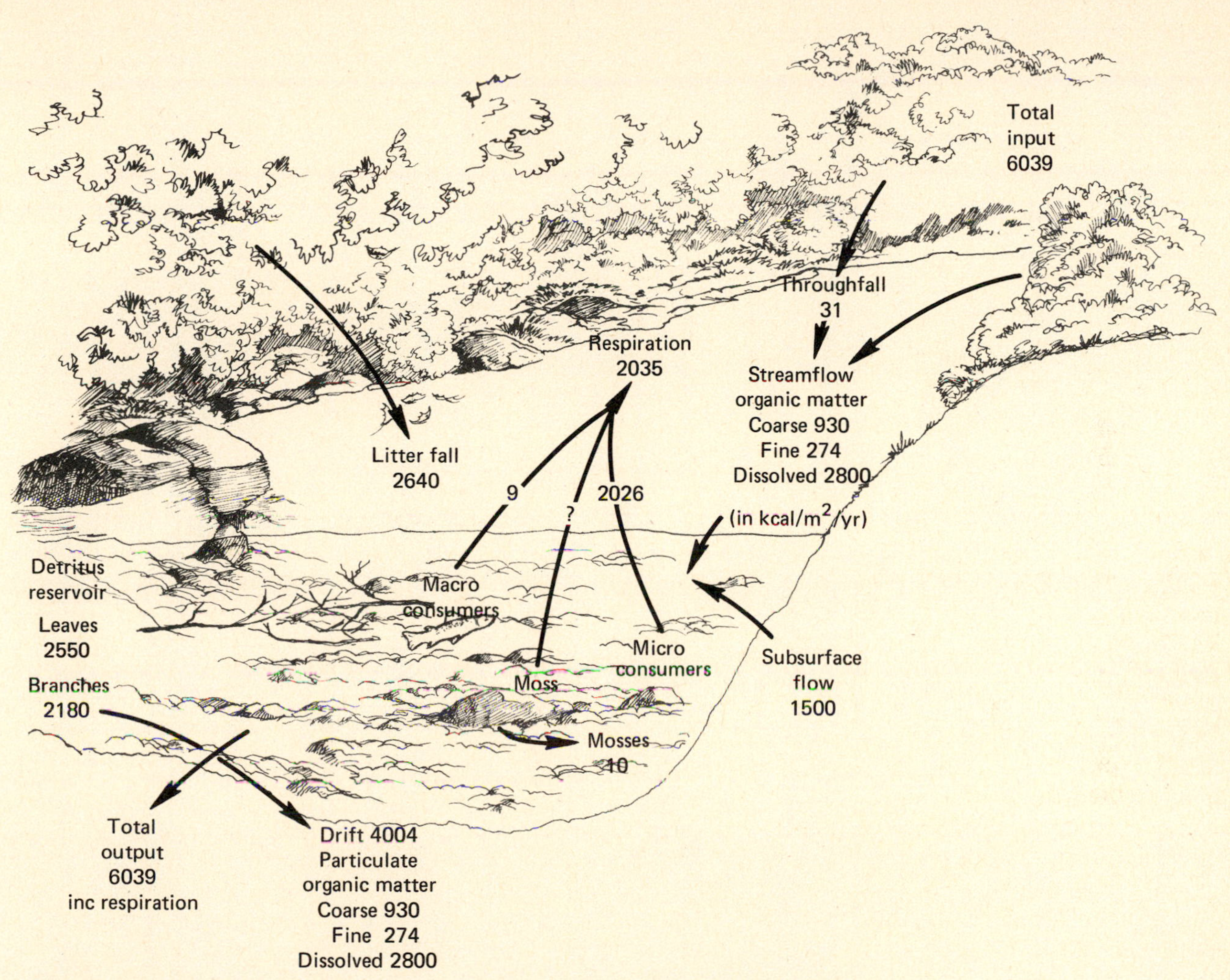

FIGURE 25.2 Functioning of a stream ecosystem. Streams are open ecosystems, with much of the energy input coming from terrestrial streamside sources. Note the great dependence on detritus and inflow from upstream and the role of coarse particulate organic matter (CPOM), fine particulate organic manner (FPOM), and dissolved organic matter (DOM). Primary production contributes little to energy flow. Energy values are based on Bear Brook, New Hampshire. (*Data from Fisher and Likens, 1973.*)

Much of the material they scrape loose enters the drift as FPOM. Behaviorally and morphologically, scrapers are adapted to maintain their position in the current, either by flattening to allow avoidance of the main force of flow or by weighting down with heavy mineral cases. Feeding on mosses and filamentous algae are the *piercers,* a feeding group made up largely of microcaddisflies.

Feeding on the detrital feeders and scrapers are predaceous insect larvae such as the powerful dobsonfly larvae (*Corydalus cornutus*) and fish such as the sculpin (*Cottus*) and trout. Even these pred-

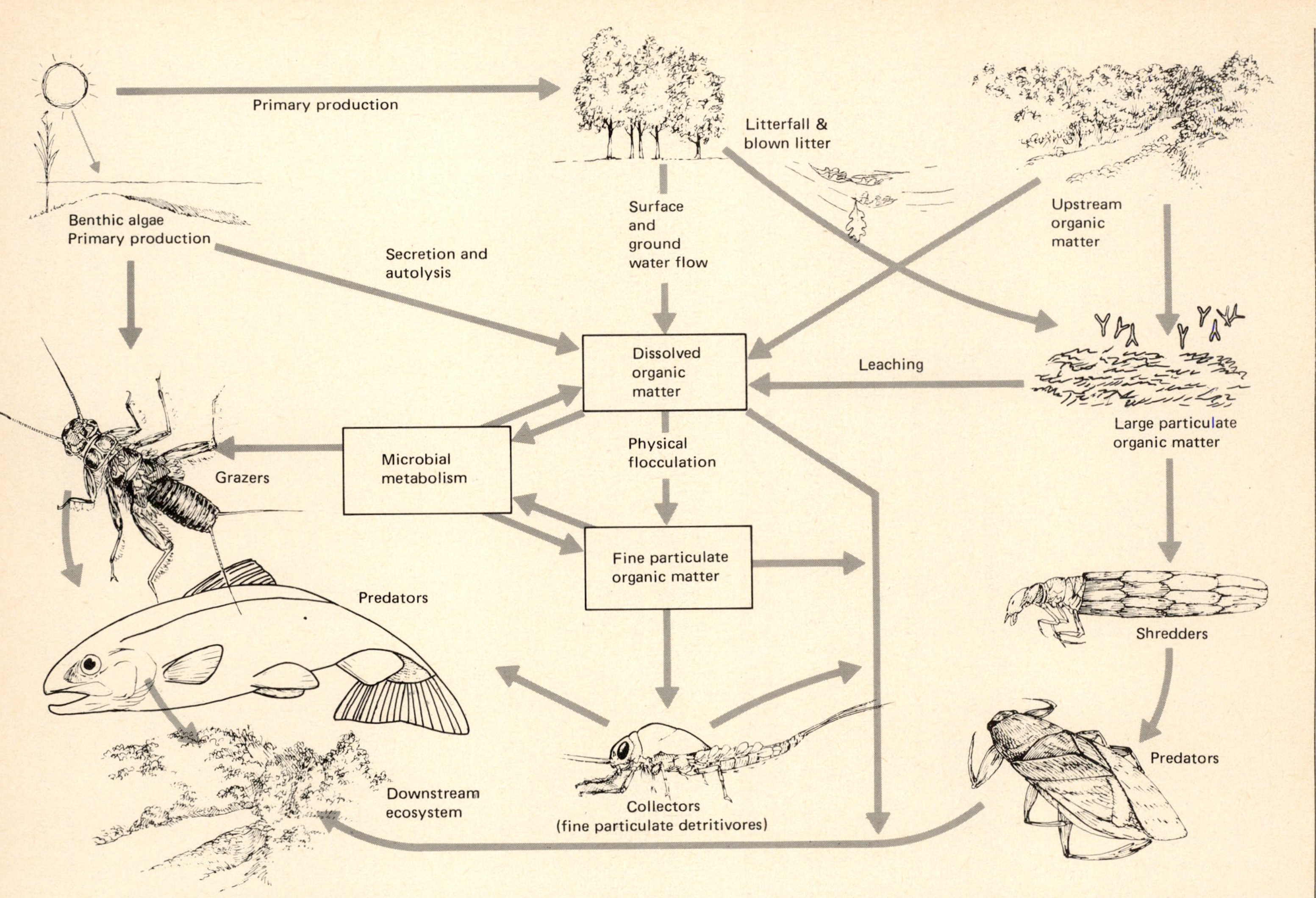

FIGURE 25.3 Model of a stream ecosystem emphasizing structure and function. It illustrates the processing of leaves and other particulate organic matter and dissolved organic matter. Leaves, branches, seeds, bark, and flowers represent coarse particulate organic matter, which is colonized by bacteria and aquatic fungi. This constitutes most of the nutrient and energy input into the stream. Algae and moss, the microproducers and macroproducers, are responsible for primary production in the stream. At any one point there is additional input of dissolved organic matter from upstream. Processing these inputs are the functional groups of organisms, mostly invertebrates: the shredders, collectors, grazers, piercers, and predators. (*Adapted from Cummins, 1974:663.*)

ators do not depend solely on aquatic insects; they also feed heavily on terrestrial invertebrates that fall or are washed into the stream.

Because of current, quantities of CPOM, FPOM, and invertebrates tend to drift downstream to form a sort of traveling benthos. This is a normal process in streams, even in the absence of high water and abnormal currents (for a detailed review see Hynes, 1970; Waters, 1972). Drift is so characteristic of streams that a mean rate of drift can serve as an index of the production rate of a stream (Pearson and Kramer, 1972).

THE RIVER CONTINUUM

Lotic ecosystems involve a continuum of physical and biological conditions from headwaters to the mouth (Vannote et al., 1980) (Figure 25.4). The upper reaches or headwaters are usually swift, cold, and in forested regions, shaded. Riparian vegetation reduces light, inhibiting autotrophic production, and contributes more than 90 percent of organic input into streams as terrestrial detritus. Even when headwater streams are exposed to sunlight and autotrophic production exceeds heterotrophic inputs, organic matter produced invariably enters detrital food chains (Minshall, 1978). Dominant organisms are shredders feeding on microbial populations on CPOM, and collectors, processors of FPOM. Populations of grazers are minimal, reflecting the small amount of autotrophic production, and predators are largely small fish—sculpins, darters, and trout. Headwater streams are accumulators, processors, and transporters of particulate organic matter from terrestrial ecosystems. As a result, the ratio of gross primary production to community respiration is less than 1. Consumer organisms utilize long- and short-chain organic compounds and leave the high–molecular weight, less soluble compounds for transport downstream. Organisms of headwater streams are adapted to a narrow temperature range, to a reduced nutrient regime, and to the maintenance of their positions in the current.

As streams increase in width to medium-sized creeks and rivers (orders 4 to 6), the importance of riparian vegetation and its detrital input decreases. The lack of shading results in an increased temperature of the stream water. As the gradient declines, the current slows and becomes more variable. The diversity of the microenvironments supports a greater diversity of organisms. An increase in light and temperature and a decrease in terrestrial input encourage a shift from heterotrophy to autotrophy, relying on algal and rooted vascular plant production. Gross primary production now exceeds community respiration. Because of the lack of CPOM, shredders disappear. Collectors, feeding on FPOM transported downstream, and grazers, feeding on autotrophic production, become the dominant consumers. Predators show little increase in biomass but shift from cold-water species to warm-water species.

As the stream order increases from 6 through 10 and higher, riverine conditions develop. The channel is wider and deeper. The volume of flow increases, and the current becomes relatively slower. Sediments accumulate on the bottom. Both riparian and autotrophic production decrease, with a gradual shift back to heterotrophy. A basic energy source is FPOM, utilized by bottom-dwelling collectors, now the dominant consumers. However, slow, deep water and dissolved organic matter (DOM) support a minimal phytoplankton and associated zooplankton population.

Throughout the downstream continuum, the lotic community capitalizes on upstream feeding inefficiency. Downstream adjustments in production and the physical environment are reflected in changes in consumer groups (Figure 25.4). Through the continuum the lotic ecosystem achieves some balance between the forces of stability, such as natural obstructions to flow that aid in the retention of nutrients upstream, and the forces of instability, such as flooding, drought, and temperature fluctuations.

ENERGY FLOW AND NUTRIENT CYCLING

Energy flow in lotic ecosystems is not widely documented. One energy budget is for the well-studied, small, forested Bear Brook in Hubbard Forest of northern New Hampshire (Fisher and Likens, 1973). That budget is summarized in Table 25.2. Over 99 percent of the energy input came from the

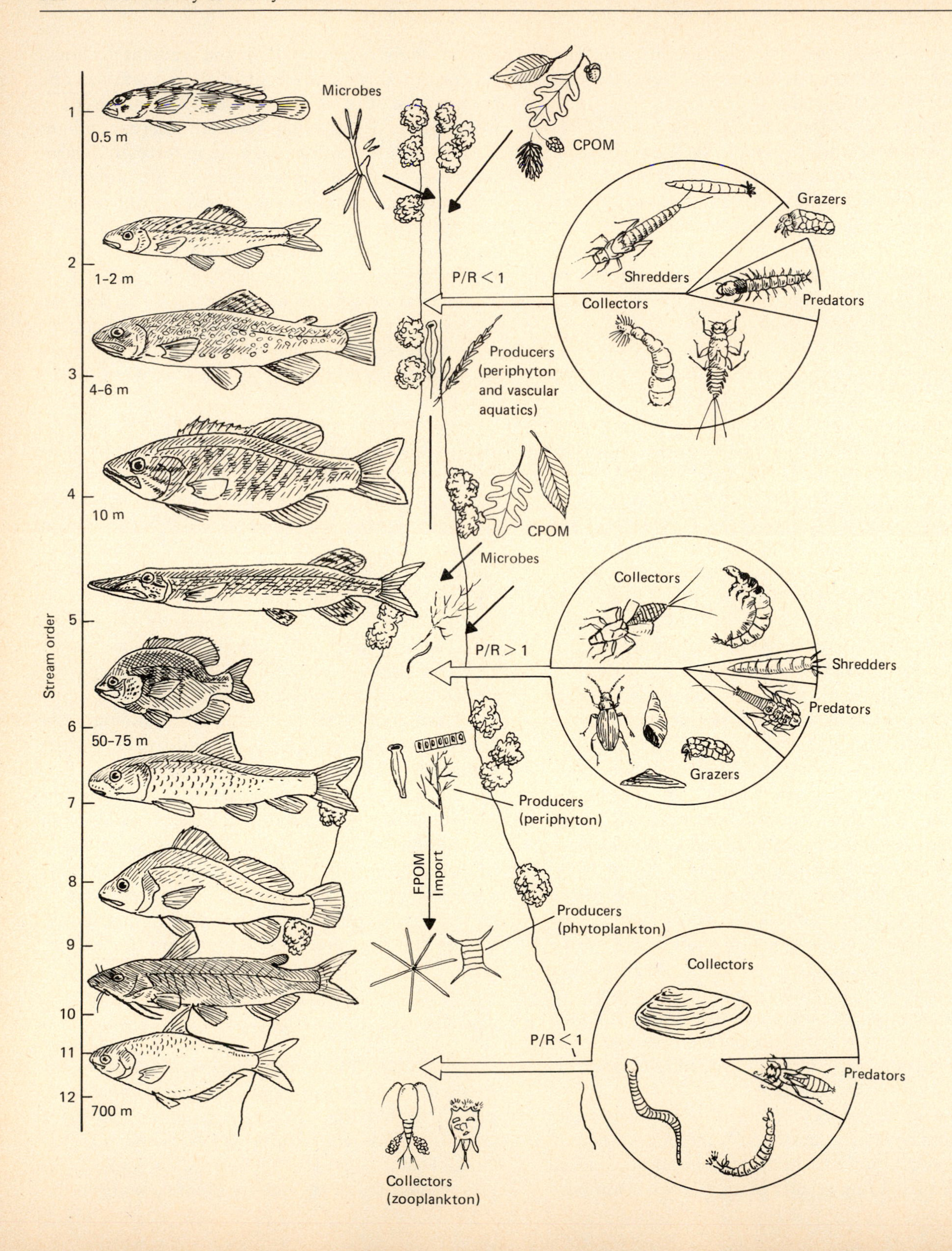
1
0.5 m
2
1-2 m
3
4-6 m
4
10 m
5
Stream order
6
50-75 m
7
8
9
10
11
12
700 m
Microbes
CPOM
Grazers
Shredders
Predators
Collectors
P/R < 1
Producers
(periphyton
and vascular
aquatics)
CPOM
Microbes
Collectors
P/R > 1
Shredders
Predators
Grazers
Producers
(periphyton)
FPOM
Import
Producers
(phytoplankton)
Collectors
P/R < 1
Predators
Collectors
(zooplankton)

surrounding forested watershed or from upstream areas. Primary production by mosses accounted for less than 1 percent of the total energy supply. Algae were absent from the stream. Inputs from litter and throughfall (meteorological) accounted for 44 percent of the energy supply, and geological inputs accounted for 56 percent. Energy was introduced in three forms: CPOM, represented by leaves and other debris; FPOM, represented by drift and small particles; and DOM. In Bear Brook 83 percent of the geologic input and 47 percent of the total energy input was in the form of DOM. Sixty-six percent of the organic input was exported downstream, leaving 34 percent to be utilized locally.

An inherent problem of flowing water ecosystems is the retention of nutrients upstream to reduce downstream losses. Nutrients in terrestrial and lentic ecosystems (see next section) are more or less recycled in place. An atom of nutrient passes from soil or water column to plants and consumers back to soil or water in the form of detrital matter or exudates and is then recycled within the same segment of the system. But in flowing water, nutrients in the form of DOM and POM undergo constant downstream displacement. Some degree of retention is accomplished by *spiraling*, which is an interdependent process of nutrient cycling and downstream transport (Webster and Patten, 1979; Newbold et al., 1982).

The process may be pictured as a spiral lying longitudinally in a stream (Figure 25.5). One cycle in the spiral involves the uptake of an atom or nutrient from DOM, its subsequent passage through

TABLE 25.2 ANNUAL ENERGY FOR BUDGET FOR BEAR BROOK

Item	Kg—whole stream[a]	Kcal/m^2	Percentage
Inputs			
Litterfall			
Leaf	1,990	1,370	22.7
Branch	740	520	8.6
Miscellaneous	530	370	6.1
Wind transport			
Autumn	422	290	4.8
Spring	125	90	1.5
Throughfall	43	31	0.5
Fluvial transport			
CPOM	640	430	7.1
FPOM	155	128	2.1
DOM, surface	1,580	1,300	21.5
DOM, subsurface	1,800	1,500	24.8
Moss production	13	10	0.2
Input total	8,051	6,039	99.9
Outputs			
Fluvial transport			
CPOM	1,370	930	15.0
FPOM	330	274	5.0
DOM	3,380	2,800	46.0
Respiration			
Macroconsumers	13	9	0.2
Microconsumers	2,930	2,026	34.0
Output total	8,020	6,039	100.2

[a] Budget in kg does not balance because of different caloric equivalents of budgetary components.

Note: CPOM—coarse particulate organic matter; FPOM—fine particulate organic matter: DOM—dissolved organic matter.
SOURCE: Fisher and Likens, 1973:425.

FIGURE 25.4 The lotic ecosystem is essentially a continuum from headwater stream to river mouth. The headwater stream is strongly heterotrophic, dependent on terrestrial input of detritus, and the dominant consumers are shredders and collectors. As stream size increases, the input of organic matter shifts from particulate organic matter to primary production from algae and rooted vascular plants. The major consumer groups are now collectors and grazers. The zone at which the shift occurs is influenced by the degree of shading. As the stream increases to a river, the lotic system shifts back to heterotrophy. It is dependent upon inputs of fine particulate organic matter and dissolved organic matter. A phytoplankton population may develop, its extent influenced by depth and turbidity. The dominant consumers are collectors, mostly bottom organisms.

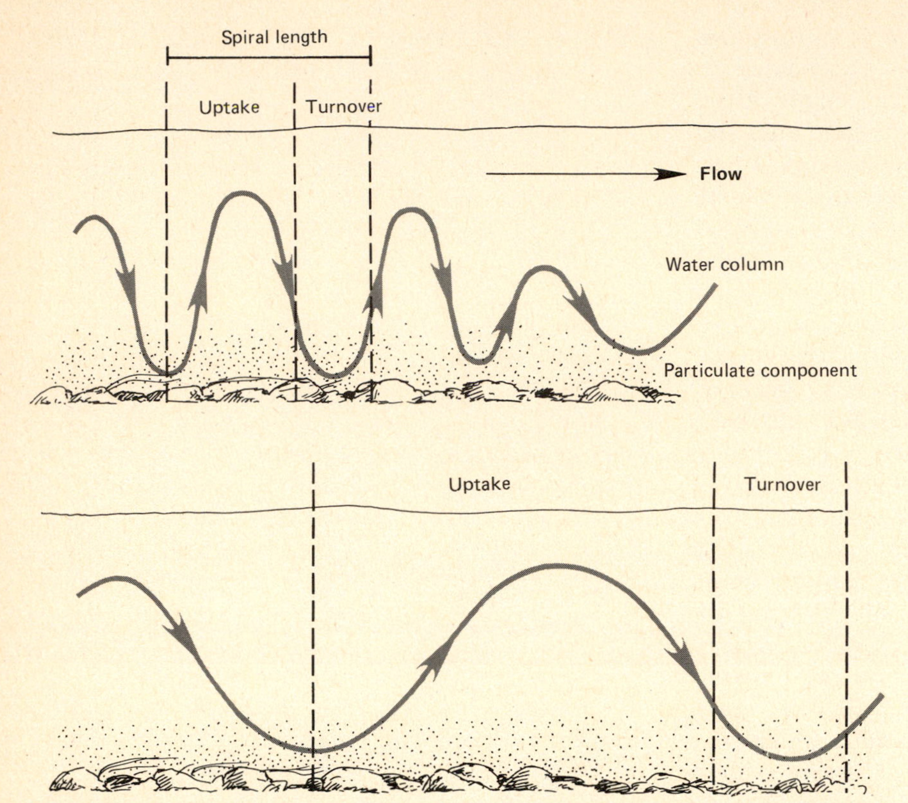

FIGURE 25.5 Nutrient spiraling between particulate organic matter, including microbes, and water column in a lotic ecosystem. Uptake and turnover take place as nutrients flow downstream. *A* represents tight spiraling and *B,* more open spiraling. The tighter the spiraling, the longer the nutrient remains in a particular segment of a stream. (*Adapted from Newbold et al., 1982:630.*)

the food chain, and its return to water, where it is available for reutilization. Spiraling is measured as the distance required for the completion of one cycle. The longer the distance required, the more open the spiral; and the shorter the distance, the tighter the spiral.

How long nutrients are retained and recycled in a given segment of a stream depends upon the tightness of the spiraling. The tighter the spiraling, the more productive is any one segment of the stream. Involved in retention is the microbial colonization of nutrients and the uptake of nutrients from CPOM as well as the subsequent consumption of both by shredders. At the same time, however, shredders enhance downstream transport by fragmenting CPOM and excreting fecal material. This FPOM joins the sloughing of algal growth and invertebrate drift being carried downstream. By trapping some of this FPOM and consuming it in place, collector organisms, especially net-spinning filter-feeders, shorten the spiral.

Newbold and his associates (1983) investigated the spiraling of exchangeable phosphorus in the form of $^{32}PO_4$ in Walker Branch at Oak Ridge, Tennessee, a first-order stream. The tagged P moved downstream at the rate of 10.4 m a day and cycled once every 18.4 days. Thus, the average downstream distance of one spiral was 190 m. In other words, one atom of P on the average completed one cycle in the ecosystem compartment for every 190 m of downstream travel. The spiraling length was partitioned into an uptake length of 165 m, associated with transport in the water column, mostly as DOM; a particulate turnover length of 25 m, associated with FPOM; and a consumer turnover length of 0.05 m, associated with consumer drift. CPOM accounted for 60 percent of the uptake; FPOM, 35 percent; and aufwuchs, 5 percent. Turnover time of P in CPOM ranged from 5.6 to 6.7 days; and FPOM, 99 days. Only 2.8 percent of the P uptake from particulate matter was transferred to consumers; the rest was transferred back to the water. About 30 percent of the consumer uptake was transferred to predators.

Still Water Ecosystems

The relatively still waters of lakes and ponds offer environmental conditions that contrast sharply with those of running water (Figure 25.6). Light penetrates only to a certain depth, depending upon turbidity. Temperatures vary seasonally and with depth. Because only a small portion of the water is in direct contact with the air and because decomposition is taking place on the bottom, the oxygen content of lentic systems is relatively low compared to that of lotic systems. In some lakes oxygen content may decrease with depth, but there are many exceptions. These gradations of oxygen, light, and temperature profoundly influence life in a lake, its distribution, and its adaptations.

PHYSICAL CHARACTERS

Each year the waters of many lakes and ponds undergo seasonal changes in temperature. In spring and early summer, surface water of ponds and lakes heats faster than deeper water. Because water reaches maximum density at 4°C, a warm, lighter layer of water floats on top of the heavier, cooler water below. The zone of steep temperature decline—approximately 1°C for every meter of depth—between the upper and lower layers is called the *thermocline* (Figure 25.7). The thermocline acts as a barrier between the upper warm layer, or *epilimnion,* and the deep cold layer, or *hypolimnion*. Because of the differences in the density of water, little circulation takes place between the surface and the lower layers. The upper well-lighted layer, where most phytoplankton grows, is well aerated because of oxygen production by plants and mixing by the wind. The lower waters, however, may be deficient in oxygen because of bacterial decomposition. Because sediments accumulate on the bottom, the deeper waters are relatively high in nutrients. The upper waters, however, become depleted of nutrients by phytoplankton uptake during the growing season.

In fall the upper waters cool, become more dense, and sink to the bottom, while the lighter, warmer water rises to the top, producing the fall overturn (Figure 25.7). This cooling continues until the temperature is uniform throughout the basin. Pond and lake waters circulate, and oxygen and nutrients are recharged. Through stirring actions of the wind, the fall overturn may last until ice forms.

As the surface water cools below 4°C, it becomes lighter, remains on the surface, and if the winter is cold enough, freezes; otherwise, it remains very close to 0°C. Now a slight inverse temperature stratification may develop, in which the water becomes warmer, up to 4°C, with depth. The water beneath the ice may be warmed by solar radiation through the ice. Because that increases its density, this water flows to the bottom, where it mixes with water warmed by heat conducted from the bottom mud. The result is a higher temperature on the bottom, although the overall stability of the water is undisturbed. As the ice melts in spring, the surface water again becomes warm, currents pass unhindered, and the spring overturn takes place.

Such seasonal changes in temperature stratification must not be considered a uniform condition in all deep bodies of water. In shallow lakes and ponds, temporary stratification of short duration may occur; in others stratification may exist, but no thermocline develops. In some very deep lakes, the thermocline may simply descend during periods of overturn and not disappear at all. In such lakes the bottom water never becomes mixed with the

(a)

(b)

FIGURE 25.6 Lakes and ponds fill basins or depressions in the land. (a) A glacial lake in a North Dakota prairie. Note the development of a littoral zone. (b) A beaver pond in a wooded landscape.

top layers. But some form of thermal stratification occurs in all very deep lakes, including those of the tropics.

The depth to which light can penetrate is limited by the turbidity of water and the absorption of light rays. On this basis, lakes and ponds can be divided into two basic layers—the *trophogenic zone,* roughly corresponding to the epilimnion, in which photosynthesis dominates; and the lower *tropholytic zone,* where decomposition is most active. The latter zone is about the same as the hypolimnion. The boundary between the two zones is the *compensation depth,* the depth at which photosynthesis balances respiration and beyond which light penetration is so low that it is no longer effective. Generally, the compensation depth occurs where light intensity is about 100 footcandles, or approximately 1 percent of full noon sunlight incident to the surface.

LAKE ZONATION

Lakes and ponds can also be subdivided into horizontal zones with differing photosynthetic activity (Figure 25.8). First is the *littoral* zone, or shallow water zone, where light penetrates to the bottom. This area is occupied by rooted plants such as water lilies, rushes, and sedges. Plants and animals found here vary with water depth, and a distinct zonation of life exists from deeper water to shore.

As sediments and organic matter accumulate at various points throughout the basin, but particularly near shore, and depth of water decreases, floating aquatics such as pond lilies and smartweeds appear. In shallow water beyond the zone of floating plants grow the *emergents,* plants whose roots

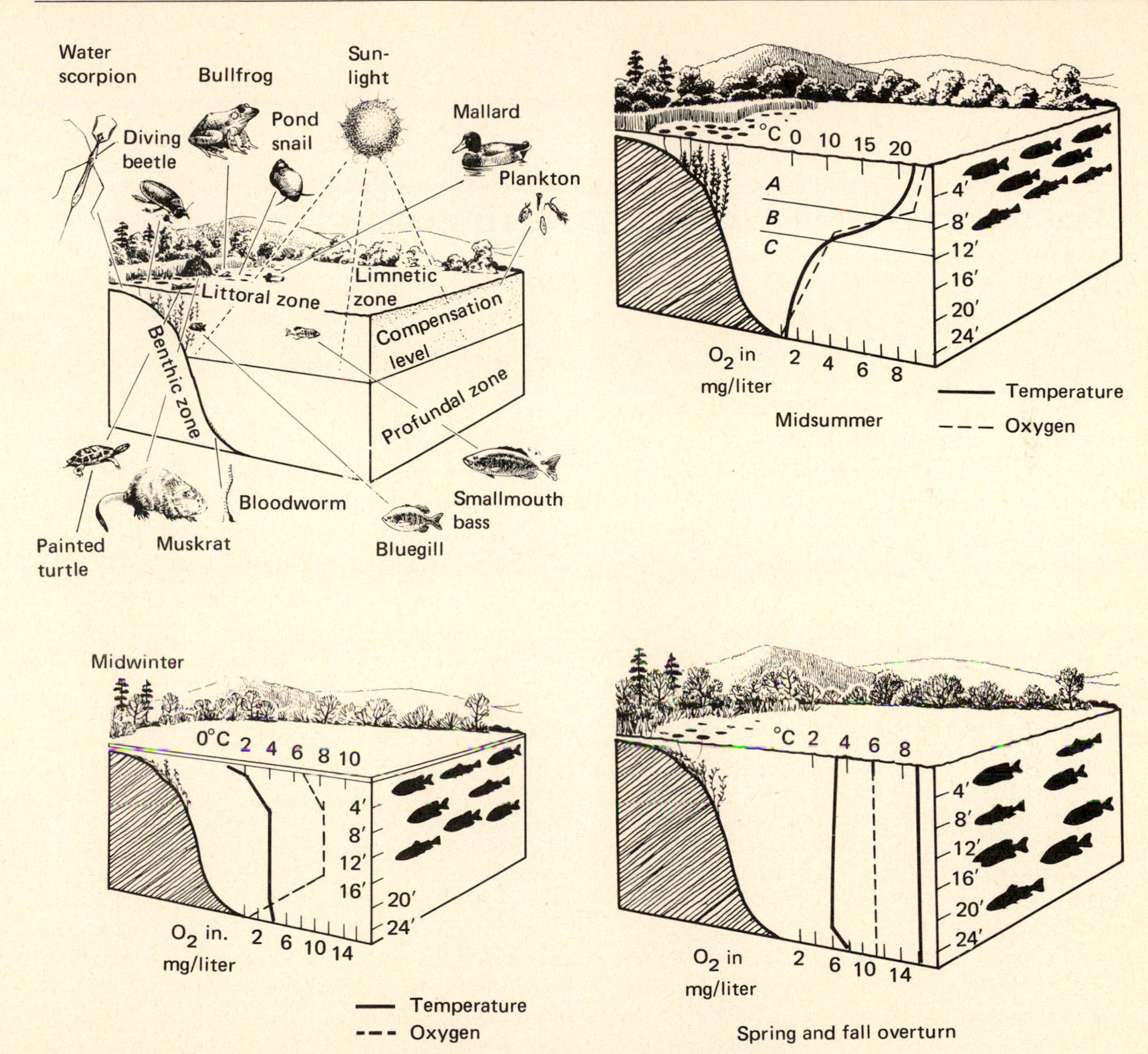

FIGURE 25.7 Seasonal variations in oxygen and temperature stratification and distribution of aquatic life in lake ecosystems. The drawing of a lake in midsummer shows the major zones—the littoral, the limnetic and the benthic—and the epilimnion, metalimnion or thermocline, and the hypolimnion. The compensation level is the depth at which light becomes too low for photosynthesis. Surrounding the lake is a variety of organisms typical of a lake community. The distribution of oxygen and temperature in a lake during the different seasons affects the distribution of fish life. The narrow fish silhouettes represent trout, or cold-water species; the wider silhouettes are bass, or warm-water species. Note the pronounced horizontal stratification in midsummer and the almost vertical oxygen and temperature curves during the spring and fall overturns.

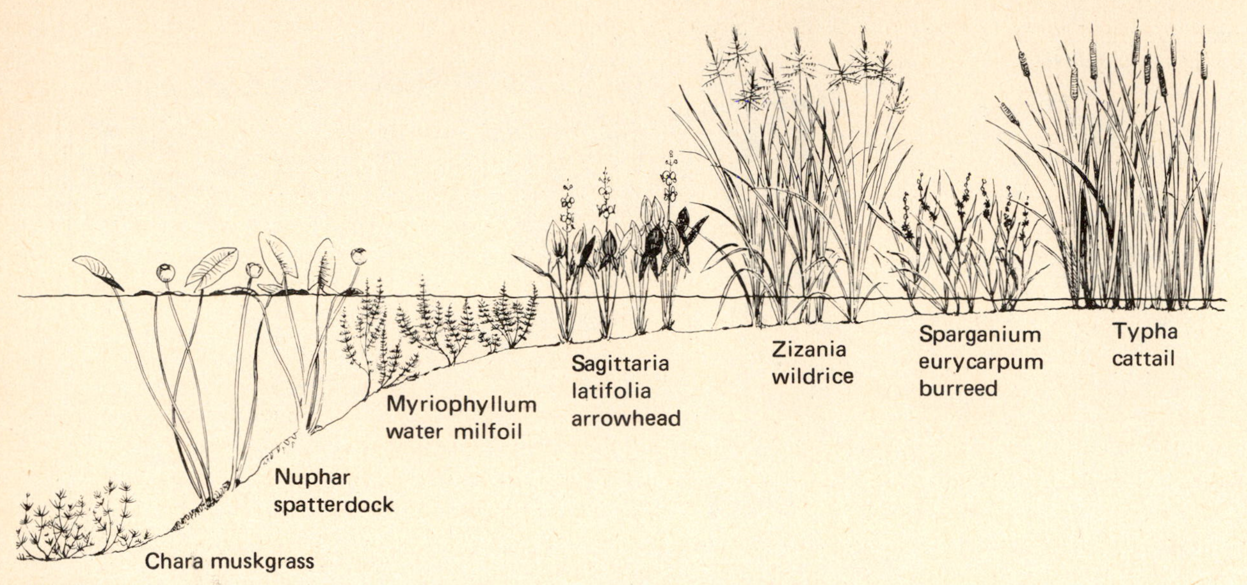

FIGURE 25.8 Zonation of emergent, floating, and submerged vegetation around a lake or pond. Note the changes of vegetation with water depth.

and lower stems are immersed in water and whose upper stems and leaves stand above water. Large areas of such plants make up a marsh. Among these emergents are plants with narrow tubular or linear leaves, such as bulrushes, reeds, and cattails. With these are associated such broadleaf emergents as pickerelweed and arrowheads. The distribution and variety of plants vary with water depth and fluctuation of water levels.

The *limnetic,* or open water, zone extends to the depth of effective light penetration. It is inhabited by suspended organisms, the *plankton,* and free-swimming organisms, the *nekton,* such as fish, capable of moving about voluntarily.

Dominating the limnetic zone is the phytoplankton, including diatoms, desmids, and filamentous algae. Because these tiny plants carry on photosynthesis in open water, they are the base on which the rest of limnetic life depends. Suspended with phytoplankton are small animals, or zooplankton, which graze on the minute plants. These animals form an important link in energy flow in the limnetic zone.

Plankton exhibits vertical distribution influenced by temperature, oxygen, light, and current. Light, of course, sets the lower limit at which phytoplanton can exist. Because animal plankton feeds on these minute plants, most of it, too, is concentrated in the trophogenic zone. Phytoplankton, by its own growth, limits light penetration and thus reduces the depth at which it can live. As the zone becomes more shallow, phytoplankton can absorb more light, and organic production is increased.

Within these limits the depth at which various species live is influenced by optimum conditions for their development. Some phytoplankton species live just beneath the water's surface; others are more abundant a few feet beneath; while those requiring colder temperatures live deeper still. Coldwater plankton, in fact, is restricted to lakes in which phytoplankton growth is scarce in the upper layer and in which the oxygen content of the deep water is not depleted by the decomposition of organic matter.

Animal plankton may exhibit stratification that changes seasonally because it is capable of independent movement. In winter some planktonic forms are distributed evenly to considerable depths; in summer they concentrate in layers most favorable to them and to their stage of development. At that season zooplankton undertakes a vertical migration during some part of the 24-hour period. Depending upon the species and their stage of development, they spend the night or day in the deep water or on the botton and move up to the surface during the alternate period to feed on phytoplankton.

During the spring and fall overturns, plankton is carried down, but at the same time, nutrients released by decomposition in the tropholytic zone are carried upward into the impoverished upper layers. In spring, when surface waters warm and stratification again develops, phytoplankton has access to both nutrients and light. A spring bloom develops, followed by a rapid depletion of nutrients and a reduction in planktonic populations, especially in shallow water.

Free to a large extent from the action of weak currents and capable of moving about at will are nekton organisms, the fish and some invertebrates. In the limnetic zone, fish make up the bulk of the nekton biomass. Their distribution is influenced mostly by food supply, oxygen, and temperature.

Common to both the tropholytic and trophogenic zones is the *benthic,* or bottom, zone. The organisms that inhabit the benthic zone are known collectively as the *benthos*. The bottom ooze is a region of great biological activity, so great that oxygen curves for lakes and ponds show a sharp drop in the water just above the bottom. Because organic muck lacks oxygen, the dominant organisms there are anaerobic bacteria. Under anaerobic conditions decomposition cannot proceed to inorganic end products. When the amounts of organic matter reaching the bottom are greater than can be utilized by bottom fauna, odoriferous muck rich in hydrogen sulfide and methane results. Thus, lakes and ponds with highly productive limnetic and littoral zones often have an impoverished fauna on and near the bottom. Life in the bottom ooze is most abundant in lakes with a deep hypolimnion in which oxygen is still available.

ENERGY FLOW AND NUTRIENT CYCLING

When compared to lotic systems, lakes and ponds are relatively closed ecosystems (Figure 25.9). Nevertheless, lentic systems are strongly influenced by inputs of nutrients from the terrestrial watersheds in which they lie. Nutrients and organic matter feed into lentic systems along biological, geological, and meteorological pathways (Likens and Bormann, 1974c).

Wind-borne particulate matter, dissolved substances in rain and snow, and atmospheric gases represent meteorological inputs to the system. Meteorological outputs are small, mainly spray aerosols and gases such as carbon dioxide and methane. Geological inputs include nutrients dissolved in groundwater, surface runoff, and inflowing streams and particulate matter washed into the basin from the surrounding terrestrial watershed. Geological outputs include dissolved and particulate matter carried out of the basin by outflowing waters and nutrients incorporated in deep sediments and possibly removed from circulation for a long time. Biological inputs and outputs, relatively small, include animals such as fish that move in and out of the basin. The lentic ecosystem also receives a hydrological input from precipitation and the drainage of surface waters. Outputs involve seepage through walls of the lake basin, subsurface flows, evaporation, and evapotranspiration. Within the lentic ecosystem, nutrients move among three compartments: dissolved organic matter, particulate organic matter, and primary and secondary minerals. Nutrients and energy move through the system by way of the grazing and detrital food chains.

Unlike lotic systems, lentic ecosystems are largely autotrophic, although the ratio of autotrophic to heterotrophic inputs varies from one lake to another. And like lotic and terrestrial ecosystems, lakes and ponds are dominated by detrital food chains even though studies of lentic ecosystems have emphasized phytoplankton-zooplankton food chains (Rich and Wetzel, 1978) (Figure 25.10). Major sources of detritus are organic materials produced within the lake or pond—particularly emergent, floating, and submerged plants of the littoral

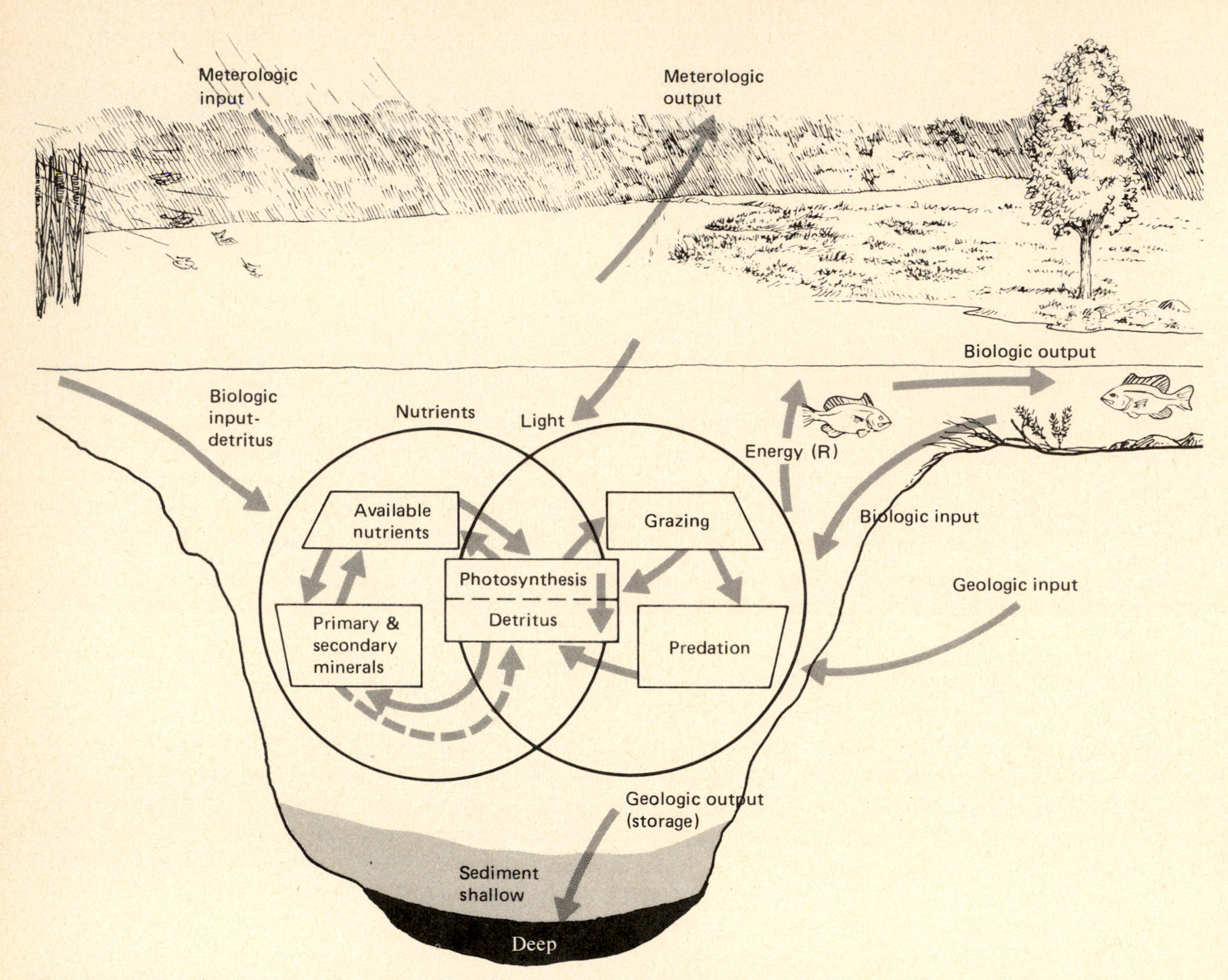

FIGURE 25.9 Model for nutrient cycling and energy flow in a lake ecosystem. Meteorological, geological, and biological inputs enter the lentic system from the watershed that contains it. The nutrient and energy inputs as well as nutrients and energy generated within the system move through a number of pathways. Part of the nutrients and energy fixed accumulates in bottom sediments. (*Based on Likens and Bormann, 1974b:447.*)

zone supplemented by phytoplankton of the limnetic zone. A secondary source is particulate and dissolved organic matter from the surrounding watershed. Most of the detrital metabolism takes place in the benthic zone, where particulate matter is decomposed, and in the open-water zone during sedimentation.

Primary production is carried out in the limnetic zone by phytoplankton and in the littoral zone by macrophytes. The ratio of the contribution of each varies among lentic systems. Phytoplankton production is influenced by nutrient availability in the water column. If nutrients are not limiting and the only losses are respiratory, the rate of net pho-

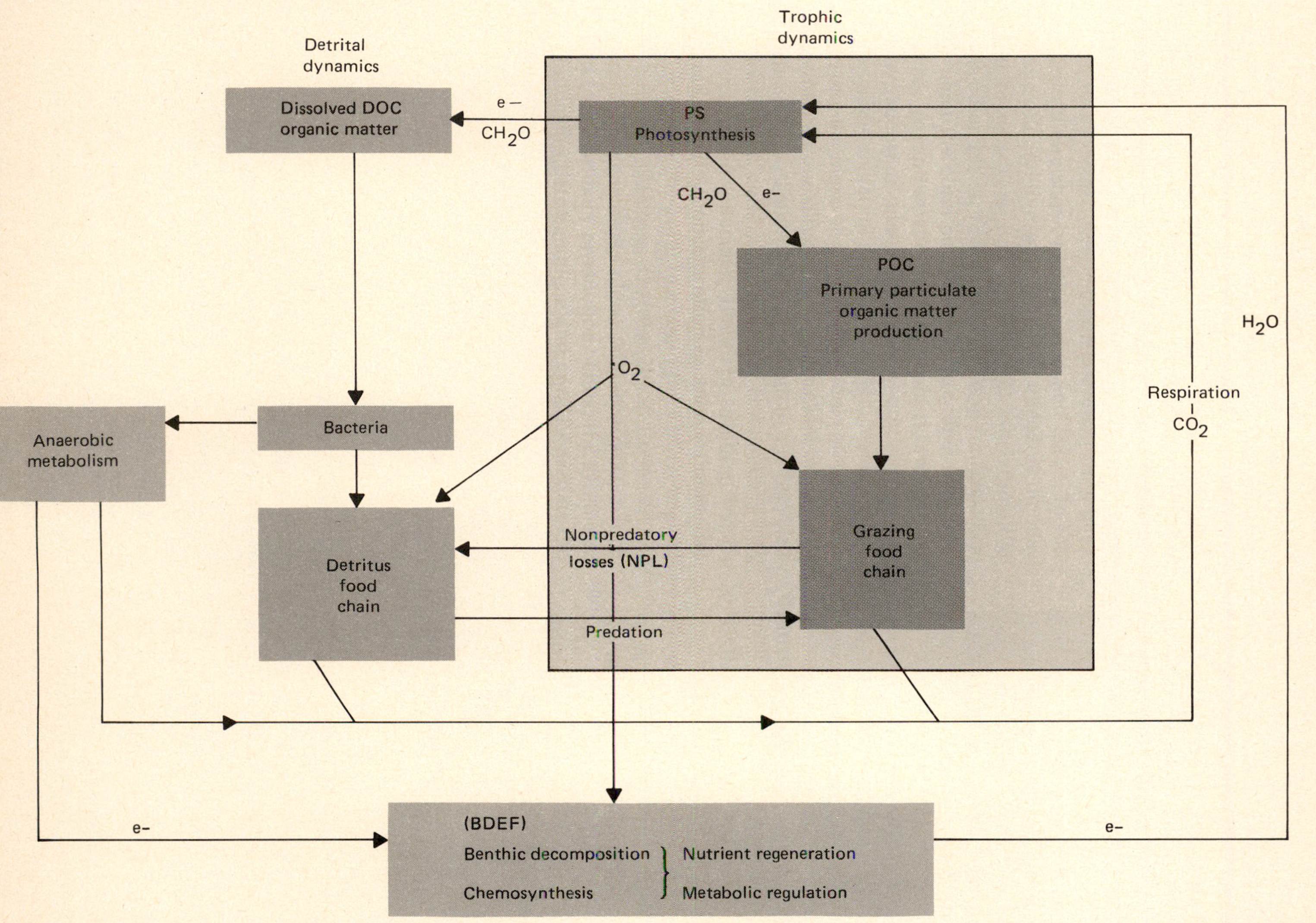

FIGURE 25.10 Energy flow and nutrient cycling in a lake ecosystem. Heterotrophic, autotrophic, and detrital pathways are necessary for operation of the whole lake ecosystem. This model shows the integration of the detrital and trophic structure involving herbivores, bacteria, and anaerobic metabolism. Herbivores and bacteria represent the grazing and detrital food chains, which depend upon each other for prey and nonpredatory losses—the movement of organic matter to decomposers. Anaerobic metabolism is a pathway for energy leaving organic substrates and the biota to interact more directly with abiotic factors. Export and uncoupled oxidation represent energetic losses, while import and net photosynthesis represent energy gains. CO_2 = oxidized carbon; H_2O = reduced oxygen; PS = photosynthesis, which reduces carbon and oxidizes water to create organic carbon, CH_2O, and molecular oxygen—O_2; DOC = dissolved organic carbon; NPL = nonpredatory loss; BDEF = benthic detrital electron flux; e = electron. (*From Rich and Wetzel, 1978:68.*)

tosynthesis and biomass accumulation is high. In fact, a linear relationship exists between phytoplankton production and phytoplankton biomass (Brylinsky, 1980) (Figure 25.11). But as phytoplankton biomass increases, shading increases, respiration per unit surface area increases, and net photosynthesis and thus production declines. When nutrients are low, respiration and mortality increase, reducing net photosynthesis and thus biomass. However, if zooplankton grazing and bacterial decomposition are high, nutrients are recycled rapidly, resulting in a high rate of net photosynthesis even though the concentration of nutrients and biomass accumulation are low.

Macrophytes also contribute heavily to lake production. Their maximum biomass is close to annual cumulative net production. The ratio of microphytic production to macrophytic production is influenced by the fertility of the lake. Highly fertile lakes support a heavy growth of phytoplankton that shades out the macrophytes and reduces their contribution. But in less fertile lakes where phytoplankton production is low, light penetrates the water and rooted aquatics can grow. Macrophytes are little affected by nutrient exchange in the water column. Rooted aquatics draw on nutrients from the sediments rather than from open water.

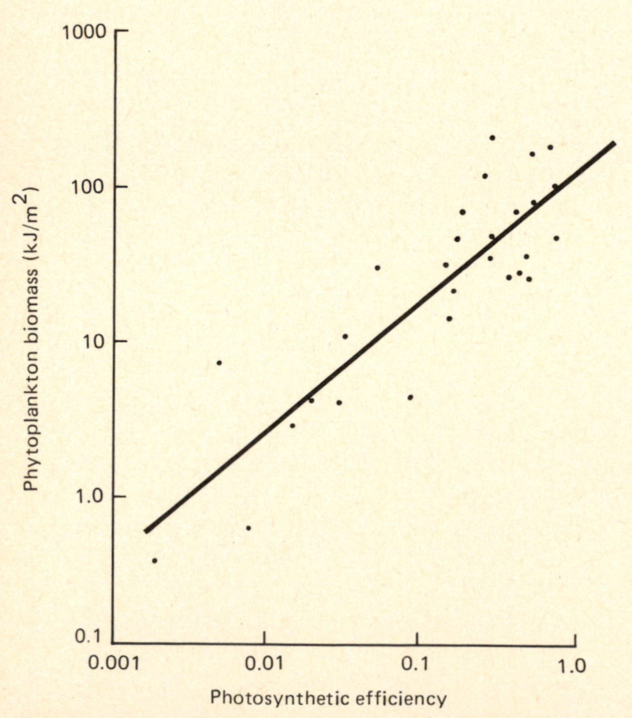

FIGURE 25.11 A strong linear relationship exists between phytoplankton biomass and photosynthetic efficiency. (*From Brylinsky, 1980:433.*)

Nutrient transfers within lentic ecosystems take place largely between the water column and sediments and involve uptake by phytoplankton, zooplankton, and bacteria and other consumers as well as sedimentation in both the water column and the benthic muds (Figure 25.12). In spring, when phytoplankton bloom is at its height, nitrogen and phosphorus become depleted in the trophogenic zone, in part because of the high rate of photosynthesis, the high rate of sinking of dead phytoplankton, and the high rate of sedimentation. The trophogenic zone experiences a decrease in particulate N and P because of decomposition, an increase in dissolved phosphorus, and a decrease in dissolved N because of denitrification.

In summmer, conditions change. Because of a decline in phytoplankton in the trophogenic zone and a slower sinking rate, as much N and P enters solution (see Chapter 21) as is taken up by phytoplankton in phtosynthesis. In the trophogenic zone, N and P increase in the dissolved and particulate pools and in bottom sediments. Phosphorus, in particular, becomes trapped in the hypolimnion and remains there unavailable to phytoplankton until the fall overturn.

NUTRIENT STATUS OF LAKES

A close relationship exists between lentic ecosystems and the watersheds in which they lie (see, for example, Wissmar et al., 1982). The nature of the watershed influences the nutrient status of lakes and ponds.

Lentic ecosystems occupying nutrient-rich watersheds such as those containing agricultural and urban areas and certain types of deciduous forests and grasslands are nutrient enriched by inflowing drainage waters. The enrichment of aquatic ecosystems is termed *eutrophication*. A nutrient-rich body

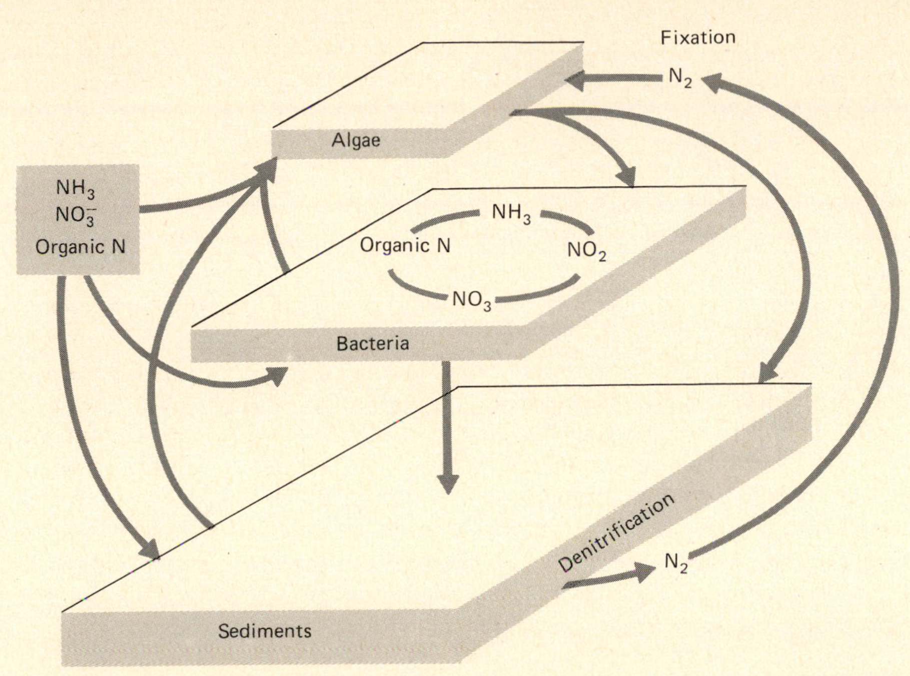

FIGURE 25.12 A model of nitrogen cycling in a lentic ecosystem showing the relationship between the water column and the sediments. Sediments are both a storehouse and a source of nutrients in the lentic system. (*Based on Golterman and Kouwe, 1980:138.*)

of water is called *eutrophic* (from the Greek *eutrophus,* "well nourished"). In general, any aquatic system in which net primary production is greater than 75 g C/m^2/year is considered eutrophic (Rohde, 1965). A typical eutrophic lake (Figure 25.13) has a high surface-volume ratio; that is, the surface area is large relative to depth. It has an abundance of nutrients that produce a heavy growth of algae and other aquatic plants. Its bottom is rich in organic sediments, and its deeper waters have continuously or seasonally low concentrations of oxygen.

Contrasting with eutrophy is *oligotrophy,* meaning "poor nourishment." Oligotrophic lakes are characterized by low surface-volume ratio, water that is clear and appears blue to blue-green in the sunlight, bottom sediments that are largely inorganic, and a high oxygen concentration that extends to the bottom. The nutrient content is low, especially in nitrogen and phosphorus. Nutrients normally added by inflow are quickly taken up by phytoplankton, whose net production is usually less than 25 g C/m^2/year. Although the number of organisms may be low, the diversity of species is high. Fish life is dominated by the salmon family.

A third type of lake is *dystrophic*. A dystrophic lake has waters that are brownish from humic materials. It is acidic; it has a reduced rate of decomposition; and its bottom consists of partially decomposed vegetation or peat. Dystrophication leads to peat bogs rich in humic materials but low in productivity.

When nutrients in moderate amounts are added to oligotrophic lakes, they are taken up rapidly and circulated. As increasing quantities are added, the lake or pond begins to change from oligotrophic to mildly eutrophic (*mesotrophic,* with net primary production between 25 and 75 g C/m^2/year) to eutrophic conditions. This has been happening at an increasing rate to oligotrophic lakes around the world.

In fact, this galloping eutrophication has been changing naturally eutrophic conditions into hyper-

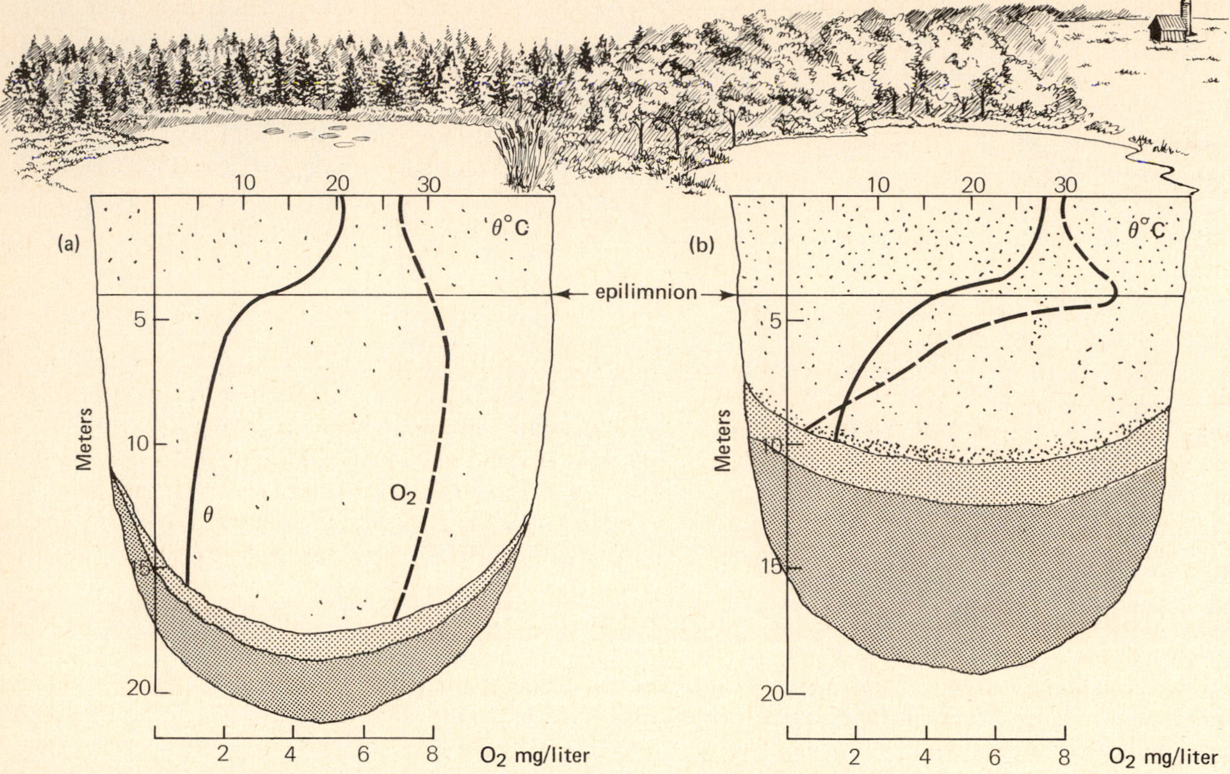

FIGURE 25.13 Comparison of oligotrophic and eutrophic lakes. (a) The oligotrophic lake is deep and has relatively cool water in the epilimnion. The hypolimnion is well supplied with oxygen. Organic matter that drifts to the bottom falls through a relatively large volume of water. The watershed surrounding the lake is largely oligotrophic, dominated by coniferous forests on thin and acid soil. (b) The eutrophic lake is shallow and warm, and oxygen in the deeper water is nearly depleted. The amount of organic detritus is large in relation to the volume of water. The watershed surrounding the lake is eutrophic, consisting of nutrient-rich farmland and deciduous forest.

trophic conditions. An excessive nutrient input results from a heavy influx of wastes, raw sewage, drainage from agricultural lands, river basin development, recreational use of water, industry, runoff from urban areas, and burning of fossil fuels. This accelerated enrichment has been called *cultural eutrophication* (Hasler, 1969). Cultural eutrophication has produced significant biological changes in many lakes. The tremendous increase in nutrients stimulates a dense growth of planktonic algae, dominated by blue-green forms, and rooted aquatics in shallow water. This upsets normal food chains. Herbivores, mostly grazing zooplankton, are unable to consume the bulk of algae as they would do normally. Abnormal quantities of unconsumed algae, as well as rooted aquatics, die and sink to the bottom. On the bottom aerobic decomposers are unable to reduce organic matter to inorganic forms and perish from the depletion of oxygen. They are replaced by anaerobic organisms that only incompletely decompose organic matter. Partially decomposed organic sediments build up on the bottom, and sulfate-reducing bacteria release hydrogen sulfide, which can poison benthic water. These chemical and environmental changes cause major shifts in plant and animal life in the affected aquatic ecosystems.

Freshwater Wetlands

Associated with lentic and lotic ecosystems are wetlands, areas where water is near, at, or above the level of the ground (Figure 25.14). Biologically, they are the richest and most interesting ecosystems. Yet they are also the least appreciated and the first to be destroyed by filling and draining. Wetlands are a halfway world between terrestrial and aquatic ecosystems and exhibit some of the characteristics of each.

TYPES OF WETLANDS

A wide variety of wetlands exists, influenced by topography and hydrological regimes (Figure 25.15). Wetlands dominated by emergent vegetation are *marshes*. Growing to reeds, sedges, grasses, and cattails, marshes are essentially wet prairies. They develop along margins of lakes (lentic), in shallow basins with an inflow and outflow of water (lotic), along slow-moving rivers (riverine), and tidal flats (tidal marshes). Wetlands in which considerable amounts of water are retained by an accumulation of partially decayed organic matter are *peatlands* or *mires*. Mires fed by water moving through mineral soil and dominated by sedges are known as *fens*. Because most of their nutrients come from mineral soil, they are called minerotrophic or *rheotrophic* mires. Mires dominated by *Sphagnum* mosses and depending largely on precipitation for their water supply are *bogs*. Mires that develop on upland situations where decomposed, compressed peat forms a barrier to the downward movement of water, resulting in a perched water table above mineral soil, are *blanket mires* or *raised bogs* (Figure 25.15). Blanket bogs are more popularly known as *moors*. Because bogs depend upon precipitation for nutrient inputs, they are highly deficient in mineral salts and low in pH. Such bogs are called *ombrotrophic*. Bogs may also develop by the filling of a lake basin from above rather than from below. Because mire vegetation often grows on a floating mat of peat over water, such bogs are often termed quaking or—perhaps more descriptively—*schwingmoor,* in German. Wooded wetlands are *swamps*. They may be dominated by trees such as cypress, tupelo, and swamp oaks or by shrubs such as alder and willow. Shrubby swamps are often called *carrs*.

STRUCTURE AND FUNCTION

Only in recent years have ecologists given serious attention to the structure and function of freshwater wetlands (Good, Whigham, and Simpson, 1978; Greeson, Clark, and Clark, 1979). Much of this work has concerned glacial prairie wetlands and freshwater tidal wetlands. Productivity is strongly influenced by hydrological regimes: groundwater, surface runoff, precipitation, drought cycles, and the like. It is also affected by the nature of the watershed in which the wetland lies, soils, nutrient availability, types of vegetation, and the life history of the plant species.

Above-ground biomass may vary with the proportionate abundance of annual and perennial species, which change in dominance through the growing season. Annual emergents exhibit a linear increase in biomass through the growing season, reaching a maximum in late summer (Figure 25.16). Perennial biomass increases during the first part of the growing season, then declines or levels off as leaves become senescent (Figure 25.16) (Whigham et al., 1978). In general, however, the average maximum standing crop of biomass matches the annual above-ground productivity. For example, a sedge wetland had a maximum standing crop biomass greater than 1,000 g/m^2 and a net primary productivity greater than 1,000 g/m^2/year. But if the life history of the plants—including mortality and regeneration, and winter activity—was considered, net productivity was 1,600 g/m^2/year (Bernard and Gorham, 1978).

Below-ground production, much more difficult to estimate, appears for some species to be highest in summer, at the same time peak above-ground biomass is achieved. For others, such as *Typha* and *Scirpus*, peak production is reached in fall, when

(a)

(b)

FIGURE 25.14 Two examples of wetlands include (a) a glacial prairie marsh (pothole) and (b) a cypress swamp of the southern United States.

nutrients are stored in the roots. Such species may have minimal root biomass in midsummer because of nutrient transfer to growing above-ground biomass.

Initial decomposition in wetlands is high as leaching removes soluble compounds that enter DOM (Figure 25.17). After initial leaching, decomposition proceeds more slowly. Permanently submerged leaves decompose more rapidly than those on the marsh surface because they are more accessible to detritivores (Smock and Harlowe, 1983) and because the stable physical environment is more favorable for microbes.

Because bog vegetation is not in contact with mineral soil and because inflow of groundwater is blocked, nutrient input is largely by way of precipitation. Availability of nutrients—especially nitrogen, phosphorus, and potassium—is low, and most of the nutrients fixed in plant tissue are removed from circulation in the accumulation of peat. But evidence suggests that some bog plants possess mechanisms to conserve nutrients.

In general, nitrogen is available from three sources: (1) precipitation, the major source; (2) nitrogen fixation by blue-green algae living in close

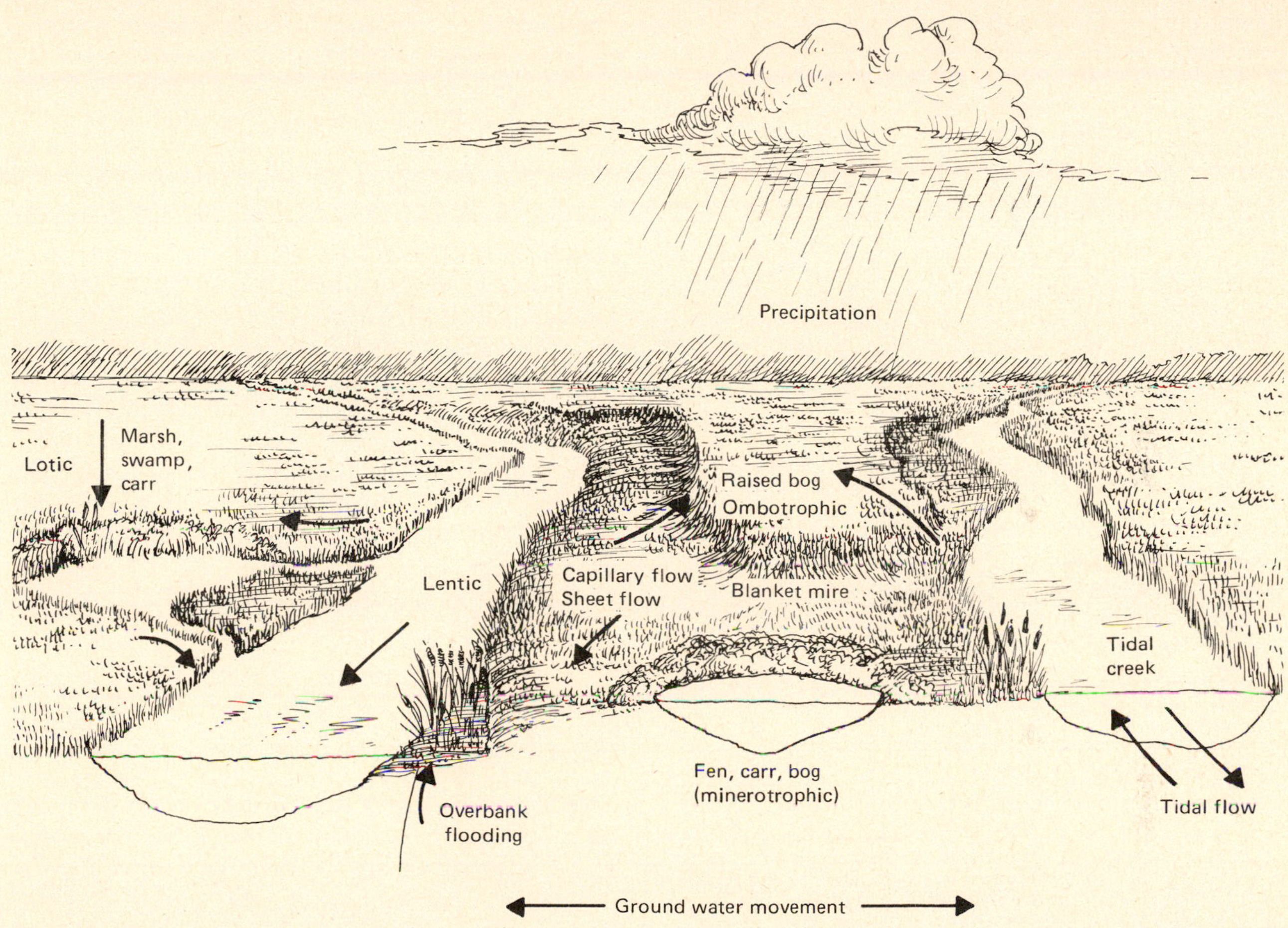

FIGURE 25.15 A schematic diagram of various types of freshwater wetlands in relation to their hydrological regimes.

association with bog mosses and by bog myrtle—if the pH of the substrate is at least 3.5; (3) carnivorous habits by which certain plants such as sundews and pitcher plants extract nitrogen from captured and digested insects. Blue-green algae may fix 0.2 to 0.9 g N/m^2/year. Rosswall et al. (1975) have calculated a nitrogen budget and nitrogen cycle for an ombrotrophic bog dominated by a trailing ground plant, *Rubus chamaemorus* (Figure 25.18). Yearly demand is about 1 to 2 g.

The phosphorus cycle is more closed than the nitrogen cycle. *Rubus chamaemorus* increases its uptake of phosphorus prior to budbreak. After budbreak the plant increases phosphorus in stem, leaf, and root. This increase correlates with a rise in peat temperature. After the plant completes shoot growth in summer, its phosphorus levels decline in the shoots as it mobilizes phosphorus in developing fruits or in roots and rhizomes. In advanced senescence the plant rapidly loses phosphorus from the shoots and accumulates it in winter buds and roots.

Energy flow in mire systems differs considerably from that in other systems because the decomposer food chain is impaired. In most ecosystems, material that enters the detrital food chain is eventually recycled, and the energy that enters the system is liberated or stored in living material. But in mire systems, material resulting from primary production accumulates in an undecomposed state, and energy is locked in peat until environmental con-

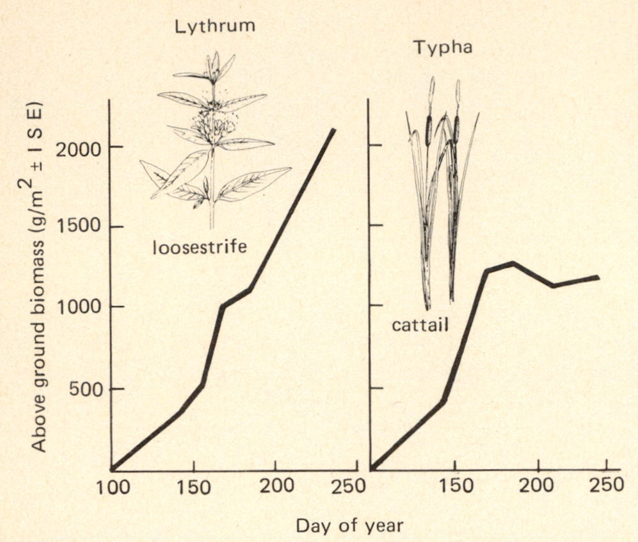

FIGURE 25.16 Pattern of above-ground biomass accumulation through the growing season for a freshwater annual, *Lythrum* (loosestrife), and a perennial, *Typha* (cattail). Note the linear increase in biomass in the annual and the sigmoid growth of the perennial. (*From Whigham et. al., 1978:12.*)

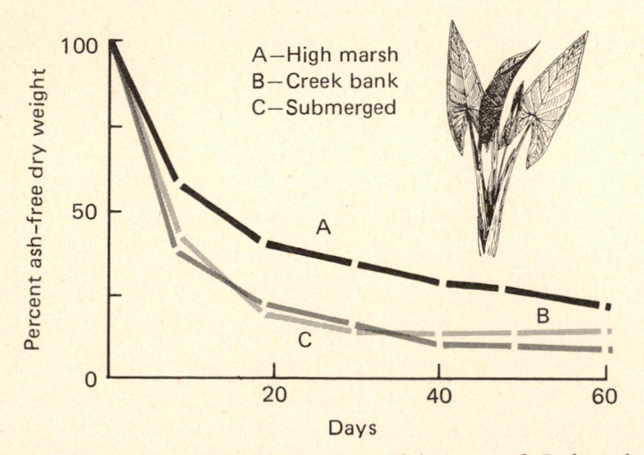

FIGURE 25.17 Decomposition of leaves of *Peltandra virginica,* arrow arum, as measured by the percentage of original ash-free dry weight remaining in litter bags under three conditions: irregularly flooded high marsh exposed to alternate wetting, creek bed flooded two times daily, and permanently submerged. Note that the detrital material consistently wet showed the highest rate of decomposition, although the overall pattern of decomposition is similar. (*From Odum and Haywood, 1978:92.*)

ditions change to favor decomposition or the material is burned.

Because of low temperatures, acidity, and nutrient immobilization, primary production in peatlands is low. For example, average production in an English blanket mire was 635 g/m^2/year, with sphagnum on wet sites contributing 300 g/m^2/year (Heal, Jones, and Whittaker, 1975).

Rates of decomposition vary among bog vegetation. Litter of *Rubus chamaemorus* may lose only 2 percent of its weight per year (Rosswall et al., 1975). Shrub litter decomposes more slowly, while sphagnum decomposes hardly at all. The rate of decomposition in an English mire is slow, with a 95 percent turnover time of 3,000 years for the system. In the top 20 cm, 95 percent turnover time is about 70 years (Heal, Jones, and Whittaker, 1975).

Summary

Two distinctive types of freshwater ecosystems are lotic, or flowing water, and lentic, or still water. Lotic ecosystems are characterized by inputs of detrital material from terrestrial sources and currents of varying velocities that carry nutrients and other materials downstream. Lotic ecosystems exhibit a continuum of physical and ecological variables from source to mouth. There is a longitudinal gradient in temperature, depth and width of the channel, velocity of the current, and nature of the bottom. Changes in physical condition are reflected in biotic structure. Headwater streams in forested regions are shaded and strongly heterotrophic, dependent upon inputs of detritus. Larger streams, open to sunlight, shift from a heterotrophic to an autotrophic condition. Primary production from algae and rooted aquatics becomes an important energy source. Large rivers return to a heterotrophic condition. They are dependent on fine particulate organic matter and dissolved organic matter as sources of nutrients and energy. Downstream systems in effect depend upon the inefficiencies of energy and nutrient processing upstream.

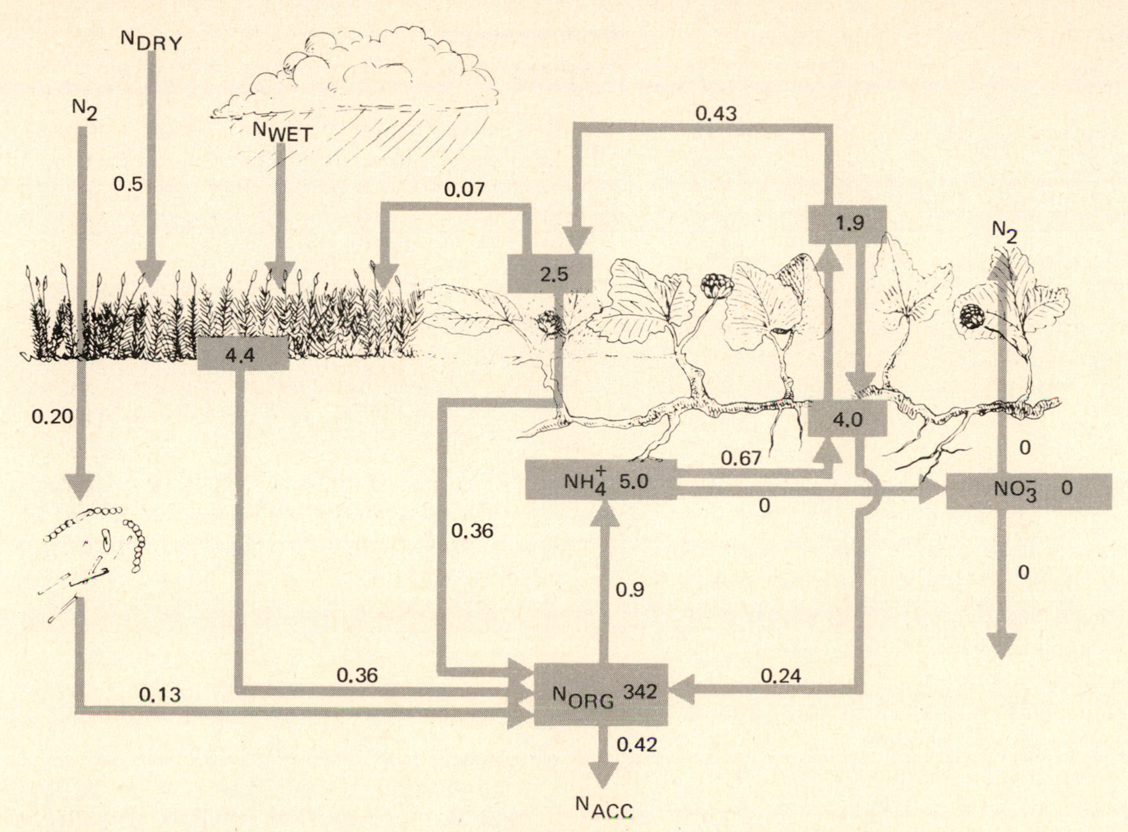

FIGURE 25.18 Nutrient budget for an ombrotrophic mire (Stordalen, Sweden). Quantities of nitrogen are expressed as g N/m^2, flows as g N/m^2/year. The vegetation has been divided into above-ground parts, illustrated by *Rubus chamaemorus* and lichens (1.6 g N/m^2); below-ground parts of vascular plants (3.0 g N/m^2); above-ground litter (3.2 g N/m^2); and byrophytes (4.4 g N/m^2). N_{org} = organic nitrogen; N_{acc} = nitrogen accumulated; N_{part} = particulate nitrogen (dry deposition). (*Revised from Rosswall, 1975:289.*)

Energy comes in three detrital fractions: coarse particulate organic matter (CPOM), fine particulate organic matter (FPOM), and dissolved organic matter (DOM), as well as in autotrophs, algae on stones and rooted aquatics. Processing this organic matter are: fungi and bacteria feeding on CPOM; shredders feeding on CPOM and its bacteria and fungi; collectors feeding on bacteria and FPOM carried by the current; scrapers working on algae; and piercers utilizing plant juices of rooted aquatics. The ratio of these feeding groups changes along the lotic continuum. A major problem in flowing water ecosystems is the retention of nutrients in any segment of the system. Nutrients cycle among particulate matter, the water column, and consumers as they move downstream. This concurrence of nutrient cycling and downstream transport has been called spiraling.

Lentic ecosystems exhibit vertical gradients in light, temperature, and dissolved gases. In summer, water in deeper lakes and ponds becomes stratified, with a layer of warm, circulating water on top, the epilimnion; the thermocline, in which temperature drops rapidly; and the hypolimnion, a bottom layer of denser water approximately 4°C, often low in oxygen but high in nutrients. In fall and spring, when surface waters cool and warm, differences in density of water between layers decrease, and water circulates throughout the lake. That is the fall and spring overturn, important in mixing bottom water with top water and circulating nutrients.

Lakes and ponds also may exhibit horizontal zonation: the littoral zone near shore, characterized by submerged, floating, and emergent vegetation; and the limnetic zone, an open-water area dominated by phytoplankton. The bottom, or benthic, zone is a place of intense biological activity, for there decomposition of organic matter takes place. Anaerobic bacteria are dominant on the bottom beneath deep water, while the benthic zone of the littoral region is rich in decomposer organisms and detritus-feeders. Although lentic ecosystems are often considered as dominated by phytoplankton and grazing food chains, lakes are actually dominated by the detrital food chain. Much of the detrital input comes from the littoral zone.

Lakes may be classified as eutrophic, or nutrient-rich; oligotrophic, or nutrient-poor; or dystrophic, acidic and rich in humic material. Most lakes are subject to cultural eutrophication, which is the rapid addition of nutrients, especially nitrogen and phosphorus, from sewage and industrial wastes. Cultural eutrophication has produced significant biological changes, mostly detrimental, in many lakes.

Closely associated with lakes and streams are wetlands, areas where water is at, near, or above the level of the ground. Wetlands dominated by grasses are marshes; those dominated by woody vegetation are swamps. Wetlands characterized by an accumulation of peat are mires. Those fed by water moving through mineral soil and dominated by sedges are fens; those dominated by sphagnum moss and dependent largely on precipitation for moisture are bogs. Bogs are characterized by blocked drainage conditions, an accumulation of peat, and low productivity. Most of the nutrients fixed in plants are removed from circulation and stored in accumulated peat; the stored energy remains locked in the peat until environmental conditions change to favor decomposition or until the material is burned.

Review and Study Questions

1. Contrast the physical differences between lotic and lentic ecosystems.
2. Compare the cycling of nutrients within a flowing water and a still water ecosystem.
3. Contrast the source and role of detritus in a lotic and lentic ecosystem.
4. What is spiraling, and how does it function in nutrient cycling in streams? What compartments and consumer groups in lotic systems are involved?
5. How do downstream lotic systems relate to upstream systems? What is the continuum concept?
6. Based on information in this chapter, what would you expect to happen to the structure and function of lotic systems when a dam is built across the channel? How does stream channelization—the straightening and deeping of a streambed—affect the structure and functioning of a lotic system?
7. Why would you not expect a well-developed phytoplankton population in a river system?
8. Discuss the vertical and horizontal zonation of a lake or pond and explain how zonation relates to the structure and functioning of a lentic ecosystem.
9. Distinguish between eutrophy and oligotrophy.
10. Compare a rheotrophic with an ombrotrophic wetland in regard to nutrient source and fate of organic production.

Suggested Readings

Brinson, M. M., A. E. Lugo, and S. Brown. 1981. Primary productivity, decomposition and consumer activity in freshwater wetlands. *Annual Review of Ecology and Systematics* 12:123–161.

Brown, A. L. 1977. *Ecology of fresh water.* Cambridge, Mass.: Harvard University Press.

Coker, R. E. 1954. *Streams, lakes, and ponds.* Chapel Hill: University of North Carolina Press.

Cowardin, L. M., V. Carter, and E. C. Golet. 1979. *Classification of wetlands and deep water habitats of the United States.* FWS/OBS–79/31. Washington, D.C.: Fish and Wildlife Service, U.S. Department of Interior.

Dykyjova, D., and J. Kvet, eds. 1978. *Pond littoral ecosystems: Structure and function.* New York: Springer-Verlag.

Forman, R. T. T., ed. 1979. *Pine barren ecosystems and landscape.* New York: Academic Press.

Good, R. E., D. F. Whigham, and R. L. Simpson, eds. 1978. *Freshwater wetlands: Ecological processes and management potential.* New York: Academic Press.

Greeson, P. E., J. R. Clark, and J. E. Clark, eds. 1979. *Wetland functions and values: The state of our understanding.* Minneapolis: American Water Resources Association.

Hutchinson, G. E. 1957–1967. *A treatise on limnology.* Vol. 1, *Geography, physics, and chemistry;* vol. 2, *Introduction to lake biology and limnoplankton.* New York: Wiley.

Hynes, H. B. N. 1970. *The ecology of running water.* Toronto: University of Toronto Press.

Lock, M. A., and D. D. Williams, eds. 1981. *Perspectives in running water ecology.* New York: Plenum.

Macan, T. T. 1970. *Biological studies of English lakes.* New York: Elsevier.

———.1973. *Ponds and lakes.* New York: Crane, Russak.

Moore, P. D., and D. J. Bellamy. 1974. *Peatlands.* New York: Springer-Verlag.

Richardson, C. J., ed. 1981. *Pocosin wetlands: An integrated analysis of coastal plain freshwater bogs in North Carolina.* New York: Scientific and Academic Editions.

Ward, J. V. 1979. *The ecology of regulated streams.* New York: Plenum.

Whitten, B. A., ed. 1975. *River ecology.* Berkeley: University of California Press.

CHAPTER 26

Marine Ecosystems

Outline

Objectives

Upon completion of this chapter, you should be able to:

1. Discuss the nature of the salinity of the oceans.
2. Discuss temperature stratification in the sea.
3. Describe the formation and types of waves and currents.
4. Explain the cause of tides.
5. Describe the characteristics of estuaries, especially as they relate to salinity.
6. Describe a tidal marsh and its relationship to the estuarine complex.
7. Explain how inhabitants of sandy and muddy beaches function in the processing of organic matter.
8. Describe the zonation of rocky shores.
9. Tell about coral reefs.
10. Discuss the role of phytoplankton, zooplankton, and nekton in nutrient cycling in open seas.

Characteristics of the Marine Environment

The marine environment is marked by a number of extreme differences from the freshwater world. It is large, occupying 70 percent of the Earth's surface. The volume of surface area lighted by the sun is small in comparison to the total volume of water involved. This and the dilute solution of nutrients limit production. It is deep—in places nearly 4 miles deep. All of the seas are interconnected by currents, dominated by waves, influenced by tides, and characterized by saline waters.

SALINITY

The salinity of the open sea is fairly constant, averaging about 35 parts per thousand (‰). This probably has not changed greatly since Earth was formed. Although quantities of salts are carried to the sea by rivers, they are removed at about the same rate by means of complex chemical reactions with sediments and particulate matter.

Salinity is due to two elements, sodium and chlorine, which make up some 86 percent of sea salt. These, along with other major elements such as sulfur, magnesium, potassium, and calcium, comprise 99 percent of sea salts. Seawater, however, differs from a simple sodium chloride solution because the equivalent amounts of cations and anions are not balanced against each other. The cations exceed the anions by 2.38 milliequivalents. As a result, seawater is weakly alkaline (pH 8.0 to 8.3) and strongly buffered, a condition that is biologically important.

The amount of dissolved salt in seawater is usually expressed as chlorinity or salinity. Because oceans are usually well mixed, sea salt has a constant composition; that is, the relative proportions of major elements change little. Thus, the determination of the most abundant element, chlorine (see Table 26.1), can be used as an index of the amount of salt present in a given volume of seawater. Chlorine expressed in ‰ is the amount of chlorine in grams in a kilogram of seawater.

TABLE 26.1 COMPOSITION OF SEAWATER OF 35‰ SALINITY,[a] MAJOR ELEMENTS (ORIGINAL)

Elements	g/kg	Millimoles/kg	Milliequivalents/kg
Cations			
Sodium	10.752	467.56	467.56
Potassium	0.395	10.10	10.10
Magnesium	1.295	53.25	106.50
Calcium	0.416	10.38	20.76
Strontium	0.008	0.09	0.18
			605.10
Anions			
Chlorine	19.345	545.59	545.59
Bromine	0.066	0.83	0.83
Fluorine	0.0013	0.07	0.07
Sulphate	2.701	28.12	56.23
Bicarbonate	0.145	2.38	—
Boric acid	0.027	0.44	—
			602.72

[a] Chlorinity can be converted to salinity, the total amount of solid matter in grams per kilogram of seawater. The relationship of salinity to chlorinity is expressed as follows:

$$S(‰) = 1.80655 \times \text{chlorinity}$$

Note: Surplus of cations over strong anions (alkalinity):2.38.

SOURCE: K. Kalle, 1971, in O. Kinne, *Marine Ecology,* vol. 2, part 1:685 (New York: Wiley-Interscience).

The salinity of parts of the ocean is variable because physical processes change the amount of water in the seas. Salinity is affected by evaporation and precipitation, by the movement of water masses, by the mixing of water masses of different salinities, by the formation of insoluble precipitates that sink to the ocean floor, and by the diffusion of one water mass to another. Salinities are most variable near the interface of sea and air. The elements affected by these physical processes are conservative elements not involved in biological processes. The most variable elements in the sea, such as phosphorus and nitrogen, are the nonconservative ones, because their concentration is related to biological activity. Taken up by organisms (see Chapter 21), these elements are usually depleted near the surface and are enriched at lower depths. In parts of the ocean, some of these nutrients are returned to the surface by upwelling.

Salinity imposes certain restrictions on the life that inhabits the oceans (see Kinne, 1970). Fish and marine invertebrates that inhabit marine, estuarine, and tidal environments have to maintain osmotic pressure under conditions of changing salinity. Most marine species are adapted to living in high salinities; the number declines as salinity is reduced.

TEMPERATURE AND PRESSURE

What has already been written about temperature and freshwater relations also applies to the sea (see Figure 26.1). The range of temperature is far less than that on land, although, as one would expect, the range of variation in temperature over the oceans is considerable. Arctic waters at −9°C are much colder than tropical waters at 27°C, and currents are warmer or colder than the waters through which they flow. Seasonal and daily temperature changes are larger in coastal waters than they are in the open sea. The surface of coastal waters is coolest at dawn and warmest at dusk. In general, seawater is never more than 2° to 3° below the freezing point of fresh water nor warmer than 27°C. At any given place the temperature of deep water is almost constant and cold, below the freezing point of fresh water. Seawater has no definite freezing point, although there is a temperature for seawater of any given salinity at which ice crystals form. Thus, pure water freezes out, leaving even more saline water behind. Eventually, it becomes a frozen block of mixed ice and salt crystals. With rising temperatures the process is reversed.

Unlike fresh water, seawater (with a salinity of 24.7‰ or higher) becomes heavier as it cools and does not reach its greatest density at 4°C. Thus, the limitation of 4°C as the temperature of bottom water does not apply to the sea. Nevertheless, the temperature of the sea bottom generally averages around 2°C even in the tropics if the water is deep enough. The temperature of the ocean floor over 1 mile deep is 3°C.

Another aspect of the marine environment is pressure. Pressure in the ocean varies from 1 atmosphere at the surface to 1,000 atmospheres at the greatest depth. Pressure changes are many times greater in the sea than in terrestrial environments, and pressure has a pronounced effect on the distribution of life. Certain organisms are restricted to surface waters, where the pressure is not so great, while others are adapted to life at great depths. Some marine organisms, such as the sperm whale and certain seals, can dive to great depths and return to the surface without difficulty.

ZONATION AND STRATIFICATION

Just as lakes exhibit stratification and zonation, so do the seas. The ocean itself has two main divisions: the *pelagic,* or whole body of water, and the *benthic,* or bottom region (see Figure 26.2). The pelagic is further divided into two provinces: the *neritic,* water that overlies the continental shelf, and the *oceanic.* Because conditions change with depth, the pelagic is divided into three vertical layers or zones: From the surface to about 200 m is the *photic* zone, in which there are sharp gradients of illumination, temperature, and salinity. From 200 to 1,000 m is the *mesopelagic* zone, where very little light penetrates and the temperature gradient is more even and gradual, without much seasonal variation. It contains an oxygen-minimum layer and often the maximum concentrations of nitrate and phosphate. Below the mesopelagic is the *bathypelagic* zone, where darkness is virtually complete except for bioluminescence, temperature is low, and pressure is very great.

The upper layers of ocean water exhibit stratification of temperature. Depths below 300 m are usually thermally stable. In high and low latitudes, temperatures remain fairly constant throughout the year (see Figure 26.1). In middle latitudes, temperatures vary with the season and are associated with climatic changes. In summer the surface waters become warmer and lighter, forming a temporary seasonal thermocline. In subtropical regions the surface waters are constantly heated, developing a

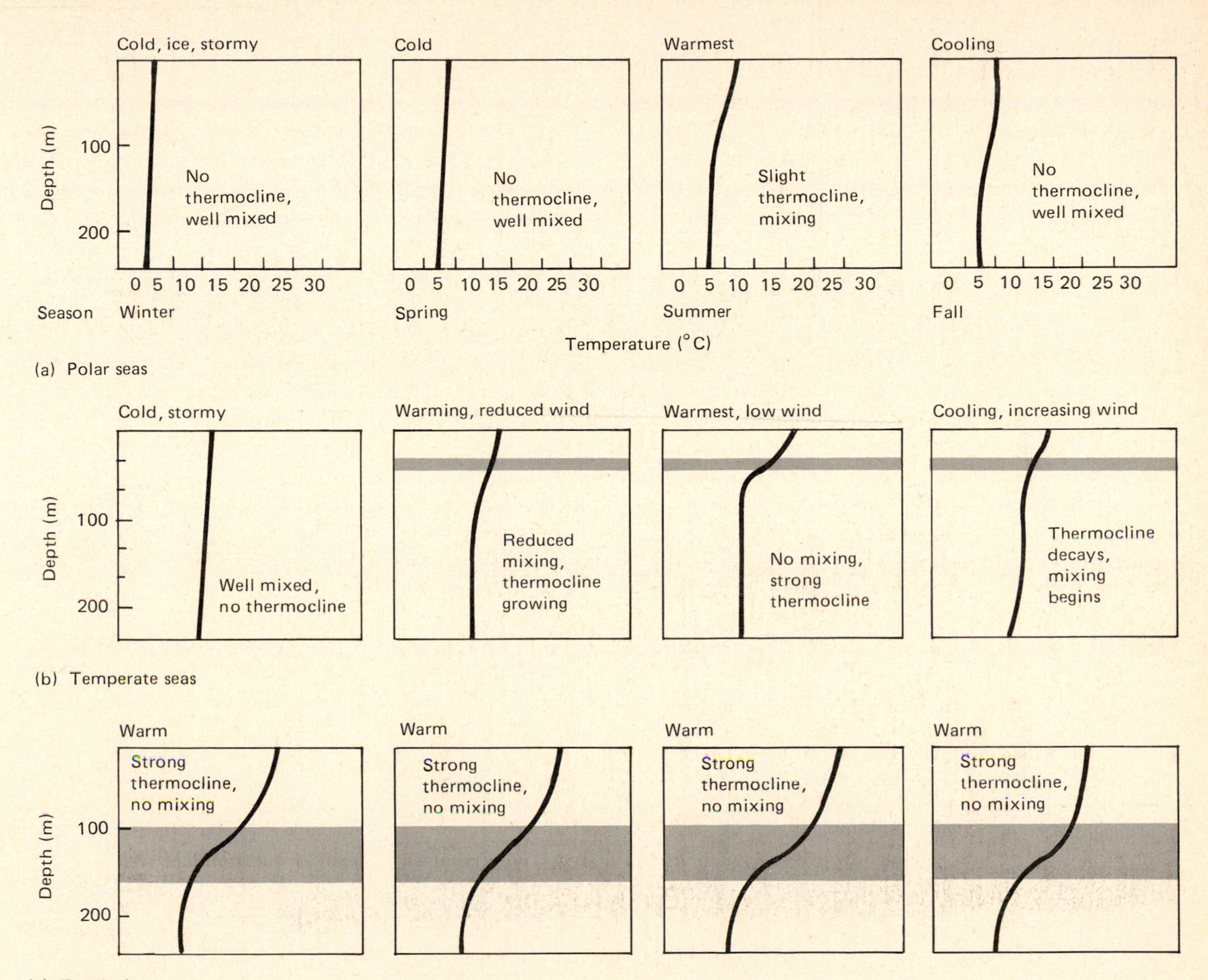

FIGURE 26.1 Thermal structure of polar, temperate, and tropical seas through the year. (a) Polar seas, covered with ice most of the year, exhibit no thermocline in winter, spring, and fall. The waters are well mixed, and nutrients are not limiting. Slight stratification occurs in polar summer (July and August). Then the ice melts, and the water warms enough and light is sufficient to support a bloom of phytoplankton. (b) In temperate seas, thermal structure changes seasonally, reflecting the amount of light and solar energy entering the water. Water in summer is thermally stratified with no mixing. In spring and fall—when the surface water warms and cools, respectively—thermal stratification decreases, and the waters become mixed to various degrees, recharging nutrients in the surface waters. (c) In tropical seas, the upper waters are well lighted, and the continuous input of energy maintains a relatively high temperature throughout the year. Light and temperature are optimum for phytoplankton production, but the waters are permanently stratified. That prevents mixing and the upward circulation of nutrients. The result is low productivity. (*Adapted from Nybakken, 1982:67.*)

marked, permanent thermocline. Between 500 and 1,500 m a permanent but relatively slight thermocline exists. Because of the contact of the deep ocean water with the cold surface waters of the high latitudes (see Figure 26.1), the deep ocean is cold and contains appreciable amounts of oxygen.

WAVES AND CURRENTS

Waves are generated by wind on the open sea. The frictional drag of the wind on the surface of smooth water ripples the water. As the wind continues to blow, it applies more pressure to the steep side of the ripple, and wave size begins to grow. As the wind becomes stronger, short, choppy waves of all sizes appear; and as they absorb more energy, they continue to grow. When the waves reach a point at which the energy supplied by the wind is equal to the energy lost by the breaking waves, they become the familiar whitecaps. Up to a certain point, the stronger the wind, the higher the waves.

The waves that break up on a beach are not composed of water driven in from distant seas. Each particle of water remains largely in the same place and follows an elliptical orbit with the passage of the wave form. As the wave moves forward with a velocity that corresponds to its length (measured from the crest to the following wave), the energy of the group of waves moves with a velocity only one-half that of individual waves. The wave at the front loses energy to the waves behind and disappears, its place taken by another. Thus, the swells that break on a beach are distant descendants of waves generated far out at sea.

As the waves approach land, they advance into increasingly shallow water. The height of each wave rises higher and higher until the wave front grows

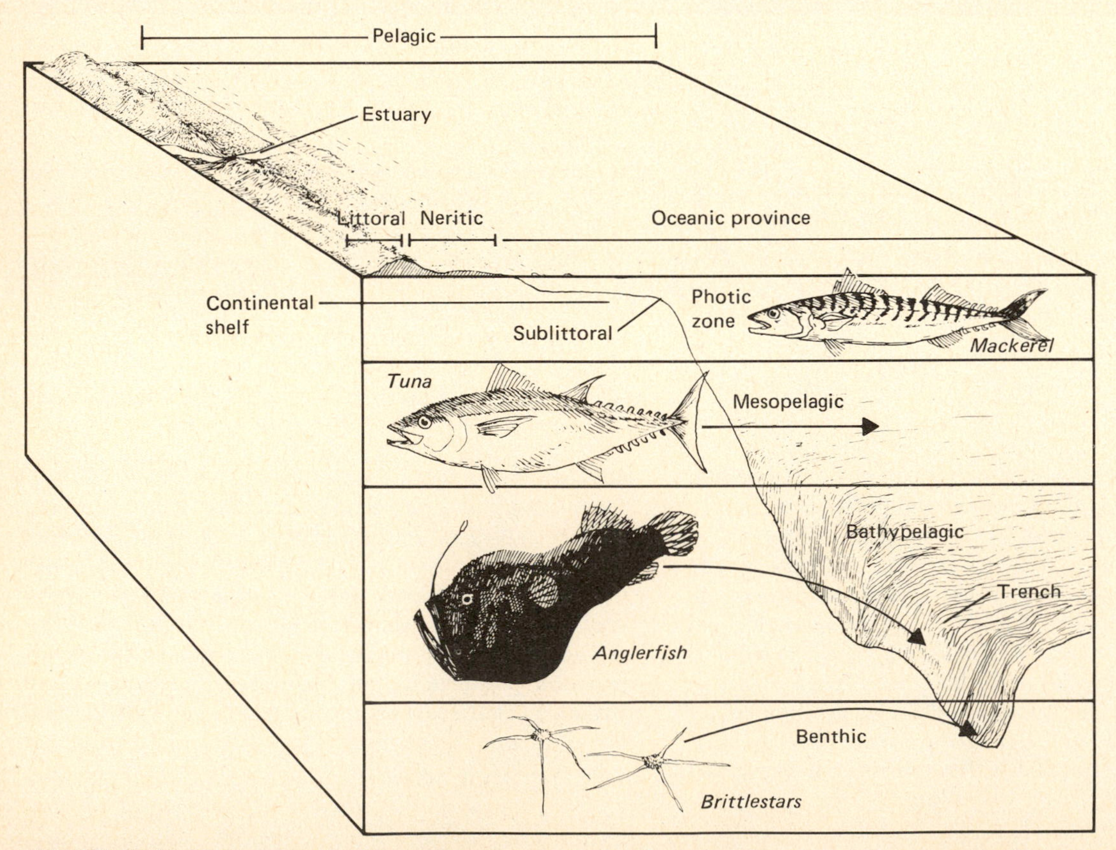

FIGURE 26.2 Major regions of the ocean.

too steep and topples over. As the waves break on shore, they dissipate their energy against the coast, pounding rocky shores or tearing away at sandy beaches at one point and building up new beaches elsewhere.

Surface waves are the most obvious ones. But in the ocean there are also internal waves. Similar to surface waves, internal waves appear at the interface of layers of waters of different densities. In addition, there are stationary waves or seiches.

Just as there are internal waves, so there are internal currents in the sea. Surface currents are produced by wind, heat budgets, salinity, and the rotation of Earth (see Chapter 4). Water moving in surface currents must be replaced by a corresponding inflow from elsewhere. Because the surface waters are cooled and salinity changes, high-density water formed on the surface, largely at high latitudes, sinks and flows toward the low latitudes. These currents are subject to the Coriolis effect—a deflection in the movement of water caused by the rotation of Earth—and are deflected or obstructed by submarine ridges and modified by the presence of other water masses. The result is three main systems of subsurface water movements: the bottom, the deep, and the intermediate ocean currents, each of which runs counter to the others.

Also influencing subsurface currents is the Eckman spiral. As the wind sets the surface layer of water in motion, that layer sets in motion a layer of water beneath, and that layer, in turn, another. Each layer moves more slowly than the layer above and is deflected to the right of it because of the Coriolis effect. At the base of the spiral, the movement of water is counter to the flow on the surface, although the average flow of the spiral is at right angles to the wind (see Strahler, 1971).

In coastal regions the Eckman transport of surface water can bring deep waters up to the surface, a process called *upwelling*. Wind blowing parallel to a coast causes surface water to be blown offshore. This is replaced by water moving upward from the deep. Although cold and containing less dissolved oxygen, upwelling water is rich in nutrients that support an abundant growth of phytoplankton. For this reason, regions of upwelling are highly productive.

TIDES

One of the fundamental laws of physics is Newton's law of universal gravitation. The law states that every particle of matter in the universe attracts every other particle with a force that varies directly as the product of their masses and inversely as the square of the distance between them. This is the reason why apples fall to Earth; it is also, in part, why the planets revolve about the sun. Earth attracts the apple, and the apple, in turn, attracts Earth; but since the mass of Earth is so much greater, the apple falls to the ground. Likewise, the sun and its planets exert an attraction on each other; but the sun, being the largest, exerts the most powerful force. There is also an attraction between Earth and the moon. Because the moon is much closer to Earth than the sun, it exerts a force twice as great as that of the sun.

The gravitational pull of the sun and the moon each cause two bulges in the waters of the oceans. The two caused by the moon occur at the same time on opposite sides of Earth on an imaginary line extending from the moon through the center of Earth. The tidal bulge on the moon side is due to gravitational attraction; the bulge on the opposite side occurs because the gravitational force there is less than at the center of Earth. As Earth rotates eastward on its axis, the tides advance westward. Thus, any given place on Earth will in the course of one daily rotation pass through two of the lunar tidal bulges, or high tides, and two of the lows, or low tides, at right angles to the high tides. Since the moon revolves in a 29½-day orbit around Earth, the average period between successive high tides is approximately 12 hours and 25 minutes.

The sun also causes two tides on opposite sides of Earth, and these tides have a relation to the sun like that of the lunar tides to the moon. Because they are less than half as high, solar tides are partially masked by lunar tides except for two times during the month—when the moon is full and when it is new. At these times Earth, moon, and sun are nearly in line, and the gravitational pulls of the sun and the moon are additive. This causes the high tides of those periods to be exceptionally large, with maximum rise and fall. These are the fortnightly *spring tides*, a name derived from the Saxon *sprun-*

gen, which refers to the brimming fullness and active movement of the water. When the moon is at either quarter, its pull is at right angles to the pull of the sun, and the two forces interfere with each other. At this time the differences between high and low tide are exceptionally small. These are the *neap tides,* from an old Scandinavian word meaning "barely enough."

Tides are not entirely regular, nor are they the same all over Earth. They vary from day to day in the same place, following the waxing and waning of the moon. They may act differently in several localities within the same general area. In the Atlantic, semidaily tides are the rule. In the Gulf of Mexico, the alternate highs and lows more or less efface each other, and flood and ebb follow one another at about 24-hour intervals to produce one daily tide. Mixed tides are common in the Pacific and Indian oceans. These are combinations of the other two, with different combinations at different places. Local inconsistencies are due to many variables. The elliptical orbit of Earth around the sun and of the moon around Earth influence the gravitational pull, as does the declination of the moon—the angle of the moon in relation to the axis of Earth. Latitude, barometric pressure, offshore and onshore winds, depth of water, contour of shore, and natural periods of oscillation, or internal waves, modify tidal movements.

Estuaries

Waters of all streams and rivers eventually drain into the sea; and the place where this fresh water joins the salt water is called an *estuary.* More precisely, estuaries are semienclosed parts of the coastal ocean where the seawater is diluted and partially mixed with water coming from the land. Estuaries differ in size, shape, and volume of water flow, all influenced by the geology of the region in which they occur. As the river reaches the encroaching sea, the stream-carried sediments are dropped in the quiet water. These accumulate to form deltas in the upper reaches of the mouth and shorten the estuary. When silt and mud accumulations become high enough to be exposed at low tide, tidal flats develop (see Figure 26.3). These flats divide and braid the original channel of the estuary. At the same time, ocean currents and tides erode the coastline and deposit material on the seaward side of the estuary, also shortening the mouth. If more material is deposited than is carried away, then barrier beaches, islands, and brackish lagoons appear.

FIGURE 26.3 A muddy tidal flat covered with blue mussels.

Current and salinity, both very complex and variable, shape life in the estuary, where the environment is neither fresh water nor salt. Estuarine currents result from the interaction of a one-way stream flow, which varies with the season and rainfall, with oscillating ocean tides and with the wind. Because of the complex nature of the currents, generalizations about estuaries are difficult to make (see Lauff, 1967).

SALINITY, TEMPERATURE, AND NUTRIENT TRAPS

Salinity varies vertically and horizontally, often within one tidal cycle. Vertical salinity may be the same from top to bottom, or it may be completely stratified, with a layer of fresh water on top and a layer of dense saline water on the bottom. Salinity is homogeneous when currents, particularly eddy currents, are strong enough to mix the water from top to bottom. The salinity in some estuaries may be homogeneous at low tide, but at flood tide a surface wedge of seawater moves upstream more rapidly than the bottom water. Salinity is then unstable, and density is inverted. The seawater on the surface tends to sink as lighter fresh water rises, and mixing takes place from the surface to the bottom. This phenomenon is known as *tidal overmixing*. Strong winds, too, tend to mix salt water with fresh water in some estuaries (Barlow, 1956); but when the winds are still, the river water flows seaward on a shallow surface over an upstream movement of seawater, only gradually mixing with the salt.

Horizontally, the least saline waters are at the river entrance and the most saline at the mouth of the estuary (see Figure 26.4). The configuration of the horizontal zonation is determined mainly by the deflection caused by the incoming and outgoing currents. In all estuaries of the Northern Hemisphere, outward-flowing fresh water and inward-flowing seawater are deflected to the right because of Earth's rotation. As a result, salinity is higher on the left side.

The salinity of seawater is about 35‰; that of fresh water ranges from 0.065 to 0.30‰. Since the concentration of metallic ions carried by rivers will vary from drainage to drainage, the salinity and chemistry of estuaries differ. The portion of dissolved salts in the estuarine waters remains about the same as that of seawater, but the concentration varies in a gradient from fresh water to sea (see Figure 26.4).

Exceptions to these conditions exist in regions where evaporation from the estuary may exceed the inflow of fresh water from river discharge and rainfall (a negative estuary). This causes the salinity to increase in the upper end of the estuary, and horizontal stratification is reversed.

Temperatures in estuaries fluctuate considerably diurnally and seasonally. Waters are heated by solar radiation and inflowing and tidal currents. High tide on the mud flats may heat or cool the water, depending on the season. The upper layer of estuarine water may be cooler in winter and warmer in summer than the bottom, a condition that, as in a lake, will result in spring and autumn overturns.

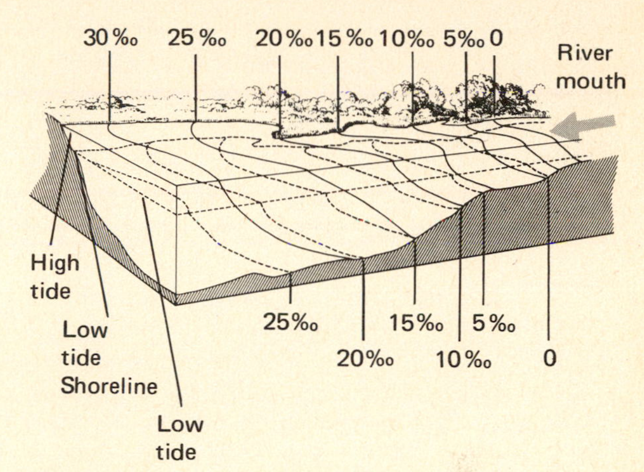

FIGURE 26.4 A generalized diagram of an estuary, showing the vertical and horizontal stratification of salinity from the river mouth to the estuary at both high and low tide. At high tide the incoming seawater increases the salinity toward the river mouth; at low tide salinity is reduced. Note also how salinity increases with depth, because the lighter fresh water flows over the denser salt water.

Mixing waters of different salinities and temperatures acts as a nutrient trap (Figure 26.5). Inflowing river waters more often than not impoverish rather than fertilize the estuary, except for phosphorus. Instead, nutrients and oxygen are carried into the estuary by the tides. If vertical mixing takes place, these nutrients are not swept back out to sea but circulate up and down between organisms, water, and bottom sediments (Figure 26.6).

SALINITY AND ESTUARINE ORGANISMS

Salinity dictates the distribution of life in the estuary. Essentially, the organisms of the estuary are marine, able to withstand full seawater. Except for anadromous fish, no freshwater organisms live there. Some estuarine inhabitants cannot withstand lowered salinities, and these species decline along a salinity gradient. Sessile and slightly motile organisms have an optimum salinity range within

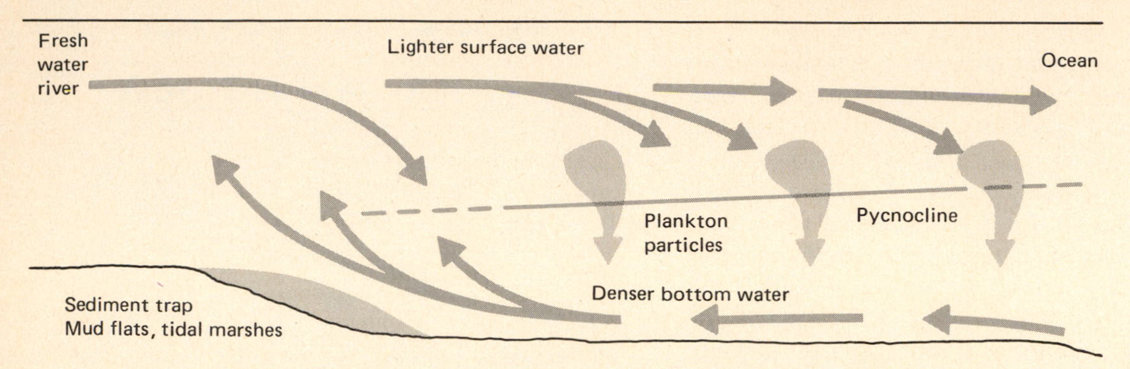

FIGURE 26.5 A diagram showing how circulation of fresh and salt water in an estuary creates a nutrient trap. Note a salt wedge of intruding seawater on the bottom, producing a surface flow of lighter fresh water and a counterflow of heavier brackish water. This countercurrent serves to trap nutrients, recirculating them toward the tidal marsh. The same countercurrent also sends phytoplankton up the estuary, repopulating the water. When nutrients are high in the upper estuary, they are taken up rapidly by tidal marshes and mud flats. These areas tend to trap particulate nitrogen and phosphorus, convert them to soluble forms, and export them back to the open water of the estuary. Plants on the tidal marsh and mud flats act as nutrient pumps between bottom sediments and surface water. (*From Correll, 1978:649.*)

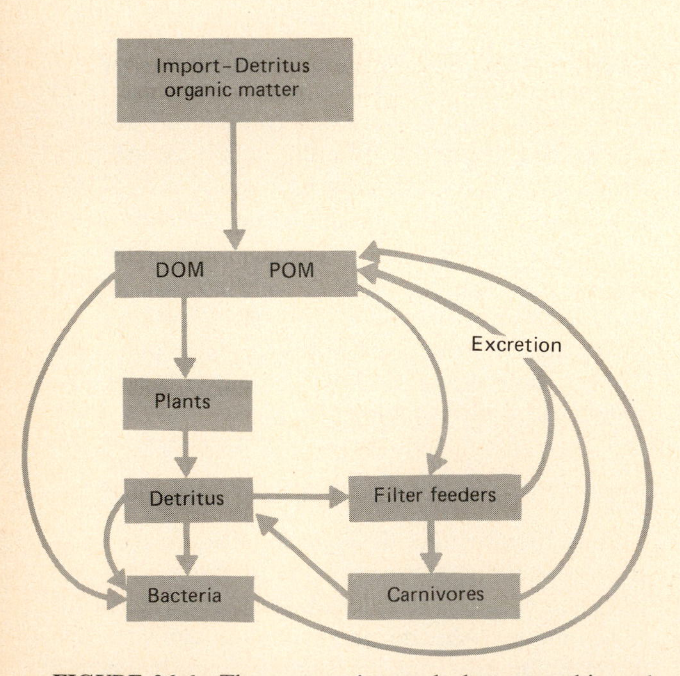

FIGURE 26.6 The estuary is mostly heterotrophic and depends upon organic material from sea and land. This diagram shows how organic matter is cycled in the estuary. (*Developed from Head, 1976.*)

which they grow best. When salinities vary on either side of this range, populations decline. Two animals, the clam worm and the scud, illustrate this situation. Two species of clam worm, *Nereis occidentalis* and *Neanthes succinea,* inhabit the estuaries of the southern coastal plains of North America. *Nereis* is more numerous at high salinities, and *Neanthes* is more abundant at low salinities. In European estuaries the scud, *Gammarus,* is an important member of the bottom fauna. Two species, *Gammarus locusta* and *G. marina,* are typical marine species and cannot penetrate very far into the estuary. They are replaced by two estuarine species—*G. zaddachi,* which lives in the landward end of the estuary, and *G. salinus,* which overlaps the distribution of *G. locusta* at the seaward end of the range.

Size influences the range of motile species within the estuarine waters. This is particularly pronounced among estuarine fish. Some, such as the striped bass, spawn near the interface of fresh and low-salinity water (see Figure 26.7). The larvae and young fish move downstream to more saline waters as they mature. Thus, for the striped bass the es-

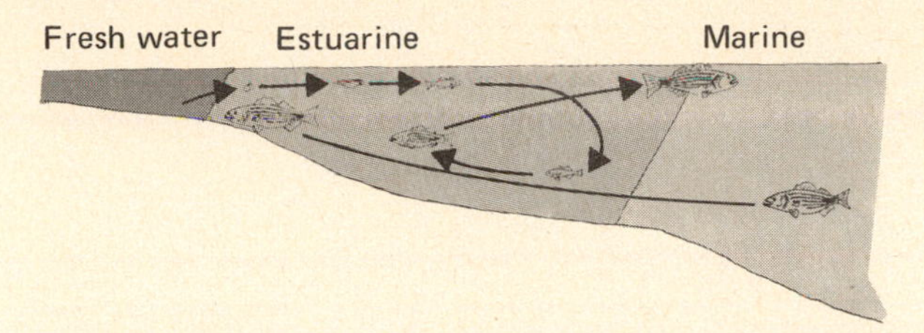

FIGURE 26.7 The relationship of a semianadromous fish, the striped bass, to the estuary. Adult striped bass enter the estuary and move up to fresh and brackish water to lay their eggs. The young move into the estuary, where they develop before moving back to the marine environment. (*From Cronin and Mansueti, 1971:32.*)

tuary serves as both a nursery and a feeding ground for the young. Anadromous species, such as the shad, spawn in fresh water, but the young fish spend their first summer in the estuary, then move out to the open sea. Other species, such as the croaker, spawn at the mouth of the estuary, but the larvae are transported upstream to feed in the plankton-rich low-salinity areas. Others, such as the bluefish, move into the estuary to feed. In general, marine species drop out toward fresh water and are not replaced by freshwater forms. In fact, the mean number of species decreases progressively from the mouth of the estuary to upstream stations (H. G. Wells, 1961).

Tidal Marshes

On the alluvial plains around the estuary and in the shelter of the spits and offshore bars and islands exists a unique community, the tidal marsh (see Figure 26.8). Although to the eye tidal marshes appear as waving acres of grass, they are actually a complex of distinctive and clearly demarcated plant associations (Figure 26.9). The reasons for this complex are, again, tides and salinity. The tides perhaps play the most significant role in plant segregation, for two times a day the salt marsh plants on the outermost tidal flats are submerged in salty water and then exposed to the full insolation of the sun. Their roots extend into poorly drained, poorly aerated soil in which the soil solution contains varying concentrations of salt. Only plant species with a wide range of salt tolerance can survive such conditions. Thus, from the edge of the sea to the highlands, zones of vegetation, each recognizable by its own distinctive color, develop.

FIGURE 26.8 View of a tidal marsh on the Virginia coast showing the pattern of salt marsh vegetation. The mosaic is influenced by both water depth at high tide and salinity. The three dominant grasses are the coarse-leaved salt marsh cord grass, *Spartina alterniflora;* the fine-leaved salt marsh hay grass, *S. patens;* and spike grass, *Distichlis spicata.*

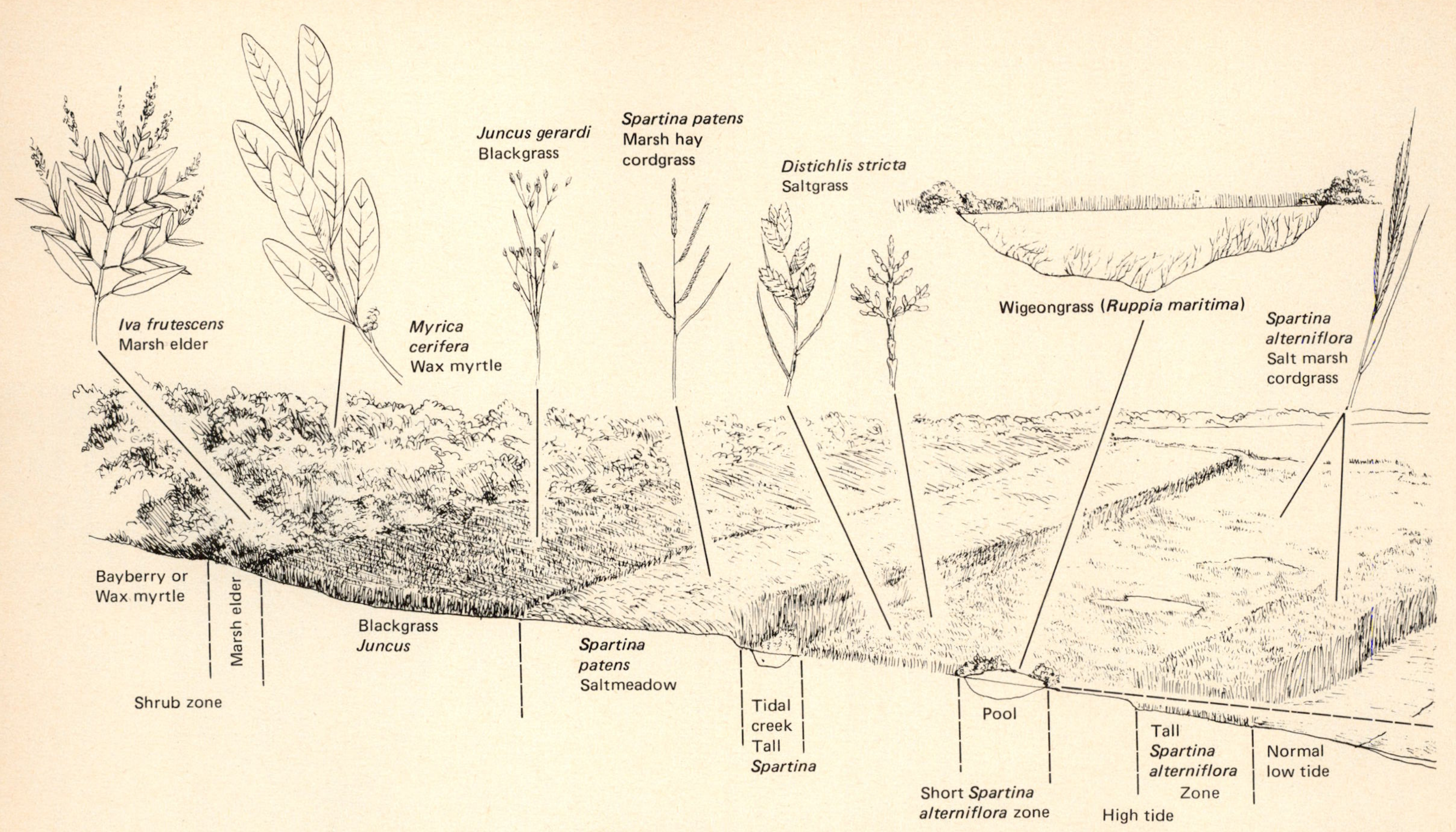

FIGURE 26.9 A stylized transect across a salt marsh showing the relationship of plant distribution to microrelief and tidal submergence.

STRUCTURE

Tidal salt marshes begin in most cases as mud or sand flats, first colonized by algae and, if the water is deep enough, by eelgrass. As organic debris and sediments accumulate, eelgrass is replaced by the first salt marsh colonists—the sea poa, *Puccinellia,* on the European coast and salt-marsh cord grass, *Spartina alterniflora,* on the coast of eastern North America. Stiff, leafy, up to 10 feet tall, and submerged in salt water at every high tide, saltwater cord grass forms a marginal strip between the open mud flat to the front and the higher grassland behind. No litter accumulates beneath the stand. Strong tidal currents sweep the floor of the *Spartina* clean, leaving only thick black mud.

As fine organic debris carried in and deposited by the tides is buried by further deposition on top, an anaerobic environment is created. Here bacteria and nematodes live on organic matter, utilizing it by means of parallel oxidizing and reducing reactions (Teal and Kanwisher, 1961). This results in the accumulation of such end products as methane, hydrogen sulfide, and ferrous compounds. Increasing degrees of reduction suppress biological activity. In fact, if the bacteria of the mud are supplied with oxygen, their rate of energy degradation increases 25 times. Thus, the tidal marsh is a vertically stratified system in which free oxygen is abundant in the surface and absent in the mud. Between these two extremes is a zone of diffusion and mixing of oxygen.

The exposed banks of the tidal creeks that braid through the salt marshes support a dense population of mud algae, the diatoms and dinoflagellates, which are photosynthetically active all year. In summer photosynthesis is highest during high tides; in winter it is highest during low tides, when the sun warms the sediments (Pomeroy, 1959). Some of the algae are washed out at ebb tide and become a temporary part of the estuarine plankton, available to such filter-feeders as oysters.

FUNCTION

Two studies of salt marsh ecosystems, one done in Georgia by Teal (1962) and the other done in New England by Nixon and Oviatt (1973), provide contrasting insights regarding the functioning of salt marshes.

Teal considered only the salt marsh, of which 42 percent was in short *Spartina alterniflora* and 58 percent was in tall *Spartina*. Average annual net primary production of tall *Spartina* was 8,470 kcal/m^2/year and of short *Spartina* 2,570 kcal/m^2/year. Average net production for the total salt marsh was 6,580 kcal/m^2/year, which amounted to 19 percent of gross production. In addition to *Spartina,* algae contributed 1,620 kcal/m^2/year to net production.

Feeding on *Spartina* were two groups of organisms: grazing herbivores, represented by the salt marsh grasshopper *Orchelimun,* which eats the plant tissues, and the plant hopper *Prokelisis,* which sucks plant juices. Neither harvested significant portions of the standing crop. Grasshoppers ingested about 3 percent of net production and assimilated 1 percent. Total annual energy flow through the grasshoppers was 28 kcal, of which production was 10.8 kcal. Plant hoppers accounted for an energy flow of 275 kcal/m^2/year, of which 70 kcal was production. Thus, most of *Spartina* production went to the detrital food chain.

Before *Spartina* is available to most detrital-feeders, it must be broken down by bacterial action. Part of the *Spartina,* particularly that in the short growth, decomposes in place in the marsh. Most of the detritus from the tall *Spartina* is washed out to sea by tidal currents. Bacterial respiration accounted for 3,890 kcal/m^2/year, or 59 percent of the available *Spartina*.

Feeding on decomposing detrital material and the bacteria it supports are fiddler crabs. Where population densities are high, fiddler crabs can sweep over the surface of a marsh between successive tides. They accounted for an assimilation of 206 kcal/m^2/year. Mussels accounted for an assimilation of 56 kcal. Feeding on the herbivores were the secondary consumers, which went largely unstudied. The mud crab *Eurytium* assimilated 27.2 kcal/m^2/year, of which 5.3 kcal was production.

Organisms of the salt marsh, like those of other ecosystems, function not only to channel energy flow but also to circulate nutrients. Many organisms may play an important role in nutrient cycling. One of the detritus feeders, the ribbed mussel, plays a major role in the turnover of phosphorus through the salt marsh ecosystem (Kuenzler, 1961). To obtain its food, which consists of small organisms as well as particles rich in phosphorus suspended in the tidal waters, the mussels must filter great quantities of seawater. Some of the particles they ingest, but most are rejected and deposited as sediment on the intertidal mud. These particles, rich in phosphorus, are retained in the salt marsh instead of being carried out to sea. Each day mussels remove from the water one-third of the phosphorus found in suspended matter. Or, to state it more precisely, every 2.6 days the mussels remove as much phosphorus as is found, on the average, in particles in the water. The particulate matter deposited on the mud is utilized by deposit-feeders, which release the phosphate back to the ecosystem. Thus, the ribbed mussel although of little economic importance to humans and relatively unimportant as an energy consumer in the salt marsh, plays a major role in the cycling and retention of phosphates in this ecosystem. The worth of an animal cannot always be measured in terms of economic values or ignored because it contributes little to energy flow.

Nixon and Oviatt concentrated on the whole system, the marsh (16,800 m^2) and its embayment, Bissel Cove (6,680 m^2). Tall *Spartina* made up about 7 percent of the marsh, short *Spartina* 78 percent, and high marsh 15 percent. The New England marsh was considerably less productive than the Georgia marsh. Production for the tall *Spartina* was 840 kcal/m^2/year and 432 g dry weight/m^2 for the short *Spartina*. Efficiency of production for the tall *Spartina* was 0.51 percent, about half that of the tall *Spartina* of the Georgia marsh.

Growth in the New England marsh ceased in the late fall—the Georgia marsh had some production in the winter—and dead grass remained until it was broken up by ice in late winter. Thus, in New England marshes ice performed the task that is carried out in part by fiddler crabs in the Georgia marsh. Fiddler crabs are rare in New England salt marshes, as are other tidal marsh inhabitants, the ribbed mussel and the marsh snail; nor is there a significant population of grasshoppers. Thus, the entire production of the marsh went to the detrital food chain.

The associated embayment supported dense but patchy populations of wigeongrass (*Ruppia*) and sea lettuce (*Ulva*), which went through several periods of rapid growth, death, decay, and export. Maximum biomass of wigeongrass was twice that of tall *Spartina alterniflora*. Major consumers in the embayment were grass shrimp (*Palaemonetes pugio*) and fish—some 20 species, including mummichogs (*Fundulus*), silversides (*Menidia*), bluefish (*Pomatomus saltatrix*), winter flounder, and stickleback. The fish were largely predatory; the grass shrimp were detritus-feeding herbivores. Feeding on the fish were terns, gulls, black ducks, and mallards. The ducks fed extensively on *Ruppia* and *Ulva*. The major fish predator was the herring gull, which consumed less than 0.5 percent of the standing crop of fish.

The New England salt marsh is characterized by sharp seasonal changes in light, temperature, and salinity, all of which lower production. Most of the production of the salt marsh goes to the embayment as detritus to enter a large sedimentary organic storage compartment as an energy source. Although the embayment is capable of considerable primary production over a short period, the marsh embayment is a semiheterotrophic system that depends on the input of organic matter produced by the marsh grasses and algae. The annual energy budget for the embayment shows that consumption exceeds production.

Just as the ribbed mussel plays an important role in the Georgia salt marsh, so the grass shrimp plays an important role in the salt marsh embayment system at Bissel Cove. The grass shrimp feeds on detrital material washed in from the salt marsh. It plucks away at the surface of dead leaves and stems, breaking them into smaller pieces that are colonized by bacteria and diatoms. Inefficient assimilation repackages the detrital material into fecal pellets and excretes large quantities of ammonia and phosphates into the water. The grass shrimp also prevents the buildup of detritus from *Ruppia* and *Ulva* by feeding on it and reducing it to fine sediment. By its action on the detrital material, the grass shrimp makes nutrients and biomass available for other trophic levels. Although the grass shrimp itself is eaten by the mummichogs, the shrimp's ability to live in a low-oxygen environment limits predation and competition and allows its population to reach the high levels necessary to function as the major detritivore.

The salt marsh and its associated estuary together make up one of the most productive of all ecosystems, including intensive agriculture. The reasons for this high productivity are several. Tides carry out wastes and are continually bringing in nutrients. The meeting of fresh and salt water traps and concentrates nutrients in the marsh, increasing natural fertility. Nitrogen-fixing blue-green algae in the marsh mud increase the supply of available nitrogen. And few nutrients are tied up in biomass for very long. Detrital material from the marsh plants is rapidly broken up and decomposed, and algal and bacterial populations turn over rapidly. This production of the marsh is continually being fed into the coastal waters and estuaries, making them the most productive waters in the world.

Mangrove Swamps

Mangrove swamps replace tidal marshes in tropical regions, and in North America they reach their best development along the southwest coast of Florida and southward (see Figure 26.10). As with the salt marsh, the vegetation is influenced by salinity and the tides. The pioneering red mangrove colonizes the submerged soft-bottomed shoals protected from the full beat of the surf. It has a peculiar system of branching prop roots that extend downward like stilts from the trunks and lower branches.

At the upper limits of high tide, the black or honey mangrove replaces the red mangrove. This mangrove does best in a sandy or less organic soil and tolerates a shorter period of flooding. It has pencil-like pneumatophores rising through the substrate from shallow horizontal roots. Behind the black mangrove and on firm ground at the edge of the tide line may be a zone of white mangrove. The

FIGURE 26.10 Mangrove swamps replace tidal marshes in tropical regions.

white mangrove also possesses pneumatophores, but they are fewer in number and smaller. Above the normal high tides grows a relative of the white mangrove, the button mangrove, with loose bark and a twisted, often prostrate trunk.

Intertidal Ecosystems

Where the edge of the land meets the edge of the sea we find the fascinating and complex world of the seashore. Rocky, sandy, muddy, protected, exposed to the pounding of incoming swells, all shores have one feature in common: they are alternately exposed and submerged by the tides. Roughly, the region of the seashore is bounded on one side by the height of the extreme high tides and on the other by the height of the extreme low tides. Within these confines conditions change from hour to hour with the ebb and flow of the tides. At flood tide the seashore is a water world; at ebb tide it belongs to the terrestrial environment, with its extremes in temperature, moisture, and solar radiation. In spite of all this, the seashore inhabitants are essentially marine, adapted to withstand some degree of exposure to the air for varying periods of time.

THE ROCKY SHORE

Zonation

As the sea recedes at ebb tide, rocks, glistening and dripping with water, begin to appear. Life hidden by tidal water emerges into the open air layer by layer. The uppermost layers of life are exposed to air, wide temperature fluctuations, intense solar radiation, and desiccation for a considerable period, while the lowest fringes on the intertidal shore may be exposed only briefly before the flood tide submerges them again. These varying conditions result in one of the most striking features of the rocky shore, the zonation of life (see Figure 26.11). Although this zonation may be strikingly different from place to place as a result of local variations in aspect, nature of the substrate, wave action, light

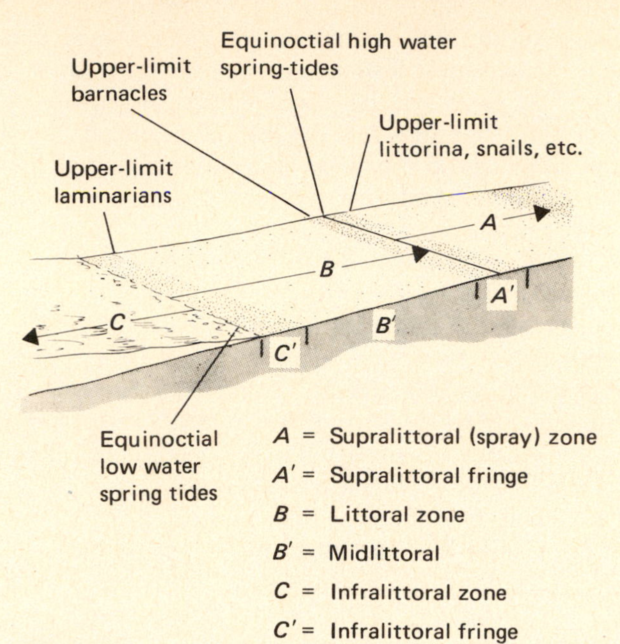

FIGURE 26.12 A diagram of the basic, or "universal," zonation on rocky shores. Use this as a guide when studying the subsequent drawings of shore zonation. (*Adapted from Stephenson, 1949.*)

FIGURE 26.11 Broad zone of life exposed at low tide on a rocky shore. Note the heavy growth of bladder wrack or rockweed on the lower portion and the distinctive white zone of barnacles above.

intensity, shore profile, exposure to prevailing winds, climatic differences, and the like, the same general features are always present. All rocky shores have three basic zones, characterized by the dominant organisms occupying them (see Figure 26.12).

The point where the land ends and the seashore begins is hard to determine. The approach to a rocky shore from the landward side is marked by a gradual transition from lichens and other land plants, the maritime zone, to marine life, dependent at least partly on the tidal waters. The first major change from land shows up on the supralittoral fringe (see Figure 26.13), where the salt water comes only every fortnight on the spring tides. It is marked by the black zone, a patchlike or beltlike encrustation of lichens of the *Verrucaria* type and *Myxophyceae* algae such as *Calothrix* and *Entrophsalis*. Capable of existing under conditions so difficult that few other plants could survive, these blue-green algae, enclosed in slimy, gelatinous sheaths, and their associated lichens represent an

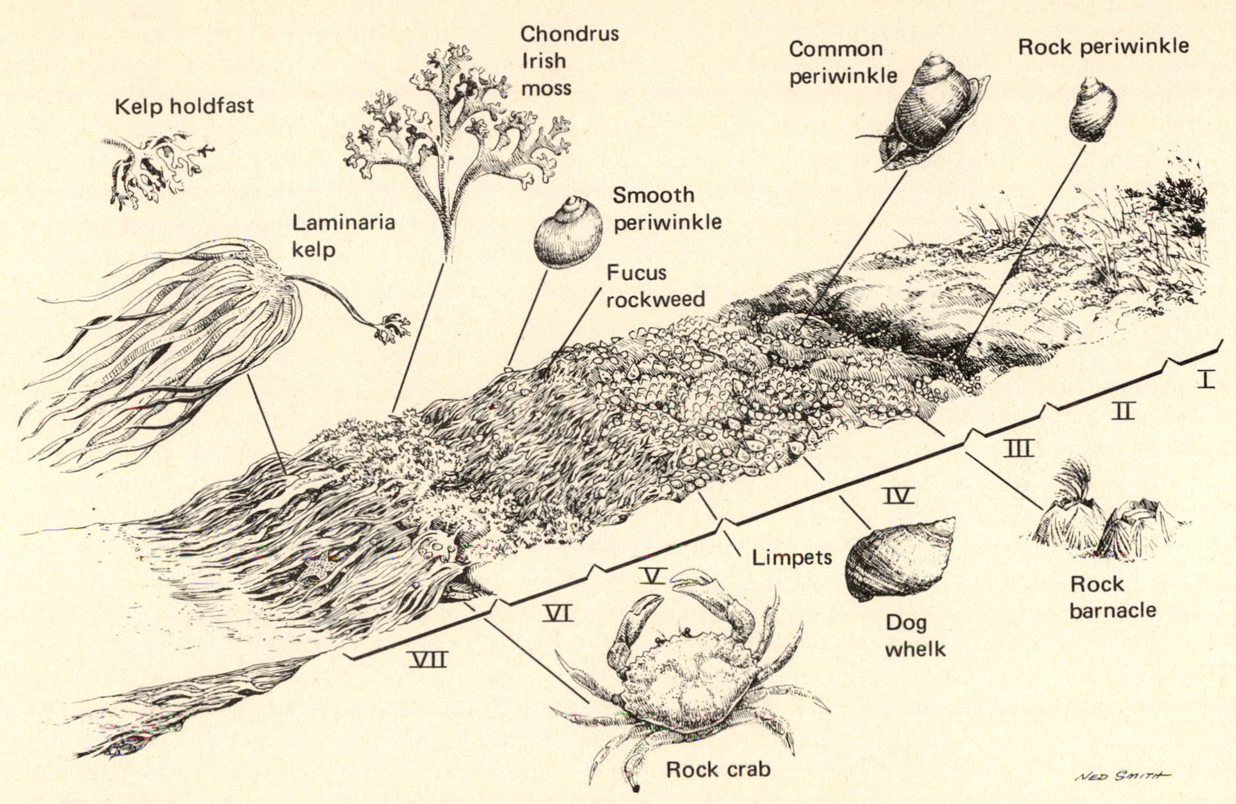

FIGURE 26.13 Zonation on a rocky shore along the North Atlantic coast. Compare this with the generalized diagram in Figure 26.12. (I) Land: lichens, herbs, grasses, and so on; (II) bare rock; (III) zone of black algae and rock periwinkles; (IV) barnacle zone: barnacles, dog whelk, a predaceous species, common periwinkles, and mussels; (V) fucoid zone: rockweed and smooth periwinkles; (VI) *Chondrus* zone: Irish moss; (VII) Laminarian zone: kelp. (*Zonation drawing based on data from Stephenson and Stephenson, 1954, and author's photographs and sketches.*)

essentially nonmarine community, on which graze the periwinkles, basically marine animals (Doty, in Hedgpeth, 1957). Common to this black zone is the rough periwinkle, which grazes on the wet algae covering the rocks. On European shores lives a similarly adapted species, the rock periwinkle, the most highly resistant to desiccation of all the shore animals.

Below the black zone lies the littoral, a region covered and uncovered daily by the tides. It is universally characterized by the barnacles, although often they are hidden under a dense growth of fucoid seaweeds or kelp (a brown alga, Phaeophyceae) and, in the more northern reaches of the North American and European coasts, by the red algae (Rhodophyceae). The littoral tends to be divided into subzones. In the upper reaches the barnacles are most abundant. The oyster, the blue mussel, and the limpets appear in the middle and lower portions of the littoral, as does the common periwinkle.

Occupying the lower half of the littoral zone of colder climates and in places overlying the barnacles is an ancient group of plants, the brown algae, more commonly known as rockweeds or wrack (*Fucus*). Rockweeds attain their finest growth on protected shores, where they may reach a length of 7 feet; on wave-whipped shores they are considerably shorter. The rockweeds that live farthest up on the tidelands are channeled rockweeds, or wrack, a species found on the European but not the American shore. It is replaced here by another, the spiral rockweed, a low-growing, orange-brown alga whose short, heavy fronds end in turgid, roundish swellings. The spiral rockweeds on the northeastern coast form only a very narrow band, if present at all, for they are replaced by the more abundant bladder rockweed and, in sheltered waters only, the knotted rockweed. The bladder rockweed can best withstand the heavy surf, since it is shorter, has a strengthening midrib, and possesses great tensile strength. Both the bladder and knotted rockweeds have gas-filled bladders that tend to buoy them up at high tide. Rockweeds have no roots—they draw their nutrients from the seawater surrounding them—but they cling to the rocks by means of disk-shaped holdfasts.

The lower reaches of the littoral zone may be occupied by blue mussels (see Figure 26.3) rather than rockweeds. This is particularly true on shores where the hard surfaces have been covered in part by sand and mud. No other shore animal grows in such abundance; the blue-black shells packed closely together may blanket the area. Mussels may grow in association with a red alga, *Gigartina,* a low-growing, carpetlike plant. The algae and the mussels together often form a tight mat over the rocks. Here, too, is found another seaweed, Irish moss, which grows in carpets some 6 inches deep. Its color is variable, ranging from purple to yellow and green; its fronds are branched and tough (Figure 26.3), and when covered by water the plant has an iridescent sheen.

Variations in Zonation

The zonation just described is essentially that of the North Atlantic coast north of Cape Cod. The coast is duplicated in part by the rocky coasts of the British Isles, for 85 percent of the plants and 65 percent of the animals are shared in common (Stephenson and Stephenson, 1949). Although adhering to the same general pattern, the details of zonation vary on other rocky shores wherever they exist.

This variation in eastern North America is due to temperature, influenced by latitude, and to that great river of the sea, the Gulf Stream. The intertidal regions of the Atlantic Coast of North America fall into tropical, warm temperate, and cold temperate regions. On the intertidal Pacific Coast, these regions are obscured by complex arrangements of currents and water masses. As a result, the intertidal flora and fauna possess many common features from Alaska to the outer coast of California. Typical zonation, suggestive of the cold temperate Atlantic Coast, exists.

Perhaps the most striking feature of the Pacific Coast intertidal region is the contrast between north-facing and south-facing shores (Stephenson, 1961). The supralittoral fringe on the north-facing shores is narrower than on the south-facing shores and is located much higher on the shore. There is also a correspondingly higher shift in the barnacle zone, which is much deeper on the north-facing shores. The barnacle zone on the south-facing shores involves two genera, an upper zone of *Chthamalus* and a lower *Balanus* zone. *Chthamalus* is missing entirely on north-facing shores. The main zone of kelp is up to 4 feet higher on the north-facing coast, and its sublittoral has a somewhat different faunal composition than the south-facing shores. Particularly characteristic of the north is the tube worm *Serpula vermicularis,* a weak fringe of kelp, and a rich group of echinoderms and anemones, all of which seemingly replace the band of short, carpetlike algae of the south-facing shores.

SANDY AND MUDDY SHORES

Sandy and muddy shores appear barren of life at low tide, in sharp contrast to the life-studded rocky shores. But the sand and black mud are not as dead as they seem, for beneath them life exists, waiting for the next high tide.

The sandy shore is in some ways a harsh environment; indeed, the very matrix of this seaside environment is a product of the harsh and relentless weathering of rock, both inland and along the shore. Through eons the ultimate products of rock weathering are carried away by rivers and waves to be deposited as sand along the edge of the sea. The size of the sand particles deposited influences the nature of the sandy beach, water retention during low tide, and the ability of animals to burrow through it. Beaches with relatively steep slopes are usually made up of larger sand grains and are subject to more wave action. Beaches exposed to high waves are generally flattened, for much of the material is transported away from the beach to deeper water and fine sand is left behind (Figure 26.14). Sand grains of all sizes, especially the finer particles in which capillary action is greatest, are more or less cushioned by a film of water, reducing further wearing action. The retention of water by the sand at low tide is one of the outstanding environmental features of the sandy shore.

In sheltered areas of the coast, the slope may be so gradual that the surface appears flat. Because the outgoing tidal currents are slow, organic material settles out and remains behind. And because of the flatness, the tide goes out a long way and the surface water fails to drain off completely, leaving the surface wet. In such situations one finds mud flats.

Characteristics

Existence of life on the surface of the sand is almost impossible. It provides no surface for attachment of seaweeds and their associated fauna; and the crabs, worms, and snails so characteristic of rock crevices find no protection here. Life is forced to exist beneath the sand (see Figure 26.15).

The life of the sandy beach does not experience the same violent fluctuations in temperature as that of the rocky shores. Although the surface temperature of the sand at midday may be 10 or more degrees centigrade higher than the returning seawater, the temperature a few inches below remains almost constant throughout the year. Nor is there a violent fluctuation in salinity, even when fresh water runs over the surface of the sand. Below 10 inches salinity is little affected.

FIGURE 26.14 A long stretch of sandy beach. Although the beach appears barren, life is abundant just beneath the sand.

For life to exist on the sandy shore, some organic matter has to accumulate. Most sandy beaches contain a certain amount of detritus from seaweeds, dead animals, feces, and material blown in from the shore. This organic matter accumulates within the sand, especially in sheltered areas. In fact, there is an inverse relationship between the turbulence of the water and the amount of organic matter on the beach, which reaches its maximum on the mud flats. In the sand, organic matter clogs the spaces between the grains and binds them together.

As water moves down through the sand, it loses its oxygen, both from the respiration of bacteria and from the oxidation of chemical substances, especially ferrous compounds. At some point within the sand, the water loses all of its oxygen, and a region of stagnation and oxygen deficiency results; this region is characterized by anaerobic bacteria and by the formation of ferrous sulfides. The iron sulfides cause a zone of blackening whose depth varies with the exposure of the beach. On mud flats such conditions exist almost to the surface.

Zonation

Zonation of animal life exists on the beach but is hidden and must be discovered by digging. Based on animal organisms, sand and mud shores can be divided roughly into supralittoral, littoral, and sublittoral zones, but a universal pattern similar to that of rocky shores is lacking (Dahl, 1953).

Life on sandy and muddy beaches consists of an *epifauna,* organisms that live on the surface, and an *infauna,* organisms living within the substrate. Most infauna either occupy permanent or semipermanent tubes within the sand or mud or can burrow rapidly into the substrate when they need to do so. The multicellular invertebrate infauna obtain oxy-

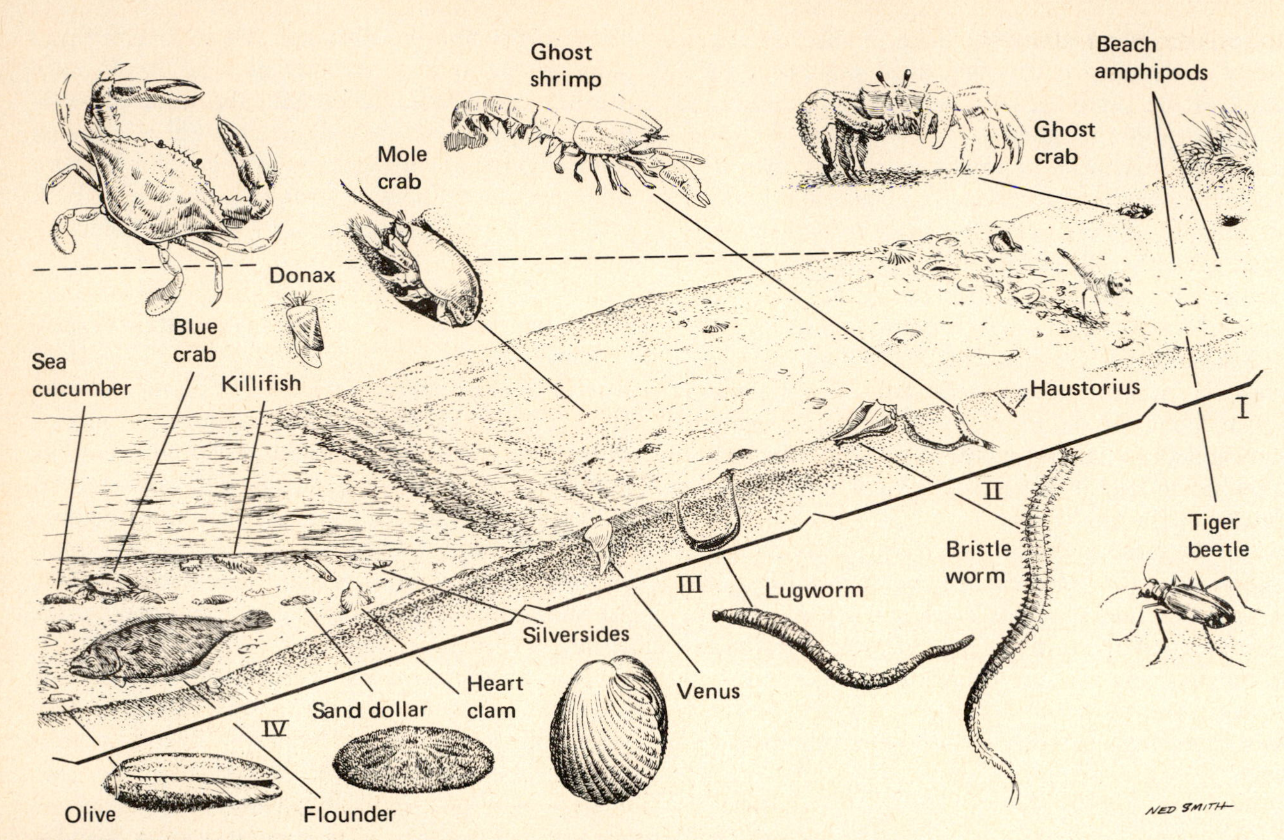

FIGURE 26.15 Life on a sandy ocean beach along the Atlantic Coast. Although strong zonation is absent, organisms still change on a gradient from land to sea. (I) Supratidal zone; ghost crabs and sand fleas; (II) flat beach zone: ghost shrimp, bristleworms, clams; (III) intratidal zone: clams, lugworms, mole crabs; (IV) subtidal zone. The dotted line indicates high tide. The lugworm is a deposit-feeder. It lies in a U-shaped burrow into which the substrate flows to the worm. The ghost shrimp excavates a vertical burrow several feet deep into the sand, where it stays at low tide. At high tide the shrimp moves to the mouth of the burrow and sifts through sand grains for organic debris. The bivalve *Donax* moves up and down with the surf, digs into the substrate, and extends its divided siphon to the surface to pick up detritus and discharge wastes as pseudofeces. The bristleworm is a dominant resident of sandy shores. It feeds on a wide range of items, from worms to nematodes and copepods. In turn, it is a prey item for fish. Other residents on the lower beach include detrital-feeders such as sand dollars and predators such as the blue crab.

gen either by gaseous exchange with the water through the outer covering or by breathing through gills and elaborate respiratory siphons.

The most conspicuous epifauna of the sandy beaches are the pale, sand-colored ghost crabs (see Figure 25.15) and sand hoppers, scavengers of the drift line. For the most part they occupy the upper beach, or supralittoral.

On the mud flat these organisms are replaced by the small gastropod mollusk *Hydrobia* and the ubiquitous *Corophium*. *Hydrobia* looks like a very small periwinkle. Although it can burrow into the mud, it spends a great deal of time on the surface. At low tide it appears by the millions on the mud. Crawling across the surface, it feeds on fine particulate organic matter, which it ingests for the bacteria it contains.

The amphipod *Corophium* crawls with a looping motion across the mud at low tide, but it never leaves the intertidal zone. It lives in a permanent to semipermanent semicircular burrow open at both ends. To feed, the amphipod excavates a small horseshoe-shaped burrow; once settled, it sets up small currents into the burrow with its walking legs. As the muddy water moves through the burrow, the amphipod sifts out quantities of large mud particles with its head appendages and crushes them with its mandibles.

Structure and Function

The intertidal beach, the littoral, is the zone where true marine life really appears. Although it lacks the variety found on the intertidal rocky shores, the populations of individual species are often enormous. The energy base for sandy-beach fauna is organic matter, much of it made available by bacterial decomposition, which goes on at the greatest rate at low tide. The bacteria are concentrated around the organic matter in the sand, where they escape the diluting effects of water. The products of decomposition are dissolved and washed into the sea by each high tide, which, in turn, brings in more organic matter for decomposition. Thus, the sand beach is an important site of biogeochemical cycling, supplying the offshore waters with phosphates, nitrogen, and other compounds.

Virtually all fauna of the sandy and muddy shore live beneath the surface, a much more hospitable place to live than on the surface. The infauna can be divided into the macrofauna, the meiofauna, and the microfauna. These organisms feed on organic matter. At this point the energy flow in sandy beaches and mud flats differs from that in terrestrial and aquatic ecosystems, for the basic consumers are bacteria. In more usual energy-flow systems, bacteria act largely as reducers responsible for the conversion of dead organic matter into a nutrient form that can be utilized by the producer organisms. But in the sandy beaches and mud flats, bacteria not only break down organic matter but are a major source of food for other consumers. In effect, the bacteria function as grazing herbivores.

A number of deposit-feeding organisms ingest organic matter largely as a means of obtaining bacteria. Prominent among these are the numerous nematodes and copepods (*Harpacticoida*). One of the most important consumers is the polychaete worm *Nereis*. Other sandy-beach animals are filter-feeders, obtaining their food by sorting particles of organic matter from the tidal waters. Two of these are "surf fishers," who advance and retreat up and down the beach with the flow and ebb of the tide. One is the mole crab, which lies partially buried within the sand and allows the surf to roll over it. The other is a bivalve, *Donax,* the coquina clam, found on the Atlantic, Gulf, and Pacific coasts. Within the beach sand and mud live vast numbers of meiofauna, species 50 to 500 mm in size, including copepods, ostracods, nematodes, and gastrotrichs, all making up the interstitial life. Interstitial fauna are generally elongated forms with setae, spine, or tubercles greatly reduced. The great majority do not have pelagic larval stages. These animals feed mostly on the microfauna, algae, bacteria, and detritus. Interstitial life shows seasonal variations, reaching maximum development in the summer months.

Because of their dependence on organic matter imported to them and their essentially heterotrophic nature, the sandy beaches and mud flats perhaps cannot be considered ecosystems without including them as part of the whole coastal environment (Figure 26.16). Except in the cleanest of sands, some primary production does take place in the intertidal zone. The major primary producers are the diatoms, confined mainly to fine-grained deposits of sand containing a high portion of organic matter. Productivity is low; one estimate places productivity of moderately exposed sandy beaches at 5 g/m^2/year. Production may be increased temporarily by phytoplankton carried in on the high tide and left stranded on the surface. Again, the majority of these are diatoms. More important as producers are the sulfur bacteria in the black sulfide or reducing layer. These are chemosynthetic bacteria that use the energy released when ferrous oxide is reduced.

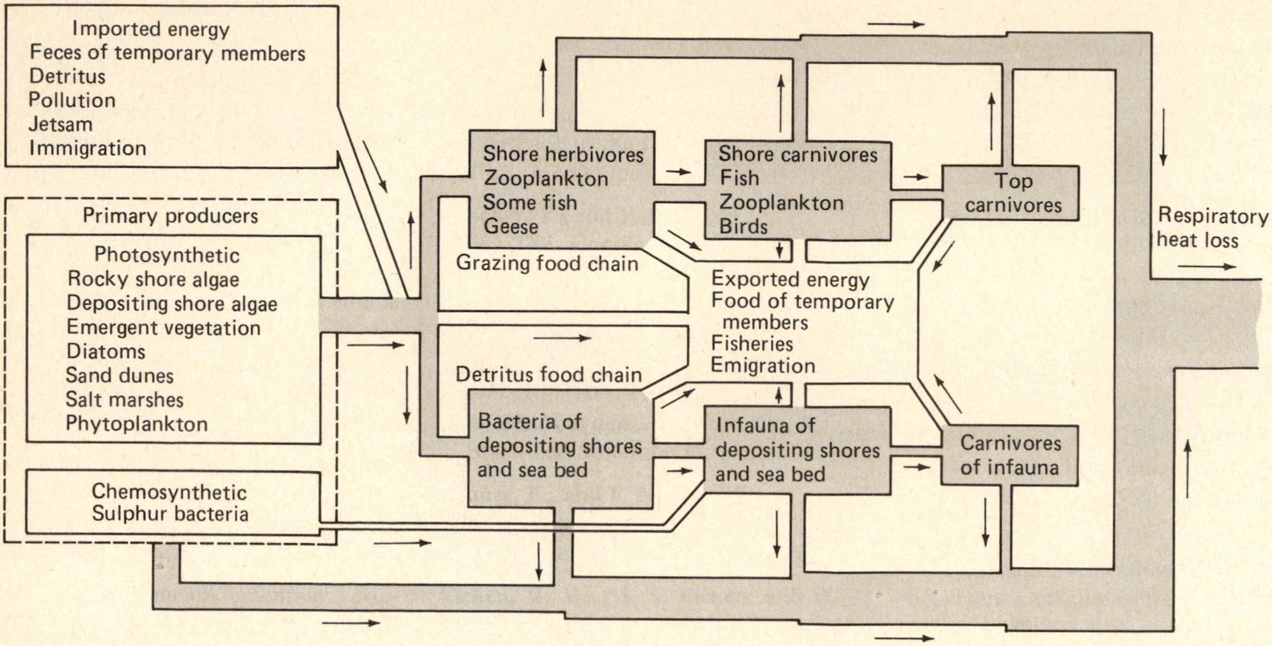

FIGURE 26.16 The coastal ecosystem is, in effect, a superecosystem, consisting of the shore, the fringing terrestrial regions, and the sublittoral zone. It encompasses two food chains: the grazing food chain, involving rocky-shore herbivores and zooplankton, and a detrital food chain, involving bacteria of the depositing shore and sublittoral muds and the dependent detrital-feeders and carnivores. Because energy import exceeds export, the system is continuously gaining energy—a reason why coastal ecosystems are so productive. (*From Eltringham, 1971:203.*)

Some mud flats may be covered with the algae *Enteromorpha* and *Ulva,* whose productivity can be substantial.

In effect, the shore and the mud flats are part of a larger coastal ecosystem involving the salt marsh, the estuary, and coastal waters. They act as sinks for energy and nutrients, because the energy they utilize comes not from primary production but from organic matter that originates outside of the area. And many of the nutrient cycles are only partially contained within the borders of the shore.

The Coral Reef

In warm, sunlit subtropical and tropical waters, one finds the coral reef, a structure of biological rather than geological origin. Coral reefs are built by carbonate-secreting organisms, of which coral (Coelenterata) may be the most conspicuous but not always the most important. Also contributing heavily are the corralline red algae (*Porolithon*) as well as foraminifera and mollusks. Reefs are formed by intergrown skeletons that withstand both the action of waves and the attack of coral-eating animals. Built only under water, usually to a maximum depth of tens of meters below low tide, coral reefs need a stable foundation to permit them to grow. Such foundations are provided by shallow continental shelves and submerged volcanoes.

There are three kinds of coral reefs, with many gradations among them: *fringing reefs,* which grow along the rocky shores of islands and continents; *barrier reefs,* which parallel shorelines along continents; and *atolls,* horseshoe-shaped reefs surrounding lagoons. Such lagoons are about 40 m deep and are usually connected to the open sea by breaks in the reef. Reefs build up to sea level. To become islands or atolls, the reefs have to be exposed by a lowering of the sea level or built up by the accumulation of sediments and the piling up of reef material by the action of wind and waves.

Ecologically, coral reefs are interesting for the life they support and the ecosystems they represent. Coral reefs are complex ecosystems involving close relationship between coral and algae. In the tissue of coral live zooxanthellae, endozoic dinoflagellate algae, and on the calcareous skeletons live still other kinds, both filamentous and calcareous. At night the coral polyps feed, extending their tentacles to capture zooplankton from the water and thus securing phosphorus and other elements needed by the coral and its symbiotic algae. During the day the algae absorb sunlight and carry on photosynthesis and directly transfer organic material to coral tissue. Thus, nutrients tend to be recycled between the coral and the algae (Johannes, 1968; Pomeroy and Kuenzler, 1969). In addition, by altering the CO_2 concentration in animal tissue, the algae enable the coral to extract the $CaCO_3$ needed to build the skeletons (Goreau, 1963). Living on the coral are filamentous algae (Chlorophyta), which add to the primary production. Thus, the symbiotic relationship between algae and coral as well as the nutrient recharging by tidal waters make the coral reef one of the most productive ecosystems.

Reflecting this high productivity and the wide variety of habitats provided by the coral structures is the great diversity of life near the coral reef. Swarming here are thousands of kinds of exotic invertebrates, some of which feed on the coral animals and algae; hundreds of highly colored herbivorous fish; and a large number of predatory fish that lie in ambush for prey in the caverns that honeycomb the reef. In addition, there is a wide array of symbionts such as the cleaning fish and crustaceans that pick parasites and detritus from larger fish and invertebrates.

The Open Sea

Beyond the estuaries and the rocky and sandy shores, beyond the mangrove swamps and the coral reefs, lies the open sea (Figure 26.17). As in lakes, the dominant form of plant life, or primary producers, is phytoplankton, small microscopic plants drifting with the currents. There is a reason for the smallness of the plants. Surrounded by a chemical medium that contains in varying quantities the nutrients necessary for life, they absorb nutrients directly from the water (see Chapter 21). The smaller the organism, the greater the surface area exposed for the absorption of nutrients and the utilization of the energy of the sun. The density of seawater is such that there is no need for well-developed supporting structures. In coastal waters and areas of upwelling, phytoplankton 100 microns or more in diameter may be common, but in general, plant life is much smaller and widely dispersed.

FIGURE 26.17 The open sea, birthplace of waves that crash against rocky and sandy shores hundreds and thousands of miles away.

PHYTOPLANKTON

Because of its requirements for light, phytoplankton is restricted to the upper surface waters, which, determined by depth of light penetration, may vary from tens to hundreds of feet. Because of seasonal, annual, and geographic variations in light, temperature, and nutrients as well as grazing by zooplankton, the distribution and species composition of phytoplankton vary from ocean to ocean and from place to place.

Each ocean or region within an ocean appears to have its own dominant forms. Littoral and neritic waters and regions of upwelling are richer in plankton than mid-oceans. In regions of downwelling, the dinoflagellates, a large, diverse group characterized by two whiplike flagellae, concentrate near the surface in areas of low turbulence. They attain their greatest abundance in warmer waters. In summer they may concentrate in the surface waters in such numbers that they color it red or brown. Often toxic to other marine life, such concentrations of dinoflagellates are responsible for red tides. In regions of upwelling, the dominant forms of phytoplankton are diatoms. Enclosed in a silica case, diatoms are particularly abundant in arctic waters. Smaller than diatoms are the Coccolithophoridae, so small that they pass through plankton nets (and so are classified as *nannoplankton*). Their minute bodies are protected by calcareous plates or spicules embedded in a gelatinous sheath. Universally distributed in all waters except the polar seas, the Coccolithophoridae have the ability to swim. They are characterized by droplets of oil that aid in buoyancy and serve as a means of storing food. In equatorial currents in shallow seas, the concentration of phytoplankton is variable. Where both lateral and vertical circulation of water is rapid, the composition reflects in part the ability of the species to grow, reproduce, and survive under local conditions.

ZOOPLANKTON

Grazing on the phytoplankton is the herbivores zooplankton, largely copepods, planktonic arthropods (the most numerous animals of the sea), and the shrimplike euphausiids, commonly known as krill. Other planktonic forms are the larval stages of such organisms as gastropods, oysters, and cephalopods. Feeding on the herbivorous zooplankton is the carnivorous zooplankton, which includes such organisms as the larval forms of comb jellies (Ctenophora) and arrowworms (Chaetognatha).

Like phytoplankton, the composition of zooplankton varies from place to place, season to season, and year to year. In general, zooplankton falls into two main groups, the larger forms characteristic of shallow coastal waters and, generally, the smaller forms characteristic of the deeper open ocean. Zooplankton of the continental shelf contains a large portion of the larvae of fish and benthic organisms. It includes a greater diversity of species, reflecting a greater diversity of environmental and chemical conditions. The open ocean, being more homogeneous and nutrient-poor, supports a less diverse zooplankton. Zooplanktonic species of polar waters, having spent the winter in a dormant state in the deep water, rise to the surface during short periods of diatom blooms to reproduce. In temperate regions distribution and abundance depend on temperature conditions. In tropical regions, where temperature is nearly uniform, zooplankton is not so restricted, and reproduction occurs throughout the year.

Also like phytoplankton, zooplankton lives mainly at the mercy of the currents. But possessing sufficient swimming power, many forms of zooplankton exercise some control. Most species migrate vertically each day to arrive at a preferred level of light intensity. As darkness falls, zooplankton rapidly rises to the surface to feed on phytoplankton. At dawn, the forms move back down to preferred depths.

NEKTON

Feeding on zooplankton and passing energy along to even higher trophic levels are the small fish, squids, and ultimately the large carnivores of the sea. Some of the predatory fish, such as herring and tuna, are more or less restricted to the photic zone. Others are found in the deeper mesopelagic and bathypelagic zones.

Living in a world that lacks any sort of refuge as a defense against predation or as a site for ambush, inhabitants of the pelagic zone have evolved various means of defense and of securing prey. Among them are the stinging cells of the jellyfish, the remarkable streamlined shapes that allow speed both for escape and for pursuit, unusual coloration, advanced sonar, a highly developed sense of smell, and social organization involving schools or packs. Some animals, such as the baleen whale, have specialized structures that permit them to strain krill and other plankton from the water. Others, such as the sperm whale and certain seals, have the ability to dive to great depths to secure food. Phytoplankton lights up darkened seas, and fish take advantage of that bioluminescence to detect their prey.

The dimly lit regions of the mesopelagic and the dark regions of the bathypelagic depend on a rain of detritus as an energy source. Such food is limited. The rate of descent of organic matter, except for larger items, is so slow that it is either consumed, decayed, or dissolved before it reaches the deepest water or the bottom. Other sources include saprophytic plankton, which exist in the darker regions; particulate organic matter; and such imported material as wastes from the coastal zone, garbage from ships, and large dead animals. Because food is so limited, species are few and the populations low.

Residents of the deep also have special adaptations for securing food. Some, like the zooplankton, swim to the upper surface to feed by night; others remain in the dimly lit or dark waters. Darkly pigmented and weak bodied, many of the deep-sea fish depend on luminescent lures, mimicry of prey, and extensible jaws and expandable abdomens (which enable them to consume large items of food) as a means of obtaining sustenance. Although most of the fish are small (usually 15 cm or less in length), the region is inhabited by rarely seen large species such as the giant squid. In the bathypelagic region bioluminescence reaches its greatest development—two-thirds of the species produce light. Bioluminescence is not restricted to fish. Squid and euphausiids possess searchlightlike structures complete with lens and iris; and squid and shrimp discharge luminous clouds to escape predators. Fish have rows of luminous organs along their sides and lighted lures that enable them to bait prey and recognize other individuals of the same species.

THE BENTHIC REGION

The term *benthal* refers to the floor of the sea, and *benthos* refers to plants and animals that live there. There is a gradual transition of life from the benthos that exists on the rocky and sandy shores to that which exists in the ocean's depths. From the tide line to the abyss, organisms that colonize the bottom are influenced by the nature of the substrate. Where the bottom is rocky or hard, the populations consist largely of organisms that live on the surface of the substrate, the *epifauna* and the *epiflora*. Where the bottom is largely covered with sediment, most of the inhabitants, chiefly animals, live within the deposits and are known collectively as the *infauna*. The kind of organism that burrows into the substrate is influenced by the particle size, since the mode of burrowing is often specialized and adapted for a certain type of substrate.

The substrate varies with the depth of the ocean and with the relationship of the benthic region to land areas and continental shelves. Near the coast, bottom sediments are derived from the weathering and erosion of land areas along with organic matter from marine life. The sediments of deep water are characterized by fine-textured material that varies with depth and with the types of organisms in overlying waters. Although these sediments are termed *organic*, they contain little decomposable carbon, consisting largely of skeletal fragments of planktonic organisms. In general, with regional variations, organic deposits down to 4,000 m are rich in calcareous matter. Below 4,000 m, hydrostatic pressure causes some forms of calcium carbonate to dissolve. At 6,000 m and lower, sediments contain even less organic matter and consist largely of red clays rich in aluminum oxides and silica.

Within the sediments are layers that relate to oxidation reduction reactions. The surface, or oxidized layer, yellowish in color, is relatively rich in oxygen, ferric oxides, nitrates, and nitrites. It supports the bulk of the benthic animals, such as polychaete worms and bivalves in shallow water; flatworms, copepods, and others in deeper water; and throughout, a rich growth of aerobic bacteria. Below this is a grayish transition zone to the black layer, which is characterized by lack of oxygen, iron in the ferrous state, nitrogen in the form of ammonia, and hydrogen sulfide. It is inhabited by anaerobic bacteria, chiefly reducers of sulfates and methane.

In the deep benthic regions, variations in temperature, salinity, and other conditions are negligible. In a world of darkness, no photosynthesis takes place, so the bottom community is strictly heterotrophic, depending entirely on what organic matter finally reaches the bottom as a source of energy. As already suggested, this is greatly limited. It is estimated that the quantity of such material amounts to 0.5 g/m^2/year (Moore, 1958). The bodies of dead whales, seals, and fish may contribute another 2 or 3 g/m^2/year.

Among the bottom organisms there are four feeding strategies. Such organisms may filter suspended material from the water, as the stalked coelenterates do; they may collect food particles that settle on the surface of the sediment, as the sea cucumbers do; they may be selective or unselective deposit-feeders such as the polychaetes; or they may be predatory, like the brittle stars and the spiderlike pycnogonids.

Important in the benthic food chain are the bacteria of the sediments. Common where large quantities of organic matter are present, bacteria may reach several tenths of grams per square meter in the topmost layer of silt. Bacteria synthesize protein from dissolved nutrients and, in turn, become a source of protein, fat, and oils for deposit-feeders.

PRODUCTIVITY

The oceans occupy 70 percent of Earth's surface, and their average depth is around 4,000 m; yet their primary production is considerably less than that of Earth's land surface. They are less productive because only a superficial illuminated area up to 100 m deep can support plant life; and that plant life, largely phytoplankton, is patchy because most of the open sea is nutrient-poor. Much of this nutritional impoverishment results from a limited to almost nonexistent nutrient reserve that can be recirculated. Phytoplankton, zooplankton, and other organisms and their remains sink below the lit zone into the dark benthic water. While this sinking supplies nutrients to the deep, it robs the upper layers.

This depletion of nutrients is most pronounced in tropical waters. There, a permanent thermal stratification—with its layer of warmer, less dense water lying on top of a colder, denser layer of deep water—prevents an exchange of nutrients between the surface and the deep. Thus, in spite of high light intensity and warm temperatures, tropical seas are the lowest in production, which amounts to an estimated 3.79×10^9 tons of carbon per year.

The temperate oceans are more productive, largely because a permanent thermocline does not exist. During the spring and to a limited extent during the fall, temperate seas, like temperate lakes, experience a nutrient overturn (see Chapter 25). This recirculation of phosphorus and nitrogen from the deep stimulates a surge of spring phytoplankton growth. As spring wears on, the temperature of the water becomes stratified and a thermocline develops, preventing a nutrient exchange. The phytoplankton growth depletes the nutrients, and the phytoplankton population suddenly declines. In the fall a similar overturn takes place, but the rise in phytoplankton production is slight because of decreasing light intensity and low winter temperatures. Reduced production in winter holds down the annual productivity of temperate seas to a level a little above that of tropical seas, 4.22×10^9 tons of carbon per year.

Most productive are coastal waters and regions of upwelling, whose annual production may amount to up to 6.90×10^9 tons of carbon per year. Major areas of upwelling are largely on the western sides of continents: off the southern California coast, Peru, northern and southwestern Africa, and the Antarctic. Upwellings result from the differential heating of polar and equatorial regions that produces the equatorial currents (see Chapter 4) and the winds. For example, as the water is pushed northward to the equator by winds blowing out of the south, it is deflected from the coast by the Coriolis force. As the deflected surface waters move away, they are replaced by an upwelling of colder, deeper water that brings a supply of nutrients into the warm, sunlit portions of the sea. As a result, regions of upwelling are highly productive, supporting an abundance of life. Because of their high productivity, upwellings support important commercial fisheries such as the tuna fishery off the California coast, the anchovy fishery off Peru, and the sardine fishery off Portugal. Other zones of high production are coastal waters and estuaries, where productivity may run as high as 1,000 $g/m^2/year$. Turbid, nutrient-rich waters are major areas of fish production.

Thus, between upwellings and coastal waters, the most productive areas of the sea are the fringes of water bordering the continental land masses. A great deal of the measured productivity of the coastal fringes comes from the benthic as well as the surface waters, since the seas are shallow. Benthic production, largely unavailable, is not considered in the productivity of the open sea. Recent estimates (Koblentz-Mishke et al., 1970) of the total production for marine plankton is 50×10^9 metric tons of dry matter per year; if benthic production is considered, total production may be 55×10^9 metric tons of dry matter per year.

Carbohydrate production by phytoplankton, largely diatoms, is the base on which the life of the seas exists (Figure 26.18). Conversion of primary production into animal tissue is accomplished by zooplankton, the most important of which are the copepods. To feed on the minute phytoplankton, most of the grazing herbivores must also be small, measuring between 0.5 and 5.0 millimeters. In the oceans most of the grazing herbivores are members of the genera *Calanus, Acartia, Temora,* and *Metridia,* probably the most abundant animals in the world. The single most abundant copepod is *Calanus finmarchicus,* with its close relative *C. helgolandicus*. In the Antarctic the shrimplike euphausiids, or krill, fed upon by the blue whale, are the dominant herbivores.

Summary

The marine environment is characterized by salinity, waves, tides, and vastness. Salinity is due largely to sodium and chlorine, which make up 86 percent of sea salt. Although sea salt has a constant composition, salinity varies throughout the oceans. It is affected by evaporation, precipitation, movement of water masses, and mixing of water masses of different salinities. Because of its salinity, seawater does not reach its greatest density at 4°C but becomes heavier as it cools.

Like lakes, the marine environment experiences stratification and zonation. The marine environment also exhibits surface waves produced by winds, internal waves at the interface of layers of water, and subsurface currents that run counter to the surface.

Estuaries, where fresh water meets the sea, and their associated tidal marshes and swamps, are units in which the nature and distribution of life are determined by salinity. In the estuary itself salinity declines from the mouth back up the river. This decrease in salinity is accompanied by a decline in estuarine fauna because the fauna consists chiefly of marine species. The estuary serves as a nursery for marine organisms, for here the young can develop, protected from predation and competing species unable to withstand the lower salinity.

Tidal marshes may add to the production of the estuary. Composed of salt-tolerant grasses and flooded by daily tides, marsh vegetation is either utilized within the marsh by detritivores or carried into the estuary, where it is utilized by bacteria and detrital-feeders. If vertical mixing between fresh and salt water takes place in the estuary, nutrients are circulated between organisms and bottom sediments.

The drift line marks the farthest advance of tides on the sandy shore. On the rocky shore the tide line is marked by a zone of black algal growth on the stone. The most striking feature of the rocky shore, its zonation of life, results from the alternate exposure and submergence of the shore by the tides. The black zone marks the supralittoral, the

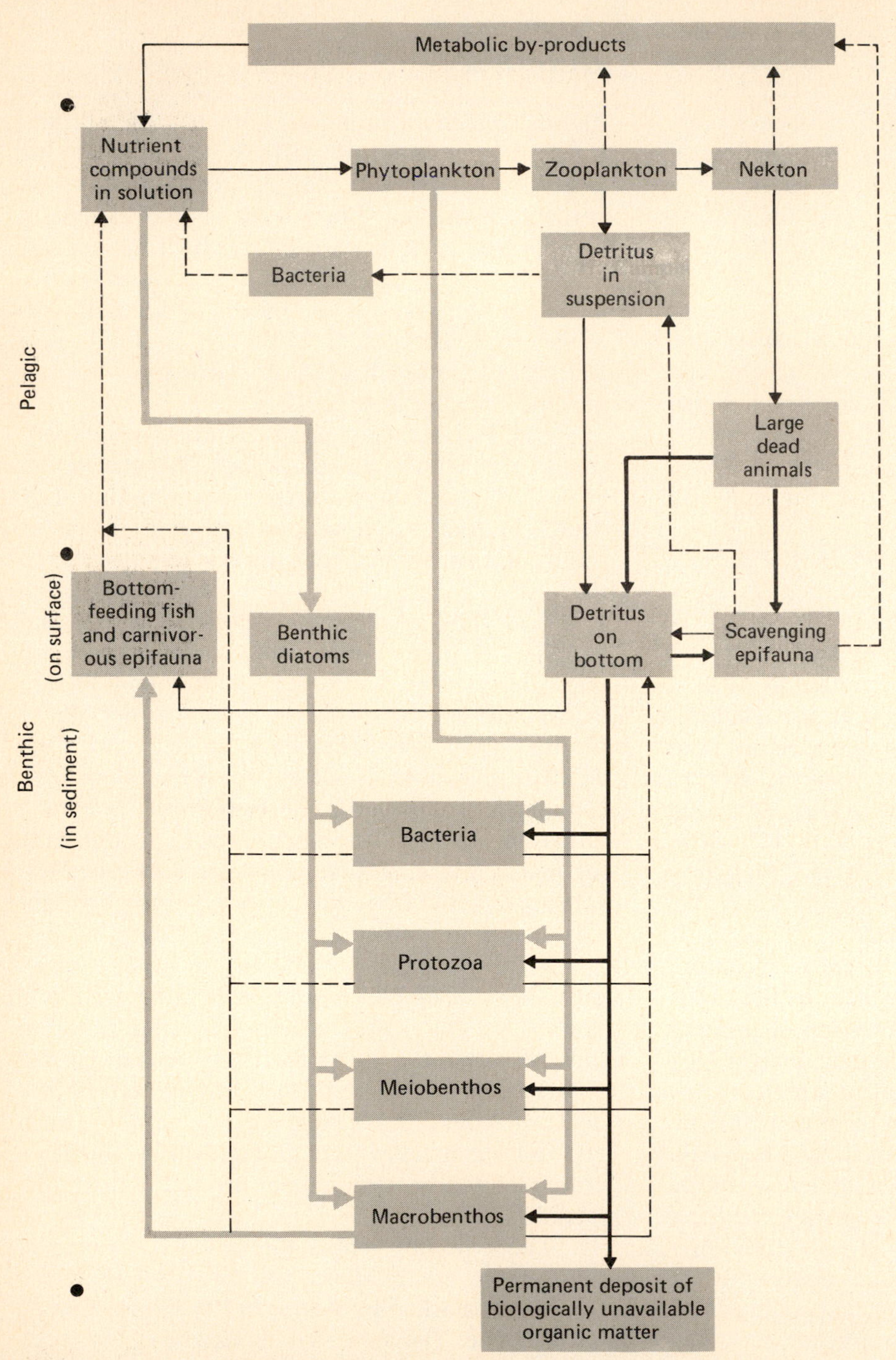

FIGURE 26.18 A simplified marine food chain. The upper portion represents the pelagic zone; below that are the epifauna and the bottom-feeding nekton. The bottom of the diagram includes the infauna and marine humus. (*After Raymont, 1963; and from M. Gross, 1972:457,* Oceanography *[Englewood Cliffs, N.J.: Prentice-Hall]*.)

upper part of which is flooded only every 2 weeks by spring tides. Submerged daily by the tides is the littoral, characterized by barnacles, periwinkles, mussels, and fucoid seaweeds.

Sandy and muddy shores, in contrast to rocky ones, appear barren of life at low tide; but beneath the sand and mud, conditions are more amenable to life than on the rocky shore. Inhabitants of mud flats and sandy beaches occupy permanent or semi-permanent tubes in the substrate and feed on organic matter carried to the shore by tides. Zonation of life is hidden beneath the surface, where the variety of life is less than on the rocky shore. But whether hidden beneath the sand or exposed on rocks, all forms of life at the edge of the sea exhibit a remarkable adaptation to the ebb and flow of tides.

Beyond the estuary and the rocky shore lies the open sea, which covers most of Earth. Because of currents and tides, the open sea can be considered one vast interconnected ecosystem that has its variants in sandy bottoms and rocky shores below the high-tide mark. Its dominant plant life is phytoplankton, and the chief consumers are zooplankton. Depending upon these for an energy base are the nekton organisms, dominated by fish.

The open sea can be divided into three main regions. The bathypelagic is the deepest, void of sunlight and inhabited by darkly pigmented, weak-bodied animals characterized by luminescence. Above it lies the dimly lit mesopelagic region, inhabited by its own characteristic species such as certain sharks and squid. Both the bathypelagic and the mesopelagic regions depend upon a rain of detritus from the upper region, the epipelagic, for their energy source.

Because of the impoverished nutrient status of the ocean water, productivity is low. That comes about because nutrient reserves in the upper layer of water are limited; phytoplankton and other life sink to the deep water; and a thermocline, permanent in deep water, prevents the circulation of deep water to the upper layer. Most productive are shallow coastal waters and upwellings, where nutrient-rich deep water comes to the surface.

Review and Study Questions

1. Why is the ocean salty?
2. Characterize the major regions of the ocean.
3. What causes tides? What are spring tides? Neap tides?
4. What types of currents, other than surface currents, exist in the oceans?
5. What is an estuary?
6. What is the relationship between vertical and horizontal stratification of salinity in the estuary?
7. What is tidal overmixing, and what influences govern it?
8. How is the estuary a nutrient trap?
9. How does a tidal marsh relate to an estuary?
10. What is the major functional difference between northern tidal marshes and southern tidal marshes?
11. What is happening to tidal marshes? What are the implications of this destruction? (You will have to go to the literature to answer this question.)
12. What are the three major zones of the rocky shore, and what are the outstanding features of each?
13. Contrast the energy source of a sandy or muddy shore with that of a rocky shore. What group of organisms are the basic consumers of sandy shores?
14. What are the outstanding features of the coral reef?
15. What regions of the seas are richest in phytoplankton? Why?
16. What is the major food source of the nekton of the deep?
17. Distinguish between epifauna and infauna.

18. What factors influence the productivity of the sea?

19. Contrast productivity of the open sea with that of coastal waters. Explain the difference.

Suggested Readings

Amos, W. H. 1966. *The life of the seashore*. New York: McGraw-Hill.

Barnes, R. S. K., and J. Green, eds. 1972. *The estuarine environment*. London: Applied Science Publishers.

Boje, R., and M. Tomezak, eds. 1978. *Upwelling ecosystem*. New York: Springer-Verlag.

Carson, R. 1955. *Edge of the sea*. Boston: Houghton Mifflin.

———. 1961. *The sea around us*. New York: Oxford University Press.

Chapman, V. J., ed. 1977. *Wet coastal ecosystems*. Amsterdam, Holland: Elsevier.

Clark, J. 1974. *Coastal ecosystems: Ecological considerations for the management of the coastal zone*. Washington, D.C.: Conservation Foundation.

Coker, R. E. 1947. *This great and wide sea*. Chapel Hill: University of North Carolina Press.

Cushing, D. H., and J. J. Walsh, eds. 1976. *The ecology of the seas*. Philadelphia: Saunders.

Dyer, K. R. 1973. *Estuaries: A physical introduction*. New York: Wiley.

Eltringham, S. K. 1971. *Life in mud and sand*. New York: Crane, Russak.

Gross, M. G. 1972. *Oceanography: A view of the earth*. Englewood Cliffs, N.J.: Prentice-Hall.

Hardy, A. 1971. *The open sea: Its natural history*. Boston: Houghton-Mifflin.

Jefferies, R. L., and A. J. Davy, eds. 1979. *Ecological processes in coastal environments*. Oxford: Blackwell.

Jones, O. A., and R. Endean, eds. 1973, 1976. *Biology and geology of coral reefs*. Vols. II, III. New York: Academic Press.

Kennedy, V. S., ed. 1980. *Estuarine perspectives*. New York: Academic Press.

Kinne, O., ed. 1978. *Marine ecology*. 5 vols. New York: Wiley.

Lauff, G., ed. 1967. *Estuaries*. Pub. no. 83. Washington, D.C.: American Association for the Advancement of Science.

Lugo, A. E., and S. C. Snedaker. 1974. The ecology of mangroves. *Annual Review of Ecology and Systematics* 5:39–64.

McLusky, D. S. 1971. *Ecology of estuaries*. London: Heinemann Educational.

———. 1981. *The estuarine ecosystem*. New York: Halsted Press, Wiley.

Mann, K. H. 1982. *Ecology of coastal waters: A systems approach*. Berkeley: University of California Press.

Marshall, N. B. 1980. *Deep sea biology: Developments and perspectives*. New York: Garland STPM Press.

Newell, R. C. 1970. *Biology of intertidal animals*. New York: Elsevier.

Nybakken, J. W. 1982. *Marine biology: An ecological approach*. New York: Harper & Row.

Odum, H. T., B. J. Copeland, and E. A. McMahan, eds. 1974. *Coastal ecological systems of the United States*. Washington, D.C.: Conservation Foundation.

Perkins, E. J. 1974. *The biology of estuaries and coastal waters*. New York: Academic Press.

Steele, J. H. 1974. *The structure of marine ecosystems*. Cambridge, Mass.: Harvard University Press.

Stephenson, T. A., and A. Stephenson. 1973. *Life between the tidemarks on rocky shores*. San Francisco: Freeman.

Teal, J., and M. Teal. 1969. *Life and death of the salt marsh*. Boston: Little, Brown.

Warner, W. W. 1976. *Beautiful swimmers: Watermen, crabs, and Chesapeake Bay*. Boston: Little, Brown.

Wiley, M., ed. 1976. *Estuarine processes*. New York: Academic Press.

GLOSSARY

abiotic Nonliving component of the environment, including soil, water, air, light, nutrients, and the like.

abyssal Relating to the bottom waters of oceans, usually below 1,000 m.

acclimation Alteration of physiological rate or other capacity to perform a function through long-term exposure to certain conditions.

acclimatization Changes or differences in a physiological state that appear after exposure to different natural environments.

acid rain Precipitation with an extremely low pH. It is brought about by a combination of water vapor in the atmosphere with hydrogen sulfide and nitrous oxide vapors released to the atmosphere from the burning of fossil fuels. The result is sulfuric and nitric acid in rain, fog, and snow.

active transport Movement of ions and molecules across a cell membrane against a concentration gradient involving an expenditure of energy. The movement of the ion or molecule is in a direction opposite to that which it would take under simple diffusion.

adaptation Genetically determined characteristic (behavioral, morphological, physiological) that improves an organism's ability to survive and successfully reproduce under prevailing environmental conditions.

adaptive radiation Evolution from a common ancestor of divergent forms adapted to distinct ways of life.

adiabatic cooling A decrease in air temperature that results when a rising parcel of warm air cools by expansion (which uses energy) rather than losing heat to the outside surrounding air. The rate of cooling is approximately 1°C/100 m for dry air and 0.6°C/100 m for moist air.

adiabatic lapse rate Rate at which a parcel of air loses temperature with elevation if no heat is gained from or lost to an external source.

aerobic Living or occurring only in the presence of free uncombined molecular oxygen either as a gas in the atmosphere or dissolved in water.

aggressive mimicry Resemblance of a predator or parasite to a harmless species to deceive potential prey.

agonistic behavior All types of hostile response to other organisms, ranging from overt attack to overt escape.

A horizon Surface stratum of mineral soil characterized by maximum accumulation of organic matter, maximum biological activity, and loss of such materials as iron, aluminum oxides, and clays.

alfisol Soil characterized by an accumulation of iron and aluminum in the lower or B horizon.

allele One of two or more alternative forms of a gene that occupies the same relative position or locus on homologous chromosomes.

allelopathy Effect of metabolic products of plants (excluding microorganisms) on the growth and development of other nearby plants.

Allen's rule Trend among homoiotherms for limbs to become longer and extremities (such as ears) to become less compact in warmer climates than in colder ones.

allochthonous Food material reaching an aquatic community in the form of organic detritus.

allogenic succession Ecological change or development of species structure and community composition brought about by some externally generated force such as fire or storms.

allopatric Having different areas of geographical distribution; possessing nonoverlapping ranges.

allopatric speciation Separation of a population into two or more species by reproductive isolation brought about by geographical separation of the subpopulations.

allopolyploidy Condition in which the several nonhomologous sets of chromosomes are derived from the union of two gametes from distantly related species.

alluvial soil Soil developing from recent alluvium (material deposited by running water); exhibits no horizon development; typical of floodplains.

alpha diversity The variety of organisms occupying a given place or habitat.

altruism Form of behavior in which an individual increases the welfare of another at the expense of its own welfare.

ambient Refers to surrounding, external, or unconfined conditions.

amensalism Relationship between two species in which one is inhibited or harmed by the presence of another.

ammonification Breakdown of proteins and amino acids, especially by fungi and bacteria, with ammonia as the excretory by-product.

anadromous Refers to fish that typically inhabit seas or lakes but ascend streams to spawn; for example, salmon.

anaerobic Adapted to environmental conditions devoid of oxygen.

anagenesis Evolutionary process that produces entirely new levels of structural organization.

antibiotic Substance produced by a living organism that is toxic to organisms of different species.

aposematism Possession of warning coloration; conspicuous markings on animals that are poisonous, distasteful, or possess some unpleasant defensive mechanism.

apparent plants Large, easy-to-locate plants possessing quantitative defenses not easily mobilized at the point of attack; for example, tannins.

assimilation Transformation or incorporation of a substance by organisms; absorption and conversion of energy and nutrient uptake into constituents of an organism.

association Natural unit of vegetation characterized by a relatively uniform species composition and often dominated by a particular species.

aufwuchs Community of plants and animals attached to or moving about on submerged surfaces; also called *periphyton,* but that term more specifically applies to organisms attached to submerged plant stems and leaves.

autecology Ecology of individual species in response to environmental conditions.

autotrophy Ability of an organism to produce organic material from inorganic chemicals and some source of energy.

balanced polymorphism Maintenance of more than one allele in a population; results from the selective superiority of the heterozygote over the homozygote.

basal metabolism Energy expenditure of an organism at rest, fasting, and in a thermally neutral environment.

Batesian mimicry Resemblance of a palatable species, the mimic, to an unpalatable species, the model.

bathyal Pertaining to anything, but especially organisms, in the deep sea, below the photic or lighted zone and above 4,000 m.

benthos Animals and plants living on the bottom of a lake or sea from the high-water mark to the deepest depths.

Bergman's rule States that populations of homoiotherms living in cooler climates tend to have a larger body size and a smaller surface-volume ratio than related populations living in warmer climates.

beta diversity Variety of organisms occupying a number of different habitats over a region; regional diversity compared to very local, or alpha, diversity.

B horizon Soil stratum beneath the A horizon characterized by an accumulation of silica, clay, and iron and aluminum oxides and possessing blocky or prismatic structure.

biennial Plant that requires 2 years to complete its life cycle, with vegetative growth the first year and reproductive growth (flowers and seeds) the second.

biochemical oxygen demand (BOD) Measure of the oxygen needed in a specified volume of water to decompose organic materials. The greater the amount of organic matter in water, the higher the BOD.

biogeocenosis Translated Germanic- and Slavic-language equivalent of *ecosystem.* Perhaps more precise than *ecosystem* because the term emphasizes the biological and geological aspects of the ecosystem.

biogeochemical cycle Movement of elements or compounds through living organisms and nonliving environment.

biological magnification Process by which pesticides and other substances become more concentrated in each link of the food chain.

biological species Group of potentially interbreeding populations reproductively isolated from all other populations.

bioluminescence Production of light by living organisms.

biomass Weight of living material, usually expressed as dry weight per unit of area.

biome Major regional ecological community of plants and animals; usually corresponds to plant ecologists' and European ecologists' classification of plant formations and classification of life zones.

biophage Organisms that feed on living material.

biosphere Thin layer about Earth in which all living organisms exist.

biotic community Any assemblage of populations living in a prescribed area or physical habitat.

blanket mire Large areas of upland dominated by sphagnum moss and dependent upon precipitation for a water supply; a moor.

bog Wetland ecosystem characterized by an accumulation of peat, acid conditions, and dominance of sphagnum moss.

boreal forest Needle-leaved evergreen or coniferous forest bordering subpolar regions; also called taiga.

bottleneck An evolutionary term for any stressful situation that greatly reduces a population.

brackish Water that has salt concentration greater than fresh water (> 0.5‰) and less than seawater (35‰).

browse Part of current leaf and twig growth of shrubs, woody vines, and trees available for animal consumption.

bryophyte Member of the division in the plant kingdom of nonflowering plants, comprising mosses (Musci), liverworts (Hepaticae), and hornworts (Anthocerotae).

buffer Chemical solution that resists or dampens change in pH upon addition of acids or bases; also used in population biology to refer to any prey species that tends to dampen the impact of predation on another.

calcification Process of soil formation characterized by accumulation of calcium in the lower horizons.

caliche Alkaline, often rocklike salt deposit on the surface of soil in arid regions; it forms at the level where leached Ca salts from the upper soil horizons are precipitated.

calorie Amount of heat needed to raise 1 gram of water 1°C, usually from 15°C to 16°C.

cannibalism Killing and consumption of one's own kind; intraspecific predation.

carnivore Organism that feeds on animal tissue; taxonomically, a member of the order Carnivora (Mammalia).

carr Vegetation dominated by alder and willow and occupying eutrophic peat.

carrying capacity (K) Number of individual organisms that the resources of a given area can support, usually through the most unfavorable period of the year. The term has acquired so many meanings that it is almost useless.

catadromous fish Fish that feed and grow in fresh water but return to the sea to spawn.

catastrophic extinction Major episode of extinction involving many taxa and occurring fairly suddenly in the fossil record.

cation Part of a dissociated molecule carrying a + electrical charge.

cation exchange capacity Ability of a soil particle to absorb + charged ions.

chamaephyte Plants characterized by perennial shoots or buds on the surface of the ground to about 25 cm above the surface.

chaparral Vegetation consisting of broad-leaved evergreen shrubs found in regions with a mediterranean climate of hot, dry summers and mild, wet winters.

character convergence Evolution of similar appearance or behavior in unrelated species.

character displacement Divergence of characteristics in two otherwise similar species occupying overlapping ranges; brought about by selective effects of competition.

character divergence Evolution of behavioral, physiological, or morphological differences among species occupying the same area brought about by the selective pressure of competition.

C horizon Soil stratum beneath the solum (A and B horizons) relatively little affected by biological activity and soil-forming processes.

circadian rhythm Endogenous rhythm of physiological or behavioral activity of approximately 24 hours duration.

cladogenesis Process of evolution that produces a series of branching events.

cleistogamy Self-pollination within a flower that does not open.

climax Stable end community of succession that is capable of self-perpetuation under prevailing environmental conditions.

climograph Diagram describing a locality based on the annual cycle of temperature and precipitation.

cline Gradual change in population characteristics over a geographical area, usually associated with changes in environmental conditions.

coarse-grained Referring to qualities or aspects of the environment that occur over large patches with respect to the activities of organisms involved.

coevolution Joint evolution of two or more noninterbreeding species that have a close ecological relationship; through reciprocal selective pressures the evolution of one species in the relationship is partially dependent on the evolution of the other.

coexistence Two or more species' living together in the same habitat, usually with some form of competitive interaction.

colluvium Mixed deposits of soil material and rock fragments accumulated near the base of steep slopes through soil creep, landslides, and local surface runoff.

commensalism Relationship between species that is beneficial to one species but is neutral or of no benefit to the other.

community Group of interacting plants and animals inhabiting a given area.

compensation point Point at which photosynthesis and respiration balance each other so that net production is zero; in aquatic systems, usually the depth of light penetration at which oxygen utilized in respiration equals oxygen produced by photosynthesis.

competition Any interaction that is mutually detrimental to both participants; occurs between species that share limited resources.

competitive exclusion Hypothesis that states that when two or more species coexist using the same resource, one must displace or exclude the other.

continuum Gradient of environmental characteristics or changes in community composition.

continuum index Measure of the position of a community on a gradient defined by species composition.

convergent evolution Development of similar characteristics in different species living in different areas but under similar environmental conditions.

Coriolis effect Physical consequences of the law of conservation of angular momentum. As a result of Earth's rotation, a moving object veers to the right in the Northern Hemisphere and to the left in the Southern Hemisphere.

countercurrent circulation Anatomical and physiological arrangement by which heat exchange takes place between outgoing warm arterial blood and cool venous blood returning to the body core. It is important in maintaining temperature homeostasis in many vertebrates.

critical thermal maximum Temperature at which an animal's capacity to move is so reduced that it cannot escape from thermal conditions that will lead to death.

cryptic coloration Coloration of organisms that makes them resemble or blend into their habitat or background.

cryptophyte Plants characterized by buds buried in the ground on a bulb or rhizome.

cyclic replacement Type of succession in which the sequence of seral stages is repeated by the imposition of some disturbance so that the sere never arrives at a climax or stable sere.

death rate Percentage of individuals in a population dying in a specified time interval.

deciduous (Of leaves) shed during a certain season (winter in temperate regions; dry seasons in the tropics); (of trees) having deciduous parts.

decomposer Organism that obtains energy from the breakdown of dead organic matter to more simple substances; most precisely refers to bacteria and fungi.

deme Local population or interbreeding group within a larger population.

density-dependent Varying in relation to population density.

density-independent Unaffected by population density.

detritivore Organism that feeds on dead organic matter; usually applies to detritus-feeding organisms other than bacteria and fungi.

detritus Fresh to partly decomposed plant and animal matter.

diameter at breast height (DBH) Diameter of a tree measured 4 feet, 6 inches (1.4 m) from ground level.

diffuse competition Type of competition in which a species experiences interference from numerous other species that deplete the same resources.

dimorphism Existing in two structural forms, two color forms, two sexes, and the like.

diploid Having chromosomes in homologous pairs, or twice the haploid numbers of chromosomes.

directional selection Selection favoring individuals at one extreme of the phenotype in the population.

dispersal Leaving an area of birth or activity for another area.

dispersion Distribution of organisms within a population over an area.

disruptive selection Selection in which two extreme phenotypes in the population leave more offspring than the intermediate phenotype that has lower fitness.

diversity Abundance in number of species in a given location.

dominance (Ecological) control within a community over environmental conditions influencing associated species by one or several species, plant or animal, enforced by number, density, or growth form; (social) behavioral, hierarchical order in a population that gives high-ranking individuals priority of access to essential requirements; (genetic) ability of an allele to mask the expression of an alternative form of the same gene in a heterozygous condition.

dominant Population possessing ecological dominance in a given community and thereby governing type and abundance of other species in the community.

dormant State of cessation of growth and suspended biological activity during which life is maintained.

drought resistance Sum of drought tolerance and drought avoidance.

drought tolerance Ability of plants to maintain physiological activity in spite of the lack of water or to survive the drying of tissues.

dystrophic Term applied to a body of water with a high content of humic organic matter, often with high littoral productivity and low plankton productivity.

ecological efficiency Percentage of biomass produced by one trophic level that is incorporated into biomass of the next highest trophic level.

ecological release Expansion of habitat or increase in food availability resulting from release of a species from interspecific competition.

ecotone Transition zone between two structurally different communities; often termed edge. (See **edge.**)

ecotype Subspecies or race adapted to a particular set of environmental conditions.

ectothermy Determination of body temperature primarily by external thermal conditions.

edaphic Relating to soil.

edge Place where two or more vegetation types meet.

edge effect Response of organisms, animals in particular, to environmental conditions created by the edge.

egestion Elimination of undigested food material.

emigration Movement of part of a population permanently out of an area.

endemic Restricted to a given region.

endothermy Regulation of body temperature by internal heat production; allows maintenance of an appreciable difference between body temperature and external temperature.

energy The capacity to do work.

entropy Transformation of matter and energy to a more random, more disorganized state.

environment Total surroundings of an organism, including other plants and animals and embracing those of its own kind.

ephemeral Organisms with a short life cycle; lasting only a season or a few days.

epifauna Benthic organisms that live on or move over the surface of a substrate.

epilimnion Upper, warm, oxygen-rich zone of a lake that lies above the thermocline.

epiphyte Organism that lives wholly on the surface of plants, deriving support but not nutrients from the plants.

equilibrium species Species whose population exists in equilibrium with resources and at a stable density.

equilibrium turnover rate Change in species composition per unit of time when immigration equals extinction.

equitability Evenness of distribution of species abundance patterns; maximum equitability is the same number of individuals in all species.

estuary Partially enclosed embayment where fresh water and seawater meet and mix.

ethology Study of animal behavior.

eukaryotic Cell that has membranous organelles, notably the nucleus.

euphotic zone Surface layer of water to the depth of light penetration where photosynthetic production equals respiration.

eutrophic Term applied to a body of water with high nutrient content and high productivity.

eutrophication Nutrient enrichment of a body of water; called cultural eutrophication when accelerated by the introduction of massive amounts of nutrients by human activity.

evapotranspiration Sum of the loss of moisture by evaporation from land and water surfaces and by transpiration from plants.

evolution Change in gene frequency through time resulting from natural selection and producing cumulative changes in characteristics of a population.

evolutionary time Period during which a population evolves and becomes adapted to an environment by means of genetic change.

exon Segment of a gene that codes for the amino acids.

exponential growth Instantaneous rate of population growth expressed as proportional increase per unit of time.

fecundity Potential ability of an organism to produce eggs or young.

fen Wetlands dominated by sedges in which peat accumulates.

field capacity Amount of water held by soil against the force of gravity.

fine-grained Qualities or characteristics of an environment that occur in patches so small relative to the activity of organisms that they do not distinguish among them.

fitness Genetic contribution by an individual's descendants to future generations.

fixation Process in soil by which certain chemical elements essential for plant growth are converted from a soluble or exchangeable form to a less soluble or nonexchangeable form.

floating reserve Individuals in a population of a territorial species that do not hold territories and remain unmated but are available to refill territories vacated by death of an owner.

flux Flow of energy from a source to a sink or receiver.

food chain Movement of energy and nutrients from one feeding group of organisms to another in a series that begins with plants and ends with carnivores, detrital feeders, and decomposers.

food web Interlocking pattern formed by a series of interconnecting food chains.

forb Herbaceous plant other than grass, sedge, or rush.

formation Classification of vegetation based on dominant life forms.

founder principle Concept stating that when a new population started by a small number of colonists isolated from the parent population contains only a small and often biased sample of genetic variation of the parent population, a markedly different, new population may result.

free-running cycle Length of a circadian rhythm in the absence of an external time cue.

frugivore Organism that feeds on fruit.

fugitive species Species characteristic of temporary habitats.

functional response Change in rate of exploitation of a prey species by a predator in relation to changing prey density.

fundamental niche Total range of environmental conditions under which a species can survive.

fynbos Sclerophyllous shrub community occurring in regions of South Africa with a Mediterranean climate.

gamma diversity Diversity differences between similar habitats in widely separated geographic regions.

gap-phase replacement Successional development in small disturbed areas within a stable plant community; filling in of a space left by a disturbance, but not necessarily by the species eliminated by the disturbance.

gene Unit material of inheritance; more specifically, a small unit of DNA molecule coded for a specific protein to produce one of the many attributes of a species.

gene flow Exchange of genetic material between populations.

gene frequency Actually allele frequency; relative abundance of different alleles carried by an individual or a population.

genetic drift Random fluctuation in allele frequency over time due to chance occurrence alone without any influence by natural selection. Important in small populations.

genetic feedback Evolutionary response of a population to adaptations of predators, parasites, or competitors.

genotype Genetic constitution of an organism.

geometric rate of increase Factor by which size of a population increases over a period of time.

gley soil Soil developed under conditions of poor drainage resulting in reduction of iron and other elements and in gray colors and mottles.

globally stable System that returns to initial conditions after all disturbances.

granivore Organism that feeds on seeds.

greenhouse effect Selective energy absorption by carbon dioxide in the atmosphere that allows short-wavelength energy to pass through but absorbs longer wavelengths and reflects heat back to Earth.

group selection Elimination of one group of individuals by another group of individuals possessing superior genetic traits. (Not a widely accepted hypothesis.)

growth form Morphological category of plants, such as tree, shrub, and vine.

guild Group of populations that utilizes a gradient of resources in a similar way.

habitat Place where a plant or animal lives.

halophyte Plant able to survive and complete its life cycle at high salinities.

haploid Having a single set of unpaired chromosomes in each cell nucleus.

hemicryptophyte Plants characterized by perennial shoots or buds close to the surface of the ground; often covered with litter.

herbivore Organism that feeds on plant tissue.

heterogeneity State of being mixed in composition; can refer to genetic or environmental conditions.

heterotrophic Requiring a supply of organic matter or food from the environment.

heterozygous Containing two different alleles of a gene, one from each parent, at the corresponding loci of a pair of chromosomes.

hibernation Winter dormancy in animals characterized by a great decrease in metabolism.

histosol Soil characterized by high organic matter content.

homeostasis Maintenance of nearly constant conditions in function of an organism or in interaction among individuals in a population.

home range Area over which an animal ranges throughout the year.

homoiohydric Ability to maintain a stable internal water balance independent of the environment.

homoiotherm Animal with a fairly constant body temperature (also spelled "homeotherm" and "homotherm").

homoiothermy Regulation of body temperature by physiological means.

homozygous Containing two identical alleles of a gene at the corresponding loci of a pair of chromosomes.

horizon Major zones or layers of soil, each with its own particular structure and characteristics.

host Organism that provides food or other benefits to another organism of a different species; usually refers to an organism exploited by a parasite.

humus Organic material derived from partial decay of plant and animal matter.

hybrid Plant or animal resulting from a cross between genetically different parents.

hyperthermia Rise in body temperature to reduce thermal differences between an animal and a hot environment, thus reducing the rate of heat flow into the body.

hypha Filament of a fungus thalli or vegetative body.

hypolimnion Cold, oxygen-poor zone of a lake that lies below the thermocline.

immigration Arrival of new individuals into a habitat or population.

immobilization Conversion of an element from inorganic to organic form in microbial or plant tissue, rendering the nutrient relatively unavailable to other organisms.

importance value Sum of relative density, relative dominance, and relative frequency of a species in a community.

imprinting Type of rapid learning at a particular and early stage of development in which the individual learns identifying characteristics of another individual or object.

incipient lethal temperature Temperature at which a stated fraction of a population of poikilothermic animals (usually 50 percent) will die when brought rapidly to it from a different temperature.

inclusive fitness Sum of the total fitness of an individual and the fitness of its relatives, weighted according to the degree of relationship.

infaunal Organisms living within a substrate.

infralittoral Region below the littoral region of the sea.

innate capacity for increase (*r*) Intrinsic growth rate of a population under ideal conditions without inhibition from competition.

instar Form of insect or other arthropod between successive molts.

interdemic selection Group selection of populations within a species.

interference competition Competition in which access to a resource is limited by the presence of a competitor.

interspecific Between individuals of different species.

introgression Incorporation of genes of one species into the gene pool of another.

introns Intervening, noncoding sequences in the structure of a gene that separate segments that code for amino acids (exons).

inversion (Genetic) reversal of part of a chromosome so that genes within that part lie in reverse order; (meteorological) increase rather than decrease in air temperature with height caused by radiational cooling of Earth (radiational inversion) or by compression and consequent heating of subsiding air masses from high-pressure areas (subsidence inversion).

isolating mechanism Any structural, behavioral, or physiological mechanism that blocks or inhibits gene exchange between two populations.

iteroparous Multiple-brooded.

karotype Characteristic chromosomes of a species determined by morphological appearance of chromosomes in the first metaphase of meiosis.

kin selection Differential reproduction among groups of closely related individuals.

krumholtz Stunted form of trees characteristic of transition zone between alpine tundra and subalpine coniferous forest.

***K*-selection** Selection under carrying-capacity conditions.

laterization Soil-forming process in hot, humid climates characterized by intense oxidation resulting in loss of bases and in a deeply weathered soil composed of silica, sesquioxides of iron and aluminum, clays, and residual quartz.

lentic Pertaining to standing water, as lakes and ponds.

life table Tabulation of the mortality and survivorship schedule of a population.

life zone Major area of plant and animal life equivalent to a biome; transcontinental region or belt characterized by particular plants and animals and distinguished by temperature differences; applies best to mountainous regions where temperature changes accompany changes in altitude.

limiting resource Resource or environmental condition that most limits the abundance and distribution of an organism.

limnetic zone Shallow-water zone of a lake or sea in which light penetrates to the bottom.

lithosol Soil showing little or no evidence of soil development and consisting mainly of partly weathered rock fragments or nearly barren rock.

littoral Shallow water of a lake in which light penetrates to the bottom, permitting submerged, floating, and emergent vegetative growth; also, shore zone of tidal water between high-water and low-water marks.

locally stable System that tends to return to its initial conditions following a small disturbance.

locus Site on a chromosome occupied by a specific gene.

logistic curve S-shaped curve of population growth that increases slowly at first, steepens, and then flattens out at asymptote, determined by carrying capacity.

logistic equation Mathematical expression for the population growth curve in which rate of increase decreases linearly as population size increases.

log normal distribution Frequency distribution of species abundance in which the x-axis is expressed in a logarithmic scale; the x-axis represents the number of individuals and the y-axis the number of species.

lotic Pertaining to flowing water.

maquis Sclerophyllous shrub vegetation in the Mediterranean region.

marl Earthy, unconsolidated deposit formed in freshwater lakes that consists chiefly of calcium carbonate mixed with clay and other impurities.

marsh Wetland dominated by grassy vegetation such as cattails and sedges.

mattoral Sclerophyllous shrub vegetation in regions of Chile with a mediterranean climate.

mediterranean-type climate Semiarid climate characterized by hot, dry summers and wet, mild winters.

meiofauna Benthic organisms within the size range of 1 to 0.1 mm; interstitial fauna.

meiosis Two successive divisions by gametic cells, with only one duplication of chromosomes so that the number of chromosomes in the two resulting nuclei is one-half the diploid or original number.

mesic Moderately moist habitat.

metalimnion Transition zone in a lake between hypolimnion and epilimnion; region of rapid temperature decline.

micelle Soil particle of clay or humus carrying + electrical charges at the surface.

microclimate Climate on a very local scale that differs from the general climate of the area; influences the presence and distribution of organisms.

migration Intentional, directional, usually seasonal movement of animals between two regions or habitats; involves departure and return of the same individual; a round-trip movement.

mimicry Resemblance of one organism to another or to an object in the environment; evolved to deceive predators.

mineralization Microbial breakdown of humus and other organic matter in soil to inorganic substances.

mire Wetland characterized by an accumulation of peat.

mitosis Division of the cell nucleus into two nuclei identical to the original one with a full complement of chromosomes genetically the same as the original ones.

moder Type of forest humus layer in which plant fragments and mineral particles form loose, netlike structures held together by a chain of small arthropod droppings.

mollisol Soil formed by calcification characterized by accumulation of calcium carbonate in lower horizons and high organic content in upper horizons.

monoecious In plants, occurrence of reproduction organs of both sexes on the same individual, either as different flowers (hermaphroditic) or in the same flower (dioceous).

monogamy Mating of an animal and maintenance of a pair bond with only one member of the opposite sex at a time.

montane Pertaining to mountains.

mor Type of forest humus layer consisting of unincorporated organic matter usually matted or compacted, or both, and distinct from mineral soil; low in bases and acid in reaction.

morphology Study of the form of organisms.

mull Humus that contains appreciable amounts of mineral bases and forms a humus-rich layer of forested soil consisting of mixed organic and mineral matter; blends into the upper mineral layer without abrupt changes in soil characteristics.

Mullerian mimicry Resemblance of two or more conspicuously marked distasteful species, which increases predator avoidance.

mutation Transmissible changes in structure of gene or chromosome.

mutualism Relationship between two species in which both benefit.

mycelium Mass of hyphae that make up the vegetative portion of a fungus.

mycorrhizae Association of fungus with roots of higher plants that improves the plants' uptake of nutrients from the soil.

natural selection Differential reproduction and survival of individuals that result in elimination of maladaptive traits from a population.

negative feedback Homeostatic control in which an increase in some substance or activity ultimately inhibits or reverses the direction of the processes leading to the increase.

nekton Aquatic animals that are able to move at will through the water.

neritic Marine environment embracing the regions where land masses extend outward as a continental shelf.

net above-ground production Accumulation of biomass in above-ground parts of plants over a given period of time.

net production Accumulation of total biomass over a given period of time; energy left over after respiration is deducted from gross production.

net reproductive rate Number of young a female can be expected to produce during a lifetime.

niche Functional role of a species in the community, including activities and relationships.

niche breadth Range of a single niche dimension occupied by a population.

numerical response Change in size of a population of predators in response to change in density of its prey.

oligotrophic Term applied to a body of water low in nutrients and in productivity.

ombrotrophic Condition in bogs or mires in which water is highly deficient in mineral salts and is low in pH.

omnivore Animal that feeds on both plant and animal matter.

opportunistic species Organisms able to exploit temporary habitats or conditions.

optimal foraging Tendency of animals to harvest food efficiently—to select food sizes or food patches that will result in maximum food intake for energy expended.

optimum yield Amount of material that can be removed from a population that will result in production of maximum amount of biomass on a sustained yield.

oscillation Regular fluctuations with a fixed cycle in a population.

osmosis Movement of water molecules across a differentially permeable membrane in response to a concentration or pressure gradient.

overturn Vertical mixing of layers in a body of water brought about by seasonal changes in temperature.

oxisol Soil developed under humid semitropical and tropical conditions characterized by silicates and hydrous oxides, clays, residual quartz, deficiency in bases, and low plant nutrients; formed by the process of laterization.

paleoecology Study of ecology of past communities by means of fossil record.

parapatric Having ranges coming into contact but not overlapping by much more than the dispersal range of an individual in its lifetime.

parapatric speciation Separation into different species of a parent population and a previously isolated subpopulation with a fixed gene mutation; if the subpopulation expands its range and comes in contact with the parent population, hybrids are selected against, producing a tension zone that prevents each population from penetrating the other, ultimately resulting in formation of a new species.

parasitism Relationship between two species in which one benefits while the other is harmed (although not usually killed directly).

parasitoid Insect larva that kills its host by completely consuming the host's soft tissues before pupation or metamorphosis into an adult.

peat Unconsolidated material consisting of undecomposed and only slightly decomposed organic matter under conditions of excessive moisture.

pelagic Referring to the open sea.

permafrost Permanently frozen soil.

perturbation Another word for disturbance; borrowed from physics to suggest an event that alters the state of or direction of change in a system.

phagioclimax Climax community maintained by a continuous human activity.

phanerophyte Plants characterized by perennial buds carried high up in the air; trees, shrubs, and vines.

phenology Study of seasonal changes in plant and animal life and the relationship of these changes to weather and climate.

phenotype Physical expression of a characteristic of an organism as determined by genetic constitution and environment.

phenotypic plasticity Ability to change form under different environmental conditions.

pheromone Chemical substance released by an animal that influences behavior of others of the same species.

photic Lighted water column of a lake or ocean.

photoperiodism Response of plants and animals to changes in relative duration of light and dark.

photosynthesis Synthesis of carbohydrates from carbon dioxide and water by chlorophyll using light as energy and releasing oxygen as a by-product.

phreatophyte Type of plant that habitually obtains its water supply from the zone of groundwater.

phyletic gradualism Evolutionary process in which a species is gradually transformed over time into a different species.

physiological longevity Maximum lifespan of an individual in a population under given environmental conditions.

phytoplankton Small, floating plant life in aquatic ecosystems; planktonic plants.

plankton Small, floating or weakly swimming plants and animals in freshwater and marine ecosystems.

podzolization Soil-forming process resulting from acid leaching of the A horizon and accumulation of iron, aluminum, silica, and clays in the lower horizon.

poikilohydric Condition in an organism in which its internal water state matches that of the environment.

poikilothermy Variation of body temperature with external conditions.

polyandry Mating of one female with several males.

polygyny Mating of one male with several females.

polymorphic Having several distinct forms in a population.

polymorphism Occurrence of more than one distinct form of a species.

polyploid Having three or more times the haploid number of chromosomes.

positive feedback Control in a system that reinforces a process in the same direction.

potential evapotranspiration Amount of water that would be transpired under constantly optimal conditions of soil moisture and plant cover.

predation Act of one living organism consuming another living organism.

primary production Production by green plants.

primary succession Vegetational development starting from a new site never before colonized by life.

production Amount of energy formed by an individual, population, or community per unit of time.

profundal Deep zone in aquatic ecosystems below the limnetic zone.

prokaryotic Any organisms whose cells lack a true membrane about the nucleus or other organelles.

punctuated equilibrium Evolution occurring not as a series of continuous sequences of changes but rather as a series of rapid changes followed by long periods in which few if any characters evolve.

pycnocline Layer of water that exhibits a rapid change in density.

rain shadow Dry area on the lee side of mountains.

raised bog Bog in which the accumulation of peat has raised its surface above both the surrounding landscape and the water table; it develops its own perched water table.

realized niche Portion of fundamental niche space occupied by a population in the face of competition from populations of other species; environmental conditions under which a population survives and reproduces in nature.

recombination Exchange of genetic material resulting from independent assortment of chromosomes and their genes during gamete production followed by a random mix of different sets of genes at fertilization.

recruitment Addition of reproduction of new individuals to a population.

red queen hypothesis States that a species must continually evolve in order to keep pace with a continually changing biotic and abiotic environment.

regolith Mantle of unconsolidated material below the soil from which soil develops.

relict Surviving species of a once widely dispersed group.

reproductive isolation Separation of one population from another by the inability to produce viable offspring when the two populations are mated.

resilience Ability of a system to absorb changes and return to its original condition.

resistance Ability of a system to resist changes from a disturbance.

resource Environmental component utilized by a living organism.

resource allocation Action of apportioning the supply of a resource to a specific use.

respiration Metabolic assimilation of oxygen accompanied by production of carbon dioxide and water, release of energy, and breaking down of organic compounds.

rheotrophic Applies to wetlands, especially bogs, that obtain much of their nutrient input from groundwater.

rhizobia Bacteria capable of living mutualistically with higher plants.

richness Component of species diversity; the number of species present in an area.

riparian Along banks of rivers and streams; river bank forests are often called gallery forests.

***r*-selection** Selection under low population densities; favors high reproductive rates under conditions of low competition.

saprophage Organism that feeds on dead plant and animal matter.

sclerophyll Woody plant with hard, leathery, evergreen leaves that prevent moisture loss.

secondary production Production by consumer organisms.

semelparity Having only one, terminal reproductive effort in a lifetime.

semiarid Region of fairly dry climate with precipitation between 25 and 60 cm a year and with an evapotranspiration rate high enough so the potential loss of water to the environment exceeds inputs.

senescense Process of aging.

seral Series of stages that follow one another in succession.

serotinous cones Cones of pine that remain on the tree several years and require the heat of fire to open them and release the seeds.

sessile Not free to move about; permanently attached to a substrate.

sibling species Species with similar appearance but unable to interbreed.

sigmoid curve S-shaped curve of logistic growth.

site Combination of biotic, climatic, and soil conditions that determine an area's capacity to produce vegetation.

soil association Group of defined and named soil taxonomic units occurring together in an individual and characteristic pattern over a geographic region.

soil horizon Developmental layer in the soil with its own characteristics of thickness, color, texture, structure, acidity, nutrient concentration, and the like.

soil profile Distinctive layering of horizons in the soil.

soil series Basic unit of soil classification consisting of soils that are essentially alike in all major profile characteristics except texture of the A horizon. Soil series are usually named for the locality where the typical soil was first recorded.

soil texture Relative proportions of the three particle sizes—sand, silt, and clay—in the soil.

soil type Lowest unit in the natural system of soil classification, consisting of soils that are alike in all characteristics, including texture of the A horizon.

speciation Separation of a population into two or more reproductively isolated populations.

species diversity Measurement that relates density of organisms of each type present in a habitat to the number of species in the habitat.

species packing Increase in species diversity within a comparatively narrow range of resource variation.

species selection Form of group selection in which sets of species with different characteristics increase by speciation or decrease by extinction at different rates because of differences in adaptive characteristics.

specific heat Amount of energy that must be added or removed to raise or lower the temperature of a substance by a specific amount.

spiraling Mechanism of retention of nutrients in flowing water ecosystems involving the interdependent processes of nutrient recycling and downstream transport.

spodosol Soil characterized by the presence of a horizon in which organic matter and amorphous oxides of aluminum and iron have precipitated; includes podzols.

stability Ability of a system to resist change or to recover rapidly after a disturbance; absence of fluctuations in a population.

stabilizing selection Selection favoring the middle in the distribution of phenotypes.

stable age distribution Constant proportion of individuals of various age classes in a population through population changes.

stand Unit of vegetation that is essentially homogenous in all layers and differs from adjacent types qualitatively and quantitatively.

stochastic Patterns arising from random factors.

stratification Division of an aquatic or terrestrial community into distinguishable layers on the basis of temperature, moisture, light, vegetative structure, and other such factors, creating zones for different plant and animal types.

sublittoral Lower division of the sea from about 40 to 60 m to below 200 m.

subspecies Geographical unit of a species population distinguishable by certain morphological, behavioral, or physiological characteristics.

succession Replacement of one community by another; often progresses to a stable terminal community called the climax.

sustained yield Yield per unit of time from an exploited population equal to production per unit of time.

swamp Wooded wetland in which water is near or above ground level.

switching A predator's changing its diet from a less abundant to a more abundant prey species.

symbiosis Living together of two or more species.

sympatric Living in the same area; usually refers to overlapping populations.

sympatric speciation Speciation without geographical isolation.

synecology Study of groups of organisms in relation to their environment; community ecology.

taiga Needle-leaved coniferous forest bordering the northern subpolar regions; boreal forest.

territory Area defended by an animal; varies among animals according to social behavior, social organization, and resource requirements of different species.

thermal conductance Rate at which heat flows through a substance.

thermocline Layer in a thermally stratified body of water in which temperature changes rapidly relative to the remainder of the body.

thermogenesis Increase in production of metabolic heat to counteract the loss of heat to a colder environment.

therophyte Annual plant that survives unfavorable periods as seeds.

threshold of security Point in local population density at which the predator turns its attention to other prey (see **switching**) because of harvesting efficiency; the segment of prey population below the threshold is relatively secure from predation.

time lag Delay in response to change.

torpidity Temporary condition of an animal involving a great reduction in respiration; results in loss of power of motion and feeling; usually occurs in response to some unfavorable environmental condition, such as heat or cold, to reduce energy expenditure.

translocation Transport of materials within a plant; absorption of minerals from soil into roots and their movement throughout the plant.

transpiration Loss of water vapor by land plants.

trophic Related to feeding.

trophic level Functional classification of organisms in an ecosystem according to feeding relationships from first-level autotrophs through succeeding levels of herbivores and carnivores.

trophogenic zone Upper layer of the water column in lakes and ponds in which light is sufficient for photosynthesis.

tundra Areas in arctic and alpine (high-mountain) regions characterized by bare ground, absence of trees, and growth of mosses, lichens, sedges, forbs, and low shrubs.

turnover rate Rate of replacement of a substance or a species when losses to a system are replaced by additions.

univoltine Breeding only once a season.

upwelling Vertical movement of water that brings nutrients from the ocean depths to the surface, usually near coastal regions.

vagile Free to move about.

Wallace's line Biogeographic line between the islands of Borneo and the Celebes that marks the eastward boundary of many landlocked Eurasian organisms and the boundary of the Oriental region.

watershed Entire region drained by a waterway that empties into a lake or reservoir; total area above a given point on a stream that contributes water to the flow at that point; the topographic dividing line from which surface streams flow in two different directions.

wilting point Moisture content of soil on an oven-dry basis at which plants wilt and fail to recover their turgidity when placed in a dark, humid atmosphere.

xeric Dry conditions, especially relating to soil.

xerophyte Plant adapted to life in a dry or physiologically dry (saline) habitat.

zoogeography Study of the distribution of animals.

zooplankton Floating or weakly swimming animals in freshwater and marine ecosystems; planktonic animals.

BIBLIOGRAPHY

Abele, L. G., and E. F. Connor
1979. Application of island biogeographic theory to refuge design: Making the right decisions for the wrong reasons. In *Proceedings of First Conference on Scientific Research in the National Parks,* ed. R. M. Linn, 89–94. Vol. 1. Washington, D.C.: U.S. Dept. Interior.

Aber, J. S., D. B. Botkin, and J. M. Melillo
1978. Predicting the effects of different harvesting regimes on forest floor dynamics in northern hardwoods. *Can. J. For. Res.* **8**:305–315.

Adkisson, P. L.
1964. Action of the photoperiod in controlling insect diapause. *Amer. Natur.* **98**:357–374.
1966. Internal clocks and insect diapause. *Science* **154**:234–241.

Ahmadjian, V., and J. B. Jacobs
1982. Algal-fungal relationships in lichens: Recognition, synthesis, and development. In *Algal symbiosis: A continuum of interaction strategies,* ed. L. J. Goff. New York: Cambridge University Press.

Alcock, J.
1973. Cues used in searching for food by red-winged blackbirds (*Agelaius phoeniceus*). *Behaviour* **46**:174–188.

Alexander, R. D., and G. Borgia
1978. Group selection, altruism, and levels of organization of life. *Ann. Rev. Ecol. Syst.* **9**:449–474.

Alexander, R. D., and T. E. Moore
1962. The evolutionary relationships of 17-year and 13-year cicadas and three new species (Homoptera, Cicadidae, *Magicicada*). *Museum Zoo. Mscl. Publ., Univ. Michigan,* **121**:1–59.

Alexander, R. D., and D. W. Tinkle, eds.
1981. *Natural selection and social behavior: Recent research and new theories.* New York: Chiron.

Alm, G.
1952. Year class fluctuations and span of life of perch. *Rept. Inst. Freshwater Res. Drottningholm* **33**:17–38.

Amadon, D.
1947. Ecology and evolution of some Hawaiian birds. *Evolution* **1**:63–68.
1950. The Hawaiian honeycreepers (Aves, Drepanididae). *Bull. Amer. Mus. Nat. Hist.* **100**:397–451.

American Chemical Society
1969. *Cleaning our environment: The chemical basis for action.* Washington, D.C.: American Chemical Society.

Anderson, J. M., and P. MacFadyen, eds.
1976. *The Role of Terrestrial and Aquatic Organisms in the Decomposition Process.* Seventeenth Symposium, British Ecological Society. Oxford, England: Blackwell.

Anderson, M., and M. O. G. Eriksson
1982. Nest parasitism in goldeneyes, *Bucephala clangula:* Some evolutionary aspects. *Amer. Natur.* **120**:1–16.

Anderson, N. H., and K. W. Cummins
1979. Influences of diet on the life histories of aquatic insects. *J. Fish. Res. Bd. Can.* **36**:335–342.

Anderson, R. C., and A. K. Prestwood
1981. Lungworms. In *Diseases and parasites of white-tailed deer,* ed. W. R. Davidson, 266–317. Mscl. Publ. no. 7, Tall Timbers Research Station. Tallahassee, Fla.: Tall Timbers Research Station.

Andrewartha, H.G., and L. C. Birth
1954. *The distribution and abundance of animals.* Chicago: University of Chicago Press.

Andrews, R., D. C. Coleman, J. E. Ellis, and J. S. Singh
1974. Energy flow relationships in a short-grass prairie ecosystem. *Proc. 1st Inter. Cong. Ecol.,* 22–28. The Hague: W. Junk Publishers.

Andrews, R., and A. S. Rand
1974. Reproductive effort in anoline lizards. *Ecology* **55**:1317–1327.

Andrzejewska, L., and G. Gyllenberg
1980. Small herbivore subsystem. In *Grasslands, systems analysis and man,* ed. A. I. Breymeyer and G. M. Van Dyne, 201–268. International Biological Programme no. 19. Cambridge, England: Cambridge University Press.

Antonovics, J.
1971. The effects of a heterogenous environment on the genetics of natural populations. *Amer. Sci.* **59**:593–599.

Arno, S. F.
1976. *The historical role of fire in the Bitterroot National Forest.* U.S.D.A. For. Serv. Res. Paper **INT 187.** Ogden, Utah: U.S. Forest Service.

Arno, S. F., and K. M. Sneck
1977. *A method for determining fire history in the coniferous forests of the mountain West.* U.S.D.A. For. Serv. Gen. Tech. Rept. **INT 42.** Ogden, Utah: U.S. Forest Service.

Aschoff, J.
1958. Tierische Periodik unter dem Einfluss von Zietgebern. *Z. F. Tierpsychol.* **15**:1–30.

Ash, J. E., and J. F. Barkham
1976. Changes and variability in the field layer of a coppiced woodland in Norfolk, England. *J. Ecol.* **64**:697–712.

Auclair, A. N., and F. G. Goff
1971. Diversity relationships of upland forests in the western Great Lakes area. *Amer. Natur.* **105**:499–528.

Avery, R. A.
1975. Clutch size and reproductive effort in the lizard *Lacerta vivipara* Jaquin. *Oecologica* **19**:165–170.

Baer, J. C.
1951. *Ecology of animal parasites.* Urbana: University of Illinois Press.

Baes, C. F., Jr., H. E. Goeller, J. S. Olson, and R. M. Rotty
1977. Carbon dioxide and climate: The uncontrolled experiment. *Amer. Sci.* **65**:310–320.

Bailey, R. W.
1976. *Ecoregions of the United States.* Ogden, Utah: U.S.D.A. Forest Service Intermountain Region.

Baker, M. C., L. R. Mewaldt, and R. M. Stewart
1981. Demography of white-crowned sparrows (*Zonitrichia leucophrys nuttali*). *Ecology* **62**:636–644.

Bakker, R. T.
1983. The deer flees, the wolf pursues: Incongruencies in predator-prey coevolution. In *Coevolution,* ed. D. Futuyma and M. Slaktin, 350–382. Sunderland, Mass.: Sinauer Associates.

Barber, H. S.
1951. North American fireflies of the genus *Photuris. Smithsonian Inst. Misc. Collections* **117**:1–58.

Barbour, M. G., J. H. Burk, and W. D. Pitts
1980. *Terrestrial plant ecology.* Menlo Park, Calif.: Benjamin/Cummings Publishing Company.

Barlow, J. P.
1956. Effect of wind on salinity distribution in an estuary. *J. Marine Res.* **15**:193–203.

Barlow, N., ed.
1958. *The autobiography of Charles Darwin 1809–1882.* 1969 reprint. New York: W. W. Norton Co.

Barrett, J. A.
1983. Plant-fungus symbioses. In *Coevolution,* ed. D. Futuyma and M. Slatkin, 137–160. Sunderland, Mass.: Sinauer Associates.

Barsdale, R. J., and V. Alexander
1975. Nitrogen balance of arctic tundra: Pathways, rates and environmental implications. *J. Env. Quality* **4**:111–117.

Barsdale, R. J., T. Fenchel, and R. T. Prentki
1974. Phosphorus cycle of model ecosystems: Significance for decomposer food chains and effect of bacterial grazers. *Oikos* **25**:239–251.

Basler, D. S., H. H. Dill, and H. K. Nelson
1968. Effect of predator reduction on waterfowl nesting success. *J. Wildl. Manage.* **32**:669–682.

Bassam, J. A.
1965. Photosynthesis: The part of carbon. In *Plant biochemistry,* ed. J. Bonner and J. E. Varnea. New York: Academic.

Batchelder, R. B.
1967. Spatial and temporal patterns of fire in the tropical world. *Proc. 6th Tall Timbers Fire Ecology Conf.,* 171–208.

Batzli, G. O., and F. A. Pitelka
1970. Influence of meadow mouse populations on California grassland. *Ecology* **51**:1027–1039.

Batzli, G. O., R. G. White, and F. L. Bunnell
1981. Herbivory: A strategy of tundra consumers. In *Tundra ecosystems: A comparative analysis,* ed. L. C. Bliss, O. W. Heal, and J. J. Moore, 359–375. International Biological Programme no. 25. Cambridge, England: Cambridge University Press.

Batzli, G. O., R. G. White, S. F. MacLean, F. A. Pitelka, and B. D. Collier
1980. The herbivore-based trophic system. In *An arctic ecosystem: The coastal tundra at Barrow, Alaska,* ed. J. Brown, P. C. Miller, L. L. Tieszen, and F. L. Bunnell, 335–410. US/IBP Synthesis Series, no. 12. Stroudsburg, Pa.: Dowden, Hutchinson & Ross.

Bawa, K. S.
1980. Evolution of dioecy in flowering plants. *Ann. Rev. Ecol. Syst.* **11**:15–39.

Bazzaz, F. A.
1975. Plant species diversity in old field successional ecosystems in southern Illinois. *Ecology* **56**:485–488.

Beacham, T. D.
1980. Dispersal during population fluctuations of the vole *Microtus townsendii. J. Anim. Ecology* **49**:867–877.

Beals, E.
1968. Spatial pattern of shrubs on a desert plain in Ethiopia. *Ecology* **49**:744–746.

Beck, S. D.
1980. *Insect photoperiodism.* 2d ed. New York: Academic Press.

Bekoff, M.
1977. Mammalian dispersal and ontogeny of individual behavioral

phenotypes. *Amer. Natur.* **111**:715–732.

Bell, G.
1980. The costs of reproduction and their consequences. *Amer. Natur.* **116**:45–76.

Bendell, J. F.
1974. Effects of fire on birds and mammals. In *Fire and ecosystems,* ed. T. Kozlowski and C. E. Ahlgren, 73–138. New York: Academic Press.

Bentley, P. J.
1977. Extrafloral nectaries and protection by pugnacious bodyguards. *Ann. Rev. Ecol. Syst.* **8**:407–427.

Berger, P. J., N. C. Negus, E. H. Sanders, and P. D. Gardner
1981. Chemical triggering of reproduction in *Microtus montanus. Science* **214**:69–70.

Bergerud, A. T.
1971. The population dynamics of Newfoundland caribou. *Wildl. Monogr.* no. **25**:1–55.

Bernard, J. M., and E. Gorham
1978. Life history aspects of primary production in sedge wetlands. In *Freshwater wetlands,* ed. R. E. Good, D. F. Whigham, and R. L. Simpson, 39–51. New York: Academic Press.

Bernier, B.
1961. Forest humus, a consequence and cause of local ecological conditions. Mimeo, Northeast Forest Soils Conference.

Berry, J., and O. Bjorkman
1980. Photosynthetic response and adaptation to temperature in higher plants. *Ann. Rev. Plant Physiol.* **31**:391–453.

Berthold, P.
1974. Circannual rhythms in birds with different migratory habits. In *Circannual clocks: Annual biological rhythms,* ed. E. T. Pengelley, 55–94. New York: Academic Press.

Bertram, B. C. R.
1975. Social factors influencing reproduction in wild lions. *J. Zool. Lond.* **177**:463–482.

Biel, E. R.
1961. Microclimate, bioclimatology, and notes on comparative dynamic climatology. *Amer. Sci.* **49**:326–357.

Billings, W. D., and H. A. Mooney
1968. The ecology of arctic and alpine plants. *Bio. Rev.* **43**:481–529.

Birch, L. C., and D. P. Clark
1953. Forest soil as an ecological community with special reference to the fauna. *Quart. Rev. Biol.* **28**:13–36.

Bird, R. D.
1930. Biotic communities of the aspen parkland of central Canada. *Ecology* **11**:356–442.

Bitman, J.
1970. Hormonal and enzymatic activity of DDT. *Ag. Sci. Rev.* **7 (4)**: 6–12.

Bjorkman, O.
1973. Comparative studies on photosynthesis in higher plants. In *Photobiology,* ed. A. C. Giese, 1–63. Vol. 8. New York: Academic Press.

Black, C. C., Jr.
1971. Ecological implications of dividing plants into groups with distinct photosynthetic production capabilities. *Adv. Ecol. Res.* **7**:87–114.
1973. Photosynthetic carbon fixation in relation to net CO_2 uptake. *Ann. Rev. Plant Physiol.* **24**:253–286.

Blackman, F. F.
1905. Optima and limiting factors. *Ann. Bot.* **19**:281–298.

Blair, A. P.
1942. Isolating mechanisms in a complex of four species of toad. *Biol. Symp.* **6**:235–249.

Blair, W. F.
1955. Mating call and stage of speciation in the *Microhyla olivaceae–M. carolinensis* complex. *Evolution* **9**:469–480.

Bliss, L. C.
1975. Devon Island, Canada. In *Structure and function of tundra ecosystems,* ed. T. Rosswall and O. W. Heal, 17–60. Stockholm: Swedish Natural Science Research Council.

Bliss, L. C., G. M. Courtin, D. L. Pattie, R. R. Wiewe, D. W. A. Whitfield, and P. Widden
1973. Arctic tundra ecosystems. *Ann. Rev. Ecol. Syst.* **4**:359–399.

Bliss, L. C., O. W. Heal, and J. J. Moore, eds.
1981. *Tundra ecosystems: A comparative analysis.* International Biological Programme no. 25. Cambridge, England: Cambridge University Press.

Blum, M. S., and N. A. Blum, eds.
1979. *Sexual selection and reproductive competition in insects.* New York: Academic Press.

Boag, P. T., and P. R. Grant
1981. Intense natural selection in a population of Darwin's finches. *Science* **214**:82–85.

Bock, W.
1970. Microevolutionary sequences as a fundamental concept in macroevolutionary models. *Evolution* **24**:704–722.

Bodola, A.
1966. Life history of the gizzard shad *Dorosoma cepedianum* (Le Sueur) in western Lake Erie. *U.S. Fish and Wildlife Service Fish. Bull.* **65**:391–425.

Bogert, C. M.
1960. The influence of sound on the

behavior of amphibians and reptiles. In *Animal sound and communications,* ed. W. E. Lanyon and W. N. Tagvolga, 137–320. AIBS Symposium Series, publ. no. 7.

Bolin, B.
1970. The carbon cycle. In *Scientific American: The biosphere,* 47–56. San Francisco: Freeman.

Bolin, B., and C. D. Keeling
1963. Large scale atmospheric mixing as deduced from seasonal and meridianal variations of carbon dioxide. *J. Geophys. Res.* **68:**3899–3920.

Bonnell, M. L., and R. K. Selander
1974. Elephant seals: Genetic variation and near extinction. *Science* **184:**908–909.

Boring, L. R., C. D. Monk, and W. T. Swank
1981. Early regeneration of a clearcut southern Appalachian forest. *Ecology* **62:**1244–1253.

Bormann, F. H., and G. E. Likens
1979. *Pattern and process in a forested ecosystem.* New York: Springer-Verlag.

Bormann, F. H., G. E. Likens, D. W. Fisher, and R. S. Pierce
1968. Nutrient loss accelerated by clearcutting of a forest ecosystem. In *Symposium on primary productivity and mineral cycling in natural ecosystems,* ed. H. E. Young. Orono, Maine: Maine University Press.

Bormann, F. H., G. E. Likens, and J. M. Melillo
1977. Nitrogen budget for an aggrading northern hardwood forest ecosystem. *Science* **196:**981–982.

Bormann, F. H., G. E. Likens, T. G. Siccama, R. S. Pierce, and J. S. Eaton
1974. The export of nutrients and recovery of stable conditions following deforestation at Hubbard Brook. *Ecol. Monogr.* **44:**255–277.

Botkin, D. B., J. F. Janak, and J. R. Wallis
1972. Some ecological consequences of a computer model of forest growth. *J. Ecol.* **60:**948–972.

Botkin, D. B., and C. R. Malone
1968. Efficiency of net primary production based on light intercepted during the growing season. *Ecology* **49:**438–444.

Boucher, D. H., S. James, and K. H. Keeler
1982. The ecology of mutualism. *Ann. Rev. Ecol. Syst.* **13:**315–347.

Bourliere, F., ed.
1983. Tropical savannas. Ecosystems of the World no. 13. Amsterdam: Elsevier Scientific Publishing.

Bourliere, F., and M. Hadley
1983. Present-day savannas: An overview. In *Tropical savannas,* ed. F. Bourliere, 1–19. Ecosystems of the World no. 13. Amsterdam: Elsevier Scientific Publishing.

Bowers, M. A., and J. M. Brown
1982. Body size and coexistence in desert rodents: Chance or community structure? *Ecology* **63:**391–400.

Box, T. W., J. Powell, and D. I. Drawe
1967. Influence of fire on south Texas chaparral. *Ecology* **48:**955–961.

Bradshaw, A. D.
1972. Some of the evolutionary consequences of being a plant. *Evol. Biol.* **5:**25–44.

Bragg, A. N.
1950. Observations on *Microhyla* (Salientia: Mycrohylidae). *Wasmann J. Biol.* **8:**113–118.

Bray, J. R., and E. Gorham
1964. Litter production in the forests of the world. *Adv. Ecol. Res.* **2:**101–157.

Breymeyer, A. I.
1980. Trophic structure and relationships. In *Grasslands, systems analysis and man,* ed. A. I. Breymeyer and G. M. Van Dyne, 799–819. International Biological Programme no. 19. Cambridge, England: Cambridge University Press.

Breymeyer, A. I., and A. Kajah
1976. Drawing models of two grassland ecosystems, a mown meadow and a pasture. *Polish Ecol. Stud.* **2:**41–49.

Breymeyer, A. I., and G. M. Van Dyne, eds.
1980. *Grasslands, systems analysis and man.* International Biological Programme no. 19. Cambridge, England: Cambridge University Press.

Briand, F.
1983a. Biogeographic patterns in food web organization. In *Current trends in food web theory,* ed. D. L. DeAngelis, W. M. Post, and G. Sugihara, 41–44. ORNL 5983. Oak Ridge, Tenn.: Oak Ridge National Laboratory.
1983b. Environmental control of food web structure. *Ecology* **64:**253–263.

Broecker, W. S.
1970. Man's oxygen reserves. *Science* **168:**1537–1538.
1974. *Chemical oceanography.* New York: Harcourt Brace Jovanovich.

Brooks, M. G.
1955. An isolated population of the varying hare. *J. Wildl. Manage.* **19:**59–61.

Brower, J. E., and J. H. Zar
1977. *Field and laboratory methods for general ecology.* 3d ed. Dubuque, Iowa: Brown.

Brower, J. V. Z.
1958. Experimental studies of mimicry in some North American butterflies: I, The monarch, *Danaus plexippus,* and Viceroy, *Limenitis*

archippus; II, *Battus philenor* and *Papilio troilus, P. polyxenes* and *P. glaucus;* III, *Danaus gilippus berenice* and *Limenitis archippus floridensis. Evolution* **12:**32–47, 123–136, 273–285.

Brower, J. V. Z., and I. P. Brower
1962. Experimental studies of mimicry: 6, The reaction of toads (*Bufo terrestris*) to honeybees (*Apis mellifera*) and their drone mimics (*Eristalis vinetorum*). *Amer. Natur.* **97:**297–307.
1969. Ecological chemistry. *Sci. Amer.* **220** (2): 22–29.

Brown, F. A.
1959. Living clocks. *Science* **130:**1535–1544.

Brown, J., P. C. Miller, L. L. Tieszen, and F. L. Bunnell, eds.
1980. *An arctic ecosystem: The coastal tundra at Barrow, Alaska.* US/IBP Synthesis Series, no. 12. Stroudsburg, Pa.: Dowden, Hutchinson & Ross.

Brown, J. C., and E. R. Brown
1981. Kin selection and individual selection in babblers. In *Natural selection and social behavior: Recent research and new theories,* ed. R. D. Alexander and D. W. Tinkle, 244–256. New York: Chiron.

Brown, J. H., and D. W. Davidson
1977. Competition between seed-eating rodents and ants in desert ecosystems. *Science* **196:**880–882.

Brown, J. H., O. J. Reichman, and D. W. Davidson
1979. Granivory in desert ecosystems. *Ann. Rev. Ecol. Syst.* **10:**201–227.

Brown, J. L.
1964. The evolution of diversity in avian territorial systems. *Wilson Bull.* **76:**160–169.
1969. Territorial behavior and population regulation in birds: A review and reevaluation. *Wilson Bull.* **81:**292–329.

Brown, W. L., and E. O. Wilson
1956. Character displacement. *System. Zool.* **5:**49–64.

Bryant, J. P., and P. J. Kuropat
1980. Selection of winter forage by subarctic browsing vertebrates: The role of plant chemistry. *Ann. Rev. Ecol. Syst.* **11:**261–285.

Bryant, J. P., G. O. Wieland, P. B. Reicherdth, V. E. Lewis, and M. C. McCarthy
1983. Pinosylvin methyl ether deters snowshoe hare feeding on green alder. *Science* **222:**1023–1025.

Brylinsky, M.
1980. Estimating the productivity of lakes and reservoirs. In *The functioning of freshwater ecosystems,* ed. E. D. LeCren and R. H. Lowe-McConnell, 411–418. International Biological Programme no. 22. Cambridge, England: Cambridge University Press.

Buell, M. F., and R. E. Wilbur
1948. Life form spectra of the hardwood forests of the Itasca Park region, Minnesota. *Ecology* **29:**352–359.

Bunning, E.
1960. Circadian rhythm and the time measurement in photoperiod. *Cold Spring Harbor Symposium* **25:**249–256.
1964. *The physiological clock.* 2d ed. New York: Academic Press.

Burgess, R. L., and D. M. Sharpe, eds.
1981. *Forest island dynamics in man-dominated landscapes.* Ecological Studies no. 41. New York: Springer-Verlag.

Burt, W. V., and J. Queen
1957. Tidal overturning in estuaries. *Science* **126:**973–974.

Butler, R. G., and S. Janes-Butler
1982. Territoriality and behavioral correlates of reproductive success of great black-backed gulls. *Auk* **99:**58–66.

Bush, G. L.
1974. The mechanism of sympatric host race formation in the true fruit flies (Tephritidae). In *Genetic mechanisms of speciation in insects,* ed. M. J. D. White, 3–23. Sydney: Australian and New Zealand Book Co.
1975. Modes of animal speciation. *Ann. Rev. Ecol. Syst.* **6:**339–364.

Campanilla, P. J.
1975. The evolution of mating systems in temperate zone dragonflies (Odonata: Anisoptera): II, *Libellula luctuosa. Behaviour* **54:**276–310.

Campanilla, P. J., and L. L. Wolf
1974. Temporal leks as a mating system in a temperate zone dragonfly (Odonata: Anisoptera): I, *Plathernis lydia* (Drury). *Behaviour* **51:**49–87.

Carey, F. G.
1982. A brain heater in the swordfish. *Science* **216:**1327–1329.

Carey, F. G., and J. M. Teal
1966. Heat conservation in tuna fish muscle. *Proc. Nat. Acad. Sci.* **56:**1464–1469.

Carlisle, A., A. H. F. Brown, and E. J. White
1966. The organic matter and nutrient elements in the precipitation beneath a sessile oak canopy. *J. Ecol.* **54:**87–98.

Carr, A.
1962. Orientation problems in the high seas travel and terrestrial movements of marine turtles. *Amer. Scient.* **80:**359–374.

Carrick, R.
1963. Ecological significance of territory size in the Australian magpie *Gymnorhina tibiten. Proc. Int. Orn. Cong.* **13:**740–753.

Carson, H. L.
1968. The population flush and its genetic consequences. In *Population biology and evolution,* ed. R. C. Lewontin, 123–137. Syracuse, N.Y.: Syracuse University Press.

Carson, H. L., and K. Y. Kaneshiro
1976. *Drosophila* of Hawaii: Systematics and ecological genetics. *Ann. Rev. Ecol. Syst.* **7:**311–345.

Casey, T. M., and J. R. Hegel
1981. Caterpillar setae: Insulation for an ectotherm. *Science* **214:**1131–1133.

Caughley, G.
1966. Mortality patterns in mammals. *Ecology* **47:**907–916.
1970. Eruption of ungulate populations, with an emphasis on Himalayan thar in New Zealand. *Ecology* **51:**154–172.
1976a. Plant and herbivore systems. In *Theoretical ecology: Principles and application,* ed. R. M. May, 94–113. Philadelphia: Saunders.
1976b. Wildlife management and the dynamics of ungulate populations. *Appl. Biol.* **1:**183–246.

Cernusa, A.
1976. Energy exchange within individual layers of a meadow. *Oecologica* **23:**141–149.

Chabot, B. F., and J. A. Bunce
1979. Drought-stress effects on leaf carbon balance. In *Topics in plant population biology,* ed. O. T. Solbrig, S. Jain, G. B. Johnson, and P. H. Raven, 338–355. New York: Columbia University Press.

Changnon, S. A.
1968. La Porte weather anomaly: Fact or fiction? *Bull. Amer. Meteorol. Soc.* **49:**4–11.

Chamie, J. P. M., and C. J. Richardson
1978. Decomposition in northern wetlands. In *Freshwater wetlands,* ed. R. E. Good, D. F. Whigham, and R. L. Simpson, 115–130. New York: Academic Press.

Chapin III, F. S., P. C. Miller, W. D. Billings, and P. I. Coyne
1980. Carbon and nutrient budgets and their control in coastal tundra. In *An arctic ecosystem: The coastal tundra at Barrow, Alaska,* ed. J. Brown, P. C. Miller, L. L. Tieszen, and F. L. Bunnell, 458–482. US/IBP Synthesis Series, no. 12. Stroudsburg, Pa.: Dowden, Hutchinson & Ross.

Chapin III, F. S., L. L. Tieszen, M. C. Lewis, P. C. Miller, and B. H. McCown
1980. Control of tundra plant allocation patterns and growth. In *An arctic ecosystem: The coastal tundra at Barrow, Alaska,* ed. J. Brown, P. C. Miller, L. L. Tieszen, and F. L. Bunnell, 140–185. US/IBP Synthesis Series, no. 12. Stroudsburg, Pa.: Dowden, Hutchinson & Ross.

Chapman, V. J.
1960. *Salt marshes and salt deserts of the world.* London: Leonard Hill.

Chappel, M. A.
1978. Behavioral factors in the altitudinal zonation of chipmunks (*Eutamias*). *Ecology* **59:**565–579.

Chappel, M. A., H. C. Heller, and A. V. Calvo
1978. Hypothalmic thermosensitivity and adaptation for heat storage behavior in three species of chipmunks (*Eutamias*) from different thermal environments. *J. Comp. Physiol B.* **126:**158–164.

Chevalier, J. R.
1973. Cannibalism as a factor in the first year survival of walleye in Oneida Lake. *Trans. Amer. Fish. Soc.* **102:**739–744.

Chew, R. M., and A. E. Chew
1970. Energy relationships of the mammals of a desert shrub (*Larrae tridentatia*) community. *Ecol. Monogr.* **40:**1–21.

Chitty, D.
1960. Population processes in the vole and their relevance to general theory. *Can. J. Zool.* **38:**99–113.

Chitty, D., and E. Phipps
1966. Seasonal changes in survival in mixed populations of two species of voles. *J. Anim. Ecol.* **35:**313–331.

Christensen, N. L., and R. K. Peet
1981. Secondary forest succession on the North Carolina piedmont. In *Forest succession: Concepts and applications,* ed. D. C. West, H. H. Shugart, and D. B. Botkin, 230–245. New York: Springer-Verlag.

Christian, J. J.
1971. Fighting, maturity, and population density in *Microtus pennsylvanicus. J. Mammal.* **52:**556–567.
1978. Neurobehavioral endocrine regulation of small mammal populations. In *Populations of small mammals under natural conditions,* ed. D. P. Snyder, 143–158. The Pymatuning Symposia In Ecology, Vol. 5 Special Publication Series, Pymatuning Laboratory of Ecology. Pittsburgh, Pa.: University of Pittsburgh Press.

Christiansen, F. B., and T. M. Fenchel
1977. *Theories of population in biological communities.* New York: Springer-Verlag.

Clack, L.
1975. Subspecific intergradation and zoogeography of the painted turtle (*Chrysemys picta*) in northern West Virginia. MS thesis, West Virginia University.

Clark, C.
1967. *Population growth and land use.* London: Macmillan; New York: St. Martin's Press.

Clark, F. E.
1969a. Ecological associations among soil microorganisms. In *Soil biology,* 125–161. Paris: UNESCO.
1969b. The microbial flora of grassland soil and some microbial influences on ecosystem functions. In *The grassland ecosystem,* ed. R. L. Dix and R. G. Beidleman. Range Sci. Dept. Ser., no. 2. Fort Collins: Colorado State University.

Clark, F. E., C. Cole, and R. A. Bowman
1980. Nutrient cycling. In *Grasslands, systems analysis and man,* ed. A. I. Breymeyer and G. Van Dyne, 659–712. International Biological Programme no. 19. Cambridge, England: Cambridge University Press.

Clarke, B., and J. Murray
1969. Ecological genetics and speciation in land snails of the genus *Partula. Biol. J. Linn. Soc.* **1**:31–42.

Clarke, J. F.
1969. Nocturnal urban boundary layer over Cincinnati, Ohio. *Monthly Weather Rev.* **97**:582–589.

Clausen, J. D., D. D. Keck, and W. M. Hiesey
1948. Experimental studies on the nature of species: 3, Environmental responses of climatic races of *Achillea. Carnegie Ins. Wash. Pub.* **581.**

Clements, F. E.
1916. *Plant succession.* An analysis of the development of vegetation. *Carnegie Inst. Wash. Pub.* **242.**

Cloudsley-Thompson, J. L.
1956. Studies in diurnal rhythms: VII. Humidity responses and nocturnal activity in woodlice (Isopoda). *J. Exp. Biol.* **33**:576–582.
1960. Adaptive functions of circadian rhythms. *Cold Spring Harbor Symp. Quant. Biol.* **255**:345–355.

Cochrane, G. R.
1968. Fire ecology in southeastern Australian sclerophyll forests. *Proc. Tall Timbers Fire Ecology Conf.* **8**:15–40.

Cody, M. L.
1966. A general theory of clutch size. *Evolution* **20**:174–184.
1969. Convergent characteristics in sympatric species: A possible relation to interspecific competition and aggression. *Condor* **71**:222–239.

Cody, M. L., and J. H. Brown
1970. Character divergence in Mexican birds. *Evolution* **24**:304–310.

Cody, M. L., and J. Diamond, eds.
1975. *Ecology and evolution of communities.* Cambridge, Mass.: Harvard UniversityPress.

Cohen, J. E.
1978. *Food webs and niche space.* Princeton, N.J.: Princeton University Press.

Coker, R. E.
1947. *This great and wide sea.* Chapel Hill: University of North Carolina Press.

Cole, D. W., S. D. Gessel, and S. F. Dice
1968. Distribution and cycling of nitrogen, phosphorus, potassium, and calcium in a second-growth Douglas-fir ecosystem. In *Symposium on Primary Productivity and Mineral Cycling in Natural Ecosystems.* Orono: University of Maine Press.

Cole, D. W., and M. Rapp
1981. Elemental cycling in forest ecosystems. In *Dynamic properties of forest ecosystems,* ed. D. E. Reichle, 341–409. New York: Cambridge University Press.

Cole, L. C.
1946. A study of the cryptozoa of an Illinois woodland. *Ecol. Monogr.* **16**:49–86.

Connell, J. H.
1961. The influence of interspecific competition and other factors on the distribution of the barnacle *Chthamalus stellatus. Ecology* **42**:710–723.

Connell, J. H., and R. O. Slatyer
1977. Mechanisms of succession in natural communities and their role in community stability and organization. *Amer. Natur.* **111**:1119–1144.

Cook, C. W., L. A. Stoddard, and L. E. Harris
1954. The nutritive value of range plants in the Great Basin. *Utah Agr. Exp. Stat. Bull.* No. **372.**

Cook, R. E.
1979. Patterns of juvenile mortality and recruitment in plants. In *Topics in plant population biology,* ed. O. T. Solbrig, S. Jain, G. B. Johnson, and P. H. Raven, 207–231. New York: Columbia University Press.

Cook, R. M., and B. J. Cockerell
1978. Predator ingestion rate and its bearing on feeding time and the theory of optimal diets. *J. Anim. Ecol.* **47**:529–548.

Cooke, G. D.
1967. The pattern of autotrophic succession in laboratory microcosms. *Bioscience* **17**:717–721.

Cope, O. B.
1971. Interaction between pesticides and wildlife. *Ann. Rev. Entomol.* **16**:325–364.

Cornwallis, R. K.
1969. Farming and wildlife conservation in England and Wales. *Biol. Cons.* **1**:142–147.

Correll, D. L.
1978. Estuarine productivity. *Bioscience* **28:**646–650.

Cott, H. B.
1940. *Adaptive coloration in animals.* London: Oxford University Press.

Coupland, R. T.
1950. Ecology of mixed prairie in Canada. *Ecol. Monogr.* **20:**217–235.

Coupland, R. T., ed.
1979. *Grassland ecosystems of the world: Analysis of grasslands and their uses.* International Biological Programme no. 18. Cambridge, England: Cambridge University Press.

Cowan, R. L.
1962. Physiology of nutrition as related to deer. In *Proc. 1st Natl. White-tailed Deer Symposium,* 1–8. Athens: Center for Continuing Education, University of Georgia.

Cox, C. B., I. N. Healey, and P. D. Moore
1973. *Biogeography: An ecological and evolutionary approach.* Oxford, England: Blackwell.

Cox, C. R., and B. J. Le Boeuf
1977. Female incitation of male competition: A mechanism of male selection. *Amer. Natur.* **111:**317–355.

Cox, G. W.
1976. *Laboratory manual of general ecology.* Dubuque, Iowa: Wm. C. Brown.

Cox, G. W., and R. E. Ricklefs
1977. Species diversity and ecological release in Caribbean land bird fauna. *Oikos* **28:**113–122.

Crawford, R. M. M.
1978. Metabolic adaptations to anoxia. In *Plant life in anaerobic environments,* ed. D. Hook and R. M. M. Crawford, 119–136. Ann Arbor, Mich.: Ann Arbor Science.

Crawford, R. M. M., and M. Barnes
1977. Tolerance of anoxia and ethanol metabolism in tree roots. *New Phytologist* **79:**519–526.

Crocker, R. L., and J. Major
1955. Soil development in relation to vegetation and surface age at Glacier Bay, Alaska. *J. Ecol.* **43:**427–448.

Cronin, E. L., and A. J. Mansueti
1971. The biology of the estuary. In *A Symposium on the Biological Significance of Estuaries,* ed. P. A. Stroud and R. H. Stroud, 14–39. Washington, D.C.: Sport Fishing Institute.

Crowley, J. M.
1967. Biogeography. *Can. Geog.* **11:**312–326.

Cummins, K. W.
1974. Structure and function of stream ecosystems. *Bioscience* **24:**631–641.
1979. The multiple linkages of forests to streams. In *Forests: Fresh perspectives from ecosystem analysis,* Proceedings 40th Ann. Biol. Colloquium, ed. R. H. Waring, 191–198. Corvallis: Oregon State University Press.

Cummins, K. W., and M. J. Klug
1979. Feeding ecology of stream invertebrates. *Ann. Rev. Ecol. Syst.* **10:**147–172.

Curio, E.
1976. *The ethology of predation.* New York: Springer-Verlag.

Curtis, J. T.
1959. *The vegetation of Wisconsin.* Madison: University of Wisconsin Press.

Dahl, E.
1953. Some aspects of the ecology and zonation of the fauna of sandy beaches. *Oikos* **4:**1–27.

Darlington, P. J.
1957. *Zoogeography: The geographical distribution of animals.* New York: Wiley.

Darwin, C.
1844. Essay of 1844. In C. Darwin and A. R. Wallace, *Evolution by natural selection,* 1958 reprint ed., 91–254. Cambridge, England: Cambridge University Press.
1859. *The origin of species.* London: Murray.

Dash, M. C., and A. K. Hota
1980. Density effects on the survival, growth rate, and metamorphosis of *Rana tigrina* tadpoles. *Ecology* **61:**1025–1028.

Daubenmire, R. F.
1968a. *Plant communities: A textbook of plant synecology.* New York: Harper & Row.
1968b. Soil moisture in relation to vegetation distribution in the mountains of northern Idaho. *Ecology* **49:**431–438.

Davidson, D. W.
1977a. Foraging ecology and community organization in desert seed-eating ants. *Ecology* **58:**725–737.
1977b. Species diversity and community organization in desert seed-eating ants. *Ecology* **58:**711–724.

Davidson, W. R., ed.
1981. *Diseases and parasites of white-tailed deer.* Southeastern Cooperative Wildlife Disease Study, Mscl. Publ. no. 7, Tall Timbers Research Station. Tallahassee, Fla.: Tall Timbers Research Station.

Davies, N. B.
1977. Prey selection and social behavior in wagtails (Aves: Motacillidae). *J. Anim. Ecol.* **46:**37–57.
1978. Ecological questions about territorial behaviour. In *Behavioural ecology: An evolutionary approach,* ed. J. R. Krebs and N. B. Davies, 317–350. Oxford, England: Blackwell.

Davis, C. B., and A. G. Van der Valk
1978. Litter decomposition in prairie glacial marshes. In *Freshwater wetlands,* ed. R. E. Good, D. F. Whigham and R. L. Simpson, 99–144. New York: Academic Press.

Davis, D. E.
1978. Physiological and behavioral responses to the social environment. In *Populations of small mammals under natural conditions,* ed. D. P. Snyder, 84–91. Vol. 5 Special Publication Ser., Pymatuning Laboratory of Ecology. Pittsburgh: University of Pittsburgh Press.

Dean, R., L. E. Ellis, R. W. Rice, and R. E. Bemeret
1975. Nutrient removal by cattle from a short-grass prairie. *J. App. Ecol.* **12**:25–29.

DeAngelis, D. L., W. M. Post, and G. Sugihara, eds.
1983. *Current trends in food web theory.* Report on a Food Web Workshop, ORNL 5983. Oak Ridge, Tenn.: Oak Ridge National Laboratory.

DeBano, L. F., P. H. Dunn, and C. E. Conrad
1977. Fire's effect on physical and chemical properties of chaparral soils. In *Proc. Symp. Environmental Consequences of Fire and Fuel Management in Mediterranean Ecosystems,* ed. H. A. Mooney and C. E. Conrad, 65–74. U.S.D.A. For. Serv. Gen. Tech. Rept. **WO-3.** Washington, D.C.: U.S. Department of Agriculture.

DeCoursey, P. J.
1960a. Daily light sensitivity rhythm in a rodent. *Science* **131**:33–35.
1960b. Phase control of activity in a rodent. *Cold Spring Harbor Symp. Quant. Biol.* **25**:49–54.
1961. Effect of light on the circadian activity rhythm of the flying squirrel, *Glaucomys volans. Z. Vergleich. Physiol.* **44**:331–354.

Deevey, E. S.
1947. Life tables for natural populations of animals. *Quart. Rev. Biol.* **22**:283–314.

DeLong, K. T.
1966. Population ecology of feral house mice: Interference by *Microtus. Ecology* **47**:481–484.

Detling, J. K.
1979. Processes controlling blue grama production in short-grass prairies. In *Perspectives in grassland ecology,* ed. N. French, 25–42. New York: Springer-Verlag.

Devlin, R. M., and A. V. Barber
1971. *Photosynthesis.* New York: Van Nostrand Reinhold.

Diamond, J.
1969. Avifaunal equilibria and species turnover rates on the Channel Islands of California. *Proc. Nat. Acad. Sci.* **64**:57–63.
1971. Comparison of faunal equilibrium turnover rates on a tropical island and a temperate island. *Proc. Nat. Acad. Sci.* **68**:2742–2745.
1975. The island dilemma: Lessons of modern biogeographic studies for the design of natural preserves. *Biol. Cons.* **7**:129–146.

Diamond, J., and R. M. May
1976. Island biogeography and conservation: Strategy and limitations. In *Theoretical ecology: Principles and application,* ed. R. M. May. Philadelphia: Saunders.

Dice, L. R.
1943. *The biotic provinces of North America.* Ann Arbor: University of Michigan Press.

Dix, R. L.
1960. The effects of burning on the mulch structure and species composition of grassland in western North Dakota. *Ecology* **41**:49–56.

Dix, R. L., and F. E. Smeins
1967. The prairie, meadow, and marsh vegetation of Nelson County, North Dakota. *Can. J. Bot.* **45**:21–58.

Dolbeer, R. A., and W. R. Clark
1975. Population ecology of snowshoe hares in the central Rocky Mountains. *J. Wildl. Manage.* **39**:535–549.

Doust, L. L., and P. B. Cavers
1982. Sex and gender dynamics in the jack-in-the-pulpit *Arisaema triphyllum* (Araceae). *Ecology* **63**:797–808.

Dowding, P., F. S. Chapin III, F. E. Wielgolski, and P. Kilfeather
1981. Nutrients in tundra ecosystems. In *Tundra ecosytems: A comparative analysis,* ed. L. C. Bliss, O. W. Heal, and J. J. Moore, 647–683. International Biological Programme no. 25. Cambridge, England: Cambridge University Press.

Dowdy, W. W.
1944. The influence of temperature on vertical migration of invertebrates inhabiting different soil types. *Ecology* **25**:449–460.

Downs, A. A., and W. E. McQuilkin
1944. Seed production of southern Appalachian oaks. *J. For.* **42**:913–920.

Doyle, T. W.
1981. The role of disturbance in the dynamics of a montane rain forest: An application of a tropical forest succession model. In *Forest succession: Concepts and Applications,* ed. D. C. West, H. H. Shugart, and D. B. Botkin, 56–73. New York: Springer-Verlag.

Drake, J. A.
1983. Invasability in Lotka-Volterra interaction webs. In *Current trends in food web theory,* ed. D. C. DeAngelis, W. M. Post, and G. Sugihara, 83–90. ORNL 5983. Oak Ridge, Tenn.: Oak Ridge National Laboratory.

Drury, W. H., and C. T. Nisbet
1972. Succession. *J. Arnold Arboretum* **54**:331–368.

Duffey, E.
1974. *Grassland ecology and wildlife management*. London: Chapman and Hall.

Duvigneaud, P., and S. Denaeyer-DeSmet
1970. Biological cycling of minerals in a temperate deciduous forest. In *Analysis of temperate forest ecosystems,* ed. D. Reichle, 199–225. New York: Springer-Verlag.

Dybas, H. S., and M. Lloyd
1962. Isolation by habitat in two synchronized species of periodical cicadas (Homoptera, Cicadidae, *Magicada*). *Ecology* **43**:444–459.

Dyer, M. I.
1980. Mammalian epidermal growth factor promotes plant growth. *Proc. Nat. Acad. Sci.* **77**:4836–4837.

Eckardt, F. E., G. Heim, M. Methy, B. Saugier, and R. Sauvezon
1971. Fonctionnement à un écosystème au niveau de la production primaire measures effectuées dans une culture d'*Helianthus annus*. *Oecol. Plant*. **6**:51–100.

Edmonds, R. L., ed.
1982. *Analysis of coniferous forest ecosystems in the western United States*. Stroudsburg, Pa.: Hutchinson Ross.

Edwards, C. A.
1973. *Environmental pollution by pesticides*. New York: Plenum.

Edwards, C. A., and A. R. Thompson
1973. Pesticides and the soil fauna. *Residue Review* **41**:1–65.

Edwards, N. T., H. H. Shugart, S. B. McLaughten, W. F. Harris, and D. E. Reichle
1981. Carbon metabolism in terrestrial ecosystems. In *Dynamic properties of forest ecosystems,* ed. D. E. Reichle, 500–536. International Biological Programme no. 23. Cambridge, England: Cambridge University Press.

Egler, F. E.
1953. Vegetation management for rights-of-way and roadsides. *Smithsonian Inst. Ann. Rept.—1953,* 299–322.
1954. Vegetation science concepts. I, Initial floristic composition—a factor in old field vegetation development. *Vegetatio* **4**:412–417.

Ehleringer, J., H. A. Mooney, S. L. Gulman, and P. Rundel
1980. Orientation and its consequences for *Copia poa* (Cactaceae) in the Atacama Desert. *Oecologica* **46**:63–67.

Ehrlich, P. R., O. D. Murphy, M. C. Singer, C. B. Sherwood, R. R. White, and I. L. Brown
1980. Extinction, reduction, stability, and increase: The responses of the checkerspot butterfly (*Euphydryas*) populations to California drought. *Oecologica* **41**:101–105.

Ehrlich, P. R., and P. H. Raven
1964. Butterflies and plants: A study in coevolution. *Evolution* **18**:586–608.

Eldredge, N., and S. J. Gould
1972. Punctuated equilibria: An alternative to phyletic gradualism. In *Models in paleobiology,* ed. T. M. J. Schopf, 82–115. San Francisco: Freeman, Cooper and Co.

Eisner, T.
1970. Chemical defense against predation in arthropods. In *Chemical ecology,* ed. E. Sondheimer and J. Simeone. New York: Academic Press.

Eisner, T., and J. Meinwald
1966. Defensive secretions of arthropods. *Science* **153**:1341–1350.

Elner, R. W., and R. N. Hughes
1978. Energy maximization in the diet of the shore crab *Carcinus maenas*. *J. Anim. Ecol.* **47**:103–116.

Elton, C.
1927. *Animal ecology*. London: Sidgwick and Jackson.

Eltringham, S. K.
1971. *Life in mud and sand*. New York: Crane, Russak.

Emanuelsson, A., E. Erikisson, and H. Egner
1954. Composition of atmospheric precipitation in Sweden. *Tellus* **3**:261–267.

Emlen, S. T., and L. W. Oring
1977. Ecology, sexual selection, and the evolution of mating systems. *Science* **197**:215–223.

Engle, L. G.
1960. Yellow-poplar seedfall pattern. *Central States For. Exp. Stat. Note* **143**.

Erikisson, E.
1952. Composition of atmospheric precipitation: I, Nitrogen compounds. *Tellus* **4**:214–232.
1963. The yearly circulation of sulfur in nature. *J. Geophys. Res.* **68**:4001–4008.

Errington, P. L.
1946. Predation and vertebrate populations. *Quart. Rev. Biol.* **22**:144–177, 221–245.
1963. *Muskrat populations*. Ames: Iowa State University Press.

Estes, R. D., and J. Goddard
1967. Prey selection and hunting behavior of the African wild dog. *J. Wildl. Manage*. **31**:52–70.

Etherington, J. R.
1982. *Environmental and plant ecology,* 2d ed. New York: Wiley.

Evans, F. C., and S. A. Cain
1952. Preliminary studies on the vegetation of an old-field community in

southeastern Michigan. *Contrib. Lab. Vert. Biol.*, University of Michigan **51**:1–17.

Eyre, S. R.
1963. *Vegetation and soils: A world picture*. Chicago: Aldine.

Farner, D. S.
1959. Photoperiodic control of annual gonadal cycles in birds. In *Photoperiodism and related phenomena*, ed. R. B. Withrow, 717–758. Washington, D.C.: American Association for the Advancement of Science.
1964a. The photoperiodic contol of reproductive cycles in birds. *Amer. Scient.* **52**:137–156.
1964b. Time measurement in vertebrate photoperiodism. *Amer. Natur.* **98**:375–386.

Farner, D. S., and R. A. Lewis
1971. Photoperiodism and reproductive cycles in birds. In *Photophysiology*, ed. A. C. Giese, 325–364. Vol. 6. London: Academic Press.

Farnsworth, E., and F. Golley
1973. *Fragile ecosystems*. New York: Springer-Verlag.

Federer, C. A., and C. B. Tanner
1966. Spectral distribution of light in the forest. *Ecology* **47**:555–560.

Feeney, P.
1975a. Biochemical coevolution between plants and their insect herbivores. In *Coevolution of animals and plants*, ed. L. E. Gilbert and P. H. Raven, 3–19. Austin: University of Texas Press.
1975b. Plant apparency and chemical defenses. In *Biochemical Interactions between Plants and Insects, Rec. Adv. Phytochem.*, ed. J. Wallace and R. Mansell, **10**:1–40.

Feinsinger, P.
1983. Coevolution and pollination. In *Coevolution*, ed. D. J. Futuyma and M. Slatkin, 282–310. Sunderland, Mass.: Sinauer Associates.

Fenchel, T.
1974. Character displacement and coexistence in mud snails (Hydrobridia). *Oecologica* **20**:19–32.

Fenchel, T., and P. Harrison
1976. The significance of bacterial grazing and mineral cycling for the decomposition of particulate detritus. In *The role of terrestrial and aquatic organisms in decomposition processes*, ed. J. Anderson and A. MacFayden, 285–299. Oxford: Blackwell Scientific Publications.

Fenner, F., and F. N. Ratcliffe
1965. *Myxamatosis*. Cambridge, England: Cambridge University Press.

Ficken, R. W., M. S. Ficken, and D. H. Morse
1968. Competition and character displacement in two sympatric pine-dwelling warblers (*Dendroica*, Parulidae). *Evolution* **22**:307–314.

Fischer, A. G.
1960. Latitudinal variation in organic diversity. *Evolution* **14**:64–81.

Fisher, R. A.
1930. *The genetical theory of natural selection*. Oxford: Clarendon Press.

Fisher, S. G., and G. E. Likens
1973. Energy flow in Bear Brook, New Hampshire: An integrated approach to stream ecosystem metabolism. *Ecol. Monogr.* **43**:421–439.

Fitzpatrick, L. C.
1973. Energy allocation in the Allegheny Mountain salamander *Desmognathus ochrophaeus*. *Ecol. Monogr.* **43**:43–58.

Flanagan, P. W., and F. L. Bunnell
1980. Microflora activities and decomposition. In *An arctic ecosystem: The coastal tundra at Barrow, Alaska*, ed. J. Brown, P. C. Miller, L. L. Tieszen, and F. L. Bunnell, 291–334. US/IBP Synthesis Series, no. 12. Stroudsburg, Pa.: Dowden, Hutchinson & Ross.

Flint, R. F.
1970. *Glacial and quaternary geology*. New York: Wiley.

Flowers, J. J., P. F. Troke, and A. R. Yeo
1977. The mechanism of salt tolerance in halophytes. *Ann. Rev. Plant Physiol.* **28**:89–121.

Forbes, S. A.
1887. The lake as a microcosm. *Bull. Peoria Sci. Assoc.*: 77–87; reprinted 1925, *Ill. Natural History Surv. Bull.*, **15**:537–550.

Forcier, L. K.
1975. Reproductive strategies and co-occurrence of climax tree species. *Science* **189**:808–810.

Forel, F. A.
1901. *Handbuch der seenkunde, allgemeine limnologie*. Stuttgart: J. Engelhorn.

Forman, R. T., A. E. Galli, and C. F. Leck
1976. Forest size and avian diversity in New Jersey woodlots with some land use implications. *Oecologica* **26**:1–8.

Fortney, J. L.
1974. Interaction between yellow perch abundance, walleye predation, and survival of alternate prey in Oneida Lake, New York. *Trans. Amer. Fish. Soc.* **105**:15–24.

Fortney, R. H.
1975. The vegetation of Canaan Valley: A taxonomic and ecological study. Ph.D. dissertation, West Virginia University.

Fowells, H. A., ed.
1965. Silvics of forest trees of the United States. *U.S.D.A. Agricultural Handbook* **271**. Washington, D.C.: U.S. Department of Agriculture.

Fowler, C. W.
1981. Density dependence as related to life history strategy. *Ecology* **62**:602–610.

Fowler, C. W., and J. MacMahon
1982. Selective extinction and speciation: Their influences on the structure of communities and ecosystems. *Amer. Natur.* **119**:480–498.

Fox, B. J.
1981. Niche parameters and species richness. *Ecology* **62**:1415–1425.

Fox, L. R.
1975a. Cannibalism in natural populations. *Ann. Rev. Ecol. Syst.* **6**:87–106.
1975b. Factors influencing cannibalism, a mechanism of population limitation in the predator *Notonecta hoffmanni*. *Ecology* **56**:933–941.
1975c. Some demographic consequences of food shortage for the predator *Notonecta hoffmanni*. *Ecology* **56**:868–880.

Fox, M. W.
1980. *The soul of the wolf*. Boston: Little, Brown and Company.

Franklin, J. F., and C. T. Dyrness
1973. Natural vegetation of Oregon and Washington. *U.S.D.A. For. Serv. Gen. Tech. Rept. PNW* **8**. Corvallis, Oreg.: USDA Forest Service.

Franklin, J. F., and M. A. Hemstrom
1981. Aspects of succession in the coniferous forests of the Pacific northwest. In *Forest succession: Concepts and applications*, ed. D. C. West, H. H. Shugart, and D. B. Botkin, 212–229. New York: Springer-Verlag.

Franz, H.
1950. *Bodenzoologie als Grundlage der Bodenpflege*. Berlin: Akademie-Verlag.

Franzblau, M. A., and J. P. Collins
1980. Test of hypotheses of territory regulation in an insectivorous bird by experimentally increasing prey abundance. *Oecologica* **46**:164–170.

Freeman, C. L., K. T. Harper, and E. L. Charnov
1980. Sex change in plants: Old and new observations and new hypotheses. *Oecologica* **47**:222–232.

French, N.
1979. Principal subsystem interactions in grasslands. In *Perspectives in grassland ecology*, ed. N. French, 173–190. New York: Springer-Verlag.

French, N., ed.
1979. *Perspectives in grassland ecology*. New York: Springer-Verlag.

French, N. R., R. K. Steinhorst, and D. M. Swift
1979. The grassland biomass trophic pyramids. In *Perspectives in grassland ecology*, ed. N. French, 59–87. New York: Springer-Verlag.

Frenzel, G.
1936. *Untersuchungen uber die Tierwelt des Weisenboden*. Jena, E. Germany: Gustav Fischer.

Fretwell, S. D., and H. L. Lucas
1969. On territorial behavior and other factors influencing habitat distribution in birds. *Acta Biotheoretica* **19**:16–36.

Friche, H., and S. Friche
1977. Monogamy and sex change by aggressive dominance in coral reef fish. *Nature* **266**:830–832.

Fritts, S. H., and L. D. Mech
1981. Dynamics, movements, and feeding ecology of a newly protected wolf population in northwestern Minnesota. *Wildl. Monogr.* **80**:1–89.

Fryer, G., and T. D. Iles
1972. *The cichlid fishes of the Great Lakes of Africa*. Neptune City, N.J.: T. F. H. Publications.

Futuyma, D. J., and M. Slatkin, eds.
1983. *Coevolution*. Sunderland, Mass.: Sinauer Associates.

Gadgil, M., and O. T. Solbrig
1972. The concept of r- and K- selection: Evidence from wild flowers and some theoretical considerations. *Amer. Natur.* **106**:14–31.

Galli, A. E., C. F. Leck, and R. T. T. Forman
1976. Avian distribution patterns on forest islands of different sizes in central New Jersey. *Auk* **93**:356–364.

Gant, R. E., and E. E. C. Clebsch
1975. The allelopathic influences of *Sassafras albidum* in old field succession in Tennessee. *Ecology* **56**:604–615.

Gates, D.
1962. *Energy exchange in the biosphere*. New York: Harper & Row.

Gaunt, S. L.
1980. Thermoregulation in doves (Columbidae): A novel esophageal heat exchanger. *Science* **210**:445–447.

Gause, G. F.
1934. *The struggle for existence*. Baltimore: Williams & Wilkins.

Gersper, P. L., V. Alexander, S. A. Barkley, R. J. Barsdale, and P. S. Flint
1980. The soils and their nutrients. In *An arctic ecosystem: The coastal tundra at Barrow, Alaska*, ed. J. Brown, P. C. Miller, L. L. Tieszen, and F. L. Bunnell, 219–254. US/IBP Synthesis Series, no. 12. Stroudsburg, Pa.: Dowden, Hutchinson & Ross.

Ghent, A. W.
1958. Studies of regeneration in forest stands devastated by the spruce

budworm. II, Age, height growth, and related studies of balsam fir seedlings. *For. Sci.* **4**:135–146.

Ghilarov, M. S.
1970. Soil biocoenosis. In *Methods of study in soil ecology,* ed. J. Phillipson, 67–77. Paris: UNESCO.

Ghiselin, J.
1981. Applied ecology. In *Handbook of contemporary developments in world ecology,* ed. E. J. Kormondy and J. F. McCormick. Westport, Conn.: Greenwood Press.

Gilbert, L. E.
1975. Ecological consequences of a coevolved mutualism between butterflies and plants. In *Coevolution of animals and plants,* ed. L. E. Gilbert and P. H. Raven, 210–240. Austin: University of Texas Press.

Gilbert, L. E., and P. H. Raven, eds.
1975. *Coevolution of animals and plants.* Austin: University of Texas Press.

Gill, A. M.
1977. Plant traits adaptive to fires in Mediterranean land ecosystems. In *Proc. Symp. Environmental Consequences of Fire and Fuel Management in Mediterranean Ecosystems,* ed. H. A. Mooney and C. E. Conrad, 17–20. U.S.D.A. For. Serv. Gen. Tech. Rept. **WO-3.** Washington, D.C.: U.S. Department of Agriculture.

Gill, F. B., and L. L. Wolf
1975. Economics of feeding territoriality in the golden-winged sunbird. *Ecology* **56**:333–345.

Gillon, Y.
1983. The invertebrates of the grass layer. In *Tropical savannas,* ed. F. Bourliere, 289–311. Ecosystems of the World no. 13. Amsterdam: Elsevier Scientific Publishing.

Gilpin, M. E., and J. M. Diamond
1980. Subdivision of nature reserves and the maintenance of species diversity. *Nature* **285**:567–568.

Gist, C. S., and D. A. Crossley, Jr.
1975. A model of mineral element cycling for an invertebrate food web in a southeastern hardwood forest litter community. In *Mineral cycling in southeastern ecosystems,* ed. F. G. Howell, J. B. Gentry, and M. H. Smith, 84–106. ERDA Symposium Series. Springfield, Va.: National Technical Information Service, U.S. Department of Commerce.

Gleason, H. A.
1917. The structure and development of the plant association. *Bull. Torrey Bot. Club* **44**:463–481.
1926. The individualistic concept of the plant association. *Bull. Torrey Bot. Club* **53**:7–26.
1927. Further views on the succession concept. *Ecology* **8**:299–326.

Godshalk, G. L., and R. G. Wetzel
1978. Decomposition in the littoral zone of lakes. In *Freshwater wetlands,* ed. R. E. Good, D. F. Whigham, and R. L. Simpson, 131–143. New York: Academic Press.

Goff, L. J.
1982. Symbiosis and parasitism: Another viewpoint. *Bioscience* **32**:255–256.

Golley, F. B.
1960. Energy dynamics of a food chain of an old-field community. *Ecol. Monogr.* **30**:187–206.
1972. Energy flux in ecosystems. In *Ecosystem structure and function,* ed. J. S. Wiens, 69–90. Biol. Colloquium 31st. Corvallis: Oregon State University Press.
1975. Productivity and mineral cycling in tropical forests. In *Productivity of world ecosystems.* Washington, D.C.: National Academy of Science.

Golley, F. B., and H. Leith
1972. Basis of organic production in the tropics. In *Tropical ecology with an emphasis on organic production,* ed. P. M. Golley and F. B. Golley, 1–26. Athens: University of Georgia Press.

Golley, F. B., K. Petrusewicz, and L. Ryszkowski, eds.
1975. *Small mammals: Their productivity and population dynamics.* Cambridge, England: Cambridge University Press.

Golley, P. M., and F. B. Golley, eds.
1972. *Tropical ecology with an emphasis on organic production.* Athens: University of Georgia Press.

Golterman, H. L., and F. A. Kouwe
1980. Chemical budgets and nutrient pathways. In *The functioning of freshwater ecosystems,* ed. E. D. LeCren and R. H. Lowe-McConnell, 88–140. International Biological Programme no. 22. Cambridge, England: Cambridge University Press.

Gomez-Pompa, A., and C. Vazquez-Yanes
1981. Successional studies of a rain forest in Mexico. In *Forest succession: Concepts and applications,* ed. D. C. West, H. H. Shugart, and D. B. Botkin, 246–266. New York: Springer-Verlag.

Good, R. E., D. F. Whigham, and R. L. Simpson, eds.
1978. *Freshwater wetlands.* New York: Academic Press.

Goreau, T. F.
1963. Calcium carbonate deposition by coralline algae and corals in relation to their roles as reef builders. *Ann. N.Y. Acad. Sci.* **109**:127–167.

Goss, R. J., C. E. Dinsmore, L. N. Grimes, and J. K. Rosen
1974. Expression and suppression of circannual antler growth cycle in

deer. In *Circannual clocks: Annual biological rhythms,* ed. E. T. Pengelley, 393–422. New York: Academic Press.

Goss-Custard, J. D.
1970. The response of redshank *Tringa totanus* (L) to spatial variation in the density of their prey. *J. Anim. Ecol.* **39:**91–113.
1977a. The energetics of prey selection by redshank *Tringa totanus* (L) in relation to prey density. *J. Anim. Ecol.* **46:**1–19.
1977b. Optimal foraging and size selection of worms by redshank *Tringa totanus. Anim. Behav.* **25:**10–29.

Gosz, J. R., G. E. Likens, and F. H. Bormann
1976. Organic matter and nutrient dynamics of the forest and forest floor in the Hubbard Brook Forest. *Oecologica* **22:**305–320.

Gould, S. J.
1982. The meaning of punctuated equilibrium and its role in validating a hierarchical approach to microevolution. In *Perspectives on evolution,* ed. R. Milkman, 83–104. Sunderland, Mass.: Sinauer Associates.

Gould, S. J., and N. Eldredge
1977. Punctuated equilibria: The tempo and mode of evolution reconsidered. *Paleobiology* **3:**115–151.

Gould, S. J., and R. F. Johnston
1972. Geographic variation. *Ann. Rev. Ecol. Syst.* **3:**457–498.

Gould, S. J., and E. S. Vrba
1982. Exaptation—A missing term in the science of form. *Paleobiology* **8:**4–15.

Graetz, P. D.
1981. Plant–animal interactions. In *Arid-land ecosystems: Structure, functioning and management,* Volume II, ed. D. W. Goodall and R. A. Perry, 85–103. International Biological Programme no. 17. Cambridge, England: Cambridge University Press.

Graham, D., and B. D. Patterson
1982. Responses of plants to low, nonfreezing temperatures: Protein metabolism and acclimation. *Ann. Rev. Plant Physiol.* **33:**347–372.

Grant, P. R.
1981. Speciation and adaptive radiation of Darwin's finches. *Amer. Scient.* **69:**653–663.

Grant, P. R., and B. R. Grant
1980. The breeding and feeding characteristics of Darwin's finches on Isla Genovesa, Galapagos. *Ecol. Monogr.* **50:**381–410.

Grant, P. R., B. R. Grant, J. N. M. Smith, I. J. Abbott, and L. K. Abbott
1976. Darwin's finches: Population variation and natural selection. *Proc. Nat. Acad. Sci.* **73:**257–261.

Grant, P. R., and N. Grant
1983. The origin of a species. *Nat. Hist.* **92** (9): 76–80.

Greenwood, P. H.
1974. The cichlid fishes of Lake Victoria, East Africa: The biology and evolution of a species flock. *Bull. Brit. Mus. (Nat. Hist.) Zool.* **6:**1–134.

Greeson, P. E., J. R. Clark, and J. E. Clark, eds.
1979. *Wetland functions and values: The state of our understanding.* Minneapolis: American Water Resources Association.

Grieg-Smith, P.
1964. *Quantitative plant ecology.* 2d ed. Washington, D.C.: Butterworth.

Grier, C. E., and R. S. Logan
1977. Organic matter distribution and net production in plant communities of a 400-year-old Douglas-fir ecosystem. *Ecol. Monogr.* **47:**373–400.

Grime, J. P.
1966. Shade avoidance and shade tolerance in flowering plants. In *Light as an ecological factor,* ed. F. Bainbridge et al., 187–207. British Ecological Society Symposium no. 6. New York: Wiley.
1977. Evidence for the existence of three primary strategies in plants and its relevance to ecological and evolutionary theory. *Amer. Natur.* **111:**1169–1194.
1979. *Plant strategies and vegetative processes.* New York: Wiley.

Grinnell, J.
1917. The niche relationships of the California thrasher. *Auk* **34:**427–433.
1924. Geography and evolution. *Ecology* **5:**225–229.
1928. Presence and absence of animals. *Univ. Cal. Chron.* **30:**429–450.

Gross, M.
1972. *Oceanography.* Englewood Cliffs, N.J.: Prentice-Hall.

Gruell, G. E.
1983. *Fire and vegetative trends in the northern Rockies: Interpretations from 1871–1982 photographs.* U.S.D.A. Forest Service Gen. Tech. Rept. **INT-158.**

Gutierrez, J. R., and E. R. Fuentes
1979. Evidence for intraspecific competition in the *Acacia cazen* (Leguminosae) savanna of Chile. *Oecol. Plant.* **14:**151–158.

Gwinner, E.
1978. Effects of pinealectomy on circadian locomotor activity rhythms in European starlings *Sturnus vulgaris. J. Comp. Physiol.* **126:**123–129.

Gyllenberg, G.
1980. Bioenergetic parameters of the main group of herbivores. In

Grasslands, systems analysis, and man, ed. A. I. Breymeyer and G. M. Van Dyne, 238–251. International Biological Programme No. 19. Cambridge, England: Cambridge University Press.

Haarlov, N.
1960. Microarthropods from Danish soils. *Oikos.* Supp. no. 3, 1–176.

Hadley, E. B., and B. J. Kieckhefer
1963. Productivity of two prairie grasses in relation to fire frequency. *Ecology* **44**:389–395.

Hairston, N. G., and C. H. Pope
1948. Geographic variation and speciation in Appalachian salamanders (*Plethodon jordani* group). *Evolution* **2**:266–278.

Haldane, J. B. S.
1932. *The causes of evolution.* London: Longmans, Green and Co.
1954. *The biochemistry of genetics.* London: G. Allen.

Halle, F., R. A. A. Oldeman, and P. B. Tomlinson
1978. *Tropical trees and forests: An architectural analysis.* New York: Springer-Verlag.

Halliday, T. R.
1978. Sexual selection and mate choice. In *Behavioural ecology: An evolutionary approach,* ed. J. R. Krebs and N. B. Davies, 180–213. Oxford, England: Blackwell.

Hamilton, W. D.
1964. The genetical evolution of social behavior I, II. *J. Theoret. Biol.* **82**:1–16, 17–52.
1972. Altruism and related phenomena, mainly in insects. *Ann. Rev. Ecol. Syst.* **3**:193–283.

Hammond, E. H.
1964. Classes of land surface form in the forty-eight states, U.S.A. *Annals. Assoc. Amer. Geog.,* v. 54, map supplement no. 4.

Hanson, A. D.
1982. Metabolic responses of mesophytes to plant water deficits. *Ann. Rev. Plant Physiol.* **33**: 163–203.

Hanson, W. D., D. G. Watson, and R. W. Perkins
1967. Concentration and retention of fallout radionuclides in Alaskan arctic ecosystems. In *Radioecological concentration processes,* ed. B. Aberg and F. P. Hungate, 233–245. London: Pergamon Press.

Harborne, J. B.
1979. Flavonid pigments. In *Herbivores: Their interaction with secondary metabolites,* ed. G. A. Rosenthal and D. H. Janzen, 619–655. New York: Academic Press.

Harden, G.
1960. The competitive exclusion principle. *Science* **131**:1292–1297.

Harper, J. L.
1961. Approaches to the study of plant competition. In *Mechanisms in biological competition: Symp. Soc. Exp. Biol.,* ed. F. L. Milthorpe, **15**:1–39.
1977. *Population biology of plants.* New York: Academic Press.
1982. After description. In *The plant community as a working mechanism,* ed. E. I. Newman, 11–25. Oxford, England: Blackwell Scientific Publications.

Harper, J. L., and A. D. Bell
1979. The population dynamics of growth form in organisms with modular construction. In *Population dynamics, 20th Symposium British Ecological Society,* ed. R. M. Anderson, 29–52. Oxford, England: Blackwell Scientific Publications.

Harper, J. L., and J. White
1974. The demography of plants. *Ann. Rev. Ecol. Syst.* **5**:419–463.

Harrington, R. W., Jr.
1959. Photoperiodism in fishes in relation to the annual sexual cycle. In *Photoperiodism and related phenomena in plants and animals,* ed. R. B. Witherow, 651–667. Publ. 55. Washington, D.C.: American Association for the Advancement of Science.

Harris, L. D.
1984. *The fragmented forest.* Chicago: University of Chicago Press.

Harris, T. M.
1958. Forest fire in the Mesozoic. *J. Ecol.* **46**:447–453.

Harris, W. F., P. Sollins, N. T. Edwards, B. E. Dinger, and H. H. Shugart
1975. Analysis of carbon flow and productivity in a temperate deciduous forest ecosystem. In *Productivity of world ecosystems.* Washington, D.C.: National Academy of Science.

Harrison, J. L.
1962. The distribution of feeding habits among animals in tropical rain forests. *J. Anim. Ecol.* **31**:53–63.

Hartesveldt, R. J., and H. T. Harvey
1967. The fire ecology of sequoia regeneration. *Proc. Tall Timbers Fire Ecol. Conf.* **7**:65–77.

Hartshorn, G. S.
1978. Tree falls and tropical forest dynamics. In *Tropical trees as living systems,* ed. P. B. Tomlinson and M. H. Zimmerman. Cambridge, England: Cambridge University Press.

Harvey, P. H., and P. J. Greenwood
1978. Antipredator defence strategies: Some evolutionary problems. In *Behavioural ecology: An evolutionary approach,* ed. J. R. Krebs and N. B. Davies, 129–151. Oxford, England: Blackwell.

Hasler, A. D.
1954. Odour perception and orientation in fishes. *J. Fisheries Res. Board Can.* **11**:107–129.
1960. Guideposts of migrating fishes. *Science* **132**:785–792.
1969. Cultural eutrophication is reversible. *Bioscience* **19**:425–431.

Hasler, A. D., ed.
1975. *Coupling of land and water systems.* New York: Springer-Verlag.

Hassell, M. P.
1966. Evaluation of predator or parasite response. *J. Anim. Ecol.* **35**:65–75.

Hassell, M. P., J. L. Lawton, and J. R. Beddington
1976. The components of arthropod predation. *J. Anim. Ecol.* **45**:135–164.

Hassell, M. P., and R. M. May
1973. Stability in insect-host parasite models. *J. Anim. Ecol.* **42**:693–726.
1974. Aggregation in predators and insect parasites and its effect on stability. *J. Anim. Ecol.* **43**:567–594.

Hawthorn, W. R., and P. B. Cavers
1978. Population dynamics of the perennial herbs *Plantago major* and *P. rugelii* Decne. *J. Ecol.* **64**:511–527.

Head, P. C.
1976. Organic processes in estuaries. In *Estuarine chemistry,* ed. J. D. Burton and P. S. Liss, 54–91. London: Academic Press.

Heal, O. W., P. W. Flanagan, D. P. French, and S. F. Maclean, Jr.
1981. Decomposition and accumulation of organic matter. In *Tundra ecosystems: A comparative analysis,* ed. L. C. Bliss, O. W. Heal, and J. J. Moore, 587–633. International Biological Programme no. 25. London: Cambridge University Press.

Heal, O. W., H. E. Jones, and J. B. Whittaker
1975. Moor House, U.K. In *Structure and function of tundra ecosystems,* ed. T. Rosswall and W. O. Heal, 295–320. Stockholm: Swedish Natural Science Research Council.

Hedgpeth, J. W.
1957. *Treatise in marine ecology. I. Ecology.* Memoir 67, Geological Society of America.

Hedrick, P. W., M. W. Ginevan, and E. P. Ewing
1976. Genetic polymorphism in heterogenous environments. *Ann. Rev. Ecol. Syst.* **7**:1–32.

Heichel, G. H., and N. C. Turner
1976. Phenology and leaf growth of defoliated hardwood trees. In *Perspectives in forest entomology,* ed. J. F. Anderson and H. K. Karpas, 31–40. New York: Academic Press.

Heinrich, B.
1976. Heat exchange in relation to bloodflow between thorax and abdomen in bumblebees. *J. Exp. Biol.* **64**:567–585.
1979. *Bumblebee economics.* Cambridge, Mass.: Harvard University Press.

Heinrich, B., and P. H. Raven
1972. Energetics and pollination ecology. *Science* **176**:597–603.

Heinselman, M. L.
1981a. Fire and succession in the conifer forests of northern North America. In *Forest succession: Concepts and applications,* ed. D. C. West, H. H. Shugart, and D. B. Botkin, 374–405. New York: Springer-Verlag.
1981b. Fire intensity and frequency as factors in the distribution and structure of northern ecosystems, 7–57. In *Fire regimes and ecosystem properties.* U.S. For. Serv. Gen. Tech. Rept. **WO-26.** Washington, D.C.: U.S. Department of Agriculture.

Heller, H. C.
1971. Altitudinal zonation of chipmunks (*Eutamias*): Interspecific aggression. *Ecology* **52**:312–319.

Heller, H. C., and D. Gates
1971. Altitudinal zonation of chipmunks (*Eutamias*): Energy budgets. *Ecology* **52**:424–433.

Henderson, L. J.
1913. *The fitness of the environment.* New York: Macmillan.

Hendriksson, E., and B. Simu
1971. Nitrogen fixation by lichens. *Oikos* **22**:119–121.

Herrera, C. M.
1982. Defense of ripe fruit from pests: Its significance in relation to plant-disperser interactions. *Amer. Natur.* **120**:218–241.

Hett, J., and O. L. Loucks
1976. Age structure models of balsam fir and eastern hemlock. *J. Ecology* **64**:1029–1044.

Hibbs, D. E.
1983. Forty years of forest succession in central New England. *Ecology* **64**:1394–1401.

Hickey, J. J.
1952. Survival studies of banded birds. *U.S. Fish and Wildlife Serv. Spec. Sci. Rept.* **15.**

Hickman, J. C.
1975. Environmental unpredictability and plastic energy allocation strategies in annual *Polygonum cascadensis* (Polygonaceae). *J. Ecology* **63**:689–701.

Higgs, A. T.
1981. Island biogeography theory and nature reserve design. *J. Biogeogr.* **8**:117–124.

Hill, R. W.
1976. *Comparative physiology of animals: An environmental approach.* New York: Harper & Row.

Hirst, E.
1974. Food related energy requirements. *Science* **186**:134–138.

Hoffman, K.
1965. Clock mechanisms in celestial orientation of animals. In *Circadian clocks,* ed. J. Aschoff. Amsterdam: North Holland Publishing Co.

Hogstedt, G.
1980. Prediction and tests of the effects of interspecific competition. *Nature* **283**:64–66.

Holbrook, S. H.
1943. *Burning an empire.* New York: Macmillan.

Holdridge, L. B.
1947. Determination of wild plant formations from simple climatic data. *Science* **105**:367–368.

Holdridge, L. R., W. C. Grenke, W. H. Hatheway, T. Liang, and J. A. Tosi, Jr.
1971. *Forest environments in tropical life zones: A pilot study.* New York: Pergamon Press.

Holling, C. S.
1959. The components of predation as revealed by a study of small mammal predation of the European pine sawfly. *Can. Entomol.* **91**:293–320.
1961. Principles of insect predation. *Ann. Rev. Entomol.* **6**:163–182.
1964. The analysis of population processes. *Can. Entomologist* **96**:335–337.
1966. The functional response of invertebrate predators to prey density. *Mem. Entomol. Soc. Canada* **48**:1–86.
1973. Resilience and stability of ecological systems. *Ann. Rev. Ecol Syst.* **4**:1–23.

Holmes, J. C.
1983. Evolutionary relationships between parasite helminths and their hosts. In *Coevolution,* ed. D. J. Futuyma and M. Slatkin, 161–185. Sunderland, Mass.: Sinauer Associates.

Holmgren, R. C.
1956. Competition between annuals and young bitterbrush (*Purshia tridentata*) in Idaho. *Ecology* **37**:370–377.

Hood, L., J. H. Campbell, and S. C. R. Elgin
1975. The organization, expression, and evolution of antibody genes and other multigene families. *Ann. Rev. Genet.* **9**:305–353.

Hopkins, H. H.
1954. Effects of mulch on certain factors of the grassland environment. *J. Range Manage.* **7**:255–258.

Horn, H. S.
1966. Measurement of "overlap" in comparative ecological studies. *Amer. Natur.* **100**:419–424.
1975. Markovian properties of forest succession. In *Ecology and evolution of communities,* ed. M. L. Cody and J. Diamond, 196–211. Cambridge, Mass.: Harvard University Press.
1981. Some causes of variety in patterns of secondary succession. In *Forest succession: Concepts and applications,* ed. D. C. West, H. H. Shugart, and D. B. Botkin, 24–35. New York: Springer-Verlag.

Horsley, S. B.
1977. Allelopathic inhibition of black cherry by fern, grass, goldenrod, and aster. *Can. J. For. Res.* **7**:205–216.

Howard, H. E.
1920. *Territory in bird life.* London: Murray.

Howard, R. D.
1978a. The evolution of mating strategies in bullfrogs *Rana catesbeiana. Evolution* **32**:850–871.
1978b. The influence of male defended oviposition sites on early embryo mortality in bullfrogs. *Ecology* **59**:789–798.
1981. Male age-size distribution and male mating success in bullfrogs. In *Natural selection and social behavior: Recent research and new theories,* ed. R. D. Alexander and D. W. Tinkle. New York: Chiron Press.

Howell, F. G., J. B. Gentry, and M. H. Smith, eds.
1975. *Mineral cycling in southeastern ecosystems.* ERDA Symposium Series. Springfield, Va.: National Technical Information Service, U.S. Department of Commerce.

Hubbard, S. F., and R. M. Cook
1978. Optimal foraging by parasitoid wasps. *J. Anim. Ecol.* **47**:593–604.

Huffaker, C. B.
1958. Experimental studies on predation: Dispersion factors and predator-prey oscillations. *Hilgardia* **27**:343–383.

Humphrey, R. R.
1958. The desert grassland: A history of vegetational changes and an analysis of causes. *Botan. Rev.* **24**:193–252.

Hunkapiller, T., H. Huang, L. Hood, and J. H. Campbell
1982. The impact of modern genetics on evolutionary theory. In *Perspectives on evolution,* ed. R. Milkman, 164–189. Sunderland, Mass.: Sinauer Associates.

Hurbert, S. H., M. S. Mulla, and H. R. Wilson
1972. Effects of an organophosphate insecticide on phytoplankton, zooplankton, and insect populations of fresh water ponds. *Ecol. Monogr.* **42**:269–297.

Huston, M.
1979. A general hypothesis of species diversity. *Amer. Natur.* **113**:81–101.

Hutchinson, B. A., and D. R. Matt
1977. The distribution of solar radiation within a deciduous forest. *Ecol. Monogr.* **47**:185–207.

Hutchinson, G. E.
1957. *A treatise on limnology.* Vol. 1, *Geography, physics, and chemistry.* New York: Wiley.
1969. Eutrophication: Past and present. In *Eutrophication: Causes, consequences, correctives.* 12–26. Washington, D.C.: National Academy of Science.

Huxley, J. S.
1934. A natural experiment on territorial instinct. *Brit. Birds* **27**:270–277.

Hyder, D. N.
1969. The impact of domestic animals on the structure and function of grassland ecosystems. In *The grassland ecosystem: A preliminary synthesis,* ed. R. L. Dix and R. G. Beidleman, 243–260. Fort Collins: Colorado State University Range Science Department, Sci. Ser. no. 2.

Hynes, H.
1970. *The ecology of running water.* Toronto, Ontario: University of Toronto Press.

Jackson, R. C.
1976. Evolution and systematic significance of polyploidy. *Ann. Rev. Ecol. Syst.* **7**:209–234.

Jacob, F.
1977. Evolution and tinkering. *Science* **196**:1161–1166.

Janzen, D. H.
1970. Herbivores and the number of tree species in tropical forests. *Amer. Natur.* **104**:501–528.
1971. Seed predation by animals. *Ann. Rev. Ecol. Syst.* **2**:465–492.
1976. Why bamboos wait so long to flower. *Ann. Rev. Ecol. Syst.* **7**:347–391.
1977. Why fruits rot, seeds mold, and meat spoils. *Amer. Natur.* **111**:691–713.
1978. Complications in interpreting the chemical defenses of trees against tropical arboreal plant-eating vertebrates. In *Ecology of arboreal folivores,* ed. G. G. Montgomery, 73–84. Washington, D.C.: Smithsonian Institution Press.
1980. When is it coevolution? *Evolution* **34**:611–612.

Janzen, D. H., and P. S. Martin
1982. Neotropical anachronisms: The fruits the gomphotheres ate. *Science* **214**:19–27.

Jenny, H.
1933. Soil fertility losses under Missouri conditions. *Missouri Agri. Exp. Stat. Bull.* **324.**
1980. *The soil resource.* New York: Springer-Verlag.

Johannes, R. E.
1964. Uptake and release of phosphorus by a benthic marine amphipod. *Limnol. Oceanogr.* **9**:235–242.
1968. Nutrient regeneration in lakes and oceans. In *Advances in the microbiology of the sea,* ed. M. R. Droop and E. J. Ferguson, 203–213. Vol. 1. New York: Academic Press.

Johnson, D. W.
1968. Pesticides and fishes: A review of recent literature. *Trans. Amer. Fish. Soc.* **97**:398–424.

Johnson, D. W., D. W. Cole, C. S. Bledsoe, K. Cromack, R. L. Edmonds, S. P. Bessell, C. C. Grier, B. N. Richards, and K. A. Vogt
1982. Nutrient cycling in forests of the Pacific Northwest. In *Analysis of coniferous forest ecosystems in the western United States,* ed. R. L. Edmonds, 186–232. Stroudsburg, Pa.: Hutchinson Ross.

Johnson, F. S.
1970. The oxygen and carbon dioxide balance in the Earth's atmosphere. In *Global effects of environmental pollution,* ed. S. F. Singer, 4–11. Dordrecht, Holland: D. Reidel Publishing Company.

Johnson, P. L., and W. O. Billings
1962. The alpine vegetation of the Beartooth Plateau and its relation to cryopedogenic processes and patterns. *Ecol. Monogr.* **32**:105–135.

Johnston, J. W.
1936. The macrofauna of soils as affected by certain coniferous and hardwood types in the Harvard Forest. Ph.D. dissertation, Harvard University Library, Cambridge, Mass.

Johnston, R. F., and R. K. Selander
1964. House sparrows: Rapid evolution of races in North America. *Science* **144**:548–550.

Jones, M. G.
1933. Grassland management and its influence on the sward. *Emp. J. Exp. Agric.* **1**:43–57, 122–128, 223–234, 360–366, 366–367.

Jordan, C. F.
1971. Productivity of a tropical forest and its relation to a world pattern of energy storage. *J. Ecol.* **59**:127–142.
1981. Do ecosystems exist? *Amer. Natur.* **118**:284–287.
1982. Amazon rain forest. *Amer. Scient.* **70**:394–401.

Jordan, C. F., F. Golley, J. D. Hall, and J. Hall
1979. Nutrient scavenging of rainfall by the canopy of an Amazonian rain forest. *Biotropica* **12**:61–66.

Jordan, C. F., and R. Herrera
1981. Biogeochemical cycles and tropical forests. *Amer. Natur.* **117**:167–180.

Jordan, C. F., R. L. Todd, and Q. Escalante
1979. Nitrogen conservation in a tropical rain forest. *Oecologica* **39:**123–128.

Kabat, C., and D. R. Thompson
1963. Wisconsin quail, 1834–1962: Population dynamics and habitat management. *Tech. Bull.* no. **30.** Madison: Wisconsin Conser. Dept.

Kajah, Z., G. Bretschko, F. Schiemer, and C. Levegue
1980. Zoobenthos. In *The functioning of freshwater ecosystems,* ed. E. D. LeCren and R. H. Lowe-McConnell, 285–307. International Biological Programme no. 22. Cambridge, England: Cambridge University Press.

Kalinin, G. R., and V. D. Bykov
1969. The world's water resources, present and future. *Impact of Science on Society* **19:**135–150.

Kanwisher, J. W.
1955. Freezing in intertidal animals. *Biol. Bull. Woods Hole* **109:**56–63.
1959. Histological metabolism of frozen intertidal animals. *Biol. Bull. Woods Hole* **116:**285–264.

Kays, S., and H. Harper
1974. The regulation of plant and tiller density in a grass sward. *J. Ecology* **62:**97–105.

Keeley, J. E.
1981. Reproductive cycles and fire. In *Proc. fire regimes and ecosystem properties,* 231–277. U.S. For. Serv. Gen. Tech. Rept. **WO-26.** Washington, D.C.: U.S. Department of Agriculture.

Keeling, C. D.
1968. Carbon dioxide in surface ocean waters. *J. Geophys. Res.* **73:**4543.

Keever, C.
1950. Causes of succession in old fields of the Piedmont, North Carolina. *Ecol. Monogr.* **20:**229–250.

Keith, L. B.
1974. Some features of population dynamics in mammals. *Proc. Cong. Game. Biol., Stockholm* **11:**17–58.

Keith, L. B., and L. A. Windberg
1978. A demographic analysis of the snowshoe hare cycle. *Wildl. Monogr.* **58.**

Keith, T. P., and R. H. Tamarin
1981. Genetic and demographic differences between dispersers and residents in cycling and non-cycling vole populations. *J. Mammal.* **62:**713–725.

Kellogg, C. E.
1936. *Development and significance of the great soil groups of the United States.* U.S.D.A. Misc. Publ. **229.**

Kellogg, W. W., R. D. Cadle, E. R. Allen, A. L. Lazrus, and E. A. Martell
1972. The sulfur cycle. *Science* **175:**587–599.

Kettlewell, H. B. D.
1961. The phenomenon of industrial melanism in Lepidoptera. *Ann. Rev. Entomol.* **6:**245–262.
1965. Insect survival and selection for pattern. *Science* **148:**1290–1296.
1973. *The evolution of melanism.* New York: Oxford University Press (Clarendon Press).

Kevan, D. E. McE
1955. *Soil zoology.* Washington, D.C.: Butterworth.
1962. *Soil animals.* New York: Philosophical Library.

Key, K. H. L.
1974. Speciation in the Australian morabine grasshoppers—taxonomy and ecology. In *Genetic mechanisms of speciation in insects,* ed. M. J. D. White, 43–56. Sydney: Australian and New Zealand Book Co.

Kiester, A. R.
1971. Species density of North American amphibians and reptiles. *Syst. Zool.* **20:**127–137.

Kimmins, J. P.
1972. Relative contributions of leaching, litterfall, and defoliation by *Neodiprion sertifera* (Hymenoptera) to the removal of ^{139}Cesium from red pine. *Oikos* **23:**226–234.

Kincaid, W. B., and G. N. Cameron
1982. Effect of species removal on resource utilization in a Texas rodent community. *J. Mamm.* **63:**229–235.

King, C. E., and W. A. Anderson
1971. Age-specific selection: II, The interaction between *r* and *K* during population growth. *Amer. Natur.* **105:**137–156.

King, C. E., and E. E. Gallagher, and D. A. Levin
1975. Equilibrium diversity in plant pollination systems. *J. Theoret. Biol.* **53:**263–275.

Kinne, O., ed.
1970. *Marine biology: A comprehensive integrated treatise on life in oceans and coastal waters.* Vol. 1, *Environmental factors.* New York: Wiley.

Klein, D. R.
1970. Food selection by North American deer and their response to overutilization of preferred plant species. In *Animal populations in relation to their food resources,* ed. A. Watson, 25–46. Oxford, England: Blackwell.

Knox, K. G.
1952. Jefferson County (N.Y.) soils and soil map. New York State College of Agriculture, Cornell University, Ithaca.

Knutson, R. M.
1974. Heat production and tempera-

ture regulation in eastern skunk cabbage. *Science* **186**:746–747.

Koblentz-Mishke, J. J., V. V. Volkovinsky, and J. G. Kabanova

1970. Plankton primary production of the world ocean. In *Scientific exploration of the South Pacific,* ed. W. S. Wooster. Washington, D.C.: National Academy of Science.

Kok, B.

1965. Photosynthesis: The pathway of energy. In *Plant biochemistry,* ed. J. Bonner and J. E. Varner, 904–960. New York: Academic Press.

Komarek, E. V., Sr.

1962–1976. *Annual Tall Timbers Fire Ecology Conference,* vols. 1-15. Tallahassee, Fla.: Tall Timbers Research Station.

1964. The natural history of lightning. *Proc. Tall Timbers Fire Ecol. Conf.* **3**:139–183.

1966. The meteorological basis for fire ecology. *Proc. Tall Timbers Fire Ecol. Conf.* **5**:85–126.

1968. Lightning and lightning fires as ecological forces. *Proc. Tall Timbers Fire Ecol. Conf.* **8**:169–197.

Konishi, M.

1973. How the owl tracks its prey. *Amer. Scient.* **61**:414–424.

Kopec, R. J.

1970. Further observations of the urban heat island of a small city. *Bull. Amer. Meteorological Soc.* **51**: 602–606.

Kowal, N. E.

1966. Shifting agriculture, fire, and pine forest in the Cordillera Central, Luzon, Philippines. *Ecol. Monogr.* **36**:389–419.

Kramer, G.

1950. Weitere analyse der Faktoren, welche die Zugaktivitat des gekafigten Vogels orientieren. *Naturwissenschaften* **37**:377–388.

Krebs, C. J.

1963. Lemming cycle at Baker Lake, Canada, during 1959–62. *Science* **146**:1559–1560.

1964. The lemming cycle at Baker Lake, Northwest Territories, during 1959–1962. *Tech. paper* no. **15.** Arctic Institute of North America.

1978. *Ecology: The experimental analysis of distribution and abundance.* 2d ed. New York: Harper & Row.

Krebs, C. J., M. S. Gaines, B. L. Keller, J. H. Myers, and R. H. Tamarin

1973. Population cycles in small rodents. *Science* **179**:35–41.

Krebs, C. J., and J. H. Myers

1974. Population cycles in small mammals. *Adv. Ecol. Res.* **8**:267–399.

Krebs, C. J., I. Wingate, J. LeDuc, J. A. Redfield, M. Taitt, and R. Hilborn

1976. *Microtus* population biology: Dispersal in fluctuating populations of *M. townsendii. Can. J. Zool.* **54**:79–95.

Krebs, J. R.

1971. Territory and breeding density in the great tit *Parus major. Ecology* **52**:2–22.

Krebs, J. R., and N. B. Davies, eds.

1978. *Behavioural ecology: An evolutionary approach.* Oxford, England: Blackwell.

Krebs, J. R., A. Kacelnik, and P. Taylor

1978. Optimal sampling by foraging birds: An experiment with great tits (*Parus major*). *Nature* **275**:27–31.

Kruuk, H.

1972. *The spotted hyena.* Chicago: University of Chicago Press.

Kucera, C. L., R. C. Dahlman, and M. R. Koelling

1967. Total net productivity and turnover on an energy basis for tallgrass prairie. *Ecology* **48**:536–541.

Kuchler, A. W.

1964. Potential natural vegetation of the conterminous United States (map and manual). *Amer. Geog. Soc. Spec. Publ.* No. **36.**

Kuenzler, E. J.

1958. Niche relations of three species of Lycosid spiders. *Ecology* **39**:494–500.

1961. Phosphorus budget of a mussel population. *Limno. Oceanogr.* **6**:400–415.

Kuhnelt, W.

1950. *Bodenbiologie mit besonderer Brerücksichtigung der Tierwelt.* Vienna: Herold.

1970. Structural aspects of soil-surface-dwelling biocoenoses. In *Methods of study in soil biology,* ed. J. Phillipson, 45–56. Paris: UNESCO.

Lack, D.

1947a. *Darwin's finches.* Cambridge, England: Cambridge University Press.

1947b. The significance of clutch size. *Ibis* **89**:30–52; **90**:25–45.

1954. *The natural regulation of animal numbers.* Oxford, England: Oxford University Press (Clarendon Press).

1971. *Ecological isolation in birds.* Oxford, England: Blackwell Scientific Publications.

Lamotte, M.

1975. The structure and function of a tropical savanna ecosystem. In *Trends in tropical ecology,* ed. F. Golley and E. Medina, 179–222. New York: Springer-Verlag.

Lamotte, M., and F. Bourliere

1983. Energy flow and nutrient cycling in tropical savannas. In *Tropical savannas,* ed. F. Bourliere, 583–604. Ecosystems of the World no. 13. Amsterdam: Elsevier Scientific Publishing.

Landahl, J., and R. B. Root
1969. Difference in the life tables of tropical and temperate milkweed bugs, genus *Oncopeltus* (Hemiptera: Lygaeidae). *Ecology* **50**:734–737.

Landsberg, H. E.
1970. Man-made climatic changes. *Science* **170**:1265–1274.

Lang, G. E., and D. H. Knight
1983. Tree growth, mortality, recruitment, and canopy gap formation during a 10-year period in a tropical moist forest. *Ecology* **64**:1075–1080.

Lang, G. E., W. A. Reiners, and R. H. Heier
1976. Potential alterations of precipitation chemistry by epiphytic lichens. *Oecologica* **25**:229–241.

Lange, O. L., L. Kappen, and E. E. Schulze, eds.
1976. *Water and plant life: Problems and modern approaches.* New York: Springer-Verlag.

Lanyon, W. E., and W. N. Tavolga, eds.
1960. *Animal sounds and communications,* AIBS Symposium Series, publ. no. 7.

Larcher, W.
1975. *Physiological plant ecology.* New York: Springer-Verlag.

Lauff, G., ed.
1967. *Estuaries.* Washington, D.C.: American Association for the Advancement of Science.

Lavenroth, W. K.
1979. Grassland primary production: North American grasslands in perspective. In *Perspectives in grassland ecology,* ed. N. French, 3–24. New York: Springer-Verlag.

Lawlor, L. R.
1976. Molting, growth, and reproductive strategies in the terrestrial isopod Armadillidum vulgare. *Ecology* **57**:1179–1194.

Lawton, J. H.
1981. Moose, wolves, *Daphnia* and *Hydra:* On the ecological efficiency of endotherms and ectotherms. *Amer. Natur.* **117**:782–783.

LeCren, E. D., and R. H. Lowe-McConnell, eds.
1980. *The functioning of freshwater ecosystems.* International Biological Programme no. 22. New York: Cambridge University Press.

Lee, R.
1978. *Forest microclimatology.* New York: Cambridge University Press.

Lees, A. D.
1973. Photoperiod time measurement in the aphid *Megoura viciae. J. Insect. Physiol.* **19**:2279–2316.

Leith, H.
1960. Patterns of change within grassland communities. In *The biology of weeds,* ed. J. L. Harper, 27–39. Oxford, England: Oxford University Press.
1963. The role of vegetation in the carbon dioxide content of the atmosphere. *J. Geophys. Res.* **68**: 1887–1898.

Leopold, A.
1933. *Game management.* New York: Scribner.

Levenson, J. B.
1981. Woodlots as biogeographic islands in southeastern Wisconsin. In *Forest island dynamics in man-dominated landscapes,* ed. R. L. Burgess and D. M. Sharpe, 13–39. Ecological Studies no. 41. New York: Springer-Verlag.

Levin, D. A., and H. W. Kerster
1978. Rings and age in *Liatris. Amer. Natur.* **112**:1120–1122.

Levins, R.
1968. *Evolution in changing environments.* Princeton, N.J.: Princeton University Press.

Levyns, M. R.
1966. *Haemanthus canaliculatus,* a new fire lily from the western Cape Province. *J. S. Afr. Bot.* **32**:73–75.

Lewin, R. A.
1982. Symbiosis and parasitism—definition and evaluation. *Bioscience* **32**:254, 257–259.
1983a. Extinction and the history of life. *Science* **221**:935–937.
1983b. What killed the giant mammals? *Science* **221**:1036–1037.

Lewis, T.
1969. The diversity of insect fauna in hedgerows and neighboring fields. *J. Appl. Ecol.* **6**:453–458.

Lidicker, W. Z., Jr.
1966. Ecological observations on a feral house mouse population declining to extinction. *Ecol. Monogr.* **36**:27–50.
1975. The role of dispersal in the demography of small mammals. In *Small mammals: Their productivity and population dynamics,* ed. F. B. Golley, K. Petrusewics, and L. Ryszkowski, 103–128. Cambridge, England: Cambridge University Press.

Liebig, J.
1840. *Chemistry in its application to agriculture and physiology.* London: Taylor and Walton.

Ligon, J. D.
1968. Sexual differences in foraging behavior in two species of *Dendrocopos* woodpeckers. *Auk* **85**:203–215.
1981. Demographic patterns and communal breeding in the green hoopoo *Phoeniculus purpureus.* In *Natural selection and social behavior: Recent research and new theories,* ed. R. D. Alexander and D. W. Tinkle, 231–246. New York: Chiron.

Likens, G. E., and F. H. Bormann
1974a. Acid rain: A serious regional environmental problem. *Science* **184**:1176–1179.
1974b. Effects of forest clearing on the northern hardwood forest ecosystem and its biogeochemistry, 330–335. *Proc. First Int. Cong. Ecology*. The Hague: W. Junk Publisher.
1974c. Linkages between terrestrial and aquatic ecosystems. *Bioscience* **24**:447–456.
1975. Nutrient-hydrologic interactions (eastern United States). In *Coupling of land and water systems,* ed. A. D. Hasler, 1–5. New York: Springer-Verlag.

Lillywhite, H. B.
1970. Behavioral temperature regulation in the bullfrog, *Rana catesbeiana*. *Copeia* **1970**:158–168.

Lindquist, B.
1942. Experimentelles Untersuchungen über die Bedeutung einiger Landmollusken für die Zersetgung der Waldstreu. *Kgl. Fysiograf. Sallskap Lund Forh.* **11**:144–156.

Linhart, Y. B.
1973. Ecological and behavioral determinants of pollen dispersal in hummingbird-pollinated *Heliconia*. *Amer. Natur.* **107**:511–523.
1974. Intra-population differentiation in annual plants. I, *Veronica peregrina* L. raised under non-competitive conditions. *Evolution* **28**:232–243.

Lister, B. C.
1981. Seasonal niche relationships of rain forest anoles. *Ecology* **62**:1548–1560.

Little, E. L., Jr.
1971. *Atlas of United States trees*. Vol. 1, *Conifers and important hardwoods*. U.S.D.A. Misc. Publ. no. **1146**.

Lloyd, J. A., J. J. Christian, D. E. Davis, and F. H. Bronson
1964. Effects of altered social structure on adrenal weights and morphology in populations of woodchucks (*Marmota monax*). *Gen. Comp. Endocrinol.* **2**:271–276.

Lloyd, J. E.
1966. Studies on the flash communication system in *Photinus* fireflies. *Misc. Publ. Mus. Zool. Univ. Mich.* **130**:1–95.

Lloyd, M., and H. S. Dybas
1966. The periodical cicada problem: I, Population ecology. *Evolution* **20**:133–149.

Lombardi, J. R., and J. B. Vandenbergh
1977. Pheromonially induced sexual maturation in females: Regulation by the social environment of the male. *Science* **196**:545–546.

Loomis, R. S.
1967. Community architecture and the productivity of terrestrial plant communities. In *Harvesting the sun,* ed. A. San Pietro, F. A. Greer, and T. J. Army, New York: Academic Press.

Lord, R. D.
1960. Litter size and latitude in North American mammals. *Amer. Midl. Natur.* **64**:488–499.

Lotan, J. E.
1974. Cone serotiny-fire relationships in lodgepole pine. *Proc. Tall Timbers Fire Ecol. Conf.* **14**:267–278.

Lotka, A. J.
1925. *Elements of physical biology.* Baltimore: Williams & Wilkins.

Loucks, O.
1970. Evolution of diversity, efficiency, and community stability. *Amer. Zool.* **10**:17–25.

Loumer, C. G.
1980. Age structure and disturbance history of southern Appalachian virgin forest. *Ecology* **61**:1169–1184.

Lowe, V. P. W.
1969. Population dynamics of red deer (*Cervus elaphus*) on Rhum. *J. Anim. Ecol.* **38**:425–457.

Lubchenco, J.
1983. *Littorina* and *Fucus:* Effects of herbivores, substratum heterogeneity, and plant escapes during succession. *Ecology* **64**:1116–1124.

Lull, H. W., and W. E. Sopper
1969. *Hydrologic effects from urbanization of forested watersheds in the Northeast*. U.S.D.A. For. Serv. Res. Paper **NE-146.**

Lutz, H. J.
1956. *Ecological effects of forest fires in the interior of Alaska*. U.S.D.A. Tech. Bull. no. **1133.**

Lutz, H. J., and R. F. Chandler
1954. *Forest soils*. New York: Wiley.

Lyman, C. P., J. S. Willis, A. Malan, and L. C. H. Wang
1982. *Hibernation and torpor in mammals and birds*. New York: Academic Press.

Lynch, J. F., and R. F. Whitcomb
1977. Effects of insularization of the eastern deciduous forest and avifaunal diversity and turnover. In *Classification, inventory, and analysis of fish and wildlife habitat*. FWS/OBS-78/76, 461–489. Washington, D.C.: U.S. Fish and Wildlife Service.

Note: Names beginning with the prefix **Mc–** are alphabetized as if spelled **Mac–**.

McArdle, R. E., and W. H. Meyer
1949. *The yield of Douglas-fir in the Pacific Northwest*. U.S.D.A. Tech. Bull. no. **201**.

MacArthur, R. H.
1958. Population ecology of some

warblers of northeastern coniferous forests. *Ecology* **39**:599–619.
1960. On the relative abundance of species. *Amer. Natur.* **94**:25–36.
1961. Population effects of natural selection. *Amer. Natur.* **95**:195–199.
1972. *Geographical ecology.* New York: Harper & Row.

MacArthur, R. H., and R. Levins
1964. Competition, habitat selection, and character displacement in a patchy environment. *Proc. Nat. Acad. Sci.* **51**:1207–1210.
1967. The limiting similarity convergence and divergence of coexisting species. *Amer. Natur.* **101**:377–385.

MacArthur, R. H., and J. W. MacArthur
1961. On bird species diversity. *Ecology* **42**:594–598.

MacArthur, R. H., and E. O. Wilson
1967. *The theory of island biogeography.* Princeton, N.J.: Princeton University Press.

McBee, R. H.
1971. Significance of intestinal microflora in herbivory. *Ann. Rev. Ecol. Syst.* **2**:165–176.

McBrien, H., R. Harmsen, and A. Crowder
1983. A case of insect grazing affecting plant succession. *Ecology* **64**:1035–1039.

McCalls, T. M.
1943. Microbial studies of the effects of straw used as mulch. *Trans. Kansas Acad. Sci.* **43**:52–56.

McCann, G. D.
1942. When lightning strikes. *Sci. Amer.* **49**:23–25.

McGinnes, W. G.
1972. North America. In *Wildland shrubs: Their biology and utilization,* ed. C. M. McKella, J. P. Blaisdell, and J. R. Goodin, 55–56. U.S.D.A. For. Ser. Gen. Tech. Rept. **INT-1.**

McIlroy, R. J.
1972. *An introduction to tropical grassland husbandry.* London: Oxford University Press.

McIntosh, R. P.
1958. Plant communities. *Science* **128**:115–120.
1967. An index of diversity and the relation of concepts to diversity. *Ecology* **48**:392–403.
1976. Ecology since 1900. In *Issues and ideas in America,* ed. B. J. Taylor and T. J. White. Norman: Oklahoma University Press.
1980a. The background of some current problems of theoretical ecology. *Synthese* **43**:195–255.
1980b. The relationship between succession and the recovery process. In *The recovery process in damaged ecosystems,* ed. J. Cairns, 11–62. Ann Arbor, Mich.: Ann Arbor Science Publishers.
1981. Succession and ecological theory. In *Forest succession: Concepts and applications,* ed. D. West, H. Shugart, and D. Botkin, 10–23. New York: Springer-Verlag.

McKella, C. M., J. P. Blaisdell, and J. R. Goodin, eds.
1972. *Wildland shrubs: Their biology and utilization.* U.S.D.A. For. Ser. Gen. Tech. Rept. **INT-1.**

McKey, D.
1975. The ecology of coevolved seed dispersal systems. In *Coevolution of animals and plants,* ed. L. E. Gilbert and P. H. Raven, 159–191. Austin: University of Texas Press.

MacLean, S. F., Jr.
1980. The detritus-based trophic system. In *An arctic ecosystem: The coastal tundra at Barrow, Alaska,* ed. J. Brown, P. C. Miller, L. L. Tieszen, and F. L. Bunnell, 411–457. US/IBP Synthesis Series, no. 12. Stroudsburg, Pa.: Dowden, Hutchinson and Ross.

MacLulich, D. A.
1947. Fluctuations in the numbers of varying hare (*Lepus americanus*). *Univ. Toronto Biol. Ser.,* no. **43.**

MacMahon, J. A.
1979. Ecosystems over time: Succession and other types of change. In *Forests: Fresh perspectives from ecosystem analysis,* ed. R. H. Waring, 27–58. Proc. 40th Biology Colloquium. Corvallis: Oregon State University Press.

McMillan, C.
1959. The role of ecotypic variation in the distribution of the central grassland of North America. *Ecol. Monogr.* **29**:285–308.

McNab, B.
1978. The evolution of endothermy in the phylogeny of mammals. *Amer. Natur.* **112**:1–22.
1980. Food habits, energetics, and the population biology of mammals. *Amer. Natur.* **116**:106–124.

McNaughton, S. J.
1979. Grazing as an optimization process: Grass-ungulate relationships in the Serengeti. *Amer. Natur.* **113**:691–703.

McPherson, J. K., and C. H. Muller
1969. Allelopathic effects of *Adenostoma fascicuatum* "chamise" in the California chaparral. *Ecol. Monogr.* **39**:177–199.

Madgwick, H. A. L., and J. D. Ovington
1959. The chemical composition of precipitation in adjacent forest and open plots. *Forestry* **32**:14–22.

Mahmoud, A., and J. P. Grime
1976. An analysis of competitive ability in three perennial grasses. *New Phytol.* **77**:431–435.

Maio, J. J.
1958. Predatory fungi. *Sci. Amer.* **199**:67–72.

Malthus, T.
1978. *An essay on the principles of population.* London: J. Johnson.

Malvin, R. L., and M. Rayner
1968. Renal function and blood chemistry in Cetacea. *Amer. J. Physiol.* **214:**187–191.

Manaut, J.
1983. The vegetaton of African savannas. In *Tropical savannas,* ed. F. Bourliere, 109–149. Ecosystems of the world no. 13. New York: Elsevier Scientific Publishing Co.

Manuwal, D.
1974. The natural history of Cassin's auklet (*Ptychoramphus aleuticus*). *Condor* **76:**421–431.

Marchand, D. E.
1973. Edaphic control of plant distribution in the White Mountains, eastern California. *Ecology* **54:**233–250.

Margalef, R.
1968. *Perspectives in ecological theory.* Chicago: University of Chicago Press.

Marks, P. L.
1974. The role of pin cherry (*Prunus pensylvanica* L.) in the maintenance of stability in northern hardwood ecosystems. *Ecol. Monogr.* **44:**73–88.

Marks, P. L., and F. H. Bormann
1972. Revegetation following forest cutting: Mechanisms for return to steady state nutrient cycling. *Science* **176:**914–915.

Marquis, D. A.
1974. *The impact of deer browsing on Allegheny hardwood regeneration.* U.S.D.A. Forest Serv. Res. Paper **NE-308.**

Marsden, H. M., and N. R. Moller
1964. Social behavior in confined populations of the cottontail and swamp rabbit. *Wildl. Monogr.* no. 13.

Martin, P. S.
1973. The discovery of America. *Science* **179:**969–974.

Marx, D. H.
1971. Ectomycorrhizae as biological deterrents to pathogenic root infections. In *Mycorrhizae,* ed. E. Hacskalo, 81–96. U.S.D.A. Misc. Publ. **1189.**

Massey, A., and J. G. Vandenbergh
1980. Puberty delay by a urinary cue from female house mice in feral populations. *Science* **209:**821–822.

Mattson, W. J., ed
1977. *The role of arthropods in forest ecosystems.* New York: Springer-Verlag.

May, R. M.
1973. *Stability and complexity in model ecosystems.* Princeton, N.J.: Princeton University Press.
1976. Models for two interacting populations. In *Theoretical ecology: Principles and application,* ed. R. M. May, 49–70. Philadelphia: Saunders.

May, R. M., ed.
1976. *Theoretical ecology: Principles and application.* Philadelphia: Saunders.

May, R. M., and R. M. Anderson
1983. Parasite-host coevolution. In *Coevolution,* ed. D. J. Futuyma and M. Slatkin, 186–206. Sunderland, Mass.: Sinauer Associates.

Mayr, E.
1942. *Systematics and the origin of species.* New York: Columbia University Press.
1963. *Animal species and evolution.* Cambridge, Mass.: Harvard University Press.
1974. The definition of the term disruptive selection. *Heredity* **32:**404–406.

Means, J. E., ed.
1982. *Forest succession and stand development research in the Northwest.* Corvallis: Forest Research Laboratory, Oregon State University.

Mech, L. D.
1970. *The wolf: The ecology and behavior of an endangered species.* Garden City, N.Y.: Doubleday.

Menant, J. C.
1983. The vegetation of African savannas. In *Tropical savannas,* ed. F. Bourliere, 109–150. Ecosystems of the World no. 13. Amsterdam: Elsevier Scientific Publishing.

Menzie, C. M.
1969. Metabolism of pesticides. *U.S.D.I. Fish and Wildlife Ser. Sp. Sci. Rept. Wildl.* No. **127.**

Merriam, C. H.
1898. Life zones and crop zones of the United States. *Bull U.S. Bureau Biol. Survey* **10:**1–79.

Meslow, E. C., and L. B. Keith
1968. Demographic parameters of a snowshoe hare population. *J. Wildl. Manage.* **32:**812–835.

Mettler, L. E., and T. G. Gregg
1969. *Population genetics and evolution.* Englewood Cliffs: N.J.: Prentice-Hall.

Michod, R. E.
1982. The theory of kin selection. *Ann. Rev. Ecol. Syst.* **13:**23–25.

Milkman, R., ed.
1982. *Perspectives on evolution.* Sunderland, Mass.: Sinauer Associates.

Miller, P. C., P. J. Webber, W. C. Oechel, and L. L. Tieszen
1980. Biophysical processes and primary production. In *An arctic ecosystem: The coastal tundra at Barrow, Alaska,* ed. J. Brown, P. C. Miller, L. L. Tieszen, and F. L. Bunnell, 66–101. US/IBP Synthesis Series, no. 12. Stroudsburg. Pa.: Dowden, Hutchinson & Ross.

Miller, P. R., J. R. Parmeter, O. C. Taylor, and E. A. Cardiff
1963. Ozone injury to the foliage of *Pinus ponderosa. Phytopathology* **53**:1072–1076.

Miller, R., and J. Rusch
1960. Zur Frange der Kohlensaure-versorgung des Waldes. *Forstwiss Cbl.* **79**:42–64.

Minshall, G. W.
1967. The role of allochthonous detritus in the trophic structure of a woodland spring brook community. *Ecology* **48**:139–149.
1978. Autotrophy in stream ecosystems. *Bioscience* **28**:767–771.

Mishustin, E. N., and V. E. Shilnikova
1969. The biological fixation of atmospheric nitrogen by free-living bacteria. In *Soil biology,* 65–124. Paris: UNESCO.

Mobius, K.
1877. An oyster-bank is a bioconose, or social community. In *Die Aiusterund die Austerwirthschaft.* Berlin: Wiegurdt, Hempel, and Parey. Translation printed in *Report of the U.S. Commission of Fisheries,* trans. H. J. Rice, 1880: 683–751. Reprinted in *Readings in Ecology,* ed. E. J. Kormondy, 1965: 121–124. Englewood Cliffs, N.J.: Prentice-Hall.

Moir, W. H.
1969. Steppe communities in the foothills of the Colorado front range and their relative productivities. *Amer. Midl. Natur.* **81**:331–340.

Monk, C. A.
1967. Tree species diversity in the eastern deciduous forest with particular reference to northcentral Florida. *Amer. Natur.* **101**:173–187.

Monsi, M.
1968. Mathematical models of plant communities. In *Functioning of terrestrial ecosystems at the primary production level,* ed. F. E. Eckardt, 131–149. Proceedings of the Copenhagen Symposium, Natural Resources Research V. Paris: UNESCO.

Mook, L. J.
1963. Birds and spruce budworm. In *The dynamics of epidemic spruce budworm populations,* ed. R. F. Morris, 244–248. Mem. Entomol. Soc. Can. no. **31.**

Mooney, H. A., and C. E. Conrad, eds.
1977. *Proc. Symp. Environmental Consequences of Fire and Fuel Management in Mediterranean Ecosystems.* U.S.D.A. For. Serv. Gen. Tech. Rept. **WO-3.** Washington, D.C.: U.S. Department of Agriculture.

Mooney, H. A., and E. L. Dunn
1970a. Convergent evolution of Mediterranean climate evergreen sclerophyll shrubs. *Evolution* **24**:292–303.
1970b. Photosynthetic systems of Mediterranean shrubs and trees of California and Chile. *Amer. Midl. Natur.* **104**:447–453.

Mooney, H. A., and D. J. Parsons
1973. Structure and function of the California chaparral—an example from San Dimas. In *Mediterranean type ecosystems: Origin and structure,* ed. F. di Castri and H. A. Mooney, 83–112. New York: Springer-Verlag.

Moore, H. B.
1956. *Marine ecology.* New York: Wiley.

Moore, J. A.
1949a. Geographic variation of adaptive characters in *Rana pipiens* Schreber. *Evolution* **3**:1–24.
1949b. Patterns of evolution in the genus *Rana.* In *Genetics, paleontology, and evolution,* ed. G. Jepsen, E. Mayr, and G. Simpson, 315–338. Princeton, N.J.: Princeton University Press.

Morris, R. F.
1963. Predictive population equations based on key factors. *Entomol. Soc. Can. Mem.* **32**:16–21.

Morris, R. F., ed.
1963. The dynamics of epidemic spruce budworm populations. *Mem. Entomol. Soc. Can.* No. **31.**

Morris, R. F., W. F. Cheshire, C. A. Miller, and D. G. Mott
1958. The numerical response of avian and mammalian predation during a gradation of the spruce budworm. *Ecology* **39**:487–494.

Moulder. B. C., and D. E. Reichle
1974. Significance of spider predation in the energy dynamics of forest floor arthropod communities. *Ecol. Monogr.* **42**:473–498.

Moulder, B. C., D. E. Reichle, and S. I. Auerbach
1970. *Significance of spider predation in the energy dynamics of forest floor arthropod communities.* Oak Ridge National Laboratory Rept. ORNL 4452.

Moynihan, M.
1968. Social mimicry: Character divergence versus character displacement. *Evolution* **22**:315–331.

Mueller-Dombois, D., and H. Ellenberg
1974. *Aims and methods of vegetation ecology.* New York: Wiley.

Muir, J.
1878. The new sequoia forests of California. *Harpers* **57**:813–827.

Muller, C. J., R. B. Hanawalt, and J. K. McPherson
1968. Allelopathic control of herb growth in the fire cycle of California chaparral. *Bull. Torrey Bot. Club* **95**:225–231.

Muller, K.
1974. Steam drift as a chronobiological phenomenon in running water ecosystems. *Ann. Rev. Ecol. Syst.* **5**:309–373.

Muller, P. E.
1889. Recherches sur les formes naturelles de l'humus. *Ann. Sci. Agron.* **6**:85–423.

Muller-Schwarze, D.
1971. Pheromones in black-tailed deer. *Anim. Behav.* **19**:141–152.

Munger, J. C., and J. H. Brown
1981. Competition in desert rodents: An experiment with semipermeable exclosures. *Science* **211**:510–512.

Munn, R. E.
1966. *Descriptive micrometeorology.* New York: Academic Press.

Murdoch, W. W.
1969. Switching in general predators: Experiments on predator specificity and stability of prey populations. *Ecol. Monogr.* **39**:335–354.
1973. The functional response of predators. *J. Appl. Ecol.* **10**:335–342.

Murdoch, W. W., and A. Oaten
1975. Predation and population stability. *Adv. Ecol. Res.* **9**:1–131.

Mutch, R. W.
1970. Wildland fires and ecosystems—a hypothesis. *Ecology* **51**:1046–1051.

Muul, I.
1965. Daylength and food caches. *Nat. Hist.* **74** (3): 22–27.

Myers, J. H.
1980. Is the insect or the plant the driving force in the Cinnabar moth-tansy ragwort system? *Oecologica* **47**:16–21.

Myers, J. H., and C. J. Krebs
1971. Genetic, behavioral, and reproductive attributes of dispersing field voles *Microtus pennsylvanicus* and *Microtus ochrogaster. Ecol. Monogr.* **41**:53–71.

Nace, R. L.
1960. Human uses of ground water. In *Water, earth, and man,* ed. R. J. Chorley, 285–294. London: Methuen.

Nagy, K. A., D. K. Odell, and R. S. Seymour
1972. Temperature regulation by inflorescence of Philodendron. *Science* **178**:1195–1197.

National Academy of Science
1974. *U.S. participation in the International Biological Program.* Rept. no. 6. U.S. Committee for the International Biological Program. Washington, D.C.: National Academy of Science.
1975. *Productivity of world ecosystems.* Washington, D.C.: National Academy of Science.

Neel, J. K.
1951. Interrelations of certain physical and chemical features in headwater limestone streams. *Ecology* **32**:368–391.

Negus, N. C., P. J. Berger, and L. G. Forslund
1977. Reproductive strategy of *Microtus montanus. J. Mammal.* **58**:347–353.

Nelson, D. J., and D. C. Scott
1962. Role of detritus in the productivity of a rock-outcrop community in a Piedmont stream. *Limnol. Oceanogr.* **3**:396–413.

Neuwirth, R.
1957. Some recent investigations into the chemistry of air and of precipitation and their significance for forestry. *Allg. Forst-v. Jagdztg* **128**:147–150.

Nevo, E.
1969. Mole rat *Spalax ehrenbergi* mating behavior and its evolutionary significance. *Science* **163**:484–486.

Nevo, E., and S. A. Blondheim
1972. Acoustic isolation in the speciation of mole crickets. *Ann. Entomol. Soc. Amer.* **65**:980–981.

Nevo, E., Y. J. Kim, C. R. Shaw, and C. S. Thaler
1974. Genetic variation, selection, and speciation in *Thomomys talpoides* pocket gophers. *Evolution* **28**:1–23.

Nevo, E., and C. R. Shaw
1972. Genetic variation in a subterranian mammal *Spalax ehrenbergi. Biochem. Genet.* **7**:235–241.

Newbold, J. D., J. W. Elwood, R. V. O'Neill, and A. L. Sheldon
1983. Phosphorus dynamics in a woodland stream: A study of nutrient spiralling. *Ecology* **65**:1249–1265.

Newbold, J. D., R. V. O'Neill, J. W. Elwood, and W. Van Winkle
1982. Nutrient spiraling in streams; implications for nutrient and invertebrate activity. *Amer. Natur.* **120**:628–652.

Newell, R. C.
1965. The role of detritus in the nutrition of two marine deposit feeders, the prosobranch *Hydrobia ulvae* and the bivalve *Macome balthica. Proc. Zool. Soc. London* **144**:25–45.

Nicholson, A. J.
1957. The self-adjustment of populations to change. *Cold Spring Harbor Symp. Quant. Biol.* **22**:153–173.

Nielsen, A.
1950. The torrential invertebrate fauna. *Oikos* **2**:176–196.

Niering, W. A., and F. E. Egler
1955. A shrub community of *Viburnum lentago:* Stable for twenty-five years. *Ecology* **36**:356–360.

Niering, W. A., and R. Goodwin
1974. Creation of relatively stable shrubland with herbicides: Arresting "succession" on rights-of-way and pastureland. *Ecology* **55**:784–795.

Nixon, S. W., and C. A. Oviatt
1973. Ecology of a New England salt marsh. *Ecol. Monogr.* **43**:463–498.

Noble, J. R., and R. O. Slatyer
1980. The use of vital attributes to predict successional changes in plant communities subject to recurrent disturbances. *Vegetatio* **43**:5–21.

Noy-mier, I.
1973. Desert ecosystems: Environment and producers. *Ann. Rev. Ecol. Syst.* **4**:25–51.
1974. Desert ecosystems: Higher trophic levels. *Ann. Rev. Ecol. Syst.* **5**:195–214.

Nybakken, J. W.
1982. *Marine biology: An ecological approach.* New York: Harper & Row.

Odum, E. P.
1959. *Fundamentals of ecology.* 2d. ed. Philadelphia: Saunders.
1964. The new ecology. *Bioscience* **14**:14–16.
1969. The strategy of ecosystem development. *Science* **164**:262–270.
1971. *Fundamentals of ecology.* 3d ed. Philadelphia: Saunders.
1983. *Basic ecology.* Philadelphia: Saunders.

Odum, E. P., C. E. Connell, and L. B. Davenport
1962. Population energy flow of three primary consumer components of old field ecosystems. *Ecology* **43**:88–96.

Odum, H. T.
1970. Summary: An emerging view of the ecological system at El Verde. In *A tropical rain forest: A study of irradiation and ecology at El Verde, Puerto Rico,* ed. H. T. Odum and R. F. Pigeon, I-191–I-281. Washington, D.C.: U.S. Atomic Energy Commission.

Odum, W. E., and M. A. Heywood
1978. Decomposition of intertidal freshwater marsh plants. In *Freshwater wetlands,* ed. R. E. Good, D. F. Whigham, and R. L. Simpson, 89–97. New York: Academic Press.

Oka, I. N., and D. Pimental
1974. Corn susceptibility to corn leaf aphids and common corn smut after herbicide treatment. *Environ. Entomol.* **3**:911–915.

Oke, T. R., and C. East
1971.The urban boundary layer in Montreal. *Boundary-layer Meteorol.* **1**:411.

Oliver, C. D., and E. P. Stephens
1977. Reconstruction of a mixed species forest in central New England. *Ecology* **58**:562–572.

Olson, J. S.
1958. Rates of succession and soil changes in southern Lake Michigan sand dunes. *Botan. Gaz.* **119**:125–170.
1963. Energy storage and balance of decomposers in ecological systems. *Ecology* **44**:322–332.
1970. Carbon cycles and temperate woodlands. In *Analysis of temperate forest ecosystems,* ed. D. E. Reichle, 226–241. New York: Springer-Verlag.

O'Neill, R. V.
1976. Ecosystem persistence and heterotrophic regulation. *Ecology* **57**:1244–1253.

O'Neill, R. V., W. F. Harris, B. S. Ausmus, and D. E. Reichle
1975. A theoretical basis for ecosystem anaylsis with particular reference to element cycling. In *Mineral cycling in southeastern ecosystems,* ed. F. G. Howell, J. B. Gentry, and M. H. Smith, 28–40. ERDA Symposium Series. Springfield, Va.: National Technical Information Service, U.S. Department of Commerce.

Oosting, H. J., and L. E. Anderson
1937. The vegetation of a bare-faced cliff in western North Carolina. *Ecology* **18**:280–292.
1939. Plant succession on a granite rock in eastern North Carolina. *Botan. Gaz.* **100**:750–768.

Opler, P. A.
1974. Oaks as evolutionary islands for leaf-mining insects. *Amer. Scient.* **62**:67–73.

Orians, G. H.
1969a. The number of bird species in some tropical forests. *Ecology* **50**:783–801.
1969b. On the evolution of mating systems in birds and mammals. *Amer. Natur.* **103**:589–603.

Orians, G. H., and M. F. Willson
1964. Interspecific territories of birds. *Ecology* **45**:736–745.

Ovington, J. D.
1961. Some aspects of energy flow in plantation of *Pinus sylvestris* L. *Ann. Bot. London,* n.s., **25**:12–20.
1965. Organic production, turnover, and mineral cycling in woodlands. *Biol. Rev. Cambridge Phil. Soc.* **40**:295–336.

Ovington, J. D., and D. B. Laurence
1967. Comparative chlorophyll and energy studies of prairie savanna, oak wood, and field maize ecosystems. *Ecology* **48**:515–524.

Owen, D. F.
1980. How plants may benefit from animals that eat them. *Oikos* **35**:230–235.

Owen, D. F., and R. G. Wiegert
1981. Mutualism between grasses and

grazers: An evolutionary hypothesis. *Oikos* **36**:376–378.

Ozoga, J. J., L. J. Verme, and C. S. Bienz
1982. Parturition behavior and territoriality in white-tailed deer: Impact on neonatal mortality. *J. Wildl. Manage.* **46**:1–11.

Pacher, P. E.
1974. *Rehabilitation potentials and limitations of surface-mined land in the northern Great Plains.* U.S.D.A. For. Serv. Gen. Tech. Rept. **INT-14.**

Paine, R. T.
1966. Food web complexity and species diversity. *Amer. Natur.* **100**:65–76.
1974. Intertidal community structure: Experimental studies on the relationship between a dominant competitor and its principal predator. *Oecologica* **15**:93–120.

Paloheimo, J. E., and L. M. Dickie
1970. Production and food supply. In *Marine food chains,* ed. J. H. Steel, 499–527. Berkeley: University of California Press.

Parenti, R. L., and E. L. Rice
1969. Inhibitional effects of *Digitaria sanguinalis* and possible role in old field succession. *Bull. Torrey Bot. Club* **96**:70–78.

Park, O. W. C., V. E. Allee, and W. Shelford
1939. *A laboratory introduction to animal ecology and taxonomy.* Chicago: University of Chicago Press.

Park, T.
1948. Experimental studies of interspecific competition. I. Competition between populations of flour beetles *Tribolium confusum* Duval and *T. castaneum* Herbst. *Ecol. Monogr.* **18**:265–307.
1954. Experimental studies of interspecific competition. II. Temperature, humidity, and competition in two species of *Tribolium. Physiol. Zool.* **27**:177–238.

Parker, J.
1981. Effects of defoliation on oak chemistry. In *The gypsy moth: Research toward integrated pest management,* ed. C. C. Doane and M. L. McManus. *For. Serv. Tech. Bull.* **1584.** Washington: U.S. Department of Agriculture.

Paris, O. H.
1969. The function of soil fauna in grassland ecosystems. In *The grassland ecosystem: A preliminary synthesis,* ed. R. L. Dix and R. L. Beidleman. Range Sci. Dept. Sci. Ser., no 2. Fort Collins: Colorado State University.

Parsons, J.
1971. Cannibalism in herring gulls. *Br. birds* **64**:528–537.

Pate, V. S. L.
1933. Studies on fish food in selected areas: A biological survey of the Raquette Watershed, N.Y. *N.Y. State Conserv. Dept. Biol. Surv.* no. **8**, 136–157.

Paulik, G. J.
1971. Anchovies, birds, and fishermen in the Peru Current. In *Environment, pollution, and society,* ed. W. M. Murdoch, 156–185. Sunderland, Mass.: Sinauer Associates.

Peakall, D. B.
1970. Pesticides and the reproduction of birds. *Sci. Amer.* **222**:72–78.

Pearce, R. B.
1967. Photosynthesis in plant communities as influenced by leaf angle. *Crop Sci.* **7**:321–326.

Pearce, R. B., R. H. Brown, and R. E. Blaser
1965. Relationship between leaf area index, light interception, and net photosynthesis in orchard grass. *Crop Sci.* **5**:553–556.

Pearson, O. P.
1966. The prey of carnivores during one cycle of mouse abundance. *J. Anim. Ecol.* **35**:217–233.
1971. Additional measurements of the impact of carnivores on California voles. *J. Mammal.* **52**:41–49.

Pearson, W. D., and R. H. Kramer
1972. Drift and production of two aquatic insects in a mountain stream. *Ecol. Monogr.* **24**:365–385.

Pease, J. L., R. H. Vowles, and L. B. Keith
1979. Interaction of snowshoe hares and woody vegetation. *J. Wildl. Manage.* **43**:43–60.

Peet, R. K.
1981. Changes in biomass and production during secondary forest succession. In *Forest succession: Concepts and applications,* ed. D. C. West, H. H. Shugart, and D. B. Botkin, 324–338. New York: Springer-Verlag.

Peet, R. K., and N. L. Christensen
1981. Succession: A population process. *Vegetatio* **43**:131–140.

Pengelley, E. T., ed.
1974. *Circannual clocks: Annual biological rhythms.* New York: Academic Press.

Pengelley, E. T., and S. J. Asmundson
1974. Circannual rhythmicity in hibernating mammals. In *Circannual clocks: Annual biological rhythms,* ed. E. T. Pengelley, 95–160. New York: Academic Press.

Peters, R. P., and L. D. Mech
1975. Scent marking in wolves. *Amer. Scient.* **63**:628–673.

Petersen, R. C., and K. W. Cummins
1974. Leaf processing in a woodland stream. *Freshwater Biol.* **4**:343–368.

Petrides, G. A., and W. G. Swank
1966. Estimating the productivity and energy relations of the African elephant. *Proc. Inter. Grassland Conf.* **9:**831–842.

Petrinovich, L., and T. L. Patterson
1982. The white-crowned sparrow: Stability, recruitment, and population structure in the Nuttall subspecies (1975–1980). *Auk* **99:**1–14.

Pfeiffer, W.
1962. The fright reaction of fish. *Biol. Rev.* **37:**495–511.

Phillips, D. L., and J. A. MacMahon
1978. Gradient analysis of a Sonoran Desert bajada. *Southwestern Natur.* **23:**669–680.

Phillips, J.
1965. Fire as master and servant: Its influence in bioclimatic regions of trans-Saharan Arica. *Proc. Tall Timbers Fire Ecol. Conf.* **4:**70–109.

Philpot, C. W.
1977. Vegetation features as determinants of fire frequency and intensity. In *Environmental consequences of fire and fuel management in Mediterranean ecosystems,* ed. H. A. Mooney and C. E. Conrad, 12–16. U.S.D.A. For. Serv. Gen. Tech. Rept. **WO-26.** Washington, D.C.: U.S. Department of Agriculture.

Pianka, E. R.
1967. On lizard species diversity: North American flatlands desert. *Ecology* **48:**333–351.
1970. On *r* and *K* selection. *Amer. Natur.* **104:**592–597.
1972. *r* and *K* selection or *b* and *d* selection? *Amer. Natur.* **100:**65–75.
1975. Niche relations of desert lizards. In *Ecology and evolution of communities,* ed. M. Cody and J. M. Diamond, 292–314. Cambridge, Mass.: Harvard University Press.
1976. Competition and niche theory. In *Theoretical ecology: Principles and application,* ed. R. M. May, 114–141. Philadelphia: Saunders.
1983. *Evolutionary ecology.* 3d ed. New York: Harper & Row.

Pickett, S. T. A.
1976. Succession: An evolutionary interpretation. *Amer. Natur.* **110:** 107–119.

Pielou, E. C.
1974. *Population and community ecology.* New York: Gordon and Breach.
1977. *Mathematical ecology.* New York: Wiley.
1981. The usefulness of ecological models: A stock-taking. *Quart. Rev. Biol.* **56:**17–31.

Pimentel, D.
1961. Animal population regulation by the genetic feedback mechanism. *Amer. Natur.* **95:**65–79.
1968. Population regulation and genetic feedback. *Science* **159:**1423–1437.
1971. *Ecological effects of pesticides on non-target species.* Washington, D.C.: Executive Office of the President, Office of Science and Technology.

Pimm, S. L.
1980. Food web design and the effect of species deletion. *Oikos* **35:**139–149.
1982. *Food webs.* London: Chapman and Hall.

Pimm, S. L., and J. H. Lawton
1977. The number of trophic levels in ecological communities. *Nature* **268:**329–331.
1982. The cause of food web structure: Dynamics, energy flow, and natural history. In *Current trends in food web theory,* ed. D. C. DeAngelis, W. M. Post, and G. Sugihara, 45–49. Report on a Food Web Workshop, ORNL 5983. Oak Ridge, Tenn.: Oak Ridge National Laboratory.

Pitelka, L. F.
1977. Energy allocation in annual and perennial lupines (*Lupinus*) Leguminosae. *Ecology* **58:**1055–1065.

Pittendridgh, C. S.
1966. The circadian oscillation in *Drosophila pseudoobscura* pupae: A model of the photoperiodic clock. *Z. Pfanzenphysiol.* **54:**275–307.

Policansky, D.
1982. Sex change in plants and animals. *Ann. Rev. Ecol. Syst.* **13:**471–495.

Polis, G.
1981. The evolution and dynamics of intraspecific predation. *Ann. Rev. Ecol. Syst.* **12:**225–251.

Polis, G. A., and R. D. Farley
1980. Population biology of a desert scorpion: Survivorship, microhabitat, and the evolution of life history strategy. *Ecology* **61:**620–629.

Polunin, N.
1955. Aspects of arctic botany. *Amer. Scient.* **43:**307–322.

Pomeroy, L. R.
1959. Algae production in salt marshes of Georgia. *Limnol. Oceanogr.* **4:**386–397.

Pomeroy, L. R., and E. J. Kuenzler
1969. Phosphorus turnover by coral reef animals. In *Symposium on Radioecology, conf. 670503,* ed. D. Nelson and F. Evans, 478–482. Springfield, Va.: National Technical Information Services.

Pomeroy, L. R., H. M. Mathews, and H. Shik Min
1963. Excretion of phosphate and soluble organic phosphorus compounds by zooplankton. *Limnol. Oceanogr.* **4:**50–55.

Poole, R. W.
1974. *An introduction to quantitative ecology.* New York: McGraw-Hill.

Post, W. M., C. C. Travis, and D. L. DeAngelis
1980. Evolution of mutualism between species. In *Differential equations and applications in ecology, epidemics, and population problems,* ed. C. L. Cooke and S. Brisenberg, 183–201. New York: Academic Press.

Price, P. W., C. E. Bouton, P. Gross, B. A. McPheron, J. N. Thompson, and A. E. Weis
1980. Interactions among three trophic levels: Influence of plants on interactions between insect herbivores and natural enemies. *Ann. Rev. Ecol. Syst.* **11**:41–65.

Primack, R. B.
1979. Reproductive effort in annual and perennial species of *Plantago* (Plantaginaceae). *Amer. Natur.* **114**:51–62.

Putwain, P. D., and J. L. Harper
1970. Studies of dynamics of plant populations: 3, The influence of associated species on populations of *Rumex acetosa* L. and *R. acetosella* L. in grassland. *J. Ecol.* **58**:251–264.

Putwain, P. D., D. Machin, and J. L. Harper
1968. Studies in the dynamics of plant populations: II, Components and regulation of a natural population of *Rumex acetosella* L. *J. Ecol.* **56**:421–431.

Rabatnov, T. A.
1974. Differences between fluctuations and successions. In *Vegetation dynamics;* Part 8, *Handbook of vegetation science,* ed. R. Knapp, 21–24. The Hague, Netherlands: Junk.

Rafes, P. M.
1970. Estimating the effects of phytophagous insects on forest production. In *Analysis of temperate forest ecosystems,* ed. D. E. Reichle, 100–106. New York: Springer-Verlag.

Raikow, R. J.
1976. The origin and evaluation of Hawaiian honeycreepers (Drepanidae). *Living Bird* **15**:95–117.

Ralls, K.
1977. Sexual dimorphism in mammals: Avian models and unanswered questions. *Amer. Natur.* **111**:917–938.

Ramensky, L. G.
1926. Die grundgesetzmassigkeiten im aufbau der vegetationsdech. *Botanisches Centralblatt N. F.* **7**:453–455. Translation in *Readings in ecology,* ed. E. J. Kormondy, 151–152. Englewood Cliffs, N.J.: Prentice-Hall.

Ranney, J. W., M. C. Bruner, and J. B. Levenson
1981. The importance of edge in the structure and dynamics of forest islands. In *Forest island dynamics in man-dominated landscapes,* ed. R. L. Burgess and D. M. Sharpe, 67–95. Ecological Studies no. 41. New York: Springer-Verlag.

Raup, D. M., and J. J. Sepkoski, Jr.
1982. Mass extinction in the marine fossil record. *Science* **215**:1501–1502.

Raymont, J. E. G.
1963. *Plankton and productivity in oceans.* Elmsford, N.Y.: Pergamon Press.

Reader, R. J.
1978. Contribution of overwintering leaves to the growth of three broad-leaved evergreen shrubs belonging to the Ericacea family. *Can. J. Bot.* **56**:1248–1261.

Rees, W. A.
1978. The ecology of the Kafue lechwe: As affected by the Kafue Gorge hydroelectric scheme. *J. App. Ecol.* **15**:205–217.

Regier, H. A., and E. B. Cowell
1972. Application of ecosystem theory, succession, diversity, stability, stress, and conservation. *Biol. Cons.* **4**:83–93.

Reichle, D. E.
1967. Radioisotope turnover and energy flow in terrestrial isopod populations. *Ecology* **48**:351–366.
1971. Energy and nutrient metabolism of soil and litter invertebrates. In *Productivity of forest ecosystems,* ed. P. Duvigneaud, 465–477. Paris: UNESCO.
1975. Advances in ecosystem analysis. *Bioscience* **25**:257–264.

Reichle, D. E., ed.
1970. *Analysis of temperate forest ecosystems.* New York: Springer-Verlag.
1981. *Dynamic properties of forest ecosystems.* England: Cambridge University Press.

Reichle, D. E., B. E. Dinger, N. T. Edwards, W. F. Harris, and P. Sollins
1973. Carbon flow and storage in a forest ecosystem. In *Carbon and the Biosphere, conf. 72501,* ed. G. M. Woodwell and E. V. Pecan, 345–365. Springfield, Va.: National Technical Information Service.

Reichle, D. E., P. B. Dunaway, and D. J. Nelson
1970. Turnover and concentration of radionuclides in food chains. *Nuclear Safety* **11**:43–56.

Reichle, D. E., R. A. Goldstein, R. I. Van Hooke, Jr., and G. J. Dotson
1973. Analysis of insect consumption in a forest canopy. *Ecology* **54**:1076–1084.

Reichman, O. J., and D. Oberstein
1977. Selection of seed distribution types by *Dipodomys merriami* and

Perognathus amplus. *Ecology* **58**:636–643.

Reifsnyder, W. E., and H. W. Lull
1965. *Radiant energy in relation to forests*. U.S.D.A. Tech. Bull. no. **1344.**

Reitmeier, R. F.
1957. Soil potassium and fertility. In *Yearbook of Agriculture*, 101–106. Washington, D.C.: U.S. Department of Agriculture.

Renner, M.
1960. The contribution of the honey bee to the study of the time sense and astronomical orientation. *Cold Spring Harbor Symp. Quant. Biol.* **25**:361–367.

Rhoades, D. F., and R. G. Cates
1976. A general theory of plant antiherbivore chemistry. *Recent Adv. Phytochem* **10**:168–213.

Rice, E. L.
1964. Inhibition of nitrogen-fixing and nitrifying bacteria by seed plants. *Ecology* **45**:824–837.
1972. Allelopathic effects of *Andropogon virginicus* and its persistence in old fields. *Amer. J. Bot.* **59**:752–755.

Rice, E. L., and L. Parenti
1967. Inhibition of nitrogen-fixing and nitrifying bacteria by seed plants: V, Inhibitors produced by *Bromus japonicus* Thunb. *Southwest Natur.* **12**:97–103.

Rice. W. R.
1982. Acoustical location of prey by the marsh hawk: Adaptation to concealed prey. *Auk* **99**:403–413.

Rich, P. H., and R. G. Wetzel
1978. Detritus in the lake ecosystem. *Amer. Natur.* **112**:57–71.

Richards, P. W.
1952. *The tropical rain forest*. London: Cambridge University Press.

Ricklefs, R.
1979. *Ecology*. New York: Chiron Press.

Riechert, S. E.
1981. The consequences of being territorial: Spiders, a case study. *Amer. Natur.* **117**:871–892.

Rigler, F. H.
1964. The phosphorus fractions and the turnover time of inorganic phosphorus in different types of lakes. *Limno. Oceanogr.* **9**:511–518.

Riley, G. A.
1973. Particulate and dissolved organic carbon in the oceans. In *Carbon and the Biosphere, conf. 72501*, ed. G. M. Woodwell and E. V. Pecan, 204–220. Springfield, Va.: National Technical Information Service.

Ringler, N.
1979. Selective predation by drift-feeding brown trout (*Salmo trutta*). *J. Fish. Res. Bd. Can.* **26**:392–403.

Rixon, A. J.
1970. Cycling of nutrients in a grazed *Atriplex vesicaria* community. In *The biology of Atriplex*, ed. R. Jones, 87–95. Canberra: CSIRO Division of Plant Industry.

Rodin, L. E., and I. Bazilevic
1967. *Production and mineral cycling in terrestrial vegetation*. Translated from the Russian by Scripta Technica: ed. G. E. Fogg. London: Oliver and Boyd.

Rohde, W.
1965. Standard correlations between photosynthesis and light. *Memorie dell'Istituto Italiano di Idrobiologia* **18** (supplement): 365–381.

Rohwer, S.
1978. Parent cannibalism of offspring and egg raiding as a courtship strategy. *Amer. Natur.* **112**:429–440.

Root, R. B.
1967. The niche exploitation pattern of the blue-gray gnatcatcher. *Ecol. Monogr.* **37**:317–350.

Rorison, J. H., ed.
1969. *Ecological aspects of mineral nutrition of plants*. Oxford, England: Blackwell.

Rosenthal, G. A., and D. H. Janzen, eds.
1979. *Herbivores: Their interaction with secondary metabolites*. New York: Academic Press.

Ross, M. D.
1982. Five evolutionary pathways to subdioecy. *Amer. Natur.* **119**:297–318.

Rosswall, T., and O. W. Heal, eds.
1975. *Structure and function of tundra ecosystems*. Stockholm: Swedish Natural Sciences Research Council.

Rosswall, T., J. G. K. Flower-Ellis, L. G. Johansson, S. Johsson, B. E. Ryden, and M. Sonesson
1975. Stordalen (Abisko), Sweden. In *Structure and function of tundra ecosystems*, ed. T. Rosswall and O. W. Heal, 265–294. Stockholm: Swedish Natural Science Research Council.

Rotenberry, J. T., and J. A. Weins
1980. Temporal variation in habitat structure and shrubsteppe bird dynamics. *Oecologica* **47**:1–9.

Roughgarden, J.
1974. Niche width: Biogeographic patterns among *Anolis* lizard populations. *Amer. Natur.* **108**:429–442.
1979: *Theory of population genetics and evolutionary theory: An introduction*. New York: Macmillan.
1983. The theory of coevolution. In *Coevolution*, ed. D. J. Futuyma and M. Slatkin, 33–64. Sunderland, Mass.: Sinauer Associates.

Rudd, R. L., and E. B. Genelly
1956. Pesticides: Their use and toxicity in relation to wildlife. *Calif. Fish and Game Bull.* no. **7.**

Rusch, D. H., E. C. Meslow, P. D. Doerr, and L. B. Keith
1972. Response of great horned owl populations to changing prey densities. *J. Wildl. Manage.* **36:**282–296.

Rutherford, M. C.
1978. Primary production ecology in Southern Africa. In *Biogeography and ecology of Southern Africa,* ed. M. J. A. Wegner, 621–659. The Hague, Netherlands: Junk.

Ryszkowski, L.
1980. Ecosystem synthesis. In *Grasslands, systems analysis and man,* ed. A. I. Breymeyer and G. M. Van Dyne, 327–334. International Biological Programme no 19. Cambridge, England: Cambridge University Press.

Sagar, G. A., and A. M. Mortimer
1976. An approach to the study of population dynamics of plants with a special reference to weeds. *Appl. Biol* **1:**1–47.

Salt, G. W.
1957. An analysis of avifauna in the Teton Mountains and Jackson Hole, Wyoming. *Condor* **59:**373–393.

Sandon, H.
1927. *The composition and distribution of the protozoan fauna of the soil.* London: Oliver and Boyd.

Sarukhan, J., and J. L. Harper
1973. Studies on plant demography: *Ranuculus repens* L., *R. bulbosus* L., and *R. acris* L.—1, Population flux and survivorship. *J. Ecol.* **61:**676–716.

Saunders, D. S.
1982. *Insect clocks.* 2d ed. Elmsford, N.Y.: Pergamon Press.

Saunders, G. W.
1976. Decomposition in fresh water. In *The role of terrestrial and aquatic organisms in the decomposition process,* ed. J. M. Anderson and P. MacFayden, 341–373. Seventeenth Symposium, British Ecological Society. Oxford, England: Blackwell.

Saunders, G. W., K. W. Cummins, D. Z. Gak, E. Pieczynska, V. Straskrabova, and R. G. Wetzel
1980. Organic matter and decomposition. In *The functioning of freshwater ecosystems,* ed. E. D. LeCren and R. H. Lowe-McConnell, 341–392. International Biological Programme no. 22. Cambridge, England: Cambridge University Press.

Scachak, M., U. N. Safriel, and R. Hunum
1981. An exceptional event of predation on desert snails by migratory thrushes in Negev Desert, Israel. *Ecology* **62:**1441–1449.

Schaefer, R.
1973. Microbial activity under seasonal conditions of drought in Mediterranean climates. In *Mediterranean type ecosystems: Origin and structure,* ed. F. di Castri and H. Mooney, 191–198. New York: Springer-Verlag.

Schaller, G. B.
1972. *Serengeti: A kingdom of predators.* New York: Knopf.

Scharitz, R. R., and J. R. McCormick
1973. Population dynamics of two competing plant species. *Ecology* **54:**723–740.

Scheffer, V.
1951. The rise and fall of a reindeer herd. *Sci. Month.* **73:**356–362.

Schmid, W. D.
1982. Survival of frogs at low temperature. *Science* **215:**697–698.

Schmidt-Nielsen, K.
1960. The salt-secreting gland of marine birds. *Circulation* **21:**955–967.
1970. *Animal physiology.* 3d ed. Englewood Cliffs, N.J.: Prentice-Hall.
1979. *Animal physiology: Adaptation and environment.* New York: Cambridge University Press.

Schouw, J. F.
1822. *Grundtraek til en almindelig plantegeographie.* Copenhagen, Denmark: Glydendal.
1823. *Grunzuge einer allgemeinen pflanzengeographie.* Berlin.

Schowalter, T. D.
1981. Insect-herbivore relationship to the state of the host plant: Biotic regulation of ecosystem nutrient cycling through succession. *Oikos* **37:**126–130.

Schroeder, M. S., and C. C. Buck
1970. *Fire weather. Agricultural handbook* no. **360.** Washington, D.C.: U.S. Department of Agriculture.

Schultz, J. C., and I. T. Baldwin
1982. Oak leaf quality declines in response to defoliation by gypsy moth larvae. *Science* **217:**149–151.

Scott, J. A., N. R. French, and J. W. Leetham
1979. Patterns of consumption in grasslands. In *Perspectives in grassland ecology,* ed. N. French, 89–105. New York: Springer-Verlag.

Scott, T. C.
1943. Some food coactions of the northern plains red fox. *Ecol. Monogr.* **13:**427–479.
1955. An evaluation of the red fox. *Ill. Nat. Hist. Surv. Biol. Notes* No. **35,** 1–16.

Searcy, W. A.
1979. Female choice of males: A general model for birds and its application to red-winged blackbirds

Agelaius phoeniceus. Amer. Natur. **114:**77–100.

Sears, P. B.
1964. Ecology—a subversive subject. *Bioscience* **14:**11.

Selleck, G. W.
1960. The climax concept. *Bot. Rev.* **26:**534–545.

Sette, O. E.
1943. Biology of the Atlantic mackerel *Scomber scombus. U.S. Fish and Wildl. Serv. Fishery Bull.* **38.**

Severinghaus, C. W.
1972. Weather and the deer population. *The Conservationist* **27**(2):28–31.

Shapiro, A. M.
1977. Photoperiod and temperature in phenotype determination of Pacific slope Pieroni: Biosystematic implications. *J. Res. Lepid.* **16:**193–200.
1981. The pierid red-egg syndrome. *Amer. Natur.* **117:**276–294.

Shapiro, D. Y.
1980. Serial female sex changes after simultaneous removal of males from social groups of a coral reef fish. *Science* **209:**1136–1137.

Sheppard, D. H.
1971. Competition between two chipmunk species (*Eutamias*). *Ecology* **52:**320–329.

Sheppard, P. M.
1959. *Natural selection and heredity.* London: Hutchinson.

Sherman, P. W.
1977. Nepotism and the evolution of alarm calls. *Science* **197:**1246–1253.
1981. Reproductive competition and infanticide in Belding's ground squirrel and other animals. In *Natural selection and social behavior: Recent research and new theories,* ed. R. D. Alexander and D. W. Tinkle, 311–331. New York: Chiron Press.

Shervis, L. H., G. M. Bush, and C. F. Koval
1970. Infestation of sour cherries by apple maggot: Confirmation of a previously uncertain host status. *J. Econ. Entomol.* **63:**993–1006.

Shine, R.
1980. "Costs" of reproduction in reptiles. *Oecologica* **46:**92–100.

Shorey, H. H.
1976. *Animal communication by pheromones.* New York: Academic Press.

Shugart, H. H., Jr., T. R. Crow, and J. M. Hett
1973. Forest succession models: A rationale and methodology for modeling forest succession over large regions. *For. Sci.* **19:**203–212.

Shure, D. J., and H. S. Ragsdale
1977. Patterns of primary succession on granite outcrop surfaces. *Ecology* **58:**993–1006.

Sibley, C. G.
1957. The evolutionary and taxonomic significance of sexual dimorphism and hybridization in birds. *Condor* **59:**166–191.
1960. The electrophoretic patterns of avian egg-white proteins as taxonomic characters. *Ibis* **102:**215–284.

Simberloff, D. S.
1969. Experimental zoogeography of islands: A model for insular colonization. *Ecology* **50:**296–314.
1974. Equilibrium theory of island biogeography and ecology. *Ann. Rev. Ecol. Syst.* **5:**161–182.

Simberloff, D. S., and L. G. Abele
1976. Island biogeographic theory and conservation practice. *Science* **191:**285–286.
1982. Refuge design and island biogeographic theory: Effects of fragmentation. *Amer. Natur.* **120:**41–50.

Simons, S., and J. Alcock
1971. Learning and foraging persistence of white-crowned sparrows *Zonotrichia leucophrys. Ibis* **111:**477–482.

Simpson, G. G.
1964. Species density of North American recent mammals. *Syst. Zool.* **13:**57–73.

Sims, P. L., and J. S. Singh
1971. Herbage dynamics and net primary production in certain grazed and ungrazed grasslands in North America. In *Preliminary analysis of structure and function in grasslands,* ed. N. R. French, 59–124. Range Sci. Dept. Sci. Ser. no. 10. Fort Collins: Colorado State University.

Sinclair, A. R. E.
1977. *The African buffalo: A study of resource limitation of populations.* Chicago: University of Chicago Press.

Singh, J. S., and M. S. Joshi
1979. Primary production. In *Grassland systems of the world,* ed. R. T. Coupland, 197–218. International Biological Programme no. 18. Cambridge, England: Cambridge University Press.

Singh, J. S., K. D. Singh, and P. S. Yadava
1979. Ecosytem synthesis. In *Grasslands systems of the world,* ed. R. T. Coupland, 231–240. International Biological Programme no. 18. Cambridge, England: Cambridge University Press.

Skellam, J. S.
1972. Some philosophical aspects of mathematical modelling in empherical science with special reference to ecology. In *Mathematical models in ecology,* ed. J. N. R. Jeffers, 13–28. Oxford: Blackwell Scientific Publications.

Slobodkin, L. B.
1974. Comments from a biologist to a mathematician. In *Proc. SIAM-SIMS Conference, Alta, Utah,* ed. S. A. Levin, 318–329.

Smalley, A. E.
1960. Energy flow of a salt marsh grasshopper population. *Ecology* **41**:672–677.

Smith, F.
1975. Ecosystem and evolution. *Bull. Ecol. Soc. Amer.* **56**(1):2–6.

Smith, H.
1982. Light quality, photoreception, and plant strategy. *Ann. Rev. Plant Physiol.* **33**:481–518.

Smith, J. N. M.
1974a. The food searching behaviour of two European thrushes: 1, Description and analysis of search paths. *Behaviour* **48**:276–302.
1974b. The food searching behaviour of two European thrushes: 2, The adaptiveness of the search patterns. *Behaviour* **49**:1–61.

Smith, J. N. M., and H. P. A. Sweatman
1974. Food searching behaviour of titmice in patchy environments. *Ecology* **55**:1316–1232.

Smith, N. G.
1968. The advantage of being parasitized. *Nature* **219**:690–694.

Smith, R. L.
1959. Conifer plantations as wildlife habitat. *N.Y. Fish Game J.* **5**:101–132.
1976a. *The ecology of man: An ecosystem approach*. New York: Harper & Row.
1976b. Socio-ecological evolution in the hill country of southwestern West Virginia. In *Hill lands, Proc. International Symposium,* ed. J. Luchok, J. D. Cawthon, and M. J. Breslin, 198–202. Morgantown: West Virginia University Books.
1980. *Ecology and field biology*. 3d ed. New York: Harper & Row.

Smith, S. M.
1978. The "underworld" in a territorial adaptive strategy for floaters. *Amer. Natur.* **112**:570–582.

Smith, T. M., and B. H. Walker
1983. The role of competition in the spacing of savanna trees. *Proc. Grassld. Soc. S. Afr.* **18**:159–164.

Smock, L. A., and K. L. Harlowe
1983. Utilization and processing of freshwater wetland macrophytes by the detritivore *Asellus forbesi*. *Ecology* **64**:1556–1565.

Snyder, D. P., ed.
1978. *Populations of small mammals under natural conditions*. The Pymatuning Symposia in Ecology. Vol. 5, Special Publication Series, Pymatuning Laboratory of Ecology. Pittsburgh: University of Pittsburgh Press.

Soil Survey Staff
1975. *Soil taxonomy: A basic system of soil classification*. Agricultural Handbook **436**. Washington, D.C.: U.S. Department of Agriculture.

Solbrig, O. T., S. Jain, G. B. Johnson, and P. H. Raven, eds.
1979. *Topics in plant population biology*. New York: Columbia University Press.

Solbrig, O. T., and B. B. Simpson
1974. Components of regulation of a population of dandelion in Michigan. *J. Ecol.* **62**:473–486.

Solomon, M. E.
1949. The natural control of animal populations. *J. Anim. Ecol.* **18**:1–32.

Soriano, A.
1972. South America. In *Wildland shrubs: Their biology and utilization,* ed. C. M. McKella, J. P. Blaisdell, and J. R. Goodwin, 51–54. U.S.D.A. For. Ser. Gen. Tech. Rept. **INT-1**.

Soule, M. E., and B. A. Wilcox, eds.
1980. *Conservation biology: An evolutionary-ecological perspective*. Sunderland, Mass.: Sinauer Associates.

Southwood, T. R.
1978. *Ecological methods*. 2d ed. London: Chapman and Hall.

Specht, R. L.
1973. Structure and functional response of ecosystems in the Mediterranean climate of Australia. In *Mediterranean-type ecosystems: Origin and structure,* ed. F. di Castri and H. A. Mooney, 113–120. New York: Springer-Verlag.

Spight, T. M., and J. Emlen
1976. Clutch size of two marine snails with a changing food supply. *Ecology* **57**:1162–1176.

Sprugel, D. G.
1976. Dynamic structure of wave generated *Abies balsamea* forests in north-eastern United States. *J. Ecol.* **64**:889–911.

Spurr, S. H.
1956. Natural restocking of forests following the 1938 hurricane in central New England. *Ecology* **37**:443–451.

Stanley, S. M.
1979. *Macroevolution: Pattern and process*. San Francisco: Freeman.

Stack, N., and C. F. Jordan
1978. Nutrient retention by the root mat of an Amazonian rain forest. *Ecology* **59**:434–437.

Starr, M. P.
1975. A generalized scheme for classifying organismic associations. *Symp. Soc. Exp. Biol.* **29**:1–20.

Stearns, S. C.
1976. Life history tactics: A review

of the ideas. *Quart. Rev. Biol.* **51**:3–47.
1977. The evolution of life history traits. *Ann. Rev. Ecol. Syst.* **8**:145–171.

Stebbins, G. L.
1950. *Variation and evolution in plants.* New York: Columbia University Press.
1970. Adaptive radiation of reproductive characteristics in angiosperms: I, Pollination mechanisms. *Ann. Rev. Ecol. Syst.* **1**:307–326.
1972. Evolution and diversity of arid-land shrubs. In *Wildland shrubs: Their biology and utilization,* ed. C. M. McKella, J. P. Blaisdell, and J. R. Goodwin, 111–116. U.S.D.A. For. Serv. Gen. Tech. Rept. **INT-1.**

Stephenson, T. A., and A. Stephenson
1949. The universal features of zonation between tide marks on rocky coasts. *J. Ecol.* **37**:289–305.
1954. Life between the tide-marks in North America: IIIA, Nova Scotia and Prince Edward Island—description of the region; IIIB, Nova Scotia and Prince Edward Island—the geographical features of the region. *J. Ecol.* **42**:14–45, 46–70.
1961. Life between the tide marks in North America: IVA, IVB, Vancouver Island, I, II. *J. Ecol.* **49**:1–29, 229–243.
1971. *Life between the tide-marks on rocky shores.* San Francisco: Freeman.

Stern, J. T., Jr.
1970. The meaning of adaptation and its relation to the phenomenon of natural selection. *Evol. Biol.* **4**:38–66.

Stern, W. L., and M. F. Buell
1951. Life form spectra of New Jersey pine barren forest and Minnesota jack pine forest. *Torrey Botan. Club. Bull.* **78**:61–65.

Stiles, E.
1980. Patterns of fruit presentation and seed dispersal in bird disseminated woody plants in the eastern deciduous forest. *Amer. Natur.* **116**:670–688.

Stiles, F. G.
1975. Ecology, flowering phenology, and hummingbird pollination of some Costa Rican *Heliconia* species. *Ecology* **56**:285–301.

Stout, J. D., K. R. Tate, and L. F. Molloy
1976. Decomposition processes in New Zealand soils with particular respect to rates and pathways of plant degradation. In *The role of terrestrial and aquatic organisms in the decomposition process,* ed. J. M. Anderson and P. MacFayden, 97–144. Seventeenth Symposium, British Ecological Society. Oxford, England: Blackwell.

Strahler, A.
1971. *The earth sciences.* New York: Harper & Row.

Strickland, J. D. H.
1965a. Phytoplankton and marine primary production. *Ann. Rev. Microbiol.* **19**:127–162.
1965b. Production of organic matter in the primary stages of marine food chains. In *Chemical oceanography,* ed. J. P. Riley and G. Skirrow. Vol. 1. New York: Academic Press.

Study of Critical Environmental Problems
1970. *Man's impact on the global environment.* Cambridge, Mass.: MIT Press.

Stuiver, M.
1978. Atmospheric carbon dioxide and carbon reservoir changes. *Science* **199**:253–258.

Swain, T.
1977. Secondary compounds as protective agents. *Ann. Rev. Plant Physiol.* **28**:479–501.

Swank, W. T., J. B. Waide, D. A. Crossley, Jr., and R. L. Todd
1981. Insect defoliation enhances nitrate export from forest ecosystems. *Oecologica* **51**:297–299.

Tamarin, R. H.
1978. Dispersal, population regulation, and K-selection in field mice. *Amer. Natur.* **112**:545–555.

Tamm, C. O.
1951. Removal of plant nutrients from tree crowns by rain. *Physiol. Plant.* **4**:184–188.

Tanner, J. T.
1966. Effects of population density on growth rates of animal populations. *Ecology* **47**:733–745.
1975. The stability and intrinsic growth rates of prey and predator populations. *Ecology* **56**:855–867.
1978. *A guide to the study of animal populations.* Knoxville: University of Tennessee.

Tansley, A. G.
1935. The use and abuse of vegetational concepts and terms. *Ecology* **16**:284–307.
1939. *The British Isles and their vegetation.* 1965 reprint. Cambridge, England: Cambridge University Press.

Tauber, M. J., and C. A. Tauber
1976a. Developmental requirements of the univoltine species *Chrisopa downesi:* Photoperiodic stimuli and sensitive stage. *J. Insect Physiol.* **22**:331–335.
1976b. Environmental control of univoltism and evolution in an insect species. *Can. J. Zool* **54**:260–265.

Taylor, C. R.
1970a. Dehydration and heat effects on temperature regulation of East

African ungulates. *Amer. J. Physiol.* **219:**1136–1139.

1970b. Strategies of temperature regulation: Effect of evaporation on East African ungulates. *Amer. J. Physiol.* **219:**1131–1135.

1972. The desert gazelle: A paradox resolved. In *Comparative physiology of desert animals,* ed. G. M. O. Malory, 215–227. Symp. Zool. Soc. London no. 31. London: Academic Press.

Teal, J. M.

1957. Community metabolism in a temperate cold spring. *Ecol. Monogr.* **27:**283–302.

1962. Energy flow in the salt marsh ecosystem of Georgia. *Ecology* **43:**614–624.

Teal, J. M., and J. Kanwisher

1961. Gas exchange in a Georgia salt marsh. *Limno. Oceanogr.* **6:**388–389.

Terborgh, J.

1974. Preservation of natural diversity: The problem of extinction prone species. *Bioscience* **24:**715–722.

Terri, J. A., and L. G. Stowe

1976. Climatic patterns and distribution of C_4 grasses in North America. *Oecologica* **23:**1–12.

Tester, J. R., and W. H. Marshall

1961. A study of certain plant and animal interrelations on a native prairie in northwestern Minnesota. *Minn. Mus. Nat. Hist., Occasional Paper* no. **8.**

Thailer, C. S.

1974. Four contacts between ranges of different chromosome forms of the *Thomomys talpoides* complex (Rodentia, Geomyidae). *Syst. Zool.* **23:**343–354.

Thieneman, A.

1931. Der produktionsbergriff in der biologie. *Arch. Hydrobiol.* **22:**616–622.

Thompson, D. Q., and R. H. Smith

1970. The forest primeval in the northeast: A great myth? *Proc. Tall Timbers Fire Ecology Conf.* **10:**255–265.

Tinbergen, N.

1960. Comparative studies of the behaviour of gulls (Laridae): A progress report. *Behaviour* **15:**1–70.

Tinkle, D. W., J. W. Condon, and P. C. Rosen

1981. Nesting season and success: Implications for the demography of painted turtles. *Ecology* **62:**1426–1432.

Townsend, C. R., and P. Calow

1981. *Physiological ecology: An evolutionary approach to resource use.* Oxford, England: Blackwell.

Tramer, E. J.

1975. The regulation of plant species diversity on an early successional old-field. *Ecology* **56:**905–914.

Trewartha, G. T.

1943. *An introduction to weather and climate.* 2d. ed. New York: McGraw-Hill.

Trimble, G. R., Jr.

1973. The regeneration of central Appalachian hardwoods with emphasis on the effects of site quality and harvesting practice. *U.S.D.A. For. Serv. Res. Paper* **NE-282.**

Van Cleve, K., and L. A. Viereck

1981. Forest succession in relation to nutrient cycling in the boreal forest in Alaska. In *Forest succession: Concepts and applications,* ed. D. C. West, H. H. Shugart, and D. B. Botkin, 185–211. New York: Springer-Verlag.

Van Den Bergh, J. P.

1969. Distribution of pasture plants in relation to chemical properties of the soil. In *Ecological aspects of mineral nutrition of plants,* ed. J. H. Rorison, 11–23. Oxford, England: Blackwell.

Van Den Bergh, J. P., and C. T. Dewit

1960. Concurrentie tussen Timothee en Reukgras. *Meded. Inst. biol. scherk Onderg Lanlb. Gewass.* **121:**155–165.

Van Hooke, R. I.

1971. Energy and nutrient dynamics of spider and orthropteran populations in a grassland ecosystem. *Ecol. Monogr.* **41:**1–26.

Vannote, R. L., G. W. Minshall, K. W. Cummins, J. R. Sedell, and C. E. Cushing

1980. The river continuum concept. *Can. J. Fish Aquat. Sci.* **37:**130–137.

Van Valen, L.

1973. A new evolutionary law. *Evol. Theory* **1:**1–30.

Varley, G. C., and G. R. Gradwell

1970. Recent advances in insect population dynamics. *Ann. Rev. Entomol.* **15:**1–24.

Vaughan, T. A.

1978. *Mammalogy.* Philadelphia: Saunders.

Veale, P. T., and H. L. Wascher

1956. Henderson County soils. *Ill. Univ. Agr. Exp. Stat. Soil Rept.* no. **77.**

Verhulst, P. F.

1883. Notice sur la loi que la population pursuit dans son accroissement. *Corresp. Math. Phys.* **10:**113–121.

Viereck, L. A.

1970. Forest succession and soil development adjacent to the Chena River in interior Alaska. *Arct. Alp. Res.* **2:**1–26.

Vogl, R. J.

1967. Fire adaptations of some southern California plants. *Proc. Tall Timbers Fire Ecology Conf.* **7**:79–109.

1969. One hundred and thirty years of plant succession in a southeastern Wisconsin lowland. *Ecology* **50**:248–255.

1974. Effect of fire in grassland. In *Fire and ecosystems,* ed. T. F. Kozlowski and C. E. Ahlgren, 139–194. New York: Academic Press.

Voigt, G. K.

1971. Mycorrhizae and nutrient immobilization. In *U.S.D.A. For. Serv. Misc. Pub.* no. **1189**, ed. E. Hasckaylo, 122–131.

Volterra, V.

1926. Variazione e fluttazioni de numero d'individiu in specie animali conviventi. *Mem. Accad. Lincei* **2**:32–113. Translated in R. N. Chapman, *Animal ecology.* New York: McGraw-Hill, 1931.

von Frisch, K.

1954. *The dancing bees.* London: Methuen.

Vose, R. N., and D. G. Dunlap

1968. Wind as a factor in the local distribution of small mammals. *Ecology* **49**:381–386.

Vrba, E. S.

1980. Evolution, species, and fossils: How does life evolve? *S. Afr. J. Sci.* **76**:61–84.

Wagner, F. H., C. D. Besadny, and C. Kabat

1965. Population ecology and management of Wisconsin pheasants. *Wisconsin Cons. Dept. Tech. Bull.* **34.** Madison: Wisconsin Conservation Department.

Wahlenberg, W. G.

1946. *Longleaf pine: Its use, ecology, regeneration, protection, growth, and management.* Washington, D.C.: Charles Lathrop Pack Forestry Foundation.

Walker, B. H.

1981. Is succession a viable concept in African savanna ecosystems? In *Forest succession: Concepts and applications,* ed. D. C. West, H. H. Shugart, and D. B. Botkin, 431–437. New York: Springer-Verlag.

Walker, J., C. H. Tompson, J. F. Fergus, and B. R. Tunstall

1981. Plant succession and soil development in coastal sand dunes of subtropical eastern Australia. In *Forest succession: Concepts and applications,* ed. D. C. West, H. H. Shugart, and D. B. Botkin, 107–131. New York: Springer-Verlag.

Wallace, A. R.

1856. On the tendency of varieties to depart indefinitely from the original type. In *Evolution by natural selection,* C. Darwin and A. R. Wallace, 1958 ed., 268–279. Cambridge, England: Cambridge University Press.

1876. *The geographical distribution of animals.* London: Macmillan.

1878. *Tropical nature and essays.* London: Macmillan.

Waller, D. M.

1982. Jewelweed's sexual skills. *Nat. Hist.* **91**(5):32–38.

Wallwork, J. A.

1973. *Ecology of soil animals.* New York: McGraw-Hill.

Walter, E.

1934. Grundlagen der allegmeinen fisherielichen Produktionslehre. *Handb. Binnenfisch. Miteleur.* **14**:480–662.

Walter, H.

1971. *Ecology of tropical and subtropical vegetation.* Edinburgh: Oliver and Boyd.

1973. *Vegetation of the earth.* New York: Springer-Verlag.

Walter, H.

1977. Effects of fire on wildlife communities. In *Proc. symp. environmental consequences of fire and fuel management in Mediterranean ecosystems,* ed. H. A. Mooney and C. E. Conrad, 183–192. U.S.D.A. For. Serv. Gen. Tech. Rept. **WO-3.** Washington, D.C.: U.S. Department of Agriculture.

Wargo, P. M.

1981. Defoliation and tree growth. In *The gypsy moth: Research toward integrated pest management,* ed. C. C. Doane and M. L. McManus, 225–240. For. Serv. Tech. Bull. **1584.** Washington, D.C.: U.S. Department of Agriculture.

Waring, R. H., ed.

1979. *Forests: Fresh perspectives from ecosystem analysis.* Biology Colloquium, 40th. Oregon State University. Corvallis: Oregon State University Press.

Warner, R. E.

1968. The role of introduced diseases in the extinction of the endemic Hawaiian avifauna. *Condor* **70**:101–120.

Wassink, E. C.

1959. Efficiency of light energy conversion in plant growth. *Plant Physiol.* **34**:356–361.

Waters, T. F.

1972. The drift of stream insects. *Ann. Rev. Entomol.* **17**:253–272.

Watson, A., and R. Moss

1971. Spacing as affected by territorial behavior and nutrition in red grouse (*Lagopus l. scoticus*). In *Behavior and environment,* ed. A. E. Esser, 92–111. New York: Plenum.

Watt, A. S.

1947. Pattern and process in the plant community. *J. Ecol.* **35**:1–22.

Weaver, J. E.

1954. *North American prairie.* Lincoln, Nebr.: Johnson.

Weaver, J. E., and F. W. Albertson
1956. *Grasslands and the Great Plains: Their nature and their use.* Lincoln, Nebr.: Johnson.

Webber, P. J., P. C. Miller, F. S. Chapin III, and B. H. McCown
1980. The vegetation pattern and succession. In *An arctic ecosystem: The coastal tundra at Barrow, Alaska,* ed. J. Brown, P. C. Miller, L. L. Tieszen, and F. L. Bunnell, 186–218. US/IBP Synthesis Series, no. 12. Stroudsburg, Pa.: Dowden, Hutchinson & Ross.

Webster, J. R., and B. C. Patten
1979. Effects of watershed perturbation on stream potassium and calcium dynamics. *Ecol. Monogr.* **49:**51–72.

Webster, J. R., J. B. Warde, and B. C. Patten
1975. Nutrient recycling and the stability of ecosystems. In *Mineral cycling in southeastern ecosystems,* ed. F. G. Howell, J. B. Gentry, and M. H. Smith, 1–27. ERDA Symposium Series. Springfield, Va.: National Technical Information Service, U.S. Department of Commerce.

Weis-Fogh, T.
1948. Ecological investigations of mites and collembola in the soil. *Nat. Jutland* **1:**135–270.

Weller, M. W., and L. H. Fredrickson
1973. Avian ecology of a managed glacial marsh. *Living Bird* **12:**269–291.

Wellington, W. G.
1964. Qualitative changes in populations in unstable environments. *Can. Entomol.* **96:**436–451.

Wells, H. G.
1961. The fauna of oyster beds with special reference to the salinity factor. *Ecol. Monogr.* **31:**239–266.

Wells, K. D.
1977. The social behavior of anuran amphibians. *Anim. Behav.* **25:**666–693.

Werner, E. E., and J. D. Hall
1974. Optimal foraging and size selection of prey by the bluegill sunfish. *Ecology* **55:**1042–1052.
1976. Niche shifts in sunfishes: Experimental evidence and significance. *Science* **191:**404–406.
1979. Foraging efficiency and habitat switching in competing sunfish. *Ecology* **60:**256–264.

Werner, P. A.
1978. On the determination of age in *Liatris aspera* using cross sections of corms: Implications for past demographic studies. *Amer. Natur.* **112:**1113–1120.

Wertz, W. A., and J. F. Arnold
1972. *Land system inventory.* Ogden, Utah: U.S.D.A. Forest Service, I Intermountain Region.

West, D. C., H. H. Shugart, and D. B. Botkin, eds.
1981. *Forest succession: Concepts and applications.* New York: Springer-Verlag.

West, N. E., N. K. H. Rea, and R. O. Harniss
1979. Plant demographic studies in sagebrush-grass communities of southeastern Idaho. *Ecology* **60:** 376–402.

West, N. E., and P. T. Tueller
1972. Special approaches to studies of competition and succession in shrub communities. In *Wildland shrubs: Their biology and utilization,* ed. C. M. McKella, J. P. Blaisdell, and J. R. Goodwin, 172–181. U.S.D.A. For. Ser. Gen. Tech. Rept. **INT-1.**

Westlake, D. F., coordinator
1980. Primary production. In *The functioning of freshwater ecosystems,* ed. E. D. LeCren and R. H. Lowe-McConnell, 141–246. International Biological Programme no. 22. Cambridge, England: Cambridge University Press.

Wetzel, R.
1975. *Limnology.* Philadelphia: Saunders.

Wheelwright, N. T., and G. H. Orians
1982. Seed dispersal by animals: Contrasts with pollen dispersal, problems with terminology and constraints on evolution. *Amer. Natur.* **119:**402–413.

Whigham, D. F., J. McCormick, R. E. Good, and R. L. Simpson
1978. Biomass and primary production in freshwater tidal wetlands of the Middle Atlantic Coast. In *Freshwater wetlands,* ed. R. E. Good, D. F. Whigham, and R. L. Simpson, 3–20. New York: Academic Press.

Whitcomb, R. F., J. F. Lynch, P. A. Opler, and C. S. Robbins
1976. Island biogeography and conservation: Strategy and limitations. *Science* **193:**1030–1032.

Whitcomb, R. F., C. F. Robbins, J. F. Lynch, B. L. Whitcomb, M. K. Klimkiewicz, and D. Bystrak
1981. Effects of forest fragmentation on avifauna of the eastern deciduous forest. In *Forest island dynamics in man-dominated landscapes,* ed. R. L. Burgess and D. M. Sharpe, 125–205. Ecological Studies no. 41. New York: Springer-Verlag.

White, E. J., and F. Turner
1970. A method of estimating income of nutrients in a catch of airborne particles by a woodland canopy. *J. Appl. Ecol.* **7:**441–461.

White, J.
1979. The plant as a metapopulation. *Ann. Rev. Ecol. Syst.* **10:**109–145.

White, M. J. D.
1968. Models of speciation. *Science* **159:**1065–1070.
1973. *Animal cytology and evolution.* Cambridge, England: Cambridge University Press.
1974. Speciation in Australian morabine grasshoppers—the cytogenic evidence. In *Genetic mechanisms of speciation in insects,* ed. M. J. D. White, 1–42. Sydney: Australian and New Zealand Book Co.
1978. *Modes of speciation.* San Francisco: Freeman.

White, M. J. D., ed.
1974. *Genetic mechanisms of speciation in insects.* Sydney: Australian and New Zealand Book Co.

Whitfield, P. J.
1979. *The biology of parasitism.* London: Edward Arnold.

Whitmore, T. C.
1975. *Tropical rain forests of the Far East.* Oxford, England: Clarendon Press.

Whittaker, R. H.
1953. A consideration of the climax theory: The climax as population and pattern. *Ecol. Monogr.* **23:**1–44.
1960. Vegetation of the Siskiyou Mountains. *Ecol. Monogr.* **30:**279–338.
1962. Classification of natural communities. *Botan. Rev.* **28:**1–239.
1963. Net production of heath balds and forest heaths in the Great Smoky Mountains. *Ecology* **44:**176–182.
1965. Dominance and diversity in land plant communities. *Science* **147:**250–260.
1967. Gradient analysis of vegetation. *Biol. Rev.* **42:**207–264.
1970. *Communities and ecosystems.* New York: Macmillan.

Whittaker, R. H., F. H. Bormann, G. E. Likens, and T. G. Siccama
1974. The Hubbard Brook ecosystem study: Forest biomass and production. *Ecol. Monogr.* **44:**233–252.

Whittaker, R. H., and P. R. Feeney
1971. Allelochemics: Chemical interactions between species. *Science* **171:**757–770.

Whittaker, R. H., S. A. Levin, and R. B. Root
1973. Niche, habitat, and biotope. *Amer. Natur.* **107:**321–338.

Whittaker, R. H., and G. E. Likens
1973. Carbon in the biota. In *Carbon and the biosphere conf. 72501,* ed. G. M. Woodwell and E. V. Pecan, 281–300. Springfield Va.: National Technical Information Service.

Whittaker, R. H., R. B. Walker, and A. R. Kruckeberg
1954. The ecology of serpentine soils. *Ecology* **35:**258–288.

Whittaker, R. H., and G. M. Woodwell
1969. Structure, production, and diversity of the oak-pine forest at Brookhaven, New York. *J. Ecol.* **57:**155–174.

Whyte, R. O.
1968. *Grasslands of the monsoons.* New York: Praeger.

Wiegert, R. G.
1964. Population energetics of meadow spittlebugs (*Philaenus spumarius* L.) as affected by migration and habitat. *Ecol. Monogr.* **34:**217–241.
1967. Investigation of secondary productivity in grasslands. In *Secondary productivity of terrestrial ecosystems,* ed. K. Petrusewicz, 499–518. Warsaw: Polish Academy of Sciences.

Wiegert, R. G., and D. F. Owen
1971. Trophic structure, available resources, and population density in terrestrial vs. aquatic ecosystems. *J. Theoret. Biol.* **30:**69–81.

Wieland, N. K., and F. A. Bazzaz
1975. Physiological ecology of three codominant successional annuals. *Ecology* **56:**681–688.

Wielgolaski, F. E., L. C. Bliss, J. Svoboda, and G. Doyle
1981. Primary production of tundra. In *Tundra ecosystems: A comparative analysis,* ed. L. C. Bliss, O. W. Heal, and J. J. Moore, 187–227. International Biological Programme no. 25. London: Cambridge University Press.

Wielgolaski, F. E., and S. Kjelvik
1975. Energy content and use of solar radiation of Fennoscandian tundra plants. In *Fennoscandian tundra ecosystems.* Part I. *Plants and microorganisms,* ed. F. E. Wielgolaski, 201–207. New York: Springer-Verlag.

Wiens, J. A.
1973. Pattern and process in grassland bird communities. *Ecol. Monogr.* **43:**247–270.
1976. Population responses to patchy environments. *Ann. Rev. Ecol. Syst.* **7:**81–120.
1977. On competition and variable environments. *Amer. Scient.* **65:**590–597.

Wilbur, H. M., D. I. Rubenstein, and L. Fairchild
1978. Sexual selection in toads: The role of female choice and male body size. *Evolution* **32:**264–270.

Williams, E. C.
1941. An ecological study of the forest floor fauna of the Nanama rain forest. *Bull. Chicago Acad. Sci.* **6:**63–124.

Williams, G. C.
1975. *Sex and evolution.* Princeton, N.J.: Princeton University Press.

Williamson, M.
1981. *Island populations.* Oxford, England: Oxford University Press.

Williamson, P.
1971. Feeding ecology of the red-eyed vireo (*Vireo olivaceous*) and associated foliage-gleaning birds. *Ecol. Monogr.* **41**:129–152.

Williamson, P.
1976. Above-ground primary production of chalk grassland allowing for leaf death. *J. Ecol.* **64**:1059–1075.

Willis, E. O.
1974. Populations and local extinctions of birds on Barro Colorado Island, Panama. *Ecol. Monogr.* **44**:153–169.

Willson, M. F.
1979. Sexual selection in plants. *Amer. Natur.* **113**:777–790.
1983. *Plant reproduction physiology.* New York: Wiley.

Willson, M. F., and N. Burley
1983. *Mate choice in plants: Tactics, mechanisms, and consequences.* Princeton, N.J.: Princeton University Press.

Wilson, D.
1975. A theory of group selection. *Proc. Nat. Acad. Sci.* **72**:143–146.

Wilson, E. O.
1971. Competitive and aggressive behavior. In *Man and beast: Comparative social behavior,* ed. J. Eisenberg and W. Dillon. Washington, D.C.: Smithsonian Institution Press.
1975. *Sociobiology.* Cambridge, Mass.: Harvard University Press.

Wilson, E. O., and E. O. Willis
1975. Applied biogeography. In *Ecology and evolution of communities,* ed. M. L. Cody and J. Diamond, 522–533. Cambridge, Mass.: Harvard University Press.

Wing, L. D., and I. D. Buss
1970. Elephants and forests. *Wildl. Monogr.* **15.**

Winston, F. W.
1956. The acorn microsere with special reference to microarthropods. *Ecology* **37**:120–132.

Wirtz, W. O.
1977. Vertebrate post-fire succession. In *Proc. symp. environmental consequences of fire and fuel management in Mediterranean ecosystems,* ed. H. A. Mooney and C. E. Conrad, 47–57. U.S.D.A. For. Serv. Gen. Tech. Rept. **WO-3.** Washington, D.C.: U.S. Department of Agriculture.

Wissmar, R. C., J. E. Richey, A. H. Devol, and D. M. Eggers
1982. Lake ecosystems of the Lake Washington drainage basin. In *Analysis of coniferous forest ecosystems in the western United States,* ed. R. L. Edmonds, 333–386. Stroudsburg, Pa.: Dowden, Hutchinson & Ross.

Witherspoon, J. P., S. I. Auerbach, and J. S. Olson
1962. Cycling of cesium-137 in white oak trees on sites of contrasting soil type and moisture. *Oak Ridge Nat. Lab. Rept.* **3328**:1–143.

Withrow, R. B., ed.
1959. *Photoperiodism and related phenomena in plants and animals.* Publ. no. **55.** Washington, D.C.: American Association for the Advancement of Science.

Witkamp, M.
1963. Microbial populations of leaf litter in relation to environmental conditions and decomposition. *Ecology* **44**:370–377.

Wojcik, V. H.
1968. Possible divergent evolution in *Helenium amarum* populations. MA Thesis, Dept. Botany, University of North Carolina, Chapel Hill.

Wolf, L. L., F. A. Hainsworth, and F. G. Stiles
1972. Energetics of foraging: Rate and efficiency of nectar extraction by hummingbirds. *Science* **176**:1351–1352.

Wolfe, J. N., R. T. Wareham, and H. T. Scofield
1949. Microclimates and macroclimates of Neotoma, a small valley in central Ohio. *Ohio Biol. Surv. Bull* no. **41.**

Wolff, J. O.
1980. The role of habitat patchiness in the population dynamics of snowshoe hares. *Ecol. Monogr.* **50**:111–130.

Wolfson, A.
1959. The role of light and darkness in the regulation of spring migration and reproductive cycles in birds. In *Photoperiodism and related phenomena in plants and animals,* ed. R. B. Withrow, 679–716. Publ. no. **55.** Washington, D.C.: American Association for the Advancement of Science.
1960. Regulation of annual periodicity in the migration and reproduction of birds. *Cold Spring Harbor Symp. Quant. Biol.* **25**:507–514.

Woodwell, G. M., and D. B. Botkin
1970. Metabolism of terrestrial ecosystems by gas exchange techniques, the Brookhaven approach. In *Analysis of temperate forest ecosystems,* ed. D. E. Reichle, 73–85. New York: Springer-Verlag.

Woodwell, G. M., and W. R. Dykeman
1966. Respiration of a forest measured by carbon dioxide accumulations during temperature inversions. *Science* **154**:1031–1034.

Woodwell, G. M., and T. G. Marples
1968. The influence of chronic gamma radiation on the production and decay of litter and humus in an oak-pine forest. *Ecology* **49**:456–465.

Woodwell, G. M., and E. V. Pecan, eds.
1973. *Carbon and the biosphere, conf. 72501.* Springfield, Va.: Na-

tional Technical Information Service.

Woodwell, G. M., C. F. Wurster, Jr., and P. A. Isaacson
1967. DDT residues in an East Coast estuary: A case of biological concentration of a persistent pesticide. *Science* **156**:821–823.

Woolfenden, G. E.
1975. Florida scrub jay helpers at the nest. *Auk* **92**:1–15.
1981. Selfish behavior by Florida scrub jay helpers. In *Natural selection and social behavior: Recent research and new theories,* ed. R. D. Alexander and D. W. Tinkle, 257–260. New York: Chiron Press.

Woolfenden, G. E., and J. W. Fitzpatrick
1977. Dominance in the Florida scrub jay. *Condor* **79**:1–12.

Woolpy, J. H.
1968. The social organization of wolves. *Nat. Hist.* **77** (5): 46–55.

Woolpy, J. H., and B. E. Ginsburg
1967. Wolf socialization: A study of temperament in a wild social species. *Amer. Zool.* **7**:357–364.

Wright, R. T.
1970. Glycollic acid uptake by plankton bacteria. In *Organic matter in natural waters,* ed. D. Wood. Occ. Publ. No. **1**. Fairbanks: Institute of Marine Science, University of Alaska.

Wright, S.
1931. Evolution in Mendelian populations. *Genetics* **16**:97–159.
1935. Evolution in a population in approximate equilibrium. *J. Genetics* **30**:243–256.

Wynne-Edwards, V. C.
1962. *Animal dispersion in relation to social behavior.* New York: Macmillan (Hafner Press).
1963. Self-regulating system in populations of animals. *Science* **147**:1543–1548.

Yarranton, M., and G. A. Yarranton
1975. Demography of a jack pine stand. *Can. J. Bot.* **53**:310–314.

Yeaton, R. I.
1978. A cyclic relationship between *Larrea tridentata* and *Opuntia leptocaulis* in the northern Chihuahuan Desert. *J. Ecol.* **66**:651–656.

Yodzis, P.
1980. The connectance of real ecosystems. *Nature* **284**:544–545.
1981. The structure of assembled communities. *J. Theoret. Biol.* **92**:103–117.
1982. The compartmentation of real and assembled communities. *Amer. Natur.* **120**:551–570.
1983. Community assembly, energy flow, and food web structure. In *Current trends in food web theory,* ed. D. L. DeAngelis, W. M. Post, and G. Sugihara, 41–44. ORNL 5983. Oak Ridge, Tenn.: Oak Ridge National Laboratory.

Yonge, C. M.
1949. *The sea shore.* London: Collins.

Zack, R., and J. B. Falls
1976a. Do ovenbirds (Aves: Parulidae) hunt by expectation? *Can. J. Zool.* **54**:1894–1903.
1976b. Foraging behavior, learning, and exploration by captive ovenbirds (Aves: Parulidae). *Can. J. Zool.* **54**:1880–1893.
1976c. Ovenbird (Aves: Parulidae) hunting behavior in a patchy environment: An experimental study. *Can. J. Zool.* **54**:1863–1879.

Zak, B.
1964. Role of mycorrhizae in root disease. *Ann. Rev. Phytopathol.* **2**:377–392.

Zamaro, B. A.
1982. Understory development in forest succession: An example from the inland Northwest. In *Forest succession and stand development research in the Northwest,* ed. J. E. Means, 63–69. Corvallis: Forest Research Laboratory, Oregon State University.

Zeller, D.
1961. Certain mulch and soil characteristics of major range sites in western North Dakota as related to range conditions. MA thesis, North Dakota State University, Fargo.

Zimen, E.
1978. *The wolf: A species in danger.* 1981 translation. New York: Dell (Delacourt Press).

Zimmerman, E. C.
1960. Possible evidence of rapid evolution in Hawaiian moths. *Evolution* **14**:137–138.

Zimmerman, J. L.
1971. The territory and its density dependent effect in *Spiza americana.* *Auk* **88**:591–612.

INDEX